RENEWALS 458-4574

WITHDRAWN
UTSA Libraries

Levantine Archaeology
1

Executive Editors

Russell Adams & Yuval Goren

Editorial Board

Ghazi Bisheh, Raphael Greenberg, Julian Henderson,
David Ilan, Naama Goren-Inbar, Zeidan Kafafi,
Thomas E. Levy, Jodi Magness, Mohammed Najjar,
Graham Philip, Hamdan Taha, Paola Villa

www.SheffieldAcademicPress.com

THE ARCHAEOLOGY OF JORDAN

Edited by
**Burton MacDonald, Russell Adams
and Piotr Bienkowski**

Copyright © 2001 Sheffield Academic Press

Cover Photograph
Petra, rock-cut façade of ad-Dayr
('the monastery'), c. AD 100
Photograph by S. Schmid

Published by Sheffield Academic Press Ltd
Mansion House
19 Kingfield Road
Sheffield S11 9AS
England

Copies of this volume and a catalogue of other archaeological
publications can be obtained from the above address or from our home page

www.SheffieldAcademicPress.com

Printed on acid-free paper in Great Britain
by Bookcraft,
Midsomer Norton, Bath

A catalogue record for this book is available from the British Library

ISBN 1-84127-136-5

Contents

List of Figures	vii
List of Tables	xii
List of Contributors	xiii
Preface	xv

Major Chapters

1. Evolving Landscape and Environment in Jordan — 1
   Phillip G. Macumber
2. The Palaeolithic Period, Including the Epipalaeolithic — 31
   Deborah I. Olszewski
3. The Neolithic Period — 67
   Gary O. Rollefson
4. The Chalcolithic Period — 107
   Stephen J. Bourke
5. The Early Bronze I–III Ages — 163
   Graham Philip
6. The Early Bronze Age IV — 233
   Gaetano Palumbo
7. The Middle Bronze Age — 271
   Steven Falconer
8. The Late Bronze Age — 291
   John Strange
9. The Iron Age — 323
   Larry G. Herr and Muhammad Najjar
10. The Persian Period — 347
    Piotr Bienkowski
11. The Nabataeans: Travellers between Lifestyles — 367
    Stephan G. Schmid
12. Roman Jordan — 427
    Philip Freeman
13. The Byzantine Period — 461
    Pamela Watson
14. Umayyad and Abbasid Periods — 503
    Donald Whitcomb

15. Fatimid, Ayyubid and Mamluk Jordan and the Crusader Interlude   515
  Alan Walmsley
16. The Ottoman Period   561
  Alison McQuitty

Topical Summaries

17. Climatic Changes in Jordan through Time   595
  Burton MacDonald
18. Water Supply in Jordan through the Ages   603
  John Peter Oleson
19. Pastoralism   615
  Alison Betts
20. Traditional Agriculture   621
  Carol Palmer
21. Archaeological Survey in Jordan   631
  Edward Banning
22. The Analysis of Chipped Stone   639
  Douglas Baird
23. Pottery Technology   653
  Henk Franken
24. Writing and Texts   659
  Alan Millard
25. The Bible, Archaeology and Jordan   663
  Burton MacDonald
26. The Mosaics of Jordan   671
  Michele Piccirillo
27. Crusader Castles in Jordan   677
  Denys Pringle
28. Ottoman Hajj Forts   685
  Andrew Petersen

Index of Place Names and Sites Mentioned in the Text   693

List of Figures

Figure	Description	
1.1	Principal physiographic provinces in Jordan	3
1.2	Central Jordan Valley and Dead Sea	4
1.3	Topography and Palaeolithic site distribution at Tabaqat Fahl	6
1.4	Schematic cross-section through Wadi al-Hammeh	6
1.5	Stratigraphic section beneath the plateau	7
1.6	Geology of al-Azraq Basin, eastern Jordan	9
1.7	Geology of southwestern Jordan	16
1.8	Geological section through the Judayid Basin	17
1.9	Terraces of the Upper Wadi az-Zarqa' region	20
1.10	The maximum extent of Late Pleistocene Lake al-Hasa	23
1.11	Upper Wadi al-Hasa	24
2.1	Major areas of the western and eastern Levant	32
2.2	Lower and Middle Palaeolithic sites in Jordan	37
2.3	Upper Palaeolithic sites in Jordan	43
2.4	Epipalaeolithic sites in Jordan	49
3.1	Map showing Neolithic sites in Jordan	68
3.2	'Ayn Ghazal projectile points	72
3.3	Western room of a two-room MPPNB house at 'Ayn Ghazal	74
3.4	An MPPNB 'haltered' cattle figurine	77
3.5	A female statue and two busts from 'Ayn Ghazal	78
3.6	Large LPPNB building from 'Ayn Ghazal	80
3.7	'Twin' LPPNB shrines from 'Ayn Ghazal	83
3.8	View to the east of the eastern room of the upper LPPNB temple at 'Ayn Ghazal	84
3.9	View to the west of the eastern room of the upper LPPNB temple at 'Ayn Ghazal	85
3.10	The lower PPNB temple at 'Ayn Ghazal	86
3.11	A PPNC courtyard work area at 'Ayn Ghazal	88
3.12	A PPNC corridor building and paved ramp at 'Ayn Ghazal	89
3.13	The large Yarmoukian dwelling at 'Ayn Ghazal	90
3.14	The stepped, walled street at 'Ayn Ghazal	90
3.15	Yarmoukian bowls from 'Ayn Ghazal	92
3.16	Yarmoukian jars from 'Ayn Ghazal	93
4.1	Plan of Late Chalcolithic areas at Tulaylat al-Ghassul	121
4.2	The 'Star frieze' wall painting from Tulaylat al-Ghassul	122
4.3	Wall painting from Tulaylat al-Ghassul	122
4.4	Chalcolithic building at Abu Hamid	124
4.5	Chalcolithic architecture at Pella	126
4.6	Chalcolithic buildings at ash-Shuna, Area D	127
4.7	Chalcolithic buildings at Sahab	128
4.8	Layout of Chalcolithic Abu Snesleh	129
4.9	Chalcolithic architecture at Jabal al-Jill (J14)	130
4.10	Chalcolithic architecture at Tall Magass	131
4.11	Sanctuary area at Tulaylat al-Ghassul	132
4.12	Late Chalcolithic cornets and cups	134
4.13	Late Chalcolithic bowls	135
4.14	Late Chalcolithic storage jars and jars	136

4.15	Late Chalcolithic jars and spouted jars	137
4.16	Late Chalcolithic fenestrated stands, churns and spoons	138
4.17	Late Chalcolithic basalt vessels	140
4.18	Late Chalcolithic basalt figurines	141
4.19	Late Chalcolithic mace heads	143
4.20	Late Chalcolithic violin figurines	148
5.1	Map showing location of Early Bronze Age sites in Jordan	164
5.2	Plan of excavated area of Khirbat az-Zaraqun	173
5.3	Tall as-Sa'idiyya Area BB 700, Stratum L2 (EBII)	177
5.4	Tall ash-Shuna, circular EB I structure, Building 9	178
5.5	Tall ash-Shuna, later EB I rectilinear structure, Building 3	179
5.6	Jawa, subcircular domestic structure in Square UTI	180
5.7	Tall Abu al-Kharaz Area 2, EB II Main Phase II	181
5.8	Bab adh-Dhra' Tomb A76, an EB IA shaft grave	198
5.9	Bab adh-Dhra' Tomb A55, an EB II–III charnel house	199
5.10	Dolmen, Jabal Mutawwaq	201
5.11	Early EB I pottery from Tall ash-Shuna	203
5.12	Tall ash-Shuna, EB I bowl forms	204
5.13	EB I pottery	205
5.14	EB I and EB II pottery	206
5.15	Khirbat Kerak ware vessels from Tall ash-Shuna	209
5.16	Khirbat az-Zaraqun, EB III storage jar with loop handles	209
5.17	Khirbat az-Zaraqun, large EB III storage jar	210
5.18	EB III pottery from Bab adh-Dhra'	210
5.19	Flint blades of Canaanean type mounted in wooden haft, Tall Abu al-Kharaz, EB II	211
5.20	Metal and groundstone artifacts from Bab adh-Dhra'	213
5.21	Sealing practices from EBA Jordan	215
5.22	Seal impressions bearing geometric designs	216
5.23	Clay figurine, EB I, Bab adh Dhra', Tomb A 5E	217
6.1	Map of sites mentioned in the text	235
6.2	Map of all EB IV sites in Jordan	237
6.3	Pottery chronology according to Dever	239
6.4	Tall Abu en-Ni'aj: surface pottery scatter	241
6.5	General view of Jabar ar-Rahil, looking east from Jabal al-'Asi	242
6.6	Aerial view of Jabal ar-Rahil	242
6.7	General view of Khirbat Iskandar, looking north	243
6.8	Menhir alignment at Lejjun	244
6.9	Enclosure or terrace wall at Jabal ar-Rahil	246
6.10	Shaft tomb at al-'Umayri	248
6.11	Dolmen at Khirbat Umm al-Ghozlan	249
6.12	'Family AZ' pottery from a tomb at al-'Umayri	252
6.13	Weapons from a tomb at al-'Umayri	256
7.1	Map of Jordan showing Middle Bronze Age sites	272
7.2	Arboreal pollen frequencies from the Sea of Galilee sediment core	275
7.3	Excavation areas on the main mound of Pella and adjoining hillsides	277
7.4	Temple and domestic architecture excavated in phases 5–2 at Tall al-Hayyat	279
7.5	Rank-size distributions for Early Bronze III and Middle Bronze IIB–C settlements	282
7.6	Rank-size distributions for Early Bronze II, Middle Bronze IIA and Middle Bronze IIB–C settlements	283
8.1	Map of Jordan in the Late Bronze Age	296

8.2	Glass pendant in form of a fertility figurine, Tall al-Fukhar	300
8.3	Bronze lamp, Tall al-Fukhar	301
8.4	The panel from the Balu' stela	301
8.5	Late Bronze Age pottery 1	302
8.6	Late Bronze Age pottery 2	305
8.7	Glazed 'knob', possibly wall decoration, Tall al-Fukhar	306
8.8	Late Bronze Age building plans	308
9.1	Schematic section of the western defences at Tall al-'Umayri	325
9.2	Plan of the Late Bronze/Iron I transitional period architecture at Tall al-'Umayri	326
9.3	Isometric drawing of the Late Bronze/Iron I transitional four-room house at Tall al-'Umayri	327
9.4	Comparison of collared pithos rims	328
10.1	Map of sites with material that has been dated to the Persian period	350
10.2	Pottery from stratified contexts at Tall al-'Umayri	353
10.3	Sausage-shaped torpedo jar from Tall as-Sa'idiyya	354
10.4	Mortarium from Tall as-Sa'idiyya	354
10.5	Bag-shaped perfume juglet from Tall as-Sa'idiyya	354
10.6	Bottle from Tall Dayr 'Alla Phase IV	355
10.7	Stratum III building at Tall as-Sa'idiyya Stratum III	355
10.8	Building at al-Dreijat	355
10.9	Final phase of Area C building at Busayra	356
10.10	Final phase of Area A building at Busayra	357
11.1	Plain and painted pottery of phase 1	369
11.2	Reconstruction of possible temporary 'tent-house' installation	370
11.3	Plain, rouletted and painted pottery of phase 2	372
11.4	Map of the city of Petra	375
11.5	Schematic plan of Nabataean villa on the south slope of az-Zantur, Petra	376
11.6	Petra, Nabataean villa on az-Zantur south terrace	377
11.7	Petra, offering place on top of Zib Atuf	377
11.8	Nabataean temples	378
11.9	Petra 'South Temple' and Ptolemaic temples of Edfu and Kalabsha	380
11.10	South Arabian temples	381
11.11	Petra, relief slab with head of Gorgo Medusa as sign shield	382
11.12	Petra, relief slab with representation of a veiled muse or goddess	383
11.13	Petra, relief slab with cupid holding a garland	384
11.14	Petra, relief slab with a winged female torso	384
11.15	Petra, relief slab from Wadi Musa	385
11.16	Petra, Al-Khazna Faraoun	385
11.17	'Iraq al-Amir (near 'Amman), Qasr al-'Abd	387
11.18	The riverboat of Ptolemy IV	388
11.19	Petra, Al-Khazna Faraoun	389
11.20	Petra, tendril capital from the area around Qasr al-Bint	389
11.21	Petra, row of tomb façades south of the theatre	390
11.22	Plain, rouletted, stamped and painted pottery of phase 3a-b; painted bowl of second/third century AD	392
11.23	Painted bowl of phase 3a	393
11.24	Painted bowl of phase 3b	393
11.25	'Amman, Atargatis panel from Khirbat at-Tannur	394
11.26	'Amman, architectural element with a bust of Hermes/al-Khutbay	394
11.27	'Amman, relief of a winged Nike supporting a zodiac from Khirbat at-Tannur	395
11.28	'Amman, relief of the so-called dolphin goddess from Khirbat at-Tannur	395

11.29	Petra, 'winged head', probably Hermes/al-Khutbay	396
11.30	'Amman, aniconic *baitylos* found in the temple of the winged lions (Petra)	396
11.31	Petra, temenos gate	397
11.32	Petra, west slope of al-Khubtha with row of tomb façades	397
11.33	Petra, tomb façade of ad-Dayr	398
11.34	(a) Petra, Roman soldier tomb and triclinium; (b) Ptolemais, Palazzo delle Colonne; (c) Reconstruction of Vitruvius's description of the Greek house; (d) Jericho, Herod's first winter palace	399
12.1	Map of Roman sites in Jordan mentioned in the text	429
12.2	Gerasa of the Decapolis (Jarash)	430
12.3	Philadelphia of the Decapolis ('Amman)	431
12.4	Gadara of the Decapolis (Umm Qays)	432
12.5	Abila of the Decapolis (Quwayliba)	433
13.1	Map showing sites mentioned in text	462
13.2	Map showing Byzantine provincial divisions	464
13.3	Khirbat Faynan, ancient Phaeno	471
13.4	View across Umm al-Jimal	472
13.5	Qasr Burqu'	473
13.6	Byzantine pottery	475
13.7	Byzantine pottery and glass	476
13.8	Petra Church mosaic detail	477
13.9	Umm al-Jimal	483
13.10	Pella, Tall al-Husn	483
13.11	Umm Qays	485
13.12	Qasr Bshir	486
13.13	Umm ar-Rasas	496
14.1	Map of sites in Jordan during the Umayyad and Abbasid periods	504
14.2	Chronological periodization of the early Islamic periods in Jordan	505
14.3	Comparison of plans of Qasr al-Mushatta and 'Amman Citadel	507
15.1	The geo-political features of tenth-century Jordan	517
15.2	The pilgrimage route and other major roads of Jordan in the tenth and eleventh centuries	518
15.3	Crusader holdings in Jordan in the twelfth century	519
15.4	Major centres, provincial divisions and the pilgrimage route in Mamluk Jordan	522
15.5	Ajlun castle: plan showing the original structure	531
15.6	Ajlun castle: the inner gateway to the additions of 1214-15	532
15.7	Azraq: entrance to the fort, 1236-37	532
15.8	Azraq: plan of the fort	533
15.9	Kahf: view of the Mamluk mosque, looking south	535
15.10	Kahf: measured sketch of the Mamluk mosque, south to top	535
15.11	Plan and section of the shrine of Abu Sulayman al-Dirani near Shawbak	537
15.12	Tabaqat Fahl: plane of the Mamluk mosque	538
15.13	Gharandal, near Tafila	540
15.14	'Amman Citadel: pottery of second half of eleventh century	546
15.15	Ayla: pottery of the eleventh century from the Pavilion Building	547
15.16	Ayla, lamps; Shawbak, coarse wares; Tabaqat Fahl (Pella), jars and bowls	548
15.17	Tabaqat Fahl, pottery of the second half of the fourteenth to fifteenth centuries	551
16.1	Map of Jordan showing sites mentioned in the text	562
16.2	Map showing economic variety in the sixteenth century	566
16.3	Al-Qatrana railway station	570
16.4	The *serai* (governor's residence) at Irbid	570

16.5	Nineteenth-century house of Haj Guweida Suleiman Obeidat in Hartha	571
16.6	Construction method of barrel-vaulted structure	572
16.7	Plan of nineteenth-century house at Khirbat Faris	574
16.8	Interior of nineteenth-century house at al-Qasr	574
16.9	Diagram to show watermill technology	576
16.10	Sixteenth-century pottery from Khirbat Faris	578
16.11	Ottoman pottery from ash-Shawbak	579
16.12	Twentieth-century *habia* from north Jordan	580
16.13	Ottoman lamp and tobacco pipe fragments from Khirbat Faris	582
16.14	Nineteenth-century grave	586
17.1	Rainfall map	596
18.1	'Ayn Shelaleh conduit	604
18.2	Gerasa nymphaeum, view	605
18.3	Humayma: run-off field with stone piles	607
18.4	Humayma: cistern with arch and slab roof	607
18.5	Petra: reservoir	609
18.6	Humayma: dam	609
18.7	Humayma: aqueduct conduit	610
18.8	Humayma: Roman bronze stopcock	611
18.9	Qasayr'Amra: reconstruction of *saqiya* arrangement	611
19.1	Modern Bedouin encampment	616
19.2	Safaitic carving from eastern Jordan	617
20.1	Adzing an oak beam to form the stilt/sole of an *ard*	623
20.2	Close-up of an *ard* during spring tillage	623
20.3	Dual animal tillage in the spring using a traditional *ard* near the village of al-Mazar, northern Jordan	624
20.4	Field fed by runoff water on a break of slope in the Wadi Siagh below Petra	625
20.5	Broadcast-sowing wheat at Mu'ta, near al-Karak	626
20.6	Harvesting wheat by uprooting in northern Jordan	626
20.7	An up-turned threshing sledge at rest	627
21.1	Map of Wadi Ziqlab's drainage basin	633
21.2	Map of pedestrian transects	633
21.3	Edge effects on site discovery by transects and square quadrats	634
21.4	The effect of site size on the probability of site detection by pedestrian transects	634
21.5	Estimates of survey coverage as a function of site diameter	635
22.1	Chipped stone types from Jordan	641
26.1	Mosaic of Hippolythus, Madaba	672
26.2	Mosaic map of the Holy Land from Madaba, with Jerusalem at its centre	673
26.3	Mosaic medallion with the personification of the Sea	674
26.4	Mosaic floor of the Church of St Stephen, Umm ar-Rasas	675
27.1	Ash-Shawbak (Montreal): the castle from the west	679
27.2	Al-Karak: the castle from the southwest	679
27.3	Al-Karak: the northeast tower of the barbican	680
27.4	Al-Habis, Petra: plan of the castle	682
28.1	Location of Hajj forts in Jordan	687
28.2	Qal'at al-Qatrana: ground floor plan	688
28.3	Qal'at al-Qatrana: first/upper floor plan	688
28.4	Qal'at al-Hasa: ground floor plan	689
28.5	Qal'at al-Hasa: first/upper floor plan	689
28.6	Qal'at al-Fassu'a: entrance and projecting corner towers	690

List of Tables

1.1	U–Th dates from the Central Plateau Region	11
1.2	Summary of major Pleistocene events from selected sites	25
2.1	Epipalaeolithic lithic assemblages by researcher	50
4.1	Radiocarbon dates for the Jordanian Chalcolithic	112
5.1	Estimate of absolute dates for EB I–III	169
5.2	Threshold area for defining an urban site	182
5.3	Excavated walled Early Bronze Age sites in Jordan	182
5.4	Percentages of identified bone fragments for main species represented at Tall ash-Shuna	186
5.5	Percentages of identified bone fragments for main species represented at Khirbat az-Zaraqun	186
5.6	Early Bronze Age sites on the east side of the Jordan Valley	191
7.1	Middle Bronze Age chronology and nomenclature	273
7.2	Early and Middle Bronze Age site counts and densities in Jordan	284
15.1	Total number of Ayyubid-Mamluk sites and percentage of total number of historical-period sites by JADIS sectors	528
16.1	Historical sources for the Ottoman period in Jordan	563
16.2	Ottoman sultans	563
16.3	Grave types	587

List of Contributors

Russell B. Adams	Department of Anthropology, University of California, San Diego, USA
Douglas Baird	School of Archaeology, Classics and Oriental Studies, University of Liverpool, PO Box 147, Liverpool L69 3BX, UK
Edward Banning	Department of Anthropology, University of Toronto, Toronto, Ontario M5S 1A1, Canada
Alison Betts	Department of Archaeology, University of Sydney, Sydney 2006, NSW, Australia
Piotr Bienkowski	Liverpool Museum, William Brown Street, Liverpool L3 8EN, UK
Stephen J. Bourke	Department of Archaeology, University of Sydney, Sydney 2006, NSW, Australia
Steven Falconer	Department of Anthropology, Arizona State University, Main Campus, PO Box 872402, Tempe, AZ 85287-2402, USA
Henk Franken	v. Essenstraat 42, 7203 DM Zutphen, The Netherlands
Philip Freeman	School of Archaeology, Classics and Oriental Studies, University of Liverpool, PO Box 147, Liverpool L69 3BX, UK
Larry Herr	Canadian University College, 235 College Avenue, College Heights, AB, T4L 2E5, Canada
Burton MacDonald	Department of Religious Studies, PO Box 5000, St Francis Xavier University, Antigonish, Nova Scotia B2G 2W5, Canada
Phillip G. Macumber	School of Earth Sciences, University of Melbourne, Melbourne, Australia
Alison McQuitty	Bij de Berg 4, Blaricum 1261 XN, The Netherlands
Alan Millard	School of Archaeology, Classics and Oriental Studies, University of Liverpool, PO Box 147, Liverpool L69 3BX, UK
Muhammad Najjar	Department of Antiquities of Jordan, Amman, Jordan
John Peter Oleson	Department of Greek and Roman Studies, University of Victoria, Victoria, B.C., Canada V8W 3P4
Deborah I. Olszewski	Department of Anthropology, Bishop Museum, 1525 Bernice Street, Honolulu, HI 96817, USA
Carol Palmer	School of Archaeological Studies, University of Leicester, University Road, Leicester LE1 7RH, UK
Gaetano Palumbo	Institute of Archaeology, University College London, 31-34 Gordon Square, London WC1H 0PY, UK
Andrew Peterson	School of History and Archaeology, Cardiff University, PO Box 909, Cardiff CF1 3XU, Wales
Graham Philip	Department of Archaeology, University of Durham, South Road, Durham DH1 3LE, UK
Michele Piccirillo	Studium Biblicum Franciscanum, Flagellation (via Dolorosa), PO Box 19424, 91193 Jerusalem, Israel

Denys Pringle	School of History and Archaeology, Cardiff University, PO Box 909, Cardiff CF1 3XU, Wales
Gary O. Rollefson	Department of Anthropology, Whitman College, Walla Walla, WA 99362, USA
Stephan G. Schmid	Ecole Suisse d'archèologie en Gréce, Odos Scaramanga 4b, GR 104 33 Athens, Greece
John Strange	Department of Biblical Exegesis, University of Copenhagen, Kobmagergade 44-46, DK-1150 Copenhagen K, Denmark
Alan Walmsley	Classics and Ancient History, University of Western Australia, WA 6907, Australia
Pamela Watson	Department of Archaeology and Palaeontology, University of New England, Armidale, NSW, Australia
Donald Whitcomb	The Oriental Institute, 1155 East 58th St., Chicago, IL 60637, USA

Preface

This volume was originally conceived by the three editors in spring 1994, during a dinner discussion at Karak Rest House in Jordan. At that time we envisaged it as a general text book, an up-to-date overview of the archaeology of Jordan aimed at students and specialists in other areas and disciplines. Such a volume did not exist, and we felt that one would be useful given the huge quantity of archaeological field work and research in Jordan in recent years.

We appreciate that the final product is rather different from that originally envisaged, and is more a collection of papers on the current state of research on the different periods and aspects of Jordanian archaeology, by scholars at the forefront of recent field work. This change in focus was not deliberate, and there are several explanations for why it occurred. These include the natural writing style of some contributors, and the nature of the overview they felt it was important to present, but perhaps the most important single factor is the impact of the changing world of academic archaeology. Most (though not all) of the contributors are based in universities; particularly in England and the USA, but also elsewhere, university departments are under pressure to publish high-quality research that can be measured and compared in assessment exercises that affect the level of funding of departments. As a result, university personnel are urged to maximize the specialist academic papers they publish, and this is to the detriment of more general, less specialized writing for wider audiences that, rightly or wrongly, does not have the same academic prestige. Put bluntly, for a university archaeologist to spend time writing a chapter or paper, it must be specialist, original research that contributes to his or her department's research profile, and not a synthesis aimed at a non-specialist audience. The present book is an inevitable reflection of this new world.

The book is divided into 16 major chapters (on the landscape and environmental background of Jordan, and 15 chronological surveys from the Palaeolithic to the Ottoman periods), followed by 12 topical summaries on particular aspects of the archaeology of Jordan that fit less easily into purely chronological treatments. Each chapter is followed by a bibliography, and there is an index of place names and sites at the back of the book. Although there is no separate chapter on the Hellenistic period, much of the Hellenistic evidence is treated within the Nabataean chapter and in the conclusion to the Persian-period chapter.

The spelling of Jordanian place names follows the system of transliteration recommended by the Royal Jordanian Geographic Centre and adopted by the Department of Antiquities of Jordan in its most recent publications. It follows a system adopted by the United Nations, by the Board on Geographic Names (a division of the United States Defense Department), and by the British Permanent Committee on Geographic Names, whose aim is a consistent method of writing Arabic names in English based on formal Arabic. The most frequent examples of this practice are 'Tall' for 'Tell' and 'Khirbat' for 'Khirbet'. Diacritics are not used, except in the titles of Arabic publications.

Dating conventions follow standard practice, using capitalized 'BC' and 'BP' ('Before Present') for calibrated dates, and lower case 'bc' and 'bp' for raw uncalibrated radiocarbon dates. No overall chronological table has been included since there are still several possible chronologies and variant terminologies. The editors did not feel it appropriate to choose one system over others, thus implying that it was in some way more 'correct'. Instead, the problems of dating and terminology within each period are discussed in the chronological chapters.

With such a large book, editorial work was inevitably divided: Russell Adams initially coordinated with Sheffield Academic Press, approached contributors and prepared the index; Burton MacDonald and Piotr Bienkowski edited the manuscripts, Burton MacDonald while on sabbatical leave at Clare Hall College, University of Cambridge; Piotr Bienkowski made the final choice of illustrations and coordinated the work at proof stage.

The editors wish to thank all the contributors for their assistance and patience during the long gestation of this book—some of the manuscripts were initially submitted in 1997—and Philip Davies at Sheffield Academic Press for his support and interest. Don Shewan of London Guildhall University produced the maps of sites at the beginning of most of the chronological chapters.

Burton MacDonald
Russell Adams
Piotr Bienkowski

1. Evolving Landscape and Environment in Jordan

Phillip G. Macumber

Introduction

The following contribution examines aspects of the physiography of Jordan as a basis for understanding processes and events that influence population distribution and occupational environment, both at present and in the past. When attempting to examine landscape alone, it becomes clear that in any archaeological context, landscape cannot be removed from the broader physical setting and must be considered alongside hydrological and climatic factors. Of special significance is the role of (palaeo)hydrology, since the availability of water is the one essential element without which ongoing occupation cannot exist, and which strongly influences the nature of that occupation, where it does exist.

This paper provides a description of the physiography of Jordan based largely on the physiographic provinces initially erected by Bender (1968, 1974, 1975), but with some modifications. For each province it examines and compares events from a small selection of well-documented sites, where the nature of long-term occupation reflects strongly on the physical evolution of Jordan through time. In dealing with palaeoenvironmental reconstructions there are many difficulties in comparing the palaeohydrology and palaeoclimatology from different areas, and this is especially the case using archaeological evidence. One reason for this is that Palaeolithic man did not live in an average geographic setting, but instead chose those areas where he could best survive. Attempts at landscape and environmental reconstruction for a region based largely on geoarchaeological data are thus strongly biased by the favourable hydrological settings of the long-term occupation sites from which the most complete records come.

Landscape, palaeoenvironmental interpretation and regional correlation

Throughout much of Jordan, landscape evolution has been highly dynamic, the region having undergone numerous physiographic and climatic changes during the period of human occupation. From the hominid occurrences at 'Ubeidiya in the northern Rift Valley, this is known to have lasted for almost 1.5 m.yr (Tchernov 1988). Much of our understanding of palaeoenvironmental change over this time stems from environment-related investigations during archaeological studies. However, one factor that strongly complicates landscape and climatic interpretations arising from studies of Levantine prehistory is the degree of geographic and climatic diversity occurring throughout what is essentially a rather small region. In addition, for much of the region, archaeological sites are located in thin surficial sequences where aeolian, alluvial and colluvial sediments rapidly interchange, both vertically and horizontally, reflecting fine-scale temporal instability.

The theoretical and disciplinary bases for palaeoenvironmental interpretations are of a diverse nature; they include pollen curves, faunal analyses, geomorphic and sedimentological studies, hydrological studies including lake level changes, etc. Ideally, for best results, such studies should contain elements of all these approaches. However, this is not normally the case, and in many instances, given limited site objectives, they cannot be justified or afforded. For instance, while detailed pollen analyses have been carried out on a number of sites, insufficient sequences have adequate absolute dating. This is especially the case in the Middle and Lower Palaeolithic. Where little or no absolute dating exists, the erection of a local temporal framework is made using stone tool industries, stratigraphic interpretation, pollen curves, etc. This framework is then related to similar but better documented/dated and understood sequences elsewhere. Where absolute chronologies are not available, these types of analyses remain the only effective way of determining and comparing evolving palaeogeographic settings for prehistoric occupation.

This is not always simple, however. Problems of local interpretation in the first instance, followed by regional correlation, are very real and cannot be expediently discounted: for example, how to interpret regionally the occurrence of erosion or sedimentation at a site within a drainage basin, given that erosion in one part of the catchment implies sedimentation in another part. The difficulties inherent in broader sequence matching are obvious, and, while there may be strong similarities in overall character, the precise correlation of the various events often remains problematical. Therefore, there must often be concern about how really representative a specific site is to the regional synthesis.

The problems of erecting regional syntheses in the light of the individuality of specific sites are obvious. However, where insufficient well-documented and dated sequences exist, it may inevitably result, in time, in the gradual reinforcement of a few better-established sequences into one or occasionally two, often opposing, structures. In many respects, the proposed reconstructions may be valid, but there may equally be times when the final monolithic presentation is far removed from reality, relying on an underpinning of many inadequate interpretations, themselves mirroring the supposed broader understanding.

Whatever the case, landscape and palaeoenvironmental interpretations must be regarded as evolving inductive hypotheses, which gain strength when subjected to rigorous examination and the accumulation of more information. One test of the worth of the monolith is its ability to encompass not simply the large number of fragmentary records, which may be inadvertently distorted to fit the hypothesis, but the fewer more complete sequences that may occasionally arise.

Water, landscape and occupation

In arid and semiarid regions, such as large parts of Jordan today, occupation is synonymous with water. Without water there is no occupation. However, surface water supplies, both fluvial and lacustrine, are at the mercy of seasonality and drought. It is the groundwater-based spring-fed sites that provide the best short- and long-term buffers to aridity. Through time, the ongoing presence of man and animals is commonly associated with major spring systems. This is the case where extended periods of occupation occur at Pella (Pottery Neolithic to present-day occupation) in the Jordan Valley. It has been the case in the surrounding region of Tabaqat Fahl/Wadi al-Hammeh for at least the last 0.5 m.yr, and also in the otherwise arid al-Azraq Basin since Lower Palaeolithic times. Both areas, one in eastern and one in western Jordan, are associated with regional groundwater discharge, with large-scale continuous spring outflow through long periods of time.

This account examines three different landscapes where occupation is closely linked to spring activity—al-Azraq Basin, Tabaqat Fahl and Wadi al-Hasa. These well-documented settings are also compared and contrasted with sites from several other areas having significant input into an understanding of the basis and relevance of palaeoenvironmental and palaeoclimatic reconstructions in Jordan. Additional sites are the upper az-Zarqa' region, Wadi Hisma of southwestern Jordan and the Badia of northeastern Jordan.

Physiographic framework of Jordan

The Hashemite Kingdom of Jordan covers an area of 96,500 km^2 at the eastern end of the Arabian Peninsula. The geology and physiography have been previously detailed in the work of Bender (1968, 1974, 1975). Given the widespread acceptance of Bender's *Geology of Jordan* as the principal authority on the underlying structural and stratigraphic framework on which the geomorphology is overlain, this paper follows closely, with some minor modification, his geomorphologic subdivisions. Accordingly, Jordan is divided into six principal geomorphologic provinces (Figure 1.1):

1. Jordan Rift Valley (includes Wadi 'Arabah and the Dead Sea graben).
2. Central Plateau (includes al-Jafr and al-Azraq/Wadi as-Sirhan Basins).
3. Northern Basalt Plateau and Northeastern Limestone Plateau.
4. Southern Mountain Desert.
5. Western Highlands (Bender's highlands east of the Rift).

The following discussion examines the physiographic framework of Jordan using examples from major geoarchaeological studies carried out in each geomorphologic province. The history of landscape evolution at the sites is compared and contrasted to provide a broad understanding of the relationship existing between physiography and occupation across Jordan.

Jordan Rift Valley Province

Jordan lies at the western edge of the Arabian plate, which extends eastwards across Saudi Arabia to Oman. Undoubtedly, the most striking physiographic feature of Jordan is the Jordan Rift Valley at the western extremity of the country (Figure 1.1). The Rift Valley

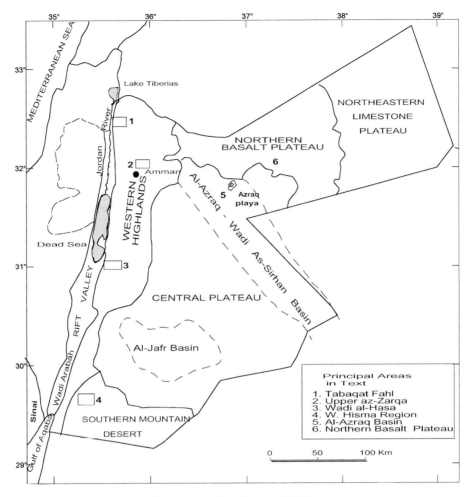

Figure 1.1. Principal physiographic provinces in Jordan (modified from Bender 1975).

is a northward extension of the Great Rift Valley system, which extends northwards from East Africa, and includes the Red Sea, the Gulf of Aden, Gulf of Suez and Gulf of al-'Aqaba. The Jordan Rift Valley marks the boundary between the Arabian and African plates with a suggested 106 km lateral strike slip movement between the plates of 0.4-0.9 cm/yr (Quennel 1958). The valley reaches a width of about 32 km near Jericho but narrows to less than 10 km farther north. It is bounded on either side by steep fault-controlled edges. Close to the Dead Sea, the floor of the Jordan Valley is some 400 m below sea level, but the landscape rises precipitously to the east on to the Western Mountain Province. Southwards, between the Dead Sea and the Gulf of al-'Aqaba, the Wadi 'Arabah consists of a series of shallow interconnected basins.

In Jordan, the Rift Valley may be divided into three sub-provinces (Figure 1.2):

1. The Dead Sea and adjacent plain containing the Lisan Peninsula that extends southwards to the Khanazir Fault that is expressed as a 50 m-high escarpment.
2. The Wadi 'Arabah extends from the Gulf of al-'Aqaba northwards to the Khanazir Fault.
3. The Jordan River north of the Dead Sea.

The rifting in western Jordan commenced about 25–30 m.yr ago in the late Oligocene or early to middle Miocene period. The resulting sedimentary pile in the Dead Sea Basin is up to 10 km deep, with subsidence probably commencing in the Pliocene accompanied by the accumulation of thick deposits of evaporites and some marine sequences (Ginzberg and Kashai 1981). Subsidence is still continuing today. Much of the later fill reflects a number of cycles of lacustrine and fluvial sedimentation when alternating lake and river systems occupied the depression. The Late Pleistocene Lake Lisan and its precursor Lake Samra were the last megalakes occupying the Rift Valley.

With continued subsidence in the centre, concomitant uplifting and arching of the edges of the Rift

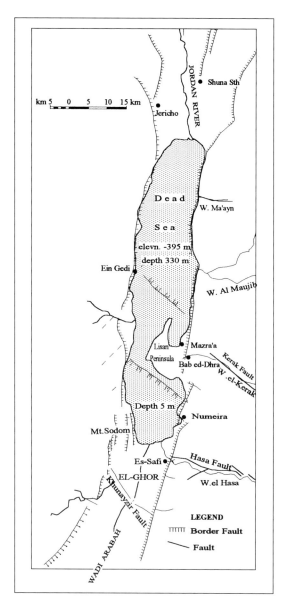

Figure 1.2. Central Jordan Valley and Dead Sea.

(Tchernov 1988). This is from 'Ubeidiya, to the south of Lake Tiberias, where steeply dipping conglomerates and clays representing former fluvial and lacustrine sequences, now overturned, contain human and animal faunal remains and Oldowan and early Acheulean stone tool industries (Stekelis et al. 1969). Wadis, such as al-Hasa and Hisma, which were important areas of Pleistocene occupation, are associated with major lateral faults, one of which led to the development of the Late Pleistocene al-Hasa Lake. Farther north in the Rift Valley, zones of spring outflow in the Tabaqat Fahl region are tectonically controlled, and neo-tectonic activity has had a significant impact upon the landscape (Macumber and Head 1991). Early phases of uplift of the Rift Valley sides led to landscape incision and the development of large ancestral wadis. These ancestral valleys underwent a phase of extensive alluviation in Late Pleistocene times in response to the filling of Lake Lisan.

Volcanic activity is associated with rifting. It began in the early Miocene period and continued to early recent times with phases of activity dated from about 18.9 m.yr to less than 10 m.yr. The basalts were commonly extruded along major fault zones; examples occur in the Dhra' area west of al-Karak, the Wadi Hisma and at Ghor al-Katar.

The development of the Rift Valley resulted in the capture of regional groundwater systems that previously passed eastwards to the Mediterranean and westwards to the inland depressions of the Central Plateau (Salameh 1985). Groundwater from inland Jordan now flows eastwards into the Rift Valley where it emerges as springs. The importance of the Rift Valley groundwater outflow zone to continuing occupation in the Jordan Valley cannot be overstated. The springs are conduits for the naturally regulated release of regional groundwater that is fed from large recharge catchments, refilled over millennia and extending over thousands of square kilometres as far eastwards as al-Azraq Basin. Flow times for water infiltrating within the highland catchments as far away as al-Azraq and discharging into the Rift Valley are put at about 3,500 years (Salameh and Udluft 1985). Unlike the rivers and streams, the springs and their associated discharge zones are strongly buffered against seasonality and short term to long term periods of aridity that are known to have affected the region since Early Pleistocene times. As a consequence, the well-watered Jordan Valley has been a corridor for migration between Africa and Europe for more than 1.5 m.yr.

Valley created the high mountainous regions lying lateral to the rift graben. The mountainous area in Jordan was termed by Bender (1968, 1974) 'the Mountain Ridge and Northern Highlands East of the Rift'. For simplicity, it is here referred to as the *Western Mountain Province*. The accumulation of thick sedimentary deposits in the graben accelerated during the Pleistocene with the deposition of alternating continental fluvial and lacustrine sequences as large volumes of sediment eroded from the steep valley flanks poured into the Rift Valley.

Tectonic activity during Rift Valley development occurred over the period of human occupation, for which the earliest evidence is about 1.4 m.yr old

Lake Lisan

Throughout Quaternary times the Jordan Rift Valley was aggraded by cycles of fluvial and lacustrine deposition, such as those that contain the Early Pleistocene occupation sequences at 'Ubeidiya (Stekelis *et al.* 1969). During the Late Pleistocene period the Jordan Valley from the Dead Sea to the Sea of Galilee was occupied by Lake Lisan (Neev and Emery 1967; Begin *et al.* 1974, 1985), in which were deposited thick sequences of chemical and detrital calcareous sediments. An earlier lake, Lake Samra, for which the ages are less certain, preceded Lisan. Lake Lisan existed for a considerable time ranging from 80,000–60,000 bp up to 18,000–15,000 bp (Neev and Emery 1967), or up to 13,000–11,000 bp (Neev and Hall 1976; Macumber and Head 1991). It reached an uppermost level of about minus 180 m before finally drying up at about 11,000 bp.

A large number of archaeological sites occur on the floor of the Rift Valley (e.g. Bab adh-Dhra' to the east of the Lisan Peninsula [Figure 1.2]). These sites postdate the drying of Lake Lisan and reflect environmental changes occurring only over the Holocene period, a very small albeit important part of the time in which humans occupied the Rift Valley. A broader understanding of long-term landscape evolution affecting human occupation can be obtained by examining areas marginal to the Jordan Valley that have undergone considerable modification through time, and exhibit a long record of environmental change. These well-watered riftside landscapes are associated with intense human occupation, and geoarchaeological studies of such areas have permitted the reconstruction of palaeoenvironmental changes over long periods of time.

Four well-documented case histories covering different physiographic provinces where occupation spans the lifetime of Lake Lisan are discussed and compared with respect to landscape evolution in the north, central and southern parts of the Rift Valley. A feature of three of the areas is the strong influence of tectonics, which clearly plays an important part in the geomorphic and, hence, environmental history of archaeological sites occurring adjacent to the Rift Valley. While all four regions have recorded histories spanning the existence of Lake Lisan, the Tabaqat Fahl region has the thickest and most complete sedimentary sequences (spanning 170 m over two contiguous vertical sequences) representing perhaps 0.5 m.yr of occupation. The areas chosen, passing from north to south, are Tabaqat Fahl/Pella, Upper Wadi az-Zarqa', Wadi al-Hasa and al-Quwayra Depression/Wadi Hisma. These areas represent the Rift Valley, the Western Mountain and the Southern Mountain Desert provinces respectively.

Example: landscape and occupation at Tabaqat Fahl and Wadi al-Hammeh

While not directly situated on the Rift Valley floor, the Tabaqat Fahl region protrudes westwards into the Jordan Valley, and geomorphic and sedimentary processes reflect events occurring in the abutting Rift Valley environment. The Tabaqat Fahl/Wadi al-Hammeh area lies opposite the junction of the Valley of Jezreel and the Jordan Valley about 30 km south of the Sea of Galilee. The 10 km^2 of the Tabaqat Fahl region and its two bounding wadis, al-Hammeh and Jirm al-Moz, provides one of the longest recorded histories of semi-continuous occupation anywhere in the Middle East, going back perhaps 500,000 years (Macumber 1992a; Edwards and Macumber 1995). This area contains the Lower Palaeolithic Masharia sites, the Middle Palaeolithic to Epipalaeolithic Wadi al-Hammeh sites, and the Pella tell sites on Wadi Jirm al-Moz, where occupation was from Pottery Neolithic virtually up to the present (Figure 1.3). The intense occupation in the Tabaqat Fahl region is associated with the tectonically uplifted Tabaqah Block composed largely of tufaceous limestone, deposited within a major zone of regional groundwater discharge at the edge of the Rift Valley (Macumber 1992b). The presence of thick tufaceous deposits (Tabaqat Fahl Formation, 100 m; Knob Limestone, 35 m) is attributed to structural control, linked to Jezreel and Jordan Valley rifting, which through time has focused regional groundwater discharge into the area. The present active major springs feeding Wadi Jirm al-Moz and Wadi al-Hammeh represent a continuation of this process.

The long geomorphic, sedimentologic and archaeological record from the Tabaqat Fahl region is undoubtedly the clearest obtained from any continental open-air sequence spanning the Lower, Middle, Upper and Epipalaeolithic periods in Jordan. While a number of *in situ* Lower Palaeolithic sites have been recorded in the 120 m-thick Middle Pleistocene Tabaqat Fahl Formation, the work here is just beginning. The 60 m-thick Late Pleistocene Wadi al-Hammeh deposits are continuous, clearly visible sequences, showing ongoing sedimentation covering a number of discrete geologic settings with equally clear stratigraphic relationships (Figure 1.4). These observations are supported by an exhaustive suite of radiocarbon dates and

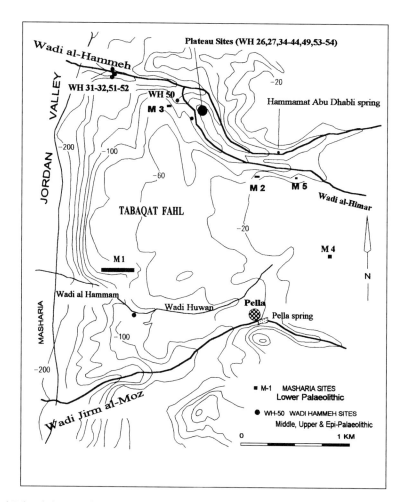

Figure 1.3. Topography and Palaeolithic site distribution at Tabaqat Fahl (from Macumber 1992b).

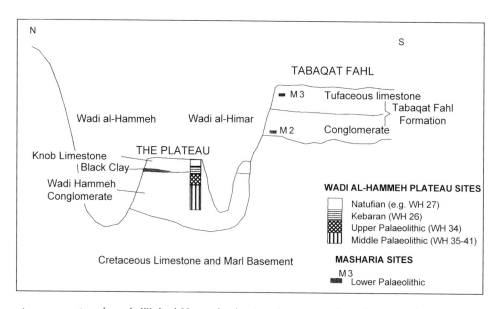

Figure 1.4. Schematic cross-section through Wadi al-Hammeh, showing the stratigraphic relationship between the Masharia and Wadi al-Hammeh Palaeolithic sites (from Macumber 1992b).

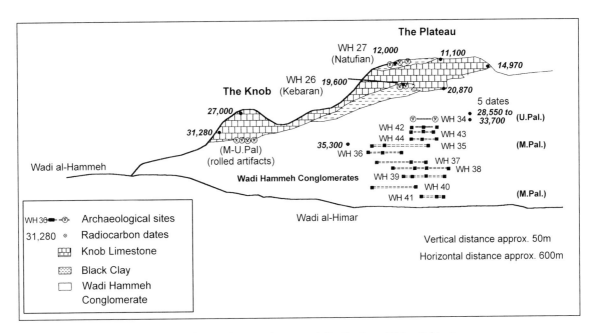

Figure 1.5. Stratigraphic section beneath the plateau showing the vertical distribution of Palaeolithic sites.

by a very large number of excavated *in situ* Middle, Upper and Epipalaeolithic sites occurring both vertically within the sequences and laterally down the wadi system. The radiocarbon dates from the sites are consistent with their accompanying stone tool industries.

Physical setting

Near the Rift Valley, the Middle Pleistocene Tabaqat Fahl Formation overlies the steeply dipping Plio-Pleistocene Ghor al-Katar Formation. The lowermost unit in the Tabaqat Fahl Formation sequence is a conglomerate, which is, in turn, overlain by thick sequences of tufaceous silt and limestone. The conglomeratic unit thickens on passing eastward away from the Rift Valley, at the expense of the tufaceous sediments. At the Rift Valley edge, the conglomerate merges with and is replaced by dense white uniform paludolacustrine limestone. The overall pattern of facies variation on approaching the Rift Valley represents a change from a fluvial setting (represented by the conglomerate) to a zone of massive spring deposition in a zone of groundwater outflow (tufaceous limestone) adjacent to a large lake (dense uniform limestone) occurring within the Rift Valley.

A similar sedimentary sequence, containing a comparable lateral facies change from conglomerate to tufaceous limestone, occurs in the Late Pleistocene Wadi al-Hammeh sequence, indicating essentially the same physical setting (Figure 1.5). The Wadi al-Hammeh sequence occupies an ancestral valley carved into the Tabaqat Fahl Formation. Infilling of the ancestral valley by fluvial and tufaceous sequences was in response to rising levels of Lake Lisan from c. 65,000 (80,000?) bp to 11,100 bp. Unbroken sedimentation continued throughout the upper Middle, Upper and Epipalaeolithic times. The resulting 60 m-thick valley-fill sediments are the terrestrial lateral time equivalents of the lacustrine Lisan Formation.

In the Wadi al-Hammeh sequence, a basal conglomerate containing red clay bands (Wadi al-Hammeh Conglomerate) is overlain by tufaceous limestone (Knob Limestone) with an intervening black humic clay (Black Clay) variously present. The conglomerate represents a fluvial wadi fill deposit; the tufaceous limestones are derived from spring outflow in a zone of regional groundwater discharge in the lower parts of the wadi, and the black clay is a fluviopaludal deposit. All three units contain bands of the fluviatile freshwater snail *Melanopsis praemorsa*, which is only found in fresh perennial streams like Wadi Jirm al-Moz. The presence of travertine horizons and *Melanopsis praemorsa* throughout the calcareous sequences indicates permanent spring-fed fresh water in the lower Wadi al-Hammeh during the period of deposition of the Knob Limestone. Archaeological sites occur scattered through all sedimentary units, and their common association with *Melanopsis* bands is seen as reflecting occupation alongside flowing streams.

The boundary between the conglomerate and tufaceous limestone units is diachronous as coarse clastic wadi-type sedimentation is gradually replaced by low-energy flood plain organic-rich paludal and spring-derived chemical deposits (Figure 1.5). The diachroneity is due to the gradual up-wadi migration of the springhead in response to rising base levels as Lake Lisan filled. The highest level reached by wadi aggradation occurred at 11,100 bp, when the uppermost sediments were deposited. It is equated with the time of highest Lake Lisan levels. Drying of Lake Lisan at about 11,000 bp was quickly followed by 30–60 m incision of the wadi-fill sequence, to form the present day topography, thereby exposing numerous *in situ* archaeological sites in the sediments of the valley walls.

A similar, albeit thinner record comes from Wadi Hisban at the northern end of the Dead Sea. Here the extensive silt terraces of Wadi Hisban contain Epipalaeolithic sites towards the top of the sequence. The final sedimentation on the terrace was dated to 11,100 bp and this is underlain by an *in situ* Natufian archaeological site dated to c. 12,000 bp.

Archaeological sites at Tabaqat Fahl

Three major cycles of sedimentation and occupation are recognized at Tabaqat Fahl: a Lower Palaeolithic (Acheulean) phase; a Middle Palaeolithic to Epipalaeolithic (Natufian) phase; and the final ongoing Holocene phase. The latter period saw the long-term occupation of the Pella tell, from Pottery Neolithic times up to the present (Edwards and Macumber 1995). Five Lower Palaeolithic sites (Masharia sites, Figure 1.3) containing bifaces occur on the Tabaqat Fahl Plateau. Of these, two sites (Masharia 1 and 3) are situated in the upper limestone member of the Tabaqat Fahl Formation towards the western end of the Tabaqat Fahl plateau (Macumber 1992b). The most extensive site (Masharia 1) occurs within tufaceous limestone some 30 m below the top of the Tabaqat and extends for perhaps 400 m around the cliff face. These artifacts are Upper Acheulean with an age range of between 200,000 and 250,000 bp and are archaeologically *in situ*. A further three sites occur in the lower conglomeratic member, in the north-central and eastern areas of the plateau (Masharia 2, 4 and 5). It is possible that certain of these sites were *in situ*, especially in the case of Masharia 4 where two hand axes were found together. No age can be given. However, sites in the conglomerate are generally stratigraphically lower in the sequence and older than those in the limestone.

A large number of *in situ* Middle Palaeolithic to Epipalaeolithic (Kebaran and Natufian) sites occur within well-exposed valley fill sediments along the Wadi al-Hammeh for more than 3 km upstream from the Rift Valley (Edwards and Macumber 1995). In a single 60 m-thick vertical sequence beneath the 'Plateau' are many *in situ* sites spanning the period from the upper Middle Palaeolithic to the Natufian period (Figure 1.5). At the top of the sequence is the very extensive 12,000-year-old WH 27 Natufian site. Vertically beneath the Natufian site are the Kebaran sites Wadi al-Hammeh 33 and 26 (19,500 bp), which have an extensive stone tool industry and faunal remains (Edwards 1984). Below WH 26 is the extensive *in situ* Upper Palaeolithic WH 34 site, which extends for 50 m along the cliff face. Radiocarbon dates from a dense band of *Melanopsis* shells immediately above the WH 34 site gave an average age of 30,000 bp. Farther down the sequence, a second dense shell band, situated above the *in situ* Middle Palaeolithic WH 35 site, gave an age of c. 35,300 bp. Spot excavations on three intervening clay bands between the WH 34 and 35 sites demonstrated a vertical continuity of occupation in the WH 42, 43 and 44 sites. Below WH 35, a further six Middle Palaeolithic sites (WH 36–41) have been excavated spanning the section to the base of the sequence.

Occupational environment at Tabaqat Fahl/Wadi al-Hammeh

During Quaternary times, upwelling carbonate-rich groundwater emerged as outseepage in the lower parts of the wadi system(s) and around the margins of Lake Lisan and an earlier lake that occupied the Jordan Valley at the time of deposition of the Tabaqat Fahl Formation. Thick tufaceous sedimentary sequences accumulated in an actively aggrading, fluvio-paludal embayment(s) across which the sluggish spring-fed water flowed gradually to build the Tabaqat Fahl Formation and, later, the 60 m-thick Wadi al-Hammeh sequences. The main occupation occurred on spring-fed river flats in the lower reaches of the wadis and around the margins of ancient lakes into which they flowed. The prime factor for long-term occupation has been the existence of a highly supportive microenvironment developed around large permanent springs, which provide a buffer against seasonal, short-term, and even long-term aridity.

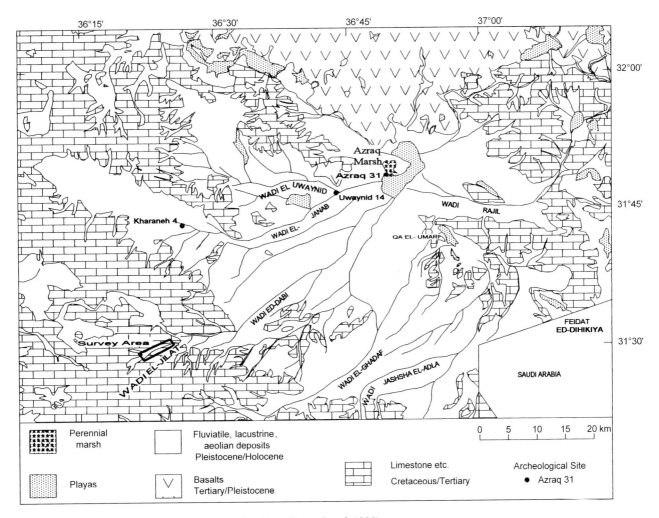

Figure 1.6. Geology of al-Azraq Basin, eastern Jordan (from Garrard et al. 1985).

The Central Plateau

In eastern Jordan there are four major physiographic features: as-Sirhan Depression, including al-Azraq Basin; al-Jafr Basin; the Black Desert; and the Limestone Desert (Figure 1.1). Of the four, as-Sirhan Depression, al-Azraq Basin and al-Jafr Basin are part of the Central Plateau Province, while the Black Desert and Limestone Deserts of the Jordan 'panhandle' form the Northern Basalt Plateau and Northeastern Limestone Plateau Province. The Central Plateau is the largest physiographic province and occupies much of central Jordan. It extends from the catchment of the broad al-Azraq/Wadi as-Sirhan depressions on the east and north, to the Western Mountain Province to the west. The southern boundary is the Ras an-Naqab escarpment and the rising ground of the Tubeiq Highlands. It includes al-Azraq Basin with its spring-fed al-Azraq marshes, and the depression of the al-Jafr Basin, a broad tectonically formed basin of internal drainage about 100 km wide and 150 km long, developed at the foot of the elongated leeward slope of the upwarped Western Highlands. Al-Azraq Basin (Figure 1.6) contains a long record of occupation from Lower Palaeolithic times.

The floor of the Central Plateau is cut across Upper Cretaceous to early Tertiary marine sequences of limestone, marl, phosphorites and chert. Shallow dipping to flat-lying, more resistant beds result in a landscape that in places consists of tableland scarps, mesas and cuestas surrounded by weathering debris. The elevation at times reaches 1000 m above sea level, falling to about 850 m in the south. The lowest part of the Central Plateau occurs in al-Jafr Basin. Chert fragments derived from the underlying marine sequences are scattered across the landscape often covering hundreds of square kilometres and forming the familiar *hammad* desert.

Drainage of the Central Plateau is eastwards towards al-Jafr Basin in the south, while in the northeast it is towards al-Azraq–as-Sirhan depression. In the west of the Central Plateau, headward erosion from the deeply incised al-Karak, al-Hasa and al-Mujib wadis has captured much of the drainage that previously flowed eastwards, and it now passes towards the Rift Valley. In wadis like al-Hasa there is a 1200 m drop from the plateau to the Rift Valley floor. In a similar setting, at the boundary between the Central Plateau and Western Mountain Province is the very large late Acheulean site at Fjaje near Ash-Shawbak Castle overlooking Wadi al-Bustan that drains towards Wadi 'Arabah. Advantage was taken by Acheulean hunters of the view from the plateau rim across the wadi system to intercept herds of animals migrating between the Rift Valley and grasslands of the Central Plateau (Rollefson 1985).

Al-Jafr and al-Azraq Basins, lying within deserts of limestone, chalk and flint, owe their origin to subsidence, commencing in Cretaceous times. In al-Jafr Basin a Pleistocene lake covered 1000–1800 km^2 at its maximum (Huckriede and Wiesemann 1968) resulting in the deposition of a 25 m-thick limestone and marl unit. During the Quaternary period, there were alternating wet and dry phases, leading to the present situation represented by a mud flat, Qa' al-Jafr. Middle and Upper Palaeolithic artifacts overlie the lacustrine limestone.

The central part of al-Azraq Basin contains perennial springs and pools, al-Azraq springs being an important source of water for cities of western Jordan. The al-Azraq region has the most complete archaeological record so far obtained in eastern Jordan, with sites ranging from the Acheulean to the present. The basin, which covers 13,000 km^2, is an irregularly shaped depression with an elevation ranging from 1800 m along its northern boundary with Jabal al-Druze to about 500 m at its centre in Qa' al-Azraq. Its eastern and western flanks rise to about 600–900 m. The southern portion of the basin is underlain by Cretaceous and Tertiary limestone, chalk and marl capped by a flint *hammad*, while the northern part is covered by Miocene to Pleistocene basalt and tuff. Al-Azraq Basin is an extension of, albeit tectonically separated from, the Wadi as-Sirhan Depression, which passes into Saudi Arabia.

Central to al-Azraq Depression is the largely fault-controlled Qa' al-Azraq. It is bounded on the northeast and southwest by al-Bayda Fault and as-Sirhan Fault respectively, and to the southeast and northwest by the Baqawiyya and Qaisiyeh Faults. To the north lies the Basalt Plateau with a 5–15 m scarp rising from the limestone plain. On the plateau the scoria cone Tall Hassan marks one of the many eruption points for basalt flows.

From late Tertiary times, lakes and marshes have been a feature of the al-Azraq region. The earliest recognized lacustrine sequence in al-Azraq Basin is the Late Miocene calcareous sandstone of the Qirma Formation (Ministry of Energy and Mineral Resources 1993), which outcrops alongside the Wadi Rajil in the eastern parts of the region. Al-Azraq Depression contains a thick Quaternary suite, which includes the Plio-Pleistocene al-Azraq Formation, thick Pleistocene alluvial sequences, and the Dashsha Silt—a widespread largely aeolian silt, observed during archaeological surveys and excavations in the area. Al-Azraq Formation reaches a maximum exposed thickness of 15 m, and consists of evaporites, claystone and marl. Gastropods and diatoms show it to be lacustrine. This is clearly seen at Feidhat ad-Dihikiya on the Saudi Arabian border at the margins of al-Azraq Basin. Here, a gorge-like depression, with cliffed 30–50 m-high wadi sides incised into Cretaceous–Lower Tertiary limestone and marl, has eroded back from Wadi as-Sirhan into al-Azraq Basin. The sediments consist of a 55 m-thick marly sequence of brackish to saline lacustrine deposits that contain gypsum and halite evaporites. The sequence includes a 5 m-thick layer of brackish water molluscs—*Cardium edule paladosa*.

In the mid-1980s, Garrard and I collected a number of limestone and shell samples from al-Jafr and al-Azraq Basins for Uranium–Thorium (U–Th) dating. These came from a range of lacustrine settings, which reflect wetter periods within al-Azraq and al-Jafr Basins and provide a few initial dates on eastern Jordan's Pleistocene lacustrine sequences. For instance, a single U–Th date on a lacustrine limestone sample from al-Jafr gave an age of 83,000 ± 8000 bp. In al-Azraq Basin, limestone samples were collected 100 m south of al-Azraq Playa at 'Ayn al-Bayda, from dense limestone beneath the Neolithic site at al-Azraq 31, and from a brackish lacustrine sequence containing a band of *Cardium edule* north of al-'Umayri. The samples, dated at the Australian National University in Canberra, gave the following ages:

In both al-Jafr and al-Azraq Basins, lacustrine influences have been active for a considerable time. The earliest age obtained is from dense limestone or travertine beneath al-Azraq 31, which was dated on U–Th

Locality	U–Th Age
al-Jafr limestone	83,000 ± 8,000 bp
al-Azraq-limestone below al-Azraq 31	220,000 ± 30,000 bp
al-Azraq–al-'Umayri	146,000 ± 15,000 bp
al-Azraq–'Ayn al-Bayda	44,000 ± 5,000 bp

Table 1.1. U–Th dates from the Central Plateau Region.

at 220,000 ± 30,000 bp. The dense travertine horizon passes northwestward and occurs along the canals near Crab Spring and at C-Spring (Besançon *et al.* 1989), where Lower Palaeolithic artifacts occur towards the base of a 3.5 m sequence. If, as it seems, it is younger than this sequence, it provides a minimal date of 220,000 ± 30,000 bp for the site.

Example: al-Azraq Basin

The central area of al-Azraq oasis, lying at an elevation of 506 m asl, is a 6–8 km² area of essentially permanent wetlands, marshes and pools, fed by permanent springs in two major areas of regional groundwater outflow. A number of large and small freshwater springs and seeps emerge from the base of the low scarp at the periphery of the basalt flow where it peters out along the northern and western edge of the basin. Near their outlets, the larger springs form shallow pools and marshes. Outflow from the springs prior to the commencement of large-scale extraction for town supply purposes was about 250,000 m³/d in the southern (al-Azraq Sheshan) area and about 90,000 m³/d from the northern (al-Azraq Druze) spring zone (Nelson 1973). The central area is surrounded by a larger seasonally wet sabkha or playa lake, the Qa' al-Azraq, covering an area of about 50–90 km². During wetter years, the Qa' al-Azraq can flood to a depth of over 2 m, with the floodwater sitting on shallow saline groundwater occurring normally at a depth of 1–2 m beneath the otherwise dry sabkha surface.

The present-day climate of al-Azraq Basin is essentially arid, with rainfall in the western areas ranging from between 100 and 200 mm/yr, while that in the east is bordering on hyper-arid with precipitation less than 50 mm. A number of normally dry wadis with associated terraces feed into the central al-Azraq region. They only flow after storms and the water either infiltrates or is lost by evaporation from small playas.

Silt dunes surround the marshy area and outcrop within it. Similar silts contain and also underlie a number of archaeological sites. Their probable source is deflation of the playa/lake floor or from the periphery of the permanently wet area. Deflation occurs wherever water tables are within capillary distance of the groundsurface, permitting salt efflorescence and break-up of the fine-grained surface sediments into sand and silt-sized aggregates prior to deflation as loess (Macumber 1970). This is likely to have been an ongoing process through time, given that saline water tables are within 1–2 m of the surface beneath much of al-Azraq sabkha. Indeed, silt is the major component of the shallow sedimentary sequences beneath the lake/marsh (Gilbertson *et al.* 1985; Garrard *et al.* 1988a).

Both the wadi systems and the central playa/lake/spring area have a long history of occupation. For instance, Copeland and Hours (1988) have shown that Wadi Rattama, Wadi Butm and the Qasr al-Kharrana wadi complex was rich in Lower Palaeolithic (Desert Acheulean) and, to a lesser degree, Middle Palaeolithic material. Other Acheulean occurrences are found within the swamp deposits of the main al-Azraq marsh areas at C-Spring and Lion Spring (Rollefson 1982).

As Copeland (1988) noted, al-Azraq Basin is in a climatically transitional zone with small changes in climate capable of having large impacts upon the physical environment. However, while this applies to the basin as a whole, it is less applicable to the spring-fed marsh region that receives regional groundwater flow, recharged on Jabal al-Druze. With a long flow path, the regional groundwater may be fossil to subfossil with origins reflecting wetter periods in the past, as may have been the case in the early Holocene. The groundwater emerging as springs around the perimeter of the basalt is largely buffered against adverse climatic change and provides a hydrologically stable setting in a region of otherwise environmental instability. In addition to the spring flow, locally recharged groundwater allows ephemeral base flow in certain of the lower wadis (Rattama, lower Butm and Enoqiyya) that permits them to maintain flow longer than would otherwise be the case.

Overall, the al-Azraq region provides a wide range of environmentally different settings, in which in some instances the full impact of deteriorating physical conditions through time would be felt, while elsewhere microenvironmental regimes provide well-watered asylum conditions largely immune to seasonal and longer term regional aridity.

Palaeoenvironment

The sediments of al-Azraq Basin have a long stratigraphic, biological and archaeological record, ranging from Acheulean times up to the present, that permits the construction of its environmental and climatic history from Middle to Late Pleistocene times. For instance, a detailed geomorphological study of al-Azraq Basin, together with palaeoclimatic interpretation, has been carried out by Besançon et al. (1989) who note the stratigraphic complexity of the region, which, together with the thinness and discontinuity of the deposits, leads to 'its interpretation being unfortunately problematic'. This is a common feature of many inland arid regions, especially where aeolian processes continue to be strongly active and the sedimentary sequences condensed. Despite this limitation, Besançon et al. (1989) divide the events into a number of 'morphoclimatic' cycles. Garrard et al. (1988b), with a more restricted archaeological perspective, recognize that five terraces are present along the wadis, perhaps representing wet and dry phases.

Given the clear complexity of the region, the intention here is to address only certain of the processes and events occurring during the latter part of the period in late Glacial and early Holocene times, covered by the more recent archaeological studies. This covers only the last 4 of the 17 episodes recorded by Besançon et al. (1989). A feature of this period is the presence of the very extensive in situ sites, such as al-Kharrana 4 and Jilat 6. While there is some uncertainty about why these sites were favoured (Garrard and Byrd 1992), it is difficult to see how they could have existed without permanent water, probably fed from active springs.

Despite a common practice of splitting the palaeoenvironmental record into a plethora of wets, dries, hots and colds, this has not been done by Garrard et al. (1988a). Instead, recognizing the difficulties of interpretation, they limit the division of Late Pleistocene palaeoclimate into two phases, 23,000–15,000 bp, when it was deemed wetter (reflected by zones of soil formation within silt sequences), and post 15,000 bp when it was drier (reflected by aeolian silts without palaeosols). Garrard et al. (1988a) note that there is much work still to be done in al-Azraq Basin before a detailed statement about the palaeoenvironment can be made. However, even for their simplified climatic interpretation, much rests on the origins of the silt.

From an environmental perspective, perhaps the most important feature of the Late Pleistocene and Holocene landscape are the ubiquitous silt deposits found throughout the region. Many excavated and dated archaeological sites in the wadis occur within aeolian silt that sometimes contains palaeosols. This is best seen in the Wadi al-Jilat study (Garrard et al. 1988b). For instance, at the Jilat 6 site, aeolian silts underlay a sequence containing two palaeosols with archaeological sites dated at 19,000 bp and > 14,500 bp respectively. The palaeosols are deemed to be short-term humid interruptions to an otherwise uniform phase of silt accumulation (Garrard et al. 1988b). Similarly, at Jilat 9, a palaeosol developed on aeolian silt beneath fluvial deposits of 'aggradation unit B' was interpreted as indicating a more humid event interrupting an otherwise arid setting. Two additional soils occur within the aggradational unit B sequence, the upper being dated at $21,250 \pm 400$ bp. Finally, at the Jilat 10 site (dated at 14,790–12,700 bp and also situated in aeolian silts), silty marsh deposits recorded from the base of the aeolian sequence are tentatively correlated to the humid event corresponding to the palaeosol between the aeolian silt and fluvial sediments in Jilat 9 (above).

At sites Jilat 8, 10 and 6, Late Epipalaeolithic assemblages younger than 14,500 bp are contained within unmodified aeolian silts. This was also the case for Late Epipalaeolithic, Natufian and Neolithic sites at the Central Playa where a series of copious springs occur around the sides of the playa. Here, the mid–late Natufian al-Azraq 18 and PPNB al-Azraq 31 sites are in aeolian silts, and silts are still being deposited.

A similar record to that of Wadi al-Jilat comes from sites at Wadi Uwaynid. At Wadi Uwaynid 18, silts occur throughout the sequence with humic silts at the base overlain by aeolian silt containing a hearth dated at 23,200 bp. The silts contain a number of poor soil horizons prior to a hiatus at c. 19,800–19,500 bp, during which there was strong pedogenesis. In a similar sedimentary setting is the nearby Uwaynid 14 site with dates of 18,900–18,400 bp. The industry is similar to that from the lower parts of Jilat 6, also contained within a palaeosol. More silts accumulated, before a final fluvial sequence was deposited containing a further palaeosol. The sequence was then incised and remains as a terrace.

Discussion

The correlation between the palaeosols in different silt sequences remains tentative. The sequences overall suggest that the period from well before 21,250 bp up to the present was one of an otherwise uniform silty depositional environment containing a number of occupation sites, which in the earlier part of the record sometimes related to periods of pedogenesis. It is clear from this data that aeolian activity was not restricted to the latter part of the Epipalaeolithic and the Holocene period but was semi-continuous back into Upper Palaeolithic times, beyond the limits of the dated archaeological sequences.

While aeolian activity is so prevalent, despite the long history of lacustrine activity in al-Azraq Basin, there is no unequivocal record of a late Pleistocene phase of expanded lake levels commensurate with the significantly wetter conditions reported from western Jordan (Tabaqat Fahl, al-Hasa, the Hisma) at the time. Earlier, Garrard et al. (1985b) considered that a caliche layer in the silt occurring at a level above that of the Uwaynid 18 and 14 horizons and found within the 530 m contour, represented a lacustrine deposit following lake expansion. This interpretation was later abandoned and the caliche deemed to be non-lacustrine and due to capillary emplacement by high water tables, or of aeolian origin (Garrard et al. 1988a). These alternative origins are not mutually exclusive, but instead may simply reflect different source areas undergoing essentially the same processes. If formed at the site (as a $Ca\ SO_4/Ca\ CO_3$ capillarity-induced evaporite), it indicates higher water tables and, therefore, wetter conditions than at present.

If the caliche is aeolian, it is probably deflated from a nearby source having high water tables where capillarity permits the break-up of surface crusts by salts into deflatable components (cf. the formation of copi deposits and clay/gypsum lunettes in Australia, Macumber 1970, 1991). The most likely ongoing source of silt is deflation from al-Azraq playa and others such as Qa' al-'Umayri. If this is the case, the presence of silt throughout much of the recorded history from Upper Palaeolithic times indicates that, while water tables at al-Azraq may have undergone fluctuations, they were always close to the surface. Silt deflation occurs under a range of climatic conditions from semi-arid to hyper-arid.

During the periods of aeolian activity, whatever the climatic situation, the local hydrological setting around al-Azraq oasis cannot have been significantly different than at present. It follows that the one essential reason for the lack of strong evidence for more clearly defined climatic fluctuations in a region, shown by Copeland (1988) as being climatically in a state of disequilibrium, is the hydrological buffering provided by al-Azraq oasis throughout the Late Pleistocene. This dampens the climatic extremes and locally limits the impacts of physiographic and climatic processes, which continue to affect the surrounding landscape. Given that the silts reflect not increased regional aridity but deflation from nearby areas underlain by high water tables, the association of archaeological sites with the silt sequences at al-Azraq underpins the notion that in arid settings humans live in (micro) climatically favourable sites not necessarily representative of the region as a whole.

Yet the period from Late Glacial times to the present is a time when major climatic change was under way. It is clearly seen in changing occupation patterns in deserts across the Middle East and North Africa. Of special significance, across much of this broad region, has been the period of increased temperatures and monsoonal activity from 10,000 to 6000 bp. This followed the cold of the Glacial Maximum and the aridity of the Younger Drias from 11,000 to 10,000 bp, a period represented by the drying of the megalake Lake Lisan. The very dry/cold phase of the Younger Drias would have obliged populations to seek refuge in such areas as al-Azraq oasis and Tabaqat Fahl region where the groundwater-fed systems provide buffering from the climatic extremes. After the aridity of the Younger Drias, there followed an amelioration of climate in early to mid Holocene times in North Africa and parts of Arabia extending from the Nile River to Oman, during which many presently dry lake systems were full.

While there are no major environmental fluctuations observed in the Holocene al-Azraq record, where groundwater buffering limits or even overrides the effects of climatic change, a clearer indication of regional environmental/climatic change as reflected in its impact on occupation would be expected from areas peripheral to the basin. Such a situation exists in the nearby northeastern basalt and limestone steppe deserts of the Badia.

Northern Basalt Plateau and Northeastern Limestone Plateau

The area referred to by Bender (1968, 1974, 1975) as the Northeastern Deserts of Jordan, or more commonly, the Badia, with its basalt (*harra*) and limestone (*hammad*) deserts, lies to the northeast of al-Azraq Basin–Wadi as-Sirhan Depression. The deserts comprise two distinct geomorphological units:

the (Northern) Basalt Plateau (often referred to as the Black Desert); and the (Northeastern) Limestone Plateau farther to the east. Both units are treated here together.

The Basalt Plateau/Black Desert is a southwards extension of the Druze Basalts, and covers about 11,000 km^2 in Jordan or about 25% of the total basalt plateau (Ash-Shammah Plateau) most of which lies in Syria. It is an extensive and rather inaccessible undulating plateau, a lava plain covered in places with a surface of basalt cobbles and boulders, forming the upper surface of lava flows that issue from basaltic and scoriaceous volcanic cones that now dot the landscape. The cones form linear patterns suggestive of fissure eruptions. Volcanic activity occurred from late Eocene to Pleistocene times and the most recent eruptions may have continued into historical times. Six different phases of volcanic activity have been recognized and the overall thickness in places reaches 270 m. The fissure eruptions are thought to correlate with the sixth stage of activity and are believed to be Pleistocene in age. The flows continue into Saudi Arabia forming an eastern limit to the Wadi as-Sirhan–al-Azraq Basin.

Eastward of the Black Desert is the Northeastern Limestone Plateau, consisting of a flat stony surface formed across early Tertiary marine siliceous and calcareous sequences. Resistant limestone beds form low scarps. Features on the Limestone Plateau are subdued, but include a number of northwest–southeast-trending depressions filled with Quaternary sediments, their shape being dictated by structural patterns in the limestone. In places, the depressions appear to suggest remnants of palaeodrainage patterns. The direction of the depressions is dictated by the principal direction of structural deformation in eastern Jordan.

The present-day climate of the eastern Badia is arid with 50–150 mm rainfall. A feature of such arid regions is great rainfall variability in which there may be months without rainfall, yet when it rains it may be intense with a single storm producing as much as the mean annual precipitation. The presence of mudflats shows that there is at times sufficient local runoff and perhaps groundwater outflow to permit occasional visits by pastoralists and in places limited occupation. The only reliable surface water is the distant al-Azraq oasis but rainpools and limited groundwater outflow provide a number of other temporary potential water sources. Such is the case with Ibn al-Ghazzi situated on the periphery of al-Azraq Basin some 75 km from al-Azraq (Betts and Helms 1989).

Occupation of the Black Desert

Despite the present inhospitable environment, occupation in the Black Desert has gone on, perhaps sporadically, for a considerable time. Little is known about the history prior to the twelfth millennium bp, other than for small scatters of Middle Palaeolithic and Epipalaeolithic artifacts. Betts (1992, 1993) notes that sites representing the first part of the early Neolithic (PPNA) are rare. By the second half of the ninth millennium bp, occupation and knapping sites are common, and hunters extensively used the Badia and especially the *harra*. Betts (1993) considers this expansion in terms of similar expansions elsewhere in Syria and the Negev. This, however, is not an explanation but a comparison, and the expansion probably reflects a more generally hospitable environment than previously existed (or followed).

Example: Neolithic settlements in the Black Desert

Helms (1982) similarly notes that, while humans probably lived in the Black Desert as early as 12,000 bp, the bulk of the physical evidence comes later. The abundance of manmade structures in the desert, principally stone hut circles and desert 'kites', implies a once intensive occupation with suggestions of relatively dense semi-permanent settlements and a highly organized system of hunting technology. Stone tool assemblages associated with the structures represent the PPNB and Late Neolithic periods dating from the early sixth (or perhaps seventh) to late fifth millennium BC. A large 'palaeo-Bedouin' population is hypothesized as existing up to the Late Neolithic period followed by a decline in which, until the late fourth millennium BC, there was no artifactual, datable evidence of occupation. The explanation postulated by Helms is that migration occurred to more verdant areas to the east, north and west, and apart from a relatively short-lived period represented by the Jawa occupation (Helms 1977), there is no archaeological evidence for significant occupation of the Black Desert up to the third millennium BC. The reason for the migration is unclear.

The number of Neolithic sites dated between 8350 and 6900 bp (Betts 1993), and their subsequent absence, raises the question of the extent of climatic influence on the landscape and occupation in eastern Jordan during the early Holocene. The early–mid-Holocene phase of occupation in the Badia desert coincides with the time when higher rainfalls resulting from increased monsoonal activities were strongly affecting nearby areas of Arabia and North Africa—areas that today have arid or even hyper-arid climates (McClure 1976, 1978). Strongest flooding occurred along the Nile River from 9000 to 6000 bp, a period seen as being one of climatic optimum throughout the Eastern Mediterranean region by Rossignol-Strick (1993). Cores from the Arabian Sea show that strong monsoonal conditions were present, and in what is now arid eastern Oman, the interdune corridors in the Wahiba Sands had seasonal ponds; nearer the coast, silts with thick zones of rhyzoconcretions were deposited (Gardner 1988). There was occupation of the Wahiba Sands during this early–mid-Holocene moist period, but this ceased once aridity returned, and has only recommenced in relatively recent times (Eden 1988).

In the Levant, increased monsoonal activity resulting in higher rainfall in the more arid southern regions has previously been suggested by a number of authors (Gat and Carmi 1987; Goldberg and Rosen 1987). However, while wetter conditions are recorded in the southern Levant during this period, they contradict drier conditions recorded in the more northern areas of the Rift Valley (Weinstein-Evron 1989) and perhaps southwestern areas at the same time. The explanation provided is that palaeoclimatic difference across the Levant may reflect different rain sources, with the southern and eastern areas being influenced by the monsoonal activity, while the more northerly areas receive rains during glacial periods from an Atlantic source.

Whatever the case, there is a clear indication that conditions were significantly wetter for much of the Middle East and North Africa than today in early–mid-Holocene times. In these latter areas there is a widespread appearance of Neolithic sites across the landscape attesting to a significant expansion of occupation lasting until mid-Holocene times. While the timing of the early–mid-Holocene moist period seems to be slightly different for different areas, it is likely that this same impact was also experienced in eastern Jordan and recorded in the events in the eastern deserts up to the onset of a more arid phase that has continued up to the present.

Finally, Rollefson (1992) has noted that major increases in population in Jordan and Palestine between 9600 and 8500 bp can be attributed to the growth of agriculture and goatherding. This is undoubtedly true, but this argument does not explain the expansion of populations into the more arid areas, nor their retreat following the demise of the early–mid-Holocene moist phase, which caused a reversion to a more poorly watered setting. Clearly both climatic and cultural factors influenced occupation in different ways.

Southern Mountain Desert Province

At its southern edge, the Central Plateau terminates against the fault-controlled Ras an-Naqab scarp that marks the boundary between the Central Plateau and the Southern Mountain Desert provinces (Figures 1.1, 1.7). The geomorphic boundary roughly follows the lithologic boundary between Mesozoic–early Tertiary limestone/marl and Palaeozoic (Ordovician–Silurian) sandstone and shale, which give the province much of its physical character. The sandstone–shale sequences overlie an acid igneous basement of granites, granodiorites, quartz porphyry and diorites, which outcrop in the southwest to form mountainous regions at the southern limit of the adjacent Western Highland Province. The highest points of the escarpment reach 1700 m above sea level. To the southeast are the Tubeiq Highlands, separated from the Southern Mountain Desert Province by outcropping Cretaceous sediments, forming an extension of the Central Plateau. South of Ras an-Naqab the topography falls abruptly almost 600 m and slopes towards the Hisma–al-Quwayra Depression developed at the western end of Wadi Hisma. Three steps or levels (pediments, Figure 1.8) have been recognized on the Ras an-Naqab escarpment, at 1600 m, 1400 m and about 1200 m, each representing erosional cycles in the development of the scarp (Hassan 1995).

The sandstone landscapes of the Southern Mountain Desert Province are deeply incised especially along the scarps. As in the case of the limestone Central Plateau, more resistant beds strongly influence the landscape, with the development of inselbergs, mesas and buttes (table mountains), and cuestas. In the Tubeiq area, a tabletop landscape of inselbergs interspersed with large flat depressions predominates, with extensive mudflats developed over thick alluvial deposits, representing deposition during fluvial phases during the Quaternary period.

The strong influence of tectonics on drainage systems bordering the Rift Valley (also evident in the case of Wadi al-Hasa) is visible in the southern Wadi 'Arabah. Faults paralleling the northwest–southeast-trending al-Hasa Fault mark the edge of the Ras an-Naqab Scarp. Faulting determines the course of Wadi Hisma, which follows

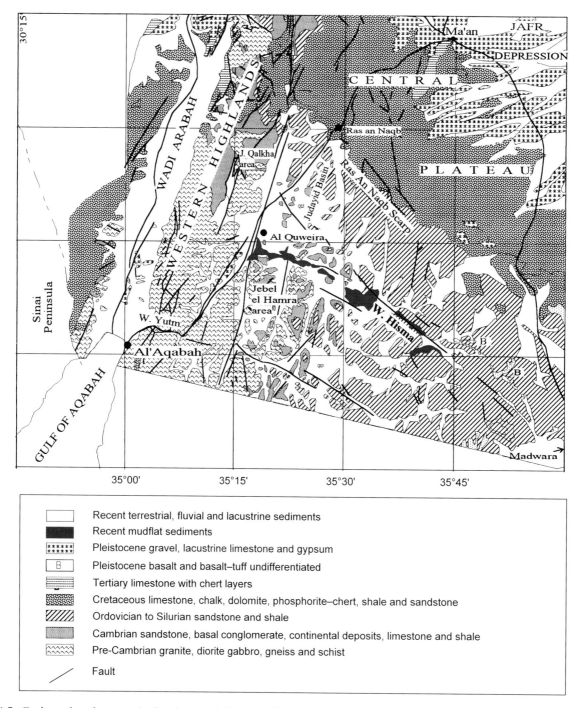

Figure 1.7. Geology of southwestern Jordan showing al-Quwayra–Wadi Hisma region (Geology after Bender 1975).

a fracture zone created by tension fissures, along which basalt dykes have intruded (marked as 'B' in Figure 1.7).

Faulting has formed the Wadi Hisma–al-Quwayra Depression into which ephemeral wadis, such as Judayid, Qalkha and Thallaga, drain from the north, and Wadis Ramman and Ramm from the south. Lying less than 20 km from the Rift Valley, the western edge of al-Quwayra Depression is bounded by a north–south-trending fault passing a little to the west of al-Quwayra where Precambrian basement outcrops. This fault has in the past dammed Wadi Yutm, which drains westward towards the Gulf of al-'Aqaba. Eastwards of the fault there has been extensive alluviation and playa development with a line of mudflats occurring along Wadi Hisma between al-Quwayra and

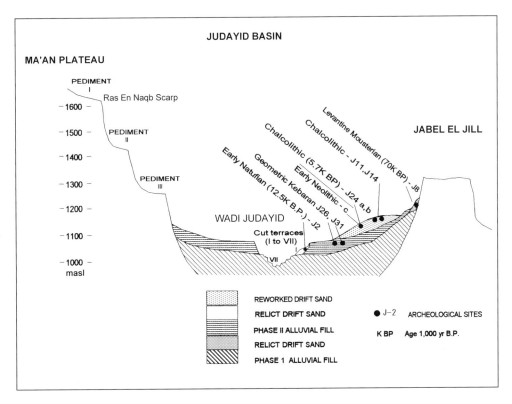

Figure 1.8. Geological section through the Judayid Basin, southwestern Jordan (from Hassan 1995).

al-Mudawwara. In the uppermost parts of the region near al-Mudawwara, lacustrine sequences of the Halat Ammar Formation contain a brackish to saline molluscan fauna. It is uncertain whether this lacustrine sequence relates to similar deposits from al-'Umayri near al-Azraq dated at 146,000 bp or whether it represents a more recent event, such as that recorded across North Africa, eastern Arabia and perhaps in the Badia of northeastern Jordan. Terrace evidence at lower Hisma, however, suggests that incision predominated during the Holocene, and it is more likely to equate with earlier wet phases.

Poking up through alluvium are Ordovician sandstone buttes and inselbergs, with Jabal Ram and Jabal Sahm occurring to the south. The Hisma region, from the edge of the Ras an-Naqab scarp southwards to the lowland area of inselbergs and sand plains/mud flats, has been extensively studied (Henry 1979, 1995). Over 100 archaeological sites, ranging in age from Lower Palaeolithic to Chalcolithic, were observed scattered across the deeply incised sandstone landscape of wadis and inselbergs. Detailed investigations carried out in the Judayid Basin and at Jabal Qahlka, Jabal Mueisi and Jabal Hamra recorded sites from a number of settings, namely alluvial fans, river terraces and rock shelters.

Example: the Hisma region of southwestern Jordan

A feature of the Hisma area is the number of sites spanning a range of different wadi and jabal settings that provide detailed accounts of environmental changes during the Middle–Late Pleistocene period. Many of the sites are, however, in stratigraphic isolation, for instance, in rock shelters with very few radiocarbon dates. While providing data on localized climatic conditions via pollen, fauna, etc., they are less useful in the reconstruction of a more general environmental framework, as would be the case if they occurred in more easily related landscapes, such as terraces, lake shorelines, etc. Nevertheless a detailed account of landscape evolution has been established (Hassan 1995), best seen in the Judayid Basin immediately south of Ras an-Naqab.

Two phases of fan-alluviation are present in the Judayid Basin (Figure 1.8). The main phase (Phase I) was aggraded prior to the Middle Palaeolithic period. The earliest artifacts obtained from the region are an assemblage containing Late Acheulean bifaces (Henry 1995) coming from near the top of a 30 m-thick section of early alluvial fill (Qalkha Formation of Hassan) in a deeply incised wadi draining Jabal Qalkha. The Phase I alluviation was followed by a period of downcutting and dune development (Hassan 1995). Remnants of dune sand overlying Phase I alluviation contain the c. 69,000-year-old Tor Sabiha Mousterian site (Henry 1982, 1995), the date being one of two obtained by amino acid racemization on ostrich shell (Henry and Miller 1992). While conditions were generally cold and arid, pollen and faunal evidence from Layer C at Tor Sabiha, and a similarly dated Mousterian deposit at Tor Faraj, suggest that conditions may have been moister than today.

The next significant period of occupation is represented by six Upper Palaeolithic sites, found at shallow depth in rock shelters in the vicinity of Jabal Qalkha, at elevations of 940–980 m above sea level. They occur in pinkish silt similar to that occurring at the nearby Mousterian site of Tor Faraj. However, their location makes it difficult to place the sites with any certainty into any local geostratigraphic framework or regional stratigraphic context (Coinman and Henry 1995).

A second phase of fan-alluviation (Phase II) commenced under warm and moister conditions. The age of commencement is unclear. Its termination, however, was put at c. 14,000–13,000 bp based on the presence of Geometric Kebaran (Middle Hamran) sites J26 and J31 eroding from alluvial fans comprising the upper units of the alluvium. There are no absolute dates, but instead the age was determined by comparison of the tool assemblages with dated Geometric Kebaran sequences elsewhere in the Levant. The period covering Phase II alluviation incorporates the early Epipalaeolithic 'Qalkhan' period represented by sites older than 14,000–15,000 bp occurring along Wadi Qalkha tributaries in the vicinity of the Jabal Qalkha (Henry 1995). Here again, dates are scarce and a suggested age for the Qalkhan of 20,000–15,500 bp is based on similarities of the material with stratified assemblages from the Uwaynid 14 and 18 sites at al-Azraq radiocarbon dated at 19,800–15,500 bp (Garrard et al. 1988a).

The second period of alluviation was in turn followed by further erosion, during which phases of wadi incision produced minor terracing along the Wadi Judayid. As downcutting of the wadi proceeded, dune drift sand accumulated on the Phase II fan-alluvium sequence (Figure 1.8). The uppermost terrace (Terrace I) is the largest, and contains an early Natufian site (J2) radiocarbon dated to c. 12,500 bp. This indicates that the initial stage of wadi incision occurred prior to 12,500 bp. The Natufian date from the terrace is similar to the two dates (c. 12,680 and 12,320 bp) from upper Layer C in the Tor Hamar rock shelter (site J431) covering the period of introduction of Helwan lunates. In the shelter, Layers B, C and D contain Epipalaeolithic (Mushabian) artifacts, while rubble Layer E contained a Qalkhan industry.

Elsewhere, dune sequences occur overlying Phase II alluvium. These contain early Neolithic and Chalcolithic sites, showing that aeolian processes were active during early to mid-Holocene times.

The 12,500 bp radiocarbon age for the J2 Natufian site, and the 12,680 and 12,320 bp dates for the Tor Hamar site, are the next oldest absolute dates in the Judayid sequence after that of the amino acid racemization dates from the Mousterian sites at c. 70,000 bp (Figure 1.8). The lack of absolute dates over this intervening period emphasizes the uncertainties inherent in relating local chronologies of rock shelter and scattered open-air stratigraphies to any more general regional geomorphic/stratigraphic framework. This is especially the case when fine subdivisions are adopted and then dated by reference to broad regional understandings or distant sites, many of which have their own uncertainties.

Hisma summary

The Hisma region has a long history of landscape evolution and human occupation. For the former, a broad relative sequence of erosional and depositional events can be constructed and tentatively compared with similar major events elsewhere in Jordan. However, at a more detailed level, correlations within the local sequences and with other regional systems are made difficult by the discontinuous nature of the sequences and lack of datable material, even though there is a surfeit of archaeological sites. The most complete stratigraphic sections come from isolated rock shelters, such as site J431 at Tor Hamar, where Natufian/Geometric Kebaran sequences overlie Kebaran sequences, which in turn overlie Upper Palaeolithic horizons. Yet even here only two dates exist (12,680 and 12,320 bp), from upper Layer C. As a consequence, the localized stratigraphies are largely artifact-based and their absolute ages obtained by comparison with similar sequences elsewhere. Because of the richness of artifactual material, the sequences are finely divided, and where fine subdivisions are erected there is always the problem of typological diachroneity. Perhaps this explains an apparent rapid climatic shift from the 'marked humid phase' from 13,000–12,500 bp from Qa' Salab to 'very cold and dry conditions' at J2 Wadi Judayid site from 12,780 to 12,090 bp (Emery-Barbier 1995). Whatever the case, it is clear that detailed climatic syntheses based on diversely erected chronologies from widely different physical settings (in this case, a rock shelter and a river terrace) must always be regarded as tentative.

Western Highlands/Western Mountain Province

The Western Highlands/Western Mountain Province border the eastern side of the Jordan Rift Valley and extend for about 370 km from Lake Tiberias to the Gulf of al-'Aqaba (Figure 1.1). In the north, the Western Mountain Province is largely composed of Cretaceous and Early Tertiary limestone, sandstone and shale sequences deposited in the Tethys Ocean, which in Mesozoic and Cenozoic times covered large areas of northeast and eastern Jordan, extending westwards across the Arabian Peninsula. In the southwest, however, the Western Mountain Province consists essentially of pre-Cambrian crystalline rocks overlain in part by red Palaeozoic sandstone (Figure 1.7). The crystalline basement forms part of the Nubo-Arabian shield that was an important source of sediment supply for the Cambrian and Ordovician sandstone landscapes of southern Jordan, and is well known from Wadi Ramm in the Hisma Region. Extensive work in this latter area at Jabal Qalkha, Judayid Basin, Jabal Mueisi and Jabal Hamra (Henry 1995) shows occupation ranged from the Lower Palaeolithic up to the Chalcolithic period.

Uplift of the Western Mountains has different expressions in the north and south of the country. In the north, tilting and arching dominate, producing such features as the broad flat upwarp of the 'Ajlun region. To the south, block faulting is more prominent (Bender 1975). Here the courses of major river valleys, such as Wadi al-Karak, Wadi al-Hasa (downthrown to the south) and Wadi Hisma between al-Quwayra and al-Mudawwara are strongly influenced by northwest–southeast-trending lateral faults, a trend that is also parallel to that of the Red Sea Rift.

The Western Highlands reflect many of the attributes of mountainous landscapes bordering rift valleys, with long gentle slopes away from the rift towards the depressions of the Central Plateau, and a steep sharp fall into the Rift Valley graben. The highest areas in Jordan are in the southern mountains to the west of Ma'an where the elevation reaches over 1800 m. Drainage is to the west to the Jordan Valley and Wadi 'Arabah, and to the southwest to the Gulf of al-'Aqaba. Farther east, drainage is inland, towards al-Jafr and al-Azraq Basins and Wadi as-Sirhan.

As a consequence of rifting and lowered base levels, the Western Highlands are deeply incised by wadis, such as the Yarmuk River, River az-Zarqa', Wadi al-Karak, Wadi al-Hasa and Wadi al-Mujib. Many of the deeply incised wadis intersect the saturated zones of Upper Cretaceous limestone aquifers, such as those of the 'Ajlun and Belqa group, and receive water as base flow and from springs, sufficient to permit perennial flow and provide water for many small settlements. Parker (1970) records a high permeability zone that occurs along the course of Wadi al-Hasa, arising from secondary porosity caused by intense jointing. He estimates that the base flow and spring flow to Wadi al-Hasa as being about 25 m. m^3/year.

The wadis are variously terraced, reflecting periods of incision, backfilling and further incision in response to changes in base level brought about by tectonism and by lake level changes in the adjacent Rift Valley to which they are graded. Upstream of their junction with the Rift Valley, thick sequences of Late Pleistocene Lisan Formation or its time-equivalent alluvium commonly infill canyons previously cut by ancestral wadis, such as al-Hasa, al-Mujib, al Hammeh and Jirm al-Moz. In some cases, tectonics have resulted in localized blocking of major wadis, resulting in the development of temporary lake systems until further downcutting has re-established the previous fluvial landscape, leaving a suite of terraces to record the various stages in the development of the lake/wadi. Such was the case in Wadi Yutm/Hisma (Osborn 1985; Hassan 1995), Wadi al-Hasa (Clark et al. 1987; Schuldenrein and Clark 1994) and at Gharandal (Bender 1974).

Example 1: Upper az-Zarqa'/Khirbat as-Samra region of west-central Jordan

One important geoarchaeological study was that carried out in the upper az-Zarqa' catchment, north of 'Amman (Besançon et al. 1984). The area includes Wadi Dhulail, a tributary of upper az-Zarqa', the zone of its confluence with Wadi az-Zarqa', and the area 5 km downstream from the junction (Figure 1.9).

The az-Zarqa' sequence is one of the better terrace sequences in Jordan where a stratigraphical framework has been clearly related to middle Pleistocene archaeological content. Besançon et al. (1984) compare the terraces with terrace sequences in Syria. This assumes, however, a more generalized and uniform evolution of the landscape than has yet been firmly established, especially given the role of tectonics as well as climate in

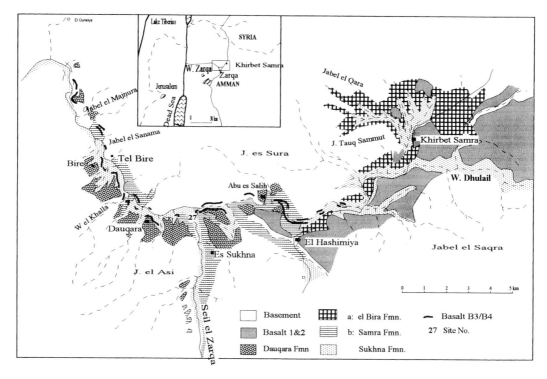

Figure 1.9. Terraces of the Upper Wadi az-Zarqa' region (from Besançon *et al.* 1984).

erosion/deposition cycles. The az-Zarqa' system tectonics have been varyingly active, and an early downcutting phase is linked to deepening of the Ghor. Similarly, tectonics have been suggested as being responsible for significant differences in the terrace system above and below al-Hashimiya, clearly shown in Figure 1.9.

The upper az-Zarqa' region was strongly influenced by basalt flows, with an early phase, which now caps the surrounding hills, dated at c. 7–5 m.yr bp. Following deep incision by the wadis, a second phase of basalt flows dated at 2.92–3.35 m.yr bp flowed along the wadis. Four (or in some instances five) alluvial or colluvial terraces were recognized, the oldest overlying the younger basalts (Figure 1.9). These terraces were termed:

Qf3   Dauqara Formation
Qf2   Bire Formation
Qf2-1 Bire-Samra
Qf1   Khirbat as-Samra Formation—last Glacial/Pluvial
Qfo   Sukhna Formation

The high terrace Qf3 (Dauqara Formation) backfills and overtops a trench carved into the basalt. It now occurs at 70–80 m above az-Zarqa' valley floor. It is composed largely of cemented conglomerates and contains traces of likely Middle Acheulean occupation. The next oldest terrace, the Bire Formation, containing large amounts of weathered basalt cobbles and pebbles, overlies the Dauqara Formation. It contains rolled Late Acheulean assemblages of bifaces, cores and flakes, with a suggested age of about 200,000 years bp (Besançon *et al.* 1984; Copeland and Hours 1988). The third terrace (Qf1), the Khirbat as-Samra Formation (together with an intermediate Bire-Samra unit, which may be a different facies containing reworked deposits), consists more of colluvium than alluvium but includes aeolian material. It is a low terrace consisting of gravels with a silty matrix and contains Middle Palaeolithic Levallois flakes and cores. It has a suggested age of 'Last Glacial/Pluvial, the early Wurm'. Following a further phase of incision of the earlier terraces, the Final Wurm to Holocene Qfo terrace (Sukhna Formation) was deposited. It contains Epipalaeolithic to Bronze Age material. For instance, at site 27 (Figure 1.9) there is multi-period occupation with a Kebaran site overlain by a possible Neolithic site. Copeland and Hours (1988) consider that the terrace is Holocene and that the Kebaran material may be reworked.

The two earliest documented phases of incision along Wadi az-Zarqa' occur prior to and following the 3.35 m.yr bp basalt flow, the latter preceding the deposition of the Qf3 (Middle Acheulean) sequence. These events are much earlier and cannot be readily correlated with most other sites under discussion. However, they may have equivalents in the Ras an-Naqab escarpment facing the Wadi Hisma catchment where three steps (pediments) at 1600 m, 1400 m and c. 1200 m respectively represent erosional cycles in the development of the scarp (Hassan 1995). Subsequently, a number of phases of downcutting and backfilling are represented by the terrace system, reflecting major base level changes affecting the Wadi az-Zarqa' system. From Middle Pleistocene times, the evolution of the terraces has been interpreted in terms of climatic change, with aggradation linked to colder or drier phases (Besançon et al. 1984).

Artifact assemblages were associated with the different terraces and four age groups for the artifacts were determined based on the geomorphologic data, but patina was also an important consideration:

a. Those geologically *in situ* within the Dauqara Formation (considered Middle Pleistocene).
b. Those of the Bire Formation: Qf2 (equated with the Late Acheulean of the Levant and considered to represent the end of the Middle Pleistocene c. 200,000 bp and the penultimate Glacial/Pluvial).
c. Those in the Bire-Samra and Khirbat as-Samra Formations. The Qf1 terrace contains Levallois flakes; it is seen as representing Last Glacial (early Wurm).
d. The most recent terraces, such as the Sukhna Formation contain Epipalaeolithic to Bronze Age material and even pottery. There may be several phases present.

Example 2: the Wadi al-Hasa region of southwestern Jordan

In the southwestern area of Jordan, there is a steep rise from the Dead Sea (elevation c. minus 400 m) to the Western Highlands and then on to the Central Plateau (at 1200 m). The Western Highlands are deeply incised by a number of major wadis draining westwards to the Jordan Valley. Drainage on the Central Plateau in most instances slopes gently to the east, with an important exception being Wadi al-Hasa. Occupation in westwards-flowing valleys has occurred since Lower Palaeolithic times and the well-documented Wadi al-Hasa sequence demonstrates the close relationship between landscape evolution and archaeology in these areas.

Wadi al-Hasa drains to the Jordan Rift Valley near the southern end of the Dead Sea. It is the only perennial stream between the Dead Sea and the Gulf of al-'Aqaba to rise in the Central Plateau and pass westwards across the Western Highlands to enter the Rift Valley. The catchment area is 1740 km^2 and the rainfall lies between 100 and 200 mm/year. In its upper catchment region, Wadi al-Hasa receives perennial base flow from the Upper Cretaceous limestone aquifer. Like Wadi al-Hammeh, the course of Wadi al-Hasa has been varyingly influenced by base level changes caused by tectonism associated with Rift Valley development, and by cyclic lacustrine and fluvial activity. The Wadi al-Hasa lies to the southern (downthrown) side of the east–west-trending al-Hasa Fault, which has diverted the course of the wadi from a former northwards flow, westwards into the Rift Valley (Clark et al. 1987).

As a consequence of the ongoing subsidence of the Dead Sea graben, the lower tract of Wadi al-Hasa is very deeply incised (as much as 1200 m) into Cretaceous and late Pre-Cambrian sandstone and limestone. Upstream, the gradient flattens, and the course of the wadi traverses a landscape that consists of a remnant plain formed across a dissected former lake bed situated at an elevation of c. 815 m. The incised valleys range from 0.2 to 1.0 km in width. The boundary between the upper (lake-plain) and lower (gorge) tract is seen as being where the wadi passes across the Wadi al-Hasa Fault that created the barred Wadi al-Hasa upper basin.

Al-Hasa wadi terraces–upper Wadi al-Hasa

At Wadi al-Hasa a well-developed set of terraces spans a period ranging from Middle Palaeolithic times to Late Holocene. Terraces occur both upstream and downstream from a fault that was responsible for the damming of the wadi in late Pleistocene times.

Vita-Finzi (1964) recognized three terraces in the upper Wadi al-Hasa. The highest is a 10–15 m-high fluvial terrace consisting of a white calcareous silt with zones of grey, green and khaki, which on the basis of Kebaran artifacts was regarded as having formed by 16,000 bp (Copeland and Vita-Finzi 1978). This terrace is the same as the one that Schuldenrein and Clark (1994) later interpreted as a lake-plain sequence developed across lacustrine marl. A second terrace (and alluvial fill) of silty gravel is also present. Distribution of the alluvial fill is somewhat localized. It was dated to 11,000–8,000 bp (Vita-Finzi 1964, 1966).

A 4–6 m-high third terrace that forms the high valley floor of upper Wadi al-Hasa is very extensive. It consists of two discrete units. At the base of the sequence is 2–3 m of silts and fine sands overlain by a 1–2 m-thick layer of cross-bedded and banded silts and fine sands (Schuldenrein and Clark 1994). Artifacts from near the base of the sequence led Vita-Finzi to conclude that aggradation commenced about 10,000 bp. Dates of 3950 ± 150 bp (Vita-Finzi 1966; Copeland and Vita-Finzi 1978) and 990 ± 120 bp (Schuldenrein and Clark 1994) have been obtained from the interval between the two sequences, suggesting a significant Middle to Late Holocene hiatus. This terrace probably corresponds with the low terrace (LT) of Wadi al-'Ali (see below).

Remnants of a fourth terrace were also recognized by Vita-Finzi and he identified this as a late Roman phase of wadi aggradation. A similar phase of alluviation is present elsewhere in the Jordan Valley, for instance, in the upper Wadi al-Hammeh and in upper Wadi Jirm al-Moz of the Tabaqat Fahl region.

In the lower Wadi al-Hasa and along the Wadi al-'Ali tributary, a three-tiered cut and fill terrace system has also been identified. The uppermost terrace is some 12–30 m above the present wadi floor, the intermediate terrace lies 5–6 m above the channel floor, and a lowermost terrace (modern flood plain) is only 0.5–1.5 m above the channel. The uppermost (high) terrace consists mainly of fluvial coarse sand and gravel but locally contains tufa, representing palaeospring deposits. At the WHS 634 site, situated some 11 km downstream of the western limit of the al-Hasa palaeolake, the high terrace surface passes laterally into a rock shelter, with the sequence containing Middle Palaeolithic occupation. An age of 105 ± 15 k yr. was obtained for near the top of the high terrace.

Lake Hasa

Clark *et al.* (1987) and Schuldenrein and Clark (1994) indicate that the upper part of Wadi al-Hasa was occupied during Late Pleistocene times by a lake (Lake al-Hasa) that reached a maximum level of 815 m (Figure 1.10). Lake al-Hasa was, therefore, at an elevation almost 1000 m above the highest level reached by the Late Pleistocene Lake Lisan (minus 180 m). Sediments from Lake al-Hasa consist of lacustrine marly sequences interdigitated with fluvial deposits, across which a number of terraces have developed.

Archaeological sites ranging in age from the Lower Palaeolithic are scattered along Wadi al-Hasa; a number of these are associated with the shoreline(s) of the lake. A lake-plain in the upper Wadi al-Hasa was eroded across a suite of marly lacustrine/fluvial sediments, at times capped with spring tufa. The palaeolake had an area of about 18 km long by 4 km wide and reached a maximum altitude of 815 m (Clark 1984). The western limit of the lake is about 40 km from the edge of the Dead Sea graben. Most archaeological sites occur nearer the western end of the lake.

The marly deposits consist of two distinct lacustrine facies. Finely laminated chemically deposited gypseous and calcareous marls abutting the outer flanks of the lake basin at elevations of 810–815 m represent the highest levels reached by the lake. They are in isolated instances capped by thin (<2 m thick) spring-derived tufa deposits, which represent deposition during the high lake stand. At slightly lower levels (<810 m) the lake sediments consist of blocky to laminar sandy silts that grade laterally into the contemporary al-Hasa terraces. They are considered to represent a paludal or ponded facies. The picture is one of an evolving environment in which the former lake represented by the highly calcareous and gypseous suite gives way to a marshy landscape of pools and swamps in which marly sandy silts accumulate. Occupation occurred around the margins of the lake and in tributaries entering the lake.

Middle Palaeolithic lithic sites occur in both the upper and lower regions of Wadi al-Hasa. For instance, the WHS 621 site in the upper al-Hasa region (Figure 1.11) is linked to an early phase of Lake al-Hasa. Site WHS 634, situated 10 km to the west, is Wadi al-'Ali, in an entirely fluvial section well below the limits of the lake. At WHS 621, artifacts cluster around an oval of calcreted rubble, which may be the remains of a former spring. Middle Palaeolithic artifacts also occur in adjacent lacustrine marls causing Clark *et al.* (1987) to note that they are not *in situ* but instead were washed into the lake. Therefore, their origin either predates the lake or at the latest was contemporaneous with it. On this basis, Clark *et al.* (1987) and Schuldenrein and Clark (1994) suggest that the earliest phase of lake activity may date to the Middle Palaeolithic period, and estimate an age of 70,000 years.

Much clearer evidence for the former existence of a lake is seen in the Upper Palaeolithic site WHS 618 with radiocarbon dates that range in age from c. 26,000–19,000 bp. Nearby, Epipalaeolithic sites WHS 1065 and 618 occupy high spurs of the

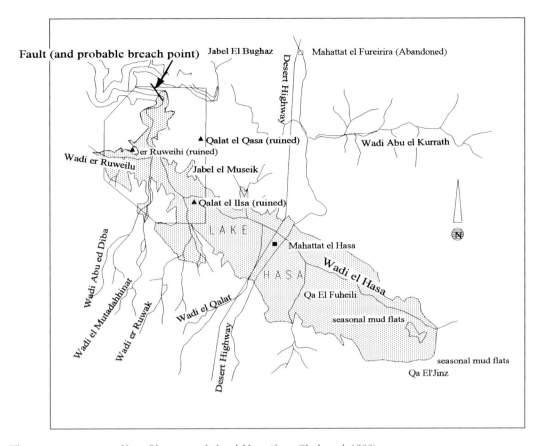

Figure 1.10. The maximum extent of Late Pleistocene Lake al-Hasa (from Clark *et al.* 1988).

dissected lake plain. At WHS 618, calcite-rich lacustrine sequences grade laterally into wadi alluvium and colluvium emanating from adjacent wadi systems. In the western locus of the site, tufaceous spring deposits and organic rich marshy deposits containing large numbers of microliths unconformably overlie laminated lacustrine marls. Northwards, the organic sediments are, in turn, overlain by alluvium. The spring activity is deemed to have occurred at peak lake levels with groundwater discharge outseeping around the margins of the lake. The high organic content of the sediments probably reflects vegetation growing within the spring zone. A date on spring tufa gave an age of $20,300 \pm 600$ bp.

At WHS 1065, situated only 200 m from 618, stratified marly sandy silts and clays, which contain extensive silicified root casts, overlie the lacustrine marls. They are seen as being marshy paludal deposits that developed after 20,000 bp, following the demise of Lake al-Hasa. Six dates on the occupation complex cluster between 16,900 bp and 15,600 bp, which agrees well with Kebaran stone tool industries present. Occupation in the paludal (at least seasonally) settings may have lasted until later than 12,000 bp as additional dates occur younger than 12,000 bp, which may represent a Natufian component. Pollen from the site suggested a Mediterranean open woodland/steppe environment, moister than that occurring at the older WHS 618.

Schuldenrein and Clark (1994) comment that the dated sequences of Wadi al-Hasa neither represent all time periods, nor cover the full extent of occupation. Instead, the clustering of dates, for example, in intervals of between 26,000 and 19,000 bp and 16,900 and 15,600 bp, is seen as representing phases of landscape stability, while the absence of dates during the intervening period is deemed as indicating landscape instability and erosion. Here the long interval prior to the Upper Palaeolithic date (26,000 bp) would imply erosion, at a time when it has been suggested that Lake al-Hasa was full. Indeed, it has been suggested that the lake may have been initiated as long as 70,000 years ago and lasted until about 20,000 bp (Schuldenrein and Clark 1994). However, the actual time when it originated is still problematical, as the Middle Palaeolithic stone tools in lake marl sequences at site WHS 621 are not *in situ* but instead were derived from adjacent areas prior

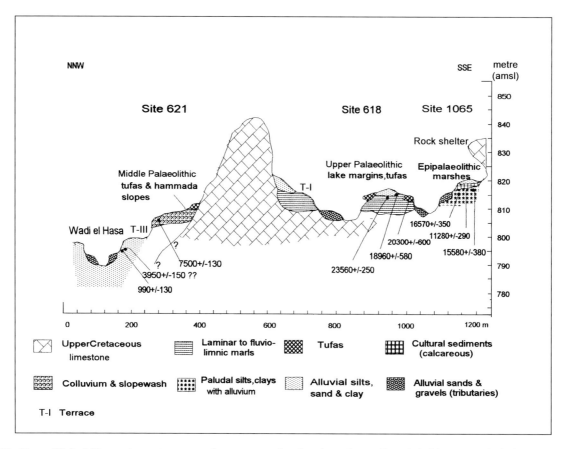

Figure 1.11. Upper Wadi al-Hasa: schematic stratigraphic section and archaeological sites (from Schuldenrein and Clark 1994).

to being deposited within the marls. Therefore, dating of the lake marls by the tools depends upon firstly clearly demonstrating that the lake and the tools were contemporaneous.

Whatever the case, it seems unlikely that with the limited catchment area (1740 km^2) there would have been sufficient runoff, other than during exceptionally wet periods, to maintain a permanent lake high in the landscape for 50,000 years, during which time there were a number of climatic oscillations. Furthermore, the lake is tectonically initiated and does not occupy a closed terminal lake basin, as is the case with other long-enduring palaeolake systems, such as Lake Lisan/Dead Sea, al-Azraq, al-Jafr, etc. Instead it occurs within the upper reaches of a deeply incised wadi system, lying 1000 m above the nearby Rift Valley. Its presence in this position is unusual and is due to the fault barrier separating the upper and lower al-Hasa regions. The lake's survival and its history will therefore be closely linked to the history of the downstream barrier. In this setting, it is most unlikely that closure of the lake could be maintained over extremely long time periods given recurrent tectonic effects and the relatively short period for which the lake basin could remain unsilted. Silting, leading in turn to overtopping, would quickly destroy the barrier and hence the lake. A lake lasting from Upper Palaeolithic times (26,000 bp) into the Epipalaeolithic period is more easily accepted.

Palaeoenvironmental summary for Wadi al-Hasa

The archaeological sites occurring along Wadi al-Hasa and Wadi al-'Ali range from the Middle Palaeolithic (with an age greater than 105,000 bp) up to the Chalcolithic. However, from an environmental perspective, the most significant aspect of the Wadi al-Hasa record is the presence during the Late Pleistocene of Lake al-Hasa. While now forming a wadi terrace, the marly lacustrine sequences associated with tufaceous and organic-rich sediments suggest a lake/spring-zone setting from about 26,000 to c. 19,000 bp. The moist conditions prevailed into Kebaran times and possibly into Natufian times when pools and swamps replaced the lake.

Site	Landform	*	Time (yr bp)	Occupation and examples of Sites
W. az-Zarqa'	Qf3 Terrace-Dauqara Formation	A	> 600,000 to >?250,000	Middle Acheulean
Tabaqat Fahl	Lower Tabaqat Fahl Formation	A	?400,000 to ~250,000	?Middle Acheulean: Sites M2, and M4
W. az-Zarqa'	Qf3–Qf2 interval	E	>250,000 to ~200,000	
al-Azraq	Lacustrine carbonate	A	220000 ± 30,000	Travertine overlies C-spring Lower Palaeolithic site
W. az-Zarqa'	Qf2 Bire Formation	A	200,000	Late Acheulean
Tabaqat Fahl	Upper Tabaqat Fahl Formation	A	250,000–150,000	Late Acheulean. Sites M1 and M3
Hisma	Phase 1 alluvium (Qalkha Formation)	A	> 150,000	Late Acheulean, from near top of 30 m section of Phase 1 fill
al-Azraq	*Edule* lacustrine deposits	A	146.000 ± 15,000	
Tabaqat Fahl	W. al-Hammeh palaeovalley cut	E	~150,000 to 80/65,000	
W. al-Hasa	W. al-'Ali, near top of high terrace	A	105,000 ± 15,000	Middle Palaeolithic WHS 634 (rock shelter)
W. al-Hasa	Calcreted rubble-spring tufa?			Middle Palaeolithic
W. Hisma	Dune sands in Phase 1 and 2 hiatus	E	> 70,000 to > 20,000	Tor Faraj and Tor Sabiha Mousterian sites
W. az-Zarqa'	Qf2-1/Qf-1 terraces. Bire-Samra and Kh. as-Samra Formation	A	'Early Wurm'	Levallois flakes in Qf1 terrace
Tabaqat Fahl	W. al-Hammeh fill of palaeovalleys	A	80/65,000–11,100	Middle Palaeolithic to Natufian sites
W. Hisban	Wadi Hisban terrace	A	?80/65,000–11,000	Near top, Epipalaeolithic–Geom. Kebaran/Natufian
W. Hisma	Phase 2 fill/rock shelters	A	> 20,000–12,300	Upper Palaeolithic and Epipalaeolithic, including Qalkhan, Madamaghan, and Hamran sites
al-Azraq Basin	Wadi al-Jilat: soils in aeolian silt	A	21,250 to >14,790	Upper Palaeolithic and Epipalaeolithic. Jilat 9, 6
al-Azraq Basin	Wadi al-Jilat: aeolian silts	A	14,790–present	Late Epipalaeolithic, Jilat 6, 8 and 10; Natufian, al-Azraq 18; PPNB, al-Azraq 31
al-Azraq Basin	W. Uwaynid: aeolian silt with palaesols	A	23,200–15,500	Uwaynid 14, 18

Table 1.2. Summary of major Pleistocene events from selected sites.

Site	Landform	*	Time (yr bp)	Occupation and examples of Sites
W. al-Hasa	Lake edge = part Terrace 1 Vita-Finzi (1964)	A	c. 26,000–19,000	Upper Palaeolithic WHS 618
W. al-Hasa	Swamp/marsh deposits postdate lake = part Terrace 1 Vita-Finzi	A	16,900–15,600	Kebaran: WHS 1065
W. al-Hasa	Data hiatus between Terrace 1 and 2	E	15,600–11,280	
W. al-Hasa	Marsh setting = Terrace 2 Vita-Finzi	?A	11,280–9,010	Natufian
W. Hisma	Incision of Phase 2 fill, terraces formed	E	>12,500–present	Natufian J2 site in terrace sediments
Tabaqat Fahl	Incision of W. al-Hammeh/ Jirm al-Moz terraces	E	<11,100–	
W. Hisban	Incision of Hisban terraces	E	<11,000–	
W. Hisma	Dune accretion	A	11,000–9,000	

* Aggradation (A) or erosion (E).

Table 1.2. (continued).

It is likely that the favourable occupation environment ended with the onset of the Younger Drias cold/dry phase, which was followed by postglacial warming. This period saw the drying of Lake Lisan, at about 11,000 bp. This would have led to a further phase of incision of Wadi al-Hasa, causing the breaching of any remaining barriers and the opening of the upper wadi to the lower wadi and hence to the Jordan Valley. By 9000–8000 bp, extensive dissection had cut tributaries down to bedrock.

Conclusion

Landscape evolution in a region as physically diverse as Jordan is not a uniform process. On the one hand, there is a tectonically active rift valley in the west, which provides the basis for large changes in base level, resulting in accelerated erosion and sedimentation both in the valley and on adjacent catchments. By contrast, in the east are the large shallow al-Azraq and al-Jafr Basins of internal drainage where base levels are relatively static. Erosion is commonly by deflation, resulting in condensed sedimentary sequences.

Any discussion on landscape evolution will therefore reveal as many differences as similarities, and no single monolithic pattern can be expected. Furthermore, while climate plays an important role in many of the evolutionary processes, other factors, such as groundwater and tectonics, will at times be locally dominant and act independently and varyingly across the region. Given that the basis for this analysis is geoarchaeological data, then aggradational features will have more relevance than erosional features. This is born out in Table 1.2, which is a résumé of Pleistocene events from the various sites.

The earliest landscape features discussed above are Pliocene, and predate human occupation. They include the development of the Jordan Rift Valley, the three erosive levels or pediments along the Ras an-Naqab scarp, and the Pliocene basalt flows that now cap the hills and infill ancient valleys in the upper az-Zarqa' region. However, the main focus in this paper is the time range spanned by occupation, from the Middle Pleistocene period to the present. Of the examples chosen, there are broad similarities in the evolution of the Jordanian landscape between some of the sites, especially during Early Palaeolithic times. For instance, the time span for the aggradation of the Qf2 Bire Formation at Wadi az-Zarqa', the Phase I fill at Hisma and the upper

Tabaqat Fahl Formation. From Middle Palaeolithic times to the present, a number of broad cycles of incision and deposition are recorded across these sites, all the catchments of which border the Jordan Valley.

However, while there are many additional similarities and coincidences of events, there are also many major differences, resulting in no clearly defined ongoing uniformity in the pattern of geomorphic evolution across the country. The problem stems partially from the lack of dating and partially because of the great diversity between the sites. For instance, the nature of al-Azraq sites is dominated by the position of the area at the lower extremes of a basin of internal drainage, and the presence of al-Azraq (and other) playas through time. The resulting strong aeolian influences reflecting deflationary processes, coupled with the thin discontinuous sedimentary sequences, make it difficult to slot landscape evolution into a regional (countrywide) framework. External climatic influences, although having real and identifiable impacts, are to a large extent muted by the hydrological and geomorphologic influence of the playas.

Similarly, Lake al-Hasa, while an important centre of occupation throughout its history, is a fault-controlled landform lying high up on the side of the Jordan Valley. The history of the creation and destruction of the barrier behind which the lake formed therefore largely determines the landscape history of the lake. Its base level is determined by the fault and not by the level of the Dead Sea. The wadi is deeply incised and a base flow system. Any barrier could lead to temporary lake formation, lasting until the barrier was breached, either through further tectonic activity or to overtopping during times of very high lake levels, or maybe headward erosion of the truncated Wadi al-Hasa. There is no requirement that the evolution of this system should bear any close relationship to the regional changes affecting other areas under discussion.

Of the sites examined, az-Zarqa', Pella and Hisma sites are best suited to reflect major regional trends in geomorphic evolution. Erosion and deposition in their catchments are strongly influenced by Rift Valley lake levels, which in turn are influenced by climate and tectonics. Of all the sites discussed, the Badia sites in the Eastern Basalt Desert Province, although limited in time, best reflect purely climatic influences.

Finally, when examining the archaeological/geomorphologic record, it seems there is an unintentional interplay between a relatively small number of environmental specialists, who commonly overlap on sites and carry with them to new sites understandings that greatly influence their interpretations, especially where limited data is present. This is brought out clearly in the large amount of site cross-referencing used to support interpretations in cases where there is otherwise too little data on which to base a conclusion. Despite this cross-pollination of ideas, there have still arisen significantly different interpretations on the nature of palaeohydrologic events occurring across Jordan since Middle Pleistocene times.

The environmental reconstructions discussed here seem to be inherently valid within the limits of their data, yet separately they have many differences. It follows that in the case of a number of the better-documented sites, attempts to reach consensus on a broad regional syntheses are not easy. It is even less appropriate, therefore, to carry out detailed regional reconstructions and correlations on the basis of less well-documented, more fragmentary sites, in widely different geographic and climatic settings. Perhaps then, before trying to tie in with any broader framework, it must be asked how appropriate is a site to meet such objectives? Clearly there will often be a stand-alone story, with or without a requirement that it be regionalized. Indeed where regionalization occurs, originality can be lost for the dubious gain of appearing to conform and by doing so add legitimacy. The absence of such questioning raises the spectre that the 'science' or 'art' of regional synthesis may at times survive simply because the fragmentary nature of the evidence lends itself to the construction of discrete schools whose doctrines prosper as much by the lack of evidence as by its availability.

References

Begin, Z.B., A. Ehrlich and Y. Nathan
   1974   Lake Lisan: The Pleistocene Precursor of the Dead Sea. *Geological Survey of Israel Bulletin* 63: 1-30.

Begin, Z.B. *et al.*
   1985   Dead Sea and Lake Lisan Levels in the Last 30,000 Years. *Geological Survey of Israel Report* 29/85: 1-30.

Bender, F.
   1968   *Geologie von Jordanie*. Berlin: Borntraeger.
   1974   *Geology of Jordan*. Berlin: Borntraeger.
   1975   Geology of the Arabian Peninsula–Jordan. *United States Geological Survey Professional Papers* 56: I.

Besançon, J., B. Gayer and P. Sanlaville
   1989   Contribution to the Study of the Geomorphology of the Azraq Basin, Jordan. In L. Copeland and F. Hours (eds.),

The Hammer on the Rock, 7-63. British Archaeological Reports, International Series 540(i). Oxford: British Archaeological Reports.

Besançon, J. *et al.*
1984 Lower and Middle Palaeolithic in the Upper Zarqa/Khirbet Samra Area of Northern Jordan: 1982–83 Survey Results. *Annual of the Department of Antiquities of Jordan* 28: 91-142.

Betts, A.V.G.
1985 Black Desert Survey, Jordan: Third Preliminary Report. *Levant* 17: 29-35.
1988 The Black Desert Survey: Prehistoric Sites and Subsistence Strategies in Eastern Jordan. In A.N. Garrard and H.G. Gebel (eds.), *The Prehistory of Jordan: The State of Research in 1986*, 369-91. British Archaeological Reports, International Series 396. Oxford: British Archaeological Reports.
1992 Eastern Jordan: Economic Choices and Site Locations in the Neolithic Periods. In M. Zaghloul *et al.* (eds.), *Studies in the History and Archaeology of Jordan*, IV: 111-14. Amman: Department of Antiquities.
1993 The Neolithic Sequence in the East Jordan Badia: A Preliminary Overview. *Paléorient* 19.1: 43-53.

Betts, A.V.G. and S.W. Helms
1989 A Water Harvesting and Storage System at Ibn El-Ghazzi in Eastern Jordan: A Preliminary Report. *Levant* 21: 3-11.

Clark, G.A.
1984 The Negev Model for Palaeoclimatic Change and Human Adaption in the Levant. *Annual of the Department of Antiquities of Jordan* 28: 225-48.

Clark, G.A., N. Coinman and J. Lindly
1986 Palaeolithic Site Placement in the Wadi Hasa, West-Central Jordan. *Annual of the Department of Antiquities of Jordan* 30: 23-29.

Clark, G.A. *et al.*
1987 Palaeolithic Archaeology in the Southern Levant: A Preliminary Report of Excavations at Middle, Upper and Epipalaeolithic Sites in Wadi el-Hasa, West-Central Jordan. *Annual of the Department of Antiquities of Jordan* 31: 19-77.
1988 Excavations at Middle, Upper and Epipalaeolithic Sites in the Wadi Hasa, West-Central Jordan. In A.N. Garrard and H.G. Gebel (eds.), *The Prehistory of Jordan: The State of Research in 1986*, 209-85. British Archaeological Reports, International Series 396(i). Oxford: British Archaeological Reports.
1992 Wadi Hasa Palaeolithic Project, 1992: A Preliminary Report. *Annual of the Department of Antiquities of Jordan* 36: 13-23.

Coinman, N. and D.O. Henry
1995 The Upper Palaeolithic Sites. In D.O. Henry, *Prehistoric Cultural Ecology and Evolution: Insights from Southern Jordan*, 123-213. New York: Plenum.

Copeland, L.
1988 Environment, Chronology, and Lower-Middle Palaeolithic Occupations of the Azraq Basin, Jordan. *Paléorient* 14.2: 66-75

Copeland, L. and F. Hours
1988 The Palaeolithic in North Central Jordan: An Overview of Survey Results from the Upper Zarqa and Azraq 1982–1986. In A.N. Garrard and H.G. Gebel (eds.), *The Prehistory of Jordan: The State of Research in 1986*, 287-309. British Archaeological Reports, International Series 396(i). Oxford: British Archaeological Reports.

Copeland, L. and C. Vita-Finzi
1978 Archaeological Dating of Geological Deposits in Jordan. *Levant* 10: 10-25.

Eden, C.
1988 Archaeology of the Sands and Adjacent Portions of the Sharqiyah. In The Scientific Results of the Royal Geographic Society's Oman Wahiba Sands Project 1985–1987. *Journal of Oman Studies Special Report 3*: 113-30.

Edwards, P. C.
1984 Two Epipalaeolithic Sites in the Wadi Hammeh. In A. McNicoll *et al.*, Preliminary Report of the University of Sydney's Fifth Season of Excavation at Pella in Jordan. *Annual of the Department of Antiquities of Jordan* 28: 55-86.

Edwards, P. C. and P. G. Macumber
1995 The Last Half Million Years at Pella. In S. Bourke and J.-P. Descoeudres (eds.), *Trade, Contact and Movement of Peoples in the Eastern Mediterranean*. Mediterranean Archaeological Supplement 3: 1-14. Sydney: Meditarch, University of Sydney.

Edwards, P. C., P. G. Macumber and M.J. Head
1996 The Early Epipalaeolithic of Wadi al-Hammeh. *Levant* 28: 115-30.

Emery-Barber, A.
1995 Pollen Analysis: Environmental and Climatic Implications. In D.O. Henry, *Prehistoric Cultural Ecology and Evolution: Insights from Southern Jordan*, 375-84. New York: Plenum.

Gardner, R.A.M.
1988 Aeolianite and Marine Deposits of the Wahiba Sands: Character and Palaeoenvironments. *Journal of Oman Studies Special Report 3*: 75-95.

Garrard, A.N. and B. Byrd
1992 New Dimensions to the Epipalaeolithic of the Wadi El Jilat in Central Jordan. *Paléorient* 18.1: 47-62.

Garrard, A.N. *et al.*
1985a Prehistoric Environment and Settlement in the Azraq Basin: A Report on the 1982 Survey Season. *Levant* 17: 1-52
1985b The Environmental History of the Azraq Basin. In A. Hadidi (ed.), *Studies in the History and Archaeology of Jordan*, II: 109-15. Amman: Department of Antiquities.
1988a Summary of Palaeoenvironmental and Prehistoric Investigations in the Azraq Basin. In A.N. Garrard and H.G. Gebel (eds.), *The Prehistory of Jordan: The State of Research in 1986*, 311-37. British Archaeological Reports, International Series 396(i). Oxford: British Archaeological Reports.
1988b Environment and Subsistence during the Late Pleistocene and Early Holocene in the Azraq Basin. *Paléorient* 14: 40-49.

Gat, J.R. and I. Carmi
1987 Effect of Climatic Change on the Precipitation Patterns and Isotopic Composition of Water in a Climate Transition Zone: Case of the East Mediterranean Sea Area. *IAHS Publication* 168: 513-23.

Gilbertson, D.D., C.O. Hunt and S. Bradley
1985 Micropalaeontological and Palaeoecological Studies of Recent Sediments from the Azraq Marshes in the Jordanian Desert. In A. Hadidi (ed.), *Studies in the History and Archaeology of Jordan*, II: 347-52. Amman: Department of Antiquities.

Ginzberg, A. and E. Kashai
1981 Seismic Measurements in the Southern Dead Sea. In R. Freund and Z. Garfunkel (eds.), *The Dead Sea Rift*: Tectonophysics 80: 67-80.

Goldberg, P. and A.M. Rosen
1987 Early Holocene Palaeoenvironments of Israel. In T.L. Levy (ed.), *Shiqmim I: Studies Concerning Chalcolithic Sites in the Northern Negev Desert, Israel*, 23-32. British Archaeological Reports, International Series 356. Oxford: British Archaeological Reports.

Goodfriend, G.A., M. Magaritz and I. Carmi
1986 A High Stand of the Dead Sea at the End of the Neolithic Period: Palaeoclimatic and Archaeological Implications. *Climatic Change* 9: 349-56.

Goren-Inbar, N.
1988 The Lower Palaeolithic-Synthesis. *Paléorient* 13.2: 109-11.

Hassan, F.A.
1995 Late Quaternary Geology and Geomorphology of the Area in the Vicinity of Ras en Naqb. In D.O. Henry, *Prehistoric Cultural Ecology and Evolution: Insights from Southern Jordan*, 23-31. New York: Plenum.

Helms, S.W.
1977 Jawa Excavations 1975: Third Preliminary Report. *Levant* 9: 21.
1982 Paleo-Bedouin and Transmigrant Urbanism. In A. Hadidi (ed.), *Studies in the History and Archaeology of Jordan*, I: 97-112. Amman: Department of Antiquities.

Helms, S.W. and A.V.G. Betts
1987 The Desert 'Kites' of the Badiyat esh-Sham and North Arabia. *Paléorient* 12.1: 41-67.

Henry, D.O.
1979 Palaeolithic Sites in the Ras en Naqb Basin, Southern Jordan. *Palestine Exploration Quarterly* 111: 79-85.
1982 The Prehistory of Southern Jordan and Relationships with the Levant. *Paléorient* 2.2: 389-93.
1995 *Prehistoric Cultural Ecology and Evolution: Insights from Southern Jordan*. New York: Plenum.

Henry, D.O. and G.H. Miller
1992 The Implications of Amino Acid Racemization Dates of Levantine Mousterian Deposits in Southern Jordan. *Paléorient* 18.2: 45-52.

Huckriede, R.
1966 Das Quartar des Arabischen Jordan-Thales und Beobachtungen über 'Pebble Culture' und 'Prae-Aurignac'. *Eizeitalter Genenwart* 17: 211-12.

Huckriede, R. and G. Wiesemann
1968 Der Jungpleistozane Pluvial-Von Al Jafr und Weitere Daten zum Quartar Jordaniens. *Geologica et Palaeontologica* 2: 73-79.

MacDonald, B.
1992 Settlement Patterns along the Southern Flank of Wadi al-Hasa: Evidence from 'the Wadi Al Hasa Archaeological Survey'. In M. Zaghloul *et al.* (eds.), *Studies in the History and Archaeology of Jordan*, IV: 73-76. Amman: Department of Antiquities.

MacDonald, B. *et al.*
1983 The Wadi el Hasa Archaeological Survey 1982. *Annual of the Department of Antiquities of Jordan* 27: 31-34.
1988 *The Wadi el-Hasa Archaeological Survey 1979-1983, West-Central Jordan*. Waterloo, Ontario: Wilfrid Laurier University.

McClure, H.A.
1976 Radiocarbon Chronology of Late Quaternary Lakes in the Arabian Desert. *Nature* 263: 755-56.
1978 Ar Rub Al Khali. In S.S. al-Sayari and J.G. Zotl (eds.), *Quaternary of Saudi Arabia* 1: 252-63. New York/Wien: Springer.

Macumber, P.G.
1970 Lunette Initiation in the Kerang District. *Victorian Mining and Geological Journal* 6: 16-18.
1984 Geology and Geomorphology of the Lower Wadi Hammeh Sites. In A.W. McNicholl *et al.*, Preliminary Report on the University of Sydney's Fifth Season of

Excavation at Pella. *Annual of the Department of Antiquities of Jordan* 28: 81-86.

1986 Environmental Reconstruction of the Wadi Hammeh Region in the Late Pleistocene. In A.W. McNicholl *et al.*, Preliminary Report on the University of Sydney's Seventh Season of Excavation at Pella (Tabaqat Fahl) in 1985. *Annual of the Department of Antiquities of Jordan* 30: 156-57.

1991 *Interaction between Groundwater and Surface Systems in Northern Victoria, Australia.* Melbourne: Ministry of Environment

1992a The Geology and Geomorphology of the Wadi Hammeh-Wadi Jirm Region, Northwestern Jordan. In A. McNicholl *et al.* (eds.), *Pella in Jordan* 2: 205-14. Canberra: Australian National Gallery.

1992b The Geological Setting of Palaeolithic Sites at Tabaqat Fahl, Jordan. *Paléorient* 18.2: 31-44.

Macumber, P.G. and H.J. Head

1991 Implications of the Wadi al-Hammeh Sequences for the Terminal Drying of Lake Lisan, Jordan. *Palaeogeography, Palaeoclimatology, Palaeoecology* 84: 163-73.

Ministry of Energy and Mineral Resources

1993 *Al 'Azraq 1:50000.* Amman, Jordan.

Neev, D. and K.O. Emery

1967 The Dead Sea, Depositional Processes and Environments of Evaporites. *Geological Survey of Israel Bulletin* 41.

Neev, D. and J.K. Hall

1976 Dead Sea Geophysical Survey 19 July–Aug. 1974. *Proceedings of the Annual Meeting of the Geological Survey of Israel, Ashqelon,* 8-10.

Nelson, B.

1973 *Azraq, Desert Oasis.* London: Lane.

Osborn, G.

1985 Evolution of Late Cenozoic Inselberg Landscape of Southwestern Jordan. *Palaeogeography, Palaeoclimatology, Palaeoecology* 49: 1-23.

Parker, D.H.

1970 *The Hydrogeology of the Mesozoic–Cainozoic Aquifers of the Western Highlands and Plateau of Eastern Jordan.* UNDP/FAO, Ag 2: SF/JOR 9, Technical Report No. 2, 424pp. Rome.

Quennel, A.M.

1958 The Structural and Geomorphic Evolution of the Dead Sea Rift. *Quarterly Journal of the Geological Society of London* 114: 1-24.

Rollefson, G.O.

1982 Late Acheulian Artifacts from Ain el-Assad (Lion's Spring), near Azraq, Eastern Jordan. *Bulletin of the American Schools of Oriental Research* 240: 1-20.

1985 Late Pleistocene Environments and Seasonal Hunting Strategies: A Case Study from Fjaje, near Shobak, Southern Jordan. In A. Hadidi (ed.), *Studies in the History and Archaeology of Jordan*, II: 103-108. Amman: Department of Antiquities.

1992 Neolithic Settlement Pattern in Northern Jordan and Palestine. In M. Zaghloul *et al.* (eds.), *Studies in the History and Archaeology of Jordan*, IV: 123-27. Amman: Department of Antiquities.

Rollefson, G.O. and A.H. Simmons

1988 The Neolithic Settlement at 'Ain Ghazzal. In A.N. Garrard and H.G. Gebel (eds.), *The Prehistory of Jordan: The State of Research in 1986*, 393-405. British Archaeological Reports, International Series 396(i). Oxford: British Archaeological Reports.

Rossingnol-Strick, M.

1993 Late Quaternary Climate in the Eastern Mediterranean Region. *Paléorient* 19.1: 135-52.

Salameh, E.

1985 The Development of the Hydrodynamic System in Connection with the Formation of the Jordan Graben and its Effects on the Generation of Oil. *Beiträge zur Hydrologie* 5.1: 49-66.

Salameh, E. and P. Udluft

1985 The Hydrodynamic Pattern of the Central Part of Jordan. *Geologie Jahrbuch* 38: 39-53.

Schuldenrein, J. and G.A. Clark

1994 Landscape and Prehistoric Chronology of West-Central Jordan. *Geoarchaeology* 9.1: 31-35.

Stekelis, M., O. Bar-Yosef and T. Schick

1969 *Archaeological Excavations at 'Ubeidiya, 1964-1966.* Jerusalem: Israel Academy of Sciences and Humanities.

Tchernov, E.

1988 The Age of 'Ubeidiya Formation (Jordan Valley, Israel) and the Earliest Hominids in the Levant. *Paléorient* 14.2: 63-65.

Vita-Finzi, C.

1964 Observations on the Late Quaternary of Jordan. *Palestine Exploration Quarterly* 95: 19-31.

1966 The Hasa Formation: An Alluvial Deposition in Jordan. *Man* 1: 386-90.

1982 The Prehistory and History of the Jordanian Landscape. In A. Hadidi (ed.), *Studies in the History and Archaeology of Jordan*, I: 23-27. Amman: Department of Antiquities.

Weinstein, M.

1976 The Late Pleistocene Vegetation and Climate of the Northern Golan. *Pollen et Spores* 17.4: 553-62.

Weinstein-Evron, M.

1989 Palynological History of the Last Peniglacial in the Levant. Les Industries à Pointes Foliacées du Paléolithique Superior Européen, Krakow 1989. Liège: Études et Recherches Archéologiques de l'Université de Liège No. 42.

2. The Palaeolithic Period, Including the Epipalaeolithic

Deborah I. Olszewski

Introduction

The modern country of Jordan is within the eastern or inland Levant, a region that has undergone striking climatic and ecological change during the course of the approximately 450,000 years of its Palaeolithic prehistory. Its story of ancient humans is one told largely through the stone tools they discarded, their campsites, the remains of their food, and the reconstruction of the landscapes they inhabited. More rarely, this story also includes evidence from burials and of artistic expression. The study of these Palaeolithic groups and their lifeways is part of a research tradition with a lengthy history, one that has accelerated at a dramatic pace since the 1970s.

Prior to the 1970s, only a few researchers had recorded and excavated Palaeolithic sites in Jordan. Notable among these early efforts are surveys by Rees (1929), Rhotert (1938), Zeuner *et al.* (1957) and Field (1960), and excavations by Waechter *et al.* (1938) and Kirkbride (1958, 1966). These studies documented the existence of a substantial Palaeolithic presence in Jordan, but pursuit of Palaeolithic research was limited, as most researchers focused their work on the sites of later time periods. In the 1970s, however, several individuals began extensive archaeological research programmes that were centred both on resurvey of areas explored in earlier decades or on vast tracts of land for which little archaeological documentation existed. Many of these programmes carried over into the 1980s and 1990s, and have been augmented by additional projects.

Palaeolithic research has concentrated on four regions explored by these projects. These are al-Azraq Basin (Garrard *et al.* 1977, 1985, 1986, 1987), Wadi al-Hasa (MacDonald *et al.* 1980, 1982, 1983; Clark *et al.* 1988, 1992, 1994), the Ras an-Naqab region (Henry 1979; Henry *et al.* 1983), and the Tabaqat Fahl area (Edwards *et al.* 1988; Macumber and Edwards 1997). Surveys and excavations in the Petra area (Schyle and Uerpmann 1988) and the Black Desert (Betts 1983, 1984, 1985) have also added sites to the Palaeolithic roster, as have dozens of smaller surveys and single site excavations. This research on the Palaeolithic landscape of Jordan has resulted in a significant transformation of our understanding of ancient lifeways, not only because there are now hundreds of documented sites but, more importantly, because the archaeological record of the eastern Levant has shown that the adaptations of these hunter-gatherer groups were much more varied, flexible and dynamic than predicted by archaeological models developed in the western Levant (Figure 2.1).

Terminology

The Palaeolithic is a period of time that is commonly known as the 'Old Stone Age'. The use of the term Palaeolithic refers to the tradition of manufacturing tools from stone, a tradition that began with the ancestors of modern human groups and continued well beyond the end of the Palaeolithic era. Human groups during the Palaeolithic were hunter-gatherers, most of whom were highly mobile. The majority of the Palaeolithic time period coincides with the Pleistocene, although the first stone tools are earlier and date to the Pliocene. During the latter part of the Pleistocene, glaciers across the northern and southern latitudes advanced and retreated numerous times. These advances (stadials) and retreats (interstadials), as well as many smaller oscillations within stadials and interstadials, had a significant effect on worldwide climate and environment. Although glaciers were not present in the Middle East, except at the tops of some mountain peaks, there were notable long-term changes in temperature, the amount of rainfall and the seasonal distribution of rainfall. These types of changes meant that the geographic extent of vegetation communities and the diversity of plant species that characterized them were subject to fluctuations over time. In turn, both the types and the abundance of animal species that depended on particular vegetation communities were also influenced, as were the human groups who hunted these animals and made use of plant foods.

Stone tools were an important adaptive technology for ancient peoples. Early groups developed methods

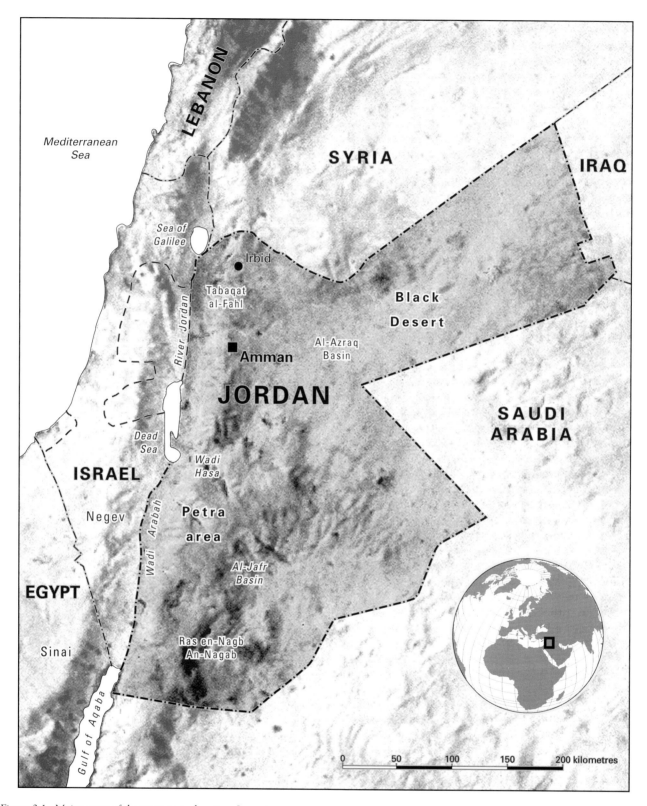

Figure 2.1. Major areas of the western and eastern Levant.

to remove effectively sharp-edged pieces of stone from larger blocks of stone. These could be used as they were or modified into special forms of tools. Removing pieces of stone is a process known as knapping (flint knapping) in which a core, such as a nodule or cobble, is struck with a stone hammerstone, a hard piece of wood or an antler segment (percussion technique). Pieces of stone are sometimes pressed off a core using pressure technique. With proper control and practice, a flintknapper can remove blades (long, parallel-edged pieces) or bladelets (small versions of blades) or flakes (wide, non-parallel-edged pieces). Blades and flakes serve as blanks that can be modified into such tools as endscrapers, sidescrapers and burins. Bladelets are often retouched or backed, becoming a tool form known as a microlith. There are also special forms of core tools, such as bifaces, that are retouched on both sides; those known as handaxes can vary in shape from pointed to oval, while cleavers have a flat edge at one end. Early flintknappers also developed specialized core techniques, such as Levallois technology, where cores were carefully prepared so that particular shapes and sizes of blanks could be removed.

The earliest archaeological evidence of Palaeolithic groups is from eastern Africa during the Pliocene, about 2.5 m.yr bp. It was not until about 1 m.yr bp or slightly earlier, during the Pleistocene, however, that ancient humans of the Palaeolithic began to migrate from Africa into the Middle East. Archaeological work in Jordan has not yet documented sites of this age, although they have been found in the western Levant (Tchernov 1988), and it is probably only a matter of time before such sites are located in the eastern Levant. Currently, the oldest sites in the archaeological record of Jordan appear to date somewhat earlier than 450,000 bp.

The Palaeolithic has been subdivided by archaeologists working in Western Eurasia (Europe and the Middle East) into four major periods based on chronology and the technology and typology of stone artifact assemblages. These periods are the Lower, Middle and Upper Palaeolithic, and the Epipalaeolithic. Although this classification system was designed to provide a set of distinct analytical units, many decades of research have now shown that there is considerable technological overlap between the Lower and Middle Palaeolithic, and substantial chronological overlap between the late Upper and the early Epipalaeolithic. Additional difficulties are raised by the lack of dates from many of the relevant sites and the advent of newer dating methodologies, for example, electron spin resonance (ESR) and thermoluminescence (TL), that have yielded dates for sites that are considerably older than expected. These are key issues that have not always been satisfactorily resolved, and which can have an important bearing on how Palaeolithic behaviours are interpreted.

Stone artifact assemblages form the primary basis of the delineation of the Palaeolithic periods. In the discussions that follow, lithic assemblages are grouped into phases, industries, and complexes, following Henry (1995c: 35-36). Briefly, an assemblage is a set of lithic artifacts recovered from a specific context within an archaeological site, generally from a particular layer or level. When assemblages share specific technological and typological characteristics, they are grouped into a phase designation, such as 'Early' or 'Late'. An industry consists of one or more phases that are similar in technology and in typology at the level of tool classes and tool types; this is an intermediate category of classification, for example, the 'Early Natufian'. A complex is composed of one or more industries that share a general technological pattern, but which have less overall similarity in specific tool types. The Early and Late Natufian, for example, contain different forms of lunate microliths; both phases/industries are components of the 'Natufian Complex'.

The Lower Palaeolithic is the oldest period (beginning in Africa about 2.5 m.yr bp). In Jordan, the oldest sites of this period are somewhat earlier than approximately 450,000 bp; this period continues until about 150,000 bp. The stone artifact assemblages of the Lower Palaeolithic in Jordan are usually classified into the Acheulean Complex. The most distinctive Acheulean stone tools are bifaces—sometimes referred to as handaxes—and cleavers; other tools include scrapers, notches and retouched flakes. A prominent technological attribute of some Acheulean assemblages is the use of the proto-Levallois and Levallois techniques of core reduction. The Acheulean is subdivided into four phases: Early, Middle, Late and Final. Of these, it is the Middle, Late and Final phases that are found in Jordan. Other distinctions noted are differences between the 'Desert Wadi Acheulean' and the 'Acheulean of Azraq' phases (Copeland 1988; Copeland and Hours 1988). These are further discussed later.

Sites of the Middle Palaeolithic span the interval from about 150,000 to 45,000 bp. Stone artifact assemblages of this period are representative of the Levantine Mousterian Complex. The industries of this complex are named for the sequence discovered at the site of Tabun in the western Levant (Mt Carmel area). At that site, the earliest industry is classified as Tabun D-type; this is followed by Tabun C-type, and then by Tabun B-type. Although these types of industries have been identified elsewhere in the Levant, they do not appear to delineate chronologically distinct entities except at the site of Tabun. Many researchers believe that the differences between the industries are more likely related to the exploitation of the resources of differing local ecologies. (The sequence at Tabun may, therefore, represent changes in local ecology over time.) For example, steppic and desert zones characterize many areas of Jordan; assemblages here are primarily of the Tabun D-type (Clark et al. 1997; Henry 1995c: 37). Typical features of these assemblages are the use of the Levallois technique to produce laminar blanks, such as blades and points. In most assemblages, these points are elongated in size; they are the primary tool type, along with tools on blades, such as burins. There are, however, also rare examples of Tabun C/B-type assemblages, characterized by the use of Levallois technique to produce Levallois flakes and other blanks that are used to manufacture various sidescrapers (Clark et al. 1988: 214-25; 1997). The distinctions between Tabun D and Tabun C/B assemblages are explored in further detail in the Middle Palaeolithic section below.

The Upper Palaeolithic period lasts from approximately 45,000 to 20,000 bp in the eastern Levant and possibly as late as 16,000 bp in the western Levant. Two major complexes of this period, found in the Middle East, are the Ahmarian and the Levantine Aurignacian. The former assemblages are characterized by blade and bladelet technology, with many of the bladelets modified into microlithic tools, such as Ouchtata bladelets and al-Wad points. There are also many endscrapers on blades and burins (Coinman 1998; Henry 1995c: 38). Currently, the evidence from Jordan indicates that sites of the Ahmarian complex are present, but the Levantine Aurignacian has not been found. Instead, there are stone artifact assemblages that occasionally include characteristic Levantine Aurignacian tools, such as endscrapers and burins on thick blades, but these are technologically more similar to the Ahmarian (Coinman 1998). Although researchers in Jordan have used the term Levantine Aurignacian to describe these assemblages, it is possible that they simply represent variability within the Ahmarian complex itself. These issues are further discussed below in the Upper Palaeolithic section.

The final period is the Epipalaeolithic, which begins about 20,000 bp and terminates at the end of the Pleistocene, about 10,300 bp. This portion of the Palaeolithic is the most complicated with respect to complexes of stone artifact assemblages, reflecting the large amount of technological and typological variability found within these assemblages. The major complexes include the Qalkhan, Kebaran, Geometric Kebaran, Mushabian and Natufian (Henry 1995c: 36), although some researchers have preferred to use terminology that is more generic, that is, the 'Non-Natufian Microlithic' (Byrd 1994: 210-11; Neeley et al. 1998). Each complex has been further subdivided into industries (Qalkhan in the Qalkhan Complex; Madamaghan in the Mushabian Complex; Early Hamran in the Kebaran Complex; Middle, Late and Final Hamran in the Geometric Kebaran Complex; Natufian in the Natufian Complex) and phases (Early, Middle, Late and Final) within particular industries. The major lithic assemblage characteristics of these complexes are blade and bladelet technology, with bladelets fashioned into a variety of retouched and/or backed microliths. There is a trend for microliths to be nongeometric in form during the early part of the Epipalaeolithic and geometric in morphology during the later portion of this period. Aside from the documentation of extensive assemblage variability, one of the most interesting developments revealed by research on Jordanian Epipalaeolithic lithic assemblages has been the recognition of the early occurrence of the microburin technique, a specialized method to segment bladelets to manufacture various types of microliths. This occurs as early as 20,000 bp (Byrd 1988: 260). The Epipalaeolithic section below examines the complexity of this period of the Palaeolithic in more detail.

Although the majority of the discussion in this chapter focuses on the Palaeolithic of Jordan, it is important to remember that the research themes that characterize it are common to research programmes in many regions of the Levant and the greater Middle East. It is virtually impossible in some instances to discuss these themes without reference to such areas as the western Levant, where the archaeological record has been intensively studied for many decades and various scholars have generated important models of Palaeolithic settlement, subsistence and other adaptive strategies. Research in Jordan, in fact, has often used these western Levantine models as initial approximations of past lifeways that can then be modified and refined using the data from archaeological sites in Jordan.

Chronology

The temporal framework of the Palaeolithic period in Jordan has been developed from both relative and chronometric sources. The relative sequence of complexes and industries resulted from the excavation of sites with stratigraphically separated assemblages. In general, materials from the lower levels of sites are older than those from upper levels. In some cases, chronometric dates have been obtained for some of these levels, allowing archaeologists to assign particular assemblages to certain ranges of time. Numerous sites, especially those that are surface lithic scatters, are relatively placed within the Palaeolithic framework on the basis of the techno-typological characteristics of their assemblages and other attributes. For example, heavily patinated surface assemblages that include numerous bifaces are likely to be attributed to the Lower Palaeolithic, while more lightly patinated surface assemblages yielding various forms of microliths usually are either Upper or Epipalaeolithic in age, depending on the specific morphologies of the microliths. Because suitable samples for chronometric dating are not always recovered from all levels of excavated sites, many of the assemblages from *in situ* contexts are also relatively dated using the characteristics of the lithic assemblages.

Until recently, the major chronometric technique used for sites in Jordan and many other areas of the Levant was radiocarbon. This method requires organic samples, such as wood charcoal, burnt animal bone or burnt seeds; it can be used to date materials from approximately the past 45,000 years. Materials that are older than this are recorded as 'greater than' dates, for example, >42,000 bp. In the Palaeolithic framework, radiocarbon is thus most useful for Upper and Epipalaeolithic sites, and is not very informative for the earlier ranges of Palaeolithic time. Middle Palaeolithic sites are often dated using one or more of several relatively new dating methodologies. These include thermoluminescence (TL), carried out on burnt lithic artifacts, electron spin resonance (ESR) used on burnt animal teeth, uranium series (Th/U) determinations on samples of travertine, and amino acid racemization on such materials as ostrich egg shell. The application of these techniques to sites from both the western and eastern Levant has yielded dates in the range from about 245,000 to 40,000 bp. No Lower Palaeolithic sites from Jordan have yet been dated using chronometric techniques; potential methods include potassium-argon (K/Ar) and other types of uranium series. Both of these techniques require finding sites that are stratigraphically positioned either immediately above or below volcanic rock layers, such as tuff or lava.

Although chronometric dates provide an absolute determination of age, they are not infallible methodologies. Great care must be taken to collect noncontaminated samples or to remove contaminants that affect the accuracy of the dates. For example, the lower end of the range of dates at 40,000 bp listed for chronometrically dated Middle Palaeolithic sites in some sources (e.g. Clark *et al.* 1997) is almost certainly too young, indicating that the sample that yielded this date is probably not reliable. All chronometric dates are cited with a standard deviation (ó), for instance, 58,400 ± 4,000 bp. This reflects the fact that the true age of the sample dated has a 68% probability of being within one standard deviation of the mean age (e.g. 54,400–62,400 bp) and a 95% probability of lying within two standard deviations of the mean age (e.g. 50,400–66,400 bp). Because standard deviations can be large, the range of dates for any given level of a site often can overlap with the range for layers below and above it. Chronometric tables for Middle, Upper and Epipalaeolithic sites can be found in Byrd (1994), Clark *et al.* (1997) and Phillips (1994).

Research themes

The main research themes of the Palaeolithic concern settlement patterning and settlement systems, subsistence and technology, primarily because the archaeological remains of this remote period of time best lend themselves to these issues. In order to examine settlement patterning and settlement systems, for example, archaeologists recover information about the geographic and topographic location of sites, past global and local environments (from oxygen isotope ratios in deep-sea cores, palynology and geoarchaeology), and aspects of lithic technology that can be used to infer the level of mobility of past groups. They also consider how landscapes have changed over time, usually to obtain a rough idea about the degree to which sites have disappeared due to erosional processes or destruction from later cultural modification of regions. Archaeological assemblages of animal bones (vertebrate fauna) and shellfish (invertebrate fauna), which are primarily found in late sites (Epipalaeolithic), as well as plant remains, which survive only rarely, are used to reconstruct the subsistence of ancient groups. Because animal bones are the most commonly found subsistence item, much of this research has focused on hunting patterns, although, based on analogy with remnant modern hunter-gatherer groups, archaeologists hypothesize that most ancient groups probably had a diet composed principally of plant foods.

One controversial research theme that is temporally restricted to the Middle Palaeolithic involves the origin of modern humans. The Middle Eastern region has yielded fossil humans of both the modern type (*Homo sapiens sapiens*) and an ancient type (*Homo sapiens neanderthalensis*), both of which are associated with Middle Palaeolithic stone tool industries. Currently, however, none of the sites of this time period in Jordan have produced human remains, and thus this research theme has been only indirectly pursued here.

Archaeological sites that date to the end of the Palaeolithic period (late Upper Palaeolithic and Epipalaeolithic) can contain a more diverse array of artifacts and structures than sites of the earlier Palaeolithic, including occupation floors, stone-walled or semi-subterranean dwellings, hearths, ornaments, art and human burials. Burials and their grave goods, ornaments and art can sometimes be used to infer social status, levels of inequality within human groups, and identity as a social grouping. There have also been attempts to isolate attributes of lithic tools that may serve as indicators of social group identity or of cultural traditions of manufacturing particular types of stone tools. One reason that the Epipalaeolithic period in particular has been the centre of these research themes is because many archaeologists believe that environmental and ecological conditions during the latter part of the Epipalaeolithic resulted in conditions favourable to the development of less mobile hunter-gatherer societies. Some of these groups apparently began to rely more heavily on cereal grasses as a food item, which eventually resulted in the deliberate cultivation of cereal grasses. Ultimately, a reliance on cultivation sparked the development of settled agricultural societies with complex social organizations in the Neolithic period that followed (see Chapter 3).

The research themes briefly outlined in this section are discussed more fully within the divisions of the Palaeolithic period that follow below.

The Lower Palaeolithic: 450,000+ to c. 150,000 bp

There are numerous sites and isolated finds of the Lower Palaeolithic recorded in Jordan. The majority of these are derived rather than *in situ* surface scatters of lithic artifacts. A small number of *in situ* sites, however, have been located and several have undergone test excavations, yielding not only lithics but also fossil animal bones. Sites discussed in this section include examples from al-Azraq Basin, Wadi al-Bustan (near ash-Shawbak), the Tabaqat Fahl area of the east Jordan Valley, and the Ras an-Naqab region of southern Jordan. All known sites of this period in Jordan are open-air rather than cave or rockshelter sites (Figure 2.2).

Although the Lower Palaeolithic spans a chronologically lengthy period of time, the understanding of its archaeological record is limited for several reasons. Principal among these are the destruction of sites over time through erosional processes, the transport of artifacts from *in situ* contexts to their current derived locations either on deflated land surfaces or in wadi deposits, poor chronology, and the limited number of associated assemblages, such as faunal remains. Reconstructing ancient settlement systems has thus proven difficult, although there are a few examples of efforts to predict why some localities appear to have been favoured for repeated visits.

Environment

The environment of the Middle East during this period was considerably different from today. Climatic conditions fluctuated between cool and moist to cool and dry. Many areas of Jordan experienced a higher water table, which resulted in numerous springs throughout many regions of the upland plateau (Transjordanian Plateau) and in Pleistocene lakes and marshes, such as those found in al-Azraq Basin and near Lake Lisan, the precursor to the Dead Sea (Copeland and Hours 1988: 296-99; Macumber and Edwards 1997). Some of these areas would have been heavily vegetated with riparian species, such as reeds (*Phragmites*) and willow (*Salix*). Areas away from the springs and lakes may have been somewhat less hospitable, but it is likely that vegetation zones like the Mediterranean forest and the steppe grasslands were much greater in geographic extent than they are today (e.g. Rollefson 1985: 105).

Research themes

Because lithic artifacts are durable and have survived in a variety of contexts, the most developed research theme for this period of the Palaeolithic has been the intensive study of the technology and typology of lithic artifacts. This has produced a relative chronology of

The Palaeolithic Period, Including the Epipalaeolithic

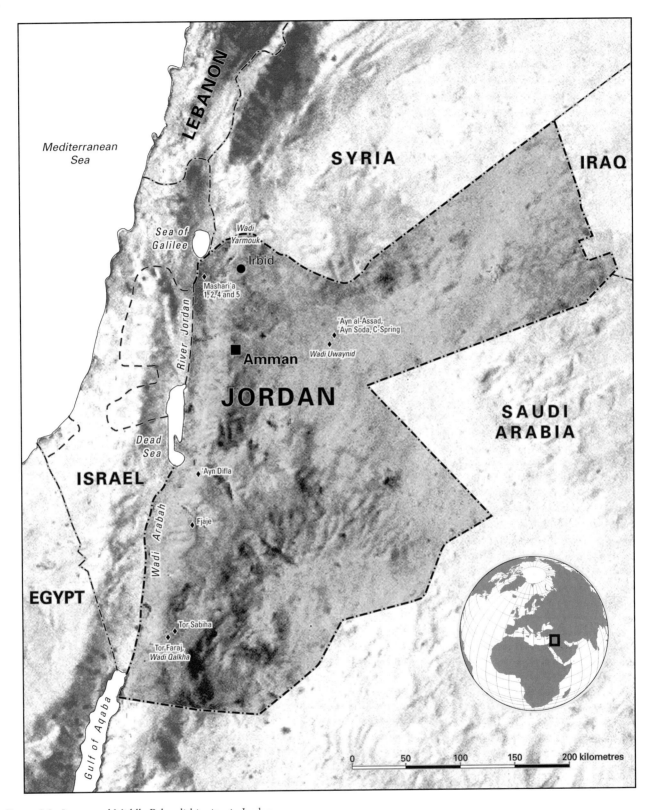

Figure 2.2. Lower and Middle Palaeolithic sites in Jordan.

Middle, Late and Final Acheulean industries, which can also be separated into spatial contexts, such as the Desert Wadi Acheulean (DWA) and the Late Acheulean of al-Azraq phases (e.g. Copeland 1988; Copeland and Hours 1988). The major distinguishing typological and technological criteria for the division of assemblages into chronological periods within the Acheulean are the classes and types of bifaces and the average length of bifaces (Gilead 1970; Macumber and Edwards 1997; Rollefson 1981, 1984). In general, ovate-shaped and Abbevillian types of bifaces typify earlier assemblages, while later assemblages contain more numerous cordiform- and lanceolate-shaped types. Average biface length appears to decrease over time. Typology and technology are also used to separate relatively contemporary assemblages from various geographic locales. For example, cleavers are particularly abundant in assemblages that lack Levallois technology; these are found in mesic settings, that is, when the lakes and marshes were highly productive resource locales. Other assemblages are characterized by use of Levallois technology and biface types other than cleavers; these are found in xeric contexts, that is, when marshes and lakes were much reduced in extent or absent. It is difficult to assess the degree to which this variability is due to chronology, over which there is relatively poor control for the Acheulean period, or to other factors, such as different sets of activities in differing ecological contexts.

Sites

One of the earliest known lithic assemblages from Jordan is a Middle Acheulean surface scatter from Wadi Uwaynid in al-Azraq Basin (Rollefson 1984). The assemblage is derived rather than *in situ* because the artifacts are heavily abraded and rolled, and there are few small–medium-sized flakes present. The scatter occurs along an approximately 2 km length of the wadi, with the majority of the bifaces from a 750 m-long stretch near the centre of the distribution (Rollefson 1984: 128). The high frequency of the class of ovate bifaces and Abbevillian types is the distinguishing trait that favours a designation of this assemblage as Middle Acheulean. Because the assemblage has been transported from an unknown location to its present place, knowledge of the geographic location of this Wadi Uwaynid site is informative only as an indicator that Middle Acheulean groups were present in this area of Jordan sometime earlier than about 450,000 bp.

Other probable Middle Acheulean sites include Mashari'a 2, 4 and 5 in the Tabaqat Fahl area of the east Jordan Valley. They are designated Middle Acheulean based on a small number of metrically large bifaces and on their stratigraphic context, which underlies deposits containing sites that are Late Acheulean in age (Macumber 1992; Macumber and Edwards 1997).

In al-Azraq Basin, the next groups to exploit the area are those identified as the Desert Wadi Acheulean (DWA), associated with the Late Acheulean. Finds are all derived, being heavily patinated and rolled; they consist of numerous bifaces, some cleavers, large-sized cores, and use of a proto-Levallois technique of core reduction (Copeland 1988: 68-71; Copeland and Hours 1988: 294-96). They may date to the interval between about 500,000 and 300,000 bp (Copeland 1988: 69). Although all of these materials were transported from their original locations, it is thought that these assemblages represent exploitation of the steppe grassland areas characteristic of many portions of al-Azraq Basin and its perimeters.

Marshes, lakes and springs were also present during certain intervals in al-Azraq Basin. These were attractive resource areas that served as focal points for both animals and the human groups who preyed upon them. Rapid burial of archaeological remains by water-deposited sediments resulted in good preservation of artifacts and fauna *in situ*. Many of the sites were found during periods of lowered water levels at the springs that exposed deposits otherwise inaccessible. The best-known sites are 'Ayn al-Assad (Copeland 1989a, 1989b; Rollefson 1982, 1983), 'Ayn Soda (Rollefson *et al.* 1997) and C-Spring (Copeland 1991). Unusually dense concentrations of lithic artifacts have been recovered from these sites—thousands of lithics, of which hundreds are classified as bifaces. Cleavers comprise close to a third of the biface assemblage in some levels of the sites, with the majority of the other bifaces classified as ovate or discoid forms. Other lithics are cores, various flake tools and the debris from knapping; there is no use of the Levallois technique of core reduction. The high concentrations of bifaces, particularly the cleavers, are thought to be associated with butchering activities; fossil bone representative of prey species includes camel (*Camelus*), hartebeest (*Alcelaphus*), wild boar (*Sus scrofa*), rhinoceros (*Dicerorhinus hemitoechus*), aurochs (*Bos primigenius*), equid (*Equus*), and elephant (*Elephas*) (Clutton-Brock 1970; Rollefson *et al.* 1997). Copeland (1988: 71-73; Copeland and Hours 1988: 296-99, 303) has classified these lithic assemblages as the Late Acheulean of al-Azraq phase because of the high number of cleavers present in the assemblages, and suggests that they may date to about 250,000–200,000 bp.

Another region of Jordan characterized by freshwater marsh and spring conditions during the Late Acheulean period is the Tabaqat Fahl region of the east Jordan Valley (Macumber and Edwards 1997). The *in situ* site of Mashari'a 1 is associated with these deposits. Recent excavations here yielded a large lithic assemblage characteristic of a biface reduction site—numerous biface thinning flakes and small flakes, shatter and debris. The recovery of very few bifaces suggests that most bifaces were curated to other locations. There are, however, many notched flakes, invasively retouched flakes and other tools, indicating that Mashari'a 1 was used for more activities than just biface manufacture and reduction.

Surface scatters of Late Acheulean artifacts have also been found in the highland plateau areas away from marsh and lake locales. Like the DWA, these are derived assemblages and their information potential is somewhat limited compared to *in situ* sites. One example of a highland plateau site is Fjaje in the Wadi al-Bustan, near ash-Shawbak. The assemblage is from a deflated land surface and thus represents a palimpsest of numerous occupations (Rollefson 1981: 6-7). Unlike the DWA, where assemblages have been transported over unknown distances, the assemblage from Fjaje is in its approximate original locale on the landscape because its artifacts have been deflated rather than transported. The topographic position of the Fjaje assemblage can thus be used to address the question of why early groups returned repeatedly over time to this locale. Rollefson (1985) hypothesizes that hunters favoured this spot because it lay along a seasonal migration route for herd animals. Animals travelling through the Wadi al-Bustan from the low areas of Wadi 'Arabah to the highlands during the spring could be relatively easily ambushed by hunters waiting along the rim at Fjaje. Prevailing winds from the west and northwest during the spring placed hunters in a downwind position, and the long climb from Wadi 'Arabah up to the highlands may have tired the animals. Both of these factors increased the success rate of hunters at Fjaje and led to their repeated use of this site area.

Another highland site is Wadi Qalkha in the Ras an-Naqab area of southern Jordan. The artifacts are lightly patinated and have sharp edges, indicating that they have not been transported very far nor been exposed for long periods of time (Henry 1995f: 43). The assemblage includes examples of bifaces, sidescrapers and retouched pieces, as well as flakes and blades; use of Levallois technology is documented. This assemblage is similar to that from Fjaje. Henry (1995f: 47) suggests this industry represents a Late Acheulean adaptive strategy common throughout steppe and light woodlands environments that were situated near plains and plateaus.

The Middle Palaeolithic: c. 150,000–45,000 bp

Sites of the Middle Palaeolithic in Jordan, like those of the Lower Palaeolithic, have been most frequently recorded in derived contexts, primarily as deflated surface scatters of lithic artifacts. Unlike the Lower Palaeolithic, however, there are examples of *in situ* Middle Palaeolithic sites from rockshelters, some of which have been test excavated. Discussed in this section are sites from the Wadi al-Hasa region and the Ras an-Naqab area (see Figure 2.2). All assemblages identified to date are attributed to the Levantine Mousterian Complex that is characterized by the use of Levallois technology. New dates that extend the lower boundary of the Middle Palaeolithic much farther back into the past, as well as shared technotypological characteristics with the Lower Palaeolithic, have prompted a number of researchers to regard a strict boundary between the Lower and Middle Palaeolithic as no longer tenable (e.g. Clark *et al.* 1997).

Environment

Environmental conditions during the 100,000 or more years of the Middle Palaeolithic were subject to considerable short- and long-term fluctuations, as well as regional variation. Pollen from lake cores, and pollen and phytoliths from archaeological sites across the Levant, indicate that north–south and west–east climatic clines existed (Clark *et al.* 1997). In general, vegetation zones like the Mediterranean forest and open steppe extended farther to the south, for example, into the Negev highlands, and to the east, for example, the Wadi al-Hasa area, during certain intervals. Lakes and marshes or seasonal ponds, resulting in small patches of trees including oak (*Quercus*) and riparian vegetation such as rushes (*Cyperus*) also characterized many locales. The early part of the Middle Palaeolithic, based on pollen evidence from 'Ayn Difla in Wadi al-Hasa, appears to have been relatively cool and dry, although more moist than modern conditions in this region (Clark *et al.* 1997). This area was primarily an open steppe or desert/steppe grassland with scattered trees along the wadi. Later in the Middle Palaeolithic, pollen and phytolith evidence from Tor Sabiha and Tor Faraj in the Ras an-Naqab region indicates a warm, dry interval, ameliorated in this area by the presence of seasonal ponds (Emery-Barbier 1995: 377–79; Rosen 1995: 401–402). Although warmer than the earlier Middle Palaeolithic, the local landscape remained dominated by open steppe grasslands and punctuated by trees along drainages.

Research themes

As during earlier portions of the Palaeolithic, research themes of the Middle Palaeolithic in Jordan are focused on lithic typology and technology, and on settlement and subsistence systems, as well as some aspects of the debate over the origin of modern humans and the implications this holds for interpretation of various of the archaeological assemblages that have been recovered. With one exception, the Levantine Mousterian lithic assemblages of the southern and inland Levant, including Jordan, are classified as Tabun D-type (e.g. Clark *et al.* 1997; Henry 1995c: 37; Lindly and Clark 1987), and represent a long-lived complex that is slow to change typologically. As discussed previously, numerous unretouched Levallois points characterize Tabun D-type; in the arid regions, these tend to be elongate in form. Recently obtained TL and ESR dates from the site of 'Ayn Difla in Jordan suggest that these assemblages are much earlier than previously believed, perhaps in excess of 135,000 bp (Clark *et al.* 1997); the Tabun D-type continues to appear at sites as late as about 45,000 bp in the Negev, although the latest dated sites in Jordan (Tor Faraj and Tor Sabiha) are about 69,000 bp.

No hominid remains have yet been recovered from Middle Palaeolithic sites in Jordan to add to the debate in the western Levant regarding the possible existence of two sympatric hominid species (*Homo sapiens neanderthalensis* and *Homo sapiens sapiens*) who apparently made and used similar stone artifacts. This debate, which is polarized into replacement versus continuity viewpoints, has a lengthy history and has not been resolved. The two major viewpoints have been extensively published elsewhere (e.g. Clark and Lindly 1988; Stringer and Andrews 1988) and are only briefly summarized here. Replacement advocates postulate that both hominid species overlapped temporally; archaic humans (*Homo sapiens neanderthalensis*) represent an evolutionary side branch that eventually became extinct. Evidence to support this viewpoint is cited from several sources including molecular genetics (mitochondrial DNA), palaeoanthropology (morpho-metric attributes of hominid fossils) and the archaeological record (chronometric dates, food acquisition techniques). Adherents of this paradigm point out that Neanderthals could not have evolved into modern humans because they survive as a species well after the appearance of moderns, based on both 'early' and 'late' dates of assemblages associated with Neanderthal fossils (around 122,000 bp at Tabun and 60,000 bp at Kebara in the western Levant) and 'early' dates for modern human fossils (about 92,000 bp or earlier at Qafzeh in the western Levant). On the basis of chronometrics alone, this evidence would seem to suggest that the two species coexisted for some tens of thousands of years. Some researchers have further argued that, although both species made and used Levantine Mousterian stone artifacts, there are differences in their subsistence patterns, indicating that they occupied somewhat different environmental niches (Lieberman 1993).

Continuity advocates, on the other hand, argue that the fossils of archaic (Neanderthal) and modern humans from the Levant represent a single evolving population. They point to several lines of evidence, including biological factors, aspects of chronological resolution and behavioural patterning. They attribute variability in skeletal morphology to changes over time, as well as to variation present within the species at any given moment in time (analogous to morphological variation visible in human populations today). Thus, they see the fossil specimens as variable, isolated individuals that must be studied within a population context rather than as individuals. Additionally, they argue that the TL, ESR and other dates associated with various sites are not comparable and can vary greatly depending on which technique is used and what type of sample is dated (e.g. Clark *et al.* 1997). In other words, while the dates obtained from these techniques indicate a relatively long span of time overall, it is difficult to assess the comparison of specific TL dates from one site to specific ESR dates from a different site. The standard deviations associated with dates from different layers at the same site also often overlap considerably; layers that have older or younger mean ages potentially can be virtually 'contemporary' when the standard deviations are taken into account. The end result is that neither 'species' can be said to predate the other. Finally, based on the competitive exclusion principle of evolutionary ecology, continuity advocates emphasize that it would have been impossible for two sympatric hominid species in the same small region to have coexisted for such a lengthy period of time without inhabiting completely different environmental niches. The association of the same lithic industry with both hominid 'species', as well as the same prey animals (similar faunal assemblages), and the use of neighbouring site locales in some areas, are all suggestive of a similar adaptation and, therefore, a single hominid line rather than sympatric species.

Another research theme is settlement patterning. Henry's (1995e: 107-27) settlement model for the Middle Palaeolithic of southern Jordan is based on the idea of seasonal transhumance. Human groups here during the fall/winter months appear to have utilized a strategy of radial settlement (or logistical mobility)—longer-term, seasonal base camps with associated ephemeral task camps in nearby localities. During the spring/summer months, they engaged in a circulating settlement pattern (or residential mobility)—a series of successively occupied short-term base camps. In the Ras an-Naqab area, Tor Faraj represents a low-elevation, long-term base camp during the fall/winter months, and Tor Sabiha one of the series of short-term camps during the spring/summer months. This approach represents a significant departure from settlement models developed for the Middle Palaeolithic of the Negev and Sinai, where researchers have argued that this period was characterized by a year-round radial settlement system that persisted for millennia (e.g. Marks and Friedel 1977). It is clear that the archaeological record of areas of the eastern Levant, such as southern Jordan, has a great deal to contribute towards the understanding of the adaptive flexibility of these early groups of highly mobile hunter-gatherers.

Sites

The earliest chronometrically dated Middle Palaeolithic site in Jordan is the *in situ* deposits from the rockshelter of 'Ayn Difla in the Wadi al-'Ali tributary drainage to Wadi al-Hasa. A series of TL, ESR and Th–U dates from samples from several levels indicates that the Tabun D-type artifacts were deposited sometime between about 180,000 to 90,000 bp (Clark *et al.* 1997). Several seasons of excavation have yielded a relatively large lithic assemblage, containing numerous elongated Levallois points, as well as many small flakes and shatter elements. Retouched tools are few, but include examples of burins, notches and denticulates (Lindly and Clark 1987: 285; Clark *et al.* 1997). The small size of many of the cores suggests that they were discarded when close to the end of their usefulness, and the presence of a wide variety of debitage items, such as the small flakes, shatter and primary elements (pieces with cortex), indicates that lithic reduction was one activity at this site. Associated faunal remains are quite fragmentary and sparse, but include equids (*Equus*), goat or ibex (*Capra*), and gazelle (*Gazella*) (Clark *et al.* 1988: 235; Lindly and Clark 1987: 289-90). It is likely that the groups who used 'Ayn Difla intermittently over thousands of years were highly mobile and provisioned themselves by carrying lithic artifacts with them as they travelled from locale to locale (Clark *et al.* 1997).

Two other *in situ* sites are the rockshelters at Tor Faraj and Tor Sabiha in the Ras an-Naqab region. These sites are archaeologically contemporary and contain Tabun D-type assemblages; both are chronometrically dated to the latter part of the Middle Palaeolithic, about 69,000 bp (Henry 1995c: 37). They are believed to represent different aspects of a Middle Palaeolithic settlement system, with a larger, winter-occupied base camp at Tor Faraj, and a smaller, summer short-term camp at Tor Sabiha (Henry 1994; 1995e: 107-32). Evidence for this settlement model is derived from a variety of sources, including the topographic location of the two sites and characteristics of the lithic assemblages. The high elevation location of Tor Sabiha, its limited extent of sheltered area and its eastern exposure are suggestive of locational attributes chosen during summer months; Tor Faraj, in contrast, has a much larger sheltered area, faces south and is at lower elevation (Henry 1995e: 123). Artifacts from Tor Faraj tend to be larger and heavier, and elements from the entire process of core reduction are found, including the production of tools, their use and resharpening (Henry 1995e: 112-14). These are characteristics of longer-term base camp occupations. Tor Sabiha, on the other hand, yielded lighter weight and smaller artifacts, with lithic reduction evidence pointing to final stage manufacture, use and resharpening of artifacts.

Faunal assemblages from Tor Faraj are limited and poorly preserved; they include gazelle (*Gazella*), equid (*Equus*) and ostrich (*Struthio camelus*) eggshell fragments (Henry 1995g: 54). A similar assemblage, with the addition of aurochs (*Bos primigenius*), was recovered from Tor Sabiha. Most of these animals are representative of open steppe-like conditions, while aurochs indicate the presence of locally available water sources (Henry 1995g: 58). The faunal species are consistent with the pollen and phytolith data from these sites that also indicate open steppe grasslands interspersed with locally mesic situations.

Microwear studies of lithic artifacts from Tor Faraj and Tor Sabiha were also conducted. At Tor Faraj, the most frequently used artifacts were Levallois points, Levallois debitage and retouched tools (Shea 1995: 97). Comparison of the microwear from these artifacts with traces on experimentally used artifacts indicates that the Tor Faraj assemblage documents evidence of impact, butchering activities and contact with bone and hafts. A small percentage of the artifacts appear to have been used on soft plant materials and on wood. It is possible that the more xeric, steppe setting of the Ras an-Naqab region contained fewer plant species of use to Middle Palaeolithic humans occupying the Tor Faraj rockshelter than plants near sites in the western Levant that were situated in a Mediterranean forest setting. This may mean that the Tor Faraj groups concentrated more exclusively on hunting and butchering activities than their contemporaries elsewhere (Shea 1995: 97). Microwear on artifacts from Tor Sabiha revealed that similar activities occurred at this high elevation site, with perhaps a greater emphasis on working of animal materials (Lee cited in Henry 1995e: 119-20).

The Upper Palaeolithic: c. 45,000–20,000 bp

The archaeological record of the Upper Palaeolithic in Jordan, which includes many surface scatters, is relatively well known partly because there are many examples of *in situ* sites compared to earlier periods of the Palaeolithic. Both open-air and rockshelter contexts have been tested in such areas as Wadi al-Hasa and Ras an-Naqab, as well as in al-Azraq Basin. Lithic assemblages are identified as belonging to one of three major groupings—an early Upper Palaeolithic that may be transitional from the Middle Palaeolithic; or the Ahmarian Complex; or a non-Ahmarian grouping that bears some similarity to both the Ahmarian and to the Levantine Aurignacian (Figure 2.3).

The early Upper Palaeolithic Complex contains Upper Palaeolithic tool types and a technology that features some attributes of Middle Palaeolithic core reduction techniques. The Ahmarian is an elongated blade- and bladelet-based complex, with endscrapers, burins and retouched bladelets. In the early phase, retouched bladelets, such as al-Wad points, are common; in the late phase, Ouchtata retouched bladelets are characteristic. Non-Ahmarian assemblages are typified by a less-refined blade technology that yields rather large, thick debitage, few retouched bladelets and few true Aurignacian tool types. It is also dissimilar to the Levantine Aurignacian as defined from rockshelter and cave contexts in the northern and western Levant, which has use of Aurignacian scalar retouch and production of thick blades and flakes, Aurignacian endscrapers, and carinated and nosed endscrapers. The non-Ahmarian grouping potentially represents variability within the Ahmarian Complex of industries; this is a research question that is in its initial stages of development (Coinman 1998; Coinman and Henry 1995: 194-95).

Environment

The environment during the Upper Palaeolithic underwent several significant fluctuations. The earlier part of this period (Early Upper Palaeolithic and Early Ahmarian) appears to have been slightly more moist than during the latter Middle Palaeolithic (Emery-Barbier 1995: 382); for example, there is evidence of significant Early Ahmarian occupation associated with marsh settings in such areas as the southern Sinai (Phillips 1988). Sites of this period are also found in southern Jordan. Conditions likely became drier throughout the duration of the Upper Palaeolithic, especially in the period following about 25,000 bp as worldwide temperatures began to decline with the onset of the last glacial maximum of the Pleistocene. In the Wadi al-Hasa region, for instance, pollen analyzed from Late Ahmarian sites documents low arboreal pollen, which decreases in frequency through time, while steppic vegetation becomes increasingly dominant (Coinman 1993: 19-20; Olszewski *et al.* 1990: 46). However, a higher water table, which led to increased spring flow and more numerous springs throughout highland areas, helped create many locally ameliorated settings in Jordan where lakes, marshes and seasonal ponds continued in existence from earlier times. These appear to have remained favoured locales for Upper Palaeolithic groups. Discussion below concentrates on sites of the latter portion of the Upper Palaeolithic because the archaeological record is better documented for this interval in Jordan.

The Palaeolithic Period, Including the Epipalaeolithic

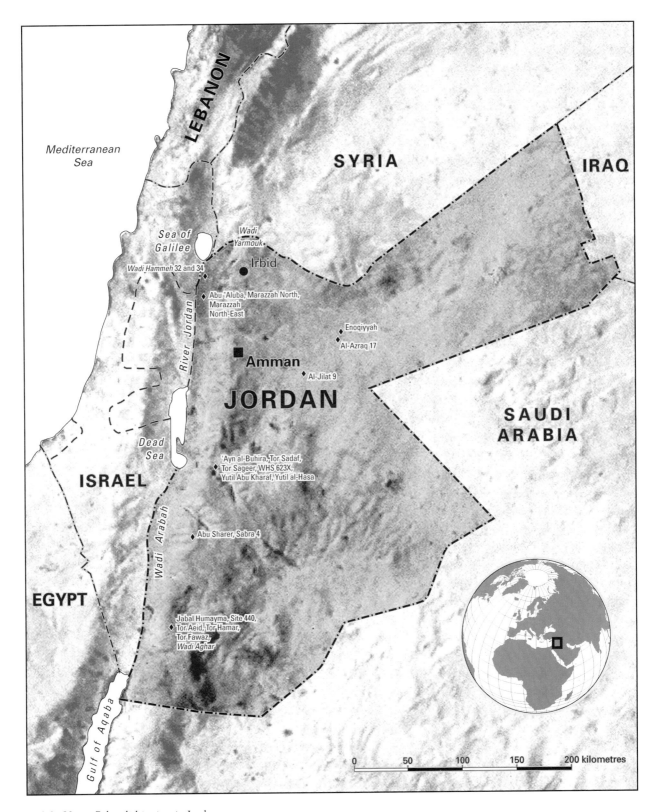

Figure 2.3. Upper Palaeolithic sites in Jordan.

Research themes

Settlement and subsistence systems are again a key research theme. Currently, the majority of the known Upper Palaeolithic sites in Jordan appear to be late assemblages, that is, Late Ahmarian or related industries. These Late Upper Palaeolithic sites have been most frequently found in association with marsh, lake and seasonal pond ecological settings. Henry (1995a: 426-27) applies his model of seasonal transhumance between low elevation, fall/winter base camps, and higher elevation, spring/summer transitory camps—initially documented in the Middle Palaeolithic—to this period of the Upper Palaeolithic in the Ras an-Naqab region as well. Because his model is dependent on great topographic relief in an area typified by seasonal ponds, however, it is probable that this type of settlement system is not applicable to all areas of Jordan during the Late Upper Palaeolithic. For example, Olszewski and Coinman (1998) have recently suggested that the Late Upper Palaeolithic of the Wadi al-Hasa area may have been a year-round radial settlement system, with settlement tethered to the lake and marshes present in the eastern part of the drainage. Such areas would have contained a number of attractive resources: freshwater springs, materials for manufacturing baskets, fish, shellfish, waterfowl, and game animals, such as gazelle, equids and aurochs. Radial settlement patterning may also have been present in such areas as al-Azraq Basin where analogous lake/marsh conditions existed.

One of the more difficult aspects has been discriminating between the end of the Upper and the beginning of the Epipalaeolithic. In part, this is because radiocarbon determinations have suggested that some of the industries of the Late Upper Palaeolithic overlap chronologically with early industries of the Epipalaeolithic. To what extent this simply represents overlap created by the standard deviation age range around the mean age obtained at various sites is unknown. It is possible, of course, that groups characterized by Upper Palaeolithic lithic assemblages coexisted for some time alongside groups that had developed Epipalaeolithic lithic assemblages. It is important to remember in this context that morphologically different tools could have been used in similar manners or for similar purposes. Thus, two groups manufacturing and using archaeologically distinct tools may have been engaged in very similar overall lifeways. Interpreting the meaning underlying the coexistence of different lithic industries is more problematic, and is a research theme studied more intensively for the Epipalaeolithic period (see p. 52).

Sites

One mesic area was al-Azraq Basin, where both *in situ* and deflated sites dating between about 25,000 and 20,000 bp have been recorded (Byrd 1988). The assemblages are quite diverse: the Trench 2 materials from al-Azraq 17 contain many thin bladelets with fine retouch (Byrd 1988: 259), analogous to Ouchtata bladelets of the Late Ahmarian (Coinman 1998); lithics from al-Jilat 9 are non-microlithic and include numerous endscrapers and retouched pieces (Garrard *et al.* 1988a: 319), perhaps comparable to the non-Ahmarian assemblages of southern Jordan (Coinman 1998). All of these Upper Palaeolithic assemblages are blade or bladelet based; therefore, it is inaccurate to classify them as Levantine Aurignacian. Associated fauna are primarily open steppic types, such as gazelle (*Gazella*) and equids (*Equus*); there are also hares (*Lepus capensis*) and scutes of land tortoise (*Testudo graeca*), which may have served as containers (Garrard *et al.* 1988b: 47). Aurochs (*Bos primigenius*), which favours a more mesic setting, is also present, indicating the availability of wetter settings. Geological studies in Wadi al-Jilat indicate that soil erosion suggestive of a drier interval at about 21,000 bp may have typified the period during which al-Jilat 9 was occupied (Garrard *et al.* 1994: 181). This corresponds to general environmental evidence cited above that indicates drier intervals towards the end of the Upper Palaeolithic period.

Wadi al-Hasa is a second region where the reconstructed palaeolandscape included springs, a lake and marshes (Schuldenrein and Clark 1994). The Pleistocene lake here probably attained its peak levels about 40,000 bp, during the Early Upper Palaeolithic, and had begun to recede beginning about 20,000 bp. The eastern part of Wadi al-Hasa forms a plain, greater than 50 km^2 in extent, representing the former bed of the lake. Archaeological surveys of the south (Wadi Hasa Survey) and north (Wadi Hasa North Bank Survey) banks of the Hasa system between 1979 and 1993 yielded 48 Upper Palaeolithic and 19 Upper to Epipalaeolithic sites (Clark *et al.* 1992, 1994; MacDonald *et al.* 1980, 1982, 1983). A number of these are *in situ* and include both rockshelters and open-air sites. The distribution of these sites is highly patterned, with most situated within about 4 km of the lake. The density of Upper Palaeolithic sites in the immediate vicinity of Lake Hasa is indicative of the importance of this ecological context. Two Late Upper Palaeolithic sites have been tested, the open-air site of 'Ayn al-Buhira and the southeast-facing rockshelter at Yutil al-Hasa.

At 'Ayn al-Buhira, a spatially large site on the ancient lakeshore, there is an extensive surface scatter of lithics as well as deeply stratified deposits (greater than 2 m) that span the interval from about 25,000 to 20,000 bp. At least two stratigraphically and spatially differentiated occupations have been identified. The earlier occupation (Unit C) is characterized by a blade and bladelet technology that yielded retouched pieces, truncations, endscrapers and microliths, and, more rarely, other tools. Most cores are quite small, suggesting that they were intensively reduced. This assemblage is not a typical Late Ahmarian industry because it contains few Ouchtata bladelets. Marshy sediments were recorded in two of the excavation units; the organics in Unit F were radiocarbon dated to $25,950 \pm 440$ bp (Beta-55928) and are presumably correlated with the marsh deposit in Unit C, which yielded the lithic assemblage (Coinman 1993: 19). The later occupation (Units H–I), which is associated with a spring deposit (tufa), yielded a radiocarbon date of $20,300 \pm 600$ bp (UA-4395) (Coinman 1993: 17). The assemblage is a blade and bladelet industry, with microliths, retouched pieces and endscrapers, as well as other tools. It is a typical Late Ahmarian assemblage with numerous Ouchtata bladelets (Coinman 1993, 1998). The fauna consist primarily of equids (*Equus*) and aurochs (*Bos primigenius*), with limited numbers of ovicaprines and tortoise (*Testudo graeca*). Among the faunal assemblage were two examples of worked bone and antler/horn (Coinman 1998). One of these tools may have been a device used as a punch to remove blades and flakes from cores; the other has notches along its edge and a groove at its base. The characteristics of the site's assemblages—a full range of tool manufacturing activities and the diversity of tool types—suggest that it functioned as a longer-term base camp.

Yutil al-Hasa is a small collapsed rockshelter situated in the narrower reaches of the drainage where the wadi begins to constrict; it is approximately 4 km north of 'Ayn al-Buhira. There is a spring tufa in deposits across the wadi from the rockshelter, and evidence at the site of seeps. Two test units (A and B) yielded Late Ahmarian lithic assemblages and a radiocarbon date of $19,000 \pm 1300$ bp (UA-4396) (Olszewski 1997; Olszewski *et al.* 1990: 38). Pollen evidence indicates a change during the Ahmarian from drier conditions dominated by amaranth and chenopod types to a more mesic setting with trees like oaks (*Quercus*). The lithic assemblage appears to be somewhat similar to the industry from the earlier occupation (Unit C) at 'Ayn al-Buhira, however, the Yutil al-Hasa assemblage contains Ouchtata bladelets similar to the later occupation at 'Ayn al-Buhira. Cores are limited in frequency and there is minimal diversity in tool types (mainly microliths, retouched pieces and notch/denticulates), suggesting relatively specialized activities, such as game hunting and processing. Freshwater springs and the site's location in a narrow section of the al-Hasa drainage would have served to attract and constrict the movement of game. The generally well-preserved faunal assemblages document the procurement of gazelle (*Gazella*), equid (*Equus*), aurochs (*Bos primigenius*) and tortoise (*Testudo graeca*).

In addition to these tested sites, there are numerous examples of open-air lithic scatters and additional rockshelters. The lithic scatters range in size from about 25,000 to 15,000 m^2. Some of the smaller sites, such as WHS 623X, are knapping stations where a limited number of cores were systematically reduced and blanks removed for use elsewhere (Lindly *et al.* 2000). Recently located rockshelters, such as Tor Sadaf, and open-air sites, such as Thalab al-Buhira, may represent other specialized task camps analogous to Yutil al-Hasa. Upper Palaeolithic sites on the plateaus away from Wadi al-Hasa and its major tributary drainages are rare; this suggests that use of the plateau areas was incidental compared to the areas in the immediate vicinity of the lake and marshes.

The ecological stability of locally ameliorated situations in the eastern al-Hasa allowed Upper Palaeolithic groups to exercise alternative settlement options compared to such areas as the Negev or the Ras an-Naqab region (Olszewski and Coinman 1998). Overall, the productivity of lake/marsh resources is high year-round for foraging and collecting activities, although hunting during the fall period may be somewhat more productive in upland areas. Upper Paleolithic groups in the eastern al-Hasa likely utilized the immediate vicinity of Lake Hasa as a base for winter, spring and early summer occupation. Supplementary hunting and gathering trips into the eastern deserts, where gazelle and other large animals could be procured, were undertaken during the winter. In the late summer and fall, hunter-gatherers followed major tributary drainages south into highland areas characterized by Mediterranean forest where different sets of resources could be exploited. The association of sites with the major tributary drainages suggests that these drainages, due in part to the presence of springs and game, served as travel routes.

In Ras an-Naqab, somewhat different ecological conditions pertained than in al-Azraq Basin and the Wadi al-Hasa regions. As during earlier periods of the Palaeolithic, the Ras an-Naqab area during the Pleistocene was characterized by the presence of seasonal ponds and pools at lower elevations rather than a lake or marshes. The type of settlement patterning here was influenced by this situation, as well as the great topographic relief in the area. Surveys in the Jabal Qalkha section located one Early Upper Palaeolithic and five Early Ahmarian and non-Ahmarian sites (Henry *et al.* 1983). All of these are *in situ* rockshelter sites located at lower elevations, analogous to the site settings of Middle Palaeolithic sites in the area.

The earliest known site of the Upper Palaeolithic in the Ras an-Naqab is Wadi Aghar, a south-facing rockshelter at the base of a cliff. The sparsity of cultural materials from the excavations suggests that this site was a short-term camp where a limited range of activities occurred. The cultural deposits are relatively shallow and include a rock-lined hearth. The industry from this site has been classified as Early Upper Palaeolithic. Many of the assemblage characteristics are reminiscent of earlier Middle Palaeolithic technology, suggesting that Wadi Aghar is an example of an industry that is transitional between the Middle and Upper Palaeolithic periods (Coinman 1998; Coinman and Henry 1995: 182-91). Lithic manufacture at the site involved the production of relatively large-sized debitage that is typified by elongated blades, many of which are pointed in morphology. Very few bladelets are present, and none are retouched. Cores are relatively numerous and generally are not intensively reduced. Tools are limited in their diversity compared to later Upper Palaeolithic assemblages, but include endscrapers and burins, as well as retouched pieces and notches. The morphology of the elongated pointed blades and technology used to manufacture them strongly suggests that the assemblage from Wadi Aghar represents a local transition from the assemblages of the Ras an-Naqab Middle Palaeolithic sites of Tor Faraj and especially Tor Sabiha.

Early Ahmarian sites are represented by the rockshelters at Tor Hamar (Layers F–G), Tor Aeid and Jabal Humayma. No radiocarbon dates are available for these sites; all have been relatively dated on the basis of the characteristics of their respective lithic assemblages. At Tor Hamar, the southwest-facing rockshelter is near the confluence of the main wadi channel and several smaller tributaries. Possible attractions include the presence of pools in the wadi that hold water during the winter months, and the fact that the wadi is a route west to the lower elevation locales in the Rift Valley (Coinman and Henry 1995: 138). The lithic assemblage is oriented to the production of elongated debitage, particularly blades and bladelets (Coinman 1998; Coinman and Henry 1995: 162-69). Cores and tools are few in number. The tools consist primarily of burins, retouched flakes, endscrapers, al-Wad points and other retouched bladelets. A small faunal assemblage includes examples of gazelle (*Gazella*), caprines (*Capra*), equids (*Equus*) and jackal (*Canis aureus*) (Klein 1995: 411). Other items of note are a few shells—*Dentalium*, *Pecten*, *Ancilla*, and a gastropod; shells are rarely found at Upper Palaeolithic sites in the Levant (Reese 1995: 385).

Tor Aeid faces south and is situated above the Wadi Aghar drainage, with a view over the Wadi Qalkha drainage. There are nearby seasonal pools and access to the top of the cliff (Coinman and Henry 1995: 142). The site contains a lithic assemblage similar to that from Tor Hamar (Layers F–G), although the al-Wad points tend to be smaller. Cores are more numerous than at Tor Hamar, but tool frequency is comparable. Scrapers and burins are the most common tool types, followed by al-Wad points and retouched blades. Distinctive traits include the use of flakes as blanks for many endscrapers and numerous burins on core-like blanks. The low frequency of tools at both Tor Hamar and Tor Aeid suggest that manufacture of tools, their resharpening and use were not significant activities at either site (Coinman and Henry 1995: 169-78).

The southwest-facing rockshelter of Jabal Humayma was initially described as a 'Levantine Aurignacian' or non-Ahmarian site (Coinman and Henry 1995: 152-60). The assemblage is characterized by large blades and flakes, and a tool component comprised of notches and retouched blades and flakes. A second season of excavation yielded a significant number of al-Wad points, indicating that this site is actually Early Ahmarian, albeit with some differences when compared to the 'typical' Early Ahmarian assemblage (Coinman 1998). As Coinman points out, the earlier Upper Palaeolithic assemblages of the Ahmarian from sites in southern Jordan are quite variable in the details of their technology and typology. This suggests that the definitions used by archaeologists for these industries will need to be refined as more studies and analyses are completed. The meaning of this variability is undoubtedly attributable to a complex combination of activity, spatial and chronological differences between the sites of the Early Ahmarian.

Two sites in the Ras an-Naqab have non-Ahmarian assemblages, Tor Fawaz and Site 440. Tor Fawaz is a south-facing rockshelter overlooking the plain to the east of Wadi Qalkha (Coinman and Henry 1995: 137). The debitage is much larger than debitage from Early Ahmarian sites in the area, and there are a significant number of tools. The most common tools are retouched pieces, followed by notches, retouched bladelets, endscrapers and burins. The retouched bladelet component may indicate that this industry is later than the Early Ahmarian, but earlier than the Late Ahmarian (Coinman 1998; Coinman and Henry 1995: 144-51). Site 440 is across the wadi from Tor Hamar and thus also situated near the seasonal winter pools and along the access route to the Rift Valley (Coinman and Henry 1995: 143). The lithic assemblage was collected from the surface; no excavation units were dug. The surface sample appears to be unrepresentative. It contains a range of core types, blades, flakes and largish bladelets. Burins are the most common tool type, and usually are on core-like blanks, much like those from Tor Aeid (Coinman and Henry 1995: 179-81).

The settlement strategies of these early Upper Palaeolithic groups in Ras an-Naqab are thought to continue the seasonal transhumance pattern documented for the Middle Palaeolithic of this region. All six of the Upper Palaeolithic sites, however, are situated at low elevation locales, thus representing only the fall/winter portion of the settlement round (Henry 1995a: 426-27). Additionally, all six sites are small rockshelters, with few indications of long-term base camp occupation. Instead, each of these sites appears to be a shorter-term camp, perhaps analogous to a task camp associated with yet undiscovered larger base camps. Alternatively, it is tempting to think that perhaps these sites represent spring/summer higher elevation sites associated with a seasonal round that incorporated movement into the lower elevations of Rift Valley during the summer. This is a pattern that is proposed for some later Epipalaeolithic groups in the region, for example, the Madamaghan (Henry and Shen 1995: 311).

There are a number of other documented Upper Palaeolithic sites in Jordan. For a variety of reasons, including modern disturbance, lack of excavation and lack of study, however, the information from these sites is less complete. Sites include Marazzah North, Marazzah Northeast and Abu 'Aluba in the North Jordan Valley, north of the Dead Sea (Muheisen 1988c: 515). These are open-air sites that have been disturbed by agricultural use of the terraces, as well as natural processes. In the Tabaqat Fahl area, open-air Upper Palaeolithic sites are exemplified by Wadi Hammeh 32 and 34 (Edwards *et al.* 1988: 532-35; Macumber and Head 1991: Table 1). Only Wadi Hammeh 32 has been briefly reported; it is a severely deflated site with a small collection of lithics. Abu Sharer and Sabra 4 are in the Petra area (Schyle and Uerpmann 1988). In addition to lithics, a small hearth was found at Sabra 4. Finally, in al-Azraq Basin, the site of Enoqiyyah is associated with sediments typical of marshes in the area north of the Pleistocene al-Azraq lake/marsh (Copeland and Hours 1988: 300). Virtually all of these assemblages have been characterized as bearing a resemblance to the Levantine Aurignacian, primarily because there is a large-sized debitage with burins and endscrapers and few bladelets or retouched bladelets. This characterization, however, must be tempered by the knowledge that most of these are surface collections or small samples. When viewed in the context of the variability known to exist within Early Ahmarian assemblages, as well as the non-Ahmarian grouping, it is just as probable that the sites of Marazzah North, Northeast, Abu 'Aluba, Wadi Hammeh 32 and 34, Abu Sharer, Sabra 4 and Enoqiyyah may represent additional examples of the Early Ahmarian/non-Ahmarian industries.

The Epipalaeolithic: c. 20,000–10,300 bp

Of all the periods of the Palaeolithic, it is this latest period, the Epipalaeolithic, that has yielded the best-known archaeological record in terms of quantity of sites and diversity of site contents. This is due, in part, to the relative recency of the Epipalaeolithic; more sites are preserved because there has been less time for them to have been subjected to erosional processes. Another factor may be increased population size; a greater number of people inhabiting the landscape will both create more sites and perhaps deposit a greater quantity of cultural materials at certain sites. *In situ* sites in rockshelters and in open-air contexts have been located, as well as numerous surface scatters. Perhaps the most striking aspects of the sites of this period, however, are the spatial and chronological variability in their lithic assemblages, the higher frequency of structures, such as 'dwellings' at sites, and the increased presence of art and ornamentation. In the less than 10,000 years that comprise the Epipalaeolithic, there are minimally nine recognizable lithic industries and phases in Jordan alone. This total rises to at least 18 industries and phases if sites from both the eastern and western Levant are combined. Some of this variability can be explained as temporal changes in lithic production techniques, and some as related to task specificity at particular sites. Additional arguments have also been made that some typological and technological diversity is due to cultural differences between groups of people or alternatively to the process of lithic reduction itself. These topics are taken up in more detail below (Figure 2.4).

In Jordan, some Epipalaeolithic lithic assemblages have been classified as Qalkhan, Hamran, Madamaghan or Natufian, although the systematics of the classification framework are still being developed as new sites and areas within Jordan are investigated in greater detail. With the exception of the Natufian industry, there is incomplete agreement between various researchers on the use of this terminology. This has resulted in several temporary compromises. For example, Byrd (1994) recognizes specifically named tool types, such as the Qalkhan point and La Mouillah point, but chooses to classify assemblages with these and other nongeometric forms of microliths from the al-Azraq region as 'Non-Natufian Microlithic'. Neeley *et al.* (1998) follow a similar approach by using 'nongeometric and geometric Epipalaeolithic' for assemblages in the Wadi al-Hasa area. Another example is the Madamaghan, which is proposed by Henry (1995c: 40) for the Ras an-Naqab area and linked by him to industries of the Mushabian Complex of the Negev and Sinai. Although he defined this industry from the site of Tor Hamar, he chose the site of Wadi Madamagh in the Petra area as the type site assemblage. Because the assemblage from the site of Wadi Madamagh is more similar to assemblages from Wadi al-Hasa and al-Azraq Basin, Olszewski (1997) chooses to use the term Madamaghan in modified form by considering it strictly an industry of the eastern Levant, rather than linked to the Mushabian Complex of the Negev and Sinai. Others, such as Edwards *et al.* (1988: 535-37) in the Tabaqat Fahl area, and Muheisen and Wada (1995) in al-Azraq Basin region, use western Levantine terminology, such as Kebaran and Geometric Kebaran. To help clarify these varying viewpoints, the various terminologies that have been applied to particular sites described below will be noted in each site discussion (Table 2.1).

The earliest lithic assemblages of the Epipalaeolithic (Qalkhan, Early Hamran, Madamaghan *sensu* Olszewski, Tabaqat Fahl and some al-Azraq Basin Non-Natufian Microlithic) contain nongeometric forms of microliths. These vary in morphology, for example, arched backed and pointed bladelets, La Mouillah points, Qalkhan points and some microburins typify the Qalkhan (Henry 1995i: 219); straight-backed bladelets with truncations opposite snaps are typical for the Early Hamran (Henry 1995d: 251); arched-backed bladelets, La Mouillah points, attenuated 'lunates', and abundant microburins are common to the Madamaghan (*sensu* Olszewski *et al.* 1994: 130-34); and arched-backed bladelets, La Mouillah points and numerous microburins characterize some of the Non-Natufian Microlithic in the al-Azraq area (Byrd 1994: 210-11). All of these assemblages are blade and bladelet technology, and all but the Early Hamran yield very narrow bladelets. Virtually all of these assemblages also contain other tools, such as endscrapers on blades, burins, perforators, notches and denticulates, and retouched pieces. Most of these assemblages are either radiometrically dated between 20,000 and 15,000 bp or are typologically and technologically similar to the dated assemblages.

Beginning c. 15,000 bp, most Epipalaeolithic assemblages begin to yield geometric forms of microliths, as well as nongeometric types. The industries include the Middle, Late and Final Hamran, as well as the Madamaghan *sensu* Henry (1995c: 39-40) and some of the Non-Natufian Microlithic of the al-Azraq area (Byrd 1994: 210-11). Typical microlithic forms consist of trapeze/rectangles and La Mouillah points, and more rarely triangles. A small number of lunates is also occasionally found. Microburin technique appears to be used in most of these assemblages except for the Middle Hamran (Henry 1995c: 39). Blade and bladelet technology continues to predominate, although most bladelets are somewhat wider on average than in earlier assemblages. Scrapers, retouched pieces and notches are included among the other types of tools in these assemblages. Unusual industries include two sites in the al-Azraq Basin region with Non-Microlithic assemblages (Byrd 1994: 211-12) and sites with very wide trapezes and/or very wide 'lunates' in the al-Azraq area (Kharaneh IV) and in Wadi al-Hasa (Tor al-Tareeq).

The period between about 12,500 and 10,300 bp is characterized by the Natufian Complex. The lunate is the primary microlithic tool, with Early Natufian assemblages (12,500–11,000 bp) containing mainly Helwan (bifacially backed) lunates, and Late/Final Natufian assemblages (11,000–10,300 bp) typified by

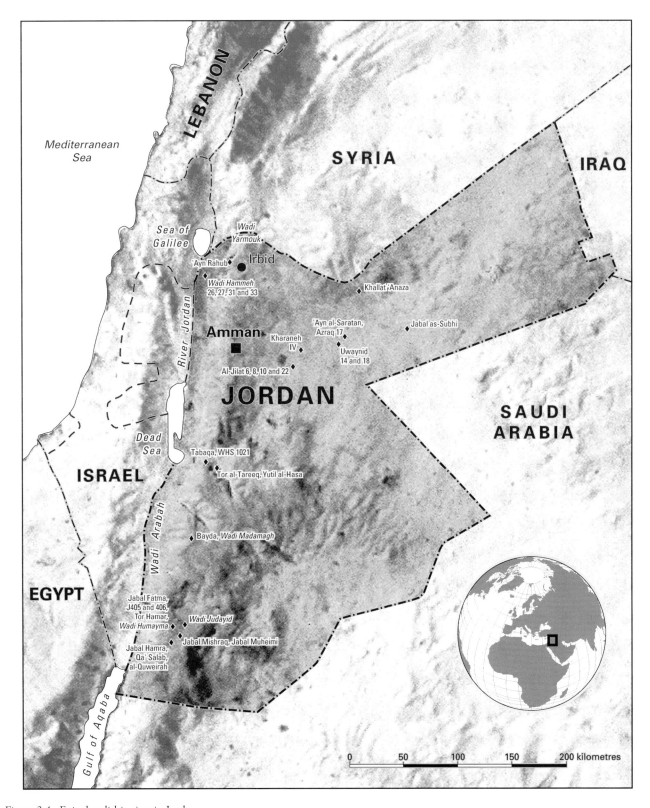

Figure 2.4. Epipalaeolithic sites in Jordan.

Qalkhan (Henry 1995)	Madamaghan (Henry 1995)	Madamaghan (Olszewski 1997)	Hamran (Henry 1995)	Non-Natufian Microlithic (Byrd 1994)	Non-Microlithic (Byrd 1994)	Unnamed Geometric (Neeley et al. 1998)	Natufian (All researchers)
Tor Hamar (E), Wadi Humayma (J406b lower), J405, J407	Tor Hamar (A-D), Jabal Fatma	Yutil al-Hasa (C, E), Tor al-Tareeq (lower)	Jabal Hamra (A-C), Jabal Misraq, Jabal Muheimi, Wadi Humayma (J406b upper), Al-Quwayra (A-C), Qa Salab (A-C), J21, J22, J26, J31	Uwaynid 14, Uwaynid 18, Al-Jilat 6, Al-Jilat 8, Al-Jilat 22 (Phase B), al-Azraq 17, Kharaneh IV (A-D)	Al-Jilat 10, Al-Jilat 22 (Phase C)	Kharaneh IV (D)	Wadi Judayid, Wadi Humayma (J406a)
Uwaynid 14, Uwaynid 18, Al-Jilat 6	Al-Jilat 8	Uwaynid 14, Uwaynid 18, Al-Jilat 6 (Phase C)					Bayda
	Wadi Madamagh	Wadi Madamagh					Tabaqa, Yutil al-Hasa (D), WHS 1021
		Wadi Hammeh 26, 31, 33		Wadi Hammeh 26, 31, 33			Wadi Hammeh 27
				Yutil al-Hasa (C,E), Tor al-Tareeq		Tor al-Tareeq (upper)	'Ayn as-Saratan
							'Khallat 'Anaza, Jabal as-Subhi
				Tor Hamar			'Ayn Rahub

Table 2.1. Epipalaeolithic lithic assemblages by researcher.

abruptly retouched lunates. Microburin technique is common to most sites of this period of the Epipalaeolithic. Other tools include scrapers, burins, notches and denticulates, and retouched pieces (Henry 1995h: 323-27). Although blade and bladelet technology is used, the blanks produced are quite short and wide. There also appears to be a higher incidence of tools on flakes compared to earlier phases of the Epipalaeolithic period.

Environment

Several significant environmental changes occurred during the span of the Epipalaeolithic period. The first of these was the global cooling that occurred as the last glacial maximum approached and peaked c. 18,000–17,000 bp. Available pollen evidence from lake cores (e.g. the Hula core) indicates that this was a cold and relatively dry interval in the Levant, characterized in many areas by an expansion of steppic vegetation communities (Baruch and Bottema 1991: 16). Certain areas of the eastern Levant, such as al-Azraq Basin and Wadi al-Hasa, probably witnessed a shrinkage in the lakes and marshes, although freshwater springs continued to flow. Sometime around 15,000 bp, climatic conditions began to ameliorate as glaciers in the northern latitudes began a relatively rapid retreat. Vegetation communities like the Mediterranean forest became much broader in distribution, as temperatures became somewhat warmer and humidity increased (Baruch and Bottema 1991: 16). A marsh ecology characterized parts of al-Azraq Basin, but this ecological setting appears to have disappeared in the Wadi al-Hasa region. This climatic optimum, characterized by increasing humidity throughout the period, continued until about 11,500 bp. From about 11,500 to 10,500 bp, a period corresponding to the Younger Dryas in the European sequence, climatic conditions worsened considerably as both temperature and humidity decreased.

This resulted in an expansion of steppic vegetation communities (Baruch and Bottema 1991: 17). The climatic sequence described here is also supported by pollen analyses derived from archaeological sites (Emery-Barbier 1995: 380-83; Leroi-Gourhan and Darmon 1991).

Research

Four major research themes characterize the study of the Epipalaeolithic: a continuing interest in reconstructing settlement and subsistence patterns; debates concerning the meaning of variability in lithic assemblages that focus primarily on whether distinct cultural groups can be identified; examination of the processes that led to early cultivation and agriculture as a subsistence strategy; and issues related to complex foraging and increased levels of social complexity. The first research theme, settlement and subsistence patterns, involves many of the models explicated for earlier periods of the Palaeolithic. The diversity of settlement strategies documented during the Epipalaeolithic, however, is a potential indication of the greater complexity of group adaptive dynamics. The steppic and desert areas of the western Levant, such as the Negev and Sinai, appear to be characterized by an abandonment of many locales during the earlier part of the Epipalaeolithic; sites of the later Epipalaeolithic (post-16,500 bp) are well represented (Goring-Morris 1987: 401-15; Marks and Freidel 1977: 146-49). Most sites in this region of the Levant are small, ephemeral camps, suggesting a highly mobile population. There are examples of Late Natufian sites, however, that contain structures, such as circular, stone-walled dwellings, that may indicate longer-term base camps. In Jordan, the closest apparent parallel to the situation in the Negev and Sinai appears to be in the Black Desert. Extensive surveys by Betts (1983, 1984) have located few traces of Epipalaeolithic sites earlier than the Late Natufian phase. One of the Late Natufian sites, Khallat 'Anaza, has structural elements, such as enclosure walls and possibly hut circles (Betts 1985: 30-32).

In the Ras an-Naqab area, Henry's (1995b: 337-43) seasonal transhumance model remains typical for settlement patterning in that region. There are, however, interesting variations compared to earlier periods of the Palaeolithic. For example, both Qalkhan and Madamaghan sites in the lower elevations appear to be winter base camps, but there is an absence of these sites in higher elevations. This suggests that the Qalkhan and Madamaghan transhumance pattern may have involved spring/summer travel to lower rather than higher elevations, for instance, to Wadi 'Arabah. The Hamran sites, on the other hand, follow a strategy of lower elevation winter base camps associated with seasonal lakes, and a higher elevation summer transitory camps. During the succeeding Early Natufian phase, multi-season base camps (e.g. Wadi Judayid J2) are established at somewhat higher elevation settings. The pattern of the Late Natufian appears to involve the use of these upper elevations, but with a return to more mobile strategies involving seasonal camps at lower elevations as well.

The Wadi al-Hasa region also documents flexibility in settlement strategies throughout much of the Epipalaeolithic period (Olszewski and Coinman 1998). During the earlier part of the Epipalaeolithic, coinciding with the apparent abandonment of many portions of the Negev and Sinai regions, the lake/marsh microhabitats of the eastern al-Hasa were subjected to a period of increased environmental fluctuation, associated with the onset, climax and retreat of the Last Glacial Maximum. Resources became more unpredictable, and, during some intervals, Lake Hasa was much reduced in size and highly saline. Despite these circumstances, the area continued to be exploited by early Epipalaeolithic (Madamaghan *sensu* Olszewski) groups. The focus of settlement, as during the Upper Palaeolithic, was within 4 km of remnant Lake Hasa. The settlement strategy is one of high residential mobility, with a probable decrease in local population size and less frequent visits to the eastern al-Hasa. Early Epipalaeolithic groups continued to establish a logistical residence (radial) pattern of relatively short duration during the spring and summer seasons of some years.

A return to more substantial occupation of the eastern al-Hasa is found beginning about 15,500 bp, broadly analogous to the reoccupation of the Negev and Sinai during this period. Local ecological conditions rebounded so that the lake and marshes achieved a resource potential comparable to the late Upper Palaeolithic period. Epipalaeolithic groups were characterized by reduced mobility (logistical mobility). As in earlier times, the immediate vicinity of Lake Hasa is the main focus of settlement. The Early Natufian phase that follows represents a distinctive shift in settlement strategy. While this does not mirror the shift witnessed in the Ras an-Naqab area during this period, it is an additional example of the flexibility of hunter-gatherer groups. Local conditions in Wadi al-Hasa remained attractive, although the lake and marsh regime characteristic of earlier periods, as well as many of the resources that typified them, disappeared. The focus of settlement shifted to areas near active springs and probably to sections of the drainage where stands of cereal grasses could be exploited. Spring/summer short-term base camps are in these locales.

The second research theme of the Epipalaeolithic concerns the interpretation of variability in lithic assemblages. There are two dimensions to this issue: contemporary but distinct assemblages found in geographically different areas (e.g. Hamran in Ras an-Naqab and 'geometric' Early Epipalaeolithic in the Wadi al-Hasa and al-Azraq areas), and contemporary but distinct assemblages found within the same region (e.g. Hamran and Madamaghan [*sensu* Henry] in the Ras an-Naqab area). Many of the distinctions that have been noted to date relate to the presence/absence of the microburin technique and to variations within the microlithic class of tools—these include variability in their morphology (i.e. different types), differences in their average length and patterning in retouch type and placement. Two research positions have crystallized—one favors an explanation based on the presence of different cultural (or 'ethnic') groups (e.g. Fellner 1995; Goring-Morris *et al.* 1996; Kaufman 1995), the other posits an explanation based on aspects of lithic reduction (e.g. Neeley and Barton 1994; Goring-Morris *et al.* 1996). Although these views appear to be polar opposites to some extent, all of these researchers emphasize the importance that such variables as raw material selection and activity focus played in the production of lithic assemblages. Perhaps the most salient point is the recognition of the fact that how raw material is exploited, including the reduction process, can be a choice that is culturally determined (e.g. Goring-Morris *et al.* 1996). This debate over cultural groups and lithic reduction processes deserves further research, in part because it is likely that elements of both explanations are responsible for the archaeological patterning that is present in Epipalaeolithic industries. Additional information from archaeological data other than the lithic assemblages will be the key to clarifying the more subtle aspects of these interpretations.

The third research theme focuses on the shift from simple to complex foraging subsistence systems during the latter part of the Epipalaeolithic. This shift appears to have involved the increased use of stone-ground plant foods—acorns and cereal grasses in the Mediterranean forest areas, and cereal grasses in the open parkland and steppic regions. Because later Neolithic groups practised an agricultural way of life, many researchers have suggested that the origins of agriculture are to be found in the archaeological record of the Late Epipalaeolithic. Concomitant with the increased emphases on certain plant foods in the Late Epipalaeolithic is the emergence of large, multi-season base camps that contain abundant examples of groundstone implements for stone grinding of acorns and cereals, such structures as circular stone-walled dwellings, bone tools, ornaments and other art, and burials. The most varied and extensive archaeological record of this socioeconomic shift is from sites in the western Levant (e.g. al-Wad and 'Ayn Mallaha), although a few examples (e.g. Wadi Hammeh 27) have also been found in Jordan. Perhaps the most intriguing aspect of this research topic lies in the knowledge that the most complex of these sites are Early Natufian in age and are located primarily in the Mediterranean forest zone. By the Late Natufian, sites appear more often to be transitory camps, except for a few sites in the Negev and Black Desert areas. The distribution of late Natufian sites suggests that increased reliance on cereal grasses, which eventually led to their domestication during the Neolithic period, was a development of the steppic areas rather than the more optimal Mediterranean forest zone (Olszewski 1993) and that it occurred in a subsistence strategy characterized by simple foraging. Previously, areas such as the Jordan Valley have been suggested as one locale of origin for early cultivation (Bar-Yosef and Belfer-Cohen 1989), but others may have been the vast open steppes that dominated much of the ancient landscape of Jordan.

The later Epipalaeolithic is also the focus for the research theme of increased social complexity, particularly during the Early Natufian phase. The large base camps, art, ornamentation and burial data have suggested to some researchers that Early Natufian societies were characterized by increased population size and density, more ceremonialism, and perhaps social ranking (e.g. Henry 1989: 197-211). For example, certain individuals appear to have been buried with more elaborate grave goods than others; there are also differences between individuals in group burials where one individual of the group is a primary burial and the others are secondary. This might indicate that people with a lower social ranking who predeceased a more important individual were removed from their original graves and re-interred when the higher ranked person died. The majority of the archaeological record utilized by researchers to investigate this topic again is from sites in the western Levant (e.g. al-Wad). The best examples of these types of sites in Jordan are Wadi Hammeh 27 and 'Ayn al-Saratan. Few would argue that the Early Natufian phase is not distinct in its archaeological signature. However, not all agree that the level of social complexity necessarily entailed social ranking (e.g. Olszewski 1991). Much of the evidence cited in support of social ranking is derived from excavations conducted in the 1930s when excavation techniques and recording methods were not as meticulous as now.

This leaves open to debate the contextual association of various burials and their contents.

Sites of the Early Epipalaeolithic

The Early Epipalaeolithic is generally used by Levantine researchers to refer to all Epipalaeolithic complexes prior to the Natufian. In Jordan, these include the Qalkhan, Hamran, Madamaghan (*sensu* both Henry and Olszewski), Non-Natufian Microlithic, and Non-Microlithic, as well as assemblages variously attributed to the western Levantine Kebaran and Geometric Kebaran. These sites are widely distributed across many areas of Jordan, for example, the al-Azraq area, Wadi al-Hasa, Ras an-Naqab, Tabaqat Fahl and the Petra region (see Figure 2.4). In the al-Azraq area, there are a number of open-air sites, including Uwaynid 14 (middle and upper phases) and 18 (upper phase), Kharaneh IV and Wadi al-Jilat 6 and 22. Others are al-Azraq 17 (Trench 1), and Wadi al-Jilat 8 and 10. The industries are described as Non-Natufian Microlithic (Byrd 1994: 210-11), although some of the assemblages, especially those from Uwaynid 14 and 18, and probably Wadi al-Jilat 6 (Phase C), could be considered Madamaghan (*sensu* Olszewski), or possibly Qalkhan (e.g. Henry 1995: 233-34).

The earliest chronometrically dated sites are Uwaynid 14 and 18, which yielded radiocarbon dates associated with diagnostic lithic industries that range from $18,400 \pm 250$ bp (OxA-866) at Uwaynid 14 to $19,800 \pm 350$ bp (OxA-864) at Uwaynid 18 (Garrard *et al.* 1988a: 325-26). The lithic assemblages are characterized by numerous arched-backed and pointed bladelets, variable quantities of La Mouillah points and truncated backed bladelets, extensive use of the microburin technique, and various larger tools, such as scrapers, burins and truncations. At Uwaynid 14 there is a temporal trend toward larger microliths and increased numbers of La Mouillah points. Faunal remains from these two sites include gazelle (*Gazella*), equids (*Equus*), aurochs (*Bos primigenius*), hare (*Lepus capensis*) and tortoise (*Testudo graeca*). A number of hearths were recorded at Uwaynid 18, and shell beads (*Dentalium*), ochre and a grinding stone were also recovered. These two sites are only 120 m apart, and may represent portions of a single large site that has been divided by erosional channels. They are both situated in palaeosols, indicating somewhat more humid conditions during the periods they were occupied (Garrard *et al.* 1987: 9-15).

At Wadi al-Jilat 6, the earliest occupation (Phase C) contains a lithic assemblage analogous to those from Uwaynid 14 (middle phase) and Uwaynid 18. Like the Uwaynid sites, the Phase C occupation is associated with a palaeosol indicating more humid climatic conditions (Garrard *et al.* 1988a: 320). The succeeding occupation at Wadi al-Jilat 6 (Phase B) is also contained within a palaeosol. This phase, like Uwaynid 14 (upper phase), yielded a lithic assemblage characterized by larger microliths, with numerous examples of La Mouillah points. Interestingly, the Phase B assemblage has a low frequency of Qalkhan points (Garrard and Byrd 1992: 50-52), which are used as a marker for the Qalkhan industry in the Ras an-Naqab area (Henry 1995: 38). The Phase B occupation predates c. 16,700–15,500 bp, based on radiocarbon dates obtained from charcoal samples in the Phase A deposit (Garrard and Byrd 1992: 50). Both the Phase B and C occupations are believed to represent short-term camps. The latest occupation at Wadi al-Jilat 6 (Phase A) is contained in aeolian silt/loess, indicating dry climatic conditions (Garrard *et al.* 1988a: 320). Somewhat contrary to expectation, it is this occupation that yields the best evidence for a longer-term base camp and/or more frequent site visits. Phase A contains three living floors, two of which are ochre stained; their edges are lipped up, suggesting that they were the floor surfaces for structures, such as huts (Garrard and Byrd 1992: 49-50). The lithic assemblage continues to contain numerous microburins, but the microliths are overwhelmingly geometric in form (primarily small triangles). This phase of occupation is radiocarbon dated by several dates between $16,700 \pm 140$ bp (AA-5494) and $15,470 \pm 130$ bp (AA-5492) (Garrard and Byrd 1992: 50). The three phases at this site yielded faunal assemblages containing gazelle (*Gazella*), equids (*Equus*), hare (*Lepus capnesis*) and tortoise (*Testudo graeca*), as well as a rarer fox (*Vulpes*), wolf (*Canis lupus*), aurochs (*Bos primigenius*), wild pig (*Sus scrofa*) and birds, such as ostrich (*Struthio camelus*) and various migratory bird species (Garrard *et al.* 1988a: 320). Other finds include shell beads (*Dentalium, Ancilla, Arcularia* and *Cerastoderma*), bone points and beads, and groundstone implements (Garrard and Byrd 1992: 52). Seeds of steppic plants, such as members of the Chenopodiaceae and Compositae, as well as various grasses, were recovered from Phase A (Garrard *et al.* 1988b: 44). All of these seeds are edible, although use of these for food in Phase A is not documented (Garrard and Byrd 1992: 52).

The site of Kharaneh IV contains several Epipalaeolithic occupations that mirror some, but not all, of the temporal changes seen at such sites as Uwaynid 14 and Wadi al-Jilat 6. Kharaneh IV is situated roughly between the Uwaynid and al-Jilat areas, but slightly farther to the west. Use of the microburin technique is not reported for Phases A–C at this site, which sets it apart from the other non-Natufian microlithic assemblages of the al-Azraq region. There are four phases of occupation; the earliest is Phase A. The Phase A lithics are reminiscent of the western Levantine Early Kebaran, with microgravettes and partially retouched bladelets. The assemblage also contains numerous endscrapers on blades along with burins and additional large tools (Muheisen 1988a: 353-54). Other significant finds include a living floor and hearth, and a faunal assemblage containing equids (*Equus*), gazelle (*Gazella*), tortoise (*Testudo graeca*) and hare (*Lepus capensis*). Phase B also yielded a living floor with a hearth, as well as two burials below the floor. The lithic assemblage contains numerous microliths that are either narrow-pointed forms or truncated and backed; it is considered to be a later Kebaran-type industry (Muheisen 1988a: 358). The Phase C occupation contains another living surface and a hearth. The microliths continue to be mainly truncated or backed and truncated types, although there are also a few geometrics, such as trapezes. Muheisen (1988a: 365) considers this assemblage to be similar to the western Levantine Geometric Kebaran. The Phase D occupation has been the most extensively studied to date (Muheisen 1988a: 362; Muheisen and Wada 1995). The lithic assemblage is somewhat unusual in that it contains very wide trapezes and some wide lunates, types known from only one other site in Jordan (Tor al-Tareeq in the Wadi al-Hasa area). Microburin technique is documented, but does not occur frequently. This assemblage is considered to represent an industry analogous to the Final Geometric Kebaran, and is radiocarbon dated to between 16,000 and 15,000 bp (Muheisen cited in Garrard and Byrd 1992: 60). A living surface with five hearths was found; these appear to be semi-encircled by probable post holes that indicate a dwelling such as a hut (Muheisen 1988a: 362). The faunal assemblage is similar to that from Phase A, with gazelle (*Gazella*), equid (*Equus*), hare (*Lepus capensis*), fox (*Vulpes*), aurochs (*Bos primigenius*) and tortoise (*Testudo graeca*). Two incised bones, shells (*Dentalium*) and ochre were also found. The density of cultural materials from each of the phases, in conjunction with the presence of living floors and the burials in Phase B, suggest that Kharaneh IV was a longer-term base camp that was repeatedly visited over the millennia.

Another somewhat atypical site in the al-Azraq area is Wadi al-Jilat 22. The earliest two occupations (Phases E and C) are radiocarbon dated to c. 13,500–12,800 bp, and are contained in marsh deposits (Garrard and Byrd 1992: 53-54). These Non-Microlithic assemblages are similar, with a low frequency of microliths and high frequencies of large tools, such as endscrapers, retouched pieces, notches and denticulates, and burins. Use of the microburin technique occurred. The most unusual tool type is the al-Jilat knife, a pointed blade with a retouched tang. Microwear analyses have determined that these tools functioned as knives rather than projectiles (Garrard and Byrd 1992: 54-56). Al-Jilat knives comprise more than 50% of the Phase C lithic assemblage, indicating that during this phase al-Jilat 22 may have been a specialized activity area. The Phase B assemblage is derived from an aeolian silt deposit suggestive of somewhat drier conditions and is microlithic in character, containing La Mouillah points, trapeze-rectangles, triangles and lunates. Use of the microburin technique is very common. A radiocarbon date of 11,920 ± 180 bp (OxA-1770) was obtained for this phase (Garrard and Byrd 1992: 53). The faunal assemblage from the phases contains a high frequency of tortoise (*Testudo graeca*) carapace fragments; there are also gazelle (*Gazella*), equid (*Equus*), hare (*Lepus capensis*), wolf/jackal (*Canis*) and fox (*Vulpes*), as well as one caprine bone (Garrard and Byrd 1992: 56). The most striking aspect of the faunal assemblage, however, is the high incidence of bird bones, which include sandgrouse, eagles and mallards (*Anas platyrhyncos*).

Early Epipalaeolithic open-air sites in the Tabaqat Fahl region include Wadi Hammeh 26, 31 and 33 (Edwards *et al.* 1988: 535-41). The largest assemblage is from Wadi Hammeh 26; it contains numerous narrow microliths of the truncated and backed type, and a small number of microburins. Faunal remains include gazelle (*Gazella*), pig (*Sus scrofa*), hare (*Lepus capensis*), fox (*Vulpus*), birds and tortoise (*Testudo graeca*). A radiocarbon date of 19,500 ± 600 bp (SUA-2101) was obtained from hearth charcoal (Edwards *et al.* 1988: 537). The lithic assemblages from these three sites have been attributed to the western Levantine Kebaran, although the presence of microburins suggests that these assemblages are more appropriately classified using eastern Levantine terminology (e.g. Non-Natufian Microlithic, or Madamaghan *sensu* Olszewski, or possibly Qalkhan).

Three sites yielding Early Epipalaeolithic assemblages are known from the Wadi al-Hasa region; these are the rockshelters at Yutil al-Hasa (Units C and E) and Tor Sageer, and the site of Tor al-Tareeq, which has two distinct Early Epipalaeolithic phases. Tor Sageer has been recently excavated and is not discussed in detail here. The predominant lake environment of the latter part of the Upper Palaeolithic was displaced by ponds and stream-fed marshes during the Early Epipalaeolithic. Yutil al-Hasa continued to be used as a task camp associated with hunting activities. The Madamaghan (*sensu* Olszewski) occupation yielded a lithic assemblage of extremely limited diversity. Tools consist almost exclusively of microliths, which are very narrow and include La Mouillah points, curved (arched-backed), backed and truncated, and 'attenuated lunate-types' (Olszewski *et al.* 1994: 131-34). A few scrapers, burins, notches and denticulates, and other large tools were also present. Faunal remains include gazelle (*Gazella*), equids (*Equus*), aurochs (*Bos primigenius*) and tortoise (*Testudo graeca*), as well as a few marine shells. A fragment of a bone awl and a polished bone also were recovered. Current evidence suggests that uses of the rockshelter were of briefer individual duration than during Late Upper Palaeolithic times. The assemblage from Tor Sageer is similar to that of Yutil al-Hasa.

The site of Tor al-Tareeq, a south-facing, possibly collapsed rockshelter, which is 4 km upstream from Yutil al-Hasa, has dense superimposed deposits on the steep terrace in front of the bedrock outcrop. The cultural deposits rest on a sediment complex underlain by lacustrine marl facies typical of a lake edge and pond environment (Schuldenrein and Clark 1994: 37). Faunal remains include gazelle (*Gazella*), equid (*Equus*), aurochs (*Bos primigenius*), hare (*Lepus capensis*), wolf (*Canis lupus*), fox (*Vulpes vulpes*), badger (*Meles meles*), tortoise (*Testudo graeca*), rock dove (*Columba livia*), fish and crab (Clark *et al.* 1988: 268), while the pollen profiles include cattail (*Typha*), riparian willow (*Salix*), alder (*Alnus*) and oak (*Quercus*), as well as a substantial Chenopodiaceae component, all of which are indicative of a mosaic of open steppe and lake/marsh environments. Tor al-Tareeq is described by Neeley *et al.* (1998) as a short-term base camp. The early phase of occupation yielded a nongeometric microlithic assemblage with frequent microburins. It is similar in composition to the assemblage from Yutil al-Hasa, and contains numerous examples of narrow arched-backed bladelets (including curved types), as well as backed and truncated varieties. This early phase occupation features low intensity of core reduction and low ratios of tools to cores, which are thought to represent relatively brief residential occupancy with curation of tools for use offsite. This assemblage has been classified as nongeometric Epipalaeolithic by Neeley *et al.* (1998) and as Madamaghan by Olszewski (1997). Radiocarbon dates place this occupation in the interval between about 16,900 and 15,500 bp. In contrast, the later geometric phase occupation has an industry analogous to that from Kharaneh IV (Phase D), with wide lunates and trapezes (classified as bi-truncated bladelets) (Neeley *et al.* 1998). Early Epipalaeolithic occupations at Tor al-Tareeq also include groundstone implements in the form of pestles, mortars and bedrock mortars, as well as several hearths and a stone alignment in the form of an arc that may represent a structure. Perforated shells of *Mitra*, *Dentalium*, *Nerita* and *Conus* were recovered; these represent ornaments. The geometric phase of occupation has evidence of more on-site reduction activity, indicative of *in situ* tool manufacture, use and discard, and potentially lengthier periods of occupation. This shift to lengthier stays has also been concluded by Stevens (1996: 154-56).

In the Petra region, the rockshelter site of Wadi Madamagh yielded a lithic assemblage containing narrow microliths in the forms of arched-backed, pointed and 'attenuated lunate type' (Kirkbride 1958). Microburin technique is common. Faunal materials, including marine shell, were also recovered, and several hearths were documented. This is a large rockshelter (c. 100 m^2); in conjunction with the density of cultural materials it may have been either a base camp or a frequently revisited task locale. New excavations in the rockshelter by Schyle and Uerpmann (1988: 47-52) and a re-examination of the lithics by Byrd (1994: 210) have confirmed its Early Epipalaeolithic character. This site was used by Henry and Garrard (1988: 17-18) as the type site for the Madamaghan industry when they reported the materials from the site of Tor Hamar in the Ras an-Naqab region to the south. The typology used by Kirkbride (1958: 57), which included the term 'microlithic lunate-type bladelets', does not refer to true lunates typical of geometric microlith assemblages. Despite the presence of very rare asymmetrical lunates and triangles, the assemblage from Wadi Madamagh is more like those from Wadi al-Hasa and the al-Azraq region than the Tor Hamar assemblage, forming the basis for Olszewski's (1997) use of the term Madamaghan *contra* Henry's (1995c: 40) use of this term.

A number of different Early Epipalaeolithic industries are found in the Ras an-Naqab area—Qalkhan, Madamaghan (*sensu* Henry) and Hamran. Many of the sites are multi-component, containing either a sequence of temporally distinct industries or phases of a particular industry. Sites with Qalkhan assemblages include J405, J407, Wadi Humayma (J406b lower), and Tor Hamar (Layer E). The Madamaghan was found at Tor Hamar (Layers A–D) and Jabal Fatma. Hamran sites are Jabal Hamra, Jabal Misraq, Jabal Muheimi, Wadi Humayma (J406b upper), J21, 22, 26 and 31.

The two most intensively investigated Qalkhan sites include one open-air locality (Wadi Humayma) and one rockshelter (Tor Hamar). Lithic assemblages are dominated by nongeometric forms of microliths (e.g. narrow arched-backed and pointed bladelets, La Mouillah points, truncations opposite snaps). Microburin technique is common, and Qalkhan points are the diagnostic element (Henry 1995i: 219). There are also a small number of geometric microliths (trapeze/rectangles). Other tools include retouched pieces, notches, scrapers and chamfered blades. No hearths or faunal remains were recovered from these sites. Because many of the Qalkhan sites are small in size and, with the exception of Tor Hamar, have shallow deposits, they are likely to represent short-term transitory camps that were revisited on an infrequent basis (Henry 1995i: 230).

Both sites yielding Madamaghan (*sensu* Henry) assemblages are rockshelters (Tor Hamar and Jabal Fatma). Nongeometric microliths are the dominant tool form and include arched-backed and straight-backed bladelets and La Mouillah points, as well as scalene bladelets. A small proportion of the microliths are geometrics, mainly Helwan lunates (Henry and Garrard 1988: 8; Henry and Shen 1995: 301). Microburin technique is used. Other tools occurring in moderate frequencies are scrapers and retouched pieces. At Tor Hamar (Layers A–D), a diverse assemblage also includes groundstone implements (a pestle, a handstone, a bowl fragment and mortars), shell (*Dentalium*, *Ancilla*, *Cerithium*, *Conus*, *Cypraea*, *Columbella*, *Gibbula*, *Glycymeris*, *Ostrea*, *Pecten* and *Arcularia*) and bone points (Henry and Shen 1995: 308-309; Henry and Garrard 1988: 13; Reese 1995: 386). The faunal assemblage from Tor Hamar includes gazelle (*Gazella*), caprines (*Capra*), equids (*Equus*), aurochs (*Bos primigenius*), hare (*Lepus capensis*), fox (*Vulpes*), jackal (*Canis aureus*), tortoise (*Testudo*) and chukar partridge (*Alectoris chukar*) (Henry and Garrard 1988: 13-14; Klein 1995). Fish scales, as yet unstudied, and ostrich (*Struthio camelus*) eggshell fragments are also reported from Tor Hamar (Klein 1995: 410). The depth of deposits at Tor Hamar suggest that it was consistently revisited over some length of time.

During the Early Hamran, there are both lowland rockshelters (Jabal Hamra Layer C, Jabal Misraq and Jabal Muheimi) and highland open-air sites (Wadi Humayma [J406b upper], J21 and 22). In these assemblages, the key lithic is nongeometric microliths of the straight-backed and truncated opposite snaps type. Bladelets, including those retouched into tool types, are somewhat wider than in other industries. Retouched pieces, scrapers and burins also occur (Henry 1995: 251-55). A small number of groundstone implements, including shallow bedrock mortars and a portable mortar (both at Jabal Misraq), were recorded. Ornamental shells include *Dentalium* and *Mitra*. Perhaps the most intriguing discovery, however, are the panels of petroglyphs at the Jabal Hamra, Jabal Mishraq and Jabal Muheima rockshelters. These include many examples of wild goats; on the basis of desert varnish that has resulted from weathering, and on superimposed ancient Thalmudic script, the petroglyphs are thought to be of great antiquity (Henry 1995d: 259-60). The lowland rockshelters have thick cultural deposits and likely represent long-term base camps that were repeatedly visited; on the other hand, the high elevation sites are small in size and have shallow deposits indicative of short-term camps.

The Middle Hamran was found at Jabal Hamra (Layers A and B), al-Quweirah (Layers B and C), J26 and 31. As during the Early Hamran, lowland sites are rockshelters (Jabal Hamra and al-Quweirah) and upland sites are open-air (J26 and 31). The nongeometric microlith component occurs in high frequency and continues to contain straight-backed bladelets with truncations opposite snaps. Geometric microliths become more common during this phase and are trapeze/rectangle in form (Henry 1995d: 262-64). The tendency toward wider bladelets and bladelet tools also continues during this phase. Microburins are extremely rare, indicating that this technique was not extensively utilized. Retouched pieces and scrapers occur among the larger tools. Groundstone implements have rarely been recovered. Ornamental shell is also rare (*Dentalium* and *Arcularia*) (Reese 1995: 386). The pattern of longer-term base camps at rockshelters at lower elevations and short-term, ephemeral camps at higher altitudes continues.

Al-Quweirah (Layer A) and Qa Salab (Layers A–C) yielded Late to Final Hamran assemblages. Both are lowland rockshelters, representing longer-term base camps. During this phase of the Hamran, trapeze/rectangles are replaced by lunates as the dominant geometric form of microlith. Nongeometric microliths, however, still dominate the tool kit and include straight-backed bladelets with truncations opposite snaps. Relatively wide bladelets continue to be manufactured. Microburin technique becomes more important during these phases, and appears to be related to the emphasis on the production of lunates (Henry 1995d: 270-75). Numerous ornamental shells (*Dentalium*, *Glycymeris*, *Pecten*, *Arcularia*, *Mitra*, *Cypraea* and *Conus*) were also recovered (Reese 1995: 386).

Sites of the Late Epipalaeolithic

Levantine researchers generally refer to all phases of the Natufian as Late Epipalaeolithic. Most of the sites that have been examined in detail in Jordan are Early Natufian in age ('Ayn al-Saratan, Wadi Hammeh 27, Tabaqa, Yutil al-Hasa, Bayda, Wadi Judayid), with a few Late Natufian sites found primarily in the Ras an-Naqab (Jabal Humayma J406a) and Black Desert (Khallat 'Anaza and Jabal as-Subhi), as well as one site ('Ayn Rahub) in the northern Jordanian highlands north of Irbid (see Figure 2.4). In the al-Azraq area, only one Late Epipalaeolithic site was recorded during the intensive surveys of the Wadi al-Jilat, Wadi Uwaynid and central al-Azraq Basin areas. This is the Early Natufian occupation at 'Ayn al-Saratan at a minor spring in the vicinity of the perennial springs at al-Azraq ash-Shishan (Garrard 1991: 235-36). The hallmark Helwan lunates are an important type in the geometric microlithic component, which also includes bipolar and abruptly backed lunates. Other tools are endscrapers, drills and notches. A few groundstone implements (handstones), shell beads (*Dentalium*), a probable bone sickle haft fragment, and a perforated horn core (aurochs) were recovered (Garrard 1991: 238). The faunal assemblage included aurochs (*Bos primigenius*), equids (*Equus*), gazelle (*Gazella*), hare (*Lepus capensis*), wolf (*Canis lupus*), fox (*Vulpes*), and such birds as pintail (*Anas acuta*) and garganey (*Anas querquedula*). Although no recognizable pit(s) could be discerned, several fragmented human skeletons were recorded. Both male and female adults were present, as well as juveniles and a subadult (Garrard 1991: 240). These appear to be secondary burials. It is probable that 'Ayn as-Saratan represents a longer-term base camp.

In the Tabaqat Fahl region, the Early Natufian site of Wadi Hammeh has also yielded a diverse assemblage, as well as structures and burials indicative of a longer-term base camp occupation. The cultural materials are contained within a humic silty clay indicative of the availability of springs in the vicinity (Edwards 1991: 124). The geometric microlith component is overwhelmingly dominated by Helwan lunates; other microlithic forms include Helwan retouched bladelets and inversely retouched bladelets. Larger tools consist of scrapers, burins and notches and denticulates. The majority of the studied materials are from Phase I, the uppermost occupation at the site; this occupation is radiocarbon dated to between c. 12,200 and 11,900 bp (Edwards 1991: 128). A very diverse array of other stone artifacts, including groundstone, was also recovered. These consist of both basalt and limestone pestles, mortars, mullers, vessel rim fragments, small bowls and shaft straighteners (Edwards 1991: 129-35). Many of these stone artifacts are decorated with crenellated motifs, grooves, incisions and zoomorphic designs. There are also three engraved slabs that were used in the construction of one of the structure walls. Two stone-walled structures are known from Phase I—one is curvilinear, the other is U-shaped. Bone artifacts include points, beads, pendants and six almost complete sickle hafts, one of which had Helwan lunates in place (Edwards 1991: 136). *Dentalium* shell beads were found in the structures as well as with one of the burials. The skeleton fragments from Phase I appear to be from at least three adults; there is also one adult male burial beneath the earliest occupation (Phase III). The faunal assemblage includes gazelle (*Gazella*), caprines (*Capra*), wild pig (*Sus scrofa*), hare (*Lepus capensis*), birds (at least seven species), equids (*Equus*), aurochs (*Bos primigenius*), red deer (*Dama dama*), roe deer (*Capreolus*), wolf (*Canis lupus*), fox (*Vulpes*), wild cat (*Felis*), tortoise (*Testudo*), freshwater crab (*Potamon potamon*) and various reptiles (Edwards 1991: 146). Identified seeds include grasses, wild barley (*Hordeum spontaneum*) and legumes.

Three sites in Wadi al-Hasa have been identified as Early Natufian. One is a rockshelter (Yutil al-Hasa Unit D), and two are open-air localities (Tabaqa and WHS 1021). As in earlier Palaeolithic times, Yutil al-Hasa appears to have functioned during the Early Natufian as a task camp associated with hunting activities. Tool diversity and core reduction processes, however, are indicative of somewhat more diverse activities during this phase of occupation than during the Madamaghan use of the site (Olszewski *et al.* 1994; Olszewski 1997). The small lithic assemblage contains Helwan lunates, Helwan bladelets, nongeometric microliths, microburins, notches and denticulates, retouched pieces and a few burins. Faunal remains consist of gazelle (*Gazella*) and aurochs (*Bos primigenius*), and a few marine shell fragments. A series of mortars is present in the bedrock that caps the ridge along which Yutil al-Hasa is located. Some of these are desert varnished, and it is tempting, on the basis of the emphasis on groundstone implements at other Early Natufian sites, to suggest that these mortars may date to the Natufian utilization of Yutil al-Hasa.

Approximately 6 km downstream is the site of Tabaqa, tested by Byrd and Colledge (1991) and later (1997) by Olszewski. The site is situated on a terrace overlooking the confluence of Wadis al-Ahmar and al-Hasa. The lithic assemblage is characterized by the microburin technique, which was used to produce Helwan lunates. Other tools include a small number of nongeometric microliths, notches and denticulates, retouched pieces, scrapers and burins. Charred plant remains yielded grasses, and the faunal assemblage contained gazelle (*Gazella*) and marine shell. A number of groundstone implements, such as handstones, bowls, pestles, mortars and querns, as well as stone beads, were also recovered (Byrd and Colledge 1991: 272). Tabaqa appears to be a base camp occupied on a seasonal basis, probably for exploiting cereal grasses. A possible associated task locale is the nearby site of WHS 1021.

Al-Bayda, a site in the Petra area best known for its Neolithic village and assemblages (see Chapter 3), yielded an Early Natufian occupation some 1.5 m below the Neolithic deposits (Byrd 1989, 1991). Radiocarbon dates indicate that this phase occurred between about 12,900 and 10,900 bp (Byrd 1991: 249). Lunates are the most frequent microlith type and are dominated by the Helwan type (Byrd 1989: 66-69). Notches and denticulates were very common; such tools as scrapers, retouched pieces, truncations and burins were also present. The faunal assemblage included caprines (*Capra*), gazelle (*Gazella*), aurochs (*Bos primigenius*) and equids (*Equus*) (Hecker 1989: 97-99). Marine shell was also recovered; several genera were present—*Dentalium, Nerita, Strombus, Pyrene, Clypeomorus, Pecten, Cypraea, Vermetus* and *Columbella* (Reese 1989: 102-103). Although no structures were recorded, there were several hearths and roasting areas (Byrd 1989: 78-81). Like Tabaqa, al-Bayda is interpreted as a seasonal base camp that was revisited on numerous occasions. Byrd (1991: 260) suggests the winter season as the most likely option.

In the Ras an-Naqab region, an Early Natufian site (Wadi Judayid) and a Late Natufian occupation (Wadi Humayma J406a) were documented. Both are open-air sites at relatively high elevation. Wadi Judayid is radiocarbon dated to c. 12,800–12,100 bp (Henry 1995h: 321). As expected, geometric microliths are a dominant tool class, containing predominantly Helwan lunates in the Wadi Judayid assemblage and bipolar and abruptly retouched lunates in the Wadi Humayma industry. Microburin technique is common at both sites. Groundstone implements include handstones and mortars (Wadi Judayid) and a grinding slab (Wadi Humayma). Ornamental shell consists of a relatively large assemblage at Wadi Judayid (*Dentalium*, *Nerita* and gastropods) and a few specimens of *Dentalium* and *Arcularia* at Wadi Humayma (Reese 1995: 387). Faunal remains at Wadi Judayid include wild goat (*Capra*) and wild sheep (*Ovis orientalis*), gazelle (*Gazella*), wild ass (*Equus*), aurochs (*Bos primigenius*), hare (*Lepus capensis*), leopard (*Panthera pardus*), fish and birds (Henry 1995a: 327). Henry (1995h: 330-31) interprets Wadi Judayid as a multi-season base camp and Wadi Humayma as a seasonal camp (mid-summer to early fall), perhaps broadly analogous to the seasonal camps at Tabaqa and al-Bayda.

Late Natufian phase sites were located in the Black Desert region (Betts 1983, 1985). These include Khallat 'Anaza and Jabal as-Subhi, both open-air locales. The chipped stone industries from both sites contain bipolar or abruptly retouched lunates, although nongeometric microlith forms are almost as common. Other tools include retouched pieces, notches and borers (Betts 1991). At Khallat 'Anaza, two enclosure walls and a circular stone-walled dwelling have been documented. Other items of note are mortars, and shell and stone beads. The faunal assemblage contains gazelle (*Gazella*), ovicaprids, equids (*Equus*), hare (*Lepus capensis*) and canid (*Canis*). Betts (1991: 230-31) suggests that Khallat 'Anaza was a seasonal base camp and Jabal as-Subhi a shorter-term occupation. Both sites may have been focused on hunting activities, particularly during the late winter and early spring.

A final Natufian settlement was recorded at 'Ayn Rahub in the Jordanian highlands (Muheisen 1988b). The site is an open-air locale on a terrace with several curvilinear and linear walls that probably represent dwellings. There is a high frequency of geometric microliths that consist mainly of bipolar and abruptly retouched lunates. Various nongeometric microliths are also present, as are retouched pieces and notches and denticulates. Other tools, such as scrapers, burins and borers are less frequent. Microburin technique was used. Two groundstone fragments and faunal remains also were recovered (Muheisen 1988b: 482). The presence of structures probably indicates that 'Ayn Rahub was a longer-term base camp.

Concluding remarks

Palaeolithic research in Jordan has both broadened and sharpened our understanding of ancient human groups during the Pleistocene in the Levant. Hundreds of open-air sites and dozens of rockshelter locales are currently documented. Many of these sites are *in situ* and have yielded lithic and faunal assemblages, and occasionally groundstone implements, structures, art, ornaments and burials. Analyses of the archaeological data from surveys and excavations have led to new or refined theoretical contributions and have raised a number of additional research questions and directions for future research.

Theoretical contributions

Investigations into several research themes in Jordan have resulted in theoretical contributions to the study of settlement systems, to the interpretation of variability in lithic assemblages and to insights into the development of agriculture. All of these contributions have begun with models and ideas developed on the basis of research in the western Levant. In the area of settlement patterning, Henry's (1995a: 417-37) seasonal transhumance model adds considerable flexibility to the Negev/Sinai model of long-term, stable radial or circulating settlement systems; each system is hypothesized to last several millennia (Marks and Freidel 1977). The seasonal transhumance model, on the other hand, posits a shift between radiating (fall/winter) and circulating (spring/summer) strategies within an annual cycle that is tied to movement between lower and higher elevational zones. For regions not characterized by great elevational contrast, and particularly for those areas with Pleistocene marsh and lake ecologies in Jordan, Olszewski and Coinman (1998) have proposed another set of settlement strategies. During more mesic periods settlement systems were likely logistical (radial) and tethered to these 'oases' for much of each year. Xeric climatic intervals witnessed residential mobility (circulating) likely tied to the availability and locations of springs in these areas. These strategies were highly flexible and could vary either within a yearly cycle or between years or decades.

The documentation and analysis of numerous lithic assemblages in Jordan, particularly those of the Upper and Epipalaeolithic periods, has led to the recognition of great variability in technology and typology. During the Upper Palaeolithic, there are indications that considerable variability exists within the Ahmarian Complex; this currently is treated within the contrast between Ahmarian and non-Ahmarian industries (Coinman 1998). During the Epipalaeolithic, the advent of the microburin technique is now known to have occurred at least as early as 20,000 bp (Byrd 1988: 263). More importantly, it would appear that this technique is not of North African origin, as previously believed, but is more probably a development of the eastern Levant. The variability between Epipalaeolithic lithic assemblages has led to the recognition of several new industries and complexes, such as the Qalkhan, Hamran and Madamaghan (Henry 1995c: 37-40). Agreement between researchers on the use of these terms is sometimes less than perfect; this is a reflection of the ongoing development of the systematics of classification in Jordan. Interpretation of this lithic variability has been seen as the result of either lithic reduction processes (Neeley and Barton 1994) or a combination of lithic reduction processes and stylistic (ethnic or cultural) traditions (Fellner 1995; Goring-Morris *et al.* 1996; Kaufman 1995).

The Late Epipalaeolithic (Natufian Complex) is widely regarded as a period of intensified cereal grass use and cultivation. This subsistence emphasis is believed to have led to the development of agriculture. The Early Natufian phase has yielded the most dramatic evidence of a complex foraging lifeway, including abundant groundstone implements, dwellings, art, ornamentation and burials with grave goods. Most sites of this phase are found within the Mediterranean forest vegetation zone. It now appears, however, that cultivation of cereal grasses may be associated with more xeric ecological settings, such as the Jordan Valley, as well as postdating the Early Natufian (Bar-Yosef and Belfer-Cohen 1989; Olszewski 1993). In addition to possible locales in the eastern Jordan Valley, Late Natufian sites in such areas as the Black Desert (Betts 1988) have the potential to address aspects relating to the shift to cultivation and eventually agriculture.

Future research directions

In many respects, Palaeolithic research programmes in Jordan are relatively recent. Work that has concentrated on research themes, such as those highlighted above, has suggested many further avenues of investigation. For example, Henry's (1995a–i) Ras an-Naqab projects have revealed the importance of a long-term seasonal transhumance strategy in this region. The most extensively documented segments, however, are the Middle Palaeolithic and the Hamran Complex within the Epipalaeolithic period. Further work at Upper Palaeolithic sites and with sites of the Qalkhan, Madamaghan and Natufian Complexes within the Epipalaeolithic will augment the strength of this model, and perhaps document additional behavioural flexibility as well. The potential of Pleistocene lake/marsh settings, on the other hand, has only barely been realized. Recent fieldwork and on-going analyses by Olszewski and Coinman in the eastern Wadi al-Hasa will lead to a better understanding of Late Upper and Epipalaeolithic hunter-gatherer use of this ecological setting. Sites of al-Azraq Basin also hold great potential to augment our knowledge of marsh/lake resource utilization, particularly during the Epipalaeolithic period (Garrard and Byrd 1992; Garrard *et al.* 1988a; Muheisen 1988a; Muheisen and Wada 1995) and the Lower Palaeolithic (Rollefson *et al.* 1997), as do sites from the Tabaqat Fahl region (Edwards *et al.* 1988; Macumber and Edwards 1997). Relatively unexplored Pleistocene lake/marsh settings include the Jafr Basin. Additional work on Late Natufian Epipalaeolithic sites in such areas as the Black Desert (Betts 1988) will provide data critical to the interpretation of inland adaptations during the period leading to the origin of agriculture.

Specialized studies are also increasingly important to the interpretation of past lifeways. Continued analysis of lithic assemblage variability is likely to result in refinement of both the systematics of classification and interpretive models of reduction processes and stylistic traditions. Documentation of subsistence choices will benefit from more emphasis on the recovery of macrobotanical remains and phytoliths, as well as microwear and residue studies of the edges of stone artifacts. Ultimately, these types of analyses, in conjunction with theoretical modelling of Palaeolithic behavioural strategies, will further our anthropological reconstruction of the archaeological record. The Jordanian Palaeolithic record has proven to be of immense value in this regard.

References

Baruch, U. and S. Bottema
1991 Palynological Evidence for Climatic Changes in the Levant ca. 17,000–9,000 B.P. In O. Bar-Yosef and F.R. Valla (eds.), *The Natufian Culture in the Levant*, 11-20. Ann Arbor: International Monographs in Prehistory.

Bar-Yosef, O. and A. Belfer-Cohen
1989 The Origins of Sedentism and Farming Communities. *Journal of World Prehistory* 3.4: 447-98.

Betts, A.
1983 Black Desert Survey Jordan: First Preliminary Report. *Levant* 15: 1-10.
1984 Black Desert Survey Jordan: Second Preliminary Report. *Levant* 16: 25-34.
1985 Black Desert Survey Jordan: Third Preliminary Report. *Levant* 17: 29-52.
1988 The Black Desert Survey: Prehistoric Sites and Subsistence Strategies in Eastern Jordan. In A.N. Garrard and H.G. Gebel (eds.), *The prehistory of Jordan*, 369–91. British Archaeological Reports, International Series 396(ii). Oxford: British Archaeological Reports.
1991 The Late Epipalaeolithic in the Black Desert, Eastern Jordan. In O. Bar-Yosef and F.R. Valla (eds.), *The Natufian Culture in the Levant*, 217-34. Ann Arbor: International Monographs in Prehistory.

Byrd, B.F.
1988 Late Pleistocene Settlement Diversity in the Azraq Basin. *Paléorient* 14.2: 257-64.
1989 *The Natufian Encampment at Beidha*. Århus: Jutland Archaeological Society Publications 23: 1.
1991 Beidha: An Early Natufian Encampment in Southern Jordan. In O. Bar-Yosef and F.R. Valla (eds.), *The Natufian Culture in the Levant*, 245-64. Ann Arbor: International Monographs in Prehistory.
1994 Late Quaternary Hunter-Gatherer Complexes in the Levant between 20,000 and 10,000 B.P. In O. Bar-Yosef and R. Vera (eds.), *Late Quaternary Chronology and Paleoclimates of the Eastern Mediterranean*, 205-26. RADIOCARBON. Tucson: Department of Anthropology, University of Arizona.

Byrd, B.F. and S.M. Colledge
1991 Early Natufian Occupation along the Edge of the Southern Jordanian Steppe. In O. Bar-Yosef and F.R. Valla (eds.), *The Natufian Culture in the Levant*, 265-76. Ann Arbor: International Monographs in Prehistory.

Clark, G.A. and J. Lindly
1988 The Biocultural Transition and the Origin of Modern Humans in the Levant and Western Asia. *Paléorient* 14.2: 159-67.

Clark, G.A. *et al.*
1988 Excavations at Middle, Upper and Epipalaeolithic Sites in the Wadi Hasa, West-Central Jordan. In A.N. Garrard and H.G. Gebel (eds.), *The Prehistory of Jordan*, 209-85. British Archaeological Reports, International Series 396(i). Oxford: British Archaeological Reports.
1992 Wadi al-Hasa Paleolithic Project, 1992: Preliminary Report. *Annual of the Department of Antiquities of Jordan* 36: 13-23.
1994 Survey and Excavation in the Wadi al-Hasa: A Preliminary Report of the 1993 Season. *Annual of the Department of Antiquities of Jordan* 38: 41-55.
1997 Chronostratigraphic Contexts of Middle Paleolithic Horizons at the Ain Difla Rockshelter (WHS 634), West-Central Jordan. In H.G. Gebel, Z. Kafafi and G. Rollefson (eds.), *The Prehistory of Jordan. II. Perspectives from 1996*, 77-100. Studies in Near Eastern Production, Subsistence, and Environment. Berlin: ex oriente.

Clutton-Brock, J.
1970 The Fossil Fauna from an Upper Pleistocene Site in Jordan. *Journal of Zoology* 161: 19-29.

Coinman, N.R.
1993 WHS 618—Ain el-Buhira: An Upper Paleolithic Site in the Wadi Hasa, West-Central Jordan. *Paléorient* 19.2: 17-37.
1998 The Upper Paleolithic of Jordan. In D.O. Henry (ed.), *The Prehistoric Archaeology of Jordan*, 39-63. British Archaeological Reports, International Series 705. Oxford: British Archaeological Reports.

Coinman, N.R. and D.O. Henry
1995 The Upper Paleolithic Sites. In D.O. Henry (ed.), *Prehistoric Cultural Ecology and Evolution*, 133-214. New York: Plenum.

Copeland, L.
1988 Environment, Chronology and Lower-Middle Paleolithic Occupations of the Azraq Basin, Jordan. *Paléorient* 14.2: 66-75.
1989a The Artifacts from the Sounding of D. Kirkbride at Lion Spring, Azraq, in 1956. In L. Copeland and F. Hours (eds.), *The Hammer on the Rock: Studies on the Early Palaeolithic of Azraq, Jordan*, 171-211. British Archaeological Reports, International Series 540. Oxford: British Archaeological Reports.
1989b The Harding Collection of Acheulean Artifacts from Lion Spring, Azraq: A Quantitative and Descriptive Analysis. In L. Copeland and F. Hours (eds.), *The Hammer on the Rock: Studies on the Early Palaeolithic of Azraq, Jordan*, 213-58. British Archaeological Reports, International Series 540. Oxford: British Archaeological Reports.
1991 The Late Acheulean Knapping-Floor at C-Spring, Azraq Oasis, Jordan. *Levant* 23: 1-6.

Copeland, L. and F. Hours
1988 The Palaeolithic in North-Central Jordan: An Overview of Survey Results from the Upper Zarqa and Azraq 1982–86. In A.N. Garrard and H.G. Gebel (eds.), *The Prehistory of Jordan*, 287-309. British Archaeological Reports, International Series 396(ii). Oxford: British Archaeological Reports.

Edwards, P.C.
1991 Wadi Hammeh 27: An Early Natufian Site at Pella, Jordan. In O. Bar-Yosef and F.R. Valla (eds.), *The Natufian Culture in the Levant*, 123-48. Ann Arbor: International Monographs in Prehistory.

Edwards, P.C. *et al.*
1988 Late Pleistocene Prehistory in the Wadi al-Hammeh, Jordan Valley. In A.N. Garrard and H.G. Gebel (eds.), *The Prehistory of Jordan*, 525-65. British Archaeological Reports, International Series 396(ii). Oxford: British Archaeological Reports.

Emery-Barbier, A.
1995 Pollen Analysis: Environmental and Climatic Implications. In D.O. Henry (ed.), *Prehistoric Cultural Ecology and Evolution*, 375-84. New York: Plenum.

Fellner, R.
1995 Technology or Typology?: A Response to Neeley and Barton. *Antiquity* 69: 381-83.

Field, H.
1960 North Arabian Desert Archaeological Survey, 1925–1950. *Papers of the Peabody Museum* 45.2.

Garrard, A.N.
1991 Natufian Settlement in the Azraq Basin, Eastern Jordan. In O. Bar-Yosef and F.R. Valla (eds.), *The Natufian Culture in the Levant*, 235-244. Ann Arbor: International Monographs in Prehistory.

Garrard, A.N. and B. Byrd
1992 New Dimensions to the Epipalaeolithic of the Wadi el-Jilat in Central Jordan. *Paléorient* 18: 47-62.

Garrard, A.N., D. Baird and B.F. Byrd
1994 Late Paleolithic and Neolithic Chronology in the Azraq Basin. In O. Bar-Yosef and R. Kra (eds.), *Late Quaternary Chronology and Paleoclimates of the Eastern Mediterranean*, 177-99. RADIOCARBON. Tucson: Department of Anthropology, University of Arizona.

Garrard, A.N., B. Byrd and A. Betts
1986 Prehistoric Environment and Settlement in the Azraq Basin: An Interim Report on the 1984 Excavation Season. *Levant* 18: 5-24.

Garrard, A.N., N. Stanley Price and L. Copeland
1977 A Survey of Prehistoric Sites in the Azraq Basin, Eastern Jordan. *Paléorient* 3: 109-26.

Garrard, A.N. *et al.*
1985 Prehistoric Environment and Settlement in the Azraq Basin. Report on the 1982 Survey Season. *Levant* 17: 1-28.
1987 Prehistoric Environment and Settlement in the Azraq Basin: An Interim Report on the 1985 Excavation Season. *Levant* 19: 5-25.
1988a Summary of Palaeoenvironmental and Prehistoric Investigations in the Azraq Basin. In A.N. Garrard and H.G. Gebel (eds.), *The Prehistory of Jordan*, 311-37. British Archaeological Reports, International Series 396(ii). Oxford: British Archaeological Reports.
1988b Environment and Subsistence during the Late Pleistocene and Early Holocene in the Azraq Basin. *Paléorient* 14.2: 40-49.

Gilead, I.
1970 Handaxe Industries in Israel and the Near East. *World Archaeology* 2.1: 1-11.

Goring-Morris, N.
1987 *At the Edge: Terminal Pleistocene Hunter-Gatherers in the Negev and Sinai*. British Archaeological Reports, International Series 361. Oxford: British Archaeological Reports.

Goring-Morris, N. *et al.*
1996 Pattern in the Epipalaeolithic of the Levant: Debate after Neeley and Barton. *Antiquity* 70: 130-47.

Hecker, H.M.
1989 Appendix C. Beidha Natufian: Faunal Report. In B.F. Byrd, *The Natufian Encampment at Beidha*, 97-101. Aarhus: Jutland Archaeological Society Publications 23: 1.

Henry, D.O.
1979 Paleolithic Sites within the Ras en Naqb Basin, Southern Jordan. *Palestine Exploration Quarterly* 111: 79-85.
1989 *From Foraging to Agriculture*. Philadelphia: University of Pennsylvania.
1994 Prehistoric Cultural Ecology in Southern Jordan. *Science* 265: 336-41.
1995a Adaptive Behaviors, Evolution, and Ethnicity. In D.O. Henry (ed.), *Prehistoric Cultural Ecology and Evolution*, 417-37. New York: Plenum.
1995b Cultural Evolution and Interaction during the Epipaleolithic. In D.O. Henry (ed.), *Prehistoric Cultural Ecology and Evolution*, 337-43. New York: Plenum.
1995c Cultural-Historic Framework. In D.O. Henry (ed.), *Prehistoric Cultural Ecology and Evolution*, 33-41. New York: Plenum.
1995d The Hamran Sites. In D.O. Henry (ed.), *Prehistoric Cultural Ecology and Evolution*, 243-93. New York: Plenum.
1995e Late Levantine Mousterian Patterns of Adaptation and Cognition. In D.O. Henry (ed.), *Prehistoric Cultural Ecology and Evolution*, 107-32. New York: Plenum.

1995f   The Lower Paleolithic Site of Wadi Qalkha. In D.O. Henry (ed.), *Prehistoric Cultural Ecology and Evolution*, 43-48. New York: Plenum.

1995g   The Middle Paleolithic Sites. In D.O. Henry (ed.), *Prehistoric Cultural Ecology and Evolution*, 49-84. New York: Plenum.

1995h   The Natufian Sites and the Emergence of Complex Foraging. In D.O. Henry (ed.), *Prehistoric Cultural Ecology and Evolution*, 319-35. New York: Plenum.

1995i   The Qalkhan Occupations. In D.O. Henry (ed.), *Prehistoric Cultural Ecology and Evolution*, 215-42. New York: Plenum.

Henry, D.O. and A.N. Garrard
1988   Tor Hamar: An Epipalaeolithic Rockshelter in Southern Jordan. *Palestine Exploration Quarterly* 120: 1-25.

Henry, D.O. and C. Shen
1995   The Madamaghan Sites. In D.O. Henry (ed.), *Prehistoric Cultural Ecology and Evolution*, 295-317. New York: Plenum.

Henry, D.O. *et al.*
1983   An Investigation of the Prehistory of Southern Jordan. *Palestine Exploration Quarterly* 115: 1-24.

Kaufman, D.
1995   Microburins and Microliths of the Levantine Epipalaeolithic: A Comment on the Paper by Neeley and Barton. *Antiquity* 69: 375-81.

Kirkbride, D.
1958   A Kebaran Rockshelter in Wadi Madamagh, near Petra, Jordan. *Man* 63: 55-58.
1966   Five Seasons at the Pre-Pottery Neolithic Village of Beidha in Jordan. *Palestine Exploration Quarterly* 98: 8-72.

Klein, R.G.
1995   The Tor Hamar Fauna. In D.O. Henry (ed.), *Prehistoric Cultural Ecology and Evolution*, 405-16. New York: Plenum.

Leroi-Gourhan, A. and F. Darmon
1991   Analyzes polliniques de stations natoufiènes au Proche-Oriente. In O. Bar-Yosef and F.R. Valla (eds.), *The Natufian Culture in the Levant*, 21-26. Ann Arbor: International Monographs in Prehistory.

Lieberman, D.
1993   The Rise and Fall of Hunter-Gatherer Seasonal Mobility: The Case of the Southern Levant. *Current Anthropology* 34: 599-631.

Lindly, J. and G.A. Clark
1987   A Preliminary Lithic Analysis of the Mousterian Site of Ain Difla (WHS Site 634) in the Wadi Ali, West-Central Jordan. *Proceedings of the Prehistoric Society* 53: 279-92.

Lindly, J., R. Beck and G.A. Clark
2000   Core Reconstructions and Lithic Reduction Sequences at WHS 623X: An Upper Paleolithic Site in the Wadi al-Hasa, West-Central Jordan. In N.R. Coinman (ed.), *The Archaeology of the Wadi al-Hasa, West-Central Jordan. II. Archaeological Excavations in the Wadi al-Hasa*, 211-26. Tempe: Anthropological Research Papers, Arizona State University.

MacDonald, B., E. Banning and L. Pavlish
1980   The Wadi el Hasa Survey 1979: A Preliminary Report. *Annual of the Department of Antiquities of Jordan* 24: 169-84.

MacDonald, B., G. Rollefson and D. Roller
1982   The Wadi el Hasa Survey 1981: A Preliminary Report. *Annual of the Department of Antiquities of Jordan* 26: 117-31.

MacDonald, B. *et al.*
1983   The Wadi el Hasa Survey, 1982: A Preliminary Report. *Annual of the Department of Antiquities of Jordan* 27: 311-24.

Macumber, P.G.
1992   The Geological Setting of Palaeolithic Sites at Tabaqat Fahl, Jordan. *Paléorient* 18.2: 31-44.

Macumber, P.G. and P.C. Edwards
1997   Preliminary Results from the Acheulian Site of Mashari'a 1, and a New Stratigraphic Framework for the Lower Palaeolithic of the East Jordan Valley. In H.G. Gebel, Z. Kafafi and G. Rollefson (eds.), *The Prehistory of Jordan. II. Perspectives from 1996*, 23-43. Studies in Near Eastern Production, Subsistence and Environment. Berlin: *ex oriente*.

Macumber, P.G. and M.J. Head
1991   Implications of the Wadi al-Hammeh Sequences for the Terminal Drying of Lake Lisan, Jordan. *Palaeogeography, Palaeoclimatology, Palaeoecology* 84: 163-73.

Marks, A.E. and D. Friedel
1977   Prehistoric Settlement Patterns in the Avdat/Aqev Area. In A.E. Marks (ed.), *Prehistory and Paleoenvironments in the Central Negev, Israel*, II: 131-58. Dallas: Southern Methodist University.

Muheisen, M.
1988a   The Epipalaeolithic Phases of Kharaneh IV. In A.N. Garrard and H.G. Gebel (eds.), *The Prehistory of Jordan*, 353-67. British Archaeological Reports, International Series 396(ii). Oxford: British Archaeological Reports.
1988b   Excavations at 'Ain Rahub: A Final Natufian and Yarmoukian Site near Irbid. In A.N. Garrard and H.G. Gebel (eds.), *The Prehistory of Jordan*, 473-502. British Archaeological Reports, International Series 396(ii). Oxford: British Archaeological Reports.
1988c   Survey of Prehistoric Cave Sites in the Northern Jordan Valley. In A.N. Garrard and H.G. Gebel (eds.), *The Prehistory of Jordan*, 503-23. British Archaeological Reports,

Muheisen, M. and H. Wada
1995 An Analysis of the Microliths at Kharaneh IV, Phase D, Square A20/37. *Paléorient* 21.1: 75-95.

Neeley, M.P. and C.M. Barton
1994 A New Approach to Interpreting Late Pleistocene Microlith Industries in Southwest Asia. *Antiquity* 68: 275-88.

Neeley, M.P. *et al.*
1998 Investigations at Tor al-Tareeq: An Epipaleolithic Site in the Wadi al-Hasa. *Journal of Field Archaeology* 25: 295-317.

Olszewski, D.I.
1991 Social Complexity in the Natufian?: Assessing the Relationship of Ideas and Data. In G.A. Clark (ed.), *Perspectives on the Past: Theoretical Biases in Mediterranean Hunter-Gatherer Research*, 322-40. Philadelphia: University of Pennsylvania.
1993 Subsistence Ecology in the Mediterranean Forest/Maquis: Implications for the Origins of Cultivation in the Epipaleolithic Southern Levant. *American Anthropologist* 95.2: 420-35.
1997 From the Late Ahmarian to the Early Natufian: A Summary of Hunter-Gatherer Activities at Yutil al-Hasa, West-Central Jordan. In H.G. Gebel, Z. Kafafi and G. Rollefson (eds.), *The Prehistory of Jordan. II: Perspectives from 1997*, 171-82. Studies in Near Eastern Production, Subsistence and Environment. Berlin: *ex oriente*.

Olszewski, D.I. and N.R. Coinman
1998 Late Pleistocene Settlement Patterns in the Wadi al-Hasa, West-Central Jordan. In N.R. Coinman (ed.), *The Archaeology of the Wadi al-Hasa, West-Central Jordan. I. Surveys, Settlement Patterns and Paleoenvironments*, 177-203. Tempe: Anthropological Research Papers, Arizona State University.

Olszewski, D.I., G.A. Clark and S. Fish
1990 WHS 784X (Yutil al-Hasa): A Late Ahmarian Activity Site in the Wadi Hasa, West-Central Jordan. *Proceedings of the Prehistoric Society* 56: 33-49.

Olszewski, D.O. *et al.*
1994 The 1993 Excavations at Yutil al-Hasa (WHS 784): An Upper/Epipalaeolithic Site in West-Central Jordan. *Paléorient* 20.2: 129-41.

Phillips, J.L.
1988 The Upper Paleolithic of the Wadi Feiran, Southern Sinai. *Paléorient* 14.2: 183-200.
1994 The Upper Paleolithic Chronology of the Levant and the Nile Valley. In O. Bar-Yosef and R. Kra (eds.), *Late Quaternary Chronology and Paleoclimates of the Eastern Mediterranean*, 169-76. RADIOCARBON. Tucson: Department of Anthropology, University of Arizona.

Rees, L.W.B.
1929 The Transjordan Desert. *Antiquity* 3: 389-407.

Reese, D.S.
1989 Appendix D: The Natufian Shells from Beidha. In B.F. Byrd, *The Natufian Encampment at Beidha*, 102-104. Århus: Jutland Archaeological Society Publications 23: 1.
1995 Shells from the Wadi Hisma Sites. In D.O. Henry (ed.), *Prehistoric Cultural Ecology and Evolution*, 385-90. New York: Plenum.

Rhotert, H.
1938 *Vorgeschichtliche Forschungen in Klein Asien und Nordafrika. I. Transjordanien*. Stuttgart: Verlag Strecker und Schröder.

Rollefson, G.O.
1981 The Late Acheulean Site at Fjaje, Wadi el-Bustan, Southern Jordan. *Paléorient* 7: 5-21.
1982 Late Acheulean Artifacts from Ain el-Assad ('Lion's Spring') near Azraq, Eastern Jordan. *Bulletin of the American Schools of Oriental Research* 240: 1-20.
1983 Two Seasons of Excavations at Ain el-Assad near Azraq. *Bulletin of the American Schools of Oriental Research* 252: 25-34.
1984 A Middle Acheulian Surface Site from Wadi Uweinid, Eastern Jordan. *Paléorient* 10.1: 127-33.
1985 Late Pleistocene Environments and Seasonal Hunting Strategies: A Case Study from Fjaje, near Shobak, Southern Jordan. In A. Hadidi (ed.), *Studies in the History and Archaeology of Jordan*, II: 103-107. Amman: Department of Antiquities.

Rollefson, G.O. *et al.*
1997 'Ain Soda and 'Ain Qasiya: New Pleistocene and Early Holocene Sites in the Azraq Shishan Area, Eastern Jordan. In H.G.K. Gebel, Z. Kafafi and G.O. Rollefson (eds.), *The Prehistory of Jordan. II. Perspectives from 1997*, 45-58. Studies in Near Eastern Production, Subsistence and Environment. Berlin: *ex oriente*.

Rosen, A.M.
1995 Preliminary Analysis of Phytoliths from Prehistoric Sites in Southern Jordan. In D.O. Henry (ed.), *Prehistoric Cultural Ecology and Evolution*, 399-403. New York: Plenum.

Schuldenrein, J. and G.A. Clark
1994 Landscape and Prehistoric Chronology of West-Central Jordan. *Geoarchaeology* 9.1: 31-55.

Schyle, D. and H.-P. Uerpmann
   1988   Paleolithic Sites in the Petra Area. In A.N. Garrard and H.-G. Gebel (eds.), *The Prehistory of Jordan*, 39-65. British Archaeological Research, International Series 396(i). Oxford: British Archaeological Reports.

Shea, J.J.
   1995   Lithic Microwear Analysis of Tor Faraj Rockshelter. In D.O. Henry (ed.), *Prehistoric Cultural Ecology and Evolution*, 85-105. New York: Plenum.

Stevens, M.N.
   1996   A Techno-Typological Analysis of Tor al-Tareeq (WHS 1065): An Epipaleolithic Site in West-Central Jordan. Unpublished M.A. thesis, University of Arizona, Tucson.

Stringer, C. and P. Andrews
   1988   Genetic and Fossil Evidence for the Origins of Modern Humans. *Science* 239: 1263-68.

Tchernov, E.
   1988   The Age of 'Ubeidiya Formation (Jordan Valley, Israel) and the Earliest Hominids in the Levant. *Paléorient* 14.2: 63-65.

Waechter, J. *et al.*
   1938   The Excavations at Wadi Dhobai 1937–1938 and the Dhobaian Industry. *Journal of the Palestine Oriental Society* 18: 172-86, 292-98.

Zeuner, F., D. Kirkbride and B. Park
   1957   Stone Age Exploration in Jordan, 1. *Palestine Exploration Quarterly* 89: 17-24.

3. The Neolithic Period

Gary O. Rollefson

Introduction

Until the beginning of the 1980s the Neolithic of Jordan east of the Rift Valley was virtually untapped. A dim outline of developments was available from unsystematic surveys (Field 1960; Kirkbride 1958; Rhotert 1938; Zeuner 1957) and a few excavations spread over a number of decades (e.g. Waechter and Seton-Williams 1938; Kirkbride 1966). Beginning in the 1970s prehistoric investigations began to intensify (Garrard and Price 1977; Henry 1979), and soon large areas of Jordan had undergone extensive systematic surveys (e.g. MacDonald *et al.* 1988; Gebel and Starck 1985). Excavations of Neolithic settlements were undertaken throughout much of the country, and the results of this research have vastly changed our views of this crucial period of human socioeconomic development, not only in Jordan itself but in the greater Near East as a whole.

The beginning and end of the Neolithic are far from clearly defined. The onset is associated with the rise of agriculture, an event that would have profound effects on human population and social organization. This development was a gradual process that had begun much earlier in the Late Epipalaeolithic period, and the resolution between non-selective cultivation and true agriculture is not easily detectable in the archaeological record. But the Neolithic period was also allied with the appearance of changes in the material culture, including the first instance of true arrowheads in the form of El Khiam and similar projectile points (Bar-Yosef 1981; Nadel *et al.* 1991), and such tangible evidence is of great assistance in deciding if an archaeological site dates to the earliest phases of the Neolithic.

Similarly, the transition from the Neolithic to the Chalcolithic is far from clear. In this instance, material correlates for the change are disputed, partly because of the scarcity of data and partly due to the varied interpretations individual scholars draw from particular aspects of the archaeological record (e.g. Goren 1990; Gopher and Gophna 1993: 337; cf. Kafafi 1998; Levy 1995: 226).

In the description and discussion below, the following chronological phases will be used as defined in earlier publications (cf. Bar-Yosef 1981; Rollefson 1989: 169):

PPNA	10,300–9,600 bp[1]
Early PPNB	9,600–9,200 bp
Middle PPNB	9,200–8,500 bp
Late PPNB	8,500–8,000 bp
PPNC	8,000–7,500 bp
Pottery Neolithic	7,500–?6,000 bp

These periods will be discussed in terms of lithic technology, architecture, subsistence economy and ritual; settlement pattern will also be addressed in so far as it is possible.

The Pre-Pottery Neolithic A

Reflecting the rare number of sites for this period throughout the Levant, only three Jordanian sites (Sabra 1, 'Iraq ad-Dubb and Dhra') are assignable to the Pre-Pottery Neolithic A (PPNA). In the case of Jordan, this mirrors more the insufficiency of surveyed landscape than of the real PPNA presence. Two of these sites occur in southern Jordan and one in the north, so no interpretation of settlement pattern during this period is possible. Another site (Jabal Queisa J-24), may in part be PPNA in age, but in view of the limited sounding here, an EPPNB age is also possible.[2]

Lithic technology and typology

Confirmed by radiocarbon dates at Dhra' and 'Iraq ad-Dubb, type fossils including Khiamian points, Hagdud truncations and unidirectional blades and blade cores all point to a PPNA time frame for the sites (Gebel 1988: 78 and fig. 6; Henry 1995: 348-49 and fig. 14.3; Kuijt *et al.* 1991: 102-04 and fig. 5; Kuijt and Mahasneh 1998: table 1 and fig. 5). Small lunates were also found at 'Iraq ad-Dubb, but since the PPNA layers occur above Late Natufian deposits, Kuijt considers that they are probably intrusive into the PPNA layers, and that lunates are not a characteristic component of the PPNA industry (Kuijt 1994b: 2-3, 1996; Kuijt and Mahasneh 1998).

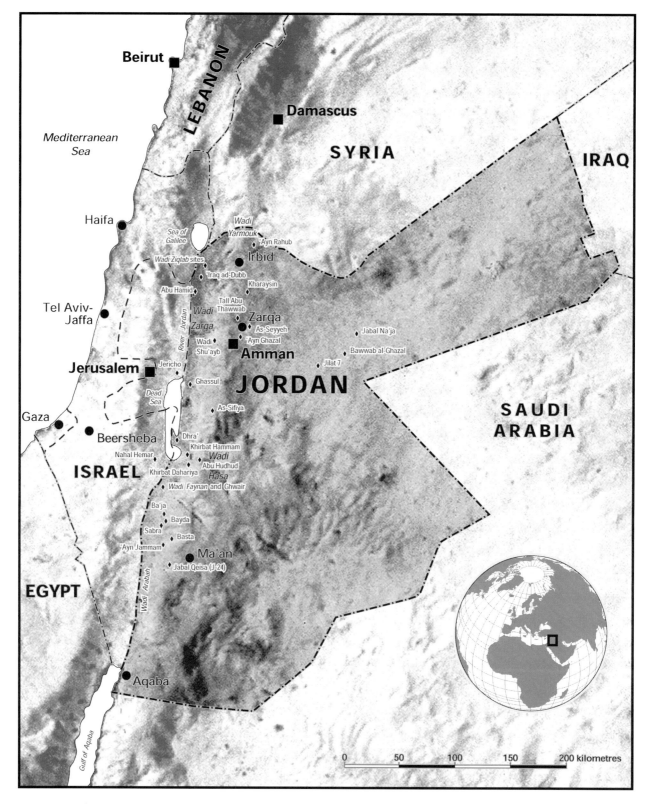

Figure 3.1. Map showing Neolithic sites in Jordan.

The temporary nature of the occupations at Sabra 1 and at Jabal Queisa (J-24) is attested by the restricted nature of the tool kits, and the relatively high percentage of points and other piercing tools indicates their status as camps associated with hunting. The inventory at 'Iraq ad-Dubb still has not been published, but the presence of sickle blades in some quantity (Kuijt et al. 1991: 102) suggests a relatively longer stay for the inhabitants of the rockshelter. Points account for more than half of the reported tool kit from Dhra', but the variety of other kinds of tools in the sample lends support to a relatively permanent occupation.

The Khiamian and Salibiya types are the most diagnostic of the PPNA period projectile point types reported from the four Jordanian sites, and they constitute virtually the only versions of arrowheads at 'Iraq ad-Dubb, Dhra' and Sabra 1 (Gebel 1988: figs. 6-9, 10). Jabal Queisa also contains Helwan points (Henry 1995: 348 and fig. 14.3), and this type is also associated with the Early PPNB period (cf. Gopher 1994: 190-99); in the absence of radiocarbon dates at Jabal Queisa, it is possible that this hunting camp may be from a time near the end of the PPNA or the beginning of the PPNB, or somewhere around 9,700–9,500 bp, and possibly even later in Kuijt's judgment (Kuijt and Chesson 1994: 36-38). That Jabal Queisa was visited in the early PPNB is supported by the presence of bi-directional naviform blade cores (Henry 1995: fig. 14.6).³

Subsistence economy

Little substantial indications of the subsistence economy practised by the inhabitants of the PPNA sites are available. Bones are mentioned for both Sabra 1 and Jabal Queisa, but no species identifications have been provided. At 'Iraq ad-Dubb, preliminary analysis has shown that all animal remains were wild, including in decreasing frequency gazelle, boar, sheep/goat and aurochs for the larger mammals; smaller animals were mostly red fox, wildcat, and lagomorphs. Birds were also preserved, including doves, grouse and pheasant, hawks and waterfowl (Mullen and Gruspier, in Palumbo et al. 1990: 114). Kuijt et al. (1991: 104) mention that botanical analysis is still in progress, but no publication of results has yet appeared. The recently concluded season at Dhra' also produced animal bones and plant remains, but analysis is still incomplete.

Architecture

In the south Jabal Queisa (J-24, Henry 1982, 1995), in the Judayid basin just south of Ras an-Naqab, and Sabra 1 (Gebel 1988), in the Petra region, are small ephemeral campsites with no evidence of architecture. Another temporary PPNA settlement occurred at 'Iraq ad-Dubb, a rockshelter in the Wadi al-Yabis in northern Jordan (Kuijt et al. 1991; Kuijt 1994a). But here, although the rockshelter is small and the terrace in front of it not very extensive, excavations revealed three oval to circular PPNA structures inside the shelter, and the labour invested in the construction indicates that the occupation was at least semi-permanent. The earliest building is known only from a partially exposed mud floor and adjacent mudbricks, a burned post (radiocarbon dated to 9950 ± 100 bp), and some pit features cut into bedrock. A second building was oval, also with a mud floor, and may have been as large as 1.5 by 3.0 m; the walls of this structure consisted of a single width of cobbles preserved to a height of 70 cm. Details of the third building are barely known since it was not extensively excavated.

The only permanent PPNA farming village known so far in Jordan appears to be the fourth site, Dhra', although results of the recent excavations remain preliminary (Kuijt and Mahasneh 1998). Located above the east shore of the Dead Sea at an elevation of c. 40 m below sea level (cf. Bennett 1980; Raikes 1980), Dhra' covers an area of less than 0.4 ha (i.e. at least 50 by 80 m) and has 'extensive evidence for oval or circular structures made of stone and mud' (Kuijt and Mahasneh 1998). A radiocarbon sample from the section produced a date of 9960 ± 110 bp. One excavated house was about 3 m in diameter and had stone walls standing to a height of 85 cm; a low stone feature, possibly a bench, was inside the building.

Ritual

Burned and unburned bones, including some from humans, were reported for Sabra 1 (Gebel 1988: 78; Roehrer-Ertl et al. 1988), but there is no clear evidence for an intentional burial. Two burials from 'Iraq ad-Dubb constitute the only evidence for ritual activity during the PPNA in Jordan; no examples of figurines or other objects associated with symbolic activity have been recovered from Jordanian PPNA sites so far.

Both 'Iraq ad-Dubb burials were flexed and appear to be complete, although one of them was not completely exposed. One skeleton was of an adult male, the other of an adolescent, and neither had any associated grave goods (Kuijt et al. 1991: 106). The two burials inside the shelter could reflect an emotional bond of the social group with the location, even if the occupation of the rockshelter was not a permanent one.

The Early Pre-Pottery Neolithic B (EPPNB)

Like the PPNA, evidence for the presence of Early PPNB (EPPNB) settlements in Jordan is relatively rare, known only from the lowest layer at the site of Jilat 7 (Garrard et al. 1994) in the eastern steppe not far from al-Azraq and from Abu Hudhud (Rollefson 1996) in Wadi al-Hasa in southern Jordan.[4]

Lithic technology and typology

The most fundamental distinction between the PPNA and PPNB lies in the shift to naviform blade production in the latter period (Cauvin 1987), and the presence of bi-directional blades from such cores characterizes the collections from both Jilat 7-I and Abu Hudhud. No naviform cores were found on the surface of Abu Hudhud, but they account for over 10% of the cores in Jilat 7-I (Baird 1994: table 2; 1995: 508-509). Reflecting its early date, about a quarter of the 45 Jilat 7-I projectile points are Khiam and related types, while almost three-quarters are Helwan points; 24 Hagdud truncations were also recovered from this phase. This mix, including the core mixture, is generally similar to the small artifact collection from Jabal Queisa (Henry 1995: 348). All eight points collected from the surface of Abu Hudhud were Helwan types, but it is possible that the generally smaller Khiamian points and Hagdud truncations may have been overlooked since the collection strategy was not systematic.[5]

The tool kit at Jilat 7-I was greatly varied (Baird 1994) and indicates that a broad range of activities was undertaken at the site; grinding stones (including mortars, pestles and stone vessels; cf. Wright in Garrard et al. 1994: table 2) demonstrate at least a minor degree of permanence for the occupation. Simple and transverse types dominated burins. A single sickle element seems odd for this arid area (Garrard et al. 1994: 89), although there is no reason that the tool was necessarily associated with agriculture. The collection of tools from Abu Hudhud, although small, is also diverse (34 pieces among 11 tool classes, in addition to the projectile points and 1 basalt pestle fragment; cf. Rollefson 1996: table 1) and supports the likelihood that Abu Hudhud was at least a semi-permanent settlement.

Architecture

One structure excavated at Jilat 7-I was rectangular (but with rounded corners) and measured approximately 3 by 3.5 m. Walls were 'built from a combination of coursed stones and uprights with rubble packing' (Garrard et al. 1994: 75), and other walls of possible contemporaneous structures were located at the edges of the excavated area.

Four small oval or subrectangular structures (c. 3 by 2.5 m) were visible at the surface of Abu Hudhud, and an erosional cut through the site revealed a corner of a building of undetermined size with walls that formed a right angle. It is not clear if the rounded alignments are remnants of house walls or if they were the foundation stones of temporary huts or tents.

Subsistence economy

Information relating to the exploitation of plant and animal resources is available only from Jilat 7-I. Despite its location in the steppe, the site yielded remains of domesticated einkorn wheat and barley; the rest of the material from flotation samples appears to represent wild seeds, fruits and nuts (Colledge in Garrard et al. 1994: 102-105). The animal bones come entirely from wild species (in decreasing frequency hare, gazelle, tortoise and fox; cf. Martin in Garrard et al. 1994: table 7). Of particular importance, perhaps, is the absence of any goat and sheep, which indicates that these animals were not available to these hunters (Martin in Garrard et al. 1994: 97).

Ritual

No burials were found at Jilat 7-I, but one engraved stone object has been described as having potential calendrical or mathematical functions (Wright in Garrard et al. 1994: 92 and fig. 10a).

The Middle Pre-Pottery Neolithic B (MPPNB)

The later centuries of the ninth millennium bp witnessed what may be taken to reflect a virtual population explosion, for in contrast to the meagre evidence for PPNA and EPPNB settlements in Jordan, later aceramic Neolithic sites are to be found throughout the countryside.

Lithic technology and typology

The definition of the PPNB in contradistinction to the PPNA relies most heavily on the widespread adoption of the naviform core-and-blade technique sometime after 9600 bp. Here again, the EPPNB evidence is critical for evaluating the expanding rate of popularity for naviform blade technology.

Technology

It appears that at MPPNB settlements there is a 'dual character' of the lithics industry: one that involves an *ad hoc* execution of the immediate satisfaction of relatively undemanding needs, and another that centres on a sophisticated blade production technique for certain applications (cf. Goring-Morris 1994: 438). Replication experiments have shown the elaborate core preparation and maintenance required of the naviform blade technique (Wilke and Quintero 1994). The narrow range of flint/chert quality used in the process, as well as the restricted distribution of naviform chipping floors, suggests the likelihood that naviform blade production was a specialization practised by relatively few members of the MPPNB 'Ayn Ghazal community (Quintero and Wilke 1995). Overall, whether dealing with *ad hoc* or specialized naviform blade production, the blade to flake ratio is almost 1 : 1, a situation that changes dramatically in later periods (cf. Rollefson 1990a, 1997b).

One topic that has received a perceptible degree of debate in the past is the presence, and even dominance, of a high-quality purple-pink flint in many assemblages throughout Jordan and Israel. At the centre of the controversy was whether this glossy material was available as a natural resource or if it represented the results of human modification of poorer quality flint through the means of painstaking heat treatment (e.g. Bar-Yosef 1981, 1991; Nadel 1989: 65-66; Edwards and Thorpe 1986: 85). This issue has been resolved for the northern and central Jordanian highlands, at least, for both outcrops of purple-pink flint and mines to extract the material have been found in the limestone hills of Wadi Huweijir, about 2 km north of 'Ayn Ghazal (Quintero 1996). It is likely that outcrops of Huweijir-type flint occur throughout the southern Levant, and that heat treatment was not resorted to in its processing.

Typology

Projectile points

As for typology, there may be some territorial differences in the popularity of some MPPNB projectile point types (Figure 3.2). Although the generalized Byblos and Amuq points are found in both northern and southern Jordan, Jericho points (Gopher's Type A4; cf. Gopher 1994: 36) with pronounced backswept barbs are absent at Bayda, for example, but they are relatively common at 'Ayn Ghazal (Eighmey 1992: 99) and Wadi Shu'ayb (unpublished data). The Abu Gosh point is a variant of the Byblos type, distinguished by a prominent and obliquely pressure-flaked tang (Lechevallier's Type 10; Lechevallier 1978: 53, fig. 17-7; also cf. Eighmey 1992: 97, fig. 8-e); they were nearly four times as popular as Jericho points at 'Ayn Ghazal (Eighmey 1992: 99, table 2). This type does not exist as a distinct pattern in Mortensen's study of the Bayda inventory, although there are illustrated examples (Mortensen 1970: fig. 16-b, c) that are grouped together in his Type A11 with points similar to the Munhata point (Eighmey 1992: 95, fig. 8-a; cf. Perrot 1966: 53, fig. 2: 1-2; Gopher 1994: fig. 5-18). In any event, Type A11 points range between 1 and 3% for levels II–VI at Bayda (Mortensen 1970: fig. 19), and a comparable grouping of Abu Gosh and Munhata points at 'Ayn Ghazal ranges from 19% (PPNC) to 30% (MPPNB) (cf. Eighmey 1992: table 2).

It is also possible that there is a tendency towards the production of slenderer, more delicate points in southern Jordan in contrast to the longer, bulkier pieces recovered from Wadi Shu'ayb and 'Ayn Ghazal, but no comparable dimensional data are available to substantiate this subjective observation that is based on published illustrations.

Burins

Data from 'Ayn Ghazal and Wadi Shu'ayb show a heavy reliance on burin manufacture, and this tool class, whose function(s) remain(s) enigmatic, shows a chronologically sensitive pattern in terms of specific type popularity; in the MPPNB period at 'Ayn Ghazal three-quarters of the burins are of the transverse type, mostly made on naviform blades (Rollefson 1995). The seriational potential of burins appears to hold for both

Figure 3.2. 'Ayn Ghazal projectile points. The lowest row is from the MPPNB; the middle row from the PPNC; and the top row from the Yarmoukian period (photo G. Rollefson).

the permanent settlements of northern Jordan and the Jordan Valley, as well as the steppe and desert sites in the eastern part of the country (Rollefson 1988, n.d.).

Sickles

MPPNB sickles appear to consist principally of long and intact naviform blades which may or may not bear a variety of retouch ranging from fine to moderate serration (Olszewski 1994; Mortensen 1970), and sickle gloss often appears on both lateral edges. The recently completed experimental work on harvesting 'green' and 'dry' cereal stands with replicated flint blades shows that it is very likely that a significant proportion of blades used as reaping tools on mature, dried-out stalks develop almost no detectable sickle gloss, and it is possible, therefore, that the reliance on the appearance of sickle gloss alone may have resulted in overlooked sickle elements in excavated assemblages (Quintero et al. 1997).

Other tools

Scrapers, borers and knives all may hold important potential for temporal seriation based on the results of sample analyses from 'Ayn Ghazal (Rollefson et al. 1994), but so far these results are not compatible with Mortensen's research (1970), so that it is not clear if differences are due to real regional variation or if the research efforts reflect personal views of typological definition, inclusion and exclusion, or a combination of both aspects. In view of the work by Quintero et al. (1997), there is now some reason to suspect that some of the blades ascribed to knives on the basis of edge damage might actually belong to the sickle category, despite the absence of visible gloss; the same observation may apply to Mortensen's results. Axes/celts/adzes, picks, and chisels are extremely rare elements in MPPNB tool kits in both northern and southern Jordan (cf. Rollefson et al. 1992: table 4; Mortensen 1970: fig. 53). To date, no tranchet axes have been noted among the MPPNB tools at 'Ayn Ghazal, and none were noted at Wadi Shu'ayb. But Mortensen shows one from Level V, which is MPPNB in date (Mortensen 1970: fig. 49-c); one from Level III at Bayda and another from Level II (Mortensen 1970: fig. 50-a, b) have been ascribed to the MPPNB, but there are some potential problems with this chronological ascription (see below); the tranchet bit is taken to reflect a tradition of axe manufacture (or at least resharpening) that reaches back to the PPNA and 'Early PPNB'[6] periods. The popularity of tranchet axes at MPPNB Jericho, Nahal Oren I–II, and Nahal Lavan 109 (Bar-Yosef 1981: 564, fig. 3) might indicate that the northern and central highlands of Jordan constitute a regional sector where this tradition of manufacture and resharpening had not been adopted.

Architecture

As is the case with so much of the information pertaining to the MPPNB in Jordan, architecture from this period is documented almost exclusively at Bayda (especially Kirkbride 1966, 1967, 1968a) and 'Ayn Ghazal (Banning and Byrd 1984, 1987; Byrd and Banning 1988) although some generalizations can be derived from unpublished plans from Wadi Shu'ayb (cf. Simmons *et al.* 1989).

Bayda Levels V–VI[7]

The earliest Neolithic layers at Bayda, whose radiocarbon dates fall comfortably into the MPPNB (cf. Mortensen 1970: 13), already show considerable architectural uniqueness in comparison with architecture from permanent MPPNB settlements anywhere else in the Levant. Circular and semi-subterranean, these one-room structures (c. 4 m in diameter) were clustered into 'honeycomb' arrangements of several units each (Kirkbride 1967: 6-8, fig. 1, pls. IB, IIA). Constructed with stone walls interspersed with wooden roof supports, these 'post-houses' resemble the circular structures of PPNA Jericho, as Kirkbride noted (1967: 6),[8] differing principally in the construction materials. The major distinctions in the houses in Levels V and VI are that the clustered arrangements that shared common walls in Level VI gave way to free-standing single-room buildings in later Level V, which Kirkbride saw as an indication of a reorganization of social structure; Kirkbride remarked that Level V buildings were also much better constructed, and the size of one of them reached 6.5 m in diameter (1967: 8). Byrd has noted that building size distribution and internal variability suggest that one of the Level V–VI/Phase A buildings may have served a public function (Byrd 1994: 649 and fig. 3).

Bayda Level IV

Continuing an architectural evolution, structures in Level IV maintained the independence and single-room nature of Level V, but they were built on the contemporaneous land surface and had changed to rectilinear shapes (sometimes with rounded corners) that would have been familiar to MPPNB architects in central and northern Jordan. Byrd describes two building sizes (small/medium and large), the latter category associated with 'non-domestic' use (Byrd 1994: 650-51, table 3 and fig. 5). The use of relatively thin rectangular stone slabs, leveled and steadied with thin chinking stones, foreshadows dressed-stone construction of LPPNB sites in Jordan.[9]

'Ayn Ghazal MPPNB

Rectangular shapes are the only forms houses assumed in permanent MPPNB settlements in north and central Jordan, although it must be admitted that the sample of excavated settlements is very small. In any event, no curvilinear structures have been found at MPPNB 'Ayn Ghazal or Wadi Shu'ayb, although the situation in the steppe/desert of eastern Jordan reveals both structural forms (and transitional expressions) erected by hunting groups at Jilat 7 and 26.

The architectural sequence at 'Ayn Ghazal covers the entire range of the MPPNB (see radiocarbon dates in Rollefson *et al.* 1992; Rollefson 1998a), and there is a clear evolution of architectural standards over the c. 700-year period. The earliest MPPNB houses were built on the surface and single-roomed (as at Level IV Bayda) and were as large as c. 5 by 8 m (cf. Byrd and Banning 1988: Rollefson and Simmons 1986), with some reaching perhaps 50 m^2 in floor area (Figure 3.3). Floors were made of a sophisticated lime plaster and massive posts up to 60 cm in diameter supported roofs, and the room had a circular hearth built as an integral part of the floor exactly at its geometric centre (Rollefson 1990b).

Through time, the single-room structures were replaced with two- and even three-roomed houses. The overall area of the buildings was essentially unchanged, although roofs were increasingly supported by stone pillars or 'piers' (Byrd and Banning 1988) in place of the thick, tree-trunk-sized roof posts, which were becoming scarcer in the immediate vicinity of the settlement (Rollefson and Köhler-Rollefson 1989). Notably, by the end of the MPPNB period, postholes rarely exceeded 15 cm in diameter, and often they were found along wall lines, not in the centre of rooms; occasionally a number of very small 'post holes' (10 cm or less in diameter) were found in some corner areas of rooms, which suggested that such installations as loom areas or storage facilities may have been allotted to some parts of the house. As was the case at Bayda Level IV, the floors were decorated with the application of pigment,

Figure 3.3. Western room of a two-room MPPNB house at 'Ayn Ghazal; the doorway connecting the two rooms is top centre. Note the size of the post holes of this c. 8100 BC lime plaster floor. Scale: 1 m; north at left (photo C. Blair).

sometimes in finger-painted designs of unknown meaning (Rollefson and Simmons 1987).

Throughout the MPPNB area excavated at 'Ayn Ghazal, the free-standing houses are often built very close together, sometimes to the point that adults would have had trouble passing between them. The slope of the hillside at 'Ayn Ghazal required that labour-intensive terracing was necessary to provide level ground for house construction, and remnants of terrace walls have been found during the excavation. The isolation of the houses suggests that the (single) families that lived in them were probably independent economic units of production and consumption, although there was probably considerable leeway for kin and friendly support in the event of particular hardship. The independence of the family units is supported by the absence of any courtyard or compound walls that could signal 'property rights' or 'social boundaries'.

MPPNB subsistence

Plant resources

Although the agricultural species inventory for the MPPNB populations in Jordan was probably not very different from PPNA times, recent evidence suggests that the situation in Jordan and elsewhere in the Levant may not have conformed to the widely held concept of a cereal-based farming economy. From the enormous stocks of charred lentils and peas in a house from MPPNB 'Ayn Ghazal, they certainly played a more important role in MPPNB diets than they are given credit for; clearly the kinds of processing of harvested crops has effected a powerful under-representation of pulses in MPPNB economies (cf. Rollefson and Simmons 1986: 152; Kislev 1992).

In an elegant experimental project, the 'domesticated' status of cereal plants in the Levant up to the end of the MPPNB has been called into question (Quintero et al. 1997). Sickle gloss was shown to form intensively and rapidly when flint blades were used to cut through 'green', partially ripened wheat stalks; blades used to reap dried wheat were less affected by the development of 'sickle sheen'. Among the implications of the experimental work, MPPNB and earlier cereal fields may have been populated by substantial proportions of mature plants whose spikes still shattered at maturity, a genetic characteristic that was not successfully controlled until the following LPPNB period (Quintero et al. 1997).

Animal resources

There was a major contrast in terms of the acquisition of meat by the beginning of the MPPNB period. In most PPNA sites in Israel and Jordan, the dominant prey for hunters was gazelle, sometimes overwhelmingly so (cf. Horwitz 1989: fig. 8); this maintained the pattern already established during the Natufian period. But, for people living in the Jordan Valley and in the Jordanian highlands, by 9200 bp there was a dramatic shift to a focus on goats. It is unfortunate that so little research has been carried out in the EPPNB, for this is when the changeover must have begun.

The prevalence of goats at MPPNB Bayda in southern Jordan may be a direct reflection of the local habitats, for gazelle accounted for only about a quarter of the bones in one Natufian sample (compared to nearly 70% for goat or ibex) and 0% in another Natufian sample from the site (cf. Hecker 1975: tables 11–12). At 'Ayn Ghazal goats accounted for half of the animal bones while gazelle reached only about 13% (Köhler-Rollefson *et al.* 1988: table 2); the situation was similar for MPPNB Wadi Shu'ayb (unpublished data, but cf. Simmons *et al.* 1989). In these last two cases, it is unlikely that the predominance of goat is explained by the immediate environmental circumstances, for the local countryside was probably more densely populated by gazelles than caprines.

The popularity of goats suggests that a different strategy for meat acquisition was in effect by the MPPNB, and that hunting had been supplemented by herding the animals, effectively isolating them from the wild caprine environment and securing them as a living, breathing meat locker. Hecker has referred to this as 'cultural control' (Hecker 1975: 306), and Köhler-Rollefson has argued that this manipulation of the species fulfils all the definitional requirements of domestication with the specific absence of genetic change (Köhler-Rollefson 1997; *contra* Ducos 1994; Horwitz 1989), an outcome of herding that may have been unintentional and unimportant to the earliest populations of goatherders (Köhler-Rollefson and Rollefson n.d.). Regardless of the terminological debate on the status of domestication for MPPNB goats, it is clear that a special relationship between human society and *Capra* sp. had developed early in the MPPNB period, a human predation on a behavioural subservience that was to have a profound effect on settled human populations long before morphologically recognizable *C. hircus* individuals are recognizable in the archaeological record (cf. Rollefson 1990b; Köhler-Rollefson and Rollefson 1990).

Beyond the new concentration on keeping and protecting goats, information from 'Ayn Ghazal (Köhler-Rollefson *et al.* 1988) and Wadi Shu'ayb reveals that the localities around these settlements consisted of lucrative ecozones that supported a rich variety of animal species; in the case of 'Ayn Ghazal, more than 50 different species reflected woodland or open woodland, *maquis*, steppe, desert and riverine gallery ecozones until the end of the MPPNB period. Although goatherding accounted for about half of the meat in the diet at 'Ayn Ghazal, hunting still provided a major part of the daily (perhaps seasonal) meat intake and afforded a probably welcome variety of food; different animals also may have contributed to a wide array of non-nutritional needs as well, such as furs, skins, bone elements for tools and other body parts for other uses.

Ritual

It is tempting to refer to the MPPNB in the Levant as the 'classical' period in view of the elaborate expressions of ritual practice. While no monumental structures can be ascribed to the MPPNB,[10] the intricate relationships of birth and renewal, death and burial, post-mortem treatment, and various paraphernalia to realize these themes are unparalleled in the Levantine Neolithic period. Space is too limited for a comprehensive treatment of the ritual aspects of the MPPNB in Jordan, but more detailed accounts are available elsewhere (Rollefson 1983, 1986, 1998b, 2000).

Burials

Kirkbride correctly concluded that the number of MPPNB burials simply did not conform to population estimates of the settlements, and that an off-site cemetery or other form of post-mortem disposal must have been the norm (Kirkbride 1968b: 272). At 'Ayn Ghazal, a total of 81 burials of all types (see below) was recovered from the MPPNB layers that spanned about 700 years, which corresponds to just over one burial per decade. Buried beneath the floors of one house that was used over a period of about 400 years (cf. Byrd and Banning 1988) were the remains of 12 individuals, or about one every 33 years (or 'long generation').

Subfloor and courtyard burials are well known from MPPNB sites in the Levant, and normally burials consisted of single interments. Wadi Shu'ayb's burials contradicted this aspect of post-mortem disposal, for multiple burials of two to four individuals were common (cf. Simmons *et al.* 1989: 36–40; K. Roler, pers. comm.)

'Ayn Ghazal has provided detailed evidence for a variety of burial types. Briefly, they include (a) subfloor or courtyard, decapitated; (b) courtyard, skull intact with the rest of the body; and (c) infant burial (cf. Rollefson 1983: 29–30). The first has been termed the 'typical' type (e.g. Rollefson *et al.* 1992: 461), but in view of the remarks in the previous paragraph, this is clearly not the case; in fact, the recovered decapitated burials are probably the exceptional kind of treatment. The second type is even rarer, and the 'trash' circumstances these burials involve support the view that in the MPPNB some people died in the community who were not afforded any ritual at all in their disposal (Rollefson 1986: 50).[11] All infants younger than c. 12–15 months were found with skulls intact with the rest of the body, a situation also noted at Bayda (Kirkbride 1967: 9). The 'casual' treatment of some infants, also consigned to rubbish pile deposition, was also probably in effect (for babies of the people associated with burial type 'b'?), although the circumstances of other infant burials indicate special ritual meaning (Rollefson *et al.* 1992: 463; Rollefson and Simmons 1986: 183). Since all infants below the age of 12–15 months were disposed of with the skull still attached to the body, there may have been a required period of physical and social maturation before public recognition of the child's entry into society, perhaps through some naming ceremony after 12–15 months of age (Rollefson 2000).

Most MPPNB burials included no 'grave goods' except the occasional personal ornament, such as a necklace of 40 small bone beads with one woman at 'Ayn Ghazal (Rollefson 1986: 51) or the cowrie-shell bracelet with a child at Bayda (Kirkbride 1967: 9). A striking exception to this paucity of grave goods was found in 1989 at Wadi Shu'ayb: in a burial pit with four individuals, a 22 cm-high human female statuette made of marl was placed on the chest of an older adult of undetermined sex; nearby were 20 plaster beads and fragments of smaller figurines (K. Roler, pers. comm.).

Skull treatment

Headless burials were found at Bayda and Wadi Shu'ayb, but so far in Jordan MPPNB skulls (e.g. Bienert 1991) have only been recovered at 'Ayn Ghazal, 13 in all.[12] Three untreated skulls were found in a pit beneath a house floor. A courtyard pit contained four skulls, two evidently untreated, but the other two bore evidence of having the face area covered with lime plaster and bitumen 'cosmetics' (eyeliner and pupils). A third plastered skull, found in a bulldozer section with little secure information concerning its context (but probably in a courtyard pit), was skilfully, delicately modeled and incised but reflected no use of bitumen (Simmons *et al.* 1990). Three more modeled skulls, older than 9100 bp, were placed in a hole dug into sterile basal clay; actually, the skulls themselves were missing, and what had been buried were only the plaster 'masks' that once covered the faces (Griffin 1993; Griffin *et al.* 1998). Three temporal and parietal fragments of a human skull were found on the floor of a house that appears to have been destroyed by fire; the pieces had been scraped clean of flesh and then painted red. Finally, the floor of one room had undergone several renewals, and in one of these the front portions of a decorated skull were shaved off, leaving only the rear portions that were coated with a dark substance, possibly bitumen (Rollefson *et al.* 1999; cf. Rollefson 1983, 1986; Butler 1989; Rollefson and Simmons 1986).[13]

There was considerable variability in treatment of skulls once they were removed from a burial: (a) no treatment except for reburial elsewhere; (b) painting and perhaps coating of portions with bitumen; (c) re-creation of faces in plaster, with no cosmetics; and (d) re-creation of faces in plaster plus the use of bitumen cosmetics. What this diversity might mean in terms of the potential status differences is conjectural; certainly there is the possibility that a part of the variability could be stylistic evolution over the 700-year span of the MPPNB.

Human and animal figurines

Bayda and Wadi Shu'ayb[14] did not produce many small clay human and animal figurines, but nearly 200 were recovered from MPPNB contexts at 'Ayn Ghazal. Of the animal figurines, some appear to be whimsical effigies made for or even by children for their amusement (cf. Rollefson 1983: pl. III-7). Other animal figurines almost certainly were involved in rites of magic, such as the 'killed' cattle recovered from a subfloor pit (Rollefson 1986: 50 and pl. II-4). Schmandt-Besserat (1997) has noted the association of fire with nearly all the figurines, and the differential firing on many of them, particularly those representing cattle, indicate that they were votive pieces used in ritual activity. It is interesting to note that around the throats of several of the cattle figurines there were depressions formed by spun thread, as if the figurines were 'haltered'

Figure 3.4. An MPPNB 'haltered' cattle figurine: note the groove behind the head, formed by a twisted, spun thread of unknown sort (photo C. Blair).

(Figure 3.4), despite the absence of any bone evidence that cattle at the time were domesticated.

MPPNB human figurines were probably mostly used as talismans for protection (perhaps at special times) from natural or spiritual forces (Rollefson 1998b), but the human form could also be used to display profane human tragedy (cf. Rollefson 1986: 47 and pl. II-3). One class of human figurines includes 'fertility figurines', representing obviously pregnant women; the rocker-stamped impressions on these figurines (but not on other types) appear to be unique to 'Ayn Ghazal (cf. Rollefson 1983: 32 and pl. II-3). Whether these figurines were used to promote fertility is a moot point; certainly it must be acknowledged that they may also have been talismans for the protection of mothers during pregnancy and especially at birth.

Lime plaster human statuary

Two caches of lime plaster human statuary were excavated at 'Ayn Ghazal in 1983 and 1985 (Rollefson 1983, 1986). The 1983 group is stratigraphically the older and consists of 25 or 26 figures (Figure 3.5); two separate radiocarbon samples from between the tightly packed statues produced radiocarbon dates of 8660 ± 80 bp and 8700 ± 80 bp. The group recovered in 1985 had been badly disturbed by bulldozer work, and only 10–11 separate pieces (seven restorable) remained in the pit; charcoal stratigraphically older than the statue pit provided a *terminus ante quem* of 8520 ± 110 bp (Rollefson et al. 1992: table 1). It is conceivable that this group actually comes from the first century or two of the LPPNB period.

Tubb and Grissom (1995) have provided technical descriptions of both caches, and no additional comments are needed here. There are close stylistic similarities with two of the plastered skulls from 'Ayn Ghazal, and it is plausible that the statues and busts represent a step higher in a hierarchy of religious expression, with the skulls representing known ancestors and the statuary symbolizing mythical ancestors that founded clans or even humanity itself (Rollefson 2000).

The two caches differ most noticeably through the presence of three busts from the 1985 group that had two heads rising from a single body (cf. Tubb and Grissom 1995: fig. 10). Interpreting such singular symbolic forms is difficult, but two developments in the southern Levant contemporaneous with the construction of the double-headed pieces include the devastating abandonment of all known permanent farming villages in the Jordan Valley and Israel by the middle of the ninth millennium bp (Rollefson 1987) and the beginning of what might have been an intensification of the exploitation of the steppe and

Figure 3.5. A female statue (with seven fingers on the left hand) and two busts from the 1983 cache from 'Ayn Ghazal (photo H. Debajah).

desert areas to the east of 'Ayn Ghazal (cf. Garrard et al. 1994; Rollefson and Köhler-Rollefson 1993). In both these situations the people of the 'Ayn Ghazal community were coming into more intimate contact with once geographically distant groups, and social bonds among them may have been depicted by a union of their respective mythical ancestors.

Settlement pattern

Numerous aceramic Neolithic sites have been located by recent surveys all across Jordan (e.g. Coinman et al. 1988: 56; table 3; Clark et al. 1994; Palumbo 1992; Mabry and Palumbo 1988; Bisheh et al. 1993; Wright et al. 1989; MacDonald et al. 1987; Betts 1992; Garrard et al. 1985), but most of the information has not been subdivided into the Middle and Late PPNB and PPNC phases. In one regard, this is because some surveys were conducted before the M/LPPNB and PPNC distinctions were made; in others, surveys included lithics analysts who were not familiar with the distinctions; finally, in many cases, it is probable that these subtle distinctions cannot be made solely on the basis of surface collections of chipped stone artifacts (cf. Rollefson 1990a: 122–23).

Despite the limited understanding of MPPNB settlement patterns in the broader sense, MPPNB settlements still show a major change of human–land relationships compared to earlier periods. The situation before 9200 bp is uncertain in terms of permanent settlements, but certainly the stone-built structures at Bayda (Kirkbride 1966), 'Ayn Ghazal (Rollefson et al. 1992), and Wadi Shu'ayb (Simmons et al. 1989) are indicative of permanent residence by a relatively large population. While detailed information is scarce, it appears that MPPNB settlements were mostly small or medium-sized, generally smaller than 4–5 ha. In the case of 'Ayn Ghazal, population estimates range from c. 500 people at the beginning to possibly more than a thousand at the end of the 750-year range of the MPPNB (cf. Köhler-Rollefson and Rollefson 1990: table 3). 'Ayn Ghazal is currently the only source for detailed evidence of population growth in Jordan for the MPPNB period.

The Late PPNB

Very little was known about the LPPNB period anywhere in the southern Levant, let alone in Jordan, until excavations began at Basta in 1986 (Gebel et al. 1988). Since then excavations have probed LPPNB settlements all along the N–S axis of highland Jordan, and they have also been sampled in the eastern steppe/desert region of Jordan. Some of the excavations are so new that little information is available at this time.

Lithic technology and typology

There is no major difference in the technological aspects of lithic manufacture in the LPPNB: the specialized naviform and ad hoc technologies continued to serve the needs, as they had in the MPPNB period, although Quintero has noted a tendency for naviform cores to lose some of their standardization in the later parts of the LPPNB (Quintero 1998). Typologically there were developments that contributed to a different 'look' of the tool kits used by the residents of the region.

Jericho and Abu Gosh points may have been relatively popular in northern Jordan in the earliest part of the LPPNB, but they drop out of the tool kit for most of the period. 'Normal' Byblos and Amuq points dominate completely, sometimes with elaborate pressure-flaked retouch. There appears to be a trend to increasingly delicate point manufacture (smaller dimensions altogether), although this has not been statistically tested, and in the south, at least, tangs are often elaborate (cf. Nissen et al. 1987: fig. 10; 1991: figs. 3-4). Projectile points from several of the temporary camps in the steppe and desert appear to retain many MPPNB characteristics (especially size) (cf. Betts 1988: fig. 11; Garrard et al. 1987: fig. 12; Baird et al. 1992: fig. 16).

The burin classes at 'Ayn Ghazal change from the highly skewed frequencies of the MPPNB to a more even distribution; nevertheless, transverse burin types still outnumber simple, dihedral and truncation burins (Rollefson 1995: table 2).

There also appears to have been a major change in sickle typology. Highly glossed sickle blades accounted for 9% of the formal tools in the MPPNB at 'Ayn Ghazal (Rollefson et al. 1992: table 4), but they made up only 2.9% of the formal inventory in the large 1993–95 LPPNB samples. Similarly, sickles at Basta were also suspiciously low at far less than 1% (Gebel, pers. comm.). In contrast, the number of 'knives' at MPPNB 'Ayn Ghazal was relatively low at 1.2%, although in the LPPNB this class rose to 29.6%, mostly due to an increase in naviform blades that showed rounded-edge damage and a slight indication of 'near gloss' developing within the first millimetre from the edge. This corresponds to the 'unglossed' sickles identified by Quintero et al. (1997) in their experimental work, and it is likely that these 'knives', then, were used for the reaping of dry, mature cereal crops; as Quintero et al. (1997) point out, this is an indication that cereal crops may not have become fully domesticated until the LPPNB period.

Architecture

The LPPNB period in Jordan saw radical changes in architecture that had held sway for several hundred years, then apparently disappeared for a millennium or more before returning as a common feature of village life in the region after the end of the Neolithic period. Houses in the MPPNB period were single family dwellings with relatively large and open rooms, although houses during this period at 'Ayn Ghazal showed a continuous increase in room number and corresponding decrease in room size (Rollefson and Köhler-Rollefson 1992). By the last three centuries of the LPPNB period, such houses may have been rare, and in their place larger, multi-family compounds appear to have been constructed to consolidate formerly independent nuclear families into larger and more efficient production and consumption units.

Basta Area B illustrates this phenomenon best (Nissen et al. 1991: fig. 1). In one compound there were 19 rooms that shared common walls around either a large central courtyard or a partially roofed (?) 'space'. The rooms varied considerably in size, some with interior dimensions of c. 1.5 by 1.5 m, others 2 by 2 m, and yet others 3 by 2 m in inside measurements. Overall, this structure covered more than 150 m^2. At least four, and perhaps five, other multi-roomed compounds shared walls with this immense building, which has been described as 'some kind of central building' (Nissen et al. 1991: 15). The compact organization reminds one to some degree of prehistoric pueblos in the American Southwest (Gebel, pers. comm.), and there is no question that the loose socioeconomic organization of earlier MPPNB settlements had undergone profound alteration.

Several new excavations reflect the change to the Basta-type architecture, including as-Sifiya in Wadi al-Mujib (Mahasneh 1997), 'Ayn Jammam, southwest of Ma'an (pers. observation; cf. Bisheh et al. 1993: 121-22), at Ba'ja north of Bayda (cf. Gebel and Bienert 1997), and in the Wadi Faynan region at Ghwair (cf. Najjar 1994; Simmons and Najjar 1996). The circumstances in northern Jordan are not so dramatically preserved, although a glimpse of something similar was noted at Wadi Shu'ayb in 1989, and buildings partially destroyed by bulldozers at 'Ayn Ghazal also show features of the architectural design seen at Basta (cf. Rollefson 1997a; Rollefson and Kafafi 1996: figs. 2-3).

LPPNB architectural changes also included the first indications of two-storied buildings in Jordan.[15] At Basta one wall is more than 4 m high and includes sockets near the centreline for floor beams; while this suggests two stories, each roughly 2 m high, it is also possible that this may be a terraced house with the rooms proceeding down the hill slope in 2 m 'steps' (Gebel, pers. comm.). Walls preserved to c. 3 m were also found at Ghwair (Najjar, pers. comm.) and 'Ayn Jammam (Waheeb and Fino 1997).

Two-storey structures are also in evidence at 'Ayn Ghazal. In one relatively large building in the North Field (partially destroyed by bulldozers), archaeological fill in three rooms recorded the LPPNB destruction of the structure due to fire at the end of the ninth millennium bp (Figure 3.6). In this fill were hundreds of thousands of charred lentils and peas, burned clay with

Figure 3.6. Large (minimum 11 rooms) LPPNB building from 'Ayn Ghazal; note circular floor hearth at right centre. A second storey existed at least over the rooms at the western end of this building (top of photo) (photo Y. Zo'bi).

ceiling beam impressions, and dense amounts of thick, red-painted floor plaster that could only have come from a floor above the ceiling (Rollefson and Kafafi 1996). Another building with room fill that includes second-storey floor plaster can be seen in the bulldozer section of the parking lot in 'Ayn Ghazal's South Field.

Overall, the architectural evidence indicates that the once-independent single-family units that produced and consumed their own food in the MPPNB were forced to adapt to the deteriorating environmental conditions of the latter half of the ninth millennium bp. As population had grown enormously by the end of the MPPNB period, increased demands for fields in the immediate vicinity of settlements could be met only by the establishment of corporate groups, composed minimally of two families (and most likely more)—probably siblings and their spouses—that pooled their resources for equitable distribution among themselves (Rollefson 1997a).

There were other stresses on the social fabric of the Jordanian townspeople of the LPPNB, and this is demonstrated in the emergence and proliferation of buildings that were directed towards religious observations. These shrines and temples are described in the following section.

One final remark on LPPNB architecture involves the presence of subfloor 'channels' in buildings in Basta, as-Sifiya and Khirbat Hammam (if this last settlement is indeed LPPNB; it is possible that it is MPPNB; cf. Rollefson and Kafafi 1985), either to isolate the floors from damp ground (Nissen et al. 1991: 14) or to provide a means of providing a level surface for house construction without the necessity of terracing a hill slope (cf. Nissen 1997). Such channels were not noted in LPPNB subflooring excavations at Wadi Shu'ayb. Subfloor channels were noted beneath the floors of an ostensibly LPPNB building (heavily remodeled in the PPNC period) in the South Field of 'Ayn Ghazal (Rollefson et al. 1990: 105 and fig. 7), but it is unlikely that channels could have been present under the large, two-storied house in the North Field, since the floors were very close to bedrock. Altogether, subfloor channels appear to be principally a characteristic of southern Jordan.

Subsistence economy

Crops planted by LPPNB farmers did not evidently change in any major way; certainly pulses remained a vital element of the diet, as more than a quarter of a million charred lentils and peas recovered from one house indicate (Neef, pers. comm.; cf. Kafafi and Rollefson 1995: 24). Hunting, which had played so prominent a role in the MPPNB subsistence economy, appears to have declined considerably in importance for some communities, perhaps because earlier habitation and exploitation had incurred severe habitat destruction in the immediate vicinity of the settlements (e.g. Köhler-Rollefson and Rollefson 1990). Currently analyzed samples from LPPNB 'Ayn Ghazal show a major reduction in animal species from the rich variety in the MPPNB (cf. Köhler-Rollefson et al. 1993: table 1), although admittedly this sample is from the 1988–89 seasons and therefore a comparatively restricted area of the site (c. 60 m^2). Nevertheless, preliminary findings of the 1993–95 LPPNB faunal remains indicate a drop in species variability from more than 50 in the MPPNB (Köhler-Rollefson et al. 1988) to just over 20 in the

LPPNB (von den Driesch, pers. comm.). It is important to note that 71% of the LPPNB animal bones are from domesticated ovicaprids (Köhler-Rollefson *et al.* 1993: table 1), so that wild species contributions had fallen by almost 50% in comparison to MPPNB times.[16]

Basta is the only other Jordanian LPPNB settlement with published results of faunal analysis, and the faunal inventory appears to be similar to that at 'Ayn Ghazal. Becker notes that both cattle and pigs appear as domesticated forms at Basta,[17] and that there are morphologically wild and domesticated forms for both sheep and goats in the faunal collection (1991: table 3). She has also remarked that domestic animals accounted for 78–82% of the animal bones at Basta, which is very close to the situation at 'Ayn Ghazal if the *Bos* and *Sus* remains are accepted as domesticates (see notes 16 and 17 and Köhler-Rollefson *et al.* 1993: table 1). Dogs also held a place in the daily life of LPPNB people (Becker 1991; Quintero and Köhler-Rollefson 1997).

There was a dramatic shift *within* the domestic animal group at 'Ayn Ghazal: a revolution in the proportions of goats and sheep. Wasse noted that in the MPPNB the ratio of goats to sheep was roughly 95 : 5%, but after 500 years by the beginning of the PPNC, this relationship had reversed to a 15 : 85% ratio, which suggests a major alteration of herd composition during the LPPNB period (Wasse 1997). Although the sample Wasse investigated had potential sampling problems *vis-à-vis* the LPPNB numbers, the larger LPPNB collection from 1993–95 (von den Driesch, pers. comm.) seems to support Wasse's prediction that the late ninth millennium bp witnessed a considerable overhaul of ovicaprid husbandry. Becker has written that the goat to sheep ratio at Basta in the LPPNB stood at 60 : 40%, which fits neatly between the MPPNB and PPNC extremes at 'Ayn Ghazal (1991: 64).

This changeover from goat to sheep dominance may have had a major effect on the development of pastoral nomadism (cf. Rollefson and Köhler-Rollefson 1993; Köhler-Rollefson 1992). Faunal remains from the Wadi al-Jilat and al-Azraq sites, in areas that are not natural habitats for either sheep or goats, include ovicaprids that appear for the first time, albeit in small numbers, during the LPPNB at c. 8200 bp (Garrard *et al.* 1987; Martin in Garrard *et al.* 1994; Baird *et al.* 1992: 27). At issue now is how incipient pastoralism developed, whether as an entirely new aspect of the subsistence economy of local steppe/desert natives (Byrd 1992: 55-56; cf. Perrot 1993; Ducos 1993) or if the herds of sheep and goats, and their herders, were seasonal visitors to the area from permanent farming settlements in the Mediterranean zones of Jordan (e.g. Köhler-Rollefson 1988, 1992; Rollefson and Köhler-Rollefson 1993; Köhler-Rollefson and Rollefson n.d.). A newly discovered LPPNB temporary camp in al-Azraq Shishan promises to offer good insight into the process of pastoral development (Wilke *et al.* 1997; Rollefson *et al.* 1999a).

Ritual

Human burials

Basta has produced the largest LPPNB mortuary population, comprising 42 individuals (Nissen *et al.* 1987: 96-97; 1991: 18; Berner and Schultz 1997). Burial practices often made use of abandoned rooms and subfloor channels beneath house floors. Skulls were often intact with the rest of the skeleton, although often the cranium was missing from a burial and skulls were found isolated from any body. It was not possible in many cases to determine if the decapitation was intentional or if it was the result of later disturbance. In many other examples, including subfloor burials, it is clear that the skull was intended to remain intact, which is a major departure from the MPPNB instances at 'Ayn Ghazal. The body was so tightly flexed in some cases that it appears the bodies may have been tied before interment (Nissen *et al.* 1991: 18). There is also evidence to show that red ochre was spread over at least some (parts of) burials at Basta.

Only one burial has been uncovered to date at 'Ayn Jammam (Waheeb, pers. comm.), but its circumstances suggest it may have been a special person. One large room in an LPPNB building included a massive, free-standing stone pillar situated towards the northeast corner that supported the roof. The dead person was placed in the northeast corner, after which walls were built to the ceiling to enclose the burial, effectively decreasing the floor area of the room by nearly 25%.

In view of the broad areas that have been excavated, the number of 'Ayn Ghazal LPPNB burials is relatively low, but this is mostly because few floors have been taken up. Several of the burials, placed inside rooms, were badly disturbed by later (PPNC?) digging activity that indicates that, as at Basta, abandoned rooms were used for interment. Courtyard burials were relatively common and often resemble MPPNB 'trash burials', in the sense that the burial pits are filled with rubbish in addition to the body, and the skull is present. One burial from the East Field showed severe treatment before interment (before death?): every long bone of the skeleton had been broken at least once. Furthermore, a fragment of a flint blade had penetrated the left temporal area, blowing a 2–3 cm section of the interior surface of the cranium into the brain (Rollefson and Kafafi 1996: 22).

Gazelles

Although gazelles had dropped in terms of their former nutritional value, they appear to have taken on some symbolic meaning in both the northern and southern parts of the country in the LPPNB period. At 'Ayn Jammam, in the same room as the free-standing pillar and the walled-in human burial, a small niche at about eye level in the eastern wall contained the skull of a gazelle arranged so that it stared into the room; on the floor beneath the niche lay a single gazelle horn core (Waheeb, pers. comm.).

There are good indications of a gazelle cult at 'Ayn Ghazal, too. In one room in the two-storey building in the North Field (Figure 3.6) were five pairs of charred gazelle horns that had fallen from the room above during the conflagration; an isolated gazelle horn was found in the fill of an adjacent room. In the courtyard, behind (uphill and to the west) the building, a small burial pit contained the articulated remains of a juvenile gazelle lying with its head turned back onto its shoulder; the strange thing about this burial is that the feet of the gazelle had been charred before interment (cf. Kafafi and Rollefson 1995: 23; von den Driesch, pers. comm.).

Human and animal figurines

While figurines have been found at LPPNB sites in Jordan, they are less remarkable than their MPPNB counterparts simply because of their relative rarity. Of note from Basta are a pendant in the shape of a human head that has some phallic implications (Nissen *et al.* 1991: fig. 6-2; Gebel, pers. comm.) and a fragment of a stone mask (Nissen *et al.* 1987: fig. 16-1) similar to the two better-preserved pieces from Nahal Hemar (Bar-Yosef and Alon 1988: 23-27).

Cultic shrines and temples

Spectacular architectural remains that undoubtedly are related to ritual activity have been exposed at 'Ayn Ghazal. Two architectural styles seem to coexist, and they probably had different functions.

Cultic shrines

Near the two-storey building in the North Field of 'Ayn Ghazal, an LPPNB building with a four-phase history certainly did not serve a domestic function, at least not in its final phase of use (Rollefson and Kafafi 1994: 20-24 and figs. 8-10). The first two phases entailed the construction of buildings with an apsidal wall at the western (uphill) end; the adjacent western courtyard consisted of a former MPPNB lime plaster house floor surrounded by the original MPPNB walls (Rollefson and Kafafi 1997). The third phase saw the construction of a straight wall across the width of the apse to counter the collapse of the curved section. The interior design of the building in the first three phases is not known due to the radical changes inflicted by the fourth phase, but it is possible that the structure was essentially 'open' and uncluttered with interior 'furniture', if the apsidal building that was excavated and re-used by Yarmoukian Pottery Neolithic inhabitants in the Central Field can be used as a model (cf. Rollefson *et al.* 1990: 110-12 and fig. 12). In this regard, there would be similarities with the T-2 and T-3 shrine or sanctuary or temple buildings described by Kirkbride outside the domestic area at Bayda (Kirkbride 1968a: 93-96 and fig. 2).

The concept of the building was completely reorganized in Phase 4 (Figure 3.7). A circular wall enclosed a space 2.5 m in diameter with a doorway leading to the east that led into an antechamber of unknown dimensions. The floors of the earlier structure were cut through in the newest renovation, and a sequence of eight red-painted lime plaster surfaces accumulated in the circular room. The centre of the room was dominated by a hole 60 cm in diameter and c. 40 cm deep. At the edges of the hole the floor plaster began to rise, just as it did along the wall, revealing that at one time there was an installation rising out of the floor at the hole. Beneath the floor were two paired channels about 20 cm wide and 15 cm high formed by limestone slabs resting across the channel sides made of flat limestone slabs set on edge (Rollefson and Kafafi 1994: fig. 10); the channels formed radiating axes in north–south and northeast–southwest directions. Presumably these channels were air ducts to bring air into the hole from outside to feed a fire in an upraised hearth, although this speculative reconstruction has not been tested by further (damaging) investigation in the field. The unique shape of the building and its

Figure 3.7. 'Twin' LPPNB shrines from 'Ayn Ghazal. The one on the right is the oldest, and the circular room is the fourth phase of construction. The one on the left (to the south) was built rapidly and poorly, and it lasted a short time (photo B. Degedeh and Y. Zo'bi).

uncommon interior arrangements are strong arguments that the building served a ritual function, and the small size and limited access indicate that the structure was probably limited to family (or lineage/clan) observances much in the nature of a cult building or shrine.

In 1996 excavations just 5 m to the south exposed a virtual twin to the cultic building. The walls of the southern room consisted of small, rounded field stones set in thin mortar, in contrast to the sturdy construction techniques in the northern structure (Figure 3.7). The thin floor, although made of red-painted lime plaster, showed that one reflooring episode had occurred, but the badly damaged surfaces of both of them revealed a very fragile substructure, and with the poor-quality wall construction, the impression one gets is that the shrine had been hastily erected, perhaps with little intention of the need for durability. It is only conjectural at the moment, but it would seem that the southern circular shrine was a successor to the northern one after the latter fell out of use for some reason (Rollefson and Kafafi 1997).

The Upper Temple

A rectangular building erected high on the slope of the central East Field at 'Ayn Ghazal is clearly distinctive from any domestic LPPNB building at the site. Measuring 4 m (N–S) by at least 5 m (E–W) (the western end was destroyed by erosion), the building is a single-room structure with a dirt floor (one of only two PPNB buildings without a lime plaster floor at 'Ayn Ghazal; see below). The carefully constructed walls, using rectangular slabs and thin leveling stones similar to the building techniques used in southern Jordan, are preserved to less than a metre high. The following 'furniture', or features, identify the structure as a temple (Figures 3.8, 3.9).

Near the centre of the southern wall two 1.0 by 0.2 by 0.2 m limestone blocks border a 0.5 m-wide patch of brick-red burnt clay about 3.0 cm thick; the eastern end of this altar-like platform is bounded with two large shapeless stones and the western (downslope) side is eroded away.

Directly to the north of this altar is a line of three 'standing stones' equidistant from each other and from the southern and northern walls of the temple. The stones stand about 70 cm high, and the central stone tumbled towards the west in antiquity.

Midway between the line of the three orthostats and the eastern wall is a red-painted lime plaster hearth, c. 50 cm in maximum dimension, set off from the surrounding floor by seven small limestone slabs.

In the centre of the eastern wall, and an integral part of it, is another orthostat of brilliant white limestone

The Lower Temple

Approximately 100 m downhill and southwest of the Upper Temple is another LPPNB structure that clearly served a ritual purpose. The lower example is larger, although erosion has removed much of the western part of both buildings (Figure 3.10). Nevertheless, it is clear that the temple once consisted of several rooms: the eastern room measured 6.5 (N–S) by 3.5 (E–W) m, the partially destroyed western room at least 6.5 (N–S) by 2.5 (E–W) metres. To the east, beyond and adjacent to the eastern room was a small storage (?) chamber excavated into sterile basal clay, measuring 3.0 (N–S) by 1.5 m (E–W) and with walls preserved to at least 1.2 m high.

The internal furniture of the eastern room demonstrates the significance of the space. The 'back wall' (east) is preserved to a height of 1.8 m, and it is clear that a westward collapse of this wall was the reason for the abandonment of the temple. The floor of the room was clay, apparently obtained from the excavation of the storage chamber east of the temple. Inside the room and against the middle of the eastern wall was an altar nearly 2 m long, consisting of two large limestone slabs c. 60 by 80 by 20 cm that had been set atop three pairs of shaped rectangular 'standing stones', each c. 40 cm on a side and roughly 45–70 cm high. In front (west) of the altar was a floor hearth of white (lime?) plaster surrounded by seven limestone slabs, which altogether had a diameter of c. 1 metre. Adjacent to the centre of the northern wall was a small (c. 45 cm on a side) square cubicle made of limestone slabs set on to the clay floor. There were no artifacts in or around this feature, so its function is conjectural.

A 1 m-wide doorway in the wall separating the eastern and damaged western rooms (the latter had, evidently, a floor of common dirt) was set off by the thin screen of a single row of stones on the south that quickly (within 60 cm) made a right-angle turn to the north, effectively blocking off all possibility of viewing activity in the east room. The situation implies an example of a 'Holy of Holies' much more ancient than ever before conceived (Figure 3.10).

The interior furniture from both these temples set them apart from the shrines in the North Field. It is interesting to note the parallel with Kenyon's PPNB 'shrine' at Jericho that incorporated the socketed orthostat in one wall niche (Kenyon 1981: 306-307 and pls. 172-73). The furniture aspect brings to mind the standing slabs in Bayda's T-1 building complex, which

Figure 3.8. View to the east of the eastern room of the upper LPPNB temple at 'Ayn Ghazal. On the floor in the centre of the photo is the 'floor altar', and to the north (left) of this are three 'standing stones' (the centre one has fallen). Near the centre of the back wall is a doorway blocked by a large anthropomorphic orthostat (top centre) and dressed stones (behind the metre stick and to the left of the point of the north arrow) (photo B. Degedeh).

c. 80 cm high and 40 cm wide. The rounded form of this 'column', plus a knob-like projection at the top, lend an anthropomorphic character to the stone.

The building's interior underwent at least one phase of renovation: a low platform of limestone slabs was built in the space between the north wall and the northernmost standing stone, and this platform was set off from the western side of the room by a thin, single-leaf wall of small limestone blocks. Although the western side of the temple is featureless (bearing in mind erosional damage), easy access to the eastern side would have been only between the central and southern standing stones.

Figure 3.9. View to the west (top) of the eastern room of the upper LPPNB temple at 'Ayn Ghazal. The 'floor altar' is at upper left, the three 'standing stones' at top centre. At right centre is a red-painted lime plaster hearth surrounded by seven stones (photo B. Degedeh and Y. Zo'bi).

also had at least two phases of use (Kirkbride 1968a: 95-96 and fig. 2).

The singular nature of these structures is most likely associated with ritual. The persistent, intense burning on the surface of the floor-level altar at 'Ayn Ghazal is unlike any other feature at the settlement, and burning (and associated smoke and fire) must have been essential elements of the rites practised in the walled surround. The absence of any associated skeletal remains excavated at the three sites, particularly plastered skulls, indicates that the ancestral cult may not have been associated with these buildings.

Settlement pattern

The lack of distinctions among the general PPNB sites found during surveys conducted in Jordan applies to the LPPNB as well as the MPPNB. Nevertheless, recent excavations have confirmed an earlier prediction that the LPPNB period was a time of population consolidation and site growth that reached 'megasite' proportions that would not be seen again until the Chalcolithic period at the earliest (Rollefson 1987, 1999). It is now apparent that large towns like 'Ayn Ghazal became established in most of the major wadi systems, including 'Ayn Jammam at the edge of Wadi Hisma, Basta, al-Baseet (Wadi Musa; Fino 1997), as-Sifiya (Wadi al-Mujib), 'Ayn Ghazal (southern Wadi az-Zarqa'), Wadi Shu'ayb, and perhaps Kharaysin[18] in the northern Wadi az-Zarqa' (Edwards and Thorpe 1986). The emerging pattern of these large centres is intriguing, and it suggests that other towns may be found in other major wadi drainages, such as Wadi Dana, Wadi at-Tafila, Wadi al-Hasa, Wadi al-Karak, etc., all along the escarpment from the north to the south of the highland spine of the country. Smaller sites, unfortunately assignable (mostly) to a specific PPNB phase only after excavation, might show dependent relationships with the larger sites in Theissen Polygon studies (cf. Levy 1986).

The reasons for the development of the megasite phenomenon probably lie in great part among the kinds of cultural degradation that slowly but steadily throttled smaller and ecologically sensitive MPPNB settlements, forcing a relocation of the affected populations, in part at least, to farming villages in more tractable environmental circumstances (Rollefson 1987; Köhler-Rollefson and Rollefson 1990). Smaller contemporaneous 'subsidiary' villages and hamlets may have existed throughout the arable countryside (and Khirbat Hammam is a possible example; cf. Rollefson and Kafafi 1985), but since small sites rarely receive the beginning or middle of the LPPNB period, a part of the environmental conscription mentioned above.[19] The abandonment tragedy also affected long-lived settlements in the Transjordanian highlands, including Bayda, which has yielded no clear indications of an LPPNB occupation;[20] perhaps many of its residents resettled in nearby Ba'ja or Basta.

But modifications in settlement pattern and adjustments in local resource exploitation were only transient remedies for the problems that had already begun a subsurface seismic disturbance to Neolithic lifeways in the MPPNB. After about 500 years of coping with social realignments, the PPNB sociocultural and socioeconomic patterns had eventually traversed a reweaving process that resulted in a new cultural fabric as distinct from the MPPNB as the MPPNB was different from the PPNA. By the end of the ninth millennium bp, new adjustments set the multidimensional PPNC rearrangement of social, cultural and economic threads that had frayed for some time.

The Pre-Pottery Neolithic C (PPNC)

In her interpretation of the stratigraphy of Jericho and its implications for the region, Kenyon viewed the ninth millennium bp abandonment of the long-lived settlement and the ensuing erosion of the site as evidence for a withdrawal from the southern Levant for up to a thousand years (Kenyon 1979: 46), a suggestion that caught a tenacious hold as the *hiatus palestinien* and that even recently has been recalled as a historical event (Nissen 1993). Investigation of the uninterrupted occupational sequence at 'Ayn Ghazal first brought the early eighth-millennium bp transitional period between the LPPNB and the Yarmoukian Pottery Neolithic to light, and, in view of its distinctive character in terms of lithic technology and typology, subsistence economy, architecture and ritual, the period was termed the 'PPNC' (Rollefson and Simmons 1986: 161). Although the validity of the PPNC as an archaeological 'entity' took time for general acceptance in the southern Levant, it has now been recognized at Khirbat Sheikh 'Ali (Garfinkel 1994) and Atlit Yam (Gopher and Gophna 1993: 307) in Israel, and it is also becoming better known in Jordan as surveys and excavations have increased the data base there. Unfortunately, most of the information that is available for the PPNC period in Jordan comes only from 'Ayn Ghazal, although there are indications that some earlier excavations may have unknowingly investigated PPNC deposits.

Figure 3.10. The lower PPNB temple at 'Ayn Ghazal. Across the top is a 20 m-plus retaining wall. The Eastern Room of the temple has an altar raised on three pairs of orthostats at the centre of the rear (east) wall, in front of which is a white plaster hearth on the floor, surrounded by seven stones. A doorway in the wall in the centre of the photo is partially blocked by a right-angle screen of stones (photo B. Degedeh and Y. Zo'bi).

intensive investigation (if any at all), the human geography of Jordan will remain obscure. A few sites in the eastern steppe and desert of Jordan have been radiocarbon dated to the LPPNB (e.g. Betts 1992; Garrard *et al.* 1994), and there was undoubtedly an eastern 'push' for part-time exploitation of the area, possibly as the initial stage of pastoral nomadism (Köhler-Rollefson 1992).

What was taking place in Jordan stands in stark contrast to contemporaneous events in the Jordan Valley and Israel. Most long-established farming villages in this western region of the southern Levant were being abandoned wholesale, so that Jericho, Abu Gosh, Munhatta, Yiftahel, etc. were mostly deserted by

Lithic technology and typology

Comparison of lithic manufacturing techniques for the LPPNB and PPNC phases show changes as striking as between the PPNA and PPNB periods (Cauvin 1987). Every major technological feature (blade : flake ratio, platform type, presence of cortex, etc.) is significantly different (Rollefson 1990a, 1997b), and these contrasts can be directly linked to the relinquishment of the naviform core-and-blade technique (Rollefson *et al.* 1994: 454). The abandonment of this technique is also reflected in the diminution of the use of the pink-purple Wadi Huweijir flint at 'Ayn Ghazal, and altogether the evidence demonstrates that specialists in flint tool production (Quintero and Wilke 1995) had lost their calling.[21] Possibly this was due, in part at least, to the decline in the importance of hunting and to the evolution of cereal grains that remained intact with the stalk at maturity.

Typologically, projectile points were much smaller than their MPPNB complements at 'Ayn Ghazal, although the statistical relationship with the LPPNB in terms of dimensions is not significant, despite a continued trend in these regards. Invasive and even covering unifacial retouch was more characteristic of the PPNC period in contrast to earlier times (Figure 3.2). Hunted game was dominated by gazelles (Köhler-Rollefson *et al.* 1993: table 1), and this may be partially responsible for the virtually exclusive use of bow-and-arrow delivery systems instead of spear-thrower reliance in the MPPNB (Eighmey 1992).

Burin production was transformed to a nearly equivalent distribution among the four burin classes (Rollefson 1995: table 2), but it should be mentioned that the pre-eminence of truncation burins in the Yarmoukian is foreshadowed by a strong representation in the PPNC. Bifacially retouched knives and tabular knives became typical elements of the tool kit in the PPNC, and scrapers changed dramatically from sidescrapers to transverse scrapers (Rollefson *et al.* 1994: 454-55). There was a trend for retouched tools of all kinds to become shorter, wider and thicker in the PPNC, undoubtedly a consequence of the switch to the increased production of flakes.

Subsistence economy

Recent research by Wasse (1997) and Von den Driesch and Wodtke (1997) reveals that by the onset of the PPNC at 'Ayn Ghazal, sheep were in overwhelming supremacy in the ovicaprid herds, and it is possible that cattle and pigs had joined the domestic inventory as well. Hunting had dropped to a minor contribution to the diet (c. 13%; cf. Köhler-Rollefson *et al.* 1993: table 1), and most of the mammals came from steppe or desert territory, including gazelle species as well as onager, which had risen to 3% of the bone sample. If the domesticates are removed from consideration, the PPNC faunal inventory at 'Ayn Ghazal mirrors a very eastern orientation, with few representatives of the Mediterranean forest/woodland or *maquis* zones. This all coincides neatly with the rise in the emphasis on sheep and the model for the intensification of incipient pastoral nomadism (Köhler-Rollefson 1992; Rollefson and Köhler-Rollefson 1993). Further validation for the emergence of pastoral nomadism is provided in the architectural and ritual sections below.

Architecture

PPNC inhabitants clearly reused and renovated standing LPPNB buildings in several areas of 'Ayn Ghazal (Rollefson *et al.* 1990), which strengthens the idea that there was a transitional period from one phase to the other instead of a replacement of one population by another. But it is also quite obvious that the LPPNB standard of two-storied, multi-family dwellings was abandoned soon in the PPNC period, and that a novel dual house plan replaced it. One house plan was used by farming families who inhabited 'Ayn Ghazal year-round; the other was the 'corridor building', a platform/storage bunker for relatives who left 'Ayn Ghazal for months at a time, returning to the town only after the harvest was in.

Smaller and simpler one-room buildings for full-time residents of 'Ayn Ghazal replaced the composite duplex/triplex houses of the LPPNB. They agreed with the reduced economic circumstances brought on with the cultural degradation of the environment for more than a thousand years at 'Ayn Ghazal, and they represented one part of a new adaptation to cope with diminished resources in the vicinity of the town. Measuring about 4 by 3 m (Figure 3.11), the houses were made with thin walls (c. 40 cm thick) and, in complete incongruity with the LPPNB period, with floors made of either dirt or an ersatz plaster made of crushed chalk and mud (*huwwar* in the local Jordanian dialect) (cf. Rollefson and Köhler-Rollefson 1993: fig. 3, upper left). Some storage facilities may have served these farmer families (cf. Rollefson 1993: pl. I-1).

'Ayn Ghazal has produced remains of four or five 'corridor buildings'—platform/storage bunkers—ranging

Figure 3.11. A PPNC courtyard work area at 'Ayn Ghazal (the house remains unexcavated to the left [north]). Near the centre of the photo is a rectangular feature that may have been a 'milling bin' (photo H. Wada).

from mere traces to one building that is substantially intact. The last is a c. 3.5 by 3.5-m semi-subterranean structure with small narrow rooms and proportionally thick walls that clearly supported an upper 'storey', although nothing in the room fills below nor in the construction of the walls themselves indicates that there was a superstructure in the sense of an upper level of rooms (Figure 3.12). Instead, the broad walls appear to have supported a wood-and-mud roof-cum-platform that could have served as a temporary tent area (Rollefson and Köhler-Rollefson 1993: 36-37 and fig. 2). A floor assemblage of artifacts, found in the cell at the southwest corner of the structure, included axes, loom weights and other heavy duty tools; nothing was found on the floors of the other cells.

Köhler-Rollefson's Incipient Migratory Pastoralism model is well supported by the two types of PPNC structures. The platform/storage bunker structures (found only in the South Field) were used by that part of the community that left the settlement with their flocks of sheep and goats for extended periods of the annual round, returning only after the harvest in early summer when the steppe and desert areas dried out and when the animals could forage on the field. During their absence, the herders could store the material not needed on the trip, and relatives could lay up a portion of the harvest in anticipation of their return, which would be compensated for by an exchange of sheep and goat products (including fertilizer on the fields) (Köhler-Rollefson 1992; Rollefson and Köhler-Rollefson 1993).

The PPNC corridor buildings at 'Ayn Ghazal and those from Levels II and III at Bayda (Kirkbride 1966: figs. 1-2; 1967: fig. 1; Phase C of Byrd 1994) are so similar that they are likely contemporaneous. There is one structural difference: several of the Bayda examples reflect a level of one or more rooms built above the semi-subterranean basement cells (Byrd 1994: fig. 8; Kirkbride 1966: 15), while this was not evident at 'Ayn Ghazal; nevertheless, this suggests local functional variability and is a weak argument against the overall congruence of the design and probable functions of the structures in both parts of Jordan. What is a problem at the moment is the MPPNB age Kirkbride assigned to Levels II and III on the basis of radiocarbon dates (Kirkbride 1966: 72).

Two radiocarbon dates of 7824 ± 240 bp (Sample AA-1165) and 7895 ± 95 bp (Sample AA-5205) uphold the PPNC ascription to the 'Ayn Ghazal corridor buildings; three other charcoal samples from the fill in and around the PPNC structure are LPPNB in age and probably reflect the disturbance caused by the PPNC builders as they excavated down to an LPPNB red-painted plaster floor for their own

Figure 3.12. A PPNC corridor building (right, south) and paved ramp (left, north) at 'Ayn Ghazal. The entrance to the semi-subterranean corridor is at the top of the photo. Notice the small size of the rooms (scale = 1 m) (photo L. Rolston).

use: four other radiocarbon dates support the general PPNC time frame at 'Ayn Ghazal (cf. Rollefson *et al.* 1992: table 1). The three dates from Levels II–III at Bayda, which ranged in assay results of 9030 ± 50 bp (GrN-5062), 8892 ± 115 bp (P-1382), and 8550 ± 160 bp (K-1085), all came from the *same* piece of charcoal, and chronologically the charcoal is contradictingly older than samples from stratigraphically earlier contexts. As Kirkbride mentioned, 'the large pits excavated to contain the semisubterranean corridor buildings of Levels II and III had largely obliterated the underlying level' (1966: 19), which undoubtedly created an enormous amount of material from mixed contexts, including an earlier charred log that ended up in Level II 'fill'. In sum, there is a strong possibility that a major PPNC occupation occurred at Bayda that was not recognized in field work or laboratory analysis, possibly due to the mixed nature of the earlier artifacts combined with contemporary discards of the early 8th millennium bp.

The large open buildings from Phase C at Bayda, to which Byrd assigns a public, non-domestic function (Byrd 1994: 656-57, fig. 9), have not been found at 'Ayn Ghazal or Wadi Shu'ayb. On the other hand, public 'architecture' of a sort has been uncovered at 'Ayn Ghazal that indicates not only non-domestic utilization, but a high degree of public cooperation in community planning and construction. One of these public works is the 'Great Wall', a relatively massive bulwark 1.4 m wide and preserved to a height of four to five courses that ran roughly east–west for more than 11 m in the Central Field;[22] this wall separated two *huwwar*-plastered courtyards, which in themselves were not characteristically domestic arrangements. The entire system remains enigmatic and only future excavation can resolve the meaning of this situation (cf. Figure 3.13).

The 'walled street', a public thoroughfare that roughly paralleled the Great Wall about 22 m to the north is the second element of public construction. Approximately 2.5 m wide, the street progressed from the roadside bulldozer cut uphill for a distance of c. 9 m before it was destroyed by later Yarmoukian activity (Figure 3.14). In this stretch there were two gateways, set off from the street by a curb of a single line of stones, that led into courtyards to the north (Rollefson and Kafafi 1994: 13-14, figs. 4 and 7).

Settlement pattern

As mentioned earlier, early eighth-millennium presence in the southern Levant was unknown prior to 1984, and it is evident that the material culture characteristics of PPNC period are still unfamiliar for some

Figure 3.13. The large Yarmoukian dwelling at 'Ayn Ghazal is at lower right, much later than the Great Wall and other structures in the upper half of the photo. The Great Wall in the centre was PPNC but used and modified by later Yarmoukian residents. Note the 'empty' courtyard top the left (north) of the house. At far left centre is an oval outhouse evidently used as a kitchen (photo Y. Zo'bi).

Figure 3.14. The stepped, walled street at 'Ayn Ghazal is in the centre of the photo. At the top is the modern highway, and to the left (north) is a Yarmoukian 'tent' structure (photo Y. Zo'bi).

recent excavators. PPNC occupations are not easily detectable on the basis of surface collections of lithic artifacts; the PPNC phenomenon is more than simply a change in lithic production (while fundamental, this change may not be easily recognizable to many surveying teams). But as the extent of PPNC identifiers becomes better known, the distribution of PPNC sites will probably increase accordingly.

PPNC occupations are already documented at 'Ayn Ghazal and Wadi Shu'ayb, and the newly discovered site of as-Seyyeh in the Wadi az-Zarqa' drainage has revealed a PPNC horizon beneath a Yarmoukian deposit (Z. Kafafi and G. Palumbo, pers. comm.). PPNC layers are also probable at Basta, and the architecture at Bayda now suggests a major PPNC occupation there in the early eighth millennium bp. Architectural characteristics at as-Sifiya (personal observation), Ghwair (personal observation) and 'Ayn Jammam (Waheeb and Fino 1997) also suggest possible PPNC presence near the end of each local occupation. Smaller farming sites in the western highlands of Jordan or pastoral sites in the steppe and desert are problematic, since they would have produced decreased numbers of distinctive artifacts; their smaller sizes would be easier to miss on surveys or, if detected, they would have been more difficult to assign to a post-PPNB ascription because of transitional characteristics in many cases.

Human burials

At 'Ayn Ghazal later cultural activity had disturbed many of the 37 burials excavated from PPNC layers; in addition, numerous scattered human bones have also turned up in PPNC contexts, and these might represent many LPPNB and even MPPNB interments dislodged by PPNC pit-digging activity.

PPNC burials are distinctive from earlier norms in several ways: (a) multiple interments in the same pit occur commonly; (b) secondary burials, previously very rare at 'Ayn Ghazal at least, became as popular as primary burials; (c) decapitation of skeletons was no longer practised, which argues that the ancestor cult had changed considerably, if it had not been abandoned altogether; and (d) grave offerings, in the form of pig bones, were common elements in burial pits.

The situation involving multiple interments in the same burial pit may have its origin in the LPPNB, for some of the subfloor channels at Basta contained several individuals; on the other hand, it is also possible that these circumstances represented repeated reopening of the channels in successive burial rites (Nissen *et al.* 1991: 18). The situation for multiple burial pits in pre-PPNC times will remain ambiguous until more complete field reports are forthcoming. At 'Ayn Ghazal, the earliest (?) PPNC burials were placed beneath house floors, a carryover from LPPNB times, but in these instances the differences clearly pointed to PPNC alterations of earlier practices. First, two pits contained two and three individuals, respectively, and, as they were primary burials, it is likely that they had died at about the same time. One implication might be that diseases had taken hold of the large settlements.[23]

The burial of people who had died elsewhere and were transported back for interment became a routine rite at 'Ayn Ghazal in the PPNC period, and roughly half of the interments are of this type. Characteristically, many of the smaller bones (wrist, ankle, ribs, etc.) are missing in such interments; on the other hand, in one case the secondary burial was represented simply by a skull with no other skeletal elements present. A common interpretation of secondary burials holds that people who died outside the permanent village were brought back to the settlement, perhaps after a considerable period of time, and in the process some bones were lost along the way. When viewed from the perspective of Köhler-Rollefson's Incipient Migratory Pastoralism model, the secondary burials could represent the part-time herder segment of the community, an interpretation supported by an especially high incidence of secondary burials in the corridor–building complex in the South Field.[24]

The presence of the isolated skull as one secondary burial in the South Field is unique at 'Ayn Ghazal; there is no evidence from the other burials that skull decapitation was still practised by the first half of the eighth millennium bp. In view of the number of subfloor burials at LPPNB Basta, where skulls were routinely included with the rest of the skeleton, the decline in the importance of the skull cult may have already begun by the last quarter of the ninth millennium bp.

A novel PPNC change in the post-mortem treatment of the dead was the inclusion of pig bones with some people. So far this appears to be known only at 'Ayn Ghazal. A detailed correlation has not yet been undertaken, but there is some indication that the pig bone offerings are principally associated with the corridor–building complex in the South Field, although there are some instances in the 'permanent village' area in the Central Field as well. If the distribution of pig bone grave offerings is restricted in both the 'herder' (South Field) and 'permanent farmer' (Central Field) sectors at 'Ayn Ghazal, we may have the earliest solid evidence of clan-based totemism in the Levant.

Human and animal figurines

PPNC deposits in Jordan have not been rich in human and animal figurines. Mention should be made of an elegant carved limestone figurine, painted red, of a stylized pregnant woman 14 cm long from 'Ayn Ghazal (Kafafi and Rollefson 1995: fig. 7). It is unique in its form, size, and material, and there is some reason to believe it was placed on a special platform in a courtyard (Schmandt-Besserat 1998). Besides this, other examples (in clay) are rare and not otherwise exceptional.

The Aceramic–Ceramic Neolithic transition

Kenyon proposed that the return of farmers to the southern Levant was a recolonization by people who, after perhaps a thousand years, had invented a fired-pottery technology (Kenyon 1979: 45-46). This entire scenario has now been shown to be of interest in the history of archaeological investigation, but it is also wrong. The stratigraphic evidence at 'Ayn Ghazal alone (and also at Wadi Shu'ayb) is strong testimony to a consistent transition from the LPPNB through the PPNC to the Yarmoukian period in terms of architectural use (Rollefson and Köhler-Rollefson 1993; Rollefson 1993).

The development of lime plaster technology satisfied all the prerequisites for the emergence of a pottery technology, and this pre-adaptation had been in place in the southern Levant since at least the beginning of the MPPNB (Rollefson 1990b). Indeed, as early as the mid-ninth millennium well-fired potsherds were found at Tall Assouad in northern Syria (Cauvin 1974) and at 'Ayn Ghazal (Rollefson and Simmons 1986). But these examples were exceptional; the tradition of pottery manufacture is certainly a characteristic that began near the latter half of the ninth millennium bp. Several instances of the transition from the Pre-Pottery to the Pottery Neolithic phases are shown in architectural use at 'Ayn Ghazal (Rollefson 1993). The stratigraphic and architectural evidence is strong testimony that there was a consistent occupation and evolution from the PPNC into the Yarmoukian period.

The Pottery Neolithic

The situation concerning the character of the Pottery Neolithic has become very complex over the past decade, and the simple scheme devised on the basis of the Jericho sequence in the 1950s is no longer tenable, not only for Jordan but for the entire southern Levant. As described above, the emergence of a ceramic technology was a local development in the area, and in view of the size and geographic diversity of the southern Levant, it should not be surprising that the trajectories of ceramic traditions did not follow a unilineal succession.

As was the case for so much of the aceramic Neolithic of Jordan, the ceramic Neolithic period remains poorly documented in the sense that the information at hand comes from only a few excavations, and these have been very different in terms of how intensively the Pottery Neolithic layers were investigated and reported. Among the most basic issues that remain to be resolved are how many distinct ceramic traditions there were, how they are distinguished, how long they lasted and how many may have been contemporaneous traditions.

The Yarmoukian

The most widely and firmly attested Pottery Neolithic entity in Jordan is the Yarmoukian tradition, a direct outgrowth of the earlier PPNC (Rollefson 1993). This ceramic tradition is defined by its distinctive ceramics, including the typical banded herringbone impression (Figures 3.15 and 3.16; Kafafi 1993). At present it appears that the Yarmoukian is restricted to the northern part of Jordan, having been reported from 'Ayn Rahub, near Irbid (Gebel and Muheisen 1985; Muheisen et al. 1988), Jabal Abu Thawwab between 'Amman and Jarash (Kafafi 1988); Wadi Shu'ayb (Simmons et al. 1989), 'Ayn Ghazal (e.g. Kafafi 1990), and as-Seyyeh near Zarqa' (Kafafi and Palumbo, pers. comm.). The only radiocarbon date from the Pottery

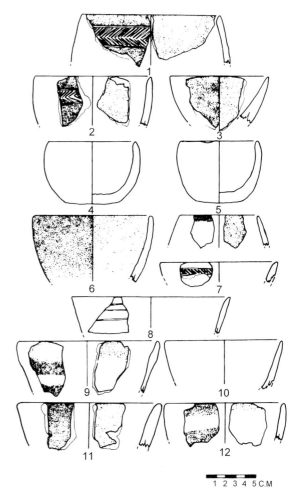

Figure 3.15. Yarmoukian bowls from 'Ayn Ghazal (drawing D. Obeidat).

Neolithic in Jordan comes from 'Ayn Rahub, which at 7480 ± 90 bp (Muheisen et al. 1988: 498) falls near the beginning of the Yarmoukian period; there is absolutely no indication of when the Yarmoukian ended.

Lithic technology and typology

There was little change in lithic technology from the PPNC to the Yarmoukian: flakes dominated the assemblage, and what blades were produced (usually short and relatively broad) came from relatively irregular unidirectional blade cores. The use of high-quality Wadi Huweijir-type flint with pink-to-purple hues disappeared altogether, although some of the projectile points and sickle blades appear to have been made on pink-purple flint naviform blades scavenged from earlier cultural deposits at 'Ayn Ghazal.

Projectile points became very small in general (Figure 3.2), and complete bifacial retouch was

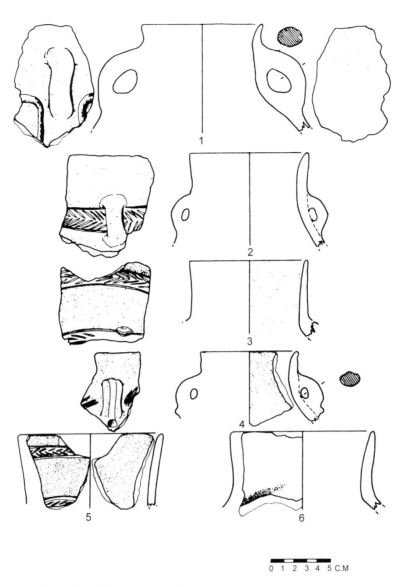

Figure 3.16. Yarmoukian jars from 'Ayn Ghazal (drawing D. Obeidat).

commonly used to shape them. Sickle blades have a deeply denticulated cutting edge formed by wide adjacent notches, and usually the tools are blade fragments that were backed and truncated at one or both ends. The high gloss on these sickle elements indicates they were not used on mature (dry) cereals, and although more research is necessary to determine their use, the cutting of reeds for mats and basketry should be considered.

Burins are heavily dominated by truncation types, particularly burins on concave truncations (Rollefson 1995). The similarities with burin sites from the steppe and desert in eastern Jordan suggest a general similarity in function throughout the region, although the use of burins of any kind remains enigmatic (cf. Finlayson and Betts 1990).

Architecture

The variability in Yarmoukian architecture is difficult to assess in view of the limited exposures of structures in Jordan. No coherent architecture was encountered in the restricted trenches at Wadi Shu'ayb and 'Ayn Rahub, and no preliminary reports have yet appeared for as-Seyyeh. Reliance must be placed, then, on the limited information from Abu Thawwab and the more extensive finds at 'Ayn Ghazal. Only very brief preliminary accounts of the architecture at Jabal Abu Thawwab have appeared so far, and little more can be said except that the Yarmoukian layers had both rectilinear and curvilinear structures; of the latter, one building had a diameter projected to be c. 4 m (Kafafi 1985: 125; cf. Kafafi 1986).

Permanent structures at 'Ayn Ghazal are mostly rectilinear, and the largest building discovered so far has three or four rooms and measures at least 4.3 by 8.8 m (Figure 3.16; Kafafi and Rollefson 1995: 14-15). The sturdy walls were built of undressed stone set in mud mortar, and one of the dirt floors was coated with a large square area of *huwwar*. Hearths and firepits were located outside in the spacious courtyards, sometimes associated with *arishas* (shaded areas). A small (2 by 1 m) oval 'kitchen' outhouse (Figure 3.13) was uncovered in one courtyard (Kafafi and Rollefson 1995: 16).

A circular building was only partially exposed in the 1996 season of excavation at 'Ayn Ghazal, and consequently there are doubts as to any interior features (Rollefson and Kafafi 1997). The wall of the exposed arc was extraordinarily thick (c. 80 cm), and the diameter would have been c. 5.5 to 6.0 m with a floor area of c. 12.5 m^2: one is not dealing with a flimsy outhouse. A doorway 80 cm wide opened to the south, complete with a threshold of flat limestone slabs and, just inside, a stone with a door socket almost 20 cm in diameter. The function of the structure remains uncertain, but since there were circular domestic structures at Jabal Abu Thawwab, it is possible this is simply a robust house with a floor plan that developed later in the Yarmoukian occupation of the settlement. It appears that the round 'house' is stratigraphically later than the rectangular house described above.

Other architectural elements used by the Yarmoukian inhabitants reflect the continuous use of the settlement across the transitional period from the PPNC. One of these is the reuse and slight modifications of the 'Great Wall' and the Walled Street originally constructed in the first half of the eighth millennium bp (Figure 3.14). Stretching back to much earlier horizons in their persistent zeal for digging pits, the Yarmoukians also encountered the lime plaster floor of an LPPNB apsidal building, which they cleared and used for their own purposes, ostensibly associated with some public function in view of the restricted presence of Yarmoukian fine ware inside and immediately outside the former late ninth-millennium building (Rollefson *et al.* 1990: 110-12).

The final kind of architecture associated with the Yarmoukian occupation at 'Ayn Ghazal is also the latest in the sequence. Foreshadowed, perhaps, by the solid round house of the later Yarmoukian levels at the site, at least two examples of circular temporary buildings have been exposed in the latest Yarmoukian layers (Rollefson and Kafafi 1994: 16-17; Rollefson and Simmons 1987: 105). The structures were outlined with a single row of stones that appear to have acted as anchors for walls made of animal skins or woven mats/cloth; the interior space had a diameter of between 4.5 and 5.0 m. It is possible that these structures are forerunners of the Bedouin tent, erected by the last Yarmoukian visitors to 'Ayn Ghazal, who brought their herds of sheep and goats to the sweet water of the spring after the settlement had become a ghost town (Figure 3.14).

Yarmoukian subsistence economy

Faunal resources are well represented and reflect a heavy dominance by sheep and goat herding. Pig and cattle were likely additions to domesticated species, although hunting was still practised (albeit at a much lower intensity compared to the MPPNB). Wild animals were primarily from steppe and desert species, including gazelle and *Equus* sp. (Köhler-Rollefson *et al.* 1988, 1993; als-Shiyab 1997; Wasse 1997), although notably the hunting of wild goat appears to have increased (Von den Driesch and Wodtke 1997), possibly since the domestic animals were away from the settlement for long periods.

Despite intensive sampling for flotation analysis at 'Ayn Ghazal, the only palaeobotanical results reflect that animal dung appears to have been the major fuel of the time. Fortuitously, remains from 'Ayn Rahub analyzed by Neef were more productive and show that emmer (and less frequently einkorn) wheat and hulled barley were the principal cereal crops, in addition to flax as a fibre or oil source. Wood charcoal was exclusively from the deciduous oak species *Quercus ithaburensis*. Other 'weedy' species included medick, mallow, clover, milk vetch and the thistles *Centaurea* and *Carthamus* (Neef, in Muheisen *et al.* 1988: 497).

Yarmoukian settlement pattern

Trying to determine a settlement pattern based on the limited amount of available information is inappropriate, but some aspects should be brought up that may reflect reality. First, the areal extent of Wadi Shu'ayb was at its greatest (c. 14 ha?), based on surface indications and on pottery in roadcuts, during the Yarmoukian period, and at 'Ayn Ghazal Yarmoukian structures were distributed over at least 12 ha. This is in contrast to the small size of such sites as Jabal Abu Thawwab, which likely did not exceed a couple of hectares at the most.

Secondly, although Wadi Shu'ayb and 'Ayn Ghazal were larger (in both area and probably population) than other known Yarmoukian settlements, it is clear that population densities in the permanent towns had dropped considerably over the period since the MPPNB and LPPNB heydays. At 'Ayn Ghazal, for instance, spaces between houses were more than 15 m (how much more remains unknown), and perhaps the total size of the 12 ha Yarmoukian population may have been even less than the original number of inhabitants in the 2 ha MPPNB settlement.

Finally, there is no reason to assume that the regional population of northern Jordan declined during the Yarmoukian period. In his survey around Jabal Abu Thawwab, Kafafi (pers. comm.) located numerous Yarmoukian hamlets within a short distance of each other, reflecting a dispersal of former human concentrations into a more diffuse scatter across the countryside. Banning's 'both sides of the surface' survey in the Wadi Ziqlab, where he has looked not only for archaeological evidence on the ground but beneath it as well, has shown that many sites, especially small ones alongside drainages, are obscured by natural processes (e.g. Banning 1996); they normally do not show up on traditional survey projects.

The Pottery Neolithic A tradition in Jordan

There is a clear distinction between the Yarmoukian tradition and the pottery recovered from Jericho Layer IX (variously termed Pottery Neolithic A [PNA], 'Jericho IX' and 'Lodian'), and one of the easiest features to recognize is the use of slips and a red pigment applied in geometric motifs (e.g. Kenyon 1979: 62 and fig. 4). The Jericho style is clearly an entity characteristic of the area west of the Jordan River and north of the Negev (Gopher and Gophna 1993: fig. 10). This early Jericho pottery has also appeared in minor quantities among the pottery repertoires of 'Ayn Ghazal and Wadi Shu'ayb (Kafafi 1998), but at Khirbat adh-Dharih (cf. Site 524 in MacDonald et al. 1982: 121) and Dhra' (Bennett 1980) in southern Jordan it appears to have constituted the major ceramic tradition. The excavations at Dhra' were very limited, but Bennett saw close similarities to PNA pit dwellings at Jericho (Bennett 1980: 33; cf. Kenyon 1979: 60-63); no hints of the size of this permanent (?) farming village have been suggested, although it may have been relatively large (cf. Raikes 1980: 56-60). Indications of pit dwellings also occurred in the road cut at Site 524 in Wadi La'ban just north of at-Tafila, but unfortunately highway construction in 1982 destroyed the remainder of the settlement before any excavations could be undertaken.

The Wadi Rabah tradition in Jordan

Although defined as a ceramic tradition in 1958 in Israel (Kaplan 1958; cf. Gopher and Gophna 1993), recognition of this tradition east of the Jordan Valley has been a recent event. Banning (1996: 35) appears to have uncovered two farmsteads in Wadi Ziqlab whose pottery and sickle blades are similar to Wadi Rabah sites elsewhere (also cf. Banning et al. 1996: figs. 6 and 7; Banning 1995: 5, 7). Kafafi (1998) assigned the 'Middle Levels' of Abu Hamid[25] to the Wadi Rabah, although the 'Lower Levels' were described as belonging to the Ghrubba tradition. In view of the broadly similar definitions of the Ghrubba and Wadi Rabah pottery wares and types (with bow-rim jars, hole-mouth jars and painted designs; compare Kafafi (1998) and Gopher and Gophna 1993: 329-33), it might be the case that the excavations at Abu Hamid have defined two successive phases of the Wadi Rabah entity (cf. Dollfus and Kafafi 1993: 246-49, 254).

It is not yet clear how widely the Wadi Rabah levels at Abu Hamid may have extended, but in view of the large size (c. 6.6 ha; Dollfus and Kafafi 1988: 23) of the sixth-millennium bp. Chalcolithic settlement in the highest levels, Abu Hamid could have been at one end of the settlement scale as opposed to the farmstead settlements in the Wadi Ziqlab (Banning 1996).

Dollfus and Kafafi note that the hunting of animals played a 'minimal' role in the subsistence economy, with bones of wild animals reaching only 10% of the faunal remains (Dollfus and Kafafi 1993: 245). This pattern was already established as early as the beginning of the PPNC (Köhler-Rollefson et al. 1993). No quantification is available yet, but the hunting of red and fallow deer, aurochs and wild boar is also reported for the Wadi Ziqlab sites (Banning et al. 1996: 43).

Other ceramic Neolithic sites in Jordan

There are other sites with Pottery Neolithic layers (and some with layers containing pottery of an unknown period) from Jordan that cannot be assigned to one of the generally accepted ceramic Neolithic phases. In some cases the samples were too small or suspicious to make a judgment. Such was the case at Basta, for example, where the quantity and nature of the potsherds could not be assessed (Gebel, pers. comm.). At 'Ayn Jammam, pottery fragments have been found on the surface, and, while they appear to be Neolithic, the small sample does not neatly conform to any of the major definitions.

A Pottery Neolithic site in Wadi Faynan is reported in some detail in terms of its circumstances, lithics and pottery (Najjar et al. 1990), but the authors have intentionally not attempted to ascribe the ceramic assemblage to a general entity, noting that there was little decorated material. Goren (1990: 102-105) assigns a collection from 'Faynan' to the Qatifian (a late Pottery Neolithic/early Chalcolithic group), although it is not clear if the collection actually comes from the site described by Najjar et al. Based on the illustrated pottery and lithics (especially sickles), Najjar et al. may be dealing with a variant of the Wadi Rabah tradition (1990: figs. 10, 12-13).

There are other issues concerning the Pottery Neolithic, not the least of which is deciding when the Neolithic ends and the Chalcolithic begins. The 'Ghassulian culture', for example, remains enigmatic in terms of its identity (e.g. Bourke 1997; Kafafi 1998). The same problem has concerned the reality of the Pottery Neolithic B (PNB) designation, particularly in view of the similarities in ceramic repertoires with the Wadi Rabah and Ghassulian assemblages. Certainly a prominent element of uncertainty lies in the restricted (geographically, stratigraphically and statistically) nature of the various assemblages found so far, but another concern must deal with treating the transitional nature of culture change. The onset of the Neolithic itself will never be defined in the field, since the Neolithic period was a process, one that was a series of transitions of other processes set in motion long before the 'beginning' of the PPNA. In the same way, the 'end' of the Neolithic and the 'beginning' of the Chalcolithic will remain as much a matter of opinion as of any specifically identifiable feature to distinguish the two periods.

The 'aceramic Pottery Neolithic of the desert'

It is not uncommon that interest in human society—political, economic, even archaeological—has focused on areas where human population has been greatest: those areas are usually where dramatic events took place and are most recognizable in the historic and prehistoric records. But focusing a spotlight on one part of the stage automatically dims action that takes place elsewhere, even when this activity is an essential part of the entire drama.

The unfarmable steppe and desert of Jordan were occupied from the beginning of Jordan's prehistoric record, and they remained important areas of exploitation in the PPN and PN periods as well. Most of our information comes from the concerted efforts of Garrard and his team (e.g. Garrard et al. 1996, with references) and the independent investigations of Betts (e.g. Betts 1986, 1992, with references). Although the steppe and desert regions could not sustain human groups in permanent settlements, there appears to have been a 'permanent occupation' of the area by either transient herders with their sheep and goats or, in Byrd's view, of resident hunters and gatherers who adopted pastoralism in response to pressures from farming and hunting groups to the west (1992).

Most of the sites from the desert and steppe areas are temporary camps with the shallowest of accumulated deposits. Jabal Na'ja was occupied around 7430 ± 100 bp (Finlayson and Betts 1990: 13) (thus generally Yarmoukian in age), and the focus on burin production is one that is shared not only in terms of a tool class (burins), but also by a burin class (concave truncation burins), with other sites throughout the arid eastern region (Betts 1988; Rollefson 1988). The association of these camps with ephemeral structures that may have been corrals suggests they were inhabited by pastoralists, but the focus on burin production remains puzzling. Betts has suggested that the burins were not intentional tools but 'cores' for the production of small splinters ('burin spalls') to be used as drill bits for manufacturing stone and bone beads (Finlayson and Betts 1990: 19-20). There are other possible uses of burin spalls (Rollefson, Quintero and Wilke 1999) that might be associated with pastoral camps, but field research has not been undertaken so far to test these hypotheses.

Concluding remarks

The results that have been generated by the research projects over the past two decades are impressive in their scale and detail. The composition of the general evolution of Neolithic socioeconomic and sociocultural systems has taken on a fully formed outline, stretching from the earliest part of the PPNA through more than 4000 years up to the onset of metal-producing cultures. Some areas require emphatic efforts, such as the poorly known EPPNB and the later phases of the ceramic Neolithic, but at least some starting points have been identified for that research.

It is clear that, despite their obvious contacts with the Jordan Valley and Israel, the inhabitants of the highland zones and steppe/desert areas in Jordan directed their developments independently, taking advantage of local resources and opportunities to adjust continually to their own impacts on the environment. This is most clearly shown in the case of the smooth passage through the ninth millennium in Jordan, while tumultuous disruptions rocked the settlements in the Rift and areas to the west; in fact, the conditions in Jordan were comfortable enough that a sizeable proportion of the populations forced to abandon their villages in the west were able to emigrate eastwards, expanding established villages like Wadi Shu'ayb and 'Ayn Ghazal and founding new towns themselves.

At least three 'provinces' appear to have characterized Jordan during the ninth millennium (and perhaps throughout the entire Neolithic period), including the northern section (from 'Ayn Rahub to 'Ayn Ghazal), the southern part (from Ras an-Naqab up to as-Sifiya in Wadi al-Mujib) and the eastern steppe and desert areas that covered two-thirds of the country. These 'territories' are identifiable by architectural traits and, to some degree, by lithic typology (especially in projectile points) and ceramic technology/typology. There may also have been real differences in the ritual arena in these three territories as well, although in terms of the elaborate statuary from 'Ayn Ghazal, this would be an argument based on 'silence' from the south, an argument that could quickly be refuted in the next season of excavations.

The emergence of enormous settlements near the middle of the ninth millennium—at the change from the MPPNB to the LPPNB—is a phenomenon that is principally characteristic of Jordan,[26] and the population concentrations must have set in operation a chain of social stresses and organizational adaptations to cope with both external and internal pressures. Part of the response could have been the scenario described by Köhler-Rollefson in her pastoralism model, wherein a new system arose that immediately reduced stress on local ecosystems by removing both humans and animals from the area for large parts of the year, beginning a process that would come to characterize the entire Levantine region for the rest of its history.

Another element of change during the existence of the town phenomenon undoubtedly had to do with social relationships between neighbours and not-so-close neighbours. As permanent populations grew, a new surprise was in store: for the first time a resident of a town could encounter another resident who may have been more or less a stranger. As competition for farm land became keener, older methods from MPPNB times may not have been effective in settling conflicting claims for fields and resources; and these conflicts may have arisen within and not just between kin groups. The growth of multi-family houses is one reflection of this kind of stress, as is the apparent disappearance of the ancestor cult by the end of the LPPNB. The late ninth-millennium temples at 'Ayn Ghazal (and perhaps the T1 complex at Bayda) show connections with plastered skulls, burials or statues. Instead, there may have been a turning outward from kin groups, and from the community, towards an external, spiritual means of resolving problems that grew among strangers and kinfolk alike.

The concentration on later Neolithic developments is only about 10 years old, and the revelations of work in Wadi Ziqlab and Jabal Abu Thawwab have shown the directions that continued research should take. The tantalizing scatters of as yet under-researched ceramic Neolithic sites in the southern part of Jordan have some indications that the social patterns of the country—and the Levant in general—were much more complex than was imagined in the early 1980s.

The accelerated field research and theoretical ruminations in the past generation have provided prehistorians with the means to approach questions that were not even considered in the early 1970s.[27] This is true for Jordan's entire prehistoric record, of course, but the richness of the data from the Neolithic period has generated enormous anthropological excitement, and remarkable, solid progress in the understanding of our food-producing predecessors is promised in the near future.

Notes

1. Uncalibrated ages are expressed in lower case 'bp', while calibrated dates use the upper case 'BP' notation.
2. New excavations at Ghwair I, in Wadi Faynan in southern Jordan, indicate that there may be some PPNA presence, as a Khiam point and many bladelets suggest (Simmons and Najjar 1996).
3. Notable among the projectile points at Bayda, 13 are Khiam points, mostly from the lower layers (Levels VI–IV), where they reflect 2–3% of the points (Mortensen 1970: 21-22, fig. 19).
4. Kuijt (1997) has a different view concerning the presence of the EPPNB in Jordan.
5. Bayda produced seven Helwan points; in the lowermost layers (V and VI) they accounted for 1.0–1.5% (Mortensen 1970: 23, fig. 19).
6. Note that this use of the term 'Early PPNB' antedates Cauvin's (1987) tripartite division of the

PPNB period; Bar-Yosef's use of the term (1981: 564) does not necessarily reflect a reference restricted to the mid- to late tenth millennium bp.
7. Byrd has recently summarized the architectural developments at Bayda in which he uses 'phases'; essentially Byrd's Phase A is equivalent to Kirkbride's Levels V–VI, Phase B to Level IV, and Phase C to Levels I–III (Byrd 1994).
8. The potential for a PPNA or EPPNB presence at Bayda is strong in view of the occurrence of Khiam and Helwan points (see notes 3 and 5), despite the 'young' 14C dates in Level VI (cf. Weinstein 1984: 328-29).
9. The structures from Bayda Levels II and III/Phase C are discussed later in the PPNC section.
10. Incomplete information from Ghwair I, which has a series of MPPNB and M/LPPNB dates, suggests one relatively large building where there may have been one room, at least, associated with ritual activities in view of three niches and what may be an altar (Simmons and Najjar 1996).
11. Another 'type' of burial may be represented by a single example. It is a courtyard burial, but unlike 'typical' decapitated burials or 'trash' burials, the 15-year-old woman appears to be in a secondary location: the long bones were neatly stacked, next to which lay the mandible; ribs and vertebrae were scattered among the long bones and jaw.
12. This total does not include the skulls that were still attached to 'trash burials' or infants.
13. The identity of the substance has not been confirmed; there are superficial similarities with three coated skulls at Nahal Hemar (Yakar and Hershkovitz 1988; Arensburg and Hershkovitz 1989), but what had been first described as bitumen has since been identified as some other organic material in view of the high collagen content (Bar-Yosef, pers. comm.; Nissenbaum, pers. comm.).
14. The rarity of figurines at Wadi Shu'ayb is explainable, in part at least, by the restricted area of MPPNB excavations (cf. Simmons *et al.* 1989).
15. Two-storied buildings are claimed for the MPPNB levels at Cafer Höyük in southeast Turkey (Molist and Cauvin 1991: 107-109, fig. 15), although the bottom storey could be a basement storage area. Other two-storied buildings may occur in eastern Anatolia during the LPPNB period.
16. The decline in wild species in the diet may, in fact, have been much more dramatic if cattle and pigs had become domesticated during this period, possibilities that remain unverified (von den Driesch and Wodtke 1997).
17. Von den Driesch has noted that the cattle and pigs from LPPNB (and PPNC) 'Ayn Ghazal are still morphologically wild, in her opinion, although the residents at 'Ayn Ghazal may have been keeping them in corrals/stalls (von den Driesch and Wodtke 1997), which could constitute the beginning phases of domestication.
18. Edwards and Thorpe (1986) caution that the 36 ha distribution of artifacts on the steep slopes of Wadi az-Zarqa' at Kharaysin may reflect erosion and deflation. How large the actual settlement was requires extensive excavation probes.
19. Kenyon once implied that the southern Levant had been abandoned by farming people for up to a millennium after the misty end of the PPNB occupation at Jericho (Kenyon 1979: 45-46), a contention that has been popular among prehistorians for a long time since. It is unlikely that the Jordan Valley and the western areas of Israel were devoid of farming communities during the LPPNB and even the PPNC periods; see below.
20. The earlier discussion of possible LPPNB temples or shrines at Bayda does not conflict with a desertion of the site as a permanent farming settlement. The temples and shrines may have served as focal points of returning pilgrims among the emigrant populations for generations.
21. This situation is also shown by the disappearance of dense chipping floors that had characterized restricted loci in MPPNB and LPPNB courtyards.
22. The eastern end was truncated by bulldozer work connected with highway construction; the western end lies somewhere under unexcavated sediments (see Rollefson *et al.* 1990: 108-109; Rollefson *et al.* 1992: fig. 6).
23. Preliminary analysis of one PPNC skeleton has revealed bone disease associated with chronic tuberculosis, and another late LPPNB skeleton also suggests a similar pathology (Najjar *et al.* 1996). Epidemic tubercular death would probably have been too rapid for bone pathologies to develop in most cases, and it is not surprising that tuberculosis appears in the Neolithic since goats, sheep and cattle are vectors for the disease.
24. Contributing to the absence of some of the bones in secondary burials under these circumstances,

if someone died in the steppe/desert in the earlier part of the pastoral visit, the body may have been left for excarnation (including under stone cairns) for several months, and this may have attracted the attention of scavengers that could have been responsible for the disappearance of bones that were not necessarily small.

25. A radiocarbon sample from the 'very top of these levels' gave a date of 6030 ± 60 bp, while the overlying Chalcolithic levels yielded three dates between 5745 ± 35 and 5651 ± 40 bp (Dollfus and Kafafi 1993: 244-45).

26. Large (greater than 10 ha) settlements are also known from Syria (e.g. Tall Abu Hureyra) and the Jordan Valley (Beisamoun and Khirbat Sheikh 'Ali), but in these areas the towns are exceptions to the 'normal' settlement sizes (cf. Rollefson 1987, 1999; Gebel and Rollefson 1997).

27. Or, if they were considered, they certainly had precious little information from which to develop answers.

References

Arensburg, B. and I. Hershkovitz
1989 Artificial Skull 'Treatment' in the PPNB Period: Nahal Hemar. In I. Hershkovitz (ed.), *People and Culture in Change*, 115-33. British Archaeological Reports, International Series 508. Oxford: British Archaeological Reports.

Baird, D.
1993 Neolithic Chipped Stone Assemblages from the Azraq Basin, Jordan, and the Significance of the Neolithic of the Arid Zones of the Levant. Unpublished PhD dissertation, University of Edinburgh.
1994 Chipped Stone Production Technology from the Azraq Project Neolithic Sites. In H. Gebel and S. Kozlowski (eds.), *Neolithic Chipped Stone Industries of the Fertile Crescent*, 525-42. Berlin: ex oriente.
1995 Chipped Stone Raw Material Procurement and Selection in the Azraq Basin: Implications for Levantine Neolithic Cultural Developments. In K. 'Amr, F. Zayadine and M. Zaghloul (eds.), *Studies in the History and Archaeology of Jordan*, V: 505-14. Amman: Department of Antiquities.

Baird, D. et al.
1992 Prehistoric Environment and Settlement in the Azraq Basin: An Interim Report on the 1988 Season. *Levant* 24: 1-31.

Banning, E.
1995 Herders or Homesteaders? A Neolithic Farm in the Wadi Ziqlab, Jordan. *Biblical Archaeologist* 58.1: 2-13.
1996 Highlands and Lowlands: Problems and Survey Frameworks for Rural Archaeology in the Near East. *Bulletin of the American Schools of Oriental Research* 301: 25-46.

Banning, E. and B. Byrd
1984 The Architecture of PPNB 'Ain Ghazal, Jordan. *Bulletin of the American School of Oriental Research* 255: 15-20.
1987 Houses and the Changing Residential Unit: Domestic Architecture at PPNB 'Ain Ghazal. *Proceedings of the Prehistoric Society* 53: 309-25.

Banning, E. et al.
1996 The 1992 Excavations in Wadi Ziqlab, Jordan. *Annual of the Department of Antiquities of Jordan* 40: 29-50.

Bar-Yosef, O.
1981 The 'Pre-Pottery Neolithic' in the Southern Levant. In J. Cauvin and P. Sanlaville (eds.), *Préhistoire du Levant*, 551-69. Paris: C.N.R.S.
1991 Raw Material Exploitation in the Levantine Epi-Paleolithic. In A. Montet-White and S. Holen (eds.), *Raw Material Exploitation among Prehistoric Hunters and Gatherers*, 235-50. Publications in Anthropology 19. Lawrence, KS: University of Kansas.

Bar-Yosef, O. and D. Alon
1988 The Excavations. *'Atiqot* 18: 1-30.

Bar-Yosef, O., A. Gopher and D. Nadel
1987 The 'Hagdud Truncation'—A New Tool Type from the Sultanian Industry at Netiv Hagdud, in the Jordan Valley. *Mitekufat Haeven* 20: 151-57.

Becker, C.
1991 The Analysis of Mammalian Bones from Basta, a Pre-Pottery Neolithic Site in Jordan: Problems and Potential. *Paléorient* 17.1: 59-75.

Bennett, C.M.
1980 Soundings at Dhra', Jordan. *Levant* 12: 30-39.

Berner, M. and M. Schultz
1997 The Physical Anthropology of the Late PPNB Population of Basta. *Neo-Lithics* 2/97: 7.

Betts, A.
1986 The Prehistory of the Basalt Desert, Transjordan: An Analysis. Unpublished PhD dissertation, University College, London, Institute of Archaeology.
1988 1986 Excavations at Dhuweila, Eastern Jordan: A Preliminary Report. *Levant* 20: 7-21.
1992 Eastern Jordan: Economic Choices and Site Location in the Neolithic Periods. In M. Zaghloul et al. (eds.), *Studies in the History and Archaeology of Jordan*, IV: 111-14. Amman: Department of Antiquities.

Bienert, H.-D.
1991 Skull Cult in the Prehistoric Near East. *Journal of Prehistoric Religion* 5: 9-23.

Bisheh, G. et al.
1993 Archaeological Rescue Survey of the Ras an-Naqab–Aqaba Highway Alignment, 1992. *Annual of the Department of Antiquities of Jordan* 37: 119-31.

Bourke, S.
1997 The 'Pre-Ghassulian' Sequence at Teleilat Ghassul: Sydney University Excavations 1975–1995. In H.G. Gebel, Z. Kafafi and G. Rollefson (eds.), *The Prehistory of Jordan II: Perspectives from 1996*, 395-418. Berlin: *ex oriente*.

Butler, C.
1989 The Plastered Skulls of 'Ain Ghazal: Preliminary Findings. In I. Hershkovitz (ed.), *People and Culture in Change*, 141-46. British Archaeological Reports, International Series 508. Oxford: British Archaeological Reports.

Byrd, B.
1992 The Dispersal of Food Production across the Levant. In A. Gebauer and T. Price (eds.), *Transitions to Agriculture in Prehistory*, 49-61. Madison: Prehistory.
1994 Public and Private, Domestic and Corporate: The Emergence of the Southwest Asian Village. *American Antiquity* 59.4: 639-66.

Byrd, B. and E. Banning
1988 Southern Levantine Pier Houses: Intersite Architectural Patterning during the Pre-Pottery Neolithic B. *Paléorient* 14.1: 65-72.

Cauvin, J.
1974 Les débuts de la céramique sur le Moyen-Euphrate: nouveaux documents. *Paléorient* 2.1: 199-205.
1987 Chronologies relative et absolue dans le Néolithique du Levant Nord et d'Anatolie entre 10.000 et 8.000 bp. In O. Aurenche, J. Evin and F. Hours (eds.), *Chronologies du Proche Orient*, 325-42. British Archaeological Series, International Series 379. Oxford: British Archaeological Reports.

Clark, G. et al.
1994 Survey and Excavation in Wadi el-Hasa: A Preliminary Report on the 1993 Season. *Annual of the Department of Antiquities of Jordan* 38: 41-55.

Coinman, N., G. Clark and J. Lindly
1988 A Diachronic Study of Paleolithic and Early Neolithic Site Placement Patterns in the Southern Tributaries of the Wadi el Hasa. In B. MacDonald et al., *The Wadi el Hasa Archaeological Survey 1979–1983, West-Central Jordan*, 48-86. Waterloo, ON: Wilfrid Laurier University.

Dollfus, G. and Z. Kafafi
1988 *Abu Hamid: Village du 4e millénaire de la vallée du Jourdain.* Amman: Economic.
1993 Recent Researches at Abu Hamid. *Annual of the Department of Antiquities of Jordan* 37: 241-62.

Ducos, P.
1993 Proto-élevage et élevage au Levant sud au VIIe millénaire B.C.: Les données de la Damascène. *Paléorient* 19.1: 153-73.
1994 À propos du modèle du pastoralisme PPNC proposé par G. Rollefson et I. Köhler-Rollefson. *Paléorient* 20.2: 165-66.

Edwards, P. and S. Thorpe
1986 Surface Lithic Finds from Kharaysin, Jordan. *Paléorient* 12.2: 85-87.

Eighmey, J.
1992 A Functional Analysis of Projectile Points from the Neolithic Site of 'Ain Ghazal. Unpublished MA thesis, San Diego State University.

Field, H.
1960 *North Arabian Desert Archaeological Survey, 1925–1950.* Papers of the Peabody Museum of Archaeology and Ethnology, Harvard University 45.2.

Finlayson, B. and A. Betts
1990 Functional Analysis of Chipped Stone Artifacts from the Late Neolithic Site of Gabal Na'ja, Eastern Jordan. *Paléorient* 16.2: 13-20.

Fino, N.
1997 Al-Baseet, a New LPPNB Site Found in Wadi Musa, Southern Jordan. *Neo-Lithics* 3/97: 13-14.

Garfinkel, Y.
1994 The PPNC Flint Assemblage from Tel 'Ali. In H.G. Gebel and S.K. Kozlowski (eds.), *Neolithic Chipped Stone Industries of the Fertile Crescent*, 543-62. Berlin: *ex oriente*.

Garrard, A. and N.S. Price
1977 A Survey of Prehistoric Sites in the Azraq Basin, Eastern Jordan. *Paléorient* 3: 109-26.

Garrard, A., S. Colledge and L. Martin
1996 The Emergence of Crop Cultivation and Caprine Herding in the 'Marginal Zone' of the Southern Levant. In D. Harris (ed.), *The Origins and Spread of Agriculture and Pastoralism in Eurasia*, 204-26. London: University College.

Garrard, A. et al.
1985 Prehistoric Environment and Settlement in the Azraq Basin: A Report on the 1982 Survey Season. *Levant* 17: 1-28.
1987 Prehistoric Environment and Settlement in the Azraq Basin: An Interim Report on the 1985 Excavation Season. *Levant* 19: 5-26.
1994 Prehistoric Environment and Settlement in the Azraq Basin: An Interim Report on the 1987 and 1988 Excavation Seasons. *Levant* 26: 73-109.

Gebel, H.G.
　1988　Late Epipaleolithic–Aceramic Neolithic Sites in the Petra Area. In A. Garrard and H. Gebel (eds.), *The Prehistory of Jordan*, 67-100. British Archaeological Reports, International Series 396. Oxford: British Archaeological Reports.

Gebel, H.G. and H.-D. Bienert
　1997　Excavating Ba'ja, Greater Petra Area, Southern Jordan. *Neo-Lithics* 1/97: 9-11.

Gebel, H.G. and M. Muheisen
　1985　Note on 'Ain Rahub, a New Late Natufian Site Near Irbid, Jordan. *Paléorient* 11.1: 107-10.

Gebel, H.G. and G. Rollefson (eds.)
　1997　Central Settlements in Neolithic Jordan. *Neo-Lithics* 2/97.

Gebel, H.G. and J.M. Starck
　1985　Investigations into the Stone Age of the Petra Area (Early Holocene Research): A Preliminary Report on the 1984 Campaigns. *Annual of the Department of Antiquities of Jordan* 29: 89-114.

Gebel, H.G. *et al.*
　1988　Preliminary Report on the First Season of Excavations at the Late Aceramic Neolithic Site of Basta. In A. Garrard and H.G. Gebel (eds.), *The Prehistory of Jordan*, 101-34. British Archaeological Reports, International Series 396. Oxford: British Archaeological Reports.

Gopher, A.
　1994　*Arrowheads of the Neolithic Levant*. Winona Lake, WI: Eisenbrauns.

Gopher, A. and R. Gophna
　1993　Cultures of the Eighth and Seventh Millennia bp in the Southern Levant: A Review for the 1990s. *Journal of World Prehistory* 7.3: 297-353.

Goren, Y.
　1990　The 'Qatifian Culture' in Southern Israel and Transjordan: Additional Aspects for its Definition. *Mitekufat Haeven* 23: 100-12.

Goring-Morris, N.
　1994　Aspects of the PPNB Lithics Industry of Kfar HaHoresh, near Nazareth, Israel. In H.G. Gebel and S.K. Kozlowski (eds.), *Neolithic Chipped Stone Industries of the Fertile Crescent*, 427-44. Berlin: ex oriente.

Griffin, P.
　1993　*Treatment Report: Head Pit Fragments*. Unpublished report of the Smithsonian Institution CAL/MSC, CAL No. 5336 (November 1993).

Griffin, P., C. Grissom and G. Rollefson
　1998　Three Late 8th Millennium Plastered Faces from 'Ain Ghazal, Jordan. *Paléorient* 24.1: 59-70.

Hecker, H.
　1975　The Faunal Analysis of the Primary Food Animals from Pre-Pottery Neolithic Beidha (Jordan). Unpublished PhD dissertation, Columbia University. Ann Arbor: University Microfilms.

Henry, D.
　1979　Palaeolithic Sites within the Ras en Naqb Basin, Southern Jordan. *Palestine Exploration Quarterly*: 79-85.
　1982　The Prehistory of Southern Jordan and Relationships with the Levant. *Journal of Field Archaeology* 9: 417-44.
　1995　*Prehistoric Cultural Ecology and Evolution: Insights from Southern Jordan*. New York: Plenum.

Horwitz, L.
　1989　A Reassessment of Caprovine Domestication in the Levantine Neolithic: Old Questions, New Answers. In I. Hershkovitz (ed.), *People and Culture in Change*, 153-81. British Archaeological Reports, International Series 508. Oxford: British Archaeological Reports.

Kafafi, Z.
　1985　Late Neolithic Architecture from Jabal Abu Thawwab, Jordan. *Paléorient* 11.1: 125-27.
　1986　Second Season of Excavations at Jabal Abu Thawwab (er-Rumman), 1985, Preliminary Report. *Annual of the Department of Antiquities of Jordan* 30: 57-68.
　1988　Jabal Abu Thawwab, a Pottery Neolithic Village in North Jordan. In A. Garrard and H.G. Gebel (eds.), *The Prehistory of Jordan*, 451-71. British Archaeological Reports, International Series 396. Oxford: British Archaeological Reports.
　1990　Early Pottery Contexts from 'Ain Ghazal, Jordan. *Bulletin of the American Schools of Oriental Research* 280: 15-30.
　1993　The Yarmoukians in Jordan. *Paléorient* 19.1: 101-14.
　1998　The Late Neolithic in Jordan. In D. Henry (ed.), *The Prehistoric Archaeology of Jordan*, 127-38. British Archaeological Reports, International Series 705. Oxford: Archaeopress.

Kafafi, Z. and G. Rollefson
　1995　The 1994 excavations at 'Ayn Ghazal: Preliminary Report. *Annual of the Department of Antiquities of Jordan* 39: 13-29.

Kaplan, J.
　1958　Excavations at Wadi Raba. *Israel Exploration Journal* 8: 149-60.

Kenyon, K.
　1979　*Archaeology in the Holy Land*. 4th edn. London: Benn.
　1981　*Excavations at Jericho*, III. London: British School of Archaeology in Jerusalem.

Kirkbride, D.
　1958　Notes on a Survey of Pre-Roman Archaeological Sites near Jerash. *Bulletin of the Institute of Archaeology* 1: 9-20.
　1966　Five Seasons at the Pre-Pottery Neolithic Village of Beidha in Jordan. *Palestine Exploration Quarterly* 98: 8-72.

1967 Beidha 1965: An Interim Report. *Palestine Exploration Quarterly* 99: 5-14.

1968a Beidha 1967: An Interim Report. *Palestine Exploration Quarterly* 100: 90-96.

1968b Beidha: Early Neolithic Village Life South of the Dead Sea. *Antiquity* 42: 263-74.

Kislev, M.
1992 Agriculture in the Near East in the VIIth Millennium bc. In P. Anderson (ed.), *Préhistoire de l'agriculture: Nouvelles approaches expérimentales et ethnographiques*, 87-93. Monographie de CRA 6. Paris: C.N.R.S.

Köhler-Rollefson, I.
1988 The Aftermath of the Levantine Neolithic Revolution in Light of Ecologic and Ethnographic Evidence. *Paléorient* 14.1: 87-93.

1992 A Model for the Development of Nomadic Pastoralism on the Transjordanian Plateau. In O. Bar-Yosef and A. Khazanov (eds.), *Pastoralism in the Levant*, 11-18. Madison: Prehistory.

1997 Proto-Élevage, Pathologies and Pastoralism: A Post-Mortem on Goat Domestication. In H.G. Gebel, Z. Kafafi and G. Rollefson (eds.), *The Prehistory of Jordan II: Perspectives from 1996*, 557-66. Berlin: ex oriente.

Köhler-Rollefson, I. and G. Rollefson
1990 The Impact of Neolithic Man on the Environment: The Case of 'Ain Ghazal, Jordan. In S. Bottema, G. Entjes-Nieborg and W. van Zeist (eds.), *Man's Role in the Shaping of the Eastern Mediterranean Landscape*, 3-14. Rotterdam: Balkema.

1998 Brooding about Breeding: Herding at the Beginning of the Animal Domestication Process. Paper presented to the international workshop on 'The Transition from Foraging to Farming in Southwest Asia', Gröningen, September.

Köhler-Rollefson, I., W. Gillespie and M. Metzger
1988 The Fauna from Neolithic 'Ain Ghazal. In A. Garrad and H. Gebel (eds.), *The Prehistory of Jordan*, 423-30. British Archaeological Reports, International Series 396. Oxford: British Archaeological Reports.

Köhler-Rollefson, I., L. Quintero and G. Rollefson
1993 A Brief Note on the Fauna from Neolithic 'Ain Ghazal. *Paléorient* 19.2: 95-97.

Kuijt, I.
1994a Pre-Pottery Neolithic A Settlement Variability: Evidence for Sociopolitical Developments in the Southern Levant. *Journal of Mediterranean Archaeology* 7.2: 165-92.

1994b A Brief Note on the Chipped Stone Assemblage from 'Iraq ad-Dubb, Jordan. *Neo-Lithics* 2/94: 2-3.

1996 Where Are the Microlithics? Lithic Technology and Neolithic Chronology as Seen from the PPNA Occupation at Dhra, Jordan. *Neo-Lithics* 2/96: 7-8.

1997 Trying to Fit Round Houses into Square Holes: Reexamining the Timing of the South-Central Levantine Pre-Pottery Neolithic A and Pre-Pottery Neolithic B Transition. In H.G. Gebel, Z. Kafafi and G. Rollefson (eds.), *The Prehistory of Jordan II: Perspectives from 1996*, 193-202. Berlin: ex oriente.

Kuijt, I., J. Mabry and G. Palumbo
1991 Early Neolithic Use of Upland Areas of Wadi el-Yabis: Preliminary Evidence from the Excavations of 'Iraq ad-Dubb, Jordan. *Paléorient* 17.1: 99-108.

Kuijt, I. and M. Chesson
1994 Investigations at Jabal Queisa, Jordan (1993): A Reconsideration of Chronology and Occupational History. *Annual of the Department of Antiquities of Jordan* 38: 33-39.

Kuijt, I. and H. Mahasneh
1998 Dhra': An Early Neolithic Village in the Southern Jordan Valley. *Journal of Field Archaeology* 25.2: 153-61.

Lechevallier, M.
1978 *Abou Gosh et Beisamoun*. Mémoires et Travaux du Centre de Recherches Préhistoriques Française de Jérusalem No. 2. Paris: Association Paléorient.

Levy, T.
1986 The Chalcolithic Period. *Biblical Archaeologist* 49: 83-106.

1995 Cult, Metallurgy and Rank Societies—Chalcolithic Period (ca. 4500–3500 BCE). In T. Levy (ed.), *The Archaeology of Society in the Holy Land*, 226-43. London: Leicester University.

Mabry, J. and G. Palumbo
1988 The 1987 Wadi al-Yabis Survey. *Annual of the Department of Antiquities of Jordan* 32: 275-305.

MacDonald, B. *et al.*
1987 Southern Ghor and Northeast 'Arabah Archaeological Survey 1986, Jordan: A Preliminary Report. *Annual of the Department of Antiquities of Jordan* 31: 391-418.

1988 *The Wadi al-Hasa Archaeological Survey 1979–1983, West-Central Jordan*. Waterloo, ON: Wilfrid Laurier University.

MacDonald, B., G. Rollefson and D. Roller
1982 The Wadi el-Hasa Survey 1981. Preliminary Report. *Annual of the Department of Antiquities of Jordan* 26: 117-31.

Mahasneh, H.
1997 Es-Sifiya: A Pre-Pottery Neolithic B Site in the Wadi el-Mujib, Jordan. In H.G. Gebel, Z. Kafafi and G. Rollefson (eds.), *The Prehistory of Jordan II: Perspectives from 1996*, 203-14. Berlin: ex oriente.

Molist, M. and J. Cauvin
1991 Les niveaux inférieures de Cafer Höyük (Malatya, Turquie): Stratigraphie et architectures (fouilles 1984–1986). *Cahiers de l'Euphrates* 5-6: 85-114.

Mortensen, P.
1970 A Preliminary Study of the Chipped Stone Industry from Beidha. *Acta Archaeologica* (Copenhagen) 41: 1-54.

Muheisen, M. *et al.*
1988 'Ain Rahub, a New Final Natufian and Yarmoukian Site near Irbid. In A. Garrard and H.G. Gebel (eds.), *The Prehistory of Jordan*, 472-502. British Archaeological Reports, International Series 396. Oxford: British Archaeological Reports.

Nadel, D.
1989 Flint Heat Treatment at the Beginning of the Neolithic Period in the Levant. *Journal of the Israel Prehistoric Society* 22: 61-67.

Nadel, D., O. Bar-Yosef and A. Gopher
1991 Early Neolithic Arrowhead Types in the Southern Levant: A Typological Suggestion. *Paléorient* 17.1: 109-20.

Najjar, M.
1994 Ghwair I, a Neolithic Site in Wadi Feinan. *The Near East in Antiquity* 4: 75-85.

Najjar, M., A. al-Shiyab and I. al-Sarie
1996 Cases of Tuberculosis at 'Ain Ghazal, Jordan. *Paléorient* 22.2: 123-28.

Najjar, M. *et al.*
1990 Tell Wadi Feinan: The First Pottery Neolithic Tell in the South of Jordan. *Annual of the Department of Antiquities of Jordan* 34: 27-56.

Nissen, H.
1993 The PPNC, the Sheep and the 'Hiatus Palestinien'. *Paléorient* 19.1: 177-83.
1997 Proto-Urbanism: An Early Neolithic Feature. Abstract in H.G. Gebel and G. Rollefson (eds.), *Neo-Lithics* 2/97: 4.

Nissen, H., M. Muheisen and H.G. Gebel
1987 Report on the First Two Seasons of Excavation at Basta (1986–1987). *Annual of the Department of Antiquities of Jordan* 31: 79-119.
1991 Report on the Excavations at Basta 1988. *Annual of the Department of Antiquities of Jordan* 35: 13-40.

Olszewski, D.
1994 The PPN Glossed Blades from 'Ain Ghazal, Jordan (1982–85 and 1988–89 Seasons). In H.G. Gebel and S.K. Kozlowski (eds.), *Neolithic Chipped Stone Industries of the Fertile Crescent*, 467-78. Berlin: *ex oriente*.

Palumbo, G.
1992 The 1990 Wadi al-Yabis Survey Project and Soundings at Khirbat Umm el-Hedamus. *Annual of the Department of Antiquities of Jordan* 36: 25-41.

Palumbo, G., J. Mabry and I. Kuijt
1990 The Wadi al-Yabis Survey: Report on the 1989 Field Season. *Annual of the Department of Antiquities of Jordan* 34: 95-118.

Perrot, J.
1966 La troisième campagne de fouilles à Munhata. *Syria* 43: 49-63.
1993 Remarques Introductives. *Paléorient* 19.1: 9-21.

Quintero, L.
1996 Flint Mining in the Pre-Pottery Neolithic: Preliminary Report on the Exploitation of Flint at Neolithic 'Ain Ghazal in Highland Jordan. In S. Kozlowski and H.G. Gebel (eds.), *Neolithic Chipped Stone Industries of the Fertile Crescent and their Contemporaries in Adjacent Regions*, 233-42. Berlin: *ex oriente*.
1998 Evolution of Lithic Economies in the Levantine Neolithic: Development and Demise of Naviform Core Technology. Unpublished PhD dissertation, Department of Anthropology, University of California-Riverside.

Quintero, L. and I. Köhler-Rollefson
1997 The 'Ain Ghazal Dog: A Case for the Neolithic Origin of *Canis familiaris* in the Near East. In H.G. Gebel, Z. Kafafi and G. Rollefson (eds.), *The Prehistory of Jordan II: Perspectives from 1996*, 567-74. Berlin: *ex oriente*.

Quintero, L. and P. Wilke
1995 Evolution and Economic Significance of Naviform Core-and-Blade Technology in the Southern Levant. *Paléorient* 21.1: 17-34.

Quintero, L., P. Wilke and J. Waines
1997 Pragmatic Studies of Near Eastern Neolithic Sickle Blades. In H.G. Gebel, Z. Kafafi and G.O. Rollefson (eds.), *The Prehistory of Jordan II: Perspectives from 1996*, 263-86. Berlin: *ex oriente*.

Raikes, T.
1980 Notes on Some Neolithic and Later Sites in Wadi Araba and the Dead Sea Valley. *Levant* 12: 40-60.

Rhotert, H.
1938 *Transjordanien: Vorgeschichtliche Forschungen*. Stuttgart: Strecker und Schröder.

Roehrer-Ertl, O., K.-W. Frey and H. Newseley
1988 Preliminary Note on Early Neolithic Human Remains from Basta and Sabra. In A. Garrard and H.G. Gebel (eds.), *The Prehistory of Jordan*, 135-36. British Archaeological Reports, International Series 396. Oxford: British Archaeological Reports.

Rollefson, G.
1983 Ritual and Ceremony at Neolithic 'Ain Ghazal (Jordan). *Paléorient* 9.2: 29-38.
1986 Neolithic 'Ain Ghazal (Jordan): Ritual and Ceremony II. *Paléorient* 12.1: 45-52.
1987 Local and Regional Relations in the Levantine PPN Period: 'Ain Ghazal, Jordan, as a Regional Centre. In

	A. Hadidi (ed.), *Studies in the History and Archaeology of Jordan*, III: 29-32. Amman: Department of Antiquities.
1988	Stratified Burin Classes at 'Ain Ghazal: Implications for the Desert Neolithic of Jordan. In A. Garrard and H.G. Gebel (eds.), *The Prehistory of Jordan*, 437-49. British Archaeological Reports, International Series 396. Oxford: British Archaeological Reports.
1989	The Late Aceramic Neolithic of the Levant: A Synthesis. *Paléorient* 15.1: 168-73.
1990a	Neolithic Chipped Stone Technology at 'Ain Ghazal, Jordan: The Status of the PPNC Phase. *Paléorient* 16.1: 119-24.
1990b	The Uses of Plaster at Neolithic 'Ain Ghazal, Jordan. *Archaeomaterials* 4: 33-54.
1993	The Origins of the Yarmoukian at 'Ain Ghazal. *Paléorient* 19.1: 91-100.
1995	Burin Variability at Neolithic 'Ayn Ghazal, Jordan. In K. 'Amr, F. Zayadine and M. Zaghloul (eds.), *Studies in the History and Archaeology of Jordan*, V: 515-18. Amman: Department of Antiquities.
1996	Abu Hudhud (WHS 1008): An EPPNB Settlement in the Wadi el-Hasa, Southern Jordan. In S. Kozlowski and H.G. Gebel (eds.), *Neolithic Chipped Stone Industries of the Fertile Crescent and their Contemporaries in Adjacent Regions*, 159-60. Berlin: *ex oriente*.
1997a	Changes in Architecture and Social Organization at 'Ain Ghazal. In H.G. Gebel, Z. Kafafi and G. Rollefson (eds.), *The Prehistory of Jordan II: Perspectives from 1997*, 287-308. Berlin: *ex oriente*.
1997b	A Further Note on the Blade: Blade Ratio as a Neolithic Phase Discriminator. *Neo-Lithics* 1/97: 20.
1998a	Expanded Radiocarbon Chronology from 'Ain Ghazal. *Neo-Lithics* 2/98: 8-10.
1998b	Neolithic 'Ain Ghazal: Ritual and Ceremony III. *Paléorient* 24.1: 43-58.
1999	Emerging Complexity in LPPNB Social Organization. In H.G. Gebel and R. Neef (eds.), *Central Settlements in Neolithic Jordan*. Berlin: *ex oriente* (in press).
2000	Ritual and Social Organization at Neolithic 'Ain Ghazal. In I. Kuijt (ed.), *Life in Neolithic Farming Villages*, 165-90. New York: Kluwer Academic/Plenum.
n.d.	Burin Production at 'Ain Ghazal, Wadi Shu'ayb, Beidha and Jericho: A Comparative Analysis. Manuscript in preparation.

Rollefson, G. and Z. Kafafi
- 1985 Khirbat Hammam: A PPNB Village in the Wadi el-Hasa, Southern Jordan. *Bulletin of the American Schools of Oriental Research* 258: 63-69.
- 1994 The 1993 Season at 'Ain Ghazal: Preliminary Report. *Annual of the Department of Antiquities of Jordan* 38: 11-32.
- 1996 The 1995 Season at 'Ayn Ghazal: Preliminary Report. *Annual of the Department of Antiquities of Jordan* 40: 11-28.
- 1997 The 1996 Season at 'Ayn Ghazal: Preliminary Report. *Annual of the Department of Antiquities of Jordan* 41: 27-48.

Rollefson, G. and I. Köhler-Rollefson
- 1989 The Collapse of Early Neolithic Settlements in the Southern Levant. In I. Hershkovitz (ed.), *People and Cultures in Change*, 73-89. British Archaeological Reports, International Series 508. Oxford: British Archaeological Reports.
- 1992 Early Neolithic Exploitation Patterns in the Levant: Cultural Impact on the Environment. *Population and Environment* 13: 243-54.
- 1993 PPNC Adaptations in the First Half of the 6th Millennium bc. *Paléorient* 19.1: 33-42.

Rollefson, G. and A. Simmons
- 1986 The Neolithic Village of 'Ain Ghazal, Jordan: Preliminary Report on the 1984 Season. *Bulletin of the American Schools of Oriental Research Supplement* 24: 145-64.
- 1987 The Life and Death of 'Ain Ghazal. *Archaeology* 40.6: 38-45.

Rollefson, G., M. Forstadt and R. Beck
- 1994 A Preliminary Typology of Scrapers, Knives and Borers from 'Ain Ghazal. In H.G. Gebel and S. Kozlowski (eds.), *Neolithic Chipped Stone Industries of the Fertile Crescent*, 445-66. Berlin: *ex oriente*.

Rollefson, G., Z. Kafafi and A. Simmons
- 1990 The Neolithic Village of 'Ain Ghazal: Preliminary Report on the 1988 Season. *Bulletin of the American Schools of Oriental Research Supplement* 27: 95-116.

Rollefson, G., L. Quintero and P. Wilke
- 1999 Bawwab al-Ghazal: Preliminary Report on the Testing Season 1998. *Neo-Lithics* 1/99: 2-4.

Rollefson, G., D. Schmandt-Besserat and J. Rose
- 1999 A Decorated Skull from MPPNB 'Ain Ghazal. *Paléorient* 24.2: 99-104.

Rollefson, G., A. Simmons and Z. Kafafi
- 1992 Neolithic Cultures at 'Ain Ghazal, Jordan. *Journal of Field Archaeology* 19.4: 443-70.

Schmandt-Besserat, D.
- 1997 Animal Symbols at 'Ain Ghazal. *Expedition* 39.1: 48-58.
- 1998 A Stone Metaphor for Creation. *Near Eastern Archaeology* 61.2: 109-17.

al-Shiyab, A.H.
- 1997 Animal Remains from 'Ain Rahub. In H.G. Gebel, Z. Kafafi and G. Rollefson (eds.), *Prehistory of Jordan II: Perspectives from 1996*, 593-600. Berlin: *ex oriente*.

Simmons, A. and M. Najjar
- 1996 Current Investigations at Ghwair I, a Neolithic Settlement in Southern Jordan. *Neo-Lithics* 2/96/: 6-7.

Simmons, A. et al.
- 1989 Test Excavations at Wadi Shu'ayb, a Major Neolithic Settlement in Central Jordan. *Annual of the Department of Antiquities of Jordan* 33: 27-42.
- 1990 A Plastered Human Skull from Neolithic 'Ain Ghazal, Jordan. *Journal of Field Archaeology* 17.1: 107-10.

Tubb, K. and C. Grissom
- 1995 'Ayn Ghazal: A Comparative Study of the 1983 and 1985 Statuary Caches. In K. 'Amr, F. Zayadine and M. Zaghloul (eds.), *Studies in the History and Archaeology of Jordan*, V: 515-18. Amman: Department of Antiquities.

Von den Driesch, A. and U. Wodtke
- 1997 The Fauna of 'Ain Ghazal, a Major PPN and Early PN Settlement in Central Jordan. In H.G. Gebel, Z. Kafafi and G. Rollefson (eds.), *Prehistory of Jordan II: Perspectives from 1996*, 511-56. Berlin: *ex oriente*.

Waechter, J. and V. Seton-Williams
- 1938 The Excavations at Wadi Dhobai 1937–38 and the Dhobaian Industry. *Journal of the Palestine Oriental Society* 18: 172-86.

Waheeb, M. and N. Fino
- 1997 'Ayn el-Jammam: A Neolithic Site near Ras e-Naqb, Southern Jordan. In H.G. Gebel, Z. Kafafi and G. Rollefson (eds.), *The Prehistory of Jordan II: Perspectives from 1997*, 215-20. Berlin: *ex oriente*.

Wasse, A.
- 1997 Preliminary Results of an Analysis of Sheep and Goat Bones from 'Ain Ghazal, Jordan. In H.G. Gebel, Z. Kafafi and G.O. Rollefson (eds.), *The Prehistory of Jordan II: Perspectives from 1997*, 575-92. Berlin: *ex oriente*.

Weinstein, J.
- 1984 Radiocarbon Dating in the Southern Levant. *Radiocarbon* 26.3: 297-366.

Wilke, P. and L. Quintero
- 1994 Naviform Core-and-Blade Technology: Assemblage Character as Determined by Replicative Experiments. In H.G. Gebel and S.K. Kozlowski (eds.), *Neolithic Chipped Stone Industries of the Fertile Crescent*, 33-60. Berlin: *ex oriente*.

Wilke, P., L. Quintero and G. Rollefson
- 1997 Bawwab el-Ghazal: A Temporary Station of Hunting Pastoralists in the Eastern Jordanian Desert. *Neo-Lithics* 3/97: 12-14.

Wright, K., R. Schick and R. Brown
- 1989 Report on a Preliminary Survey of the Wadi Shu'eib. *Annual of the Department of Antiquities of Jordan* 33: 345-50.

Yakar, R. and I. Hershkovitz
- 1988 Nahal Hemar. The Modeled Skulls. *'Atiqot* 18: 59-63.

Zeuner, F.
- 1957 Stone Age Exploration in Jordan, I. *Palestine Exploration Quarterly* 89: 17-54.

4. The Chalcolithic Period

Stephen J. Bourke

Introduction

Attempting an overview of Chalcolithic material culture from a specifically Jordanian perspective is made difficult by a relatively uneven coverage. Intensive investigation of the Jordan Valley in general, and the site of Tulaylat al-Ghassul in particular, is to be contrasted with a still superficial knowledge of settlement and land use in the western half of the verdant north/central uplands and the more marginal zones to the east and south. Anything like an equal consideration of all regions is not yet possible, as intensive exploration of upland and marginal zones is still in its infancy.

Nevertheless, there is value in an assessment of the Chalcolithic from a Jordanian perspective since much of the current debate on economic and cultural development is sourced almost exclusively to material from sites located in the northern Negev, which has skewed debate to the detriment of central and southeastern Levantine data. A Jordanian perspective will only highlight the need for a more even distribution of research effort, and the limited value to be placed on generalizations sourced to a single region of what is both geomorphologically and culturally a sharply dissected landscape.

This being said, any assessment of the Jordanian Chalcolithic will still be dominated by Jordan Valley data in general and that from Tulaylat al-Ghassul in particular (although this has recently begun to change). This skewing of the Jordanian data impacts unfavourably on synthetic analysis, but is presently unavoidable. After introductory remarks on history of research and chronology, my procedure will be to outline settlement patterns, economic and subsistence developments, and material cultural remains (including trade patterns), initially with reference to the evidence from Tulaylat al-Ghassul, then from a Jordan Valley perspective, before integrating these with the more limited data from upland/marginal areas to form a (necessarily provisional) synthesis.

History of research

Pre-war (1929–1967)

Concerted study of the Chalcolithic period in Jordan began with the first excavations at Tulaylat al-Ghassul. The site has remained central to all subsequent research. The eight seasons of excavation by the Pontifical Biblical Institute (1929–38) uncovered spectacular remains of a hitherto little-known Chalcolithic civilization. Mallon and Koeppel distinguished four main phases of occupation (Ghassul I–IV), with the latest (Ghassul IV) spread over some 20 ha. The failure to investigate the earlier three phases in any detail and the importance of this omission went largely undetected in the enthusiastic description of the newly discovered Ghassul IV culture, which rapidly became the exemplar for the entire south Levantine Chalcolithic, thereafter known as the Ghassulian culture.

The next major development in Jordanian Chalcolithic studies was the publication of Nelson Glueck's extensive wartime surveys east of the Jordan. These surveys illustrated the great spread of prehistoric settlement throughout the uplands and out into the marginal zones (Glueck 1951). Overconfident acceptance of Glueck's field identifications, however, has led to much subsequent confusion, and a probable over-estimation of the importance of upland Chalcolithic settlement intensity. In 1952, shortly after Glueck's publication, Henri de Contenson and James Mellaart carried out a much more intensive survey in the Jordan Valley as part of the Point IV Irrigation Project, and sounded seven of the more important prehistoric mounds discovered therein (Leonard 1992). Soundings at ash-Shuna, Abu Habil, Sa'idiyya Tahta, Umm Hammad and Ghrubba uncovered a string of well-established Chalcolithic settlements along the eastern edge of the valley, and revealed the existence of both 'Ghassulian' and 'non-Ghassulian' Chalcolithic assemblages in the northern part of the valley (de

Contensen 1960a; Mellaart 1962). The publication of these important discoveries was much delayed, greatly truncated and presented piecemeal when finally published, which led to many important observations being ignored.

As the 1950s unfolded, Mellaart's publication (1956) of the Early Chalcolithic Ghrubba material, with apparent Halafian links and implied population movements down Wadi az-Zarqa' and de Contensen's (1961) suggestion that the Early Chalcolithic of ash-Shuna ash-Shamaliyya was no more than 'Ghassulian influenced', underlined the need to know more about the earliest occupation (Ghassul I–III) at Ghassul. Responding to this need, North's single season of renewed excavations at Tulaylat al-Ghassul in 1959–60 focused on exploring the early horizons at the site (North 1961). His conclusions appeared unequivocal. He reaffirmed the fourfold division of occupation at the site, and, notwithstanding the great depth of occupation across the site, claimed no significant difference in culture between the first three phases and the last, discounting significant development from one stage to the next.

Hennessy's return to Ghassul in 1967 marks a turning point in the study of the Chalcolithic period in Jordan. Planned as a long-term project, excavations were interrupted after a single season. Even so, Hennessy's findings were revolutionary, and totally unexpected in the light of North's then recent pronouncements. Hennessy (1969) described a nine-stage sequence, and outlined a clear cultural progression, with basal deposits characterized as a local manifestation of the Jericho ceramic Neolithic. He concluded that the Ghassulian culture developed smoothly out of a local Late Neolithic precursor, passing through readily distinguished developmental stages before culminating in the well-known, Classic Ghassulian culture revealed by past excavators.

Post-war (1968–1988)

In the years immediately after the Six-Day War, Hennessy's conclusions were rapidly assimilated into the doctoral literature of a 'younger brigade' of scholars (Moore, Lee and Elliot), although established researchers (Perrot, Kenyon, Lapp and Mellaart) continued to view the Ghassulian as a short-lived, foreign-inspired culture. A growing knowledge of Late Neolithic culture at Jericho allowed Moore (1973) to establish convincing links between the stone tool traditions of Jericho and Ghassul, outweighing vague ceramic links with the Syro-Mesopotamian northeast. At the same time, the 'northern Chalcolithic' was codified in assemblage and distribution by de Vaux (1966). Soon after, Kaplan (1969) reinforced de Vaux's advocacy of an indigenous origin for the 'northern Chalcolithic', and nominated the newly codified (Halaf-influenced) Rabah Neolithic as its likely parent, an impression Kirkbride (1969) and Copeland's (1969) survey work in the Lebanese Beqa'a, and Dunand's research at Byblos (1973) was held to support. As Kaplan has suggested that the Jericho VIII assemblage (loosely equated with Kenyon's Pottery Neolithic B) was a degenerate variety of the Rabah Neolithic, once Moore (1973) had linked the basal Ghassul assemblage to the Jericho Pottery Neolithic B, an indigenous origin for the Ghassulian seemed inescapable.

With the great change in the political landscape after the Six-Day War, renewed interest in regional survey occurred. Although opportunistic small-scale surveys had begun to exploit Glueck's magnificent database in the late fifties and early sixties, it was Mittmann's North Jordan Survey that began the systematic checking of Glueck's database in the uplands of the North Jordan Plateau (Mittmann 1970; Boling 1988). This revealed a limited Chalcolithic presence in the western uplands and along wadi systems. The documentation of numbers of settled agricultural communities on the plateau threw into sharp focus the need to determine the relationship (if any) between the Jordan Valley and Upland Plateau settlements.

As conditions normalized in the Jordan Valley by 1975, the intensive Joint Survey Project (Ibrahim et al. 1976; Yassine et al. 1988) did for the Valley floor what Mittmann had done for the plateau, doubling the number of sites known from earlier Point IV days. This survey began to illustrate the close relationship of Chalcolithic site clusters and the lateral wadi systems linking the valley and the plateau, while at the same time contrasting Chalcolithic patterns with those of the succeeding Early Bronze Age. From this time forward, intensive survey of the Jordan Valley lateral wadi systems began to 'knit together' the often large valley-base settlements, smaller foothill sites and the little-known upland extensions into what are arguably regional associations (Betts 1992).

At the same time, survey work in the West/Central (Hisban 1973–76: Ibach 1987) and Southern Plateau (Central/Southern Moab 1978–82: Miller 1991) regions and in the 'Arabah (North 'Arabah 1979–81: Jobling 1989; East 'Arabah/Faynan 1978–79: Raikes 1980) intensified, revealing scattered Chalcolithic settlements throughout the uplands, and along the lateral wadis running into the north 'Arabah and south Jordan Valleys. Of particular importance here was intensified work in the Fidan/Faynan wadi systems, focusing on the enormous copper ore sources in this region, and much expanded work in the South East Dead Sea plain (Expedition to the Dead Sea Plain 1973–74: Rast and Schaub 1976; Arab Potash Company Townships: McCreery 1978–79). Surveys along the south plateau (Wadi al-Hasa Survey 1979–83: MacDonald *et al.* 1988), in the 'Arabah (Southern Ghors and Northeast 'Arabah Survey 1985–86: MacDonald *et al.* 1992) and in the southern lowlands (al-'Aqaba-Ma'an 1980–92: Munro *et al.* 1997) intensified during the early eighties and continues to the present day.

While expanded survey work gradually began to reveal the distribution of Chalcolithic settlements throughout the verdant uplands and some form of presence out into the marginal zones, renewed excavations began to collect the physical evidence needed to address the strengthening economic/environmental concerns of the seventies and early eighties. The single most important excavation was Hennessy's return to Tulaylat al-Ghassul.

Hennessy led three seasons of renewed excavations at Tulaylat al-Ghassul between 1975 and 1977. It was during these seasons that he expanded the size of entrenchments into the earliest phases of occupation at Ghassul. This work strengthened first impressions of a Neolithic ancestry for the Ghassulian, outlined a meticulously stratified 'pre-Classic Ghassulian' sequence, and obtained the radiometric dates that confirmed the long occupational prehistory of the Ghassulian culture (Bourke 1997b). However, the Sydney expedition is probably best known for the discovery of the Late Chalcolithic Sanctuary complex, in the southwest of the site, and the famous polychrome-figured 'Processional Frieze' wall painting from an early level (Hennessy E/F), in the central mound area (Hennessy 1982; Cameron 1981). Hennessy's view of the Neolithic origins of the Ghassulian culture is now largely accepted (Mazar 1990; Gilead 1990; Stager 1992; Levy 1993).

The decade after Hennessy's work at Ghassul ended saw a flowering of new projects researching Chalcolithic settlements in the Jordan Valley. The most prominent were Hanbury-Tenison's at Jabal Sartaba (1980–83) as part of the Pella Excavation Project, Leonard's at Katarit as-Samra' (1982), Helms' at Umm Hammad (1982–84), Gustavson-Gaube's problematic excavations at ash-Shuna (1984–85), and, most importantly, Dollfus and Kafafi's excavations at Tall Abu Hamid (1986–92). The first two projects (Jabal Sartaba and Umm Hammad) concentrated on latest Chalcolithic deposits. Most published material from Abu Hamid is of similar date, although there is undoubtedly a very long sequence at the site (Dollfus and Kafafi 1993; Lovell *et al.* 1997). Gustavson-Gaube's material from ash-Shuna was difficult to interpret, but should probably be seen as part of the 'Early Chalcolithic', pre-Ghassulian horizons discussed by de Contenson (Baird 1987; de Contenson 1964; Leonard 1992; Baird and Philip 1994). Together, these projects greatly expanded the Chalcolithic database, allowing, for the first time, a more nuanced view of Jordan Valley material culture.

At the same time, intensified survey work in the lateral wadis of the east Jordan Valley enriched understanding of Chalcolithic settlement patterns, underscoring the relationship of site location and water utilization, with winter (and spring?) wadi floodwater diversion strategies employed in tandem with permanent spring exploitation. Surveys in Wadi al-'Arab (1978: Kerestes *et al.* 1977/78; 1984: Hanbury-Tenison 1987), Wadi Ziqlab (1981–95: Banning *et al.* 1989, 1996, 1998), Wadi al-Yabis (1987–92: Mabry and Palumbo 1992, Wadi Sarar (1987–88: Mabry 1989), Wadi Kufrinjeh (1985: Greene 1986), Wadi az-Zarqa' (1978: Kerestes *et al.* 1977/78; 1980–82: Gordon and Villiers 1986), Wadi Nimrin (Simmons *et al.* 1989), Wadi Hisban (Iktanu Survey 1966–67: Prag 1974; Hisban Survey 1973–74: Ibach 1987), and the Naur-Dead Sea/Wadi Kufrein area developed an increasingly comprehensive view of the relationship of geography and Chalcolithic site location.

While intensified survey in the valley and foothill regions progressed, the first intensive work in the northern uplands occurred. The most notable for early remains are: Lenzen and McQuitty's Bayt Ras (1983–84: Lenzen and McQuitty 1983) survey of the Irbid-ar-Ramtha region; Green's 'Ajlun-Kufrinjeh (1986); Leonard's Jarash-Husn (1984: Leonard 1989); Sapin (1982–88: Sapin 1992) and Hanbury-Tenison's (1984: Hanbury-Tenison 1987) Jarash Region surveys; McGovern's Baqa'a Valley (1977–87: McGovern 1989); Kafafi and Simmon's 'Ayn Ghazal Region (1987: Simmons and Kafafi, 1992); the Cultural Resource Management Survey of Greater 'Amman (1987–88: Abu Dayyah *et al.* 1991); and Ibrahim's Sahab Regional Survey (1983–86: Gustavson-Gaube and Ibrahim 1986). On the eastern periphery, survey work included Helms's (1972–76: Betts 1991) pioneering Wadi Rajil–Jawa area survey; Betts's more extensive Black Desert Survey (1979–86: Betts 1986); and Garrard's Wadi al-Jilat Survey (1981–88: Garrard et al. 1987). This work covered areas that are far into the marginal zones.

The 10 years after Hennessy's excavations at Ghassul were marked by several major syntheses important for Chalcolithic studies. Among these were the final publication of the Jericho Neolithic (Kenyon and Holland 1982, 1983); Kafafi's (1982) synthesis of the Jordanian Ceramic Neolithic; the collection (and eventual publication) of much of Mellaart's material from Point IV soundings by Leonard (complete 1985, published 1992); Hanbury-Tenison's wide-ranging synthesis of the Jordanian Late Chalcolithic (1986); and Levy's (1987) first synthetic study of the excavations at Shiqmim. Together, these represent a sea-change in knowledge, for Kafafi and Hanbury-Tenison's syntheses of Jordanian Neolithic and Chalcolithic material respectively created the first thoroughgoing consensus on the extent and nature of the Jordanian Neolithic/Chalcolithic, with which the ever-increasing body of survey and excavation data could be integrated, and against which Palestinian Neolithic and Chalcolithic discoveries could be viewed.

The present day (1988–1998)

Excavations in the Jordan Valley over the last decade have focused on relatively small-scale, problem-oriented studies at previously investigated sites. Among these, renewed work at ash-Shuna (1991–94), deep soundings on the main tall of Pella (1992–97) and renewed work at Tulaylat al-Ghassul (1994–97) may be highlighted. Work on the plateau has been more restricted but grows steadily. Ibrahim's pioneering soundings at Sahab, southeast of 'Amman (1972–77), began detailed work on the Chalcolithic in the region. This has been strengthened recently by 'Amr and Najjar's work northeast of 'Amman at Wadi al-Qattar (1989), Kerner's work farther to the east at Abu Snesleh (1990–92), and Ji's work in the northwest at 'Iraq al-Amir (1996–97). Najjar's excavations at Tall Wadi Faynan (1988) began exploration of the Neolithic/Chalcolithic strata at what is probably the central site in the crucial Faynan copper ore source, although Hauptmann's important metallurgical analyses had been going on for some time (1984–88).

Exploration of the Faynan region has intensified throughout the nineties, with several large projects currently working in the region (Barker et al. 1997, 1998; Wright et al. 1998; Simmons et al. 1997; Levy et al. in press). Henry's soundings in the Judayid Basin sites of Queisa and al-Jill (1979–83), Hart's at Khirbat Qurrein North (1985) and Khalil's more substantial investigations at Magass (1985–95) began concentrated work on the variety of Chalcolithic cultures in the south, while Betts's soundings at al-Hibr (1989) did the same for the eastern marginal zones.

Survey work in the nineties was more tightly focused, and increasingly constituted urgent rescue work ahead of ever increasing roadworks and industrial/housing developments. Surveys along the Wadi al-Ajib in the far northeast (Betts et al. 1995), the Pella Hinterland (Watson and O'Hea 1996), the az-Zarqa' headwaters (Palumbo et al. 1996; Kafafi et al. 1997), the Hisban uplands (Prag 1991), the eastern Dead Sea shore ('Amr et al. 1996), the central/east al-Karak plateau (Mattingly 1996), and the southern lowlands (Bisheh et al. 1993; Waheeb 1996) work in a wide range of environments, many under imminent threat, inevitably resulting in an ever-more uneven coverage. Surveys typically sample the urban fringe and the pristine far horizons, while ignoring the perhaps compromised but less threatened middle distance. The institution of a long-term Cultural Resource Management programme in 1987, along with the development of the Jordan Antiquities Database and Information System (JADIS) (Palumbo 1994), will be crucial in counterbalancing the flood of emergency rescue work with a long-term programme of intensive regional survey.

Major trends in the nineties have seen intensified use of technical/analytical studies. Ceramic studies have been revolutionized by determined efforts to source clays and inclusions, classify manufacturing techniques and quantify production regimes. Research on ground stone and flint materials also seeks to determine quarry sources, manufacturing techniques and function. After much preliminary survey, detailed investigation of the Wadi Faynan/Fidan copper ore sources will soon provide a reliable archaeological background to the Bochum Museum's long-term metallurgical study. A heightened awareness of the need to document the economic underpinnings of Chalcolithic society has given a new prominence to geochemical, climatic, zoological and botanical studies. Comparative studies are now more commonly sourced to a suite of radiometric dates rather than a comprehensive typological analysis. Sharp subregional assemblage variability and differential rates of culture change are recognized factors in the increasingly dissected cultural landscapes of Chalcolithic Jordan.

Even though research efforts are still dominated by Jordan Valley data, excavation of upland and marginal zone settlements steadily grows, and will eventually allow for a more balanced summary of Chalcolithic settlement data than currently possible. Adequate coverage of the Jordan Valley is certainly possible today, but even with the much-increased upland survey efforts, no more than an impression of settlement patterns can be offered. The summary offered below is neither definitive nor synthetic but more a report on work very much in progress.

Relative and absolute chronology

There is no demonstrable link between the historical chronologies of Egypt or Mesopotamia and the Chalcolithic of Jordan. There are a variety of local objects that some feel reflect knowledge of the predynastic cultures of Egypt, and even more tenuous parallels with north Syrian/north Mesopotamian prehistoric assemblages have been drawn from time to time. None of these very general (at best) associations supply anything like the ability to construct even the most simple regional, far less a detailed subregional relative chronology. For this, radiocarbon dating needs to be employed, but until very recently the number of dates available for the south Levantine Chalcolithic was too concentrated on the Negev for widespread application outside that area. However, the situation has changed significantly over the last 10 years, as the steady rise in the number of analytical studies sourced to a radiocarbon-based chronology illustrates (Levy 1992; Gilead 1994; Joffe and Dessel 1995).

Table 4.1 is a list of Jordanian radiocarbon determinations and a restricted number from adjacent regions for comparative purposes. It is now generally accepted that the Jordanian Chalcolithic develops relatively smoothly out of the preceding Late Neolithic. Nominating an artificial point at which various transitions can be said to have occurred serves little purpose beyond delimiting parameters for a general review of the evidence such as this. Broadly speaking, the vast majority of the cultural material generally considered to belong to the Jordanian Chalcolithic falls within the late seventh/sixth millennium BP or the fifth/early fourth millennium BC. The cultural phasing is roughly delimited by the large suite of Late Neolithic dates from Wadi Ziglab 200 (Banning et al. 1996) and a comparable suite of early EB I dates from ash-Shuna (Chapter 5, this volume).

From the radiometric determinations listed in Table 4.1, it can be seen that any attempt to divide the Jordanian Chalcolithic into either chronological or regional subphases would be premature at present. There is currently little radiometric support for discerning a 'post-Ghassulian' phase within the Jordanian sequence (contra Hanbury-Tenison 1986: 117-18), although there is some (albeit controversial) support for a 'post-Ghassulian' afterglow in the northern Negev and West Dead Sea area (Gilead 1994; Joffe and Dessel 1995). Whether a 'pre-Ghassulian' phase can be adequately defined radiometrically will hinge on new assays at ash-Shuna ash-Shamaliyya, Pella, Abu Hamid and Tulaylat al-Ghassul, all currently in progress. There is some radiometric support for such a view, and cultural sequencing from Tulaylat al-Ghassul and Abu Hamid in the south, and ash-Shuna and Pella in the north seem to warrant it (Bourke 1998b). From a south Levantine perspective, the Jordanian Chalcolithic should be seen as later than the Qatifian and the early phases of the Munhata/Rabah complex, and contemporaneous with the later Rabah ('Tsafian') and the Beersheban Besoran complexes, although the Negevite Chalcolithic may extend slightly beyond the time frame of the Jordanian Ghassulian. In the wider world, at least partly contemporary are the Merimda/al-Omari (Delta) cultures and the Badarian/Naqada I (Valley) cultures in Egypt, the poorly differentiated Syrian Chalcolithic and the northern Mesopotamian varieties of the later Ubaid.

Settlement patterns

In spite of the great increase in survey work over the last generation, the record is still largely deficient in detail, published in a most preliminary form, largely uncorrected for geomorphologic processes, and biased towards large sites and more visible remains. Even so, a general pattern to the settlement data can be discerned. Highest settlement densities occur in the Jordan Valley, with a more restricted spread into the foothills and the western margins of the upland plateau. Plateau settlements are concentrated either close by the head of the many lateral wadis leading into the lowlands or along the better-watered upland valleys. Coverage becomes ever more sparse as one moves farther east and south along the seasonal wadis into the marginal zones.

Site	Site phase	Lab	Date bp	Date BC	Period
Ziglab 200	E33	TO-3409	6900 ± 70	5740	Late Neo
Nizzanim	Early Neo	Hv-8509	6740 ± 90	5620	Late Neo
Tsaf	I	RT-508A	6720 ± 460	5590	Late Neo
Ziglab 200	F34	TO-3411	6670 ± 60	5560	Late Neo
Chagar Bazar	Level 11/12	P-1487	6665 ± 77	5550	Late Halaf
Tall Leilan	Period VIa	UM-1817	6580 ± 100	5480	Early Ubaid
Ghassul	G/(A:xi-x)	SUA-732	6550 ± 160	5450	Neo/Early Chal
Ghassul	G/(A:xi-x)	SUA-736	6430 ± 180	5360	Neo/Early Chal
Wadi Faynan	Prof. B	HD-10567	6410 ± 115	5350	Late Neo
Ziglab 200	G34	TO-3412	6380 ± 70	5320	Late Neo
Ghassul	G/(A:xi-x)	SUA-734	6370 ± 105	5310	Neo/Early Chal
Wadi Faynan	Prof. B	HD-12335	6360 ± 45	5300	Neo/Early Chal
Ziglab 200	E33	TO-3410	6350 ± 70	5275	Late Neo
Ghassul	G?/(E:vi)	SUA-738/1	6300 ± 110	5240	Neo/Early Chal
Abu Hamid	Lower	Ly-6174	6200 ± 80	5165	Neo/Early Chal
Ziglab 200	E33	TO-3408	6190 ± 70	5135	Neo/Early Chal
Abu Hamid	Lower	Ly-6254	6190 ± 55	5135	Neo/Early Chal
Abu Hamid	Lower	Ly-6255	6160 ± 70	5100	Neo/Early Chal
Abu Hamid	Lower	Ly-6259	6135 ± 80	5075	Neo/Early Chal
at-Turkman	Period IVB	GrN-13038	6110 ± 80	5060	Late Ubaid
Wadi Faynan	Baulk	HD-12338	6110 ± 75	5060	Middle Chal
Ras Shamra	Period IIIC	P-389	6098 ± 173	5020	Post-Halaf
Ghassul	G?/(E:vi-v)	SUA-739	6070 ± 130	4950	Middle Chal
Qatif Y3	Late Neo	PTA-2968	6040 ± 80	4930	Middle Chal
Abu Hamid	Middle	GrN-16357	6030 ± 60	4920	Middle Chal
Sukas	G11: layer 58	K-936	5910 ± 100	4860	Syrian Chal
Ghassul	A-C/(E:iii)	SUA-511b	5796 ± 115	4690	Late Chal
et-Turkman	Period IVB	GrN-13041	5760 ± 80	4670	Late Ubaid
Shiqmim	Upper Vill.	RT-649B	5750 ± 180	4660	Late Chal
Merimde	Early	W-4355	5750 ± 100	4660	Late Chal
Abu Hamid	Upper	GrN-16358	5745 ± 35	4650	Late Chal
Ziglab 200	Area A	TO-1086	5740 ± 110	4625	Late Chal
Wadi Faynan	Sq. A loc 23	HD-12337	5740 ± 35	4625	Late Chal
Jabal Queisa	J24	SMU-804	5720 ± 150	4560	Late Chal
Shiqmim	Phase IV	OxA-2523	5710 ± 140	4550	Late Chal
Abu Hamid	Upper	GrN-14623	5670 ± 40	4500	Late Chal
Golan	Site 21	RT-1864	5565 ± 60	4485	Late Chal
Ghassul	A-C/(E:iii)	SUA-511c	5661 ± 120	4480	Late Chal
Abu Hamid	Middle	Ly-6249	5655 ± 210	4465	Late Chal
Abu Hamid	Upper	GrN-17496	5651 ± 40	4460	Late Chal
Abu Hamid	Middle	Ly-6248	5650 ± 75	4460	Late Chal
Abu Hamid	Middle	Ly-6251	5580 ± 95	4400	Late Chal
Shiqmim	Phase IV	OxA-2526	5540 ± 150	4360	Late Chal
Thawwab	'Post-Neo'	GrN-15192	5540 ± 110	4360	Late Chal
Golan	Faras Silo	RT-718	5540 ± 110	4360	Late Chal
Shiqmim	Phase III	OxA-2521	5530 ± 130	4350	Late Chal
Ghassul	A-C/(E:iii)	SUA-511a	5507 ± 120	4340	Late Chal
Ghassul	IV?/(G:vi-v)	RT-390A	5500 ± 110	4340	Late Chal
Abu Snesleh	Chal	Hv-20791	5445 ± 195	4295	Late Chal
Gilat	Phase III	RT-860A	5440 ± 180	4280	Late Chal
Nahal Qanah	Gold Dep.	RT-861E	5440 ± 100	4270	Late Chal
Pella	Area XIV	SUA-2391	5430 ± 60	4270	Late Chal
Safadi	Lower	M-864A	5420 ± 350	4260	Late Chal
Shiqmim	Phase III	OxA-2525	5385 ± 130	4245	Late Chal
al-Mudiq	Layer IV	IRPA-168	5380 ± 210	4240	Late Ubaid

Table 4.1. Radiocarbon dates for the Jordanian Chalcolithic.

Site	Site phase	Lab	Date bp	Date BC	Period
Wadi Faynan	Sq.A loc. 8	HD-12336	5375 ± 30	4235	Late Chal
Shiqmim	Phase II	RT-859D	5370 ± 180	4230	Late Chal
Ghassul	A-B/(A:iii-i)	GrN-15194	5330 ± 25	4190	Late Chal
Abu Hamid	Middle	Ly-6257	5325 ± 140	4180	Late Chal
Hemamieh	Badarian	WSU-1728	5290 ± 130	4120	Badarian
Horvat Beter	Early	W-254	5280 ± 150	4080	Late Chal
Safadi	Middle	M-864B	5270 ± 300	4060	Late Chal
Harbush	Golani	RT-525	5270 ± 140	4060	Late Chal
Ghassul	A/(E:ii-i)	GrN-15195	5270 ± 100	4060	Late Chal
Merimda	Late	WSU-1846	5260 ± 90	4045	Early Delta
Shiqmim	Phase I	RT-554A	5250 ± 140	4030	Late Chal
Abu Hamid	Upper?	Ly-6258	5205 ± 95	3995	Late Chal
Ash-Shuna	E II 43	GrN-15200	5125 ± 25	3960	Late Chal
Safadi	Upper	M-864C	5120 ± 350	3950	Late Chal
Ash-Shuna	E I 12/3	GrN-15199	5115 ± 25	3945	Late Chal
Ghassul	A/(A:ii-i)	GrN-15196	5110 ± 90	3940	Late Chal
Horvat Beter	Late	PTA-4312b	5100 ± 130	3920	Late Chal
Hab.Kabira	Uruk	GrN-00000	5085 ± 65	3900	Early Uruk
Shiqmim	Phase I	OxA-2520	5060 ± 140	3860	Late Chal
Naq. KH1	Early Naq.	SMU-360	5030 ± 100	3820	Naqada I
Naq. NT	Early Naq.	WSU-2257	4990 ± 80	3800	Naqada I
Naq. KH3	Early Naq.	TX-2340	4970 ± 70	3780	Naqada I
Harbush	Golani	RT-1862	4945 ± 65	3755	Late Chal
Shiqmim	Phase I	RT-1339	4940 ± 70	3750	Late Chal
Maadi	Early	KN-3574	4940 ± 60	3750	Late Delta
Mishmar	Hoard Mat	W-1341	4880 ± 250	3670	Late Chal
Maadi	Early	R-1425	4860 ± 70	3650	Late Delta
Harbush	Golani	RT-1866	4810 ± 90	3600	Late Chal
Gilat	Phase II	RT-860B	4800 ± 135	3600	Late Chal
Hierakonpolis	Site 29	WIS-1183	4800 ± 75	3600	Naqada II
Mishmar	Hoard Mat	I-285	4780 ± 100	3560	Late Chal
Hierakonpolis	Site 29	WIS-1169	4760 ± 80	3545	Naqada II

Table 4.1. (continued).

The Jordan Valley

Major Chalcolithic settlements are positioned at the point where side wadis enter the valley from the uplands to the east. The largest settlements are positioned beside the major wadis, with exploitation of seasonal wadi floodwaters central to the agricultural practices of at least the Later Chalcolithic period. Potential settlement size seems tightly correlated with size of adjacent floodplain. Smaller sites are found along the upper courses of each wadi system, larger and more closely packed in the western half and less numerous and smaller in size the farther east into the uplands one goes. There are no significant settlements between Wadis az-Zarqa' and Nimrin, which, given the absence of lateral wadis, strengthens the correlation of settlement size and associated wadi/floodplain. There are major Chalcolithic settlements at ash-Shuna (Wadi al-'Arab), Tall Fendi (Wadi Ziglab), Pella/Jabal Sartaba (Wadi Malawi/al-Jirm), Tall Abu al-Kharaz/Tall Abu Habil (Wadi al-Yabis), Abu Hamid (Wadi 'Ajlun), Tall al-Handaquq (Wadi Serar), Sa'idiyya Shimali/Tahta (Wadi Kufrinji), Katarit as-Samra'/Umm Hammad (Wadi az-Zarqa'), Ghrubba/Tall Ghanam (Wadi Nimrin), Tall al-Malih (Wadi Kefrein), Tulaylat al-Ghassul (Wadi ar-Rama/Djaffara) and Tall al-Azeimeh (Wadi Azeimeh). This pattern continues south along the eastern shore of the Dead Sea with az-Zara 25 (Wadi 'Ayn Umm Hudayb), Bab adh-Dhra' APC (Wadi al-Karak) and Tall Wadi Faynan (Wadi Faynan) the more prominent settlements.

Overall, there are more than a hundred Chalcolithic period sites known for the East Jordan Valley alone. Less than 10% are larger than 2 ha. While several sites are very large (>10 ha) and some are mid-sized (c. 5 ha), any suggestion that the variation in site size can be related to a rank-size distribution and a more formal integration into subregional economic associations is premature at best and improbable on current indicators. Given the correlation between site magnitude and associated floodplain size, and the absence of any clear evidence for marked economic specialization, growth is likely to be correlated with irrigated agricultural potential. However, where at least some form of stratification can be inferred (Ghassul, Abu Hamid), it is associated exclusively with the largest sites, which may imply a limited role as a 'cultural central place', even if characterization as an 'economic central place' cannot yet be sustained.

Of the major settlements, Chalcolithic deposits have been intensively investigated at ash-Shuna, Pella, Abu Hamid, Umm Hammad and Tulaylat al-Ghassul, and more limited soundings have been made at Tall Fendi, Abu al-Kharaz, Abu Habil, Sa'idiyya Tahta and Ghrubba. Material from surface surveys, restricted soundings and large-scale area excavations all suggest that the az-Zarqa'/Nimrin 'settlement gap' marks an effective border between relatively distinct Early Chalcolithic assemblages of the north and south valley, a distinctiveness much reduced (if not entirely removed) in the Late Chalcolithic period. It is possible that the Early Chalcolithic sites of the northern valley are more closely associated with the late Rabah/Tsaf complexes than sites in the south. Early Chalcolithic sites of the southern valley have their precursors in the still poorly characterized early assemblages at Tulaylat al-Ghassul and Ghrubba, and may ultimately prove to derive from the latest phases of the Jericho Ceramic Neolithic.

The Jordanian uplands and the marginal zones

Once the recalibration of Glueck's pottery reading is noted (for Middle Chalcolithic read Ghassulian, for Late Chalcolithic read EB I), the non-intensive format of most early (and some modern) survey projects factored in, and the tendency to characterize most fourth-millennium upland sites as 'Chal/EB' (with most turning out to be EB I when scrutinized closely) noted, a very superficial survey of the upland/inland Chalcolithic would suggest a rough count of around 120 sites reported in some form or another. Most seem quite small (>1 ha), with the vast majority associated with wadi systems and/or the western half of the upland plateau.

There is a cluster of around 20 sites in the northern uplands, with concentrations on the Irbid/Ar-Ramtha Plain (Lenzen and McQuitty 1983) around the largest site of Sal (Kamleh 1998), some located along the western escarpment, and a few far into the eastern Harra (Betts 1992). A variety of surveys (Leonard 1987; Hanbury-Tenison 1987; Sapin 1992) have identified another 10 between al-Husn and Jarash, 10 more between Jarash and 'Amman, and a cluster of around 20 in the environs of 'Amman, with Sahab the largest (Ibrahim 1984), and Wadi Qattar/Abu Snesleh ('Amr et al. 1993) smaller sites to the northeast and southwest respectively. All sites of any size are positioned either beside wadis or at the confluence of two systems. The Hisban/Madaba Plains survey area has produced perhaps another 20, most small and clustered along the western escarpment wadis (Ibach 1987; Mortensen and Thuesen 1998), or located in and around the later Early Bronze tulul sites (Harrison 1997). The various al-Karak Plateau surveys (Miller 1991; Mattingly 1996) account for another 10, mostly small and more definitely associated with the western escarpment and the lateral wadis. The East Dead Sea Plains (Rast and Schaub 1974; McCreery 1978/79), Wadi al-Hasa and northeast 'Arabah (MacDonald et al. 1988, 1992) and the Wadi Faynan regions (Barker et al. 1997, 1998; Levy et al., in press; Simmons et al. 1998) are linked to another 10–15 sites. Most are small, concentrated in the Faynan/Fidan region, dominated by Tall Wadi Faynan, and even more closely wedded to the eastern wadi systems. The foothills and the lowlands of the Hisma (Henry 1995) reveal a similar number of noticeably smaller sites (such as Wadi Queisa), probably correlated with specialist pastoral activities. Even so, Henry's association of these sites with an eastern 'Timnian' seems unwise given the radiocarbon dating for Queisa, and the lack of material cultural parallels with the Negev. Another 10 sites can be documented along the Edomite plateau (Hart and Faulkner 1985), with the spring-fed site of Khirbat Qurrein North the largest in the region (Hart 1987). A further 10 sites can be identified in the al-'Aqaba/Ma'an survey region (Jobling 1989), although all are small and similar in type to those of the Hisma region. Close by the south coast, the surprisingly large site of Tall Magass (Khalil 1995) dominates the coastal region and may well be better associated with the copper-mining north 'Arabah cluster (or that of Timna in the south) than the pastoral encampments to the east (Khalil 1992).

Few upland sites have been investigated in any detail, making material cultural characterization perilous and diachronic analysis impossible. The sites of Sal (Irbid) and al-Hibr (eastern Harra) allow no more than a glance at the northern plain, although Sahab (particularly), Wadi Qattar and Abu Snesleh provide reasonable coverage of the 'Amman region. Farther to the south, the more isolated activity at Tall Wadi Faynan and Tall Magass, Hart's sounding at Khirbat Qurrein North, and Henry's detailed characterization of the Hisma Basin sites represent small islands of knowledge in a notoriously difficult to interpret sea of survey results. Comments below are the barest of preliminary impressions of an extremely variegated data set.

North plateau sites

These are reported to be a mixture of small settled villages (if close by water sources) or pastoral encampments (if removed from water sources). It is unlikely that site function can be so easily determined, although a shifting mosaic of sedentary, pastoral and seasonal lifestyles is likely. Optimal locations are all occupied, and even minor wadi systems are utilized. Some marginal zones are exploited to a limited extent. The impression is one of an even, if sparsely occupied, landscape, but with no obviously neglected econiches. Material cultural associations are very poorly understood. Ceramics are reported to be predominantly coarse, red-slipped wares, with painted decoration quite rare. Associations with the Golani Chalcolithic culture may be hinted at by similarities in ground stone inventories (Ibrahim and Mittmann 1998), but a comprehensive description of the North Jordanian Upland culture is yet to be developed.

Epstein (1978b) views the Golani Chalcolithic culture as a distinct subregional grouping, with few links to the better-known Ghassulian of the Jordan Valley. The Golani subsistence regime is dominated by olive cultivation and transhumant pastoralism. Olive cultivation became increasingly important as the Chalcolithic economy developed (Epstein 1993) and settlement studies (Finkelstein and Gophna 1993) relate a Late Chalcolithic spread into the Palestinian uplands with spreading olive cultivation. A similar scenario may well be at least partly applicable to the north Jordanian uplands (Kamleh 1998). It is possible that the Golani and adjacent Irbid/ar-Ramtha plain settlements are to be associated in one subregional unit, with one segment of the society practising transhumant pastoralism, journeying from the central Jawlan (summer) to the eastern *harra* (winter), and the balance (particularly in the larger sites northeast of Irbid) practising a sedentary, mixed-farming regime, enriched by a developing emphasis on horticulture.

Radiocarbon dating for the Golani Chalcolithic (Carmi *et al.* 1995) would see it confined to the second half of the fifth and first half of the fourth millennium BC. As latest dates for Rasm Harbush are as late as any south Levantine Chalcolithic settlement site, it seems unlikely that the northern Chalcolithic ended significantly earlier than southern Beersheban settlement. If the Irbid/ar-Ramtha complex can be related to the Golani Chalcolithic, then a similar relatively Late Chalcolithic chronology may be proposed for the majority of the sites in the region.

Central plateau sites

The site of Sahab, southeast of 'Amman, would appear to be a mature, settled, agricultural community. The material culture is said to have particular links with the south Jordan Valley Ghassulian sites, with painted pottery a more common occurrence than in northern assemblages. The presence of numerous massive storage silos (Ibrahim 1987) implied the existence of a substantial grain surplus, and the discovery of a seal fragment suggested regulated interregional trade (Ibrahim 1984). Although the excavator favours immigration into the uplands from the Jordan Valley as the preferred explanation for the origin of the Sahab area settlements (Ibrahim 1987), Late Neolithic settlement is known in the region (Kafafi 1982; Kerner, pers. comm.). The Wadi Qattar Chalcolithic site may contain an early Chalcolithic component ('Amr *et al.* 1993). Kerner's work at Late Chalcolithic Abu Snesleh (Kerner *et al.* 1992) contrasts somewhat with Sahab, perhaps illustrating the more restricted assemblage of a small community in a more marginal (northeastern) niche. Given the relatively low frequency of sickle blades and the high frequency of tabular scrapers (Lehmann *et al.* 1991), it is possible that the pastoral component plays a more central role at Snesleh. Material culture and radiometric data favour association with the larger Sahab community.

The south

The Chalcolithic settlements in Wadi Faynan have only recently begun to be explored intensively. Najjar's work at Tall Wadi Faynan has identified the likely main focus of Neolithic/Chalcolithic occupation in the region. Radiometric data would suggest a more extensive Chalcolithic period occupation at the site (particularly in Area A) than first indicated. Neolithic/Chalcolithic material culture is still poorly known, but published Neolithic ceramic material (Najjar *et al.* 1990) has reasonably close form parallels in the early assemblages at

Tulaylat al-Ghassul (Hennessy 1982), and the lithic assemblage suggests a mixed-farming regime, probably based on floodwater irrigation (Barker *et al.* 1997). A surface figurine fragment may suggest, albeit tenuous, links with Yarmoukian contemporaries (Najjar *et al.* 1990: 32–33), although radiometric data would suggest a later association, not inconsistent with early Ghassul parallels (Hennessy 1969). Kafafi (1993) suggests an association with the PNA-related sequences at Dhra' and adh-Dharih, which seems more likely than Goren's (1990) posited links with the Negevite Qatifian.

Khalil's work at the well-established, mixed-farming village of Tall Magass, slightly inland from the southern coast, stands in high contrast to the seasonal 'Timnian' pastoral encampments, if Henry's detailed analysis of the Hisma occupation (Henry 1995) is an accurate reflection of settlement trends in the far south. Magass is a relatively large village site with architectural and ceramic links to the more established settlements of the north (Khalil 1992). The most common explanation for the apparently anomalous developments at Magass is (presumably Timnan) copper exploitation, although the local microenvironment seems capable of supporting mixed farming, perhaps with a high pastoral component (Khalil 1995). Indeed, if Henry's view of the Hisma occupation as part of an (early?) Timnian complex is sustained, then it may well be that earliest-phase Magass stands in a similar relationship to its pastoral hinterland as Finkelstein (1990) suggested for Arad. Indeed, the apparently large very late Chalcolithic settlement of Arad V could have stood in a similar relationship to its hinterland (Gilead and Goren 1986; Avner *et al.* 1994) as that Finkelstein suggested for the subsequent Early Bronze Age settlement. A more speculative observation might see both Magass and Arad as similar 'pastorally enhanced', mixed-farming villages, serving as 'copper stations' along routes to the important copper-using settlement concentrations to the north and west.

Summary

Chalcolithic settlement is widespread throughout Jordan. A major concentration in the Jordan Valley, directly associated with floodwater management strategies, is suggested by surveys old and new. This follows on from a less-intensive Late Neolithic exploitation of valley floor niches, and a much less pronounced settlement of the foothill region, implying considerably less emphasis on floodwater irrigation and horticulture in earlier times. Upland occupation is widespread, although still tending to concentrate along wadi systems, the western escarpment and optimal farmlands, suggesting floodwater management is still an important factor in upland settlement location. Both settlement number and site size increase (on average) from Late Neolithic levels and a greater variety of upland habitats are exploited, suggesting a more flexible combination of pastoral/agricultural/horticultural segments. Even so, settlement numbers are small in comparison to the subsequent Early Bronze Age, where a considerable increase in settlement number (if not necessarily population) occurs. In the south, permanent settlement numbers are much reduced, small in scale, and more certainly restricted to the western escarpment wadi systems, although small seasonal encampments are numerous in the east. Permanent resettlement of the far south probably occurs for the first time during the Chalcolithic, although a pastoral element was certainly present throughout the Neolithic. Due to the well-known uncertainties of the survey data base, we cannot yet speak of changing settlement patterns over the course of the Chalcolithic. However, Wadi Qattar hints at an early Chalcolithic occupation of the central plateau, and isolated ceramic elements from the al-Karak plateau sites would seem to support this.

Economic activity

Introduction

The developmental status of the Chalcolithic economy is controversial. Until recent times, there was a tendency to dichotomize the Chalcolithic and Early Bronze Age economic modes, with many innovations claimed for the latter period, yielding clear differences in economic potential. However, few of the alleged differences (niche exploitation of fruits, product intensification through irrigation and deep plough agriculture, and long-distance trade in staples facilitated by equid domestication) stand up to close analysis. None of these features are exclusive to the Early Bronze Age.

There is considerable evidence for olive harvesting, both in the valley floor (Meadows 1998, 1999) and in the hill regions (Willcox 1992; Bourke *et al.* 1999b) from Late Neolithic times. This develops into full-blown cultivation by the Late Chalcolithic period

(Meadows 1999), probably throughout the eastern foothills (Mabry and Palumbo 1992). Floodwater irrigation systems are well documented in the Negev (Levy 1987, 1995), and are likely to have existed in the south Jordan Valley as well (Glueck 1951; Zohary and Speigel-Roy 1975). Settlement pattern and (particularly in the south) rainfall data provide compelling evidence in favour of widespread employment of simple, floodwater-management schemes. It is more difficult to determine the timing for the introduction of deep-plough agriculture, although zoopathological evidence would suggest this dates from the Late Chalcolithic period in the western Negev (Grigson 1995). At Tulaylat al-Ghassul, cattle numbers approach the 20% 'deep-plough' threshold (Wapnish and Hesse 1991: 26–27) from Middle Chalcolithic times (Mairs 1997). It seems that limited use of the deep plough occurred in the Jordanian uplands by the Late Chalcolithic ('Amr *et al.* 1993: 274). The domestication of the donkey is now well documented in Late Chalcolithic times (Grigson 1995; Mairs 1997), as has been implied for some time by Late Chalcolithic clay models (Epstein 1985). With the probable exception of the grape (Stager 1985; Grigson 1995), most fruits were domesticated by the Late Chalcolithic (Grigson 1995; Meadows 1998, 1999). Even if this latter is a genuine Early Bronze Age innovation, it is likely to have taken some time to have had a significant economic impact.

It is difficult to develop a distinctive picture of the Jordanian Chalcolithic economy that differs noticeably from that of the succeeding Early Bronze Age. It is undeniable that the intensity of certain aspects of economic activity (deep plough, tree crops, copper metallurgy, trade) alters over the course of the Early Bronze Age. This may be due to contrasting social regulatory mechanisms and heightened levels of interregional interaction in the latter period. Contrasting economic potentials will no longer serve as an explanation for the manifest changes between the two periods.

Case study: husbandry and agricultural practices at Tulaylat al-Ghassul

The 6 m-deep, 1500-year long occupational sequence at Tulaylat al-Ghassul is grouped into five major archaeological horizons of unequal length (Bourke 1997a, 1997b). The entire sequence is to be characterized by relatively slow material cultural progression from one mosaic of attributes to another, with two main instances of relatively more pronounced cultural change. The first occurs between the Late Neolithic and Early Chalcolithic phases, Hennessy Phases H/I and G (Hennessy 1969, 1982; Bourke 1997a, 1997b), witnessing major changes to architectural forms and ceramic and lithic tool kits towards something more definitely ancestral to the later 'Classic Ghassulian' assemblage (Bourke 1997b). The second major change is that between the Middle 'pre-Ghassulian' and Late 'classic Ghassulian' phases, Hennessy Phases F/D and C/A respectively (Hennessy 1969, 1982; Bourke 1997a, 1997b). This may be characterized as the development of specialized lithic tool kits, enriched ceramic repertoires, distinct architectural forms, large storage facilities, formalized cult activity and bureaucratic regulatory devices (Bourke 1998b).

These two main alterations in the material cultural assemblage (discussed in more detail in appropriate sections below) are probably indicative of fairly significant economic and social change. The first may well be characterized as the time when intensified agricultural practices are adopted. The second is probably the time at which significant societal change towards a more stratified economically diverse society becomes visible in the material record. More speculatively, this socioeconomic change may be the by-product of diversifying and intensified irrigation-assisted agricultural practices, which in turn encouraged the development of more specialized production and managerial elites and more elaborate social regulatory devices, on the one hand, while producing the significantly enhanced agricultural surpluses needed to sustain and buttress any such developing elite.

Zoological (Mairs 1997) and botanical (Meadows 1998) evidence

Over the course of the two-thousand-year sequence, major changes in husbandry practices occur. The marked increase in the retention of aged cattle and absence of neonates from Early Chalcolithic times onwards, suggests some form of milk-production regime. The sudden change in pig-harvesting patterns in the Early Chalcolithic suggests adoption as a (if not the) main domestic meat-protein source. The gradual reduction in neonate and aged ovicaprines from the Middle Chalcolithic period onwards may well suggest an increasing emphasis on milk production over time. This harvest pattern might seem to deny significant specialized wool production in the Chalcolithic. However, in the later Chalcolithic phases, where numbers are sufficient to allow reliable separation of sheep and goat, sheep numbers outweigh goat by a considerable margin (7 : 1 NISP and 4 : 1 MNI), perhaps to be correlated with increased numbers of tabular scrapers at this time (Henry 1995). One possible interpretation might see flock pasturing off-site for a significant part of the year (thus removing neonatal sheep remains from the bone assemblage) as more and more fields on the nearby floodplain were given over to intensified agriculture, only returning during the shearing season.

There is a clear impression of increasing agricultural intensification and the exploitation of more productive strains of wheat and barley over time. There is a tendency towards crop diversification over time, with legumes and fruits steadily increasing in importance. As well, there is the suggestion of a major alteration in field management strategies in the Late Chalcolithic, with the development of green-field fallowing (Meadows 1998). Distinct spatially segregated crop-processing regimes develop by the Late Chalcolithic. The exploitation of tree crops from an early stage, and evidence for domestication of olive (and possibly fig) from Middle Chalcolithic times underlines the growing importance of fruit yield over time (Meadows 1999).

Given the probable annual rainfall for the Chalcolithic south Jordan Valley, even allowing for a (still-controversial) modest increase from changing summer monsoon patterns, it is difficult to see how the increasingly sophisticated production regime at Ghassul was achieved in the absence of some form of floodwater management/irrigation. Evidence is only suggestive and admittedly largely circumstantial at present, although specific irrigation weed types and seed morphology vectors are currently under investigation. Irrigation is likely to have begun as a simple exploitation of the winter wadi floodwaters, perhaps supplemented by springwaters in other seasons, although increased summer rainfall early in the Chalcolithic is a distinct possibility if the monsoonal rain systems did shift north in this period. However, the sophisticated mixed farming regime of Late Chalcolithic Ghassul would seem to demand a more structured management of competing demands for what may have become an increasingly scarce resource, if the monsoonal systems shifted south towards the end of the Chalcolithic.

Ghassul in context

The picture of agriculture and husbandry at Tulaylat al-Ghassul is significantly more complete than for most other Chalcolithic sites in Jordan, although many share features with Ghassul. All major Jordan Valley sites display a mixed farming regime by the Late Chalcolithic. The contemporaneous botanical and zoological assemblages at Pella, Abu Hamid and Ghassul can be viewed as essentially similar. Olive cultivation is fundamental to all, as is a mixed wheat/barley cereal crop. The more productive (if more environmentally fragile) varieties of wheat (emmer) and barley (six-row) dominate cereal assemblages. Legumes are an important source of protein, which may allow one to speculate on the general introduction of green-field fallowing in all Jordan Valley sites by this time. Flax is grown as an important specialized crop, and fruit crops are now a significant part of the assemblage.

At Abu Hamid (Dollfus and Kafafi 1988), the increase in pig numbers in comparison with those at Ghassul is more a consequence of the wetter conditions in the mid-valley than any significant variation in husbandry practice. This point is reinforced by the even greater importance of pig in the swampy conditions at ash-Shuna (Baird and Philip 1994). Not surprisingly, the slightly drier conditions in the foothill regions around Pella result in reduced pig numbers in comparison with ash-Shuna (Bourke *et al.* 1998). Environmental variability rather than differential husbandry practice is the likely explanation for slight assemblage variability up and down the valley (Grigson 1995). Equids are a consistent presence, as are small numbers of deer and gazelle, but the main food animals are cattle, ovicaprines and pig. The age–class harvest patterns at Ghassul, Abu Hamid and Pella are very similar, with an implied emphasis on cattle and ovicaprine milk production, and pig as an important domestic, meat-protein source. The absence of ovicaprine neonatal classes at Abu Hamid implies winter pasturing at some remove from the main site (Dollfus and Kafafi 1988: 28), and provides further evidence for the existence of a well-established pastoral component by Late Chalcolithic times.

Although evidence from the Late Chalcolithic uplands is very slight, the Golani assemblages include cereals, legumes and olives and a mixed husbandry regime, which is likely to be at least similar to that of the north Jordan plateau sites. The Sahab assemblage includes cereals, olives and legumes, and a mixed husbandry regime. The material from Wadi Qattar includes cereals and fruits, and a husbandry regime in which ovicaprines predominate. Of particular interest are the observations on cattle age classes, which suggest the employment of aged animals for secondary milk production and possible deep-plough agriculture ('Amr et al. 1993: 274). A similar state of affairs exists at Abu Snesleh, with the zoological assemblage suggesting a greater reliance on ovicaprines and hunted species (Kerner, pers. comm.), a state of affairs more in keeping with the transhumant assemblages of the Hisma (Henry 1995) and the Eastern Desert margins (Betts 1992).

It is more difficult to determine if the Early Chalcolithic assemblages at Ghassul are in any way representative of the country as a whole, due mainly to the absence of contemporary comparative data. There is a limited amount of evidence from ash-Shuna and Pella that hints at more variation in earlier Chalcolithic assemblages. While olive seems to have been a constant presence within the plant assemblage at Ghassul from Neolithic times, the evidence for domestication only manifests itself from Late Chalcolithic times onwards. In the north valley sites, olive is said to be absent from Early Chalcolithic deposits at ash-Shuna (Chapter 5, this volume), and occurs in much-reduced numbers in Early Chalcolithic phases at Pella. The relatively late introduction of domesticated olive (and other tree crops?) into the north Jordan Valley may well suggest that horticulture in general came later to the north valley than to the south. However, the presence of tree crops in the Golan, the eastern foothill and north/central upland zones by the Late Chalcolithic (Epstein 1993; Willcox 1992; 'Amr et al. 1993) suggests that any spread from a putative south Jordan Valley 'heartland' must have been thoroughgoing and relatively rapid, given the ubiquity of olive cultivation during the Late Chalcolithic.

Interpretation

The move from simple economic description to socio-economic reconstruction is likely to remain hazardous for some time to come. Any attempt at societal reconstruction from the current data set might seem closer to fraud than legitimate speculation. However, all but extreme minimalists should acknowledge that significant socioeconomic change does occur at some stage during the Chalcolithic period (Levy 1995, 1996). Practitioners will continue to differ over the economic significance of observed changes (Levy 1987; Gilead 1988), the potential and actual social change that need be inferred from such economic changes (Levy 1995; Yoffee 1993), and the degree of continuity between Late Chalcolithic and Early Bronze Age systems (Hanbury-Tenison 1986; Joffe 1993).

The relatively undifferentiated Late Neolithic and Early Chalcolithic economic systems of the Jordan Valley sites would suggest little economic and/or social differentiation, a tribal society in which autonomous, household-production units are the norm (Kafafi 1993; Gopher and Gophna 1993). Over time, intensified and specialized agricultural practices, buttressed by an increasing reliance on labour-intensive and elite-regulated exploitation of floodwater irrigation systems, may have led to the development of distinct (and perhaps competing) social classes, sourced to the redistribution of agricultural surplus (staple finance). It may be that the exploitation of high-risk, high-yield cereals, alongside more traditionally exploited cultivars, and the development of specialized husbandry and horticultural activities alongside simple, mixed-farming regimes mark a crucial diversification of economic activity during the Late Chalcolithic period.

The evidence from Ghassul would suggest a gradual and piecemeal transformation of a relatively undifferentiated society (Early Chalcolithic) into one in which specialized economic activity existed alongside a more undifferentiated economic mode (Late Chalcolithic). The very large grain storage systems at Ghassul, Abu Hamid, Pella and Sahab illustrate the significance of centrally stored, agricultural produce to at least a segment of society by the Late Chalcolithic. It may well be that the much-discussed 'restructuring' of Early Bronze Age society represents no more than the triumph of one pre-existing mode of socioeconomic competition over another, or in anthropological terms, the movement from group-oriented to an individualizing chiefdom (Earle 1991; Finkelstein 1995; Stein 1998).

Architecture

Settlement layout and domestic architecture

Few settlements have been extensively excavated. Very large surface areas were excavated by the first Pontifical Biblical Institute mission at Tulaylat al-Ghassul (over $12,000\,\text{m}^2$) but not carefully, so that the association of architectural units across large and apparently featureless courtyards and their combination into general phase plans is rarely as secure as the published settlement plans imply. However, without the evidence from Ghassul, there is virtually no information on any but the latest settlements at most sites (ash-Shuna, Pella, Sahab and Magass), and with the exception of Abu Hamid, even these exposures rarely extend beyond one or two dwelling units. As with many other aspects of the Jordanian Chalcolithic, a close consideration of the evidence from Tulaylat al-Ghassul is central to any attempt at synthesis.

Late Chalcolithic domestic architecture and settlement layout at Tulaylat al-Ghassul

Most reviews of Ghassulian architecture (Lee 1973; Mazar 1990; Gonen 1992; Porath 1992) seem agreed on a number of points. Houses are rectilinear in form, and simply constructed of handformed, sun-dried bricks, sometimes set upon a single course of undressed fieldstone foundations, or otherwise built directly on to prepared, rammed-clay surfaces. Dwellings were flat roofed and constructed from wood beams, reed battens and mud plaster, with no second storey apparent. Walls were carefully plastered (often many times over), and not uncommonly painted, mostly in a monochrome buff but occasionally in brilliant polychrome. As generally envisaged, the basic architectural unit consisted of a roofed broadroom dwelling (5 × 5 m) with an attached (10 × 5 m) open or partly roofed courtyard, together forming one unit on average 15 by 5 m in extent. The roofed dwelling was thought to have been used primarily for sleeping, with the majority of food preparation, cooking and craft activities taking place in the courtyards, in which the majority of hearths and storage pits were generally located. These simple housing units (assumed to be the dwellings of nuclear families) were arranged in irregular clumps around larger open more public spaces ('plazas'), which were linked together by narrow winding alleyways. The implication of this traditional account of architectural form and settlement plan is that there was little or no architectural differentiation, reflecting the conditions existing in a large but undifferentiated farming village.

A glance at Mallon's Tulayl 1 settlement plan (Mallon *et al.* 1934) would seem to reinforce this impression of an undifferentiated village (Figure 4.1). However, a close examination of plans and photographs of individual dwellings suggests variation not apparent from a consideration of the 'sanitised' settlement top-plan alone. There is considerable variation in the size, shape and elaboration of construction in individual dwellings, and in the size and layout of attached courtyards. Some of the units appear quite substantial, particularly those with stonewall footings. These are often equipped with stone and plaster-lined storage pits, brick benches, free-standing mudbrick bins, and centrally positioned, large, flat-stone anvils. Other structures appear relatively small and undistinguished in their fittings. Further support for this suggested architectural differentiation is drawn from a consideration of Koeppel's little studied top-plans from the better-documented excavations on Tulayl 3 (Koeppel *et al.* 1940).

Even the most cursory view of the Tulayl 3 Phase IVA/B top-plans illustrates the substantial variation in unit type, with several large, multi-room complexes associated with numerous one and two room storage units (Koeppel *et al.* 1940; Lee 1973). The multi-room complexes often have brick-built, free-standing bins, stone-lined storage pits and large ceramic pithoi placed along the interior walls and in the corners of generally trapezoidal courtyards, or within the simple 'row house' storage and work rooms, these latter arranged in lines and opening into the courtyards. A number of the larger 'plazas' have small structures at their centre, perhaps for the feeding of livestock. Nowhere can one discern the idealized simple courtyard units. Large (perhaps agglomerative), multi-room complexes dominate the Tulayl 3 settlement plan, with variability in the size and elaboration of individual-dwelling units, courtyard size and storage capacity. Overall, the impression is one of considerable variability, certainly not consistent with the traditional view of standardized Chalcolithic architectural form and settlement plan.

The generally accepted view that elaborate figured polychrome wall paintings do not correlate with any artefactual differentiation should be reconsidered (Figures 4.2, 4.3). Whenever it is possible to determine context with some accuracy, both building form and artifact types do suggest a functional and qualitative difference associated with the presence of wall paintings (Koeppel *et al.* 1940: 82; Lee 1973: 339; Elliot 1977; Cameron 1981), whether this be measured by elaboration of construction, quantity and quality of artifacts, storage capacity or number of activity areas (Lee 1973; Elliot 1977; Bourke 1998a, 1998b). All else being equal, the presence of wall paintings probably does imply a special function for the structure in question, with cultic-ritual use probable (Joffe 1998; Garfinkel 1998). It should be remembered that Hennessy did find large quantities of polychrome wall painting fragments in association with his Sanctuary buildings, although these were not recoverable.

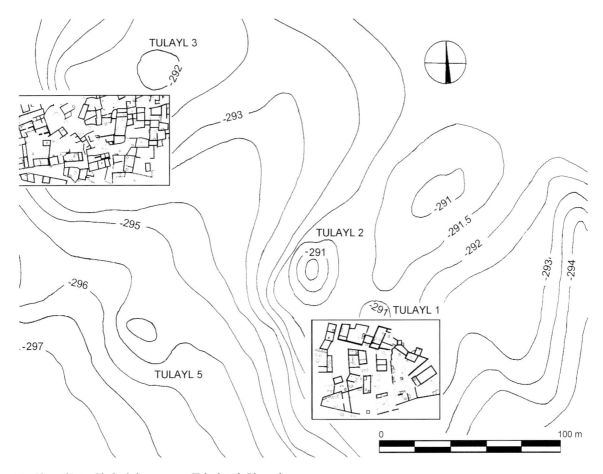

Figure 4.1. Plan of Late Chalcolithic areas at Tulaylat al-Ghassul.

Current excavations at Ghassul (1994–1997): the Area H complex

Current investigations on the westernmost hillock of Tulayl 6 (Area H) have explored an area approximately 30 × 15 m in extent (H II–IV), and fully excavated two phases of a Late Chalcolithic broadroom dwelling (5 × 7 m) and courtyard (13 × 7 m) complex (Bourke et al. 1995, 1999a; Bourke 1997a). The two-room complex is well constructed of handmade bun-shaped bricks, built on a single course of fieldstone foundations. Interior walls are thickly plastered and contain fragments of polychrome on white-painted plaster. Small plaster-lined pits exist in two corners, and a collection of stone tools (tabular scrapers and flint chisels) were recovered from a third corner. Rammed-clay pathways ran around the exterior north and west walls of the dwelling, and a doorway opened directly on to the western 'street', while a second opened into the northwestern corner of the attached courtyard. The courtyard was probably roofed along its northern end and features a large, centrally placed hearth, a number of stone and brick bins built against the western courtyard wall, a number of shattered storage vessels placed in small depressions against the eastern wall, a paved (and perhaps roofed) work area in the southeast corner, and an exterior doorway in the southwestern corner make up the main features of the complex. Outside this entrance, more disturbed features indicate the presence of a series of stone and brick fodder bins and a moderately sized, shallow plaster-lined water trough, set around the north and western edges of a small pebble-paved 'plaza'. The cobbled surface of the 'plaza' leads up to the courtyard entrance, and runs south to an adjoining east/west alleyway, which runs along the south wall of the complex, separating this 'unit' from others to the south and west. The courtyard area seems to have been the main work and storage area. A collection of tabular scrapers were found together, along with several chisels, small hammerstones, and what may be resharpening debitage, all in a discrete area beside the northern wall. Open

Figure 4.2. The 'Star frieze' wall painting from Tulaylat al-Ghassul.

Figure 4.3. Wall painting from Tulaylat al-Ghassul, showing three masked figures in procession.

areas immediately beside the central stone-lined hearth contain a rich collection of animal bones, including all major domesticates, but with sheep outnumbering goats by seven to one. Botanical samples hint at the storage of cereals and flax in the northeastern area and fodder crops in the south and in the exterior 'plaza' bins. Overall, the impression is one of a well-designed, solidly built and well-maintained, multi-purpose farmhouse, with discrete-sheltered, interior work and storage areas and external, stock-feeding facilities, all linked together by well-maintained cobbled or rammed-earth pathways. Current excavations in the northern (Area G), central (Area N), southern (Area A) and eastern (Area Q) areas of the site further reinforce this impression of well-ordered, multi-purpose dwellings, displaying non-standardized ground plans, a variety of construction techniques and numerous, special-purpose work and storage areas (Bourke et al. 1999a).

Late Chalcolithic domestic architecture at Ghassul: summary

This review of Late Chalcolithic architectural evidence from Tulaylat al-Ghassul reveals a rather different picture to that generally portrayed. While much architecture is rectilinear, curved walls, 'corrals' and windbreaks are known. Rectangular, trapezoidal and triangular courtyards are all found together. Single, double and multi-room complexes are all known, with various combinations of small, medium and large rooms. Some floor surfaces and passageways are cobbled, but most employ prepared rammed clay surfaces for interiors, and less-prepared, compacted surfaces for alleyways. The majority of floors occur at approximately the same level as exterior surfaces, but some are semi-subterranean, and these are generally cobbled. Distinct activity and storage areas are demonstrable from both the material cultural and environmental data. Thick mudplaster was widely employed to seal walls and roofs, line pits and bins. It was mostly unpainted, but occasionally walls were painted in brilliant mineral polychromes, often displaying an elaborate complex symbolism, much complicated by multiple overpainting and differential wear.

Early Chalcolithic domestic architecture at Tulaylat al-Ghassul

Far less is known about the architecture of the early phases at Tulaylat al-Ghassul. Although a large area of Tulayl 3 was excavated into Stratum III layers, little coherent architecture or settlement layout can be recovered from the fragmentary plans produced by Koeppel and Lee (Koeppel 1938; Lee 1973). Hennessy excavated about half of two 5 × 7 m trenches (A II and A III) from the early Late Chalcolithic levels reached at the end of the 1967 season (Hennessy 1969), down to sterile sand by the end of his second 1977 season (Hennessy 1982; Bourke 1997b; Lovell 1999). In these two trenches, he uncovered four coherent phases of multi-room, rectilinear-mudbrick and stone architecture. These were associated with plastered pits and bins, distinct activity areas (including a complete flint-knapping and/or food preparation area), one with an elaborate polychrome wall painting. Only with the earliest two phases (Hennessy's Late Neolithic Phases H and I) does the architecture change to single-room, semi-subterranean round dwellings.

The current excavations have explored two small areas of Early Chalcolithic architecture (Bourke 1997b; Bourke et al. 1999a). In 5 × 5 m trench G II, a neatly constructed complex of storage bins (two rows of 2 × 1 m brick bins) was uncovered. They were built against the northeastern corner of a large room, or perhaps a roofed courtyard. One row of three bins contained quantities of flax, legume and cereal residues respectively. Two in the second row produced immature pig teeth (perhaps suggesting use as farrowing pens), and the last held a quantity of beads and a collection of bone and stone awls (Bourke et al. 1999a). Excavations in 5 × 5 m trench A XI (the eastern half of Hennessy's original trench A I), uncovered two phases of multi-room, rectilinear architecture equivalent to Hennessy Phase E. Along with portions of two well-constructed mudbrick rooms (one in the northwest, the other in the southeast corner), numerous thickly plastered firepits in the southwest corner were found to contain a rich assortment of cereal and fruit remains.

Current excavations have explored meaningful areas of two main phases of rectilinear architecture below the 'classic' Late Chalcolithic Ghassulian horizon (Bourke 1997b; Bourke et al. 1999a). Observations match those from Hennessy's deep probes (Hennessy 1977, 1982, 1989). Carefully constructed rectilinear, multi-room (probably multi-purpose) units seem to be a feature of the architecture of Ghassul from the beginning of the Chalcolithic period.

Chalcolithic domestic architecture at other sites

Abu Hamid

Over 2000 m^2 of Late Chalcolithic architecture has been uncovered in two main excavation fields at Abu Hamid (Dollfus and Kafafi 1988, 1989, 1993) (Figure 4.4). The main architectural themes reflect those at Ghassul. Multi-room, rectilinear architecture, built of handmade mudbricks, set on a single course of fieldstone foundations, are most common. The settlement seems well laid out, with considerable distances (large 'plazas') existing between dwelling units.

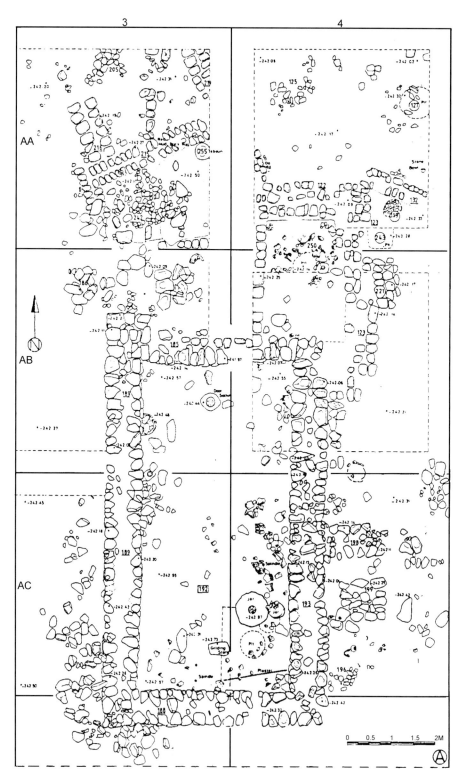

Figure 4.4. Chalcolithic building at Abu Hamid.

Most units display the 'dwelling and courtyard' design, although rows of contiguous smaller rooms become more common over time, and are associated with the larger open 'plazas'. Most work and bulk storage areas are located in the enclosed courtyard areas or small 'row houses' associated with them. There does seem to be differentiation between complexes, with more substantial 'longhouse' units associated with lines of much smaller 'row houses', as at Ghassul. The famous very large storage jar/grain silo (Dollfus and Kafafi 1988: 46) was found in a small, semi-subterranean 'row house' associated with one of the more substantial complexes. Other small, 'row house' rooms are associated with storage jars, food preparation and specialized flint working (Dollfus and Kafafi 1993: 244-45). Thick plaster-lined bins and pits are common, and distinct work areas have been detected within structures. Overall, this seems a very similar settlement landscape to that reflected in Koeppel's Level IV plans from the Tulayl 3 excavations at Ghassul. Details of internal layout and function can be matched by the findings of the more recent Sydney University work at Ghassul.

The appropriate stratigraphic relationship of the Abu Hamid Early Chalcolithic 'Middle Phase' architecture to the Tulaylat al-Ghassul sequence remains unclear, although association with Ghassul Phase D/G assemblages is likely. The information gained from the relatively large-scale 'Middle Phase' excavations is uniquely important, as it provides something of a backdrop against which the Late Chalcolithic norms at both sites may be viewed. The main 'Middle Phase' variations would seem to be less evidence for plaza 'row houses' and more for 'stand-alone' multi-cellular complexes, organized in a more dispersed built environment. The 'Middle Phase' complexes are built of well-formed, plano-convex brick superstructures set on fieldstone foundations. Plans consist of large, open-rectangular living rooms with defined work areas, linked by narrow passageways to smaller rectangular rooms associated with storage or special activities. Mudbrick benches, partition walls, plastered stone-lined subterranean pits and above-ground built storage bins, plastered polychrome painted walls and plastered pillar bases, pebble-paved zones and semi-circular boundary walls all feature in architectural units that were constantly remodelled, built over and subdivided as needs altered. As well, there is the suggestion that special-purpose industrial areas exist in the north and east of the southern sector in 'Middle Phase' times.

The existence of stand-alone, multi-room, multi-purpose complexes in 'Middle Phase' times (c. 4500 BC) would seem to imply at least some tendency towards diversified economic activity, centralized redistribution of agricultural products and craft specialization. Comparison of Middle and Upper Phase settlements would suggest a strengthening of this tendency over time, with more provision for centralized grain storage, more easily identified (and more numerous), activity areas and more distinctive-craft specialization, all features of the Upper Phase (c. 4000 BC) assemblages.

Pella

Late Chalcolithic architecture has been exposed in two main locations at Pella: (1) an isolated farmstead on the western slopes of the Jabal Sartaba hillside, about a kilometre southeast of the main settlement (Area XIV); and (2) a series of wall lines and silos set around a large stone and mudbrick platform located on the central-south side of the main mound of Khirbat Fahl (Area XXXII) (Figure 4.5). The first settlement (Area XIV) seems likely to have been the site of a dispersed agricultural complex, with at least three distinct units, one (at least) of which was strongly geared towards the production and storage of olives as fruit and oil. Very shallow occupation deposits and uneven, sharply sloping bedrock resulted in the recovery of fragmentary plans and virtually no superstructure, but enough was preserved to reveal the familiar rectilinear multi-room, enclosed courtyard ground plan, featuring small 'row houses' given over to storage and special activity areas (olive processing and storage), with many stone-lined pits, plaster-lined bins and silos, and stone and mudbrick benches located in the courtyards. Material culture and radiometric dating suggests a Late Chalcolithic date. Hanbury-Tenison drew parallels between the Jabal Sartaba ground plans and the upland 'row house' layouts of the Golani Chalcolithic (McNicoll et al. 1992: 24) and those discovered in his survey work east of Jarash (Hanbury-Tenison 1986).

On the main mound of Khirbat Fahl, approximately 200 m^2 of a multi-use area has been exposed, dominated by a massive, grain storage complex (Bourke 1997c; Bourke et al. 1998). The central southern area of the exposure (trench XXXIID, 5 × 5 m in extent) contains a network of six (two rows of three) storage silos. Each is 1 m wide and 1–2 m deep. Several are thickly plastered and three are ceramic-lined. They were all enclosed within a partly roofed courtyard. Immediately to the northeast is a rectangular stone and mudbrick platform some 2.5 m thick, raised approximately 0.5 m above the surrounding surfaces, and exposed over 5 m of its length (the rest is obscured by a Bronze Age temple). The structure was originally thought to be a major wall, but is now viewed as the stone foundations and mudbrick flooring of the east end of a massive,

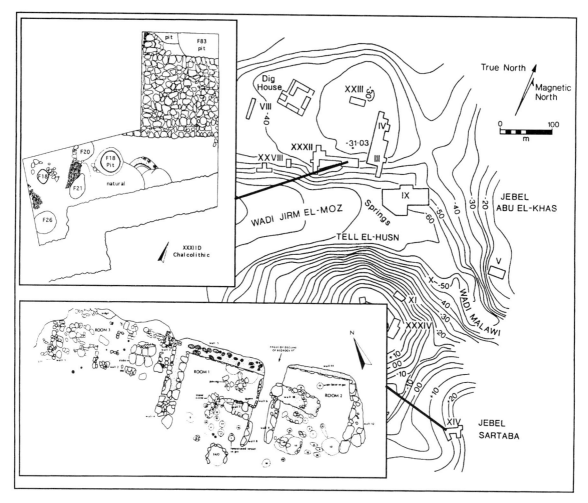

Figure 4.5. Chalcolithic architecture at Pella.

raised grain storage facility. Excavations in the area of the platform recovered a massive burnt deposit of pre-sorted (99% clean) wheat-grain product (Bourke *et al.* 1999b). The platform storage complex also featured areas of neat, stone slab paving (along the interior north face of the platform), stone stairs (built into the eastern end), stretches of pebble paving (around the plaster-lined silos), neatly constructed small rectilinear rooms (to the west), a gently curving, courtyard border wall (to the east), and numbers of small, thickly-plastered bins and pits (to the south). All of this material dates to the Late Chalcolithic period and is to be compared with the classic Ghassulian phases at Tulaylat al-Ghassul, the Upper Phase assemblages at Abu Hamid, and the 'Ghassulian-related' assemblage from Neve Ur. Overall, the Late Chalcolithic evidence from Pella indicates intensive cultivation of the olive and centralized storage of grain, in a prosperous, well-regulated, Late Chalcolithic settlement.

The Early Chalcolithic presence at Pella is less well known. A limited sounding (5 × 7 m) in the western edge of the Area XXXII exposure was taken down to bedrock during the 1997 field season (Bourke *et al.* 1999b). The majority of the Early Chalcolithic exposure is dominated by a courtyard, much disturbed by multi-phase plaster-lined storage pits. It features several stone and mudbrick bins built against the northern courtyard wall, which has an entrance in its centre. Two stone-lined silos fill the centre of the courtyard, surrounded by several smaller, relatively shallow, plaster-lined bins, areas of hard, orangey-yellow, plastered floor and at least two plaster-lined, flat-circular, pillar supports, perhaps suggesting that at least the northern half of the courtyard was roofed. Associated fragmentary architecture in the north and west consists of small mudbrick, rectilinear multi-cellular structures, perhaps used for food storage (in the west) and preparation (in the north). Little more can be added at

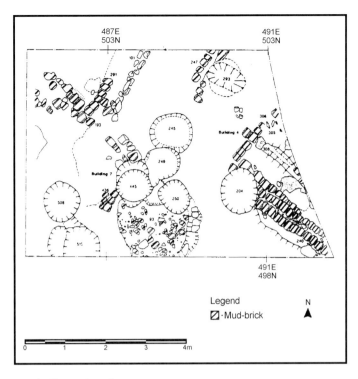

Figure 4.6. Chalcolithic buildings at ash-Shuna, Area D.

present as much material remains to be processed, but the assemblage has more affinities with the 'Northern Chalcolithic' assemblages at Tall al-Far'a North and ash-Shuna than with the Early Chalcolithic assemblages at Abu Hamid and Tulaylat al-Ghassul.

Ash-Shuna

Three expeditions (de Contensen/Mellaart, Gustavson-Gaube, Baird and Philip) have excavated Chalcolithic period remains at ash-Shuna and together they have exposed approximately 100 m^2 of pre-Bronze Age deposits. The recent detailed investigations by Gustavson-Gaube (Area E 1984–85) and Baird (Area D 1991–94), working on the western edge of the central tell, close by one another, uncovered approximately 85 m^2 of a tightly packed, rectilinear, multi-cellular complex of large and small rooms built around a roughly square courtyard (Baird and Philip 1994) (Figure 4.6). Structures were built from hand-formed mudbricks laid on fieldstone foundations, with beaten earth floors. Plaster-lined pits, semi-circular brick silos and patches of cobbled surface, several associated with discrete storage or activity areas, are positioned within the courtyard. The complex was rebuilt several times, with walls and installations added as required, and seems to have been in use for some considerable period of time.

Appropriate comparative stratigraphy is unclear at present, but the presence of 'Tsafian' Lattice Painted pottery (Leonard 1992; Baird and Philip 1994: 123-24) within the assemblage suggests an Early Chalcolithic date. Architectural layout and assemblages from Gustavson-Gaube and Baird's 'Shuna Chalcolithic' are broadly consistent with the Abu Hamid 'Middle Phase' assemblages, and quite similar to the recently excavated Early Chalcolithic material from Pella. Most importantly, the Shuna Early Chalcolithic seems likely to be the first reliably excavated exemplar of de Vaux's 'Northern Chalcolithic' assemblage east of the Jordan River. Many isolated, generally poorly stratified assemblage 'scraps' have been assigned to this 'Northern Chalcolithic' culture over the years (de Vaux 1966), but hitherto the combination of poor description and uncertain context robbed these observations of much significance, even though the likely status as indigenous successor to local Ceramic Neolithic cultures had long been recognized (Kaplan 1968; Perrot 1968). Baird's meticulous investigations give promise of the first reliable description and evaluation of this potentially widespread, Early Chalcolithic,

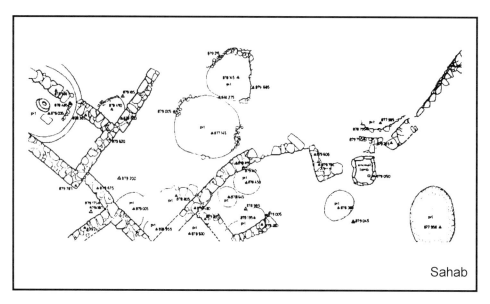

Figure 4.7. Chalcolithic buildings at Sahab.

pre-Ghassulian culture (Baird 1987; Baird and Philip 1994).

The central uplands: Sahab and Snesleh

Ibrahim's pioneering investigations at Sahab in the early 1970s provided the first reliable excavated sequence of the upland Late Chalcolithic culture, which was suggested as deriving from earlier Chalcolithic Jordan Valley exemplars, with some Damascene elements (Ibrahim 1984: 257-58; 1987: 73-75). Approximately 150 m^2 of Late Chalcolithic architecture was exposed in Ibrahim's Area E excavations (Figure 4.7). Architectural form and constructional technique is close to Ghassulian exemplars (Ibrahim 1987: 73). Constructions are primarily of hand-formed mudbrick walls set on fieldstone foundations. Layout displays the familiar central courtyard with associated-rectilinear, small 'row house' flanking rooms, not unlike Golani and Jabal Sartaba layouts (Hanbury-Tenison 1986). Several of the small rooms are closely packed with storage pits, and a number of very large stone-lined storage silos are located centrally in the courtyard. This provides reasonable evidence for centralized control of what seems to have been a considerable agricultural surplus. The Area E complex was modified a number of times, suggesting some time depth to the occupation. Ibrahim believes that the layout of the complex implies occupation by an extended family unit (1987: 73). The material assemblage has links to the late phases of the classic Ghassulian (Hennessy Phase A-D) and Abu Hamid 'Upper Phase' assemblages.

Closely associated with survey work in the region around Sahab southeast of 'Amman, the more recent Abu Snesleh excavations of the early 1990s (Lehmann et al. 1991; Kerner et al. 1992) have uncovered approximately 100 m^2 of a similarly dated Late Chalcolithic settlement (Figure 4.8). Fragmentary remains of a rectangular broadroom structure were recovered. Construction utilizes quite large, roughly square stone blocks, but otherwise suggest a very similar technique to that employed at Sahab. The rectilinear broadroom, associated courtyard and centrally positioned storage installation suggest a similar layout (and function) as well. Overall, the architectural layout and material cultural homogeneity of the central upland region in the Late Chalcolithic period favours the existence of a thinly spread (Gustavson-Gaube 1986; Bernbeck et al. 1995) system of extended family-run farmsteads in the countryside surrounding the larger central settlement of Sahab.

The contrasting south: Jabal al-Jill and Tall Magass

Henry's investigations in the Hisma Basin have documented a number of the semi-permanent campsites of the transhumant pastoralists (Henry 1995: 356) who

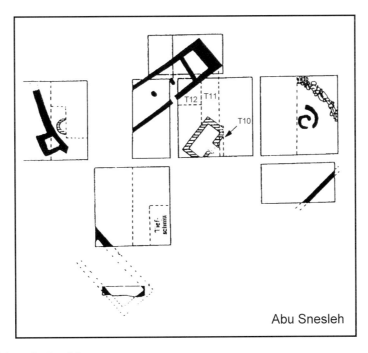

Figure 4.8. Layout of Chalcolithic Abu Snesleh.

occupied the region during the fourth millennium BC. The architectural form and layout differs sharply from that of the permanent farming communities on the plateau. Henry divides the transhumant pastoral sites of the Judayid Basin into two main categories, ephemeral camps and 'long-term' camps. Architectural remains of the latter at Jabal al-Jill (J14) (Figure 4.9), and less coherent traces at Jabal Queisa (J24), consist of loosely associated groups of circular structures stretching over an excavated area of approximately 200 m². Constructions feature semi-subterranean, stone-lined, single-room roundhouses, which contain stone-lined hearths, storage pits and discrete work and food preparation areas (Henry 1995: 358-61). They are associated with large circular 'corrals', smaller circular storage facilities and patches of pebble pavement. Individual architectural form and settlement layout contrast sharply with the architecture of the permanent farming villages of the north plateau and the south coast. This may be attributed in some degree to variant subsistence regime and cultural associations with either the southwestern Timnian complex (Henry 1995; Rothenberg 1999) or southeastern Hijazi cultures.

Khalil's recently renewed excavations at the south coastal site of Tall Magass, now 4 km north of al-'Aqaba but perhaps closer to the shore in antiquity, have uncovered approximately 150 m² of a four-phase (I–IV) agricultural/industrial settlement in two distinct fields (A/C and B) of excavation (Khalil 1987, 1992, 1995) (Figure 4.10). The architectural technique and layout are typical of the permanent settlements of the north/central uplands and sharply in contrast with the transhumant camps immediately to the north and east. Layout duplicates the familiar plan of central courtyard with multiple, tightly packed clusters of associated small rooms, storage bins and work areas. Fieldstone foundations, hand-formed mudbricks, stone and plaster-lined storage pits, large mudbrick and stone storage silos, and what may well be ceramic firing pits and small-scale, copper-smelting facilities, all feature in a picture of densely packed industrial (?) activity areas, storage facilities and domestic food preparation areas (Khalil 1995). Some structures include numbers of small storage bins, very similar in form to those recently uncovered in Tulaylat al-Ghassul Area G II (Bourke et al. 1999a).

The south Jordanian material assemblage is still little known (Genz 1997) and the relative chronology of the Magass 'culture' assemblage remains controversial (Hanbury-Tenison 1986; Kerner 1998). However, it seems likely that at least some of the ill-defined and diverse range of domestic, storage and industrial/manufacturing facilities at Magass date to the Late Chalcolithic period, although what proportion is currently unclear. The overall pattern revealed by Khalil's excavations is one of a permanent mixed farming and small industrial settlement, set in a hinterland dominated by semi-permanent pastoral camps

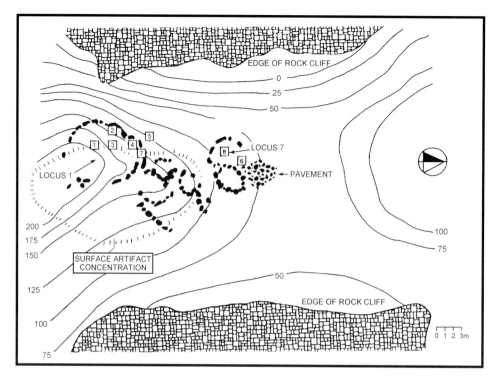

Figure 4.9. Chalcolithic architecture at Jabal al-Jill (J14).

(Jobling 1989: 16-24), although Khalil's soundings at Hujayrat Ghuzlan (Khalil 1995), slightly over 1 km farther east up Wadi al-Yutum, suggest that at least some of the hinterland settlements are agricultural villages exploiting the winter wadi floodwaters.

Cultic architecture case study: the Tulaylat al-Ghassul sanctuary area

Introduction

Only one building complex in Jordan has ever been identified as having been purpose-built for Chalcolithic cultic activities, and that is the Tulayl 5 (Hennessy Area E) Sanctuary Complex at Tulaylat al-Ghassul. Over the course of three field seasons (1975–77), Hennessy opened 16 trenches in and around the Sanctuary Complex, fully excavating the two nominated sanctuary buildings (Sanctuaries A and B) (Figure 4.11), stretches of the northern and western Temenos Enclosure wall, and a 10 × 5 m area 15 m to the north of the temenos wall (Hennessy 1982, 1989). The current Sydney University expedition has carried out three further seasons (1994–97) of renewed investigations, opening three additional trenches in and around the Sanctuary Complex, for the most part in areas unexplored by Hennessy. A major structure linked to Sanctuary A (Installation C), a shallow gateway in the northwest corner of the Stone Temenos phase wall, and an earlier phase Mudbrick Temenos wall were added to Hennessy's discoveries, and a number of internal phasing issues clarified (Bourke et al. 1999a).

Architectural form and cult practice: some contextual remarks

Debate over the form and variety of Chalcolithic cult practice has been dominated by both architecture and assemblages drawn from the Negev (Gilat/Shiqmim) and the west Dead Sea (En Gedi and Nahal Mishmar). Initial suggestions that the Ghassul wall paintings were indicators of cultic activity were hamstrung by Koeppel's (1940) insistence that associated architectural contexts were unexceptional and associated assemblages standard (although fine). North (1961) suggested that Koeppel's apparently authoritative denial of unusual architectural context for the famous wall paintings had more to do with poor attention to detail than any clear determination of standardized architectural forms. North made a case for an unusual architectural context

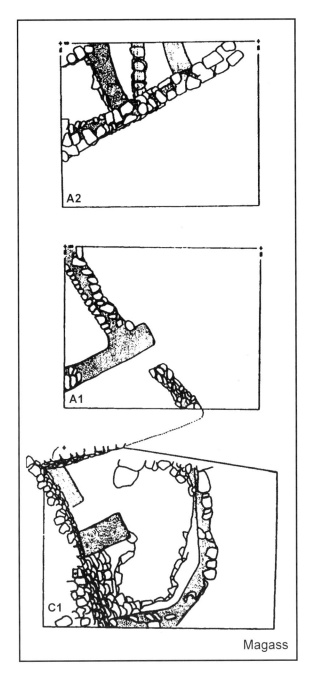

Figure 4.10. Chalcolithic architecture at Tall Magass.

for his Tiger Fresco, suggesting that it be associated with a semi-subterranean, stone-lined 'throne' feature (North 1961: 6-16). Although North's observations may still be valid, a close consideration of the published photographs make it more likely that his two linked features (wall painting and stone-lined silo) are associated with unrelated superimposed structures (Blackham 1999).

Nonetheless, there are distinct clusters of what may reasonably be interpreted as cultic or ritual paintings in separate building complexes on Tulayl 3 (Koeppel *et al.* 1940; Elliot 1977: 19-20), and on Tulayl 1 (Hennessy 1978, 1989; Cameron 1981). The most spectacular of these is the multi-roomed, broadroom complex (Building 78) uncovered by Mallon in Square B 7 in the centre of Tulayl 3 (Mallon *et al.* 1934). It contained the famous 'Bird' and 'Notables' paintings in a context including a built altar, offering pits and a small storage room (Room 81) off to the east. Most importantly, this last contained over 30 broken cornet cups (Koeppel *et al.* 1940; Cameron 1995). The recent association of fourth-millennium alcohol consumption, cult practice, developing social stratification and social control is of particular relevance here (Joffe 1998). Further support for a cultic function is drawn from the subject matter of the 'Notables' painting, which illustrate an explicitly ritual and hierarchical event (Stager 1992). An associated, semi-circular, horseshoe-shaped structure (Installation 75a) is positioned in the centre of an enclosed courtyard (Area 77) to the west of Building 78. This layout of enclosed courtyard and associated multi-room complex remained largely unaltered from earlier Stratum II/III times. Given the central role of a large star in the subject matter of the 'Notables' painting, it seems probable that the building in which the famous 'Star' painting was discovered (Building 10) had a cultic function as well, although much of this latter complex lay north of the Pontifical Biblical Institute's excavation area and was not investigated. A third, and rather more tentative cultic complex, may be associated with broadroom Buildings 51 and 53, stretches of exterior paving, and perhaps the semi-circular Installation 44f.

Hennessy's 'Processional' painting (Cameron 1981) was discovered in 1975, during the excavation of a multi-roomed broadroom and courtyard complex not unlike Mallon's Building 78. Hennessy's Area A broadroom structure is, however, less well preserved and only partly excavated. However, the discovery of a cornet cup deposit within the structure strengthens the link between ritual drinking and/or libation offerings and cult practice (Hennessy 1977; Bourke 1998b). The architectural features loosely associated with the Pontifical Biblical Institute's wall paintings—a central, internally subdivided broadroom with ancillary storage chambers, exterior paved areas, enclosed courtyards with semi-circular offering installations—were sufficiently irregular in arrangement to prevent one from nominating a cult architectural template with any confidence. This changed with Hennessy's discoveries in Area E (Hennessy 1982) and the full publication of

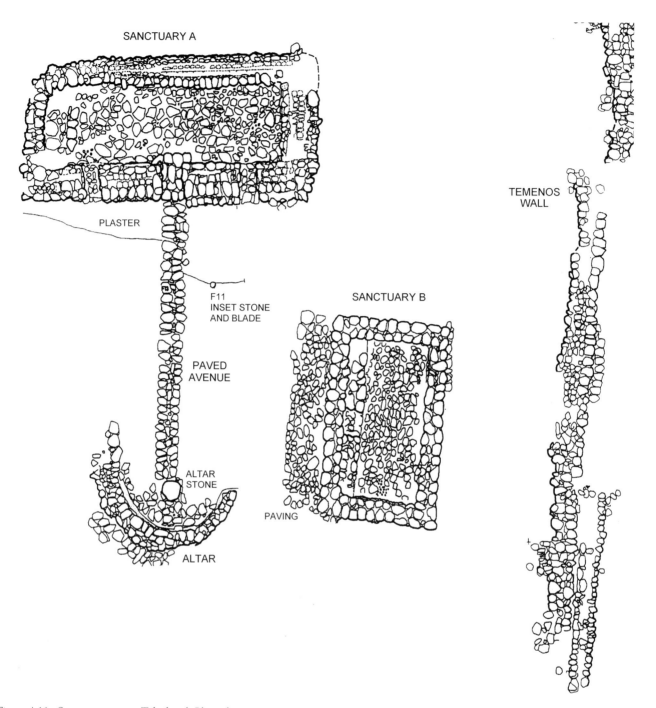

Figure 4.11. Sanctuary area at Tulaylat al-Ghassul.

the previously excavated En Gedi sanctuary (Ussishkin 1980). This made it possible to define just such a template—central broadrooms, enclosed courtyard with semi-circular offering installation (Ottosson 1980; Hennessy 1982, 1989). It is possible that the more formalized arrangement of the Area E sanctuary complex may reflect its later date and architectural codification over time.

Conclusions

Cultic installations with a variety of architectural forms coexisted within the Late Chalcolithic, Ghassulian-cultural complex. These variant architectural contexts may point to differing cult practice (Amiran 1989; Fox 1997), although the quite similar cultic assemblages (anthropomorphic and zoomorphic figurines and vessels, fenestrated stands, mortars and pestles and numerous cornet cups) from En Gedi, Gilat and Ghassul would imply an essentially similar cult practice at each site. Minor variations may have a chronological explanation and may be due to later codification of earlier, less-formalized practice (Bourke 1998b). Equally, changing economic (Joffe 1998) or environmental (Prag 1995; Hole 1997) conditions may have elicited different responses from autonomous religious communities as circumstances changed.

Material remains

Ceramics

It is not yet possible to provide an adequate description of ceramic assemblages spanning the entire Chalcolithic period. Much is known about the latest Chalcolithic assemblages, generally (but not always) to be related in various degrees of association with the familiar, classic-Ghassulian assemblage (Mallon et al. 1934; Koeppel et al. 1940; Lee 1973; Elliot 1978), recently summarized by form (Gonen 1992) and chronological positioning (Stager 1992). Early Chalcolithic assemblages are far less well known at present, although recent work at Ghassul (Lovell 1999), Abu Hamid (Lovell et al. 1997), ash-Shuna ash-Shamaliyya and Pella (Lovell 2000) begins to redress the balance. The present summary will confine itself to highlighting major trends visible at this stage in what are very much ongoing analyses, and comment on the potential significance of variations over time and space (Figures 4.12–4.16).

The classic-Ghassulian, Late Chalcolithic assemblage is dominated numerically by medium-sized holemouth or short-necked storage jars and hole-mouth cooking pots. Distinctive forms, such as cornet cups, fenestrated stands, necked churns, ceramic spoons and very large storage containers are not unknown in earlier assemblages (Garfinkel 1992b; Lovell 1999), but become more common as time progresses. Most ceramic classes can be divided into forms associated with the two basic activities of food preparation/consumption and storage. Large-scale feasting and other forms of conspicuous consumption (Joffe 1998), and the centralized bulk storage and redistribution of agricultural produce (Earle 1991), are recognized as important indicators of emerging stratified societies. Defining ceramic assemblages that may be associated with such activities provides an important line of evidence in ongoing analyses of Chalcolithic societal development. Identifying specialized types (large storage vessels) and mass-produced cooking (hole-mouth cooking pots) and feasting vessels (cornets) is one way to advance analysis. In the Early Chalcolithic assemblages (as with Late Neolithic predecessors), there are few ceramic indicators of such activity (Garfinkel 1992a). However, with the onset of the Late Chalcolithic horizons at Ghassul, Abu Hamid and Pella in the lowlands, and at Sahab in the highlands, such trends within the ceramic assemblages become more pronounced, and are reasonably to be interpreted as indicating developing social stratification.

Another indicator of developing specialized activity within the ceramics industry may be discerned in the changing manufacturing processes and expanding distribution networks evident over time. Early Chalcolithic assemblages display little evidence of standardized form/fabric repertoires, manufacturing techniques or distribution networks. However, recent work on the Late Chalcolithic Ghassul (Lovell 1999), Abu Hamid (Roux and Courty 1997) and Pella (Lovell 2000, pers. comm.) assemblages would suggest the development of distinctive form-specific manufacturing and firing (Edwards and Segnit 1984) techniques. While the incidence of such specialized potting activities within the ceramic assemblage as a whole remains relatively small, when coupled with limited chemical-analytical evidence from Pella and Tulaylat al-Ghassul (Bourke 1998a; Bourke et al. 1999a) for expanding regional distribution networks over time, the picture of a slowly diversifying ceramic industry may be developed without too much risk of overstatement.

The presence of unusual ceramic forms may also be an indicator of specialized activity. Obvious examples of such material are the extraordinary anthropomorphic and zoomorphic vessels from the Gilat and En Gedi sanctuaries. Similar vessels have been found within the sanctuary complex at Ghassul (Hennessy 1989; Seaton 1999; Bourke et al. 1999a), and Abu Hamid (Dollfus and Kafafi 1995). Such vessels indicate distinctive cult practice (Amiran 1989; Fox 1997) as well as distinctive potting traditions.

Most variation between Chalcolithic ceramic assemblages is confined to decorative regimes. Surface decoration varies widely both temporally and spatially (Lovell 1999) and is often employed to differentiate assemblages chrono-spatially. While it is likely that culturally significant information is encoded

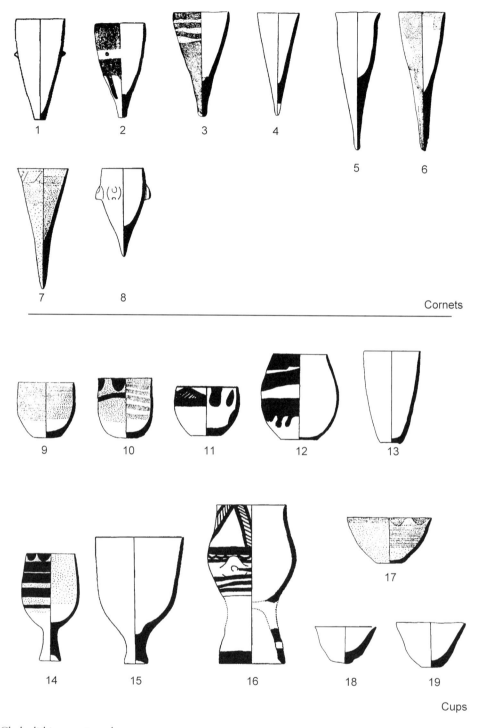

Figure 4.12. Late Chalcolithic cornets and cups.

within distinct decorative regimes, undue emphasis should not be given to minor ceramic variations between cultures (Mellaart 1975) as ceramic variation need not translate into broader cultural differences.

Stone tools

Earlier Chalcolithic lithic assemblages are poorly defined, although current work on the long sequences from ash-Shuna, Pella, Abu Hamid and Tulaylat

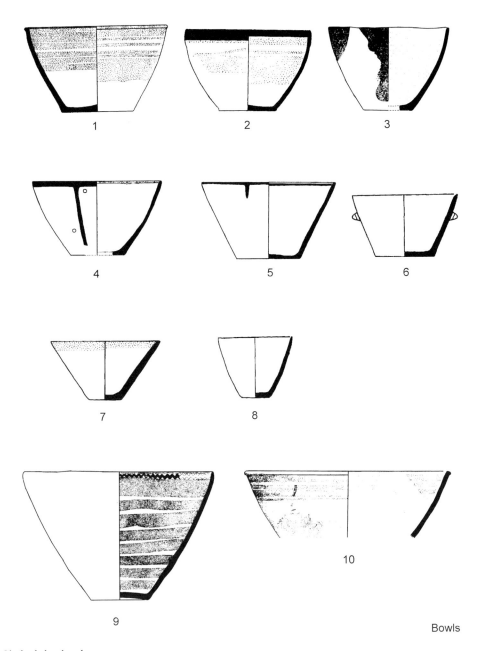

Figure 4.13. Late Chalcolithic bowls.

al-Ghassul will eventually produce tightly defined assemblages, which will fill a large technological gap between the relatively well-known Late Neolithic and Late Chalcolithic assemblages. The Late Chalcolithic Ghassulian assemblage is often characterized by the presence of distinctive tool forms, such as the chisel, the fan-scraper and the backed blade (Rosen 1997). Although related and probably ancestral forms of many typical Late Chalcolithic artifacts can be traced back to a variety of Neolithic precursors, they characterize, in combination, the Ghassulian culture effectively. However, the degree of relatedness of the various Late Chalcolithic lithic assemblages remains debatable. While we can successfully characterize the lithic assemblage of Late Chalcolithic Ghassul, for example, it remains unclear whether it is justified to generalize a 'Ghassulian' or 'Late Chalcolithic' lithic assemblage from a single, even if widely known assemblage. While we will highlight prominent elements of the Ghassul lithic assemblage below, contemporaneous regional variations in the type and frequency of individual elements (tabular scrapers, holed disks)

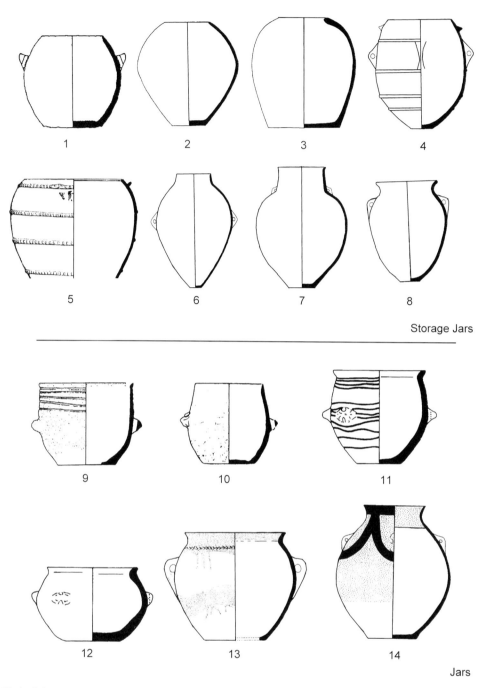

Figure 4.14. Late Chalcolithic storage jars and jars.

or manufacturing techniques (Baird 1987) remind us that it may well be unwise to extrapolate from any one assemblage to the Jordanian Chalcolithic in general.

At Tulaylat al-Ghassul, the generic chisel form includes a number of distinctive subclasses defined by the presence or absence of hafting, hafting format, edge shape, length and preparation and butt condition (Lee 1973: 247-50). Hennessy's first season at Ghassul produced a large lithic assemblage, but a very restricted sample of early phase material. Stockton's report on the 1967 material (Hennessy 1969: 17) should be regarded as a detailed statement on the Late Chalcolithic assemblage only, with comments on incidence by phase treated with caution. Habgood's detailed study of Ghassul chisels (Habgood 1983) is

The Chalcolithic Period

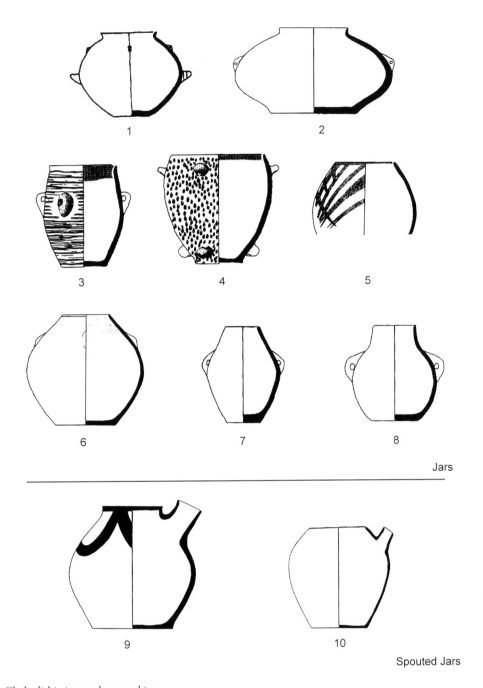

Figure 4.15. Late Chalcolithic jars and spouted jars.

based on Hennessy's 1975–77 Early Chalcolithic lithic assemblages as well as Stockton's 1967 study. Habgood concludes that chisels are present throughout the sequence and that they are both more abundant and more typologically variable in Late Chalcolithic times. As well, a tendency towards greater overall body length, reduced body thickness and increased blade width is observed during Late Chalcolithic times (Habgood 1983). Typological (Habgood 1983) and technological (Lee 1973: 250) studies are consistent with a variety of chopping, cutting and manufacturing uses.

Fan scrapers are another distinctive tool type found within the Late Chalcolithic tool kit (Lee 1973: 253-54). While rare in earlier Chalcolithic assemblages, they come to dominate later tool kits (Hennessy 1969: 17-18; Habgood 1983), often displaying distinct intra-site distributional patterning

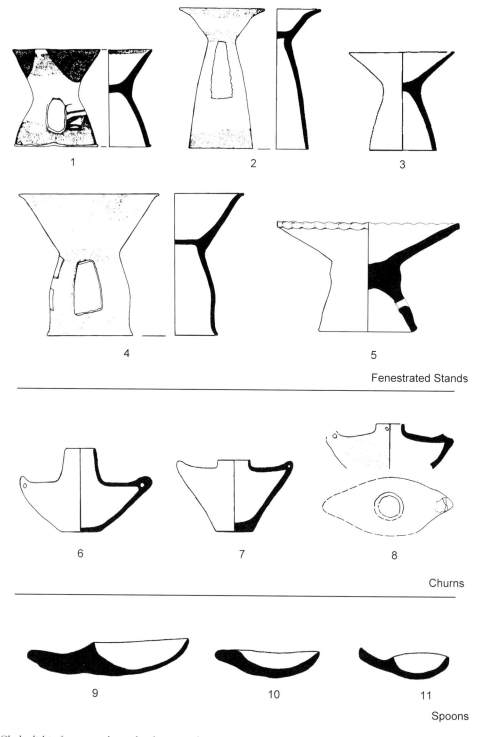

Figure 4.16. Late Chalcolithic fenestrated stands, churns and spoons.

(Bourke *et al.* 1999a). In south Levantine assemblages, they are made from a distinctive, brown-cortical flint, generally sourced to the western Negev (Rosen 1983: 80; 1989: 109; 1997: 75) and exported from there throughout the south/central Levant. Rosen's suggested geometric fall-off distribution is attractive (Rosen 1983: 80-81; Baird 1987: 476), although the recent Ghassul data reinforces earlier observations (Helms 1987) that Chalcolithic assemblages do not fit the logarithmic fall-off model nearly so well as subsequent Early Bronze Age assemblages do (Rosen 1997: 75, 105-106). Exploitation of additional raw material sources in southern Jordan (Hanbury-Tenison 1986; Henry 1995) may be a factor in the Late Chalcolithic, as may the changing nature of nomad/sedentary interaction over time. Both factors may account for the contrasting distribution patterns of Chalcolithic and Early Bronze Age scrapers. The fan scraper has a relatively limited typological variation (Rosen 1997: 74) and its function is likely to have been multi-purpose, with use as butchering knives (McConaughy 1979), hafted cutting tools (Lee 1973) and sheep shears (Henry 1995: 372-73).

The backed blade is another hallmark of the later Chalcolithic assemblages. Ghassulian blade-manufacturing techniques tended to produce small-to-medium prismatic blades, fashioned from single-platform, local-wadi cobbles (Rosen 1997: 45-48). Backing is usually a unifacial-abrupt, hard-hammer retouch, while working edge retouch is often very slight ('nibbling') to a light serration, when present at all. Retouch is general only on sickle blades, which commonly display silica sheen (Hennessy 1969: 18; Lee 1973: 250–51; Rosen 1997). Sickle blades tend to be between 5 and 8 cm long, rectangular in shape, truncated and reworked at each end, and triangular in section (Rosen 1997: 50).

There is some suggestion that the proto-Canaanean, blade-manufacturing technology begins towards the end of the Chalcolithic period (Hanbury-Tenison 1986: 146-50; Baird 1987: 476-78; Rowan and Levy 1994: 167-68), so that a limited presence of Canaanean technology is consistent with a Late Chalcolithic date. However, Canaanean technology remains rare in pre-Bronze Age assemblages (Rosen 1997: 50-55) and concentrated in the latest Chalcolithic horizons at ash-Shuna (Baird 1987: 478-79) and Magass (Hanbury-Tenison 1986: 147; Rowan and Levy 1994: 168).

The perforated disc and/or star-shaped tool is one of the more enigmatic Chalcolithic tool types. It is made from a large flake, similar in initial production regime to the tabular scraper. Shapes range from semi-circular to rounded, with a multi-pronged or star-shaped variant. The primary flake is often retouched bifacially to thin the body, and abruptly retouched to shape the blunt side of asymmetrical tools. Most feature a central hole pecked through the body from one side. Distribution is concentrated in southern Syria, northern Palestine and upland-north Jordan, with isolated examples found in northern Jordan Valley sites (Hanbury-Tenison 1986: 142-43; Epstein and Noy 1988; Rosen 1997: 85).

Most well-dated examples are concentrated in the latest phases of the Late Chalcolithic period. Function remains unclear, but suggestions range from the prosaic use as a cutting tool (Epstein and Noy 1988), a wool carding device (Nasrallah 1948) to the ceremonial/symbolic (Hanbury-Tenison 1986; Rosen 1997). The pronged or star-shaped disc may have functioned as an actual weapon or as a symbol of authority. Parallels with the famous Nahal Mishmar star-shaped copper standards (Bar-Adon 1980: 47-50), or the more familiar mace/standards with holed, disc-shaped tops (Bar-Adon 1980: 84-87), may well be relevant.

The bulk of the Chalcolithic lithic assemblage is made up of 'ad hoc' tools (notches and denticulates, awls and borers, side scrapers and retouched flakes) and debitage. Chronological and typological distinctiveness of such tool categories within Chalcolithic assemblages remains unclear. It is probable that larger flake tools, abruptly retouched steep side scrapers, end scrapers, burins and bladelet tools are better associated with early Chalcolithic assemblages, although this remains speculative (Hennessy 1969: 18; Baird 1987: 476-78; Rosen 1997: 67). The 'ad hoc' elements may represent less-specialized tool elements fashioned as needed for a range of tasks. At Tulaylat al-Ghassul, they show no particular spatial patterning, although bulk lithic clustering in one area (F II) may indicate warehousing of primary materials or a lithic 'workshop' area.

Groundstone

A feature of the Chalcolithic groundstone assemblage is the basalt bowl forms (Hanbury-Tenison 1986: 162-63; Gonen 1992: 57-58; Gilead 1995: 310-11), although other stone materials (phosphorite and limestone) are occasionally utilized. The most common products are the wide, flaring rim, flat-based V-shaped bowl form (Figure 4.17). These are often decorated with shallow-incised, diagonally filled, triangular motifs, pendent from the rim interior. A less common variation on this bowl form features the addition of a fenestrated foot or bar-ring stand, which together form a chalice-like vessel. These fenestrated vessels occasionally display elaborate, horizontal-ridged decoration at the body/base join, and cross-hatched light incision on leg exteriors. Many vessels (both simple and elaborate) are highly polished on interior surfaces.

A second distinctive product of the Chalcolithic groundstone industry is the basalt-sculptured, figurative vessels (Figure 4.18). These can be both

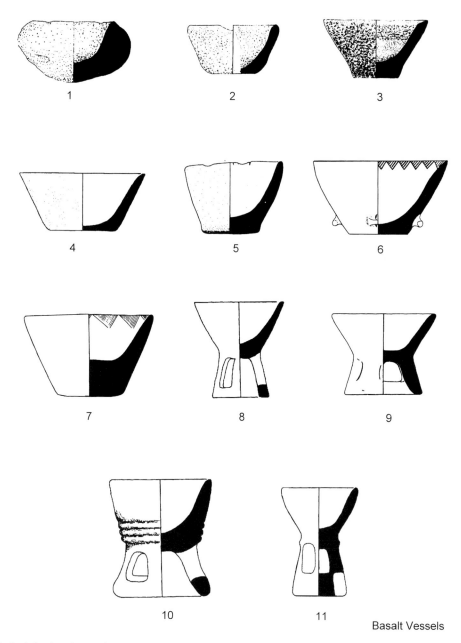

Figure 4.17. Late Chalcolithic basalt vessels.

anthropomorphic and zoomorphic, although the cylindrical, anthropomorphic 'pillar figures' are the better known. Pillar figures (Epstein 1975, 1988) feature prominent facial features (ears, eyes, nose) and usually include a shallow bowl at the top of the figure. Zoomorphic figures (Ibrahim and Mittmann 1998), horned bovines or caprines with a shallow bowl on the back, are less well known but equally spectacular. Pillared figures (anthropomorphic and zoomorphic) are generally considered products of the 'Golani' Chalcolithic culture (Epstein 1988; Ibrahim and Mittmann 1998), which spans the lowlands of the eastern Galilee, the Golan and the Damascus basin south into upland-north Jordan (Epstein 1978b; Hanbury-Tenison 1986).

Ghassul has been suggested as a major basalt vessel manufacturing centre in the past (Gilead and Goren 1989; Gilead 1995). However, recent petrological and chemical studies of a selection of Late Chalcolithic basalt tools and fine-stone vessels from Ghassul suggest that while utilitarian tools derived from the local Sweimeh outcrops, the fine bowls were made from north Jordanian upland basalts (Philip and Williams-Thorpe 1993; Philip 1998). Large (and virtually unexplored) north Jordanian sites, such as Sal, are

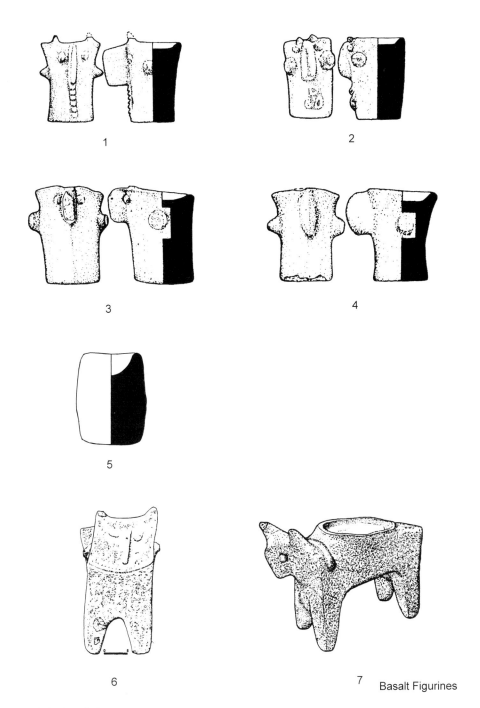

Figure 4.18. Late Chalcolithic basalt figurines.

littered with basalt vessels made from outcrops close by the site (Philip and Williams-Thorpe 1993) and at least some of the distinctive figurative vessels are probably to be sourced to this region (Kamleh 1998; Ibrahim and Mittmann 1998). Philip's most recent chemical analyses (1998) suggest that a major production centre for fine-basalt bowls and figurative pieces is to be located on the north Jordan plateau. While the fine-bowl forms seem to have been widely traded, the figured-basalt pieces are rarely found outside the Central Levantine core area. However, one non-figurative cylindrical vessel, a solid concave-sided pillar with characteristic shallow bowl on top, was recently discovered in the Ghassul sanctuary area (Seaton 1999; Bourke *et al.* 1999a). Although the Ghassul pillar may be unrelated to the Golani sculptures (Hanbury-Tenison 1986: 163), the similarity to rare undecorated (early?) pillar figures (Epstein 1985: 55) suggests that it is. Given

Epstein's suggested cultic context for the Golani pillar figures (Epstein 1978a), the sanctuary context of the Ghassul pillar may be significant (Bourke et al. 1999a).

Ground and polished stone maceheads are a characteristic feature of the Chalcolithic ground stone assemblage (Bar-Adon 1980: 116; Hanbury-Tenison 1986: 164; Gonen 1992: 58) (Figure 4.19). They have been found at Magass (Khalil 1992: 144-45), Ghassul (Mallon et al. 1934: 71; Lee 1973: 276-77), Abu Hamid (Dollfus and Kafafi 1989: 107 and fig. 9), Pella (Bourke et al. 1994: 91-93) and Tubna (Banning et al. 1998: 153). Most maceheads are either spherical or pear-shaped, with drilled central perforations. A variety of stones are employed in their manufacture, although limestone, calcite and hematite are the most common. While the limestone is likely to derive from local sources (Bender 1974), and the hematite may be sourced to the 'Ajlun region (Hauptmann, pers. comm.), the calcite is probably imported from the Eastern Desert of Egypt (Lucas and Harris 1962).

The majority of Chalcolithic maceheads are made from high prestige (copper), rare (hematite) or imported (calcite) materials, making them far more likely to be markers of status (Blackham 1999: 62) than simple weapons of war. Their presence in assemblages need not imply militarization of society (Levy 1995: 243), but probably serves as further evidence for social stratification (Cialowicz 1989). At Ghassul, the recent discovery of a miniature hematite macehead, a siltstone stamp seal, a mother of pearl pectoral and several miniature ceramic vessels closely associated within a large multi-purpose complex, is a further indication of the link between maceheads (particularly miniatures) and elite activity (Bourke et al. 1999a).

Miscellaneous groundstone objects bulk large within the Chalcolithic assemblage and range from quite large, coarsestone mortars, through various-sized discoid spindle whorls, to a variety of shaped semi-precious polished and bored-stone beads (Lee 1973: 263-83). The majority of the roughly fashioned stone mortars, pestles, hoes, picks and hammerstones derive from local limestone, basalt and flint sources. Stone spindle whorls are generally discoid shaped with a bored central hole, and derive from local limestone. Beads are made from a wide variety of materials, ranging from locally available limestone, through semi-precious hematite, carnelian and greenstone, to exotic imported serpentinite and calcite examples (Bourke et al. 1995: 49-53).

Worked bone, ivory and shell

At Ghassul and similarly at other valley sites, the bone tool assemblage seems to derive from local domesticates (Lee 1973: 284-93; Hanbury-Tenison 1986: 164-65), with ovicaprine metapodials the predominant material used (Bourke et al. 1995: 49-50). Most are fashioned into a variety of more or less keenly sharpened points, needles, burnishers and spatulae. The majority of such points, awls and needles were probably used in hide, fibre and food preparation. A minority of the more finely worked and highly polished bone points were more likely used as hair or clothing pins, perhaps in association with numerous small, flat-disk beads and occasionally polished, trapezoidal bone pendants as jewellery items. To be included among jewellery items are several highly polished, miniature, violin-shaped pendants known from Pontifical Biblical Institute (Scham 1997: 108) and recent Sydney University excavations (Bourke et al. 1999a). More enigmatic, possibly mnemonic pieces, are known (Levy and Golden 1996: 155-57; Scham 1997: 108), although such challenging interpretations remain contentious. Several small, highly polished and skilfully worked, hippopotamus tusk, ivory pendants were recovered from Pontifical Biblical Institute's excavations (Lee 1973: 288-89), and another in more recent Sydney University seasons (Bourke et al. 1999a). Tortoise shell was occasionally used for bead and pendant manufacture (Lee 1973: 285), although both ivory and tortoise are rare in the bone assemblage.

At Ghassul, both worked and unworked shell occurs within an assemblage dominated by local Melanopsis and Unio species (Lee 1973: 305-16; Hanbury-Tenison 1986: 165; Reese 1998). The Pontifical Biblical Institute's excavators found several pieces of Red Sea mother of pearl (*Maleagrina margaritifera*) shaped into crescent pectorals with holes at each end, and a further pectoral was found in recent Sydney University excavations (Bourke et al. 1999a). Numbers of the Nilotic *Aspatharia* (found by both expeditions) may have been employed as both tools and ornaments (Reese et al. 1986). Examples of the less frequently found *Glycymeris*, *Cardium*, *Conus* and *Tridacna* were probably also employed as jewellery (Lee 1973). In southern Jordan, Tall al-Magass profits from its proximity to the Red Sea sources of several shell families, with considerable evidence for shell bead manufacture recovered from the site (Khalil 1992: 144).

The Chalcolithic Period

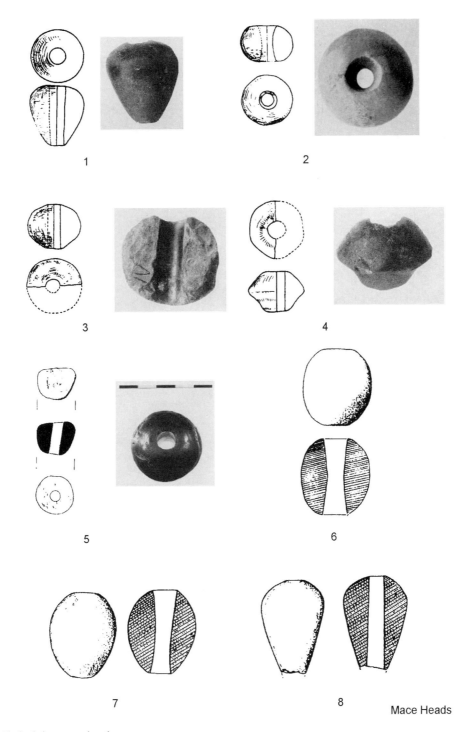

Figure 4.19. Late Chalcolithic mace heads.

Metallurgy and faience

Copper tools are very rare and occur in Late Chalcolithic assemblages only. At Ghassul, two complete and several fragmentary chisel-like, flattened axes were found during Mallon's excavations on Tulayl 1 (Mallon *et al*. 1934: 75). They seem quite similar in overall dimensions to the polished, narrow-pointed flint chisels and probably served a similar function (Rosen 1997). As well, several quite large fish-hook-shaped implements were recovered. Their function remains unclear, although it is probably architectural. Mallon's early chemical analysis (Mallon *et al*. 1934: 77) returned the surprising result of a high tin

143

content. Recent re-examination of two of the axes by Hauptmann has found the original analyses to have been in error. The axes were made from a nearly pure copper, suggesting an unsurprising association with Faynan ores (Hauptmann, pers. comm.). A single axe-head fragment recovered from the surface of Tulayl 3 in 1995 is currently undergoing analysis. Otherwise, with the exception of a number of simple pin fragments, copper tools remain very rare at Ghassul.

Copper of any sort is equally rare at Abu Hamid (Dollfus and Kafafi 1989: 106), Pella (Bourke *et al.* 1999b) and ash-Shuna (Rehren *et al.* 1997) and generally confined to pin fragments. However, the situation at Tall al-Magass in the south of Jordan is noticeably different. Here copper ore, slag in small quantities and metallic copper finds are considerably more common than elsewhere (Khalil 1992: 144-47). The site is likely to have benefited from its proximity to Timnan ores, the source of some (if not all) of Magass's copper (Khalil 1992: 147). Association with Timnan ores might imply an Early Bronze Age date for the Magass levels in which the copper appears (Rothenberg 1999).

If copper distribution is accepted as a useful indicator of interregional interaction, then the rarity of copper tools and the absence of any of the more specialized arsenic/antimony copper alloy objects in Jordan would seem to imply that Jordanian sites played only a marginal role in western networks. However, as the arsenic/antimony objects are regarded by many as special status objects (Tadmor *et al.* 1995; Gopher and Tsuk 1996), their absence from Jordan could equally be accounted for by variant cult practices or differing, elite status networks. A chronological element may also exist. Copper production in general, and especially the production of arsenic/antimony alloy objects, would seem to be confined to the latest phases of the Beersheban Chalcolithic (Joffe and Dessel 1995). If new dates for the end of significant occupation at Ghassul have any more general application, then many southern sites may well have been in sharp decline or have ceased to be occupied altogether by late Beersheban times (Bourke 1998a, 1999b).

A number of pale-brown, faience-disk beads have been recovered in recent excavations at Ghassul. Faience remains very rare in all but the latest Chalcolithic contexts, and the small quantities known to exist at Ghassul (Bourke *et al.* 1995: 52-53) may well derive from Egypt (Peltenburg 1987: 11-12; 1995: 36-37), although local production associated with small-scale copper smelting should not be discounted.

Seals, tokens and potmarks

Stamp seals are very rare elements in Late Chalcolithic assemblages (Ben-Tor 1995: 365-66) and their significance in a south Levantine context remains unclear (Hanbury-Tenison 1986: 100). In Syro-Mesopotamia, similar seals are viewed as unambiguous indicators of centralizing tendencies in the economic sphere and growing complexity in the social sphere (Schmandt-Besserat 1992). A detailed analysis of the spiral-incised stamp seal from Grar (Ben-Tor 1995) has identified it as phyllite, a non-local, grey-green soapstone equivalent (Ben-Tor 1995: 361-62), of likely Syrian origin. Further, Ben-Tor has suggested that most other spiral-incised stamp seals found west of the Jordan are made of this material, implying all are imports (Ben-Tor 1995: 366). While this provocative suggestion needs to be confirmed by petrographic analysis, it holds the implication that, as occasional imports into the late (perhaps very late) southern Levantine Chalcolithic, such seals need not hold any significance for local socioeconomic development.

At Ghassul, recent excavations discovered a beautifully made, square, cross-hatch incised, siltstone-stamp seal, in a reliable Late Chalcolithic context, in association with a hematite macehead and mother-of-pearl jewellery (Bourke *et al.* 1999a). This is quite unlike the small conical, circle-incised stamp seals found in Mallon's excavations (Mallon *et al.* 1934; Lee 1973; Elliot 1978), but is very close to the seal fragment found at Sahab (Ibrahim 1983: 44). Helms's (1991: 113-17, 333) synthesis of the early Jordan glyptic evidence would suggest that cross-hatch seal findspots are concentrated in the south Jordan Valley, as products of that region. The cross-hatch motif on the Sahab seal might link adjacent eastern upland regions with the south Jordan Valley, consistent with Sahab's general cultural affiliations (Ibrahim 1987: 73-75). The distribution of the round spiral-incised seals may indicate the existence of a second, coastal association (Mellaart 1975: 158; Keel 1989: 3-4; Ben-Tor 1995).

However one interprets the significance of the spiral and cross-hatch incised seals, they remain very rare occurrences in the south Levantine Chalcolithic, and if they are to be seen as first indicators of a nascent bureaucratic system, further signs of increasing concern for record keeping might be expected. In Syro-Mesopotamia, a variety of small geometric counters (Schmandt-Besserat 1992) and an increased incidence of incised and stamped potmarks (Helms 1991) are two of the more common indicators of such a tendency, and both are present in the Late Chalcolithic assemblages of the Jordan Valley (Lee 1973; Helms 1987a, 1987b, 1987c).

The small ceramic and stone tokens that have recently been documented as precursors to writing in Syro-Mesopotamia (Schmandt-Besserat 1992) are easily misidentified as gaming pieces or miscellaneous jewellery products. Even so, rather more have been discovered in the Jordanian Chalcolithic than have generally been credited to date. Two stone-spiral, grooved-conoid tokens (Schmandt-Besserat 1992: 204, Cone Type 1:38) were recovered off a floor in recent work at Ghassul (Bourke et al. 1999a). These may be associated with several stone conoids with incised circles around their bases (Lee 1973: 278-79, LB 61), quite close to Schmandt-Besserat's Punctated Base Conoids (1992: 204, 1:25). These very small pieces have often been identified as stamp seals (Elliot 1978), but are perhaps better seen as tokens. In addition, a number of small limestone flattened spheroids, with incised points set around their rim (Lee 1973: 276-77, LB 513 h-i) and identified as small loom weights, may possibly be seen as Spheroid Punctated tokens (Schmandt-Besserat 1992: 206, 2:6A). At Abu Hamid, a number of animal 'horns' published with ceramic figurine fragments (Dollfus and Kafafi 1989: 106, 1995: 451, fig. 1.2) may be ceramic conoid tokens (Schmandt-Besserat 1992: 203, 1.3, 14), and the small partly perforated ceramic ball (Dollfus and Kafafi 1989: fig. 9.3), may be a Plain Sphere token (Schmandt-Besserat 1992: 206, 2:1). At Pella, a ceramic Bent Conoid token (Schmandt-Besserat 1992: 203, Cone Type 1:14) was recently discovered inside a large grain silo (Bourke 1997c: 99).

Detecting Levantine tokens farther south will not be a straightforward procedure, as an element of subjectivity must remain until an agreed-upon corpus is constructed. This is unlikely to be identical to Schmandt-Besserat's Syro-Mesopotamian corpus, although it probably will emulate essential features. It remains to be seen if greater vigilance and increased awareness in the wake of Schmandt-Besserat's synthetic study produces more tokens than the few examples offered above, but this seems quite likely. While the isolated finds documented above do not allow us to demonstrate a greatly increased concern for record keeping at present, they do suggest a first movement in this direction.

Incised potmarks are a common occurrence on Late Chalcolithic pottery vessels at Ghassul. These marks are generally very simple devices, such as crosses of various sorts, circles in isolation or with a variety of linear ornament, diagonal and upright cross-hatched patches or simple tree motifs (Mallon et al. 1934: 121-22; Lee 1973; Helms 1987c). On rare occasions, they can include zoomorphic figures (North 1964). Perhaps to be associated with the simple potmarked vessels are a number of docket-like small geometric stone plaques, generally with a suspension hole at their centre-top, known from Pontifical Biblical Institute's excavations (Lee 1973: 278-79, LB 62; Schmandt-Besserat 1992: 219, 7: 20-23). All potmarks, plaques and zoomorphic-incised patterns are confined to the latest phases at Ghassul. With caution, one may associate such tendencies with the first predynastic Egyptian steps towards hieroglyphic recording practice (Baines 1988; Schmandt-Besserat 1992: 167-83; Schulman 1992: 412-15). Again, one must be careful not to overstate the case for heightened levels of bureaucratic practice in Late Chalcolithic Ghassul, as the illustrations offered above remain isolated cases. Although incised potmarks are a typical occurrence at Late Chalcolithic Abu Hamid and Pella (Lovell 2000), a comprehensive analysis of motif incidence is yet to take place. Nonetheless, the weight of evidence (seal, token and potmark) is consistent with a first tentative move towards heightened levels of bureaucratic recording practice in Late Chalcolithic Jordan. The indicators are fitful at present, but the corpus of bureaucratic recording devices steadily grows.

Wall paintings

Wall painting is one of the most distinctive, and artistically brilliant features of the Chalcolithic period in Jordan (Cameron 1981). At Ghassul, when first discovered by Mallon in 1931, the polychrome figured wall paintings seemed a uniquely early manifestation of artistic brilliance (Mallon et al. 1934: 129-41) (Figures 4.2, 4.3). Mellaart's subsequent magnificent discoveries at Neolithic Çatal Hüyük (Mellaart 1967, 1975) and the full publication of the Hierakonpolis T.100 material (Payne 1973; Kemp 1973, 1989) demonstrated that such endeavour had a very long history. Rather than being a spectacular anomaly, the Chalcolithic paintings from Ghassul represent a brilliant episode within a long developmental process. However, one must guard against the tendency to relate the earlier (Çatal Hüyük) and later (Hierakonpolis) manifestations via the medium of Ghassul (Cameron 1981: 32-35), as there is no demonstrable direct link between the three artistic traditions. New discoveries, such as those at Arslantepe (Frangipane1997), make a nonsense of sweeping associations.

Numerous wall painting fragments and assorted pieces of painted plaster have been found over the course of the four phases of field excavation at Ghassul. In her detailed consideration of the Ghassul paintings, Cameron (1981: 3) isolates seven paintings for detailed consideration, although it must be emphasized that many more fragments of painted wall decoration are known from across the site. Of the seven most complete paintings, three (the Notables, Star and Bird friezes) were found during the first Pontifical Biblical Institute's excavations (Mallon *et al.* 1934: 129-40) and another two (the Geometric and Tiger friezes) in subsequent Pontifical Biblical Institute work (North 1961: 32-36; Blackham 1999). Hennessy found the large polychrome geometric Zig-Zag fragment in his 1967 season (Hennessy 1969: 7) and the figured Procession Frieze in the 1975 season (Hennessy 1977).

The Ghassul wall paintings are normally painted on to brilliant white through yellow-buff backgrounds of fine lime wash laid over more substantial mudplaster backing. Paints are generally mineral based, and probably applied to a still-damp surface. Colours include deep scarlet, dark blue, black, green, yellow and a range of browns, all typically well preserved. As each wall shows evidence of multiple overpaintings, and each individual mural began with a new fine-lime wash, the adhesion of each overpainting to the previous murals is quite poor, and flakes of various depths easily detach. It is quite common to lift a fragment of painting that displays elements of up to five overpaintings on the preserved surface. This complicates interpretation and much confusion has resulted from a failure to differentiate between the various overpainting phases, often with hilarious results—the 'walrus fangs' reconstructed on many of the 'spook' masks are the result of earlier cornet cup motifs combined with later 'spook' masks (Cameron 1981: 10). Much laboured interpretation of apparently related motifs on paintings is misguided, as preserved motifs often represent a palimpsest of between 5 and 20 unrelated overpaintings.

The first painting, the 'Notables frieze' (Cameron 1981: 4), is drawn from the bottom half-metre strip of an approximately 5 m-long painting. The painting illustrates the feet of a series of large seated figures facing a much smaller figure that is standing in front of a large painted star. The second, the famous 'Star frieze' (Figure 4.2; Cameron 1981: 5), consists of an approximately 2 m wide, eight-pointed polychrome-decorated star, apparently associated with a selection of architectural, zoomorphic, geometric and masked anthropomorphic motif fragments, many of which are overlapped by the central Star motif. Much ingenuity has been invested in the interpretation of these isolated motifs, but as most (if not all of them) belong to earlier, overpainted scenes, it is unlikely that many are related to the Star motif. The third painting, the 'Bird frieze' (Cameron 1981: 6), was immediately recognized as being painted in a noticeably different style to the previous finds. The rendering of the plumage on a partridge-like fowl is naturalistic (almost impressionistic) in style and sharply different to the crisp linear technique of the preceding murals. Unfortunately, little associated contextual motifs were recovered from the 5 m-long stretch of painting.

North's second-phase Pontifical Biblical Institute's work first discovered the large 'Geometric frieze' (Cameron 1981: 7-8). Important points are North's observations on the unique combination of raised, hollow-plastic ornament (craters) with geometric-lined decoration (North 1961: 35) and Cameron's controversial suggested parallel with the famous Vulture Frieze at Çatal Hüyük (Cameron 1981: 20-21), with all that this implies for shared funerary excarnation rituals (Cameron 1995: 118). North's second discovery, the famous polychrome 'Tiger frieze' (North 1961: 6; Cameron 1981: 9), consists of a confusing array of overlapping segments of unrelated motifs, apparently dominated by a large striped animal. Here Cameron's suggested reconstruction (Cameron 1981: 28-31) of an elaborately dressed male figure in place of the 'tiger' is more plausible.

Hennessy's polychrome 'Zig-Zag' fragment came from the large broadroom discovered in 1967 (Hennessy 1969: 7; Cameron 1981: 12). It was part of a larger painting, but only a small metre-square piece was recoverable. It may illustrate a textile pattern or possibly an architectural fragment similar to those found in the 'Processional' frieze (Cameron 1981: 31). Hennessy discovered the famous 'Processional frieze' in 1975, lifted it during the 1977 field season, and had it conserved in 'Amman in 1978. This scene, also much complicated by multiple overpaintings, appears to show a procession of elaborately attired masked figures, the largest carrying a sickle-shaped object, promenading towards a walled complex shown in plan to the left (Figure 4.3; Cameron 1981: 13). It only remains to be noted that Hennessy found large amounts of shattered, wall painting fragments in both his sanctuary buildings, along with what can plausibly be reconstructed as a painter's workshop outside the complex to the northeast (Hennessy 1978). However, no substantial wall painting fragments were recoverable during excavations in the sanctuary precinct.

Recently renewed excavations at Ghassul have recovered small pieces of painted plaster in a number of locations, but only one substantial half-metre square fragment, found in 1995 in newly opened trench N II, northeast of Hennessy's Area A (Bourke *et al.* 1999a). The 'Garlanded Sickle' fragment shows a red sickle-shaped object, with white garlands threaded through circular holes in the object. It is vaguely similar to the perforated, sickle-like objects discovered at Nahal Mishmar (Bar-Adon 1980: 16-21), one of which retained traces of red colour and fragments of fibre. It is possible that the Garlanded Sickle painting and the Mishmar Perforated sickles relate in some way to the sickle-like object held by the lead figure in Hennessy's 'Processional' frieze (Cameron 1981: 27).

For 50 years, the Ghassul wall paintings were unique in the Chalcolithic of Jordan (Cameron 1981: 3), but a recent find from Abu Hamid has demonstrated that polychrome wall painting was not confined to one site in the south Jordan Valley (Dollfus and Kafafi 1993: 251-53). The Abu Hamid fragment consists of a series of linear and filled geometric motifs painted on to an off-white background. It is similar in design, technique and execution to several of the smaller geometric fragments from Ghassul.

The wall paintings from Ghassul are a uniquely important record of ancient sociocultic behaviour, as well as representing an important landmark in the history of art. Art speaks with an immediacy that is matched by few other media, and the quality of the painting and the apparent naturalism of the Ghassulian subject matter serves only to emphasize a feeling of accessibility. However, interpretation of subject matter and evaluations of cultural significance are far more troubled territory than earlier analysts acknowledged (Cameron 1981: 32-35; Mellaart 1998: 37-39; Hodder and Matthews 1998: 50-51). While it remains possible that a number of the more famous Ghassulian wall paintings are connected with cultic activity, this is by no means the only way to account for the paintings.

Figurines

Stylized human and animal figurines are a feature of the Jordanian Chalcolithic and both have a long history in the region (Mellaart 1975; Garfinkel 1995; Gopher and Orelle 1996; McAdam 1997). The most distinctive is the 'violin-shaped' figurine (Figure 4.20). The small–medium-sized figurines are made predominantly of ground and polished limestone, although a variety of stone types, including calcite, marble, greenstone and sandstone have been recorded (Levy and Golden 1996: 153-56). Miniature violin pendants in ivory and bone are known from Ghassul (Lee 1973: 279, e-f; Bourke *et al.* 1999a) and in greenstone from Abu Hamid (Dollfus and Kafafi 1988: 49, fig. 80). The extreme stylization of most violin figurines is in contrast with more naturalistic, but much rarer basalt (Shiqmim) and ivory (Abu Matar/Safadi) figurines from the northern Negev (Levy and Golden 1996: 154-55).

The majority of violin figurine find contexts west of the Jordan have been interpreted as cultic (Perrot 1959; Alon and Levy 1989; Levy and Golden 1996). The sheer concentration ($n = 59$) of such figurines at Gilat would seem to be support for such a contention (Levy and Golden 1996: 155). Violin figurines are known from Ghassul (Lee 1973: 278, LB 72), Abu Hamid (Dollfus and Kafafi 1995: 454-55) and Pella (Bourke *et al.* 1999b). While none are from an explicitly cultic context, at Ghassul Lee's (1973) Tulayl 3 findspot coordinates are consistent with a relationship to the Building 78 cultic complex (Elliot 1977; Bourke 1998b). The Pella findspot lies directly below later (Middle Bronze/Late Bronze) cultic establishments, and is contemporary with the platform storage facility discovered in 1995 (Bourke *et al.* 1998: 180-81). The context of the Abu Hamid figurines remains to be clarified, although association with the zoomorphic-bull rhyta and the macehead cache should not be ruled out (Dollfus and Kafafi 1988; 1995: 455). However, no violin figurines were found associated with Hennessy's Sanctuary complex (Seaton 1999), none are reported from upland Chalcolithic sites, and no Jordan Valley site has recorded anything like the concentration of figurines reported at Gilat. The presence of violin figurines in Jordan Valley assemblages may represent items of exchange rather than indigenous products (Alon and Levy 1989: 209; Dollfus and Kafafi 1995: 455), perhaps indicating some form of cult association between a Gilat central place and Jordan Valley peripheries.

Animal figurines are very common within early and later Chalcolithic assemblages and both technique and subject matter hark back to Neolithic exemplars (Garfinkel 1995; McAdam 1997). The majority are small and fashioned by hand, although bone tools are occasionally used to pick out finer details. Whereas many Neolithic figurines are unfired

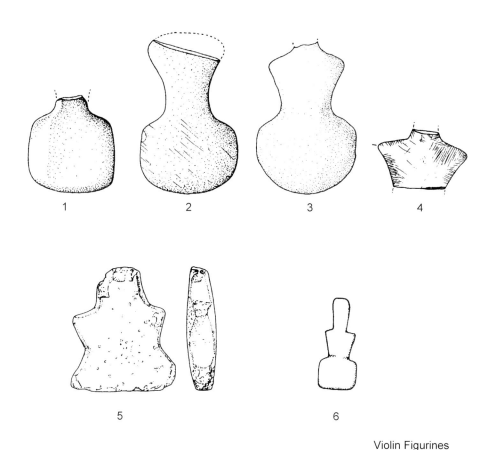

Figure 4.20. Late Chalcolithic violin figurines.

(McAdam 1997: 135), most Chalcolithic examples are generally well fired. Although identifying types is a manifestly subjective endeavour, it would appear that the majority of Ghassul's figurines represent domestic ovicaprines and cattle, and a variety of deer, dog and bird species. Animal and bird figurines are also known from Abu Hamid (Dollfus and Kafafi 1995, 451-55) and animal figurines from Pella (Bourke et al. 1999b).

Function is not obvious beyond simple representation, but the fact that many of the figurines found at Ghassul are broken at the neck or mid-body may be significant, given the apparent 'ritual killing' of many of the 'Ayn Ghazal figurines (McAdam 1997: 131-35). While damage to the generally unfired Neolithic figurines from 'Ayn Ghazal may be due to their original fragility, the well-fired Chalcolithic examples (Mallon et al. 1934: 83-87; Bourke et al. 1999a) suggest something more than simple breakage may be occurring. However, at Ghassul, contextual details do not support any specialized function, with most figurine fragments deriving from rubbish deposits (Bourke et al. 1999a). Many show traces of burning, but this may relate to the often ashy depositional contexts. Function as ornaments (Levy 1993), talleys (Schmandt-Besserat 1992; McAdam 1997) or protective talismans are all possible. Whatever function they may have served, they document a concern for a relatively limited number of species, in a lively plastic art tradition.

Burial customs

Subfloor, shallow-pit burials are the norm for neonatal and young child burials in the Chalcolithic southern Levant (Hanbury-Tenison 1986: 205-31; Gonen 1992: 74-78). Around 40 neonatal and child burials are known from Tulaylat al-Ghassul (Lee 1973: 332-38; Elliot 1977: 22; Hennessy 1969: 7; Blackham 1999: 37-38), some buried in jars, and others roughly contained by large, storage-jar sherds. They are generally to be located beneath floors, in room corners. However, early suggestions that this practice documents child sacrifice (Mallon et al. 1934: 49-50; Lee 1973: 334-36) are overstated and likely to be incorrect. Similar child burials are known from Pella (Hanbury-Tenison 1986) and ash-Shuna. Burial-custom analysis is commonly employed to infer socioeconomic developmental status. Child burials with disproportionate amounts of grave goods at Shiqmim (Levy and Alon 1982: 52-53) and Nahal Qanah (Gopher and Tsuk 1996: 226) are claimed as evidence for an ascribed status regime in late Chalcolithic times, but such evidence is equivocal at best (Hanbury-Tenison 1986: 220-24).

Intramural adult burials are very rare in Jordanian Chalcolithic sites (Elliot 1977: 21-22; Hanbury-Tenison 1986: 224) and extramural cemeteries are both rare and difficult to define. It is commonly observed that select elements of the ruinfields at Adeimeh, located 2 km southeast of Ghassul, constitute the most likely cemetery for Ghassul (Mallon *et al.* 1934; Stekelis 1935; Hanbury-Tenison 1986). Adeimeh was excavated by Stekelis in 1933 (Stekelis 1935) concurrently with Pontifical Biblical Institute's work at Ghassul. Confusion has arisen from the intermixture of a number of distinct and perhaps unrelated elements within the one 3 ha ruinfield. The first component is a large, cist-grave cemetery, consisting of several hundred metre-long, stone-lined rectangular pits, containing either tightly flexed, primary inhumations or secondary burials, consisting of long-bone and skull clusters. The cist cemetery may be associated with a series of c. 5 m-wide stone and rubble tumuli, as some overlay cist graves, while others do not. A relationship is generally assumed, but the evidence is defective. To the south of the cist/tumuli complex lies a 30 m-wide, roughly rounded stone enclosure, and farther to the south a series of dolmen clusters (Hanbury-Tenison 1986: 218-19).

Attempting to assess functional relationships across the ruinfield is made difficult by the limitations of 60-year-old evidence, and present inability to re-examine the primary data, which lies within a military zone. Prag (1995: 76-77) differentiates Late Chalcolithic cist burials from EB I dolmen fields, and follows Stekelis (1935: 67-69), Elliot (1977: 19) and Hanbury-Tenison (1986: 218) in dating the large enclosure to the Chalcolithic as well. However, the ceramic evidence for association between Adeimeh and Ghassul remains slight. The possibility remains that the cist, tumuli and enclosure complex could equally well have served a semi-nomadic (perhaps even post-Ghassulian) component of Chalcolithic society in the same way that the dolmen complexes are thought to have served a later transhumant people (Prag 1995: 83-84).

The few other Chalcolithic cemeteries known in Jordan appear to share features with Shiqmim (Levy and Alon 1985; Hanbury-Tenison 1986), and Adeimeh (McCreery 1978/79; Clarke 1978/79). Around 30 roughly rounded cairn tombs were discovered c. 1 km south of Bab adh-Dhra' in 1977 (McCreery 1978/79) and investigated shortly thereafter as a rescue excavation (Clarke 1978/79). The cairn tombs appear to be larger versions of the Adeimeh cists, except that each one appears to be covered with a stone and rubble tumulus, somewhat similar to the Adeimeh tumuli (Clarke 1978/79: 60-69). As well, Clarke investigated two large 25 m-wide enclosures, similar to that discovered at Adeimeh by Stekelis (Clarke 1978/79: 69-73). These appear to be directly related to the cairn tombs and can be associated with funerary activities, strongly implying that the Adeimeh enclosure was used for a similar activity (Clarke 1978/79: 75-76). Date and cultural affiliation was difficult to determine due to the paucity of finds and their generally poor condition. However, Clarke felt that the best parallels lay with the Ghassulian Chalcolithic (Clarke 1978/79: 73-74).

Very little is known about Chalcolithic funerary activity. However, if the few fragmentary examples have been interpreted correctly, then it would appear that funerary culture may have included open-air enclosures (perhaps for primary excarnation) in association with cist and tumulus graves (for ultimate secondary burial). This culture seems broadly consistent around the Dead Sea periphery, as an enclosure similar to those at Adeimeh and Bab adh-Dhra' is known from the Nahal Mishmar region (Bar-Adon 1980: 12-13). Moorey (1989: 179) has raised the possibility that the Nahal Mishmar 'crowns' be viewed as model funerary enclosures, functioning as excarnation areas. It may be that the assumed link between the Nahal Mishmar hoard and En Gedi is unnecessary (Bar-Adon 1980: 202) and that the hoard relates to funerary activities in the Nahal Mishmar region itself. Alternatively, if the link is a valid one, then perhaps both the En Gedi and Ghassul Area E sanctuaries functioned as funerary enclosures, which might account for their isolated positioning. Either way, the existence of a Dead Sea regional-funerary culture seems likely and should be viewed in contrast to the coastal/central upland cave and ossuary funerary culture (Hanbury-Tenison 1986: 205-12; Perrot and Ladiray 1980; Gopher and Tsuk 1996; Gal *et al.* 1997).

Trade and foreign relations

Much has been written on traded materials and their apparent illustration of far-flung foreign relations during the Chalcolithic period (Hanbury-Tenison 1986: 95-103; Bourke 1998b). Little is known about the earlier phases of the Chalcolithic in Jordan and comments are generally confined to vague observations on ceramic-motif parallels with the Halafian north (Kaplan 1960; Mellaart 1975; Gophna and Sadeh 1988/89) or the Badarian south (Kaplan 1959; Kantor 1992). Such statements rarely document actual imports, far less method or motivation. If interregional trade did exist in this early period, it was a very minor component of material variability and not developmentally significant. In the later Chalcolithic period, evidence does exist for relatively widespread materials exchange. Care should be taken to distinguish relatively large-scale within-region trade (hard stone, copper, ceramics) from small-scale interregional exchange of exotics. Regional trade in basalt, alabaster and hematite is well documented, and likely to be of some significance. However, the economic importance of interregional trade in exotic items of adornment (shell, faience, ivory and semi-precious stones) should not be over-emphasized. The vast majority of the Jordanian Late Chalcolithic material assemblage derives from the immediate hinterland of each site, and is locally manufactured.

While the bulk of all ceramic assemblages seem to be locally made, distinctive manufacturing technique at Abu Hamid (Roux and Courty 1997) and physico-chemical analysis at Ghassul (Edwards 1997; Bourke 1998b) suggest that a small percentage of Late Chalcolithic ceramics derive from the north Jordan Valley and the upland regions to the north and east of the Jordan Valley. The presence of the distinctive Cream Ware at Ghassul (Bourke 1998b; Blackham 1999) suggests a small but distinctive interregional element in this exchange. While basalt suitable for utilitarian tools was exploited from a number of locations throughout Jordan (Philip and Williams-Thorpe 1993), basalt suitable for more elaborate vessel manufacture is sourced to the northern Jordan uplands (Philip 1998). Although lithic assemblages were overwhelmingly crafted from locally available resources (Baird 1987; Rollefson 1992), a small percentage of fine flint tools probably derives from the southwestern Negev (Rosen 1997). Hematite is sourced to the upland regions of central Jordan, and alabaster, marble and serpentinite all probably derive from the Eastern Desert of Egypt. Virtually all Jordan Valley copper comes from the Faynan region (Hauptmann, pers. comm.), although south Jordanian sites may draw some copper from Timna (Rothenberg 1999). Much exotic shell derives from the Red Sea, and a smaller percentage from Mediterranean and Nilotic sources (Lee 1973; Reese *et al.* 1986; Bourke 1998b).

There is considerable evidence for regional movement of relatively bulky materials (basalt, hematite, copper and flint) and interregional movement of fine items of adornment (semi-precious stone, faience, ivory and shell). However, there is little reason to posit systematic, long-distance, reciprocal trade between Jordanian Chalcolithic sites and centres in lower Egypt, northern Syria or southern Mesopotamia. A low-intensity, 'down the line' exchange is more likely, but such cannot be held to have played an important role in the local Jordanian Chalcolithic economies. However, the bulk of the evidence for interregional exchange is concentrated in the latest Chalcolithic phase at most sites (Bourke 1998b). It would thus seem reasonable to suggest that intensity of contact (both regional and interregional) increases over time, to reach its greatest heights at the very end of the Chalcolithic period.

Chalcolithic society

Much has been written on the nature of social organization in the Chalcolithic period in the southern Levant over the last 15 years (Levy 1995, 1996; Yoffee 1993, 1995). Arguments over the archaeological correlates of social transformation have intensified as the sophisticated field projects of the 1980s move from preliminary description to final publication (Gilead 1995; Alon and Levy, 1989). At one level, discussion is strictly anthropological and concerns itself with identifying the unsustainable assumptions in over-rigid, developmental typologies (Yoffee 1993; Rothman 1994). At another level, prehistorians continue to interpret archaeological cultures with reference to anthropological theories (Yoffee 1995; Levy 1996) because identifying significant vectors in the development of later state-level societies remains a priority research concern (Stein 1998).

Current research agendas in the fifth-millennium BC Middle East acknowledge the need to abandon over-rigid typologies in organizational states, such as chiefdoms, while preserving the broad concept of mid-range societies with distinctive developmental features (Stein 1998: 4-7). Refinements to originally monolithic-organizational classes (simple/complex or individualizing/group-oriented chiefdoms) have been generally accepted (Stein 1998: 8-10) and much attention is now focused on the mechanisms through which subchiefly lineage groups compete with each other for power (Stein 1998). Researchers recognize the need to evaluate the varying significance of enhanced agricultural surpluses (Stein 1994b), rural/urban interaction (Falconer and Savage 1995), interregional exchange (Savage 1997), craft specialization (McCorriston 1997), writing and literacy (Baines 1988) and cult practice (Alon and Levy 1989), in this competition for influence and power. This is never straightforward, and in the relatively impoverished landscapes of the southern Levant, particularly challenging.

It would be premature to do more than sketch the barest outline of Jordanian Chalcolithic social organization at present, as many important field projects are currently in publication phase (ash-Shuna, Pella, Abu Hamid, Tulaylat al-Ghassul). We can do no more than describe changes from Neolithic norms (Gopher 1995; Gopher and Gophna 1993) and (with less confidence) begin to identify potential agencies in such change. Our ignorance of Early Chalcolithic norms is almost total and this hamstrings thoroughgoing processual analysis. Our treatment will concentrate on the better known later Chalcolithic cultures in general and the site of Ghassul in particular, because this is where the majority of our evidence is concentrated. Contrasting individual features of the Late Neolithic and the Late Chalcolithic assemblages is feasible, but an accurate assessment of when each individual change occurs remains elusive. This is an important caveat because it renders difficult (if not impossible) the evaluation of the degree of interrelatedness of individual changes. The Late Chalcolithic cultures are the endpoint of a millennium-long process, during which individual strands of socioeconomic organization may have altered at differing times and at differing rates. Without the long perspective of a well-documented Early Chalcolithic developmental sequence, we risk viewing change in many individual factors as related when they are not, and may fail to identify the key factors through an inadequate time perspective (Bourke 1998a).

Economy and society

Increasing differentiation, specialization and intensification of economic activity from early in Late Chalcolithic times (c. 4500 BC) is consistent with the environmental evidence recently documented at Tulaylat al-Ghassul (Bourke 1997b, 1998a; Bourke *et al.* 1999a; Meadows 1998; Mairs 1997). One interpretation of the Ghassul evidence would see the Late Chalcolithic economy as one in which a number of distinctive subsistence strategies exist in parallel, some generalist and focused on subsistence, and others more specialized in high yield (but environmentally fragile) monocultures (Mairs 1997; Meadows 1998; Bourke 1998a). Given that high-yield cereals, pulses and fruits have a relatively short storage life, the motivation to move away from traditional subsistence regimes towards the production of specialized agricultural surpluses would suggest heightened internal demand for these products. The motivation to produce larger surpluses remains a matter of debate, although it is consistent with the mobilization of staple surpluses in support of developing elites (D'Altroy and Earle 1985; Earle 1987). The architectural evidence at Ghassul for much greater storage capability in the later phases of the Chalcolithic, with specialized storage facilities (for grain and fodder) located within enlarged multi-room courtyard complexes, is not in doubt (Bourke 1998a; Blackham 1999; Bourke *et al.* 1999a). This tendency towards large storage facilities is also seen at Abu Hamid (Dollfus and Kafafi 1988), Pella (Bourke *et al.* 1998), north Jordan Valley sites (Tzori 1958, 1967; Garfinkel 1993), and Sahab (Ibrahim 1984).

While increased ability to store agricultural surpluses is widespread, the absence of many of the additional trappings of a 'typical' chiefdom must be acknowledged. There are few signs of overt architectural differentiation (Blackham 1999), although creation of distinctive cultic establishments does occur (Bourke 1998a; Bourke *et al.* 1999b). Differential mortuary practice is not documented east of the Jordan. There is little evidence for marked concentrations of elite wealth or symbols of authority. Interregional traded goods remain a minor factor in the local economy, with long-distance imports confined to minor jewellery items consistent with only modest levels of wealth display (Bourke 1998b). In short, if one seeks to view Late Chalcolithic society as some form of chiefdom (Levy 1995), then it must be acknowledged to be a chiefdom with few of the traditional trappings. This circumstance need not be the impediment it is often said to be (Gilead 1988), as contemporaneous Ubaid-period, southern Mesopotamia provides an instructive parallel in type and circumstance (Stein 1994a). Stein views Ubaid society as consisting of a series of small group-oriented 'egalitarianizing' chiefdoms (Stein 1994a: 42-43), which derived much of their wealth and power through the ability to mobilize labour from extended kinship groups to generate a reliable agricultural surplus. Such a developmental scenario has a number of striking material cultural parallels with the Late Chalcolithic of Jordan (Bourke 1998a). In the Jordan Valley, the early domestication and subsequent intensive exploitation of olive fruit and oil (Meadows 1999) and the early development of a significant pastoral component (Mairs 1997) were key factors in the development of a reliable agricultural surplus.

Several additional factors may have played a part. Landform and climate combine to circumscribe settlement distribution in the southern Jordan Valley, suggesting a variant on Carneiro's Circumscription Hypothesis (Carneiro 1970; Falconer and Savage 1995). A review of settlement patterns in the south Jordan Valley reveals a series of wadi-related segmentary distributions (Stein 1994b), with the largest settlement typically on the floodplain adjacent to winter floodwater-producing wadis (Falconer 1994). If the wadi systems between Nimrin and the Dead Sea are viewed as a region, then Ghassul enjoys an acknowledged primacy over all other settlements. Such primacy is typically accounted for through the concentration of unique services (economic, cultic and trade) on the primate centre (Falconer and Savage 1995: 40). Ghassul may well have enjoyed an enhanced regional-cultic function (Cameron 1981; Levy 1995; Joffe 1998). Intensified regional and more limited interregional exchange is well documented at Late Chalcolithic Ghassul (Philip 1998; Reese 1998; Bourke 1998b). The present state of the evidence is in keeping with Stein's (1994b: 11-12) segmentary state model for the Jeziran countryside, with the largest settlements exercising a limited selective control over an effectively autonomous hinterland. There is sufficient evidence to suggest that the Jordan Valley settlements in general, and Tulaylat al-Ghassul in particular, enjoyed heightened levels of internal organization in comparison with upland settlements.

Conclusion: the demise of the Ghassulian Chalcolithic

The view of the Jordanian Chalcolithic offered above relies heavily on evidence drawn from Jordan Valley sites. Emphasis on the relatively enhanced circumstance at Ghassul recurs throughout this overview. This emphasis is due in part to Ghassul being the most intensively investigated Chalcolithic site in Jordan, but it also reflects the apparent absence of similar data from most other sites. Looking ahead to the imminent publication of several major field projects, future overviews should be able to achieve a considerably more balanced assessment of the countryside as a whole. As it is, this overview largely reflects the concerns and biases of the last 50 years, perspectives that are changing rapidly with the onset of an ever-more comprehensive coverage of the landscape, and a greatly enhanced analytical toolkit.

The demise of the Jordanian Chalcolithic and the related question of the degree of continuity with the successor cultures of the Early Bronze Age will continue to bedevil scholars for some time to come. At present, the fashion in Early Bronze Age studies is to downplay the traditional view of a relatively sophisticated Early Bronze Age 'walled town culture' in comparison with the preceding Chalcolithic 'village-based society'. Arguments advanced herein for a modest complexity in later Chalcolithic society (if accepted) result in a further narrowing of the distance between Chalcolithic and Early Bronze Age socioeconomic norms. Modest stratification, based on elite control of agricultural surpluses, was not unique to the Early Bronze Age, as this was one important element in the Late Chalcolithic agricultural transformation. Even so, significant change across the periods does occur. The rich symbolic landscape of the Late Chalcolithic vanishes forever.

It may be that Jordanian Late Chalcolithic society can be viewed as one in which the 'traditional' symbolic/religious ruling elite gradually came into competition with an emerging elite based on the control of agricultural surpluses. While there is no evidence for heightened levels of conflict or destruction, it is possible that in the increasingly harsh environmental regime of the latest Chalcolithic leadership in society shifted from those who manifestly could not control the environment to those who had at least developed a means of coping with it. If this is so, the gradual sociopolitical marginalizing of the old ritual elite may have resulted in an actual physical displacement of these religious elites and their symbolic paraphernalia to the marginal environments of the Dead Sea region.

All evidence currently available suggests that the Ghassulian Chalcolithic had its origins in the small village societies that grew up along the northeastern shores of the Dead Sea early in the Chalcolithic period. It seems likely that a sharp climatic alteration intervened to halt the gradual economic progression evidenced in the long sequence at Ghassul. While hardier economic elements survived to rebuild a new society in the succeeding Early Bronze Age, a great many of the most brilliant artistic elements were lost forever. If radiometric data is an accurate guide, the Chalcolithic civilization that grew up by the shores of the Dead Sea probably came to an end in much the same environment. Today it is a harsh landscape, but still strangely beautiful.

References

Abu Dayyah, A. *et al.*
1991 Archaeological Survey of Greater Amman, Phase I: Final Report. *Annual of the Department of Antiquities of Jordan* 35: 361-95.

Alon, D. and T. Levy
1989 The Archaeology of Cult and the Chalcolithic Structures at Gilat. *Journal of Mediterranean Archaeology* 2: 163-221.

Algaze, G.
1993 Expansionary Dynamics of Some Early Pristine States. *American Anthropology* 95: 304-33.

Amiran, R.
1955 The 'Cream Ware' of Gezer and the Beersheba Late Chalcolithic. *Israel Exploration Journal* 4: 240-45.
1989 The Gilat Goddess and the Temples of Gilat, En-Gedi and Ai. In P. de Miroschedji (ed.), *L'urbanisation de la Palestine a l'age du Bronze Ancien*, 53-61. Oxford: British Archaeological Reports, International Series 527.

Amr, K. *et al.*
1993 Wadi al-Qattar Salvage Excavation 1989. *Annual of the Department of Antiquities of Jordan* 38: 263-78.

Amr, K. *et al.*
1996 Archaeological Survey of the East Coast of the Dead Sea Phase I: Suwayma, az-Zara and Umm Sidra. *Annual of the Department of Antiquities of Jordan* 40: 429-50.

Avner, U., I. Carmi and D. Segal
1994 Neolithic to Bronze Age Settlement of the Negev and Sinai in Light of Radiocarbon Dating: A View from the Southern Negev. In O. Bar-Yosef and R. Kra (eds.), *Late Quaternary Chronology and Paleoclimates of the Eastern Mediterranean*, 265-300. RADIOCARBON. Tucson: Department of Geoscience, University of Arizona.

Baines, J.
1988 Literacy, Social Organisation and the Archaeological Record: The Case of Early Egypt. In J. Gledhill, B. Bender and M. Larsen (eds.), *State and Society: The Emergence and Development of Social Hierarchy and Political Centralisation*, 192-214. London: Unwin–Hyman.

Baird, D.
1987 A Preliminary Analysis of the Chipped Stone from the 1985 Excavations at Tell esh-Shuna North. *Annual of the Department of Antiquities of Jordan* 31: 461-80.

Baird, D., and G. Philip
1994 Preliminary Report on the Third (1993) Season of Excavations at Tell esh-Shuna North. *Levant* 26: 111-33.

Banning, E., M. Blackham and D. Lasby
1998 Excavations at WZ 121, a Chalcolithic Site at Tubna in the Wadi Ziglab. *Annual of the Department of Antiquities of Jordan* 42: 141-59.

Banning, E. *et al.*
1989 Wadi Ziglab Project 1987: A Preliminary Report. *Annual of the Department of Antiquities of Jordan* 33: 43-58.
1996 The 1992 Season of Excavations in Wadi Ziqlab, Jordan. *Annual of the Department of Antiquities of Jordan* 40: 29-50.

Bar-Adon, P.
1980 *The Cave of the Treasure*. Jerusalem: Israel Exploration Society.

Barker, G. *et al.*
1997 The Wadi Faynan Project, Southern Jordan: A Preliminary Report on the Geomorphology and Landscape Archaeology. *Levant* 29: 19-40.
1998 Environment and Landuse in the Wadi Faynan, Southern Jordan: The Second Season of Geoarchaeology and Landscape Archaeology. *Levant* 30: 5-25.

Bender, F.
1974 *Geology of Jordan*. Berlin: Contributions to the Regional Geology of the Earth.

Ben-Tor, A.
1995 A Stamp Seal and a Seal Impression of the Chalcolithic Period from Grar. In I. Gilead (ed.), *A Chalcolithic Site in the Northern Negev*, 361-75. Beersheva: Ben-Gurion University.

Bernbeck, R. *et al.*
1995 Wasserspeicherung in der jordanischen Steppe. *Das Altertum* 40: 163-74.

Betts, A.
1986 The Prehistory of the Black Desert Transjordan. Unpublished PhD dissertation, University of London.
1992 Tell el-Hibr: A Rock Shelter Occupation of the Fourth Millennium BCE in the Jordanian Badiya. *Bulletin of the American Schools of Oriental Research* 287: 5-23.

Betts, A. (ed.)
1991 *Excavations at Jawa 1972–1986*. Edinburgh: Edinburgh University Press.

Betts, A. et al.
  1995 Archaeological Survey of the Wadi al-Ajib, Mafraq District. *Annual of the Department of Antiquities of Jordan* 39: 149-68.

Bisheh, G. et al.
  1993 Archaeological Rescue Survey of the Ras an-Naqb–Aqaba Highway Alignment, 1992. *Annual of the Department of Antiquities of Jordan* 37: 119-34.

Blackham, M.
  1999 Teleilat Ghassul: An Appraisal of Robert North's Excavations (1959–1960). *Levant* 31: 19-64.

Boling, R.
  1988 The Early Biblical Community in Transjordan. Sheffield: Almond Press.

Bourke, S.
  1997a The Urbanisation Process in the South Jordan Valley: Renewed Excavations at Teleilat Ghassul 1994/1995. In M. Zaghloul et al. (eds.), *Studies in the History and Archaeology of Jordan*, VI: 249-59. Amman: Department of Antiquities.
  1997b The 'Pre-Ghassulian' Sequence at Teleilat Ghassul. In: H-G. Gebel et al. (eds.), *The Prehistory of Jordan. II. Perspectives from 1997*, 395-417. Berlin: *ex oriente*.
  1997c Pre-Classical Pella in Jordan: A Conspectus of Ten Year's Work (1985-1995). *Palestine Exploration Quarterly* 129: 94-115.
  1998a A Chalcolithic Origin for the Bronze Age State Systems in the Southern Levant: Evidence from Teleilat Ghassul, Jordan. In P. Matthiae et al. (eds.), *First International Congress on the Archaeology of the Ancient Near East*. Rome: Universita di Roma La Sapienza.
  1998b Teleilat Ghassul: Foreign Relations in the Late Chalcolithic Period. In T. Levy and E. van den Brink (eds.), *Egyptian–Canaanite Interaction from the Fourth through Third Millennium BCE*. Jerusalem: Israel Exploration Society.

Bourke, S. J. et al.
  1994 Preliminary Report on the Fourteenth Season of Excavation by the University of Sydney at Pella in Jordan. *Annual of the Department of Antiquities of Jordan* 38: 81-126.
  1995 Preliminary Report on a First Season of Renewed Excavations at Teleilat Ghassul by the University of Sydney, 1994. *Annual of the Department of Antiquities of Jordan* 39: 31-64.
  1998 Preliminary Report on the University of Sydney's Sixteenth and Seventeenth Seasons of Excavations at Pella (Tabaqat Fahl) in 1994/95. *Annual of the Department of Antiquities of Jordan* 42: 179-211.
  1999a Preliminary Report on a Second and Third Season of Renewed Excavations at Teleilat Ghassul by the University of Sydney, 1995/1997. *Annual of the Department of Antiquities of Jordan* 44.
  1999b Preliminary Report on the University of Sydney's Eighteenth and Nineteenth Seasons of Excavations at Pella (Tabaqat Fahl) in 1996/97. *Annual of the Department of Antiquities of Jordan* 45.

Cameron, D.
  1981 *The Ghassulian Wall Paintings*. London: Kenyon–Deane.
  1995 The Ghassulian Wall Paaintings. *Boston College Studies in Philosophy* 28: 114-20.

Carmi, I., C. Epstein and D. Segal
  1995 Radiocarbon dates from Chalcolithic Sites in the Golan. *Atiqot* 27: 207-209.

Carneiro, R.
  1970 A Theory for the Origin of the State. *Science* 69: 733-78.

Cialowicz, K.
  1989 *Les têtes de massues des périodes Prédynastique et Archaique dans la Vallèe du Nil*. Krakow: Krakow University.

Clark, V.
  1978/79 Investigations in a Prehistoric Necropolis near Bab edh-Dhra. *Annual of the Department of Antiquities of Jordan* 22-23: 57-77.

Contenson, H. de
  1960a Three Soundings in the Jordan Valley. *Annual of the Department of Antiquities of Jordan* 3-4: 12-98.
  1960b La Chronologie relative du Niveau le plus Ancien de Tell esh Shuna. *Mélanges de l'Université Saint-Joseph* 37: 57-75.
  1961 Remarques sur le Chalcolithique Recent de Tell esh-Shuna. *Revue Biblique* 68: 546-56.
  1964 The 1953 Survey in the Yarmouk and Jordan Valleys. *Annual of the Department of Antiquities of Jordan* 8-9: 30-46.

Copeland, L.
  1969 Neolithic Village Sites in the South Beqaa, Lebanon. *Mélanges de l'Université Saint-Joseph* 45: 85-114.

Coughenour, R.
  1976 Preliminary Report on the Exploration and Excavation of Mugharet el Warda and Abu Thawab. *Annual of the Department of Antiquities of Jordan* 21: 71-76.

D'Altroy, T. and T. Earle
  1985 Staple Finance, Wealth Finance and Storage in the Inka Political Economy. *Current Anthropology* 26: 187-206.

Dollfus, G. and Z. Kafafi (eds.)
  1988 *Abu Hamid: Village du 4ème Millénaire de la vallée du Jordain*. Amman: Centre culturel français.
  1989 Abu Hamid. In D. Homès-Fredericq and J.B. Hennessy (eds.), *Archaeology of Jordan. II.1. Field Reports: Surveys and Sites A–K*, 102-13. Leuven: Peeters.
  1993 Recent Researches at Abu Hamid. *Annual of the Department of Antiquities of Jordan* 37: 241-62.

1995 Representations Humains et Animales sur le site d'Abu Hamid. In K. 'Amr, F. Zayadine and M. Zaghloul (eds.), *Studies in the History and Archaeology of Jordan*, V: 449-57. Amman: Department of Antiquities.

Donahue, J., B. Peer and R.T. Schaub
1997 The Southeastern Dead Sea Plain: Changing Shorelines and their Impact on Settlement Patterns through Historical Periods. In M. Zaghloul *et al.* (eds.), *Studies in the History and Archaeology of Jordan*, VI: 127-36. Amman: Department of Antiquities.

Dunand, M.
1973 *Fouilles de Byblos*, V. Paris.

Earle, T.K.
1987 Chiefdoms in Archaeological and Ethnohistorical Perspective. *Annual Review of Anthropology* 16: 279-308.

Earle, T.K. (ed.)
1991 *Chiefdoms: Power, Economy and Ideology*. New York: Cambridge University.

Edwards, W.I.
1997 *Report on Teleilat Ghassul Petrography 1995–96*. Unpublished Manuscript.

Edwards, W.I. and E.R. Segnit
1984 Pottery Technology at the Chalcolithic Site of Teleilat Ghassul. *Archaeometry* 26: 69-77.

Elliot, C.
1977 The Religious Beliefs of the Ghassulians. *Palestine Exploration Quarterly* 109: 3-25.
1978 The Ghassulian Culture in Palestine: Origins, Influences and Abandonment. *Levant* 10: 37-54.

Epstein, C.
1975 Basalt Pillar Figures from the Golan. *Israel Exploration Journal* 25: 193-201.
1978a Aspects of Symbolism in Chalcolithic Palestine. In P. Moorey and P. Parr (eds.), *Archaeology in the Levant: Essays for Kathleen Kenyon*, 23-35. Warminster: Aris & Phillips.
1978b A New Aspect of Chalcolithic Culture. *Bulletin of the American Schools of Oriental Research* 229: 27-45.
1985 Laden Animal Figurines from the Chalcolithic Period in Palestine. *Bulletin of the American Schools of Oriental Research* 258: 53-62.
1988 Basalt Pillar Figures from the Golan and the Huleh Region. *Israel Exploration Journal* 38: 205-23.
1993 Oil Production in the Golan Heights during the Chalcolithic Period. *Tel Aviv* 20: 133-46.

Epstein, C. and T. Noy
1988 Observations Concerning Perforated tools from Chalcolithic Palestine. *Paléorient* 14: 133-44.

Falconer, S.E.
1994 Village Economy and Society in the Jordan Valley: A Study of Bronze Age Rural Complexity. In G. Schwartz and S. Falconer (eds.), *Archaeological Views from the Countryside: Village Communities in Early Complex Societies*, 121-42. Washington.

Falconer, S.E. and S.H. Savage
1995 Heartlands and Hinterlands: Alternative Trajectories of Early Urbanisation in Mesopotamia and the Southern Levant. *American Anthropology* 60: 37-58.

Finkelstein, I.
1990 Early Arad-Urbanism of the Nomads. *Zeitschrift des Deutschen Palästina-Vereins* 106: 34-50.
1995 Two Notes on Early Bronze Age Urbanisation and Urbanism. *Tel Aviv* 22: 47-69.

Finkelstein, I. and R. Gophna
1993 Settlement, Demographic, and Economic Patterns in the Highlands of Palestine in the Chalcolithic and Early Bronze Periods and the Beginning of Urbanism. *Bulletin of the American Schools of Oriental Research* 289: 1-22.

Fox, N.
1997 The Striped Goddess from Gilat: Implications for the Chalcolithic Cult. *Israel Exploration Journal* 47: 212-25.

Frangipane, M.
1997 A 4th-Millennium Temple/Palace Complex at Arslantepe-Malatya. North–South Relations and the formation of Early State Societies in the northern regions of Greater Mesopotamia. *Paléorient* 23: 45-73.

Gal, Z., H. Smithline and D. Shalem
1997 A Chalcolithic Burial Cave in Peqi'in, Upper Galilee. *Israel Exploration Journal* 47: 145-54.

Galili, E. and J. Sharvit
1994/95 Evidence of Olive Oil Production from the Submerged Site at Kfar Samir, Israel. *Mit. Haev.* 26: 122-33.

Garfinkel, Y.
1992a The Material Culture in the Central Jordan Valley in the Pottery Neolithic and Early Chalcolithic Periods. Unpublished PhD dissertation, Hebrew University, Jerusalem.
1992b *The Pottery Assemblages of the Sha'ar Hagolan and Rabah Stages of Munhatta (Israel)*. Paris.
1993 Tel 'Eli. *Excavations and Surveys in Israel* 12: 19.
1995 *Human and Animal Figurines from Munhatta (Israel)*. Paris: Association Patéorient.
1998 Dancing and the Beginning of Art Scenes in the Early Village Communities of the Near East and Southeast Europe. *College Art Journal* 8: 207-37.

Garrard, A.
1998 Environment and Cultural Adaptations in the Asraq Basin: 24000-7000 BC. In D. Henry (ed.), *British Archaeological Reports, International Series* 705. Oxford: Archaeopress.

Garrard, A. et al.
　1987　Prehistoric Environment and Settlement in the Azraq Basin: An Interim Report on the 1985 Excavation Season. *Levant* 19: 5-25.

Genz, H.
　1997　Problems in defining a Chalcolithic for Southern Jordan. In H.G. Gebel et al. (eds.), *The Prehistory of Jordan. II. Perspectives from 1997*, 441-48. Berlin: *ex oriente*.

Gilead, I.
　1988　The Chalcolithic Period in the Levant. *Journal of World Prehistory* 2.4: 397-443.
　1990　The Neolithic–Chalcolithic Transition and the Qatifian of the northern Negev and Sinai. *Levant* 22: 47-63.
　1994　The History of the Chalcolithic Settlement in the Nahal Beer Sheva Area: The Radiocarbon Aspect. *Bulletin of the American Schools of Oriental Research* 296: 1-13.
　1995　*Grar: A Chalcolithic Site in the Northern Negev*. Beersheva: Ben-Gurion University.

Gilead, I. and Y. Goren
　1986　Stations of the Chalcolithic Period in Nahal Sekher, Northern Negev. *Paléorient* 12: 83-90.
　1989　Petrographic Analyses of Fourth Millennium BC Pottery and Stone Vessels from the Northern Negev, Israel. *Bulletin of the American Schools of Oriental Research* 275: 5-14.

Glueck, N.
　1951　*Explorations in Eastern Palestine IV*. Annual of the American Schools of Oriental Research (1945–49). New Haven, CT: American Schools of Oriental Research.

Gonen, R.
　1992　The Chalcolithic Period. In A. Ben-Tor (ed.), *The Archaeology of Ancient Israel*, 40-80. New Haven, CT: Yale University.

Gopher, A.
　1995　Early Pottery Bearing Groups in Israel—the Pottery Neolithic Period. In T. Levy (ed.), *The Archaeology of Society in the Holy Land*, 205-25. London: Leicester University.

Gopher, A. and R. Gophna
　1993　Cultures of the Eighth and Seventh Millennia BP in the Southern Levant: A Review for the 90s. *Journal of World Prehistory* 7: 297-353.

Gopher, A. and E. Orelle
　1996　An Alternative Interpretation for the Material Imagery of the Yamukian, a Neolithic Culture of the Sixth Millennium BC in the Southern Levant. *College Art Journal* 6: 255-79.

Gopher, A. and T. Tsuk (eds.)
　1996　*The Nahal Qanah Cave*. Tel Aviv: Tel Aviv University.

Gophna, R. and S. Sadeh
　1988/89　Excavations at Tel Tsaf: An Early Chalcolithic Site in the Jordan Valley. *Tel Aviv* 15-16: 3-36.

Gordon, R. and L. Villiers
　1986　Telul edh Dhahab and its Environs Surveys of 1980 and 1982: A Preliminary Report. *Annual of the Department of Antiquities of Jordan* 30: 275-89.

Goren, Y.
　1990　The Qatifian Culture in southern Israel and Transjordan. *Mitukot Haeven* 23: 100-12.

Greene, J.
　1986　*Ajlun-kufrinji Archaeological Survey*. Amman: Department of Antiquities.

Grigson, C.
　1995　Plough and Pasture in the Early Economy of the Southern Levant. In T. Levy (ed.), *The Archaeology of Society in the Holy Land*, 245-68. London: Leicester University.

Gustavson-Gaube, C.
　1986　Tall as Shuna North 1984–85. *Archiv für Orientforschung* 31: 218-22.

Gustavson-Gaube, C. and M. Ibrahim
　1986　Sahab Survey 1983. *Archiv für Orientforschung* 31: 283-86.

Habgood, P.
　1983　Report on the Lithic Assemblage from the 1975–77 Excavations at Teleilat Ghassul. Unpublished manuscript.

Hanbury-Tenison, J.
　1986　*The Late Chalcolithic to Early Bronze I Transition in Palestine and Transjordan*. Oxford: British Archaeological Report, International Series 311.
　1987　Jerash Region Survey 1984. *Annual of the Department of Antiquities of Jordan* 31: 129-57.

Hanbury-Tenison, J. et al.
　1984　Wadi Arab Survey 1983. *Annual of the Department of Antiquities of Jordan* 28: 385-424.

Harrison, T.
　1997　Shifting Patterns of Settlement in the Highlands of Central Jordan during the Early Bronze Age. *Bulletin of the American Schools of Oriental Research* 306: 1-37.

Hart, S.
　1987　Five Soundings in Southern Jordan. *Levant* 19: 33-47.

Hart, S. and R. Falkner
　1985　Preliminary Report on a Survey in Edom. *Annual of the Department of Antiquities of Jordan* 29: 255-77.

Hauptmann, A.
1989   The Earliest Periods of Copper Metallurgy in Feinan/Jordan. In A. Hauptmann, E. Pernicka and G. Wagner (eds.), *Old World Archaeometallurgy*, 119-35. Bochum: Selbstverlag des Deutschen Bergbau-Museums.

Helms, S.
1987a A Question of Economic Control during the Proto-Historical Era of Palestine/Transjordan. In A. Hadidi (ed.), *Studies in the History and Archaeology of Jordan*, III, 41-51. Amman: Department of Antiquities.
1987b Jawa, Tell Um Hammad and the EB I/Late Chalcolithic Landscape. *Levant* 19: 49-82.
1987c A Note on Some 4th Millennium Stamp Seal Impressions from Jordan. *Akkadica* 52: 29-31.
1991   Stamped, Incised and Painted Designs. In A. Betts (ed.), *Excavations at Jawa 1972–1986*, 113-24. Edinburgh.

Hennessy, J.B.
1969   Preliminary Report on a First Season of Excavations at Teleilat Ghassul. *Levant* 1: 1-24.
1977   *Teleilat Ghassul: An Interim Report*. Sydney: Sydney University.
1978   City Below the Sea. *Hemisphere* 22.7: 30-35.
1982   Teleilat Ghassul and its Place in the Archaeology of Jordan. In A. Hadidi (ed.), *Studies in the History and Archaeology of Jordan*, I: 55-58. Amman: Department of Antiquities.
1989   Ghassul, Teleilat. In D. Homès-Fredericq and J.B. Hennessy (eds.), *Archaeology of Jordan*. II.1. *Field Reports: Surveys and Sites A-K*, 230-41. Akkadica Supplementum 7. Leuven: Peeters.

Henry, D.
1995   *Prehistoric Cultural Ecology and Evolution: Insights from Southern Jordan*. New York: Plenum.

Hodder, I. and R. Matthews
1998   Çatalhöyük: The 1990s Seasons. In R. Matthews (ed.), *Ancient Anatolia*, 43-51. London: British Institute of Archaeology at Ankara.

Hole, F.
1997   Evidence for Mid-Holocene Environmental Change in the Western Khabur Drainage, Northeastern Syria. In H. Dalfes *et al.* (eds.), *Third Millennium* BC *Climate Change and Old World Collapse*, 39-66. Berlin.

Hourani, F. and M.-A. Courty
1997   L'évolution morpho-climatique de 10 500 à 5 500 BP dans la vallée du Jourdain. *Paléorient* 23: 95-105.

Ibach, R.
1987   *Archaeological Survey of the Hesban Region: Catalogue of Sites and Characterisation of Periods*. Hesban 5. Berrien Springs, MI: Institute of Archaeology and Andrews University.

Ibrahim, M.
1983   Siegel und Siegelabdrucke aus Sahab. *Zeitschrift des Deutschen Palästina-Vereins* 99: 43-53.
1984   Sahab. *Archiv für Orientforschung* 29: 256-60.
1987   Sahab and its Foreign Relations. In A. Hadidi (ed.), *Studies in the History and Archaeology of Jordan*, III: 73-81. Amman: Department of Antiquities.

Ibrahim, M. and S. Mittmann
1998   Eine chalkolithische Stierskulptur aus Nordjordanien. *Zeitschrift des Deutschen Palästina-Vereins* 114: 101-105.

Ibrahim, M., J. Sauer and K. Yassine
1976   The East Jordan Valley Survey. *Bulletin of the American Schools of Oriental Research* 222: 41-66.

Ji, C.
1997   The Chalcolithic and Early Bronze Cemeteries near Iraq al-Amir and the Preliminary Report on Salvage Excavations. *Annual of the Department of Antiquities of Jordan* 41: 49-68.

Jobling, W.
1989   Aqaba-Ma'an Archaeological and Epigraphic Survey. In D. Homès-Fredericq and J.B. Hennessy (eds.), *Archaeology of Jordan*. II.1. *Field Reports: Surveys and Sites A–K*, 16-24. Akkadica Supplementum 7. Leuven: Peeters.

Joffe, A.
1993   *Settlement and Society in Early Bronze Age I and II Southern Levant: Complementarity and Contradiction in a Small-Scale Complex Society*. Sheffield: Sheffield Academic Press.
1998   Alcohol and Social Complexity in Ancient Western Asia. *Current Anthropology* 39: 297-322.

Joffe, A. and J.-P. Dessel
1995   Redefining Chronology and Terminology for the Chalcolithic of the Southern Levant. *Current Anthropology* 36: 507-18.

Kafafi, Z.
1982   The Neolithic of Jordan. Unpublished PhD dissertation, Freie Universität, Berlin.
1992   Pottery Neolithic Settlement Patterns in Jordan. In M. Zaghloul, K. 'Amr, F. Zayadine and N.R. Tawfiq (eds.), *Studies in the History and Archaeology of Jordan*, IV: 115-23. Amman: Department of Antiquities.
1993   The Yarmoukians in Jordan. *Paléorient* 19: 101-14.

Kafafi, Z. *et al.*
1997   The Wadi az-Zarqa'/Wadi ad-Dulayl Archaeological Project: Report on the 1996 Fieldwork Season. *Annual of the Department of Antiquities of Jordan* 41: 9-26.

Kamleh, J.
1998 Der Zeiraqon-Survey (1989–1994): Mit Beiträgen zur Methodik und zur geschichtlichen Auswertung archäologischer Oberflächenuntersuchungen in Palästina. Unpublished PhD dissertation, University of Tubingen.

Kantor, H.
1992 The Relative Chronology of Egypt and its Foreign Correlations before the First Intermediate Period. In R. Ehrich (ed.), *Chronologies in Old World Archaeology*, 3-21. Chicago: University of Chicago.

Kaplan, J.
1959 The Connections of the Palestinian Chalcolithic Culture with Prehistoric Egypt. *Israel Exploration Journal* 9: 134-36.
1960 The Relation of the Chalcolithic Pottery of Palestine to Halafian Ware. *Bulletin of the American Schools of Oriental Research* 159: 32-36.
1969 'Ein el-Jarba: Chalcolithic Remains in the Plain of Esdraelon. *Bulletin of the American Schools of Oriental Research* 194: 2-39.

Keel, O.
1989 *Studien zu der Stamp Siegeln aus Palästina*, II. Göttingen: Orbis Biblicus et Orientalis.

Kemp, B.
1973 Photographs of the Decorated Tomb at Hierakonpolis. *Journal of Egyptian Archaeology* 59: 36-43.
1989 *Ancient Egypt: Anatomy of a Civilisation*. London: Routledge.

Kempinski, A.
1992 Chalcolithic and Early Bronze Age Temples. In A. Kempinski and R. Reich (eds.), *The Architecture of Ancient Israel*, 53-59. Jerusalem.

Kenyon, K. and T. Holland
1982 *Excavations at Jericho*, IV. London: British School of Archaeology in Jerusalem.
1983 *Excavations at Jericho*, V. London: British School of Archaeology in Jerusalem.

Kerestes, T. *et al.*
1977/78 An Archaeological Survey of the Three Reservoir Areas in Northern Jordan, 1978. *Annual of the Department of Antiquities of Jordan* 22: 108-35.

Kerner, S.
1998 Entwicklung handwerklicher Spezialisierung im 6. bis 4 vorchristl. Jts. in der südlichen Levante. Unter Berücksichtigung des Zusammenhangs zwischen vertikaler und horizontaler Organisation. Unpublished PhD dissertation, Freie Universität, Berlin.

Kerner, S. *et al.*
1992 Excavations in Abu Snesleh: Middle Bronze Age and Chalcolithic Architecture in Central Jordan. In S. Kerner (ed.), *The Near East in Antiquity*, III, 43-54. Amman: German Protestant Institute of Archaeology.

Khalil, L.
1987 Preliminary Report on the 1985 Excavation at el-Maqass-Aqaba. *Annual of the Department of Antiquities of Jordan* 31: 481-85.
1992 Some technological features from a Chalcolithic site at Magass-Aqaba. In M. Zaghloul, K. 'Amr, F. Zayadine and N.R. Tawfiq (eds.), *Studies in the History and Archaeology of Jordan*, IV: 143-48. Amman: Department of Antiquities.
1995 A Second Season of Excavations at al-Magass-Aqaba. *Annual of the Department of Antiquities of Jordan* 39: 65-80.

Kirkbride, D.
1969 Early Byblos and the Beqa'a. *Mélanges de l'Université Saint-Joseph* 45: 45-60.

Koeppel, R.
1938 Die achte Grabung in Ghassul. *Biblica* 19: 260-66.

Koeppel, R., R. Mallon and R. Neuville
1940 *Teleilat Ghassul*, II. Rome: Pontifical Biblical Institute.

Lee, J.
1973 Chalcolithic Ghassul: New Aspects and Master Typology. Unpublished PhD dissertation, Hebrew University, Jerusalem.

Lehmann, G., R. Lamprichs, S. Kerner and R. Bernbeck
1991 The 1990 Excavations at Abu Snesleh: Preliminary Report. *Annual of the Department of Antiquities of Jordan* 35: 41-63.

Lenzen, C.
1986 Tall Irbid and Beit Ras. *Archiv für Orientforschung* 31: 164-66.

Lenzen, C. and A. McQuitty
1983 A Preliminary Survey of the Irbid–Beit Ras Region, Northwestern Jordan. *Annual of the Department of Antiquities of Jordan* 27: 656.

Leonard, A.
1987 The Jarash–Tell el-Husn Highway Survey. *Annual of the Department of Antiquities of Jordan* 31: 343-90.
1989 A Chalcolithic Fine Ware from Katuret es-Samra in the Jordan Valley. *Bulletin of the American Schools of Oriental Research* 276: 3-14.
1992 *The Jordan Valley Survey, 1953: Some Unpublished Soundings Conducted by James Mellaart*. Winona Lake, WI: Eisenbrauns.

Levy, T.
1992 Radiocarbon Dating of the Beersheba Culture and Predynastic Egypt. In E.C.M. van den Brink (ed.), *The Nile Delta in Transition: 4th–3rd Millennium* BC, 345-36. Jerusalem: Israel Exploration Society.

1993 Ghassul, Tuleilat el. In E. Stern (ed.), *New Encyclopaedia of Archaeological Excavations in the Holy Land*, 506-11. New York: Simon & Schuster.

1995 Cult, Metallurgy and Rank Societies—Chalcolithic Period (ca. 4500–3500 BCE). In T. Levy (ed.), *The Archaeology of Society in the Holy Land*, 226-244. London: Leicester University.

1996 Anthropological Approaches to Protohistoric Palestine: A Case Study from the Negev Desert. In J. Seger (ed.), *Retrieving the Past: Essays on Archaeology and Research in Honor of Gus van Beek*, 163-78. Winona Lake, WI: Eisenbrauns.

Levy, T. (ed.)

1987 *Shiqmim I: Studies Concerning Chalcolithic Societies in the Northern Negev Desert, Israel*. British Archaeological Reports. Oxford: BAR.

Levy, T. and D. Alon

1982 The Chalcolithic Mortuary Site near Mezad Aluf, Northern Negev Desert: A Preliminary Study. *Bulletin of the American Schools of Oriental Research* 248: 37-59.

Levy, T. and J. Golden

1996 Syncretistic and Mnemonic Dimensions of Chalcolithic Art: A New Human Figurine from Shiqmim. *Biblical Archaeologist* 59: 150-59.

Levy, *et al.*

in press Eearly Metallurgy, Interaction and Societal change: The Jabul Hamrat Fidan (Jordan) Archaeological Survey 1998. In G. Bisheh (ed.) *Studies in the History and Archaeology of Jordan* VII. Amman: Department of Antiquities.

Lovell, J. L.

1999 The Late Neolithic and Chalcolithic Periods in the Southern Levant: New Data from Teleilat Ghassul, Jordan. Unpublished PhD dissertation, University of Sydney, Australia.

2000 Pella in Jordan in the Chalcolithic Period. In G. Philip and D. Baird (eds.), *Breaking with the Past: Ceramics and Change in the Early Bronze Age of the Southern Levant*, 59-71. Sheffield: Sheffield Academic Press.

Lovell, J.L., Z. Kafafi and G. Dollfus

1997 A Preliminary Note on the Ceramics from the Basal Levels at Abu Hamid. In H.G. Gebel *et al.* (eds.), *The Prehistory of Jordan II: Perspectives from 1997*, 361-70. Berlin: ex oriente.

Lucas, A. and J.R. Harris

1962 *Ancient Egyptian Materials and Industries*. 4th edn. London: Edward Arnold.

Mabry, J.

1989 Investigations at Tell el-Handaquq, Jordan (1987–88). *Annual of the Department of Antiquities of Jordan* 33: 59-95.

Mabry, J. and G. Palumbo

1988 The 1987 Wadi el-Yabis Survey. *Annual of the Department of Antiquities of Jordan* 32: 275-305.

1992 Environmental, Economic, and Political Constraints on Ancient Settlement Patterns in the Wadi al-Yabis Region. In M. Zaghloul *et al.* (eds.), *Studies in the History and Archaeology of Jordan*, IV: 67-72. Amman: Department of Antiquities.

MacDonald, B. *et al.*

1988 *The Wadi el Hasa Archaeological Survey 1979–1983, West-Central Jordan*. Waterloo, ON: Wilfrid Laurier University.

1992 *The Southern Ghors and Northeast 'Arabah Archaeological Survey*. Sheffield Archaeological Monographs 5. Sheffield: J.R. Collis.

Mairs, L.

1997 Ghassul Archaeozoological Report 1995 and 1997 Seasons. Unpublished Manuscript.

Mallon, A., R. Koeppel and R. Neuville

1934 *Teleilat Ghassul I*. Rome: Pontifical Biblical Institute.

Mattingly, G.

1996 Al-Karak Resources Project 1995: A Preliminary Report on the Pilot Season. *Annual of the Department of Antiquities of Jordan* 40: 349-68.

Mazar, A.

1990 *Archaeology of the Land of the Bible*. New York: Doubleday.

McAdam, E.

1997 The Figurines from the 1982–5 Seasons of Excavations at Ain Ghazal. *Levant* 29: 115-52.

McConaughy, M.

1979 Formal and Functional Analysis of Chipped Stone Tools from Bab esdh Dhra. Unpublished PhD dissertation, Ann Arbor, MI.

McCorriston, J.

1997 The Fibre Revolution: Textile Extensification, Alienation and Social Stratification in Ancient Mesopotamia. *Current Anthropology* 38: 517-49.

McCreery, D.

1978/79 Preliminary Report of the APC Township Archaeological Survey. *Annual of the Department of Antiquities of Jordan* 22-23: 150-62.

McGovern, P.

1989 Baq'ah Valley Project-Survey and Excavation. In D. Homès-Fredericq and J.B. Hennessy (eds.), *Archaeology of Jordan. II.1. Field Reports: Surveys and Sites A–K*, 25-44. Akkadica Supplementum 7. Leuven: Peeters.

McNicoll, A.W. *et al.*

1992 *Pella in Jordan 2*. Sydney: Mediterranean Archaeology.

Meadows, J.

1998 Ghassul Archaeobotany Report: 1997 Season. Unpublished Manuscript.

1999 Olive Domestication in the Jordan Valley. In L. Hopkins and A. Parker (eds.), *The Ancient Near East: An Australian Postgraduate Perspective.* Sydney: Archaeological Computing Laboratory.

Mellaart, J.
1956 The Neolithic Site of Ghrubba. *Annual of the Department of Antiquities of Jordan* 3: 24-40.
1962 Preliminary Report of the Archaeological Survey in the Yarmouk and Jordan Valley. *Annual of the Department of Antiquities of Jordan* 6-7: 126-58.
1967 *Çatalhöyük: A Neolithic Town in Anatolia.* London: Thames & Hudson.
1975 *The Neolithic of the Near East.* London: Thames & Hudson.
1998 Çatalhöyük: The 1960s Seasons. In R. Matthews (ed.), *Ancient Anatolia*, 35-41. London: British Institute of Archaeology at Ankara.

Miller, J. (ed.)
1991 *Archaeological Survey of the Kerak Plateau.* Atlanta, GA: Scholars Press.

Mittmann, S.
1970 *Beiträge zur Siedlungs und Territorialgeschichte des Nördlichen Ostjordanlandes.* Wiesbaden: Harrassowitz.

Moore, A.
1973 The Late Neolithic in Palestine. *Levant* 5: 36-68.

Moorey, P.R.S.
1989 The Chalcolithic Hoard from Nahal Mishmar, Israel, in Context. *World Archaeology* 20: 171-89.

Mortensen, P. and I. Thuesen
1998 The Prehistoric Periods. In M. Piccirillo and E. Alliath (eds.), *Mount Nebo: New Archaeological Excavations 1967–1997*, 84-99. Jerusalem: Franciscan.

Munro, R., R. Morgan and W. Jobling
1997 Optical Dating and Landscape Chronology at ad-Disa, Southern Jordan, and its Potential. In M. Zaghloul et al. (eds.), *Studies in the History and Archaeology of Jordan*, VI: 97-104. Amman: Department of Antiquities.

Najjar, M. et al.
1990 Tell Wadi Feinan: The First Pottery Neolithic Tell in the South of Jordan. *Annual of the Department of Antiquities of Jordan* 34: 27-51.

Nasrallah, J.
1948 Une station ghassoulienne du Hauran. *Revue Biblique* 55: 81-103.

Neef, R.
1990 Introduction, Development and Environmental Implications of Olive Culture. In S. Bottema et al. (eds.), *Man's Role in the Shaping of the Eastern Mediterranean Landscape*, 295-306. Rotterdam.

North, R.
1961 *Ghassul 1960 Excavation Report.* Rome: Pontifical Biblical Institute.
1964 Ghassul's New-Found Jar-Incision. *Annual of the Department of Antiquities of Jordan* 8-9: 68-74.

Ottosson, M.
1980 *Temples and Cult Places in Palestine.* Uppsala: Uppsala University.

Palumbo, G.
1994 *JADIS: The Jordan Antiquities Database and Information System: A Summary of the Data.* Amman: Department of Antiquities and the American Center of Oriental Research.

Palumbo, G., J. Mabry and I. Kuijt
1990 The Wadi el-Yabis Survey: Report on the 1989 Field Season. *Annual of the Department of Antiquities of Jordan* 34: 95-118.

Palumbo, G. et al.
1996 The Wadi az-Zarqa/Wadi ad-Dulayl Excavation and Survey Projects. *Annual of the Department of Antiquities of Jordan* 40: 375-429.

Payne, J.
1973 Tomb 100: The Decorated Tomb at Hierakonpolis Confirmed. *Journal of Egyptian Archaeology* 59: 31-35.

Peltenburg, E.
1987 Early Faience: Recent Studies, Origins and Relations with Glass. In M. Binson and I. Freestone (eds.), *Early Vitreous Materials*, 5-29. London: British Museum.
1995 Kissonerga in Cyprus and the Appearance of Faience in the East Mediterranean. In S. Bourke and J.-P. Descoeudres (eds.), *Trade, Contact and the Movement of Peoples in the Eastern Mediterranean: Studies in Honour of J. Basil Hennessy*, 31-41. Sydney: Mediterranean Archaeology.

Perrot, J.
1959 Statuettes en ivoire et autres objets en ivoire et os provenent des gisements préhistoriques de la région de Beershéba. *Syria* 36: 8-19.
1968 La préhistoire Palestinienne. *Supplément au Dictionnaire Archéologique de la Bible* 8: 286-446.
1972 La préhistoire Palestinienne. In L. Pirot et al. (eds.), *Supplément au Dictionnaire de la Bible* 8: 286-446. Paris: Letouzey & Ané.

Perrot, J. and D. Ladiray
1980 *Tombes à ossuaires de la région cotière Palestinienne au IVe millenaire l'ère chrétienne.* Paris: Association Patéorient.

Philip, G.
1998 The Basalt Industry in the Southern Levant. In P. Matthiae et al. (eds.), *First International Congress on the Archaeology of the Ancient Near East.* Rome: Universita di Roma La Sapienza.

Philip, G. and O. Williams-Thorpe
1993 A Provenance Study of Jordanian Basalt Vessels of the Chalcolithic and Early Bronze Age I Periods. *Paléorient* 19: 51-63.

Porath, Y.
1992 Domestic Architecture of the Chalcolithic Period. In A. Kempinski and R. Reich (eds.), *The Architecture of Ancient Israel*, 40-50. Jerusalem: Israel Exploration Society.

Prag, K.
1974 The Intermediate Early Bronze–Middle Bronze Age: An Interpretation of the Evidence from Transjordan, Syria and Lebanon. *Levant* 6: 69-116.
1991 A Walk in the Wadi Hesban. *Palestine Exploration Quarterly* 123: 48-61.
1995 The Dead Sea Dolmens: Death and the Landscape. In S. Campbell and A. Green (eds.), *The Archaeology of Death in the Ancient Near East*, 75-85. Oxford: Oxbow.

Raikes, T.
1980 Notes on Some Neolithic and Later Sites in Wadi Araba and the Dead Sea Valley. *Levant* 12: 40-60.

Rast, W. and R.T. Schaub
1974 Survey of the Southeastern Plain of the Dead Sea, 1973. *Annual of the Department of Antiquities of Jordan* 19: 5-53.

Reese, D.S.
1998 A First Study of Teleilat Ghassul Shell Remains 1994–97. Unpublished Manuscript.

Reese, D.S., H.K. Mienis and F.R. Woodward
1986 On the Trade of Shells and Fish from the Nile River. *Bulletin of the American Schools of Oriental Research* 264: 79-84.

Rehren, T., K. Hess and G. Philip
1997 Fourth Millennium BC Copper Metallurgy in Northern Jordan: The Evidence from Tell esh-Shuna. In H.G. Gebel *et al.* (eds.), *The Prehistory of Jordan II: Perspectives from 1997*, 625-40. Berlin: ex oriente.

Rollefson, G.
1992 Chipped Stone Tools from Pella. In A.W. McNicoll *et al.*, *Pella in Jordan*, II: 231-41. Sydney: Mediterranean Archaeology.

Rosen, S.A.
1983 Tabular Scraper Trade: A Model for Material Culture Dispersion. *Bulletin of the American Schools of Oriental Research* 249: 79-86.
1989 The Origins of Craft Specialisation: Lithic Perspectives. In I. Hershkovitz (ed.), *People and Culture in Change*, 107-14. Oxford: British Archaeological Reports International Series.
1997 *Lithics after the Stone Age*. Walnut Creek: Altamira.

Rothenberg, B.
1999 Archaeometallurgical Researches in the Southern Arabah 1959–1990. Part 1: Late Pottery Neolithic to Early Bronze IV. *Palestine Exploration Quarterly* 131: 68-89.

Rothman, M.
1994 Evolutionary Typologies and Cultural Complexity. In G. Stein and M. Rothman (eds.), *Chiefdoms and Early States in the Near East: The Organisational Dynamics of Complexity*, 1-22. Madison.

Roux, V. and M.-A. Courty
1997 Les bols elabores au tour d'Abu Hamid: Rupture technique au 4e millenaire avant J.-C. dans le Levant-Sud. *Paléorient* 23: 25-43.

Rowan, Y. and T. Levy
1994 Proto-Canaanean Blades of the Chalcolithic Period. *Levant* 26: 167-74.

Sapin, J.
1992 De l'occupation à l'utilisation de l'espace à l'aube de l'age du bronze dans la région de Jérash et sa périphérie orientale. In M. Zaghloul *et al.* (eds.), *Studies in the History and Archaeology of Jordan*, IV: 169-74. Amman: Department of Antiquities.

Savage, S.H.
1997 Descent Group Competition and Economic Strategies in Predynastic Egypt. *Journal of Anthropological Archaeology* 16: 226-68.

Scham, S.
1997 Shiqmim's Violin-Shaped Figurines and Ghassulian Bone Artifacts. *Biblical Archaeologist* 60: 108.

Schmandt-Besserat, D.
1992 *Before Writing*. I. *From Counting to Cuneiform*. Austin, TX.

Schulman, A.
1992 Still More Egyptian Seal Impressions from 'En Besor. In E.C.M. van den Brink (ed.), *The Nile Delta in Transition: 4th–3rd Millennium BC*, 395-417. Jerusalem: Israel Exploration Society.

Seaton, P.
1999 Recent Research in the Chalcolithic Sanctuary Precinct at Teleilat Ghassul. In P. Matthiae (ed.), *First International Congress on the Archaeology of the Ancient Near East*. Rome: Universita' di Roma 'La Sapienza'.

Simmons, A. and Z. Kafafi
1992 The Ain Ghazal Survey: Patterns of Settlement in the Greater Wadi az-Zarqa Area, Central Jordan. In M. Zaghloul *et al.* (eds.), *Studies in the History and Archaeology of Jordan*, IV: 77-82. Amman: Department of Antiquities.

Simmons, A. and M. Najjar
1998 Al-Ghuwayr I, a Pre-Pottery Neolithic Village in Wadi Faynan, Southern Jordan: A Preliminary Report on the

1996 and 1997/98 Seasons. *Annual of the Department of Antiquities of Jordan* 42: 91-101.

Simmons, A. et al.
1989 Test Excavations at Wadi Shu'eib, a Major Neolithic Settlement in Central Jordan. *Annual of the Department of Antiquities of Jordan* 33: 27-42.

Stager, L.
1985 The First Fruits of Civilisation. In J. Tubb (ed.), *Palestine in the Bronze and Iron Age*, 172-88. London.
1992 The Periodisation of Palestine from the Neolithic through Early Bronze Times. In R. Ehrich (ed.), *Chronologies in Old World Archaeology*, 22-60. Chicago: University of Chicago.

Stein, G.
1994a Economy, Ritual and Power in 'Ubaid Mesopotamia. In G. Stein and M. Rothman (eds.), *Chiefdoms and Early States in the Near East: The Organisational Dynamics of Complexity*, 35-46. Madison, WI: Prehistory Press.
1994b Segmentary States and Organisational Variation in Early Complex Societies. In G. Schwartz and S. Falconer (eds.), *Archaeological Views from the Countryside: Village Communities in Early Complex Societies*, 10-18. Washington, DC: Smithsonian Institution.
1998 Heterogeneity, Power and Political Economy: Some Current Research Issues in the Archaeology of Old World Complex Societies. *Journal of Anthropological Research* 6: 1-44.

Stein, G. and M. Rothman (eds.)
1994 *Chiefdoms and Early States in the Near East: The Organisational Dynamics of Complexity*. Madison, WI: Prehistory Press.

Stekelis, M.
1935 *Les Monuments Mégalithiques de Palestine*. Paris: Archives de l'Institut de Paléontologie Humaine.

Tadmor, M. et al.
1995 The Nahal Mishmar Hoard from the Judean Desert: Technology, Composition and Provenance. *Atiqot* 27: 95-148.

Tzori, N.
1958 Neolithic and Chalcolithic Sites in the Valley of Beth Shan. *Palestine Exploration Quarterly* 90: 44-51.
1967 On Two Pithoi from the Beth Shan Region of the Jordan Valley. *Palestine Exploration Quarterly* 99: 101-103.

Ussishkin, D.
1980 The Ghassulian Shrine at En-Gedi. *Tel Aviv* 7: 1-44.

Vaux, R. de
1966 Palestine during the Neolithic and Chalcolithic Periods. In I. Edwards et al. (eds.), *Cambridge Ancient History*, I: 498-538. Cambridge: Cambridge University Press.

Waheeb, M.
1996 Archaeological Excavation at Ras en-Naqb–Aqaba Road Alignment: Preliminary Report (1995). *Annual of the Department of Antiquities of Jordan* 40: 339-48.

Wapnish, P. and B. Hesse
1991 Faunal Remains from Tel Dan: Perspectives on Animal Production at a Village, Urban and Ritual Center. *Archaeozoologia* 4: 9-86.

Watson, P. and M. O'Hea
1996 Pella Hinterland Survey 1994: Preliminary Report. *Levant* 28: 63-76.

Wilcox, G.
1992 Preliminary Report on the Plant Remains from Pella. In A.W. McNicoll et al., *Pella in Jordan*, II: 253-56. Sydney: Mediterranean Archaeology.

Wright, K. et al.
1998 The Wadi Faynan 4th and 3rd Millennia Project, 1997: Report on the First Season of Test Excavations at Wadi Faynan 100. *Levant* 30: 33-60.

Yassine, K., J. Sauer and M. Ibrahim
1988 The East Jordan Valley Survey, 1976: Part 2. In K. Yassine (ed.), *Archaeology of Jordan: Essays and Reports*, 189-207. Amman: Department of Archaeology, University of Jordan.

Yoffee, N.
1993 Too Many Chiefs? (or, Safe Texts for the 90s). In N. Yoffee and A. Sherratt (eds.), *Archaeological Theory: Who Sets the Agenda?*, 60-78. New York: Cambridge University Press.
1995 Conclusion: A Mass in Celebration of the Conference. In T. Levy (ed.), *The Archaeology of Society in the Holy Land*, 542-48. London: Leicester University.

Zohary, D. and P. Spiegel-Roy
1975 The Beginnings of Fruit Growing in the Old World. *Science* 187: 319-27.

5. The Early Bronze I–III Ages

Graham Philip

Introduction

The following discussion is not intended as a history of Early Bronze Age studies in Jordan (for this see Geraty and Willis 1986). Neither is it a review of the Early Bronze Age in the southern Levant as a whole. For example, the influence of Egypt upon developments in the southern Levant is not discussed because the relevant evidence comes almost entirely from Palestine (van den Brink 1992; Harrison 1993; Levy 1995a). I have tried to go beyond a purely descriptive account, and have drawn upon a range of current theoretical models to challenge some traditional interpretations, and highlight problematic areas, in the hope that others may feel ready to approach these issues from new directions.

Early Bronze Age studies in the southern Levant

Origins of the city-state model

There is no published discussion of the Early Bronze Age from a specifically Jordanian perspective. All published accounts dealing with the southern Levant are based upon the data from Palestine, with Jordan regarded simply as its eastern extension. This is in spite of clear geographical and environmental contrasts between the two areas.

Scholars have generally considered the Early Bronze Age within a broadly neo-evolutionary framework (Richard 1987; Mazar 1990; Ben-Tor 1992; Esse 1989a, 1991; Finkelstein 1995a). The traditional interpretation sees increasing complexity during EB I, leading to the appearance of a stratified society during EB II–III. In regional terms, this takes the form of a series of broadly homologous EB II–III city-states, each based around a walled urban centre. These are sometimes viewed as 'peer-polities' (Finkelstein 1995a). This phase of complexity is followed by a period of collapse at the transition from EB III to EB IV (see Esse 1989a: table 1; Palumbo, this volume).

It is important to understand that the notion of Early Bronze Age city-states in the southern Levant became entrenched in the literature (e.g. Albright 1949: 74) many years before scholars began to debate the material correlates of urbanism and stratified societies (Adams 1966; Flannery 1972; Service 1962, 1975; Wright 1977). As a result, the presence at these sites of the classic characteristics of state societies—regional polities, social stratification, elite control of the economic base, administrative systems—was never actually demonstrated; it was simply assumed.

In fact, the idea of Early Bronze Age city-states appears to have derived from two sources. Firstly, Wright's (1961: 81) remark that the Early Bronze Age represented 'the beginning of the city-state system' makes it clear that Early Bronze Age urbanism was really a projection back in time of the situation in the Late Bronze Age as revealed by documentary sources that indicate the existence of small city-states, based around the main urban centres (Moran 1992; Bunimovitz 1995: 326). This model also appears to fit the Middle Bronze Age data (Ilan 1995: 301), and scholars have assumed a broad equivalence between the Early Bronze Age walled settlements and the urban communities of the second millennium BC (Finkelstein 1995a: 55; de Miroschedji 1999: 12). However, the EB III is separated from the Middle Bronze Age by a gap of several centuries, characterized by a very different form of socioeconomic organization (Palumbo, this volume). There is, therefore, no reason to *assume* continuity of social or political organization between the two periods.

The second source was the widely held belief that developments in the southern Levant should comprise a parallel, if smaller scale, version of processes of urbanization and state development taking place in contemporary Mesopotamia and Egypt (Wright 1961: 81; Kenyon 1979: 84-86). In this regard, Ben-Tor's (1992: 86) analogy that places the southern Levant at the edge of a pond, receiving the outer ripples of a stone (urbanism) thrown into the south Mesopotamian centre, is particularly revealing. There existed, therefore, an expectation that urban settlements existed,

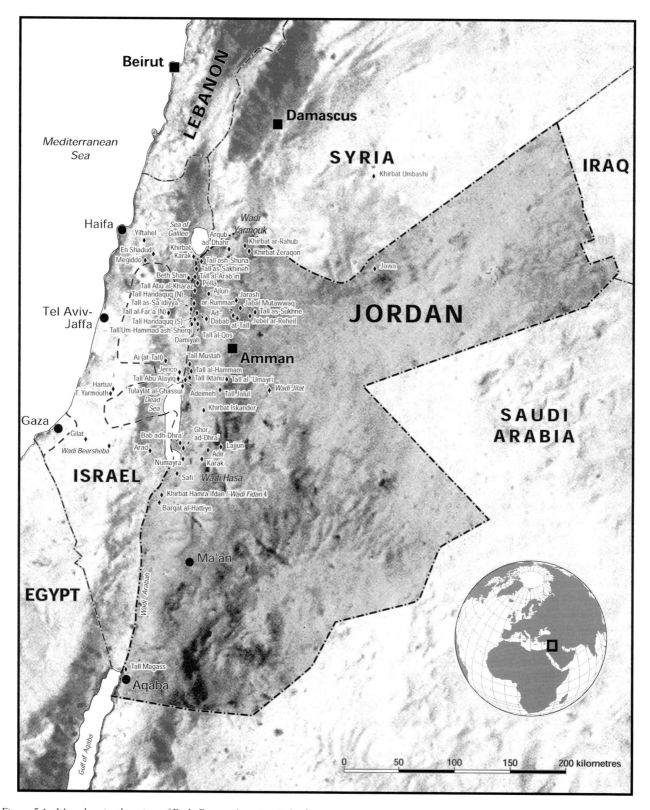

Figure 5.1. Map showing location of Early Bronze Age sites in Jordan.

one that was readily satisfied by the presence of walled sites.

Evidence for the existence of city-states

While scholars working within the framework of traditional Palestinian 'historical' or 'biblical' archaeology (e.g. Kenyon 1979; Mazar 1990) appear to have taken for granted the 'fact' of Early Bronze Age city-states, others have sought to situate Palestinian city-states within a broader theoretical debate, and to identify specific correlates of complexity and urbanism within the regional data (Esse 1989a: fig. 1; de Miroschedji 1989: fig. 1). These include the presence of public architecture, including defensive and administrative structures, evidence for the growth of social, political and settlement hierarchies, and various forms of economic specialization including both crafts and subsistence practices.

Esse (1989a), for example, argued that Early Bronze Age Palestine represented an instance of Secondary State Formation (Price 1978). Thus, the development of small-scale polities controlled by local elites took place in response to connections with the more developed state of Egypt, which presented opportunities for incipient local elites to enhance their power by exploiting strategic economic niches. The 'urban' centres thus constituted regional foci of specialization and control. Such a coincidence of political and economic power is widely assumed in neo-evolutionary literature (e.g. Flannery 1972). The image of the Early Bronze Age city-states as peer polities (Finkelstein 1995a, after Renfrew and Cherry 1986) paints a picture of hierarchical, territorial political and economic units each organized around a walled central place. Once the reality of such polities was accepted, their existence provided a ready-made explanation for a host of socioeconomic developments, such as agricultural intensification, the construction of major public architecture and decreasing regionalism in material culture, topics which have, therefore, rarely been examined individually.

The reality is less clear-cut, however. De Miroschedji (1989: 73-74) has stressed the diversity of developmental trajectories in the different regions of Palestine, a point amplified in a study of settlement data by Finkelstein and Gophna (1993). Secondly, de Miroschedji (1989: 70-71) observes that the patterning of some aspects of material culture, the production and distribution of pottery and chipped stone in particular, suggests that some areas of the economy operated at a spatial level far more extensive than that of individual polities. Both points raise doubts about the reality of the existence throughout the southern Levant of a series of homologous city-states each acting as a regional focus for political, economic and ritual power. This review of the evidence from Jordan presents an ideal opportunity to devise an alternative analytical framework for the study of the Early Bronze Age in the southern Levant.

Critique of the city-state model

There are clear discrepancies between the empirical data from the Early Bronze Age southern Levant and the criteria taken as indicative of complex urban-based polities elsewhere in western Asia. Major problem areas include:

a. *Urbanism*. The small size of many walled settlements indicates concentrations of population well below those generally understood as indicative of urban communities. Territorial populations too appear significantly below the absolute levels generally associated with the existence of states (Falconer 1987, 1994; Joffe 1993).

b. *Settlement structure*. The presence of regional settlement hierarchies is generally seen as a necessary condition for the existence of states (Flannery 1972: 412; Wright 1977: 383). These cannot be demonstrated convincingly using the settlement data from the Early Bronze Age southern Levant (Falconer and Savage 1995; Harrison 1997).

c. *Administrative and control systems*. Many models of state formation have stressed the importance of administration to complex organizations (Flannery 1972; Wright and Johnson 1975). However, in the southern Levant material indications for bureaucratic organization, such as writing or sealing systems designed to track commodity flows, are conspicuous by their absence (Joffe 1993).

d. *Elites*. Sociopolitical hierarchies are viewed as an essential component of states (Service 1975; Wright 1977). However, the forms of evidence considered indicative of such hierarchies elsewhere in southwest Asia—large public buildings with reception facilities and service areas, a highly differentiated burial record, the presence and uneven distribution of 'wealth' items—are absent in the southern Levant. (A large public building, 'Palace B', recently excavated at Tel Yarmouth has been compared to 'palaces' from other regions of western Asia [de Miroschedji 1999: 10-11]). However, as demonstrated by the manner in which cylinder seals were used (see [p. 214], the existence of a degree of formal similarity between material from the southern Levant and that from Mesopotamia is not in itself evidence of functional equivalence. Moreover, this building dates to a late stage of EB III, and cannot be used to argue for the existence of either city-states or palace-based elites throughout the EB II–III period.)

Recent studies (e.g. Joffe 1993) have sought to contextualize the model by developing concepts of complexity and urbanism suited to small-scale societies, arguing that, seen in terms of local environmental and subsistence conditions, Early Bronze Age walled settlements represent significant concentrations of both population and human energy. In this usage, the local 'urban' units are seen as agglomerated, nucleated and differentiated settlements, representing not a distinct 'tier' in a hierarchy, but rather lying at one end of a continuum of forms of settlement and organization (Joffe 1993: 64-65). 'Palaces' are re-interpreted as elite residences, although features such as defensive walls and cultic structures are still accorded a high priority in terms of understanding the political and economic roles of these sites. While this may well prove a fruitful approach, it requires us to posit a quantitatively different form of 'urbanism', for which material correlates and conceptual frameworks have yet to be established. We may do better simply to abandon notions of cities and states altogether and approach the data from other perspectives.

Alternative interpretations

The chiefdom model

The obvious alternative is the chiefdom, which has proved a useful heuristic device for the analysis of ancient societies, and which has recently been applied to the data from Chalcolithic Palestine and a range of fifth through third-millennium BC societies from other parts of the Near East (Levy 1986, 1995a; Stein and Rothman 1994). This form of political organization is characterized by a modest degree of social and economic stratification and involves regional populations in the thousands or tens of thousands (Johnson and Earle 1987; Earle 1987, 1991), figures that appear appropriate for Early Bronze Age Jordan. However, the concept originated as one stage in a neo-evolutionary sequence of societal 'types' (Service 1962). As a distinct set of organizational characteristics was attributed to each stage, this can result in a lack of sensitivity to the subtleties of individual historical trajectories (Crumley 1987; Paynter 1989; Yoffee 1993), suggesting that a less rigid analytical tool may be required.

Middle-range societies

The notion of middle-range societies, developed by Feinman and Nietzel (1984) on the basis of ethnographic data, was designed to conceptualize a spectrum of organizational forms intermediate between mobile gatherer-hunter groups and bureaucratic states, and thus subsumes the chiefdom. Many of the characteristics of middle-range societies appear germane to the Early Bronze Age data from Jordan; political organization is highly variable and there is a clear correlation between the total population of a society and the number of levels of political decision-making. Ethnographic data indicated that no groups including less than 4000 members (a level unlikely to have been attained by any of the walled sites in Jordan, see below) revealed more than two levels of decision-making (Feinman and Neitzel 1984: 69). Methods of transferring leadership positions vary considerably, although neither purely achieved power nor inflexible hereditary forms are common. There is, however, a tendency for positions to remain within certain family lines, a point perhaps connected with inherited economic advantage. However, the evidence revealed a high degree of administrative flexibility, and economic inequalities did not always equate to political rank (Feinman and Neitzel 1984; Hastorf 1990: 148). Recent work in western Asia and elsewhere indicates that significant areas of economic activity could exist outside the operational arena of political power (Stein and Blackman 1993; Wattenmaker 1994; Potter and King 1995; Levy 1995), and there is now a trend towards approaches that seek to 'unpack' the various components of complexity, political and economic structures in particular (Netting 1990). Given the variety of organizational forms characteristic of middle-range societies, Feinman and Neitzel (1984) suggest that the model can be best employed as a means to explore the interrelationships between different aspects of the given society, by using the archaeological data to investigate key dimensions of social organization.

Staple finance

D'Altroy and Earle (1985) have defined two organizational modes that describe the economic basis of political power in chiefdoms, wealth finance and staple finance. In the former, political power is based upon the control and manipulation of access to valued substances or products, something for which Early Bronze Age Jordan has produced little evidence. More relevant perhaps is staple finance in which power derives from the control and manipulation of the products of subsistence production, an organizational feature generally associated with societies characterized as agricultural, collective and territorial, and which may prove of value in terms of the southern Levant. Archaeological criteria by which staple finance might be recognized include evidence for intensified agricultural production, and the availability of storage and transport facilities (Schwartz 1994; Stein 1994).

Heterarchy

Crumley (1995: 3) observes that the 'almost unconscious assumption of hierarchy as order... has made it difficult to imagine, much less recognize and study patterns of relations that are complex but not hierarchical'. The idea of hierarchy as order has underpinned the city-state model, and has structured most discussions of the nature of Early Bronze Age societies in the southern Levant, although the existence of hierarchy cannot be readily demonstrated through the data (see below). The problem may lie not with the evidence, but in the appropriateness of the assumptions integral to a hierarchical model. In fact, significant elements of social and economic complexity can arise from non-hierarchical relationships, for example, those between groups differentiated on the basis of household, age or gender (Paynter 1989). Building upon Crumley's ideas, instead of seeking to explain the evidence in terms of a single organizational principle, we should think in terms of overlapping, and at times contradictory, organizational forms, a situation that she has termed *heterarchy* (Crumley 1987, 1995).

Crumley (1979: 144) defines a heterarchical system as one in which 'each element is either unranked (relative to other elements) or possesses the *potential* for being ranked in a number of different ways, depending on systemic requirements'. In this light, we should view intergroup relations, including structures of power and domination, not as fixed within an overarching framework but as transient and contingent. Thus, we must reject the simple unifocal perspective implied by elite-dominated regional centres in favour of altogether more complex forms of organization in which different types of relationships (e.g. cultic, exchange, kinship) may cross-cut each other and/or be organized along quite different lines.

The idea of heterogeneous communities has found favour in recent research on Mesopotamia (Pollock 1992; Yoffee 1993), but the concept appears particularly well suited to regions where there is clear evidence for the exercise of power on a considerable scale, but where the data does not provide the traditional indicators of elite-driven state societies, such as Cyprus (Keswani 1996) or the Levant. In contrast to the unifocal perspectives implied by city-state or chiefdom models, this chapter will treat Early Bronze Age Jordan as heterogeneous, marked by multiple coexisting sources of power.

Complexity and the corporate village

Archaeologists working in the Levant have traditionally contrasted 'urban' sites, characterized as large diverse centres for the provision of specialist services, with villages, seen as small homogenous units supplying primary products, but dependent upon urban centres for specialist goods and services. However, as Schwartz and Falconer (1994: 2) observe, villages are both more complex, and socially and economically diverse, than has generally been assumed. The notion of village complexity appears to offer a potentially useful way of comprehending Early Bronze Age Jordan.

A potentially valuable insight into the nature of Early Bronze Age communities is offered by the corporate village, an organizational form identifiable in Late Bronze Age documentary sources from the Levant dealing with agricultural communities (Magness-Gardiner 1994, with references; Schwartz and Falconer 1994). In such communities land was held in a variety of communal forms—within a single family, between several families, as village lands—and external demands for tax and labour obligations were met jointly. The apparent inconsistency between the clear evidence for coordinated large-scale projects (e.g. defensive walls) and limited indications for the existence of institutionalized elites in the Early Bronze Age can be explained by the proposition that investment activities requiring cooperation beyond the level of individual households were undertaken on a corporate basis by a variety of organizational units, including kinship groups and the entire community. It is worth making the point that, in the absence of elites focused upon conspicuous consumption, village communities may have been able to retain sufficient surplus production to finance investments in valuable communal infrastructure, such as agricultural and defensive works and collective burial monuments. This model, which shares certain features with Renfrew's (1974) idea of 'group-oriented chiefdoms', appears to offer the basis of a new approach to the evidence from Early Bronze Age Jordan.

The key to maintaining such a system over time, and avoiding a fatal transition towards institutionalized inequality, appears to lie in what has recently been termed a 'corporate power strategy' (Blanton et al. 1996), through which power is distributed across different groups and sectors within society in a heterarchical fashion. An appropriate 'group-centred' ideology combined with effective networks of resistance to potential attempts by particular individuals or interests to monopolize power would contribute to the long-term stability of the system. This is not to argue that such systems were egalitarian (Sweet 1960, cited in Schwartz and Falconer 1994: 6). However, disparities take the form of differential ownership of land, livestock and personal reputation rather than conspicuous consumption, all hard to detect in the archaeological record, and reminiscent of staple rather than wealth finance.

Following Renfrew (1974) and Blanton et al. (1996: 6-7), archaeological correlates of a corporate village organization might include defences, irrigation works, certain categories of public building (but not palaces), group-centred burial monuments, a lack of 'prestige' goods in spite of technological innovation in the production of utilitarian artifacts, and minimal evidence for the systematic differentiation of individuals. These criteria, which centre on the deployment of wealth in projects of value to the community, rather than its consumption as a facet of elite lifestyles, appear highly relevant to the Early Bronze Age data from the southern Levant.

Nor is the corporate village model a 'static' phenomenon, unable to account for diachronic change. Evidence from later periods (Joffe 1993: 48 with references) suggests that a village would have included elements from several different kinship-based groups, showing differing degrees of wealth, prestige and extra-village connections. The shifting pattern of cooperative and competitive interaction between these groups would have provided an important internal dynamic for change. In addition, Magness-Gardiner (1994: 44-45) has noted that successful units will tend to expand at the expense of the less successful ones, leading to the development of villages of greater and lesser importance, and a diversity of village forms. The suggestion that walled settlements constituted a fairly heterogeneous group is consistent with Finkelstein's (1995b: 79-86) suggestion that the site of Arad in southern Palestine should be attributed to the settlement of local pastoralist groups.

Clearly, damage to the agricultural base, or other localized problems, would directly affect a community's ability to 'reinvest', and could result in potentially rapid changes in fortune. Thus, in a system in which the individual corporate village constituted the highest order organizational unit, we might expect a relatively stable region-wide structure, but a high degree of instability in the case of its separate components. In this light, the evidence for significant discontinuities in the occupational record of individual sites and regions (Portugali and Gophna 1993; Gophna 1995a) conforms to the expectations of a dynamic corporate village model.

Review of the evidence

Chronology

The establishment of absolute chronologies

An Egyptian historical chronology based upon the sequence of royal dynasties exists from the end of the fourth millennium BC. In theory, this permits the accurate dating of artifacts from certain Egyptian contexts, often the graves of named individuals. Thus, material of Egyptian origin, when recovered at sites outside Egypt, or Levantine artifacts from reliable Egyptian contexts, provide synchronisms that allow the correlation of Levantine archaeological deposits with Egyptian historical chronology. Limitations of this method include the rarity of finds of Egyptian origin in Jordan, and the possibility that these may have entered the archaeological record long after the date of their production.

Historical dating provides good evidence only for the end of EB I. A growing corpus of Egyptian material, including pottery and *serekhs*, several of which can be read as Narmer, first king of Dynasty 1, have been recovered from late EB I contexts in southern Palestine (Brandl 1992; Levy 1995a: 31; van den Brink 1996). These provide a firm correlation between the beginning of Egyptian Dynasty 1 and the end of EB I.

The beginning of EB II can be equated with a relatively early point within the Egyptian First Dynasty, as vessels of forms diagnostic of EB II first appear in the tomb of the Pharaoh Djer. However, Egyptologists' dates for the beginning of the First Dynasty have ranged between c. 3100 and 2900 BC (see Kantor 1992; Wilkinson 1996: 9-15 for details). Current Egyptological opinion appears to favour a date between 3100 and 3000 BC for the beginning of Dynasty 1 (Hendrickx 1996: table 9; Wilkinson 1999: 27), which would place the beginning of EB II at the very end of the fourth millennium BC, in good agreement with the radiocarbon evidence (see below).

EB III, 2700–2350 BC

It is generally believed that the beginning of EB III equates with the Egyptian Third Dynasty, so the EB II/III transition has been traditionally placed c. 2700–2650 BC (Stager 1992: 41; Joffe 1993: 68), although this should be regarded as no more than an estimate (see below). The linkage with Egyptian chronology is looser than in the case of EB II because many of the Levantine ceramic forms found in Egyptian Third Dynasty and later contexts have a relatively wide duration in the Levant (see Esse 1991: 103-16; Stager 1992: 37-39 for discussion). Stager gives correlations between EB II–III Palestine and the Egyptian dynastic sequence (1992: fig. 16).

Radiocarbon dating

Despite some initial scepticism, radiocarbon dating is assuming an ever greater importance in the construction of chronologies in the region (Levy 1992a; Gilead 1994; Joffe and Dessel 1995). It has two key advantages over a chronology built upon historical synchronisms:

1. The method allows the extension of absolute dating to a period well before the upper limit of the Egypt historical chronology; absolute dates for the EB I period are dependent upon radiocarbon.

2. Radiocarbon dates are taken on decayed organic material, which, unlike Egyptian imports, can be recovered from most sites.

In combination, the two methods now provide a reasonably secure absolute chronology for the main phases of the Early Bronze Age.

A growing body of radiocarbon dates indicates that the Chalcolithic period terminates in the first few centuries of the fourth millennium BC (Joffe and Dessel 1995; Carmi *et al.* 1995). Most studies have placed the beginning of the Early Bronze Age around 3500 BC (Esse 1991; Stager 1992: 40; Joffe 1993; Gophna 1995), although unpublished radiocarbon dates from recent excavations at Tall ash-Shuna suggest that in northern Jordan at least the EB I period began no later than 3600 BC.

Radiocarbon dates from Jordan

Although the corpus of Early Bronze Age I–III radiocarbon dates from Jordan is growing rapidly, many dates suffer from poor documentation of the samples and the archaeological contexts from which they were obtained, which can render them hard to interpret. Therefore, the selected groups here have well-documented dates from clear chronological horizons.

A series of dates from early EB I deposits from Tall ash-Shuna fall between 3650 and 3350 BC; those from the later EB I deposits fall between 3400 and 3000 BC, while a group of late EB I/EB II dates from Tall Abu al-Kharaz (Fischer 1998: 219-20) cluster between 3350 and 2900 BC. Five dates, including four from charcoal and one from grape seeds, collected from the exclusively EB III site of Numayra (Rast and Schaub 1980: Table 3) fall between 2900 and 2500 BC. All dates discussed are expressed at two standard deviations and were calibrated using version 2.18 of the OxCal calibration programme (Bronk Ramsey 1995) (Table 5.1).

The establishment of relative ceramic chronologies

Many sites lack both reliable radiocarbon dates and Egyptian material. In these cases, pottery is the prime dating indicator. On the whole, Early Bronze Age pottery from Jordan is comparable to that from Palestine, the ceramic chronology of which has been the subject of regular discussion, founded upon the studies of Wright (1937) and Amiran (1969); a more recent summary has been provided by Stager (1992).

Period	Absolute dates BC
Early EB I	3600–34/3300
Late EB I	34/3300–31/3000
EB II	31/3000–2850/2750
EB III	2850/2750–24/2300

Table 5.1. Estimate of absolute dates for EB I–III, based upon calibrated radiocarbon measurements.

Developments in EB I and II Palestine are described by Joffe (1993), while more detailed treatments of pottery from northern (Esse 1991; Greenberg 1996) and southern areas (Fargo 1979; Seger 1989) are also available. Early Bronze Age ceramics from the southern Levant have been treated extensively in a recent collection of papers (Philip and Baird 2000). The discussion that follows concentrates upon those aspects of the material directly relevant to an understanding of developments in Jordan, or where the Jordanian evidence may require modification of conventional wisdom.

EB I

While the definition and nomenclature of the EB II and III periods have been reasonably uncontentious, the terminology and definition of EB I has been a source of scholarly disagreement. The underlying reason for this is that EB I pottery differs markedly from region to region. The traditional two-phase terminology of EB IA and IB for this period (Stager 1992) is avoided here for two reasons:

1. Material tends to be assigned to one or other of these categories on the basis of its place within a localized ceramic sequence. However, in the absence of radiocarbon dates it is not clear that material termed EB IA or IB in one area is necessarily contemporary with that described by the same term elsewhere.
2. Increasingly refined chronological control has meant that in southern Palestine some scholars now recognize three (Stager 1992: 31-32) and others four phases of EB I (Amiran and Gophna 1992: Table 1; Levy 1995a). Until reliable correlations have been established between regional pottery sequences, it seems better to simply allocate material to an earlier or later position within its own regional sequence.

Two main phases of EB I have been recognized in north Jordan. Early EB I is defined on the basis of 'Grey Burnished Ware' and what have been termed 'Impressed-Slash Wares' by Stager (1992: 29-30), a term that appears to be extended (wrongly) by some to include the later Tall Umm Hammad Ware (e.g. Joffe 1993: 39). Late EB I is characterized by vessels bearing what has been termed 'Band Slip' or 'Grain Wash' decoration.

In southern Palestine, early EB I is marked by the use of a red paint on white wash, which appears to develop later into the so-called 'Line Painted Group' or B-Tradition, sometimes also termed 'Line Group Painted Ware' (LGPW), characteristic of late EB I in central Palestine and Jordan (Schaub 1982; Joffe 1993). While LGPW is well known from sites in south-central Jordan, where it appears in the later EB I, early EB I material, as known from Bab adh-Dhra', forms a distinct regional group (Schaub 1987). At present, there is no Jordanian counterpart to the large quantities of Egyptian pottery that are reported from late EB I assemblages from southwestern Palestine (Stager 1992: 32-33; Brandl 1992).

A refined picture of regional ceramic chronology is dependent upon long stratified sequences, such as those from Tall Umm Hammad (Betts 1992) and Tall ash-Shuna (Baird and Philip 1994) in the Jordan Valley. These sites have demonstrated the length of the EB I period and the existence of two chronologically distinct ceramic phases at each site, while the contrasts between the two assemblages highlight the degree of ceramic regionalism. Additional good EB I material has been excavated at Bab adh-Dhra' in the southern Ghor, and recently at nearby as-Safi (Schaub and Rast 1989; Waheeb 1995), revealing the existence of at least one distinct ceramic region located east and south of the Dead Sea. In this case, however, the bulk of the published data comes from tombs rather than stratified deposits, and so cannot provide detailed material for construction of an interregional ceramic chronology.

EB II

This period witnesses a decline in the earlier ceramic regionalism, with pottery becoming increasingly homogenous throughout the southern Levant. While this makes it relatively easy to recognize EB II material, it is not yet possible to correlate EB II sequences at different sites to provide an internal ceramic chronology for the period. Thus, there exists no clear basis for recognizing internal subdivisions within EB II, other than the relative stratigraphic positions of groups of material at individual sites. In Palestine, two main EB II ceramic regions, north and south, have been defined (Stager 1992; Joffe 1993). In general, the northern region is characterized by the presence of vessels in a wide variety of shapes made in highly fired 'Metallic' Ware, and the southern by red-slipped and burnished vessels made from softer fabrics (Greenberg and Porat 1996). Provisional evidence suggests that these divisions are broadly applicable to Jordan, although we currently have no way of determining whether the appearance of Metallic Ware was synchronous at all sites. Lacking the independent verification of radiocarbon dates, the assignation of deposits to EB II on the basis of the presence of Metallic Ware has a certain circularity. In particular, we have little idea as to how long Metallic Ware remained in production, or how it was distributed, and it is not impossible that the key indicator of EB II was a short-lived, or unevenly distributed, style of pottery.

Recently, the validity of traditional ceramic distinctions between EB I and II has been called into question. Work at Tall Abu al-Kharaz has revealed that jars bearing Band Slip decoration and vessels in Metallic Ware, hitherto believed diagnostic of late EB I and EB II respectively, occurred together on the same floors (Fischer and Toivonen-Skage 1995: 587, figs. 3, 4; Fischer 2000), indicating their contemporaneous usage. The additional observation of storage vessels bearing a style of painted decoration very similar to that known as Band Slip, in EB III contexts at Khirbat az-Zaraqun (Genz 2000), confirms that this style of pottery can no longer be taken as diagnostic of late EB I occupation alone. This may require a reconsideration of settlement pattern data, as the presence of Band Slip decoration is likely to have been a major factor in the classification of surface-collected material to late EB I.

EB III

The beginning of EB III in the north Jordan Valley is generally believed to be indicated by the appearance of the distinctive Khirbat Kerak ware (Stager 1992: 39), although there may be an element of circular reasoning in this reconstruction (Philip 1999: 34). Whatever the case, its restricted spatial distribution (Esse 1991: 138-40, fig. 25) renders Khirbat Kerak ware unsuitable as a universal indicator of EB III. While abundant at Tall ash-Shuna (Leonard 1992), it is rare on the Jordanian plateau (Mittmann 1994: 10), and even in the Jordan Valley becomes sporadic south of Beth Shan (Esse 1991: 138-39). However, sites located as far south as Bab adh-Dhra' show particular bowl forms that appear to have been influenced by features of Khirbat Kerak ware, which should indicate an EB III date (Schaub and Rast 1989: 439).

Khirbat Kerak ware apart, substantial groups of (typologically) EB III pottery come from three sites, namely, Bab adh-Dhra', Tall al-'Umayri and Khirbat az-Zaraqun (Schaub and Rast 1989; London 1991; Mittmann 1994; Ibrahim and Mittmann 1994), with a smaller quantity published from Tall Handaquq (S) (Chesson 1998). However, no EB III stratigraphic sequence from Jordan has yet been published in detail, and ceramic developments within EB III are poorly understood (but see now Harrison 2000). While it is possible to distinguish between EB II and EB III assemblages in many cases, a more refined periodization cannot be undertaken without close reference to the evidence from Palestine (Callaway 1978; de Miroschedji 1988: 70-83; Seger 1989; Esse 1991; Greenberg 1996).

It should be clear from the above that the dating of sites through ceramic criteria is far less precise than is often assumed. In fact, material can rarely be placed more precisely than within a phase of several centuries' duration. Assuming that received wisdom in ceramic chronology is broadly correct (and it may not be), it is clear that many sites were occupied during only part of the EB I-III. Furthermore, the presence at a site of material dating to any one period does *not* imply continuous occupation during the whole of that period, although this appears to be the assumption in much of the literature. The implication is that occupations measured in terms of a few generations, or one or two centuries, may have been common, even at fortified sites. If correct, it is hard to see individual sites as playing a critical role in structuring long-term regional social and economic relationships, implying a more fluid settlement and political universe than has generally been envisaged. Such a concept can be better related to a flexible, heterarchical world of corporate villages and middle-range societies than to the institutionalized power structures generally associated with city states.

The material remains: architecture

Archaeological field work has been focused mainly upon walled sites; far less is known about the material culture and spatial organization of smaller settlements. The inherent bias within the available data has, therefore, conditioned the following discussion of architectural remains, the aim of which is to summarize the available information, before returning to consider its significance for our overall interpretation of the nature of the local Early Bronze Age.

Walls

With the exception of Jawa, the earliest fortification walls presently known in Jordan appear to date to the EB II period, although more information is available on examples dating to EB III. As in Early Bronze

Age Palestine (Kempinski 1992a: 72), fortifications grew by accretion, with walls subject to successive phases of thickening or other forms of elaboration. For example, the EB III stone wall at Bab adh-Dhra' appears to have replaced an earlier mudbrick structure (Rast 1995: 126), while the thickening of walls over time by the addition of extra layers of stone is well illustrated at Khirbat az-Zaraqun. Building materials generally reflect what was locally available, for example, sandstone at Numayra, and limestone at Khirbat az-Zaraqun.

The main wall at Bab adh-Dhra', which dates to EB III, was at least 7 m wide at the base, and bore traces of a decayed mudbrick superstructure. It was built in sections some 15–20 m long separated by transverse walls, a technique widely paralleled in Palestine (e.g. Amiran *et al.* 1978: 11-12, pl. 178), with the number of stone foundation courses varying according to local topography (Rast 1995: 126). In some instances, the stone used for the core differs between adjacent sections (Rast and Schaub 1978: 12), perhaps indicating a division of the work among different groups of workers, or the wall's construction in stages, a possible contrast to the more unified building operations hypothesized for key Palestinian sites (de Miroschedji 1989: 68). At Numayra (Coogan 1984), two parallel walls constituted the inner and outer wall-faces, while the space in between was filled with rubble. Here, too, there is evidence for transverse sections, but placed at c. 7 m intervals.

Semi-circular towers of the kind known from several EB II sites in Palestine (Kempinski 1992a: 72; Braun 1996) are as yet unknown in Jordan. A number of walled sites do, however, exhibit rectangular towers or bastions. Bab adh-Dhra' has evidence for at least two such structures. One located in the highest, northeastern part of the site revealed two phases of construction (Rast and Schaub 1978: 14) and appears to have been built in mudbrick founded upon wooden beams set between small stones (Rast and Schaub 1980: 26). An area of packed marl at the western edge of the site may indicate the presence of another tower (Rast and Schaub 1981: 14). The wall was 'strengthened' by the addition of further layers of stone adjacent to the external face of the inner wall. Regularly spaced rectangular towers formed part of the enclosure system at Tall Handaquq (N) in the Jordan Valley (Mabry *et al.* 1996: 123), while a rectangular bastion some 30 m in length protected the main entrance at Khirbat az-Zaraqun (Ibrahim and Mittmann 1994: 13). This was constructed by the addition of extra layers of stone, resulting in a final width of more than twice that of other parts of the circuit wall. The presence of a rectangular tower at the 1 ha site of Numayra indicates that these were not just a feature of the larger settlements.

At Pella, a 20 m-long stretch of walling located at the eastern edge of the tell, just inside the Middle Bronze Age town wall, has been interpreted as part of an Early Bronze Age perimeter wall (Bourke 1997: 99). This construction, which shows two distinct phases, is dated between late EB I and the end of EB II. The presence of what appear to be two rectangular towers or revetments would appear to negate the idea of generalized shift from EB II semi-circular to EB III rectangular towers that has been claimed on the basis of the Palestinian evidence (Kempinski 1992a: 72). What may be a continuation of this wall has been uncovered in Area XXXII along the southern edge of the tell (Bourke 1997: 99-100, figs. 4-5).

The complexity of the Early Bronze Age occupation at Pella is exemplified by the presence of a large stone platform on the summit of Tall al-Husn, a natural hill located across Wadi Jirm from the main tell (Bourke *et al.* 1994: 98, figs. 8-9; 1997: 100-103, fig. 8), and that appears to represent some sort of defended complex. The extensive Early Bronze Age remains include two rubble platforms with dressed stone margins; each platform measured at least 15 by 15 m. Evidence for processing and storage functions occurs in the form of plaster-lined pits and grinding equipment. Within the destruction debris of the final use of the structure, a hoard of copper objects included axes, chisels and a spearhead. The whole complex is dated to EB I/II, and its floruit would appear broadly in line with that of the rest of the site. In concept at least, this unit is reminiscent of EB III stone platforms known at such sites as Tel Yarmouth (de Miroschedji 1988: 52-53).

We should be wary of assuming, however, that all walls were intended for defence. A case in point is Jabal Mutawwaq (Hanbury-Tenison 1989a: 137) where analysis revealed that the enclosure wall consisted of a single course only of long, undressed stone blocks, set end-to-end. Both its construction and the presence of domestic units outside the circuit (Hanbury-Tenison 1989b: 57) argue against a defensive function.

Gates

The EB III gate at Bab adh-Dhra' was of direct-entry design, without towers or other defensive elaboration, prior to the construction of a blocking wall in its final phase (Schaub and Rast 1984: 43-46, figs. 4-5). At Jawa

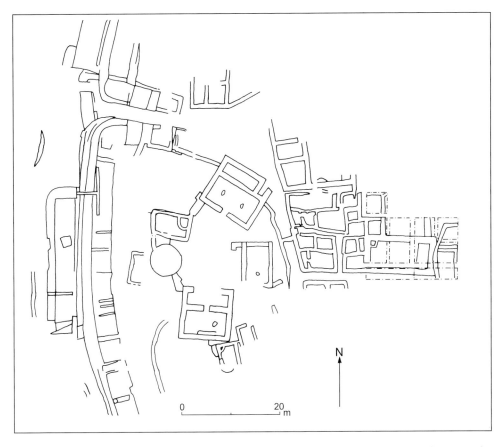

Figure 5.2. Plan of excavated area of Khirbat az-Zaraqun, showing walls, main gate, streets and cultic complex. Note the contrast between the latter and the more densely built-up area to the east (redrawn by Y. Beadnell).

in the steppe, however, the EB I Upper Fortifications included gate G 1, which featured a pair of small piers projecting slightly beyond the exterior face of the wall to create a small chamber (Helms 1981: 103-105, fig. 38; Betts 1991: 34-35, fig. 37), while a paved gateway opening between two rubble buttresses has recently been encountered within the EB II platform structure at Tall al-Husn near Pella (Bourke 1997: 102, figs. 8, 9). However, the sole instance of a large Early Bronze Age gate complex that has been investigated in detail is the main gate at Khirbat az-Zaraqun (Figure 5.2). This consists of a corridor 15 m long by 2 m broad, which passes between two towers and two rectangular rooms, before opening into a triangular open area within the upper settlement. This gate revealed three constructional phases, while another gate in the southwestern wall of the site revealed two separate phases of use prior to its being built over (Mittmann 1994; Ibrahim and Mittmann 1994). Such frequent modification of defensive structures is characteristic of EB III building in Palestine (Kempinksi 1992a: 73). It is not clear whether the successive stages represented real technical improvements, necessary repairs and refurbishment, or were more connected with embellishment for the purposes of display and prestige.

In addition to the main gates, the circuit wall of Khirbat az-Zaraqun is periodically interrupted by small gaps, some 1 m in width, which are interpreted as posterns, a feature familiar from Palestinian sites (Helms 1975). While these might have a military function, Herzog (1986: 30) sees posterns as allowing easy access to the exterior, avoiding what would otherwise have been frequent, lengthy detours via the major entrances. The most obvious explanation is that posterns were included to facilitate access to cultivated areas and the movement of livestock. Thus, defensive concerns appear to have been tempered with regard to the demands of daily agricultural life.

Water systems

Water storage structures are a well-documented feature of Early Bronze Age walled sites in Palestine (Ben-Tor 1992: 104). The capture of seasonal precipitation would have been even more vital in parts of the Jordanian plateau where some settlements were located close to the limits of dry farming.

A series of tunnels cut into the limestone bedrock deep below the site of Khirbat az-Zaraqun allowed access to an underground water source. While there is no conclusive evidence for an Early Bronze Age date for this system, the physical relationship of the system to the walled Early Bronze Age settlement, and the importance of water storage facilities at other Early Bronze Age sites in the region, are highly suggestive. A more elaborate water management system has been documented at Jawa, located in the arid basalt *harra* some 160 km east of the Jordan Valley (Helms 1981; Betts 1991), where there is evidence for the use of stone-built dams designed to channel water into storage pools via canals. The main catchment at Jawa is Wadi Rajil, which carries winter run-off from the Jabal al-'Arab lying to the north. Despite the lack of direct stratigraphic connections between it and the settlement, the excavator is surely correct in dating the water system to the Early Bronze Age (Betts 1991: 54). However, the exact nature and duration of the occupation remains a matter of debate (Betts 1991; Braemer 1993a; McClelland and Porter 1995).

Helms (1982: 109) draws attention to common construction techniques linking the Jawa water system to the fortification walls found at various Early Bronze Age settlements, and would view the site as an example of 'transmigrant urbanism', that is, a short-term occupation by incomers from more developed areas. Rather than interpreting Jawa as an isolated engineering marvel in the steppe, however, what is remarkable about the site is the excellent preservation of the structural remains. This is attributable to its location in the steppe where it has been relatively little affected by subsequent colluviation processes, later settlement or stone robbing. Thus, Jawa is a uniquely well-preserved instance of water management techniques that may have been quite widely practised during the later fourth millennium BC.

While parallels are rare, Mabry *et al.* (1996: 124) have argued that two stone-built dams spanning Wadi Sarar near the large Early Bronze Age site of Tall Handaquq (N) date no later than the fourth millennium BC, and are connected with efforts to retain water and silt brought down by seasonal floods. It is probably no coincidence that Early Bronze Age settlements in the Jordan Valley are concentrated along the major side wadis (Ibrahim *et al.* 1988: 171), the places most suitable for water capture techniques.

While physical evidence for water management systems in the valley remains limited, this may reflect the impact of post-Early Bronze Age geomorphological processes—the downcutting of wadi beds (Donahue 1985; Mabry 1989: 59) and the build-up of colluvium along the valley floor (Banning 1996)—in addition to later building and agricultural activities. Other topographical features that may indicate water storage activities include a large depression within the site of Tall Handaquq (N), and a substantial basin detected at Tall Jalul near Madaba (Mabry *et al.* 1996: 124, fig. 2; Harrison 1997). There appears to be a limited but suggestive body of evidence for both on-site water storage and for major structures designed to exploit seasonal water surpluses.

Internal organization of walled settlements

A number of Palestinian Early Bronze Age sites appear to show a division into an acropolis area, containing key public buildings, and a lower town, comprising domestic structures (Kempinski 1992a: 79). The implied difference has been seen as indicative of social differentiation. A similar arrangement of upper and lower town appears to have characterized a number of walled sites in Jordan, such as Tall al-Hammam and Tall Handaquq (S) in the Jordan Valley, and al-Lajjun on the al-Karak plateau (Prag 1991: 60; Chesson 1998; Miller 1991: 102). However, this was not universal since at Numayra the entire central area was occupied by domestic structures (Coogan 1984). Clearly, all walled settlements were not functionally equivalent.

Khirbat az-Zaraqun is the only site where a sufficiently large contiguous area (5000 m^2, see Figure 5.2) has been excavated to permit discussion of internal organization (Mittmann 1994). The site shows good evidence for systematic planning and appears to have been divided into an upper town, located east of the main gate, containing cult structures and other public buildings, and a lower town to the south. The latter revealed a regular network of streets, with dwelling units separated by narrow passageways. Of particular interest is the location within the lower town of what would appear to be a substantial non-domestic building, which may have been connected with the gate to the lower town, and appears to have gone out of use when the latter was blocked.

There is little to suggest that Early Bronze Age walled settlements in Jordan were intended to function as centres of specialist manufacturing activity. No walled site has yet produced a clearly defined industrial installation, or even a marked concentration of manufacturing debris, suggesting that much craft activity was undertaken outside the walled areas. However, soundings at Bab adh-Dhra' have revealed good evidence for extramural occupation, including material both predating and contemporary with the walled settlement (Rast and Schaub 1980: 23), raising the possibility that excavators' concentration upon intramural occupation may have skewed the data.

Public buildings

Cult structures from settlements

While Chalcolithic cult structures appear both within settlements and as isolated units (Levy 1995a: 235-36), their concentration within settlements in EB I and within the central areas of walled sites (that is, differentiated from domestic occupation) by EB II–III has been taken as indicative of increasing centralized control over religious activity (de Miroschedji 1989: 69).

Khirbat az-Zaraqun has revealed an excellent example of such a cult area (Figure 5.2), located just southeast of the main entrance (Mittmann 1994; Ibrahim and Mittmann 1994). This consisted of an enclosure composed of four 'broadroom' structures, both elements (enclosure and broadroom layout) characteristic of Early Bronze Age cultic buildings in Palestine (Ben-Tor 1992: 101; Kempinski 1992b: 57-58, fig. 1). These buildings were of different sizes, and formed a rough circle around a courtyard. The southern structure had a shallow anteroom placed before the 'cella' and is interpreted as a temple by excavators. On the floor were two flat stone slabs, which would have functioned as the bases for timber roof supports, partly embedded in the limestone-marl floor. Benches of plastered stone lined the walls. Three other 'broadroom' structures surrounded the courtyard; that in the northeast corner revealed indications of a possible staircase giving access to the roof or an upper storey in an annex against the rear wall. At the eastern side of this area, a circular stone-built platform was located. Some 6.5 m in diameter and bearing traces of lime plaster rendering, the platform was preserved to between 0.5 m and 1.0 m in height, with access gained via several steps on its eastern side. The cultic nature of this area is confirmed by the existence of a very similar structure (altar 4017) in the EB III cultic area at Megiddo in northern Palestine (Ben-Tor 1992: 103: figs. 14.16, 14.17). The presence of two such similar structures might suggest not only shared architectural styles, but even a degree of common cult practices between the two sites.

However, what have been interpreted as 'cult' structures are not restricted to upper towns. Building 1.3 located in the lower town at Khirbat az-Zaraqun, in architectural terms a domestic structure, has also been viewed as 'cultic' because it produced a number of figurines. The excavators have suggested that this structure represented the focus of some sort of domestic cult, in contrast to the 'official' practices represented by the main 'cultic' area in the upper city.

A rectangular structure some 12 by 6 m in size and located within Field XII at Bab adh-Dhra' has also been interpreted as 'cultic' (Rast and Schaub 1981: 27-31). This revealed two building phases similar in size and alignment, but separated by a layer of gravel. An area of flagstones is located at the north end of the interior, and the lower phase has preserved three large stone slabs each preserving *in situ* the stump of a wooden beam. The later reconstruction included a mudbrick stairway and the brick pavement of an open courtyard to the west. This paving abutted a semi-circular stone structure (Locus 13), which has been interpreted as an 'altar'. An unusual assemblage—five flint tabular scrapers and a seal impression—was recovered from its vicinity. Preserved traces of internal decoration include possible fragments of ivory and a large wooden beam that had been inlaid with squares of plaster (Rast and Schaub 1981: fig. 25). If these structures are allowed at least some kind of cult function, it appears that even within walled settlements certain aspects of cultic practice lay outside the direct physical control of elites.

Standing stones

Groups of free-standing stone monoliths, often viewed as of cultic significance, are known from the Chalcolithic site of Gilat (Alon and Levy 1989: 182-84) and in arid areas of southern Palestine (Avner 1984, 1990). However, several examples from south-central Jordan are located within, or close to, Early Bronze Age walled settlements.

A short way from the walled Early Bronze Age site of al-Lajjun, located east of al-Karak, Glueck (1934: 45, fig. 19) reported a group of 16 uncut limestone blocks each around 1.5 m high and erected on end to form a gentle curve. At Adir, some 10 km to the southwest, Glueck (1934: 45-47, figs. 19-21) recorded three much larger monoliths. One, 4.5 m high, remained erect, while two others, each 3.8 m in length, were lying on the ground. Albright (1924: 10) recorded the presence of monoliths at Bab adh-Dhra', although these have since been destroyed, while others exist at Khirbat Iskandar (Richard and Borass 1984: fig. 5). Several groups of standing stones from southern Jordan appear to reveal a spatial association with Early Bronze Age settlements. A monolithic pillar standing upon an artificial stone terrace in Ghor adh-Dhra' east of the Dead Sea has been dated to the Early Bronze Age (Korber 1993), although the associated material remains unpublished. However, it is not clear whether the apparent concentration of monolithic sites in south-central Jordan reflects their original distribution, or a lower rate of destruction there compared with areas that have seen more intensive subsequent land use.

Standing stones in Jordan have traditionally been dated to EB IV, a view apparently based upon Glueck's early surveys in 'Moab' (see Mattingly 1983: 482-83), and it has often been assumed that these 'open-air' monuments were associated with nomadic groups (see Graesser 1972 for general discussion). However, the location of examples close to, or within, walled EBA II–III sites would suggest an earlier date, while the alleged association of stone monoliths with mobile groups would appear qualified by their presence in cult structures located in the heart of major second-millennium BC towns (Ottosson 1980; Kempinski 1992b: 174-75). Moreover, the incorporation of a pre-existing row of free-standing worked stone slabs within the wall of a substantial EB I building at Hartuv in Palestine (Mazar and de Miroschedji 1996: 6-10, figs. 5, 6) underscores the connection between superficially very different architectural forms, namely roofed rectangular structures, and open-air monoliths. Given the ability of humans to transform and reinterpret structures, attempts to associate either type of monument with a particular social group would appear overly simplistic.

Discussion of cult structures

Individual units appear relatively small, and there is no indication of storage or administrative facilities directly associated with cultic installations. Many structures take the form of broadrooms, with roofs supported on wooden pillars resting upon stone bases. The discovery of a large cult installation at EB III Khirbat az-Zaraqun appears to support Joffe's (1993: 83) argument that there was, at least by EB III, a tendency to locate certain aspects of cult activity within central areas of larger sites, as at Khirbat az-Zaraqun and Megiddo. This could be interpreted as indicative of growing control over these agricultural communities by emergent elites, who sought to ulitize cult connections as part of a strategy of legitimation.

However, this may only be one part of the story, and alternative interpretations are possible. The variety of potential Early Bronze Age 'cultic' structures, ranging from single rooms to large complexes, as well as a range of open-air sites of a very different physical nature, suggests the simultaneous existence of multiple spheres of cult activity, not all of which would have been equally well integrated with systems of political control. Moreover, while one building from Khirbat az-Zaraqun has offered some evidence for cult furnishings, it remains the case that the majority of so-called cult structures have produced far less in the way of distinctive paraphernalia than have their second-millennium BC equivalents. The absence from Early Bronze Age cult areas of the offering pits, and large quantities of portable artifacts so characteristic of second-millennium BC cult sites (Philip 1988; Ilan 1992), is interesting, and suggests that the practices conducted within and around cultic buildings differed between the two periods. Given the close relationship between Middle Bronze Age grave goods and contemporary cultic offerings (Philip 1988: 193), the rarity of small non-ceramic objects in both Early Bronze Age graves and cult sites may not be fortuitous (see further discussion under burial evidence, [p. 199-200]).

Administrative structures

In comparison to their defences, the preserved public architecture from Early Bronze Age sites in the southern Levant is relatively unimpressive (Kempinski 1992a: 78; Joffe 1993: 84). Despite the evidence of a large late EB III architectural complex with significant storage facilities at Tel Yarmouth (de Miroschedji 1999: 10-12), there is nothing from either Palestine or Jordan that resembles the administrative complexes seen in a contemporary Syrian centre, such as Tall Mardikh. This has revealed evidence for reception suites, bureaucratic archives, specialized craft workshops, and stockpiles of valuable imports (Archi 1991; Mazzoni 1991; Pinnock 1991). The absence of these features from sites in the southern Levant, despite the evidence for 'public works' in the form of defensive and water management systems, suggests a sociopolitical organization in the south that was qualitatively different from that of Syria.

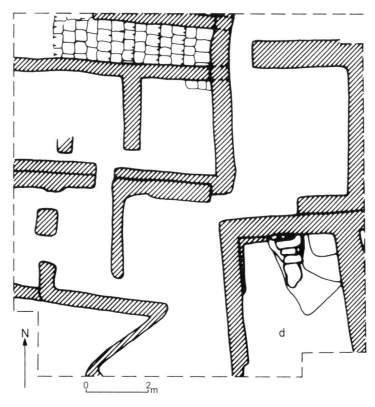

Figure 5.3. Tall as-Sa'idiyya Area BB 700, Stratum L2 (EB II). An example of agglomerative rectilinear architecture; the sunken room 'd' was apparently used for storing oil (redrawn by Y. Beadnell).

In the southern Levant, large, non-domestic structures regularly take the form of rectangular halls with a central row of pillars, a design that persists right through EB I–III (Marquet-Krause 1949: fig. C; de Miroschedji 1988: 38, fig. 2; Mazar and de Miroschedji 1996; Mazar 1997: 148). Only one excavated Early Bronze Age structure from Jordan falls into this category, a building complex located to the east of the cultic area at Khirbat az-Zaraqun (Figure 5.2). This consists of a large 'broadroom' with thick walls, perhaps indicative of the presence of an upper storey. Adjoining it is a complex of small rooms, which produced many small finds and medium-sized storage vessels. While provisionally termed a 'palace' (Mittmann 1994), the finds suggest that the structure functioned more in connection with work and storage than with official reception and government. Another large building was located in the lower town, connected with the gate. Measuring 11.5 m long by 5.0 m broad, and with a row of four stone pillar bases on the floor, this building appears not unlike the so-called 'Palace' at Ai, now generally interpreted as a temple of some kind (Callaway 1993: 41).

Pillared halls have been variously interpreted as temples, palaces and storage complexes, and it may be that this was no more than an architectural form employed for structures fulfilling a range of functions. Perhaps the notion of a clear distinction between buildings having cultic, administrative or elite residential functions is inappropriate in an Early Bronze Age context. These structures might be better termed 'non-domestic' or 'special purpose' complexes (Mazar and de Miroschedji 1996: 13). As such, they may have taken the form of individual buildings able to serve a variety of the community's needs, an interpretation that appears in accord with the cross-cutting power structures envisaged under a heterarchical model.

A connection between large structures and storage is seen in a late EB I pillared hall recently excavated at Beth Shan (Mazar 1997), which contained large amounts of carbonized grain and lentils and many broken jars. The pattern is reinforced by recent evidence from Tall as-Sa'idiyya. Here, a substantial EB II multi-roomed installation included a sunken room measuring some 4 by 3 m and entered via steps (Figure 5.3), which contained the remains of 12–13 large storage jars, suggesting the need to keep a commodity cool, most likely olive oil (Tubb and Dorrell 1993, 1994). The presence of almost 200 pierced bivalve shells within one room has led the excavator to suggest their employment within some

Figure 5.4. Tall ash-Shuna, circular EB I structure, Building 9; note stone base for central roof support (G. Philip).

sort of recording system connected to storage activities (Tubb and Dorrell 1994: 63). While evidence for specialized craft production is scant, the data does argue that certain non-domestic structures within Early Bronze Age walled settlements played an important role vis-à-vis the storage of agricultural products, such as grain, lentils and liquids. In this light, the fortifications would have offered the critical elements of both external security and internal control, points perhaps significant on both physical and ideological levels.

To summarize, evidence for large complexes of public architecture appears only during EB III, be these 'cultic' or administrative. If these are indicative of growing inequality, and or even of an emergent elite (Joffe 1993: 85-87), then the creation of defensive walls appears substantially to predate this development. The evidence of non-palatial public architecture, central storage facilities for agricultural products and of defensive walls, all of which predate the emergence of local elites, appears in harmony with the models of staple finance and the corporate village community.

Domestic structures

EB I

In contrast to the wide distribution of 'rectangular halls', Early Bronze Age domestic structures assumed a variety of forms. A tradition of free-standing curvilinear structures is documented in the early EB I of northern Palestine (Ben-Tor 1992: 62-64). A distinctive feature of those from Yiftahel in Galilee was the presence of an internal area of stone paving (Braun 1989b: 4, fig. 4). An arrangement of slabs observed in the partly exposed early EB I Building 5 at Tall ash-Shuna (Baird and Philip 1994: 116, fig. 5) may indicate the presence of similar units at sites in northern Jordan.

While small rural sites such as Yiftahel may well have been composed largely of low-density, free-standing, curvilinear structures (Braun 1989a: fig. 4), larger settlements have revealed greater architectural diversity. In addition to Building 5 described above, the early EB I phase at Tall ash-Shuna has revealed a circular structure some 5.5 m in diameter, built of mudbrick on stone foundations (Figure 5.4), with a stone slab in the centre forming the base for a timber roof support. Building 9 appears to have been one of several circular structures, grouped closely in the central area of the site, revealing a greater density of structures than that seen in many 'rural' sites. There is an interesting correspondence between the form of this structure and circular built tombs at Bab adh-Dhra' ascribed to EB IB (e.g. Schaub 1982: 73).

Also dating to the early EB I period at Tall ash-Shuna was Building 13, a rectilinear multi-roomed structure, more akin to the later EB I buildings from the site (Figure 5.5) and EB I structures from nearby sites like Tall Kitan in Palestine

Figure 5.5. Tall ash-Shuna, later EB I rectilinear structure, Building 3; note stone bases for uprights supporting roofed areas (G. Philip).

(Eisenberg 1993: 879-80). Among the noteworthy features of Building 13 are plastered hearths positioned in the corners of rooms and the presence in the floor of irregular stone pillar bases, indicative of substantial roofed areas—illustrating a connection between local domestic and public architectural forms. The evidence from Tall ash-Shuna, therefore, argues against a 'normative' view of EB I domestic structures as dominated by a single architectural style. Furthermore, the variety noted at Tall ash-Shuna is in sharp contrast to the apparently homogenous architecture seen at smaller sites, such as Yiftahel and 'Ayn Shadud, suggesting that there was a real difference in both architectural diversity and the density of occupation between the larger, long-lived EB I sites, such as Tall ash-Shuna and the shorter occupations evidenced at small rural sites. This fact emphasizes the pivotal role played by the larger northern EB I settlements in the restructuring of post-Chalcolithic society (Joffe 1993: 46). The location of Tall ash-Shuna and other large early EB I sites, such as Tall Umm Hammad (Betts 1992), on major wadi systems offering considerable irrigation potential is also likely to form part of this picture.

Occupation at a far lower density is seen at the fourth-millennium BC site of Jabal Mutawwaq east of Jarash, where surface remains suggest the presence of several hundred ovoid structures dispersed over some 28 ha. Some lie within and others outside the so-called enclosure wall. Individual units are 6–10 m in length with foundations of long stone slabs and a doorway in one of the long sides flanked by upright pillars, a feature also seen in the entrance to one of the rectangular halls at Hartuv in Palestine (Mazar and de Miroschedji 1996: 9, fig. 5). The contrast between such low-density arrangements of relatively homogenous units and the more heterogeneous makeup of higher density occupations may well indicate the greater functional diversity and socioeconomic complexity of the latter, which would appear to anticipate the role of the later EB II–III walled settlements.

The walled site of Jawa in the steppe presents yet another contrast. Here, small subcircular structures with walls coated in yellow clay plaster have been interpreted as dwellings. A structure in square UT1 consisted of a subcircular main room, with two smaller compartments (Figure 5.6). Flat stones in the centre of floor may have provided a base for roof supports (Betts 1991: 35-36, figs. 39, 40). However, there is little in the evidence from Jawa to suggest that this architecture represented an indigenous steppe tradition, and it is better seen as an adaptation of familiar western techniques to suit locally available building materials and environmental conditions. The walled EB II settlement of Arad in the arid northern Negev has recently been reinterpreted as a community of sedentarized pastoralists, facilitating interaction between the populations of the south Palestinian steppe and the Mediterranean zone (Finkelstein 1990; 1995b: 80-82).

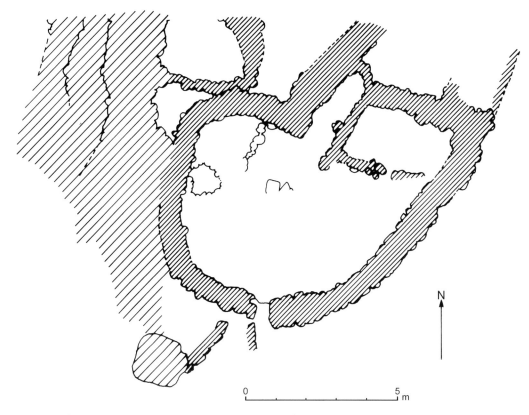

Figure 5.6. Jawa, subcircular domestic structure in Square UT1; structure abuts main site wall to the west (drawn by Y. Beadnell).

The architecture of this site, with its compounds, rectilinear structures and evidence of planning (Ben-Tor 1992: 62-63), is substantially different from that so far revealed at Jawa, suggesting that there may be significant differences between the ways in which southern and eastern areas of the steppe were incorporated into the wider regional system.

EB II–III

The domestic architecture of EB II–III is characterized by rectilinear, multi-roomed structures generally built of mudbrick on stone foundations, sometimes with internal benches. The influence of locally available materials on building practices is neatly illustrated by the fact that buildings at Numayra are made in local sandstone, while those at nearby Bab adh-Dhra' are built almost entirely of mudbrick, reflecting its position on marl deposits (Rast 1995: 127). This procedure extended to the site's local funerary architecture as well, with stone foundations restricted to major architectural units.

Typical domestic installations include a variety of ovens and hearths and large jars sunk into floors (Herr et al. 1991: 11). At Tall Abu al-Kharaz (Fischer 1993, 1994), structures are terraced to deal with the sloping topography, and are separated by narrow pebbled streets. Overall, the evidence appears to suggest that multi-roomed, rectilinear structures, separated by narrow alleyways were the norm for EB II–III sites (Figure 5.7), although the picture would be more satisfactory were more undefended 'rural' sites to be excavated to complement the evidence from the fortified settlements. While general accounts focus attention upon the predominance of 'broadroom' architecture during EB II–III (Ben-Tor 1992: 62-64), many domestic structures do not appear to conform to this category. The main common feature of domestic architecture throughout the Mediterranean zone appears to be the rectilinear plan, within which there exists a considerable degree of variety.

Tall al-'Umayri has revealed multi-roomed EB III domestic complexes that might be described as consisting of groups of 'long' rooms. Artefacts reflect subsistence-related activities—grinding stones, storage jars and debris from chipped-stone working, while a range of spindle whorls indicate textile production (Herr et al. 1991: 10-11) within domestic structures. Multi-roomed rectilinear complexes also characterized the central area at Numayra where several such units

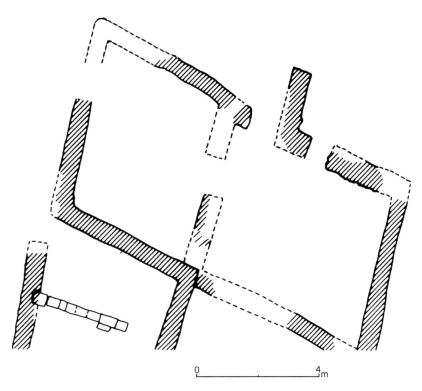

Figure 5.7. Tall Abu al-Kharaz Area 2, EB II Main Phase II; two-roomed, rectilinear architecture (drawn by Y. Beadnell).

were separated by open areas. This site also revealed evidence for roofing techniques using wooden beams covered in reeds and grasses tied together with ropes to produce a kind of thatch (Coogan 1984). Installations such as hearths, stone mortars and storage jars were set into the ground, while the large assemblage of storage vessels, and a number of pits and unfired clay silos, indicate that a significant amount of agricultural produce was stored within individual dwelling complexes. Viewed as a whole, the evidence suggests that many aspects of production and storage were undertaken at the level of individual households.

Numayra, at 1 ha, is small for a walled site, and, despite the emphasis on storage seen within domestic units, has not produced any structure equivalent to the pillared halls seen at some other sites. There is clearly no direct functional equivalence between a small, walled, arid-zone site, such as Numayra, and much larger walled settlements, such as Khirbat az-Zaraqun, located in areas with far greater agricultural potential.

Early Bronze Age walled settlements: organization, significance and diversity

While some walled sites were clearly more than simple agglomerations of domestic units, there is no evidence for the palace complexes or specialist workshops seen in Syria. Rather, the major architectural units appear to have been multi-purpose pillared halls, a form derived from domestic architectural traditions, and which may have been connected with the storage of agricultural products. The architectural evidence suggests a society that, while sufficiently corporate to make substantial investments in common facilities, remained heterogeneous and fragmented in many aspects of subsistence, residence and cultic activity.

The traditional focus upon the presence of defensive walls as a criterion for indicating 'significance' has distracted attention from the fact that even the largest Early Bronze Age walled sites in Jordan encompass no more than 10–20 ha, which is small by the standards of 'urbanism' present in Syria and Mesopotamia (Falconer 1987; Joffe 1993). A consideration of the population levels of Jordanian walled settlements should provide some indication of their likely sociopolitical organization. Falconer (1994: 312) argues that a settlement should only be defined as 'urban' when its population lies beyond that which can be supported from its immediate agricultural sustaining area. A common method of reckoning the populations of ancient sites is the use of estimates based upon site areas. However, calculation is complicated by the contrasting figures

Sustaining area (radius in km)	Sustaining area (ha)	Sustainable population	Threshold 100 persons/ha	Threshold 250 persons/ha
3	2850	1900	19.0	7.6
4	5000	3350	33.5	13.4

Table 5.2. Threshold area for defining an urban site, using sustaining areas of both 3 and 4 km radius and population estimates of 100 and 250 persons/ha.

Site name	Site area (ha)	Periods of occupation
Jawa	c. 5 (upper town)	EB I (Betts 1991: fig. 5)
Khirbat az-Zaraqun	8	EB III (Mittmann 1994)
Pella/Husn	8	EB I–II (Bourke 1997)
Tall Abu al-Kharaz	4	EB I–II (Fischer 1994)
Tall al Handaquq (N)	7 (fortified), 25–30 (scatter)	EB I–II (Mabry 1996)
Tall as-Sa'idiyya	8	EB I–II (Tubb and Dorrell 1994)
Tall al-Handaquq (S)	15	EB IB–III (Chesson 1998: 20)
Tall al-Hammam	15	EB III (Prag 1991: 65)
Tall al-'Umayri	4.25	EB III (Harrison 1997)
Bab adh-Dhra'	4 +	EB I–III (Rast and Schaub 1981)
Numayra	1	EB III (Coogan 1984)

Table 5.3. Excavated walled Early Bronze Age sites in Jordan (site areas and periods of occupation) listed from north to south.

employed by scholars to estimate population per unit area, which range between 100 and 250 persons per ha (Esse 1991: 162; Harrison 1997). Using Falconer's figure of 1.5 ha of land per person, and an agricultural sustaining area of 3–4 km radius from the site, both derived from ethnographic sources (see Falconer 1994: 312 for details), it is possible to estimate where the urban/rural boundary would lie under population densities of 100 and 250 persons per ha (Table 5.2).

Given the difficulty of relating site area to population (Esse 1991; Harrison 1997) and the very different agricultural potentials of the hinterlands of individual sites, these figures are of heuristic value only. However, it is immediately apparent that only by using the maximum estimate for population density and the minimum agricultural sustaining area (3 km radius) does the limit for 'urban' fall below 8 ha. In reality, the limit probably lies some way above this, with the result that few if any of the Early Bronze Age walled sites in Jordan would be classed as 'urban' on this basis (for site areas see Table 5.3). In fact, they should be seen as non-urban agricultural communities, a situation aptly summed up in Falconer's (1987) phrase 'Heartland of Villages', a conclusion in accord with the architectural evidence treated above.

Using the figures quoted above, population estimates for a 10 ha settlement would lie in the range of 1000–2500 persons. A number of cross-cultural studies suggest that communities with populations exceeding 2500 seem to be associated with a significant degree of organizational complexity (Feinman 1995: 260). At the other end of the spectrum, figures quoted by Upham (1990: 12) suggest 400 individuals as providing the upper limit for the population of most non-hierarchical sedentary societies. Under most of the estimates cited above, the population of nearly all Jordanian Early Bronze Age walled sites would fall between these thresholds, and would be classed as 'middle range' societies according to the criteria of Feinman and Neitzel (1984). Of critical importance for our understanding of the Early Bronze Age data is the point that, with groups of this size, there is no automatic connection between *political* leadership and control over strategic *economic* resources, a position in keeping with the ideas of kinship-based groups and heterarchy.

EB II–III

These two are frequently treated together in settlement studies (Gophna and Portugali 1988; Finkelstein and Gophna 1993), and perceived as the floruit of the Early Bronze Age city-state system (Esse 1989: table 1). However, the tendency to treat EB II and III as a unit may have concealed very real changes taking place within this lengthy period, a point that has already been made by Joffe (1993). In practice, sites showing settlement continuity through EB II–III are the exception. Many sites show only partial Early Bronze Age sequences, often relatively short periods of florescence punctuated by significant phases of decline or abandonment (see Table 5.3). Some of the largest EB I settlements, such as Tall ash-Shuna and Tall Umm Hammad, did not develop into major EB II centres, while many walled EB II sites—Pella, Tall Abu al-Kharaz, Tall as-Sa'idiyya, for example—show little sign of significant EB III occupation. On the other hand, some sites with substantial EB III remains, for example Bab adh-Dhra' and Khirbat az-Zaraqun, reveal little evidence for an EB II occupation of comparable scale (see Table 5.3). Major Palestinian settlements, such as Megiddo and Beth Shan (Esse 1991: 83; Finkelstein 1995: 50; Mazar 1997) appear to lack significant EB II occupations, despite the presence of substantial EB I and EB III remains, while other important sites, such as Arad and Tall al-Far'a (N) (Amiran and Ilan 1996: 1; de Miroschedji 1993a: 437), were abandoned before the end of EB II.

There has been a tendency in the literature to assume that the presence of EB II or EB III occupation means that the site was occupied throughout the entire EB II or EB III periods. Implicit in this view is the notion that walled sites were fundamental structural components of the political and economic organization of their local regions. However, to my knowledge, no walled site in Jordan has yet produced the large number of successive stratigraphic phases of domestic occupation that would be expected to result from the continuous intensive use of mudbrick architecture over several centuries. Long continuous occupations require to be demonstrated on stratigraphic grounds, rather than simply assumed on the basis of our still inadequate knowledge of local ceramic sequences.

The large number of sites that do not show smooth transitions between periods should cast doubt on assumptions of continuity of occupation within periods. This provides further evidence for viewing individual walled settlements as something less than permanent structural features of the Early Bronze Age landscape. It would perhaps be better to view their expansion, continuance and contraction as contingent upon particular, and perhaps short-lived, combinations of circumstances within individual localities.

Unstable hierarchies and highly fluid systems of political control are common features of middle-range societies (Paynter 1989: 381-83; Earle 1991: 4). Thus, regional polities, where they existed, may have been short-lived political phenomena, deriving material support from the local subsistence base, but not constituting deeply rooted features of local socioeconomic organization, and certainly not indicative of institutionalized, elite-dominated resource extraction. A constantly mutating system of this kind might explain some of the apparent oddities of site location noted by Finkelstein (1995a) in his discussion of theoretical site territories in EB II–III Palestine. It might also account for the significant number of large EBA II–III sites that did not reappear as important settlements in the Middle and Late Bronze Ages.

Settlement and subsistence

Subsistence economy

The fourth millennium BC saw a number of changes to the subsistence economy of the southern Levant. These innovations, when combined with the intensification of practices already documented in the Chalcolithic, appear to have permitted a significant increase in agricultural production. The major areas of change were:

a. Irrigation agriculture, which, although practised during the Chalcolithic, appears to have been more widely adopted during the Early Bronze Age (Rosen 1995: 33-35).
b. Extensive cultivation, using the ox-drawn plough (Kolska-Horwitz and Tchernov 1989; Grigson 1995).
c. A substantially increased emphasis on tree crops, the olive and vine in particular (Stager 1985; Lipschitz et al. 1991; Esse 1991: 119-25; Finkelstein and Gophna 1993).
d. The large-scale employment of the donkey as a pack animal (Ovadia 1992; Grigson 1995).
e. The greater availability of metal tools for agricultural maintenance and general construction purposes. Evidence in this case comes as much from a decline in the frequency of heavy chipped-stone tools such as axes, adzes and chisels at the beginning of the Early Bronze Age, as from the corpus of surviving metalwork (Rosen 1997a: 161-62).

The effective exploitation of most of the above innovations would have required a degree of long-term investment. The will to commit resources in this manner has recently been linked to the stimulus provided by Egyptian demand for the products of the Levant

(Esse 1989; Joffe 1993; Gophna and Finkelstein 1993; Levy 1995a). The argument is that the appropriation of produce and, thus, control over investment decisions was increasingly concentrated in the hands of emergent local political elites, who functioned as coordinating mechanisms for production, and as channels through which products were collected and transported to the Egyptian market. According to this view, changes in subsistence practices were linked to local political change.

Increasing agricultural production to a level some way beyond that required for immediate subsistence needs would have provided an opportunity for surpluses to be deployed in a variety of projects. However, the manner in which such surpluses were deployed would have been a factor of sociopolitical organization, the social relations of production in Marxian terms. The unique archaeological record of the Early Bronze Age constitutes the material results of these choices. Once more, the lack of evidence for conspicuous consumption and the emergence of elites contrasts with the substantial evidence for investment in a variety of community-centred projects—defences, irrigation works, storage facilities, etc. The manner in which surpluses were deployed is consistent with the anticipated outcome of decision-making on a corporate basis (Schwartz and Falconer 1994; Blanton et al. 1996), in which investment in community-level projects was favoured over individual consumption.

The above is, of course, simply a generalized scheme. In fact, the environmental diversity of Jordan was such that irrigation, tree crops and extensive plough agriculture were not equally suited to all areas. Regional differentials are also seen in Early Bronze Age Palestine where there existed a contrast between the coastal plain, suited for cereal growing, and the upland areas of the West Bank, seen as more suited to tree crops and seasonal grazing (Rosen 1995: 38; Falconer and Savage 1995: 51). It can be argued, therefore, that the supraregional changes in settlement and economy that are documented (see below) could only have resulted from the effective integration of the productive capacities of different regions.

Irrigation

In a region in which rainfall was highly seasonal, and in which livestock raising was likely to be an important component of subsistence strategies, structural evidence for the collection of water cannot automatically be taken as indicative of irrigation agriculture.

The structural indications from Tall Handaquq (N) at present constitute the only physical evidence for the existence of major water catchment systems outside the basalt *harra*. Support is provided by the recovery of grains of naked bread wheat of a sufficiently large size to suggest irrigation agriculture (Mabry et al. 1996: 142). In addition, the botanical record from Bab adh-Dhra', located on Wadi al-Karak in the arid southern Ghor, indicates the cultivation of cereals and pulses, suggesting some sort of irrigation. Additional evidence in the form of linseeds, of a size too great to have been cultivated without irrigation (McCreery 1981: 167), provides further support for this view. Although no structural remains of Early Bronze Age water systems were found, the existence of a system perhaps involving the channelling of spring flows or captured run-off water would have considerably increased the agricultural potential of the area, and would go a long way towards explaining the growth of a walled settlement in that particular location.

Simple floodwater irrigation systems are documented from earlier Chalcolithic settlements located along watercourses such as Wadi Bi'r as-Saba' (Levy 1995b: 230; A. Rosen 1989, 1991, 1995), indicating that the necessary technology was already available at the beginning of the Early Bronze Age. In geographical terms, the east side of the Jordan Valley, with its large number of tributary wadi systems would appear especially suitable for simple irrigation systems. The site of Jawa in the steppe has revealed both structural and botanical evidence (e.g. the presence of six-row barley that could not have been grown locally without irrigation) for the practice of small-scale irrigation (Helms 1982: 105; Wilcox 1981). Regardless of whether the water captured at Jawa was used for cultivation or livestock raising alone, the presence of dams and pools at a fourth-millennium BC site far out in the steppe is strong evidence for the employment of similar techniques elsewhere in Jordan.

Another argument in favour of irrigation agriculture in the Jordan Valley is the sheer number and size of Early Bronze Age settlements and their concentration along the major tributary wadis. Many of these sites are located in areas where annual precipitation is below 300 mm, which, when interannual variability is taken into account (Esse 1991: 11), would render reliable cultivation of most cereals and pulses problematic (Lipschitz 1989: 274). Within these regions, it is hard to see how the population concentrations implied by sites covering 10 ha or more could have been sustained without the increased agricultural production that irrigation would have provided (A. Rosen 1995: 33). In fact, the investment and long-term commitment to a location implied by irrigation may have played a part in the development of walled communities in these key locations.

A dependence upon irrigation agriculture might also provide a clue to the lack of significant post-EB III occupation at many of the larger Early Bronze Age sites in the Jordan Valley. Although too little geomorphological work has yet been undertaken to establish a general pattern, significant episodes of post-Early Bronze Age erosion, observed at Bab adh-Dhra' and Numayra in the southern Ghor (Donahue 1985: 136-37; 1984: 86-87) and at Tall Handaquq (N) farther north (Mabry 1989: 59), may have reduced irrigation potential through the downcutting of the nearby water courses. A. Rosen (1995: 36) has observed that erosional processes towards the end of the Early Bronze Age appear to have rendered floodwater irrigation difficult at certain Palestinian Early Bronze Age sites, perhaps significantly reducing the carrying capacity of the local agricultural base in these locations. Were it established that the two sets of processes were related, this would constitute important evidence for our understanding of both the origins, and eventual decline, of Early Bronze Age subsistence and settlement patterns in the region.

Tree crops

Not all large Palestinian Early Bronze Age sites were in locations offering the possibility of irrigation agriculture (A. Rosen 1995: 36). Perhaps as much as half of the 'built-up' area in EB II–III Palestine was concentrated in the upland areas of Galilee and the West Bank (Broshi and Gophna 1984; Palumbo 1990: 44), a pattern strikingly different from that of the Middle and Late Bronze Ages. This shift appears to be connected with the expansion of upland olive and grape cultivation (Stager 1985; Finkelstein and Gophna 1993). In Jordan, the upland areas along the northwestern margins of the plateau, where precipitation was highest, would have had good potential for the cultivation of tree crops. Published survey data is scarce, but while few large sites have been identified in this zone, a number of defended hilltop sites have been recorded in the uplands around 'Ajlun (Braemer, pers. comm.), perhaps indicative of a subsistence regime similar to that posited for the Palestinian uplands.

Turning to the botanical evidence, while cultivated olives have a larger fruit and higher oil content than the wild variety (Lipschitz *et al.* 1991: 442), the latter would have grown in this area, and it is difficult to distinguish the two on the basis of archaeobotanical evidence. While Lipschitz *et al.* (1991: 449) argue that Chalcolithic data cannot go beyond gathering of wild olives, Epstein (1993) has drawn attention to evidence suggesting Chalcolithic olive cultivation in the Jaulan: olive stones, a significant proportion of olive wood, and artifactual evidence possibly indicative of processing—basalt basins and ceramic spouted vats (see Epstein 1993: table 1). One alternative indicator for deliberate cultivation is the presence of olives outside environmental zones in which wild plants would be expected to occur. Examples include Chalcolithic specimens from Tulaylat al-Ghassul and Tall ash-Shuna in the Jordan Valley (Neef 1990), arguing for initial olive cultivation during the Chalcolithic. Evidence from recent excavations at Tall ash-Shuna indicates that olives are absent from early Chalcolithic deposits, dating around 5000 BC calibrated, but are present in some quantity in pits dating to the very beginning of the fourth millennium BC (calibrated). A second line of evidence is the presence of significant amounts of olive wood at EB I Tall ash-Shuna and EB II Tall as-Sa'idiyya (Cartwright and Clapham 1993). This argues for local cultivation, on the grounds that the wood is likely to have originated from the pruning and felling of trees located at no great distance from these sites. Less is known about wine production. However, the grape makes its first appearance in early EB I contexts at Bab adh-Dhra' (McCreery 1979) and Tall ash-Shuna, in keeping with the evidence from Palestine (Lipschitz 1989: 274). As the native habitat of the grape lies far to the north (Zohary and Hopf 1993), its appearance should indicate its deliberate introduction to Jordan.

How do tree crops fit into the development of the economy as a whole? Unlike cereals, these require an element of long-term investment, with olive trees requiring six to seven years between planting and fruiting (Zohary and Hopf 1993). Furthermore, cultivated fruit trees are propagated by vegetative reproduction, that is, cuttings and grafting (Lipschitz *et al.* 1991: 441), techniques differing from those required by cereals and pulses. In order to maximize the return on such crops, the ability to invest on a delayed return basis is required, as are a guarantee of sufficient stability to render the investment worthwhile and a commitment to regular maintenance (Zohary and Hopf 1993: 134-37). In the absence of evidence for an elite-directed economy, communities making corporate investment decisions appear to offer a plausible alternative mechanism.

In the Jaulan, the fact that 'equipment used for the production of olive oil was present in most households' (Epstein 1993: 143) would imply a domestic level of production. However, in the Early Bronze Age there is evidence for a marked intensification of production, underlined by storage facilities for liquid products at EB II Tall as-Sa'idiyya (Tubb and Dorrell 1993, 1994) and EB III Khirbat az-Zaraqun (Ibrahim and Mittmann 1994), suggesting that some walled sites may have acted as focal points for the collection and storage of a variety of agricultural products. Equally, new ceramic forms characteristic of the Early Bronze Age include jugs, juglets, narrow-necked storage jars and 'vats' interpreted as separators, all of which indicate an increased emphasis upon the handling of liquid commodities (Stager 1985: 176-77; Esse 1991: 119-24). Oils and fats are also likely to have provided the basis for a variety of scents. Recent analytical work has revealed the presence of vegetable oils in jars of Levantine forms found in First Dynasty cemeteries in the Nile Valley (Serpico and White 1996) and would appear to indicate some sort of trade in fine oils. Possible evidence for the presence of wood resins, such as cedar or pine, in at least one imported one-handled jug (Serpico and White 1996: 136) argues for the production and export of scents, thus combining two local specialities (oil and scented wood species) to create an added-value product.

Large-scale production of olives in upland areas, however, presupposes that a proportion of the product was destined for consumption elsewhere. In this case, there would have been value in having collection points either in or en route to likely areas of consumption. The presence of a major oil storage structure at Tall as-Sa'idiyya in the Jordan Valley (Tubb and Dorrell 1993, 1994) may indicate that a proportion of the produce of upland areas was ultimately destined for storage in, and distribution from, the larger walled sites in the lowlands. It is possible that this greater level of organization relates to a combination of improved transport and storage capabilities, stimulated by external demand from Egypt, at least during EB I (Finkelstein and Gophna 1993). However, firm evidence for large-scale Egyptian demand for such products during EB II–III is limited.

More relevant may be Manning's (1993: 47) point that emergent elites may seek to produce their own high value resources for exchange or for display and consumption. Within corporate village communities, the units in competition are more likely to have been the individual kinship-based components that together constituted the corporate group. Thus, given the lack of evidence for 'wealth' in the form of imports and fine craft goods, it is possible that competition was centred around the production, stockpiling and consumption of valued liquids. The inclusion of a large number of jugs and juglets in Early Bronze Age burials (see burials) might be regarded as an act of conspicuous consumption, both providing for the deceased and simultaneously maintaining the value of such products by removing substantial quantities from circulation (Meillassoux 1968; Gregory 1982). It is worth noting that Early Bronze Age jugs and juglets are ubiquitous, occurring in tombs at sites both within and outside those areas best suited for the cultivation of tree crops, suggesting that their significance was recognized on a regional rather than local scale. Even were these vessels placed in tombs empty, the symbolism may still have been as described above.

Plough agriculture

The faunal assemblages from both EB I Tall ash-Shuna and EB III Khirbat az-Zaraqun revealed a significant cattle component (Tables 5.4, 5.5), suggesting that developments in Jordan were in line with those in Palestine where introduction of the ox-drawn plough during the Early Bronze Age is now widely accepted (Grigson 1995: 267-68). As part of a general trend towards increased output, the productivity of areas of the plateau suitable for dry farming, but offering little potential for irrigation agriculture or the intensive cultivation of tree crops, could have been transformed by plough cultivation. The adoption of the ox-drawn plough has been suggested as one reason for Early Bronze Age settlement expansion in the Madaba Plains (Harrison 1997), and would surely have been equally valuable in other upland regions, such as the

Period	Cattle %	Pig %	Caprines %	Total bones
Chalcolithic	15.0	46.3	38.7	313
Early EB I	13.0	37.2	49.7	223
Late EB I	28.1	19.1	52.8	335

Table 5.4. Percentages of identified bone fragments for main species represented at Tall ash-Shuna (after Croft 1994).

Cattle %	Pig %	Caprines %
20	0.7	70

Table 5.5. Percentages of identified bone fragments for main species represented at Khirbat az-Zaraqun (after Dechert 1995).

Hauran and the al-Karak Plateau, which also show a significant expansion of settlement at this time (see settlement, p. 190, 193).

Conditions in these regions would have suited the planting of relatively large areas of low-density cereal crops. Such extensive agriculture, perhaps employing a fallowing strategy rather than the more labour-intensive rotation with pulses (Halstead 1995: 13), would have provided sufficient output to support the local population, without requiring such high levels of labour input as to render the whole system difficult to maintain. Periodic demand for additional labour for harvesting, for example, might have been met through drawing on the labour of part-time specialists, such as potters or metalsmiths, or those engaged mainly in animal herding, to whom stubble grazing may have been highly attractive.

The adoption of the ox-drawn plough permits both increased output and the growth of wealth disparities within a community (Sherratt 1981; Gilman 1991). Furthermore, the training and maintenance requirements of oxen impose a significant 'entry cost', which effectively places ownership out of the reach of small land-holders (Halstead 1995: 17). Thus, plough agriculture appears to benefit larger landowners disproportionately through increased outputs combined with the ability to control others' access to the critical resource of cattle. It appears, then, that the very strategy that seems to have permitted settlement expansion over significant areas of the Jordanian plateau was structured so as to result in growing inequality. However, this assumes that access to plough animals was mediated through patron–client relationships. Once more, the notion of corporate communities, composed of a variety of kinship-based units, would provide an alternative way of gaining access to and meeting the costs of draught animals. However, as with irrigation agriculture, this strategy holds considerable potential for exacerbating initial differences between groups in terms of size and resources, and thus wealth and prestige, and may hold the seeds of important social changes.

Animal raising

The key to interpreting livestock-raising strategies lies in understanding the overall herd structure, that is, the proportions of different species. In Jordan, however, animal-raising strategies would have been partly determined by local environmental conditions (Kolska-Horwitz and Tchernov 1989: 287-88; Grigson 1995: 251-55, figs. 6a-c), with sites in the more arid areas generally showing low frequencies of cattle and pig. Detailed information is currently available from only two sites, EB I Tall ash-Shuna ash-Shamaliyya and EB III Khirbat az-Zaraqun (Croft 1994; Dechert 1995).

Remarks

The marked presence of pig at Tall ash-Shuna is probably related to the distinctive environment of the Jordan Valley, a point reinforced by the presence of significant quantities of pig bone at Chalcolithic Pella (Bourke, pers. comm.), although the decline from Chalcolithic into later EB I is in line with the general regional pattern (Grigson 1995: 254). The other striking characteristic is the increase in the percentage of cattle bone from early to late EB I, which would render bovids the major contributor to meat production. Particularly important is the fact that the late EB I period sees cattle exceed 20%, which has been suggested as a threshold figure indicative of the utilization of the ox-drawn plough (Hesse and Wapnish 1991). This shift appears to coincide with a significant increase in settlement size.

EB III Khirbat az-Zaraqun, located close to the steppe edge, provides a very different faunal sample.

The low percentage of pig at Khirbat az-Zaraqun reflects the site's location on the plateau where the water and shade favoured by pigs were not readily available. Here, too, cattle are present at a level consistent with plough agriculture (Hesse and Wapnish 1991). The predominance of sheep and goat is in line with the general picture reported from Bab adh-Dhra' (Rast and Schaub 1978: 51), reinforcing their critical role in the Early Bronze Age subsistence economy, in particular in the more arid areas.

The presence of roughly equal proportions of male and female sheep might suggest a herd kept mainly to provide wool (Dechert 1995) rather than dairy products. A contrasting pattern is seen in the goat bones, with adult females outnumbering males by 15 to 1. The shortage of evidence for the slaughter of young males, normally indicative of a dairy herd (Grigson 1995: 256-57), is striking. While this might in part result from the difficulty of differentiating between the sex of immature animals (Dechert 1995: 85), it is also possible that caprid bone from Khirbat az-Zaraqun represents but one part of a more complex, and perhaps geographically dispersed, stock-raising system.

At Khirbat az-Zaraqun, the sheep to goat ratio is 55 to 45, in marked contrast to the ratio of 75 to 25 documented at Tel Yarmouth in Palestine (Davis

1988: 145). While this may reflect herding strategy, it is possible that the higher proportion of goat bone at Khirbat az-Zaraqun reflects its proximity to the eastern steppe. Wapnish and Hesse (1988; Hesse and Wapnish 1991) have stressed that faunal remains are directly indicative of animal consumption patterns rather than local herd structures. Caprines are both valuable and mobile, and may be produced and consumed within different economies. The complex patterning of faunal remains seen at Khirbat az-Zaraqun may indicate the consumption on-site of mature female animals deriving from herds kept by groups exploiting the steppe, and that arrived at Khirbat az-Zaraqun through exchange. The existence of a steppe connection of some sort is confirmed by the presence of the remains of both ostrich and wild equid at the site (Dechert 1995).

The evidence from Jawa (Betts 1991) demonstrates clearly the existence of connections between the steppe and areas to the west. Furthermore, recent research in the Syrian Hauran (Braemer et al. 1993; Braemer and Échallier 1995) suggests a role for Early Bronze Age sites in that area in connection with animal-herding strategies. Although caprine herding in the Jordanian steppe has a long history (Garrard et al. 1996), the degree of integration between steppe and farming groups remains unclear. However, one effect of agricultural intensification in the moister areas of Jordan may have been to encourage the development of increasingly close economic links between the west and a steppe economy focused on livestock raising, thus adding an additional block to an increasingly integrated regional economy, and one more element to a heterarchical organization.

The notion of a regionally specialized economy integrated through upland–lowland relationships has been advanced for Palestine (Esse 1991: 156-58; Finkelstein and Gophna 1993), and interactions between the larger sites in the Jordan Valley and the uplands and perhaps steppic areas to the east may reflect a similar structure of regional economic specialization. Physical evidence includes the presence of upland crops at Bab adh-Dhra', but more significant may be the generalized east–west distribution of Early Bronze Age ceramic style zones (see pottery, p. 204-5, 209).

Equids and transport

The evidence discussed above suggests an intensified, multi-component Early Bronze Age subsistence economy, involving new products and technology, the growth of centralized storage and regional specialization within an increasingly integrated economy. The effective operation of such a system would have required a reliable and economical means of transport, the lack of which would have acted as a significant constraint upon interregional connections. Pack animals, the donkey in particular, would have offered just such a mechanism. While animal transport may well have begun during the Chalcolithic (Epstein 1985; Grigson 1987; 1995: 258), the evidence (and requirement) for pack donkeys becomes clear during EB I (Ovadiah 1992: 20-22). In addition to transporting produce, donkeys would also have offered a means of transporting water, tools and building materials to both settlements and to agricultural land, and their employment on a substantial scale would surely have constituted a significant enabling factor in wider economic developments. Actual faunal evidence is limited, although domestic equids were present at Khirbat az-Zaraqun (Dechert 1995). This, however, is the case in many periods (Ovadiah 1992: table 1), and probably results from the limited role of equids as sources of meat (Grigson 1995: 258), with the result that their remains were rarely incorporated within on-site domestic refuse deposits.

Discussion

Although the existence of major structures with a significant storage function is documented (see administrative structures, p. 176), there is also ample evidence for storage of cereals and pulses at household level. Examples include a plaster-lined bin at EB II Tall as-Sa'idiyya (Tubb and Dorrell 1993), both jars and wooden containers at Tall Abu al-Kharaz (Fischer 1993: 285), and several thousand chick-peas in an EB III domestic structure at Tall al-'Umayri (Herr et al. 1991: 10). Many aspects of Early Bronze Age agriculture offered scope for increasing production through the investment of human labour—irrigation and water storage, terracing, extensive weeding. In the absence of clear evidence for elite control over agricultural production or storage, we should perhaps view kinship-based corporate groups as the means of coordinating economic activity on a supra-household level.

Recent reviews have seen the Chalcolithic–Early Bronze Age transition as a restructuring rather than a complete break (Joffe 1991; 1993: 41-48). One of the most striking features of these changes is the disappearance of familiar Chalcolithic prestige items, for which no Early Bronze Age equivalents have been identified. The disappearance from the record of a variety of fine artifacts, the 'ritual paraphernalia' of Joffe (1993: 37), may reflect a shift from wealth to staple finance, a move away from symbolically charged artifacts as a source of prestige (Levy 1993, 1995b), and towards the basing of status upon access to physical resources. The shift may well have resulted from the realization that an alternative path to power existed through the exploitation of new agricultural technologies. These were able to provide the material capital that would enable even higher levels of production and accumulation to be obtained and thus further increase status. The result would have been a new focus upon investment in physical resources and the production of storable products. Egyptian interest in particular commodities that may have given these a 'cachet' in the eyes of local communities perhaps reinforced the process.

The changing economy and power potentialities

A degree of social inequality has been posited for the Chalcolithic with power based upon the control of symbolic artifacts, rare imported commodities and specialist craft production (Joffe 1993; Levy 1993, 1995b). Reconstructions favouring a 'gap' between the Chalcolithic and EB I periods have been contradicted by recent radiocarbon evidence (Joffe and Dessel 1995; Carmi et al. 1995; unpublished early EB I dates from Tall ash-Shuna cited above). However, the Early Bronze Age as outlined here suggests a society organized rather differently from its Chalcolithic predecessor. Changes in the nature and scale of subsistence activities offered means to power (through access to land, labour and surpluses for reinvestment) that were within the control of individual communities, and that would have diminished the significance of participation in the circulation of prestige goods. This point alone would explain some of the most striking differences between the material remains of the two periods.

Equally important to the intensification of agricultural production is the ability to deploy human labour (Knapp 1990, 1993: 88). In the absence of clear evidence for an EB I–II administrative elite, corporate groups could have provided units sufficiently large to underwrite tasks requiring substantial initial investment while offering an increased, but delayed, return, such as the construction of water storage facilities, or the rearing and training of cattle for the plough. Such groups may have assumed the functions of planning and coordination hitherto attributed to elites.

Many have argued that such investment-led intensification of subsistence activities results in an increased potential for the development of inequality (Sherratt 1981; Gilman 1991; Halstead 1995). However, the importance of such groups, and the possibility of inherent intergroup conflict, may have acted to hinder the development of monopoly control over production by maintaining a distinction between structures of political and economic power. In this view, economic production may have remained rooted in such corporate groups, while political power may have proved a more fluid and transitory affair, shifting between groups and between settlements according to a host of local factors. Under such circumstances, walled settlements might be understood not only as a means of protection for a community and its stored agricultural produce, but also as a key element in transitory political strategies concerned with the competitive expression of power through the creation and embellishment of built monuments. Many of the apparently sharp contrasts between the archaeology of the Chalcolithic and Early Bronze Age periods can be accounted for through their different subsistence possibilities and related settlement organization, and the contrasting modes through which power was expressed.

Settlement evidence

The subsistence data reviewed above provides a basis for reaching an understanding of Early Bronze Age settlement evidence. The new subsistence possibilities are likely to have had a significant and highly varied impact upon the economic potential of different regions of Jordan, with the result that regional settlement organization should be expected to differ quite substantially from that of previous periods.

Limitations of the data

Before examining regional settlement data in detail, it is appropriate to discuss the limitations of the available evidence.

a. Good survey data is only available for a small percentage of Jordan: in many areas basic evidence for settlement distribution is lacking. As a result, certain areas for which information is available must be asked to 'stand for' larger, unsurveyed regions.

b. In many cases, evidence is published in preliminary form only. Furthermore, this comes from surveys undertaken with varied aims, resources and field methodologies. The result is that data from different areas is not always directly comparable. Thus, the interregional comparisons attempted below are no more than broad interpretative statements, based on the sketchy information available. Our knowledge of the distribution of small sites in particular is skewed towards those few areas that have been subject to systematic, intensive survey coverage (Harrison 1997: figs. 2, 3). Elsewhere, only the largest sites are generally documented.

c. Many of the larger settlements represent multi-period occupations, which present familiar problems (Joffe 1993: 13-14), such as the over-representation of material from the latest or most extensive phase of occupation, and a tendency to quote a single figure for site area regardless of fluctuations through time.

d. The dating of survey evidence is dependent upon the interpretation of surface pottery. Chronological precision is thus limited both by our knowledge of the local ceramics and the size and composition of the available sample. Many sites can be characterized only as EB II–III, which presents problems for the analysis of settlement organization, all the more so given the possibility that individual sites may have been occupied for quite short spells within these long periods. Re-evaluation of dates is rarely possible as few researchers have published ceramic evidence on a site-by-site basis.

e. In Jordan's highly dissected terrain, geomorphological processes are likely to have had a significant and uneven impact upon the nature of the visible settlement record (Joffe 1993: 12; Banning 1996). For example, heavy post-Early Bronze Age sedimentation occurring in the vicinity of Tall al-'Umayri southwest of modern 'Amman (Schnurrenberger 1991: 372-74) is likely to have considerably reduced the visibility of Early Bronze Age settlement in this area. However, few published surveys appear to have attempted to counter this by the use of appropriate sampling techniques, and so their results may not reflect closely original settlement distributions.

Regional settlement data

North Jordan Plateau

Although systematic survey data is sparse, it is clear that northwestern Jordan supported a number of relatively large Early Bronze Age settlements, reflecting its favourable environment, with Mittmann (1970: 256-64) recording a number of substantial Early Bronze Age sites, including the walled settlements of Khirbat az-Zaraqun and Khirbat ar-Rahub located only 5 km apart. Detailed survey of the area around Khirbat az-Zaraqun (Kamlah n.d.) has revealed several quite extensive Early Bronze Age settlements on the plateau northwest of Irbid, some of which were occupied during EB I as well as EB II–III.

The Jordan Valley

Low rainfall and the infrequency of reliable springs ensured that major Early Bronze Age sites were primarily located at the edge of the valley floor at the points where the major side wadis enter the valley from the eastern escarpment (Ibrahim *et al.* 1988: 171). Most of the wadi systems reveal one major settlement, frequently walled (Table 5.6). A. Rosen's (1995) suggestion that Early Bronze Age sites on the coastal plain of Palestine were positioned to exploit seasonal wadi floodwaters appears highly suggestive, and the positioning of major settlements within the Jordan Valley might imply the use of similar irrigation techniques.

The lack of intensive surveys renders it hard to reconstruct local settlement organization in detail. Most field projects have concentrated upon the excavation of individual sites, paying little attention to the regional settlement context. Exceptions include survey work in the 'az-Zarqa' triangle', and in the area immediately north of the Dead Sea (Helms 1992; Prag 1992). In the former, EB I occupation is dominated by the extensive, apparently unwalled site of Tall Umm Hammad alongside which existed a number of smaller dispersed communities. However, EB II–III saw a marked concentration of population upon the walled site of Tall Handaquq (S). This was followed in EB IV by a return to smaller dispersed settlements. A similar trend was reported from a survey of the lower az-Zarqa' Basin to the east (Gordon and Villiers 1983). Work in the southern part of the Jordan Valley (Prag 1992: 155) reveals a similar picture with EB II–III witnessing the appearance of walled sites and a

Site	Wadi system	Periods of occupation EB I–III	Nature of occupation	Subsequent occupation
Tall ash-Shuna	W. al-'Arab	EB I, EB III	Large unwalled EB I, smaller EB III	Hell
Tall ash-Sakhineh*	Small wadi S. of W. al-'Arab	EBA	?	MB, LB, Iron
Tall al-'Arab in*	W. Ziqlab	EBA	?	MB, LB, Iron
Pella/al-Husn	W. Jirm	EB I–II	Walled EB I–II	Major MBA and LBA
Tall al-Husn	W. Jirm	EB II ?	Major structure, EBII	Small MB/LB
Tall Abu al-Kharaz	W. al-Yabis	EB I–II	Walled 4 ha	Smaller LBA
Ad-Debab*	W. az-Zarqa'	Chalco-EB	?	EB IV, Classical
Tall al-Handaquq (N)	Wadi Sarar	EB I–II	Walled 7 ha, with larger sherd scatter	None
Tall as-Sa'idiyya	W. Kufrinjeh	EB II	Walled? 8 ha	Small, end of LBA
Tall al-Handaquq (S)	W. az-Zarqa'	EB IB–III ?	Walled, 15 ha	None
Tall Umm Hammad	W. az-Zarqa'	EB I	Unwalled, 20 ha scatter	EB IV
Tall al-Qos*	W. Rajib	EB I–II ?	?	Iron Age
Tall al-Hammam	W. Kafrain	EB III	Walled, 15 ha ?	MB/LB
Tall Mustah*	W. Kafrain	?	?	LBA, Byz
Bab adh-Dhra'	W. al-Karak	EB I–III	Walled, 4–8 ha	None
Numayra	W. Numayra	EB III	Walled, 1 ha	None

*The information is derived from Ibrahim et al. (1988), with no subsequent confirmation.

Table 5.6. Early Bronze Age sites on the east side of the Jordan Valley (listed from north to south).

corresponding decline in the actual number of settlements compared to EB I. These areas appear to demonstrate the pattern of population agglomeration familiar from the Palestinian data (Esse 1991; Joffe 1993). However, a contrasting pattern is seen in some areas of the valley. In the north, for example, at Tall ash-Shuna, an early EB I settlement estimated at c. 5–10 ha, was succeeded by a larger late EB I settlement, which, covering c. 15–20 ha, was larger than many EB II–III walled sites. However, there is little evidence for a substantial EB II occupation in this area. Here, then, the agglomeration process (if such there was—no reliable survey data is available) took place rather earlier, and there appears to have been local settlement disruption during EB II. However, the site revealed an EB III occupation, characterized by the presence of Khirbat Kerak ware, no less extensive than the 5–10 ha estimated for the early EB I. This is all the more surprising in the light of the paucity of EB II evidence and may raise questions concerning the reliability of the ceramic criteria used to define occupational periods (see Philip and Baird 2000).

With the exception of Tall ash-Shuna however, there is an apparent absence of major EB III settlements on the east side of the northern part of the valley. Tall ash-Shuna apart, the most northerly fortified EB III settlement is Tall Handaquq (S), a marked change in settlement distribution compared to EB II, which saw the presence of several fortified sites in this part of the valley (Table 5.6). A similar contraction in settlement numbers between EB II and EB III is also documented in the Coastal Plain and Jezreel regions (Palumbo 1990: 47, fig. 19), where it has been interpreted as indicating the concentration of population into a small number of ever-larger centres. However, the southern part of the valley sees an increase in the number of walled settlements during EB III, a situation more in line with that seen in southern Palestine (Finkelstein 1995a: 63). The different patterns of settlement development within the Jordan Valley may reflect not just subsistence factors, but may be related to those processes governing the appearance of Khirbat Kerak ware (Philip 1999) (see pottery, p. 208).

Upland areas

'Ajlun

The highly localized variations in geomorphology and ground cover conditions characteristic of dissected upland territories can have a striking effect upon site recovery rates (Banning 1996: 29-31). While this renders many details of settlement development in the 'Ajlun highlands uncertain, the broad outlines of a settlement history are now becoming clear. Although Neolithic occupation is known in this region (Kuijt et al. 1991; Banning et al. 1994), settlement in the uplands appears to have undergone a major expansion during EB I–II, including the appearance of defensive walls at some sites (Sapin 1985, 1992; Mabry and Palumbo 1988). While Sapin (1992: 171) argues for an abatement of settlement with EB III, others note that the ceramic criteria for distinguishing the local EB II and III are uncertain (Mabry and Palumbo 1988: 289). A phase of Chalcolithic settlement recently documented in Wadi al-Yabis has been connected with the cultivation of tree crops, for which this area was particularly well suited (Mabry and Palumbo 1988: 189). Thus, settlement increase in EB I would appear to relate to the expansion, rather than the initial adoption of arboriculture. The broad contemporaneity of this development with the growth of large Early Bronze Age settlements in the lowlands appears to support the view that the two processes were both aspects of the development of an increasingly integrated regional economy.

East of Jarash

In contrast to the Mediterranean environment of the western uplands, the more arid areas east of Jarash offer rather different subsistence possibilities (Sapin 1985). Early Bronze Age exploitation of this terrain is represented by settlements, enclosures, reservoirs, tumuli and rock art and shows an extent and intensity far exceeding that seen in the second millennium BC, a fact that Sapin (1992) attributes to soil degradation resulting from over-exploitation of this fragile environment during the Early Bronze Age.

A large number of EB I sites have been reported from the Wadi az-Zarqa', ar-Rumman and Jarash areas, at least some of which appear to show ceramic links with both the central Jordan Valley and the steppe site of Jawa (Gordon and Villiers 1983: 288; Hanbury-Tenison 1987; Helms 1987: 55-60; Betts 1991: 74-75; 1992: 47-48). While it is difficult to distinguish between agricultural and pastoralist sites on the basis of surface-collected data, proximity to water sources would appear to offer one possible criterion (Hanbury-Tenison 1987; Sapin 1992) and it is very likely that some of the extensive physical remains in this area relate to the activities of EB I animal-herding groups (or the animal management activities of those groups responsible for the agricultural sites in the region). In contrast, EB II–III sedentary occupation, at least in the lower az-Zarqa' Basin, appears less extensive. Agglomeration of population within the single enclosed site of at-Tall at 2.4 ha cannot account for the general decline in overall settled area.

One might suggest that the apparent high visibility of mobile groups during the fourth millennium BC reflects some particular aspect of the manner in which the stock rearing and agricultural components of the population were integrated during the EB I period. One possible explanation is that the two were relatively little distinguished during the earlier fourth millennium with economic activity in this region constituted at the level of relatively small, autonomous groups. Recent survey on the fringe of the arid zone in the middle az-Zarqa' Basin has identified a striking range of Early Bronze Age occupation, elements dating to EB II in particular (Chesson et al. 1995: 122). This includes both walled and open settlements, as well as a series of stone circles associated with basalt-tempered pottery of a type found at Jabal ar-Reheil in clear EB II contexts. In this area, a number of different settlement forms may well have coexisted, while the range of variability within individual forms (village, for example) remains to be explored. Moreover, there is no indication as yet that the walled settlements lay at the apex of a regional settlement hierarchy; the concept of heterarchy seems more useful.

The ceramic parallels between the az-Zarqa' and Jordan Valleys noted in EB I are continued into EB II (Chesson et al. 1995: 120, fig. 3). Tall as-Sukhne, at 3.5 ha, represented a walled EB II settlement of some significance, but was apparently occupied only intermittently (Chesson et al. 1995: 116), reinforcing our suggestion that walled settlements should be seen as but one element that could be drawn upon as part of the flexible construction of sociopolitical units, and need not have been associated exclusively with sedentary agricultural groups.

If we accept the idea that human groups may alter their respective emphases upon mobile and sedentary subsistence strategies as circumstances require (LaBianca 1990), we might view the apparent decline in settlement numbers in the lower az-Zarqa' and Jarash areas during EB II–III (Gordon and Villiers 1983; Gordon and Knauf 1987; Hanbury-Tenison 1987) as indicative of the emergence of a less visible, more fully mobile sector within the population. Such a development might be viewed as a response to a demand from more sedentary communities for animal products, which had grown to such a level as to encourage the development of a distinctive, specialist pastoralist sector. Moreover, the appearance of walled settlements, such as Tall as-Sukhne (and in southern Syria [Braemer 1993b; Braemer et al. 1995]), suggests that groups adopting flexible, resilient subsistence strategies were able to appropriate and reinterpret a variety of contemporary settlement forms characteristic of the more fertile regions. While the above reconstruction is no more than a suggestion, the critical points are the existence of contrasting regional patterns of development within Jordan and that formal similarities between settlement types should not be equated with equivalence of function.

Central Jordan Plateau

Madaba region

Surveys in the Madaba Plain (Harrison 1997) reveal a tendency for EB I settlement clusters in wadi systems along the western edge of the plateau to be replaced by single larger settlements during EB II–III, the classic agglomeration pattern observed elsewhere in the southern Levant (Gophna and Portugali 1988; Esse 1991; Finkelstein and Gophna 1993; Joffe 1993; Falconer 1994). Tall Madaba, with an *estimated* Early Bronze Age area of 16 ha, would have constituted a regional centre as large as all but the biggest of those known from Palestine (Finkelstein 1995a). In this case, the absence of Early Bronze Age settlement beyond the 250 mm isohyet (Harrison 1997: 16) confirms the association of sedentary activity with the limits of dry farming, although it appears likely that settlements in these areas would, through necessity, have practised a more resilient subsistence strategy than those in the side wadis of the Jordan Valley. Once again, it is interesting to note that quantitative studies of settlement data (Harrison 1997: fig. 9) indicate little intersite economic differentiation, and betoken a low level of settlement integration. While large sites exist, they do not form the upper tier of a clearly structured hierarchy, but may be better viewed as possibly short-lived epiphenomena.

Several of the larger EB II–III sites on the plateau developed from smaller Late Chalcolithic/EB I occupations (Harrison 1997: table 3), suggesting settlement expansion from the wadi systems on to the plateau during the later fourth millennium. Given the nature of the plain, such sites would almost certainly have been connected with extensive plough-based cereal cultivation. However, cattle are heavily water dependent, and the rarity of springs and reliable streams on the plateau would have rendered water management techniques, such as dams and reservoirs, necessary in order to exploit the plain as an agricultural rather than a grazing resource (note the existence of enclosed depressions at Tall Jalul [Harrison 1997: 17]). However, the area has long provided summer grazing for mobile groups exploiting either the steppe to the east or the Jordan Valley to the west (Prag 1985, 1991; LaBianca 1990), and their presence during the Early Bronze Age cannot be ruled out.

Al-Karak region

While the al-Karak Plateau has been surveyed in recent years, the lack of consistently recorded data on site areas renders inappropriate the analytical techniques employed in the case of the Madaba Plains. The major site of al-Lajjun typifies the problem, with site area quoted variously as 8 ha (Steele 1990), 14 ha (Homès-Fredericq and Hennessy 1989: 360) and 17.5 ha Glueck (1934: 44).

However, despite some discrepancies between the settlement data as reported by Miller (1991) and Steele (1990), it is clear that there was a large increase in both site numbers and areas from EB I to EB II–III, and that there existed a strong association between site location and optimal positions for rain-fed agriculture (Steele 1990: 11-13, fig. 5). At least six sites, ranging between 1 ha and 8 ha, are said to reveal enclosure walls, with Steele (1990: 16) arguing for an EB II–III settlement hierarchy with fortified sites ranked above the rural component, but below a possible central place tentatively located at Adir. However, the variability in the size of defended sites must raise questions about the extent to which they can be considered as 'equivalents' in any hierarchical scheme. Furthermore, given the lack of fully quantified settlement data, the absence of evidence regarding either the duration of occupation at individual sites or the degree of contemporaneity of occupation at different sites (we are speaking of some 700 years after all), and the contrary evidence from the Madaba Plains, attempts to reconstruct settlement hierarchies on the al-Karak Plateau appear somewhat premature. It is preferable to view the al-Karak Plateau as one more example of a poorly integrated heterarchical structure.

However, another aspect of the data from the al-Karak Plateau deserves consideration. This is the fact that Early Bronze Age settlement in this area appears to extend significantly farther east than is the case during the second millennium BC (Miller 1991: 307), a pattern reminiscent of that seen in Wadi az-Zarqa' (Chesson *et al.* 1995: 122). Also striking is the degree of continuity of occupation from EB II–III into EB IV witnessed in the region (Palumbo 1990: 98-102). Parallels with the situation in the az-Zarqa' Basin, where sedentary EB IV settlement has also been recorded (Palumbo 1990: 58-59), suggest that this is attributable to the fact that neither of these areas, both located near the limits of dry farming, was particularly suited to highly intensive agricultural strategies. Low-intensity, extensive crop production in combination with ready access to the steppe would have provided the basis for a resilient food system (LaBianca 1990: 20) involving a significant animal-herding component. Such a structure would have been less inclined to collapse under pressure than the more highly specialized systems posited for areas to the west. It is worth reiterating again that the presence of walled settlements in two areas does not demand the existence of a common socioeconomic structure.

Wadi al-Hasa and settlement in the southern arid zone

Surveys in recent years have documented a considerable body of evidence for Early Bronze Age occupation in areas of southern Jordan where local annual precipitation is insufficient to support widespread intensive agricultural activity. The few large Early Bronze Age settlements in this area are restricted to alluvial terraces situated adjacent to major wadis draining the plateau to the east (Donahue 1981: 140) and which offer some potential for irrigation. In that sense, they resemble the larger settlements of the Jordan Valley to the north.

However, survey in Wadi al-Hasa and the southern Ghor indicates the presence of a significant number of Early Bronze Age 'campsites' and 'corrals' in areas less obviously suited for agriculture, which are thereby taken to indicate pastoralist activity. However, in this case there appears to be a marked predominance of EB I over EB II–III settlement (MacDonald *et al.* 1988: 155-66, tables 24-26; 1992: 61-71), the reverse of the situation documented in the al-Karak Plateau and Madaba Plains. The discrepancy may be partly methodological. For example, the reporting of lithic scatters in the Southern Ghor (Neeley 1992) contrasts with their absence from the catalogues of Chalcolithic–Early Bronze Age sites recorded in the al-Karak Plateau (Miller 1991). However, the presence of EB II–III sites in areas with good agricultural potential renders the absence of indications of contemporary evidence from smaller sites rather odd. In fact, this raises the question of the relationship between site function and the range of diagnostic material present. Of the 38 sites in Wadi al-Hasa listed as EB I (MacDonald *et al.* 1988: table 4), no less than 30 produced 10 or fewer diagnostic sherds. Given the small samples, and an environment favouring small or seasonally occupied settlements, the Early Bronze Age ceramic inventory is likely to have been impoverished and to lack the classic EB II–III diagnostic forms, a point borne out by the presence of 29 sites in the southern Ghor classed as 'Early Bronze Age' without further qualification (MacDonald *et al.* 1992: 71).

MacDonald *et al.* (1988: 166) also note that some of the EB I 'camp sites' resemble those known from Sinai and the Negev, which are traditionally dated to EB II on the basis of their alleged association with Arad (Beit-Arieh 1981). Recent radiocarbon evidence (Avner *et al.* 1994) now suggests that the Negev was occupied continuously throughout the Chalcolithic and Early Bronze Age, and that assemblages consisting of hole-mouth vessels and chipped stone alone cannot allow occupation to be placed more precisely than within a broad Chalcolithic–Early Bronze Age phase (Sebbane *et al.* 1993: 43-45). In this light, the data from southern Jordan could be interpreted as evidence for a continuous exploitation of the region throughout the fourth and third millennia, a scenario more in keeping with settlement evidence from areas to the north and west. Similarly, while Henry (1995: 354) equates 'Chalcolithic' sites in the Hisma, characterized by corrals and curvilinear structures and producing small quantities of locally manufactured thick, heavily tempered plain pottery, with 'Timnian' material from the west side of the 'Arabah, radiocarbon dates appear to fall between 5700 and 4000 bp (Henry 1995: 359, fig. 15.2), that is, well into the Early Bronze Age reinforcing the points made above.

Fourth- and third-millennium BC evidence from southern Jordan is far more striking than that of the 2nd millennium (see MacDonald *et al.* 1988: tables 168-70), with the evidence, such as pens and corrals, presumably connected with animal management activities especially prevalent. This would appear in keeping with the idea expressed above of the growth during the Early Bronze Age of a specialized animal-raising economy in the arid zone. The faunal evidence from Jabal Jill (Henry 1995: 368-69), which produced 74% caprine and 25% gazelle bones, implies an unusual combination of herding and hunting strategies, and might indicate the extent to which caprine raising for 'export' had become integrated within the subsistence strategies of diverse steppe-based communities.

The evidence from eastern Jordan

Early Bronze Age exploitation of eastern Jordan is poorly understood. Two sites in Wadi al-Jilat have been attributed to the Early Bronze Age (Garrard 1989) and fourth–third-millennium sites have been identified elsewhere (Betts 1991: 189), but there is insufficient data to describe a pattern. However, recent evidence from southern Syria appears to indicate the development there of a subsistence strategy based around a small number of larger Early Bronze Age settlements, in locations suited to the capture of seasonal run-off water. This is markedly different from the pattern observed in southern Jordan, and has recently been put forward as evidence for the periodic collecting together of animals, perhaps for counts, slaughter or shearing (Braemer et al. 1993; Braemer and Échaillier 1995; Échallier and Braemer 1995). These sites, which appear to have been occupied, perhaps intermittently, throughout most of the Early Bronze Age, may provide a context for the otherwise rather anomalous EB I site of Jawa, and may indicate the emergence of specialized animal management strategies in the eastern steppe during the fourth millennium, presumably in response to demand from communities to the west. Clear ceramic links between the larger steppe sites and their western counterparts (Betts 1991: 103–104, table 3; Braemer and Échallier 2000: 406) support this possibility.

Southern Jordan

Evidence from southern Jordan is both limited in quantity and difficult to interpret. The importance of the region is clear from the presence of major copper deposits, apparently exploited as early as the Chalcolithic period to supply ores to sites in southern Palestine (Hauptmann et al. 1992; Shalev 1994). Recent research points towards a largely continuous exploitation of the Faynan ores throughout the Early Bronze Age, with evidence for copper-working now known from sites assigned to EB I, EB II and EB III–IV periods (Fritz 1994; Adams and Genz 1995: 14; Adams 2000). However, the critical questions concerning the organizational basis of copper production, and the means by which control was exercised over this important resource, cannot yet be answered.

Tall Magass, located 4 km north of al-'Aqaba, has also produced evidence for copper working, and while a significant Chalcolithic occupation is clear (Khalil 1987), the presence of Canaanean sickle blades suggests some Early Bronze Age activity (Khalil 1988: 93, e.g. fig. 11.4). The difficulty of identifying the transition between the Chalcolithic and Early Bronze Age in southern Jordan is exemplified by the site of Wadi Fidan 4 which, while originally classed as 'Chalcolithic' (Adams and Genz 1995: 19). In contrast to the rather localized ceramic assemblage from this site, pottery from the early third-millennium BC site of Barqat al-Hatiye shares many features of shape and surface treatment with more northerly EB II forms (Fritz 1994), and suggests that the south was drawn further into regional interaction systems as the Early Bronze Age progressed. There is clearly much yet to be understood concerning the nature of Early Bronze Age developments in southern Jordan and their connection with the exploitation and control of local copper resources.

Major patterns in settlement data

Patterns of settlement development in Early Bronze Age Jordan are regionally differentiated. This applies not simply to the differences between the arid and fertile zones, but between different areas within these zones. However, the apparently disparate developments may reflect regional specialization within an increasingly integrated interregional economy.

The distribution of Early Bronze Age settlement is in striking contrast to patterns seen during the second millennium BC, when major sedentary occupations appear to be more heavily concentrated in the north and west. In the Jordan Valley, the few Middle Bronze Age sites located south of Wadi az-Zarqa' (Najjar 1992) offer a sharp contrast to the more southerly spread of walled EB II–III settlements (Ibrahim et al. 1988; Prag 1992: 155). In the case of sites in the Jordan Valley, this may be attributable to the effects of erosional regimes on irrigation possibilities, as described above, but a similar pattern is also seen in areas where irrigation was less feasible. The scale of Early Bronze Age settlement in Wadi az-Zarqa' and the al-Karak Plateau, for example, contrasts with the settlement record of the second millennium BC (Dornemann 1983; Miller 1991), while in the area south of 'Amman, as a whole, it is only with the Iron II period that we see the re-emergence of a major sedentary component resembling that documented for EB II–III (MacDonald et al. 1988; Miller 1991).

While this may reflect environmental changes towards the end of the third millennium BC, the data for the Holocene is patchy (Goldberg 1994: 99), and the various strands of palaeoenvironmental evidence are by no means in agreement (see papers in Bar-Yosef and Kra 1994 and discussion in A. Rosen 1995). It is worth noting that any particular climatic shift would have had a differential impact upon the various regions of Jordan. Increased rainfall may cause erosion of agricultural soils in one area and better crop growth in another. On the whole, environmental models operate at a scale well beyond the level at which such developments were perceived, and responded to, in human terms.

In fact, it may be useful to consider the reason for the development of the distinct Early Bronze Age settlement pattern in the first place. It can be argued that this is indicative of a form of Early Bronze Age socioeconomic organization, within the more arid regions in particular, that is not repeated during the Middle and Late Bronze Ages. In Palestine, too, a significant number of large Early Bronze Age centres were not reoccupied in the second millennium BC (Broshi and Gophna 1986; Falconer 1987; Finkelstein and Gophna 1993), suggesting a real qualitative difference between the settlement landscapes of the third and second millennia BC, and highlighting the scale and complexity of Early Bronze Age settlement development in the region. Settlement data appear to confirm the initial suggestion that the Middle and Later Bronze Ages were not simply the revival of Early Bronze Age 'urbanism'.

Analyses of settlement data (Falconer 1994: 315-17; Falconer and Savage 1995; Harrison 1997) suggest that the Early Bronze Age is characterized by poorly integrated settlement systems, showing significant regional differentiation. The existence of regional settlement hierarchies focused around the largest sites has not been demonstrated, suggesting that analyses framed in terms of 'central places' (e.g. Finkelstein 1995a) may be inappropriate for Early Bronze Age Jordan. Rather, the settlement landscape has been characterized as a patchwork in which the largest units were superimposed upon a much broader resilient network of smaller sites following their own courses of development (Falconer and Savage 1995: 55). This would suggest that interregional integration was to a great extent constituted at a day-to-day level between individual communities and their members, and not channelled through a few major sites.

At present, we lack settlement data that is sufficiently fine-grained to allow a comparison of individual regional settlement patterns before, during and subsequent to the florescence of large sites, but the settlement data, in combination with the limited evidence for institutionalized control structures within walled sites, argues against their being assigned a strictly causal role in local socioeconomic developments, and that the subsistence base could function with or without the presence of major local settlements. The settlement data argues for a flexible system, free to integrate in a variety of ways, rather than supporting a rigid hierarchical structure. Such a construct would appear consistent with the notion of walled sites in some regions as representing temporary expressions of *political* concerns, rather than economic central places. The idea of walling a site would, therefore, have formed an ideological resource, open to reworking according to specific local concerns. Thus, the presence of walled sites in different locations may need not imply similar underlying socioeconomic structures. The concentration of Palestinian Middle Bronze Age walled sites in a more restricted number of locations than was the case in the Early Bronze Age (Falconer 1994: 326) appears to support the notion that the placing of such sites in the Early Bronze Age was not purely a function of economics.

Mobile groups

A form of mobile animal herding, closely integrated with a sedentary agricultural economy, described as village-based transhumance, has been documented during the fifth millennium BC in southwestern Palestine (T. Levy 1992b: 73; 1995b: 232). However, settlement evidence in both southern and eastern arid zones suggests the emergence during the Early Bronze Age of significant, if regionally differentiated, systems of specialized livestock management providing secondary animal products for sedentary communities.

Given the location of many walled settlements in the Rift Valley and along the edge of the steppe, there are grounds for suggesting that the inhabitants of these sites too were involved to some degree in seasonal movement, with some perhaps quite extensively involved with mobile groups. (Note Finkelstein's [1990, 1995b] suggestion that Arad represented a gateway community for interaction between mobile and sedentary groups.) However, this would have constituted a rather different system from the more specialized animal management posited for groups exploiting the arid zone. These more localized processes of nomadization and sedentarization, perhaps fluctuating year-to-year, region-to-region, would have provided a relatively fine-tuned response to local environmental, economic and sociopolitical factors. The on-site storage of agricultural staples, combined with the possibility of seasonal variations in the size of the resident population, may have given added emphasis to wall construction as a means of protecting valuable community resources.

Highland–lowland–steppe connections and the growth of a regional economy

There is both ceramic (east–west style zones) and economic evidence (e.g. lowland oil storage) to argue for a connection between settlement development in upland and lowland areas. I have suggested that different regions followed agricultural strategies best described as 'subsistence plus' with each developing a specialist component suited to the local environment. Each would, therefore, have contributed specific resources to an increasingly integrated regional economy. Intensive horticulture in the valley would have exploited the potential of floodwater irrigation while requiring a significant labour force. Extensive plough agriculture would have rendered possible the production of agricultural surpluses in suitable areas of the plateau. The cultivation of fruit trees in well-watered upland zones would have supplied a product suitable for exchange, perhaps with the larger valley bottom sites, while the development of a mobile pastoralist component would have catered for sedentary communities' need for animal products, as well as providing a pool of additional labour that could be drawn upon as required. Critically, these changes appear to have taken place in the absence of convincing evidence for the existence of institutionalized elite power.

The material remains: burial data

A case study: graves from Bab adh-Dhra'

The sole Jordanian cemetery producing material from all periods of the Early Bronze Age is located just outside the walled settlement of Bab adh-Dhra'. Despite some local characteristics, examples selected from this site will be used as the basis for discussion.

EB IA burials occur in shaft and chamber graves (Figure 5.8). Tomb A 78 (Schaub 1981a) consisted of a vertical circular shaft opening into four chambers, the entrance to each of which was closed by a blocking stone. The disarticulated skeletal remains of several individuals formed a pile in the centre of each chamber. Skulls were usually placed to the left of the pile, while grave goods, mainly ceramic vessels, were placed around the edge of the chamber or to the right of the entrance. The graves contained both adult and child burials, and are assumed to represent family groups. Finds were dominated by pottery, mostly bowls (50–60%), jars and juglets with wide necks (Figure 5.13.1-2). Other artifacts, such as limestone maceheads, shell bracelets, basalt bowls and clay figurines (Figures 5.20.2, 5.23), occur less frequently.

EB IB tombs (Schaub and Rast 1989) are characterized by shafts opening into a single chamber, with entrances that include a threshold stone flanked by stone orthostats. A similar practice has been observed at the EB I pillared hall at Hartuv (Mazar and de Miroschedji 1996), suggesting an overlap between structures used by the living and the dead. EB IB graves contain more articulated burials, and disarticulated remains when found are not placed in the deliberate piles characteristic of EB IA. Some EB IB graves are circular, mudbrick structures with a semi-circular forecourt; the apparent resemblance to the entrance shaft of contemporary rock-cut tombs may be intentional.

EB II–III funerary architecture consists of rectangular mudbrick structures 7–11 m long and 4–7 m broad, termed 'charnel houses' (Figure 5.9). The entrance was in the long side and had a stone threshold, giving the tombs a close resemblance to the layout of the so-called 'broadroom' house. In many cases, two monoliths flanked the entrance, a feature with parallels in both stone-built tombs and other architectural units. Some EB II charnel houses were circular, with walls corbelled inwards and a central slab, presumably the base for a roof support. While these carry over certain design features from shaft and chamber graves, parallels with circular architecture at northern sites, such as Tall ash-Shuna (Figure 5.4) are also close.

Charnel houses could contain more than 100 disarticulated burials, in some cases forming distinct layers, suggesting an extended period of use. One large, long-lived tomb, A 22 (Rast and Schaub 1980), included six articulated burials, three of those associated with the remains of what may have been a pallet or bed made of poles and matting, possibly connected with the transport or laying out of the deceased. All other burials, including 161 skulls, were disarticulated and clustered in piles along the external and internal walls.

Discussion

EB IA disarticulated burials have been interpreted as indicating periodic visits to the site by non-sedentary groups (Schaub 1981b: 81; Khazanov 1984; Dever 1987), with the dead transported in a decomposed state. However, Palumbo (1990: 122) makes the point that many sedentary communities also practise disarticulated burial. In fact, discrepancies between

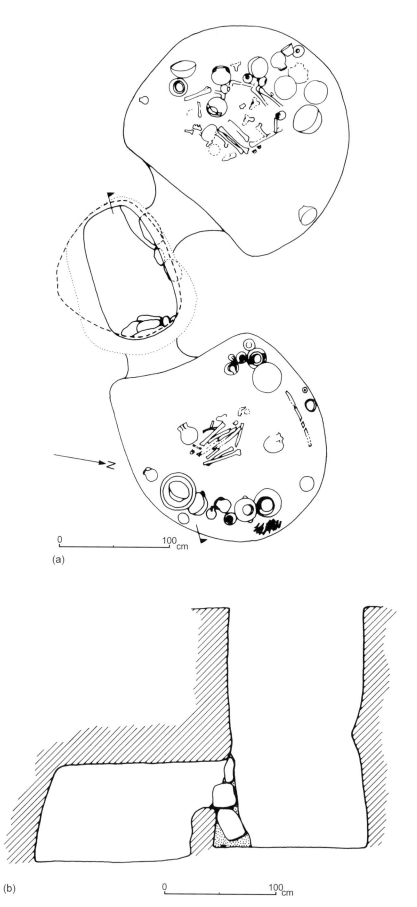

Figure 5.8. Bab adh-Dhra' Tomb A76, an EB IA shaft grave: a. plan; b. section (drawn by Y. Beadnell).

Figure 5.9. Bab adh-Dhra' Tomb A55, an EB II–III charnel house used for multiple successive interment over an extended period; doorway to left of picture, blackened area beyond scale produced burnt linen cloth covering pottery and bones (photo by R.T. Schaub).

numbers of skulls and long bones recovered from individual tombs at Bab adh-Dhra', and the stylized arrangement of both skeletal material and artifacts (see also Jericho Tomb A 94 [Kenyon 1960: 20] and an EB I tomb at 'Ayn al-Assawir in Palestine [Mazar 1990:98]), appear suggestive not of mobility, but specific practices involving human remains. The handling, movement, even veneration of ancestral remains area phenomena well documented in both the ethnographic and archaeological literature from other parts of the world (Barrett 1988). Anthropological research (Van Gennep 1960; Turner 1969; Bloch 1971) suggests that structures designed for long term reuse, allowing regular access to the community of the deceased, would permit the symbolic presence of the ancestors to be drawn into the world of the living. Thus, tombs would provide foci for links with the ancestral group, suggesting that the interment of an individual was understood as part of a process, a link in a chain. The maintenance of an ancestral unit would help explain the sheer number of individuals interred within single funerary structures, and it seems reasonable to suggest that multiple successive burial was connected to the developing importance of kin-based groups as corporate units.

Burial equipment

This suggestion finds support in the stereotypical nature of the grave goods, which are dominated by ceramics, the forms and fabrics of which overlap to a great extent with those of domestic pottery. The rarity of metal goods is particularly striking in the light of evidence for Early Bronze Age copper exploitation at Faynan (Hauptmann et al. 1985, 1992; Adams and Genz 1995), and for copper working at individual sites (Rehren et al. 1997). The grave repertoire appears to have excluded many forms of artifact. Of particular note is the rarity of those things most closely associated with the marking of individuals in EB IV and Middle Bronze Age graves—pins, jewellery, weapons, meat offerings—items placed close to the body and presumably associated with specific individuals (Hallotte 1996; Baxevani 1995; Ilan 1995; Philip 1995b). This might suggest that EBA burial sought deliberately to subsume individuals within the group.

Cooking and large storage vessels are also rare in graves. They do, however, contain a significant number of small bowls and narrow-necked jugs and juglets (Schaub 1996: 234, table 1), best characterized as vessels for food consumption, and containers for such valuables as oil, wine or scents. Given the apparent significance of oil production to the wider economy, it may be that these vessels and their contents expressed in burial contexts messages concerning access to the critical agricultural resources of land, labour and investment. The attribution of symbolic value to locally produced substances certainly appears consistent with arguments presented earlier concerning a new emphasis upon the mobilization of agricultural products.

The presence of numerous walled settlements alongside 'collective' burial practices lacking obvious wealth differences supports the view that kinship-based groups were the core component of Early Bronze Age social and economic organization. In the absence of convincing evidence for institutionalized elites, it would appear that these groups provided the basis for undertaking major delayed return investments, and working in cooperation could have created material remains superficially akin to those produced by stratified state societies.

The stereotypical form and long-term continuity of collective tombs argue that they were closely involved in the reproduction of the social structure (Barrett 1988), and by providing an ideological context for living corporate groups they may have served to maintain a system of rights and obligations and to provide a legitimating 'past' for current social realities. Mortuary practices may, therefore, have served the interests of an incipient elite by reproducing structures of authority through the promotion of an 'egalitarian' or 'group-centred' ideology, akin to Renfrew's (1974) concept of group-orientated chiefdoms. Such an ideology may have served to play down the reality of growing inequality of access to power and resources, a point perhaps reflected in Chesson's (1999) distinction between 'Greater' and 'Lesser' charnel houses. Differentiated in terms of size, number of burials and range of grave offerings, these appear to provide evidence for growing disparities between the power of different corporate groups during EB II–III. This situation appears entirely compatible with the internal dynamics of heterarchically organized corporate village communities.

Built-stone tombs

Alongside the well-known shaft and chamber graves and the charnel houses of Bab adh-Dhra', there existed a second major category of mortuary structures, built-stone tombs. Those most commonly associated with the Early Bronze Age are the so-called dolmens (Figure 5.10), which consist of a chamber built from roughly worked stone uprights, supporting a capstone: chambers can be up to 2–3 m in length.

Dating stone burial monuments is difficult as many have been subject to later disturbance and reuse. While it is now clear that cist graves excavated by Stekelis (1935) at Adeimeh are of Chalcolithic date (Prag 1995: 76), several dolmens from Damiyeh, others from Wadi Hisban, and one near Tall al-'Umayri on the plateau south of 'Amman are all of EB I date (Yassine 1988: 51; Dajani 1968; Dabrowski and Krug 1994). A similar date has been also proposed for dolmens located in the Jarash region (Hanbury-Tenison 1986: 245; Sapin 1992: 173). While no Jordanian stone burial monuments have yet been dated to EB II–III, Vinitsky's (1992) observation of the clear spatial association between EB II–III settlements and dolmens in both the Jaulan and Galilee argues that the possibility should not be excluded. Well-documented groups of Early Bronze Age shaft and chamber graves are relatively infrequent in Jordan, in particular when compared to the greater number of such tombs known from later periods, a point which should give pause for thought in the light of Vinitsky's remarks.

The appearance of above-ground mudbrick charnel houses at Bab adh-Dhra', and of underground stone-lined chambers at as-Safi featuring slab-roofs and entered via a short flight of steps (Waheeb 1995: figs. 1, 2), appears to indicate that the simple dichotomy of underground shaft and chamber graves and above-ground built stone tombs is no longer tenable. Further confirmation comes from the presence east of Jarash of tombs in which the chamber is partially rock-cut, but roofed over by a large slab or slabs in the manner of a dolmen (Sapin 1992: 174). Moreover, rock-cut tombs at Tall Handaquq (N) have a distinctive squared recessed opening (Mabry et al. 1996: 126) resembling the doors carved on certain dolmens (Yassine 1985; Hanbury-Tenison 1986: 244). The sheer variety of burial monuments brings into question the value of the traditional typological approach to the study of structures 'whose form may not be fixed but rather embody ideas which themselves may be variable and changing' (Bradley 1993: 71-72). The possibility that such potentially long-lived and accessible monuments may have been subject to reinterpretation suggests that rather than defining discrete classes prior to analysis we might do better to emphasize those points held in common among various monument types.

Figure 5.10. Dolmen, Jabal Mutawwaq (G. Philip).

Traditionally, scholars have connected 'megalithic' monuments in the Levant with pastoralist groups (Zohar 1992: 52-55). The monuments tend to appear in clusters that have been argued to show an association with land of limited agricultural value but offering good grazing potential (Webley 1969; Prag 1995). However, Stekelis (1961) believed the large dolmens at ad-Damiyeh, located on a rocky ridge overlooking the az-Zarqa' triangle from the east, were built by agricultural groups exploiting the valley floor below. This case finds some support in the association between the distribution of built stone monuments and of EB I settlements in the area northeast of the Dead Sea (Prag 1995: 79), a point reinforced by the locations of surveyed dolmen fields in Wadi al-Yabis (see Palumbo *et al.* 1990: 480).

The main concentrations of built stone monuments are located along the escarpment and eastern slopes of the Jordan Valley. Monuments extend from the region west of Madaba and Wadi Hisban, and other drainage systems northeast of the Dead Sea northwards through Adeimeh to ad-Damiyeh (Stekelis 1935; 1961; Neuville 1930; Swauger 1965) above Wadi az-Zarqa'. Northwards again, they occur in Wadi al-Yabis (Palumbo *et al.* 1990: 480) and in the hills east of Pella (Watson 1996). However, their distribution is not restricted to the Jordan Valley. On the plateau proper, numerous monuments are located in the area east of Jarash (Sapin 1992; Hanbury-Tenison 1989a) and have been recorded in and around modern 'Amman (see older literature summarized by Zohar [1992] and Prag [1995]). Distribution is wide and encompasses a variety of environmental zones. However, it is important to remember that we may be observing the pattern of monument survival, rather than an accurate reflection of their original distribution, and their apparent association with 'pastoralist' territory may reflect a far higher destruction rate within the more heavily cultivated areas.

Neither the spatial nor structural distinctions between shaft graves and stone monuments appear as clear as might have been expected, with both dolmens and shaft tombs occurring together at Jabal Mutawwaq (Hanbury-Tenison 1989a: 137-39). The use of monolithic door jambs in Hall 134 at Hartuv in Palestine (Mazar and de Miroschedji 1996) has clear parallels with the design of the entrances to charnel houses at Bab adh-Dhra', stone-lined cists at as-Safi (Waheeb 1995: figs. 1, 2) and houses at Jabal Mutawwaq (Hanbury-Tenison 1989a: 137). The close connections between stone-working techniques employed on superficially different structures raise questions concerning their ascription to distinct sections of the population. Further doubt has been cast upon the simple association of shaft tombs with sedentary groups and dolmens with mobile groups through the excavation of a largely undisturbed late EB I dolmen near Tall al-'Umayri, southwest of 'Amman (Dabrowski and Krug 1994: 241-42). This tomb contained the remains of more than 20 individuals, the earlier burials pushed towards the rear to accommodate later additions. Thus, the quality of accessibility was common to both stone monuments and shaft and chamber graves, while the grave assemblage compares closely to those of contemporary shaft tombs. A further intriguing aspect of this dolmen was the presence of an area of plaster surface immediately outside the chamber (Herr *et al.* 1997: 153), which might indicate the regular performance of activities outside or even involving the chamber, strengthening possible parallel practices posited for shaft tombs. In fact, the shared features between these two burial forms appear at least as strong as the contrasts.

Before built stone monuments could be securely ascribed to long-range pastoralist groups we would require much better information on the nature of the burial record in the arid zone, an area that would surely have featured prominently in patterns of seasonal movement (Köhler-Rollefson 1992; Garrard *et al.* 1996). In fact, 'megalithic monuments are so widely distributed and so numerous as to suggest that they represented the 'normal' mode of burial for the dead of a large part of the population over an extended period' (Prag 1995: 83). For example, we should not rule out a possible connection between stone burial monuments along the edge of the north Jordan Valley and hilltop settlements in the uplands around 'Ajlun. The use of similar burial monuments by various groups would appear in keeping with the notion of flexible subsistence strategies, and might account for the common features of the two monument types.

Early Bronze Age social and cultural landscapes

The Early Bronze Age saw the incorporation of various novel features within the topography of Jordan: walled human settlements, dams and irrigated fields, orchards, extensive new areas of ploughed arable land, numerous built stone tombs, and presumably a variety of paths and routes whereby people, animals and commodities could move across the landscape. It is important, therefore, to consider the sheer physical impact of Early Bronze Age activity, and the degree to which the pre-existing landscape would have been modified. Recent theoretical discussions (Thomas 1991; Bradley 1993; Tilley 1994) have stressed that humans not only create and modify landscapes but that these, in turn, constitute social and ideological symbols that play a role in influencing people's comprehension and experience of the world.

Thus, while the growth of large Neolithic settlements and the domestication of caprines may have impacted significantly upon the environment (Rollefson 1993, this volume; Garrard *et al.* 1996), this would have been the cumulative, perhaps imperceptible, result of short-term subsistence choices. The highly visible Early Bronze Age modifications were deliberate constructions, and created a landscape that had been formed by human choices and action, and the development of which was open to day-to-day perception and judgment. From the later fourth millennium BC onwards, the inhabitants of Jordan would have moved within a landscape much of which originated in the Early Bronze Age. Recognizing the way in which the built environment alters, even forms, people's perceptions (Thomas 1991), it can be suggested that the Early Bronze Age sees the inscription into the landscape of both living groups and their past, through the creation of a world in which space was structured by highly visible human creations.

Material remains: portable artifacts

The underlying view of craft specialization held in Levantine archaeology is that it evolved concomitant with increasingly complex and centralized political development (Rosen 1997b: 82-83). However, it is not clear that such a reconstruction can be supported on the evidence from the southern Levant during the Early Bronze Age (Rosen 1993, 1997b). Various criteria have been suggested to distinguish between specialization and basic household-level production (e.g. Feinman *et al.* 1984; Costin 1991). On the basis of such criteria as standardization of form and technique and restricted availability of raw materials, a number of products, such as Canaanean blades and Metallic Ware pottery appear likely to have been produced by specialists for wide distribution, a point reinforced by their suitability for transportation (see below, p. 208, 211).

Specialization in what are in essence utilitarian products, however, generally indicates production for domestic consumption, with certain households or communities producing for other units (Costin 1991: 13). Such a system can function independently of urban communities or elites, and need not be indicative of stratification or inequality. For example, flint blades and Metallic Ware storage vessels were important features of the economy and appear to have been produced in restricted locations. However, they were utilized on a spatial scale far greater than that of any individual polity, suggesting that their distribution lay outside political control.

While craft specialization has generally been seen as an important component of hierarchical organizations, with the growth of attached specialists working for elites and producing high-status goods seen as a typical facet of complex societies (e.g. Blumfiel and Earle 1987), there is little evidence for high-status manufactured goods from Early Bronze Age Jordan. Nor is there much to indicate large concentrations of manufacturing activities within major settlements. Thus, the data indicate that various segments of the economy were organized quite differently, with significant areas of production and procurement apparently outside the control of political authorities. Yet again, we appear to be witnessing potentially heterarchical relationships rather than a single organizational principle, the city-state.

Pottery

The basic sequence of Early Bronze Age pottery from the southern Levant has been summarized by Stager (1992). To avoid repetition, this section will concentrate upon specific issues raised by the Jordanian material. Much discussion of Early Bronze Age pottery is framed in terms of a traditional set of ceramic categories. The difficulty is that, while these groups have been defined by twentieth-century archaeologists, they are frequently treated as though these same categories had some kind of living reality in the past, rather than as abstractions created by present-day scholarship (on this problem see Barrett's [1994] comments on the status of the 'beaker' phenomenon in European prehistory). Chronology apart, much of the traditional literature has concentrated upon pottery as a 'cultural' indicator (e.g. Kenyon 1979), or as evidence for interregional contacts (Hennessy 1967; Ben-Tor 1986; Stager 1992: 40-41). Most studies have worked with typologies based upon shape and decoration, with scholars appearing reluctant to move beyond traditional questions and methodologies (Philip and Baird 2000: 3-4). With the exception of the pioneering work of Franken (1974), studies of fabric and manufacturing techniques are a relatively recent innovation, while researchers are only beginning to consider the socioeconomic implications of patterns of ceramic production and distribution. However, the various dimensions of pottery as an investigative tool are becoming apparent, and recent work in Jordan has highlighted the accretive multi-component nature of the ceramic assemblages recovered from many sites (Beynon et al. 1986; Schaub 1987; London 1991; London et al. 1991; Betts 1991: 103-107) and which cannot, therefore, be directly equated with particular local 'cultures'.

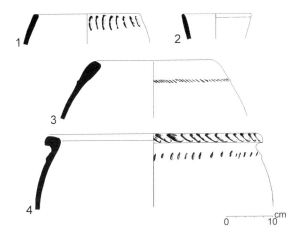

Figure 5.11. Early EB I pottery from Tall ash-Shuna (Baird and Philip 1994: fig. 10).

EB I

The internal development of the EB I period is particularly well documented in Jordan, with three recently excavated sites having produced clear evidence for two distinct ceramic subphases (Tall ash-Shuna, Tall Umm Hammad and Bab adh-Dhra'), the first two from long stratified sequences of occupational deposits, the latter from tombs.

Tall ash-Shuna

The bulk of the early EB I assemblage consists of handmade vessels in a coarse, poorly levigated fabric, bearing a light orange wash. Shapes are simple conical bowls with straight sides and simple jar forms; decoration is restricted to heavy thumb impressions. Rounded ledge handles, thumb impressed around the edge, are characteristic. Hole-mouth cooking pots show a distinctive calcite temper, beneficial in the control of thermal shock, and may bear impressed ovoids or slashes on the rim (Figure 5.11).

The most distinctive component of the pottery assemblage of the earlier EB I phase in the north is Grey Burnished Ware (GBW), also called Esdraelon Ware, although it comprises but a small percentage of the total repertoire. Vessels have a grey-black, highly burnished surface and are restricted to large bowl forms, sometimes bearing plastic decoration (Figure 5.12.1-2, 4). While traditionally associated with Anatolian emigrants (Hennessy 1967; Stager 1992: 29), more recent interpretations have seen the pottery as a skeuomorphic form linked to basalt bowls (Ben-Tor 1992; Joffe 1993), or perhaps produced in imitation of tarnished silver vessels (Philip and Rehren 1996). Petrographic

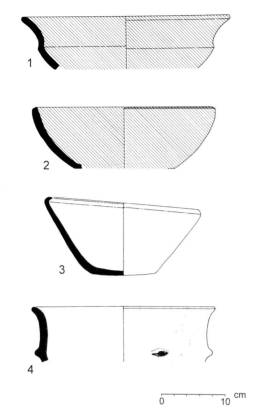

Figure 5.12. Tall ash-Shuna, EB I bowl forms: 1, 2 and 4 are Grey Burnished Ware, 3 is Crackled Ware (Baird and Philip 1994: fig. 11).

evidence suggests that GBW was a specialized product, but manufactured at more than one centre (Goren and Zukermann 2000: 169-70). The vessels are best seen as tablewares connected with the presentation of food and drink, and one might ask whether their appearance is indicative of a new collective dimension to food consumption, perhaps periodic dining in groups larger than that of the individual household. GBW is characteristic of northern Palestine and Transjordan. Its distribution extends to northern parts of the plateau (Kamlah n.d.) but does not reach Jawa. In the Jordan Valley, it occurs as far south as Tall Umm Hammad and at Tall Abu Alayiq near Jericho (Betts 1992: 76-77), but was not reported from the extensive EB I occupations recorded in the Jarash and ar-Rumman regions (Hanbury-Tenison 1987; Gordon and Knauf 1987).

The later EB I pottery from Tall ash-Shuna consisted of large vessels, such as hole-mouth jars and pithoi, bearing bands of streaky red wash known as Band Slip decoration, and small vessels, mainly bowls and closed containers, with red-slipped and burnished surfaces. Hemispherical bowls replaced straight-sided forms, and impressed decoration is less common. New developments include Crackled Ware (Figure 5.12.3), which appears to represent a late descendant of GBW (Esse 1989b; Rowan 1994). The early and late EB I ceramic repertoires from Tall ash-Shuna have good parallels respectively at Yiftahel (Braun 1997) and 'Ayn Shadud (Braun 1985) in northern Palestine.

In comparison to the industry of the earlier EB I, which is characterized by handmade vessels using a limited range of fabrics, the later EB I material shows a much greater incidence of wheel finishing, a generally higher standard of vessel finish and firing, and a wider range of fabrics employed in a more selective fashion for the production of different vessel forms. Such a technically superior, more diversified industry appears indicative of an increased level of craft specialization, although there is no evidence to connect this with any element of political control.

Band Slip pottery is often described as a 'ware' (e.g. Stager 1992: 30), although it is better understood as a decorative technique. It is distributed across a broad east–west band extending from the north Jordan plateau and the Syrian Hauran (Glueck 1946: 5; Braemer and Échallier 2000: 406) to the Mediterranean coast (Kempinski and Niemeier 1990: x). On the east side of the Jordan Valley it occurs at Tall Abu al-Kharaz and Tall Handaquq (N). While its southern limit appears to lie in the area of Tall Umm Hammad, there are sporadic occurrences at more southerly sites, such as Tall Iktanu (Prag 1993: 269). There is a marked association between this decorative technique and vessels used for storage. However, the technique is not specific to any one vessel form, and in fact continues despite significant changes in vessels styles through time. It appears on hole-mouth vessels in EB I contexts at Tall ash-Shuna and Tall Umm Hammad (Betts 1992: 51-53, 66), and additionally on necked jars with everted and rolled rims in EB II contexts at Tall Abu al-Kharaz and Tall Handaquq (N) (Fischer 1993: 287, fig. 13.5; Mabry et al. 1996: 136-38, figs. 8.7-8, 14.4-6). The exact significance of this connection remains to be explored.

Tall Umm Hammad

A long EB I sequence from Tall Umm Hammad in the central part of the Jordan Valley (Betts 1992: 17-21) also reveals two distinct ceramic phases. The earlier part (stage 2) includes a large number of hole-mouth jars bearing impressed or slashed decoration along or just below the rim, as seen at Tall ash-Shuna. One particular variant is a group of hole-mouth jars

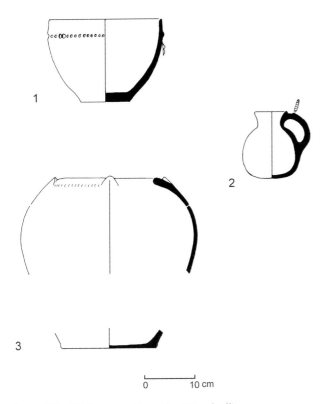

Figure 5.13. EB I pottery (drawn by Y. Beadnell).
1. Bab adh-Dhra', EB IA bowl with conical projections below rim.
2. Bab adh-Dhra', EB IA jug.
3. Tall Umm Hammad, EB IA hole-mouth jar with pushed-up lug handles of a type occurring at sites in Wadi Zarqa' and Jawa in the steppe.

with four pushed-up ledge handles (Figure 5.13.3), apparently produced in a single fabric (Betts 1992: 47). This distinctive form has been recognized elsewhere in the central Jordan Valley (Mabry 1989: 61), in the uplands to the east where it has been used as a chronological marker for early EB I (Hanbury-Tenison 1987; Betts 1992: 48) and as an import at Jawa in the basalt *harra* (Betts 1991: 72), where it provides clear evidence for a connection between the central Jordan Valley and the *badia* during EB I. The physical transport of pottery from the central Jordan Valley to the steppe may well be indicative of patterns of seasonal mobility between these areas (Helms 1987). In this light, the absence of such vessels from Tall ash-Shuna to the north hints at patterns of interaction framed in east–west rather than north–south terms.

Red-slipped and burnished surface treatment appears restricted to loop-handled juglets and small hemispherical bowls, and appears to increase in frequency with time (Betts 1992: 99). Also characteristic of this phase are large GBW bowls (Betts 1992: 76-77, genre 45). Also noteworthy is the presence of what may be a related form, that is, deep bowls with everted rim (Betts 1992: 78-79, Genre 48). These share certain characteristics of size and shape with GBW, but lack the distinctive surface treatment. This raises the possibility that the scholarly concentration upon the colour of GBW may have diverted attention from their real significance, their size and shape. It is possible that scholars have not recognized the functional equivalence of large open bowls generally, including those from such sites as Bab adh-Dhra' where GBW is lacking (see below), and have, thus, underestimated the significance of the widespread adoption of vessels designed for communal food consumption. Perhaps this is the origin of the large platter bowls so characteristic of EB II–III assemblages (see below).

Stage 3 produced a later EB I assemblage, although the exact relationship between stages 2 and 3 remains unclear, owing to the limited excavation area and the possibility of a local stratigraphic break (Philip 1995b: 166). Traits distinctive of stage 3 include a decline in the proportion of hole-mouth jars bearing impressed decoration, with many now undecorated, and the appearance on others of Band Slip decoration. Red-slipped and burnished surface treatments become increasingly common and a number of new shapes appear, including high-necked jars, platters with inverted rims and spouted bowls.

This stage also sees the appearance of the distinctive Tall Umm Hammad ash-Sharqiyya (TUH) ware (Leonard 1992: 83; Betts 1992: 107, repertoire 6), which appears in a heavy red fabric as large bowls and jars bearing multiple applied bands with thick thumb-impressed and moulded decoration (Figure 5.14.1). The distribution of this material during the EB I period is restricted to the central Jordan Valley (see Betts 1992: 56 *inter alia*) with an extension westwards to Tall al-Far'a (N) where it was termed Pré-Urbaine-D (de Miroschedji 1971: 38-40, fig. 14). It was not detected by surveys in districts east of the valley (Hanbury-Tenison 1987; Gordon and Knauf 1987) despite earlier ceramic connections between these areas, and is absent from Pella (Bourke, pers. comm.). Both the appearance and limited distribution of TUH ware suggest that it represents a localized ceramic product, while the range of shapes in which it occurs indicates a connection with storage. The contrast between the limited distribution of TUH ware and the far wider extent of Band Slip decoration highlights the complexity of EB I ceramic patterning

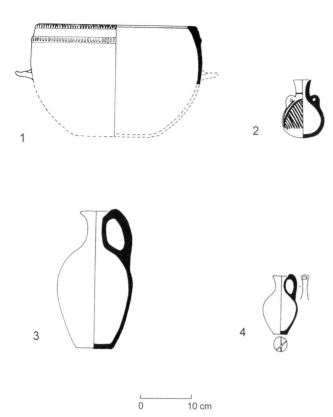

Figure 5.14. EB I and EB II pottery (drawn by Y. Beadnell).
1. Tall Umm Hammad, EB IB, bowl in Tall Umm Hammad ash-Sharqiyya ware.
2. Bab adh-Dhra', EB IB, amphoriskos bearing Line Group Painted decoration.
3. Bab adh-Dhra', EB II, jug, red-slipped, EB II.
4. Tall Abu al-Kharaz, EB II, juglet in metallic ware.

and the need to accommodate the simultaneous operation of a variety of processes at contrasting scales.

Tall Umm Hammad ware proper, and hole-mouth jars bearing impressed and slashed decoration, have at times been conflated by scholars more familiar with the data from Palestine to form a loosely defined group of 'Impressed Slashed Wares' (Stager 1992: 29; Joffe 1993: 39). Hopefully, the discussion above has removed some of the confusion surrounding this issue, and has identified Tall Umm Hammad ware as a distinct ceramic entity.

Bab adh-Dhra' and the southern Ghor

The cemetery at Bab adh-Dhra' in the southern Ghor also reveals two distinct phases of EB I use, referred to as EB IA and IB. In contrast to the position at ash-Shuna, the material from the early phase at Bab adh-Dhra' (Figure 5.13.1, 2) frequently bears a distinctive red slip. Ledge handles occur in a variety of forms and many vessels bear punctured or slashed bands, or horizontal arrangements of small clay projections. The most distinctive northern traits, such as GBW and Band Slip decoration, are absent, and although the pottery shares a number of individual features with that of other EB I sites, the overriding impression is that this represents a local assemblage (Schaub 1987).

The pottery of the later phase shows some similarities to EB IA, although red slips, punctured and slashed bands are all less common. Large jars now have large 'duck bill' ledge handles, while bowls with a 'trumpet' spout appear. Everted-rim bowls, so characteristic in the north, are absent, although deep bowls occur with small ledge handles below the rim. Juglets have round bases and short flaring necks drawn up from body, with handles pulled up higher, and attached higher up the rim than in EB IA examples, although juglets and amphoriskoi with narrow necks do not appear until EB II.

This phase sees the appearance of vessels bearing Line Group painted decoration, which consists of parallel lines of red or brown paint applied directly to the surface of small vessels (Figure 5.14.2). In Palestine, pottery bearing Line Group decoration extends from Ai and Jericho, as far south as Arad (Schaub 1982; Stager 1985; 1992: 29), while its distribution in Jordan embraces the south Jordan Valley (Prag 1989: 38-40), southern Ghor and the south-central plateau (Brown 1991: 173; Dabrowski and Krug 1994). However, the exact significance of this style is not clear, and its very distinctiveness may have accorded it greater prominence in scholarly discussions than it really warrants.

The south

The material from southern Jordan suffers from the fact that the non-sedentary subsistence strategies most common in this area tend to produce impoverished ceramic repertoires, although current work on settlements associated with Early Bronze Age copper production in the Wadi Faynan area is sure to revolutionize our understanding of the area. Despite substantial evidence for direct Egyptian involvement in southern Palestine (papers in van den Brink 1992; Gophna 1995b; Levy 1995a), there is at present no indication of a similar degree of contact with southern Jordan, a point that may be relevant to patterns of sociopolitical development in the two regions.

Summary

While ceramic regionalism is generally taken as characteristic of EBI, the term as traditionally employed amounts to the recognition of a few distinctive features within a complex of cross-cutting ceramic assemblages; some are styles of painting, while others appear to be associated with specific combinations of vessel form and fabric. In reality, it appears that ceramic 'regions' are constituted at different spatial scales. While the material from Bab adh-Dhra' is very distinctive and apparently of highly localized distribution, that from Tall ash-Shuna conforms to a widely distributed 'northern' sphere. Equally, the limited range of fabrics occurring at the first site and the far wider repertoire from Tall ash-Shuna indicate assemblages resulting from rather different patterns of production and procurement. The varied constitution of EB I assemblages may provide a context for the greater, if somewhat uneven, homogenization of ceramic assemblages during EB II.

Certain broad patterns are already clear, however. The main elements of traditional 'regionalism'—GBW, Band Slip and Line Group decoration—are manifest spatially as extensive east–west distributions, but their north–south limits are more restricted. Even pottery of more localized distribution appears to follow this pattern with the four lugged jars known from Jawa and the central Jordan Valley absent at sites both to the north and south, while Tall Umm Hammad ash-Sharqiyya ware appears concentrated along the axis of Wadis az-Zarqa' and Far'a. These ceramic distributions might appear counter-intuitive, in that they cut across natural topographic boundaries, while evidence for the north–south communications along the valley floor or plateau is less apparent. This appears to argue for some wider role for an east–west axis, one that might be connected to growing economic integration between the lowland, upland and steppe regions with their different subsistence potentials. It is also worth observing that, despite the differences in specific styles, many of the key *functional* ceramic categories are present at agriculture-based settlements throughout Jordan, suggesting that a rich and complex network of shared values and practices underpinned day-to-day activities among Early Bronze Age communities.

EB II

Pottery of this period from Palestine appears to fall into two broad groups, namely, Metallic Wares in the north and 'local' softer, red-slipped and burnished pottery in central and southern areas (Greenberg 1996:137), a characterization that is broadly applicable to Jordan (Figure 5.14.3-4).

Metallic ware

Metallic Ware (Greenberg and Porat 1996), though mostly red, can appear buff or grey in thin vessels, and takes its name from the distinctive ringing sound it emits when struck. Vessels were fired to a temperature between 850 and 950 °C, which produces the ware's distinctive appearance and matt surface. Greenberg and Porat (1996) argue, on petrographic grounds, that Metallic Wares from sites in northern Palestine were made from a single clay body originating in the Lower Cretaceous formation, which outcrops at the foot of Jabal Sheikh (Hermon). Metallic Ware appears in a wide range of vessel forms, and contributes 50–70% of the ceramic assemblage at EB II sites as far south as the Jezreel Valley. That a major component of the pottery occurring at northern EB II sites was imported from this source indicates both large-scale production and a significant degree of economic integration. Cooking pots, however, do not appear in Metallic Ware, and were presumably manufactured locally.

Greenberg and Porat (1996: 20) argue that the extensive distribution of Metallic Ware during EB II is connected with changing political structures in northern Palestine. However, there are good arguments against this view. Metallic Ware was utilized on a spatial scale far greater than that of any posited individual polity, while there is now growing evidence from western Asia indicating that significant areas of specialized economic activity were organized independently of the existence of political hierarchies (Stein and Blackman 1993; Wattenmaker 1994; Rosen 1997b). Costin (1991: 13) has suggested that items manufactured by independent specialists tend to be utilitarian, used by most households and available without restriction, while specialists attached to elites tend to produce luxury items and wealth-generating goods. Metallic Ware belongs to the first rather than the second of these categories, while the distinctive concentration of production is simply explained by the restricted physical distribution of the required raw materials, a factor that Costin (1991: 13-15) has acknowledged as playing a key role in determining the basis of production. The distribution of Metallic Ware is probably better explained as the summative outcome of numerous small-scale transactions undertaken at community or household level, rather than as an indication of any overarching political organization. In this sense, it represents an incremental advance along the trajectory of increasing regional economic interaction that was already well under way during EB I.

Metallic Ware has been reported in Jordan from Tall Abu al-Kharaz (Fischer 1994, 2000; Fischer and Toivonen-Skage 1995), where it appears as platters, jugs and juglets and large storage jars with pattern-combed decoration, Tall Handaquq (N) (Mabry 1989: 79, fig. 10.17), and Tall as-Sukhne in Wadi az-Zarqa' (Chesson et al. 1995), revealing the eastward extent of this ceramic form. In the case of Abu al-Kharaz, for which most information is available, the presence of platters suggests that the pottery itself was a commodity rather than simply a container, a point reinforced by the relatively low weights of individual vessels (Greenberg and Porat 1996: 19). Local production remained important at these sites, however, and included a range of bowls, jugs and amphoriskoi as well as large storage vessels, bearing red-painted decoration executed in a variety of styles.

Non-metallic wares

In southern Palestine, Metallic Ware vessels are rare (Greenberg 1996: 137) and they have not yet been reported from sites in south-central Jordan, where red-slipped and burnished vessels made in softer fabrics predominate. Here, despite similarities with Metallic Ware in terms of vessel shape, production appears to have been almost entirely localized as exemplified by the material from the southern Ghor (Beynon et al. 1986). At Bab adh-Dhra', the EB II assemblage included an increased frequency of narrow-necked jugs and juglets, presumably an indication of the importance of specialized liquid products, and wide shallow platters. High-shouldered jars with short flaring necks and ledge handles appear, a form with parallels at other southern sites, such as Jericho. To some extent the very decline in ceramic regionalism characteristic of EB II is the result of the degree of overlap between vessel forms seen in Metallic Ware and other fabrics, and is presumably indicative of an increasing commonality of economic and social practices. The presence at Barqat al-Hatiye in Wadi Faynan of red-slipped pottery including spouted bowls, necked jars with ledge handles and platters (Fritz 1994) reveals the extent to which a particular set of ceramic forms had become established in domestic contexts throughout the region by the EB II period.

Once more, it is the case that, despite superficial similarities of vessel style, individual assemblages were generated according to specific local circumstances. The degree of flexibility existing within strategies of ceramic procurement suggests that pottery production and distribution was not an arena for political action and played little part in structuring mechanisms of political and economic power.

EB III

In general terms, the EB III period sees a continuation of the trend towards greater convergence of regional ceramic assemblages, with one exception.

Khirbat Kerak ware

This highly distinctive form of pottery has a soft fabric, resulting from firing at a relatively low temperature. Vessels bear a highly burnished red or black slip, with relief decoration in some cases, and occur in shapes foreign to the ceramic traditions of the southern Levant (Philip 1999: 38-40, figs. 4, 5). While generally seen as the hallmark of the EB III period in northern Palestine, the distribution of Khirbat Kerak ware is concentrated in the northern part of the Jordan Valley (Esse 1991; Philip 1999). The only substantial excavated assemblage from Jordan is that from Tall ash-Shuna, where the wide range of vessel forms, including pot-stands, vessel lids and portable hearths, as well as shapes more typical of the local industry, such as bowl and platter forms (Figure 5.15), suggests the presence of a production centre (Philip 1999: 43).

Esse (1991: table 4, fig. 25) notes that occurrences of Khirbat Kerak ware outside the 'core' area of the north Jordan Valley tend to be restricted to the smaller bowl forms. While this appears to hold in the case of southern sites, the presence of a one-handled vessel at Khirbat az-Zaraqun (Mittmann 1994: 10), part of a stand and a piece of an incised vessel from a tomb at Arqub ad-Dhahr (Parr 1956, nos. 208, 214), suggests that we have much to learn concerning the distribution of Khirbat Kerak ware on the north Jordan plateau.

Scholars have generally interpreted Khirbat Kerak ware in terms of the arrival of new population groups, ultimately of east Anatolian origin (Hennessy 1967: 79; Esse 1991: 139-40). However, there are difficulties with this view (Philip 1999: 35-36), and it is noteworthy that Khirbat Kerak ware lacks many of the ceramic forms most closely associated with the 'classic' Early Bronze

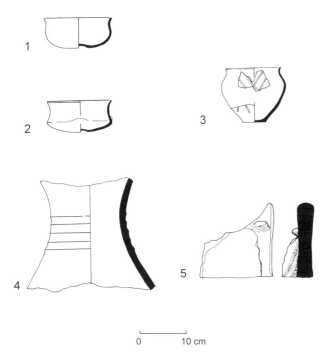

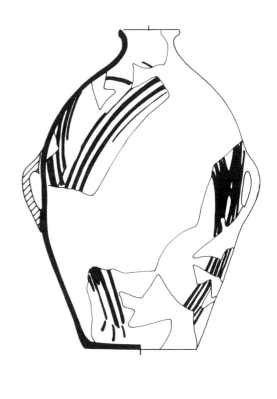

Figure 5.15. Khirbat Kerak ware vessels from Tall ash-Shuna (drawn by Y. Beadnell).
   1–2. Bowls.
   3. Jar.
   4. Pot-stand.
   5. Part of portable horseshoe-shaped hearth.

Age agricultural economy—jugs and juglets, storage vessels, large platter bowls—and it has recently been argued (Philip 1999) that scholars seeking to explain the appearance of this pottery should focus upon the functional and socioeconomic status of the user communities.

Other aspects of EB III ceramics

EB III pottery in the local tradition is represented throughout Jordan: on the plateau at Khirbat az-Zaraqun near Irbid; at Tall al-'Umayri south of 'Amman; in the Jordan Valley at Tall al-Handaquq (S); and at Bab adh-Dhra' and Numayra in the southern Ghor. Specific forms that appear to be diagnostic of EB III include deep bowls with flattened inverted rims (hammer-rimmed), red-brown slipped and burnished platters with inverted or vertical rims, bearing a small external groove below the rim, and juglets with piriform (pear-shaped) bodies and narrow stump bases, all clearly developments from familiar EB I–II forms.

EB III material from north Jordan is typified by that from the final occupation phase at Khirbat az-Zaraqun. Storage vessels were common, occurring as pithoi around 1 m high (Figures 5.16, 5.17). Some were

Figure 5.16. Khirbat az-Zaraqun, EB III storage jar with loop handles. The style of painted decoration is reminiscent of earlier Band Slip decoration (drawn by Y. Beadnell).

in a fabric resembling what appears to be Metallic Ware and bore vertical combing and horizontal applied rope decoration at the neck. These may represent the continuation in a narrower range of forms of the Metallic Ware of the EB II period. Loop handles are characteristic, with ledge handles better represented at southern sites. The continued appearance in clear EB III contexts of Band Slip decoration on large storage vessels (Genz 2000: 280; see Figure 5.16) has important implications for its use as a dating criterion. Deep-combed inverted rim bowls with spout and loop handles (Figure 15.18.6) may be connected with olive oil production (Stager 1985: 176-77; Esse 1991: 119-24), while the recovery of juglets suggests an association with precious liquid commodities. In what would appear to be a distinctive northern feature, round-based hole-mouth jars appear to have functioned as cooking pots.

In general, EB III material from south-central Jordan (Figure 5.18.1-5) compares well typologically with that from central and southern Palestine (Harrison 2000:

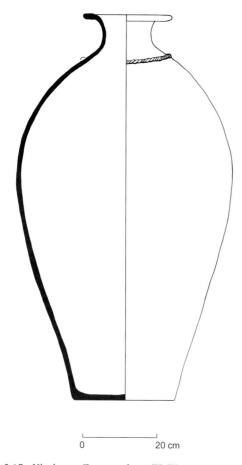

Figure 5.17. Khirbat az-Zaraqun, large EB III storage jar; note that the applied rope-like clay band at the neck lies in the place where a seal impression would be placed (drawn by Y. Beadnell).

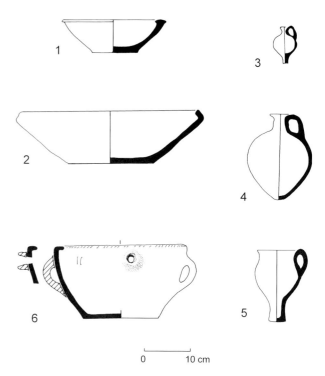

Figure 5.18. EB III pottery from Bab adh-Dhra' (1–5) (drawn by Y. Beadnell).
1–2. Platter-bowls with inverted rim.
3–5. Jug forms with characteristic pointed and stump bases.
6. Khirbat az-Zaraqun, vat, possibly used in the processing of olive oil.

360). The EB III repertoire of the central-southern plateau is typified by that from Tall al-'Umayri, and consists of a variety of jars including examples of both flaring and hole-mouth forms, a range of platter bowls, some with inturned rims, and small bowls decorated with vertical bands of red paint (Daviau 1991: 102). Small–medium-sized bowls and hole-mouth jars are dominant numerically, while the use of ledge handles on storage jars, and the presence of jars with a 'combed' exterior bearing a white chalky slip are characteristic of southern sites (Harrison 2000: 353–55). Overall, both northern and southern assemblages appear to represent direct continuations of local EB II traditions, rendering Khirbat Kerak ware all the more distinctive in comparison.

A technical study by London et al. (1991) has revealed that at Tall al-'Umayri a range of pot-forming techniques and fabrics were employed for different vessel types, leading London (1991: 394) to suggest that different parts of the ceramic repertoire were produced by different potters, possibly working on different bases. Petrographic studies of EB II–III pottery from sites in the southern Ghor (Beynon et al. 1986) indicate a mainly local production, with some movement of specific products between Bab adh-Dhra' and Numayra, but with little evidence for a significant input of material from outside the southern Ghor area. It is striking that assemblages, which in the case of individual sites develop from the coalescence of a number of different strands of ceramic production, appear quite similar when viewed on a site-to-site comparative basis. This reinforces the idea that the underlying pattern of demand reflected a fairly uniform set of requirements, itself indicative of quite similar social and economic practices.

Stone

Chipped stone

Studies of Early Bronze Age chipped stone have focused upon the most distinctive artifact types, the so-called Canaanean blades, and tabular scrapers.

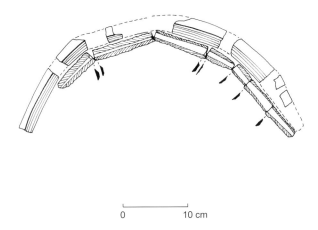

Figure 5.19. Flint blades of Canaanean type mounted in wooden haft, Tall Abu al-Kharaz, EB II (drawn by Y. Beadnell).

Sickles

The sickle elements known as Canaanean blades were long parallel-sided blades of distinctive trapezoidal cross-section, struck from a prepared, single platform core (Rosen 1997a: 46-49). The resulting long blades could be snapped to produce shorter segments 4–6 cm in length, which were mounted in groups of five to six, set into notches on a wooden handle; a partly preserved example was recovered from destruction debris at Tall Abu al-Kharaz (Figure 5.19; Fischer 1994: fig. 11.4). EB I examples from Tall Iktanu revealed polish on one edge that is believed to be indicative of the reaping of grasses, while the opposite edge frequently bore traces of bitumen intended to fix the blade to a haft (McCartney 1996: 145).

This blade form does not derive from earlier local flint-working traditions, but is related to a technological complex spanning the Levant, north Syria and Anatolia. In the southern Levant, blades are usually made from brown, fine-grained, Eocene flint nodules, and caches of unretouched examples found at several sites in Palestine and at Tall Iktanu in Jordan provide evidence for their distribution as unfinished blades (McCartney 1996: 145; Rosen 1997a: 141). This fact, plus the lack of manufacturing debris recovered from most excavations, suggests specialized production. The absence of evidence for production at major tell sites, with the exception of EB III Tall Halif (Futato 1996), argues, however, that manufacture was not organized by political elites, and that these blades constitute a good example of a utilitarian specialist product produced and distributed independently of political power structures (Hayden 1994: 200). The predominant use of Eocene flint for blade production appears to confirm Costin's (1991: 13-15) suggestion that (as with Metallic Ware) crafts organized on such a basis may be based upon raw materials of uneven distribution. Rosen's (1997a: 108) point that production was largely based within rural settlements is consistent with such an interpretation, and is supported by their continued occurrence during EB IV (Rosen 1997a: 143), at which point major centres had effectively disappeared.

The site of Bab adh-Dhra', however, produced sickles of a different type, backed blades related to Chalcolithic traditions (McConaughty 1980; Rosen 1997a: 50-51). Yet again, however, evidence for on-site production is absent. Similar tools are found in the arid areas of southern Palestine (Rosen 1997a: 141), suggesting the existence of a separate industry in the south, perhaps a steppe tradition. Despite the evidence from Bab adh-Dhra', Canaanean blades are present at EB III Numayra (McConaughty 1980), highlighting the diversity which underlies the superficial similarity of walled settlements.

Tabular scrapers

Tabular scrapers (Rosen 1997a: 71-79) are a feature of Chalcolithic assemblages and consist of large retouched flakes, showing the deliberate retention of cortex over most of the dorsal surface. They vary in shape and measure on average 15–20 cm long by 1 cm thick (Futato 1996: 61). The distribution of tabular scrapers appears concentrated at sites in the arid regions of southern Palestine, while known quarry sites are concentrated in the southern Negev and Sinai, although they also occur at more northerly sites (Rosen 1997a: 75). A connection between the long-range distribution of such tools and the activities of pastoral groups has been proposed (Rosen 1997a: 107).

Suggested functions include butchering tools, a fact that may find support in their occurrence in what are interpreted as cult contexts at Bab adh-Dhra' and several Palestinian sites (Rosen 1997a: 74). Others have seen them as tools for shearing wool-bearing sheep (Henry 1995: 372). If correct, this would have important implications for the development of large-scale textile production in the region, and for the nature of the EB I–III exploitation of the arid zones. Whatever the case, it appears that the tabular scrapers and Canaanean blades were produced and distributed on rather different bases, with neither providing evidence for centralized economic direction.

Ad hoc production

In the case of Umm Hammad (Betts 1992: 122-23), on-site flint working appears to have been limited to the fashioning of 'irregular' tools from flakes struck from local cherts obtained as wadi cobbles. Thus, there existed a separation between 'imported' specialized and *ad hoc* tools, produced locally as required, presumably at a household level. The apparent lack of an equivalent element in late fourth-millennium BC, chipped stone assemblages from Mesopotamia (Pope and Pollock 1995) highlights the organizational differences between the economies of the two areas.

Groundstone

The groundstone industry of the Early Bronze Age has not been subject to a comprehensive study and is often inadequately reported in archaeological publications. However, by analogy with earlier periods (Hanbury-Tenison 1986; Wright 1993), one would expect this to include a substantial body of heavy food-processing tools, mortars, rubbers, pounders and so on. The stone chosen appears to vary according to local availability. There also existed a range of finely manufactured basalt vessels, generally with flat bases and flaring walls that tapered towards the rim (Beebe 1989; Braun 1990). These appear in both grave and settlement contexts (Figure 5.20.2).

However, groundstone appears to offer considerable potential for further analysis, as shown by investigations of the petrology and geochemistry of Early Bronze Age basalt artifacts from Jordan (Philip and Williams-Thorpe 1993). These have revealed, as with pottery, the multi-component nature of the assemblages of basalt artifacts occurring at individual sites, raising the possibility that vessels and grinding tools may have been acquired through different sets of socioeconomic relationships. As far as bowls are concerned, the material found at sites in the southern Ghor appears to originate in basalt flows located around al-Karak, while material from sites in many other areas of the southern Levant appears to have derived from the extensive basalt outcrops in northern Palestine and Transjordan. There existed, therefore, more than one centre of bowl production, the products of which reveal rather different spatial patterning.

Metals

Copper

The most frequently used metal was copper; the repertoire of artifacts is similar to that from Palestine. Weapons occur in a few EB III tombs, and include narrow-bladed daggers with a wooden handle attached by rivets (Philip 1989: 103-104, type 2) and crescent-shaped axes from Bab adh-Dhra' (Figure 5.20.1; Philip 1989: 45-46, type 1). With the exception of a single tin–bronze dagger from Bab adh-Dhra' (Maddin *et al.* 1980: 115), they are all made from unalloyed copper. Characteristic metal tools include flat axes and chisels, along with a variety of awls and other smaller items. Recent finds include an EB II 'hoard' of metal objects from Pella (Bourke 1997), an axe and chisel from domestic contexts at Tall Abu al-Kharaz (Fischer 1993: 285), and a group of EB III copper adzes from Numayra (Rast and Schaub 1980: 44).

Jordan contains major copper deposits in the Faynan area, east of Wadi 'Arabah, ores that were employed at Chalcolithic settlements in Wadi Bi'r as-Saba' (Shalev 1994). Recent work by German researchers has produced evidence for ancient mining and smelting from a variety of periods, including the Early Bronze Age (Hauptmann *et al.* 1985; Hauptmann 1989; Fritz 1994). On-site metalworking is attested at several settlements in the area, including the late fourth-millennium sites of Wadi Fidan 4 (Adams and Genz 1995; Levy *et al.* 1999), while evidence from Khirbat Hamra Ifdan appears to indicate the continuation of local copper working into the late EB III and EB IV periods (Adams 2000: 393). Recent archaeometallurgical research (Hauptmann *et al.* 1992, 1999) suggests that, contrary to previous assumptions, the main source of the copper used at the EB II site of Arad in the northern Negev was not Sinai, but Faynan. This would give a new twist to reconstructions that stress the role of Arad as supplying arid-zone products to settlements of the Mediterranean zone (Finkelstein 1990, 1995b), in that copper from deposits in Jordan would have been a major factor in the development of the regional economy.

Despite intensive archaeometallurgical research (Hauptmann 1989; Hauptmann *et al.* 1992), we still lack a real understanding of the organizational basis of copper extraction at Faynan, although new investigations in Wadi Fidan may change this situation (Levy *et al.* 1999). At present, there is no evidence

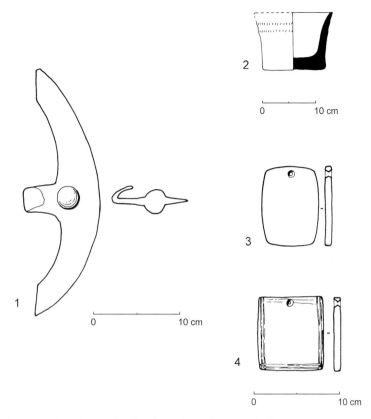

Figure 5.20. Metal and groundstone artifacts from Bab adh-Dhra' (drawn by Y. Beadnell).
  1. Crescentic Axe from Bab adh-Dhra', Tomb A 44, EB III.
  2. Basalt vessel from Bab adh-Dhra', Tomb A 70, EB IA.
  3–4. Siltstone palettes from EB II–III graves at Bab adh-Dhra'.

for a major EB II–III centre in Faynan in any way comparable to those known from the north and west. While it is possible that, following the demise of Arad, the role of intermediary between Faynan and the major agricultural areas might have been assumed by centres such as Bab adh-Dhra' in the southern Ghor, this cannot be demonstrated at present. On present evidence, it would be possible to argue a case against any form of clear political control over copper production in southern Jordan and view this as one more instance of interregional economic links, constituted at the level of intercommunity relationships. However, the aridity of the Faynan area may have constrained the development there of the kind of walled settlements characteristic of the Early Bronze Age elsewhere in Jordan, and it is possible that, given the inherent vulnerability of mining, smelting and transportation operations, control over copper resources was exercised in a manner that left little archaeological evidence.

Whatever the role of Faynan as a supplier of copper, it is clear that copper working, as opposed to extraction, was widely dispersed throughout Jordan, and the notion of a concentration of artifact production in the vicinity of the mining sites (Ilan and Sebbane 1989: 144) must be rejected. In the south, evidence for metalworking occurs at Tall Magass near al-'Aqaba (Khalil 1992: 144), a site that appears to include an Early Bronze Age as well as a Chalcolithic settlement component (Khalil 1992: 143, see illustration in Hanbury-Tenison [1986: fig. 21.3]). At the opposite end of Jordan, a late EB I midden deposit at Tall ash-Shuna has produced metalworking debris including round-based crucibles, small rectangular moulds and metal prills. As would be expected, the evidence seems to indicate the melting and casting of copper, rather than smelting of ore (Rehren et al. 1997). Later copper-working evidence comes from Room 15 in the EB III settlement of Numayra (Coogan 1984: 77) and, taken together, the dispersed nature of artifact production and the consistent artifact repertoire argues for widely held culturally determined notions of appropriate forms, that is, the communication and sharing of concepts at a spatial level well beyond that of potential political units.

It would be misleading, however, to assume a single monolithic copper industry deriving all raw material from Faynan. While most published Early Bronze Age artifacts appear to be produced in an unalloyed copper low in impurities, which is generally held to be compatible with the likely smelting products of Faynan ores (Hauptmann *et al.* 1992; Shalev 1994), there are exceptions. The copper recovered from Tall ash-Shuna appears to contain both arsenic and nickel at around 2%, a composition incompatible with any ore yet investigated at Faynan (Rehren *et al.* 1997). Typically, moulds bear matrices formed to cast tools in near-final form, thus reducing the need for subsequent working by smiths. However, those from Tall ash-Shuna appear designed to cast small rectangular copper blocks or thin ingots, perhaps designed for subsequent hammering to form copper sheet, or to produce small billets of standard size for subsequent redistribution. This cannot be attributed simply to north–south differences, as the analysis of copper artifacts from the EB I site of Yiftahel in Galilee revealed the low impurity unalloyed copper traditionally associated with Chalcolithic and Early Bronze Age tools (Shalev and Braun 1997: 95, table 11.3; Shalev 1994). Early Bronze Age copper procurement appears as a more complex and varied process than might have been expected in the light of the ready availability of ores in southern Jordan.

Gold and silver

Both gold and silver artifacts are rare in the southern Levant. Neither metal occurs locally; the nearest gold sources are in Egypt, silver in Anatolia (Moorey 1994). Several pieces of gold leaf jewellery come from EB II–III charnel house A 22 at Bab adh-Dhra', while a group of gold beads and tiny gold spacers appears to belong to a necklace (Rast and Schaub 1980: fig. 14). Silver artifacts are known from late EB I contexts at Tall ash-Shuna (a pin and a fragment of silver sheet) and Bab adh-Dhra' (Philip and Rehren 1996).

Analysis (Rehren *et al.* 1996) has revealed both silver artifacts from Tall ash-Shuna to be artificial alloys, containing a few per cent each of gold and copper, the latter presumably added to harden the metal (Moorey 1994: 238). The presence of gold appears to indicate the use of mixed gold–silver alloys by ancient craftsmen, suggesting that smiths had little idea of the exact nature of their material, and perhaps indicating the presence within the region of a supply of partly recycled precious metal 'stock'. The presence of silver or gold artifacts does not, therefore, constitute evidence for direct contacts with either Egypt or Anatolia.

Seals and sealings

In recent years, the corpus of Early Bronze Age sealings from Jordan has grown to more than 160, recovered from 15 sites (Vieweger 1997: 151). These belong to a tradition, widespread in the southern Levant, of placing sealings on the shoulders of vessels, just above the point where the neck was joined to the body (Figure 5.21.2). Impressions were formed by rolling a cylinder seal repeatedly to form a continuous band, a practice not so common in contemporary Mesopotamia, but shared with western Syria (Collon 1987: 113).

The sealing of jars would appear congruent with a system of staple finance, in that they would provide evidence for administrative control over commodity storage and transport. In Mesopotamia, however, seals were employed in such a manner as to allow their easy removal and subsequent resealing as required, a process clearly connected with administrative activities (Collon 1987: 113). There is no such evidence from Jordan. Vessels were only impressed once, and this had to take place prior to firing, that is, within the potter's workshop, a far less flexible system from an administrative point of view.

In Jordan, the largest single group comes from Khirbat az-Zaraqun with 143 seal impressions from 118 different seals (Mittmann 1994: 15), the bulk of these found on Metallic Ware vessels. The high ratio of impressions to vessels would seem to argue against an inherent logic within the system. We urgently need more information on the fabrics of the vessels bearing seal impressions in order to assess the strength of the relationship between sealing and Metallic Ware vessels: it is possible that in north Jordan the occurrence of sealings reflects little more than the distribution of Metallic Ware pithoi. Given recent petrographic work (Greenberg and Porat 1996) suggesting that Metallic Ware vessels were manufactured and presumably sealed in a single centre, it is hard to see how such sealings could have been responsive to the specific requirements of administrative practices at individual sites. In fact, while jar sealings may indicate some quality of the vessels themselves (Mazzoni 1984: 32-33), the number of different designs and the long duration of seal usage, with examples spanning the entire EB II–III period, argues against any universal system of 'meaning'.

EB II–III sites in south-central Palestine have produced fewer seal impressions than those in the north

The Early Bronze I–III Ages

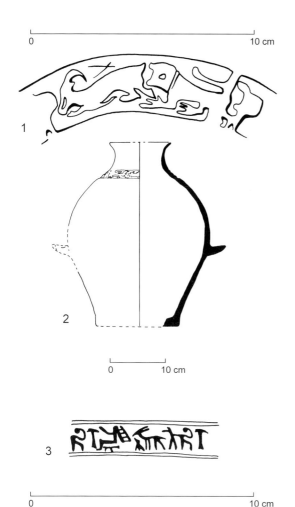

Figure 5.21. Sealing practices from EBA Jordan (drawn by Y. Beadnell).
1. Sealing employing zoomorphic motif, Bab adh-Dhra', No. 1301.
2. Position of seal impression on storage vessel; note how seal has been rolled around the neck of the vessel.
3. Impression of cylinder seal bearing design which includes a seated female figure, Bab adh-Dhra', No. 2823.

(Ben-Tor 1992: 117). In south-central Jordan, however, impressions are known from Tall al-'Umayri (Lapp 1991: 242, fig. 6.23.17) and the al-Karak Plateau (Brown 1991: 179, no. 64), plus a substantial corpus from the southern Ghor (Lapp 1995). The contrast with the situation in Palestine is striking. Even here, where Metallic Ware appears to be absent and pottery vessels may have been manufactured closer to their intended locus of use (see pottery), there is no evidence to suggest the existence of a coherent system of meaning. (For a recent discussion of this topic, see Flender 2000.)

While sharing many general stylistic features with Palestinian sealings (Figure 5.22.1), those from Bab adh-Dhra' suggest additional connections as well. Three seals (one each in ceramic, alabaster and chlorite) and 28 clay impressions have been recovered. One seal is made from pink alabaster of a kind that Lapp (1995: 44) considers of Egyptian origin, although locally carved. A black chlorite seal features a female figure seated before a table (Figure 5.21.3), and shows affinities with Egyptian First Dynasty cylinder seals rather than northern forms (Lapp 1989: 9-10). Other examples have better northern parallels, including one in 'cultic' style from the EB III 'Sanctuary' (Lapp 1989: 6-7). Sealing in Early Bronze Age Jordan cannot be deployed as convincing evidence for complex administrative systems but represents no more than an idea adopted from what Joffe (1993: 56) has termed 'a distant and undifferentiated Syro-Mesopotamian world', stripped of its original meaning and subsequently reinterpreted in a local context.

Other artifacts

Although a number of additional artifact forms are reported from Early Bronze Age sites in Jordan, the range and variety is quite limited. However, there do exist intersite distinctions in the material, presumably related to the kind of diversity of local practices that might underlie distinctive regional ceramic and architectural styles.

Bone, clay, shell and stone

Two particular classes of artifact stand out, both of which are frequently treated as *objets d'art* (Ben-Tor 1992: 121). The function and significance of decorated bone tubes, distributed widely throughout Syria and the Levant (Zarzecki-Peleg 1993), remains uncertain. Examples from Jordan, not cited in the aforementioned study, include one from Tall ash-Shuna, probably of EBA date, and several EB III instances from Khirbat az-Zaraqun (Genz, pers. comm.), a largely northern distribution. An EB III bull's head from Charnel House A 21 at Bab adh-Dhra' (Wilkinson 1989: 458, fig. 262. 2) belongs to a group of small carved bovine heads known from EB II–III contexts in Palestine (de Miroschedji 1993b). One might speculate that, lacking other types of animal figurine in the Early Bronze Age, the choice of bovids indicates an association with draft animals and, thus, with agricultural productivity.

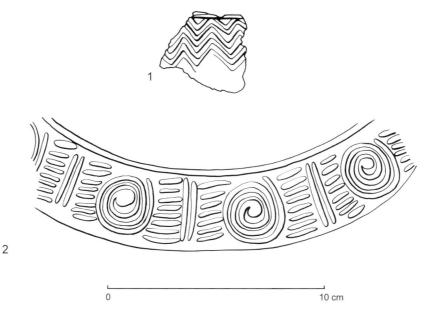

Figure 5.22. Seal impressions bearing geometric designs (drawn by Y. Beadnell).
1. Zig-zag pattern, Bab adh-Dhra', No. 2817.
2. Tall as-Sukhne, EB II; composition: central motif consisting of a row of spirals placed within groups of short horizontal lines, and bordered by lines top and bottom. The motif is framed on each side by a vertical line, thus its repetition as the seal is rolled over the vessel produces pairs of vertical lines, D. 1.8 cm, Ht 4.1 cm (Chesson *et al.* 1995: 120 n. 25).

Unbaked clay figurines with upraised arms are so far unique to graves at Bab adh-Dhra' (Wilkinson 1989) and may represent a phenomenon of particular significance at that site. Contacts with the Red Sea area are revealed through shell bracelets that occur in EB I tombs at Bab adh-Dhra' and at Tall Magass near al-'Aqaba (Khalil 1992: 1.3, 1.6-8). Similar artifacts are found in Palestine, Sinai and Egypt (Wilkinson 1989) and highlight contacts with the arid zone to the south.

Seven 'schist' or 'grey stone' palettes, mostly perforated and of rectangular shape, were recovered from EB II–III graves at Bab adh-Dhra' (Figure 5.20.3-4; Wilkinson 1989: 453-55, fig. 261). These resemble the Egyptian stone cosmetic palettes discussed by Petrie (1921), made from grey siltstone (Klemm and Klemm 1993: 369). Similar palettes are known from Chalcolithic and Early Bronze Age sites in Palestine and are believed to be Egyptian imports (Ben-Tor 1992: 94), although 'local' examples occur in such materials as limestone and granite (Hennessy 1967: 32; Brandl 1992: 447). At Bab adh-Dhra', the apparent lack of palettes in local stone suggests that it may have been important to produce palettes in stones with particular characteristics of colour and texture. Additional examples are reported from the Early Bronze Age occupation at Tall al-'Umayri (Geraty *et al.* 1986: 135).

Jacobs (1996: 130) notes that a number of such palettes found in the southern Levant come from EB III contexts, which would place them several centuries later than their closest stylistic parallels from Egypt, a pattern that extends to those from Bab adh-Dhra'. This apparent chronological discrepancy receives support from recent work at Minshat Abu Omar (Kroeper 1996: 72), where palettes do not occur after MAO IV, that is, First to mid-Second Dynasty (Kroeper 1996: 81-82, fig. 8). Although reducing the gap somewhat, these are still considerably earlier than the EB III dates of some examples from the Levant. Palettes found in the southern Levant are frequently perforated (Wilkinson 1989: fig. 269; Jacobs 1996: figs. 1, 3, 6), which is rarely the case with Egyptian examples, suggesting that this type of artifact may have been reinterpreted in a Levantine context, and may not simply represent the direct adoption of an Egyptian custom.

Jordan during Early Bronze Age I–III: an image

This chapter has put forward an alternative to the traditional city-state model for the Early Bronze Age. In the absence of evidence for institutionalized elites, it is argued that the distinctive archaeological record of the Early Bronze Age indicates prosperous village communities, in which a highly developed agricultural base provided the resources necessary for undertaking major projects on a corporate basis. Within these communities, wealth was understood in terms of land and productive facilities, a significant change in the symbolism of power compared to the preceding Chalcolithic.

Kinship-based groups are seen as constituting the basic units of action at subsistence level, with many aspects of the craft production focused upon independent specialists. There is no evidence to indicate that economic and political power were systematically linked in such a way as to offer a basis for sustained monopolistic control. Greater regional interaction over time is attributed to demand resulting from increasing economic specialization, intended to produce additional resources for investment rather than for elite consumption.

There is some evidence from Palestine for growing centralization and inequality during EB III (Joffe 1993: 86-87; de Miroschedji 1999), although this may relate to increasing disparities in wealth and power between different kin groups. However, the outstanding characteristic of this period is the relatively slow rates of change apparent in most aspects of material culture. This may be attributable to the absence of the elite-driven, competitive dynamic that appears so influential in the transformation of society, economy and material culture in later periods (see Sherratt and Sherratt 1991). One might suggest that the evidence from the southern Levant, Jordan in particular, provides a rare glimpse of the form taken by local societies in the absence of the exploitative, conspicuously consuming elites, the presence of whom shaped the material record of later periods in the region.

Occupation at many walled sites appears to have been relatively short, or punctuated, suggesting that they did not represent basic structuring elements of the local economic system. The extent of defended sites along the eastern fringes of the arid zone indicates that walled settlements were not simply associated with communities practising intensive agriculture, but became a symbolic statement, a resource capable of being reworked in a variety of contexts. Walled sites alone cannot be equated with any particular form of political organization.

Acknowledgments

I am grateful to the Director and staff of the Department of Antiquities of Jordan, who have supported

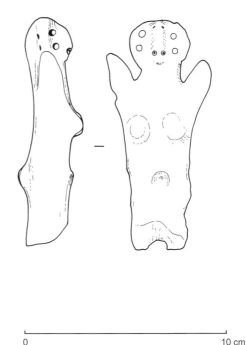

Figure 5.23. Clay figurine, EB I, Bab adh-Dhra', Tomb A 5E (drawn by Y. Beadnell).

and facilitated my research in many ways over the years, and to the British Institute at 'Amman for Archaeology and History (now the Council for British Research in the Levant), which provided both the opportunity and means to investigate many of the issues raised here through archaeological field work at the EB I site of Tall ash-Shuna. Drs Frank Braemer, Herrmann Genz, Kay Prag and Stephen Bourke were kind enough to read and comment upon a draft of this contribution, but bear no responsibility for the final content. I am grateful also to Mrs Yvonne Beadnell, who prepared the line drawings. Finally, I should acknowledge the supportive role of my family, who have shown immense forbearance in the face of sustained archaeological provocation.

References

Adams, R.
 2000 The Early Bronze Age III-IV Transition in Southern Jordan: Evidence from *Khirbet Hamra Ifdan*. In G. Philip and D. Baird (eds.), *Ceramics and Change in the Early Bronze Age of the Southern Levant*, 379-401. Sheffield: Sheffield Academic Press.

Adams, R. and H. Genz
 1995 Excavations at Wadi Fidan 4, a Chalcolithic Village Complex in the Copper Ore District of Feinan, Southern Jordan. *Palestine Exploration Quarterly* 127: 8-20.

Adams, R. McCormick
1966 *The Evolution of Urban Society*. Chicago: Aldine.

Albright, W.F.
1924 The Archaeological Results of an Expedition to Moab and the Dead Sea. *Bulletin of the American Schools of Oriental Research* 14: 2-12.
1949 *The Archaeology of Palestine*. Harmondsworth: Penguin.

Alon, D. and T.E. Levy
1989 The Archaeology of Cult and the Chalcolithic Sanctuary at Gilat. *Journal of Mediterranean Archaeology* 2: 163-221.

Amiran, R.
1969 *Ancient Pottery of the Holy Land*. Jerusalem: Masada.
1970 The Beginnings of Urbanization in Cannan. In J.A. Sandars (ed.), *Near Eastern Archaeology in the Twentieth Century: Essays in Honor of Nelson Glueck*, 83-100. New York: Doubleday.
1978 *The Chalcolithic Settlement and Early Bronze Age City: First–Fifth Seasons of Excavations 1962–1966*. Jerusalem: Israel Exploration Society.

Amiran, R. and R. Gophna
1992 The Correlation between Lower Egypt and Canaan during the Early Bronze I Period. In E.C. van den Brink (ed.), *The Nile Delta in Transition: 4th–3rd Millennium BC*, 357-60. Jerusalem: Israel Exploration Society.

Amiran, R. and O. Ilan
1996 *Early Arad II: The Chalcolithic and Early Bronze IB Settlements and Early Bronze II City: Architecture and Planning. Sixth to Eighteenth Seasons of Excavations, 1971–1978, 1980–1984*. Jerusalem: Israel Exploration Society.

Amiran, R. *et al.*
1978 *Early Arad: The Chalcolithic Settlement and Early Bronze Age City I: First–Fifth Seasons of Excavations 1962–1966*. Jerusalem: Israel Exploration Society.

Archi, A.
1991 Ebla: la formazione di uno stato del III millennio a.C. *La Paroloa del Passato* 46: 195-219.

Avner, U.
1984 Ancient Cult Sites in the Negev and Sinai Deserts. *Tel Aviv* 11: 115-31.
1990 Ancient Agricultural Settlement and Religion in the Uvda Valley in Southern Israel. *Biblical Archaeologist* 53: 125-41.

Avner, U., I. Carmi and D. Segal
1994 Neolithic to Bronze Age Settlement of the Negev and Sinai in the Light of Radiocarbon Dating: A View from the Southern Negev. In O. Bar-Yosef and R. Kra (eds.), *Late Quaternary Chronology and Paleoclimates of the Eastern Mediterranean*, 265-300. RADIOCARBON. Tucson: University of Arizona.

Baird, D. and G. Philip
1994 Preliminary Report on the Third (1993) Season of Excavations at Tell esh-Shuna North. *Levant* 26: 111-33.

Banning, E.B.
1996 Highlands and Lowlands: Problems and Survey Frameworks for Rural Archaeology in the Near East. *Bulletin of the American Schools of Oriental Research* 301: 25-46.

Banning, E.B., D. Rahimi and J. Siggers
1994 The Late Neolithic of the Southern Levant: Hiatus, Settlement Shift and Observer Bias. *Paléorient* 20.2: 151-64.

Bar-Yosef, O. and R. Kra (eds.)
1994 *Late Quaternary Chronology and Paleoclimates of the Eastern Mediterranean*. RADIOCARBON. Tucson: The University of Arizona Press.

Barrett, J.
1988 The Living, the Dead and the Ancestors: Neolithic and Early Bronze Age Mortuary Practices. In J. Barret and I. Kinnes (eds.), *The Archaeology of Context in the Neolithic and Bronze Age: Recent Trends*, 30-41. Sheffield: Department of Archaeology and Prehistory, University of Sheffield.
1994 *Fragments from Antiquity: An Archaeology of Social Life in Britain, 2900–1200 BC*. Oxford: Blackwell.

Baxevani, E.
1995 The Complex Nomads: Death and Social Stratification in EB IV Southern Levant. In A.C. Green and S. Campbell (eds.), *The Archaeology of Death in the Ancient Near East*, 85-95. Oxbow Monograph 51. Oxford: Oxbow.

Beebe, H.K.
1989 Basalt Bowls. In R.T. Schaub and W.E. Rast (eds.), *Bab edh-Dhra: Excavations in the Cemetery Directed by Paul W. Lapp (1965–67)*. Reports of the Expedition to the Dead Sea Plain, Jordan, vol. I. Winona Lake, IN: Eisenbrauns.

Beit-Arieh, I.
1981 A Pattern of Settlement in Southern Sinai and Southern Canaan in the Third Millennium B.C. *Bulletin of the American Schools of Oriental Research* 243: 31-55.

Ben-Tor, A.
1986 The Trade Relations of Palestine in the Early Bronze Age. *Journal of the Economic and Social History of the Orient* 29: 1-27.
1992 The Early Bronze Age. In A. Ben-Tor (ed.), *The Archaeology of Ancient Israel*. New Haven, CT: Yale University Press.
1994 Early Bronze Age Cylinder Seal Impressions and a Stamp Seal from Tell Qashish. *Bulletin of the American Schools of Oriental Research* 295: 15-30.

Betts, A.V.G. (ed.)
- 1991 *Excavations at Jawa 1972–1986*. Edinburgh: Edinburgh University Press.
- 1992 *Excavations at Tell Um Hammad: The Early Assemblages (EB I–II)*. Edinburgh: Edinburgh University Press.

Beynon, D.E. *et al.*
- 1986 Tempering Types and Sources for Early Bronze Age Ceramics from Bab edh-Dhra'. *Journal of Field Archaeology* 13: 297-305.

Blanton, R.E. *et al.*
- 1996 A Dual-Processual Theory for the Evolution of Mesoamerican Civilization. *Current Anthropology* 37: 1-31.

Bloch, M.
- 1971 *Placing the Dead*. London: Seminar.

Blumfiel, E. and T.E. Earle
- 1987 Specialization, Exchange and Complex Societies: An Introduction. In E. Blumfiel and T.E. Earle (eds.), *Specialization, Exchange and Complex Societies*, 1-10. Cambridge: Cambridge University Press.

Bourke, S.J.
- 1997 Pre-Classical Pella in Jordan: A Conspectus of Ten Years' Work (1985–1995). *Palestine Exploration Quarterly* 129: 93-115.

Bourke, S.J. *et al.*
- 1994 Preliminary Report on the University of Sydney's Fourteenth Season of Excavations at Pella (Tabaqat Fahl) in 1992. *Annual of the Department of Antiquities of Jordan* 38: 81-126.

Bradley, R.
- 1993 *Altering the Earth: The Origins of Monuments in Britain and Continental Europe*. Society of Antiquaries of Scotland Monograph Series No. 8. Edinburgh: Society of Antiquaries of Scotland.

Braemer, F.
- 1993a Review of A.V.G. Betts, *Excavations at Jawa 1972–1986*. *Syria* 70: 283-86.
- 1993b Prospections archéologiques dans le Hawran (Syria). *Syria* 70: 117-70.

Braemer, F. and J.-C. Échaillier
- 1995 La marge désertique en Syrie du Sud au IIIe Millénaire: Éléments d'appréciation de l'évolution du milieu. In *L'homme et la dégradation de l'environment*, 345-56. XVe Recontres Internationales d'Archéologie et d'Histoire d'Antibes. Juan-les-Pins: Éditions APDCA.
- 2000 A Summary Statement on the Early Bronze Age Ceramics from Southern Syria, and the Relationship of this Material with that of Neighbouring Regions. In G. Philip and D. Baird (eds.), *Ceramics and Change in the Early Bronze Age of the Southern Levant*, 403-10. Sheffield: Sheffield Academic Press.

Braemer, F., J.-C. Échallier and A. Taraqji
- 1993 Khirbet el Umbashi (Syrie): rapport préliminaire sur les campagnes 1991 et 1992. *Syria* 70: 415-30.

Brandl, B.
- 1992 Evidence for Egyptian Colonization in the Southern Coastal Plain and Lowlands of Canaan during the EB I Period. In E.C.M. van den Brink (ed.), *The Nile Delta in Transition: 4th–3rd Millennium BC*, 441-76. Jerusalem: Israel Exploration Society.

Braun, E.
- 1985 *En Shadud: Salvage Excavations at a Farming Community in the Jezreel Valley, Israel*. British Archaeological Reports, International Series 249. Oxford: British Archaeological Reports.
- 1989a The Transition from the Chalcolithic to the Early Bronze Age. In P. de Miroschedji (ed.), *L'Urbanisation de la Palestine à l'age du Bronze Ancien*, 7-29. British Archaeological Reports, International Series 527. Oxford: British Archaeological Reports.
- 1989b The Problem of the Apsidal House. *Palestine Exploration Quarterly* 2: 1-43.
- 1990 Basalt Bowls of the EB I Horizon in the Southern Levant. *Paléorient* 16.1: 87-96.
- 1996 Salvage Excavations at the Early Bronze Age Site of Me'ona: Final Report. *'Atiqot* 28: 1-39.
- 1997 *Yiftah'el: Salvage and Rescue Excavations at a Prehistoric Village in Lower Galilee, Israel*. Jerusalem: Israel Antiquities Authority Reports No. 2.

Bronk Ramsey, C.
- 1995 Radiocarbon Calibration and Analysis of Stratigraphy: The OxCal Program. In *RADIOCARBON— Proceedings of the 15th International Radiocarbon Conference, Glasgow*. Tucson: The University of Arizona.

Broshi, M. and R. Gophna
- 1984 The Settlements and Population of Palestine during the Early Bronze Age II–III. *Bulletin of the American Schools of Oriental Research* 253: 41-53.
- 1986 Middle Bronze Age II Palestine: Its Settlements and Population. *Bulletin of the American Schools of Oriental Research.* 261: 73-90.

Brown, R.
- 1991 Ceramics from the Kerak Plateau. In J.M. Miller (ed.), *Archaeological Survey of the Kerak Plateau*. American Schools of Oriental Research Archaeological Monographs 1, 169-280. Atlanta, GA: Scholars Press.

Bunimovitz, S.
- 1995 On the Edge of Empires: The Late Bronze Age 1500–1200 B.C. In T.E. Levy (ed.), *The Archaeology of Society in the Holy Land*, 320-31. London: Leicester University Press.

Callaway, J.A.
1978   New Perspectives on Early Bronze III Canaan. In P.R.S. Moorey and P.J. Parr (eds.), *Archaeology in the Levant: Essays for Kathleen Kenyon*, 46-58. Warminster: Aris & Phillips.
1993   Ai. In E. Stern (ed.), *The New Encyclopedia of Archaeological Excavations in the Holy Land*, I: 39-45. Jerusalem: Israel Exploration Society.

Carmi, I., C. Epstein and D. Segal
1995   Radiocarbon Dates from Chalcolithic Sites in the Golan. *'Atiqot* 27: 207-209.

Cartwright, C.
1994   The Archaeobotanical Remains from the 1993 Season. In J.N. Tubb and P.G. Dorrell, Tell es-Sa'idiyeh 1993: Interim Report on the Seventh Season of Excavations. *Palestine Exploration Quarterly* 126: 52-67.
1996   The Archaeobotanical Remains from the 1995 Season. In J.N. Tubb and P.G. Dorrell, Interim Report on the Eighth (1995) Season of Excavations at Tell es-Sa'idiyeh. *Palestine Exploration Quarterly* 128: 16-40.

Cartwright, C. and A. Clapham
1993   The Archaeobotanical Remains from Tell es-Sa'idiyeh, Jordan. In J.N. Tubb and P.G. Dorrell, Interim Report on the Sixth Season of Excavations. *Palestine Exploration Quarterly* 125: 50-74.

Chesson, M.S.
1998   Preliminary Result of Excavations at Tell el-Handaquq South (1993–1996). *Palestine Exploration Quarterly* 130: 20-34.
1999   Libraries of the Dead: Early Bronze Age Charnel Houses and Social Identity at Urban Bab edh-Dhra', Jordan. *Journal of Anthropological Archaeology* 18: 137-64.

Chesson, M.S. *et al.*
1995   Tell es-Sukhne North: An Early Bronze Age II Site in Jordan. *Paléorient* 21.1: 113-23.

Collon, D.
1987   *First Impressions: Cylinder Seals in the Ancient Near East*. London: British Museum.

Coogan, M.D.
1984   Numeira 1981. *Bulletin of the American Schools of Oriental Research* 255: 75-82.

Costin, C.L.
1991   Craft Specialization: Issues in Defining, Documenting and Explaining the Organization of Production. *Archaeological Method and Theory* 3: 1-56.

Croft, P.
1994   Some Preliminary Comments on the Animal Remains from the First Three Seasons at Shuna. In D. Baird and G. Philip, Preliminary Report on the Third (1993) Season of Excavations at Tell esh-Shuna North. *Levant* 26: 130-31.

Crumley, C.L.
1979   Three Locational Models: An Epistemological Assessment for Anthropology and Archaeology. In M.B. Schiffer (ed.), *Advances in Archaeological Method and Theory* 2: 141-73. New York: Academic Press.
1987   A Dialectical Critique of Hierarchy. In T.C. Patterson and C.W. Gailey (eds.), *Power Relations and State Formation*, 155-69. Washington, DC: American Anthropological Assocation.
1995   Heterarchy and the Analysis of Complex Societies. In R.M. Ehrenreich, C.L. Crumley and J.E. Levy (eds.), *Heterarchy and the Analysis of Complex Societies*, 1-5. Archaeological Papers of the American Anthropological Association. Arlington, VA: American Anthropological Association.

D'Altroy, T. and T.E. Earle
1985   State Finance, Wealth Finance and Storage in the Inka Political Economy. *Current Anthropology* 26: 187-206.

Dabrowski, B. and H. Krug
1994   Megalithic Tomb at Tell el-'Umeiri, Jordan. *Biblical Archaeologist* 57: 241-42.

Dajani, R.W.
1968   Excavations in Dolmens. *Annual of the Department of Antiquities of Jordan* 12.13: 56-64.

Daviau, P.M.M.
1991   Field D: The Lower Southern Terrace. In L.G. Herr *et al.* (eds.), *Madaba Plains Project: The 1987 Season at Tell el-'Umeiri and Vicinity and Subsidiary Studies*, 87-155. Berrien Springs, MI: Andrews University.

Davis, S.J.M.
1988   The Mammal Bones: Tel Yarmouth 1980-1983. In P. de Miroschedji (ed.), *Yarmouth I: Rapport sur les trois premières campagnes de fouilles à Tel Yarmouth (Israël)*, 143-49. Paris: Éditions Recherche sur les Civilisations.

Dechert, B.
1995   The Bone Remains from Hirbet ez-Zeraqon. In H. Buitenhuis and H.-P. Uerpmann (eds.), *Archaeozoology of the Near East*, II, 79-87. Leiden: Bachhuys.

Dever, W.G.
1987   Funerary Practices in EB IV (MBI) Palestine: A Study in Cultural Continuity. In J.H. Marks and R.M. Good (eds.), *Love and Death in the Ancient Near East: Essays in Honor of Marvin H. Pope*, 9-19. Guildford, CT: Four Quarters.

Donahue, J.
1981   Geologic Investigations at Early Bronze Sites. In W.E. Rast and R.T. Schaub (eds.), *The Southeastern Dead Sea Plain Expedition: An Interim Report of the 1977 Season*, 137-54. Annual of the American Schools of Oriental Research 46. Cambridge, MA: American Schools of Oriental Research.

1984 Geologic Reconstruction of Numeira. *Bulletin of the American Schools of Oriental Research* 255: 83-88.

1985 Hydrologic and Topographic Change during and after Early Bronze Age Occupation at Bab edh-Dhra and Numeira. In A. Hadidi (ed.), *Studies in the History and Archaeology of Jordan*, II: 131-40. Amman: Department of Antiquities.

Dornemann, R.
1983 *The Archaeology of the Transjordan in the Bronze and Iron Ages*. Milwaukee, WI: Milwaukee Public Museum.

Earle, T. E.
1987 Chiefdoms in Archaeological and Historical Perspective. *Annual Review of Anthropology* 16: 279-308.

Earle, T. E. (ed.)
1991 *Chiefdoms, Power, Economy and Ideology*. Cambridge: Cambridge University Press.

Échallier, J.-C. and F. Braemer
1995 Nature et fonctions des 'desert kites': Données et hypothèses nouvelles. *Paléorient* 21.1: 35-63.

Eisenberg, E.
1993 Tell Kitan. In E. Stern (ed.), *The New Encyclopedia of Archaeological Excavations in the Holy Land*, II: 878-81. Jerusalem: Israel Exploration Society.

Epstein, C.
1985 Laden Animal Figurines from the Chalcolithic Period in Palestine. *Bulletin of the American Schools of Oriental Research* 258: 53-62.

1993 Oil Production in the Golan Heights during the Chalcolithic Period. *Tel Aviv* 20: 133-46.

Esse, D.L.
1989a Secondary State Formation and Collapse in Early Bronze Age Palestine. In P. de Miroschedji (ed.), *L'urbanisation de la Palestine à l'age du Bronze Ancien*, 81-96. British Archaeological Reports, International Series 527. Oxford: British Archaeological Reports.

1989b Village Potters in Early Bronze Age Palestine: A Case Study. In A. Leonard Jr. and B.B. Williams (eds.), *Essays in Ancient Civilization Presented to Helene Kantor*, 77-92. Studies in Ancient Oriental Civilization 47. Chicago: University of Chicago Press.

1991 *Subsistence, Trade and Social Change in Early Bronze Age Palestine*. Studies in Ancient Oriental Civilization 50. Chicago: University of Chicago Press.

Falconer, S.E.
1987 Heartland of Villages: Reconsidering Early Urbanism in the Southern Levant. Unpublished PhD dissertation, The University of Arizona, Tucson.

1994 The Development and Decline of Bronze Age Civilization in the Levant: A Reassessment of Urbanism and Ruralism. In C. Mathers and S. Stoddart (eds.), *Development and Decline in the Mediterranean Bronze Age*, 305-33. Sheffield Archaeological Monographs 8. Sheffield: J.R. Collis.

Falconer, S.E. and S. Savage
1995 Heartlands and Hinterlands: Alternative Trajectories of Early Urbanization in Mesopotamia and the Southern Levant. *American Antiquity* 60: 37-58.

Fargo, V.
1979 Early Bronze Age Pottery at Tell el-Hesi. *Bulletin of the American Schools of Oriental Research* 236: 23-40.

Feinman, G.M.
1995 The Emergence of Inequality: A Focus on Strategies and Processes. In T.D. Price and G.M. Feinman (eds.), *Foundations of Social Inequality*, 255-80. New York: Plenum.

Feinman, G.M. and J. Neitzel
1984 Too Many Types: An Overview of Sedentary Prestate Societes in the Americans. In M. Schiffer (ed.), *Advances in Archaeological Method and Theory* 7, 39-102. New York: Academic Press.

Feinman, G.M., S.A. Kowalewski and R.E. Blanton
1984 Modelling Ceramic Production and Organizational Change in the Pre-Hispanic Valley of Oaxaca. In S.E. van der Leuw and A.C. Pritchard (eds.), *The Many Dimensions of Pottery*, 297-333. Amsterdam: Universiteit van Amsterdam.

Finkelstein, I.
1990 Early Arad: Urbanization of the Nomads. *Zeitschrift des Deutschen Palästina-Vereins* 106: 34-50.

1995a Two Notes on Early Bronze Age Urbanization and Urbanism. *Tel Aviv* 22: 47-69.

1995b *Living on the Fringe: The Archaeology and History of the Negev, Sinai and Neighbouring Regions in the Bronze and Iron Ages*. Monographs in Mediterranean Archaeology. Sheffield: Sheffield Academic Press.

Finkelstein, I. and R. Gophna
1993 Settlement, Demographic and Economic Patterns in the Highlands of Palestine in the Chalcolithic and Early Bronze Periods and the Beginning of Urbanism. *Bulletin of the American Schools of Oriental Research* 289: 1-22.

Fischer, P.M.
1993 Tell Abu al-Kharaz: The Swedish Jordan Expedition 1991, Second Season Preliminary Excavation Report. *Annual of the Department of Antiquities of Jordan* 37: 279-306.

1994 Tell Abu al-Kharaz: The Swedish Jordan Expedition 1992, Third Season Preliminary Excavation Report. *Annual of the Department of Antiquities of Jordan* 38: 127-46.

1995 Tell Abu al-Kharaz: The Swedish Jordan Expedition 1993, Fourth Season Preliminary Excavation Report. *Annual of the Department of Antiquities of Jordan* 39: 93-119.

1998 Tall abu al-Kharaz. The Swedish Jordan Expedition 1997, Eighth Season Preliminary Excavation Report. *Annual of the Department of Antiquities of Jordan* 42: 213-23.

2000 The Early Bronze Age at Tell Abu al-Kharaz, Jordan Valley: A Study of Pottery Typology and Provenance, Radiocarbon Dates and Synchronization of Palestine and Egypt. In G. Philip and D. Baird (eds.), *Ceramics and Change in the Early Bronze Age of the Southern Levant*, 201-32. Sheffield: Sheffield Academic Press.

Fischer, P.M. and E. Toivonen-Skage
1995 Metallic Burnished Early Bronze Age Ware from Tall Abu al-Kharaz. In K. 'Amr, F. Zayadine and M. Zaghloul (eds.), *Studies in the History and Archaeology of Jordan*, V: 587-98. Amman: Department of Antiquities.

Flanagan, J.G.
1989 Hierarchy in Simple 'Egalitarian' Societies. *Annual Review of Anthropology* 18: 245-66.

Flannery, K.V.
1972 The Cultural Evolution of Civilizations. *Annual Review of Ecology and Systematics* 3: 399-426.

Flender, M.
2000 Cylinder Seal Impressed Vessels of the Early Bronze Age III in Northern Palestine. In G. Philip and D. Baird (eds.), *Ceramics and Change in the Early Bronze Age of the Southern Levant*, 295-314. Sheffield: Sheffield Academic Press.

Franken, H.J.
1974 *In Search of the Jericho Potters: Ceramics from the Iron Age and from the Neolithicum*. Amsterdam: North-Holland.

Fritz, V.
1994 Vorbericht über die Grabungen in Barqa el-Hetiye im Gebeit von Fenan, Wadi el-'Arabah, Jordanien. *Zeitschrift des Deutschen Palästina-Vereins* 110: 125-50.

Futato, E.M.
1996 Early Bronze III Canaanean Blade/Scraper Cores from Tell Halif, Israel. In J.D. Seger (ed.), *Retrieving the Past: Essays on Archaeological Research and Methodology in Honor of Gus W. Van Beek*, 61-74. Winona Lake, WI: Eisenbrauns.

Garrard, A.N.
1989 Jilat Survey. In D. Homès-Fredericq and J.B. Hennessy (eds.), *Archaeology of Jordan. II.1. Field Reports: Surveys and Sites A–K*, 60-66. Akkadica Supplementum 7. Leuven: Peeters.

Garrard, A.N., S. Colledge and L. Martin
1996 The Emergence of Crop Cultivation and Caprine Herding in the 'Marginal Zone' of the Southern Levant. In D. Harris (ed.), *The Origins and Spread of Agriculture and Pastoralism in Eurasia*, 204-26. London: UCL.

Genz, H.
1993 Zur bemalten Keramik der Frühbronzezeit II–III in Palästina. *Zeitschrift des Deutschen Palästina-Vereins* 109: 1-19.

2000 Grain-wash Decoration in Early Bronze Age III? The Evidence from Khirbet ez-Zeraqon. In G. Philip and D. Baird (eds.), *Ceramics and Change in the Early Bronze Age of the Southern Levant*, 279-86. Sheffield: Sheffield Academic Press.

2001 The Organization of Early Bronze Age Metalworking in the Southern Levant. *Paléorient* 26: 55-65.

Geraty, L.T. and L.A. Willis
1986 The History of Archaeological Research in Transjordan. In L.T. Geraty and L.G. Herr (eds.), *The Archaeology of Jordan and Other Studies Presented to Siegfried Horn*, 3-74. Berrien Springs, MI: Andrews University.

Geraty, L.T., L.G. Herr and Ø.S. LaBianca
1986 Madaba Plains Project. A Preliminary Report of the 1984 Season at Tell el-'Umeiri and Vicinity. *Bulletin of the American Schools of Oriental Research, Supplementary Series* 24: 117-19.

Gilead, A.
1994 The History of the Chalcolithic Settlement in the Nahal Beer Sheva Area: The Radiocarbon Aspect. *Bulletin of the American Schools of Oriental Research* 296: 1-14.

Gilman, A.
1991 Trajectories towards Social Complexity in the Later Prehistory of the Mediterranean. In T.E. Earle (ed.), *Chiefdoms, Power, Economy and Ideology*, 146-68. Cambridge: Cambridge University Press.

Glueck, N.
1934 *Explorations in Eastern Palestine I*. Annual of the American Schools of Oriental Research 14: 1-113. New Haven, CT: American Schools of Oriental Research.

1946 Band Slip Ware in the Jordan Valley. *Bulletin of the American Schools of Oriental Research* 101: 3-20.

Goldberg, P.
1994 Interpreting Late Quaternary Continental Sequences in Israel. In O. Bar-Yosef and R.S. Kra (eds.), *Late Quaternary Chronology and Paleoclimates of the East Mediterranean*, 89-103. RADIOCARBON. Tucson: The University of Arizona.

Gophna, R.
1995a Early Bronze Age Canaan: Some Spatial and Demographic Observations. In T.E. Levy (ed.), *Archaeology of Society in the Holy Land*, 269-80. London: Leicester University Press.

1995b *Excavations at 'En Besor*. Tel Aviv: Ramot.

Gophna, R. and Y. Portugali
1988 Settlement and Demographic Processes in Israel's Coastal Plain from the Chalcolithic to the Middle

Bronze Age. *Bulletin of the American Schools of Oriental Research* 269: 11-28.

Gophna, R., N. Lipschitz and S. Lev-Yadun
1986–87 Man's Impact on the Natural Vegetation of the Central Coastal Plain of Israel during the Chalcolithic Period and the Bronze Age. *Tel Aviv* 13-14: 71-84.

Gordon, R. and L. Villiers
1983 Telul edh-Dhahab and its Environs: Survey of 1980 and 1982—Preliminary Report. *Annual of the Department of Antiquities of Jordan* 27: 275-89.

Gordon, R.L. Jr and E.A. Knauf
1987 Er-Rumman Survey 1985. *Annual of the Department of Antiquities of Jordan* 31: 289-98.

Goren, Y.
1995 Shrines and Ceramics in Chalcolithic Israel: The View through the Petrographic Microscope. *Archaeometry* 37: 285-305.

Goren, Y. and S. Zukermann
2000 An Overview of the Typology, Provenance and Technology of the Early Bronze Age I 'Grey Burnished Ware'. In G. Philip and D. Baird (eds.), *Ceramics and Change in the Early Bronze Age of the Southern Levant*, 165-82. Sheffield: Sheffield Academic Press.

Graesser, C.F.
1972 Standing Stones in Ancient Palestine. *Biblical Archaeologist* 35: 34-64.

Greenberg, R.
1996 The Early Bronze Age Levels. In A. Biran (ed.), *Dan I: A Chronicle of the Excavations, the Pottery Neolithic, the Early Bronze Age and the Middle Bronze Age Tombs*, 83-160. Jerusalem: Nelson Glueck School of Biblical Archaeology.

Greenberg, R. and N. Porat
1996 A Third Millennium Levantine Pottery Production Center: Typology, Petrography and Provenance of the Metallic Ware of Northern Israel and Adjacent Regions. *Bulletin of the American Schools of Oriental Research* 301: 5-24.

Gregory, C.
1982 *Gifts and Commodities*. London: Academic Press.

Grigson, C.
1987 Shiqmim: Pastoralism and Other Aspects of Animal Management in the Chalcolithic of the Northern Negev. In T.E. Levy (ed.), *Shiqmim I: Studies Concerning Chalcolithic Societies in the Northern Negev Desert, Israel (1982–1984)*, 219-41. British Archaeological Reports, International Series 356. Oxford: British Archaeological Reports.
1995 Plough and Pasture in the Early Economy of the Southern Levant. In T.E. Levy (ed.), *Archaeology of Society in the Holy Land*, 245-68. London: Leicester University Press.

Hallotte, R.S.
1996 Mortuary Archaeology and the Middle Bronze Age Southern Levant. *Journal of Mediterranean Archaeology* 8: 93-122.

Halstead, P.
1995 Plough and Power: The Economic and Social Significance of Cultivation with the Ox-drawn Ard in the Mediterranean. *Bulletin on Sumerian Agriculture* 8: 11-22

Hanbury-Tenison, J.W.
1986 *The Late Chalcolithic–Early Bronze I Transition in Palestine and Transjordan*. British Archaeological Reports, International Series 311. Oxford: British Archaeological Reports.
1987 Jarash Region Survey 1984. *Annual of the Department of Antiquities of Jordan* 31: 129-57.
1989a Jebel Mutawwaq 1986. *Annual of the Department of Antiquities of Jordan* 33: 137-44.
1989b Desert Urbanism in the Fourth Millennium. *Palestine Exploration Quarterly* 121: 55-63.

Harrison, T.
1993 Economics with an Entreprenurial Spirit: Early Bronze Age Trade with Late Predynastic Egypt. *Biblical Archaeologist* 56: 81-92.
1997 Shifting Patterns of Settlement in the Highlands of Central Jordan during the Early Bronze Age. *Bulletin of the American Schools of Oriental Research* 306: 1-38.
2000 The Early Bronze III Ceramic Horizon for Highland Central Jordan. In G. Philip and D. Baird (eds.), *Ceramics and Change in the Early Bronze Age of the Southern Levant*, 347-64. Sheffield: Sheffield Academic Press.

Hartal, M.
1989 *Northern Golan Heights: The Archaeological Survey as a Source for Regional History*. Jerusalem: Israel Department of Antiquities.

Hastorf, C.
1990 One Path to the Heights: Negotiating Political Inequality in the Sausa of Peru. In S. Upham (ed.), *The Evolution of Political Systems: Sociopolitics in Small-Scale Sedentary Societies*, 146-75. Cambridge: Cambridge University Press.

Hauptmann, A.
1989 The Earliest Periods of Copper Metallurgy in Feinan. In A. Hauptmann, E. Pernicka and G.A. Wagner (eds.), *Old World Archaeometallurgy*, 119-35. Der Anschnitt Beiheft 7. Bochum: Deutsches Bergbau-Museum.

Hauptmann, A., F. Begemann and S. Schmidt-Strecker
1999 Copper Objects from Arad: Their Composition and Provenance. *Bulletin of the American Schools of Oriental Research* 314: 1-17.

Hauptmann, A., G. Weisgerber and E.A. Knauf
1985 Archäometallurgische und bergbauarchäologische Untersuchungen im Gebeit von Fenan, Wadi Arabah (Jordanien). *Der Anschnitt* 37: 163-95.

Hauptmann, A. *et al.*
1992 Early Copper Produced at Feinan, Wadi Araba, Jordan: The Composition of Ores and Copper. *Archaeomaterials* 6: 1-33.

Hayden, B.
1994 Village Approaches to Complex Societies. In G.M. Schwartz and S.E. Falconer (eds.), *Archaeological Views from the Countryside: Village Communities in Early Complex Societies*, 198-206. Washington, DC: Smithsonian Institution.

Helms, S.W.
1975 Posterns in Early Bronze Age Fortifications of Palestine. *Palestine Exploration Quarterly* 107: 133-50.
1981 *Jawa: Lost City of the Black Desert*. London: Methuen.
1982 Paleo-Beduin and Transmigrant Urbanism. In A. Hadidi (ed.), *Studies in the History and Archaeology of Jordan*, I: 97-113. Amman: Department of Antiquities.
1987 Jawa, Tell um Hamad and the E.B.I/Late Chalcolithic Landscape. *Levant* 19: 49-81.
1992 The 'Zarqa Triangle': A Preliminary Appraisal of Protohistoric Settlement Patterns and Demographic Episodes. In M. Zaghloul *et al.*, *Studies in the History and Archaeology of Jordan*, IV: 129-35. Amman: Department of Antiquities.

Hendrickx, S.
1996 The Relative Chronology of the Naqada Culture: Problems and Possibilities. In J. Spencer (ed.), *Aspects of Early Egypt*, 36-69. London: British Museum.

Hennessy, J.B.
1967 *The Foreign Relations of Palestine during the Early Bronze Age*. London: Colt Archaeological Institute.

Henry, D.O.
1995 *Prehistoric Cultural Ecology and Evolution: Insights from Southern Jordan*. New York: Plenum.

Herr, L.G. *et al.* (eds.)
1991 *Madaba Plains Project 2: The 1987 Season at Tell el-'Umeiri and Vicinity and Subsidiary Studies*, 87-155. Berrien Springs, MI: Andrews University.

Herr, L.G. *et al.*
1997 Madaba Plains Project 1996: Excavations at Tall el-'Umayri, Tall Jalul and Vicinity. *Annual of the Department of Antiquities of Jordan* 41: 145-68.

Herzog, Z.
1986 *Das Stadttor in Israel und in den Nachbarländern*. Mainz am Rhein: Von Zabern.

Hesse, B. and P. Wapnish
1991 Faunal remains from Tel Dan: Perspectives on Animal Production at a Village, Urban and Rural Centre. *ArchaeoZoologia* 4.2: 9-86.

Homès-Fredericq, D. and J.B. Hennessy (eds.)
1989 *Archaeology of Jordan*. II.1. *Field Reports: Surveys and Sites A–K*, 230-41. Akkadica Supplementum 7. Leuven: Peeters.

Ibrahim, M. and S. Mittmann
1994 Excavations at Khirbet ez-Zeraqon 1993. *Newsletter of the Institute of Archaeology and Anthropology, Yarmouk University* 16: 11-15.

Ibrahim, M., K. Yassine and J.A. Sauer
1988 The East Jordan Valley Survey 1975 (Parts 1 and 2). In K. Yassine (ed.), *The Archaeology of Jordan: Essays and Reports*, 159-207. Amman: Department of Archaeology, University of Jordan.

Ilan, D.
1992 A MBA Offering Deposit from Tel Dan. *Tel Aviv* 19: 247-66.
1995 The Dawn of Internationalism: The Middle Bronze Age. In T.E. Levy (ed.), *Archaeology of Society in the Holy Land*, 297-319. London: Leicester University Press.

Ilan, O. and M. Sebbane
1989 Metallurgy, Trade and the Urbanization of Southern Canaan in the Chalcolithic and Early Bronze Age. In P. de Miroschedji (ed.), *L'Urbanisation de la Palestine à l'age du Bronze Ancien*, 139-62. British Archaeological Reports, International Series 527. Oxford: British Archaeological Reports.

Jacobs, P.F.
1996 A Cosmetic Palette from Early Bronze III at Tell Halif. In J.D. Seger (ed.), *Retrieving the Past: Essays on Archaeological Research and Methodology in Honor of Gus W. Van Beek*, 123-34. Winona Lake, WI Eisenbrauns.

Joffe, A.H.
1991 Early Bronze I and the Evolution of Social Complexity in the Southern Levant. *Journal of Mediterranean Archaeology* 4.1: 3-58.
1993 *Settlement and Society in the Early Bronze I and II Southern Levant: Complementarity and Contradiction in Small-Scale Complex Society*. Sheffield: Sheffield Academic Press.

Joffe, A.H. and J.-P. Dessel
1995 Redefining Chronology and Terminology for the Chalcolithic of the Southern Levant. *Current Anthropology* 33.3: 507-17.

Johnson, G. and T. Earle
1987 *The Evolution of Human Society: From Forager Group to Agrarian State*. Stanford: Stanford University Press.

Kamlah, J.
   n.d.   Zeraqon Survey. Unpublished manuscript.

Kantor, H.
   1992   The Relative Chronology of Egypt and its Foreign Correlations before the First Intermediate Period. In R.W. Ehrich (ed.), *Chronologies in Old World Archaeology*, 3-21. Chicago: University of Chicago Press.

Kempinski, A.
   1992a Fortifications, Public Buildings and Town Planning in the Early Bronze Age. In A. Kempinski and R. Reich (eds.), *The Architecture of Ancient Israel from the Prehistoric to the Persian Periods*, 68-80. Jerusalem: Israel Exploration Society.
   1992b The Middle Bronze Age. In A. Ben-Tor (ed.), *The Archaeology of Ancient Israel*, 159-210. New Haven, CT: Yale University Press.

Kempinski, A. and W.-D. Niemeier
   1990   *Excavations at Kabri: Preliminary Report of 1989 Season*. Tel Aviv: University of Tel Aviv.

Kenyon, K.M.
   1960   *Excavations at Jericho*, I. London: British School of Archaeology in Jerusalem.
   1979   *Archaeology in the Holy Land*. 4th edn. London: Benn.

Keswani, P.S.
   1996   Hierarchies, Heterarchies and Urbanization Processes: The View from Bronze Age Cyprus. *Journal of Mediterranean Archaeology* 9: 211-50.

Khalil, L.
   1987   A Preliminary Report on the 1985 Excavations at Tell Magass-Aqaba. *Annual of the Department of Antiquities of Jordan* 31: 481-83.
   1988   Excavations at Magass-Aqaba, 1985. *Dirasat* 15.7: 71-117.
   1992   Some Technological Features from a Chalcolithic Site at Magass-Aqaba. In M. Zaghloul *et al.* (eds.), *Studies in the History and Archaeology of Jordan*, IV: 143-48. Amman: Department of Antiquities.

Khazanov, A.M.
   1984   *Nomads and the Outside World*. Cambridge: Cambridge University Press.

Klemm, R. and D. Klemm
   1993   *Steine und Steinbrüche im Alten Ägypten*. Berlin: Springer.

Knapp, A.B.
   1990   Production, Location and Integration on Bronze Age Cyprus. *Current Anthropology* 31: 147-76.
   1993   Social Complexity: Incipience, Emergence, and Development in Bronze Age Cyprus. *Bulletin of the American Schools of Oriental Research* 292: 85-108.

Köhler-Rollefson, I.
   1992   A Model for the Development of Nomadic Pastoralism on the Transjordanian Plateau. In O. Bar-Yosef and A.M. Khazanov (eds.), *Pastoralism in the Levant: Archaeological Materials in Anthropological Perspectives*, 11-18. Madison, WI: Prehistory.

Kolska-Horwitz, L. and E. Tchernov
   1989   Animal Exploitation in the Early Bronze Age of the Southern Levant: An Overview. In P. de Miroschedji (ed.), *L'urbanisation de la Palestine à l'age du Bronze Ancien*, 279-96. British Archaeological Reports, International Series 527. Oxford: British Archaeological Reports.

Korber, C.
   1993   Edh-Dhra' Survey 1992. *Annual of the Department of Antiquities of Jordan* 37: 550-53.

Kramer, C.
   1982   *Village Ethnoarchaeology: Rural Iran in Archaeological Perspective*. New York: Academic Press.

Kristiansen, K.
   1991   Chiefdoms, States and Systems of Political Social Evolution. In T.E. Earle (ed.), *Chiefdoms, Power, Economy and Ideology*, 16-43. Cambridge: Cambridge University Press.

Kroeper, K.
   1996   Minshat Abu Omar—Burials with Palettes. In J. Spencer (ed.), *Aspects of Early Egypt*, 70-92. London: British Museum.

Kuijt, I., J. Mabry and G. Palumbo
   1991   Early Neolithic Use of Upland Areas of Wadi el-Yabis: Preliminary Evidence from the Excavations of 'Iraq ed-Dubb, Jordan. *Paléorient* 17.1: 99-108.

LaBianca, Ø.S.
   1990   *Sedentarization and Nomadization: Food System Cycles at Hesban and Vicinity in Transjordan*. Berrien Springs, MI: Andrews University.

Lapp, N.
   1989   Cylinder Seals and Impressions of the Third Millennium BC from the Dead Sea Plain. *Bulletin of the American Schools of Oriental Research* 273: 1-15.
   1991   EB III Seal Impression. In L.G. Herr *et al.* (eds.), *Madaba Plains Project 2: The 1987 Season at Tell el-'Umeiri and Vicinity and Subsidiary Studies*, 242-43. Berrien Springs, MI: Andrews University.
   1995   Some Early Bronze Age Seal Impressions from the Dead Sea Plain and their Implications for Contacts in the Eastern Mediterranean. In S.J. Bourke and J.-P. Descoeudres (eds.), *Trade, Contact and the Movement of Peoples in the East Mediterranean*, 43-52. Mediterranean Archaeology Supplements. Sydney: Meditarch.

Leonard, A.
   1992   *The Jordan Valley Survey, 1953: Some Unpublished Soundings Conducted by James Mellaart*. Annual of the American Schools of Oriental Research 50. Winona Lake, WI: Eisenbrauns.

Levy, J.E.
1995 Heterarchy in Bronze Age Denmark: Settlement Pattern, Gender and Ritual. In R.M. Ehrenreich, C.L. Crumley and J.E. Levy (eds.), *Heterarchy and the Analysis of Complex Societies*. Archaeological Papers of the American Anthropological Association 6: 41-53. Arlington, VA: American Anthropological Association.

Levy, T.E.
1986 Archaeological Sources for the History of Palestine: The Chalcolithic Period. *Biblical Archaeologist* 49: 82-108.
1992a Radiocarbon Chronology of the Beersheba Culture and Predynastic Egypt. In E.C.M. van den Brink (ed.), *The Nile Delta in Transition 4th–3rd Millennium B.C.*, 345-56. Jerusalem: Israel Exploration Society.
1992b Transhumance, Subsistence and Social Evolution in the Northern Negev Desert. In O. Bar-Yosef and A. Khazanov (eds.), *Pastoralism in the Levant: Archaeological Material in Anthropological Perspectives*, 65-82. Madison, WI: Prehistory.
1993 Production, Space and Social Change in Protohistoric Palestine. In A. Holl and T.E. Levy, *Spatial Boundaries and Social Dynamics: Case Studies from Food-Producing Societies*, 63-82. Ann Arbor, MI: International Monographs in Prehistory.
1995a Cult, Metallurgy and Rank Societies: Chalcolithic. In T.E. Levy (ed.), *Archaeology of Society in the Holy Land*, 226-43. London: Leicester University Press.
1995b New Light on King Narmer and the Protodynastic Egyptian Presence in Canaan. *Biblical Archaeologist* 58: 26-36.

Levy, T.E., R.B. Adams and M. Najjar
1999 Early Metallurgy and Social Evolution: Jabal Hamrat Fidan. *ACOR Newsletter* 11.1: 1-3.

Lipschitz, N.
1989 Plant Economy and Diet in the Early Bronze Age in Israel: A Summary of Present Research. In P. de Miroschedji (ed.), *L'urbanisation de la Palestine à l'Age du Bronze Ancien*, 269-78. British Archaeological Reports, International Series 527. Oxford: British Archaeological Reports.

Lipschitz, N. et al.
1991 The Beginning of Olive (*Olea europaea*) Cultivation in the Old World: A Reassessment. *Journal of Archaeological Science* 18: 441-53.

London, G.A.
1991 Aspects of Early Bronze and Late Iron Age Ceramic Production at Tell el-'Umeiri. In L.G. Herr et al. (eds.), *Madaba Plains Project 2: The 1987 Season at Tell el-'Umeiri and Vicinity and Subsidiary Studies*, 383-419. Berrien Springs, MI: Andrews University.

London, G.A., H. Plint and J. Smith
1991 Preliminary Petrographic Ananlysis of Pottery from Tell el-'Umeiri and Hinterland Sites 1987. In L.G. Herr et al. (eds.), *Madaba Plains Project 2: The 1987 Season at Tell el-'Umeiri and Vicinity and Subsidiary Studies*, 423-39. Berrien Springs, MI: Andrews University.

Mabry, J.
1989 Investiations at Tell el-Handaquq, Jordan (1987–88). *Annual of the Department of Antiquities of Jordan* 33: 59-95.

Mabry, J. and G. Palumbo
1988 The 1987 Wadi el-Yabis Survey. *Annual of the Department of Antiquities of Jordan* 32: 275-306.

Mabry, J. et al.
1996 Early Town Development and Water Management in the Jordan Valley: Investigations at Tell el-Handaquq North. *Annual of the American Schools of Oriental Research* 53: 115-54. Cambridge, MA: American Schools of Oriental Research.

MacDonald, B. et al.
1988 *The Wadi el Hasa Archaeological Survey, 1979–1983, West-Central Jordan*. Waterloo, ON: Wilfrid Laurier University.
1992 *The Southern Ghors and Northeast 'Arabah Archaeological Survey*. Sheffield Archaeological Monographs 5. Sheffield: J.R. Collis.

Maddin, R.A., J.D. Muhly and T. Stech-Wheeler
1980 Research at the Centre for Ancient Metallurgy. *Paléorient* 6: 111-19.

Magness-Gardiner, B.
1994 Urban–Rural Relations in Bronze Age Syria: Evidence from Alalakh Level VII Palace Archives. In G.M. Schwartz and S.E. Falconer (eds.), *Archaeological Views from the Countryside: Village Communities in Early Complex Societies*, 37-47. Washington, DC: Smithsonian Institution.

Manning, S.W.
1993 Prestige, Distinction and Competition: The Anatomy of Socioeconomic Complexity in Fourth to Second Millennium B.C.E. Cyprus. *Bulletin of the American Schools of Oriental Research* 292: 35-58.

Marquet-Krause, J.
1949 *Les fouilles de Ai (et-Tell) 1933–1935*. Paris: Geuthner.

Mattingly, G.L.
1983 Nelson Glueck and Early Bronze Age Moab. *Annual of the Departent of Antiquities of Jordan* 27: 481-89.

Mazar, A.
1990 *Archaeology of the Land of the Bible 10,000–586 B.C.E.* New York: Doubleday.
1997 The Excavations at Tel Beth Shean during the Years 1989–94. In N.A. Silberman and D.B. Small (eds.), *The Archaeology of Israel: Constructing the Past, Interpreting the Present*, 144-65. Sheffield: Sheffield Academic Press.

Mazar, A. and P. de Miroschedji
 1996 Hartuv, an Aspect of the Early Bronze I Culture of Southern Israel. *Bulletin of the American Schools of Oriental Research* 302: 1-40.

Mazzoni, S.
 1984 Seal Impressions on Jars from Ebla in EB IVA-B. *Akkadica* 37: 18-40.
 1991 Ebla e la formazione della cultura urbana in Siria. *La Paroloa del Passato* 46: 163-94.

McCartney, C.
 1996 A Report on the Chipped Stone Assemblage from Tell Iktanu, Jordan Valley. *Levant* 1996: 131-56.

McClelland, T.L. and A. Porter
 1995 Jawa and North Syria. In K. 'Amr, F. Zayadine and M. Zaghloul (eds.), *Studies in the History and Archaeology of of Jordan*, V: 49-65. Amman: Department of Antiquities.

McConaughty, M.A.
 1980 Chipped Stone Tools from Bab edh-Dhra'. *Bulletin of the American Schools of Oriental Research* 240: 53-58.

McCreery, D.W.
 1981 Flotation of the Bab edh-Dhra' and Numeira Plant Remains. In W.E. Rast and R.T. Schaub (eds.), *The Southeastern Dead Sea Plain Expedition: An Interim Report of the 1977 Season*, 165-70. Annual of the American Schools of Oriental Research 46. Cambridge, MA: American Schools of Oriental Research.

Meillassoux, C.
 1968 Ostentation, Destruction, Reproduction. *Economie et Societé* 2: 760-72.

Miller, J.M. (ed.)
 1991 *Archaeological Survey of the Kerak Plateau*. American Schools of Oriental Research Archaeological Monographs 1. Atlanta, GA: Scholars Press.

Miroschedji, P. de
 1971 *L'époque préurbaine en Palestine*. Cahiers de la Revue Biblique 13. Paris: Gabalda.
 1988 *Yarmouth I: Rapport sur les Trois Premières Campagnes de Fouilles à Tel Yarmouth (Israël) (1980–1982)*. Paris: Éditions Recherche sur les Civilisations.
 1989 Le processus d'urbanisation en Palestine au Bronze Ancien: chronologie et rythmes. In P. de Miroschedji (ed.), *L'urbanisation de la Palestine à l'age du Bronze Ancien*, 63-80. British Archaeological Reports, International Series 527. Oxford: British Archaeological Reports.
 1993a Tell el-Far'ah North: Neolithic to Middle Bronze Age. In E. Stern (ed.), *The New Encyclopedia of Archaeological Excavations in the Holy Land*, 1: 433-38. Jerusalem: Israel Exploration Society.
 1993b Notes sur les têtes de taureau en os, en ivoire et en pierre du Bronze Ancien de Palestine. In M. Heltzer, A. Segal and D. Kaufman (eds.), *Studies in the Archaeology and History of Ancient Israel in Honour of Moshe Dothan*, *29-*39. Haifa: Haifa University.
 1999 Yarmouth: The Dawn of City-States in Southern Canaan. *Near Eastern Archaeology* 62: 2-19.

Mittmann, S.
 1970 *Beitrage zur siedlungs-und territorialgeschichte des nordlichen Ostjordanlandes*. Wiesbaden: Harrassowitz.
 1994 Hirbet ez-Zeraqon: Eine Stadt der frühen Bronzezeit in Nordjordanien. *Archaeologie in Deutschland*, 10-15.

Moorey, P.R.S.
 1994 *Ancient Mesopotamian Materials and Industries*. Oxford: Clarendon.

Moran, W.
 1992 *The Amarna Letters*. Baltimore: Johns Hopkins University.

Najjar, M.
 1992 The Jordan Valley (East Bank) during the Middle Bronze Age in the Light of New Excavations. In M. Zaghloul et al. (eds.), *Studies in the History and Archaeology of Jordan*, IV: 149-54. Amman: Department of Antiquities of Jordan.

Neef, R.
 1990 Introduction, Development and Environmental Implications of Olive Culture. In S. Bottema, G. Entjes-Nieborg and W. Van Zeist (eds.), *Man's Role in the Shaping of the Middle Eastern Landscape*, 295-306. Rotterdam: Balkema.

Neeley, M.P.
 1992 Lithic Period Sites. In B. MacDonald et al., *The Southern Ghors and Northeast 'Arabah Archaeological Survey*, 25-52. Sheffield Archaeological Monographs 5. Sheffield: J.R. Collis.

Netting R.McC.
 1990 Population, Permanent Agriculture, and Polities: Unpacking the Evolutionary Portmanteau. In S. Upham (ed.), *The Evolution of Political Systems: Sociopolitics in Small-Scale Sedentary Societies*, 21-61. Cambridge: Cambridge University Press.

Neuville, R.
 1930 La nécropole megalithique d'el-Adeimeh (Transjordania). *Biblica* 11: 249-65.

Ottosson, M.
 1980 *Temples and Cult Places in Palestine*. Uppsala: Almquist & Wiksell International.

Ovadia, E.
 1992 The Domestication of the Ass and Pack Transport by Animals: A Case of Technological Change. In O. Bar-Yosef and A.M. Khazanov (eds.), *Pastoralism in the Levant: Archaeological Materials in Anthropological Perspectives*, 19-28. Madison, WI: Prehistory.

Palumbo, G.
1990 *The Early Bronze IV in the Southern Levant: Settlement Patterns, Economy and Material Culture of a 'Dark Age'*. Contriibuti e materiali di archeologica orientale III. Rome: Universita di Roma 'La Sapienza'.

Palumbo, G., J. Mabry and I. Kuijt
1990 Survey in the Wadi el-Yabis. *Syria* 67: 479-81.

Parr, P.
1956 A Cave at Arqub El Dhahr. *Annual of the Department of Antiquities of Jordan* 3: 61-73.

Paynter, R.
1989 The Archaeology of Equality and Inequality. *Annual Review of Anthropology* 18: 369-99.

Petrie, W.F.
1921 *Corpus of Prehistoric Pottery and Palettes*. London: British School of Archaeology in Egypt.

Philip, G.
1988 Hoards of the Early and Middle Bronze Ages in the Levant. *World Archaeology* 20: 190-208.
1989 *Metal Weapons of the Early and Middle Bronze Ages in Syria-Palestine*. British Archaeological Reports, International Series 526 (i) and (ii). Oxford: British Archaeological Reports.
1995a Warrior Burials in the Ancient Near Eastern Bronze Age: The Evidence from Mesopotamia, Western Iran and Syria-Palestine. In A.C. Green and S. Campbell (eds.), *The Archaeology of Death in the Ancient Near East*, 140-54. Oxbow Monograph 51. Oxford: Oxbow.
1995b Jawa and Tell Um-Hammad: Two EBA Sites in Jordan. *Palestine Exploration Quarterly* 127: 161-70.
1999 Complexity and Diversity in the Southern Levant during the Third Millennium BC: The Evidence of Khirbet Kerak Ware. *Journal of Mediterranean Archaeology* 12: 26-57.

Philip, G. and D. Baird
2000 Early Bronze Age Ceramics in the Southern Levant: An Overview. In G. Philip and D. Baird (eds.), *Ceramics and Change in the Early Bronze Age of the Southern Levant*, 3-30. Sheffield: Sheffield Academic Press.

Philip, G. and T. Rehren
1996 Fourth Millennium BC Silver from Tell esh-Shuna, Jordan: Archaeometallurgical Investigation and Some Thoughts on Ceramic Skeuomorphs. *Oxford Journal of Archaeology* 15: 129-50.

Philip, G. and O. Williams-Thorpe
1993 A Provenance Study of Jordanian Basalt Vessels of the Chalcolithic and Early Bronze Age I Periods. *Paléorient* 19.2: 51-63.

Pinnock, F.
1991 Considerazioni sul sistema commerciale di Ebla proto-siriana. *La Parola del Passato* 46: 270-84.

Pollock, S.
1992 Bureaucrats and Managers, Peasants and Pastoralists, Imperialists and Traders: Research on the Uruk and Jemdet Nasr Periods in Mesopotamia. *Journal of World Prehistory* 6: 297-326.

Pope, M. and S. Pollock
1995 Trade, Tools and Tasks: A Study of Uruk Chipped Stone Industries. *Research in Economic Anthropology* 16: 227-65.

Porat, N.
1996 Petrography of the EB Pottery at Tell Dan. In A. Biran (ed.), *Dan I: A Chronicle of the Excavations, the Pottery Neolithic, the Early Bronze Age and the Middle Bronze Age Tombs*, 134-35. Jerusalem: Nelson Glueck School of Biblical Archaeology.

Portugali, J. and R. Gophna
1993 Crisis, Progress and Urbanization: The Transition from Early Bronze I to Early Bronze II in Palestine. *Tel Aviv* 20: 164-86.

Potter, D.R. and E.M. King
1995 A Heterarchical Approach to Lowland Maya Socioeconomics. In R.M. Ehrenreich, C.L. Crumley and J.E. Levy (eds.), *Heterarchy and the Analysis of Complex Societies*, 17-32. Archaeological Papers of the American Anthropological Association No. 6. Arlington, VA: American Anthropological Association.

Prag, K.
1985 Ancient and Modern Pastoral Migration in the Levant. *Levant* 17: 81-88.
1989 Preliminary Peport on Excavations at Tell Iktanu, Jordan. *Levant* 21: 33-45.
1991 Preliminary Peport on Excavations at Tell Iktanu, and Tell el-Hammam, Jordan 1990. *Levant* 23: 55-66.
1992 Bronze Age Settlement Patterns in the South Jordan Valley: Archaeology, Environment and Ethnology. In M. Zaghloul et al., *Studies in the History and Archaeology of Jordan*, IV: 155-60. Amman: Department of Antiquities.
1993 The Excavations at Tell Iktanu 1989 and 1990: The Excavations at Tell al-Hammam 1990. *Syria* 70 (1.2): 269-73.
1995 The Dead Sea Dolmens: Death and the Landscape. In S. Campbell and A. Green (eds.), *The Archaeology of Death in the Ancient Near East*, 75-84. Oxbow Monographs 51. Oxford: Oxbow.

Price, B.J.
1978 Secondary State Formation: An Explanatory Model. In R. Cohen and E.R. Service (eds.), *The Origins of the State*, 161-86. Philadelphia, PA: Institute for the Study of Human Issues.

Rast, W.E.
1995 Building on Marl: The Case of Bab adh-Dhra'. In K. 'Amr, F. Zayadine and M. Zaghloul (eds.), *Studies in the*

History and Archaeology of Jordan, V: 123-28. Amman: Department of Antiquities.

Rast, W.E. and R.T. Schaub
- 1978 A Preliminary Report of the Excavations at Bab edh-Dhra 1975. In D.N. Freedman (ed.), *Preliminary Excavation Reports: Bab edh-Dhra', Sardis, Meiron, Tell el-Hesi, Carthage (Punic)*. Annual of the American Schools of Oriental Research 43: 1-32. Cambridge, MA: American Schools of Oriental Research.
- 1980 Preliminary Report of the 1979 Expedition to the Dead Sea Plain, Jordan. *Bulletin of the American Schools of Oriental Research* 240: 21-63.
- 1981 *The South Eastern Dead Sea Plain Expedition: An Interim Report of the 1977 Season*. Annual of the American Schools of Oriental Research 46. Cambridge, MA: American Schools of Oriental Research.

Rehren, T., K. Hess and G. Philip
- 1996 Auriferous Silver in Western Asia: Ore or Alloy? *Journal of the Historical Metallurgy Society* 30.1: 1-10.
- 1997 Fourth Millennium BC Copper Metallurgy in Northern Jordan: The Evidence from Tell esh-Shuna. In H.-G. Gebel, Z. Kafafi and G.O. Rollefson (eds.), *Prehistory of Jordan. II. Perspectives from 1997*. Studies in Early Near Eastern Production, Subsistence and Environment 4, 625-40. Berlin: ex oriente.

Renfrew, A.C.
- 1974 Beyond a Subsistence Economy: The Evolution of Social Organizations in Prehistoric Europe. In C.M. Moore (ed.), *Reconstructing Complex Societies: An Archaeological Colloquium*, 69-95. Bulletin of the American Schools of Oriental Research Supplements 20. Cambridge, MA: American Schools of Oriental Research.

Renfrew, A.C. and J.F. Cherry
- 1986 *Peer Polity Interaction and Socio-Political Change*. Cambridge: Cambridge University Press.

Richard, S.L.
- 1987 The Early Bronze Age: The Rise and Collapse of Urbanism. *Biblical Archaeologist* 50: 22-43.

Richard, S.L. and R.S. Boraas
- 1984 Preliminary Report of the 1981–82 Season of the Expedition to Khirbet Iskander and its Vicinity. *Bulletin of the American Schools of Oriental Research* 254: 63-87.

Rollefson, G.O.
- 1993 PPNC Adaptations in the First Half of the Sixth Millennium B.C. *Paléorient* 19.1: 33-42.

Rosen, A.M.
- 1989 Environmental Change at the End of the Early Bronze Age. In P. de Miroschedji (ed.), *L'urbanisation de la Palestine à l'age du Bronze Ancien*, 247-55. British Archaeological Reports, International Series 527. Oxford: British Archaeological Reports.
- 1991 Early Bronze Age Tel Erani: An Environmental Perspective. *Tel Aviv* 18: 192-204.
- 1995 The Social Response to Environmental Change in Early Bronze Age Canaan. *Journal of Anthropological Archaeology* 14: 26-44.

Rosen, S.A.
- 1989b The Analysis of Early Bronze Age Chipped Stone Industries: A Summary Statement. In P. de Miroschedji (ed.), *L'urbanisation de la Palestine à l'age du Bronze Ancien*, 199-222. British Archaeological Reports, International Series 527. Oxford: British Archaeological Reports.
- 1993 Metals, Rocks, Specialization and the Beginning of Urbanism in the Northern Negev. In A. Biran and J. Aviram (eds.), *Biblical Archaeology Today 1990: Pre-Congress Symposium Supplement—Population, Production and Power*, 41-56. Jerusalem: Israel Exploration Society.
- 1997a *Lithics after the Stone Age: A Handbook of Stone Tools from the Levant*. London: Altamira.
- 1997b Craft Specialization and the Rise of Secondary Urbanism: A View from the Southern Levant. In W.E. Aufrecht, N.A Mirau and S.W. Gauley (eds.), *Urbanism in Antiquity: From Mesopotamia to Crete*, 82-91. Journal for the Study of the Old Testament Supplement Series 244. Sheffield: Sheffield Academic Press.

Rowan, D.
- 1994 The Early Bronze Age Ceramics from Area A. In D. Baird and G. Philip, Preliminary Report on the Third (1993) Season of Excavations at Tell esh-Shuna North. *Levant* 26: 124-29.

Sapin, J.
- 1985 Prospection géo-archéologique de l'Ajlûn 1981–82: example de recherche intégrante. In A. Hadidi (ed.), *Studies in the History and Archaeology of Jordan*, II: 217-27. Amman: Department of Antiquities.
- 1992 De l'occupation à l'utilisation de l'espace à l'aube de l'age du Bronze dans le région de Jerash et sa péripherie Orientale. In M. Zaghloul et al. (eds.), *Studies in the History and Archaeology of Jordan*, IV: 169-74. Amman: Department of Antiquities.

Schaub, R.T.
- 1981a Patterns of Burial at Bab edh-Dhra. In W.E. Rast and R.T. Schaub (eds.), *The Southeastern Dead Sea Plain Expedition: An Interim Report of the 1977 Season*, 45-68. Annual of the American Schools of Oriental Research 46. Cambridge, MA: American Schools of Oriental Research.
- 1981b Ceramic Sequences in the Tomb Groups at Bab edh-Dhra. In W.E. Rast and R.T. Schaub (eds.), *The Southeastern Dead Sea Plain Expedition: An Interim Report of the 1977 Season*, 69-118. Annual of the

American Schools of Oriental Research 46. Cambridge, MA: American Schools of Oriental Research.

1982 The Origins of the Early Bronze Age Walled Town Culture of Jordan. In A. Hadidi (ed.), *Studies in the History and Archaeology of Jordan*, II: 67-75. Amman: Department of Antiquities.

1987 Ceramic Vessels as Evidence for Trade and Communication during the Early Bronze Age in Jordan. In A. Hadidi (ed.), *Studies in the History and Archaeology of Jordan*, III: 247-50. Amman: Department of Antiquities.

1992 A Reassessment of Nelson Glueck on Settlement on the Jordan Plateau in Early Bronze III and IV. In M. Zaghloul et al. (eds.), *Studies in the History and Archaeology of Jordan*, IV: 161-68. Amman: Department of Antiquities.

1996 Pots as Containers. In J.D. Seger (ed.), *Retrieving the Past: Essays on Archaeological Research and Methodology in Honor of Gus W. Van Beek*, 231-43. Winona Lake, WI: Eisenbrauns.

Schaub, R.T. and W.E. Rast
1984 Preliminary Report of the 1981 Expedition to the Dead Sea Plain, Jordan. *Bulletin of the American Schools of Oriental Research* 254: 35-60.

1989 *Bab edh-Dhra: Excavations in the Cemetery Directed by Paul W. Lapp (1965-67)*. Reports of the Expedition to the Dead Sea Plain, Jordan, vol. I. Winona Lake, WI: Eisenbrauns.

Schnurrenberger, D.W.
1991 Preliminary Comments on the Geology of the Tell el-'Umeiri Region. In L.G. Herr et al. (eds.), *Madaba Plains Project 2: The 1987 Season at Tell el-'Umeiri and Vicinity and Subsequent Studies*. Berrien Springs, MI: Andrews University

Schwartz, G.M.
1994 Rural Economic Specialization and Early Urbanization in the Khabur Valley, Syria. In G.M. Schwartz and S.E. Falconer (eds.), *Archaeological Views from the Countryside: Village Communities in Early Complex Societies*, 19-36. Washington, DC: Smithsonian Institution.

Schwartz, G.M. and S.E. Falconer
1994 Rural Approaches to Social Complexity. In G.M. Schwartz and S.E. Falconer (eds.), *Archaeological Views from the Countryside: Village Communities in Early Complex Societies*, 1-9. Washington, DC: Smithsonian Institution.

Sebbane, M. et al.
1993 The Dating of the Early Bronze Age Settlements in the Negev and Sinai. *Tel Aviv* 20: 41-54.

Seger, J.
1989 Some Provisional Correlations in EB III Stratigraphy in Southern Palestine. In P. de Miroschedji (ed.), *L'urbanisation de la Palestine à l'age du Bronze Ancien*, 117-35. British Archaeological Reports, International Series 527. Oxford: British Archaeological Reports.

Serpico, M. and R. White
1996 A Report on the Analysis of the Contents of a Cache of Jars from the Tomb of Djer. In J. Spencer (ed.), *Aspects of Early Egypt*, 128-39. London: British Museum.

Service, E.
1962 *Primitive Social Organization*. New York: Random House.
1975 *Origins of the State and Civilizations*. New York: Norton.

Shalev, S.
1994 The Change in Metal Production from the Chalcolithic Period to the Early Bronze Age in Israel and Jordan. *Antiquity* 68: 630-37.

Shalev, S. and E. Braun.
1997 The Metal Objects from Yiftah'el. In E. Braun (ed.), *Yiftah'el: Salvage and Rescue Excavations at a Prehistoric Village in Lower Galilee, Israel*, 92-96. Jerusalem: Israel Antiquities Authority Reports No. 2.

Sherratt, A.
1981 Plough and Pastoralism: Aspects of the Secondary Products Revolution. In I. Hodder, G. Isaac and N. Hammond (eds.), *Pattern of the Past*, 261-305. Cambridge: Cambridge University Press.

Sherratt, A. and E. Sherratt
1991 From Luxuries to Commodities: The Nature of Mediterranean Bronze Age Trading Systems. In N.H. Gale (ed.), *Bronze Age Trade in the Mediterranean*, 351-86. Studies in Mediterranean Archaeology 90. Jonsered: Åströms Förlag.

Simmons, A.H. and Z. Kafafi
1992 The 'Ain Ghazzal Survey: Patterns of Settlement in the Greater Wadi az-Zarqa Area, Central Jordan. In M. Zaghloul et al. (eds.), *Studies in the History and Archaeology of Jordan*, IV: 77-81. Amman: Department of Antiquities.

Stager, L.
1985 The First Fruits of Civilization. In J.N. Tubb (ed.), *Palestine in the Bronze and Iron Ages: Papers in Honour of Olga Tufnell*, 172-88. London: Institute of Archaeology.

1992 The Periodization of Palestine from Neolithic through Early Bronze Times. In R.W. Ehrich (ed.), *Chronologies in Old World Archaeology*, 22-41. Chicago: University of Chicago Press.

Steele, C.S.
1990 Early Bronze Age Socio-Political Organization in Southwestern Asia. *Zeitschrift des Deutschen Palästina-Vereins* 106: 1-33.

Stein, G.
1994 Segmentary States and Organizational Variations in Early Complex Societies: A Rural Perspective. In G.

Schwartz and S.E. Falconer (eds.), *Archaeological Views from the Countryside: Village Communities in Early Complex Societies*, 10-18. Washington, DC: Smithsonian Institution.

Stein, G. and M.J. Blackman
1993 The Organizational Context of Specialised Craft Production in Early Mesopotamian States. *Research in Economic Anthropology* 14: 29-59.

Stein, G. and M. Rothman (eds.)
1994 *Chiefdoms and Early States in the Near East: The Organizational Dynamics of Complexity*. Madison, WI: Prehistory.

Stekelis, M.
1935 *Les monuments mégalithiques de Palestine*. Archives de l'Institute de Paléontologie Humaine Mémoire 15. Paris: Masson.
1961 *La Nécropolis Magalítica di Ala Safat Transjordania*. Barcelona.

Swauger, J.L.
1965 Study of Three Dolmen Sites in Jordan. *Annual of the Department of Antiquities of Jordan* 10: 3-36.

Sweet, L.
1960 *Tell Touqan: A Syrian Village*. Anthropological Papers of the Museum of Anthropology, University of Michigan No. 14. Ann Arbor: University of Michigan.

Thomas, J.
1991 *Rethinking the Neolithic*. Cambridge: Cambridge University Press.

Tilley, C.
1994 *A Phenomenology of Landscape: Places, Paths and Monuments*. Oxford: Berg.

Tubb, J.N. and P.G. Dorrell
1993 Tell es-Sa'idiyeh: Interim Report on the Sixth Season of Excavations. *Palestine Exploration Quarterly* 125: 50-74.
1994 Tell es-Sa'idiyeh 1993: Interim Report on the Seventh Season of Excavations. *Palestine Exploration Quarterly* 126: 52-67.

Turner, V.
1969 *The Ritual Process*. New York: Cornell University Press.

Upham, S. (ed.)
1990 *The Evolution of Political Systems: Sociopolitics in Small-scale Sedentary Societies*. Cambridge: Cambridge University Press.

Van den Brink, E.
1996 The Incised Serekh-Signs of Dynasties 0-1, Part 1: Complete Vessels. In J. Spencer (ed.), *Aspects of Early Egypt*, 140-58. London: British Museum.

Van den Brink, E.C.M. (ed.)
1992 *The Nile Delta in Transition: 4th–3rd Millennium* BC. Jerusalem: Israel Exploration Society.

Van Gennep, A.
1960 *The Rites of Passage*. Chicago: University of Chicago Press.

Vieweger, D.
1997 An Early Bronze Age Handle with Cylinder Seal Impression from Dayr Qiqub, Jordan. *Levant* 29: 147-52.

Vinitsky, L.
1992 The Date of the Dolmens in the Golan and the Galilee: A Reassessment. *Tel Aviv* 19: 100-12.

Waheeb, M.
1995 The First Season of the An-Naq' Project, Ghawr As-Safi. *Annual of the Department of Antiquities of Jordan* 39: 553-55.

Wapnish, P. and B. Hesse
1988 Urbanization and the Organization of Animal Production at Tell Jemmeh in the Middle Bronze Age Levant. *Journal of Near Eastern Studies* 47.2: 81-94.

Watson, P.
1996 Pella Hinterland Survey 1994: Preliminary Report. *Levant* 28: 63-76.

Wattenmaker, P.
1994 State Formation and the Organization of Domestic Craft Production at Third Millennium BC Kurban Höyük, Southeast Turkey. In G. Schwartz and S. Falconer (eds.), *Archaeological Views from the Countryside: Village Communities in Early Complex Societies*, 109-20. Washington, DC: Smithsonian Institution.

Webley, D.
1969 A Note on the Dolmen Field at Tell el-Adeimeh and Teleilat Ghassul. *Palestine Exploration Quarterly* 101: 42-43.

Wilcox, G.H.
1981 Plant Remains. In S.W. Helms, *Jawa: Lost City of the Black Desert*, 247-48. London: Methuen.

Wilkinson, A.
1989 Objects from the EB II and III Tombs. In R.T. Schaub and W.E. Rast, *Bab edh-Dhra: Excavations in the Cemetery Directed by Paul W. Lapp (1965–67)*, 444-72. Reports of the Expedition to the Dead Sea Plain, Jordan, vol. I. Winona Lake, WI: Eisenbrauns.

Wilkinson, T.A.H.
1996 *State Formation in Egypt: Chronology and Society*. British Archaeological Reports, International Series 651. Oxford: Tempus Reparatum.
1999 *Early Dynastic Egypt*. London: Routledge.

Wright, G.E.
1937 *The Pottery of Palestine from the Earliest Times to the End of the Early Bronze Age*. New Haven, CT: American Schools of Oriental Research.

1961 The Archaeology of Palestine. In G.E. Wright (ed.), *The Bible and the Ancient Near East: Essays in Honor of William Foxwell Albright*, 73-112. London: Routledge & Kegan Paul.

Wright, H.T.
1977 Recent Research on the Origin of the State. *Annual Review of Anthropology* 6: 379-97.

Wright, H.T. and G.A. Johnson
1975 Population, Exchange and Early State Formation in South Western Iran. *American Anthropologist* 77: 267-91.

Wright, K.I.
1993 Early Holocene Ground Stone Assemblages in the Levant. *Levant* 25: 93-111.

Yassine, K.
1985 The Dolmens: Construction and Dating Reconsidered. *Bulletin of the American Schools of Oriental Research* 259: 63-69.
1988 *Archaeology of Jordan: Essays and Reports*. Amman: Department of Archaeology, University of Jordan.

Yoffee, N.
1993 Too Many Chiefs? (or, Safe Texts for the 90's). In N. Yoffee and A. Sherratt (eds.), *Archaeological Theory: Who Sets the Agenda*, 60-78. Cambridge: Cambridge University Press.

Zarzecki-Peleg, A.
1993 Decorated Bones of the Third Millennium B.C.E. from Palestine and Syria: Stylistic Analysis. *Israel Exploration Journal* 43: 1-22.

Zohar, M.
1992 Megalithic Cemeteries in the Levant. In O. Bar-Yosef and A.M. Khazanov (eds.), *Pastoralism in the Levant: Archaeological Materials in Anthropological Perspectives*, 43-63. Madison, WI: Prehistory.

Zohary, D. and M. Hopf
1993 *Domestication of Plants in the Old World: The Origins and Spread of Cultivated Plants in Western Asia, Europe and the Nile Valley*. 2nd edn. Oxford: Clarendon Press.

6. The Early Bronze Age IV

Gaetano Palumbo

Introduction

The Early Bronze Age IV (c. 2350–2000 BC) is characterized in Jordan by an extensive occupation of both fertile and arid environments, in what seems to be an essentially rural occupation of the landscape, and with a typical absence of urban centres. This characteristic makes the entire period a 'rural interlude' between the Early Bronze II–III and the Middle Bronze Ages, periods when the urban element of those societies appeared to be much stronger. The same situation is reflected in Israel and Palestine, as well as southern Lebanon and Syria, while in northern Syria large urban centres (such as Ebla) flourished precisely during the last quarter of the third millennium BC.

The cause of this apparent sudden demise of the Early Bronze Age urban environment was sought in the past in the arrival of nomadic groups from the desert fringes and the Syrian steppes (identified with the Amorites). A combination of new archaeological data from surveys and excavations and a different interpretative perspective, centred on socioeconomic factors rather than political episodes, has modified the nomadic invasion theory in favour of an approach oriented to a model of 'ruralization' of the cultural landscape of the Levant. This 'ruralization' can be interpreted as an answer to the internal crisis of the urban structure that dominated the cultural landscape of the Levant during the first seven centuries of the third millennium BC (Dever 1995; Palumbo 1990).

Terminology and theoretical models

Early Bronze Age IV (EB IV) is only one of several 'labels' used to identify the chronological and cultural phase covering approximately the last three centuries of the third millennium BC. The adoption of the EB IV terminology stresses the idea that, especially in Jordan, this period seems to have more important links with the previous, rather than the following periods. The phase was recognized as early as 1932 by Albright (1932), who called it Middle Bronze (MB) I and placed it in absolute dates between 2100 and 1900 BC. Glueck, in his *Survey of Eastern Palestine* (1934, 1935, 1939, 1951), followed the same terminology, while Wright introduced the EB IV terminology for the first time, identifying a phase immediately preceding Albright's MB I, which he recognized at several locations, including Tomb 351 at Jericho, and the assemblage at Wadi Ghazzeh site H (Wright 1937: 78-81). None of the complexes Wright mentions are earlier than Albright's 'MB I', with the exception of Tomb 351 at Jericho, and stratum J at Tall Beit Mirsim, both rather late EB III in date, and Wadi Ghazzeh site H, which dates to Late Chalcolithic/Early Bronze I (Dever 1973: 39). With the excavations at Jericho by Kenyon came also the formulation of the nomadic invasion theory and the creation of another term to describe the period: 'Intermediate Early Bronze–Middle Bronze' (EB–MB) (Kenyon 1951: 106; 1956: 41-42). Kenyon's 'Amorite hypothesis', which sees in the arrival of populations from the Syrian steppes the cause for the collapse of urbanization at the end of EB III in Palestine, met the favour of many scholars. The theory, however, was not completely original, since already in 1938 Wright had expressed an opinion linking the Amorites with the changes observed between the EB III and the following periods. This opinion was also repeated by Albright (1940) and De Vaux (1946), but it was Kenyon who linked the archaeological evidence found at several late third-millennium cemeteries in Palestine and formulated the 'Amorite hypothesis' that was at the base of the nomadic model until recently. The core of the theory claims that the Amorites, a Semitic, non-indigenous population, entered Palestine and occupied the hill country, while the native Canaaneans, who were responsible for the urban Early Bronze Age civilization, were confined to the coastal plain (Kenyon 1956: 41). Very little is said about the regions east of the Jordan by Kenyon, but the absence of excavations, with the exception of Bab adh-Dhra', could not help her in establishing the facts for her theory. At Jericho she compiled a strict typology of tomb types that she attributed to separate ethnic groups (Kenyon 1960: 180-259; 1965: 33-161).

While some scholars adhered to Kenyon's 'pure' invasion theory (Lapp 1966: 111-16;[1] Kochavi 1963a: 141-42), many others adopted a more moderate view, where infiltrations of nomadic groups, rather than invasions, are at the base of the changes observed in the archaeological record (Amiran 1960: 224-25; Dever 1970: 145; 1971: 211-25; Prag 1974: 106-107; Tufnell 1958: 41-42).

While not clearly supporting or denying the 'Amorite hypothesis', Amiran added an interesting contribution to the debate, identifying a series of pottery groups, or 'families', which she named in a variety of ways and dated to different phases within the period. She finally defined her position in a 1974 paper where she presented three basic families: the Southern Group, the Northern Group and the Bethel Group. In contrast with previous positions, she saw these families as largely contemporary (1974: 2*).

The acceptance of this basic division by many scholars is connected with its strong emphasis on regionalism, which is still today understood as one of the most distinct characteristics of this period's material culture.

Between 1971 and 1973 Dever expanded the 'families' subdivision pioneered by Amiran, improving it by taking into consideration the entire material culture of the regions investigated, and not only the pottery typology. Dever recognized seven families: Northern (N); North-Central (NC); Jericho/Jordan Valley (J); Southern (S); Central Hills (CH); Coastal (C); and Jordan (TR). He did not believe in rigid boundaries between the groups, and thought that the families largely overlapped chronologically, even if he dated the TR to an earlier phase than the N, NC, J and CH, and put the S to the terminal phase of the period. He rejected his original MB I terminology to adopt the EB IV terminology, which he also subdivided into three phases (A, B, C) (Figure 6.3).

In 1980 Dever basically confirmed this division and chronology, formulating his theory on pastoral nomadism. In developing this theory, he adopted the fundamental studies by Liverani (1970, 1973) and Rowton (1967), where the historicity of the Amorites was put in doubt by stringent arguments denying the existence of nomadic groups originating in northern Syria, where instead a sophisticated urban culture and a high degree of sedentarization existed at the end of the third millennium BC.

Liverani showed, in fact, that our knowledge of the presence of 'savage nomads' in Syria derived from the propaganda of the Mesopotamian cities that described the MAR-TU as people 'not knowing houses, not knowing cities', 'eating raw meat, they do not have houses, and when they die, they are not buried', 'tent dwellers'.[2] But these nomads were only the pastoral groups living at the fringe of the desert, sometimes pushing their way into the fertile grounds of Mesopotamia: the population in Syria during the last quarter of the third millennium BC mainly included urban dwellers and agricultural communities. Another argument pushed forward by Liverani was that the kind of nomadism represented in Kenyon's Amorite hypothesis is of the historical Central Asian 'type' (Patai 1951; Bacon 1954; Khazanov 1978; 1984: 44-53), not applicable to Syria-Palestine, where nomadism is of a pastoral type characterized by short movements and transhumance (Fabietti 1982; Khazanov 1984: 53-59).[3]

The concept of 'dimorphic society', introduced by Liverani to describe the social structure of Syria during this period, derives from Rowton's reconstruction of Old Babylonian society at Mari. Rowton used this term to describe the characteristics of a social web where the nomadic/pastoral and the agricultural/sedentary elements are in symbiosis. In the case of Mari's society during the eighteenth century BC, and northern Syrian society at the end of the third millennium BC as analysed by Liverani, the sedentary component was urbanized, adding another variable to the complex system of social relations (Rowton 1967).[4]

The concept of dimorphic society and a more careful look at the phenomenon of pastoral nomadism in ancient Middle Eastern societies are at the basis of a revival of theoretical studies, which can be placed in the decade between the mid-1970s and the mid-1980s, concerning the Early Bronze Age IV. Dever, for example, moved from the MB I to the EB IV terminology, changing also his theoretical perspective. From a moderate version of the 'Amorite hypothesis', that looked to the Jabal al-Bishri area in central Syria for the source of the nomadic invasions, he came to support the pastoral nomadism model and the concept of a 'dimorphic society', describing its effects as 'the brief triumph of the desert over the sown' (1980: 58).

This model of pastoral nomadism sees wider mobility by part of the nomads and access to 'forbidden' areas formerly controlled by the central power as a consequence of the disintegration of the Early Bronze Age urban structure.[5] The new type of settlement, based on seasonal transhumance, prevailed throughout the EB IV according to Dever, especially in the more marginal zones of Palestine and Jordan. He conceded, however, that greater settlement stability could

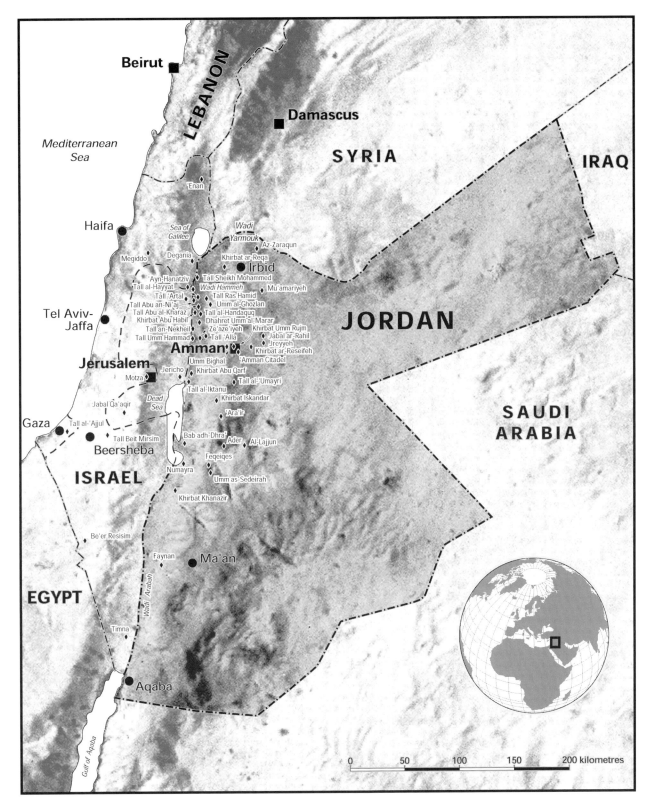

Figure 6.1. Map of sites mentioned in the text.

have existed in such areas as the Jordan Valley. Dever applied this model to an area he designated the 'Hebron Hills–Negev Complex', where he conducted a series of archaeological surveys and excavations at Jabal Qa'aqir (Dever 1972, 1981; Gitin 1975) and Khirbat Kirmil (Dever 1975a) in the Hebron Hills, and at Be'er Resisim in the Central Negev (Cohen and Dever 1978, 1979, 1981; Dever 1983, 1985). Dever interpreted the sites of the Hebron area as summer camps and cemeteries of semi-nomadic populations who spent their winter months in the Central Negev (1980: 57).

Thompson (1975, 1979) believes not only that the continuity of cultural traditions and settlement between the Early Bronze Age I–III and the EB IV are substantial, but also that most of the settlement during EB IV was sedentary and based on agriculture. His extreme position, however, was based on an acritical reading of the site lists and showed little familiarity with both the archaeological evidence and the various types of nomadism found in the region.

In the same years, Prag developed her view of the EB IV as a period of change derived not from invasions, but from infiltrations of nomadic groups or true 'pastoral migrations' from Syria toward the main valleys (Huleh, Jordan and Jezreel). For Prag, these nomadic movements were also a consequence of the decline of the Early Bronze Age urban civilization (Prag 1984: 67, 1985).

In the 1980s and 1990s, several excavations and survey data have helped to advance the understanding of the period.

Falconer and Magness-Gardiner (1984, 1989) based their observations on the data coming from two excavations they conducted in the Jordan Valley, at Tall al-Hayyat and Tall Abu an-Niʻaj. The variety of subsistence strategies identified in their excavations suggested that the model of pastoral nomadism cannot serve, alone, as an explanation for the type of mixed economic solutions adopted by the EB IV people. In his dissertation, Falconer (1987a) expanded this intuition in a model that sees pastoralism as part of a rural economic system, formed by various components, sometimes in balance, other times in conflict, but basically representing an element of stability in a region where cities could not develop into the well-organized structure visible in Mesopotamia or Syria. Dever called this the 'rural-nomadic' model (1992: 88). With different nuances, it is still the model most accepted today.

For Richard and Long (who based their work on the results of their excavations at Khirbat Iskandar, the only true 'urban' EB IV settlement excavated so far in the southern Levant[6]) the importance of pastoral nomadism was limited to local or regional boundaries. Richard sees, in the change occurring at the end of Early Bronze Age III in the society's organization and production subsystems, a reaction to an irreversible crisis in the larger urban system. The direction of change is not from a sedentary to a nomadic way of life, but from an urban to a non-urban and pastoral production system, the latter being more typical of villages and small centres (Richard and Boraas 1988; Richard 1987). Her model is based on the theory of chiefdoms as described by Service (1962), but this adaptation has been unconvincing so far.

In another contribution, Long (1986) and Richard and Long (1989) adapted a model formulated by Bates and Lees (1977: 837-39), which suggests that decline in the specialization of an economic system is followed by decline in the agricultural and pastoral modes of production. Instead of fluctuation between different types of subsistence strategies, such as agriculture and pastoralism, fluctuation should be seen in terms of different degrees of specialization. In this case, rather than shifts from one subsistence strategy to another, as implied by the pastoral nomadism model, the whole system would have been more amalgamated and less specialized.

More sophisticated than Long and Richard's contribution is the model formulated by Finkelstein (1989). He goes beyond the 'dimorphic society' approach, introducing a 'multimorphic' approach, which would better explain the causes of urban civilization's collapse. While he believes that the EB IV period in the southern Levant is characterized by the collapse of the specialization system (following, even if not explicitly, Bates and Lees), he stresses, quite rightly, the importance of the marginal nomadic or semi-nomadic pastoral groups for the understanding of this period, and identifies in the 'visibility' of these groups, from an archaeological point of view, the consequence of a partial sedentarization dictated not by a political power, but by more autonomous productive strategies, since no agricultural surplus was available any more from the sedentary communities (Finkelstein 1989: 135-36). Between 1992 and 1995 Finkelstein further refined his theory, adopting Lemche's (1985) definition of 'polymorphous society'. By doing so he also denied the chronological variations seen by Dever in his pottery 'families' and added trade as one of the factors that encouraged the sedentarization of pastoral nomads, citing the evidence provided by recent work in the copper-mining area of Faynan. He also stressed the difficulty in tracing the 'archaeological line' between agriculturalists practising herding and herders practising agriculture, especially in areas of traditional settlement (Finkelstein 1992: 134; 1995: 99).

My general work on the EB IV of the southern Levant (Palumbo 1990) introduced the concept of 'diversified levels of specialization', following

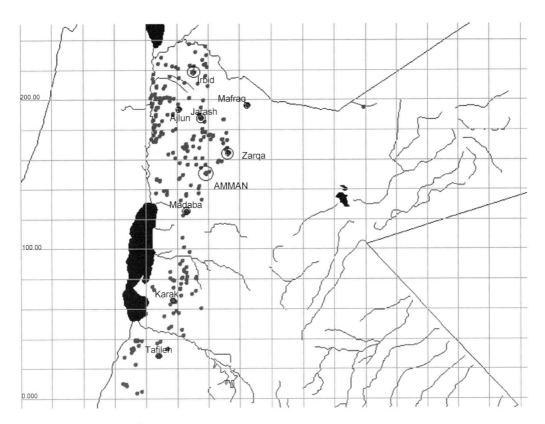

Figure 6.2. Map of all EB IV sites in Jordan.

Marfoe's approach (1979). This is an improvement over both the dimorphic/multimorphic and the specialization/despecialization models, in that it is not a 'linear' model, but explains variation in terms of 'shifts along a spectrum of available economic strategies and sociocultural roles' (Marfoe 1979: 8). In this perspective, the entire Early Bronze–Middle Bronze transition can be explained in terms of adaptations to changing physical and political conditions. Both the Early Bronze urban 'collapse' and the Middle Bronze urban rise can be explained in this context, but the model has a special significance in explaining the variety of adaptive strategies that are observed throughout the southern Levant during the last quarter of the third millennium BC. It is the rural, rather than the pastoral aspect of production that really differentiates this period from the more 'urban' EB II–III and Middle Bronze periods, and it is this 'rural response' to urban collapse that underlines the capability of 'ruralism' to survive stress and continue virtually unchanged through periods of urban control (Palumbo 1990: 130-31).

Similar conclusions were reached by Esse (1991: 175), who, in his analysis of the evolution of the cultural, economic and social landscape of northern Palestine during the Early Bronze Age, maintained that throughout this period the entire spectrum of adaptive strategies existed: what changed from one period to another was their balance, and in EB IV he sees the dominance of the pastoral component.

LaBianca (1990) observed in the data gathered during the Hisban/Madaba Plains project a recurrence of phases in which settlement is more stable against phases in which it is more dispersed, following responses to changing political or environmental circumstances. He called these 'cycles of intensification and abatement', and they can be easily adapted to explain what happened in the region between the Early and the Middle Bronze Ages.

This stress on the 'rurality' of the Early and Middle Bronze Age cultures of the southern Levant has been partially accepted by Dever in his latest contributions to this argument (1992, 1995). There, Dever admits that the data from Jordan support a more diversified economic environment, but he stresses that regionalism and pastoral nomadism are better explanations for the situation in Cisjordan. In a revision of the models concerning the EB IV, he concludes that the archaeological evidence is not sufficient to test the applicability of chiefdom models (Service 1962; Carneiro 1981) to the situation in the southern Levant during the EB IV. He rejects the model of 'multimorphic society' as formulated by Finkelstein, finding it excessively particularizing. Dever prefers to resort to a tribal model that can explain the pastoral nomadic social structure of such regions as the Hebron Hills, the regionalism evidenced in various aspects of the material culture, and the presence of separate groups in cemeteries, such as Jericho and Jabal Qa'aqir. As for the rural agricultural villages, Dever speaks about a 'rural' model that incorporates pastoral nomadism. He sees the villages as an outgrowth of tribal societies, clan-like conglomerates with a largely egalitarian social structure. According to Dever, the EB IV is a period in which the strong regional component is the reflection of a social division based on semi-nomadic tribal groups that partly sedentarize, creating new economic and social units. Dever agrees with Joffe's (1991) view of 'alternating phases of low-level and higher-level organization, a pattern of socio-political complexity spiralling upwards and downwards around the "domestic mode of production"' as being the essential characteristic of social evolution in this region (Dever 1992, 1995). Finally, while not directly related to the EB IV of the southern Levant, a series of articles written by Buccellati on the problem of the identification and role of the Amorites in third-millennium societies of Syria has some methodological relevance here. His reconstruction of the rural landscape of Terqa, Mari, and Ebla (1990a, 1990b, 1992, 1993) sees the Amorites as an anomalous rural class, independent from the urban centres, but substantially Mesopotamian in character. Defining the role of the Amorites, Buccellati identifies in them a class of peasants/herders, a consequence of 'nomadization of the peasants' that occupies a precise niche in the ecological landscape of the upper Euphrates. This approach is in contrast to Rowton's symbiotic model, and introduces a new point of view in the definition of 'ruralism' in the emerging urban societies of the third millennium BC. Even more radical is the contribution to the debate presented by Marx (1992), who disputes the existence of pastoral nomads in the Levant as a separate entity, since they are an integral part of the economy and structure of urban societies.

In summary, after 20 years of pastoral nomadic/rural models, there is still no real consensus on the historical and social picture of this period. It is clear, however, that the most promising approach in explaining the crisis of the urban element towards the end of the third millennium BC is in the reconstruction of the economic and social role of rural settlement and subsistence strategies in an urban environment.

Chronology

While the absolute date of the entire period is commonly accepted (2350/2300–2000 BC), roughly corresponding to the First Intermediate Period in Egypt, and the end of the Akkad dynasty and the Ur III period in Mesopotamia, the internal division of the period is still open to debate.

The tripartite division proposed by Dever (1980) was until recently the most accepted one. According to this division, EB IVA, dated between 2350 and 2200 BC, is only represented by Family TR, or Transjordan. EB IVB, dated between 2200 and 2100 BC, is represented by Families N (North), NC (North-Central), J (Jordan Valley), and CH (Central Hills), the latter going into EB IVC (2100–2000 BC), mostly represented by Family S (Southern) (Figure 6.3). The challenge to this view comes from several excavations recently conducted in Jordan, such as Falconer and Magness-Gardiner's Tall al-Hayyat and Tall Abu an-Ni'aj, Richard, Boraas and Long's Khirbat Iskandar, and Prag's Tall Iktanu.

Falconer and Magness-Gardiner's tentative solution to the problem of apparent overlap of Families traditions was to consider Tall al-Hayyat as on the border zone between two different but contemporary pottery traditions (Families J and NC) (Falconer and Magness-Gardiner 1984: 57-58). Helms (1986: 46) not only sees the Southern Family as contemporary to most of the other regional groups,[7] but also recommends not a tripartite but a bipartite division (Helms 1989: 32).[8] At Khirbat Iskandar red-slipped wares (typical of the Transjordan Family) were found in stratigraphic association with pottery types and decorations of the Southern Family (Richard and Boraas 1988: 124),[9] and handleless storage jars were found together with folded 'envelope' ledge-handle jars, two 'signatures' that cannot appear together in Dever's classification. For Richard and Boraas, this is due not to chronological differences, but to regional variations in contemporaneous pottery traditions.

At Tall Iktanu there are some substantial differences between the two archaeological phases (1 and 2, 1 being the most ancient) recognized by Prag. Totally absent in Phase 2 are the red-slipped ware and the large bowls with grooves under the inverted rim, the most common type in southern Transjordanian repertoires, which is missing in the repertoires of sites north of the az-Zarqa' River. In Phase 1 band-combed and painted wares are lacking (they are not common in Phase 2, either).[10]

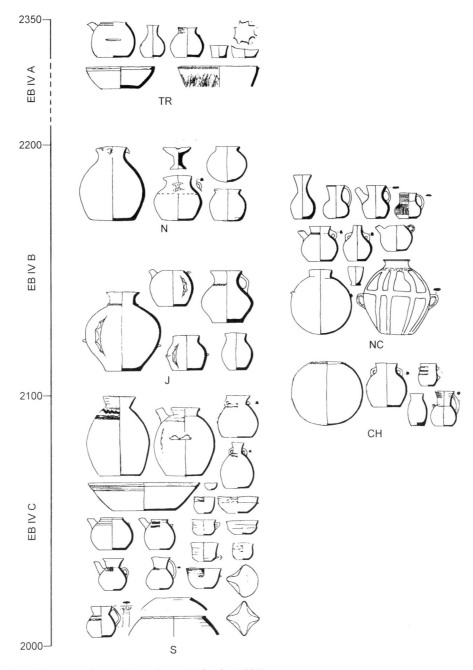

Figure 6.3. Pottery chronology according to Dever (source Palumbo, 1990).

Besides these characteristics, there are also some pottery types or decorations prevailing in one phase or in the other, explained by Prag in chronological terms. This includes the presence of incised decoration on Phase 1 jars, and the differences in the folded 'envelope' ledge-handle types, which in Phase 1 tend to have smaller and spaced flaps, while in Phase 2 they are usually larger and overlapping. Prag tried to establish some correlations with other sites in Transjordan and with the Jericho tell and tombs using this sequence (1974: 77-81; 1986). These correlations seem to be valid mainly for the southern sites, such as 'Ara'ir and Ader, and much less for the sites north of the az-Zarqa' River, whose repertoires frequently diverge from those at Iktanu.

The four stages (5–8) identified by Helms at Tall Umm Hammad can also be divided into two main periods. Stage 5 is characterized by pottery types defined by Helms as the 'regional equivalent of the red burnished, rilled, and other wares described as EB IVA', some of which are paralleled at Tall Iktanu, Phase 1 (Helms 1989: 29). In stages 6–8, the pottery is mainly of EB IVB types, but includes also EB IVC (i.e. Southern Family) types. The earlier examples of the pottery from these stages are paralleled by types known at Tall Iktanu, Phase 2 (Helms 1989: 29).

It is not clear whether the long occupational sequence (five phases) at Khirbat Iskandar is also reflected by several pottery phases, or whether the stylistic evolution at that site is more continuous than at Iktanu. Mixed pottery styles were found in the B8–B7 storerooms at this site (Richard and Boraas 1988: 124-27), including red-slipped and burnished vessels, considered among the earlier EB IV types, together with unslipped grey-ware types with incised or combed decoration, similar to Dever's Family S types.

Two phases were also identified by Olávarri at 'Ara'ir (niveaux VIb and VIa), so distinct in their characteristics that they were assigned to the 'Intermediate Bronze I and II', respectively (Olávarri 1969).[11] Levels VIb and VIa roughly correspond to Phases 1 and 2 at Tall Iktanu as shown by parallels in the pottery repertoire.

From the evidence that is now accumulating after the most recent, focused archaeological projects, it is clear that Dever's Families will turn out to be largely contemporary, or with important overlappings. The geographic 'areas of influence' of Families are also being reconsidered in the light of the contemporary presence of different Families at several sites.

Settlement patterns

One of the reasons for the reconstruction of the EB IV as a period of pastoral nomadic way of life has been the lack of clear 'settlements' to be dated to this period, and, on the other hand, a very large number of cemeteries without an apparent association with occupation layers at nearby sites, which often seem to be abandoned during EB II–III or occupied only during the beginning of the Middle Bronze Age.

This very general picture does not correspond to the facts, especially in Jordan. Over the past 15 years, a number of survey and excavation projects has tackled the problem of EB IV settlement, succeeding in the identification of a myriad of little rural sites, a few villages, and at least one 'town': Khirbat Iskandar. My general analysis of the EB IV in the southern Levant (Palumbo 1990) collected information on 258 sites in Jordan. Of these, only 29 were cemeteries, a pattern opposite to the tenet of the pastoral nomadic model, according to which the cemeteries are the only 'visible' archaeological sites of a semi-nomadic culture. Another small number of sites (nine) had both surface or stratified remains and a cemetery or tombs. Finally, only 20 of the total number of sites could be defined as a 'settlement', on the basis of either excavated remains or the importance of the material culture record found. At that time, only nine sites with consistent EB IV remains had been excavated, producing also a stratigraphic sequence.

These sites were Tall Abu an-Ni'aj, Ader, 'Ara'ir, Bab adh-Dhra' (area X), Tall al-Hayyat, Tall Iktanu, Khirbat Iskandar, Tall Umm Hammad, and az-Zaraqun.[12] Since then, work has continued at some of those sites (Tall Abu an-Ni'aj, Tall Iktanu, Khirbat Iskandar, and az-Zaraqun), and started at others (Jabal ar-Rahil).

The JADIS (Jordan Antiquities Database and Information System) database, developed from 1990 to 1994 at the Department of Antiquities of Jordan, has allowed a more thorough investigation of published and unpublished records, and the addition of information from recent archaeological projects. While only a handful of these sites have been confirmed as having occupational remains of the period, a look at the distribution of EB IV findspots in Jordan is useful to observe possible variations from the previous and following periods. As a matter of fact, this variation is quite evident in comparing both absolute numbers and the distributional charts. A general pattern of site dispersion is evident at the end of the EB II–III period: the absolute number of sites increases from EB III to EB IV, and decreases again from EB IV to MB IIA. From the more general perspective of EB II–III and the Middle Bronze Age, however, the number of sites is actually less in EB IV than in the other periods. This has led many scholars to conclude that the EB IV marked a period of site abandonment and nomadization. In reality, it is only the site size that drops drastically from EB II–III to EB IV: in the Madaba region median site size drops from 4.4 to 2.3 ha (Harrison 1997: 17), and in northern Palestine the drop is even more dramatic: from 3.2 to 0.84 ha (Esse 1991: 151-52).

On the contrary, the occupation of EB III and MB IIA sites shows a marked association with EB IV presence: both in the Jordan Valley and northern Jordan (north of the az-Zarqa' River), for example, about 45% of the EB III sites are abandoned, and 50% or more continue into the EB IV. The same proportion is also valid for the MB IIA, with 45% of the sites not showing EB IV presence, against over 50% with both EB IV and MB IIA remains. It is interesting to note that this 'smooth' transition is especially valid in Jordan, while in Cisjordan there are drastic variations from region to region.[13]

This phenomenon could be explained in two ways: 'ground visibility' or 'archaeological visibility' of pastoral sites. 'Ground visibility' of ephemeral sites, such as pastoral camps, is much higher in desert areas than in cultivated or vegetated zones. The 'archaeological visibility' of pastoral sites instead derives from a shift in economic and productive strategies, which in periods of crisis become more varied and less mobile. This means that sites that were previously occupied for brief periods, thus leaving scant or no traces, are now settled for longer periods of time, leaving a more evident record (Finkelstein 1989: 135; 1995: 98-100).

The pattern of settlement observed in Jordan shows some differences between the valleys and the plateaus, and between northern and southern regions. These differences are also reflected in the material culture, especially in the pottery types found on the surface or deposits of settlements and cemeteries.

Jordan Valley

Settlements on the east bank of the Jordan Valley are of various types, including open sites such as Tall Abu an-Ni'aj (Figure 6.4), Tall al-Hayyat, Tall Umm Hammad, Ze'aze'iyyeh (Ibrahim *et al.* 1988: 191, site 144), all situated on low ground, or Tall Iktanu (Prag 1974, 1986, 1988, 1989), located on higher ground, but probably not for defensive purposes. A site located on a hilltop, probably in a more defensive location, is Dhahret Umm al-Marar (Ibrahim *et al.* 1976: 50), where possible traces of fortifications are visible on the surface. Their common characteristic is proximity to a spring or a watercourse, and the presence of good agricultural soils in their vicinity. Traces of less permanent occupation are found at sites with more consistent remains from both earlier and more recent periods, such as Tall Abu Alubah (Mabry and Palumbo 1988: site 23), Tall Abu Habil North, Tall Abu al-Kharaz, Tall Abu Qarf, Tall 'Alla (near Tall Dayr 'Alla), Tall al-Handaquq (Mabry 1989), and others.[14]

It is not rare to find cemeteries (characterized by shaft and cist tombs) associated with 'permanent' sites: the best examples illustrating this situation are given by the sites of Tall Abu an-Ni'aj and Tall Umm Hammad (Helms 1983, 1986; Tubb 1985).

Figure 6.4. Tall Abu en-Ni'aj: surface pottery scatter.

Northern Jordan

One hundred and twenty-three sites are known from this region, but unfortunately it is not easy to evaluate the real intensity of the settlement, since almost all evidence comes from surface surveys or accidental discoveries.[15] Few sites in this region have been excavated: at Khirbat az-Zaraqun the EB II and III settlement was fortified, but during the EB IV a sedentary village with well-built houses existed on this site. At Jabal ar-Rahil, soundings revealed both EB II and EB IV phases (Palumbo *et al.* 1996). Several other sites, many of which are situated on hilltops, have substantial remains visible on the surface. Hilltop location predominates in this region: 44% of the habitation sites are located in this position and another 26% on slopes or ridges.

Some of these sites were probably even fortified, as in the cases of Jabal ar-Rahil (Palumbo *et al.* 1996) (Figures 6.5, 6.6), ar-Reseifeh (Glueck 1939), and Jreyyeh (Palumbo *et al.* 1996) or at least surrounded

Figure 6.5. General view of Jabal ar-Rahil, looking east from Jabal al-'Asi.

by enclosure walls. Ar-Reseifeh was unfortunately destroyed by the cemetery of the nearby village. Both Albright and Glueck describe the site as clearly fortified (Albright 1934: 15; Glueck 1939: 205-206), with many traces of walls visible on the surface. Their reading of the pottery as EB IV has been confirmed by the discovery of a tomb, and by the collection of surface material (Palumbo and Peterman 1993).

The presence of these sites in this area of Jordan is very interesting. They are all close to the az-Zarqa' River, a permanent watercourse. They are also very close to the 200 mm isohyet, which marks the minimum amount of rainfall for the practice of dry farming. These sites could be examples of defended settlements in an area where the relationships between sedentary and nomadic components of the society were probably less peaceful than in other regions of the country. In this same region, there are also clear examples of sedentary rural settlements in open areas (valleys, plateaus), showing that in some cases there was no need either to occupy defended positions or to compete for the land.[16]

There are 12 cemeteries or tomb locations known in northern Jordan, most in the 'Amman area. Unfortunately, the urban growth of the capital does not allow us to understand whether these cemeteries were associated with settlements. Occupational remains have been identified so far only on the 'Amman Citadel.

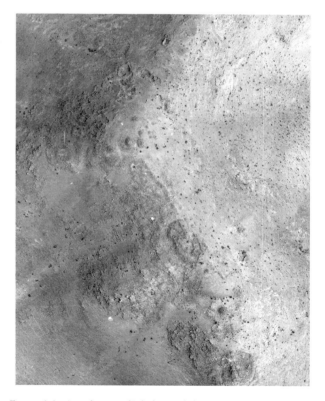

Figure 6.6. Aerial view of Jabal ar-Rahil.

In conclusion, it is clear that occupation in northern Jordan had an agricultural basis. Sedentary villages played an important role in this economy, even if the pastoral component of the society was probably very strong, especially along the eastern fringes

Figure 6.7. General view of Khirbat Iskandar, looking north.

of the plateau, where the limited amount of rainfall and the absence of permanent water courses means that the exploitation of natural resources is limited to pasture.

Central and southern Jordan

In central Jordan the continuity of settlement between the EB II/III and IV is very clear. The continuity in types of settlement is especially attested by the ongoing urban tradition in the EB IV culture. Khirbat Iskandar is the only excavated EB IV site with fortifications built or at least reused during this period (Figure 6.7). Richard believes that the fortifications at Iskandar went out of use during a later phase of the EB IV settlement.[17] It confirms that in this region certain conditions existed allowing the survival of EB II/III cultural traditions, which flourished only in the first phases of the EB IV period and died out gradually towards the end of the millennium. It is not surprising that the first and most unequivocal evidence of EB IV sedentary occupation comes from this region. Glueck had already ascertained the importance of sites like Khirbat Iskandar, 'Ara'ir, Ader, Bab adh-Dhra', Feqeiqes and al-Lajjun (Figure 6.8) for the occupational history of the region during the EB IV. The topographic location of the sites seems to favour a defensive attitude with 38% of the sites on hilltops or ridges, and another 25% on slopes. Only 24% are in a more open position on the plateau or on terraces. Only a few cemeteries (four) have been found in this region. This is probably due to the small number of accurate surveys made in the region.

Human settlement in central Jordan during the EB IV had an important agricultural and sedentary base. Khirbat Iskandar cannot be considered an exception to the rule of undefended villages in Jordan during the EB IV. New research is needed to understand both the pattern of fortified settlements and their significance in a political and economic structure based almost exclusively on local rural organization. But in this case, too, we cannot rule out the idea that Iskandar, and probably Ader and al-Lajjun, could have been centres of sedentary rural activity 'on the border line' with a more turbulent and unstable desert region. The presence of pastoral settlements in this region can probably be identified at several sites with traces of EB IV occupation but no architectural remains, found especially in south-central Jordan by Mattingly and his survey team (1984).

In southern Jordan the situation is rather different. MB II presence has been tentatively identified by MacDonald *et al.* at only two of the 20 EB IV sites known in this region (Rababeh: 1982: 44; SGNAS Site 198: 1992). Only one site, Khanazir, was occupied both in the EB II/III and EB IV periods. At Khanazir, the EB IV presence is characterized exclusively by tombs and no trace of EB IV occupation has been identified by Rast and Schaub on the tell proper (1974: 12-13).

Figure 6.8. Menhir alignment at Lejjun.

The few sites found do not permit statistics on their topographic location, but the trend in southern Jordan seems to contrast with the situation observed in central Jordan, where defence was highly valued in the choice of settlement location. In southern Jordan only two out of 13 sites are on high locations, all the others on lower slopes, terraces, plateaus or even valley bottoms. Very few surveys have been made in the southernmost areas of this region. The writer's 1990 summary work on EB IV had expressed the hope that more surveys in the southernmost regions of Jordan could clarify the apparently complete absence of EB IV occupation only a few kilometres south of Wadi al-Hasa.[18] The work of Adams in the Faynan area has started to modify this picture, confirming an active exploitation of the copper mines during the EB IV (pers. comm.). Haiman (1996) and Goren (1996) go as far as relating the EB IV settlement of the Negev and northern Sinai to the copper trade from the Faynan area to Egypt. The petrographic analysis of pottery found at sites in the Negev conducted by Goren shows two main sources: the Judaean hills south of Jerusalem and central/southern Jordan (including the Faynan area). This second group makes up 38% of the entire EB IV ceramic repertoire found in the Negev so far. With the percentage of 'foreign' imports reaching 80% in the Negev, half of which originated in Jordan, it is clear that the transhumance model proposed by Dever must be reassessed (in his key site, Be'er Resisim, the majority of the pottery has a Jordanian origin).

Besides the sites in the Faynan area, the only other important EB IV site in southern Jordan is Khirbat Khanazir (Abu Ishribsheh), a large cemetery spread out along the base of hills above the 'Arabah Valley for about 2 km (MacDonald et al. 1987: 406; MacDonald 1995; Khouri 1988: 113). Surface remains consist of exclusively EB IV pottery and of rectangular tombs with very well-preserved walls up to 1 m high. Surveys have so far failed to reveal the presence of settlements in the vicinity. This site should perhaps be interpreted as a seasonal settlement, no different from most of the Negev sites.

Material culture

Public and domestic architecture

With few exceptions, there is no real evidence of urban planning at EB IV settlements, even if layouts can be recognized at sites with more important architectural remains. Even the domestic architecture cannot be easily characterized, given the range, but also the scarcity, of architectural remains. Substantial remains, however, do exist, especially in Jordan. It is on the basis of these remains that it has been possible to recognize a rural component in EB IV societies that cannot be characterized as exclusively pastoral.

Town planning

Very few sites can provide information on the composition and structure of settlement. Excavations at Khirbat Iskandar, Tall Umm Hammad, Tall Abu an-Niʻaj, and at Bab adh-Dhraʼs extramural areas, however, have revealed elements that can perhaps be defined as an attempt at spatial organization.

At Tall Iktanu excavation and the survey of buildings visible on the surface made a reconstruction of the village plan possible. Prag observes that there is a clear order in this plan. Streets are narrow (c. 2 or 3 m wide), and they divide areas formed by several habitation units. The streets converge towards the centre of the settlement, where no excavation or survey has been made to date. When Prag was writing in the early 1970s, Tall Umm Hammad was known only for its surface pottery scatter, identified by Glueck (1951: 318-29). Prag cited this site as a typical example of an EB IV nomadic camp, on the basis of the large area covered by the sherd scatter. She also did not believe Glueck's hypothesis that the site was more sedentary than a nomadic encampment (1974: 96-97).

Recent excavations at Tall Umm Hammad, directed by Helms, have proved Glueck's interpretation correct. Tall Umm Hammad was a village formed by habitation areas separated by narrow streets (c. 2 m wide), with a pattern very similar to that observed by Prag at Iktanu (Helms 1986: 30). As at Iktanu, there is also evidence for several building phases (at least two at Tall Iktanu, and three at Tall Umm Hammad). At Tall Umm Hammad the foundation walls were made of mudbricks, and it is for this reason that no traces of architecture had been observed on the surface by previous investigators. The same situation was observed during the test excavation at Tall Abu an-Niʻaj. This tell is situated on a *qattara* hill near the *Zor*, in the Jordan Valley. Its surface is strewn with thousands of pottery sherds, but none of the mudbrick architectural remains were visible before excavation.

Another site with possible evidence of town planning is Khirbat Iskandar, the only excavated EB IV site with fortifications belonging to this period. Domestic structures found at the site seem to be organized around courtyards, but the arrangement of these structures and their relations to the streets is still not clear since the area exposed by excavation is still limited (Richard and Boraas 1988: 115). In general, there are several analogies with what has already been observed at Tall Iktanu and at Tall Umm Hammad, while the presence of fortifications adds an important element to the structure of settlements in the central and northern areas of the southern Levant.

Public buildings

This type of building, including defensive works and fortifications, is extremely rare during this period. Khirbat Iskandar's fortifications are the only ones excavated to date, while the soundings conducted at Jabal ar-Rahil (Palumbo *et al.* 1996) are of too limited extent to confirm the presence of true fortifications there (Figure 6.9).

Fortifications or enclosures may be present at a handful of other sites: Dhahret Umm al-Marar, Jreyyeh, Rusaifah, Ader, and ʻAraʻir. The date of the fortification at al-Lajjun is certainly EB II–III, but it is still possible that there was an EB IV reuse.

At Khirbat Iskandar, clear evidence of fortifications belonging to the EB IV comes from area B (Richard and Boraas 1988). The building technique observed here recalls a characteristic recorded at Bab adh-Dhraʼ and Numayra. There, the fortification wall is built by filling the space between two lines of large stone blocks with smaller stones and rubble. Every 8 m (7 m at Bab adh-Dhraʼ and Numayra; Rast and Schaub 1978: 10; 1980: 40-42) a perpendicular wall was inserted in order to strengthen the structure. In area C a structure formed by a series of parallel long rooms is interpreted as a gate, or at least as a building of a certain importance. One of the rooms is paved and provided with benches along its inner perimeter. This entire structure is very different from those domestic units found in area B.

The other 'public' structure known before the most recent excavations is the Bab adh-Dhraʼ 'cultic' structure, excavated by Rast and Schaub in Field XVI. Remains of this structure belong to four different phases of use, which could possibly be related to the four occupation phases found in the Field X domestic area. The structure is characterized by terracing and retaining walls, and by a series of small rooms with well-built walls and plastered surfaces. A pedestal stone was found reused in a wall of phase 3. This stone is very similar to those found at ʻAi in Sanctuary A (Callaway 1972: 247). According to the excavators, however, the best evidence for the cultic use of the area comes from phase IV, to which belongs an 'altar' composed of a large flat

Figure 6.9. Enclosure or terrace wall at Jabal ar-Rahil.

stone set on a platform of smaller stones inside a room, and an 'offering table' outside this building built with a flat stone slab set on a mudbrick wall five courses high and three rows wide. On this slab were more than a dozen horns of animals, some of which were identified as sheep/goat (Schaub and Rast 1984: 57-58).

The Ader temple, excavated by Albright in 1933 (Albright 1934: 13-18; Cleveland 1960: 77-97), is an isolated structure, almost completely destroyed by the foundations of a modern house. EB IV pottery was found in the area, but the stratigraphic evidence is not conclusive. In fact, an EB IV date for this structure was suggested by Albright only because of its position near some monoliths, one of which is still standing.

Domestic structures

As noted above, domestic units seem to be built in 'blocks' formed by four or more houses. These blocks are separated by narrow alleys, as at Tall Iktanu, Tall Umm Hammad, Bab adh-Dhra' (Rast and Schaub 1978: 21), and, possibly, Khirbat Iskandar. At all of the above-mentioned sites, the typical domestic structure is characterized by a series of small square or rectangular, sometimes interconnecting, rooms built around a central courtyard. The latter was probably not roofed, because the available wood beams could not span that space. At Bab adh-Dhra' the style is different. The houses are single rooms of *Breitraum* type, which finds its roots in the EB II/III construction styles.[19] The houses found at other sites are more or less rectangular, and their length often surpasses 10 m. According to Helms this architectural style finds its origins in the building traditions of the Early Bronze Age II/III, or is directly derived from Syrian influences. Prag made the same hypotheses, with reference to the temple and private houses at Megiddo (Helms 1986: 32; Prag 1971: 304). There are several building techniques reflecting the different local availability of materials. At Jericho, Tall Iktanu, Khirbat Iskandar and Bab adh-Dhra' foundation walls were built with unhewn stones, and walls were erected using mudbricks or *terre pisée*. At Tall Umm Hammad and at Tall Abu an-Ni'aj even the foundations were made of mudbrick.

Cemeteries and tombs

Only 25% of identified EB IV cemeteries are found close to habitation sites with occupations dated to the same period. This figure is drastically different from the percentage calculated for the EB II/III period, as 64% of cemeteries or single tombs have been found in proximity to a settlement. The reason for this must be found in the uneven archaeological record, and in the fact that many isolated discoveries of EB IV tombs have been made during construction works, often not followed by surveys to identify possible settlement remains.

The main EB II/III cemeteries were also in use during the EB IV, such as at Bab adh-Dhra' (Schaub and Rast 1989) and Jericho (Kenyon 1960, 1965). As for the general structure of cemeteries, the EB IV period is characterized by the first constant use of areas expressly designated for the burial of the deceased. During EB II/III the expansion of cemetery areas was very rare, due to the practice of collective burial or family tombs.[20] During EB IV the widespread reintroduction of the single burial in a shaft tomb, already known but used sporadically during the EB II/III period, and the abandonment of the practice of burials in natural caves were two factors that contributed to the modification of the basic concept of a cemetery, from burial ground to 'city of the dead'.

Typical EB IV mortuary structures are shaft tombs dug in a chalky, soft and compact rock (Figure 6.10). There are several types of shaft tombs. The most common type has a circular or ovoid shaft, about 2 m deep, with a small entrance leading to the chamber. The entrance is sometimes blocked by a stone slab. The chamber has a circular shape, 2 or 3 m in diameter, and a low ceiling. Sometimes a lamp niche is present. There are a number of variants to this type: (a) the shaft or chamber shape (oblong, rectangular, irregular); (b) the presence of one or more steps, or a short corridor leading into the chamber; (c) the dimensions can vary for the shaft's depth from 1 to 5 m; and (d) the chamber's diameter, from less than a metre to over 6 m. Multiple chamber tombs are common. In these, chambers may be separate (having just the shaft in common, such as the 'Amman Sport City tomb: Zayadine 1978: fig. 2). Construction of a shaft tomb must have been a labour-intensive enterprise, and various considerations point to the existence of professional groups of grave diggers, a sort of economic specialization that, together with other characteristics, such as pottery village production, characterizes EB IV as a period of diversified economic specializations.

Shaft tombs are the most common mortuary structure during the EB IV, but they are not the only ones. In the regions where the soft limestone is absent the tombs are above ground, built with stone slabs or piled rocks. These structures are known as dolmens and cairns. A tumulus sometimes covers the built structure. In Jordan they belong mainly to the Late Chalcolithic/EB I period, but their reuse during EB IV has been confirmed at various sites. In the Golan, the earliest date of use of dolmens seems to be EB IV (Epstein 1985). These monuments are not exclusive to the desert or to the basalt areas in Palestine-Jordan, but are also found where shaft tombs are common. For this reason, it has been suggested that tumuli and cists might have contained primary burials, which were then collected and reburied in shaft tombs (London 1987: 73). Cairns are also peculiar to the southern regions, while dolmens are more common in the north. Being very visible, these monuments have been constantly reused or robbed through time, so that today it is difficult even to date them.[21]

On the Jordanian plateau, dolmens are often found close to EB IV settlements, as at Khirbat Umm al-Ghozlan, in Wadi al-Yabis (Figure 6.11). Indirect proof of an EB IV date for the construction or reuse of at least some dolmens can also be seen in the 'built' shaft tomb (i.e. a tomb with slab lining) found at Degania, just south of the Sea of Galilee (Kochavi 1973: 51), and at Tall al-'Ajjul, near Gaza (Petrie 1932: pl. LIII: 1516, 1517).

The entire set of data concerning the megalithic monuments points to a use, if not construction, of many of these structures during the EB IV period. Unfortunately, not a single undisturbed burial has been found. For this reason, the dolmens are an interesting typological addition to the assemblage of mortuary structures in the southern Levant dating to this period, but they still do not provide useful information on the mortuary practices performed in them. The same can be said for the cairns found in the southern Jordan Valley (Tall Iktanu) and the Southern Ghors (Bab adh-Dhra'). Their EB IV date is indicated by the presence of habitation sites of the same period in their vicinities, and by the accurate dating of some objects found within the tombs.[22]

Mortuary practices

The variety and complexity of mortuary practices in EB IV might be interpreted as signs of a complex social structure. For many scholars, this variability has a chronological basis (Dever 1987), while for others it has to do with the presence of different familiar or tribal groups with different mortuary traditions in the same area (Kenyon 1960). More recent work maintains that these variations cannot be explained by chronology or differences in material culture alone, and that they reflect the deceased's social status (Palumbo 1990).

Body treatment

Three types of body treatment are found throughout Palestine and Jordan during EB IV: secondary disarticulated; primary flexed; and primary extended or

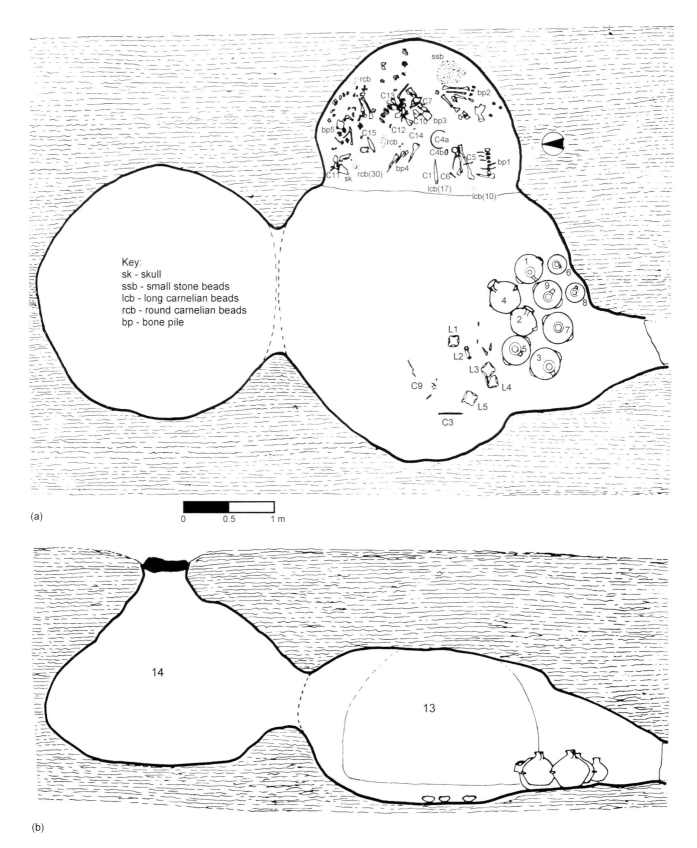

Figure 6.10. Shaft tomb at al-'Umayri (from *ADAJ* 37, 1993).

Figure 6.11. Dolmen at Khirbat Umm al-Ghozlan.

semi-flexed. Primary flexed burials were rare during the EB I–III, when extended and disarticulated burials prevailed. During the EB IV disarticulated burials prevail, followed by the flexed burial, while the extended or semi-flexed burial is quite rare.

Disarticulated burials are often the only type of body treatment found in EB IV cemeteries. Disarticulation was probably caused by decay of bodies in the open air, or temporary burial and reinterment. Bones in disarticulated burials are usually found in one or more piles in the centre of the burial chamber or at the bottom of the chamber itself. Sometimes they are found with no order in the chamber. It is very rare to find a complete disarticulated skeleton. The bones most commonly found are the long bones and the skulls. This could be considered an indirect proof of temporary burial, followed by gathering of some bones to represent the whole body in the final interment. Secondary burial has been considered by some scholars as distinctive of nomadic societies. The opposite, however, is true, since it is more common among sedentary societies in various regions of the world and throughout history.

Primary flexed burial is less frequent, but it has been recorded all over the country, in its two variants (on the left and on the right side). At Jericho more than a quarter of the EB IV burials belong to this type.[23] The body is generally turned towards the chamber entrance, with arms bent and placed against the chest, in foetal position. Based on a study conducted on tombs of this type in the Jericho cemetery, it has been suggested that burials on the right side are those of males, and those on the left side are those of females (Palumbo 1986, 1987). In the absence of anthropological studies, this interpretation is based on the association with grave goods (weapons are found in association with burials flexed on the right, no offerings or few objects, such as pins and spindle whorls, in association with burials flexed on the left), an association that has no exceptions so far in the numerous tombs and ten sites in the southern Levant where this burial type is present.

Only one tomb in Jordan (Wadi Hammeh Tomb 80E) has been found with an extended burial. Ten more tombs found in Israel and Palestine have this type of burial (Palumbo 1990). Notwithstanding its rarity, this burial is maintained as a 'type' because its constant association with 'rich' graves gives additional significance to this particular body treatment, representing, perhaps, a symbolic expression of social status.

Another very distinctive characteristic of EB IV mortuary practices, and rarely found during the EB II/III and Middle Bronze periods, are the large numbers of single interments, as well as the limited number of people buried in the same chamber (no more than four or five) in the case of multiple burials. During both the EB II/III and the Middle Bronze Age, the common practice was collective burial, containing in some cases hundreds of bodies. Not only were burials usually collective during these periods, but grave goods as well were left without association with separate burials. In the EB IV collective burials, in contrast, the single interments do not lose their individuality, since there was a clear attempt to assign grave goods to each one of the burials present in the tomb. The relation between the type of burial and the type of grave goods associated with it was certainly more important during EB IV than in the previous and following periods. As we will see, there was also a relation between grave goods and type of tomb. These three categories of analysis (burials, grave goods and tomb architecture) are, thus, important for reconstructing EB IV mortuary practices in their wider context, rather than in an isolated, typological perspective.

Grave goods

Pottery is the object most commonly found in tombs. Pottery types include lamps, small jars, amphoriskoi and large jars. Bowls, cups (more frequent in the southern regions) and spouted vessels are types found less frequently. In general it is rare to find a large number of vessels within a tomb. Sometimes weapons, pins or beads are found in the tomb. Weapons include daggers, knives, spearheads and axes (the latter have been found in very few tombs). When a dagger is present, a single specimen is usually placed close to the body of the deceased.

Animal bones are also found in tombs, and they are certainly the remains of food offerings given to the deceased. The bones are almost always of sheep and goat. Data is still scant on the association between animal offerings and 'prestige' burials, an association that has been observed by London at the cemetery of Jabal Qa'aqir, on the West Bank (London 1987).

Tombs containing human remains but totally deprived of grave goods are not uncommon. In other tombs, the only objects found are a few beads and, sometimes, a copper pin. Copper belts and 'crowns' have also been found on the bodies of primary burials in the Jericho cemetery. These were certainly 'prestige' items representing the deceased's higher social status (Shay 1983: 34).

There are no standards in the method of placing the grave goods in the tomb chamber. Pottery is usually found near or close to the chamber walls, while weapons and personal equipment (pins, combs, beads) are closer to the body of the deceased. Daggers and knives were probably tied to a leather belt worn by the deceased. This hypothesis is suggested by the position of the dagger in primary burials, commonly found at the deceased's waist or shoulder.

The clearest example of a relation between grave goods and architectural elements of the tomb is provided by the presence of a lamp niche in many tomb chambers. At Jericho, the lamp niche was often used, but sometimes no lamps were found in the tomb, even if a niche was. In other cases, lamps were among the grave goods, but they were not placed in the niche or there was no niche, so that this relation is not as secure as might be expected.

There are basic differences between EB I–III and EB IV mortuary practices. Differences are found already in the tomb types: dolmens, cairns and shaft tombs represent three different burial traditions, the first two surviving in Jordan and the Negev desert from the Late Chalcolithic and Early Bronze Age I periods, the third finding its full development in EB IV all across the southern Levant, but also originating in the Early Bronze Age cultures of this same region. In EB IV, the predominance of single depositions, of 'true' disarticulated burials (as opposed to the 'derived secondary' burials of EB I–III, above), and of a particular care in the allotment of grave goods to each burial in case of multiple burials, are all elements that are not found on this scale in EB I–III mortuary practices. The differences *within* the EB IV mortuary practices are also important, because they could be interpreted as symbolic expressions of social status. The evidence in favour of this interpretation is found in the almost constant association of primary burials with 'richer' grave goods (above), and in an even finer distinction between primary extended and primary flexed burials, with the former exhibiting the 'richest' grave goods in the Jericho cemetery (Palumbo 1987). Present evidence is not adequate to draw conclusions about the structure of EB IV society. That social status was ascribed at birth is demonstrated, in my opinion, by the presence at Jericho of differential body treatment in child burials, which are always disarticulated and without grave goods, except for two tombs,[24] where they are flexed on their right side and provided with a simple grave assemblage. The 'visibility' of children and infant burials, and, even more, their differential body treatment is commonly considered a characteristic emerging with the increase of hierarchical aspects (Brown 1982: 29), but data is still inconclusive for demonstrating that the formation of social hierarchies was a common phenomenon throughout the southern Levant.

Aspects of material culture: pottery production, styles and 'families'

Pottery typology has been for many scholars one of the keys to the understanding of the chronological problems of this period. Unfortunately, the stratigraphic sequence still missing between EB III and EB IV leaves much of this research in an abstract layer.

The division of pottery types into 'families' by Amiran and, later, by Dever, has been for the past 30 years a fixed point in the understanding of the period. This reconstruction has been used for chronological considerations, as well as for the understanding of the period as a nomadic interlude with localized craft productions. Recent research, however, is changing this view. The evidence of strong 'regionalism' in the EB IV pottery styles is unquestionable. The decline of urban-based workshops and urban-controlled regional and interregional trade is certainly one of the causes of the development of isolated pottery traditions.

A chronological revision, however, is in order to accommodate recent archaeological evidence and eliminate the gaps left by the original reconstruction. Dever was probably conscious of these limitations to his theory, since he never explicitly mentioned the existence of 'gaps' in entire regions. In 1995, however, he admitted that chronological revisions were necessary, even if he seemed to maintain his general sequence, allowing for larger overlaps between families (Dever 1995).

Three of Dever's original 'Families' are present in Jordan: Families J, NC, and TR.

Family J or Jericho/Jordan Valley is present both in Jordan and the West Bank, and is mainly found in Jordan Valley sites between the area a few kilometres south of Beth Shan and the northern shores of the Dead Sea. To the east, Dever included also the Jordanian plateau between the area south of al-Husn (which he considers of NC influence) and the 'Amman area (but see below for the suggested addition of another group for the 'Amman/az-Zarqa' area). To the west, Dever recognized J Family types in the eastern flank of the Central Hills. This family is characterized by the Jericho tomb assemblages, the materials from Tall Iktanu, and the pottery found at sites in the central and southern Jordan Valley. The most common forms are the large ovoid jar with two folded-envelope ledge-handles (Kenyon 1965: fig. 16.9), ovoid jugs with a vertical handle (Kenyon 1965: fig. 15.8), the spouted hole-mouth jar, the amphoriskos (Kenyon 1965: fig. 15.6), and the small jar without handles (Kenyon 1965: fig. 15.2). Decoration is limited to a coil applied to jar necks and decorated with nail or finger impressions.

Family NC, or North-Central, is mostly based, according to Dever, in northern Israel, covering the Jezreel Valley, the Beth Shan area and lower Galilee. In Jordan, pottery of this type has been found at al-Husn, near Irbid, and in the northern Jordan Valley. The hallmark of this pottery group is its painted decoration, consisting of red lines, sometimes grouped in series, decorating the interior and exterior of several forms. Some of the forms also exhibit a 'Syrian' flare, without doubt derived from contacts with the north. The most characteristic forms are the ovoid jar with vertical handles, basins with ledge-handles below the rim (Palumbo 1990: fig. 37, 39), small 'trickle-painted' cups and plates (Wightman 1988: fig. 12), amphoriskoi, and jugs with squat body and very high flared neck and long vertical handle (Wightman 1988: fig. 8)

Family TR, or Transjordan, was initially designated by Dever to cover the pottery found in southern Jordan as a result of the examination of Glueck's survey material and of Bab adh-Dhra' tomb A54. In 1980, however, Dever also assigned to family TR the material coming from northeastern Jordan and the eastern flank of the Jordanian plateau (Dever 1980: 48). Pottery found at Bab adh-Dhra' (Schaub and Rast 1989), Khirbat Iskandar (Richard 1983, 1990; Richard and Boraas 1984, 1988), Abu Ishribsheh (MacDonald *et al.* 1987: 407), Ader, Feqeiqes, al-Lajjun and Umm as-Sedeirah (Palumbo 1990: figs. 54-58) has widened the small repertoire of forms, initially characterized in Dever's classification by platter bowls, amphoriskoi, spouted hole-mouth jars and handleless flasks. New forms include the burnished red-slipped amphoriskos, the jar with incised decoration and large platter bowls, frequently red-slipped and burnished. Family TR types are the ones that are typologically closer to EB II–III forms.

In 1993 Palumbo and Peterman added a new family to this repertoire. The new classification derived from the analysis of pottery groups in the 'Amman and az-Zarqa' areas that exhibit homogeneous repertoire of forms and decorations, while at the same time lack the hallmarks of the other 'families' that are present in Jordan. This group was named *Family AZ* (or 'Amman/az-Zarqa'). It is characterized by strap handle jugs, storage jars with a strap handle and folded-envelope ledge-handles (both of them are the hallmark of this group) and a limited repertoire of amphoriskoi and ovoid storage jars, often with incised decoration on the neck and shoulder (Palumbo and

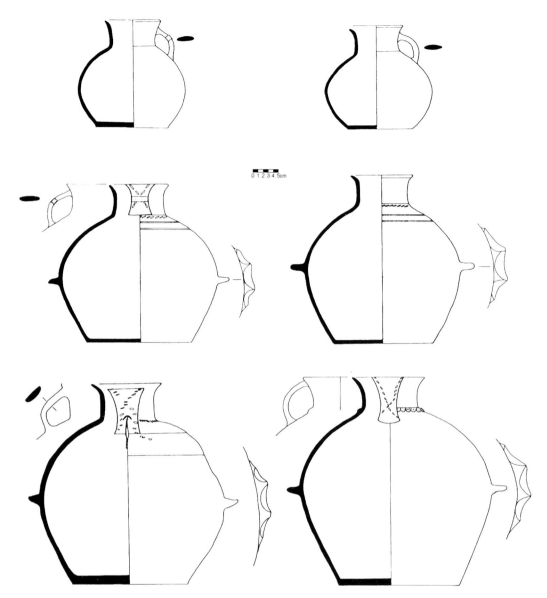

Figure 6.12. 'Family AZ' pottery from a tomb at al-'Umayri (from *ADAJ* 37, 1993).

Peterman 1993: figs. 1, 5). The boundaries of this 'family' have been so far recognized in the upper az-Zarqa' drainage and the 'Amman region, but may extend north to the vicinity of Irbid and south to the Madaba area, as the discovery of an EB IV cemetery with this kind of pottery in the vicinity of Tall al-'Umayri may demonstrate (Waheeb and Palumbo 1993) (Fig. 6.12).

Recent scientific analyses of different pottery groups in Jordan and Israel have produced interesting results that provide important insights into pottery production and trade in EB IV societies.

Falconer's neutron activation analysis of a limited sample of sherds collected on the surface and from controlled excavation at a few EB IV Jordan Valley sites (from north to south: Tall al-Hammeh, Tall al-Hayyat, Tall Abu an-Ni'aj, Dhahret Umm al-Marar and Tall Umm Hammad al-Gharbiyya) showed quite unexpected results. One of the sites, Tall Abu an-Ni'aj, emerges as a manufacturing and distributing centre of fine ware, such as 'trickle-painted' cups. The predicted pattern for these sites was that of autonomous pottery production, as a result of the very limited nature of the economic relationship between villages during the period (Falconer 1987a, 1987b). Falconer's analysis clearly showed that pottery production was primarily local. In the case of the pottery produced at Tall Abu an-Ni'aj, however, the data show that not only was this a centre for the manufacture of fine 'trickle-painted' ware, but also that this type of pottery was distributed among villages in the area. The only site among those examined by Falconer where a 'trickle-painted' bowl sherd has not been found is Tall

Umm Hammad al-Gharbiyya, which is also the farthest site from Tall Abu an-Niʻaj.

A similar experiment of neutron activation analysis has been conducted by Yellin and Perlman on a pottery tomb group from Tall ʻArtal (a site only 5 km from Tall Abu an-Niʻaj, on the west bank of the Jordan), which was compared with other vessels from ʻAyn Hanatziv and Megiddo.[25] The analyses showed that most of the ʻArtal vessels, a jug from tomb 89 at ʻAyn Hanatziv, and one amphoriskos from Megiddo tomb 1098A came from the same production centre. One may suggest that this pottery also originated in Abu an-Niʻaj, but in the absence of analyses of the clay found near ʻArtal this remains a hypothesis. Falconer (1987a) believes that both sites may have specialized in the production of fine wares. This interpretation is stimulating, being in character with the main thesis formulated by Falconer, according to which there was a differentiated rural economy prior to the advent of Middle Bronze Age urbanism, and that small villages were dependent on larger ones for exchange and trade of goods (1987b: 257-58). The presence of two fine-ware manufacturing places so close to each other cannot be considered a contradiction, but confirms instead the fundamentally disarticulated and autonomous character of EB IV intersite organization.

The petrographic analyses conducted by Goren (1996) on northern Negev and southern Jordanian pottery also show a much wider presence of 'foreign' imports in local pottery assemblages, to the extent that these imports are sometimes the majority.

At this point, the idea of strong regional boundaries should probably be modified in favour of a less abstract concept of cultural exchange between neighbouring communities. Exchange of actual goods was probably an ordinary phenomenon. Pottery vessels and their contents were perhaps among the most important items traded. From this perspective, the presence of the same form of vessel or pottery decoration at several sites could be attributed to trade from a single or a few manufacturing centres, and not to a regional style influencing the potters' production. The modified view of regionalism proposed here would include, on the one hand, a more integrated pattern of relations among communities, and, on the other, the existence of some sorts of 'regional centres', at least in pottery production. The concept of 'Central Place' as applied by Johnson to Mesopotamian and Persian situations (1972, 1973) does not find its application here, since the economic integration of EB IV communities in the southern Levant never reached a level where a Central Place distribution occurred or an administrative power emerged. The re-emergence of local production (and exchange?) centres after the EB III crisis, however, might be postulated on the basis of the latest research. When (and if) these centres were able to create around them a net of depending villages cannot yet be demonstrated, and it is still a matter of pure speculation.

Metal, flint, and stone tools

Metal sources and mining

Until recent research in the Faynan area was conducted, there was very scarce evidence confirming mining activities during the EB IV (Rothenberg 1978: 14). In the Timna area mining activities were only assumed on the basis of indirect evidence, such as the discovery of copper ingots in several Negev and northern sites, but the evidence attesting mining activities in the Faynan area during the EB IV is now clear.[26] Archaeological research in this region is revealing more intense and larger scale exploitation of the copper ores than in the Timna area (Hauptmann and Weisgerber 1987: 422). In both regions the earliest evidence for mining and copper smelting is attributed to the Late Chalcolithic period. According to Hauptmann and Weisgerber, there are no important technological changes in the copper production process in the Faynan area from the Chalcolithic to the Middle Bronze Age (422). Stone tools were used to dig the mining pits, only 1.0–1.5 m high, but up to 50 m wide; artificial pillars were also used for security (424). At Faynan, the pits follow the inclined strata of copper-bearing minerals (such as malachite and copper silicates) up to a depth of 15-20 m (424).

The discovery of several copper hoards and copper artifacts across Palestine confirms an extensive trade in this metal. In 1985, Stech, Muhly and Maddin concluded that copper and tin were imported into Palestine via Syria, possibly from Anatolia, where both these metals are found. The presence of tin in high percentages in some of the artifacts analysed is not considered natural by these scholars, who believe that these artifacts are of true tin bronze. Tin, however, is present in relatively high percentages in the Faynan copper ores.

It is difficult to understand whether small percentages of elements other than copper are part of the copper ore or were introduced during the smelting process; arsenic and lead, for example, are often present in varying quantities in the copper objects, but this might be due to the characteristics of these elements, since arsenic tends to become volatile under reducing conditions during smelting, so that the variations in arsenic content 'could be produced by different atmospheres at several points during production' (Stech et al. 1985: 78). Lead, on the other hand, tends to concentrate in different points of the copper artifact, so that there might be strong variations in lead content in the same artifact (79). For this reason, the deliberate use of arsenic as an alloy to produce bronze is not yet certain. According to Khalil, however, addition of arsenic to copper was deliberate, in order to produce a better material (1984: 168); his analyses of daggers from Tall al-'Ajjul and Jericho generally show consistent percentages of arsenic that he attributes to an alloying procedure. Moorey and Schweizer (1972) also believe in the existence of arsenical bronze during the third millennium BC in the southern Levant. The presence of tin as an alloy, on the other hand, is now certain for at least one class of EB IV artifact, the dagger.

Until a few years ago, it was commonly believed that no bronze technology was achieved in Palestine until the Middle Bronze Age (Moorey and Schweizer 1972: 193; Gerstenblith 1983: 96-98), even if the presence of a bronze artifact, the dagger from Jericho Tomb G83a, was already recognized, but treated as an isolated and possibly imported artifact (Moorey and Schweizer 1972: 195). The discovery of an EB IV tomb near 'Enan, in the Huleh Valley, and the analysis of the metal objects found in it, showed that while the artifacts other than the daggers were made of arsenical copper, the daggers were made of tin bronze (Eisenberg 1985: 65; Stech et al. 1985: 79), thus providing the Jericho dagger with a more secure context. Stech et al. (1985: 76) believe that tin bronze technology was introduced to Syria-Palestine via Anatolia, where it was known already during the EB II (also Muhly 1985: 284-85 and references there). The sporadic presence of bronze objects found in Syria, however, must be corrected by the epigraphical evidence. At Ebla, in fact, bronze was well known during the period of the state archives (Mardikh IIB1—c. 2400–2300 BC), and in several texts mentioning the import of copper and tin these two metals appear frequently in the constant proportion of nine to one (Matthiae 1989: 272). Matthiae points out that this proportion corresponds to that used in the alloying procedure in more advanced technological development of bronze production.

The introduction of bronze technology represents an advance in EB IV over the preceding periods and confirms this as a period of increasing specialization and technological advancement, even if in village rather than urban contexts.

No smelting furnaces dated to the EB IV have been found in the southern Levant to date. It might be assumed that the technology of copper smelting did not change dramatically during the third, and possibly early second millennium BC, so that Early Bronze and MB II furnaces can be taken as an example for the EB IV period as well. At Faynan, Early Bronze Age furnaces are generally shallow, made of thin pieces of stone, or thick layers of clay on the base and sides (Hauptmann 1986). There is no evidence for the use of skin or pot bellows and tuyères to allow the ventilation of the slag during smelting. Hauptmann reconstructs, for the Faynan furnaces, a technique for natural ventilation known in Mesoamerica, where stone slabs form a grating in the front of the furnace, allowing the air to reach the slag. At the same time, clay sticks (which have been found near these furnaces) were used in vertical position. Their function was probably to avoid the blockage of the air passages by the slag (Hauptmann 1986: 417; Khouri 1988: 124). Pot bellows, however, are known in Palestine from MB II contexts at Tall Beit Mirsim (Davey 1979: 106, fig. 3.3),[27] and from mid-third-millennium BC contexts in northern Syria (Beit-Arieh 1985: 116). Skin bellows were apparently in use in Egypt as early as the nineteenth century BC.[28] This evidence is important not only for its early date, but also because of the context in which the possible bellows are represented, probably representing a group of itinerant metalsmiths. That metalsmithing could be a specialized activity belonging to nomadic groups is a current opinion not only for this early period, but also for late second millennium BC situations (Beit-Arieh 1985: 115). Stech et al. (1985) explain the typological uniformity of metal artifacts during the EB IV by the presence of nomadic metalsmiths, while Gerstenblith, (1983: 97) referring to the Middle Bronze Age situation, would rather attribute this uniformity, which is also evident in metal types of that period, to the spread of ideas along trade routes, rather than to actual itinerant metalworkers.

Metal artifacts

There are only a few basic categories of metal artifacts, stylistically similar to the EB II–III metal types: weapons, tools and ornamental/functional artifacts. The first category is the most common, and includes daggers, spearheads and axes. Tools are much rarer to find, being almost exclusively simple adzes and chisels. Pins are the most common artifact that can be included in the ornamental/functional category, besides bracelets and rings. Copper belts and 'crowns', more rare, may be added to this category.

Weapons

These are usually found in funerary contexts. The dagger is the most common type of weapon associated with a burial. The relationship between grave goods and position of the body in the tomb has been treated in another section.

Several types of daggers have been identified so far, but the basic one is clearly related to the EB II/III type, consisting of a flat, elongated blade, not ribbed, with four to six rivets at the hilt attachment. Dagger shapes show a greater variation than in the previous periods. Besides the daggers with rounded tip, no mid-rib, and almost parallel sides (Kenyon 1965: 52, fig. 24.7), there are also pointed daggers with tapering sides and narrow hilt attachment (Kenyon 1956: 51, fig. 10.5-9; Oren 1973: 174, fig. 20.17), daggers with a pronounced rib (Waheeb and Palumbo 1993: 160, fig. 7.1) (Figure 6.13) and even short, triangular-blade daggers, which announce MB IIA types (Negbi 1966: 23, fig. 2.2; Eisenberg 1985: 67, fig. 7.39).

Spearheads are less frequently found than daggers, but they do appear in many tomb offerings across the southern Levant, sometimes as the only weapon in the grave assemblage, but more frequently found in association with a dagger, as at 'Amman (Zayadine 1978), al-'Umayri (Waheeb and Palumbo 1993) and at a number of sites in Palestine (Palumbo 1990). They have different shapes: elongated with a leaf-shaped point and round section (Kenyon 1965: 84, fig. 41.11, 13), very narrow, with quadrangular section (Figure 6.13) (Waheeb and Palumbo 1993: 160, fig. 7.3, Kenyon 1965: 84, fig. 41.15), all with curled tangs, or similar to daggers, with tapering sides, more or less pronounced mid-rib, and both curled (Tzaferis 1968: pl. 1A, incorrectly identified as daggers) and riveted tang (Eisenberg 1985: 68, fig. 8). Spears were probably deposited in the tombs complete with their shafts; traces of olive wood have been identified attached to a spearhead found in a tomb at Motza (Bahat 1975).

Arrowheads have been found in very limited number in a few tombs, especially in northern Palestine (see list and parallels in Palumbo 1990: 110). They have a leaf-shaped point and a short tang, circular or square in section.

Fenestrated battle-axes are a rare find in EB IV tombs, especially at northern sites, but they are a particularly interesting artifact, being a typological development of the EB III lunated, or 'epsilon' axe, and preceding the classic MB II 'duckbill' axe.[29] They have not been found in Jordan so far (see Palumbo 1990: 110-11 for a list of axes found in Israel). While they are mostly found in funerary contexts, an axe found at Megiddo is particularly important, since it was found in a wall of temple 4040, stratum XIV (Loud 1948: pl. 128: 3), allowing the dating of the stratum and the temple to the EB IV. It is interesting to note that while this type of axe disappears in Palestine in the following MB IIA period, substituted by the 'duckbill' axe, in Syria both the 'fenestrated' and the 'duckbill' axe are found even side-by-side in MB II contexts.[30]

Tools

Metal objects belonging to this category are extremely rare. Flint tools were more widely used for different purposes. The only metal tools I am aware of are simple chisels, flat adzes (none of which, however, has been found in Jordan, Palumbo 1990: 111), and small awls (found at al-'Umayri, Waheeb and Palumbo 1993: 160, fig. 7.4).

Ornaments

Metal objects with an ornamental/functional purpose are not very common, if we exclude the pins, which are fairly common in grave assemblages, especially in the northern regions of the country.

Pins occur in four varieties: (1) short, pointed at both ends, round or square in section; (2) with simple curled heads and round section; (3) with broad shaft heads, a circular perforation and round section; and (4) with hemispherical head, elongated perforation and round section. All these four types were contemporary, since they have been found together in a tomb near 'Enan (Eisenberg 1985: 70, fig. 10.53, 59, 61, 62). They have various sizes, between 5 and 10 and 25 and 30 cm. Less common are longer pins, such a 49.5 cm pin from 'Enan (Eisenberg 1985: 70, fig. 10.64). Only pins of type 1 are found all across Palestine, but pins of type 2 have also been found at al-'Umayri, south of 'Amman (Waheeb and Palumbo 1993: 160, fig. 7.5-6). The distribution of the more complex types is limited almost exclusively to the north, and their close parallels with pins found in Syria, especially in the Hama region (Prag 1974: 95), witness a strong northern influence on the production of these objects, if not an actual origin in Syrian workshops.[31]

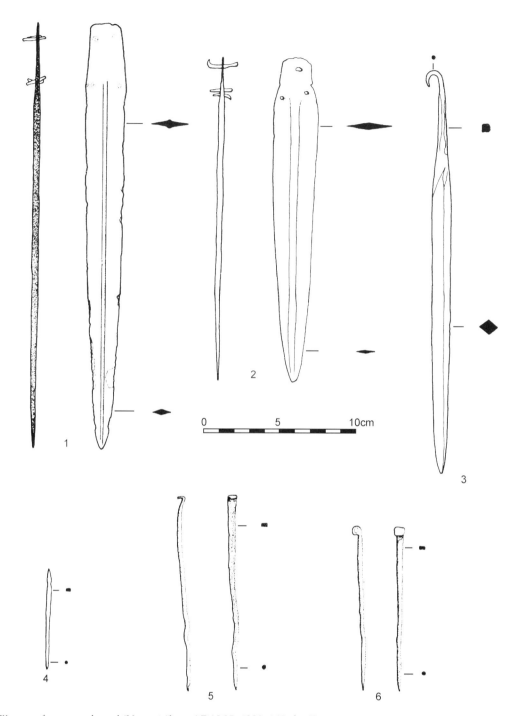

Figure 6.13. Weapons from a tomb at al-'Umayri (from *ADAJ* 37, 1993: 160, fig. 7).

Rings and bracelets have been found exclusively in grave assemblages in northern regions (Palumbo 1990: 112). They usually have open ends, sometimes overlapping. More elaborate patterns are less common. According to Epstein (1985) and Eisenberg (1985), the origin of these objects must be sought in Syria, where the best comparisons are found. At al-'Umayri (Waheeb and Palumbo 1993: 156-57, figs. 4.5, 5.7) a bronze band or ring was 'wrapped' around a jar and a jug handle, a unique find so far.

Until a salvage excavation of an EB IV tomb near Tall al-'Umayri in 1992, in which a belt and possibly a buckle were found, belts and 'crowns' made of copper were known only from Jericho. The belt (Waheeb and Palumbo 1993: 161, fig. 8) is decorated in repoussé with parallel lines of dots.

Flint tools and implements

Flint tools production in the EB IV period can be considered as a survival of Chalcolithic and Early Bronze Age traditions. The most common tool found in EB IV contexts is the so-called 'Canaanean' sickle blade, which finds its origin in the EB I period. The term 'Canaanean' identifies a technique common throughout the Early Bronze Age, which has been described in detail by Rosen (1983, 1997). Already Neuville (1930) and, in greater detail, Crowfoot Payne (1948) had identified the common characteristics of Canaanean blades, and specifically their trapezoidal cross-section, the two parallel lines on the dorsal side running near the centre of the blade itself, the prepared and narrow striking platform, a deep negative bulb of percussion and two spurs on either side. Besides technological considerations, Canaanean blades are characterized by the extremely good quality of the flint used to produce them. This, and the fact that tools made out of Canaanean blades were used and resharpened several times, mean that this type of flint product was not made at every site, and must have been an important trade item throughout the Early Bronze Age.

Early Bronze Age Canaanean blade workshops have been identified at several sites (Rosen 1997: 108), almost all of them small villages. The production and trade of this item seems to be a village-based activity throughout the Early Bronze Age, and an activity that certainly continues in the EB IV, as the presence of this tool type at several sites demonstrates. Data is still insufficient, however, to confirm the existence of a few centres of blade production during the period. The rare presence of tabular flint scrapers in EB IV contexts (Palumbo 1990: 115), an Early Bronze II/III type, confirms the substantial continuity of lithic technologies and patterns of flint tool production from the 'urban' EB II/III into the 'rural' EB IV. In other words, this type of specialization, and possibly the local trade network connected with it, was not affected by the collapse of the urban civilization at the end of the EB III, since these production and trade activities were probably independent of urban control.

Canaanean sickle blades were produced by snapping segments from longer blades. They were not retouched during first use, but sickle sheen, which often appears on one or both edges (Prag 1971; Rosen 1982; Betts 1984), and some edge sharpening and resharpening, show that the segments were reversed in the sickle haft after some use. Five to seven segments were probably used in each sickle, which was about 30 cm long on current estimates. Bitumen might have been used to secure hafting on the sickle. Bitumen traces were found in the EB IV strata of Tall Iktanu (Prag 1971: 261).

Blade width seems to increase through time, reaching maximum broadness at the end of the Early Bronze Age (Waechter 1958; Crowfoot Payne 1983; Betts 1984). According to Rosen, however, variations in blade width have a local rather than a chronological significance, in the sense that they might be due to the activity of different regional production centres (1983: 26). At Tall Umm Hammad al-Gharbiyya, EB IV types also have more pronounced trapezoidal cross-sections (Betts 1984).

Reaping knives were made out of complete, or at least longer sections, of Canaanean blades. They are usually longer than 10 cm, and their use as tools is witnessed by the sheen appearing along one or both edges. They have been found at several EB IV sites, such Tall Abu an-Ni'aj and Tall an-Nekheil (Palumbo 1990: 114), Jericho (Crowfoot Payne 1983), Tall Iktanu (Prag 1971) and sites in Israel. These blades show lustre and abrasion along both edges up to the distal end, but stopping 2–3 cm above the butt. According to Crowfoot Payne, then, this means that the blade was end hafted, or, as an alternative, held in the hand, possibly with wrappings, such as leather or other materials, around the butt (1983: 723; also Prag 1971; Rosen 1983).

There are no clear examples of complete EB IV lithic assemblages published yet, so it is not known what the proportion is at different sites between Canaanean and simple sickle blades. The presence and proportion of other tool types, such as scrapers and chisels, is also generally unknown. Besides some dissertation material (Hammond 1977), there is very little published on the technological aspect of local flint production. Such a study would provide interesting insights, since Canaanean technology is less common at some sites than expected. McConaughy (1979) and Rosen (1983) have already observed that at Bab adh-Dhra' only about 15% of the sickle blades are technologically Canaanean, and that this technology becomes very rare in the Negev. Prag seems to think that Canaanean industry is absent or very degenerate at such northern and central Jordan sites as Khirbat Umm Rujm, Mu'amariyeh, Ader and 'Ara'ir (Prag 1971).[32] Evidence from Jabal ar-Rahil (Palumbo 1990) also indicates that Canaanean technology is rare in northeastern Jordan.

In the Jordan Valley the Canaanean blade is very common, but here too with some interesting variants. At Tall Iktanu Phase 1 only a few flints have been recorded, none of which are Canaanean, while in Phase 2 the only flint tools found are sickle blades, mainly of the Canaanean type (Prag 1971). They are also common at Tall Umm Hammad (Prag 1971; Betts 1984), as they are at other central Jordan Valley sites, such as Tall an-Nekheil (Glueck 1951: pl. 112.1) and Ze'aze'iyyeh (Palumbo 1990: 114). The differences between Tall Abu an-Ni'aj and Dhahret Umm al-Marar are not limited to the pottery repertoires, but are found in the lithic assemblages as well. Tall Abu an-Ni'aj is characterized by a typical blade industry on a very fine flint. At least two-thirds of this industry is Canaanean. Particularly interesting in this surface collection is the presence of two backed blades, very uncommon elsewhere. At Dhahret Umm al-Marar surface flints are much rarer, and the assemblage is predominantly, but not exclusively, characterized by sickle blades. The flint used for the sickle blades is not as fine as at Tall Abu an-Ni'aj, and, most importantly, the technology is probably not Canaanean (Palumbo 1990: 115). This raises the question, unresolved for the pottery assemblage, whether the differences between these two neighbouring sites are of a regional or chronological character. Prag tried to reach some conclusions by comparing Iktanu's and other Palestinian and southern Jordanian assemblages, but the available evidence was not sufficient for postulating more than just preliminary thoughts. She concludes that the earlier stage of the EB IV (her EB–MB) period, characterized by a degeneration of the Early Bronze Age lithic traditions, was followed by a revival of those same traditions at a later stage of the period. Her impressions were based on the presence–absence patterns of the lithics at a few excavated sites rather than on the study of intrasite variations in the technology and in the assemblages.

Fragments of basalt or limestone querns are a common surface find at many EB IV sites, witnessing, together with the sickle blades, an important agricultural component at these settlements. Unfortunately, no precise information is given on the quantities and size of quern fragments found on the surface. At Jabal ar-Rahil, north of az-Zarqa', they are very common, but also shattered in small fragments; at Mu'amariyeh, as well, they are present in quantity, together with rubbing implements (also Prag 1971: 266). Prag found fragments of querns at several of her survey sites (1971). They are a common feature at Jordan Valley sites, such as Tall Abu an-Ni'aj, Dhahret Umm al-Marar, Tall an-Nekheil, Ze'aze'iyyeh, Tall Umm Hammad al-Gharbiyya and Tall Iktanu. They have been found at many sites in central Jordan, such as Khirbat Iskandar, Ader and Feqeiqes, together with other pounding and grinding implements (Richard 1983). Querns and rubbing stones coming from archaeological deposits are much rarer: one quern and one rubbing stone were found together in a Phase 2 room at Tall Iktanu (Prag 1971: fig. 59.1), while querns have been found at Bab adh-Dhra' Fields IX and X (Rast and Schaub 1978: 19, 21).

Mortars were found at Bab adh-Dhra' Fields IX and X (Rast and Schaub 1978: 19, 21), Tall Iktanu (Prag 1971: fig. 59.3-4) and Khirbat Iskandar (Richard 1983). Pestles are also reported from the latter site.

Limestone or basalt rings are large (10 cm in diameter or larger), doughnut shaped and their possible function has been discussed at some length by Prag (1971: 267-68). Their most probable function was as digging stick weights, and quite interestingly they have been found only in Jordan and at Jericho.[33]

Smaller perforated stones were probably spindle-whorls. Their shape is totally different from the digging stick weights, having a more flattened aspect. The hole is also much smaller in proportion to the surface. This tool also has only been found in EB IV contexts in Jordan and at Jericho (see Palumbo 1990: 117).

Stone and shell ornaments

Stone beads have been found at several sites in grave assemblages (Palumbo 1990: 117 n. 80). They are of different shapes and materials: disc, barrel-shaped, biconical and cylinder-shaped. The stone used for their production includes agate, carnelian, chalcedony, quartz, calcite and alabaster.

Frit beads are not uncommon in EB IV grave assemblages, and they are always found in association with stone beads. Bone beads are rare, but they are also very fragile, and they might not have survived intact among the grave goods. Shell beads have also been found at a few sites, including Jericho.

Shell or bone plaques of oblong shape and two perforations at their extremities have been found at several sites, mainly, but not exclusively, in grave assemblages (Palumbo 1990: 118).

Economy

Agriculture

Far from being a period of exclusive animal husbandry, EB IV sites are revealing evidence of plant cultivation. Cereals, pulses and fruits were all cultivated, especially where water was easily available. At Tall al-Hayyat and Tall Abu an-Ni'aj, Falconer and Magness-Gardiner found wheat, barley and a variety of legumes and fruits, including grape and fig. Not common, but also present, was the olive. At Jericho a similar variety was found by Kenyon: wheat and barley, lentils, peas and chick-peas, together with grapes, figs, dates and pistachios.

As a general observation, there is a smaller number of cultivated plant species found at EB IV sites than in the previous and following periods, perhaps an indication of a more limited production, concentrated on garden-like crops. The presence of grapes and fruits is significant, because these only reproduce by cloning, and only give fruit after several years. All this implies a settled human element (Grigson 1995: 259).

Pastoralism/husbandry

One of the elements used to substantiate the presence of a strong pastoral nomadic component in EB IV was the large number of sheep and goat bones found, especially in mortuary contexts. This picture may be misleading and the result of the use of these animals as food offerings in funerary rites. In reality, high percentages of cattle and pig bones have been found at several villages, including Tall Abu an-Ni'aj. Sheep/goat only prevail at sites in the desert, an environment that is more favourable to these animals. In the case of the presence of cattle and pig, clearly we cannot speak of pastoral nomads. This is also another element that has allowed scholars to look beyond the pastoral nomadic model to describe the economic basis of EB IV societies. Pastoral nomadism may have been an option at sites in the desert or marginal areas. In other cases sheep/goat husbandry may have taken the form of transhumance between winter and summer pastures conducted by specialized shepherds belonging to settled communities. In some cases, sheep/goat pastoralism may have been a purely local, village-based activity that excluded nomadic movements or even transhumance. There is an evident change from EB II–III to IV in the percentage of sheep/goat versus cattle and pig, and this is probably due to the fact that small, semi-isolated communities may tend to limit the herding of animals more dependent on humans for their care, especially if there are limited opportunities to sell or exchange these animals for other products.

Trade

Evidence of trade during the EB IV is found in almost every material culture component, from pottery to metal and flint tools. The presence at several sites in northern Israel and as far south as Khirbat Iskandar (Richard 1990: 36) of Syrian 'caliciform' grey-painted teapots, goblets and cups (Dever 1980; Mazzoni 1985) indicates that these items were treasured and often deposited in grave assemblages.

The contacts between Syria, especially the Hama region (Prag 1974; Mazzoni 1985), and Palestine are clear, even if they are limited to sites in the northern regions of Palestine. The reasons why this diffusion did not reach sites farther south than Megiddo (with very few exceptions) has probably to do with the fact that these villages were at the margins of an active trade network among the urban communities of northern Syria, and that these communities did not have interest in expanding their area of influence south of the Damascene. It certainly cannot be said that the southern Levant was involved in long-distance international trade during the EB IV. The sites in Palestine touched by the exchange of goods with northern regions were not able to start a process of *internal* diffusion of these foreign goods, probably because a true 'central market place' with redistributive functions was missing.

Much more important for the economy of the southern Levantine EB IV and for its future development into the Middle Bronze Age was the function of local trade (Falconer 1987a, 1987b; Rosen 1983, 1997). As we have seen, Falconer (1987b: 252) showed through neutron activation analysis of clays and pottery from various sites in the Jordan Valley that, while utilitarian pottery is locally produced, finer wares tend to be produced at one or a few sites and distributed to the neighbouring communities. This pattern coincides with the conclusions reached by Rosen (1983, 1997), who believes that the so-called 'Canaanean' blades were produced at a few specialized centres and then traded. It is quite interesting to note that a village in the Jordan Valley, Tall Abu an-Ni'aj, could be one of these specialized centres of *both* pottery and flint production. Copper production at Faynan also played an important role in the continuation of a trade pattern across the Negev, and perhaps into Sinai and the Nile Delta. This is demonstrated by the discovery at a number of Negev sites of pottery produced in Jordan.

While the disappearance of international trade at the end of EB III was a logical consequence of the crisis of the 'urban' system, the persistence of a village exchange system from the EB II/III (Rosen 1997) into the EB IV shows that the rural economy at its basic level did not change throughout the Early Bronze Age.

EB IV in its historical context

The emphasis on the urban characteristics of EB II and III is one of the reasons why the importance of the small rural settlement and of its relationships with the larger sites and the nomad or pastoral groups has been underestimated in the literature. The crisis that affects all urban centres during the EB III does not change the traditional village production and craft systems, which seem to survive in the EB IV, especially in Jordan. Settlement patterns, however, are disrupted, and this is probably due to a process of nucleation of the population during the EB II–III periods, which led to the abandonment of villages.

The abandonment of rural villages might also be due to a 'flight' of some segments of the population towards less controllable subsistence strategies, such as pastoral activities. The beginning and the subsequent development of the economic crisis was perhaps caused by several factors, such as drought, intersite competition, interruption of trade routes, as well as by the inadequate system of rural villages supplying the main centres. When this crisis reached its peak, the segment of population more closely linked to rural production began to move in the opposite direction, into the countryside again, reoccupying the abandoned villages or founding new ones. For the next three centuries the population reverted to a simpler subsistence economy, to a domestic mode of production that had apparently disappeared during the predominance of an urban-based economy, and instead re-emerged in all its pervasive force during the eclipse of the urban system.

From this perspective, the 'rural' EB IV and its local, domestic mode of production cannot be considered foreign to the Early Bronze Age cultures of the southern Levant. The 'domestic unit of production', as Joffe (1993) calls it, was instead the *basic* economic system of the entire Early Bronze Age. The beginnings of the urban exploitation did not affect the domestic mode of production, except, perhaps, during the EB III, when some sort of demographic and possibly political centralization occurred. But as soon as the crisis process started, it was probably natural for most of the population to revert to the basic and traditional subsistence strategies. Exaggerating the terms of the question, we may say that the Early Bronze Age urbanization was achieved only *at the expense of* the rural basis. The collapse of the system started when the needs of the 'cities' surpassed the carrying capacity of the rural villages. What may be considered as the first 'experiment' in the southern Levant in the exploitation of a population segment ended as a result of being pushed too far, too early.

According to this interpretation of collapse, the social and cultural change observed during the EB IV period is a consequence of the crisis of the urban structure and of the reaction started by the rural basis of those same structures. The main characteristics of the EB IV culture in the southern Levant are few and well defined: (a) the prevailing economy is the domestic mode of production; (b) there is no large-scale international trade; (c) regional trade is limited to a few craft/goods production specialized centres; (d) settlement is dispersed and characterized by mobility in some marginal areas and sedentism in other regions; and (e) there is a strong tendency towards regionalization of cultural traits. To some scholars, these characteristics are typical of a nomad–pastoral culture. The transhumant model (Dever's Negev–Hebron Hills Complex), limited to the region between the Negev and the South-Central Hills, can be used to describe a situation where a rural pastoral component prevailed.

In other regions there is a more balanced relationship between sedentary and pastoral elements of the society, such as in the Jordan Valley and on the plateaus of northern and central Jordan. The economic choice of the people living in more favourable areas of the country was to diversify production, in order to be able to withstand periods of crisis. The combination of agriculture and pastoralism, and the selection of one or the other system as the main production activity during certain periods are not due to the prevailing of one segment of the population over another, but rather to a well thought-out decision of the same rural population, which remains open to shifting from one system to another, as deemed necessary by the circumstances, and to maintain diversified levels of specialization within its own economic structure (Marfoe 1979).

These interpretations can be applied to the situation of the entire southern Levant, and can be used to restructure the concept of 'dimorphic society', which has been used by many to presume the success of the nomadic 'element' during periods of unbalanced relationships between the two components of the society. The dichotomy congenital in the concept of 'dimorphic society' is too rigid, as Kamp and Yoffee (1980: 93) already observed. There is no justification, then, for using a 'dimorphic' approach, when it is clear that relationships and exchanges between groups were much more complex.[34] This interaction must have been even more evident during periods of crisis; from this perspective, the 'ruralization' of some elements of the urban component, the dispersion of village dwellers into smaller productive units (farms or hamlets at a domestic level of production) and the adoption of a mode of production more oriented towards pastoralism could have been accompanied by a corresponding sedentarization of nomadic groups that could have taken advantage of a weakened and disappearing urban power, or forced by a decrease in agricultural surplus to practise their own farming.

On the basis of the available evidence, the characteristics of the EB IV period can be reconstructed with an economy based on the domestic mode of production integrating both pastoralism and agriculture, and trade on a local basis, perhaps on a larger scale in the north. This region was probably in the sphere of influence of urban centres in central and northern Syria, which were flourishing during this period. Local trade may have been favoured by the presence of large villages more active in specialized craft production.

Some trends towards labour division can be seen in the specialized production and trade of pottery, 'Canaanean' blades, metal weapons, in the mining and metalsmithing activities in the Timna and Faynan areas, and in the presence of 'professional' groups charged with the construction of the large shaft tombs at several sites.

Traces of a complex social structure can be identified in the mortuary practices. The presence of similar differentiated mortuary practices at several sites in the southern Levant might be due to the existence of a common 'ideology' and of a common social structure, besides ethnic and/or tribal differences.

As regards settlement patterns, the available evidence shows a dispersed pattern of settlements, not only in marginal areas, such as the Negev Highlands, but also where water and good agricultural soil is present. The rural landscape of EB IV in the southern Levant is one of small villages filling available ecological niches. Depending on the environmental constraints and the economy practised at these sites, some of them were certainly less stable and more ephemeral than others. Thus, villages in central Jordan or the Jordan Valley were probably occupied permanently, with a mixed economy based on agriculture and herding, perhaps integrated by local trade and specialized pottery, flint, or metal production, as might have been the case with Tall Abu an-Ni'aj in the Jordan Valley. The sites involved in this production may have acted as 'regional centres' for trade or exchange between neighbouring communities. Traces or perhaps remnants of long-distance trade may be seen in the Syrian commodities reaching Palestinian sites in the north. The presence of fortifications or enclosures at a few sites implies the existence of small towns or regional centres and may be interpreted, according to Richard and Boraas, as a survival of Early Bronze Age traditions and as a 'readaptation to a level of political autonomy probably best explained by the chiefdom model' (1988: 128). Without pushing the evidence too far, it should be noted, however, that the distribution of EB IV 'forted' sites is marginal to the main agricultural regions, such as the Huleh, the Jezreel and the Jordan Valley, and generally located between the 200 and 300 mm isohyet. It is possible that in these marginal regions there was higher competition for land between sedentary and mobile pastoral groups, which resulted in more defensive behaviour by the sedentary communities. At this point, the growth of both regional craft production and trading centres, and 'military', or at least defended agricultural outposts, created a hierarchical settlement system, which eventually was also reflected in the social structure (above). This may have been the basis for the development of a new urban culture.

The 'continuity' between EB II/III and EB IV does not mean that there was a sharp interruption in settlement patterns and cultural traits between the latter period and the Middle Bronze Age. We have seen that in some regions there is even more continuity in settlement location between EB IV and MB IIA than between EB III and IV. The importance of trade for the development of the Middle Bronze Age urban culture, in Gerstenblith's opinion (1983), is perhaps overemphasized. There is no doubt that trade may have played an important role in the development of the new urban civilization, but certainly not as a single, or even the main factor. Social and cultural change at the beginning of the second millennium BC can be seen as the result of the influence of political, climatic, economic and technological factors on a economic (the rural/domestic mode of production) and social base (where the beginnings of social stratification can be seen) that were ready to receive and develop the new trends. From this perspective, the renewed success of the urban structure in the southern Levant can be considered a local development of trends derived from the economic and social traditions of both the 'urban' EB II/III and the 'rural' EB IV societies. We have recognized these trends mainly in the adaptive ability of rural settlement and production strategies, as well as in the existence, during the EB IV, of a type of social structure that was able to develop into a more stratified component, where the administrative and, later, urban elites played a pivotal role in the formation of a new structured society. Finkelstein interprets this process as a shift back to the 'multimorphic society', and the emergence of a stronger rural system (1989: 137). The fact that Falconer (1987b) also sees in the MBA the superimposition of towns on a rural landscape shows a convergence of interpretations on the role that the rural component of society played in the economic and political development between the Early and Middle Bronze Ages in this area.

In conclusion, far from being a Dark Age, the EB IV emerges in these interpretations as a period in which its system of specialized local production and mixed agricultural strategy, as well as technological innovations (such as the widespread introduction of tin bronze technology), served to reinforce the role of the rural component of the society, and laid the basis for the development of a renewed and stronger urban culture.

Notes

1. See Palumbo (1990: 10-11) for a detailed presentation of Lapp's argument on the origin of nomadic invasion from the movements of Transcaucasian Kurgans.
2. On the relationships between Ur III urban elites and pastoral nomadic people as seen from the 'centre', see Buccellati (1966); Liverani (1970, 1973; 1988: 298-307); De Geus (1971); Haldar (1971); Kamp and Yoffee (1980).
3. On the emergence of specialized pastoral production in the southern Levant, see Lees and Bates (1974); Sherratt (1981); Levy (1983); Rosen (1988).
4. The concept of 'dimorphic society' became a central theme in a long series of articles by the same author. Through the years, he fossilized his theory into a overly rigid dichotomy between nomads and sedentary peoples (Rowton 1973, 1976, 1977). See Kamp and Yoffee (1980: 93-94) for a criticism of Rowton's position.
5. The causes of this collapse are generically listed by Dever. He mentions 'political misfortunes, economic reverses, a series of natural calamities such as drought or pestilence, over-population of urban centres and exhaustion of natural resources, or some combination of these and other factors' (Dever 1980: 58). The aim of Dever's analysis was not to look for causes, but rather to show that the priority of a pastoral economy and the dispersion of human settlement were consequences of much more important events than just an 'invasion'.
6. Excavations at Khirbat Iskandar are directed by S. Richard (Drew University, Madison, New Jersey). Preliminary reports in Richard (1990); Richard and Boraas (1988); Richard and Long (1995).
7. In his analysis of Tall Umm Hammad pottery repertoire, Helms reached the conclusion that many different styles influenced pottery production at this site.
8. Helms criticizes only the chronology proposed by Dever, judging as 'convincing and comprehensive' the typological analysis (1989: 32). See also Helms and McCreery 1988: 341.
9. Found in the B8–B7 storerooms.
10. Only little more than a dozen band-combed fragments and a few painted hole-mouth jar rims.
11. Because of the typological and stylistic characteristics of the pottery repertoire, Olávarri himself would have preferred to use the terms EB IVA and EB IVB, but he refrained from doing so to avoid confusion (1969: 259).
12. Tall Abu an-Ni'aj: Falconer and Magness-Gardiner 1984; 1989; Ader: Cleveland 1960; 'Ara'ir: Olávarri 1969; Bab adh-Dhra': Rast and Schaub 1981; Schaub and Rast 1989; Tall al-Hayyat: Falconer and Magness-Gardiner 1984, 1989; Tall Iktanu: Prag 1986, 1988, 1989; Khirbat Iskandar: Richard 1990; Richard and Boraas 1988; Richard and Long 1995; Tall Umm Hammad: Helms 1986; az-Zaraqun: Ibrahim and Mittmann 1987.
13. On the Coastal Plain, for example, only 8% of the EB III sites are also occupied in EB IV, and only 20% of the MB IIA sites also have EB IV occupation, a situation in contrast with the Hebron region, where 70% of EB III sites are reoccupied in EB IV, and 75% of the MB IIA sites have an EB IV presence (Palumbo 1990: 48-62).
14. For all of these sites: Ibrahim et al. 1976: 48-49; 1988: 191-92.
15. Glueck 1939, 1951; Mittmann 1970; Prag 1971 (Shu'eib-Hisban Survey); Ibach 1978 (Hisban Survey); Gordon and Villiers 1983; Kerestes et al. 1978; Banning 1985 (Wadi Ziqlab Survey); Mabry and Palumbo 1988; Palumbo et al. 1990 (Wadi al-Yabis Survey); Palumbo et al. 1996 (Wadi az-Zarqa' Survey).

16. See for example the sites of Muʻamariyeh and Khirbat ar-Reqaʻ. Both are quite large (about 4 ha); the first is spread on both sides of a wadi and on the hill slopes close to it, and the second is on a plateau on the edge of a valley (Glueck 1951: 82-84, 180-81).
17. The Iskandar fortifications are dated to phases C and D of occupation. In phase B they were probably abandoned, and by phase A they were certainly out of use (Richard and Boraas 1988: 112-13).
18. Surveys in the Wadi al-Hasa and Southern Ghors region were carried out by MacDonald *et al.* (1980, 1982, 1987, 1988, 1992).
19. Rast and Schaub's (1978: 22) work in Bab adh-Dhra's Areas IX and X outside of the EB II/III city wall showed that at this site there was also a consistent sedentary occupation during EB IV.
20. The only exception is the large cemetery at Bab adh-Dhraʻ, where during EB I and part of EB II, shaft tombs containing single or a few burials were the norm. But at this site, too, the typical EB II/III mortuary practice was collective burial, represented here by charnel houses, some of which contained hundreds of bodies.
21. Some of them still have a 'mysterious' character. See, for example, the large tumulus at Rujm al-Hiri, in the Golan, surrounded by a triple wall circle, which has been interpreted as a cultic site (Zohar 1989). A large tumulus surrounded by stone cairns was identified near Kufr Abil during the 1989 Wadi al-Yabis survey season (Palumbo *et al.* 1990).
22. The Bab adh-Dhraʻ cairns were empty (Rast and Schaub 1978: 29), like those dug in several Negev sites.
23. Flexed burials on the right side have been found in Jordan at ʻAmman and Bab adh-Dhraʻ. Flexed burials on the left side have been observed at Umm Bigal (near ʻAmman) and Bab adh-Dhraʻ.
24. Tombs L8 and M12 (Kenyon 1965).
25. With the exception of a lamp from Tall ʻArtal, all the specimens tested are 'trickle-painted' ware. Appendix in Hess 1984: 55-60.
26. See the jar sherd used as a lamp found 22 m below the surface in copper mine no. 43 at Wadi Khalid (Hauptmann and Weisgerber 1987: 424, 426, fig. 5.1), and the sites identified by R. Adams, with evidence of smelting activities at Khirbat Hamra Ifdan.
27. See also the reconstruction made by Beit-Arieh of pot or stone bellows operation on the basis of the Rekhmi-Re wall painting at Thebes (1985: 94, fig. 4), dated to the Eighteenth Dynasty.
28. They are represented with a group of Asiatics in the Tomb of Khnum-hotpe III at Beni Hasan (Davey 1979: 111 and references there).
29. For the typological development of this weapon, with parallels from Palestine and Syria, see Dever (1975b: 30).
30. As at Ebla, Hypogeum Q.78.B1 and Q.78.C, Tomb of the Lord of the Goats (Mardikh IIIB, c. 1750–1700 BC) (Matthiae 1980: fig. 10, 11, 12; 1984: pl. 83). A complete mould for a fenestrated axe has been found at Ebla in Tomb D3712, of the Old Syrian period (c. 1850–1750 BC) (Weiss 1985: 183, 243). The axes from Hypogeum Q.78.C belong, however, to a rather different type from the 'classic' fenestrated axes, since their blade is already quite elongated. Matthiae classifies the axes on the basis of their width–length ratio as 'large' (fenestrated), 'medium' and 'long' (duckbill) types (1980). To the 'medium' type he attributes the axes found at Ebla, which could have a possible Iranian origin (1980: 59).
31. An extensive discussion on the parallels with Syria is in Prag (1974: 93-95).
32. Richard reports the presence of 'the typical Early Bronze Age sickle blades of the characteristic trapezoidal shape' at Khirbat Iskandar (1983: 51), but whether these blades are Canaanean or not is not explicitly stated.
33. Tall Iktanu, Muʻamariyeh, Tall Umm Hammad al-Gharbiyya, Tall Sheikh Mohammed (Prag 1971); Zeʻazeʻiyyeh, Tall Ras Hamid (Palumbo 1990: fig. 59: 1, 6); Jericho Tombs N4 and M5 (Kenyon 1965: 60, fig. 29: 1-2).
34. On the same lines is Finkelstein's (1989) definition of 'multimorphic society', which is applied, however, only to the 'urban' Early Bronze and Middle Bronze societies, since he sees EB IV as a period of collapsing specializations.

References

Albright, W.F.
    1932   *The Excavation of Tell Beit Mirsim I*. Annual of the American Schools of Oriental Research 12. New Haven, CN: American Schools of Oriental Research.
    1934   Soundings at Ader: A Bronze Age City of Moab. *Bulletin of the American Schools of Oriental Research* 53: 13-18.

1940   *From the Stone Age to Christianity*. Baltimore: The Johns Hopkins University Press.

Amiran, R.
1960   Pottery of the Middle Bronze Age I in Palestine. *Israel Exploration Journal* 10: 204-25.
1974   A Tomb-Group from Geva'-Carmel. *'Atiqot* 7: 1-12 (Hebrew Series).

Bacon, E.E.
1954   Types of Pastoral Nomadism in Central and Southwest Asia. *Southwestern Journal of Anthropology* 10: 44-68.

Bahat, D.
1975   A Middle Bronze Age Tomb Cave at Motza. *Eretz Israel* 12: 18-23 (Hebrew).

Banning, E.B.
1985   Pastoral and Agricultural Land Use in the Wadi Ziqlab, Jordan: An Archaeological and Ecological Survey. Unpublished PhD dissertation, University of Toronto.

Bates, D.G. and S.H. Lees
1977   The Role of Exchange in Productive Specialization. *American Anthropologist* 79: 824-41.

Beit-Arieh, I.
1985   Serabit al-Khadim: New Metallurgical and Chronological Aspects. *Levant* 17: 89-116.

Betts, A.
1984   Chipped Stone Tools. *Levant* 16: 52.

Brown, J.A.
1982   The Search for Rank in Prehistoric Burials. In R. Chapman and K. Randsborg (eds.), *The Archaeology of Death*, 25-37. Cambridge: Cambridge University Press.

Buccellati, G.
1966   *The Amorites of the Ur III Period*. Napoli: Istituto Orientale.
1990a   'River Bank', 'High Country' and 'Pasture Land': The Growth of Nomadism on the Middle Euphrates and the Khabur. In S. Eicher, M. Wäfler and D. Warburton (eds.), *Tell al-Hamidiyah*, II: 87-117. Göttingen: Vandenhoek & Ruprecht.
1990b   The Rural Landscape of the Ancient Zor: The Terqa Evidence. In B. Geyer (ed.), *Techniques et pratiques hydro-agricoles traditionnelles en domaine irrigué*, 159-69. Paris: Geuthner.
1992   Ebla and the Amorites. *Eblaitica* 3: 83-104.
1993   Gli Amorrei e l' 'addomesticamento' della steppa. In O. Rouault and M.G. Masetti-Rouault (eds.), *L'Eufrate e il tempo: Le civiltà del Medio Eufrate e della Gezira siriana*, 67-69. Milano: Electa.

Callaway, J.A.
1972   *The Early Bronze Age Sanctuary at 'Ai (et-Tell)*. London: Quarritch.

Carneiro, R.L.
1981   The Chiefdom: Precursor of the State. In G.D. Jones and R.R. Kurtz (eds.), *The Transition to Statehood in the New World*, 37-79. Cambridge: Cambridge University Press.

Cleveland, R.L.
1960   The Excavation of the Conway High Place (Petra) and Soundings at Khirbet Ader. *Annual of the American Schools of Oriental Research* 35: 59-97. New Haven, CN: American Schools of Oriental Research.

Cohen, R. and W. G. Dever
1978   Preliminary Report of the Pilot Season of the 'Central Negev Highlands' Project. *Bulletin of the American Schools of Oriental Research* 232: 29-45.
1979   Preliminary Report of the Second Season of the 'Central Negev Highlands' Project. *Bulletin of the American Schools of Oriental Research* 236: 41-60.
1981   Preliminary Report of the Third and Final Season of the 'Central Negev Highlands' Project. *Bulletin of the American Schools of Oriental Research* 243: 57-77.

Crowfoot Payne, J.
1948   Some Flint Implements from 'Affula. *Journal of the Palestine Oriental Society* 21: 72-78.
1983   The Flint Industries of Jericho. In K.M. Kenyon and T.A. Holland (eds.), *Excavations at Jericho. V. The Pottery Phases of the Tell and Other Finds*, 682-758. London: British School in Jerusalem.

Davey, C.J.
1979   Some Ancient Near Eastern Pot Bellows. *Levant* 11: 101-11.

De Geus, G.H.J.
1971   The Amorites in the Archaeology of Palestine. *Ugarit Forschungen* 3: 41-60.

De Vaux, R.
1946   Les Patriarches Hébreux et les découvertes modernes, I. *Revue Biblique* 53: 321-48.

Dever, W.G.
1970   The 'Middle Bronze I Period' in Syria-Palestine. In J.A. Sanders (ed.), *Near Eastern Archaeology in the Twentieth Century: Essays in Honor of Nelson Glueck*, 132-63. Garden City, NY: Doubleday.
1971   The Peoples of Palestine in the Middle Bronze I Period. *Harvard Theological Review* 64: 197-226.
1972   A Middle Bronze I Site in the West Bank of the Jordan. *Archaeology* 25: 231-33.
1973   The EB IV–MB I Horizon in Transjordan and Southern Palestine. *Bulletin of the American Schools of Oriental Research* 210: 37-63.
1975a   A Middle Bronze I Cemetery at Khirbat al-Kirmil. *Eretz Israel* 12: 18*-33*.
1975b   MB IIA Cemeteries at 'Ain es-Samiyeh and Sinjil. *Bulletin of the American Schools of Oriental Research* 217: 23-36.

1980 New Vistas on the EB IV (MB I) Horizon in Syria-Palestine. *Bulletin of the American Schools of Oriental Research* 237: 35-64.
1981 Cave G26 at Jabal Qa'aqir: A Domestic Assemblage of Middle Bronze I. *Eretz Israel* 15: 22*-32*.
1983 Be'er Resisim: A Late Third Millennium B.C.E. Settlement. *Qadmoniot* 16: 52-57 (Hebrew).
1985 Village Planning at Be'er Resisim and Socio-Economic Structure in Early Bronze Age IV Palestine. *Eretz Israel* 18: 18*-28*.
1987 Funerary Practices in EB IV (MB I) Palestine: A Study in Cultural Discontinuity. In J.H. Marks and R.M. Goods (eds.), *Love and Death in the Ancient Near East: Essays in Honor of Marvin H. Pope*, 9-19. London: Guilford.
1992 Pastoralism and the End of the Urban Early Bronze Age in Palestine. In O. Bar-Yosef and A. Khazanov (eds.), *Pastoralism in the Levant: Archaeological Materials in Anthropological Perspective*, 83-92. Madison, WI: Prehistory.
1995 Social Structure in the Early Bronze IV Period in Palestine. In T.E. Levy (ed.), *The Archaeology of Society in the Holy Land*, 282-96. London: Leicester University Press.

Eisenberg, E.
1985 A Burial Cave of the Early Bronze Age IV (MB I) near 'Enan. *'Atiqot* 17: 59-74 (English Series).

Epstein, C.
1985 Dolmen Excavated in the Golan. *'Atiqot* 17: 20-58 (English Series).

Esse, D.L.
1991 *Subsistence, Trade, and Social Change in Early Bronze Age Palestine*. Chicago: Oriental Institute.

Fabietti, U.
1982 *Nomadi del Medio Oriente*. Torino: Loescher.

Falconer, S.E.
1987a Heartland of Villages: Reconsidering Early Urbanism in the Southern Levant. Unpublished PhD dissertation, University of Arizona, Tucson.
1987b Village Pottery Production and Exchange: a Jordan Valley Perspective. In A. Hadidi (ed.), *Studies in the History and Archaeology of Jordan*, III: 251-59. Amman: Department of Antiquities.

Falconer, S.E. and B. Magness-Gardiner
1984 Preliminary Report of the First Season of the Tell al-Hayyat Project. *Bulletin of the American Schools of Oriental Research* 255: 49-74.
1989 Hayyat (Tell al-). In D. Homès-Fredericq and J.B. Hennessy (eds.), *Archaeology of Jordan*. II.1. *Field Reports: Surveys and Sites A–K*, 254-61. Akkadica Supplementum 7. Leuven: Peeters.

Finkelstein, I.
1989 Further Observations on the Socio-Demographic Structure of the Intermediate Bronze Age. *Levant* 21: 129-40.
1992 Pastoralism in the Highlands of Canaan in the Third and Second Millennia B.C.E. In O. Bar-Yosef and A. Khazanov (eds.), *Pastoralism in the Levant*, 133-42. Madison, WI: Prehistory.
1995 *Living on the Fringe*. Sheffield: Sheffield Academic.

Gerstenblith, P.
1983 *The Levant at the Beginning of the Middle Bronze Age*. American Schools of Oriental Research Dissertation Series, 5. Winona Lake, IN: Eisenbrauns.

Gitin, S.
1975 Middle Bronze I 'Domestic Pottery' at Jabal Qa'aqir: A Ceramic Inventory of Cave G23. *Eretz Israel* 12: 46-62.

Glueck, N.
1934 *Explorations in Eastern Palestine I*. Annual of the American Schools of Oriental Research 14. New Haven, CN: American Schools of Oriental Research.
1935 *Explorations in Eastern Palestine II*. Annual of the American Schools of Oriental Research 15. New Haven, CN: American Schools of Oriental Research.
1939 *Explorations in Eastern Palestine III*. Annual of the American Schools of Oriental Research 18-19. New Haven, CN: American Schools of Oriental Research.
1951 *Explorations in Eastern Palestine IV*. Annual of the American Schools of Oriental Research 25-28. New Haven, CN: American Schools of Oriental Research.

Gordon, R.L. and L.E. Villiers
1983 Telul edh-Dhahab and its Environs: Surveys of 1980 and 1982: A Preliminary Report. *Annual of the Department of Antiquities of Jordan* 27: 275-89.

Goren, Y.
1996 The Southern Levant in the Early Bronze Age IV: The Petrographic Perspective. *Bulletin of the American Schools of Oriental Research* 303: 33-72.

Grigson, C.
1995 Plough and Pasture in the Early Economy of the Southern Levant. In T. Levy (ed.), *The Archaeology of Society in the Holy Land*, 245-63. London: Leicester University Press.

Haiman, M.
1996 Early Bronze Age IV Settlement Pattern of the Negev and Sinai Deserts: View from Small Marginal Temporary Sites. *Bulletin of the American Schools of Oriental Research* 303: 1-32.

Haldar, A.
1971 *Who Were the Amorites?* Leiden: E.J. Brill.

Hammond, W.M.
1977 The Raw and the Chipped: An Analysis of Correlations between Raw Materials and Tools of a Lithic Industry

from Tell el-Hesi, Israel. Unpublished PhD dissertation, Columbia University.

Harrison, T.P.
1997 Shifting Patterns of Settlement in the Highlands of Central Jordan during the Early Bronze Age. *Bulletin of the American Schools of Oriental Research* 306: 1-38.

Hauptmann, A.
1986 Archaeometallurgical and Mining-Archaeological Studies in the Eastern 'Arabah, Feinan Area, 2nd Season. *Annual of the Department of Antiquities of Jordan* 30: 415-19.

Hauptmann, A. and G. Weisgerber
1987 Archaeometallurgical and Mining-Archaeological Investigations in the Area of Feinan, Wadi 'Arabah (Jordan). *Annual of the Department of Antiquities of Jordan* 31: 419-37.

Helms, S.W.
1983 The EB IV (EB–MB) Cemetery at Tiwal esh-Sharqi, in the Jordan Valley, 1983. *Annual of the Department of Antiquities of Jordan* 27: 55-85.
1986 Excavations at Tell Um Hammad, 1984. *Levant* 18: 25-49.
1989 An EB IV Pottery Repertoire at Amman, Jordan. *Bulletin of the American Schools of Oriental Research* 273: 17-36.

Helms, S.W. and D.W. McCreery
1988 Rescue Excavations at Umm el-Bighal: The Pottery. *Annual of the Department of Antiquities of Jordan* 32: 319-47.

Hess, O.
1984 MB I Tombs at Tel 'Artal. *Bulletin of the American Schools of Oriental Research* 253: 55-60.

Ibach, R.
1978 Expanded Archaeological Survey of the Hesban Region. *Andrews University Seminar Studies* 16: 201-13.

Ibrahim, M. and S. Mittmann
1987 Tell el-Mughayyir and Khirbat Zeraqoun. *Newsletter of the Institute of Archaeology and Anthropology of Yarmouk University* 4: 3-6.

Ibrahim, M., J.A. Sauer and K. Yassine
1976 The East Jordan Valley Survey, 1975. *Bulletin of the American Schools of Oriental Research* 222: 41-66.
1988 The East Jordan Valley Survey, 1976 (Part Two). In K. Yassine (ed.), *Archaeology of Jordan: Essays and Reports*, 189-207. Amman: Department of Archaeology, University of Jordan.

Joffe, A.H.
1991 Early Bronze I and the Evolution of Social Complexity in the Southern Levant. *Journal of Mediterranean Archaeology* 4: 3-58
1993 *Settlement and Society in the Early Bronze Age I and II, Southern Levant.* Sheffield: Sheffield University.

Johnson, G.A.
1972 A Test of the Utility of Central Place Theory to Archaeology. In P. Ucko, R. Tringham and G.W. Dimbleby (eds.), *Man, Settlement, and Urbanism*, 769-85. London: Duckworth.
1973 *Local Exchange and Early State Development in Southwestern Iran.* Anthropological Papers 51. Ann Arbor, MI: Museum of Anthropology, University of Michigan.

Kamp, K.A. and N. Yoffee
1980 Ethnicity in Ancient Western Asia during the Early Second Millennium B.C.: Archaeological Assessments and Ethnoarchaeological Prospectives. *Bulletin of the American Schools of Oriental Research* 237: 85-104.

Kenyon, K.M.
1951 Some Notes on the History of Jericho in the Second Millennium B.C. *Palestine Exploration Quarterly* 83: 101-38.
1956 Tombs of the Intermediate Early Bronze–Middle Bronze at Tell el-Ajjul. *Annual of the Department of Antiquities of Jordan* 3: 41-55.
1960 *Excavations at Jericho. I. The Tombs Excavated in 1952–1954.* London: British School of Archaeology in Jerusalem.
1965 *Excavations at Jericho. II. The Tombs Excavated in 1955–1958.* London: British School of Archaeology in Jerusalem.
1966 *Amorites and Canaanites.* London: Oxford University.

Kerestes, T.M. *et al.*
1978 An Archaeological Survey of Three Reservoir Areas in Northern Jordan, 1978. *Annual of the Department of Antiquities of Jordan* 22: 108-35.

Khalil, L.
1984 Metallurgical Analyses of Some Weapons from Tell el-'Ajjul. *Levant* 16: 167-70.

Khazanov, A.M.
1978 Characteristic Features of Nomadic Communities in the Eurasian Steppes. In W. Weissleder (ed.), *The Nomadic Alternative*, 119-26. The Hague: Mouton.
1984 *Nomads and the Outside World.* Cambridge: Cambridge University Press.

Khouri, R.G.
1988 *The Antiquities of the Jordan Rift Valley.* Amman: Al-Kutba.

Kochavi, M.
1963a Har Yeruham. *Israel Exploration Journal* 13: 141-42.
1963b The Excavations at Tel Yeruham. *Yediot* 27: 284-92 (Hebrew).
1973 A Built Shaft Tomb of the MB I at Degania 'A'. *Qadmoniot* 6: 50-53 (Hebrew).

LaBianca, O.S.
1990 *Sedentarization and Nomadization: Food System Cycles at Hesban and Vicinity in Transjordan.* Hesban 1. Berrien Springs, MI: Andrews University.

Lapp, P.W.
1966 *The Dhahr Mirzbaneh Tombs.* New Haven, CN: American Schools of Oriental Research.

Lees, S.H. and D.G. Bates
1974 The Origins of Specialized Pastoral Nomadism: A Systemic Model. *American Antiquity* 39: 187-93.

Lemche, N.P.
1985 *Early Israel.* Leiden: E.J. Brill.

Levy, E.T.
1983 The Emergence of Specialized Pastoralism in the Southern Levant. *World Archaeology* 15: 15-36.

Liverani, M.
1970 Per una considerazione storica del problema amorreo. *Oriens Antiquus* 9: 5-27.
1973 The Amorites. In D.J. Wiseman (ed.), *Peoples of Old Testament Times*, 100-33. Oxford: Clarendon.
1988 *Antico Oriente: Storia Società Economia.* Bari: Laterza.

London, G.A.
1987 Homage to the Elders. *Biblical Archaeologist* 50: 70-74.

Long, J.C.
1986 Sedentism in Early Bronze IV Palestine-Transjordan: An Analysis of Sociocultural Variability in the Late Third Millennium BC. Paper presented at the Annual Meetings of the American Schools of Oriental Research, Atlanta, GA.

Loud, G.
1948 *Megiddo II, Seasons of 1935–1939* (OIP 162). Chicago: Oriental Institute.

Mabry, J.
1989 Investigations at Tell el-Handaquq, Jordan (1987–88). *Annual of the Department of Antiquities of Jordan* 33: 59-95.

Mabry, J. and G. Palumbo
1988 The 1987 Wadi el-Yabis Survey. *Annual of the Department of Antiquities of Jordan* 32: 275-305.

MacDonald, B.
1995 EB IV Tombs at Khirbat Khanazir: Types, Construction, and Relation to Other EB IV Tombs in Syria-Palestine. In K. 'Amr, F. Zayadine and M. Zaghloul (eds.), *Studies in the History and Archaeology of Jordan*, V: 129-34. Amman: Department of Antiquities.

MacDonald, B. *et al.*
1980 The Wadi el-Hasa Survey 1979: A Preliminary Report. *Annual of the Department of Antiquities of Jordan* 24: 169-83.
1982 The Wadi el-Hasa Survey 1979 and Previous Archaeological Work in Southern Jordan. *Bulletin of the American Schools of Oriental Research* 245: 35-52.
1987 Southern Ghors and Northeast 'Arabah Archaeological Survey 1986, Jordan: A Preliminary Report. *Annual of the Department of Antiquities of Jordan* 31: 391-413.
1988 *The Wadi el-Hasa Archaeological Survey 1979–1983, West-Central Jordan.* Waterloo, Ontario: Wilfrid Laurier University.
1992 *The Southern Ghors and Northeast 'Arabah Archaeological Survey.* Sheffield Archaeological Monographs 5. Sheffield: J.R. Collis.

Marfoe, L.
1979 The Integrative Transformation: Patterns of Sociopolitical Organization in Southern Syria. *Bulletin of the American Schools of Oriental Research* 234: 1-42.

Marx, E.
1992 Are There Pastoral Nomads in the Middle East? In O. Bar-Yosef and A. Khazanov (eds.), *Pastoralism in the Levant*, 255-60. Madison, WI: Prehistoric.

Matthiae, P.
1980 Sulle asce fenestrate del 'Signore dei Capridi'. *Studi Eblaiti* 3.3-4: 53-62.
1984 *I Tesori di Ebla.* Bari: Laterza.
1989 *Ebla, un Impero Ritrovato.* Torino: Einaudi.

Mattingly, G.L.
1984 The Early Bronze Age Sites of Central and Southern Moab. *Near East Archaeological Society Bulletin, New Series* 23: 69-88.

Mazzoni, S.
1985 Elements of the Ceramic Culture of Early Syrian Ebla in Comparison with Syro-Palestinian EB IV. *Bulletin of the American Schools of Oriental Research* 257: 1-18.

McConaughy, M.
1979 Formal and Functional Analyses of Chipped Stone Tools from Bab edh-Dhra. Unpublished PhD dissertation, Ann Arbor, MI.

Mittmann, S.
1970 *Beiträge zur Siedlungs- und Territorial-Geschichte des Nördlichen Ostjordanlandes.* Wiesbaden: Abhandlungen des Deutschen Palästinavereins.

Moorey, P.R.S. and F. Schweizer
1972 Copper and Copper Alloys in Ancient Iraq, Syria and Palestine: Some New Analyses. *Archaeometry* 14: 177-97.

Muhly, J.D.
1985 Sources of Tin and the Beginnings of Bronze Metallurgy. *American Journal of Archaeology* 89: 275-91.

Negbi, O.
1966 *Canaanite Burial Caves at Hanita*. Hanita: Kibbutz Hanita (Hebrew).

Neuville, R.
1930 Notes de préhistoire Palestinienne. *Journal of the Palestine Oriental Society* 10: 64-75.

Olávarri, E.
1969 Fouilles à Aro'er sur l'Arnon. *Revue Biblique* 76: 230-59.

Oren, E.D.
1973 *The Northern Cemetery of Beth Shan*. Leiden: E.J. Brill.

Palumbo, G.
1986 Per un'analisi delle sepolture contratte del Bronzo Antico IV di Gerico. *Contributi e Materiali di Archeologia Orientale* 1: 287-306.
1987 'Egalitarian' or 'Stratified' Society? Some Notes on Mortuary Practices and Social Structure at Jericho in EB IV. *Bulletin of the American Schools of Oriental Research* 267: 43-59.
1990 *The Early Bronze Age IV in the Southern Levant: Settlement Patterns, Economy, and Material Culture of a 'Dark Age'*. Contributi e Materiali di Archeologia Orientale III. Roma: Università di Roma 'La Sapienza'.

Palumbo, G. and G. Peterman
1993 Early Bronze Age IV Ceramic Regionalism in Central Jordan. *Bulletin of the American Schools of Oriental Research* 289: 23-32

Palumbo, G., J. Mabry and I. Kuijt
1990 The Wadi el-Yabis Survey: Report on the 1989 Field Season. *Annual of the Department of Antiquities of Jordan* 34: 95-118.

Palumbo, G., *et al.*
1996 The Wadi az-Zarqa/Wadi ad-Dulayil Excavations and Survey Project: Report on the October–November 1993 Fieldwork Season. *Annual of the Department of Antiquities of Jordan* 40: 375-427.

Patai, R.
1951 Nomadism: Middle Eastern and Central Asian. *Southwestern Journal of Anthropology* 7: 401-14.

Petrie, W.F.
1932 *Ancient Gaza*, II. London: British School of Archaeology in Egypt.

Prag, K.
1971 A Study of the Intermediate Early Bronze–Middle Bronze Age in Transjordan, Syria and Lebanon. Unpublished PhD dissertation, St Hugh's College, University of Oxford.
1974 The Intermediate Early Bronze–Middle Bronze Age: An Interpretation of the Evidence from Transjordan, Syria and Lebanon. *Levant* 6: 69-116.
1984 Review Article: Continuity and Migration in the South Levant in the Late Third Millennium: A Review of T.L. Thompson's and Some Other Views. *Palestine Exploration Quarterly* 116: 58-68.
1985 Ancient and Modern Pastoral Migration in the Levant. *Levant* 17: 81-88.
1986 The Intermediate Early Bronze–Middle Bronze Age Sequences at Tell Iktanu Reviewed. *Bulletin of the American Schools of Oriental Research* 264: 61-72.
1988 Kilns of the Intermediate Early Bronze–Middle Bronze Age at Tell Iktanu: Preliminary Report, 1987 Season. *Annual of the Department of Antiquities of Jordan* 32: 59-72.
1989 Preliminary Report on the Excavations at Tell Iktanu, Jordan, 1987. *Levant* 21: 33-45.

Rast, W.E. and R.T. Schaub
1974 Survey of the South Eastern Plain of the Dead Sea, 1973. *Annual of the Department of Antiquities of Jordan* 19: 5-53.
1978 A Preliminary Report of Excavations at Bab edh-Dhra', 1975. In D.N. Freedman (ed.), *Preliminary Excavation Reports: Bab edh-Dhra', Sardis, Meiron, Tell el-Hesi, Carthage (Punic)*, 1-39. Cambridge, MA: American Schools of Oriental Research.
1980 Preliminary Report of the 1979 Expedition to the Dead Sea Plain, Jordan. *Bulletin of the American Schools of Oriental Research* 240: 21-61.
1981 *The Southeastern Dead Sea Plain Expedition: An Interim Report of the 1977 Season*. Annual of the American Schools of Oriental Research 46. Cambridge, MA: American Schools of Oriental Research.

Richard, S.
1983 Report on the 1982 Season of Excavations at Khirbet Iskander. *Annual of the Department of Antiquities of Jordan* 27: 45-53.
1987 The Early Bronze Age in Palestine: The Rise and Collapse of Urbanism. *Biblical Archaeologist* 50: 22-43.
1990 The 1987 Expedition to Khirbet Iskander and its Vicinity: Fourth Preliminary Report. In W.E. Rast (ed.), *Preliminary Reports of ASOR-Sponsored Excavations 1983–1987*. Bulletin of the American Schools of Oriental Research Supplement, 26, 33-58. Baltimore, MD: American Schools of Oriental Research.

Richard, S. and R.S. Boraas
1984 Preliminary Report of the 1981–82 Seasons of the Expedition to Khirbet Iskander and its Vicinity. *Bulletin of the American Schools of Oriental Research* 254: 63-87.
1988 The Early Bronze IV Fortified Site of Khirbet Iskander, Jordan: Third Preliminary Report, 1984 Season. In W.E. Rast (ed.), *Preliminary Reports of ASOR-Sponsored Excavations 1982–1985*. Bulletin of the American Schools of Oriental Research Supplement, 25, 107-30. Baltimore, MD: American Schools of Oriental Research.

Richard, S. and J.C. Long
  1989 Specialization/Despecialization: A Model to Explain Culture Change at the End of the Early Bronze Age, ca. 2350–2000 B.C. Paper read at the Annual Meetings of the American Schools of Oriental Research, Anaheim, CA.
  1995 Archaeological Expedition to Khirbet Iskander and its Vicinity, 1994. *Annual of the Department of Antiquities of Jordan* 39: 81-92.

Rosen, S.A.
  1982 Flint Sickle-Blades of the Late Prehistoric and Early Historic Periods in Israel. *Tel Aviv* 9: 139-45.
  1983 The Canaanean Blade and the Early Bronze Age. *Israel Exploration Journal* 33: 15-29.
  1988 Notes on the Origins of Pastoral Nomadism: A Case Study from the Negev and Sinai. *Current Anthropology* 29: 498-506.
  1997 *Lithics after the Stone Age*. Walnut Creek, CA: Altamira.

Rothenberg, B.
  1978 Chalcolithic Copper Smelting. *Archaeo-Metallurgy* 1: 1-23.

Rowton, M.B.
  1967 Physical Environment and the Problem of the Nomads. In J.R. Kupper (ed.), *La civilisation de Mari*, 109-21. Paris: Les Belles Lettres.
  1973 Autonomy and Nomadism in Western Asia. *Orientalia* 42: 242-58.
  1976 Dimorphic Structure and the Tribal Elite. *Studia Instituti Anthropos* 30.
  1977 Dimorphic Structure and the Parasocial Element. *Journal of Near Eastern Studies* 36: 181-98.

Schaub, R.T. and W.E. Rast
  1984 Preliminary Report of the 1981 Expedition to the Dead Sea Plain, Jordan. *Bulletin of the American Schools of Oriental Research* 254: 35-60.
  1989 *Bab edh-Dhra: Excavations in the Cemetery Directed by Paul W. Lapp (1965–1967)*. Winona Lake, WI: Eisenbrauns.

Service, E.R.
  1962 *Primitive Social Organization*. New York: Random House.

Shay, T.
  1983 Burial Customs at Jericho in the Intermediate Bronze Age: A Componential Analysis. *Tel Aviv* 10: 26-37.

Sherratt, A.
  1981 Plough and Pastoralism: Aspects of the Secondary Products Revolution. In I. Hodder, G. Isaac and N. Hammond (eds.), *Patterns of the Past: Studies in Honor of David Clarke*, 261-305. Cambridge: Cambridge University.

Stech, T., J.D. Muhly and R. Maddin
  1985 Metallurgical Studies on Artifacts from the Tomb Near 'Enan. *'Atiqot* 17: 75-82 (English Series).

Thompson, T.L.
  1975 *The Settlement of Sinai and the Negev in the Bronze Age*. Wiesbaden: Harrassowitz.
  1978 The Background of the Patriarchs: A Reply to Dever and Clark. *Journal for the Study of the Old Testament* 9: 2-43.
  1979 *The Settlement of Palestine in the Bronze Age*. Wiesbaden: Harrassowitz.

Tubb, J.N.
  1985 Excavations in the Early Bronze Age Cemetery of Tiwal esh-Sharqi: A Preliminary Report. *Annual of the Department of Antiquities of Jordan* 29: 115-30.

Tufnell, O. et al.
  1958 *Lachish IV: The Bronze Age*. London: Oxford University.

Tzaferis, V.
  1968 A Middle Bronze I Cemetery in Tiberias. *Israel Exploration Journal* 18: 15-19.

Waechter, J.
  1958 Flint Implements. In O. Tufnell (ed.), *Lachish IV: The Bronze Age*, 325-27. London: Oxford University.

Waheeb, M. and G. Palumbo
  1993 Salvage Excavations at a Bronze Age Cemetery near Tell el-Umeiri. *Annual of the Department of Antiquities of Jordan* 37: 147-63.

Weiss, H. (ed.)
  1985 *Ebla to Damascus: Art and Archaeology of Ancient Syria*. Washington, DC: Smithsonian Institution.

Wightman, G.J.
  1988 An EBIV Cemetery in the North Jordan Valley. *Levant* 20: 139-59.

Wright, G.E.
  1937 *The Pottery of Palestine from the Earliest Times to the End of the Early Bronze Age*. New Haven, CN: American Schools of Oriental Research.
  1938 The Chronology of Palestinian Pottery in Middle Bronze I. *Bulletin of the American Schools of Oriental Research* 71: 27-34.

Zayadine, F.
  1978 An EB–MB Bilobate Tomb at Amman. In P.R.S. Moorey and P.J. Parr (eds.), *Archaeology in the Levant: Essays in Honor of Kathleen M. Kenyon*, 59-66. Warminster: Aris & Phillips.

Zohar, M.
  1989 Rogem Hiri: A Megalithic Monument in the Golan. *Israel Exploration Journal* 39: 18-31.

7. The Middle Bronze Age

Steven Falconer

Introduction

The beginning of the Middle Bronze Age (c. 2000 BC) was signalled by the rapid rejuvenation of towns and cities atop the stratified tells of the southern Levant. While some communities were founded anew, many sites that had lain abandoned during the collapse of urbanism in the last centuries of the Early Bronze Age were now reoccupied, expanded and refortified. Although the cultural heritage of the entire Bronze Age is commonly ascribed to the Canaanites of the Old Testament (Dever 1987; Joffe 1993) and other ancient texts (e.g. Matthiae 1981: 187), the archaeological record for the Middle Bronze Age represents 'a major break in terms of technology, trade, and social and political institutions from the preceding period' (Gerstenblith 1983: 123). In contrast, the technological and stylistic continuity of material culture through the Middle Bronze Age and into the Late Bronze Age is striking. Recent archaeological syntheses focused west of the Jordan River (e.g. Dever 1987; Mazar 1990; Ilan 1995) emphasize pronounced increases in regional population density, and in the number and size of Levantine cities, all of which reached levels by the end of the Middle Bronze Age (c. 1500 BC) that remained unsurpassed until the Roman and Byzantine periods more than a millennium later (Broshi 1979).

In the lands of modern Jordan, the Middle Bronze Age represents a fundamental archaeological paradox. While this period represents the high point of urbanized Canaanite society in the southern Levant generally, material evidence of Middle Bronze Age society remains relatively elusive east of the Jordan River. On the basis of his pioneering explorations, Glueck (1970: 21) surmised that between major proliferation of settlement in the Early Bronze and the Iron Ages, Jordan witnessed 'a gap in the history of permanent sedentary occupation, lasting from the end of the twentieth to the beginning of the thirteenth century BC'. In light of a growing abundance of archaeological research conducted since the 1970s, we may now view Glueck's assessment as an exercise in hyperbole. In many areas of Jordan, Middle Bronze Age settlements *are* less mboxnumerous and substantial than those that preceded or followed them. However, a variety of excavations and surveys provide tantalizing glimpses of Middle Bronze Age town and village life that augment our appreciation of Canaanite society, and may help explain the apparently modest Middle Bronze Age settlement and population densities east of the Jordan.

Chronology and nomenclature

The beginning and end dates traditionally assigned to the Middle Bronze Age derive, in large part, from the dynastic history of ancient Egypt. Between roughly 2300 and 2000 BC centralized pharaonic authority apparently collapsed during the 'First Intermediate Period' of Egypt, with serious political and economic ramifications for neighbouring societies. The contemporaneous time span in the southern Levant included periods labeled 'EB IV' and 'MB I' by Albright (1962, 1966). Alternative nomenclatures now prevail in which the entire interval is labelled 'EB IV' or 'Intermediate Early Bronze–Middle Bronze', for example (see Chapter 6).

The beginning of the Middle Bronze Age is correlated axiomatically with the end of the First Intermediate Period and the ascension of Egypt's Twelfth Dynasty c. 1991 BC, which reestablished centralized government and widespread commercial exchange in the eastern Mediterranean (see, e.g., Weinstein 1975, 1981). Albright's pioneering chronology divided MB II into sub-periods A, B and C, which spanned roughly the same time range as Egypt's Middle Kingdom and 'Second Intermediate Period' (see Table 7.1). During five centuries of considerable population increase (see Broshi and Gophna 1986), the rebirth of Levantine towns in MB IIA was followed by their expansion into fortified cities in MB IIB and C, culminating with the apex of Canaanite urban development c. 1500 BC. Middle Bronze IIC coincided with another era of political turmoil in Egypt during which the Fifteenth Dynasty of 'Asiatic' or 'Hyksos' kings ruled much of Lower Egypt. Military disruptions associated with the

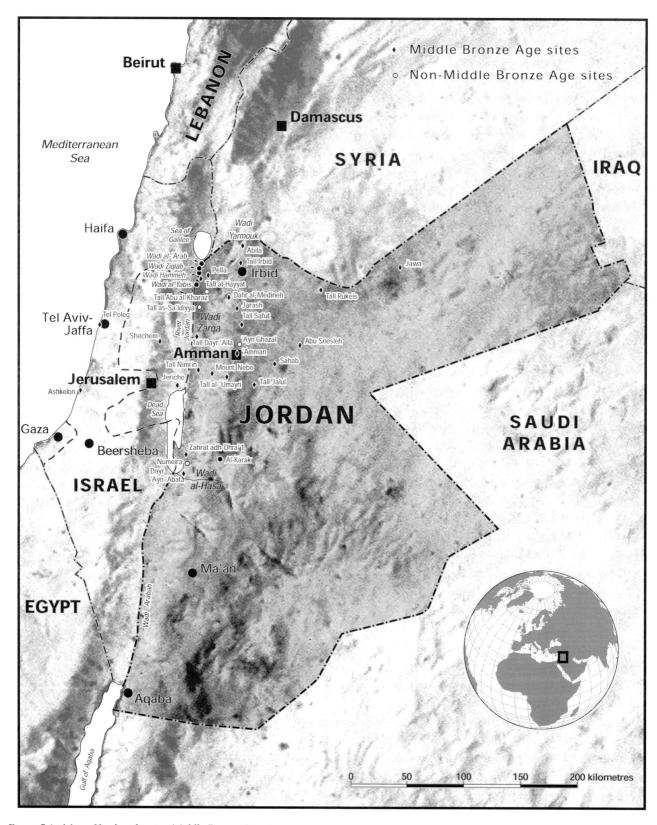

Figure 7.1. Map of Jordan showing Middle Bronze Age sites.

		Southern Levant			
Egypt	Years BC	Albright	Israeli	Kenyon	Dever
Second Intermediate (14th–17th Dynasties)	c. 1650–1500	MB IIC	MB IIB	MB II	MB III
Middle Kingdom	c. 1800–1650	MB IIB	MB IIB	MB II	MB II
(12th and 13th Dynasties)	c. 2000–1800	MB IIA	MB IIA	MB I	MB I

Source: Dever 1987: 149.

Table 7.1. Middle Bronze Age chronology and nomenclature.

collapse of Hyksos power, and possibly their expulsion into the southern Levant, provide an end point for the Middle Bronze Age that is marked archaeologically by the destruction of many Canaanite cities (Weinstein 1991).

In addition to presumed political linkages, the tripartite Middle Bronze Age chronology has hinged on selected innovations in city fortifications as implicit material reflections of the process of urbanization. For example, following their appearance at a few sites in MB IIA, the virtual ubiquity of rampart fortifications (i.e. featuring an earthen core and banked slopes constructed against town walls or tell slopes) among the cities of the southern Levant is cited as a hallmark of MB IIB (Dever 1976: 16; Mazar 1990: 198). A variant form of rampart fortification, known as a glacis, first materialized in MB IIB (a single MB IIA example is reported from Tel Poleg, Israel) (Mazar 1990: 202). These hard-packed ramparts lay at the foot of masonry fortification walls and usually had a coating of lime, creating a smooth steep impediment to ladders or siege equipment. Similarly, the construction of Cyclopean walls (i.e. characterized by huge unhewn stonework), following town destructions in the late seventeenth century BC, was used as a defining feature of MB IIC a generation ago (e.g. Mazar 1968: 91), but tends to be disregarded as a chronological marker today.

The most striking regional innovations in town architecture emerged at the beginning of MB IIB. Hence, there has been a general tendency to emphasize the more readily visible distinctions between MB IIA and B, while downplaying contrasts between MB IIB and C. For example, most Israeli archaeologists subsume Albright's MB IIB and C within 'MB IIB', citing greater continuity than disjunction in material culture and architecture. Following this reasoning, Kenyon (1973) also suggested two Middle Bronze Age sub-periods, to be relabelled 'MB I' and 'MB II'. Subsequently, Gerstenblith (1983: 2-3) and Dever (1987) have returned to a tripartite Middle Bronze chronology that includes sub-periods I, II and III. Since most literature retains Albright's nomenclature, this chapter will follow suit. In fact, the terminological controversy summarized above obscures a more important consensus that the Middle Bronze Age, however it is subdivided and labelled, witnessed the dramatic reappearance and more gradual florescence of Canaanite cities and urbanized society.

More fundamentally, Ilan (1995: 299) has called for an independent, radiocarbon-based Middle Bronze Age chronology based on the scarcity of chronologically sensitive Egyptian material culture in the Levant, the present abundance of alternative chronological schemes (as noted above) and the poorly understood geographical variability that marks some Middle Bronze Age evidence. As a chronological caveat, we should note the potential tautology of using Egyptian political transitions to define Middle Bronze Age chronology for the Levant, and subsequently explaining the major junctures within the Middle Bronze Age on the basis of Egyptian political dynamics. Unfortunately, an independent Middle Bronze Age chronology will not emerge until a sufficiently robust collection of radiocarbon dates (with the inherent uncertainties of standard deviations and recalibration) can counteract the allure of apparently firm, historically based dates from Egypt and elsewhere. Presently, data for historically independent chronologies (radiocarbon or otherwise) are embryonic at best.

Material culture

The beginning of the Middle Bronze Age featured dramatic discontinuities in material culture and settlement patterns. For example, the hand-built pottery of the Early Bronze Age was replaced by a new repertoire of wheel-thrown vessel forms, and bronze tools augmented a metal industry limited previously to copper. As Dever noted two decades ago, 'the break between [the Early and Middle Bronze ages] in terms of their material culture is one of the most abrupt and complete in the entire cultural sequence of Palestine' (1976: 5). Previous attributions of this disjunction to the arrival of new populations, especially the Amorites (Mazar 1968; Kenyon 1973; Dever 1976), in the southern Levant do not figure prominently in Middle Bronze Age research at present (but cf. Mazar 1990: 189; Ilan 1995: 300-301).

The innovation of the fast potter's wheel permitted a broadened range of technically more sophisticated vessel forms, many with Syrian parallels, but very few of which derive from EB IV ceramics. Basic vessel forms, which provide the basis for regional ceramic chronologies developed over the last century, display striking continuity through the Middle Bronze Age and into the Late Bronze Age, as an important aspect of Canaanite material culture. Middle Bronze Age fineware pottery is particularly indicative of the increasing importance of regional ceramic exchange. This aspect of ceramic technology became particularly accentuated in the ceramic trade relations between the Levant and Cyprus in the Late Bronze Age.

A notable characteristic of MB IIA pottery is burnished red slip decoration, particularly common on small vessels. Painted motifs, again often on smaller forms and modestly sized jars, often feature horizontal black or red bands. While some studies suggest derivation of Levantine painted styles from similar examples of 'Khabur Ware' from northern Syria (e.g. Dever 1976; Gerstenblith 1983: 59-64), others argue that this Levantine decorative tradition is independent (Tubb 1983).

During MB IIB, red slip decoration gradually gave way to white or cream slips, again primarily on smaller vessel forms. Tall al-Yehudiyeh Ware, a distinctive dark greyish-black ceramic with punctate decoration, was used primarily to manufacture juglets. Chemical analysis of Tall al-Yehudiyeh vessels from a variety of locales suggests its manufacture at two centres, one in the eastern Nile Delta and the second at or near Afula, in modern northern Israel (Kaplan 1980). Its widespread occurrence along the Mediterranean seaboard and on Cyprus attests to a growing regional trade in pottery and other commodities.

The pottery of MB IIC features two distinct fineware traditions: Chocolate-on-White Ware and Bichrome Ware. Chocolate-on-White vessels are found most commonly in the northern Jordan Valley, where many apparently were manufactured (Knapp 1993). Bichrome Ware, characterized by its red and black painted decoration, is found more widely distributed throughout the southern Levant. Chemical analyses suggest that most Bichrome vessels were made on Cyprus (Mazar 1990: 259-61). Both of these decorative traditions continue into the Late Bronze Age as parts of a larger Canaanite cultural continuity through the second millennium BC (Mazar 1990: 261).

Human impact on the Bronze Age environment

Before considering the social implications of regional settlement patterns, it is imperative that we consider evidence for the increasingly humanized environmental setting to which Middle Bronze Age urbanism contributed and in which it flourished. The natural environmental setting for Levantine agrarian prehistory was far from pristine. A variety of evidence from 'Ayn Ghazal attests to severe *localized* deforestation as early as the Pre-Pottery Neolithic (e.g. Rollefson and Kohler-Rollefson 1992; Rollefson et al. 1992). A second wave of agricultural innovation, featuring intensive orchard cultivation, did not become pronounced until the third and second millennia BC, roughly 5000 years after the domestication of cereals and legumes (N.F. Miller 1991). Palynological analysis of lake cores taken along the Jordan Rift (see Baruch 1990) suggests that the major impetus for the anthropogenic vegetation so characteristic of Jordan and the eastern Mediterranean stems from agricultural intensification to serve emerging urban markets. This trend was most pronounced during several centuries of Roman rule, but had deeper roots lying at least as early as the Middle Bronze Age (Falconer and Fall 1995).

The domestication of several strategic orchard taxa, including olive (*Olea europea*), grape (*Vitis vinifera*) and fig (*Ficus carica*) began in the Levant during the fourth millennium BC (Zohary and Spiegel-Roy 1975; Neef 1990). Olive, arguably the most important fruit tree of the eastern Mediterranean, was valued primarily as a producer of oil for eating and cooking, ointments and fuel for lamps (Zohary 1982). Olive oil could be stored for long periods, and consequently was a valued trade commodity and a symbol of wealth throughout the Mediterranean basin beginning in the Early Bronze Age (Zohary and Hopf 1988). Olive fruits were not pickled and eaten whole until the Hellenistic and Roman periods (Stager 1985).

The grape vine, a similarly versatile orchard taxon, produced fruits that could be eaten fresh, dried to make raisins, or pressed to render juice and wine (Zohary

The Middle Bronze Age

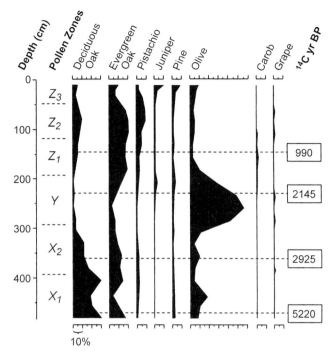

Figure 7.2. Arboreal pollen frequencies from the Sea of Galilee sediment core (after Baruch 1990).

1982). Figs, which could be dried, stored and eaten year round, supplied fresh foodstuffs rich in sugar. Olive oil, wine and dried figs could be sealed in pottery jars, stored and transported, making them particularly suitable for regional exchange (Stager 1985). By the beginning of the Middle Bronze Age, olive, grape and fig had long been established as the three main horticultural taxa in areas of rain-fed agriculture throughout the eastern Mediterranean (Zohary and Hopf 1988).

Analysis of pollen from a 5 m sediment core recovered from the southeastern portion of the Sea of Galilee permits the reconstruction of surrounding vegetation before, during and after the Middle Bronze Age (see Baruch 1990). The pollen assemblages from this core are divided into six zones that illustrate the general course of human environmental impact along the Jordan Rift.

The most striking feature of the Sea of Galilee pollen diagram (Figure 7.2) is a massive peak in olive pollen deposition in Zone Y slightly before 2000 years ago. This peak, coupled with sharply reduced amounts of oak pollen (both deciduous *Quercus boissieri* and evergreen *Q. calliprinos*), graphically reflects an era of widespread forest clearance and particularly large-scale olive oil production to serve the markets of the Roman empire. The deposition of grape pollen provides an impressive reflection of the importance of vineyards on surrounding hill slopes, since grape vines are very modest pollen producers.

Interestingly, the pronounced impact of Roman arboriculture appears to be part of a longer process of culturally induced environmental flux beginning at least 5000 years ago. The two lowest pollen zones from the Sea of Galilee show an aggregate trend of decreased arboreal pollen (primarily oak, but also pistachio [*Pistacia atlantica*] and pine [*Pinus halepensis*]), suggesting long-term forest clearance. In Zone X_1, at the core's base, pollen from dense oak forest declines, and olive pollen increases to a modest peak roughly contemporaneous with the Early and Middle Bronze Ages. Following a subsequent short-term regeneration of oak forest, dwindling oak pollen frequencies, a slight olive peak and the first appearance of grape pollen in Zone X_2 (c. 3000 bp) implicate renewed forest clearance and the cultivation of orchards and vineyards during the Iron Age.

Given its low production levels, the absence of grape pollen from Zone X_1 does not rule out significant Bronze Age vineyard horticulture. Indeed, the earliest evidence of grape cultivation in the southern Levant appears as carbonized seeds excavated from Early Bronze sites, including Numayra (McCreery 1980) and Jericho (Hopf 1983). Fig, the region's third major economic taxon, is a wasp-pollinated tree and therefore does not appear in Levantine pollen records. However, the earliest fig pips date to the Chalcolithic (Hopf 1983) and the Early Bronze Age (McCreery 1980), and evidence of fig cultivation is widespread by the Middle Bronze Age (Lines 1995).

These palynological data for economic and wild taxa alike suggest the beginnings of regional disturbance of natural taxa, including significant deforestation. Extrapolating from the Sea of Galilee pollen record, we may envision a Middle Bronze Age landscape still characterized by large tracts of oak/pistachio woodland (see also Gophna *et al.* 1986; Liphschitz *et al.* 1989). However, with the rise of urban markets and the regional commerce that they must have facilitated, the settlements of the southern Levant now actively remoulded their surrounding countrysides. Lowlands and river valleys became increasingly checkered with annual subsistence crops, while perennial fruit-bearing trees and vines encroached on neighbouring uplands (Stager 1985). The distinctly anthropogenic character of modern Middle Eastern landscapes has been achieved only after several subsequent millennia of human impacts. The *regional* transformation of that landscape, however, is one of the abiding legacies of

Levantine urbanism, which enjoyed its first heyday during the Middle Bronze Age.

Levantine towns and cities

With some sense of the environmental setting in which it developed, we may consider the dynamics of Middle Bronze Age settlement in the southern Levant generally, and in the surveyed locales of Jordan more specifically. While much of the regional populace continued to live in small rural sites, practising seasonal pastoralism and sedentary farming, vanishingly few EB IV sites remained occupied into the Middle Bronze Age. In addition, and in stark contrast to EB IV settlements, relatively large towns and cities, often fortified, appeared atop the region's stratified tells. Many of these larger communities were built on the remains of EB II or III settlements that had remained unoccupied during EB IV.

Interpretive scenarios focused west of the Jordan River argue that Middle Bronze Age towns developed first along the Mediterranean coastal plain and lowland valleys communicating with the coast (Gerstenblith 1983: 118; Mazar 1990: 176-78; Ilan 1995: 301). Indeed, MB IIA walled towns dotted the Levantine coast between Ashkelon to the south and Ras Shamra (ancient Ugarit) in Syria (Kenyon 1973: 84; Ilan 1995: 301). Traditional interpretations hold further that substantial inland towns did not emerge until later in MB IIA (e.g. Shechem) or in MB IIB (e.g. Jericho) (Gerstenblith 1983: 118-19; Mazar 1990: 197-98).

The coastal distribution of towns has been attributed to mercantile influences, especially from Egypt, delivered via maritime or overland trade (Gerstenblith 1983: 109-26; Gophna 1984). Foreign missions have been interpreted from Egyptian statuary found at Levantine coastal sites and from Egyptian documentary evidence (Mazar 1968: 73; Posener 1971: 549). More recently, Bunimovitz (1989: 47-74; see Ilan 1995: 302) has suggested a united Middle Bronze Age coastal polity. However, critical reviews point out the limited archaeological data on which these arguments are built (e.g. Weinstein 1975), and infer a series of localized polities lying along the east–west river drainages that linked the hills of Palestine with the coastal plain (Ilan 1995: 302).

Excavated settlements in Jordan

While the general coastal orientation of the largest Middle Bronze Age towns is apparent (Falconer 1994), and although evidence from MB IIA is poorly represented in Jordan (Dornemann 1983: 15), excavations at a variety of sites east of the Jordan River provide new insights on Middle Bronze Age settlements ranging from small hamlets to sizable towns, some of which date to the very beginning of the period. The overview below begins with excavated evidence from the Jordan Valley and the Dead Sea Plain, then considers the archaeologically known towns of the northern and central Transjordanian plateau, as well as the two examples of Middle Bronze Age settlement in the eastern desert of Jordan. Subsequently, a review of several major regional surveys provides a broadly painted portrait of how localized settlement trends in Jordan relate to the more comprehensive settlement patterns reported west of the Jordan River.

Jordan Valley

Only four Middle Bronze Age settlements in the Jordan Valley have been excavated to a significant extent, and many other sites show an apparent hiatus in occupation during the period (e.g. Tall Abu al-Kharaz, Tall as-Sa'idiyya). Three of the four excavated sites, Tall Dayr 'Alla, Tall Nimrin and Tall al-Hayyat, lie in the *ghor*, the rich agricultural area covering the broad Pleistocene terrace above the active floodplain of the Jordan River, which is known in Arabic as the *zor*. Pella, at present the best-known Middle Bronze Age town in Jordan, lies in the first line of foothills lying immediately overlooking the *ghor* from the east.

Pella

The site of ancient Pella lies adjacent to the modern village of Tabaqat Fahl and commands a permanent spring in the adjacent Wadi Jirm. Its settlement history extends back to the Neolithic and features extensive ruins dating to the Roman period, during which Pella was a member of the league of cities known as the Decapolis.

Given the particularly lengthy occupation of the site, Middle Bronze Age data derive from deep soundings with limited lateral exposure (see Figure 7.3). For example, before being discontinued and backfilled, the West Cut at Pella (Area VIII) produced MB IIB–C sherds at a depth of 14 m below the modern tell surface (Smith 1993: 1177). Elsewhere, Middle Bronze Age ceramics come from stratified sedimentary fill in Areas XXV and XXVIII along Pella's south slope (e.g. Smith and McNicoll 1986: 113, fig. 20). Pella's

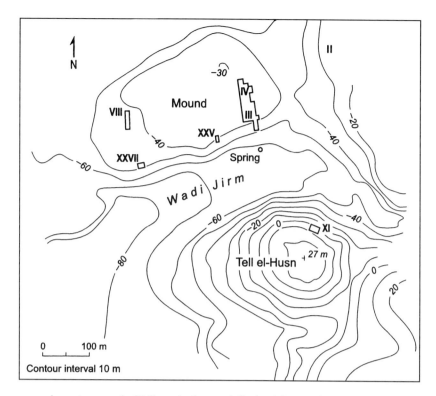

Figure 7.3. Excavation areas on the main mound of Pella and adjoining hillsides (after Smith 1993: 1174).

role as a major town early in the Middle Bronze Age is attested further by two apparently cornering sections of a massive mudbrick town wall in the East Cut (Area III, plots D and F) (Smith and Potts 1992: 35, 40-44, fig. 7). The excavators are wary of assigning a specific construction date for this architecture, the earliest known pre-classical fortification at Pella, since contemporaneous deposits have produced no reconstructible pottery thus far (Potts *et al.* 1988: 131). Currently available sherd assemblages suggest a construction date in MB IIA (Smith 1993: 1176). Excavations in 1997 revealed an additional stretch of Middle Bronze fortification wall in Area XXVIII, approximately 300 m west of Area III (Bourke, pers. comm.). Jointly, these walls hint at a Middle Bronze Age defensive system that encompassed the entire main mound at Pella, an area of perhaps 8 ha. By applying an estimated population density of 200–300 people per hectare, which is derived from a plethora of ethnographic accounts of traditional Middle Eastern communities (see especially Kramer 1982: 155-81), Pella might have housed about 2000 people. The likelihood that Pella was a substantial fortified town early in the Middle Bronze Age is strengthened by its inclusion (as Pihilum) among the presumably bothersome and ritually cursed cities of the Egyptian Execration Texts (Smith 1973: 23; see discussion below).

While Areas VIII, XXV and XXVIII have produced MB IIB and C ceramics, the best stratified evidence for these periods comes from the East Cut (Area III). Here the entire Middle Bronze Age sequence is represented in strata X–VI, and plots III C and F reveal MB IIB mudbrick domestic architecture just inside the town wall (Smith and Potts 1992: 44-47, fig. 8).

In addition to its tell deposits, Pella has produced abundant mortuary evidence from off-site tombs. While MB IIA burials are lacking, the contents of later tombs suggest significant international connections, especially at the end of the Middle Bronze Age and into the Late Bronze Age. A single infant jar burial dating to MB IIB was excavated in Area VIII on the tell (Smith 1973: 367), and a variety of tombs, most previously looted or eroded, have been surveyed west and southwest of Pella, but remain unexcavated (Smith 1993: 1177). More fruitful excavations have investigated Middle and Late Bronze Age cemeteries east of the tell (Area II) and on the north slope of neighbouring Tall al-Husn (Area XI), which lies on the south side of Wadi Jirm. These cemeteries include chamber and cist tombs, many spanning the MB IIC/LB I transition. For example, Tombs 1 (in Area II) and 62 (in Area XI) feature abundant Chocolate-on-White chalices, jugs and bowls (Smith 1973: 170, l. 46; Potts *et al.* 1988: 206), like those traded between Pella and distant markets to the west and north (Knapp 1993). The three burial chambers in Tomb 62 produced over 2000 vessels, abundant personal ornaments and the skeletal remains of more than 100 individual burials, suggesting significant material affluence and long-term use by the people of Pella (Potts *et al.* 1988: 205, figs. 9, 10).

Tall Dayr 'Alla

Tall Dayr 'Alla, which lies near the confluence of the az-Zarqa' and Jordan rivers, may be equated with ancient Succoth of the Old Testament and Talmud (van der Kooij and Ibrahim 1989: 75-76). An excavated exposure of less than 100 m² at the southeast foot of the tell reveals the earliest known Middle Bronze Age remains at the site: a substantial mudbrick town wall preserved to a height of 2.5 m (van der Kooij and Ibrahim 1989: 75-76). This wall in Dayr 'Alla's Phase IV dates, based on associated Tall al-Yehudiyeh Ware, no earlier than MB IIB. Domestic architecture, including mudbrick walls and courtyards, is also found in Phase IV and the subsequent Phase V. If the entire tell was inhabited during MB II, Dayr 'Alla would represent a small fortified town a few hectares in size, with a population in the order of 1000 inhabitants.

Tall Nimrin

Lying near the mouth of Wadi Shu'eib in the modern town of South Shuna, Tall Nimrin contains nearly 6 m of stratified remains from the Middle Bronze Age. Unlike Tall Dayr 'Alla, Nimrin provides stratified domestic architecture and material culture dating to MB IIA, apparently reflecting an unfortified village. Massive stone-founded mudbrick fortification walls date later in the Middle Bronze Age, and remain preserved up to 4 m in height, with a glacis along their exterior. This monumental architecture was constructed in two phases, the first in late MB IIB and the second in MB IIC (Flanagan *et al.* 1994: 217-18). Interestingly, this massive wall system may have protected a village perhaps only 1 ha in size (200–300 inhabitants), which was abandoned abruptly in the sixteenth or early fifteenth century BC (Flanagan *et al.* 1994: 217-19).

Tall al-Hayyat

The low mound of Tall al-Hayyat lies amid the *ghor* approximately 2 km east of the Jordan River and 7 km southwest of Pella. Hayyat embodies a Middle Bronze Age hamlet that measured about one-half hectare and would have housed no more than 150 inhabitants. Excavations have exposed 400 m² of Hayyat's surface area and proceeded through 4.5 m of cultural deposition at the mound's centre (Falconer and Magness-Gardiner 1989).

Habitation at Tall al-Hayyat began very late in EB IV and continued through six major stratigraphic/architectural phases to late MB IIC. An ephemeral EB IV stratum (Phase 6) lies at the tell's base. Hayyat's earliest architecture consists of a small, centrally located mudbrick shrine dating to early MB IIA (Phase 5), which is the first in a stratified series of mudbrick temples that are enlarged and elaborated in Phases 4–2 (MB IIA, B and C, respectively). These sanctuaries become much larger than Hayyat's dwellings, show remarkable continuity between phases and provide examples of the 'Migdal' temples known primarily from cities in the Levant, Syria and Lower Egypt (see Figure 7.4 and discussion in Magness-Gardiner and Falconer 1994).

In each stratum, a surrounding wall clearly segregates each temple compound from mudbrick household structures. Domestic architecture was not found in Phase 5, but in Phases 4, 3 and 2 the areas outside each temple compound feature houses, walled courtyards and alleyways. In addition, Hayyat revealed an intact MB IIA (Phase 4) pottery kiln on its southern slope (see Falconer 1987). Phase 1, dating to late MB IIC, produced only fragments of domestic architecture and no evidence of a temple.

Patterns of faunal, floral and ceramic deposition reveal several salient aspects of village economy at Tall al-Hayyat. In turn, these data provide a glimpse of urban–rural relations in the Jordan Valley. Hayyat's economic evidence hints at a few increased connections with external markets, which we might expect for such a community (see Falconer 1995; Fall *et al.* 1998). For example, between Phases 5 and 3 sheep became more commonly husbanded than goats, presumably because sheep generate secondary products, especially wool, with greater market value. Increased remains of marketable fruit (e.g. olive, grape and figs) and ceramic serving vessels made by craft specialists also hint at greater exchange of agricultural and manufactured products.

However, these trends are overshadowed by data suggesting accentuated economic autonomy at Tall al-Hayyat. For instance, despite the market potential of their wool, sheep remained important sources of meat for domestic consumption. Hayyat also featured

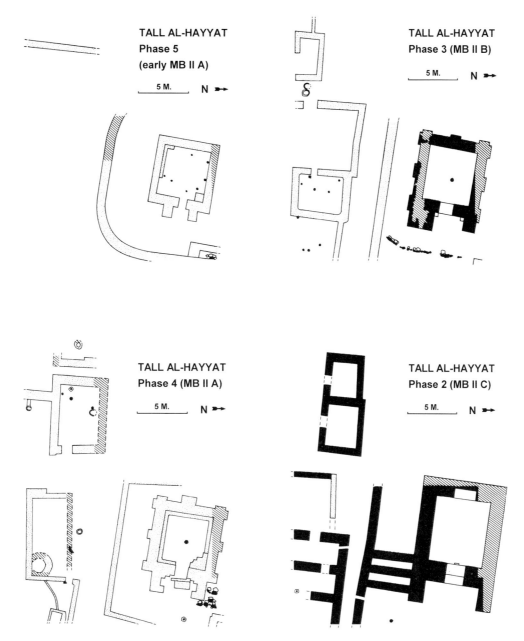

Figure 7.4. Temple and domestic architecture excavated in phases 5–2 at Tall al-Hayyat (after Magness-Gardiner and Falconer 1994: Fig. 5).

a surprisingly robust element of household pig herding that increased markedly through time. Finally, the ceramic repertoire shows that the manufacture of cooking pots became increasingly localized, storejars became more commonly suitable as short-term containers, rather than transport of marketable goods, and the storage functions of Hayyat's temples increasingly resembled those of households (Falconer 1995; Fall et al. 1998).

In sum, most of the evidence from Tall al-Hayyat implies increasing economic self-reliance throughout the Middle Bronze Age. Growing local investment is reflected in the increased size and elaboration of Hayyat's temple and domestic architecture (see discussion in Magness-Gardiner and Falconer 1994; Falconer 1995). Likewise, this village performed activities otherwise concentrated in cities, including the manufacture of copper alloy tools, weapons and figurines (Falconer and Magness-Gardiner 1989). The farmers of Tall al-Hayyat apparently exploited some opportunities presented by town and city markets, but simultaneously ensured community autonomy in the face of the inevitable liabilities of urbanism.

Dead Sea Plain

The only Middle Bronze Age sites known in the Jordan Rift south of Tall Nimrin are found at Zahrat adh-Dhra' and Dayr 'Ain 'Abata southeast of the Dead Sea. The settlement at Zahrat adh-Dhra' 1 is marked by visible stone architecture along a northwest–southeast-trending ridge in the Plain of Dhra', east of Bab adh-Dhra' (Edwards *et al.* 1998). The remains of numerous rectangular structures and curved enclosures are spread widely over approximately 6 ha. Excavations directed by me in the winter of 1999/2000 suggest intermittent occupation in MB II A and B. Severe downcutting followed this occupation (or may have ended it), as shown by the erosion of Middle Bronze Age structures into the steep-sided wadis to the northeast and southwest. This evidence suggests the site originally extended over a much larger area.

At least 19 Middle Bronze Age single- and double-chamber cairn tombs lie north of the Byzantine monastery at Dayr 'Ayn 'Abata, on the steep slopes overlooking the modern town of Ghor as-Safi (Politis 1990, 1993, 1995, 1997). Although these tombs have been looted heavily, their assemblages suggest a date late in MB II.

Northern Plateau: Tall Irbid, Abila, Jarash, Dahr al-Mudayna

Excavations along the west and northeast sides of Tall Irbid have exposed a three-tier mudbrick city wall system, the earliest of which dates to c. 2000 BC (Lenzen and McQuitty 1989: 299-300). Tall Irbid extends over approximately 20 ha and, if fully encircled by walls, embodies a Middle Bronze Age fortified town formerly inhabited by perhaps 5000 people.

Abila and Jarash, two members of the classical Decapolis in the general vicinity of Irbid, have also produced substantial amounts of Middle Bronze Age pottery (Braemer 1987; Mare 1991: 206). At Jarash, living surfaces just above bedrock include abundant Chocolate-on-White Ware, indicating a use date in MB IIC. At both sites, which extend over very large areas in later periods, the extent of Middle Bronze Age settlement is unknown.

At Dahr al-Mudayna, a modest ridgetop site northwest of Jarash, limited soundings again reveal Chocolate-on-White Ware, as well as substantial amounts of MB IIA and B pottery (Kafafi and Knauf 1989).

Central Plateau: 'Amman, Tall Safut, Abu Snesleh, Sahab, Tall Jalul, Tall al-'Umayri

A deep sounding on the lower terrace of the citadel of 'Amman reveals a stone-founded city wall dating to Middle Bronze II B and C (Greene and 'Amr 1992: 116-17). A glacis on the upper terrace apparently also dates to the latter part of the Middle Bronze Age (Dornemann 1983: 18; Zayadine *et al.* 1987: 308), although little associated material culture has been published. Although its size during the second millennium BC is difficult to estimate, the citadel may have measured 10 ha, with a population of 2000–3000 people.

More impressive remains are reported from Middle Bronze Age chamber tombs in the vicinity of 'Amman. These tombs, which feature pottery dating between late MB IIB and the beginning of the Late Bronze Age, include ornate bronze and beaded jewellery, scarabs and alabaster vessels (Piccirillo 1978; Najjar 1991). A single MB IIB–C tomb has also been reported near Mount Nebo in Wadi Abu an-Naml (Dornemann 1983: 17).

Middle Bronze IIB–C fortifications with an associated glacis are also attested at Sahab, lying southeast of 'Amman, approximately halfway between the modern capital and the edge of the eastern desert (Ibrahim 1987: 76; 1989). The extent of these walls is uncertain and, if comparable to Sahab's Late Bronze Age fortifications, would have enclosed a few hectares at most (see Ibrahim 1989: 519). Two nearby caves, used as tombs, produced human remains and ceramic assemblages dating to MB IIB or C. Sahab's modest size, fortifications and location suggest that it may have served as an outpost on the desert fringe, possibly comparable to those of the *limes* system during the Roman and Byzantine periods (Ibrahim 1987: 76).

Tall Safut, perched at the southern edge of the Baq'ah Valley northwest of 'Amman, has also revealed deposition identified as a glacis (Ma'ayah 1960: 115). Following subsequent tentative acceptance of this interpretation (Dornemann 1983: 19), these deposits most recently have been reassessed as part of the site's natural geology (Wimmer 1989: 513).

Middle Bronze Age domestic architecture is reported from Abu Snesleh, a settlement of uncertain size lying east of 'Amman at the desert margin. Limited soundings have exposed a stone-founded mudbrick house with an assemblage of domestic pottery similar to the more abundant MB IIB and C collections from sites west of the Jordan River (Lehmann *et al.* 1991).

In the vicinity of Madaba, surveys of two major *tulul* have revealed abundant Middle Bronze Age pottery. While excavations have begun only recently at Tall Jalul, several field seasons at Tall al-'Umayri have exposed a beaten earth rampart along its western slope. Pottery related to the construction of the rampart (including Chocolate-on-White Ware), as well as that recovered from the eastern portion of the site, suggests that al-'Umayri was a fortified settlement in MB IIB and C (Herr *et al.* 1991: 159, 166). The tell's full extent of approximately 6.5 ha suggests that it may represent a moderately sized town with about 1600 inhabitants.

Eastern desert: Jawa, Tall Rukeis

The only excavated Middle Bronze Age sites in Jordan's eastern desert are fortified settlements at Jawa and Tall Rukeis. The latest architecture at Jawa includes a stone 'citadel' measuring somewhat less than 1 ha, and dating to the Middle Bronze Age (Helms 1975: 22-26; 1976: 2-7; 1977: 23-27). This unusual stone structure features two stories of cell-like rooms arrayed around a central walled courtyard and flanked by three covered corridors with stone roof supports. The citadel's floor plan was readily apparent from the surface (see Helms 1976: fig. 3), necessitating only limited excavation. Following its construction, apparently in MB IIA, the citadel may have served as a caravanserai for overland trade between Syria or Mesopotamia and the Levant (Helms 1975: 26).

Tall Rukeis, near the Syrian border, is associated with several small Middle Bronze Age farmsteads along the Wadi al-'Ajib, which drains south from the Jebel Druze (Betts *et al.* 1996; Betts 1998). Excavations reveal two phases of Middle Bronze architecture, including basalt fortifications with a chambered gate and small-scale domestic structures, possibly for storage. Agricultural features on the landscapes around both Tall Rukeis and Jawa suggest intensive Bronze Age water conservation in this agriculturally marginal environment.

Regional settlement patterns in the southern Levant

The most widely published and analyzed Bronze Age settlement data presently derive from a variety of archaeological surveys that have covered approximately 14,000 km^2 between the Mediterranean and the Jordan Valley (see Ibrahim *et al.* 1976, 1988; Broshi and Gophna 1984, 1986; Gophna and Portugali 1988; Joffe 1991; Finkelstein and Gophna 1993; Falconer 1994; Falconer and Savage 1995). These studies show that the number of archaeologically detectable settlements increased substantially between MB IIA and MB IIB–C. However, this trend incorporated only modest growth in sedentary populations, and a noticeable *decrease* in average site size (to c. 2 ha in MB IIB–C) (Falconer 1994). These characteristics indicate a form of growth primarily involving smaller settlements. This inference is corroborated by two additional trends: a substantial drop in the percentage of the regional populations living in towns (i.e. sites larger than 10 ha), and a rise in the number of rural villages (i.e. sites smaller than 5 ha) (Falconer 1994). In concert, these trends describe a process of growth manifested primarily in a proliferation of villages and hamlets, especially in the hills of Palestine and in the Jordan Valley.

Rank-size analysis

Rank-size analysis provides a particularly economical method for comprehending Middle Bronze Age settlement patterns and comparing them with the Levant's first expression of urbanism in the Early Bronze Age. This method assesses regional political and economic integration by correlating the size and rank of communities in a settlement system. The 'rank-size rule' holds that for regions 'with a long tradition of urbanism, which are politically and economically complex', the size of any n-ranked place may be predicted by dividing the size of the largest place by n (Berry 1961: 582; Zipf 1949). In other words, in politically mature regions with well-integrated economic systems the largest city is expected to have twice the population of the second-ranked city, three times the population of the third-ranked city, and so on. When plotted logarithmically, the settlement sizes within such a system will describe a straight line, log-normal distribution in accordance with the rank-size rule.

Interestingly, archaeological settlement patterns seldom conform to the expectations of the rank-size rule, and their interpretation usually stems from the manner in which they depart from log-normal (e.g. Johnson 1977, 1980; Paynter 1983). For example, 'primate' rank-size distributions are characterized by a first-ranked place (a 'primate' city) that is much larger than predicted by the rank-size rule (Jefferson 1939). Primate rank-size curves, which tend to dip below a log-normal line, reflect extreme interdependence between communities. In contrast, 'convex' rank-size curves, which rise above a log-normal line, commonly

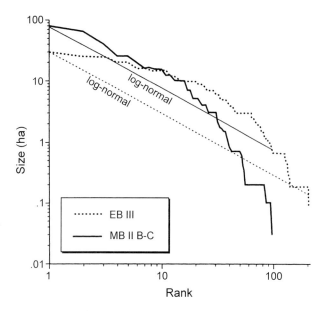

Figure 7.5. Rank-size distributions for Early Bronze III and Middle Bronze IIB–C settlements of known size in the southern Levant. For EB III: $K^- = .406$, $p<.01$; For MB IIB–C: $K^- = .479$, $p<.01$. Data from Ibrahim et al. (1976, 1988), Broshi and Gophna (1984, 1986), Gophna and Portugali (1988), Joffe (1991) and Finkelstein and Gophna (1993) (after Falconer and Savage 1995: fig. 8).

derive from the pooling of multiple settlement systems or from systems characterized by poor integration, in which community sizes are less hierarchical than predicted by the rank-size rule (Johnson 1977: 497, 1980).

In light of the pre-eminence normally ascribed to fortified Bronze Age cities in the southern Levant (e.g. Dever 1987), we might anticipate accordingly primate rank-size settlement distributions. However, Early and Middle Bronze Age survey data collected from the Jordan Valley west to the Mediterranean produce a series of *convex* departures from log-normal for both the Early and Middle Bronze Ages (Falconer and Savage 1995; Fall et al. 1998; see Figure 7.5). Although the MB IIB–C data describe a closer approximation of log-normal than seen for MB IIA, both curves suggest consistently poor political and economic integration in the southern Levant generally, the pooling of multiple localized settlement enclaves, or both.

The Egyptian Execration Texts provide an indirect, but intriguing, reflection of Levantine regional politics that may help us interpret these patterns. These texts consist of sherds (the 'Berlin Texts') and statuette fragments (the 'Brussels Texts') inscribed with lists of Asiatic localities (cities or regions) and their rulers who were cursed as adversaries of Pharaoh. The Berlin Texts, dated to the reign of Sesostris III (Posener 1971: 541), have been correlated with MB IIA in Palestine (Mazar 1968: 75). These texts enumerate 20 territories associated with 30 'rulers'. The larger collection of Brussels Texts dates approximately one generation later, and enumerates 62 territories and a similar number of rulers (Posener 1971: 541, 555). Among the Brussels Texts, perhaps dating to early MB IIB, more numerous territories suggest the spread of towns and cities, while more frequent single rulers imply slightly more centralized 'city-states' (Posener 1971: 541; Mazar 1968: 75, 81). When viewed in tandem, these documentary and survey accounts hint at the gradual coalescence of small local polities, but more fundamentally portray a highly fragmented Levantine political landscape.

Localized settlement patterns in the Jordan Valley

At present, regional survey coverage in Jordan tends to be intensive, but geographically discontinuous (see discussion below). Fortunately, Bronze Age settlement patterns in the Jordan Valley, a clearly circumscribed region of approximately 1700 km², are illuminated by archaeological surveys covering both sides of the Jordan River (see Ibrahim et al. 1976, 1988; Broshi and Gophna 1984, 1986; Joffe 1991; Falconer and Savage 1995).

Early and Middle Bronze Age rank-size curves for the Jordan Valley are effectively log-normal, showing statistically insignificant convex departures (Figure 7.6). In contrast to the southern Levant generally, this log-normality suggests more effective economic and political integration, the existence of very few localized settlement systems, or both. During the Early Bronze Age, the Jordan Valley enjoyed settlement densities comparable to, or greater than, those found anywhere in the southern Levant, including the heavily populated Mediterranean coastal plan (Joffe 1991; Falconer and Savage 1995). Following the EB IV collapse, Middle Bronze Age settlements redeveloped, but on a much-reduced scale. Middle Bronze Age sites are strikingly smaller in number and size than their Early Bronze Age predecessors. During the Middle Bronze Age, the Jordan Valley's population grew slightly, while the number of settlements roughly doubled between MB IIA and B–C (Broshi and Gophna 1986; Falconer 1994).

Archaeological survey investigations to the east and south of the Jordan Valley have tended to be discontinuous, with the coverage defined within individual wadi drainages or in the hinterlands surrounding focal sites. Although these surveys are not amenable to a rank-size

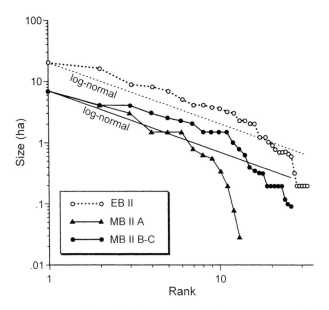

Figure 7.6. Rank-size distributions for Early Bronze II, Middle Bronze IIA and Middle Bronze IIB–C settlements of known size in the Jordan Valley. For EB II: $K^- = .219$, $p = .69$; For MB IIA: $K^- = .231$, $p = .29$; For MB IIB–C: $K^- = .308$, $p = .24$. Data from Ibrahim *et al.* (1976, 1988), Broshi and Gophna (1984, 1986), Joffe (1991).

analysis for a contiguous region, their results do show a number of common patterns, most notably a drop in settlement density between the Early and Middle Bronze Ages.

The following discussion summarizes the results from a number of surveys conducted along drainages into the Jordan Valley and Wadi 'Arabah, then considers surveys conducted on the Transjordanian plateau. This overview emphasizes settlement distinctions in the regional urbanism of the Early and Middle Bronze Ages. In light of this, and because its evidence is reported inconsistently, EB IV settlement is, regrettably, excluded. Likewise, since our focus lies on domestic sites, evidence from very ephemeral or special activity sites (e.g. sherd scatters) is omitted.

Several surveys have been conducted in the major wadis that drain the Transjordanian plateau into the Jordan River. Systematic coverage of wadi bottoms and adjacent hills is reported for Wadi al-'Arab (Hanbury-Tenison 1984) Wadi Ziqlab (Banning and Fawcett 1983), Wadi Hammeh (Petocz and Villiers 1984) and Wadi al-Yabis (Mabry and Palumbo 1988). Although field methods varied between surveys, each project utilized some form of systematic sampling in designating areas for pedestrian reconnaissance.

Likewise, these surveys use different chronological categories to report their results. For example, the Wadi al-'Arab and Wadi Ziqlab surveys combine data for EB II and III settlements, as well as for the Middle and Late Bronze Ages.

Given their comparably systematic sampling methods, it is interesting to note significant variability in site frequencies between drainages. For example, Wadis al-'Arab and al-Yabis reveal modest, but roughly comparable, Early and Middle Bronze Age settlement frequencies. However, Wadi Ziqlab produced no Early Bronze Age sites and only low-density Middle–Late Bronze Age sherd scatters (Banning and Fawcett 1983), while in Wadi Hammeh, Early and Middle Bronze Age evidence is limited to Pella, aside from a single small Early Bronze tell (Petocz and Villiers 1984).

Survey coverage of the Transjordanian Plateau

On the Transjordanian Plateau between the upper courses of the Yarmouk and az-Zarqa' Rivers, a non-systematic survey of Roman roads reports the widespread occurrence of Middle Bronze Age sites, and isolated soundings confirm the presence of stratified Middle Bronze Age remains at several of them (Mittmann 1970).

Two surveys on the central Transjordanian Plateau reveal a substantial drop in settlement between the EB I–III periods and the Middle Bronze Age. Non-random survey coverage of the lands within a 10 km radius of Hisban revealed Early Bronze Age sherds on the surface of 46 sites, while Middle Bronze Age remains were limited to 10 sites, most of which are quite small (Ibach 1987: 3, tables 3.5, 6). Sizeable Middle Bronze Age occupations were limited to Tall al-'Umayri and Tall Jalul, both of which also feature substantial evidence from the Early Bronze Age.

The Madaba Plains Survey investigated 55 visible sites within a 10 × 10 km quadrat around Tall al-'Umayri, of which 10 were occupied during EB I–III, while only two provided evidence from the Middle Bronze Age (Boling 1989). Additional reconnaissance of a 1.5% random sample within this area revealed 11 further Early Bronze Age settlements, but only two other Middle Bronze Age sites.

A survey of 875 km² on the al-Karak plateau documented 28 sites that may have been EB II–III habitations (i.e. with five or more surface sherds) and 15 such sites dating to the Middle Bronze Age (J.M. Miller 1991:

Survey	Area Surveyed (km^2)	EB Sites (no./100 km^2)	MB Sites (no./100 km^2)
East Jordan Valley	700	40[a] (5.7)	27 (3.9)
Wadi al-'Arab	25	4[a] (16)	3[c] (12)
Wadi al-Yabis	186	13[b] (7.0)	14 (7.5)
Hisban	314	46[b] (14.6)	10 (3.2)
Madaba Plains	100	11[b] (11.0)	2 (2.0)
Al-Karak Plateau	875	28[a] (3.2)	15 (1.7)

Source: Data from Ibrahim *et al.* 1976, 1988; Hanbury-Tenison 1984; Mabry and Palumbo 1988; Ibach 1987; Boling 1989; J.M. Miller 1991.
Notes: [a]EB II–III; [b]EB I–II; [c]MB–LB.

Table 7.2. Early and Middle Bronze Age site counts and densities in Jordan.

308; Brown 1991: 185). Again, these sites, especially those dating to the Middle Bronze Age, are commonly quite small.

Farther south, the Wadi al-Hasa Survey has sampled varying portions of the 600 km^2 drainage with a combination of pedestrian and vehicular coverage, documenting dozens of EB I sites, but very few from any other periods within the Bronze Age (MacDonald *et al.* 1988: 166-68; Clark *et al.* 1994).

Overview of settlement pattern data

Table 7.2 summarizes site counts and densities for six major surveys conducted during the last 20 years in Jordan, in which systematic sampling detected a significant number of Middle Bronze Age sites. When viewing these data, several caveats merit recognition. First, the essential criteria for identifying sites, especially domestic sites, vary between surveys. Miller's criterion of domestic sites having five or more diagnostic sherds is a rare exercise in explicit definition (J.M. Miller 1991). Second, Table 7.2 hints at variability in survey intensity, especially in light of the general inverse correlation of total survey coverage and site density. It appears, not surprisingly, that more localized surveys may detect smaller, more ephemeral sites that are overlooked by more ambitious, but less intensive reconnaissance. Likewise, more localized surveys may be more lenient in the criteria used to define a 'site'. Further, the temporal conflation that inevitably accompanies survey chronologies takes different forms from survey to survey. Most significantly here, an 'Early Bronze Age' site designation subsumes EB I, II and III for some surveys, but only EB II and III for others.

Even with these cautions in mind, two major settlement trends, one chronological and the other geographical, emerge fairly clearly. First, with the exception of the near-equality of sites from the Early and Middle Bronze Ages reported by the Wadi al-Yabis Survey, there is a consistent drop in settlement density between these two periods in the other major surveyed regions of Jordan. This characteristic is particularly pronounced for the surveys of the central and southern regions of the Transjordanian plateau. Secondly, Middle Bronze Age site densities, which are highest in Wadis al-'Arab and al-Yabis along the northern Jordan Valley, again dwindle to extremely low values on the central and southern plateau, with no sites reported to date south of Wadi al-Hasa.

Conclusions

The review above clarifies the relative paucity of evidence for Middle Bronze Age society in modern Jordan. However, rather than accept Glueck's premature impression of a regional settlement hiatus during this period, a number of tantalizing trends in our available data cry out for more substantial field investigations. First, although most of the excavated remains and virtually all of the mortuary evidence date to MB IIB and C, a number of sites reveal substantial occupations during MB IIA. These communities stretch from Pella in the Jordan Valley to the eastern desert at Jawa. Secondly, Middle Bronze Age communities in Jordan ranged from small cities of 10–20 ha (e.g. Pella, Tall Irbid, possibly the 'Amman Citadel), to towns with 1000–2000 inhabitants (e.g. Tall Dayr 'Alla, Tall al-'Umayri, Sahab), to villages with a few hundred occupants at most (e.g. Tall al-Hayyat, Tall Nimrin). Thirdly, following the fortification of some communities in MB IIA (Pella, Tall Irbid, 'Amman Citadel), virtually all known settlements apparently were ringed by protective walls during MB IIB–C (with the exceptions of smaller sites like Tall al-Hayyat, Abu Snesleh and Dhar al-Mudayna). Thus, the commonly cited coastal orientation for the development of Middle Bronze Age settlement and society (e.g. in Gerstenblith 1983; Mazar 1990; Falconer 1994; Ilan 1995) must be modified to accommodate the widespread, and often fortified, spectrum of communities east of the Jordan River.

Regional survey data, as discontinuous as they may be, highlight major distinctions between the Bronze Age settlement trajectories of Palestine and Israel to the west versus those of Jordan to the east. West of the Jordan River, the Middle Bronze Age may be characterized as an apex of pre-classical urbanism and population density with some justification (e.g. Dever 1987; Mazar 1990). However, to the east, its regional patterns reflect a recession of sedentary settlement that follows the advent of urbanism in the Early Bronze Age. Additionally, there is a fairly clear north–south fall-off pattern in Middle Bronze settlement densities, suggesting the possible importance of northerly economic and political ties stemming from the Beth Shan and Jezreel Valleys and the connections to the Mediterranean that they represent. In this regard, it is not altogether surprising that the northern towns of Pella and Tall Irbid potentially represent two of the most substantial centres of Middle Bronze Age society in Jordan.

In sum, while Jordan does not present evidence of settlement or population densities comparable to those in modern Palestine and Israel (with the possible exception of the northern Jordan Valley), it can no longer be characterized as a largely blank map during the Middle Bronze Age. Rather, it should be reconsidered as a thinly veiled mosaic of communities, whose larger integration within Canaanite society may not simply follow the relatively well-documented patterns of urbanization west of the Jordan River, but now must be explored on a broader stage stretching from the Mediterranean to the eastern desert.

References

Albright, W.F.
1962   The Chronology of MB I (Early Bronze–Middle Bronze). *Bulletin of the American Schools of Oriental Research* 168: 36-42.
1966   Remarks on the Chronology of Early Bronze IV–Middle Bronze II A in Phoenicia and Syria-Palestine. *Bulletin of the American Schools of Oriental Research* 184: 26-35.

Banning, E.B. and C. Fawcett
1983   Man–land Relationships in the Ancient Wadi Ziqlab. *Annual of the Department of Antiquities of Jordan* 27: 291-309.

Baruch, U.
1990   Palynological Evidence of Human Impact on the Vegetation as Recorded in Late Holocene Lake Sediments in Israel. In S. Bottema, G. Entjes-Nieborg and W. van Zeist (eds.), *Man's Role in the Shaping of the Eastern Mediterranean Landscape*, 283-93. Rotterdam: Balkema.

Berry, B.J.L.
1961   City Size Distributions and Economic Development. *Economic Development and Culture Change* 9: 573-88.

Betts, A.
1998   *The Harra and the Hamad: Excavations and Surveys in Eastern Jordan, vol. I*. Sheffield Archaeological Monographs. Sheffield: J.R. Collis.

Betts, A. *et al.*
1996   Studies of Bronze Age Occupation in the Wadi al-'Ajib, Southern Hauran. *Levant* 27: 27-39.

Boling, R.G.
1989   Site Survey in the el-'Umeiri Region. In L.T. Geraty *et al.* (eds.), *Madaba Plains Project I: The 1984 Season at Tell el-'Umeiri and Vicinity and Subsequent Studies*, 98-188. Berrien Springs, MI: Andrews University.

Braemer, F.
1987   Two Campaigns of Excavations on the Ancient Tell of Jerash. *Annual of the Department of Antiquities of Jordan* 31: 525-29.

Broshi, M.
1979   The Population of Western Palestine in the Roman–Byzantine Period. *Bulletin of the American Schools of Oriental Research* 236: 1-10.

Broshi, M. and R. Gophna
1984   The Settlements and Population of Palestine During the Early Bronze Age II-III. *Bulletin of the American Schools of Oriental Research* 253: 41-53.
1986   Middle Bronze Age II Palestine: Its Settlements and Population. *Bulletin of the American Schools of Oriental Research* 261: 73-90.

Brown, R.
1991 Ceramics from the Kerak Plateau. In J. M. Miller (ed.), *Archaeological Survey of the Kerak Plateau*, 169-279. Atlanta: Scholars.

Bunimovitz, S.
1989 The Land of Israel in the Late Bronze Age: A Case Study of Socio-Cultural Change in a Complex Society. Unpublished Ph.D. dissertation, Tel Aviv University (Hebrew with English Summary).

Clark, G.A. *et al.*
1994 Survey and Excavation in Wadi el-Hasa: A Preliminary Report of the 1993 Field Season. *Annual of the Department of Antiquities of Jordan* 38: 41-55.

Dever, W.G.
1976 The Beginning of the Middle Bronze Age in Syria-Palestine. In F.M. Cross, W.E. Lemke and P.D. Miller (eds.), *In Magnalia Dei: The Mighty Acts of God, Essays on the Bible and Archaeology in Memory of G. Ernest Wright*, 3-38. Garden City, NY: Doubleday.
1987 Palestine in the Middle Bronze Age: The Zenith of the Urban Canaanite Era. *Biblical Archaeologist* 50.3: 149-77.

Dornemann, R.
1983 *The Archaeology of the Transjordan in the Bronze and Iron Ages*. Milwaukee: Milwaukee Public Museum.

Edwards, P.C., P. Macumber and M.K. Green
1998 Investigations into the Early Prehistory of the East Jordan Valley: Results of the 1993/4 La Trobe University Survey and Excavation Season. *Annual of the Department of Antiquities of Jordan* 42: 15-39.

Falconer, S.E.
1987 Village Pottery Production and Exchange: A Jordan Valley Perspective. In A. Hadidi (ed.), *Studies in the History and Archaeology of Jordan*, III: 251-59. Amman: Department of Antiquities.
1994 The Development and Decline of Bronze Age Civilisation in the Southern Levant: A Reassessment of Urbanism and Ruralism. In C. Mathers and S. Stoddart (eds.), *Development and Decline in the Mediterranean Bronze Age*, 305-33. Sheffield: Sheffield Academic Press.
1995 Rural Responses to Early Urbanism: Bronze Age Household and Village Economy at Tell el-Hayyat, Jordan. *Journal of Field Archaeology* 22.4: 399-419.

Falconer, S.E. and P.L. Fall
1995 Human Impacts on the Environment during the Rise and Collapse of Civilization in the Eastern Mediterranean. In D.W. Steadman and J.I. Mead (eds.), *Late Quaternary Environments and Deep History: A Tribute to Paul S. Martin*, 84-101. Hot Springs, SD: The Mammoth Site.

Falconer, S.E. and B. Magness-Gardiner
1989 Bronze Age Village Life in the Jordan Valley: Archaeological Investigations at Tell el-Hayyat and Tell Abu en-Ni'aj. *National Geographic Research* 5.3: 335-47.

Falconer, S.E. and S.H. Savage
1995 Heartlands and Hinterlands: Alternative Trajectories of Early Urbanization in Mesopotamia and the Southern Levant. *American Antiquity* 60.1: 37-58.

Fall, P.L., L. Lines and S.E. Falconer
1998 Seeds of Civilization: Bronze Age Rural Economy and Ecology in the Southern Levant. *Annals of the Association of American Geographers* 88.1: 107-25.

Finkelstein, I. and R. Gophna
1993 Settlement, Demographic and Economic Patterns in the Highlands of Palestine in the Chalcolithic and Early Bronze Periods and the Beginning of Urbanism. *Bulletin of the American Schools of Oriental Research* 289: 1-22.

Flanagan, J.W., D.W. McCreery and K.N. Yassine
1994 Tell Nimrin: Preliminary Report on the 1993 Season. *Annual of the Department of Antiquities of Jordan* 38: 205-44.
1996 Tall Nimrin: Preliminary Report on the 1995 Excavation and Geological Survey. *Annual of the Department of Antiquities of Jordan* 40: 271-92.

Gerstenblith, P.
1983 *The Levant at the Beginning of the Middle Bronze Age*. American Schools of Oriental Research Dissertation Series, no. 5. Winona Lake, IN: American Schools of Oriental Research.

Glueck, N.
1970 *The Other Side of the Jordan*. 2nd edn. Cambridge, MA: American Schools of Oriental Research.

Gophna, R.
1984 The Settlement Landscape of Palestine in the Early Bronze Age II-III and Middle Bronze Age. *Israel Exploration Journal* 34: 20-31.

Gophna, R. and J. Portugali
1988 Settlement and Demographic Processes in Israel's Coastal Plain from the Chalcolithic to the Middle Bronze Age. *Bulletin of the American Schools of Oriental Research* 269: 11-28.

Gophna, R., N. Liphschitz and S. Lev-Yadun
1986 Man's Impact on the Natural Vegetation of the Central Coastal Plain of Israel during the Chalcolithic Period and the Bronze Age. *Tel Aviv* 13.14: 71-84.

Greene, J.A. and K. 'Amr
1992 Deep Sounding on the Lower Terrace of the Amman Citadel: Final Report. *Annual of the Department of Antiquities of Jordan* 36: 113-44.

Hanbury-Tenison, J.W.
1984 Wadi Arab Survey 1983. *Annual of the Department of Antiquities of Jordan* 28: 385-424.

Helms, S.W.
1975 Jawa 1973: A Preliminary Report. *Levant* 7: 20-38.
1976 Jawa Excavations 1974: A Preliminary Report. *Levant* 8: 1-35.
1977 Jawa Excavations 1975: Third Preliminary Report. *Levant* 9: 21-35.

Herr, L.G., L.T. Geraty, O.S. LaBianca and R.W. Younker
1991 Madaba Plains Project: The 1989 Excavations at Tell el-'Umeiri and Vicinity. *Annual of the Department of Antiquities of Jordan* 35: 155-80.

Hopf, M.
1983 Jericho Plant Remains. In K.M. Kenyon and T.A. Holland (eds.), *Excavations at Jericho 5*, 576-621. London: British School of Archaeology in Jerusalem.

Ibach, R.D.
1987 *Archaeological Survey of the Hesban Region*. Berrien Springs, MI: Andrews University.

Ibrahim, M.M.
1987 Sahab and its Foreign Relations. In A. Hadidi (ed.), *Studies in the History and Archaeology of Jordan*, III: 73-81. Amman: Department of Antiquities.
1989 Sahab. In D. Homès-Fredericq and J. B. Hennessy (eds.), *Archaeology of Jordan*. II.2. Field Reports: Sites L–Z, 516-20. Akkadica Supplementum 8. Leuven: Peeters.

Ibrahim, M.M., J. Sauer and K. Yassine
1976 The East Jordan Valley Survey, 1975. *Bulletin of the American Schools of Oriental Research* 222: 41-66.
1988 The East Jordan Valley Survey, 1976 (Part Two). In K. Yassine (ed.), *Archaeology of Jordan: Essays and Reports*, 189-207. Amman: Department of Archaeology, University of Jordan.

Ilan, D.
1995 The Dawn of Internationalism—The Middle Bronze Age. In T. E. Levy (ed.), *The Archaeology of Society in the Holy Land*, 297-319. London: Leicester University.

Jefferson, M.
1939 The Law of the Primate City. *Geography Review* 29: 226-32.

Joffe, A.H.
1991 Settlement and Society in Early Bronze I and II Canaan. Unpublished Ph.D. dissertation, University of Arizona, Tucson.
1993 *Settlement and Society in the Early Bronze I and II Southern Levant*. Sheffield: Sheffield Academic Press.

Johnson, G.A.
1977 Aspects of Regional Analysis in Archaeology. *Annual Review of Anthropology* 6: 479-508.
1980 Rank-size Convexity and System Integration: A View from Archaeology. *Economic Geography* 56.3: 234-47.

Kafafi, Z.A. and E.A. Knauf
1989 Dahr el-Medineh. In D. Homès-Fredericq and J.B. Hennessy (eds.), *Archaeology of Jordan*. II.1. Field Reports: Surveys and Sites A–K, 198-200. Akkadica Supplementum 7. Leuven: Peeters.

Kaplan, M.
1980 *The Origin and Distribution of Tell el-Yehudiyeh Ware*. Gothenborg: Åstrom.

Kenyon, K.
1973 Palestine in the Middle Bronze Age. In *Cambridge Ancient History*, 2, Part 1, 77-116. 3rd edn. Cambridge: Cambridge University.

Knapp, A.B.
1993 *Society and Polity in Bronze Age Pella: An Annales Perspective*. Sheffield: Sheffield Academic Press.

Kramer, C.
1982 *Village Ethnoarchaeology: Rural Iran in Archaeological Perspective*. New York: Academic Press.

Lehmann, G. et al.
1991 The 1990 Excavations at Abu Snesleh. *Annual of the Department of Antiquities of Jordan* 35: 41-65.

Lenzen, C. and A.M. McQuitty
1989 Irbid (Tell). In D. Homès-Fredericq and J. B. Hennessy (eds.), *Archaeology of Jordan*. II.1. Field Reports: Surveys and Sites A–K, 298-300. Akkadica Supplementum 7. Leuven: Peeters.

Lines, L.
1995 Bronze Age Orchard Cultivation and Urbanization in the Jordan River Valley. Unpublished PhD Dissertation, Arizona State University, Tempe.

Liphschitz, N.R. Gophna and S. Lev-Yadun
1989 Man's Impact on the Vegetational Landscape of Israel in the Early Bronze Age II–III. In P. de Miroschedji (ed.), *L'urbanisation de la Palestine à l'âge du Bronze Ancien*, 263-68. British Archaeological Reports, International Series 527. Oxford: BAR.

Ma'ayah, F.
1960 Recent Archaeological Discoveries in Jordan. *Annual of the Department of Antiquities of Jordan* 4/5: 114-16.

Mabry, J. and G. Palumbo
1988 The 1987 Wadi el-Yabis Survey. *Annual of the Department of Antiquities of Jordan* 32: 275-305.

MacDonald, B. et al.
1988 *The Wadi el-Hasa Archaeological Survey 1979–1983, West Central Jordan*. Waterloo, Ontario: Wilfrid Laurier University.

Magness-Gardiner, B. and S.E. Falconer
    1994   Community, Polity, and Temple in a Middle Bronze Age Levantine Village. *Journal of Mediterranean Archaeology* 7.2: 127-64.

Mare, H.
    1991   The 1988 Season of Excavation at Abila of the Decapolis. *Annual of the Department of Antiquities of Jordan* 35: 203-21.

Matthiae, P.
    1981   *Ebla: An Empire Rediscovered*. New York: Doubleday.

Mazar, A.
    1990   *Archaeology of the Land of the Bible*. New York: Doubleday.

Mazar, B.
    1968   The Middle Bronze Age in Palestine. *Israel Exploration Journal* 18: 65-97.

McCreery, D.W.
    1980   The Nature and Cultural Implications of Early Bronze Age Agriculture in the Southern Ghor of Jordan: An Archaeological Reconstruction. Unpublished PhD dissertation, University of Pittsburgh.

Miller, J.M. (ed.)
    1991   *Archaeological Survey of the Kerak Plateau*. Atlanta, GA: Scholars.

Miller, N.F.
    1991   The Near East. In W. van Zeist, K. Wasylikowa and K.-E. Behre (eds.), *Progress in Old World Paleoethnobotany: A Retrospective View on the Occasion of 20 Years of the International Work Group for Palaeoethnobotany*, 133-60. Rotterdam: Balkema.

Mittmann, S.
    1970   *Beiträge zur Siedlungs- und Territorial Geschichte des Nördlichen Ostjordanlandes*. Wiesbaden: Harrassowitz.

Najjar, M.
    1991   A New Middle Bronze Age Tomb at the Citadel of Amman. *Annual of the Department of Antiquities of Jordan* 35: 105-34.

Neef, R.
    1990   Introduction, Development and Environmental Implications of Olive Culture: The Evidence from Jordan. In S. Bottema, G. Entjes-Nieborg and W. van Zeist (eds.), *Man's Role in the Shaping of the Eastern Mediterranean Landscape*, 295-306. Rotterdam: Balkema.

Paynter, R.W.
    1983   Expanding the Scope of Settlement Analysis. In J. Moore and A. Keene (eds.), *Archaeological Hammers and Theories*, 233-75. New York: Academic Press.

Petocz, D. and L. Villiers
    1984   Wadi Hammeh Survey. *Annual of the Department of Antiquities of Jordan* 28: 77-81.

Piccirillo, M.
    1978   Una tomba del Bronzo Medeo ad Amman. *Liber Annuus* 28: 73-86.

Politis, K.
    1990   Excavations at Deir 'Ain 'Abata, 1990. *Annual of the Department of Antiquities of Jordan* 34: 377-87.
    1993   The 1992 Season of Excavations and the 1993 Season of Restorations at Deir 'Ain 'Abata. *Annual of the Department of Antiquities of Jordan* 37: 503-20.
    1995   Excavations and Restorations at Dayr 'Ain 'Abata, 1994. *Annual of the Department of Antiquities of Jordan* 39: 447-91.
    1997   Excavations and Restorations at Dayr 'Ain 'Abata, 1995. *Annual of the Department of Antiquities of Jordan* 41: 341-50.

Posener, G.
    1971   Syria and Palestine ca. 2160-1780 B.C.: Relations with Egypt. In *Cambridge Ancient History*, 1, part 2, 532-58. 3rd edn. Cambridge: Cambridge University.

Potts, T.F., *et al.*
    1988   Preliminary Report on the Eighth and Ninth Seasons of Excavation by the University of Sydney at Pella (Tabaqat Fahl), 1986 and 1987. *Annual of the Department of Antiquities of Jordan* 32: 115-49.

Rollefson, G. and I. Kohler-Rollefson
    1992   Early Neolithic Exploitation Patterns in the Levant: Cultural Impact on the Environment. *Population and Environment* 13: 243-54.

Rollefson, G., A. Simmons and Z. Kafafi
    1992   Neolithic Cultures at 'Ain Ghazal, Jordan. *Journal of Field Archaeology* 19: 443-70.

Smith, R.H.
    1973   *Pella of the Decapolis*, vol. I. Wooster, OH: Wooster College.
    1993   Pella. In E. Stern (ed.), *The New Encyclopedia of Archaeological Excavations in the Holy Land*, 1174-80. Jerusalem: Israel Exploration Society.

Smith, R.H. and A.W. McNicoll
    1986   The 1982 and 1983 Seasons at Pella of the Decapolis. In *Bulletin of the American Schools of Oriental Research Supplement* 24: 89-116. Philadelphia: American Schools of Oriental Research.

Smith, R.H. and T.F. Potts
    1992   The Middle and Late Bronze Ages. In A.W. McNicoll *et al.* (eds.), *Pella in Jordan 2*, 35-81. Sydney: University of Sydney.

Stager, L.E.
    1985   The Firstfruits of Civilization. In J.N. Tubb (ed.), *Palestine in the Bronze and Iron Ages: Papers in Honor of Olga Tufnell*, 172-88. London: Institute of Archaeology.

Tubb, J.N.
    1983   The MB II A Period in Palestine: Its Relationship with Syria and its Origin. *Levant* 15: 49-62.

Van der Kooij, G. and M.M. Ibrahim
- 1989 *Picking Up the Threads... A Continuing Review of Excavations at Deir Alla, Jordan.* Leiden: University of Leiden.

Weinstein, J.M.
- 1975 Egyptian Relations with Palestine in the Middle Kingdom. *Bulletin of the American Schools of Oriental Research* 217: 116.
- 1981 The Egyptian Empire in Palestine: A Reassessment. *Bulletin of the American Schools of Oriental Research* 241: 128.
- 1991 Egypt and the Middle Bronze II C/Late Bronze I A Transition in Palestine. *Levant* 23: 105-15.

Wimmer, D.
- 1989 Safut (Tell). In D. Homès-Fredericq and J. B. Hennessy (eds.), *Archaeology of Jordan*. II.2. Field Reports: Surveys and Sites L–Z, 512-15. Akkadica Supplementum 8. Leuven: Peeters.

Zayadine, F., M. Najjar and J.A. Greene
- 1987 Recent Excavations on the Citadel of Amman (Lower Terrace). *Annual of the Department of Antiquities of Jordan* 31: 299-311.

Zipf, G.K.
- 1949 *Human Behavior and the Principle of Least Effort: An Introduction to Human Ecology.* Reading, MA: Addison-Wesley.

Zohary, D. and M. Hopf
- 1988 *Domestication of Plants in the Old World.* Oxford: Clarendon.

Zohary, M.
- 1982 *Plants of the Bible.* Cambridge: Cambridge University.

Zohary, M. and P. Spiegel-Roy
- 1975 Beginnings of Fruit Growing in the Old World. *Science* 187: 319-27.

8. The Late Bronze Age

John Strange

Introduction

The Late Bronze Age in Jordan is divided into LB I, IIA and IIB. These divisions are dictated partly by the archaeological finds, to a certain extent also by political realities and, ultimately, by the terminology of the archaeology of the Levant, especially that of Palestine west of the river Jordan. Jordan itself may be divided into northern Jordan, that is, the northern plains from the Yarmuk River to the mountains south of Jarash, or to Wadi az-Zarqa'; central Jordan, the area south of this to Wadi al-Hasa; southern Jordan, the area south of Wadi al-Hasa; the Jordan Valley from Tall ash-Shuna to the Dead Sea; and, finally, the desert to the east.

The character of the evidence

The Late Bronze Age is the first period in the history and archaeology of Jordan from which we find a small body of written evidence. This is due to the fact that the Egyptians took an interest in the area. The Egyptian texts, although sporadic and accidental, give some possibility of reconstructing certain aspects of the history and archaeology of Jordan that would otherwise be impossible. The most important texts are the toponym lists from Egyptian temples, which, however, regrettably largely omit Jordanian toponyms, royal stelae that are important especially for the later part of the period, and, finally, biographical texts and accidental references in administrative documents and the like (for the character of the sources cf. Redford 1982).

The published evidence of the material culture of the period is disappointingly meagre compared to other periods in the history of Jordan. Only a few of the excavated sites are published in final form, for example, the excavation of the Late Bronze Age temple at Tall Dayr 'Alla, the Baq'ah Valley, parts of the Pella excavations, and the 'Amman Airport temple. Otherwise, only preliminary publications exist, if at all.

Origins

The Late Bronze Age is a continuation of the Middle Bronze Age. There is no real cultural break. This may be seen, for example, from the pottery forms and technology. Examples of the former are carinated and pedestalled bowls, dipper juglets, amphorae, and, not least, Chocolate-on-White Ware that overlap the transition between the periods. Regarding this transition, pottery is still produced at the beginning of the Late Bronze Age using the true potter's wheel. At the 'Amman Airport temple, imported pottery ranged from the late Middle Bronze Age through the Late Bronze Age (c. 1700–1200 BC), with many of the pieces being heirlooms, while the local pottery seems to date exclusively to the thirteenth century BC (Herr 1997). This seems to indicate continuity also in population groups. There also seems to be continuity in architecture and the arts. Pella is a good example of this. Here the transition from the Middle Bronze Age to the Late Bronze Age is smooth (McNicoll *et al.* 1992: 35, pls. 47-50, 53-59). Heirlooms, such as the ivory inlaid box from the Middle Bronze Age (McNicoll *et al.* 1992: 35, 59-63), point to continuation of both people and culture, as does Tomb 62, probably used by a family or clan in both the late Middle and early Late Bronze Ages (McNicoll *et al.* 1992: 69-81). At Irbid, moreover, the Middle Bronze Age city wall survived and was used in the Late Bronze Age (Lenzen *et al.* 1985). Finally, surveys have shown that a large proportion of settlements of the Middle Bronze Age survived into the Late Bronze Age (see, for example, Najjar 1992).

Chronology and dating evidence

Here the low Egyptian chronology (Kitchen 1987: 52) will be used wherever possible for all historical dates.

The Late Bronze Age is usually said to begin c. 1550 BC, a conventional date derived from Egyptian chronology and history. It is, however, difficult to pinpoint the beginning of the period archaeologically, as the transition from the Middle Bronze Age is not marked by a uniform destruction horizon or the like (see, however, Weinstein 1991; cf. Thompson 1992: 205-208). The urban system of the Middle Bronze Age II crumbled in western Palestine over a period of 100–120 years (Na'aman 1994: 184), and, by analogy, the same would probably also be the case east of the Jordan. The Middle Bronze/Late Bronze Age transition period lasts from c. 1600 BC to c. 1500 BC: Chocolate-on-White pottery overlaps the transition, which spans a very long period, perhaps as much as a century. It therefore seems reasonable to date the beginning of the true Late Bronze Age from the Egyptian conquest. This brought Jordan into the international system of the Late Bronze Age (see below), and prepared for the real innovation of the period, namely, the abundant trade with other areas of the Levant and the Eastern Mediterranean world as reflected in the import of Cypriot and Mycenaean pottery and, conversely, the probable export of such commodities as wine or olive oil from Jordan to the west. This, of course, again makes the date of the beginning of the period dependent on the Egyptian evidence and mixes archaeological and historical reasoning. It must be remembered, however, that the divisions of the Bronze Age are to a large extent conventional and artificial, created after a tripartite system in European prehistory. The Egyptian conquest of Palestine, including Jordan, took place just after the co-regency of Hatshepsut and Tuthmosis III (1479–1457 BC) in the twenty-second year, the year of Hatshepsut's death, or 1457 BC. Since the transition period more or less coincides with parts of the Late Bronze I period, the beginning of the Late Bronze Age can be dated between 1550 and 1450 BC. LB I would be partially the Middle Bronze Age/Late Bronze Age transition period.

The date of the end of the Late Bronze Age is also uncertain because of another transition period that is evident, for example, at Tall Dayr 'Alla (Franken 1964b: 418-20), at Tall al-Fukhar (Strange 1993, 1997) and at Tall as-Sa'idiyya (Tubb 1988b: 65). In Jordan there is, however, some solid archaeological evidence that links the end of the period to Egyptian chronology: the temple at Tall Dayr 'Alla was destroyed at the end of the period, and since nothing later than LB IIB material, for example, no Mycenaean pottery later than Mycenaean IIIb and its local imitations, was found in the destruction level (Franken 1992), this is a convenient benchmark; an important artifact from the destruction level is a faience vase with the cartouche of Queen Tewosret, or Tausert (1188–1186 BC) (Franken 1992: 30-31), giving the date 1188 BC as *terminus a quo* (Kitchen 1987: 52); Franken dates the destruction to just after the first decade of the twelfth century BC, although with some reservations (1992: 176-77.).

At Tall as-Sa'idiyya Stratum XII, the Late Bronze Age/Iron Age transition period is characterized by a possible Egyptian residence, a palace built in Egyptian style, and a cemetery with signs of very strong Egyptian influence (Tubb 1990: 26-37), including finds that may connect this cemetery with the presence of Sea Peoples (e.g. Tubb 1990: 33). After this any Egyptian presence in Jordan seems to be a thing of the past. As will be argued below, the Late Bronze Age of Jordan coincides more or less with the Egyptian domination of the country. It is possible to take the transition period together with the Late Bronze Age and claim that the Egyptian presence at Tall as-Sa'idiyya marks the end of the Late Bronze Age. It is possible, however, to take the presence of Sea Peoples as a sign of the beginning of the Iron Age. In this chapter the former alternative is preferred.

There is evidence at Tall al-Fukhar for the end of the Late Bronze Age, as the palace there was apparently abandoned and then occupied by squatters, before it was destroyed in a violent conflagration and replaced by a village of the Late Bronze Age/Early Iron Age transition period or Early Iron IA period with no Mycenaean pottery. The only Philistine sherd ever found in the highlands of Jordan was in fill in the subsequent stratum, that is, the early Iron I period. This seems to give a *terminus ante quem* of c. 1050 BC for the beginning of this period (Strange 1997: 402-403). This gives the dates c. 1550–1450 BC to c. 1200–1150 BC for the beginning and end of the Late Bronze Age respectively, with a transition period included both at the beginning and at the end of the period.

The LB I period is contemporary with part of the transition period and the LB I proper, roughly from 1550–1450 to c. 1400 BC, just before the Amarna period. The next century, the fourteenth century BC, the LB IIA period, is the period of the waning of Egyptian dominance in the area, as may be seen from the Amarna letters, which, however, refer mostly to western Palestine. The LB IIB comprises the rest of the Late Bronze Age, from c. 1300 BC to the end in the first decades of the twelfth century BC. It should be noted that the periods are defined also by the corresponding periods in western Palestine.

Radiocarbon dates

Very few radiocarbon dates from Late Bronze Age Jordan have been published. The ones that have seem useless because they are far too high for dating the period. This is due to the character of the specimens investigated, mostly carbonized wood from roof constructions. Since the timber used in construction is invariably older, and sometimes considerably older, than the building, to say nothing of its destruction, the radiocarbon date of the specimen would necessarily be expected to be older than the pottery and other artifacts in the destruction layer.

Radiocarbon dates

Tall Dayr 'Alla, carbonized roof beam
(Franken 1992: 177)

	95.4% (2 sigma)	68.3% (1 sigma)
	1292–1266 BC or 1522–1296 cal. BC	1510–1478 BC or 1460–1382 BC or 1344–1320 cal. BC

Baq'ah Valley Project
(McGovern et al. 1986: 333-34)

	calibrated dates
Upper burial layer of a LB II age cave	1590–1520 ± 60 BC
	1710–1690 ± 70 BC
LB II pit (with LB AI pottery)	1910–1780 ± 60 BC
+	2290–2190 ± 70 BC
top of pit	2090 ± 70 BC

These radiocarbon dates are too high to say anything meaningful about the chronology of the Late Bronze Age (McGovern et al. 1986: 333-34).

Tall al-Fukhar, destruction level, probable roof material
(Carbon-14 Dating Laboratory, National Museum, Copenhagen)

Sample	^{14}C b 1950	BC cal.	± stand. dev. BC cal	^{13}C promille PBD
1. Charred banches (Olea europea)	3380 ± 90	1680	1750 – 1520	–21.0
2. Part of an oakbeam	3210 ± 90	1490–1450	1530 – 1400	–22.8
3. Charred heartwood (Quercus sp.)	3610 ± 90	1940	2120 – 1790	–24.2
4. Charcoal (Olea europea L.)	3260 ± 65	1520	1610 – 1440	–21.5
Average of 1, 2 and 4	3280 ± 45	1520	1610 – 1510	

No. 3 is heartwood and could be considerably older than the outer layers reflecting the felling of the tree and the subsequent use of the beam. Here again, the dates seem much too high, placing the construction of the palace in the LB I period, but they reflect the age of the timber used in the construction of the building.

Carbonized grain from the destruction layer dated by pottery to 1300–1150 BC at Irbid (Ambers et al. 1989: 28)

cal. ^{14}C date bpcal.	^{14}C date BC	^{13}C
3040 ± HO	1395–1260	–20.0

According to McQuitty (Lenzen, Gordon and McQuitty 1985) the sample was taken from the grain cache that probably dates to within one year of the destruction of the building and places it firmly within the traditional Late Bronze Age rather than the Late Bronze Age/Iron Age transition.

Claims of radiocarbon dates falling within the Late Bronze Age come from Wadi Faynan (Bartlett 1997: 190), the largest copper production centre of the southern Levant. No contextual evidence for the dates has, as yet, been presented. This date is, again, probably too high since there are claims elsewhere that there is no proof of smelting in Wadi Faynan in the second millennium BC (Hauptmann and Weisgerber 1992: 63).

Environment

Beginning about the sixteenth century BC, the climate in Palestine seems to have been much drier than that of the preceding centuries of the Middle Bronze Age (see Chapter 17). Evidence for this relative drought comes both from a substantial drop in olive production, as well as a lowering of the water table indicated by a shift of settlement in the Iron Hills in Western Palestine, dependent on wells (Lipschitz 1986: 37-58; 1996: 139-45; Thompson 1992: 204-205). In contrast to the period after 1000 BC, this relative drought probably culminated around 1200 BC, when the period was dominated by a sub-Atlantic period of aridity with both a decrease in rainfall of approximately 20% and rising temperatures of 2–3°C. Evidence for this is drawn from coastal Syria, the Great Syrian steppe, the Dead Sea Basin and Jordan (Shehadeh 1985: 26-27; Nützel 1976: 11-24; Neuman and Parpola 1987: 161-82; van Zeist 1985: 199-204; Thompson 1992: 215-21). This relative drought would affect the marginal areas to the east and the south most severely, and would, to a large extent, account for the change in settlement pattern that can be observed in Jordan, as well as contributing to the economic decline during the Late Bronze Age.

At Pella, oak is less frequent compared to the periods before the Bronze Age; instead willow and/or poplar, *Ziziphus* sp., almond, hackberry and olive are found, perhaps indicating that degradation of forests on the lower slopes had started. This could also be due, however, to climatic influences or selection (McNicoll *et al.* 1992: 255).

Historical background

The Late Bronze Age was the first truly international age in the Eastern Mediterranean and the Levant. During this period the world was divided into spheres of interest, each dominated by a great king, and the interaction between these kings and their spheres of interest was regulated by means of a complicated system of trade relations, cultural exchanges, dynastic marriages and treaties (Liverani 1987; Merrillees 1986; Strange 1986: 2-7). The Levant and its hinterland as far north as Ugarit lay in the Egyptian sphere of interest in the early part of the Late Bronze Age. To the north, the Hittite Empire sought to expand. This resulted in a series of exhausting wars between the two states, ending in a treaty between Egypt and Hatti in 1254 BC and defining the border between the two in Lebanon. This left Egypt in control of Palestine, Transjordan and the areas in south Lebanon and around Damascus. In the Mediterranean, the Mycenaean Empire was the major power. Cyprus, which was probably independent in the early part of the Late Bronze Age but later became a colony of the Mycenaeans, played the role of trade emporium. To the east, Mitanni was a great power in the earlier part of the Late Bronze Age and, throughout the whole period, also Assyria and Babylonia, who were important trade partners but did not act directly politically in the Levant.

At the outset of the Late Bronze Age, Egypt conquered Jordan, as may be seen from the great list at Karnak where at least toponyms nos. 89–101 should be situated in Jordan, although the precise identification of names is in most cases impossible (Helck 1971: 127-28; Smith 1973: 23-32; Aharoni 1979: 173, 159; Redford 1982: 115-19; Knauf and Lenzen 1987: 49-53; Lenzen and Knauf 1987: 54-59, but see Timm 1989: 53-56). At Soleb, some names are given in a list from Amenophis III (1390–1352): *smt* in Shasu, *yhw* in Shasu, *trbr* in Shasu, *bt* ' (?). They also, however, cannot be identified with any certainty (Giveon 1971: 27).

The degree of Egyptian domination is difficult to assess. No doubt the main interest of the Egyptian kings was to control the area militarily to avoid unpleasant surprises and protect the Jordan Valley and western Palestine, and to protect the trade routes. Another interest lay in tapping the wealth of the country by trade or tribute (cf. Redford 1993: 209-13; Knapp 1993: 40-49). The southern part of the country, south of Wadi al-Hasa, was probably left alone, and at present there is no evidence of settled occupation in the Late Bronze Age. It is noteworthy, especially in view of the Egyptian activity at Timna on the other side of Wadi 'Arabah, that there was no exploitation of the copper ore deposits in Wadi Faynan in the Late Bronze Age (Hauptmann and Weisgerber 1992: 63). Southern Jordan, which was later to become the kingdoms of Moab and Edom, was inhabited by the Shasu, nomadic elements known from the Egyptian texts and reliefs (Giveon 1971; Bienkowski 1991: 14).

Even in the Amarna period Egyptian predominance continued. Evidence for this comes from the Amarna Letters mentioning several toponyms east of the Jordan, for example, Pella, controlled by King Mutbalu, the main city in a city-state bordering on the city-state of Ashtaroth (Tall Ashtara) controlled by King Ayyub (Amarna Letter no. 256; see Albright 1943: 9-15; Helck 1971: 184; Aharoni 1979: 159-60; Knapp 1993: 42-48). Here again the identification of the names is not certain. It seems, at least to judge from the evidence from western Palestine, that Egyptian influence was weakened considerably in the course of the fourteenth century BC, although the evidence of the Amarna Letters can be interpreted in different ways.

The Nineteenth Dynasty, however, already under Sethos I (1294–1279), who, after provocation from the Shasu, campaigned in northern Palestine, southern Lebanon and Syria and apparently northern Jordan (Kitchen 1982: 20-24), followed by Ramesses II (1279–1213), reasserted and strengthened Egyptian hegemony (Redford 1993: 179-91; Knapp 1993: 48; cf. the Egyptian policy in western Palestine where Weinstein assesses the evidence to be a military occupation [1982: 17-22]). A list of Asiatic names on a statue of Ramesses II from the temple at Luxor includes Moab (no. 14), together with Hatti, Naaharina and Assur. The list mixes great states with smaller ones, and the mention of Mitanni is suspicious. But it does mention Moab as a political entity outside the realm of Egypt. There is, thus, proof of a state in Jordan (cf. Timm 1989: 5-9). A relief in the court of Ramesses II of the Luxor temple shows two prisoners from Moab taken after the plunder of Butartu (Bienkowski 1991: 14).

The stelae erected by Sethos I and Ramesses II at Beth Shan testify to military expeditions across the Jordan River, again with a number of toponyms (Aharoni 1979: 166-70). The upper part of a stele erected by Sethos I was found at Tall al-Shihab on the Yarmuk, c. 28 km northeast of Irbid (Erman 1893: 100-101). Moreover, a stele with the cartouche of Ramesses II was set up at Sheikh Sa'ade, c. 30 km northwest of Dar'a (Weinstein 1982: 20). Both stelae, which are on the route to Damascus, indicate Egyptian presence. This coincides with the beginning of the LB IIB period, found for example at Tall al-Fukhar (Strange 1994: 537; 1997) and at Irbid (Lenzen et al. 1985: 152). Also the possible Egyptian governor's palace at Tall as-Sa'idiyya testifies to the importance of Jordan to the Egyptians in this period (e.g. Tubb 1988c).

At the end of the LB II period, the Egyptians withdrew from Jordan and concentrated their efforts on Cisjordan. The cities in the country east of the river were left to their own fate. Evidence for this may perhaps be seen for example at Tall al-Fukhar (Strange 1997). In the following period, the early Iron Age, there is no historical evidence concerning Jordan. Only later, in the Iron Age II, do we again find any historical evidence.

The purported biblical material on the history of Jordan in the Late Bronze Age, although it is in itself abundant and important in some respects, is in many respects manifestly contradictory to the archaeological finds (MacDonald 1994). Thus, in the present state of biblical research, this material is, from a historical and chronological point of view, so unsatisfactory that it is probably safer to disregard it completely as historical evidence pertaining to the Late Bronze Age.

Settlement patterns

Since Glueck's explorations more than 50 years ago a great deal of archaeological survey work has been conducted in Jordan (Figure 8.1). As a result of these efforts, a fairly reliable description of the pattern of settlement can be drawn. Such a description must, however, be taken with caution, as every survey in itself is often incomplete, and because the various surveys have not been conducted by the same standards and methods. Furthermore pottery, which is the main dating tool of a survey, is not always sufficiently diagnostic to give safe conclusions, and there is, as in all research, a tendency to find what is sought for. Also, very little material from excavations has been published from the Late Bronze Age except for preliminary publications. Only parts of the excavations at Pella and the temple of Tall Dayr 'Alla, and part of the Baq'ah Valley Project have been published in final form. Thus, there is very little stratigraphically dated material to check the dates of the pottery found by the surveys.

A convenient departure point may be the data collected in JADIS (Palumbo 1994), although the data are to some extent useless in the opinion of the editor, who observes (p. 1.15) that the information is biased towards the Iron Age and Classical periods. It can be seen from Palumbo's table (p. 1.14) that from the Early Bronze Age 1102 sites are listed, from the Middle Bronze Age 290 sites, from the Late Bronze Age 295, from the Iron Age I–II 1614 sites, from the Roman period 2306 sites, and from the Byzantine period 2299 sites (p. 1.14). This encourages one to think that rather than observing an explosion of human occupation, we are seeing a biased data collection strategy. More than a third of the sites listed provide inadequate information on the period(s) of occupation. Finally, the information in JADIS is biased towards the cultivated areas (p. 1.15). Palumbo concludes that the data cannot be used for settlement pattern studies. It is more useful for a sociological and historical profile of archaeological research. The voids in the research still need to be filled by intensive archaeological surveys.

JADIS divides Jordan into 15 sectors. This makes its use somewhat awkward, because the sections do not stem from the natural geographical regions of the country, but are derived from the mechanical use of the UTM coordinates and consequently cut across these regions. Thus, the Jordan Valley is treated together with the Western Hills in sections 1, 2 and 4. As a result, the sites in the valley cannot be separated easily from those in the mountains. Roughly speaking, sectors 1, 2 and 3 are situated in northern Jordan; sectors 4, 5, 6, 7, 9 and 10 in central Jordan; sectors 11, 12, and 14 in southern Jordan; and, finally, sectors 7, 8, 13 and 15 are situated in the desert to the east.

It should be kept in mind that the lists in JADIS do not give the relative size of the settlements in the individual periods. Moreover, we simply do not know the population density of Jordan, for instance, in the Late Bronze Age as compared to the Middle Bronze or Iron Ages.

Still, despite all these caveats, it may be possible to deduce some trends in the settlement from the

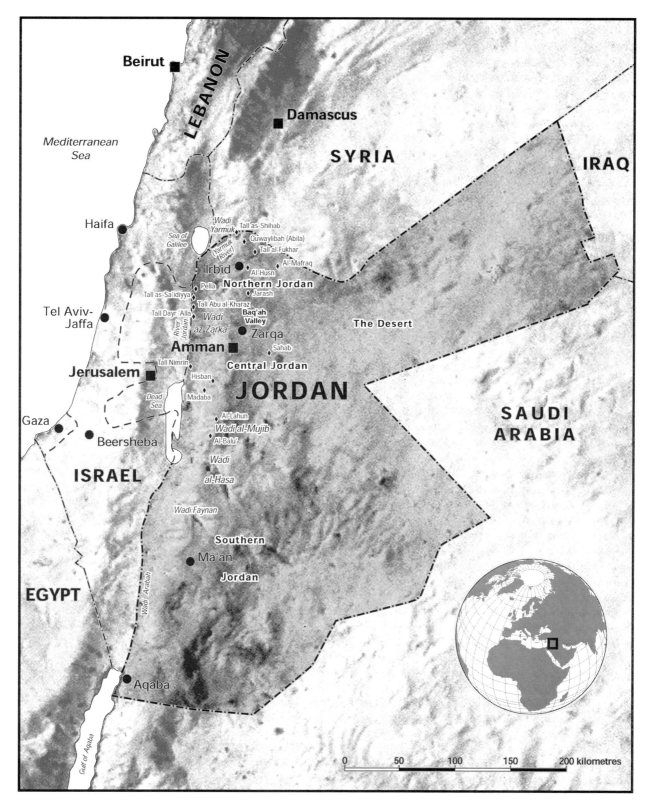

Figure 8.1. Map of Jordan in the Late Bronze Age.

JADIS data and then compare and control them with more precise surveys and knowledge gained from excavations.

First of all, as would be expected, the settlements are concentrated in the western and northern parts of the country in the Late Bronze Age: northern Jordan, 117 sites with 102 sites surviving from earlier, and 15 newly founded; central Jordan, 176 sites with 89 surviving and 87 new; southern Jordan, 18 sites with 7 surviving and 11 new; the desert, 4 sites with 1 surviving and 3 new; and the Jordan Valley, part of sectors 1, 2 and 4, part of 124 sites with 101 surviving and 23 new. The data on southern Jordan should be seriously questioned (see below).

Secondly, a decrease in the number of settlements from the Middle Bronze to the Late Bronze Age can be observed in most parts of the country. Only in sectors nos. 6, 9, 10 in central Jordan, together with sector no. 11 in southern Jordan, is an increase listed, and these data should be treated with suspicion, as will be seen below.

Finally, an enormous increase in the number of settlements may be seen from the Late Bronze to the Iron Age (both I and II) all over the country. This increase does not necessarily reflect an increase in the number of inhabitants. It may simply be a spreading of the population from the larger centres of the Late Bronze Age.

By and large, the above observations are corroborated by the various surveys and soundings that have been published over the last 25 years. Mittmann found a decrease in settlement in northern Jordan from the Middle Bronze to the LB I period and a further decrease in the LB II period (Mittmann 1970: 256-64). The main cities in the north were occupied in the Late Bronze Age. Irbid, for example, was a large town in the LB IIC period (1300–1150 BC). This is attested by a temple, a basalt city wall incorporated from the Middle Bronze Age, and major public structures and tombs. The major occupational phase in the history of the city was c. 1200 BC when the city covered the entire tell and ended in a destruction that was not manmade (Dajani 1964; Lenzen 1986, 1992; Lenzen *et al.* 1985: 153-54; Lenzen and Knauf 1986; 1987b: 60). Its name was possibly Gintot (Lenzen and Knauf 1987b).

Although sparse information has been published, it appears that Quwayliba (Abila) was occupied in the Late Bronze period (Mare 1989: 474-77). However, tomb material from the site has been published (Kafafi 1984).

At Tall al-Fukhar massive fortifications and a large public building with imported luxury wares show that the city was important in the Late Bronze Age. Moreover, a scarab from the Nineteenth Dynasty was found (Ottosson 1992; Strange 1997: 402).

There are reports of Late Bronze Age pottery, today unfortunately lost, from Tall al-Husn (Lenzen and Knauf 1987: 61; Sauer 1986: 6; cf. Leonard 1987b: 359-60, 363).

The easternmost settlement in northern Jordan seems to be reflected in a tomb at al-Mafraq, but this is from the end of the Late Bronze Age or from the Late Bronze Age/Iron Age transition period (Piccirillo: 1976: 27-30) and may be seen as a harbinger of the more dense settlement in the Iron Age.

Braemer found in the Jarash region an increase in settlements, though not accompanied by an increase in population, from the Middle Bronze to the Late Bronze Age, a spreading of settlements to the mountains (Braemer 1992). There seems, moreover, to be a major settlement phase on new sites in the middle of the second millennium BC (Braemer 1987: 529). It must be kept in mind that the Jarash region is adjacent to, or could even be considered part of, central Jordan, where the JADIS data shows an increase in settlement.

In Jarash itself, the Late Bronze Age levels were found on bedrock. There were well-preserved floors of which four belong to the end of the period. Pottery included Chocolate-on-White and White Slip Ware (Braemer 1987: 525-28), perhaps indicating that the settlement lasted the whole period.

In central Jordan, the 'Ajlun highlands and the Jordan plateau attest to only a few sites occupied (Gordon and Knauf 1987: 294-97; Mabry and Palumbo 1992). In Wadi al-Yabis, there is a decline from the Middle Bronze to the Late Bronze Age. Farther south, the plateau shows occupation in the Late Bronze Age in an abated state. The relatively rare town settlements, such as Tall Nimrin (although three seasons of excavation at Tall Nimrin have failed to produce clear evidence from the Late Bronze and early Iron Ages), and the presence of late Iron Age walls built immediately over MB II structures suggest an occupational gap during the second half of the second millennium BC (Flanagan *et al.* 1992: 535-36). Sahab and Umm ad-Dananir were small in size (Herr 1992). Moreover, sites like Umm al-Qanafah, Madaba and 'Amman were occupied (Prag 1992).

Late Bronze Age tombs are attested at 'Amman (Dajani 1966). Moreover, on the Qal'ah, the Late Bronze Age has been found on the first and second plateau, indicating that the town was possibly as large and important as during the later Iron Age.

It was perhaps walled. Associated pottery included Chocolate-on-White and white burnished wares from the beginning of the period (Dornemann 1983: 105-21; 1997: 99; Zayadine et al. 1987: 308).

A remarkable temple was found at the former 'Amman civil airport (Hennessy 1966), and at al-Mabrak, 4 km southeast of the airport, an analogous square, but probably domestic, building was found, dating from the same period (Yassine 1988; see also below).

Northwest of 'Amman, in the Baq'ah Valley, numerous settlements from the Late Bronze Age were found. Tall Umm ad-Dananir in particular must be mentioned, with a *quadratbau*-type structure, possibly cultic, and an overlapping sequence of Late Bronze Age tombs with imports (McGovern 1992; see also below).

Tall Safut, overlooking a major entrance to the Baq'ah Valley, has a perimeter wall, possibly a Middle Bronze Age defence in continued use into the LB II period (Wimmer 1987a: 280), and a probable sanctuary (Wimmer 1989: 514). In the destruction layer, a chalice with barley and a bronze figurine of a god were found (Wimmer 1987a: 280-81).

Only six of the 155 sites surveyed within 16 km of Hisban have possible Late Bronze Age pottery. Only two major sites, Tall Jalul and Tall al-'Umayri, show evidence from the period. Tall Jalul has both LB I and II sherds in fills (Herr et al. 1994: 161). The pottery from the period at Tall al-'Umayri shows close parallels to that at Tall Jalul (Ibach 1987: 157).

At al-Balu' there are Late Bronze Age remains in the area immediately east and west of the *qasr*, which may be from the Iron Age (Worschech et al. 1986: 292; Worschech 1997: 270). Furthermore, the Balu' stele (see below) should be mentioned.

There is evidence of occupation at Khirbat 'Ara'ir at the end of the Late Bronze Age (Olávarri 1997: 178).

A settlement from the Late Bronze Age lies below the Iron Age fortress at al-Lahun. Houses with clay ovens, silos, grinding stones, cooking pots and storage jars were excavated. A scarab from the Twentieth Dynasty was also found (Homès-Fredericq 1989: 354; 1997: 155).

Sedentary occupation in the area between Wadi al-Mujib and Wadi al-Hasa is perhaps corroborated by the Egyptian epigraphic evidence where we find Moab mentioned on temple walls at Luxor (cf. above).

Sahab, 12 km southeast of 'Amman, has evidence of Late Bronze Age occupation in the form of a Late Bronze/Iron Age tomb (Dajani 1970), a thirteenth-century building, and a town wall, 75 m of which has been uncovered (Ibrahim 1974: 60-61; 1975: 78-80). The town seems to have had unbroken occupation from the Middle Bronze to the Iron Age I period (Ibrahim 1989: 516), with the pottery from the Late Bronze Age covering the period up to the late thirteenth century (Ibrahim 1987: 76-77).

Late Bronze Age sites are lacking north and northeast of 'Amman in the 'Ayn Ghazal and Wadi az-Zarqa' area, but these regions are close to the desert (Simmons and Kafafi 1992: 81).

Betts's (1989) survey in the northeast desert region of Jordan shows that it was not occupied during the Late Bronze Age.

The northern half of the Jordan Valley, from the Yarmuk River to Wadi ar-Rajib, shows 6 out of 106 surveyed sites occupied in the Late Bronze Age: Tall al-Meqbereh and Tall Abu al-Kharaz (associated or twin sites); Tall as-Sakhineh; Tall as-Sa'idiyya and Tall al-Keraimeh (two more associated sites); and Tall al-Mazar (Yassine et al. 1988: 167-74). Pella must also be mentioned, although it is not in the survey.

Three of the above-listed sites have been excavated. Pella, the most important and most northerly located of the three, produced evidence of rich Late Bronze Age remains.

Tall Abu al-Kharaz has yielded buildings from LB I, transition layers from LB I and II, a temple from LB IIA and walls built during the period and reused in the Iron Age (Fischer 1991, 1993, 1994).

At Tall al-Mazar domestic installations from the Late Bronze Age were found, but no architectural remains (Yassine 1989).

Tall as-Sa'idiyya had rich Late Bronze Age deposits. Most notable is a possible Egyptian governor's residence from the thirteenth century, Stratum XII, an elaborate water system and a Late Bronze Age cemetery (Tubb 1986; Tubb et al. 1996: 24-27).

The southern half of the Jordan Valley, from Wadi ar-Rajib to the Dead Sea, had 14 out of 118 surveyed sites occupied in the Late Bronze Age: Tall an-Nkeil South; Tall Ghazaleh; Tall al-Kharabeh; Tall Qa'dan; Tall Qa'dan South; Tall Dayr 'Alla; Tall ar-Rabi'; Tall Abu Nijrah; Tall al-'Argadat; Katarit as-Samra, 'Ayn al-Bassah; Tall al-Tahune; al-Jazayir; and Tall Mustah (Yassine et al. 1988: 190-97).

Of these sites two, Katarit as-Samra and Tall Dayr 'Alla, have been examined more closely. A cemetery was found at the former and the tell has material from the Middle Bronze/Late Bronze Age transition, and much from the LB I and II, including Cypriot imports. A Late

Bronze tomb yielded material, including imports, from throughout the period, especially from the thirteenth century (Leonard 1979, 1986).

At Tall Dayr 'Alla, a temple renovated several times, existed throughout the Late Bronze Age. The site seems to have been a trading station, probably controlled by the Egyptians (Franken 1992).

It appears that the Jordan Valley south of 'Ayn al-Bassah (which is near Katarit as-Samra) was virtually uninhabited during the Late Bronze Age. Only three sites, at Wadi Nimrin and Wadi Kufrein, were found to be settled farther south (Yassine *et al.* 1988). South of the Dead Sea, in the Southern Ghors and Northeast 'Arabah, no Late Bronze Age sites are found (MacDonald 1992a: 71).

There is very little evidence of occupation on the plateau between Wadi al-Mujib and Wadi al-Hasa. The Late Bronze Age was attested at five sites by Miller (1979; but soundings at one of these have not confirmed that survey's identification of Late Bronze Age pottery [see Bienkowski *et al.* 1997; Bienkowski and Adams 1999]), revisiting sites earlier surveyed by Glueck. The most important of these sites is Balu'. The same is true in northwest Ard al-Karak, where Worschech (1992) found Late Bronze Age pottery at only four sites, after a sudden decline from the Middle Bronze Age.

In southern Jordan, south of Wadi al-Hasa, no site has yielded definite Late Bronze Age pottery in stratified sequences (Bienkowski 1996; Bienkowski *et al.* 1997; Bienkowski and Adams 1999). MacDonald *et al.* (1988: 166-70 and pl. 6) report Late Bronze Age pottery at three sites and Late Bronze/Iron Age pottery at five sites). There is no proof of smelting during the second millennium BC in Wadi Faynan (Hauptmann and Weisgerber 1992: 63). There are reports of 'Midianite pottery' only from the far south (Jobling 1981: 31). This pottery, however, should, according to some writers, perhaps be dated to the Iron Age (cf. Fritz 1994: 144-45; Rothenberg and Glass 1983; but see below).

It should not be assumed, however, that southern Jordan was empty. The area was probably inhabited by nomads who were caravaneers and lived by herding, and who left no trace (cf. above on the Egyptian evidence of the Shasu).

Summing up, it seems that the number of settlements in Jordan was small in the Late Bronze Age. It was certainly less than in the previous period. Only in confined areas, for example, around Jarash, did the number of settlements increase. This is probably due to a dissemination of the population and resettlement from other towns. The southern parts of central Jordan were apparently thinly populated, with the southern part of the Jordan Valley and Jordan south of Wadi al-Hasa completely empty of settlements, but probably inhabited by nomadic pastoralists, if not necessarily Shasu.

Material culture

Woodwork, bone and ivory

A large number of objects of ivory and bone has been found (e.g. McGovern *et al.* 1986: 243-46; Tubb 1988b: 67; Franken 1992: 33-34, 42-43, 45-46, 58, 60, 64-65; McNicoll *et al.* 1982: 43, 46-48, 102-103; 1992: pl. 72). The production is best represented by the Lion Box from Pella (Potts 1987; Bienkowski 1991: 104-105; McNicoll *et al.* 1992: 59-62). The wooden frame has perished completely, but the shape can be reconstructed from the position of the carved ivory panels when it was found. On the lid, there is a unique representation of two rampant lions roaring at each other with their front paws resting on two *uraei* whose bodies intertwine between the lions' legs forming a kind of ornament. Also on the lid is a winged sundisc with two *uraei*, and on the sides there are panels with *Djed* pillars. The box is Egyptian in inspiration, but also shows influence from other places, for example, Syria and Mesopotamia, and was probably made either as a Hyksos work from the Nile Delta or in the Levant between the eighteenth and sixteenth centuries BC, which, strictly speaking, makes it an heirloom from the Middle Bronze Age (Potts 1987: 69-70). Also from Pella comes a box decorated with a continuous frieze of *Djed* pillars with possibly wooden panels with other motifs; it was probably manufactured at the same workshop as the Lion Box, perhaps even by the same craftsman (McNicoll *et al.* 1992: 62-64 and pl. 38). Finally, an ivory cosmetic box in the shape of a fish found inside a bronze bowl in the cemetery at Tall as-Sa'idiyya must be singled out (Tubb 1988b: 64, 66; Bienkowski 1991: 104-105).

Spindle whorls and handle
(McGovern *et al.* 1986: 243-44)

Glass

Pendant of glass in form of a fertility type figurine (Strange 1997: 402 and fig. 7) (Figure 8.2).

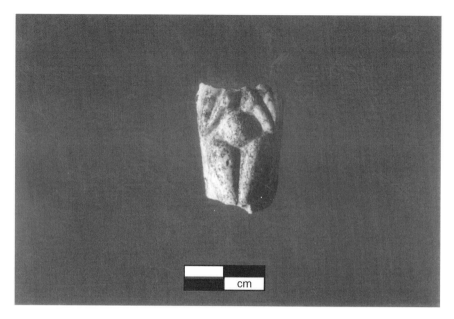

Figure 8.2. Glass pendant in form of a fertility figurine, Tall al-Fukhar.

Bronzework

McGovern *et al.* (1986: 245-57, 268; see also 278-83 for the metallurgy) give a typology of copper-base artifacts found in the Baq'ah Valley from the Late Bronze Age. This can be supplemented by published material from the temple at Tall Dayr 'Alla, Pella and several artifacts from other sites: *anklets and bracelets* (McGovern *et al.* 1986: 246, 250-55; McNicoll *et al.* 1992: 77); *earrings* (McGovern *et al.* 1986: 247, 250-53); *rings* (McGovern *et al.* 1986: 248, 250-55, 286-87; McNicoll *et al.* 1992: 77, pl. 61); *pendants* (Franken 1992: 33-34); *daggers* (McGovern *et al.* 1986: 248, 256-57); *knives* (Franken 1992: 33-34; McNicoll *et al.* 1992: 77, pl. 61); *spearpoints* (Franken 1992: 33-34; McNicoll *et al.* 1982: 43); *arrowheads* (McGovern *et al.* 1986: 249, 256-57; Franken 1992: 33-34; McNicoll *et al.* 1992: 77); *scales for armour* (Franken 1992: 31-33); *pins* (McNicoll *et al.* 1992: 77); *toggle pins* (McGovern *et al.* 1986: 249, 250-51, 254-57; McNicoll *et al.* 1982: 46-48; 1992: 77, pl. 61); *hooks* (Franken 1992: 33-34); *tweezers* (McNicoll *et al.* 1982: 43, 47; 1992: 77); *spatulae* (Franken 1992: 66-67); *rods* (McGovern *et al.* 1986: 249, 254-55); *needles* (Franken 1992: 31, 33-34, 91, 94); *figurines* (Wimmer 1987a: 280; Franken 1992: 42, 44); *lamps* (Strange 1997: 402) (Figure 8.3); *bowls* (McGovern *et al.* 1986: 268); *tableware*: from the cemetery at Tall as-Sa'idiyya came a bronze wine set, a platter, a jug and a strainer (Tubb 1988b: 69-70; Bienkowski 1991: 94-95).

Stone objects

Alabaster jars (e.g. McNicoll *et al.* 1982: pl. 109; McGovern *et al.* 1986: 270; Franken 1992: 59, 64, 91, 94); *tazza of serpentinite of Egyptian type* (Bourke *et al.* 1994: 92, 109-110); *Egyptian stone vessels* (Hankey 1974: 168-75); *objects made of Cretan limestone* (Hankey 1974: 175-76); *beads* (McGovern *et al.* 1986: 202-42); *basalt bowls, mortars, rubbing stones and querns* (Hankey 1974: 177; McGovern 1986: 269-70; Franken 1992: 30-31, 34-35, 76, 78, 88-89; McNicoll *et al.* 1992: 77); also *ostrich eggs* (McGovern *et al.* 1986: 271), *shells* etc. (Reese in McGovern *et al.* 1986: 320-32).

Sculpture

Two monuments from the end of the Late Bronze Age stand out. From Rujm al-Abd comes a basalt stela, showing a warrior god with a spear and a kilt, made in the thirteenth–twelfth centuries BC. The kilt is similar to those worn by tributaries in Egyptian private tombs from the Eighteenth and Nineteenth Dynasties (Zayadine 1991: 36-37). The relief is similar to those showing fertility and/or warrior gods from northwest Syria.

The stela from Balu' (Figure 8.4) (Ward and Martin 1964: 14; Zayadine 1991: 35, 37) is in sharp relief with three standing figures. On the left is a standing figure with a loincloth. He is wearing the double crown of Upper and Lower Egypt. He holds a *was*-sceptre in his

Figure 8.3. Bronze lamp, Tall al-Fukhar.

Figure 8.4. The panel from the Balu' stela (Ward and Martin 1964: pl. 3).

left hand and an obscure object, which he is giving to the figure in the middle, in his right hand. The middle figure is turned with lifted hands to receive the object. He is in gala dress worn by kings from the Amarna period onwards (Ward and Martin 1964: 14). His headdress is known from Egyptian scenes where it is used by Shasu Bedouin (Giveon 1971: pl. V, D-F); he is evidently the local king of Moab. A goddess is standing to the right. She wears a common dress, known from the Eighteenth Dynasty onwards, worn both by members of the royal family and by goddesses, and an *'atef*-crown, normally worn by a male god in Jordan (Daviau and Dion 1994: 162-65). In her right hand, she holds an *ankh*-sign. The scene is strongly reminiscent of Egyptian reliefs showing the living king, with a goddess in attendance, adoring the head of the pantheon (Ward and Martin 1964: 17). Two symbols are shown above the shoulders of the king in the middle. The one before the face of the king could be an orb and a crescent and should be associated with Baal-Saphon, while the one behind the king is a crescent and should be associated with the goddess who would thus be a moon goddess (Ward and Martin 1964: 16). In my opinion, the relief depicts an Egyptian scene, namely, the investiture of the local king by the two main deities of the local pantheon, probably El, Ba'al-Saphon or Kemosh and Astarte or Anat, transferred to Moab. It is executed by an artist who knew Egyptian art, but was not an Egyptian himself (Ward and Martin 1964: 14). The relief is thus evidence of an artistic tradition in Jordan at least at the end of the Late Bronze Age (for the inscription above the relief see below, p. 314). The early dating is feasible in view of the mention of Moab as a political entity in the list in the temple of Luxor (cf. above,

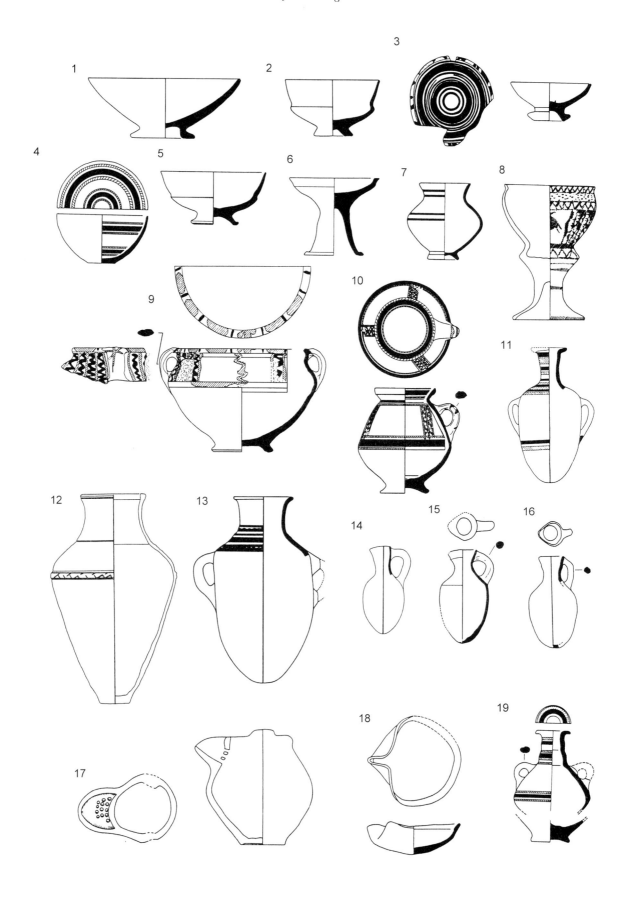

p. 294). It should, however, be mentioned that the stela is dated to the tenth century BC by Worschech (1997: 270; see also Timm 1989: 92-93 for the latest discussion).

Apart from these large pieces of art and some bronze figurines (see below), most of the sculptures known from the Late Bronze Age are miniature art, that is, a large number of scarabs and cylinder seals, or imprints, found in excavations of tells or in tombs (e.g. Ward 1964, 1966; McGovern *et al.* 1986: 284-94; Tubb 1988b: 71; Franken 1992: 28-33, 41-42, 58-59, 64-67, 76-78; McNicoll *et al.* 1992: 77-78, pl. 39).

A seal from Tall Abu al-Kharaz, showing the tree of life with antithetic ibexes, must be mentioned because of its religious significance, representing the Tree of Life (Fischer 1994: 133, 137).

Pottery

The most obvious characteristic of the pottery of the Late Bronze Age is the continuity of forms from the Middle Bronze Age, together with new developments. Chocolate-on-White Ware continues into the LB I period in the northern part of the country. It is found, for example, at Pella (Hennessy 1985), Tall Abu al-Kharaz (Fischer 1991: 86-87; 1993: 292-94), Katarit as-Samra (Leonard 1979, 1986), the Baq'ah Valley (McGovern *et al.* 1986: 76, 89 nos. 4-5, 173-74), Quwayliba (McGovern *et al.* 1986: 76), 'Amman (McGovern *et al.* 1986: 76), Tall al-Fukhar (McGovern 1997: 424; Strange 1997: 402) and was possibly produced in Katarit as-Samra (McGovern 1997: 424). The lamp, in the form of a flat bowl pinched on one side, is another example of a type of pottery that continued from the Middle Bronze Age.

New forms are also found, for example, the cooking pot with sharp carination and triangular rims, and the pilgrim flask, as well as forms that were copied from imports such as pyxides and stirrup jars. Technically, more and more pottery, not only the large pieces, but also jugs and bowls, was made by coiling instead of on a true potter's wheel in the LB I period. This method becomes predominant in the LB II period (McGovern *et al.* 1986: 168-69, 174-77). In the Baq'ah Valley in the LB I period, the fine wares were tempered with 5–10% quartz and/or calcite, and are very well fired (at c. 900°C); in the LB II period, the absolute amount of temper increases to as high as 20%, and the vessels are fired at a lower temperature; the slow wheel is still in use (McGovern *et al.* 1986: 145). More and more vessels are decorated with painting, making the Late Bronze Age pottery the most colourful pottery in the Bronze and Iron Ages. Burnishing becomes rare. The decoration is often bichrome, painted in horizontal or vertical straight and wavy bands, bowls often with concentric circles, but also with metopes or motifs; a well-known one seems to be the Tree of Life (Franken 1992: 92, 138, 184; Fischer 1991: 87) with attendant animals or worshippers (Potts *et al.* 1988: 138; Franken 1992: 82, 124, 130).

The repertoire includes the following forms (the list is not exhaustive concerning forms, and includes only the more important and published find assemblages, the Baq'ah Valley, Pella; Tall as-Sa'idiyya, Tall Dayr 'Alla, Tall Abu al-Kharaz, and for the Mycenaean imports especially the 'Amman Airport Temple'): *bowls, open and rounded, many of them carinated, and/or pedestalled* (Figure 8.5. 1-5) (McGovern *et al.* 1986: 86-89, 94-95, 98, 117, 168-70; McNicoll *et al.* 1982: pls. 105, 107, 109, 111, 117, 119; 1992: pls. 28, 43, 44, 47, 48; Pritchard 1980: 81; Tubb 1988b: 42; Fischer 1991: 88-89; 1993: 291-92; Franken 1992: 29, 39, 44, 48-49, 54, 62, 68, 77, 80-81, 85-86, 90, 99-100); *kraters* (Figure 8.5.9) (McGovern *et al.* 1986: 120-25, 148-49; McNicoll *et al.* 1992: pl. 49; Fischer 1991: 88-89; Franken 1992: 77); *chalices* (Figure 8.5.6) (McGovern *et al.* 1986: 88-89, 118-19; Franken 1992: 29); *cups* (Pritchard 1980: 81); *Goblets* (Figure 8.5.8) (Fischer 1991: 88-89; Franken 1992: 40, 54-55, 62, 68); *jars* (Figure 8.5.7, 10) (McGovern *et al.* 1986:

Figure 8.5. Late Bronze Age pottery 1. Bowl, Late Bronze Age IA (McGovern *et al.* 1986: 87, 1). 2. Bowl, Late Bronze Age IA (McGovern *et al.* 1986: 87, 2). 3. Bowl, Late Bronze Age IB (McGovern *et al.* 1986: 87, 2). 4. Bowl, Late Bronze Age II (McGovern *et al.* 1986: 101, 8). 5. Bowl, Late Bronze Age II (McGovern *et al.* 1986: 105, 2). 6. Chalice, Late Bronze Age II (McGovern *et al.* 1986: 119, 9). 7. Jar, Late Bronze Age IA (McGovern *et al.* 1986: 88, 5). 8. Goblet, Late Bronze Age II (Franken 1992: 125, 3). 9. Krater, Late Bronze Age II (McGovern *et al.* 1986: 125, 3). 10. Jar, Late Bronze Age II (McGovern *et al.* 1986: 127, 3). 11. Amphoriskos, Late Bronze Age II (McGovern *et al.* 1986: 91, 1). 12. Storage jar, Late Bronze Age II (Tall al-Fukhar). 13. Storage jar, Late Bronze Age IA (McGovern *et al.* 1986: 89, 7). 14. Juglet, Late Bronze Age IA (McGovern *et al.* 1986: 91, 8). 15. Juglet, Late Bronze Age IB/IIA (McGovern *et al.* 1986: 131, 4). 16. Juglet, Late Bronze Age II (McGovern *et al.* 1986: 131, 5). 17. Strainer jug, Late Bronze Age II (Franken 1992: 82). 18. Lamp, Late Bronze Age II (McGovern *et al.* 1986: 138, 3). 19. Jug, Late Bronze Age II (McGovern *et al.* 1986: 127, 13).

88-89, 148-49; McNicoll *et al.* 1982: pls. 105, 117, 119; 1992: pls. 28, 44, 48; Pritchard 1980: 83; Tubb 1988b: 42; Franken 1992: 39-40, 77-78, 86-87, 93, 99); *storage jars* (Figure 8.5.12-13) (McGovern *et al.* 1986: 88-89, 132-35, 167-69; Pritchard 1980: 83; Tubb 1988b: 58, 67; Fischer 1991: 94-95; Franken 1992: 94; for the *collared rim jar* and its use see Wengrow 1996); *amphoriskoi* (Figure 8.5.11) (McGovern *et al.* 1986: 89-90; Pritchard 1980: 81); *jugs and juglets* (Figure 8.5.14-16, 19) (McGovern *et al.* 1986: 90-91, 124-33, 167-69; McNicoll *et al.* 1982: pls. 105, 107, 113, 117, 119; 1992: pls. 28, 48; Pritchard 1980: 81, 83; Tubb 1988b: 42; Tubb 1990: 30; Fischer 1991: 90-91; Franken 1992: 39-40, 44, 50, 55, 62, 68, 77, 81-83, 86, 91, 99-100); *strainer jugs* (Figure 8.5.17) (McGovern *et al.* 1986: 136-37; Franken 1992: 82); *pilgrim flasks* (Figure 8.6.1) (McGovern 1986: 136-37; Pritchard 1980: 83; Tubb 1988b: 67; Franken 1992: 29, 55, 68, 99-100); *pilgrim flasks with figural painting* (Franken 1992: 55, 57: man with dog leading a goat; *pyxides* (Pritchard 1980: 81; Tubb 1988a: 42); *lamps* (Figure 8.5.18) (McGovern *et al.* 1986: 92-93, 138-45, 148-49, 168-70; McNicoll *et al.* 1982: pls. 109, 113; Pritchard 1980: 83; Tubb 1988b: 67; Fischer 1991: 92-93; Franken 1992: 29, 39, 49, 54, 62, 77, 80, 90, 99); *cooking pots* (Figure 8.6.2-3) (McGovern *et al.* 1986: 136-37, 169-70; McNicoll *et al.* 1982: pl. 119; 1992: pls. 28, 34, 45, 48; Tubb 1988b: 42-43; Tubb 1990: 30; Fischer 1991: 94-95; 1993: 291-92; Franken 1992: 81, 85, 99); *cult vessels* (Figure 8.6.6-7) (Fischer 1991: 92-93; Franken 1992: 30, 41, 56-57, 68, 83); *Chocolate-on-White* (Figure 8.6.4-5) (Leonard 1981; Hennessy 1985; McGovern *et al.* 1986: 173-74; McNicoll *et al.* 1992: 74, pls. 56-57; Fischer 1993: 292-94); *miscellaneous* (*cup-and-saucer, anthropomorphic vessel, zoomorphic vessel, storage vessel, clay tablets* [Figure 8.6.10], *funnel*) (McGovern *et al.* 1986: 268-69; Franken 1992: 29, 38, 41, 44, 51, 55, 59, 62-64, 66, 77, 83; Fischer 1994: 133, 137); *figures* (Bourke *et al.* 1992: 11; Franken 1992: 34-35; Kamlah 1993).

Mycenaean storage jars (Hankey 1974: 145-47); *Mycenaean kraters* (Hankey 1974: 147-49; Herr 1983b: 229); *Mycenaean jugs* (Pritchard 1980: 81, 62); *Mycenaean squat jars or alabasters* (Hankey 1974: 149-50), *Mycenaean stirrup jars* (Hankey 1974: 150-55; McGovern *et al.* 1986: 194-98; Pritchard 1980: 83; Tubb 1988b: 67; Franken 1992: 40); *Mycenaean flasks* (Hankey 1974: 155-56); *Mycenaean pyxides* (Pritchard 1980: 81); *Mycenaean open vessels* (Hankey 1974: 156-57). Also a few sherds of *Minoan vessels* were found (Hankey 1974: 157-58). *Cypriot White slip ware* (McGovern *et al.* 1986: 196-200; Fischer 1991: 90-91); *Cypriot Base Ring Wares* (McGovern *et al.* 1986: 196-200; Pritchard 1980: 81; Fischer 1991: 90-91). *Faience* (Franken 1992: 45-46, 58-61, 64-67); *knob*, possibly a wall decoration (Strange 1997: 404) (Figure 8.7).

For sherd typology see Dornemann (1983: 20-22); McGovern *et al.* (1986: 64-201); Franken (1992: 115-48); for technology see also Vilders (1993, 1995). For forms see also Hendrix *et al.* (1996: 146-58); for the published pottery up to 1995, see Herr and Trenchard 1996.

Social and political organization

Very little is known about the social and political organization in the Late Bronze Age. It seems, however, from the distribution of the larger settlements, in so far as it is possible to deduce this from the published material, that the city-state organization of the Middle Bronze Age survived the Egyptian conquest and was maintained during the whole period, probably under Egyptian hegemony. This can also be inferred from the Amarna Letters, for example, EA 256, concerning Pella (Albright 1943: 10-15), and the Balu' stela. The former, however, pertains only to a single incident and only to the northernmost part of Jordan. The latter may perhaps point to the beginning of the territorial states known later from the Iron Age (see below, p. 306 also).

The Egyptians had a person who was either a local king, at least seen from his own point of view, or something like a modern *mukhtar*, seen from the Egyptian point of view, in charge of each of these city-states. Some were possibly Egyptian envoys. However, the names of the petty kings in the Amarna Letters are all either Semitic or Hurrian. The kings were, therefore, probably indigenous.

The two temples at the 'Amman Airport and at Tall Dayr 'Alla in central Jordan suggest some tribal meeting points where non-sedentary people gathered, either to bury their dead or to trade with the Egyptians. This does not, however, necessarily point to any tribal confederation or amphictyonic organization, as was formerly postulated for the early Israelites, but rather to some (vague) arrangement, perhaps in the latter case under Egyptian supervision. (For the temple at Tall Dayr 'Alla see further below, p. 307) Alternatively, the temple at the 'Amman Airport, together with one found at Umm ad-Dananir, could simply be located at some distance from the community using them.

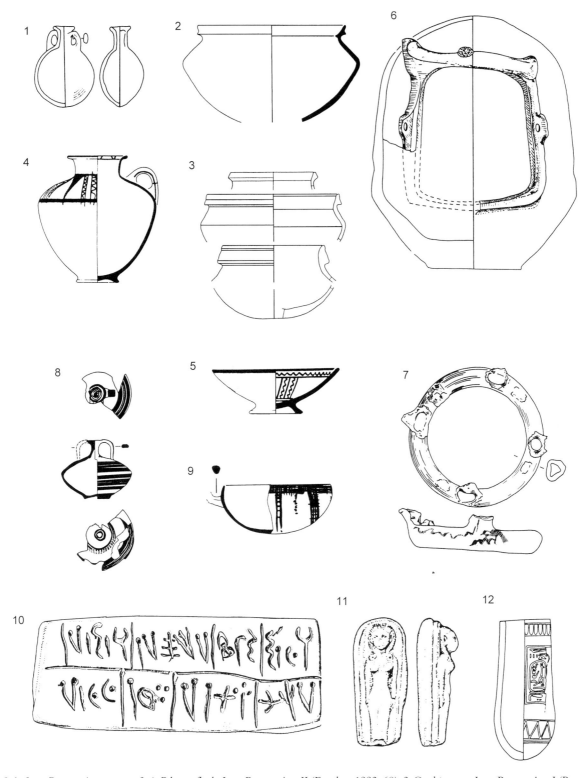

Figure 8.6. Late Bronze Age pottery 2. 1. Pilgrim flask, Late Bronze Age II (Franken 1992: 68). 2. Cooking pot, Late Bronze Age I (Bourke *et al.* 1992: pl. 45). 3. Cooking pots, Late Bronze Age II (Tall al-Fukhar). 4. Chocolate-on-White Ware, Late Bronze Age I (Hennessy 1985: 101). 5. Chocolate-on-White Ware, Late Bronze Age I (Hennessy 1985: 101). 6. Fenestrated cult vessel, Late Bronze Age II (Franken 1992: 30). 7. Cult vessel (*kernos*), Late Bronze Age II (Franken 1992: 68). 8. Mycenaean stirrup jar, Mycenaean IIIB (McGovern *et al.* 1986: 197, 2). 9. Cypriot White Slip II milk bowl (McGovern *et al.* 1986: 197, 11). 10. Inscribed clay tablet, Late Bronze Age II (Franken 1964a: 377). 11. Fertility figurine, Late Bronze Age (Bourke *et al.* 1994: 111). 12. Faience vase with cartouche of Taousert (Tewosret) (Franken 1992: 31).

Figure 8.7. Glazed 'knob', possibly wall decoration, Tall al-Fukhar.

In the later part of the period, LB IIB, the finds at Tall as-Sa'idiyya may point to a real Egyptian colonization with a garrison of mercenaries to protect an Egyptian governor (Tubb 1988c). Events in the northern part of the country, however, where, for example, the palace at Tall al-Fukhar was destroyed and not rebuilt, show that at this stage the Egyptian presence was probably confined to the Jordan Valley, leaving the rest of the country to itself. It is interesting to note that the sedentary population in cities, such as Sahab, Jarash, 'Amman, lived there simultaneously with the non-sedentary population frequenting the temple at the 'Amman Airport. This points to a situation where society was changing, with city-states and their satellites in the mountains, living together with the nomads, the Shasu, who were probably the only people living in the south. Only later in the Iron Age was this situation replaced by the kingdoms of Ammon, Moab and Edom, although the Balu' stela and the occurrence of Moab in Egyptian lists point to the beginnings of this development already in the LB IIB period.

Subsistence

Flora

Grain, *Hordeum? disticon* (two row-hulled barley) and *Triticum dicoccum* (emmer), from the Late Bronze IIB has been found at Irbid (Lenzen *et al.* 1985).

Although the period is poorly represented at Pella, two row-hulled barley and emmer, horse bean, grape, grass, bitter vetch, broad bean, chick-pea, lentils, oak, almond, willow/poplar, Christ's-thorn, olive and hackberry are attested. Proportionally, fewer weeds and grass are found than before, suggesting perhaps more sophisticated crop-processing techniques (McNicoll *et al.* 1992: 255, endplate 10-11).

Fauna

The bones of sheep/goat, dog, equid, and cattle are reported from Baq'ah Valley Cave A2 (from the LB I period). Sheep/goat, dog, cattle and donkey (also rodents, bird, snake, frog and turtle) accounted for 57.97% of the bone material and reflect grave offerings of food for the dead. From Cave B3 (from the LB II period), the bones of sheep/goat, equid and cattle are reported. Here again, the majority of bones are of sheep/goat (and rodent, bird, frog, turtle, fish). Finally, a pit from the LB II period at Khirbat Umm ad-Dananir contained the bones of cattle, equid, sheep/goat and carnivore remains (McGovern *et al.* 1986: 315-16).

Taken as a whole, the published material is too sparse to conclude anything about subsistence in the Late Bronze Age. However, the evidence from the imported pottery, as well as the evidence for exports of wine or oil (see below), makes it certain that there was a surplus of agricultural production for trade or tribute.

Technology

Pritchard discovered and partially examined a water system at Tall as-Sa'idiyya (1985: 57-59). The lowest part, a semicircular chamber fed by an underground spring, was later cleared (Miller in Tubb 1988b: 84-88). Finally, a type of gallery, paved with flat cobbles, and surrounding the top of the pool to the north with access from the northern circuit roadway on the tell, was found. The excavated pottery proved the water system belonged to Stratum XII, the Egyptian phase, to which the Egyptian residency also belongs (Tubb et al. 1996: 35-36).

Architecture

Houses

Very few complete plans of buildings from the Late Bronze Age have been published. Parts of buildings, however, are known.

A large building, probably a local ruler's palace, was found at Pella (Figure 8.8.1). It was built in the LB I period and was, according to the excavators, in use until early in the LB IIB period, that is, from the late fifteenth and during the first half of the fourteenth centuries BC (c. 1420–1350). It had trenchbuilt walls, having foundations of massive fieldstones with a superstructure of mudbricks, with thick plaster floors in the courtyard and in the adjacent rooms. The building measured c. 15 by 15 m and had a courtyard with two rows of rooms in its first phase to the north and to the south. In a later phase, characterized by a series of ashy floors, two or three more rooms with stone and mudbrick walls were built against the eastern and western walls of the courtyard making it smaller. One of the rooms was a lavatory with drainage. Two decorated Egyptian ivory boxes, two fragments of cuneiform tablets, scarabs and a seal impression were found in the building. The plan is best compared to the form called 'Governor's Residency' (Bourke et al. 1994: 104-107).

Another possible governor's residency, from a later period (area AA, Stratum XII), was found at Tall as-Sa'idiyya. The Egyptian-style plan and the construction method, with deep foundations of mudbrick, led the excavator to classify it as an Egyptian Governor's Residency (Tubb and Dorrell 1991: 69).

A palace complex (in the same Stratum XII) was found next to the outer wall at Tall as-Sa'idiyya (Figure 8.8.2). The building consisted of a series of rooms, courtyards and rooms on both sides of a passageway leading through to the city wall. A pair of interconnecting vaulted cisterns and a pool with a system of channels were found in the building, while in the pool a large number of Egyptian-type store-jars were found (Tubb and Dorrell 1993: 57, 59-61; Tubb 1997: 453-54).

A large monumental building from the LB IIA and B periods was partially uncovered at Tall al-Fukhar (Figure 8.8.3) (Strange 1997: 402). It is either a temple or a palace, probably the latter. Its length is at least 25 m, while the breadth cannot as yet be determined. The outer walls are very thick, c. 1.5 m, with a 1.5 m foundation of large stones and a mudbrick superstructure. Together both stand up to a height of 2.1 m. There was a thick white plaster floor in the rooms, with later floors above it. The destruction layer inside the building is 2 m thick, and the nature of the debris suggests that the building had an upper floor. The two rooms that have been partially excavated have entrances to other rooms to the south and to the north where the main entrance is presumably located. A staircase, only half of which has been excavated, has five steps leading down from an outer court. It is flanked by a doorpost or 'pillar' built of large boulders measuring 1 by 1 m square. The front of the building with its entrance suggests a 'Bit Hilani'-type building.

A kitchen, dated by the finds to the LB I period, was found at Tall Abu al-Kharaz. In it were found a bowl and jug, with metopic pattern, a spouted krater with 'double base ring', a large krater and a storage jar with two handles (Fischer 1993: 283).

Temples

An interesting class of buildings is the square buildings (*Quadratbau*) found at three places, all in the vicinity of 'Amman: at the 'Amman Airport from the LB IIA–B period, a little before 1300 and in use to before 1200 BC (Figure 8.8.5) (Hennessy 1966; 1989: 169-77; Herr 1983a, 1983b); at al-Mabrak of an uncertain date (not strictly a square but a rectangular building) (Yassine 1988: 61-64); and at Umm ad-Dananir from the LB IB–II period (from c. 1500–1200 BC; McGovern 1989: 35). The first two have been completely cleared, while only a quarter has been excavated of the Umm ad-Dananir building. The excavator, however, is certain it is a square building. The three buildings have a central room with a number of corridor-like rooms around it along the outer walls. The 'Amman Airport building and the Umm ad-Dananir one are temples. The use of the al-Mabrak structure is not determined. The pottery, nondescript body sherds, is probably from

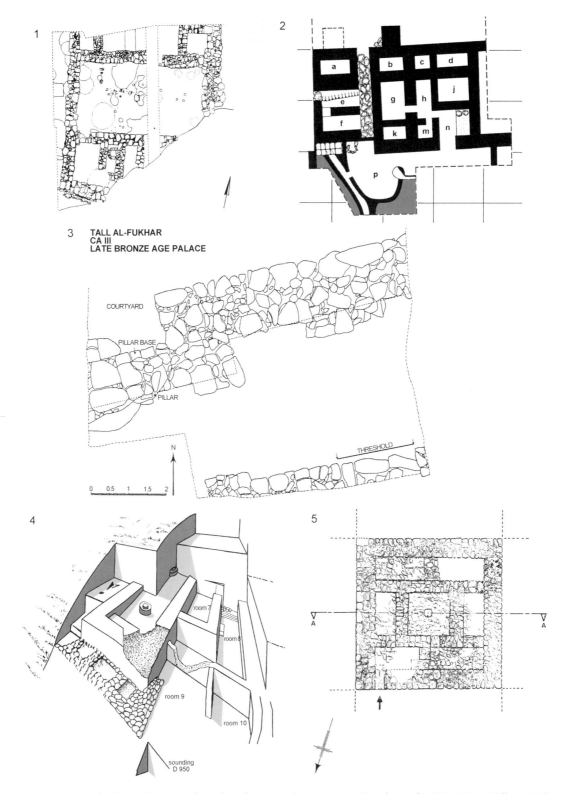

Figure 8.8. Late Bronze Age building plans. 1. Pella, palace from Late Bronze Age I (Bourke *et al.* 1992: 10). 2. Tall as-Sa'idiyya, western palace complex with aqueduct from Late Bronze Age II (Tubb and Dorrell 1994: 57). 3. Tall al-Fukhar, palace from Late Bronze Age IIB. 4. Tall Dayr 'Alla, cella with other rooms in temple from Late Bronze Age II (Franken 1992: 24). 5. 'Amman Airport, square temple from Late Bronze Age II (Hennessy 1966: 58).

the Late Bronze Age and the proximity and likeness to the other buildings make it likely that it was also intended to be a temple, although apparently never finished and used. The excavator, however, believes it to be a domestic structure for everyday use (Yassine 1988).

A temple (Figure 8.8.4) was found at Tall Dayr ʿAlla. It is not earlier than the sixteenth century and existed into the twelfth century BC when it was destroyed by an earthquake (Franken 1992: 176). It was built on a 8 m-high platform. The cella was another metre above the other rooms of the temple. The building belongs, in its latest phase, to the *Langbau* type with a non-axial or indirect entry (Franken 1992: 166). In its excavated segment, it had walls four to five bricks wide and two pillar bases on the 15 m-long axis. West of the cella, and separated from it by a brick wall, there was a complex of storerooms stacked with a workaday repertoire of large deep bowls, jugs and jars; east of the cella, there was a small courtyard, and beyond that several rooms were dug. In one of the rooms there were 12 clay tablets, indicating a temple archive, while in other rooms finer pottery, cylinder seals, scarabs, faience flasks, alabaster dishes, beads and Mycenaean imports were found. The building had been renewed regularly by adding a thick layer of fresh clay and stone to the pillar base. Because of the lack of space on the tell for ordinary houses, the temple has been interpreted as a shrine for traders who frequented the spot (van der Kooij 1993: 339-40; Franken 1961: 361-68 and pls. 2-12; Franken 1992: 23-103).

A rectangular temple from the LB IIA period (trench III, Stratum 3) measuring c. 6.8 by 5.2 m, was excavated at Tall Abu al-Kharaz. Its outer walls were c. 0.75 m wide. The building was oriented northwest–southeast, and was open to the northwest where stone slabs were found which supported a roof. A wall (wall 7) runs to the outer northwest wall. Together they create an enclosure of c. 6 m². The inner space of the larger room is 20 m², and the excavator found it surprising that no roof supports were found in the centre of the room (could it be that the room was an open space, and that the little room at the northwestern corner was the only covered part of the temple, a kind of propylaion?). A square altar was built in the eastern part of the enclosure and a bench in its northeastern part. A rich repertoire of pottery was found in the temple: fine quality vessels, pedestalled bowls, bowls and carinated bowls, decorated jugs and juglets, goblet, a stand, as well as Cypriot imports, among them a Cypriot early White Slip bowl with a possible Kalavassos provenance from c. 1375 BC (Fischer 1991: 79-80; 1993: 282-83).

Foreign relations and trade

Jordan was in intensive contact with the outside world in the Late Bronze Age. A few examples of artifacts from various excavations will suffice to demonstrate this. Mesopotamian material has been found: examples include a cylinder seal, probably made in Babylon, at the ʿAmman Airport temple (Hennessy 1989: 176), fragments of two tablets with cuneiform writing at Pella (McNicoll *et al.* 1992: 199-201), a glass pendant in the form of a fertility-type figurine (Figure 8.2), and a unique knob (Figure 8.7), possibly a wall decoration, at Tall al-Fukhar (Strange 1994; 1997: 402 and fig. 6). Moreover, Egyptian unguentaria have been found at Pella (Homès-Fredericq and Franken 1986: 135) and Egyptian pottery at Tall as-Saʾidiyya (Pritchard 1980: 7-8). Mycenaean and Cypriot pottery has been found in the Baqʿah Valley (McGovern *et al.* 1986: 194-201).

The trade followed a network that can be reconstructed with some confidence. The main gateway was Pella. Commodities came probably from Beth Shan, Megiddo and ultimately from Tall Nami (at least attested as an export emporium for Jordan), Tall Abu Hawam or Akko on the Mediterranean coast (Artzy 1994: 132; Leonard 1987a: 265; Knapp 1993: 24-28, 87). From Pella, the goods went down the Jordan Valley to Tall as-Saʾidiyya and farther to Tall Dayr ʿAlla and ʿAmman. Alternatively, there was a route to Irbid and from there south to Umm ad-Dananir and ʿAmman, from where the route forked to Sahab and to Madaba (Leonard 1987a: 265). But there must have been other gateways. The Midianite ware found at the ʿAmman Airport temple (Hennessy 1989: 171), although this piece must be seen as an isolated find (Parr 1982: 129), and in the south of Jordan (Jobling 1981: 110, pl. 31), demonstrates that there must have been a caravan route, possibly reflected in the find of a large scarab bought in Wadi al-Musa (Ward 1973: 45-46) from the reign of Amenhotep III. This route was probably operated or 'protected' by the Shasu from Arabia to ʿAmman where the goods could then be taken via the network described above north to Irbid and down to the Jordan Valley. Also from the north, namely, from Damascus, there was a gateway. This is the best explanation for the finds from the palace at Tall al-Fukhar where several items, some of them unique in Jordan, were imports from Syria or Mesopotamia (Strange 1997). Since Tall al-Fukhar lies northwest of Ramtha, just north of the later Roman road from the coast to Bosra, there must have been easy access to Irbid, with a link to Pella. Alternatively, the road could have gone directly from Ramtha to Jarash, the Baq'ah Valley and 'Amman on what was later called the King's Highway and *Via Nova Traiana* (cf. Knapp 1993: 26-27).

Other trade routes of less importance must have existed. For example, pottery of special types was imported to Tall al-Fukhar from southern Palestine in the MB II period (cooking pot and a jar/goblet), the LB II period (cooking pot), and the Iron IB period (collared rim jar and Philistine bowl) (McGovern 1997: 423-24), suggesting a route probably across the highlands of Palestine to either Tall as-Sa'idiyya or to Pella and from there to Irbid and farther east; later the Roman road from Irbid to Dera'a crossed the Wadi Shellale just south of Tall al-Fukhar. Moreover, the spread of the Chocolate-on-White Ware, probably from a single production centre in the northern Jordan Valley (McGovern 1997: 424), shows a local network of trade routes in Jordan itself.

The nature of the imports can only be guessed at, as we do not know for certain what was transported in the Mycenaean and Cypriot vessels. It has been suggested, partly because of the closed forms, that the Mycenaean vessels contained speciality oils and unguents (Leonard 1987a: 262-64). As for the Cypriot pottery, we can only guess at what was transported in the milk bowls, an open and rather shallow form. Perhaps they were imported as tableware for their beauty. At least one class of imported vessels, that is, the chariot vases, were probably imported for their intrinsic value (Hankey 1974: 147), but they are from the latest phase of the Late Bronze Age, and so far found only at the 'Amman Airport temple.

We have some idea of what was exported since wine and oil are mentioned in Egyptian texts as commodities from Asia (Helck 1971: 396-97, 398-402; Redford 1993: 212). This idea may be supported by the finds of the collared rim jar in Jordan and Palestine. The jars were probably used for the transport of these liquids, possibly on camels, from production centres in Jordan to the harbours of Tel Nami and Tel Nahariya on the Mediterranean coast (Wengrow 1996: 12-23; Artzy 1994: 132, 136-38).

Pella is mentioned in Papyrus Anastasi IV: 16, 6-17, 4 as a supplier of chariot parts in the LB II period (Smith 1973: 32). Other possible export products could have been grain and cattle (Helck 1971: 370-72).

Defence

Tall al-Fukhar had massive walls built of medium-large stones in the LB II period. Any traces of a gate were unfortunately robbed in the Late Bronze Age/early Iron Age transition period or in the early Iron Age (Ottosson 1992: 517).

Irbid was initially defended in the LB IA period by a basalt wall that was a survival from the Middle Bronze Age. This wall was incorporated into the city in the LB II period and another wall, of which 48 m was excavated, was built outside the former wall. A tower also seemed to belong to the city's defences (Lenzen 1992; Lenzen *et al.* 1985: 153-55).

A town wall from the Middle Bronze Age was found at Pella. It seems, however, that it was out of use in the Late Bronze Age and the town was no longer fortified, as it was then repeatedly mentioned in Egyptian texts as conquered (Smith 1973: 24-32; McNicoll *et al.* 1992: 40-41, 47). Perhaps it points to a status of peace under Egyptian hegemony.

Tall Safut had a perimeter wall founded on bedrock with a very narrow foundation trench (Wimmer 1987a: 280; 1987b: 165) during the Late Bronze period.

Sahab was defended by a town wall of which 75 m has been uncovered. The complete wall appears to have been symmetrical, with rounded angles and resembling a kind of super-ellipse. It existed from the fifteenth into the late thirteenth centuries BC. However, only its foundations, sunk into earlier deposits in the form of a deep trench lined on both sides with large and medium-sized stones, were found (Ibrahim 1975: 78-80; 1987: 76; 1989: 519).

In Stratum XII at Tall as-Sa'idiyya, a casemate city wall, with a very large building behind it, was found. A near-vertical trench had been cut into the slope of the mound, and in the bottom large, flat stones were placed, above which a thick layer of pisé had been poured. Into this bedding were set the lowest courses of brickwork. The casemate wall's external and internal leaves are 1.10 m and 1.20 m thick respectively and the entire structure has a width of 3.5 m. The compartments were filled with rubble and the entire complex was buttressed (Tubb 1988a: 41-45; Tubb and Dorrell 1993: 57-60). In Stratum XII the site had as its main entry an indirect access approach, namely, a pebble-paved road ascended the tell from the northeast. The road entered the city through a 4 m-wide vaulted mudbrick gate through the casemate wall. Beyond the rear of the wall, the street turned 90 degrees by means of a gate chamber that is not yet fully explored (Tubb *et al.* 1996: 31-33).

A city wall, originally built in the Early Bronze Age, but re-used in the Late Bronze Age, was found at Tall Abu al-Kharaz (Fischer 1996: 520).

The existence of fortifications around all the major cities excavated may indicate unrest or warlike conditions in the Late Bronze Age. This is corroborated by Amarna Letter 256 in which the king of Pella describes serious disturbances (Albright 1943: 10-15). If such fortifications did indeed exist, then the situation in Jordan was the opposite of that in Palestine, where sites were unfortified during the Late Bronze Age (Gonen 1984).

Burial customs

Various burial customs were in use in the Late Bronze Age in Jordan.

Cave burials (Gonen 1992: 9-15)

The usual custom was to place the bodies in a cave with their grave goods. A cave could be used over a long period of time and was often reused from an earlier period, perhaps indicating continuity.

Tomb 62 at Pella dates from the Middle Bronze II–Late Bronze I period. It consists of three rock-cut chambers (3.15 by 3.78; 2.79 by 3.30; and 1.80 by 2.04 m) with a short dromos; a step led up from chamber 1 to chamber 2. The cave, before serving as a tomb, had been used for processing food in the Middle Bronze Age. The roofs of the chambers had fallen in and had crushed most of the pottery in the tomb, but still hundreds of vessels survived intact. The individuals buried in the tomb number 100–150, probably an extended family or clan; nothing in the tomb suggests great wealth or status. The grave goods comprised more than 2000 artifacts including bowls, jars and jugs, juglets, lamps and imported Cypriot wares. The many Chocolate-on-White vessels must be singled out. Along with the pottery, gold, silver and bronze utensils, such as toggle pins, earrings, pendants, bracelets, tweezers, arrowheads, a knife, beads of various materials, scarabs, cylinder seals, stone vessels, clay artifacts, bone implements, a shell ring and a shark's tooth were found. The tomb testifies to a continuity from the Middle to the Late Bronze Age at Pella (Potts *et al.* 1985: 193, 205-208; Smith and Hennessy in McNicoll *et al.* 1992: 69-81).

Tomb 1 from the east cemetery at Pella is another tomb from the Middle Bronze/Late Bronze transition period. It is dated, according to the excavators, to c. 1600–1550/25 (Smith 1973: 170-71).

Tombs 20 and 21 at Pella come from the very beginning of the Late Bronze Age, that is, the LB IA period, or close to 1550. The original chambers seem to have been kidney shaped with a small side chamber. The main chamber was used for seven burials that were very badly preserved because of flooding and collapse. As a result, four pots floated in from an earlier Middle Bronze Age tomb. Grave goods included bowls, jars, pithoi, jugs and juglets, lamps, alabaster cosmetic jars, one with lid, a basalt axe head or hammer head, beads, a gold ring, incised bone inlays and fragments, bronze toggle pins, a javelin with tweezers and scarabs. Much of the pottery is Chocolate-on-White Ware. The side chamber of Tomb 21 contained two bowls, a jar and a bronze toggle pin (McNicoll *et al.* 1982: 40, 43-49).

Cave A2 in the Baq'ah Valley dates to the LB I period. Only a quarter of the cave was excavated. It had two chambers, an elliptical one measuring c. 5.2 by 3.4 m and, adjoined to it by a short passage, another chamber measuring c. 5.2 by 5.5 m. The entrance to the cave was through a triangular opening measuring c. 1.1 by 0.6 m. The cave had two distinct layers of burials: layer 2b, 0.5 m thick, representing a thick layer of disarticulated burials from the LB IA period; and layer 2a, representing a disturbed layer with much LB IB material. The grave goods included pottery, beads, shell-beads, cylinder seals, scarabs, spindle-whorls, anklets or bracelets, earrings, toggle pins, and ostrich egg fragments. More than two-thirds of the datable artifacts belong to the LB IA period (Brown and McGovern in McGovern *et al.* 1986: 32-44).

Tombs 13 and 14 from Quwayliba are dated to LB I–II. Only the pottery has been published. It is locally made except for one piece, a dark-grey jug that is an import (Kafafi 1984: fig. 5.6).

A tomb from LB I–II was found at Katarit as-Samra (Leonard 1981: 179-95). Other tombs from the Middle Bronze/Late Bronze period are Tombs 91, 94 and 95 from Pella (Edwards *et al.* 1990: 86).

Cave B3 in the Baq'ah Valley dates to the LB II period. It measures c. 9 by 4 m. The entrance was horizontal, elliptical, at a steep angle to the hillside. It had probably been closed by a large boulder. The grave contained two burial layers. Altogether, the minimum number of burials was 64 individuals. The upper layer contained the remains of at least eight individuals of which three had been intentionally placed around an inverted bichrome bowl, as well as a blackened cooking pot, suggesting that a funerary meal or offering was prepared in the tomb. A 1 cm-thick sterile layer separated this upper layer from the lower one. In the latter, one of the burials had a necklace of glass and copper beads with a pair of carnelian beads. The grave goods comprised about 300 vessels (Brown and McGovern in McGovern *et al.* 1986: 44-53).

Tomb D from Irbid is from the Late Bronze/Iron Age transition period (between 1350 and 1100 BC, in the excavator's opinion). Here again, the grave goods are abundant (Dajani 1964: 99-101).

A Late Bronze/Early Iron Age period tomb was found at Sahem. The tomb contained 78 ceramic vessels as well as beads, pendants, buttons, a female stone figurine and three clay figurines (Fischer 1997).

Pit burials (Gonen 1992: 15-20)

A very large cemetery with pit burials and jar burials was found on the lower segment of Tall as-Sa'idiyya. The burials date to the LB IIB period, that is, the twelfth century BC, contemporary with the Egyptian residency of Stratum XII. The total number of burials exceeds 500 individuals. The graves are simple pit burials, sometimes marked or lined with stones or mudbricks. They are cut into a 30–40 cm-thick silt deposit overlying an occupation from the Early Bronze Age. One tomb could even be described as a shaft tomb. It was not possible to determine the exact level from which the graves had been cut, and the continued use of the same area created considerable disturbance and intercutting of graves (Pritchard 1980: 10-30; Tubb 1988b: 73-80; 1990: 29-37, 38-42; 1991: 190; 1997: 453; Tubb and Dorrell 1991b: 84-86; 1993: 67-72; 1994: 65-66; Tubb *et al.* 1996: 36-39). The publication of Pritchard (1980: 32-34) gives a catalogue, with illustrations, of the grave deposits that include bowls, jars, storage jars, jugs and juglets, flasks, goblets, lamps, pyxides, imported stirrup jars, base ring jugs, together with bronzework, bracelets, anklets, beads, scarabs, seals, alabaster and ivory (Pritchard 1980: 31-103). Tubb notes further grave goods, including a very fine bronze wine set, and a surprising number of Egyptian burial customs (Tubb 1988a: 63-71; 1990: 36).

Bench burials (Gonen 1992: 23-24)

Bench burial caves have rock-cut benches or stone and earth benches constructed around the walls of the cave. The benches were used for multiple burials over a long period of time. The bones of earlier burials were removed to make room for new burials. At least one bench burial cave is known, Cave C from Sahab. It dates to the Late Bronze Age–Iron Age periods (from the fourteenth to the ninth centuries BC, according to the excavator). The cave measures c. 8.2 by 4.5 m and is c. 1.8 m high. Near its southeast corner there is a chimney-like construction reaching the surface and perhaps indicating that the cave was used earlier as a dwelling. A rock-cut bench, c. 50 cm high, is located along its south side and on its northern side there is another bench, c. 1.5 m wide and 20 cm high. A pavement runs from the entrance eastwards to a recess in which the chimney is situated. Another pavement runs from the centre of the cave to the benches. The entrance to the cave was reached by several steps. There is no information on the number of burials in the cave, but apparently it was in use throughout the whole period. Grave goods included bowls, jugs and juglets, dippers, cooking pots, flasks, pyxides, lamps, zoomorphic figurines and imported stirrup vases; other objects included ostrich eggs, daggers and knives, toggle pins, arrowheads, bracelets, anklets, earrings and finger rings, kohl sticks, pendants, stone objects and seals and stamps. These latter objects are, however, not dated (Dajani 1970: 29-32, 34, 53-64).

Structural tombs (Gonen 1992: 27-28)

A number of structural tombs were found in the cemetery at Tall as-Sa'idiyya. They were built-up chambers made of horizontally laid and bonded mudbricks and were lined up in two parallel rows in an east–west direction. The tombs were obviously planned in advance and were intended to be visible above the ground (Tubb 1988b: 60). The roof construction has gone, but it is tempting to suggest a corbelled roof like the five corbelled tombs found in western Palestine (Gonen 1992: 27).

Jar or pithos burial (Gonen 1992: 30)

A number of jar burials have been found at the Late Bronze Age cemetery at Tall as-Sa'idiyya. Single jar burials have been found with infants or foetuses, sometimes with grave goods, such as beads, amulets, bracelets and anklets; very often textiles showed that the bodies were wrapped, as also indicated by fibulae. The jars were sealed with a bowl or the base of another jar. Twenty-seven examples of double pithos burials of adults were also found: two extremely large jars were joined shoulder-to-shoulder after the necks had been removed and the junction was sealed using additional store jar sherds or stones. A variation was to cover the body with sherds from store jars and set the head inside a jar. As this type of burial, double pithos, is very rare in Palestine both east and west of the Jordan River, the excavator has connected these burials with a population of alien elements, perhaps Sea Peoples while the city was under Egyptian control (Pritchard 1980: 10-30; Tubb 1988b: 73-80; 1990: 31-42; 1991: 190; 1997: 453; Tubb and Dorrell 1991: 84-86; 1993: 67-72; 1994: 65-66; 1997: 453; Tubb et al. 1996: 36-39). Some peculiar burial customs were also attested in this cemetery. In one instance, additional skulls were re-buried with the original interment (Tubb 1988b: 63). Often, an inverted bowl, commonly of bronze, was placed on the pelvis in a position showing that it had covered the genitals of the deceased. In one case, the body had been placed face down. Bones of fish were found on the back of the skull. The hands met at the pelvis, and over them a bronze bowl with a ivory cosmetic box in the shape of a fish had been placed. In another grave, a javelin was found on the chest of the deceased with the point in the direction of the throat (Tubb 1988b: 63-66). Whatever these practices mean, they all bear witness to an extraordinary care for the deceased and probably a belief in an afterlife.

No examples have yet been found from Jordan in the Late Bronze Age for the other burial types in Gonen's typology (intramural burial, loculi burial, bilobate cave burial, open-pit burial, larnax burial, coffin burial).

Cremation

A rock pile, originally a stone-built incinerator, was found close to the square building at the 'Amman Airport temple. A number of bones, 95% of which were human, were scattered around it. All the human bones showed definite signs of burning, but not one of the animal bones had been burned. The excavator interpreted the finds as a temple for human sacrifice or a mortuary temple (Hennessy 1989: 171-77). Later excavators, however, have interpreted the place as a sanctuary for the cremation of the dead due to the finds of imported ceramics, fine ceramic vessels, bronze weapons and gold jewellery. This might be typical tomb furniture of a ritual centre connected with the dead and their cremation and burial, mainly for nomadic groups, as there was no sign of occupation in the vicinity (Herr 1983a: 9-11). Cremation did not, however, seem to be connected with the square temple at Umm ad-Dananir.

Religion and ritual

A number of representations of gods are attested in the material from the Late Bronze Age.

At Tall Safut, a male figurine made of bronze (92.8% copper, 5.45% tin and 1.75% other elements) was found. The figurine is a seated, smiling deity with a crown and with forearms extended and wrapped in gold. Two bronze pegs suggest it had been attached to some wooden structure. The eyes are hollow and may have been inlaid (Wimmer 1987a: 280; 1987b: 165 and pl. 12). Probably it is an El-type, like the seated gods from Ugarit.

A male god with a spear in his hand and wearing a kilt is carved on a basalt stela found at Rujm al-Abd. This god is probably a Ba'al type warrior god (Zayadine 1991: 36-37; see also above, p. 300).

There is the depiction of a male god wearing the Egyptian double crown along with a goddess and the local king on the Balu' stele (Ward and Martin 1964: pls. 1 and 3). The male god is possibly El, Ba'al-Saphon or Kemosh, while the goddess could be either Ashtarte or Anat (see above, p. 301).

A fertility-type goddess, with Hathor hairstyle, swollen abdomen and holding her breasts, was excavated from the destruction level of the palace at Tall al-Fukhar. The hair has been restored, from fragments found together with the figurine (Strange 1997: 402) (Figure 8.2). The goddess, probably an heirloom, conforms to a type well attested in the LB I period in the Near East and probably manufactured in Mesopotamia (McGovern 1985: 30). The goddess is probably an Egyptianized Astarte (cf. Keel and Uehlinger 1993: 74-80, 110-22).

A clay figurine, showing a woman covering her pubic hair with the right hand and holding her breast with the left, and a mould for clay figurines with pronounced pubic hair and holding two plants were found at Tall al-Fukhar (Kamlah 1993: 107-14). Both are representations of Astarte (cf. Kamlah 1993: 109; Keel and Uehlinger 1992: 74-80, 110-22). The mould perhaps indicates the popularity of the goddess, since the plaquettes seem to have been locally produced. Another example of a clay plaquette with an Astarte figurine comes from Pella (Figure 8.6.11) (Bourke et al. 1994: 110-12).

These figurines, which have countless parallels in other countries of the Ancient Near East (cf. Keel and Uehlinger 1993; 1993: 55-110), show that religion in Jordan conformed to the well-known type of fertility religion that was spread all over Syria-Palestine in the Late Bronze Age. It is, however, not possible at the present stage of research to describe the religion in detail. For example, we do not know whether the attempts above of attributing names of specific gods or goddesses to the figurines are more than guesswork since we do not know if there was continuity in the names of gods from the Late Bronze Age to the Iron Age, when we do know some of the names. Nor do we know if the names of the gods and goddesses in Jordan were the same as those of similar gods and goddesses in other parts of Syria-Palestine (e.g. the Ugaritic texts). We might expect a good deal of variation from place to place.

A glimpse of the cult can be obtained from the 'Amman Airport temple. When the temple was built, the operations commenced with digging a foundation trench, providing the site with a level surface. The outside walls were then placed on this surface in the form of a square. After the first course had been laid, the whole interior was packed with a fill (15–20 cm) of yellow clay and terra rossa. In this fill, there were frequent pockets of burnt clay, ashes and bone, evidence of a series of dedication offerings, consisting of a burnt offering together with small pieces of gold jewellery, beads, scarabs, cylinder seals, bone and ivory pieces and pottery fragments. Next, the inner walls were positioned, together with a cylindrical altar, in the centre of the building. The foundation trench was finally filled in and a number of arrow or small javelin heads were offered at the end (Hennessy 1966: 157).

A similar kind of ritual is attested at the square building at Tall Umm ad-Dananir (McGovern 1989: 35).

Writing

A number of hieroglyphic inscriptions, on monuments, on seals, and imprints from seals, have been found in Jordan from the Late Bronze Age.

Most important is the Balu' stela from the thirteenth or early twelfth century BC. It was probably executed by a local craftsman in Moab. The inscription at the top of the stela, above the relief, is so badly worn that it is not possible to read it in full. Only parts and single hieroglyphs can be read. The best publication is by Ward and Martin (1964), who conclude that the inscription was probably executed by the same artist, who was not an Egyptian, as the relief. The artist knew enough of the Egyptian culture to make a relief and inscription in Egyptian style. Timm (1989: 93) maintains that the writing is not Egyptian, but perhaps Minoan or Cypro-Minoan. But he seems to date the stela to the Iron Age, and would then presumably have to connect it with the Sea Peoples. On balance and in view of the occurrence of Moab as a political entity on an inscription in the Luxor temple, it appears to me that the text is at least an attempt to produce an Egyptian inscription.

Two fragments of cuneiform tablets have been found at Pella. On the basis of the stratigraphic context, they should be dated to the earlier part of the Late Bronze Age between 1550 and 1450 BC. They are too fragmentary to give any meaning, and it is not possible to determine where they were written (McNicoll *et al.* 1992: 299-301). They do, however, testify to knowledge of cuneiform writing at Pella, as does the later letter from Pella found at Tall al-Amarna (no. 256) (Albright 1943: 10-15).

The only examples of indigenous writing come from Tall Dayr 'Alla (Figure 8.6.10). In the debris of the temple, 11 tablets were found with a script that has until now defied all attempts at decipherment (Franken 1964a; 1964b: 421 and pl. V; 1965: 150-52; 1992: 57-59, 62-63; van den Branden 1965: 129-50; van der Kooij 1993: 339). The tablets are undoubtedly in a local script and probably served economic purposes. Perhaps the very existence of a local script shows a certain independence from the Egyptian overlords at the site at this specific time as the end of Egyptian hegemony drew to its close.

Summary of the period

The Late Bronze Age shows a waning of settlement in most areas of Jordan compared to the Middle Bronze Age. This abatement is probably due to the deterioration of the climate, which became drier over the period, but could partly be due also to Egyptian exploitation. Jordan was most heavily settled in the north and the west in the mountains along the Jordan Valley and in the valley itself. The southern part of the valley and the southern part of Jordan south of Wadi al-Hasa were virtually empty, inhabited only by nomadic people.

Jordan was more or less under Egyptian rule during the entire period. The country was in close contact with other areas, and evidence of trade in the form of imported pottery and other artifacts is found all over the country. This contact is probably due to Egyptian overlordship bringing the international trade network into contact with Jordan, first and foremost at Pella. As a result, the material culture seems heavily influenced from abroad, but it has, for example, in the pottery decoration, a colourfulness that, together with the pottery of Palestine, seems to be indigenous. At the same time, the Egyptians exploited the country heavily, and material culture suffered from it.

The country seems to have been divided into city-states, each with a kind of king or Egyptian *mukhtar*. In spite of the Egyptian hegemony, however, there seems to have been a state of warfare or at least a state of manoeuvring between the small city-states. Evidence for this situation may be the numerous defences around the cities, and in one instance at least in an historical source, EA 256. The Egyptian kings from time to time also needed to reassert their power in the area.

At the very end of the period, when Egyptian influence was waning, there seem to be indications of the emergence of new kingdoms such as Moab. Whether this is also found farther to the north cannot be said with certainty. However, in view of the string of fortified settlements in what was later Ammon—Tall Umm ad-Dananir, Tall Safut, 'Amman, Tall al-'Umayri, and Tall Sahab—it is possible that the Ammonite kingdom developed earlier than Moab and existed already early in the Late Bronze Age since settlement in Jordan has usually been from north to south (LaBianca and Younker 1995: 406-407). It seems that the central part of Jordan around 'Amman was an exception to the overall abatement of settlement in the period.

References

Aharoni, Y.
1979 *The Land of the Bible*. Tr. and ed. A.F. Rainey. London: Burns & Oates.

Albright, W.F.
1943 Two Little Understood Amarna Letters from the Middle Jordan Valley. *Bulletin of the American Schools of Oriental Research* 89: 7-17.

Ambers, J., K. Matthews and S. Bowman
1989 British Museum Natural Radiocarbon Measurements I. *Radiocarbon* 31: 15-32.

Artzy, M.
1994 Incense, Camels and Collared Rim Jars: Desert Trade Routes and Maritime Outlets in the Second Millennium. *Oxford Journal of Archaeology* 13: 121-47.

Bartlett, J.R.
1997 Edom. In E.M. Meyers (ed.), *The Oxford Encyclopedia of Archaeology in the Near East* 2: 189-90. New York: Oxford University Press.

Betts, A.V.G.
1989 Black Desert Survey. In D. Homès-Fredericq and J.B. Hennessy (eds.), *Archaeology of Jordan*. II.1. Field Reports: Surveys and Sites A–K, 45-48. Akkadica Supplementum 7. Leuven: Peeters.

Bienkowski, P.
1996 Soundings in the Wadi Hasa, 1995. *Palestine Exploration Quarterly* 128: 87.

Bienkowski, P. (ed.)
1991 *Treasures from an Ancient Land: The Art of Jordan*. Stroud: Alan Sutton.

Bienkowski, P. and R. Adams
1999 Soundings at Ash-Shorabat and Khirbat Dubab in the Wadi Hasa, Jordan: The Pottery. *Levant* 31: 149-72.

Bienkowski, P. *et al.*
1997 Soundings at Ash-Shorabat and Khirbat Dubab in the Wadi Hasa, Jordan: The Stratigraphy. *Levant* 29: 41-70.

Bourke, S.J. *et al.*
1994 Preliminary Report on the University of Sydney's Fourteenth Season of Excavation at Pella (Tabaqat Fahl) in 1992. *Annual of the Department of Antiquities of Jordan* 38: 81-126.

Braemer, F.
1987 Two Campaigns of Excavations on the Ancient Tell of Jerash. *Annual of the Department of Antiquities of Jordan* 31: 525-30.

1992 Occupation du sol dans la région de Jérash aux périodes du Bronze Récent et du Fer. In M. Zaghloul, K. 'Amr and F. Zayadine (eds.), *Studies in the History and Archaeology of Jordan*, IV: 191-98. Amman: Department of Antiquities.

Dajani, R.W.
1964 Iron Age Tombs from Irbid. *Annual of the Department of Antiquities of Jordan* 8–9: 99-101.

1966 Jabal Nuzha Tomb at Amman. *Annual of the Department of Antiquities of Jordan* 11: 48-52.

1970 A Late Bronze–Iron Age Tomb Excavated at Sahab, 1968. *Annual of the Department of Antiquities of Jordan* 15: 29-34, 53-64.

Daviau, P.M.M. and P.E. Dion
1994 El, the God of the Ammonites? The Atef-Crowned Head from Tell Jawa, Jordan. *Zeitschrift des Deutschen Palästina-Vereins* 110: 158-67.

Dornemann, R.H.
1983 *The Archaeology of Transjordan in the Bronze and Iron Ages*. Milwaukee, WI: The Milwaukee Public Museum.

1997 Amman. In E.M. Meyers (ed.), *The Oxford Encyclopedia of Archaeology in the Near East*, I: 99-102. New York: Oxford University Press.

Edwards, P.C. *et al.*
1990 Preliminary Report on the University of Sydney's Tenth Season of Excavations at Pella (Tabaqat Fahl) in 1988. *Annual of the Department of Antiquities of Jordan* 34: 57-93.

Erman, A.
- 1893 Die Denkmal Ramses' II im Ostjordanland. *Zeitschrift für Ägyptische Sprache und Altertumswissenschaft* 31: 100-101.

Fischer, P.M.
- 1991 Tell Abu Kharaz. The Swedish Jordan Expedition 1989: First Preliminary Report from Trial Soundings. *Annual of the Department of Antiquities of Jordan* 34: 67-103.
- 1993 Tell Abu Kharaz. The Swedish Jordan Expedition 1991: Second Season Preliminary Report. *Annual of the Department of Antiquities of Jordan* 37: 279-305.
- 1994 Tell Abu Kharaz. The Swedish Jordan Expedition 1992: Third Season Preliminary Report. *Annual of the Department of Antiquities of Jordan* 38: 127-45.
- 1996 Tell Abu al-Kharaz. In P.M. Bikai and V. Egan (eds.), Archaeology in Jordan. *American Journal of Archaeology* 106: 507-35.
- 1997 *A Late Bronze to Early Iron Age Tomb at Sahem, Jordan*. Abhandlungen des Deutschen Palätina-Vereins 21. Wiesbaden: Harrassowitz.

Flanagan, J.W., D.W. McCreery and K.N. Yassine
- 1992 Preliminary Report of the 1990 Excavation at Tell Nimrin. *Annual of the Department of Antiquities of Jordan* 36: 89-111.

Franken, H.J.
- 1961 The Excavations at Deir 'Alla in Jordan. *Vetus Testamentum* 11: 361-72.
- 1964a Clay Tablets from Deir 'Alla, Jordan. *Vetus Testamentum* 14: 377-79.
- 1964b Excavations at Deir 'Alla, Season 1964. *Vetus Testamentum* 14: 417-22.
- 1965 A Note on How the Deir 'Alla Tablets Were Written. *Vetus Testamentum* 15: 150-52.
- 1992 *Excavations at Tell Deir 'Alla: The Late Bronze Age Sanctuary*. Leuven: Peeters.

Fritz, V.
- 1994 Vorbericht über die Grabungen in Barqa el-Hetiye im Gebiet von Fenan, Wadi el-'Araba (Jordanien) 1990. *Zeitschrift des Deutschen Palästina-Vereins* 110: 125-50.

Giveon, R.
- 1971 *Les Bédouins Shosou des Documents Égyptiens*. Documenta et Monumenta Oriens Antiqui 12. Leiden: E.J. Brill.

Gonen, R.
- 1984 Urban Canaan in the Late Bronze Period. *Bulletin of the American Schools of Oriental Research* 253: 61-73.
- 1992 *Burial Patterns and Cultural Diversity in Late Bronze Age Canaan*. American Schools of Oriental Research Dissertation Series 7. Winona Lake, WI: Eisenbrauns.

Gordon, R.L. Jr. and E.A. Knauf
- 1987 Er-Rumman Survey 1985. *Annual of the Department of Antiquities of Jordan* 31: 289-98.

Hankey, V.
- 1974 A Late Bronze Age Temple at Amman. I: The Aegean Pottery. II: Vases and Objects made of Stone. *Levant* 6: 131-34.

Hauptmann, A. and G. Weisgerber
- 1992 Periods of Ore Exploitation and Metal Production in the Area of Feinan, Wadi 'Arabah, Jordan. In M. Zaghloul, K. 'Amr and F. Zayadine (eds.), *Studies in the History and Archaeology of Jordan*, IV: 61-66. Amman: Department of Antiquities.

Helck, W.
- 1971 *Die Beziehungen Ägyptens zu Vorderasien im 3. und 2. Jahrtausend v.Chr.* Ägyptologische Abhandlungen 5. Wiesbaden: Harrassowitz.

Hendrix, R.E., P.R. Drey and J.B. Storfjell
- 1996 *Ancient Pottery of Transjordan*. Berrien Springs, MI: Institute of Archaeology/Horn Archaeological Museum and Andrews University.

Hennessy, J.B.
- 1966 Excavation of a Late Bronze Age Temple at Amman. *Palestine Exploration Quarterly* 98: 155-62.
- 1985 Chocolate on White Ware in Pella. In J.N. Tubb (ed.), *Palestine in the Bronze and Iron Ages: Papers in Honour of Olga Tufnell*, 100-13. London: Institute of Archaeology.
- 1986 Pella. In D. Homès-Fredericq and H. J. Franken, *Pottery and Potters—Past and Present*, 134-38. Tübingen: Attempto.
- 1989 Amman Airport. In D. Homès-Fredericq and J.B. Hennessy (eds.), *Archaeology of Jordan*. II.1. *Field Reports: Surveys and Sites A–K*, 167-78 Akkadica Supplementum 7. Leuven: Peeters.

Herr, L.G.
- 1983a *The Amman Airport Excavations, 1976*. The Annual of the American Schools of Oriental Research 48. Winona Lake, WI: Eisenbrauns.
- 1983b The Amman Airport Structure and the Geopolitics of Ancient Transjordan. *Biblical Archaeologist* 46: 223-29.
- 1992 Shifts in Settlement Patterns of Late Bronze and Iron Age Amman. In M. Zaghloul, K. 'Amr and F. Zayadine (eds.), *Studies in the History and Archaeology of Jordan*, IV: 175-76. Amman: Department of Antiquities.
- 1997 Amman Airport Temple. In E.M. Meyers (ed.), *The Oxford Encyclopedia of Archaeology in the Near East*, I: 102-103. New York: Oxford University Press.

Herr, L.G. and W.C. Trenchard
- 1996 *Published Pottery of Palestine*. Atlanta, GA: Scholars.

Herr, L.G. et al.
1994   Madaba Plains Project: The 1992 Excavations at Tell el-'Umeiri, Tell Jalul, and Vicinity. *Annual of the Department of Antiquities of Jordan* 38: 147-72.

Homès-Fredericq, D.
1989   Lehun. In D. Homès-Fredericq and J.B. Hennessy (eds.), *Archaeology of Jordan*. II.2. *Field Reports: Surveys and Sites L–Z*, 349-59. Akkadica Supplementum 8. Leuven: Peeters.
1997   *Lehun*. Bruxelles: Imprimerie Cultura Wetteren.

Homès-Fredericq, D. and H.J. Franken
1986   *Pottery and Potters, Past and Present*. Tübingen: Attempto.

Homès-Fredericq, D. and J.B. Hennessy (eds.)
1989   *Archaeology of Jordan*. II. *Field Reports, Surveys and Sites*. Akkadica Supplementum 7-8. Leuven: Peeters.

Ibach, R.D.
1987   *Archaeological Survey of the Hesban Region*. Hesban 5. Berrien Springs, MI: Institute of Archaeology and Andrews University.

Ibrahim, M.M.
1974   Second Season of Excavation at Sahab, 1973. *Annual of the Department of Antiquities of Jordan* 19: 55-61.
1975   Third Season of Excavations at Sahab, 1975 (Preliminary Report). *Annual of the Department of Antiquities of Jordan* 20: 69-82.
1987   Sahab and its Foreign Relations. In A. Hadidi (ed.), *Studies in the History and Archaeology of Jordan*, III: 73-81. Amman: Department of Antiquities.
1989   Sahab. In D. Homès-Fredericq and J.B. Hennessy (eds.), *Archaeology of Jordan*. II.2. *Field Reports: Surveys and Sites L–Z*. Akkadica Supplementum 8. Leuven: Peeters.

Jobling, W.J.
1981   Preliminary Report on the Archaeological Survey between Ma'an and 'Aqaba. *Annual of the Department of Antiquities of Jordan* 26: 105-11.

Kafafi, Z.
1984   Late Bronze Pottery from Qwelbe (Jordan). *Zeitschrift des Deutschen Palästina-Vereins* 100: 12-29.

Kamlah, J.
1993   Tell el Fuhhar (Zarqu?) und die pflanzenhaltende Göttin in Palästina Ergebnisse des Zeraqon-Surveys 1989. *Zeitschrift des Deutschen Palästina-Vereins* 109: 101-27.

Keel, O. and Ch. Uehlinger
1993   *Göttinnen, Götter und Gottessymbole*. Quaestiones Disputatae 134. Freiburg: Herder.

Kitchen, K.A.
1982   *Pharaoh Triumphant: The Life and Times of Ramesses II*. Warminster: Aris & Phillips.
1987   The Basics of Egyptian Chronology in Relation to the Bronze Age. In P. Åström (ed.), *High, Middle or Low?: Acts of an International Colloquium on Absolute Chronology Held at the University of Gothenburg 20th–22nd August 1987*, 37-55. Gothenburg: P. Åström.

Knapp, A.B.
1993   *Society and Polity at Bronze Age Pella: An Annales Perspective*. JSOT/ASOR Monograph Series 6. Sheffield: Sheffield Academic Press.

Knauf, E.A. and C.J. Lenzen
1987   Notes on Syrian Toponyms in Egyptian Sources II. *Göttinger Miszellen* 98: 49-53.

LaBianca, Ø.S. and R.W. Younker
1995   The Archaeology of Society in Late Bronze/Iron Age Transjordan (ca. 1400–500 BCE). In T.E. Levy (ed.), *The Archaeology of Society in the Holy Land*, 399-415. London: Leicester University Press.

Lenzen C.J.
1986   Tell Irbid and Tell Ras. *Archiv für Orientforschung* 33: 164-66.
1992   Tell Irbid. In D.N. Freedman (ed.), *Anchor Bible Dictionary*, III, 456-57. New York: Doubleday.

Lenzen, C.J. and E.A. Knauf
1986   Tell Irbid and Beit Ras, 1983–86. *Liber Annuus* 36: 361-63.
1987   Notes on Syrian Toponyms in Egyptian Sources I. *Göttinger Miszellen* 96: 59-64.

Lenzen, C.J., R.L. Gordon and A.M. McQuitty
1985   Excavations at Tell Irbid and Beit Ras. *Annual of the Department of Antiquities of Jordan* 29: 151-59.

Leonard, A. Jr.
1979   Kataret es-Samra: A Late Bronze Age Cemetery in Transjordan? *Bulletin of the American Schools of Oriental Research* 234: 53-65.
1981   Kataret es-Samra: A Late Bronze Age Cemetery in Transjordan? *Annual of the Department of Antiquities of Jordan* 25: 179-95.
1986   Kattaret as-Samra. *Archiv für Orientforschung* 33: 166-67.
1987a The Significance of the Mycenaean Pottery Found East of the Jordan River. In A. Hadidi (ed.), *Studies in the History and Archaeology of Jordan*, III: 261-66. Amman: Department of Antiquities.
1987b The Jarash–Tell el-Husn Highway Survey. *Annual of the Department of Antiquities of Jordan* 31: 343-90.

Lipschitz, N.
1986   Overview of the Dendrochronological and Dendroarchaeological Research in Israel. *Dendrochronologia* 4: 37-58.
1996   Olives in Ancient Israel in View of Dendroarchaeological Investigations. In M. Heltzer and D. Eitam (eds.),

Olive Oil in Antiquity, 139-45. Padova: Sargon (originally issued at Haifa University 1988).

Liverani, M.
- 1987 The Collapse of the Near Eastern Regional System at the End of the Bronze Age: The Case of Syria. In M. Rowlands, M. Larsen and K. Kristiansen (eds.), *Centre and Periphery in the Ancient World*, 66-73. Cambridge: Cambridge University Press.

Mabry, J. and G. Palumbo
- 1992 Environmental, Economic and Political Constraints on Ancient Settlement Patterns in the Wadi al-Yabis Region. In M. Zaghloul, K. 'Amr and F. Zayadine (eds.), *Studies in the History and Archaeology of Jordan*, IV: 67-72. Amman: Department of Antiquities.

MacDonald, B.
- 1992a Settlement Patterns Along the Southern Flank of Wadi alHasa: Evidence from 'The Wadi al-Hasa Archaeological Survey'. In M. Zaghloul, K. 'Amr and F. Zayadine (eds.), *Studies in the History and Archaeology of Jordan*, IV: 73-76. Amman: Department of Antiquities.
- 1992b *The Southern Ghors and the Northeast 'Arabah Archaeological Survey*. Sheffield Archaeological Monographs 5. Sheffield: J.R. Collis.
- 1994 Early Edom: The Relation between the Literary and Archaeological Evidence. In M.D. Coogan, J.C. Exum and L.E. Stager (eds.), *Scripture and Other Artifacts: Essays on Archaeology and the Bible in Honor of Philip J. King*, 230-46. Louisville, KY: Westminster/John Knox Press.

MacDonald, B. *et al.*
- 1988 *The Wadi el Hasa Archaeological Survey 1979–1983, West Central Jordan*. Waterloo, Ontario: Wilfrid Laurier University.

Mare, H.
- 1989 Quweilbeh. In D. Homès-Fredericq and J.B. Hennessy (eds.), *Archaeology of Jordan*. II.2. *Field Reports: Surveys and Sites L–Z*, 472-86. Akkadica Supplementum 8. Leuven: Peeters.

McGovern, P.E.
- 1985 *Late Bronze Age Palestinian Pendants*. JSOT/ASOR Monograph Series 1. Sheffield: Sheffield Academic.
- 1989 Baq'ah Valley Survey. In D. Homès-Fredericq and J.B. Hennessy (eds), *Archaeology of Jordan*. II.1. *Field Reports: Surveys and Sites A–K*, 25-44. Akkadica Supplementum 7. Leuven: Peeters.
- 1992 Settlement Patterns of the Late Bronze and in the Greater Amman Area. In M. Zaghloul, K. 'Amr and F. Zayadine (eds.), *Studies in the History and Archaeology of Jordan*, IV: 179-83. Amman: Department of Antiquities.
- 1997 A Ceramic Sequence for Northern Jordan: An Archaeological and Chemical Perspective. In G. Bisheh, M. Zaghloul and I. Kehrberg (eds.), *Studies in the History and Archaeology of Jordan*, VI: 421-25. Amman: Department of Antiquities.

McGovern, P.E. *et al.*
- 1986 *The Late Bronze and Early Iron Ages of Central Transjordan: The Baq'ah Valley Project*. University Museum Monograph 65. Philadelphia: The University Museum.

McNicoll, A., R.H. Smith and J.B. Hennessy
- 1982 *Pella in Jordan 1: An Interim Report on the Joint University of Sydney and The College of Wooster Excavations at Pella 1979–81*. Canberra: Australian National Gallery.

McNicoll, A.W. *et al.*
- 1992 *Pella in Jordan 2*. Mediterranean Archaeology Supplement 2. Sydney: University of Sydney.

Merrillees, R.S.
- 1986 Political Conditions in the Eastern Mediterranean during the Late Bronze Age. *Biblical Archaeologist* 49: 42-50.

Miller, J.M.
- 1979 Archaeological Survey South of Wadi Mujib: Glueck's Sites Revisited. *Annual of the Department of Antiquities of Jordan* 23: 79-92.

Mittmann, S.
- 1970 *Beiträge zur Siedlungs und Territorialgeschichte des nördlichen Ostjordanlandes*. Abhandlungen des Deutschen Palästinavereins. Wiesbaden: Harrassowitz.

Na'aman, N.
- 1994 The Hurrians and the End of the Middle Bronze Age in Palestine. *Levant* 26: 175-87.

Najjar, M.
- 1992 The Jordan Valley (East Bank) during the Middle Bronze Age in the Light of New Excavations. In M. Zaghloul, K. 'Amr and F. Zayadine (eds.), *Studies in the History and Archaeology of Jordan*, IV: 149-53. Amman: Department of Antiquities.

Neuman, J. and S. Parpola
- 1987 Climatic Change and Eleventh–Tenth Century Eclipse of Assyria and Babylonia. *Journal of Near Eastern Studies* 46: 161-82.

Nützel, W.
- 1976 The Climatic Changes of Mesopotamia and Border Areas. *Sumer* 32: 11-24.

Olávarri, E.
- 1997 'Ara'ir. In E.M. Meyers (ed.), *The Oxford Encyclopedia of the Archaeology in the Near East*, I: 177-78. New York: Oxford University Press.

Ottosson, M.
- 1992 Tell el-Fukhar. *American Journal of Archaeology* 96: 516-18.

Palumbo, G. (ed.)
- 1994 *JADIS: The Jordan Antiquities Database and Information System: A Summary of the Data*. Amman: Department

Parr, P.
1982 Contacts between North West Arabia and Jordan in the Late Bronze and Iron Ages. In A. Hadidi (ed.), *Studies in the History and Archaeology of Jordan*, I: 127-33. Amman: Department of Antiquities.

Piccirillo, M.
1976 Una Tomba del Ferro I a Mafraq. *Liber Annuus* 26: 27-30.

Potts, T.F.
1987 A Bronze Age Ivory-Decorated Box from Pella (Pahel) and its Foreign Relations. In A. Hadidi (ed.), *Studies in the History and Archaeology of Jordan*, III: 59-71. Amman: Department of Antiquities.

Potts, T.F., P.C. Colledge and P.C. Edwards
1985 Preliminary Report on a Sixth Season of Excavation by the University of Sydney at Pella in Jordan (1983/84). *Annual of the Department of Antiquities of Jordan* 29: 181-210.

Potts, T.F. *et al.*
1988 Preliminary Report on the Eighth and Ninth Seasons of Excavation by the University of Sydney at Pella (Tabaqat Fahl), 1986 and 1987. *Annual of the Department of Antiquities of Jordan* 32: 115-49.

Prag, K.
1992 Bronze Age Settlement Patterns in the South Jordan Valley: Archaeology, Environment and Ethnology. In M. Zaghloul, K. 'Amr and F. Zayadine (eds.), *Studies in the History and Archaeology of Jordan*, IV: 155-60. Amman: Department of Antiquities.

Pritchard, J.B.
1980 *The Cemetery at Tell es-Sa'aidiyeh, Jordan*. University Museum Monograph 41. Philadelphia: The University Museum.
1985 *Tell es-Sa'idiyeh: Excavations on the Tell, 1964–1966*. University Museum Monograph 60. Philadelphia: The University Museum.

Redford, D.B.
1982 Contact between Egypt and Jordan in the New Kingdom: Some Comments on Sources. In A. Hadidi (ed.), *Studies in the History and Archaeology of Jordan*, I: 115-19. Amman: Department of Antiquities.
1993 *Egypt, Canaan, and Israel in Ancient Times*. Princeton: Princeton University.

Rothenberg, B. and J. Glass
1983 The Midianite Pottery. In J.F.A. Sawyer and J.A. Clines (eds.), *Midian, Moab and Edom*, 65-124. Journal for the Study of the Old Testament. Supplement Series 24. Sheffield: University of Sheffield.

Sauer, J.A.
1986 Transjordan in the Bronze and Iron Ages: A Critique of Gluek's Synthesis. *Bulletin of the American Schools of Oriental Research* 263: 1-26.

Shehadeh, N.
1985 The Climate of Jordan in the Past and Present. In A. Hadidi (ed.), *Studies in the History and Archaeology of Jordan*, II: 24-37. Amman: Department of Antiquities.

Simmons, A.H. and Z. Kafafi
1992 The 'Ain Ghazal Survey: Patterns of Settlement in the Greater Wadi az-Zarqa Area, Central Jordan. In M. Zaghloul, K. 'Amr and F. Zayadine (eds.), *Studies in the History and Archaeology of Jordan*, IV: 77-82. Amman: Department of Antiquities.

Smith, R.H.
1973 *Pella of the Decapolis I*. Wooster, OH: The College of Wooster.

Strange, J.
1986 The Transition from the Bronze Age to the Iron Age in the Eastern Mediterranean and the Emergence of the Israelite State. *Scandinavian Journal of the Old Testament* 1: 1-19.
1993 Tell el-Fukhar. *American Journal of Archaeology* 97: 484-87.
1994 Tell el-Fukhar. *American Journal of Archaeology* 98: 521-59.
1997 Tall al-Fukhar 1990–1993. In G. Bisheh, M. Zaghloul and I. Kehrberg (eds.), *Studies in the History and Archaeology of Jordan*, VI: 399-406. Amman: Department of Antiquities.

Thompson, T.L.
1992
(1994) *Early History of the Israelite People from the Written and Archaeological Sources*. Leiden: E.J. Brill.

Timm, S.
1989 *Moab Zwischen den Mächten: Studien zu historischen Denkmälern und Texten*. Ägypten und Altes Testament 17. Wiesbaden: Harrassowitz.

Tubb, J.N.
1986 Tell es-Sa'idiyeh: Interim Report of the Second Season of Excavations. *Annual of the Department of Antiquities of Jordan* 30: 119-29.
1988a Tell es-Sa'idiyeh: Third Season Interim Report. *Annual of the Department of Antiquities of Jordan* 32: 41-58.
1988b Preliminary Report on the First Three Seasons of Renewed Excavations on Tell el-Sa'idiyeh. *Levant* 20: 23-88.
1988c The Role of the Sea Peoples in the Bronze Industry of Palestine/Transjordan in the Late Bronze/Early Iron Age Transition. In J. Curtis (ed.), *Bronze-Working Centres of Western Asia c. 1000–539 B.C.* London and

New York: Kegan Paul in association with the British Museum.

1990 Preliminary Report on the Fourth Season of Excavations at Tell es-Sa'idiyeh in the Jordan Valley. *Levant* 22: 21-42.

1991 Preliminary Report on the Fifth Season of Excavations on Tell es-Sa'idiyeh. *Annual of the Department of Antiquities of Jordan* 35: 181-94.

1997 Tell es-Sa'idiyeh. In E.M. Meyers (ed.), *The Oxford Encyclopedia of Archaeology in the Near East*, IV: 452-55. New York: Oxford University Press.

Tubb, J.N. and P.G. Dorrell

1991 Tell es-Sa'idiyeh: Preliminary Report on the Fifth Season of Excavations. *Levant* 23: 67-86.

1993 Tell es-Sa'idiyeh: Interim Report on the Sixth Season of Excavation. *Palestine Exploration Quarterly* 125: 50-74.

1994 Tell es-Sa'idiyeh 1993: Interim Report on the Seventh Season of Excavation. *Palestine Exploration Quarterly* 129: 52-67.

Tubb, J.N., P.G. Dorrell and F.J. Cobbing

1996 Interim Report on the Eighth (1995) Season of Excavations at Tell es-Sa'idiyeh. *Palestine Exploration Quarterly* 128: 16-40.

Van den Branden, A.

1965 Essai de déchiffrement des inscriptions de Deir 'Alla. *Vetus Testamentum* 15: 129-50.

Van der Kooij, G.

1993 Tell Deir Allah. In E. Stern (ed.), *The New Encyclopedia of Archaeological Excavations in the Holy Land*, I, 338-42. Jerusalem: The Israel Exploration Society/Carta.

Vilders, M.

1993 Some Remarks on the Production of Cooking Pots in the Jordan Valley. *Palestine Exploration Quarterly* 125: 149-56.

1995 Some Technological Features of Tall as-Sa'idiyya Cooking Pots. In K. 'Amr, F. Zayadine and M. Zaghloul (eds.), *Studies in the History and Archaeology of Jordan*, V: 597-601. Amman: Department of Antiquities.

Ward, W.A.

1964 Cylinders and Scarabs from a Late Bronze Age Temple at 'Amman. *Annual of the Department of Antiquities of Jordan* 8-9: 47-55.

1966 Scarabs, Seals and Cylinders from Two Tombs at Amman. *Annual of the Department of Antiquities of Jordan* 11: 5-18.

1973 A Possible New Link between Egypt and Jordan during the Reign of Amenhotep III. *Annual of the Department of Antiquities of Jordan* 18: 45-46.

Ward, W.A. and M.F. Martin

1964 The Balu'a Stele: A New Transcription with Palaeographical and Historical Notes. *Annual of the Department of Antiquities of Jordan* 8: 5-29.

Weinstein, J.M.

1982 The Egyptian Empire in Palestine: A Reassessment. *Bulletin of the American Schools of Oriental Research* 241: 1-28.

1991 Egypt and the Middle Bronze IIC/Late Bronze IA Transition in Palestine. *Levant* 23: 105-15.

Wengrow, D.

1996 Egyptian Taskmasters and Heavy Burdens: Highland Exploitation and the Collared-Rim Pithos of the Bronze/Iron Age Levant. *Oxford Journal of Archaeology* 15: 307-26.

Wimmer, D.H.

1987a The Excavations at Tell Safut. In A. Hadidi (ed.), *Studies in the History and Archaeology of Jordan*, III: 279-82. Amman: Department of Antiquities.

1987b The Safut Excavations 1982–85 Preliminary Report. *Annual of the Department of Antiquities of Jordan* 31: 159-71.

1989 Safut (Tell). In D. Homès-Fredericq and J.B. Hennessy (eds.), *Archaeology of Jordan*. II.2. *Field Reports: Sites L–Z*, 512-15. Akkadica Supplementum 8. Leuven: Peeters.

Worschech, U.

1992 Ancient Settlement Patterns in the Northwest Ard al-Karak. In M. Zaghloul, K. 'Amr and F. Zayadine (eds.), *Studies in the History and Archaeology of Jordan*, IV: 83-88. Amman: Department of Antiquities.

1997 Balu'. In E.M. Meyers (ed.), *The Oxford Encyclopedia of Archaeology in the Near East*, I: 269-70. New York: Oxford University.

Worschech, U.F.C., U. Rosenthal and F. Zayadine

1986 The Fourth Season in the Northwest Ard el-Kerak and Soundings at Balu' 1986. *Annual of the Department of Antiquities of Jordan* 30: 285-301.

Yassine, K.

1988 El-Mabrak: An Architectural Analogue of the Amman Airport Building. In K. Yassine (ed.), *Archaeology of Jordan: Essays and Reports*, 61-64. Amman: University of Jordan.

1989 Tell el Mazar. In D. Homès-Fredericq and H.J. Hennessy (eds.), *Archaeology of Jordan*. II.2. *Field Reports: Sites L–Z*, 381-84. Akkadica Supplementum 8. Leuven: Peeters.

Yassine, K.H., J. Sauer and M. Ibrahim
 1988 The East Jordan Valley Survey, 1975. In K.H. Yassine (ed.), *Archaeology of Jordan: Essays and Reports*, 159-207. Amman: University of Jordan.

Zayadine, F.
 1991 Sculpture in Ancient Jordan. In P. Bienkowski (ed.), *Treasures from an Ancient Land: The Art of Jordan*, 31-61. Stroud: Alan Sutton.

Zayadine, F., M. Najjar and J.A. Greene
 1987 Recent Excavations on the Citadel of Amman (Lower Terrace): A Preliminary Report. *Annual of the Department of Antiquities of Jordan* 31: 299-310.

Zeist, W. van
 1985 Past and Present Environment of the Jordan Valley. In A. Hadidi (ed.), *Studies in the History and Archaeology of Jordan*, II: 199-204. Amman: Department of Antiquities.

9. The Iron Age

Larry G. Herr and Muhammad Najjar

Introduction

The Iron Age in Jordan saw the settlement of a variety of people groups or 'tribes' in the arable parts of the plateau of Jordan and their gradual coalescence into formal 'national' states with their own monarchies and governmental bureaucracies. Although the Jordan Valley and the northern plateau (north of Wadi az-Zarqa') had been constantly and relatively heavily settled throughout the Bronze Age, the plateau south of the az-Zarqa' had been comparatively empty during the Middle and Late Bronze Ages. Settlement began again in the 'Amman region toward the end of the Late Bronze Age; in the al-Karak region during Iron Age I; and in the at-Tafila and al-'Aqaba regions during the second half of Iron II. Settlement thus spread gradually in a southward direction throughout the Iron Age.

Stratigraphic excavations at Iron Age sites that exposed significant remains did not occur until the 1980s, except for those of Glueck at Tall al-Khalayfi (Pratico 1993) and Bennett at Umm al-Biyara (Bienkowski 1990), Tawilan (Bennett and Bienkowski 1995), and Busayra (Bienkowski 1990). Even the extensive excavations at Dhiban in the 1950s and 1960s (Tushingham 1993 and cited sources) and Hisban in the 1960s and 1970s (Merling and Geraty 1994) produced only fragmentary Iron Age remains, albeit with copious pottery in secondary contexts. Moreover, most of these excavations produced only a few finds from Iron I. Indeed, that period has only begun to be exposed significantly in the last decade (Ji 1995).

In assigning sites to specific sub-periods of the Iron Age below, we have suggested somewhat nuanced datings for several sites based on the ceramic assemblages. Much progress has occurred recently on the pottery of Jordan, allowing us to redate older excavated materials. For instance, we now have very consistent groupings of different types of collared pithoi into assemblages with equally different cooking pots, jugs, bowls and jars. Simply to find a collared pithos is no longer enough to ascribe a settlement to the early Iron I period (as did, for instance, Sauer [1986: 10]; for the typological history of the collared pithos in Jordan see Herr [forthcoming b]).

No attempt has been made to isolate all survey sites that have produced Iron Age pottery. Moreover, surveyed sites do not always tell an accurate story. Whereas one survey suggested Late Bronze to Iron Age settlement for one particular site in the south, excavation found nothing except a few examples of Iron II pottery in secondary deposits (Bienkowski *et al.* 1997).

Although not mentioned specifically as a bibliographic source, the results of many of the Iron Age sites appear in Homès-Fredericq and Hennessy (1989).

Late Bronze Age/Iron I transition (late thirteenth–early twelfth centuries BC)

Egyptian sources for the early Iron Age include Papyrus Anastasi VI during the reign of Merneptah (c. 1206 BC), which mentions the Shasu, apparently nomads from a region that included southern Jordan. On the basis of this and earlier inscriptions from the reign of Ramesses II, Kitchen has come to the conclusion that 'Moab, and especially Edom, should be considered mainly as "tented kingdoms"' (1992: 27).

The only other textual sources relevant to this period are later reports and/or remembrances of the biblical documents. These include stories of the origins of the Ammonites and Moabites (Gen 19.38), the defeat of King Og in a battle that mentions his astounding iron bed (Deut 3.1-11), and a series of wars with Israel and the Transjordanian states about who was to control the land east of Jordan (e.g. the Jephthah story in Judges 11) and even that to the west (e.g. the stories of Ehud against the Moabites in Judges 4 and Gideon against the Midianites in Judges 6).

Excavated sites with at least preliminary reports on the Late Bronze Age/Iron I transition

Abila (?)	Walls (Mare 1992 and other references there)
'Amman	Tomb (Dajani 1966a); pottery (most is Iron IIC, Hadidi 1970: Pl. 1.11)
'Ara'ir V	Houses (Olávarri 1965: fig. 1.1-4)

Baq'ah Valley	Tombs (McGovern 1986)
Umm ad-Dananir (?)	Unpublished pottery (McGovern 1986: 61)
al-Fukhar	Re-used palace?; house walls (Strange 1997)
Hisban 21	Rock-cut trench; cistern (Fisher 1994)
Irbid Phase 2	City wall; tower; cultic building (Lenzen 1988); Tomb B (Dajani 1966b)
Jarash (?)	Floor fragments; no published pottery (Braemer 1987)
al-Lahun	Fortified village; casemate wall; four houses (Homès-Fredericq 1997b)
Madaba	Tomb (Harding and Isserlin 1953)
Pella	Town with destruction (end of LB); village (McNicoll et al. 1982; Bourke 1997: 103-13)
Rujm al-Henu (?)	Unpublished pottery (McGovern 1983: 126)
Safut (?)	Unpublished pottery (Wimmer 1987: 281)
Sahab	Tomb (Dajani 1970)
Saham	Tomb (Fischer 1997)
as-Sa'idiyya XIV–XV	Tombs (Pritchard 1980); wall and cobbled surface (Tubb et al. 1996)
Tawilan	Pottery (Hart 1995: 60, fig. 6.19.11)
al-'Umayri	Fortification system; four-room house; other houses (Clark 1997)

This sub-period is defined ceramically by a mixture of early Iron I and Late Bronze Age pottery. A few sites on the plateau continued from the Late Bronze Age (Abila (?), 'Amman, the Baq'ah Valley, Umm ad-Dananir, al-Fukhar, Irbid, Jarash, Safut (?), Sahab (if the tomb reflects a contemporary settlement), and al-'Umayri. This suggests a peaceful continuity from the Late Bronze Age into the Iron Age, at least in the north where most of these sites occur. A similar picture obtains for the Jordan Valley at Pella and as-Sa'idiyya only (Dayr 'Alla seems to have been uninhabited at this point).

Most of the finds on the plateau come from tombs (Saham, 'Amman, Sahab, the Baq'ah Valley and Madaba); fragmentary architectural remains are reported for Abila?, Hisban (Sauer 1986), Irbid, and Jarash (the published collared pithos is similar to those from al-'Umayri [compare Braemer 1987: fig. 2.8 with Clark 1997: figs. 4.14-20]); isolated and/or secondary pottery finds seem to come from 'Ara'ir, Umm ad-Dananir?, Rujm al-Henu?, Safut?, Jalul (Younker, pers. comm.), and Jawa south (Daviau, pers. comm.). The finds at Hisban include a small cistern and a long trench cut into bedrock, possibly a deep, narrow moat at the edge of the hilltop (LaBianca and Ray, pers. comm.). Preliminary reports from Irbid indicate a very thick destruction level (up to 4 m) that covered the city wall, a tower, and a two-storied public building that contained cultic vessels.

Al-Lahun has produced a casemate wall and four houses, some of which are pillared. The remains seem to date to a broader time than just the transitional period, but not enough pottery has been published to make independent decisions (Homès-Fredericq 1997b). Whereas the remains are very shallow, several episodes of rebuilding in the houses suggest a long period of time for the settlement (Homès-Fredericq 1992: 190). Farther south the remains are very sparse. At Tawilan a cooking pot from this period and a Midianite potsherd were found in mixed contexts.

The best-preserved remains so far come from al-'Umayri (Clark 1994; Herr et al. 1997b). The defensive system (Figure 9.1) included a dry moat cut out of the original ridge upon which the site was founded (14); a retaining wall (12) supporting a massive rampart that repaired a crack in bedrock probably caused by an earthquake (9); and a defensive wall surrounding the site at the top of the rampart (8), which has been traced for approximately 70 m with two interruptions.

Inside the town walls were the remains of two houses (Figure 9.2). Building B, preserved over 2 m high in places by a massive brick destruction layer from the upper storey of the building, was a four-room house with post bases separating the long rooms (Room B2). The broadroom (Room B3) contained almost 40 collared pithoi lining the walls of the room and from the collapsed upper storey (Figure 9.4, top, shows the rim of one of these vessels). Six bronze weapons in the destruction debris of the room suggest that the site was destroyed by military attack. A paved annex to the east (Room B1), surrounded by a narrow wall interspersed with post bases, may have been an animal pen. Another building to the south (Building A) contained

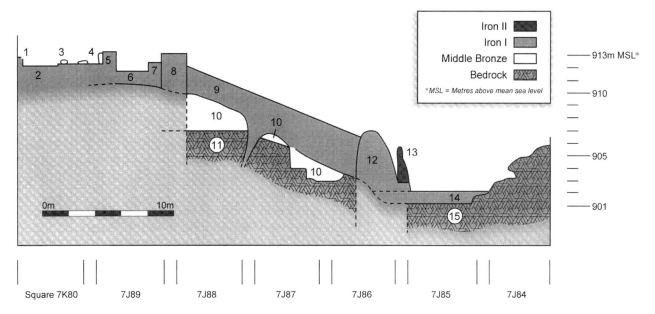

Figure 9.1. Schematic section of the western defences at Tall al-'Umayri. Iron IIC: No. 13, addition to retaining wall. Late Bronze/Iron I Transitional period: Nos. 1–7, features of Building A (compare with Figure 9.2); No. 8, defensive wall; No. 9, beaten earth rampart; No. 12, retaining wall; No. 14, bottom of moat. MB IIC: No. 10, beaten earth rampart; No. 15, bottom of moat.

more collared pithoi in the broadroom (Room A3). In the paved area of Room A2 was a standing stone with a votive altar or table in front, but in the nearby courtyard were domestic remains, suggesting a house with a small shrine.

Parallels to the objects from these structures are found mostly in the highland regions north of Jerusalem and range from pottery to potters' marks to seals. Similarities are notably less extensive in the Jordan Valley. Al-'Umayri, however, was larger and somewhat more prosperous than the sites west of the Jordan Valley. The finds probably reflect socioeconomic or lifestyle connections with the highlands of western Palestine. The limited assemblage of the finds suggests a simple economic system brought about by tribal groups beginning a lengthy sedentarization process in the highland areas of Jordan that was more-or-less free of prior occupation. A major catalyst of this initial settlement may have been the north–south trade routes.

In the Jordan Valley, the most important site is Pella, although generally the Iron Age is not very clear there. The remains published as coming from the end of the Late Bronze Age are considered to be transitional, if the pottery published in 1982 also comes from that horizon (McNicoll *et al.* 1982: 121-27); it should, however, be noted that much of the pottery seems to be from later Iron I, as well. Dayr 'Alla seems to have been uninhabited until slightly later. The large and rich cemetery at as-Sa'idiyya with LB IIB and Iron I pottery and other objects suggests a significant town or city there. Indeed, recent excavations have uncovered tantalizing hints of its existence, but the exposure is very small. The Jordan Valley sites generally produced finds of a much more varied and luxurious repertoire than the plateau sites.

Iron I (twelfth–eleventh centuries BC)

Excavated sites with at least preliminary reports on the Iron I period

Abila (?)	Walls (Mare 1992 and other references there)
Abu al-Kharaz	Citadel? (Fischer 1994: 130)
'Amman (?)	Unpublished pottery (Zayadine *et al.* 1987: 308; Najjar 1997); pottery (Dornemann 1983: 97)
'Ara'ir (?)	Houses? (Olávarri-Goicoechea 1993: 93)
Dayr 'Alla	Phases A to G or H; bronzesmith workshop; pits (Franken 1969)
Dhiban	Pottery in secondary deposits (Winnett and Reed 1964: pl. 76.11-13)

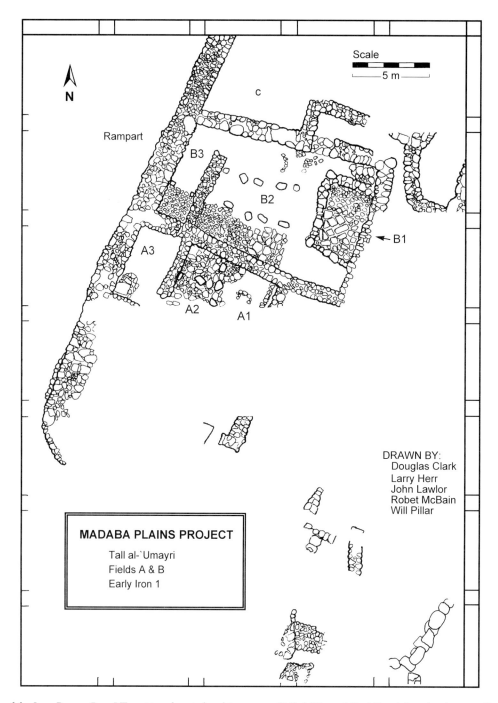

Figure 9.2. Plan of the Late Bronze/Iron I Transitional period architecture at Tall al-ʿUmayri. Building A has the shrine in Room A2; Building B is the four-room house.

al-Fukhar	Reused palace?; house walls; Philistine potsherd (Strange 1997)		fig. 9b.1-2, 20-23; the rest belong to Iron IIC and later—no. 8 is Roman)
al-Hajjar (?)	Walls (Thompson 1972)	Irbid Phase 1	City wall; houses; wine installation (Lenzen 1988)
Hisban 20-19	Pottery in rock-cut trench (Sauer 1994)	Jarash	Pottery in secondary deposits (Braemer 1986: fig. 15.9-10 [no. 8 is probably LB])
ʿIraq al-ʿAmir (?)	Fills and possible fortification wall (N. Lapp 1983: 10; 1989:		

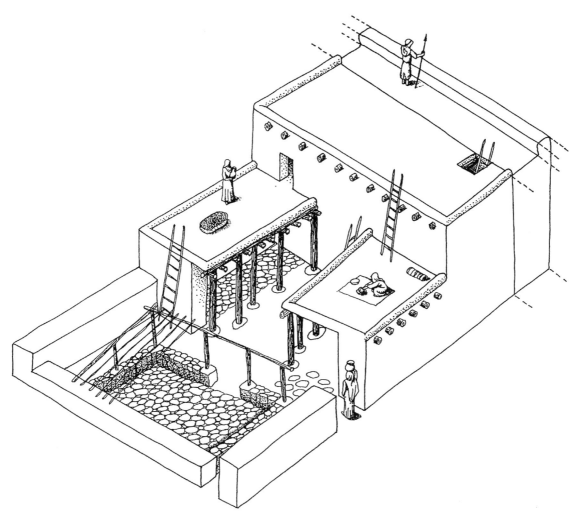

Figure 9.3. Isometric drawing of the Late Bronze/Iron I transitional four-room house at Tall al-'Umayri as reconstructed from the finds in 1996 (drawing by Rhonda Root).

al-Lahun (?)	Fortified village; casemate wall; four houses (Homès-Fredericq 1997b)	Pella VII	Village wall fragments (Bourke 1997)
		Rujm al-Malfuf S (?)	Unpublished pottery (Thompson 1973)
Madaba	Tomb (Piccirillo 1975; Thompson 1986)	Safut (?)	Mudbrick installation; unpublished pottery (Wimmer 1989)
al-Mazar (?)	Courtyard building with cultic objects (Yassine 1988: 115-35)	Sahab	Domestic house fragments; collared pithoi burials; burial cave (Ibrahim 1987: 77-78)
Mudayna 'Alia	Houses, city walls and gate; eleventh century only (C. Routledge 1995)	as-Sa'idiyya XII	Parts of the cemetery (Pritchard 1980, e.g. figs. 16.2, 3; 30.2; 31.1, etc.—brown and black juglets); administrative complex; steps to water source (Tubb et al. 1996: 24-27)
Mudayna Mu'arraja	City walls, gate, towers, houses (Olávarri 1978; 1983)		
Nimrin	Wall fragments (Flanagan et al. 1994: 212-16)		

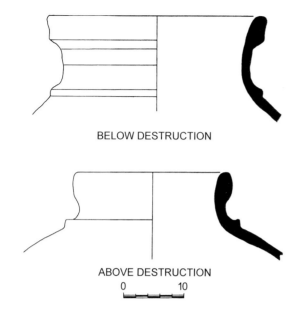

Figure 9.4. Comparison of collared pithos rims from below the destruction of the Late Bronze/Iron I transitional period (top) and those above the destruction (bottom).

Al-ʿUmayri — Storerooms and destruction layer (Clark 1989: 249-50 [Phases 5 and 4; Phase 4 was misdated to early Iron II])

New sites include, most significantly, those toward the southern edges of settlement in the regions of Madaba and al-Karak, continuing the gradual southward movement of settlements throughout Iron I. This period includes several subphases of activity that overlap throughout the period. Not all the sites listed above were contemporary.

On the northern plateau, Tall al-Fukhar seems to continue re-using the Late Bronze palace, while Abila may have extended into this time period as well. Phase 1 at Irbid, which lasted into Iron IIB, included a rebuilt city wall and domestic buildings associated with an industrial installation, which the excavators suggest was for wine. However, the published evidence for all three sites in this region is sparse.

In central areas, the sites of Tall al-ʿUmayri and Hisban continue into the middle of the Iron I period. At the former site, a storeroom was built on top of the bricky destruction of the transitional town and at least 18 collared pithoi were embedded in the fallen bricks (Herr et al. 1997b: fig. 3, bottom, shows an example of one of these vessels). At Hisban, the bedrock trench (moat?) was filled with a layer of Iron I pottery representing forms from the late twelfth and eleventh centuries BC. This would suggest the trench was no longer in use. The Iron I pottery from Sahab is virtually identical to that from al-ʿUmayri at this time (for one example among many, compare the collared pithos from ʿUmayri at the bottom of Figure 9.4 (Herr et al. 1997b) with an example from Sahab published by Ibrahim 1978: 116, 119). Of special interest are the seal impressions on the rims of many of the Sahab collared pithoi. One rim with a seal impression was found at al-ʿUmayri. None of the house plans at Sahab were complete, but enough was uncovered to characterize the rooms as rectangular and mostly paved with flagstones (Ibrahim 1974: pls. 15, 18 for the house plans).

Other central plateau sites which may have continued from the previous period were ʿAmman (incoherent walls), Madaba (if the late tomb can infer a settlement), Safut (also without published pottery), and probably al-Lahun, though here also published pottery is lacking. New sites may include al-Hajjar (a farm or tower, but no published pottery), ʿIraq al-Amir town (fills and a possible fortification wall; note that P. Lapp apparently identified Iron IIC pottery as Iron I) and Rujm al-Malfuf (S) (the pottery is unpublished). A few potsherds from the period have been published from Dhiban.

Towards the end of this period settlements were appearing in the al-Karak region at two very similar sites, Mudayna ʿAlia and Mudayna Muʿarraja. Because the ruins are visible on the surface, it is easy to describe the city walls, towers, town gates, dry moats severing the sites from neighbouring hills, and the roadways approaching the sites. Four-room houses are visible and are sometimes preserved as high as the lintels spanning the doors. In some cases large slabs of stone are still visible spanning the rooms of the houses with a corbelling technique. The pottery from the excavations seems to date to the eleventh century BC.

The Jordan Valley site of Pella continued from the earlier period, but the wall fragments are not as yet easily interpreted. Tubb et al. (1996) seem to suggest that stone terraces at as-Saʿidiyya may lead to entrances to the stepped structure, excavated by Pritchard, which descended to the water source at the foot of the site. The closest parallels to this latter feature are found in Mycenaean Greece (Mycenae and Tiryns). A large residency or administrative complex, possibly with Egyptian connections, was also found there. Three major sites began at this time: Abu al-Kharaz, Dayr ʿAlla, and al-Mazar. There may have been a citadel at Abu al-Kharaz, but the initial report has not been pursued in later seasons of excavation. At Dayr 'Alla Phases A–G or H all belong to the Iron I period. The first four phases (A–D) include a possibly itinerant bronze-smith's workshop on top of the Late Bronze sanctuary and probably date to the twelfth century BC. Nearby deposits of clay were used for moulds and the metal was fired in a large oven. Pear-shaped pits were also found around the smithy, as were small wall fragments. Some of the painted pottery has been connected with Philistine ware (Sauer 1986: 12 and fig. 12). The excavators suggest Phase C ended in an earthquake. Phases E–G/H were characterized by a much heavier settlement, but with walls often only one brick wide and founded on a layer of reeds. Phase H produced a major building of uncertain use. Towards the end of the period, an open-court sanctuary was constructed on the lower mound at al-Mazar with three rooms at the end of a large courtyard. The pottery, much of which was found together outside the door to one of the rooms, dates to the end of Iron I and the beginning of Iron II (Yassine 1988: 122-24). A few wall fragments and pottery appeared at Nimrin.

Iron II

The most extensive textual sources for Iron II comprise passages from the Deuteronomistic history of the Hebrew Bible, which were written considerably later and, therefore, reflect later historical and theological concerns, especially through the eyes of the kingdom of Judah (Hübner 1992). They chronicle a series of wars between Israel and the Transjordanian states (e.g. 2 Sam. 10.7-8; 11.1; 12.26-31; 17.27). Other references mention ambassadors of the kings of Edom, Moab, Ammon, Tyre and Sidon in Jerusalem (Jer. 27.1), probably to arrange a rebellion against Babylon. Many of the prophets castigate the Transjordanian kingdoms for social, political, economic and military offences against Judah. The only Egyptian reference comes from the triumphal reliefs of Shoshenq 1 (945–924 BC) commemorating his successful campaign into Palestine and Transjordan. Names in rows H and V of the topographical list have been identified with Transjordanian sites (Kitchen 1992: 29). From the ninth century BC onwards Egyptian influence in the area was confined to certain artistic elements in the material culture (Kitchen 1992: 30).

Among Assyrian sources the Black Obelisk of Shalmaneser III (858–824 BC) records that 'Ba'sa, son of Ruhubi from Ammon' was part of a rebellion (Pritchard 1969: 279; but cf. Millard 1992).

The fact that extant Assyrian texts do not refer to Ammon during the next 110 years suggests that there was little or no contact between them. The Building Inscription of Tiglath-Pileser III (744–727 BC) lists 18 kings who sent tribute to Assyria, including Sanipu of Bit Ammon, Salmanu of Moab, and Qaushmalaku of Edom (Pritchard 1969: 282). These three nations were again listed among the tributaries of Sennacherib (704–681 BC): 'Buduili from Bet Ammon, Kammusnadbi from Moab, and Aia'rammu from Edom' (Pritchard 1969: 287). Among the 22 kings listed on Cylinder C of Ashurbanipal (668–633 BC) were 'Amminadbi king of Bet Ammon, Qaushgabri, king of Edom, Musuri, king of Moab' (Pritchard 1969: 298). Several other texts speak of tribute from the Transjordanian countries in the seventh century BC (see Pritchard 1969: 287-300). Most of the tribute was precious metals such as silver and gold. We must assume a relatively stable economy in order to meet the demand of such tribute. On the other hand, Assyrian rule stimulated political and socioeconomic stability. It allowed development of national cultures within the broad context of the Assyrian empire and fostered close contact with a wide network of vassal states and provinces (Bienkowski 1992b: 5). The political independence of the vassal states seems to have been assured as long as the tribute was paid. The Assyrian sources apparently suggest that actual Assyrian presence in Moab and Edom did not begin until the reign of Ashurbanipal; nor did they deport anyone from (at least) Moab and Edom (Millard 1992).

After the collapse of the Neo-Assyrian empire in 612 BC and a short rule by Egypt, the Babylonians controlled the region under Nebuchadrezzar II (605-562 BC). In the 580s BC Baalis, the king of Ammon, was a member of an anti-Babylonian rebellion. Josephus states that Ammon and Moab were reconquered in 582/1 BC (*Ant.* 10.9.7). After the collapse of the Neo-Babylonian empire the Transjordanian nations probably became provinces within the Persian empire. Finds from Tall al-'Umayri suggest this was the case for Ammon, at least (Herr 1997b: 325-27).

Iron IIA (tenth century BC)

Excavated sites with at least preliminary reports on the Iron IIA period

'Amman — Tomb G? (Dornemann 1983: 49-50); pottery from soundings mixed with earlier and later forms

	(Dornemann 1983: 90, 170)
ʻAraʻir V (?)	Fortress (Olávarri 1965: 85, 87; Olávarri-Goicoechea 1993: 93)
al-Baluʻ (?)	Storage area; pottery (Worschech and Ninow 1994)
Dayr ʻAlla H–K	Tower and city wall; houses; jar with insects (Franken 1969)
Fadayn (Fedein) (?)	Fortress? (Humbert 1989)
Hisban	Pottery (Sauer 1994: 239-44)
al-Husn (?)	Casemate wall? and destruction (Leonard 1987: 359-69)
Irbid	Tomb C (Dajani 1966b); city wall; houses; wine installation (Lenzen 1988)
al-Lahun (?)	Fortress for agricultural storage (Homès-Fredericq 1997b)
Madaba	Tomb (Piccirillo 1975; Thompson 1986)
al-Mafraq	Tomb (Piccirillo 1976)
al-Mazar	Courtyard building (Yassine 1988: 115-35)
Mudayna ʻAlia (?)	Houses, city walls and gate (C. Routledge 1995)
Mudayna Muʻarraja (?)	City walls, gate, towers, houses (Olávarri 1978, 1983)
Nebo?	Tomb (Saller 1966)
Nimrin	Wall fragments, pottery (Flanagan et al. 1994: 212-16)
Pella VI (?)	Houses with plaster walls, benches, floors, and bins; destruction; tomb (Bourke 1997: 111-13)
ar-Rumeith VIII	Fortress (P. Lapp 1975:114-15)
as-Saʻidiyya X–IX	Open courtyard; re-used walls of Stratum XI tower (Tubb 1988: 34-37)
Sahab	Tomb (Dajani 1970: 33-34)

The early centuries of Iron II are presently very difficult to document in Jordan. Specifically, the tenth-century assemblage of pottery is not easy to isolate, making many of the attributions here uncertain. Some could be placed slightly earlier in Iron I. Roughly half of the material attributed to this period comes from tombs, many of which contain mixed ceramic assemblages. Because of these uncertainties and the lack of any significant, large excavations, we cannot make generalizations regarding settlement patterns or zones of material culture. In spite of this dearth of materials, it is possible that, already by this time, we can suggest incipient national groups that controlled parts of the plateau. The following suggestions of people (sociopolitical) groups are based on material culture, inscriptions, the Deuteronomist's accounts in the Bible and later geopolitical developments that reflect continuity of settlement: Israelites (possibly alternating with Aramaeans) seem to have controlled the northern plateau of Irbid and ʻAjlun; Ammonites began grouping in the region around ʻAmman; other Israelite groups were in the Madaba plain; Moabites occupied the Dhiban and al-Karak regions; and Edomites and possibly related groups (e.g. possible groups called Temanites and Midianites etc.) seem to have continued primarily a non-sedentary way of life south of Wadi al-Hasa. The Israelites probably controlled most of the Jordan Valley.

Northern Plateau (Israel/Aram-Damascus)

Already by the early ninth century BC the Bible remembers strong antagonisms between Israel and the Aramaeans of Damascus in this region. This geopolitical situation may remember earlier (tenth century BC) claims to the region by both groups. However, during the tenth century BC, the material culture is not well enough established to differentiate between the two societies. Moreover, the population (demography) of the area probably did not change when the geopolitics altered. The sites of the region are either very recently excavated or incompletely published. It is, therefore, very difficult to be sure of their dates and the precise nature of their finds.

The mudbrick 'fortress' at ar-Rumeith VIII contained finds largely of a domestic nature. The outer wall of an apparent casemate construction was only 1.25–1.50 m thick and does not inspire certainty that the structure was a fortress. More likely it was simply a small walled settlement, primarily domestic in nature. A small recessed gate pierced the north wall. Because the two subphases were each destroyed by fires it is possible to speculate, among other scenarios, that the border conflicts between Israel and Damascus may have caused them. Sauer confirms that some of the unpublished pottery was red-slipped and burnished (1986: 14). Another possible fortress was found at Fadayn, but the results are very sketchy.

Irbid Tomb C contained earlier pottery, but some seems to date to this period. The excavations at Irbid have apparently produced a continuous settlement from Iron I to Iron IIB with the same remains described above in the Iron I section. Another tomb was found at al-Mafraq that has parallels to the tenth-century BC ceramic corpus at Hazor (Sauer 1986: 14). The preliminary publication of the sounding at al-Husn tentatively suggests an early Iron II casemate wall possibly destroyed slightly later (Leonard 1987: 369); some of the pottery seems to belong to this phase (Leonard 1987: figs. 16g; 18a, c, m).

North Central Plateau (Ammon)

The Bible talks about Ammonite kings as contemporaries of David and Solomon, but tenth-century BC remains are tantalizing at best on the 'Amman Citadel where a tomb and small soundings seem to have uncovered ill-defined remains. There also appears to be another tomb at Sahab, re-used from Late Bronze/Iron I times. It is very likely that excavations at al-'Umayri will uncover Iron IIA remains. Large amounts of pottery at Hisban indicate the presence of occupation but all architectural and living remains were destroyed by later inhabitants, unless the large reservoir dates to this period (Sauer 1994: 241-42; see Iron IIB for more details on its construction).

Madaba Plateau (Israel)

The tombs at Madaba and Nebo seem to belong to this period and imply settlements at those sites. Nearby Jalul will probably produce tenth-century BC remains. The border fortresses at al-Lahun and 'Ara'ir were most likely connected geopolitically with the Israelite groups to the north because both sites seem situated ideally to protect access from the south. Dhiban may not have been settled, but we await a better understanding of Moabite pottery.

South Central Plateau (Moab)

At al-Balu', eleventh–tenth-century BC collared pithoi were discovered in a storage area (Worschech and Ninow 1994: 202; the date is Herr's).

Southern Plateau (Edom)

We are still not convinced that there was sedentary occupation in the region south of Wadi al-Hasa. Some have argued for it (Sauer 1986; Finkelstein 1992; but compare Bienkowski 1992a and Herr forthcoming b) partially by misunderstanding the ceramics of Transjordan (Finkelstein) and by the simple assumption that there must be something there (Sauer, pers. comm.). So far there is no clear published evidence that would alter the views of Bienkowski (1992b: 7-8) and others that no Iron Age settlements have been excavated in the region dating before the eighth century BC. Glueck's early publications on al-Khalayfi at the northern tip of the Gulf of al-'Aqaba misdated a type of handmade pottery called 'Negev Ware' by restricting it to the tenth century BC. A recent reappraisal has placed the earliest remains there in the eighth century BC (Pratico 1993).

Jordan Valley (Israel)

All four Jordan Valley sites in our list are in the northern part. Pella included houses with a remarkably extensive use of plaster on walls, benches, floors and bins. For the two destructions involved, the excavators suggest a military attack for the first one (Shoshenq?) and an earthquake for the second. A courtyard at as-Sa'idiyya was paved with cobbles with depressions that contained large amounts of animal bones and a substance analysed as dung; there were significant amounts of pig bones suggesting an animal pen. The presence of the pig bones could indicate a population of non-Semites or that the Semitic prescription against pigs simply was not practised consistently, or, indeed, was not well known or enforced at that time.

Yassine dates the pottery on the floor of Room 101 in the courtyard building at al-Mazar to the tenth century BC (1988: 122-24). This would represent the final stage of the building that was probably built towards the end of Iron I. The pottery seems to have fallen from shelves, while the chalices suggest a cultic function for the room. At Dayr 'Alla, the phases are somewhat difficult to ascribe to centuries because of their relatively large number. It would seem, however, that the round mudbrick tower and associated thick city wall belong here. The excavators suggest this may have been part of a gate complex (van der Kooij 1993). Nimrin has also apparently produced finds from this period.

Iron IIB (ninth–eighth centuries BC)

Excavated sites with at least preliminary reports on the Iron II B period

Abu al-Kharaz 3–4	City wall; towers; pit (Fischer 1995: 102-103; 1996: 103-104)
'Amman	Possible pottery (Dornemann 1983: 170-71); Citadel Inscription (Thompson and Zayadine 1973)
'Ara'ir V–IV	Fortress; reservoir (Olávarri 1965)
al-Balu'	Public building (Worschech 1989)
Busayra	Pottery (Finkelstein 1992—most of these are Iron IIB forms)
Dayr 'Alla IX–VI	Houses; Balaam Inscription (Franken and Ibrahim 1989; van der Kooij 1993)
Dhiban	Podium for palace?; tower; Mesha Inscription (Tushingham 1972, 1990)
Faris	Pottery (Johns *et al.* 1989)
al-Hajjar	Pottery (Thompson 1972)
Hisban	Water reservoir (Sauer 1994: 244-46)
al-Husn	Pottery (Leonard 1987)
Irbid	City wall; houses; wine installation (Lenzen 1988)
Jalul	Paved road; gate (Herr *et al.* 1996: 71-74)
Jarash (?)	Pottery? (Braemer 1987: fig. 2.9)
Jawa South	Houses; casemate wall; gate (Daviau 1996a: 84-90)
al-Khalayfi IC	Casemate fortress (Pratico 1993: 71)
al-Lahun (?)	Fortress (Homès-Fredericq 1997b)
Madaba	Tomb portions (Thompson 1986)
Mudayna Thamad	City wall; six-chamber gate (Daviau, pers. comm.)
al-Mazar	Pottery (Yassine 1989)
Nebo	Pottery (Saller and Bagatti 1949)
Nimrin	Pits; pottery (Dornemann 1990: fig. 5.8-22; Flanagan and McCreery 1990: 148)
Pella VI (?)	Houses with plaster walls, benches, floors and bins; destruction; tomb (Bourke 1997; Sauer 1994: 245 and references there)
ar-Rumeith VII–V	Fortress; casemate wall; houses; copper-refining kiln (P. Lapp 1975: 115-19)
Safut	Destruction layer; pottery (Wimmer 1987: 281)
Sahab (?)	Pillared house? (Ibrahim 1987: 78-79)
as-Sa'idiyya VIII–IV	Block of houses (Pritchard 1985; Tubb and Dorrell 1991)
al-'Umayri	Wall fragments (Herr *et al.* 1994: 149-50)

With Iron IIB we are able to construct a much more precise synthesis of the people and their culture than we could for Iron IIA. Several major sites have been excavated in several parts of the country and we have more precise historical sources as well (above).

Northern Plateau (Israel/Aram)

After initially being held by Israel, this region probably went to the Aramaeans of Damascus during the ninth-century BC wars between the two nations recorded by the Deuteronomist in 2 Kings. The fortified enclosure at ar-Rumeith was bolstered with an outer wall in Stratum VII and a small gate was constructed in the east wall. The heavy destruction, dated by P. Lapp to the mid-ninth century BC, left some of the walls standing 2.5 m high. Both Strata VI and V expanded beyond the fortified enclosure of Stratum V by adding houses in the southeast but their walls were only 50 cm wide. A copper-refining kiln with copious amounts of slag was discovered here. The houses of Stratum V may have been destroyed by Tiglath-Pileser III. At Irbid the earlier houses and thick wall, which probably was part of a fortification system, apparently continued into this period. Some of the Iron II pottery excavated at al-Husn by Leonard seems early in the period (e.g. Leonard 1987: fig. 16 g, j, p). Only one of the potsherds published by Braemer could be Iron IIB (1987: fig. 2.9—the krater does not seem to have reached the full hole-mouth form of Iron IIC).

Central Plateau (Ammon/Moab)

The pottery of Moab (best seen at Dhiban) may be similar to that from Hisban at this time (Sauer 1994: 245), suggesting that Moab may have reached as far north as the southern hills around 'Amman. Certainly the Mesha Inscription indicates such a scenario. North of Hisban the Ammonites seem to have been the primary group at al-'Umayri and Jawa.

This period is very weakly witnessed in the 'Amman region, although pottery and wall fragments have been found at several sites ('Amman, al-Hajjar, Safut, and al-'Umayri). The 'Amman Citadel Inscription records a major building project at 'Amman in the second half of the ninth century BC. The statue head of a deity, probably Milkom or Il, seems to come from this time. At Sahab, Ibrahim found a 'pillared house' that he equates with the four-room houses west of the Jordan. He notes parallels to the pottery from both Iron IIB and IIC. The most extensively exposed site so far is Jawa South. A casemate wall fortified the settlement and a possible four- or six-chambered gate seems to have pierced the fortifications in the southeast. A large building was composed of four parallel long rooms, but the finds on the floor were not specific enough to suggest a clear function for any of the rooms. In another area, a large warren of at least 11 rooms contained mostly domestic finds. Some of the walls were constructed in the *a telaio* technique whereby segmented or pillared orthostats alternated with uncoursed stone fill.

In the southern part of the region there are many sites that seem to date to this time. The most important site of the area was Dhiban where the Mesha Inscription was discovered at the end of the nineteenth century; it is the longest inscription from the southern Levant and records the Moabite rebellion against Israel (also told from an Israelite perspective in the Bible, 2 Kgs 3). At the site of Dhiban itself, later construction has made the early Iron II period virtually impossible to detect outside the pottery in secondary deposits. But there may have been a podium for a palace or temple and a tower. Much of the Iron II pottery from Dhiban seems to date to this period, but Moabite pottery is not well known. The six-chamber gate recently found at Mudayna Thamad had a length of 13.7 m (Daviau 1996b). Beams from the roof were carbon dated to about 800 BC (Daviau, pers. comm.). The finds reflect storage and worship, as well as game playing. For a site this small, the gate represents considerable investment of labour. Besides the casemate wall connected with the six-chamber gate, there was a lower wall encircling the site part way down the slope, perhaps built on a natural promontory. At 'Ara'ir, Olávarri suggests that the fortress of Stratum V was destroyed by Mesha in the late ninth century BC, and then rebuilt immediately after (Stratum IV). The new fortress had a thick wall surrounding it made up of three parallel walls filled with debris. On the northern side, most vulnerable to attack, a double wall was added. Typical of the constructions Mesha records on his stela, a reservoir found at the northwestern foot of the fortress may have belonged to his works.

The northernmost extremity of Moab was apparently Hisban, which, according to Sauer (above), had pottery more akin to Dhiban than 'Amman. Certainly the biblical remembrance of Heshbon belonging to Moab (Isa. 16.8-9) may stem from this geopolitical situation. In any case, like Dhiban, the Iron IIB remains at Hisban have been destroyed by later occupants, leaving only pottery in secondary deposits. Jalul may contain some of the best remains, but they seem to be limited to the two stone-paved approach roads and outer (?) gate, preliminarily dated to the ninth and eighth centuries BC. A small fortress at al-Lahun, overlooking Wadi al-Mujib, had a main gate in the north and a smaller postern in the south; four towers stood at each of the corners. It is difficult to date the fortress, but some of the pottery appears to be early Iron II (Homès-Fredericq 1992: 196), but some goes into Iron IIC (Homès-Fredericq 1992: 197). Portions of the tombs from Madaba and Nebo seem to extend from Iron IIA into very early Iron IIB.

South Central Plateau (Moab)

Parts of a large public building were excavated at al-Balu' that produced pottery apparently from Iron IIB and C. This city is the largest site on the al-Karak plateau from the Iron II period. The excavators suggest that King Mesha began the major buildings in the ninth century BC, but much more of the site needs to be exposed and published before a clear stratigraphy can be suggested. At Khirbat Faris a few of the published Iron II sherds could be Iron IIB (Johns *et al.* 1989: fig. 23.4, 5, 8).

South Plateau (Edom)

Very few settlements from this period existed on the southern Edomite plateau. Indeed, all we have for certain are the Iron II collared pithoi published by Bennett from Busayra in various articles and then grouped by Finkelstein as 'Iron I' (1992). We now know, however, that in Transjordan this ceramic form continued throughout Iron II (Herr forthcoming b). Many of the major buildings at the site may have begun toward the end of the eighth century and continued into Iron IIC (see that period for a more detailed description).

Although it does not belong to this region, we will consider al-Khalayfi here because of its Edomite connections. The earliest building, Glueck's Casemate Fortress IC, seems to have dated to the eighth century, but the attribution of the pottery to strata was impossible to reconstruct (for renewed excavations, see Mussell 1999). It consisted of a four-room building surrounded by a casemate wall (Pratico 1993: 172), and may have functioned as a storage depot for trade on the Red Sea.

Jordan Valley (Israel, Aram, Ammon)

In the Jordan Valley were several prosperous excavated sites. Because the valley is so long, it may have been associated with several geopolitical entities. Many of the vessels from Tomb 89 at Pella belong to the early centuries of Iron II (Bourke 1997: 112), but the evidence for a contemporary settlement is vague at best. Several metres of later material usually cover Iron Age deposits. Bourke suggests a modest settlement on the northwest slopes of Jabal Abu al-Khas. At Abu al-Kharaz early Iron II pottery was found in Strata 4 and 3, which included two towers, a city wall encircling the upper part of the mound and a pit with a metal figurine of a deity. All phases of the large block of 10 houses and the streets flanking it uncovered by Pritchard at as-Sa'idiyya belong to this period. The houses contained brick pillars separating some of the rooms, which often contained loom weights arranged in rows, suggesting they were attached to looms when the houses were destroyed. At al-Mazar and Nimrin early Iron II pottery has been found, but *in situ* remains elude excavators.

The most remarkable find from Dayr 'Alla IX, dated to around 800 BC, is the Balaam Inscription found on plaster fragments that had fallen from the wall of a house. It recounts a vision seen by the same prophet Balaam mentioned in the Bible (Num. 22–24). The language and script of the inscription are problematic; scholars are divided as to whether they are Ammonite or Aramaic (Hoftijzer and van der Kooij 1991). An earthquake apparently destroyed the several houses found in this stratum. After two ephemeral phases, Dayr 'Alla VI produced the remains of several more houses and a storeroom containing many ceramic vessels from the eighth century.

Iron IIC (seventh–sixth centuries BC)

Excavated sites with at least preliminary reports on the Iron IIC period

Abila	Burial; domestic remains (Mare 1991)
Abu al-Kharaz 1–2	City wall; towers; pit (Fischer 1995: 102-103; 1996: 103-104)
Abu Nusayr	Tomb (Abu Ghanimeh 1984)
Ader (?)	Wall fragments (Cleveland 1960)
'Amman	Palace?; wall fragments; tombs (Dornemann 1983; Zayadine et al. 1987, 1989); wall fragments; storeroom, storage pits (Najjar 1997)
'Arbid	Farmstead (B. Routledge 1995)
Ba'ja III	Acro-site (Lindner and Farajat 1987)
Balu'	Houses; pottery (Worschech 1989: 202-203)
Busayra	Palace; city wall (Bienkowski 1990)
Dayr 'Alla VI	Houses; pits (van der Kooij 1993: 341)
Dhiban	Pottery (Tushingham 1972)
al-Drayjat	Fortress (Younker et al. 1990: 11-14)
Ghrareh	Fortress; four-room house (Hart 1989)
al-Hajjar	Circular tower (Thompson 1972, 1977)
Henu, Rujm al-	Fortress (Clark 1983; McGovern 1993: 146)
Hisban 16	Wall fragments; reservoir (Fisher 1994; Lugenbeal and Sauer 1972)
Khilda	Circular tower and house (Najjar 1992)
Iktanu (?)	Fortress; watchtower; silos (Prag 1989; 1990: 122)
'Iraq al-'Amir	Fortification wall; pottery (N. Lapp 1989: 285, 288)
Jabal al-Qsayr	Acro-site (Lindner et al. 1996b)
Jalul	Tripartite pillared building; pillared house (Younker et al. 1997: 232)
Jawa South	Casemate wall; house (Daviau 1996a: 90-93)

al-Khalayfi	Offset-inset wall; gate fortress (Pratico 1993: 26-28; Mussell 1999)
al-Lahun	Town settlement (Homès-Fredericq 1992)
Madaba	City wall (Harrison, pers. comm.)
Malfuf (S), Rujm al-	Farmstead? (Thompson 1973)
al-Mazar	Tombs (Yassine 1984)
Muʿallaq	Fortress (Lindner et al. 1996a)
Mudaybiʿ	Fortress; volute capitals (Mattingly 1997)
Mudayna Thamad	Casemate city wall; six-chamber gate (Daviau 1996b)
Mukhayzin	Watchtower (Thompson 1984)
Nebo	Tomb (Saller and Bagatti 1949; Saller 1966)
Nimrin	Pottery; wall fragments (Dornemann 1990)
Safut	City wall; houses (Wimmer 1987)
Sahab	Wall fragments (Ibrahim 1987: 79, here dated to the eighth and seventh centuries BC)
as-Saʿidiyya IV	Pits (Pritchard 1985)
Salama	Farmstead? (Lenzen and McQuitty 1989; Bikai 1993: 521)
as-Sela	Neo-Babylonian inscription (Dalley and Goguel 1997)
Siran	Bottle inscription (Thompson and Zayadine 1973)
Tawilan	Houses; unfortified (Bennett and Bienkowski 1995)
Thamayil	Farmstead (B. Routledge 1995)
al-ʿUmayri	Administrative buildings; houses; monumental entry (Herr et al. 1991, 1997a)
al-ʿUmayri	Five farmsteads and one fortress (Geraty et al. 1988: 226-28; Younker et al. 1990: 11-13; 1993: 209; 1996: 68)
Umm al-Biyara	Houses (Bienkowski 1990)
Umm ad-Dananir	Cobbled courtyard; ostracon (McGovern 1993: 147)

It is from the Iron IIC finds that we have the best evidence for the separation of nationalities based on material culture. As national awareness of the political, social and ideological/religious levels rose, so did its expression in the archaeological record, especially in terms of writing, language, religious art and pottery. In Jordan, the material culture of Iron IIC seems to continue into the early Persian period, perhaps as late as the late fifth century BC (Sauer 1986: 18; Herr 1995). Because there are so many sites from this period, we have decided to present a true synthesis and not mention the finds site-by-site (Herr 1997a for more details).

Northern Plateau (Assyria?)

This part of the country seems to have been largely bereft of settlement. The Assyrian destruction seems to have destroyed the local will to establish significant settlements in the area.

Central Plateau and Jordan Valley (Ammon)

This was the era of greatest prosperity for Ammon, whose geographical boundaries seem to have stretched southward through most of the Madaba Plain to the Wadi Wala drainage system. Ammonite material culture is found at al-ʿUmayri, Jawa South, Hisban 16, Jalul, and on the surface at Khirbat al-Hari on the bank of a tributary of Wadi ath-Thamad. We probably may include sites in the Jordan Valley east of the river as Ammonite, though their material culture is not as homogeneous as sites on the plateau: Dayr ʿAlla VI, al-Mazar, Nimrin and as-Saʿidiyya IV. The northern border of Ammon may also have extended beyond Wadi az-Zarqaʾ (biblical Jabbok River) to include possible sites in the ʿAjlun area. The desert was a natural boundary on the east (Herr 1992).

The settlement pattern of Ammon was centered on ʿAmman where, unfortunately, no major multi-season excavation has yet taken place on the Iron Age remains. Small fragments of city fortifications, building walls of houses and a possible palace, and collections of material culture including fine pottery and a volute (Proto-Ionic) capital, all speak of a thriving royal city. Surrounding it were smaller towns, such as Safut on the north, ʿIraq al-Amir on the west, Jawa South, Sahab, and, later, al-ʿUmayri on the south, and the Jordan Valley sites on the west. Smaller villages or agricultural sites were in the hinterland, such as the scores of farmsteads in the highlands around ʿAmman (Younker 1991b), including Mukheizin and Salama. Some of these smaller sites were fortresses (e.g. Rujm al-Henu and al-Drayjat; see Kletter 1991); that is, they were situated in strategic locations, were somewhat larger than the agricultural sites, and had no associated agricultural installations, for example, winepresses. Ammon was thus a large city-state with its major capital city surrounded by scattered towns/villages, fortresses and rural farmsteads.

Rainfall in the highlands is sufficient for dry farming. Grain was produced in the valleys and plains, while orchards and vegetables grew on the hillsides and flocks grazed in the open spaces between fields and on the eastern steppe bordering the desert. Later, when Babylon took over the area and the Ammonite king Ba'alyasha' (a seal impression of one of his officials was found at al-'Umayri [Herr 1985]) conspired against them with a prince of Judah (Jer. 40.14), Babylon conquered them and placed them under tribute (Josephus, *Ant.* 10.9.7). The administrative site of al-'Umayri was apparently built at that time to oversee the crown's investment in wine production at scores of farmsteads in the hinterland to pay this tribute (many of the rural sites in al-'Umayri's hinterland were associated with one or more winepresses).

No Ammonite site has been excavated extensively enough to gain a clear picture of an urban plan. The best glimpse is al-'Umayri, but it was not a normal residential site and was founded very late in the period. It included administrative buildings in the southwestern quarter of the site with domestic structures housing the officials to the north and east. There was no city wall, but a monumental entrance structure with a small shrine (standing stone and basin) were found facing the valley where the King's Highway most likely passed. But no streets have been found. Jawa South was a fortified residential site with houses inside a casemate wall, but here also nothing can be said about street plans.

An idea of architectural plans of Ammonite buildings is beginning to emerge. The rectangular plan of the Ammonite fortresses is best seen at Rujm al-Henu in the Baq'ah Valley and al-Drayjat south of Tall al-'Umayri. Most of these fortress sites were inhabited in later centuries, making clear descriptions difficult. The palace of the Ammonite kings or at least a major administrative building may have been found at 'Amman in the east-central part of the site by a French–Jordanian team (Zayadine *et al.* 1989: 362). Certainly the building was an important one with very large walls surrounding a courtyard paved with a high-quality plaster floor. The rich finds and their international flair (a clay mask, Phoenician ivories, a green glass goblet, lapis lazuli fragments, and perhaps four double-faced Hathor heads) suggest a palatial interpretation. The administrative buildings at al-'Umayri had very thick walls (up to 2 m) and were constructed with basements. One of the buildings was constructed in the four-room plan. A tripartite pillared building at Jalul is the first one found in Transjordan. City walls included casemate structures at Jawa and the fortress of Rujm al-Henu. A very thick solid wall was uncovered in 1998 at Madaba and another solid wall seems to have circled 'Amman; the latter also possibly included a circular tower. A city gate was discovered at Jawa, but its plan has not yet been published. A house at al-'Umayri, only partially excavated, may have had a four-room plan with a cobbled longroom and a cobbled broadroom. But the form is otherwise rare in Ammon; indeed, there does not seem to be a typical Ammonite house plan. One of the houses at Jawa South contained two stairways and monolithic pillars separating some of the 11 rooms (Daviau 1996a: 90-93). Another flight of steps was found at Khilda; it led to a platform at the level of the second floor. Other house fragments have been found at Dayr 'Alla VI, Jalul, Safut and Sahab. Several houses have over 10 interconnected rooms. It is possible that these are basements supporting a more coherent plan in the upper storey.

Of the 150 buildings that have been identified as 'Ammonite monumental structures' (35 circular 'towers', 122 rectangular 'fortresses'), only about 6% have been excavated. Much debate has centred around them regarding function (were they fortresses or towers connecting a military defence system, or were they agricultural facilities, and/or settlements?) and date (were they Iron Age, Roman, or Byzantine or multi-period?). The best suggestion seems to be that of Younker (1991b) and others who say that the buildings on hilltops with strategic views in many directions functioned militarily, while those on hillsides overlooking valleys and next to orchards were farmsteads. Many of them were initially built in the Iron Age, but were re-used in one or more later periods. Some, on the other hand, were not Iron Age at all, as was apparently the case with Rujm al-Malfuf North.

Ammonite pottery entered its most distinctive and superior phase in Iron IIC as potting technologies improved, perhaps from Assyrian inspiration. Very few of the typical vessel forms found in Ammon have been discovered outside the region. Several excavations on the Ammonite plateau have produced a great amount of Ammonite pottery in the last two decades, including Hisban (first defined by Lugenbeal and Sauer 1972), Rujm al-Henu, al-'Umayri, Jawa South and Jalul. The Jordan Valley sites of Dayr 'Alla and as-Sa'idiyya have it too, but not in the same proportions, and it is almost completely lacking from Pella (perhaps because the site was not occupied; Bourke 1997: 113). Several types of fine wares included elegant shallow bowls or plates, sometimes rivalling the much later Nabataean ware for fineness; there were also decoratively burnished bowls, some with a grey burnish made with a manganese tool; but the most distinctive development occurred with a variety of burnished black bowls called, fittingly enough, 'black-burnished ware'.

The imported items found in the palace at 'Amman, as well as the Ammonite black-burnished bowl from Batash in Judah (Kelm and Mazar 1985: fig. 16.4), indicate active trade patterns. There was a major north–south road in Transjordan traditionally called the 'King's Highway'; and at least two other roads must have crossed the Jordan Valley from 'Amman to Jerusalem and the Samaria region. Trade on the King's Highway is represented in the lists of goods on the Hisban ostraca (soon to be published in full in a book of collected essays by F.M. Cross). The sites in the Jordan Valley may have seen more traded items than those on the plateau because they were on both north–south and east–west routes. The tombs at al-Mazar illustrate this with their Assyrian, Judaean and Phoenician vessels (Yassine 1984). There are also indications from sites on the plateau for trade with Phoenicia: artistic motifs suggest Phoenician themes (Bordreuil 1973); pottery from tombs in 'Amman (Gal 1995: 90-91); and a seal written in Ammonite script mentioning Astarte of Sidon (below). The lenticular body of a New Year Flask from Egypt, made of a greenish-turquoise faience, was found in a storage cave near an agricultural site in the al-'Umayri region (Younker 1991a: fig. 12.122.15). These vessels were traded all over the Mediterranean during the Saite (26th) dynasty (seventh–sixth centuries BC) (Homès-Fredericq 1992: 198).

Ammonite scribes developed their own distinctive writing style after borrowing the Aramaic script at the beginning of Iron IIB (Cross 1975; Herr 1980). The formal script is characterized by vertical stances; the heads of some letters opened very late in the seventh century, following an Aramaic development that began a century earlier. The most important inscription of the period is the Siran Bottle, which contained eight lines of Ammonite semi-cursive writing of around 600 BC. It mentions at least three kings of Ammon (Thompson and Zayadine 1973; Cross 1973). There are also scores of seals with several found *in situ* at Tall al-'Umayri and one of its agricultural farmsteads (Aufrecht 1989). The most famous one is the seal impression of an official of an Ammonite king named Ba'alyasha (Herr 1985). There were also several ostraca found in the fill of the Hisban 16 reservoir; several of them represent receipts of trade items. The Ammonite script gave way to Aramaic in the middle of the sixth century BC (Cross 1975).

The Ammonite language belonged to the 'Canaanite' family of Northwest Semitic, but contained what appears to be an element of Arabic, especially in names, perhaps because of Ammon's proximity to the eastern desert. Ammonite differs from neighbouring Hebrew and Moabite in small but not insignificant ways (Jackson 1983: 108). One difference seems to have been a different pronunciation of sibilants (Hendel 1996).

The religion of Ammon centred around its god Milkom, who may be depicted by several male statues and figurines wearing the *atef* crown (Daviau and Dion 1994). Like Yahweh in Judah he was probably an El deity; his iconographic symbol of a bull with huge horns is ubiquitous on Ammonite seals (Aufrecht 1989). A seal written in Ammonite script that mentions Astarte of Sidon suggests that Ammonites also worshipped Astarte. No Ammonite temples have been found, but small shrines or cultic corners were found at al-'Umayri (a standing stone with a basin at the entrance to the settlement) and perhaps in the palace at 'Amman.

There is less evidence of large, monumental art than in Iron IIB, although some of the statues mentioned above may have come from Iron IIC. Tall al-'Umayri produced fragments of a life-sized ceramic statue or anthropomorphic cultic object in 1998 (unpublished as yet). Fertility goddess figurines, some with eyes bugged out and noses made by pinching the clay between thumb and forefinger, are the most frequent human types of figurines, while horses with riders, bovines and lions are the most frequent animals depicted. Seals also present iconographic scenes. One example from al-'Umayri, though it is extremely small, is so nicely carved we can suggest that the species of bird on the seal was an orange-tufted sunbird, a small nectar-feeding bird still seen today (Herr 1997b). The seal impression of the official of Ba'alyasha' contained the depiction of a flying scarab beetle, probably a royal symbol as it was in Judah on the *lmlk* jars (Younker 1985).

The several tombs found in the 'Amman region were chambers cut into bedrock cliffs, much like those from the Jerusalem area. A very large cemetery in the Jordan Valley at Tall al-Mazar, which was made up mostly of pit graves, produced a cornucopia of finds, including pottery, glass, stone and metal vessels, bronze weapons, jewellery, beads, seals and bone and shell objects (Yassine 1984).

A water system has been known at the Citadel of 'Amman as early as 1889 (Conder 1889: 34). Further investigations have been conducted since the beginning of the twentieth century (Dornemann 1983: 90; Zayadine *et al.* 1989: 357, fig. 1-2). The underground water system consists of a large plastered cistern (c. 700 m^3) that is most likely later in date and a 23 m-long passageway carved in the native rock, leading from the top of the cistern. The difference in the altitude between the entrance of the shaft and the floor of the reservoir exceeds 17 m. The occupation in the area went back to MB II and Iron Age periods (Dornemann 1983: 90), making it possible that the water system was in use during these periods. Among other water systems, the most significant was the reservoir at Hisban that continued in use from Iron IIB. It measured 17 by 17 m in size and was 7 m deep and could have held around 2000 m^3 of water (about 2 m.l). The interesting feature of the reservoir was that it was so near the top of the mound and, even though every drop of rainfall could have been channelled into it, it would have needed thousands more donkey loads of water to fill it.

South Central Plateau (Moab)

From the few excavations that have been conducted in Moab it would seem that the most significant remains come from Iron IIB. But the pottery of Moab is not well enough known for us to subdivide Iron II easily. Moreover, very recently excavated sites are beginning to produce results from Iron IIC. The Moabite remains from Mudayna ath-Thamad represent the northeastern border. Some feel that Moab was eclipsed at this time by Arab tribes (Olávarri-Goicoechea 1993: 93). However, an Egyptian New Year Flask was found at al-Lahun confirming trade on the King's Highway during Iron IIC. The houses at al-Balu' have only been partially excavated and we do not have enough to suggest coherent plans.

A monumental gate (at least four chambers) was uncovered at Mudaybi' that contained four or more large volute (proto-Ionic) capitals (Drinkard 1997). This large fortress is surrounded by a huge fortification wall and was apparently intended to guard approaches to the al-Karak region through the Fajj, a natural roadway connecting the King's Highway with the Desert Highway to the east.

A few (under 30) Moabite seals from this time, whose owners carry the theophoric element Kemosh in their names, indicate that at least some organized entity of Moab existed. In fact, one seal refers to its owner as a *mazkir* or official 'recorder', probably of the government (Israel 1987). Moabite writing was similar to Hebrew but with larger heads on some letters and a squatter look overall; some letters were also similar to Aramaic or Ammonite. A beautifully preserved papyrus contains a short inscription in Moabite writing that refers to a *marzeah* (Bordreuil and Pardee 1990), usually interpreted as a funerary feast. Some scholars, however, doubt the authenticity of the document because it was purchased on the antiquities market and is in remarkably good shape for an ancient document. Isolated farmsteads have been excavated at Thamayil and 'Arbid.

The Moabite language was very closely related to Hebrew, except for a few differences, such as *-în* for the masculine plural ending instead of Hebrew *-îm* (Jackson 1983: 108). The best inscription for Moabite is the Mesha Stela that is dated to Iron IIB; it is virtually identical to what we know of Hebrew of that period.

Southern Plateau (Edom)

Almost all Iron Age remains in Edom date to this period, when the Edomites also expanded to the west into the southern Negev. The core territory of Edom, however, was southeast of the Dead Sea around the capital city of Busayra situated on a ridge with a stunning view overlooking deep wadis. Here were the only major public architectural remains so far found in Edom, except for the fortresses at Tall al-Khalayfi and Hazeva 5–4. Tawilan was a small town, while Mu'allaq seems to have been a fortress. There were also several small settlements in out-of-the-way locations high on rugged mountain tops, such as Umm al-Biyara in Petra, as-Sela', Baja III and Jabal al-Qseir, as well as others located by surveys but not yet excavated (Lindner 1992). The difficult access to these 'acro-sites' reminds one of the rocky fastness of which the biblical texts sometimes speak of Edom (2 Chron. 25.12). Because these sites were made up of residences and not fortresses, they present a problem of interpretation. Why were they inhabited? Were conditions so bad that people simply moved there to live? How could they support themselves 300 m above the floor of (later) Petra at Umm al-Biyara? Were these sites constructed much like the *ghorfas* of the Berbers in North Africa, as places of retreat during military invasions or raids? In fact, the small rooms, low doors and lack of streets at Umm al-Biyara remind one of the Lilliputian features of the *ghorfas*.

Edom was located in a narrow strip of cultivable land where dry farming could occur in the highlands east of the Wadi 'Arabah, which reach over 1500 m in altitude. Outside this strip the climatic conditions are that of a semi-desert at best where there was probably a strong element of nomadic pastoralism. This difficult climate is probably what inhibited Edomite sedentarism in the earlier parts of the Iron Age even though we have them mentioned by name on Egyptian inscriptions of the Late Bronze Age (Kitchen 1992).

Because of the specialized nature of Edomite sites, it is difficult to draw a good picture of a typical Edomite urban plan. A solid city wall surrounded Busayra with a simple opening in the wall 2 m wide for a gate (probably a postern). In the middle of the site were two phases of a monumental building, possibly a temple in the Assyrian style. A portion of another public building was excavated to the south that may have been a palace, again in the Assyrian style. Some even suggest the Assyrians may have constructed the buildings for themselves, but there is little other evidence for a dominant physical Assyrian presence. Residential areas probably surrounded the monumental buildings. The residential town of Tawilan was unwalled but excavations were too limited to gain a good picture. No streets are visible in the plans, leaving one with the impression of a warren of walls and rooms. The acro-site of Umm al-Biyara also reminds one of a warren. The excavations were fairly widespread, but no streets were found. The plan of the fortress at Tall al-Khalayfi is clear during two phases following the eighth-century BC phase (Pratico 1993). A new thick wall enclosed the Iron IIB structure into a large area probably including a few residences.

The architecture of the possible temple at Busayra has a plan much like Assyrian buildings, with large open courts surrounded by rooms; but much of the building lies below a modern school and cannot be excavated. The thickness of the walls and the size of the rooms suggest the structure was a very important one. After entering the probable temple in the courtyard, one could turn to the right and walk up a series of broad stairs and enter a long room that may have been the cella. Two stone bases flanked the entrance at the top of the stairs and may have been to support columns or statues. Because of the presence of pottery statuary at other Edomite shrines in the west (Qitmit and Hazeva 5—below. p. 340), the latter suggestion may be correct, although no statuary has been found at Busayra. Other rooms flanked the central structure in this monumental compound. The central building in the fortress at Tall al-Khalayfi was a large version of a four-room house. When they existed, city walls tended to be solid, such as Busayra; those of fortresses could be casemate or solid (offset-inset). Too few complete or coherent houses have been excavated at Edomite sites to be certain if there was a typical Edomite house plan or not; however, one four-room example was uncovered at Ghrareh. At times, it is also difficult to separate one house from its neighbour. Several rooms at Tawilan had pillars. One feature of houses in Edom appears to be their close connection to neighbouring buildings. The warren-like nature of their residences at Tawilan and Umm al-Biyara may reflect a lack of privacy or an intense awareness of kinship. A few door lintels were preserved at Umm al-Biyara; their low height of about 1.3–1.5 m probably indicates economy or a temporary/emergency type of settlement rather than the short stature of the people.

Edomite architecture lacked ashlar masonry at Busayra. Mudbrick was used at al-Khalayfi where stone would have had to be brought from several kilometres away; flat slabs of stone were used at Umm al-Biyara because they were ready at hand.

Edomite painted pottery, found in great quantities at Busayra but also elsewhere, is characterized by multi-coloured bands and geometric patterns; it is distinctive and easily recognizable. Even the unpainted pottery such as jugs and cooking pots can be clearly identified as Edomite.

Another aspect of Iron II pottery in the southern reaches of Jordan is called Negev ware. It was handmade and is found from southern Jordan to Sinai. But rather than identifying it with Edomites, it should more properly be identified with the pastoral–nomadic lifestyle of the desert people; any group living in the area could have made it.

Trade is probably best represented at the Edomite fortress at Tall al-Khalayfi, situated at the tip of the Gulf of al-'Aqaba. Although it is presently located about half a kilometre from the coast, in antiquity the coastline was farther inland. Its main building was probably a well-fortified storehouse or emporium for receiving and sending shipped goods. Other Edomite trade probably passed through its holdings to the west. Hazeva 5–4 was located at a junction of the north–south road in the 'Arabah with an east–west route leading from the King's Highway via Wadi al-Hasa/Wadi Faynan and northern Edom to Gaza. Holding these sites would have secured the Arabian caravan trade in Edomite hands.

Like Moabite, Edomite script is characterized by a syncretistic use of some letters in a Hebrew style and others in an Aramaic style and two letters that sometimes are written upside down. Many of the names in this script use the theophoric element Qaus (or Qôs), the Edomite deity. Several seals have been found at sites in both Jordan and the Negev, and ostraca were found at al-Khalayfi and Busayra. The Edomite language, based on these and other ostraca, is closely related to Hebrew and Moabite within the 'Canaanite' family of Northwest Semitic. Indeed, one scholar has recently stated that there are no features that would suggest Edomite was an independent dialect from Hebrew (Vanderhooft 1995). One of the most astounding finds from the territory of Edom was the discovery of a Neo-Babylonian cuneiform inscription high on the cliff at as-Sela'. There was probably a major road descending from the highlands to the Wadi Faynan area at this point.

Because of the sensational sculptural finds west of the 'Arabah at Qitmit and Hazeva 5–4, we can now say more about the iconography and visual aspects of Edomite religion than any other group among the nations of the southern Levant. The religion of their national deity, Qaus (or Qôs), was probably very much like that of Yahweh in Judah, Milkom in Ammon and Kemosh in Moab. If he was the one depicted on the statues found at Qitmit (his name figures prominently in the inscriptions found there; Beit-Arieh 1995) and Hazeva, he had a barrel chest, a beard and moustache, and was dressed in a pointed headdress and possibly a fringed garment. The three rooms of the rectangular structure at Qitmit suggest a triad of deities, including no doubt the goddess Asherah (or Astarte) whose figurines were found at Qitmit. Among the animal figurines from Qitmit was a stunning example of a cherub, or sphinx, leading us to suggest a similarity to the Jerusalem cult of Yahweh whose primary iconography was the cherub (a guardian composite beast).

Water was always a problem in Edom. Its relative scarcity did not allow its inhabitants to reach the same prosperity as their northern neighbours and it limited the rate at which they could sedentarize. Springs could be found in the low valleys, but in the acro-sites the people needed to depend completely on cisterns. Indeed, the name Umm al-Biyara means 'Mother of Cisterns'.

Conclusion

The two primary aspects of settlement patterns in Jordan during the Iron Age were the gradual growth of 'national' or 'people-group' awareness among territorial segments of society and the gradual spread of occupation from north to south.

References

Abu Ghanimeh, K.
  1984 Abu Nseir Excavations. *Annual of the Department of Antiquities of Jordan* 28: 305-10.

Aufrecht, W.E.
  1989 *A Corpus of Ammonite Inscriptions*. Lewiston: Mellon.

Beit-Arieh, I.
  1995 *Horvat Qitmit: An Edomite Shrine in the Biblical Negev*. Tel Aviv: Sonia and Marco Nadler Institute of Archaeology.

Bennett, C.-M. and P. Bienkowski
  1995 *Excavations at Tawilan in Southern Jordan*. British Monographs in Archaeology 8. Oxford: Oxford University Press.

Bienkowski, P.
  1990 Umm el-Biyara, Tawilan and Buseirah in Retrospect. *Levant* 22: 91-109.
  1992a The Beginning of the Iron Age in Edom: A Reply to Finkelstein. *Levant* 24: 167-70.
  1992b The Beginning of the Iron Age in Southern Jordan: A Framework. In P. Bienkowski (ed.), *Early Edom and Moab: The Beginning of the Iron Age in Southern Jordan*, 1-12. Sheffield Archaeological Monographs 7. Sheffield: J.R. Collis.

Bienkowski, P. et al.
  1997 Soundings at Ash-Shorabat and Khirbat Dubab in the Wadi Hasa, Jordan: the Stratigraphy. *Levant* 29: 41-70.

Bikai, P.M.
  1993 Khirbet Salameh 1992. *Annual of the Department of Antiquities of Jordan* 37: 521-32.

Bordreuil, P.
  1973 Inscriptions sigillaires ouest-sémitique. *Syria* 50: 181-85.

Bordreuil, P. and D. Pardee
  1990 Le papyrus du marzeah. *Semitica* 38: 49-68.

Bourke, S.J.
  1997 Pre-Classical Pella in Jordan: A Conspectus of Ten Years' Work. *Palestine Exploration Quarterly* 129: 94-115.

Braemer, F.
  1986 Études complémentaires. In F. Zayadine (ed.), *Jerash Archaeological Project 1981–1983*, 61-66. Amman: Department of Antiquities.
  1987 Two Campaigns of Excavations on the Ancient Tell of Jarash. *Annual of the Department of Antiquities of Jordan* 31: 525-29.

Clark, D.R.
- 1989 Field B: The Western Defense System. In L.T. Geraty et al. (eds.), *Madaba Plains Project 1: The 1984 Season at Tell el-'Umeiri and Vicinity and Subsequent Studies*, 244-57. Berrien Springs, MI: Andrews University/Institute of Archaeology.
- 1994 The Iron I Defense System at Tell el-'Umeiri, Jordan. *Biblical Archaeologist* 57: 241-42.
- 1997 Field B: The Western Defensive System. In L.G. Herr et al. (eds.), *Madaba Plains Project 3: The 1989 Season at Tell el-'Umeiri and Vicinity and Subsequent Studies*, 53-98. Berrien Springs, MI: Andrews University/Institute of Archaeology.

Clark, V.A.
- 1983 The Iron IIC/Persian Pottery from Rujm al Henu. *Annual of the Department of Antiquities of Jordan* 27: 143-63.

Cleveland, R.L.
- 1960 The Excavation of the Conway High Place (Petra) and Soundings at Khirbet Ader. *Annual of the American Schools of Oriental Research* 34-35: 79-97.

Conder, C.R.
- 1889 *The Survey of Eastern Palestine: Memoirs of the Topography, Orography, Hydrography, Archaeology, etc. The 'Adwan Country*. London: Palestine Exploration Fund.

Cross, F.M.
- 1973 Notes on the Ammonite Inscription from Tell Siran. *Bulletin of the American Schools of Oriental Research* 212:12-15.
- 1975 Ammonite Ostraca from Heshbon: Heshbon Ostraca IV–VIII. *Andrews University Seminary Studies* 13: 1-20.

Dajani, R.W.
- 1966a Jabal Nuzha Tomb at Amman. *Annual of the Department of Antiquities of Jordan* 11: 48-51.
- 1966b Four Iron Age Tombs from Irbid. *Annual of the Department of Antiquities of Jordan* 11: 88-101.
- 1970 A Late Bronze–Iron Age Tomb Excavated at Sahab, 1968. *Annual of the Department of Antiquities of Jordan* 15: 29-34.

Dalley, S. and A. Goguel
- 1997 The Sela' Sculpture: A Neo-Babylonian Rock Relief in Southern Jordan. *Annual of the Department of Antiquities of Jordan* 41: 169-76.

Daviau, P.M.M.
- 1996a The Fifth Season of Excavations at Tall Jawa (1994): A Preliminary Report. *Annual of the Department of Antiquities of Jordan* 40: 83-100.
- 1996b Wadi eth-Themed. *American Center of Oriental Research Newsletter* 8.1: 5-6.

Daviau, P.M.M. and P.E. Dion
- 1994 El, the God of the Ammonites? The Atef-Crowned Head from *Tell Jawa*, Jordan. *Zeitschrift des Deutschen Palästina-Vereins* 110: 158-67.

Dornemann, R.H.
- 1983 *The Archaeology of the Transjordan in the Bronze and Iron Ages*. Milwaukee: Milwaukee Public Museum.
- 1990 Preliminary Comments on the Pottery Traditions at Tell Nimrin, Illustrated from the 1989 Season of Excavations. *Annual of the Department of Antiquities of Jordan* 34: 153-81.

Drinkard, J.F.
- 1997 New Volute Capital Discovered. *Biblical Archaeologist* 60: 249-50.

Finkelstein, I.
- 1992 Edom in the Iron I. *Levant* 24: 159-66.

Fischer, P.M.
- 1994 Tell Abu al-Kharaz: The Swedish Jordan Expedition 1992: Third Season Preliminary Excavation Report. *Annual of the Department of Antiquities of Jordan* 38: 127-45.
- 1995 Tall Abu al-Kharaz: The Swedish Jordan Expedition 1993: Fourth Season Preliminary Excavation Report. *Annual of the Department of Antiquities of Jordan* 39: 93-119.
- 1996 Tall Abu-Kharaz: The Swedish Jordan Expedition 1994: Fifth Season Preliminary Excavation Report. *Annual of the Department of Antiquities of Jordan* 40: 101-10.
- 1997 *A Late Bronze to Early Iron Age Tomb at Sahem, Jordan*. Wiesbaden: Otto Harrassowitz.

Fisher, J.R.
- 1994 Hesban and the Ammonites during the Iron Age. In D. Merling and L.T. Geraty (eds.), *Hesban after 25 Years*, 81-96. Berrien Springs, MI: Andrews University/Institute of Archaeology.

Flanagan, J.W. and D.W. McCreery
- 1990 First Preliminary Report of the 1989 Tell Nimrin Project. *Annual of the Department of Antiquities of Jordan* 34: 131-52.

Flanagan, J.W. et al.
- 1994 Tell Nimrin Preliminary Report on the 1993 Season. *Annual of the Department of Antiquities of Jordan* 38: 205-44.

Franken, H.J.
- 1969 *Excavations at Tell Deir 'Alla I*. Leiden: E.J. Brill.
- 1992 *Excavations at Tell Deir 'Alla: The Late Bronze Age Sanctuary*. Louvain: Peeters.

Franken, H.J. and M.M. Ibrahim
- 1989 Deir 'Alla (Tell). In D. Homès-Fredericq and J.B. Hennessy (eds.), *Archaeology of Jordan*. II.1. *Field Reports: Surveys and Sites A–K*, 201-205. Akkadica Supplementum 7. Leuven: Peeters.

Gal, Z.
1995 The Diffusion of Phoenician Cultural Influence in Light of the Excavations at Hurvat Rosh Zayit. *Tel Aviv* 22: 89-93.

Geraty, L.T. et al.
1988 The Joint Madaba Plains Project: A Preliminary Report on the Second Season at Tell el-'Umeiri and Vicinity. *Andrews University Seminary Studies* 26: 217-52.

Hadidi, A.
1970 The Pottery from the Roman Forum at Amman. *Annual of the Department of Antiquities of Jordan* 15: 11-15.

Harding, G.L. and B.S.J. Isserlin
1953 An Early Iron Age Tomb at Madeba. *Palestine Exploration Fund Annual* 4: 27-33.

Hart, S.
1989 Ghrareh. In Homès-Fredericq and J.B. Hennessy (eds.), *Archaeology of Jordan*. II.1. *Field Reports: Surveys and Sites*, 241-45. Akkadica Supplementum 7. Leuven: Peeters.
1995 The Pottery. In C.-M. Bennett and P. Bienkowski (eds.), *Excavations at Tawilan in Southern Jordan*, 53-66. British Monographs in Archaeology 8. Oxford: Oxford University Press.

Hendel, R.S.
1996 Sibilants and *sibbolet* (Judges 12:6). *Bulletin of the American Schools of Oriental Research* 301: 69-76.

Herr, L.G.
1980 The Formal Scripts of Iron Age Transjordan. *Bulletin of the American Schools of Oriental Research* 238: 21-34.
1985 The Servant of Baalis. *Biblical Archaeologist* 48: 169-72.
1992 Shifts in Settlement Patterns of Late Bronze and Iron Age Ammon. In S. Tell et al. (eds.), *Studies in the History and Archaeology of Jordan*, IV: 175-78. Amman: Department of Antiquities.
1995 The Late Iron II–Persian Ceramic Horizon at Tall al-'Umayri. In K. 'Amr, F. Zayadine and M. Zaghloul (eds.), *Studies in the History and Archaeology of Jordan*, V: 617-19. Amman: Department of Antiquities.
1997a The Iron Age II Period: Emerging Nations. *Biblical Archaeologist* 60: 114-83.
1997b Epigraphic Finds from Tall al-'Umayri during the 1989 Season. In L.G. Herr et al. (eds.), *Madaba Plains Project 3*: 323-29. Berrien Springs, MI: Andrews University/Institute of Archaeology.
forthcoming a. The Iron Age Pottery. In J.A. Sauer and L.G. Herr (eds.), *Hesban: The Ceramic Remains*. Berrien Springs, MI: Andrews University/Institute of Archaeology.
forthcoming b. The History of the Collared Pithos at Tell el-'Umeiri, Jordan. In S.R. Wolff (ed.), *Memorial Festschrift for Douglas Esse*.

Herr, L.G. et al.
1991 *Madaba Plains Project 2: The 1987 Season at Tell el-'Umeiri and Vicinity and Subsequent Studies*. Berrien Springs, MI: Andrews University/Institute of Archaeology.
1994 Madaba Plains Project: The 1992 Excavations at Tell el-'Umeiri, Tell Jalul, and Vicinity. *Annual of the Department of Antiquities of Jordan* 38: 147-72.
1996 Madaba Plains Project 1994: Excavations at Tall al-'Umayri, Tall Jalul, and Vicinity. *Annual of the Department of Antiquities of Jordan* 40: 63-81.
1997a *Madaba Plains Project 3: The 1989 Season at Tall al-'Umayri and Vicinity and Subsequent Studies*. Berrien Springs, MI: Andrews University/Institute of Archaeology.
1997b Madaba Plains Project 1996: Excavations at Tall al-'Umayri, Tall Jalul and Vicinity. *Annual of the Department of Antiquities of Jordan* 41: 145-68.

Hoftijzer, J. and G. van der Kooij (eds.)
1991 *The Balaam Text from Deir 'Alla Re-Evaluated—Proceedings of the International Symposium Held at Leiden, 21–24 August 1989*. Leiden: E.J. Brill.

Homès-Fredericq, D.
1992 Late Bronze and Iron Age Evidence from Lehun in Moab. In P. Bienkowski (ed.), *Early Edom and Moab: The Beginning of the Iron Age in Southern Jordan*, 187-202. Sheffield Archaeological Monographs 7. Sheffield: J.R. Collis.
1997a Lehun. In E.M. Meyers (ed.), *The Oxford Encyclopedia of Archaeology in the Near East*, III, 340-41. New York: Oxford University Press.
1997b *Lehun et la voie royale*. Brussels: Comité Belge de fouilles en Jordanie/Belgisch comité voor Opqravingen in Jordanie.

Homès-Fredericq, D. and J.B. Hennessy (eds.)
1989 *Archaeology of Jordan*. II. *Field Reports: Surveys and Sites*. Akkadica Supplementum 7-8. Leuven: Peeters.

Hübner, U.
1992 *Die Ammoniter: Untersuchungen zur Geschichte, Kultur und Religion einer Transjordanischen Volkes im 1. Jahrtausend V. Chr.* Wiesbaden: Otto Harrassowitz.

Humbert, J.-B.
1989 Fedein (el). In D. Homès-Fredericq and J.B. Hennessy (eds.), *Archaeology of Jordan*. II.1. *Field Reports: Surveys and Sites A–K*, 221-24. Akkadica Supplementum 7. Leuven: Peeters.

Ibrahim, M.M.
1974 Second Season of Excavation at Sahab, 1973. *Annual of the Department of Antiquities of Jordan* 19: 55-61.
1978 The Collared-Rim Jar of the Early Iron Age. In P.R.S. Moorey and P. Parr (eds.), *Archaeology in the Levant: Essays for Kathleen Kenyon*, 117-26. Warminster: Aris & Phillips.

1987 Sahab and its Foreign Relations. In A. Hadidi (ed.), *Studies in the History and Archaeology of Jordan*, III: 73-81. Amman: Department of Antiquities.

Israel, F.
1987 Studi Moabiti I: Rassegna di epigrafia Moabita e i sigilli. In G. Bernini and V. Brugnatelli (eds.), *Studi Camito-semitici e Indeuropei*, 101-38. Rome: Unicopli.

Jackson, K.P.
1983 *The Ammonite Language of the Iron Age*. Chico, CA: Scholars.

Ji, C.-H.C.
1995 Iron Age I in Central and Northern Transjordan: An Interim Summary of Archaeological Data. *Palestine Exploration Quarterly* 127: 122-40.

Johns, J. *et al.*
1989 The Faris Project: Preliminary Report upon the 1986 and 1988 Seasons. *Levant* 21: 63-95.

Kelm, G.L. and A. Mazar
1985 Tel Batash (Timnah) Excavations, Second Preliminary Report (1981–1983). *Bulletin of the American Schools of Oriental Research Supplement* 23: 93-120.

Kitchen, K.A.
1992 The Egyptian Evidence on Ancient Jordan. In P. Bienkowski (ed.), *Early Edom and Moab: The Beginning of the Iron Age in Southern Jordan*, 21-34. Sheffield Archaeological Monographs 7. Sheffield: J.R. Collis.

Kletter, R.
1991 The Rujm El-Malfuf Buildings and the Assyrian Vassal State of Ammon. *Bulletin of the American Schools of Oriental Research* 284: 33-50.

Lapp, N.L.
1983 Introduction. In N.L. Lapp (ed.), *The Excavations at Araq el-Emir*, 1-11. Annual of the American Schools of Oriental Research 47. Winona Lake, IN: American Schools of Oriental Research/Eisenbrauns.
1989 'Iraq el Amir. In D. Homès-Fredericq and J. B. Hennessy (eds.), *Archaeology of Jordan*. II.1. *Field Reports: Surveys and Sites A–K*, 280-88. Akkadica Supplementum 7. Leuven: Peeters.

Lapp, P.W.
1975 Excavations at Tell er-Rumeith. In N.L. Lapp (ed.), *The Tale of the Tell: Archaeological Studies*, 111-19. Pittsburg: Pickwick.

Lenzen, C.J.
1988 Tell Irbid and its Context: A Problem in Archaeological Interpretation. *Biblische Notizen* 42: 27-35.

Lenzen, C.J. and A.M. McQuitty
1989 Salameh (Khirbet). In D. Homès-Fredericq and J.B. Hennessy (eds.), *Archaeology of Jordan*. II.2. *Field Reports: Sites L–Z*, 543-46. Akkadica Supplementum 8. Leuven: Peeters.

Lenzen, C.J. *et al.*
1985 Excavations at Tell Irbid and Beit Ras, 1985. *Annual of the Department of Antiquities of Jordan* 29: 151-59.

Leonard, A.
1987 The Jarash-Tell el-Husn Highway Survey. *Annual of the Department of Antiquities of Jordan* 31: 343-90.

Lindner, M.
1992 Edom outside the Famous Excavations: Evidence from Surveys in the Greater Petra Area. In Bienkowski, P. (ed.), *Early Edom and Moab: The Beginning of the Iron Age in Southern Jordan*, 143-66. Sheffield Archaeological Monographs 7. Sheffield: J.R. Collis.

Lindner, M. and S. Farajat
1987 An Edomite Mountain Stronghold North of Petra (Ba'ja III). *Annual of the Department of Antiquities of Jordan* 31: 175-85.

Lindner, M. *et al.*
1996a An Edomite Fortress and a Late Islamic Village near Petra (Jordan): Khirbat al-Mu'allaq. *Annual of the Department of Antiquities of Jordan* 40: 111-35.
1996b Jabal al-Qseir: A Fortified Iron II (Edomite) Mountain Stronghold in Southern Jordan, its Pottery and its Historical Context. *Annual of the Department of Antiquities of Jordan* 40: 137-66.

Lugenbeal, E.N. and J.A. Sauer
1972 Seventh–Sixth Century B.C. Pottery from Area B at Heshbon. *Andrews University Seminary Studies* 10.1: 21-69.

Mare, W.H.
1991 The 1988 Season of Excavation at Abila of the Decapolis. *Annual of the Department of Antiquities of Jordan* 35: 203-21.
1992 The Abila Excavation: The Seventh Campaign at Abila of the Decapolis. *Near East Archaeological Society Bulletin* 37: 10-18.

Mattingly, G.L.
1997 A New Agenda for Research on Ancient Moab. *Biblical Archaeologist* 60: 214-21.

McGovern, P.E.
1983 Test Soundings of Archaeological and Resistivity Survey Results at Rujm al-Henu. *Annual of the Department of Antiquities of Jordan* 27: 105-41.
1986 *The Late Bronze and Early Iron Ages of Central Transjordan: The Baq'ah Valley Project, 1977–1981*. Philadelphia: The University Museum.
1993 Baq'ah Valley. In E. Stern, (ed.), *The New Encyclopedia of Archaeological Excavations in the Holy Land*, I: 338-42. Jerusalem: Israel Exploration Society.

McNicoll, A.W. *et al.*
1982 *Pella in Jordan I*. Canberra: Australian National Gallery.

Merling, D. and L.T. Geraty (eds.)
  1994 *Hesban after 25 Years.* Berrien Springs, MI: Andrews University/Institute of Archaeology.

Millard, A.
  1992 Assyrian Involvement in Edom. In P. Bienkowski (ed.), *Early Edom and Moab: The Beginning of the Iron Age in Southern Jordan,* 35-39. Sheffield Archaeological Monographs 7. Sheffield: J.R. Collis.

Mussell, M-L.
  1999 Tell el-Kheleifeh. *ACOR Newsletter* 11.1: 5-6.

Najjar, M.
  1992 Rescue Excavations at Khilda/Amman. *Annual of the Department of Antiquities of Jordan* 36: 420-09 [sic] (Arabic).
  1997 The 1990–1992 Excavations. In A. Koutsoukou et al. (eds.), *The Great Temple of Amman: The Excavations,* 1-22. Amman: American Center of Oriental Research.

Olávarri, E.
  1965 Sondages a 'Aro'er sur l'Arnon. *Revue Biblique* 72: 77-94.
  1978 Sondeo Arqueologico en Khirbet Medeineh junto a Smakieh (Jordania). *Annual of the Department of Antiquities of Jordan* 22: 136-49.
  1983 La campagne de fouilles 1982 à Khirbet Medeinet al-Mu'arradjeh pres de Smakieh (Kerak). *Annual of the Department of Antiquities of Jordan* 27: 165-78.

Olávarri-Goicoechea, E.
  1993 Aroer (in Moab). In E. Stern (ed.), *The New Encyclopedia of Archaeological Excavations in the Holy Land,* I: 92-93. Jerusalem: Israel Exploration Society.

Piccirillo, M.
  1975 Una tomba del ferro I a Madaba. *Liber Annuus* 25: 199-224.
  1976 Una Tomba del Ferro I a Mafraq (Giordania). *Liber Annuus* 26: 27-30.

Prag, K.
  1989 Iktanu (Tell). In D. Homès-Fredericq and J. B. Hennessy (eds.), *Archaeology of Jordan. II.1. Field Reports: Surveys and Sites A–K,* 275-80. Akkadica Supplementum 7. Leuven: Peeters.
  1990 Preliminary Report on the Excavations at Tell Iktanu. *Annual of the Department of Antiquities of Jordan* 34: 119-30.

Pratico, G.D.
  1993 *Nelson Glueck's 1938–1940 Excavations at Tell el-Kheleifeh: A Reappraisal.* Atlanta, GA: Scholars.

Pritchard, J.B.
  1980 *The Cemetery at Tell es-Sa'idiyeh, Jordan.* Philadelphia: The University Museum.
  1985 *Tell es-Sa'idiyeh: Excavations on the Tell, 1964–1966.* University Museum Monograph 60. Philadelphia: The University Museum/University of Pennsylvania.

Pritchard, J.B. (ed.)
  1969 *Ancient Near Eastern Texts.* 3rd edn. Princeton: Princeton University.

Routledge, B.
  1995 Archaeological Explorations in the Vicinity of Khirbat ath-Thamayil 1992. *Annual of the Department of Antiquities of Jordan* 39: 127-47.

Routledge, C.
  1995 Pillared Buildings in Iron Age Moab. *Biblical Archaeologist* 54: 236.

Saller, S.J.
  1966 Iron Age Tombs at Nebo, Jordan. *Liber Annuus* 16: 165-298.

Saller, S.J. and B. Bagatti
  1949 *The Town of Nebo (Khirbet el-Mekhayyat).* Studium Biblicum Franciscanum 7. Jerusalem: Franciscan.

Sauer, J.A.
  1986 Transjordan in the Bronze and Iron Ages: A Critique of Glueck's Synthesis. *Bulletin of the American Schools of Oriental Research* 263: 1-26.
  1994 The Pottery at Hesban and its Relationships to the History of Jordan: An Interim Hesban Pottery Report, 1993. In D. Merling and T. Geraty (eds.), *Hesban after 25 Years,* 225-82. Berrien Springs, MI: Andrews University/Institute of Archaeology.

Strange, J.
  1997 Tall al-Fukhar 1990-91: A Preliminary Report. In G. Bisheh et al. (eds.), *Studies in the History and Archaeology of Jordan VI,* 399-406. Amman: Department of Antiquities.

Thompson, H.O.
  1972 The 1972 Excavation of Khirbet al-Hajjar. *Annual of the Department of Antiquities of Jordan* 17: 47-72.
  1973 Rujm el-Malfuf South. *Annual of the Department of Antiquities of Jordan* 18: 47-50.
  1977 The Ammonite Remains at Khirbet al-Hajjar. *Bulletin of the American Schools of Oriental Research* 227: 27-34.
  1984 The Excavation of Rujm el-Mekheizin. *Annual of the Department of Antiquities of Jordan* 28: 31-38.
  1986 An Iron Age Tomb at Madaba. In L.T. Geraty and L.G. Herr (eds.), *The Archaeology of Jordan and Other Studies: Presented to Siegfried H. Horn,* 331-63. Berrien Springs, MI: Andrews University.

Thompson, H.O. and Zayadine, F.
  1973 The Tell Siran Inscription. *Bulletin of the American Schools of Oriental Research* 212: 5-12.

Tubb, J.N.
  1988 Tell es-Sa'idiyeh: Preliminary Report on the First Three Seasons of Renewed Excavations. *Levant* 20: 23-88.

Tubb, J.N. and P.G. Dorrell
- 1991 Tell es-Saʻidiyeh: Interim Report on the Fifth (1990) Season of Excavations. *Levant* 23: 67-86.

Tubb, J.N. *et al.*
- 1996 Interim Report on the Eighth (1995) Season of Excavations at Tell es-Saʻidiyeh. *Palestine Exploration Quarterly* 128: 16-40.

Tushingham, A.D.
- 1972 *The Excavations at Dibon (Dhiban) in Moab*. Annual of the American Schools of Oriental Research 40. Cambridge, MA: American Schools of Oriental Research.
- 1990 Dhiban Reconsidered: King Mesha and his Works. *Annual of the Department of Antiquities of Jordan* 34: 183-92.
- 1993 Dibon. In E. Stern (ed.), *The New Encyclopedia of Archaeological Excavations in the Holy Land*, I: 350-52. Jerusalem: Israel Exploration Society.

Van der Kooij, G.
- 1993 Deir ʻAlla, Tell. In E. Stern (ed.), *The New Encyclopedia of Archaeological Excavations in the Holy Land*, I: 338-42. Jerusalem: Israel Exploration Society.

Vanderhooft, D.S.
- 1995 The Edomite Dialect and Script: A Review of Evidence. In D.V. Edelman (ed.), *You Shall not Abhor an Edomite for he is Your Brother: Edom and Seir in History and Tradition*, 137-58. Atlanta, GA: Scholars Press.

Wimmer, D.H.
- 1987 The Excavations at Tell Safut. In A. Hadidi (ed.), *Studies in the History and Archaeology of Jordan*, III: 279-82. Amman: Department of Antiquities.
- 1989 Safut (Tell). In D. Homès-Fredericq and J.B. Hennessy (eds.), *Archaeology of Jordan*. II.2. *Field Reports: Sites L–Z*, 512-15. Akkadica Supplementum 8. Leuven: Peeters.

Winnett, F.V. and W.L. Reed
- 1964 *The Excavations at Dibon (Dhiban) in Moab*. Annual of the American Schools of Oriental Research 36-37. New Haven, CT: American Schools of Oriental Research.

Worschech, U.
- 1989 Preliminary Report on the Second Campaign at the Ancient Site of el-Baluʻ in 1986. *Annual of the Department of Antiquities of Jordan* 33: 111-21.

Worschech, U. and F. Ninow
- 1994 Preliminary Report on the Third Campaign at the Ancient Site of el-Baluʻ in 1991. *Annual of the Department of Antiquities of Jordan* 38: 195-203.

Yassine, K.
- 1984 *Tell el Mazar I: Cemetery A*. Amman: The University of Jordan.
- 1988 *Archaeology of Jordan: Essays and Reports*. Amman: University of Jordan/Department of Archaeology.
- 1989 Mazar (Tell el). In D. Homès-Fredericq and J.B. Hennessy (eds.), *Archaeology of Jordan*. II.2. *Field Reports: Sites L–Z*, 381-84. Akkadica Supplementum 8. Leuven: Peeters.

Younker, R.W.
- 1985 Israel, Judah, and Ammon and the Motifs on the Baalis Seal from Tell el-ʻUmeiri. *Biblical Archaeologist* 48: 173-83.
- 1991a The Judgment Survey. In L.G. Herr *et al.* (eds.), *Madaba Plains Project 2: The 1987 Season at Tell el-ʻUmeiri and Vicinity and Subsequent Studies*, 269-334. Berrien Springs, MI: Andrews University/Institute of Archaeology.
- 1991b Architectural Remains from the Hinterland Survey. In L.G. Herr *et al.* (eds.), *Madaba Plains Project 2: The 1987 Season at Tell el-ʻUmeiri and Vicinity and Subsequent Studies*, 335-41. Berrien Springs, MI: Andrews University/Institute of Archaeology.

Younker, R.W. *et al.*
- 1990 The Joint Madaba Plains Project: A Preliminary Report of the 1989 Season, Including the Regional Survey and Excavations at el-Dreijat, Tell Jawa, and Tell el-ʻUmeiri. *Andrews University Seminary Studies* 28: 5-52.
- 1993 The Joint Madaba Plains Project: A Preliminary Report of the 1992 Season, Including the Regional Survey and Excavations at Tell Jalul and Tell el-ʻUmeiri. *Andrews University Seminary Studies* 31: 205-38.
- 1996 Preliminary Report of the 1994 Season of the Madaba Plains Project: Regional Survey, Tall al-ʻUmayri, and Tall Jalul Excavations. *Andrews University Seminary Studies* 34: 65-92.
- 1997 Preliminary Report of the 1996 Season of the Madaba Plains Project: Regional Survey, Tall al-ʻUmayri and Tall Jalul Excavations. *Andrews University Seminary Studies* 35: 227-40.

Zayadine, F. *et al.*
- 1987 Recent Excavations on the Citadel of Amman (Lower Terrace): A Preliminary Report. *Annual of the Department of Antiquities of Jordan* 31: 299-311.
- 1989 The 1988 Excavations on the Citadel of Amman, Lower Terrace, Area A. *Annual of the Department of Antiquities of Jordan* 33: 357-63.

10. The Persian Period

Piotr Bienkowski

Chronology and historical background

The absolute chronology of the Persian period in Transjordan can be fixed at 539–332 BC. In 539 BC Cyrus II, king of Persia, entered Babylon, and it is generally accepted that Syria and Palestine submitted to him in the same year, immediately following the fall of Babylon (Dandamaev 1989: 60-61; Lemaire 1994a: 11-13). Alexander the Great arrived in Phoenicia in 332 BC and, despite some local opposition, that date can be used conveniently as marking the end of the Persian period (cf. Eph'al 1988: 147).

However, it is unclear if the Persians had effective control of Transjordan for the whole period. There are no ancient written sources that refer to Transjordan during the period. In the Persian sources, the area from the Euphrates to southern Palestine, including Transjordan, is known by the territorial term 'Beyond the River' (though the term had already been used in Neo-Assyrian and Neo-Babylonian times, Eph'al 1988: 141). During the fourth year of the rule of Cyrus in Babylonia, a united province was created consisting of Babylonia and 'Beyond the River' (Dandamaev 1989: 60-61; Eph'al 1988: 153). There appears to have been little administrative change in the transition from Neo-Babylonian to Persian rule. Since the entire Neo-Babylonian empire came under the rule of a single governor, this suggests that, for the time being, Persian rule in the Levant maintained the same administrative patterns as in the Neo-Babylonian period (Hoglund 1992: 5; Stern 1990: 221; Aharoni 1979: 411).

Unfortunately, the administrative position of Transjordan in Neo-Babylonian times, and what happened to the Iron Age kingdoms of Ammon, Moab and Edom, is not at all clear. Ammon and Moab are not mentioned in any contemporary inscriptions. According to Josephus (*Ant.* 10.9.7), Nebuchadnezzar conquered Ammon and Moab in 582/1 BC, and most scholars interpret this as meaning that Ammon and Moab were annexed at this point and were, henceforth, ruled directly from Babylon (e.g. Ahlström 1993: 801). Josephus does not explicitly state this, however; indeed, although he mentions the exile of Jews to Babylon in the same passage, he makes no mention of the exile of Ammonites or Moabites, which might be expected to have automatically followed a full annexation.

Edom may appear in the Nabonidus Chronicle for his third year, 553 BC, but the signs are broken and the exact reading is not certain (Beaulieu 1989: 166, 169; Grayson 1975: 105, 282). What is normally restored is: 'He/they encamped against the land of Edom' or 'against the city of Edom'. This is usually understood as meaning a siege of Busayra, Edom's capital, and the annexation of Edom (Bartlett 1989: 157-61; Ahlström 1993: 805), but the evidence is scarcely conclusive.

Ammon, Moab and Edom may, therefore, have been annexed by the Neo-Babylonians, but conclusive proof is lacking, although admittedly it is unlikely that they survived as independent kingdoms into the Persian period (Eph'al 1988: 142).

After 486 BC, 'Beyond the River' became a satrapy in its own right (Eph'al 1988: 153-55; Lemaire 1994a: 13).[1] A sub-unit of a satrapy was a province, ruled by a governor. The only certain provinces within the satrapy of 'Beyond the River' in the sources are Judah and Samaria in Palestine (Eph'al 1988: 158; Lemaire 1994a: 16-24, 41-46).

There is some evidence that Ammon might have been a Persian province. The book of Nehemiah mentions someone called 'Tobiah, the servant, the Ammonite' (Neh. 2.10, 19), and it has been suggested that 'servant' refers to a fuller title 'servant of the king', and that Tobiah was a Persian official, perhaps even a governor (e.g. Eph'al 1988: 152; Lemaire 1994a: 48-50), but this is uncertain. Three stamped impressions on jars from Tall al-'Umayri, written in Aramaic and dating to the late sixth or early fifth century BC, have been interpreted as the Ammonite equivalent of the *Yehud* stamps from Judah, which contained the name of the Persian province and probably served as a stamp on goods in the Persian provincial tax system (Herr 1992; Herr 1993: 35; Lemaire 1994b: 264). If this is correct, the 'Umayri stamps, two of which read 'Shuba Ammon' and one 'Aya Ammon', may name governors or treasurers of the Persian province of Ammon.

Evidence from elsewhere suggests that governors were members of the local ethnic groups (Eph'al 1988: 151-52), so there is no need to envisage a large influx of native Persians or of Persian material culture.

Of course, even if there was a Persian province of Ammon, we do not know what its borders were,[2] nor do we know the status of Edom and Moab. A common view (e.g. Lemaire 1994a: 28) is to exclude the existence of a Persian province of Edom/Idumaea on the grounds that southern Transjordan and southern Judah would have been under the control of the Qedarite Arabs mentioned by Herodotus 3.5 (Lemaire 1994a: 24-27), although Idumaea may have been created as a province later, in the fourth century BC (Eph'al 1984: 199; Graf 1990: 139-43). Nevertheless, the cuneiform tablet excavated at Tawilan, north of Petra in Edom, but written in Harran, 900 km away in north Syria, dating to the accession year of one of the Achaemenid kings named Darius (see below; Dalley, 1995: 67-68), does imply that, whatever Edom's status, there were no barriers to communication and commerce with the rest of the Persian empire.

Lemaire tentatively suggests the existence of a province called Moab situated essentially east of the Dead Sea, although he acknowledges that direct evidence is non-existent (1994a: 46-47). The proposal that the legend on a pseudo-Athenian coin, previously attributed to Gaza and read as *b'm*, now be read *m'b* and attests to the existence of a Moabite mint, is scarcely conclusive (Lemaire 1994a: 46-47; 1994b: 282). The inclusion of Moab (*m'b*), in the 'Hierodulen' texts from Ma'in in South Arabia, as a location associated with the commercial activities of the Mineans, does not necessarily imply that this was the name of a province, although Ammon (*'mn*) is also included (Lemaire 1994a: 47; Graf 1990: 143). Lemaire's (1994a: 52-53) tentative hints that there were also Persian provinces of Gilead, Qarnini and Haurina in northern Transjordan are even more speculative; they are based not on direct evidence but on the assumption that these had already existed as Assyrian provinces, which is arguable (Bienkowski 2000).

Graf (1993: 156-60) attempts to reconstruct a Persian royal road system in Syria-Palestine, including Transjordan. From Damascus, his proposed route links Busra, 'Amman, Hisban, al-Karak, Busayra, Tawilan, Fardakh and Tall al-Khalayfi. These are sites that have all allegedly yielded Persian-period finds and that are along the route of the later Via Nova Traiana, the Roman imperial road constructed between AD 111 and 114. Graf suggests that it may have had a Persian predecessor, but much of the evidence is tenuous: there are no specifically Persian-period finds from al-Karak or Fardakh (which is included only because the name seems to be of Persian derivation), and there is no evidence for an actual road, although at this period even a royal road would have been a dirt track that is hard to detect archaeologically (cf. Graf 1993: 165-66).

Eph'al has argued that at the beginning of the Persian period southern Transjordan was still viewed as part of the empire, but when it appears again in the sources in the early Hellenistic period, this region was not included in the empire created by Alexander and his successors (1984: 205; 1988: 163). In the third century BC, during the Ptolemaic rule of Coele Syria, it is clear that the region of Ammon was under Ptolemaic rule, former Edom was under Nabataean control, and the status of Moab is uncertain. Eph'al proposes that Persian rule in Transjordan collapsed after the death of Darius II in 404 BC, when there were anti-Persian activities, fomented by Egypt, in the Levant and Cyprus and rebellions and internal struggles within the central government. However, by 344 BC the Phoenician revolt against Persia had been defeated, and Persia reconquered Egypt in 342 BC (Dandamaev 1989: 270-73). It is thus likely that Transjordan was under Persian rule at that time (and it has been suggested that the Persians may have fortified their southern frontier at this point, cf. Lemaire 1994: 28-30). Nevertheless, the situation remains unclear, and it is impossible to be certain if Transjordan was under Persian control for the whole period or not. In any case, by the time of Alexander's conquest it would seem that Persian rule in the Levant was minimal and nominal: the opposition to Alexander in Tyre and Gaza came from local elements, not from Persia itself (Eph'al 1988: 147).

A relative archaeological chronology cannot yet be constructed for the Persian period in Transjordan. Some scholars refer to 'Early Persian' or 'Late Persian', but these distinctions appear to be based on the proportion of Iron II or Hellenistic ceramic types in levels judged to be Persian, and not on well stratified and dated assemblages of unambiguous Persian material.

Settlement patterns

On the basis of his surveys in Transjordan in the 1930s, Glueck concluded that there was a full gap in occupation from the end of Iron II to the Hellenistic and Nabataean periods (cf. conveniently Sauer 1986 with full bibliography). Until recently, this idea of a

Persian-period occupation gap in most of Transjordan was generally accepted and it was concluded, on very little evidence, that Edom and Moab at least were overrun by nomadic Arab tribes (Ezekiel 25.4's 'people of the East', cf. Eph'al 1988: 163; Lemaire 1994a: 51; see still Sapin 1996, who envisages an extensive semi-nomadic and pastoralist population between the fifth and third centuries BC during a phase of climatic aridity). These tribes tended to be identified with the Qedarite Arabs who were regarded as having jurisdiction over 'the sparsely populated areas that were not occupied by a settled people' (Stern 1990: 223).

Archaeological evidence for Persian-period settlement in Transjordan was not widespread enough to counter this view, and Albright (1954) seems to have been alone in recognizing the continuity of late Iron Age Ammonite material culture, especially the pottery, into later periods. Although there were isolated tomb deposits and a few other finds in Transjordan dated to the Persian period—mostly foreign imports, a few inscriptions, and tenuous parallels from Palestine—there was no proper excavated sequence of material that could be shown unambiguously to date to the period. It was suggested that the pottery from the Hisban excavations of the 1970s might continue from Iron II into the Persian period, but since no stratigraphic sub-phasing could be established there, this did not constitute convincing evidence (Herr 1995; Sauer in Merling and Geraty 1994: 246-48). There appeared to be four Persian phases at Tall Dayr 'Alla from the excavations of the mid-1980s (van der Kooij 1987), but little hard evidence has yet been published.

The ongoing excavations at Tall al-'Umayri have now provided good stratigraphic evidence for continuity from Iron II to the Persian period (cf. Herr 1995, 1999). Two continuously occupied strata bridge the sixth century BC: the second stratum reused the walls of the first and has been dated to the late sixth and early fifth centuries BC. This stratigraphic evidence from al-'Umayri for continuity between Iron II and the Persian period now allows us to draw together the relatively few other scraps of evidence for the Persian period into a more coherent and defensible framework.

Increasingly, it seems that there is continuity from Iron II through the Persian period and perhaps into the Hellenistic period (Homès-Fredericq 1996). Many surveys and excavations have identified little or no diagnostic Persian material, but in fact we should now acknowledge that in some of these at least there might be an element of continuity from Iron II into the Persian period without much of a definable change in the material culture.

The map of Persian-period sites in Transjordan now covers the whole area from north to south. Given the lack of previous syntheses of the Persian period in Transjordan (though see Homès-Fredericq 1996), and general uncertainty regarding the nature of the period, the excavated sites that have identified material dated to the Persian period (Figure 10.1) are listed below with bibliographic references, although the use of the term 'Iron II/Persian' by excavators is confusing: it either masks uncertainty as to the precise chronological span of the material, or reflects genuine continuity between the Iron II and Persian periods.

Excavations (north to south)

Quwayliba/Abila (Mare 1989: 474): Persian-period sherds identified in Area A.

Tall al-Mugayyir (Ibrahim and Mittmann 1986: 171): Persian-period sherds identified, in association with a 'fort', 'way station' or 'caravanserai' dating from the Iron Age to the Hellenistic period.

Tall al-Fukhar (Ottosson 1993: 100; Strange 1997): 'Iron IIC/Persian' silos or bins and some building remains.

Tall as-Sa'idiyya (Pritchard 1985: 60-68, 86-87; Tubb and Dorrell 1994: 52-59; Tubb, Dorrell and Cobbing 1996: 22, 25 fig. 12): Stratum III square building of the Persian period, dated primarily by an Aramaic inscription with letters characteristic of the late sixth to fourth centuries BC, plus two ostraca, with recent excavations suggesting earlier phases of the building, and some Persian-period graves.

Tall al-Mazar (Yassine 1984, 1988b, 1988c; Yassine and Teixidor 1988): cemetery, grain silos, houses; ostracon and seals dated to the Persian period.

Tall Dayr 'Alla (van der Kooij 1987; van der Kooij and Ibrahim 1989: 89-90): Phase V buildings, Phase IV, Phase III wall foundations and small pits, Phase II pit all probably Persian, dated by fifth-century BC Attic lamp (Phase V), fourth-century BC Greek 'fish-plate' (Phase III), and inscribed ostraca dated between 500 and 400 BC.

Khirbat Umm ad-Dananir (McGovern 1989: 40-42): 'Iron IIC/Persian' courtyard with oven, and inscribed storage jar.

Rujm al-Henu West (Clark 1983): 'Iron IIC/Persian tower'.

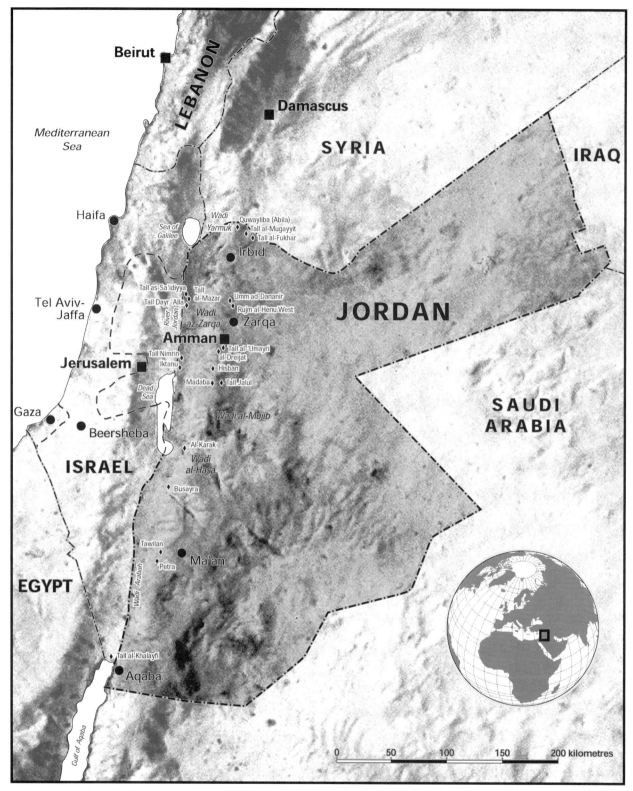

Figure 10.1. Map of sites with material that has been dated to the Persian period. 'Amman = 'Amman Citadel, Khilda, Khirbat al-Hajjar, al-Meqabelein, tomb of Adoni-Nur, Umm Udhayna.

'Amman Citadel (Greene and 'Amr 1992: 126 and fig. 7.6): 'Iron III/Persian'-early Hellenistic pottery lid in deep sounding.

Khilda (Yassine 1988a): two rock-cut tombs dating between the seventh and fifth centuries BC, possibly associated with an Ammonite tower.

Al-Meqabelein tomb (Harding 1950: pls. 13.3, 15.11; Sauer 1986: 18): sand-core glass dated to c. 500 BC, and pottery similar to Tall al-Mazar.

Umm Udhayna (Hadidi 1987): rock-cut tomb with characteristic Persian metalwork and fifth-century BC Attic lekythoi.

Tall al-'Umayri (Herr et al. 1991a: 13, 241-42; Herr 1992, 1993, 1995; Herr et al. 1994: 155; Younker et al. 1996: 77-79): three stamped jar impressions possibly of the Persian province of Ammon, dated to the late sixth century BC, and Attic sherds of the fifth century BC, found in an administrative building with evidence of continuity from late Iron II.

Al-Dreijat (Younker et al. 1990: 13): 'fort' perhaps originally dating to Iron II, cleared to bedrock and new walls constructed in the Late Persian/Early Hellenistic period.

Tall Nimrin (Flanagan, McCreery and Yassine 1992: 98-100, 1994; Dempsey 1993): one locus with two phases of occupation and partially preserved structures of the Persian period, together with eight ostraca, sixth–fourth centuries BC.

Iktanu (Prag 1989: 40-44, figs. 8-11): fort.

Hisban (Sauer 1994: 246-48; Cross and Geraty 1994: 172-74; Merling 1994: 215-16): ostraca and pottery apparently covering the whole of the Persian period, and cistern probably originally carved in the 'Iron II/Persian period'.

Madaba (Harrison 1994: 429): 'Persian-Hellenistic' sherds noted without further details.

Tall Jalul (Younker et al. 1996: 72-73; Younker and Merling 1999): parts of at least three late Iron II/Persian-period buildings were found in Fields C and D, with pottery including imported Attic ware. The Field D building seems to have consisted of rooms surrounding a central courtyard, with pillars supporting the ceiling. One of the Field C buildings sealed a burial cave containing 14 skeletons, with late Iron II/Persian pottery. In Field B the latest phase of an entrance road associated with an outer gatehouse is a repaving dating to the late Iron II/Persian period (sixth–fifth centuries BC). In Field A were found several pits, some sections of walls, and a stretch of pavement (dated to the fifth–fourth centuries BC).

Busayra (Bennett 1977: 8 and fig. 5:1201): cooking pot of Persian-period type possibly indicating that the public building in Area C continued into the Persian period. Two late fourth-century BC Attic sherds found in the last phase of the Area A public building, below the destruction level, indicate that the building continued in use to the end of the Persian period.

Tawilan (Bienkowski 1990; Bennett and Bienkowski 1995): cuneiform tablet dated to the accession year of one of the Achaemenid kings named Darius, found in an accumulation deposit overlying the main occupation.

Tall al-Khalayfi (DiVito 1993: 58-63): ostraca dated to fifth and fourth centuries BC.

A Phoenician coin found inside the tower at Khirbat al-Hajjar and dated to c. 400 BC was explained as a post-occupational deposit by the excavator (Thompson 1972: pl. 7, 1977: 31). The tomb of Adoni-Nur in 'Amman (Harding and Tufnell 1953) is essentially seventh century BC, but Attic imports and sand-core glass dating to c. 500 BC may suggest that its use continued into the Persian period (Sauer 1986: 18).

Surveys

Given the problems of identifying characteristic Persian-period material in Transjordan, surveys are unreliable for locating Persian-period settlement. Many surveys have recorded finding no diagnostic Persian-period sherds. Surveys that claim some Persian-period material are:

Baq'ah Valley Survey (McGovern 1989: 40-42): three 'Iron IIC/Persian' (c. 650–400 BC) sites located.

Dhiban Plateau Survey (Ji 1999): pottery of the 'Iron II/Persian' period was collected at 37 sites, that is, 17.96% of the sites explored. A decline is suggested for the late Persian period, before a settlement increase in the Hellenistic period to a level slightly higher than in the Iron II/Persian period.

East Jordan Valley Survey (Yassine, Ibrahim and Sauer 1988: 175-77, 198-99): thirteen sites identified, but with only small quantities of Persian-period sherds. There appears to be some continuity between Iron II, Persian and Hellenistic periods, with sherds of all three periods found at six of the sites.

Heshbon Survey (Ibach 1987: 158-68): pottery of the 'Iron II/Persian' period (c. 900–500 BC) was collected at 63 sites, that is, 43% of the sites explored by the survey. The 'Late Persian' period (c. 500–250 BC) was almost completely unrepresented in the survey. Only one site, Tall Jalul, yielded three sherds of the period, and this is now being excavated (see above).

'Iraq al-Amir and Wadi al-Kafrayn Surveys (Ji 1998; Ji and Lee 1999): pottery of the 'Iron II/Persian' period was collected at 54 sites, that is, 31.6% of the sites explored by the survey.

Karak Plateau Survey (Miller 1991): pottery of the 'Iron IIC/Persian' period (c. 540–332 BC) was collected at 20 sites (Miller 1991: 310). However, diagnostic Persian-period pottery was very rare (Brown 1991: 203-205).

Tafila-Busayra Survey (MacDonald 1999): pottery of the 'Persian/Hellenistic' period was found at some sites, but the Persian period is described as poorly represented.

Tall al-'Umayri Hinterland Survey (Herr et al. 1991a: 278): one 'Early Persian' sherd from Site 69.

Wadi al-Yabis Survey (Mabry and Palumbo 1989: 94-96): diagnostic 'Persian/Hellenistic' pottery was confirmed at four sites. Three of these showed continuity from the Iron II period.

Pottery

Local pottery, indistinguishable from Iron II pottery, has been excavated in contexts associated with imported Greek pottery of the fifth and fourth centuries BC at Tall al-'Umayri, Tall Jalul, Umm Udhayna, Khilda, Tall Dayr 'Alla (Homès-Fredericq 1996: 74) and Busayra (for Greek ceramic imports into Palestine, see Stern 1982: 136-41). There is no doubt at all, therefore, that there is continuity from Iron II into the Persian period in Transjordan; much of the Iron II pottery repertoire can thus be regarded as still being in use in the Persian period. Here, only those pottery forms that appear to be new to or characteristic of the Persian period will be highlighted, and particularly those that come from the well-stratified and dated contexts at Tall al-'Umayri and Tall Dayr 'Alla (cf. Hendrix et al. 1997: 202).[3]

At al-'Umayri, local pottery was found in stratified contexts associated with fifth-century BC imported Attic sherds, suggesting a date within the Persian period. Many of these local forms were nevertheless typical of the late Iron II corpus, but other forms appear to have Persian parallels, especially to recent, well-stratified assemblages at Tall al-Hesi and Gezer (Herr 1995): triangular jar rims on a variety of forms, necked kraters, particular bowl types, flat lamps, and chevron decoration (Figure 10.2). Totally absent were characteristic Persian types such as sausage jars, high-necked cooking pots and amphorae. Other typical Persian types were very rare, such as mortaria and shallow rounded bowls. Some of these have been found elsewhere in Transjordan. Sausage jars occur at Tall as-Sa'idiyya (Figure 10.3), Umm Udhayna and al-Meqabelein, some with pointed bases, some disc-based. Amphorae (straight-shouldered jars) are found at Umm Udhayna and as-Sa'idiyya. Mortaria have been found at as-Sa'idiyya (Figure 10.4) and at Dayr 'Alla, where the pottery is also in contexts associated with fifth and fourth-century BC imported Greek pottery. Bag-shaped perfume juglets are found at as-Sa'idiyya (Figure 10.5). At Dayr 'Alla, a form characteristic of the Persian levels appears to be a long narrow bottle, narrower than late Iron II types (Figure 10.6; cf. van der Kooij and Ibrahim 1989: 107 no.149, illustrated on p.104). The most common Persian-period form at Tall Nimrin was a storage jar with a simple, thickened, slightly everted rim (Flanagan, McCreery and Yassine 1994: 236 fig. 19.6-9). Sauer has concluded that typical features of the Hisban 'Iron IIC-Persian' assemblage, paralleled elsewhere, include widely spaced interior black wheel-burnishing, short-necked cooking pots, and offset-rimmed, red-burnished bowls (Sauer 1994: 246-48).

As in the Iron II period, there is likely to be regional variation within Transjordan, but this cannot yet be demonstrated.

Architecture

Domestic

The only identifiable domestic architecture comes from Tall al-Mazar, Dayr 'Alla, Khirbat Umm ad-Dananir and Tall al-'Umayri. The first three had architecture based on courtyards containing tabuns or silos; at al-Mazar Stratum II and Dayr 'Alla Phase V, the courtyards were surrounded by mud-brick walls, occasionally on stone foundations. Later Persian-period phases at al-Mazar and Dayr 'Alla seem to consist of pits. Al-Mazar Stratum I, the final occupation phase dated to the fourth century BC, consisted of deep, rounded storage pits and silos, many lined with bricks, stones or mud-bricks. The pits were cylindrical, barrel- or cone-shaped, and were found filled with chaff, charred grains, pottery, copper and stone vessels. Dayr 'Alla Phase III consisted of heavy wall foundations and small pits, while Phase II was virtually one huge pit.

At Tall al-Fukhar, more than twenty stone-lined silos or bins were scattered over the whole excavated

The Persian Period

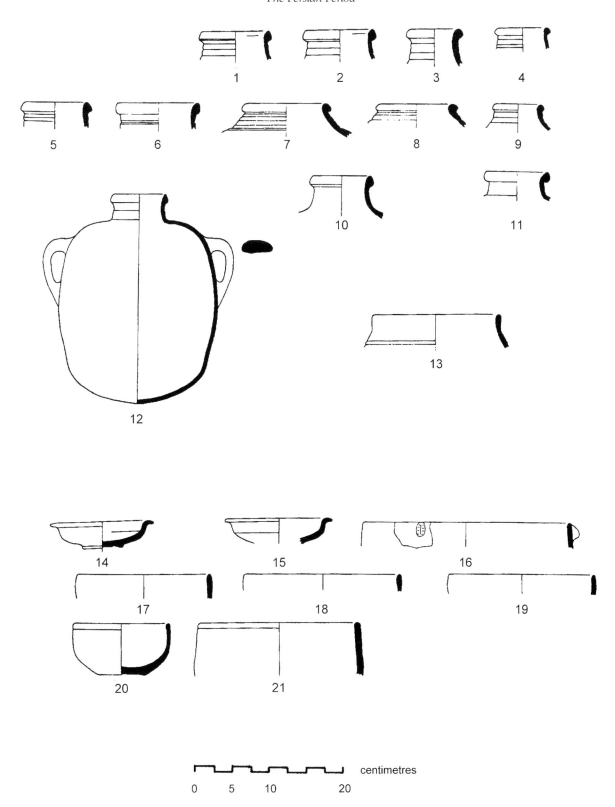

Figure 10.2. Pottery from stratified contexts at Tall al-'Umayri which has Persian-period parallels: 1-12, triangular jar rims (from Geraty *et al.* 1989: 321, nos. 9-17, 22-24); 13, necked krater (from Geraty *et al.* 1989: 325 no. 11); 14-21, bowls (from Herr *et al.* 1991: 42 nos. 24-25, 31 nos. 22-27).

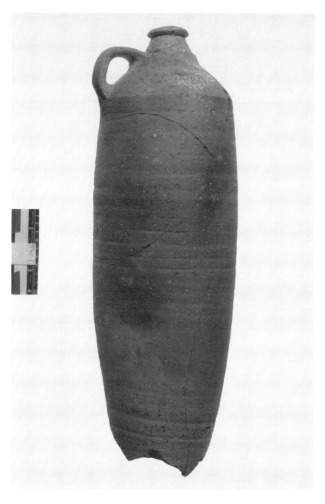

Figure 10.3. Sausage-shaped torpedo jar from Tall as-Sa'idiyya.

Figure 10.5. Bag-shaped perfume juglet from Tall as-Sa'idiyya.

Figure 10.4. Mortarium from Tall as-Sa'idiyya.

area, and possibly connected with building remains, although this could not be proved (Strange 1997). Some of the silos were paved and covered.

Building C at al-'Umayri had domestic materials such as small ash lenses, basalt grindstones, spindle whorls, jar stoppers, oven fragments and mammal ribs on the floors of seven interconnected rooms (Herr *et al.* 1991a: 17-19, 1991b: 157-58). Stairs led down into one room, and another room had two pillars. The building's position immediately to the north of the administrative Buildings A and B tempts speculation that Building C may have housed the bureaucrats (pers. comm. L.G. Herr). Dating to the very beginning of the Persian period at al-'Umayri is a possible tent or shack platform made of flimsy walls, built on the edge of the moat (Herr *et al.* 1994: 155).

Public

Buildings identified as public or administrative and possibly dating to the Persian period have been excavated at Tall as-Sa'idiyya Stratum III (Figure 10.7), Tall al-'Umayri, Tall Jalul, al-Dreijat (Figure 10.8) and Busayra (Figures 10.9-10.10). There are similarities

The Persian Period

Figure 10.6. Bottle from Tall Dayr 'Alla Phase IV (Reg. 2561).

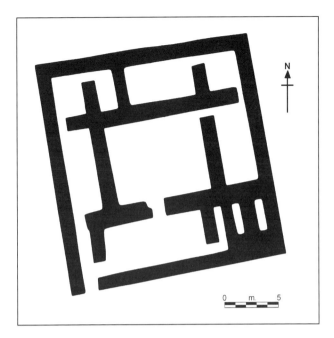

Figure 10.7. Stratum III building at Tall as-Sa'idiyya Stratum III (after Pritchard 1985: fig. 185).

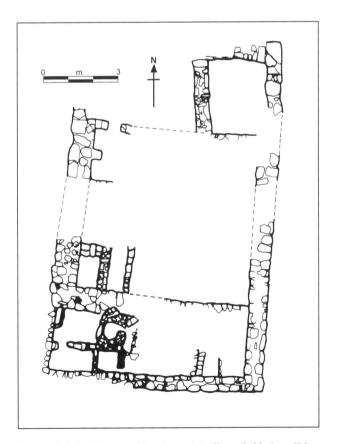

Figure 10.8. Building at al-Dreijat, originally probably Iron II but remodelled in the Late Persian/Hellenistic period (after Younker et al. 1990: 33 pl. 6).

between the Busayra and as-Sa'idiyya buildings: they appear to be in the 'open-court' style known from the Neo-Assyrian period in Mesopotamia and the Levant. At as-Sa'idiyya, the courtyard was stone-paved, at Busayra the courtyards of the Area A and C buildings were plastered. However, below the as-Sa'idiyya stone courtyard were two phases of weathered plaster associated with pits and one phase of a pebbly surface, which appear to be earlier phases of the building, dated by the excavator to the sixth century BC (Tubb and Dorrell 1994: 54-57). The as-Sa'idiyya courtyard was surrounded by rooms on all four sides; the Busayra Area C building had a series of rooms parallel to the court, the remainder of the plan being unexcavated, while the Area A building had rooms on three sides of the courtyard. The excavator concluded that the as-Sa'idiyya building was for defence, although small finds suggested activities such as spinning, weaving, sewing and milling.

Piotr Bienkowski

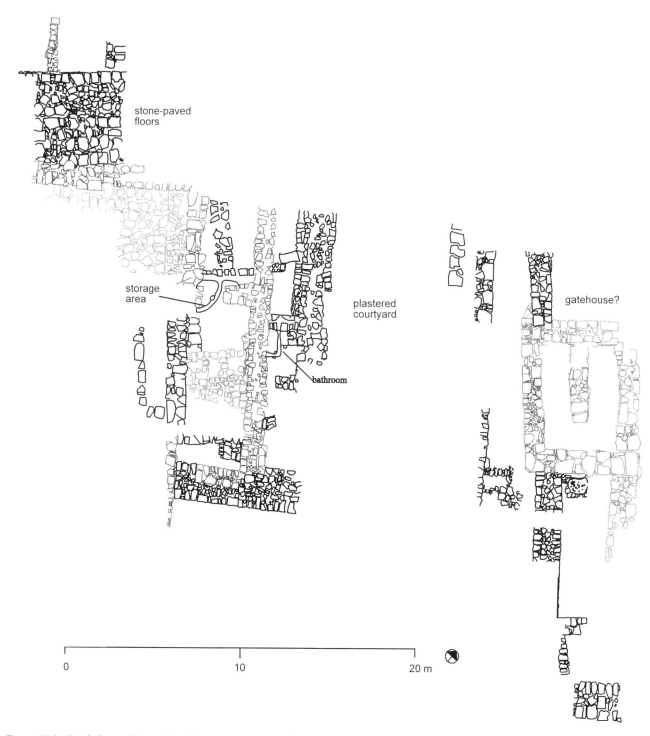

Figure 10.9. Final phase of Area C building at Busayra, probably a palace (drawing by Piotr Bienkowski).

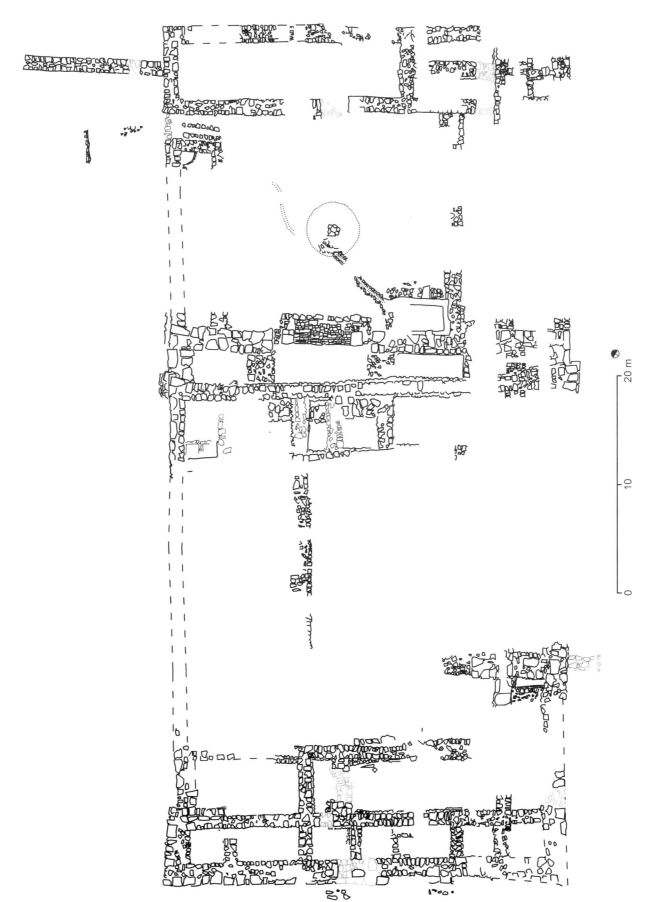

Figure 10.10. Final phase of Area A building at Busayra, originally Iron II, but which continued into the fifth century BC (drawing by Piotr Bienkowski).

It should be noted that the only evidence presented so far for dating the as-Sa'idiyya 'fortress' to the Persian period is an inscription on an incense burner from the floor of one room (see below), dated by the shape of the letters between the late sixth and fourth centuries BC (Pritchard 1985: 67). Evidence for dating the earlier phases of the building has yet to be published (Tubb and Dorrell 1994: 52-59).

The Persian date for the Busayra Area C building is even less secure: one cooking pot, with characteristic Persian-period globular sack-shaped body, short neck, and handles from rim to shoulder (Bennett 1977: 8; cf. Stern 1982: 100-102). The building was probably an Iron II palace or residence which may have continued in use into the Persian period. The earlier Iron II building was raised on a stone platform, with a large plastered courtyard or reception room with rooms on at least one side. These rooms had extensive plaster and stone-paved floors, a bathroom, and storage areas. In the later (Persian?) phase the spaces were subdivided and new doorways were added, and a possible gatehouse with a piered gate was built on the opposite side of the courtyard.

The Area A building at Busayra, probably a temple, was also originally constructed in Iron II on a stone platform, but two Attic sherds in the final pre-destruction deposit prove that it continued in use to the late fourth BC. The plan was based on two wings, one the temple proper with a possible sanctuary approached from a courtyard, the other wing perhaps a storage/administrative annexe.

A rectangular structure at al-Dreijat (Figure 10.7), originally probably Iron II but remodelled in the Late Persian/Hellenistic period, may have been a fort and is of similar size to the as-Sa'idiyya building (Younker *et al.* 1990: 13, 33 pl. 6). It was built of massive unhewn and partially hewn stones; the excavators suggested that the interior walls supported a second storey.

Two buildings at al-'Umayri have been described as administrative, and seem in their later phases to have continued from Iron II into the Persian period. Building A was apparently a single room with thick walls; Building B was planned like a four-room house but with monumental walls and a basement. The large building at neighbouring Tall Jalul has two or three rows of stone pillars supporting the roof, which had collapsed into the rooms; the associated artifacts included an incense stand/altar, a stone roof roller, basalt food preparation vessels and iron tools (Younker *et al.* 1996: 72-73, 89 pl. 16), perhaps suggesting a function beyond the strictly administrative.

Hoglund (1992: 165-205) argues that there is a distinctive type of 'fortress' dating to the mid-fifth century BC that housed Persian garrisons as a result of insecurity in the Levant because of the Egyptian revolt and the Delian league's involvement. He includes Tall as-Sa'idiyya, and amends Pritchard's date of 420 BC upwards to the mid-fifth century BC. However, the recent excavations at as-Sa'idiyya (Tubb and Dorrell 1994: 52-59) have shown that Pritchard's Stratum III square building was the culmination of several earlier levels of a courtyard building, perhaps of Persian date but maybe earlier. These earlier levels may not fit into Hoglund's suggested chronology, or his proposed construction specification for a garrison. Furthermore, the Persian-period administrative building at al-'Umayri also continues from the Neo-Babylonian period (Herr 1993), suggesting that such administrative buildings simply re-used existing structures.

Burial customs

Apart from isolated tombs, the only cemetery dated to the Persian period that has been excavated and fully published is that north of the mound at Tall al-Mazar (Yassine 1984). In general, there appears to be a continuation of Iron II burial practices (themselves little published from Transjordan) and similarities with Persian-period burials in Palestine (see Yassine 1999).

Eighty-five graves were excavated at Tall al-Mazar, dated to the sixth–fifth centuries BC (possibly, therefore, beginning still within the Neo-Babylonian period). With some exceptions, the bodies were oriented east-west, heads to the east and faces to the south, with males extended and females flexed (Yassine 1984: 8-9).

Five types of burial were found at Tall al-Mazar (Yassine 1984: 6-7):

1. The majority were simple pit burials, with the body wrapped in a cloth or mat and laid in a pit, a type also common in Palestine (Stern 1982: 86).
2. Brick-lined tombs (3).
3. Tombs lined with stone on one side, the body laid on the earth, possibly with a wooden cover (6).
4. Infant storage jar burial (1).
5. Larnax burial, i.e., a clay burial coffin shaped like a bathtub (1).

Associated with the male burials were generally pottery and metal objects (vessels, arrowheads, spearheads, swords and knives); female burials tended to have bracelets, necklaces, bronze and silver earrings, and beads of faience, glass and carnelian. Animal bones accompanying one burial were tentatively identified by the excavator as a pet or joints of meat (Yassine 1984: 36).

The Persian-period cemetery on the lower tell at nearby Tall as-Sa'idiyya (Tubb, Dorrell and Cobbing 1996: 22) generally had grave pits dug among the Early Bronze Age buildings. The excavator notes that these Persian-period burials were identical in every respect to the Late Bronze Age burials in the same area, and could be chronologically distinguished only on the basis of any diagnostic material. Unfortunately, the inventories of graves at Tall as-Sa'idiyya published in the preliminary reports do not indicate their date, so it is uncertain which graves belong to the Persian period. However, the few graves specifically noted in the preliminary reports as Persian-period are pit burials with the same orientation as those at Tall al-Mazar and similar grave goods. One of the graves contained four stamp seals (Tubb, Dorrell and Cobbing 1996: 25 fig.12, though only three are noted in the inventory on p. 38).

Four isolated rock-cut tombs have also been dated to the Persian period: al-Meqabelein (Harding 1950),[4] Umm Udhayna (Hadidi 1987), and two poorly recorded tombs at Khilda (Yassine 1988a). Al-Meqabelein, Umm Udhayna and Khilda Tomb 2 consisted of burial chambers with benches. Al-Meqabelein had a shallow central pit, while Umm Udhayna had a main chamber with 15 skeletons, probably multiple successive burials, and other burials scattered on the benches. The associated finds in each case are similar to those known from Tall al-Mazar and Tall as-Sa'idiyya.

Material culture

Very few objects found in Transjordan are diagnostically 'Persian'. There are occasional discoveries of fine pieces such as a pair of painted terracotta riders from the al-Meqabelein tomb (Bienkowski 1991: 45 pl. 46 in colour), a silver earring from the Umm Udhayna tomb (Bienkowski 1991: 103 pl. 124 in colour), or an elaborately decorated and inscribed limestone incense burner from Tall as-Sa'idiyya (Pritchard 1985: 66-68 and fig. 18.22), which arguably are Persian-period, but the most common small finds on all sites are bronze objects. The vast majority of these can only be dated broadly Iron II-Persian, since they are common or at least occur already in Iron II contexts but probably continue in use into the Persian period.

Bronze bowl types are essentially the same as those in Palestine (Stern 1982: 144-45). The two most common types at al-Mazar, Khilda, Umm Udhayna, al-Meqabelein and as-Sa'idiyya are: (a) flat, carinated, with an everted rim; and (b) round, slightly carinated, with a convex base and omphalos, normally decorated in relief with floral patterns. Both types appear first in late Iron II contexts in Palestine and continue into the Persian period, the latter being the common bowl of the Persian period according to Stern. A third type found at al-Mazar is a deep rounded bowl without an umbo, also a common Persian-period type (Amiran 1972), with parallels in Palestine and Syria. An unusual bronze dish from Umm Udhayna has a handle comprising two horses' heads (Zayadine 1986: 147).

Two types of bronze strainer are found at al-Mazar, Umm Udhayna and Khilda: (a) a possibly earlier Iron Age type with horizontal handles and deep carinated bowl; and (b) a possibly later (?Persian) type with an upright handle and deep but plain round bowl. A third type at al-Mazar has a duck-head terminal and floral stem, similar to Persian-period types in Palestine (Amiran 1972; Stern 1982: 147).

Triangular and/or bow-shaped bronze fibulae are found at Tawilan, the al-Meqabelein tomb, the Umm Udhayna tomb, al-'Umayri, as-Sa'idiyya, al-Mazar and Dayr 'Alla. These occur first in the eighth and seventh centuries BC, but continue in use through the Persian period and even later throughout the Near East.

Standard Iron II-Persian leaf-shaped bronze arrowheads are common, but 'Irano-Scythian' trilobe bronze arrowheads, often regarded as diagnostic of the Persian period, are found at Tawilan, al-'Umayri, al-Mazar and Dayr 'Alla. However, although 'Irano-Scythian' arrowheads are found throughout the limits of the Persian empire, the type first appears in Assyrian-period contexts and must be dated broadly seventh–fourth centuries BC (cf. Stern 1982: 154-56).

Circular bronze mirrors with long tapering plain tangs are found at Umm Udhayna, al-Meqabelein, as-Sa'idiyya and al-Mazar. In Palestine, these appear at the very end of the Iron Age and are popular in the Persian period (Stern 1982: 147).

Rings and bracelets formed from bent copper or copper alloy/bronze rods or occasionally sheet bronze are found at Tawilan, Tall al-Khalayfi, Dayr 'Alla, as-Sa'idiyya and al-Mazar. This type can be dated broadly Iron II-Persian, but in fact this simple shape appears in many periods, and is even found in Islamic-period tombs.

A unique piece is a bronze censer with a lid, supported by a stand in the shape of a human female figure on a tripod base, from the Umm Udhayna tomb (Khalil 1986; Zayadine 1986: 144-45; Bienkowski 1991: 96 pl. 115 in colour). The incised decoration appears to be in Achaemenid style. The hairstyle is particularly distinctive because of its 'trimmed' style in which the pattern of small curls appears at the end of wavy strands. These strands in turn radiate from the circular ring holding the stem of the incense bowl. The curls are represented with a grid pattern. This hair arrangement is depicted on the terracotta reliefs of bearded males at Susa and Persepolis.

Inscriptions and seals

The only inscribed document from Transjordan officially dated within the Persian period is the cuneiform tablet from Tawilan (Dalley 1995: 67-68). This is a legal document drawn up in Harran, dated to the accession year of one of the Achaemenid kings named Darius (i.e., 521, 423 or 335 BC), containing testimony to the disputed sale of two rams. The name and patronymic of the man responsible for the testimony are both compounded with the divine name 'Qos'; other names cited are Aramaic or Babylonian. Unfortunately, the tablet was not found in an occupation deposit at Tawilan and so cannot be used unambiguously to date settlement at the site into the Persian period (Bennett and Bienkowski 1995: 102-03).

Inscriptions from Transjordan from late Iron II, especially the sixth century BC, are difficult to date precisely, but Lemaire (1994b: 262) concludes that no Ammonite, Moabite or Edomite inscription proper dates to the Persian period. Aramaic inscriptions, on the other hand, are well attested (Lemaire 1994b: 262-66).

A limestone incense burner from Tall as-Sa'idiyya, decorated with geometric designs, a horse and a human figure, was inscribed in Aramaic *lzkwr*, '(belonging) to Zakkur', and dated palaeographically between the late sixth and fourth centuries BC (Pritchard 1985: 66-68).

Ostraca from Tall as-Sa'idiyya, Tall al-Mazar, Tall Dayr 'Alla, Tall Nimrin, Hisban, and Tall al-Khalayfi, dated palaeographically to the Persian period, have mostly short Aramaic inscriptions, some of which can be identified as records of goods, often with personal names, although generally they are not very informative. They date between the late sixth and fourth–third centuries BC, and demonstrate that Aramaic had largely replaced the national scripts in Transjordan by the end of the sixth century BC (Cross and Geraty 1994: 172).

The personal names compounded with divine names on ostraca form the only evidence for religion from this period. As well as on the Tawilan tablet, Edomite 'Qos' occurs in a name on a late sixth-century BC ostracon from Hisban (Cross and Geraty 1994: 173 [A6]), and 'Qos', Canaanite 'Baal' and Phoenician 'Eshmun' on fifth to early fourth-century BC ostraca from Tall al-Khalayfi (DiVito 1993, Reg. Nos. 2070-1). 'Milkom' and 'El' occur on an ostracon from Tall al-Mazar (Yassine and Teixidor 1988: 141), and 'El' on a fifth-century BC ostracon from Dayr 'Alla (van der Kooij and Ibrahim 1989: 69-70). 'Yahweh' may occur in a fourth to third-century BC ostracon from Tall al-Mazar (Yassine and Teixidor 1988: 137).

Three stamped jar impressions from al-'Umayri (Herr 1992) were found associated with the public buildings. They are written in Aramaic script and have been dated to the late sixth century BC. Their reading is not certain, but Herr suggests that two with the same inscription read 'Shuba Ammon'. By analogy with the *yhd/yhwd* stamps from Judah, which probably served as a stamp on goods in the Persian provincial tax system, Herr suggests that Shuba might be the name of the governor or treasurer of the Persian province of Ammon. A third stamped jar impression was found during the 1994 excavations, with a different name, 'Aya' (Lemaire 1994b: 264).

Stamp and cylinder seals from Tall al-'Umayri (Porada 1989: 381-84), Tall al-Mazar (Yassine 1984: 103-107 and fig. 9.1, 6, 1988c: 150 no. 302), Khilda (Yassine 1988a: 21, 30 fig. 8.4) and Tall as-Sa'idiyya (Tubb, Dorrell and Cobbing 1996: 25 fig. 12)[5] have good parallels to fifth-century BC examples elsewhere. Porada (1989: 381) suggests that, while the inspirations may come from genuinely Achaemenid seals, in most cases the designs and workmanship appear to be local. Although insufficient Persian-period seals have yet been found in Transjordan to define a local style, many of those identified so far have human or animal figures in Achaemenid style but with significant differences of detail unparalleled elsewhere. Other seals found in supposedly Persian-period contexts, either at these sites or at Umm Udhayna, have parallels in the late Iron II period (Neo-Babylonian, earlier sixth century BC).

Conclusion: from the Iron Age to the Hellenistic period

Until recently the evidence for Persian-period occupation in Jordan was sparse and stratigraphically unclear, allowing even the latest commentators to conclude that sedentary life had disappeared from large parts of the country. Recent evidence shows clear proof for continuity of settlements from Iron II through the Persian period and possibly even into the Hellenistic period. There is also some evidence that Ammon might have been a Persian province. There is no evidence for the status of Edom or Moab within the Persian empire, and no basis for the bald statement that Busayra and Tall al-Khalayfi were integrated into the Persian administrative system (Elayi and Sapin 1998: 155).

Virtually no research has been undertaken to synthesize evidence for the Hellenistic period in Jordan (cf. Smith 1990). Occasionally there is an element of terminological and chronological confusion, since of course the Nabataean period was coterminous with the Hellenistic, but, certainly in its earlier stages, Nabataean culture was confined to southern Jordan—and it should be stressed that Hellenistic pottery can be identified quite separately on Nabataean sites. The clearest chronological definitions are based on historical, not archaeological, criteria: the Hellenistic period beginning conventionally in 332 BC with Alexander the Great's arrival in Phoenicia, with Jordan under the rule of the Ptolemies from Egypt (c. 301–198 BC) and then the Seleucids from Syria (c. 198–63 BC), and ending with Pompey's conquest of Syria in 63 BC (for the Nabataeans, see the chapter by Schmid, this volume).

There is still no conclusive archaeological evidence concerning continuity between Edomite and Nabataean settlements (Bartlett 1990; Bienkowski 1990). Busayra appears to have been occupied from Iron II to the end of the Persian period, but Hellenistic and Nabataean pottery found at the site is unstratified and cannot be associated with the Iron II/Persian settlement. Neither Tall al-Khalayfi—apart from a surface find of a Rhodian jar handle dating to c. 200 BC (Pratico 1993: 62)—nor Ayla (Roman Aqaba, Parker 1997: 39) have yielded evidence for the Hellenistic period. Farther north it appears that there was continuity of settlement from Iron II through the Persian period and into the Hellenistic period, at least at 'Umayri (Herr 1993), Sa'idiyya (Pritchard 1985), and Fukhar (Strange 1997: 403-405). In general, there is a lack of early Hellenistic sites, whereas late Hellenistic sites (from c. 200 BC) are well attested and prosperous with numerous imports, for example at Umm Qays, Jarash, 'Amman and 'Iraq al-Amir (al-Muheisen and Villeneuve forthcoming). This pattern has been attributed to initial stagnation due to Ptolemaic domination and a possible period of drought, and then Seleucid fostering of new trade routes and a cooler, wetter climate (Smith 1990). Nevertheless, Hellenistic settlement patterns are still not clear, and recent survey work in central and northern Jordan has demonstrated differences between regions (Ji 1998; Ji and Lee 1999). On the Jordanian plateau and in the north Jordan Valley there may have been a gap in the late Persian period, but in the early and particularly the late Hellenistic period there was a considerable increase in settlements. In the south and central Jordan Valley, however, there seems to have been continuity from the Persian into the early Hellenistic period, but thereafter a gap until the Roman period. It is difficult to assess how much confidence can be placed in these survey results given the still sketchy knowledge of Hellenistic-period pottery in different regions, and the lack of a published stratified corpus of early Hellenistic pottery from either Jordan or Palestine (Smith 1990: 125), though without doubt there was some continuity from the Persian into the Roman period.

Acknowledgments

This chapter is a completely revised and updated version of a chapter written with Dianne Rowan for *Archaeology of Jordan* 3 (D. Homès-Fredericq and J. B. Hennessy [eds.], unpublished). The author is grateful to Larry G. Herr, Mohammad Najjar and Ephraim Stern for their helpful comments on an earlier draft of this chapter, Philip R. Drey for some missed references, and to Dianne Rowan for permission to amend previous joint work.

Notes

1. Lemaire 1994a supersedes his earlier basic study of the Persian administration of Palestine and Transjordan (1990).
2. Lemaire (1994a: 51) proposes that the borders of the Persian province of Ammon corresponded to the borders of the Ammonite kingdom in the seventh and sixth centuries BC, in his view the area around modern 'Amman plus the central Jordan Valley. Unfortunately, the precise borders of the

Iron Age kingdom of Ammon are uncertain (cf. Hübner 1992: 131-57).

3. Most of the pottery from Transjordan with Persian parallels is still unpublished or illustrated only by photographs, so line drawings are difficult to illustrate here. The author thanks L.G. Herr for his assistance with the material from Tall al-'Umayri and for his permission to reproduce pottery drawings in Figure 10.2 originally published by him in Geraty *et al.* 1989 and Herr *et al.* 1991a; Jonathan Tubb for his permission to use previously unpublished photographs of pottery from Tall as-Sa'idiyya (Figures 10.3-5); and H.J. Franken for Figure 10.6.

4. The date of the al-Meqabelein tomb has been much discussed (cf. Stern 1982: 79-80). It could date anytime between the late seventh and early fifth centuries BC, with Stern preferring a mid-sixth-century BC date.

5. An unpublished conical stamp seal from Tall as-Sa'idiyya Grave 27 (cf. Tubb 1988: 74), probably to be dated to the beginning of the Persian period, depicts a horned animal with a long tail, in front a leaf or a stylized (Egyptian?) feather, and above a winged scarab. I am grateful to the excavator, Jonathan Tubb, for permission to refer to this object prior to its publication.

References

Aharoni, Y.
 1979 *The Land of the Bible: A Historical Geography.* 2nd edn. Trans. and ed. A.F. Rainey. London: Burns & Oates.

Ahlström, G.W.
 1993 *The History of Ancient Palestine from the Palaeolithic Period to Alexander's Conquest.* Ed. D. Edelman. Journal for the Study of the Old Testament, Supplement Series 146. Sheffield: Sheffield Academic Press.

Albright, W.F.
 1954 Notes on Ammonite History. In *Miscellanea Biblica B. Ubach*, 131-36. Montserrat.

Amiran, R.
 1972 Achaemenian Bronze Objects from a Tomb at Kh. Ibsan in Lower Galilee. *Levant* 4: 135-38.

Bartlett, J.R.
 1989 *Edom and the Edomites.* Journal for the Study of the Old Testament, Supplement Series 77. Sheffield: JSOT Press.

 1990 From Edomites to Nabataeans: The Problem of Continuity. *Aram* 2/1, 2: 25-34.

Beaulieu, P.-A.
 1989 *The Reign of Nabonidus, King of Babylon: 556–539 BC.* New Haven: Yale University Press.

Bennett, C.-M.
 1977 Excavations at Buseirah, Southern Jordan, 1974: Fourth Preliminary Report. *Levant* 9: 1-10.

Bennett, C.-M. and P. Bienkowski
 1995 *Excavations at Tawilan in Southern Jordan.* Oxford: Oxford University Press.

Bienkowski, P.
 1990 The Chronology of Tawilan and the 'Dark Age' of Edom. *Aram* 2/1, 2: 35-44.
 2000 Transjordan and Assyria. In L.E. Stager, J.A. Greene and M.D. Coogan (eds.), *The Archaeology of Jordan and Beyond: Essays in Honor of James A. Sauer*, 44-58. Winona Lake, IN: Eisenbrauns/Semitic Museum.

Bienkowski, P. (ed.)
 1991 *Treasures from an Ancient Land: The Art of Jordan.* Stroud: Alan Sutton.

Brown, R.M.
 1991 Ceramics from the Kerak plateau. In J. M. Miller (ed.), *Archaeological Survey of the Kerak plateau*, 169-279. Atlanta, GA: Scholars press.

Clark, V.A.
 1983 The Iron IIC/Persian Pottery from Rujm al-Henu. *Annual of the Department of Antiquities of Jordan* 27: 143-63.

Cross, F.M. and C.T. Geraty
 1994 The Ammonite Ostraca from Tell Hesban. In D. Merling and C.T. Geraty (eds.), *Hesban after 25 years*, 169-74. Berrian Springs, MI: Institute of Archaeology/Siegfried H. Horn Archaeological Museum.

Dalley, S.
 1995 The Cuneiform Tablet. In C.-M. Bennett and P. Bienkowski, *Excavations at Tawilan in Southern Jordan*, 67-8. Oxford: Oxford University Press.

Dandamaev, M.A.
 1989 *A Political History of the Achaemenid Empire.* Leiden: E. J. Brill.

Dempsey, D.
 1993 An Ostracon from Tell Nimrin. *Bulletin of the American Schools of Oriental Research* 289: 55-58.

DeVito, R.A.
 1993 The Tell el-Kheleifeh Inscriptions. In G.P. Practico, *Nelson Glueck's 1938–1940 Excavations at Tell el-Kheleifeh : A Reappraisal*, 51-63. Atlanta, GA: Scholars Press.

Elayi, J. and J. Sapin
1998 *Beyond the River: New Perspectives on Transeuphratene*. Journal for the Study of the Old Testament, Supplement Series 250. Sheffield: Sheffield Academic Press.

Eph'al, I.
1984 *The Ancient Arabs: Nomads on the Borders of the Fertile Crescent 9th–5th Centuries B.C.* Jerusalem: Magnes.
1988 Syria-Palestine under Achaemenid Rule. *The Cambridge Ancient History*, IV: 139-64. 2nd edn. Cambridge: Cambridge University Press.

Flanagan, J.W., D.W. McCreery and K.N. Yassine
1992 Preliminary Report of the 1990 Excavation at Tell Nimrin. *Annual of the Department of Antiquities of Jordan* 36: 89-111.
1994 Tell Nimrin: Preliminary Report on the 1993 Season. *Annual of the Department of Antiquities of Jordan* 38: 205-44.

Geraty, L.E. et al. (eds.)
1989 *Madaba Plains Project 1: The 1984 Season at Tell el-'Umeiri and Vicinity and Subsequent Studies*. Berrien Springs, MI: Andrews University.

Graf, D.F.
1990 Arabia during Achaemenid Times. In H. Sancisi-Weerdenburg and A. Kuhrt (eds.), *Achaemenid History. IV. Centre and Periphery*, 131-48. Leiden: Nederlands Instituut voor het Nabije Oosten.
1993 The Persian Royal Road System in Syria-Palestine. *Transeuphratène* 6: 149-68.

Grayson, A.K.
1975 *Assyrian and Babylonian Chronicles*. Locust Valley, NY: Augustin.

Greene, J.A. and K. 'Amr
1992 Deep Sounding on the Lower Terrace of the Amman Citadel: Final Report. *Annual of the Department of Antiquities of Jordan* 36: 113-44.

Hadidi, A.
1987 An Ammonite Tomb at Amman. *Levant* 19: 101-20.

Harding, G.L.
1950 An Iron-Age Tomb at Meqabelein. *Quarterly of the Department of Antiquities of Palestine* 14: 44-48.

Harding, G.L. and O. Tufnell
1953 The Tomb of Adoni Nur in Amman. *Palestine Exploration Fund Annual* 6: 48-75.

Harrison, T.P.
1994 A Sixth–Seventh Century Assemblage from Madaba, Jordan. *Annual of the Department of Antiquities of Jordan* 38: 429-46.

Hendrix, R.E., P.R. Drey and J.B. Storfjell
1997 *Ancient Pottery of Transjordan: An Introduction Utilizing Published Whole Forms*. Berrien Springs, MI: Institute of Archaeology/Horn Archaeological Museum, Andrews University.

Herr, L.G.
1992 Two Stamped Jar Impressions of the Persian Province of Ammon from Tell el-'Umeiri. *Annual of the Department of Antiquities of Jordan* 36: 163-66.
1993 What Ever Happened to the Ammonites? *Biblical Archaeology Review* 19/6: 26-35, 68.
1995 The Late Iron II-Persian Ceramic Horizon at Tell el-'Umeiri. In K. 'Amr, F. Zayadine and M. Zaghloul (eds.), *Studies in the History and Archaeology of Jordan*, V: 617-19. Amman: Department of Antiquities.
1999 The Ammonites in the Late Iron Age and Persian Period. In B. MacDonald and R.W. Younker (eds.), *Ancient Ammon*, 219-37. Studies in the History and Culture of the Ancient Near East. Leiden: E.J. Brill.

Herr, L.G. et al. (eds.)
1991a *Madaba Plains Project 2: The 1987 Season at Tell el-'Umeiri and Vicinity and Subsequent Studies*. Berrien Springs, MI: Andrews University.

Herr, L.G. et al.
1991b Madaba Plains Project: The 1989 Excavations at Tell el-'Umeiri and Vicinity. *Annual of the Department of Antiquities of Jordan* 35: 155-79.

Herr, L.G. et al.
1994 Madaba Plains Project: The 1992 Excavations at Tell el-'Umeiri, Tell Jalul, and Vicinity. *Annual of the Department of Antiquities of Jordan* 38: 147-72.

Hoglund, K.G.
1992 *Achaemenid Imperial Administration in Syria-Palestine and the Missions of Ezra and Nehemiah*. Atlanta, GA: Scholars Press.

Homès-Fredericq, D.
1996 Influences diverses en Transjordanie à l'époque achéménide. *Transeuphratène* 11: 63-76.

Hübner, U.
1992 *Die Ammoniter: Untersuchungen zur Geschichte Kultur und Religion einer transjordanischen Volkes im 1. Jahrtausend v. Chr.* Wiesbaden: Otto Harrassowitz.

Ibach, R.D.
1987 *Archaeological Survey of the Hesban Region*. Berrien Springs, MI: Andrews University.

Ibrahim, M. and S. Mittmann
1986 Al-Mugayyir. *Archiv für Orientforschung* 33: 167-72.

Ji, C.-H.C.
1998 Archaeological Survey and Settlement Patterns in the Region of 'Iraq al-'Amir, 1996: A Preliminary Report. *Annual of the Department of Antiquities of Jordan* 42: 587-608.

1999 Survey of the Dhiban Plateau. *ACOR Newsletter* 11.1: 5.

Ji, C.-H.C. and J.K. Lee
1999 The 1998 Season of Archaeological Survey in the Regions of 'Iraq al-'Amir and Wadi al-Kafrayn: A Preliminary Report. *Annual of the Department of Antiquities of Jordan* 43: 521-39.

Khalil, L.A.
1986 A Bronze Caryatid Censer from Amman. *Levant* 18: 103-10.

Kooij, G. van der
1987 Tell Deir 'Alla (East Jordan Valley) During the Achaemenid Period: Some Aspects of the Culture. In H. Sancisi-Weerdenburg (ed.), *Achaemenid History. I. Sources, Structures and Synthesis*, 97-102. Leiden: Nederlands Instituut voor het Nabije Oosten.

Kooij, G. van der and M.M. Ibrahim (eds.)
1989 *Picking up the Threads...A Continuing Review of Excavations at Deir Alla, Jordan*. Leiden: University of Leiden Archaeological Centre.

Lemaire, A.
1990 Populations et territoires de la Palestine à l'époque perse. *Transeuphratène* 3: 31-74.
1994a Histoire et administration de la Palestine à l'époque perse. In E.-M. Laperrousaz and A. Lemaire (eds.), *La Palestine à l'époque perse*, 11-53. Paris: Cerf.
1994b Epigraphie et numismatique palestiniennes. In E.-M. Laperrousaz and A. Lemaire (eds.), *La Palestine à l'époque perse*, 261-87. Paris: Cerf.

Mabry, J. and G. Palumbo
1989 Wadi Yabis Survey, 1987. In D. Homès-Fredericq and J.B. Hennessy (eds.), *Archaeology of Jordan. II/1. Field Reports: Surveys and Sites A–K*, 91-97. Akkadica Supplementum 7. Leiden: Peeters.

MacDonald, B.
1999 Tafila-Busayra Survey. *ACOR Newsletter* 11.1: 4-5.

Mare, W.H.
1989 Quweilbeh (Abila). In D. Homès-Fredericq and J.B. Hennessy (eds.), *Archaeology of Jordan. II/2. Field Reports: Sites L–Z*, 472-87. Akkadica Supplementum 8. Leiden: Peeters.

McGovern, P.E.
1989 Baq'ah Valley Survey. In D. Homès-Fredericq and J.B. Hennessy (eds.), *Archaeology of Jordan. II/1. Field Reports: Surveys and Sites A–K*, 25-44. Akkadica Supplementum 7. Leiden: Peeters.

Merling, D.
1994 The 'Pools of Heshban': As Discovered by the Heshban Expedition. In D. Merliny and L.T. Geraty (eds.), *Hesban after 25 Years*. 211-23. Berrien Springs, MI: Institute of Archaeology/Siegfried H. Horn Archaeological Museum.

Miller, J.M. (ed.)
1991 *Archaeological Survey of the Kerak Plateau*. Atlanta: Scholars Press.

al-Muheisen, Z. and F. Villeneuve
forthcoming La Jordanie aux époques héllenistique, nabatéenne et romaine (fin du IVe siècle av. J.C. – fin du IIIe siècle ap. J.C.). *Studies in the History and Archaeology of Jordan VII*.

Ottosson, M.
1993 The Iron Age of Northern Jordan. In A. Lemaire and B. Otzen (eds.), *History and Traditions of Early Israel: Studies Presented to Eduard Nielsen*, 90-103. Leiden: E.J. Brill.

Parker, S.T.
1997 Preliminary Report on the 1994 Season of the Roman Aqaba Project. *Bulletin of the American Schools of Oriental Research* 305: 19-44.

Porada, E.
1989 Two cylinder seals from 'Umeiri, Nos. 49 and 363. In L.T. Geraty et al. (eds.), *Madaba Plains Project I: The 1984 Season at Tell el-'Umairi and Vicinity and Subsequent Studies*, 381-84. Berrien Springs, MI: Andrews University.

Prag, K.
1989 Preliminary Report on the Excavations at Tell Iktanu, Jordan, 1987. *Levant* 21: 33-45.

Pratico, G.D.
1993 *Nelson Glueck's 1938–1940 Excavations at Tell el-Kheleifeh: A Reappraisal*. Atlanta, GA: Scholars Press.

Pritchard, J.B.
1985 *Tell es-Sa'idiyeh: Excavations on the Tell, 1964–1966*. Philadelphia, PA: University Museum, University of Pennsylvania.

Sapin, J.
1996 Réflexions sur des stratégies et des techniques d'adaptation dans la Transjordanie du 1er millénaire. *Transeuphratène* 11: 45-61.

Sauer, J.A.
1986 Transjordan in the Bronze and Iron Ages: A Critique of Glueck's Synthesis. *Bulletin of the American Schools of Oriental Research* 263: 1-26.
1994 The Pottery at Hesban and its Relationships to the History of Jordan: An Interim Hesban Pottery Report, 1993. In D. Merling and L.T. Geraty (eds.), *Hesban after 25 years*, 225-81. Berrien Springs, MI: Institute of Archaeology/Siegfried H. Horn Archaeological Museum.

Smith, R.H.
1990 The Southern Levant in the Hellenistic Period. *Levant* 22: 123-30.

Stern, E.
1982 *Material Culture of the Land of the Bible in the Persian Period 538–332 B.C.* Warminster: Aris & Phillips/Jerusalem: Israel Exploration Society.
1990 New Evidence on the Administrative Division of Palestine in the Persian Period. In H. Sancisi-Weerdenburg and A. Kuhrt (eds.), *Achaemenid History. IV. Centre and Periphery*, 221-26. Leiden: Nederlands Instituut voor het Nabije Oosten.

Strange, J.
1997 Tall al-Fukhar 1990–1991: A Preliminary Report. In G. Bisheh, M. Zaghloul and I. Kehrberg (eds.), *Studies in the History and Archaeology of Jordan*, VI: 399-406. Amman: Department of Antiquities.

Thompson, H.O.
1972 The 1972 Excavation of Khirbet el-Hajjar. *Annual of the Department of Antiquities of Jordan* 17: 47-72.
1977 The Ammonite Remains at Khirbet el-Hajjar. *Bulletin of the American Schools of Oriental Reaearch* 227: 27-34.

Tubb, J.N.
1988 Tell es-Sa'idiyeh: Preliminary Report on the First Three Seasons of Renewed Excavations. *Levant* 20: 23-88.

Tubb, J.N. and P.G. Dorrell
1994 Tell es-Sa'idiyeh 1993: Interim Report on the Seventh Season of Excavations. *Palestine Exploration Quarterly* 126: 52-67.

Tubb, J.N., P.G. Dorrell and F.J. Cobbing
1996 Interim Report on the Eighth (1995) Season of Excavations at Tell es-Sa'idiyeh. *Palestine Exploration Quarterly* 128: 16-40.

Yassine, K.
1984 *Tell el-Mazar Cemetery A.* Amman: University of Jordan.
1988a Ammonite Fortresses, Date and Function. In K. Yassine (ed.), *Archaeology of Jordan: Essays and Reports*, 11-31. Amman: Department of Archaeology, University of Jordan.
1988b Tell El Mazar Field I: Preliminary Report of Areas G, H, L, and M. In K. Yassine (ed.), *Archaeology of Jordan: Essays and Reports*, 73-113. Amman: Department of Archaeology, University of Jordan.
1988c Ammonite Seals from Tell el-Mazar. In K. Yassine (ed.), *Archaeology of Jordan: Essays and Reports*, 143-55. Amman: Department of Archaeology, University of Jordan.
1999 Burial Customs and Practices in Ancient Ammon. In B. MacDonald and R.W. Younker (eds.), *Ancient Ammon*, 137-51. Studies in the History and Culture of the Ancient Near East. Leiden: E.J. Brill.

Yassine, K., M.M. Ibrahim and J. Sauer
1988 The East Jordan Valley Survey, 1975 and 1976. In K. Yassine (ed.), *Archaeology of Jordan: Essays and Reports*, 159-207. Amman: Department of Archaeology, University of Jordan.

Yassine, K. and J. Teixidor
1988 Ammonite and Aramaic Inscriptions from Tell el-Mazar. In K. Yassine (ed.), *Archaeology of Jordan: Essays and Reports*, 137-42. Amman: Department of Archaeology, University of Jordan.

Younker, R.W. and D. Merling
1999 Tall Jalul. *ACOR Newsletter* 11.1: 6-7.

Younker, R.W. et al.
1990 The Joint Madaba Plains Project: A Preliminary Report of the 1989 Season, including the Regional Survey and Excavations at El-Dreijat, Tell Jawa, and Tell el-'Umeiri (June 19 to August 8, 1989). *Andrews University Seminary Studies* 28/1: 5-52.

Younker, R.W. et al.
1996 Preliminary Report of the 1994 Season of the Madaba Plains Project: Regional Survey, Tall al-'Umayri, and Tall Jalul Excavations. *Andrews University Seminary Studies* 34/1: 65-92.

Zayadine, F.
1986 L'époque perse-achéménide. In *La voie royale: 9000 ans d'art au royaume de Jordanie*, 142-51. Paris: Association française d'action artistique.

11. The Nabataeans: Travellers between Lifestyles

Stephan G. Schmid

Introduction

Nabataean studies, both in the field of archaeology and history, have expanded in the last decades. Evidence for this comes from the increasing numbers of exhibitions and publications devoted to the subject.[1] However, there are still many unanswered questions including details of chronology and general interpretation.[2] The aim of this chapter, therefore, is to report on the current state of research in most of the categories usually dealt with within Nabataean culture and to propose some general thoughts about the development of Nabataean art and culture.

A people called Nabataeans are first mentioned in 312 BC (Diodorus 19.94.1; 95.1-97.6). Diodorus, writing in the first century BC, describes an attempt by Athenaios, a general of Antigonos Monophthalmos, one of the successors of Alexander the Great, to conquer the Nabataeans. He is most probably referring to Hieronymos of Kardia, who as a historian participated in this campaign and was, therefore, an eyewitness of this first reported contact between the Nabataeans and the Hellenistic world. As the Nabataeans apparently did not have any historiography of their own, we are—other than partial information given by inscriptions—strongly dependent on the picture of them as drawn by Greeks and Romans. The first account of 312 BC depicts the Nabataeans as both nomads and skilful traders between South Arabia and the Syro-Phoenician coast. The description contains all the typical characteristics of nomads, including laws forbidding them to build houses, cultivate plants, practise agriculture, and so on. Their behaviour, while engaging the Greek troops and their mercenaries, is characteristically nomadic: they left behind their goods and families on a rocky stronghold and were on their way to an annual meeting where they conducted trade and other business. This is a typical non-sedentary feature that can be witnessed in more or less the same fashion even today on market days in Arab villages, bringing together the nomads from the surrounding areas to exchange their goods.[3] Special attention is given to Diodorus's mention of the rocky stronghold with only one easily defensible access. This description would fit Umm al-Biyara, one of the characteristic rocks surrounding the city of Petra. Some scholars claim that as-Sela' and Busayra would fit better the geographical description and distances that Diodorus gives.[4]

Some scholars believe that Diodorus's description may not reflect the Nabataeans of 312 BC but rather a collection of *topoi* used by Greeks and Romans in general to describe distant peoples not well known to them. However, this position involves at least two serious problems concerning the history and the archaeology of the Nabataeans. The first problem is determining where the Nabataeans were *before* 312 BC. Present archaeological evidence does not support the presence of the Nabataeans in central or southern Jordan, where the text seems to locate them, at such an early date. Formerly, the theory of a material continuity between Edomites and Nabataeans was occasionally supported but recently has lost most of its attraction, mainly because there is no archaeological evidence from excavations or surveys.[5] It, therefore, seems that the Nabataeans were not an indigenous population of these areas but immigrated around the middle of the first millennium BC. Recent studies based on historical and linguistic evidence favour north or northeast Arabia rather than South Arabia as their place of origin.[6]

The second problem involves tracing the Nabataeans *after* 312 BC. Although there are historical sources making clear that they have to be located in central and south Jordan (Diodorus 3.42.5; 3.43.4-5; 2 Macc. 5.8; 1 Macc. 5.25; 9.35; Josephus, *Ant.* 12.8.3), there seems to be almost no material evidence that could be assigned to the Nabataeans during the third and second centuries BC.

It is only c. 100 BC that material culture identified as Nabataean appears.[7] There are some older finds, especially from Petra, reported and connected with the Nabataeans, for example, stamped amphorae handles and occasionally black glazed pottery. However, in most cases these finds are apparently intrusive and are, therefore, not of value.[8]

The earliest material culture and the process of settlement

As previously mentioned, definitive evidence of a Nabataean material culture does not occur before the end of the second or the beginning of the first century BC. The most readily available evidence may be obtained from coins because there is usually little doubt about their date and ethnic attribution.[9] The earliest, anonymous mints are attributed to Aretas II (120/10–96 BC)[10] or Aretas III (87/84–62 BC).[11] Recent finds support the earlier dating.[12] The earliest coins—clearly modelled on Hellenistic prototypes—were minted in Damascus by Aretas III from 84 BC onwards, but probably quite soon afterwards (after 72 BC?) in Petra.[13] Not only is the style of the first Nabataean coins purely Hellenistic but they also contain Greek inscriptions. This sudden initialization of a coin mint requires an explanation that will be attempted later. Here it is sufficient to point out that the minting of coins means that there must have been some kind of infrastructure. This appears to be the case since even an irregular minting requires a political and economic overlord and a superimposed organization, both of which are difficult to imagine in the case of the non-sedentary infrastructure of a people. In this context the title 'king' of the Nabataeans must be mentioned. If the hypothesis that Knauf proposes is correct and Aretas III was the first to use the title 'king', in the Hellenistic meaning of the term, in the early first century BC, this could indeed prove a structural change in Nabataean society.[14]

A similar picture is obtained by looking at characteristic Nabataean pottery. Again, none of this pottery can be dated before the late second or early first century BC.[15] Decisive chronological evidence is provided by the fact that along with the earliest Nabataean pottery, fragments of the so-called Eastern terra sigillata A (ESA) are found. Such pottery was not produced before the very end of the second century BC. The first fine-ware Nabataean pottery, with its smoothly rounded shapes and frequently occurring reddish or brownish slip, imitates the late Hellenistic pottery of the Near East (Figure 11.1).[16] In fact, if this pottery were not known to have been found in Nabataean territory, its shape and surface treatment would not be distinctive from any other pottery of the Hellenized east.[17] The most outstanding feature of this pottery, mostly in the form of bowls, is the painting, which consists of simple lines on the inner body of the vessel (Figure 11.1.5). Moreover, it occurs at a time and in an environment that was not favourable at all for painted pottery. In Greece and the Near East, painted pottery was produced to a much lesser extent in the Hellenistic Age and began to be habitually replaced by stamped or mould-made decoration, such as the so-called Megarian bowls or Eastern terra sigillata. This development was due to economic conditions. Stamped and mould-made pottery could be produced in much larger quantities, that is, in an almost industrial way in the modern sense of the word. Therefore, these categories of mould-made bowls and terra sigillata became major export articles in the second and first centuries BC and are found all over the Mediterranean. In contrast, Nabataean pottery apparently was not produced for export, even though the potters achieved a very high technical level as we shall see later. While the shapes of the Nabataean fine-ware pottery of the first half of the first century BC are entirely Hellenistic, a determination of the painting as shown in Figures 11.1 and 11.2 is much more difficult as no immediate prototypes can be found. Keeping in mind the presumed immigration of the Nabataeans from the northern or northeastern part of the Arabian Peninsula around the middle of the first millennium BC, there may, however, be a surprising explanation. It is precisely on the southern shores of the Arabian-Persian Gulf and in Iran that very similar painting on pottery is found, dating from the second millennium BC to the fourth century BC.[18] Moreover, some Nabataean pottery forms clearly seem to be based on Mesopotamian, north Arabian even Iranian prototypes.[19] This may be the first *archaeological* evidence for the origin of the Nabataeans. It must be mentioned that from this first stage of Nabataean pottery onwards there are already closed vessels like the amphora (Figure 11.1.5), indicating that this pottery was not—as believed earlier—used for cultic purposes, but for everyday life, that is, eating, drinking and storage.

The same general picture as seen in fine-ware pottery occurs in pottery lamps. Again, no lamps dating before the very late second century BC are known from any stratified Nabataean context. The earliest lamps are either imports or more or less accurate imitations of common late Hellenistic types.[20]

With the sudden introduction of these pottery categories around 100 BC, the material culture of the Nabataeans not only has its beginning but develops amazingly to include, in its later phases, huge architectural complexes, both of profane and sacral

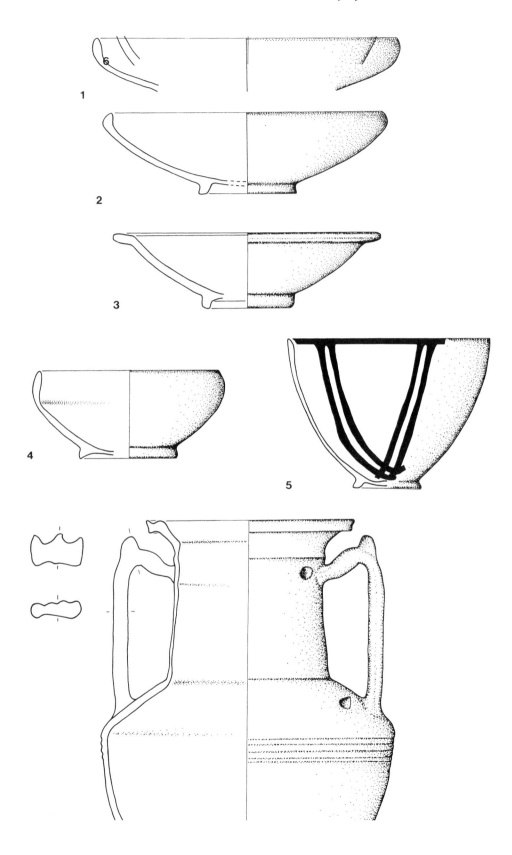

Figure 11.1. Plain and painted pottery of phase 1 (c. first half of first century BC) (drawings by S. Schmid). Scale 1:2.

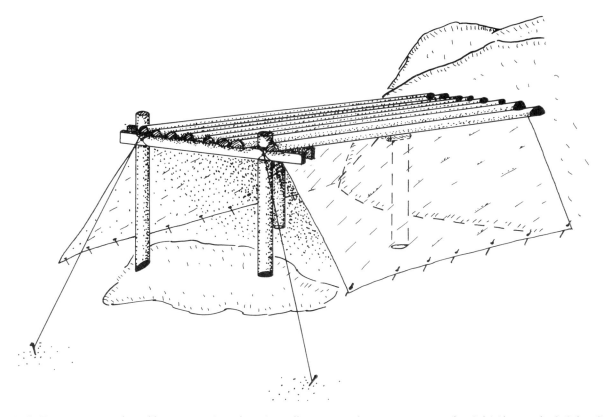

Figure 11.2. Reconstruction of possible temporary 'tent-house' installation using the carvings on top of as-Sela' (drawing by S. Schmid).

use, as well as the eye-catching rock-cut tomb façades of Petra (see below). When we try to understand the fact that the Nabataeans appear to have lived in central and southern Jordan from the time of their presumed immigration, that is, in the middle of the first millennium BC, or from their first mention in 312 BC at the latest, until the introduction of their material culture, as described above, without leaving any traces for at least 200 years, one explanation seems best.

As we have seen, Diodorus's description depicts the Nabataeans as nomads or semi-nomads. If so, then the Nabataeans would have had almost no need for a 'normal' material culture. Of course, during the entire span of this nomadic period there may have been other categories of material culture such as textiles, for example carpets and the like, or wooden tools and idols that are not preserved.[21] The absence of other categories can easily be explained by their nomadic lifestyle. For instance, nomads do not produce pottery simply because they are never long enough in one place where clay sources, water for the production of pottery, and wood for firing are available in large quantities.[22] In this case, our evidence would suggest that their settlement took place around 100 BC when they were forced to create a more permanent infrastructure in order to remain competitive in trade. The end of the second/beginning of the first century BC was marked by the decline of the Seleucid dynasty. This led to a power vacuum in the Syro-Phoenician area. This, in turn, led also to the rise of the Hasmoneans and other local powers, on the one hand,[23] and to an increase of Ptolemaic long-distance trade, due mostly to the discovery of the direct sea route from South Arabia to India, on the other.[24] Both events may have encouraged the Nabataeans to start settling on a more permanent basis in order to fill partially the gap left by the Seleucids and to challenge the Ptolemies in long-distance trade.[25] It was also during this time that historical sources report the Nabataeans as being increasingly involved in supra-regional conflicts with their neighbours. Around 93 BC the Nabataean king Obodas I defeated the Hasmonean Alexander Jannaeus.[26] The battle of Motho (c. 84 BC) between the Nabataeans and the Seleucids deserves special mention: it is not entirely clear who won the battle but the Seleucid king Antiochus XII died.[27] This resulted in the temporary takeover of Damascus by the Nabataeans under king

Aretas III[28] and had an important influence on the creation of Nabataean coinage, as indicated above, since some of the first Nabataean coins were struck in Damascus.

The presumed settlement of the Nabataeans then led to the creation of their material culture. But as there was no immediate preceding local material culture, the Nabataeans did the logical thing. They oriented their new culture according to the mainstreams of the contemporary Hellenistic world in its Near Eastern variant as it was on display in major Hellenistic cities, such as Alexandria in Egypt and Seleucia in Syria, as well as at the trading places they frequented, for example Rhodes, Kos, Priene, Milet and Delos. These and other places, according to inscriptions, are known to have been visited by Nabataeans or to have maintained contacts with the Nabataeans, as in the case of Priene.[29]

No architectural remains of this earliest phase of Nabataean settlement are presently known. It is possible that the former nomads continued living for a while in tents before they started building houses.[30] Or, for example, as in Petra, they lived in caves.[31] Since caves cannot be precisely dated, the chronological relation between houses and inhabited caves is not clear. However, some small fragments of painted stucco found in an early first-century BC context illustrate that there was already, during this early stage of settlement at Petra, a form of 'architecture'—whether built or carved in the rock is unknown—that could be decorated with paintings.[32] What a possible intermediate stage between nomadism and settlement would have looked like may be illustrated on the top of the rock of as-Sela', a candidate for the 'national stronghold' (see above). At several places at as-Sela', there are not only many huge rock-cut cisterns but also carvings in the higher rocks. These appear to be related to two postholes a few metres in front of them. Maybe the carvings and the postholes were used for the installation of temporary structures, using wooden beams, resulting in a mixture of tents and houses (Figure 11.2). To the reconstruction proposed in Figure 11.2, there should be added some kind of roofing, probably using small wooden limbs or other organic materials. But as long as there are no systematic excavations at as-Sela' (there are, however, some illicit ones), this interpretation and primarily the date of these carvings must remain hypothetical.

As the Nabataeans did not produce any historiography of their own, we are greatly restricted in our knowledge of their sociocultural organization. It should be sufficient to point out that ethnographic studies show that with the process of settlement a stronger specialization in terms of trade, crafts and administration is observed, leading, of course, to the creation of different social groups within a society.[33] In the earlier stages of settlement there were people within the community that could be called part-time sedentarized because they were involved in long-distance camel trade and, therefore, on the road for about six months of the year, while others were permanent residents.[34] With the introduction of new means of transport, most of the former camel raiders had to change their occupation and new social differentiation resulted.[35] In part, the same can be concluded in the case of the Nabataeans. With their definitive settlement c. 100 BC—at least in and around Petra—some specialization had necessarily taken place.[36] In order to maintain their economic wealth, that is, the supply of spices and other goods from South Arabia for sale in the ports of the eastern Mediterranean, some would have travelled throughout the peninsula for most of the year.[37] Others would have specialized in crafts and industry as can be seen in the case of pottery, because Nabataean pottery is certainly the product of professional potters from its first stage onwards.[38]

The monumentalization of the material culture

The chronological basis—pottery and coins

As mentioned above, no definite remains of architecture from the early period of Nabataean culture survive. The same is true for sculpture. It seems that the Nabataeans, in their first stage of settlement and production of material culture, adopted only 'smaller' categories, such as coins, pottery and terracotta. The Nabataean pottery is probably the best indicator of change and continuity since it is found in abundance and reflects easily recognizable fashions.

According to this view, the first period or phase of Nabataean pottery, as represented in Figure 11.1, lasts for roughly two generations, from c. 100 BC until the middle of the first century BC. After this sudden initiation, a development started, leading to a distinctive Nabataean style. However, external influences are not uncommon. These can be observed in fine-ware pottery in the late first century BC when some characteristic prototypes from the Roman thin-walled

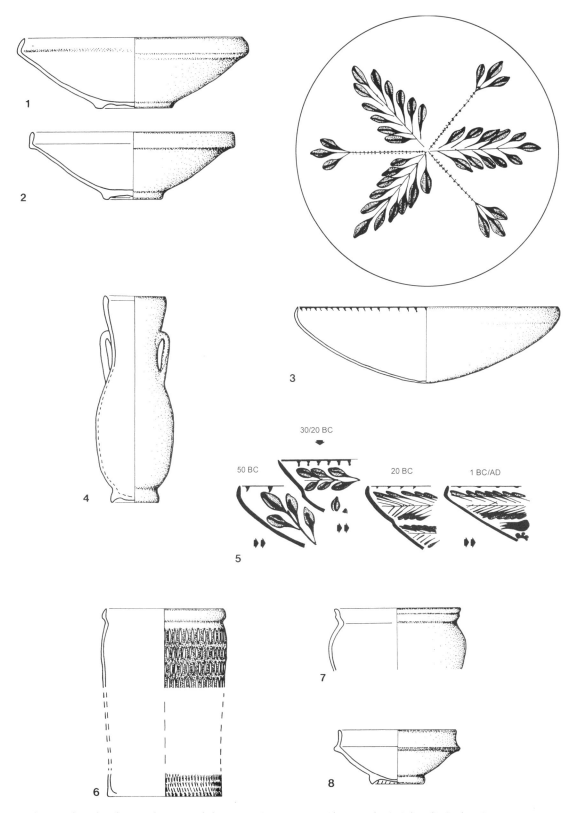

Figure 11.3. Plain, rouletted and painted pottery of phase 2 (c. 50 BC–AD 20) (drawings by S. Schmid). Scale 1:2.

and sigillata pottery are adopted, but only temporarily (Figure 11.3.6-8).[39] The main shapes, primarily open drinking bowls and plates, probably used for eating, of the second half of the first century BC and the beginning of the first century AD, are in most cases developed versions of the initial vessels of the earlier phases, but showing sharper forms and thinner walls (Figure 11.3.1, 2). However, there is a decisive change in the decorative patterns around 50 BC. The simple lines of the first phase are replaced by floral motifs, especially rows of leaves and fine limbs, across the inner body of the vessel (Figure 11.3.3). One of the characteristics of these bowls is the absence of a foot or any other standing device. As there are almost no stands found in Nabataean pottery, this means that such painted bowls were held in the hand when being filled. This is a clear indication of their function as drinking bowls, as can be seen in Greek pottery types as well, for instance in the case of *rhyta*. In general terms, the spectrum of Nabataean fine ware covers aspects of daily life well. There is a huge quantity of plates and drinking bowls (about 75%) and a smaller, but representative, quantity of flasks, jugs and other shapes that were used for serving and storage (Figure 11.3.5). Of course, cooking pots and other vessels that were used for the preparation and storage of food must be added to these forms of 'table ware'.

The occasional adoption of foreign pottery shapes also leads to one of the characteristic changes in the decoration of Nabataean bowls. Around 25 BC a flat bowl with a stepped wall is introduced into the pottery spectrum. It is clearly borrowed from Iranian or Mesopotamian prototypes, where this form was already common for centuries.[40] Since this addition to the vessel wall made it difficult to apply the radial decoration as shown in Figure 11.3.3, the decorative system changes into concentrically applied floral motives (Figure 11.3.5). Interestingly, even when the stepped wall disappears a few years later, the concentric decoration remains. A further development sees an increasing *horror vacui* of the painters, trying to cover completely the vessel's body with increasingly smaller motifs, such as dots, 'eyes', small leaves, and so on (Figure 11.3.5 right; cf. below Figure 11.23). This characteristic decoration, no longer in the bright red of the earlier painting, but in rather brownish and violet colours, marks the turning of the centuries, that is, about the first 30 years of Aretas IV's reign (9 BC–AD 40).

As observed, Nabataean pottery of the second half of the first century BC is quite sensitive to what could be called 'international fashion'. It shows at the same time, nevertheless, the first tendencies towards a distinctive Nabataean style.

Similar features can be observed in Nabataean coin minting. On the occasion of the Nabataeans' first reported clash with the western world in 312 BC, Diodorus reports that the Nabataeans were using 'Syrian' letters. This certainly means that they were writing in Aramaic, the *lingua franca* of the entire Near East at the time. The fact that their first coins bear Greek inscriptions underlines how strong and complete must have been the sudden input of Hellenistic culture at the time of their settlement. Therefore, the most remarkable step towards a typical Nabataean coin minting is certainly the change from Greek to Aramaic legends, already during the first century BC. The international component, however, is shown by the style of the coins, especially the portraits of Nabataean kings. For instance, the coins of Aretas IV (9 BC–AD 40) show the king's head in a style similar to contemporary Roman coins, although he wears the traditional Nabataean haircut with long curls at the neck.[41] A further sign of a certain Romanization in Nabataean coinage may be seen by the fact that from the year 14/13 BC on, coins of Obodas III and Aretas IV show the king crowned with a (laurel?) wreath. Earlier all the Nabataean kings were shown exclusively diademed according to the customs of Hellenistic kings. Obodas III minted coins with both types of portraits during the last five years of his reign,[42] as did Aretas IV for the first four years of his reign.[43] However, after 5 BC Nabataean kings are shown exclusively with the wreath on their coins.

Although the chronological coincidence is rather striking, it is difficult to decide whether or not the above-described situation may be more than just fashion: the years from 14/13–10/9 BC were marked by an intense struggle between the Nabataeans and Herod the Great.[44] Augustus consistently supported Herod with one exception, namely, when the Nabataean 'prime minister' Syllaios made him believe that Herod had been the aggressor.[45] Therefore, the overall situation for the Nabataeans in this conflict was rather uncomfortable during the last years of Obodas III and the early reign of Aretas IV, because Rome was on the side of their Jewish opponents. The situation was so bad that Augustus was apparently thinking about donating the Nabataean kingdom to Herod. It is said that only the continuing problems within Herod's family kept Augustus from doing so (Josephus, *Ant.* 16.10.9 [353-55]). Considering their situation, it must have seemed appropriate for the Nabataean kings to show their devotion to Rome and the *princeps*—at least initially. It could very well be that Obodas III and Aretas IV did so by replacing the traditional Hellenistic diadem with the typical Roman wreath. In light of this, it is certainly no coincidence that the ambassadors sent to Rome by Aretas IV brought Augustus an expensive golden wreath, symbol of the ruler's legitimization. It is also significant that Augustus did not accept the gift because he was angry at the Nabataean ruler for ascending to the throne without asking for his approval. This puzzling event shows that, indeed, there must have been a kind of dependence on Rome at the time (Josephus, *Ant.* 16.9.4 [296]).

Architecture: houses and temples

Pottery and coin minting, therefore, clearly show the reception of international fashion on the one hand and the development towards a distinct Nabataean style on the other. The most striking feature in Nabataean material culture of the late first century BC and the early first century AD, however, is the tendency towards monumentalization. As mentioned above, there are hardly ever any traces of architecture and sculpture in the previous periods. Again, as in the case of pottery and coins, we witness a sudden beginning of these categories apparently falling in the last quarter of the first century BC, with an additional boost in the early years of Aretas IV.

It is from this period that the first houses so far excavated at Petra date.[46] From examples presently known, it becomes clear that there is a kind of social differentiation, as houses clearly differ in size and ornamentation. If we try to trace back the structure and ground plan of these early houses we witness a phenomenon similar to that of the earliest Nabataean pottery and coins. Here again, there are strong indications that the Nabataeans were inspired by the current type of (late) Hellenistic houses in the Near East. One of the most characteristic features in this context is certainly the peristyle courtyard, as seen in the houses at az-Zantur and al-Khatute (see Figure 11.4 for their locations). While the house at al-Khatute gives an almost perfect example of a Hellenistic peristyle house, including the huge cistern under the main courtyard, such as are on display, for instance, on the island of Delos,[47] the case seems to be somewhat different for the house at az-Zantur. There, the open areas allowing light and fresh air to circulate into the inner rooms behind the columns are not in the centre of the house but rather pushed towards the periphery. This, as well as the not very systematic alignment of the smaller rooms, may well be the original Nabataean contribution to their house types.[48] However, very prominent examples like the famous villa at Boscoreale (Italy) or some houses from Pergamon show that even in such cases the courtyards could be decentrated.[49] In general terms, it seems that living in Petra during the early imperial years was not much different from life in the big centres of the eastern Mediterranean or the Near East.[50]

One of the best examples of this may be the huge dwelling recently investigated on the southern terrace of az-Zantur (Figure 11.5).[51] Its position covering the entire natural terrace, and thus offering a splendid view over the suburbs and the rows of rock-cut façades of the entire southern part of Petra, brings to mind the rich houses of Delos or Ephesus.[52] Beside its impressive architectural remains, this dwelling will become famous first of all because of its interior decoration, especially the wall paintings. Before taking a closer look at the paintings it should be mentioned that they come from the lower level of the house, that is, the ground floor. But as the house is built partially into the natural slope, the ground floor in some parts becomes the substructure for the upper floor. Therefore, most rooms of the upper floor would probably have belonged to a splendid reception area. This is supported by the finds such as parts of very fine mosaic floors belonging to the first floor![53] For the time being, we can only imagine what the upper floor may have looked like, but the finds from the ground floor more than compensate for this lack. Well-preserved wall paintings were found (Figure 11.6) in room 1 (cf. the plan Figure 11.5). The paintings illustrate illusionistic architecture similar to the stucco decoration from the Qasr al-Bint (cf. below) and are topped off by a decoration in the form of the so-called masonry style.[54] Room 6 contained a great mass of richly decorated stucco fragments that once covered the entire area. The semicolumn with floral elements and gilded painting as well as the fragment of a moulded dentil cornice with painting in the form of fine tendrils illustrate the high living standards of the villa of az-Zantur. Relative to the house's date, a fragment of painted pottery of phase 3a was found in one of the stucco fragments, thus giving a *terminus post quem* of c. AD 20 for the entire decoration.[55] This date, together with the fact that different styles of wall painting were apparently used in the building, along with the so-called masonry[56] and second Pompeiian (Figure 11.6) styles, at a time when they were both no longer in use in the main centres, may seem a problem. However, similar out-of-fashion elements were observed in such prominent buildings

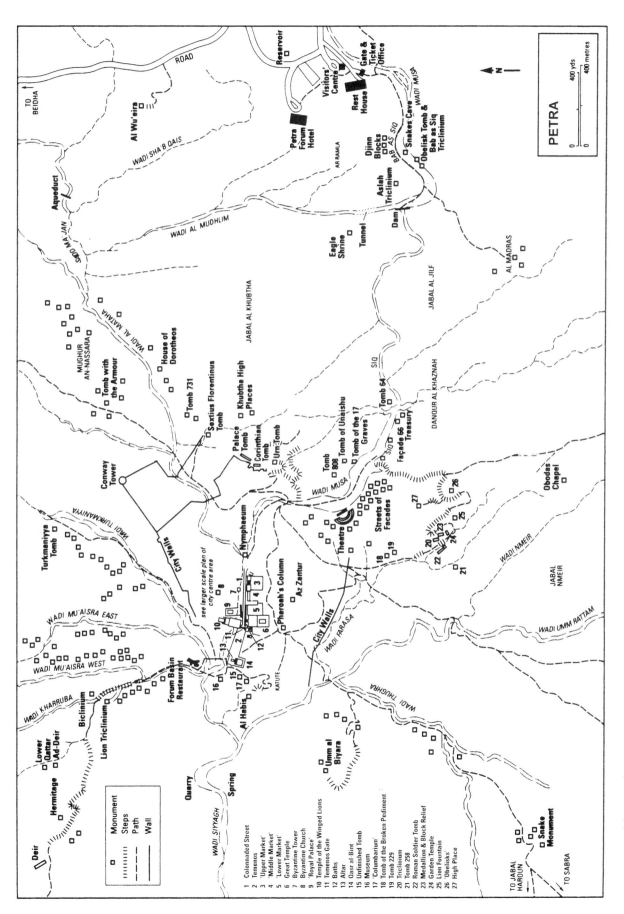

Figure 11.4. Map of the city of Petra (taken from Blue Guide Jordan, by Robert Smith © A & C Black [Publishers] Limited; Rollin and Streetly 1996: 212)

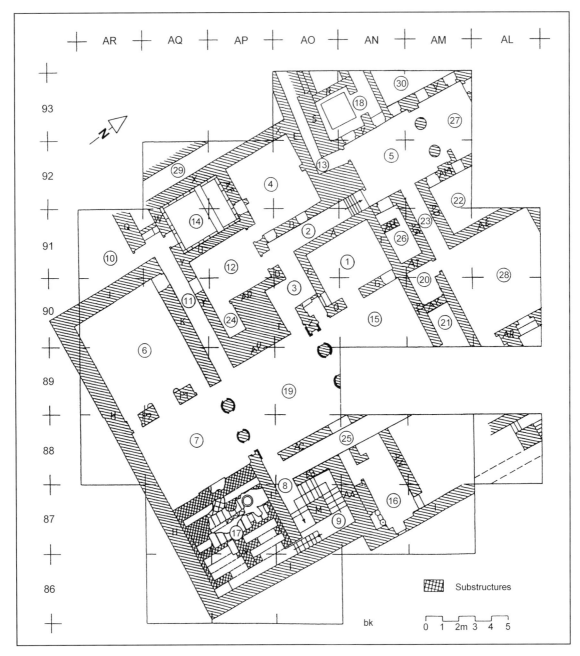

Figure 11.5. Schematic plan of the Nabataean villa on the south slope of az-Zantur, Petra (drawing by B. Kolb).

as the palaces of Herod the Great.[57] According to the overall picture obtained from Nabataean art in this general period, it seems we should grant the Nabataeans enough cultural autonomy to have deliberately selected such features, rather than labelling them provincial retards.[58]

To draw conclusions about the cultural heritage of a population in general, it may not be a good idea to begin with their cultic buildings, because these may very well contain some special characteristics that are related to religious purposes. So far, we have seen how the general development of Nabataean material culture in the areas of coins, pottery and private architecture seems to follow some general rules or models. The case of temples is, however, somewhat special.

Like the first private dwellings, the first built temples do not seem to appear before the late first century BC. There is, however, another aspect of Nabataean religion that may find its first manifestation somewhat earlier.

Figure 11.6. Petra, Nabataean villa on az-Zantur south terrace. Paintings on walls C/A in room 1 (cf. Figure 11.5), representing illusionistic architecture topped off by so-called masonry style (photo by O. Jaeggi, courtesy of B. Kolb).

Figure 11.7. Petra, offering place on top of Zib Atuf, so-called High Place (photo by S. Schmid).

The Nabataeans were known to worship their gods on so-called high places that could be located on any natural elevation. In fact, their principal deity—Dusares or Dushara, 'the one from Shara'—owes his name to a mountainous region.[59] Especially at Petra, many structures have been found on the tops of almost every rock. This shows an active religious life and the worship of gods in a kind of natural setting.[60] These structures usually consist of a kind of altar, a cistern and an entire system of water basins and channels, probably related to a procedure, that is, the pouring of a liquid such as wine, water or the blood of animals offered to the deity. One of the best-preserved examples of this specific aspect of Nabataean religion is located on top of Zib Atuf (Figure 11.7).

Concerning the temples, as in all the other aspects of material culture investigated so far, we witness a sudden flowering. The first huge buildings seem to appear out of nowhere, that is, we do not see any small predecessors but the development seems to start directly with such large scale buildings as Qasr al-Bint at Petra (Figure 11.8a.6).[61] Research on Nabataean temples is progressing but, to date, it is difficult to draw an overall picture.[62] At an early stage in the history of Nabataean research, temples were roughly divided into two groups, a northern and a southern one, as they are grouped in Figure 11.8a and b. Recently, a much clearer differentiation is proposed and future research may change that picture.[63]

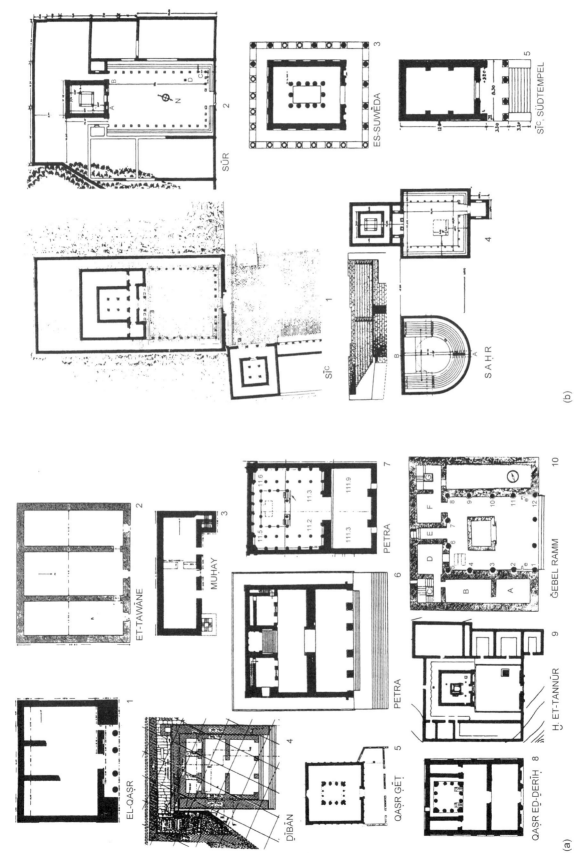

Figure 11.8. (a) Nabataean temples, 'southern' group (after Wenning 1990: fig. 6.4); (b) 'northern' group (after Wenning 1990: fig. 6.2).

A common feature of most typical Nabataean temples (Figure 11.8a) is the fact that they are more square than rectangular as are, for instance, Greek and Roman ones. This is especially true for the innermost part of the building, the *cella*, that in some cases is even wider than it is long (Figure 11.8a.3, 4, 6, 8), although it is sometimes subdivided into smaller units.[64] A further non-Greek or Roman characteristic is a kind of *podium*, the so-called *motab*, in the temple (Figure 11.8a.5, 7-10). This may be connected to the preference of the Nabataeans to worship their gods on the tops of mountains, as illustrated by the so-called high places (above). In fact, the temple of Khirbat at-Tannur, situated on the top of a steep hill, without any other related structure, is a kind of built high place. The aspect of the elevated platform for religious purposes may, therefore, stand behind the *podia* in the above-mentioned temples. This aspect appears even clearer at Qasr al-Bint at Petra (Figure 11.8a.6) where on both sides at the rear steps lead to real platforms. Although we do not know how the cult was practised in Nabataean temples, the steps leading to platforms or even onto the top of the building readily recall much older Mesopotamian temples. In fact, Strabo states that the Nabataeans 'worship the sun, building an altar on the top of the house, and pouring libations on it daily and burning frankincense' (Strabo, *Geogr.* 16.4.27). Such installations in private houses are yet to be found, but the practice described by Strabo is well confirmed by the structures at high places and at Qasr al-Bint.

As mentioned above, Syrian and Mesopotamian forerunners have been suggested to explain the above.[65] Recently, however, other possible prototypes for Nabataean temples have been discussed. Beginning with the aspect of the 'shrine in a shrine', that is, a smaller structure within a larger one, as is best demonstrated at Qasr al-Get and in Wadi Ramm (Figure 11.8a.5, 10), but elsewhere as well, Egyptian and, more specifically, Ptolemaic temples are compared to the Nabataean ones.[66] We may mention here that many Egyptian temples, both from Pharaonic times and the Hellenistic period, have stairs and huge ramps leading to the roof (cf. Figure 11.9.c, d).

Before returning to the possible Egyptian influence on Nabataean temples, the other main group, that is, the so-called northern group (Figure 11.8b). In contrast to the southern group, these temples show a much stronger long-rectangular ground plan. A further characteristic in most cases are huge courtyards with porticoes in front of the temple proper. That the geographical subdivision is rather artificial is shown by the recently investigated 'South Temple' at Petra (Figure 11.9.a, b), belonging rather to this group than to the former.[67] Mainly for the latter group, prototypes from Syria have been cited so far. One may also add that the strong tendency to emphasize the building's front by huge steps (Figure 11.8b.5) and courtyards in front of the temple is a typical Roman feature. Indeed, most of the parallels quoted from Syria already date from a time of considerable Roman influence in the Near East.[68] Since the above-mentioned theory about Ptolemaic influence on Nabataean temples seems to be presently *en vogue*, we may add a few more thoughts on the subject. The large courtyard surrounded by porticoes in front of the temple and the general installation of the temple building itself in a huge architectural complex can easily be found in almost every Egyptian temple from the Ptolemaic period, for example, the one from Edfu (Figure 11.9.c, d).[69]

Since the development of Ptolemaic temples follows a strong pharaonic tradition, it has nothing in common with Nabataean cult buildings. Therefore, we should be careful about pointing out too strong a connection between them. However, it is worth mentioning that all characteristics of Nabataean temples, that is, the corridor or passageway in the inner building, the steps leading to a platform or to the roof, and the courtyard in front of the temple, can be found in prototypes from Hellenistic Egypt. In general, the manifold influence from Ptolemaic Egypt, especially on the huge tomb façades of Nabataean Petra (see below), shows that a cultural interchange existed, regardless of the economic and political differences and quarrels.

Not only Egyptian, but also possible South Arabian influences have been suggested for Nabataean temples.[70] Since then, more work on these monuments has been done and, indeed, there are common features, such as the tripartite backside of the buildings, the inner courtyard (Figure 11.10.a, b), the additional shrine (Figure 11.10.b), and the general quadratic aspect (Figure 11.10.c). The problem with the South Arabian temples as possible prototypes is mainly a chronological one, since they are dated c. 700 and in the seventh century BC[71] and are therefore considerably earlier than the Nabataean ones. However, there is evidence in some cases for their use until the fourth century BC and there are also considerably later buildings in the same tradition.

As for the dating of the Nabataean temples considered so far, in most cases the situation is far from clear. However, it seems that a significant number of these buildings were constructed at the end of the

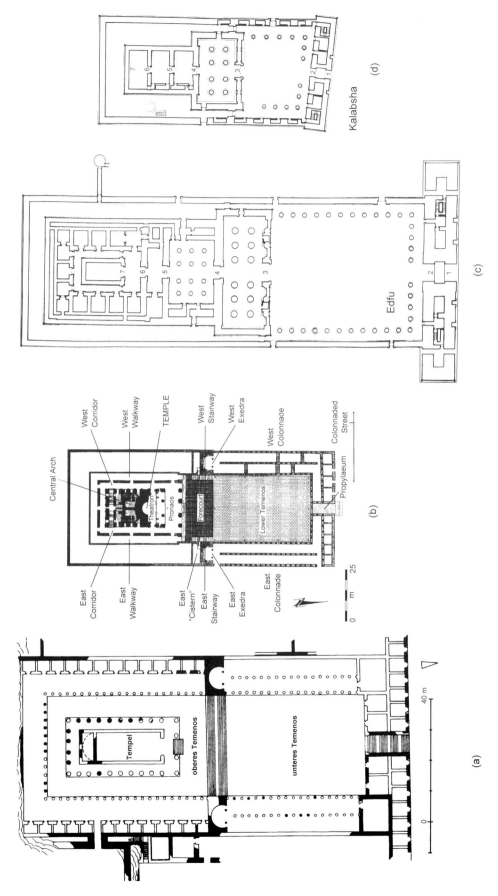

Figure 11.9 (a) Petra, 'South Temple' according to Bachmann *et al.* 1921 (after Weber and Wenning 1997: 81); (b) Petra, 'South Temple' according to recent excavations (after Joukowsky 1999: 196 fig. 1; reproduced with permission); (c) and (d) Ptolemaic temples of Edfu and Kalabsha (after Siegler 1969: fig. 8).

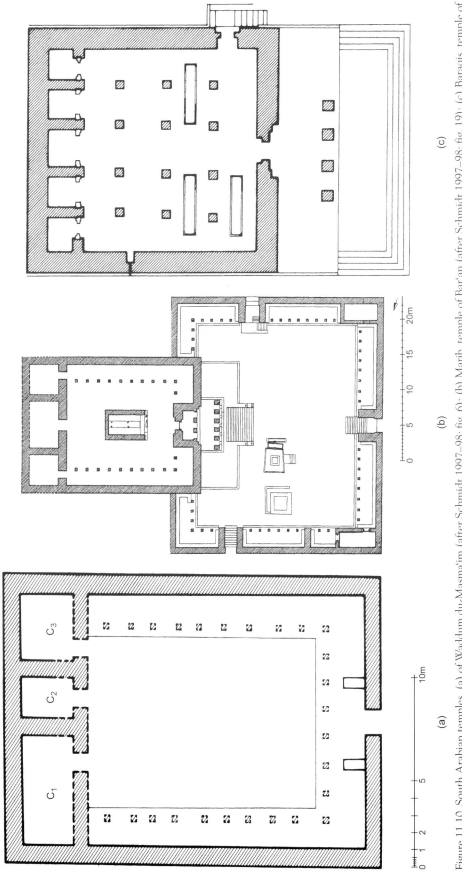

Figure 11.10 South Arabian temples, (a) of Waddum du-Masma'im (after Schmidt 1997–98: fig. 6); (b) Marib, temple of Bar'an (after Schmidt 1997–98: fig. 19); (c) Baraqis, temple of Nakrah (after Schmidt 1997–98: fig. 26).

Figure 11.11 Petra, relief slab with head of Gorgo Medusa as shield sign (photo by S. Schmid).

first century BC and the beginning of the first century AD. The most prominent case may be the Qasr al-Bint at Petra (Figure 11.8a.6), where we have some indications for a date in the late first century BC.[72] The same date is proposed for the first construction phase of the temple in Wadi Ramm (Figure 11.8a.10).[73] The 'temple of the winged lions' was built, according to the archaeological evidence, during the early years of Aretas IV (Figure 11.8a.7).[74] As for the 'South Temple' (Figure 11.9a, b), one will have to wait for a final publication. It is worth mentioning, however, that under the pavement of the temple forecourt no pottery later than phase 2b was found, giving a *terminus post quem* of 1 BC for the initial construction of the monument.[75] The excavators dated the temple from Dhiban to c. AD 10. This is confirmed by the pottery analysis of the construction layers.[76]

Architectural sculpture

Because of the available evidence, in a consideration of the development in sculpture we may focus mainly on reliefs rather than on sculpture in the round. Since the territory of the Nabataeans did not provide any marble quarries, only a very small number of marble sculptures in the round are known. Since these may well be imported, they can hardly be studied for any stylistic development in Nabataean art. On the other hand, we do possess an important number of relief slabs, made of sandstone and limestone, from different places that allow us to trace their development.[77]

For the study of chronological development, two groups of sculpture are of particular interest, one from Petra and the other from Khirbat at-Tannur. In accordance with the organization of this chapter, these will be treated separately in two subsections. We will, however, have to compare them directly in order to work out the stylistic and chronological differences. From near the temenos gate at Petra comes a complex of relief slabs that is known, after the date of its discovery, as 'the 1967 group of sculpture'.[78] To this group belongs a series of representations of weapons and sea creatures from a triumphal procession (Figure 11.11) and a series of blocks with representations of busts, mainly of gods (Figure 11.12). To these can be added, on the basis of stylistic comparisons, a third that shows cupids and garlands (Figure 11.13). They have not, however, been found with the others. Before we come to the absolute date of these sculptures, some common features must be stressed. All show a very realistic style in representing humans (or human-like gods), animals and floral decoration. Particular attention is paid to small details as in the case of the cricket on the garland of Figure 11.13. Also, the very fine cloth and hair on Figure 11.12 make the entire group comparable to the best examples of late Hellenistic and early imperial

Figure 11.12. Petra, relief slab with representation of a veiled muse or goddess (?) (photo by S. Schmid).

sculpture. On the basis of such characteristics, two more pieces may be added: another slab from near the temenos gate with the representation of a winged female torso (Nike?) on Figure 11.14, and the bust of a god (Dusares? Baalshamin?) in a medallion (Figure 11.15) that was found near the mosque in Wadi Musa.[79]

It is generally accepted that a date in the late first century BC, or at the turn of the century at the latest, would fit these reliefs. A confirmation for this dating can be found in one slab with the representation of the god Helios that reportedly was found in debris belonging to the frieze of Qasr al-Bint.[80] Most of these sculptures show rather fleshy faces, as can be seen, for example, on Figure 11.11. This is a characteristic feature of late Hellenistic sculpture, especially of the Near East, where such fleshy forms were considered a sign of richness and wealth leading even to *epitetha* of kings such as Tryphon and Physkon.[81] On the other hand, there are some parallels to the sculptural fashion of the middle Hellenistic period. Examples include the deep set eyes and carvings of the head (Figure 11.15), the particular naturalistic representation of the dress (Figure 11.14) that seems to be pressed against the body by the wind of the quick moving forward Nike (?), or the 'wild' hair of the Gorgo (Figure 11.11).[82] The eclectic character of the different groups, but also within single groups, combining features of different phases, that is, high or middle Hellenistic as well as late Hellenistic elements, can be seen as further confirmation of their date. It is precisely during the very late first century BC or the beginning of the first century AD that such features are very prominent in the centres of the ancient world, when a certain playful handling of the different styles was *en vogue*.[83] It seems, once again, that the Nabataeans were very well familiar with the international trends and features and—this being the most important aspect—willing to adopt them.

Although we do not know what was the exact context of most of these fragments, a few thoughts on their original function may be added. As seen with the bust of Helios, the panels with representation of humans or gods (Figure 11.12) may probably come from the decoration of Qasr al-Bint. For the group with weapons, there is good evidence that they once belonged to another richly decorated monument, because one slab is the edge of a deeply broken pediment. To this piece, the one with the head of Medusa (Figure 11.11) and a few others may also be joined.[84] This pediment may have been part of the front of either a predecessor of the present temenos gate (see below) or of another building. Of special interest is the fact that with this evidence we witness the use of a *built* broken pediment such as that represented on the huge tomb façades (below) at a very early date. The Triton and the weapons point towards a political alliteration of the original monument that could have been connected to a kind of naval victory by a Nabataean king, although most scholars think that it has to be seen as a reference to the victory of Octavian, the later *princeps* Augustus, at the battle of Actium in 30 BC.[85] Whatever the case, we see that during this period at Petra not only does the style of sculpture definitely follow international outlines, especially late Hellenistic and early imperial stylistic tendencies, but also thematic alliterations. The same is true for the slabs with cupids and garlands, as they are undoubtedly referring to a kind of golden age that supposedly flourished in the reign of a Nabataean king, probably Aretas IV (9 BC–AD 40).[86]

Figure 11.13. Petra, relief slab with cupid holding a garland (photo by S. Schmid).

Figure 11.14. Petra, relief slab with a winged female torso (Nike?) (photo by S. Schmid).

Tomb façades

For the modern visitor, the most eye-catching monuments of the ancient Nabataean capital of Petra are without doubt the large tomb façades hewn out of red sandstone. They can be roughly divided into two groups: 'more oriental' and 'more Hellenistic-western'.[87] Their attraction certainly existed in antiquity as well, maybe even in a more forceful way, because they were most probably once covered with painted stucco and, therefore, must have offered a very spectacular sight.[88]

The date of most of the tomb façades of Petra is difficult to establish. Such is only possible on the

Figure 11.15. Petra, relief slab from Wadi Musa with a medallion of a god (?) (photo by S. Schmid).

Figure 11.16. Petra, Al-Khazna Faraoun (photo by S. Schmid).

basis of stylistic comparisons since only very few are dated by inscriptions. The only other place where Nabataean tomb façades occur, Hegra in Saudi Arabia, offers a much better chronological frame, as inscriptions date many of these monuments. Unfortunately, at Hegra only the simple monuments occur and, therefore, the huge two- or more-storey, richly decorated monuments like al-Khazna Faraoun (Figure 11.16) or ad-Dayr (Figure 11.33) remain somewhat enigmatic. An attempt is made here to establish a chronological sequence for the tomb façades as well as placing them in the context of Nabataean material culture.

In discussions on Nabataean tomb façades, one of the earliest monuments, the now destroyed tomb of Hamrath at Souweida, although its Nabataean origin may be doubtful, is usually not considered.[89] On the basis of present evidence, this rectangular monument, divided by semicolumns and adorned with weapon reliefs, should be dated to the early first century BC, and it may very well be that such monuments provide the missing link to the Hellenistic world in Nabataean sepulchral architecture. For instance, on the island of Rhodes, several monuments show similar characteristics.[90] An example of the same rectangular construction on a stepped base, subdivided by semicolumns, is seen in the so-called Ptolemaion in the Rhodini park.[91] As the most probable reconstruction of its upper part is a stepped pyramid, it becomes clear that the 'Ptolemaion' stands in the tradition of such monuments as the Mausoleum at Halikarnassos.[92] This may also be the explanation for the tomb of Hamrath and other similar constructions. They represent a mixture between traditional funeral architectural elements and a strong Greek influence.[93] Another funeral monument from Rhodes even shows further parallels to the tomb of Hamrath as it was originally decorated with weapon reliefs very similar to the ones from Souweida.[94]

Several other monuments from Petra, although not as spectacular as the huge main façades and, therefore, not as often considered, belong to the same category of eclectic monuments, as Zayadine has pointed out.[95] Recent investigations at Jarash show that such free-standing monuments—in this case a two-storied round monument in the shape of a *tholos*, quite similar to the upper part of al-Khazna, Corinthian tomb and ad-Dayr—existed elsewhere in Jordan too.[96]

Al-Khazna Faraoun[97] (Figures 11.16, 11.19), the treasury of the Pharaoh, is the façade with the richest decoration and one of the best preserved, due to its location at the end of as-Siq, where it was well protected from corrosion by wind and rain over the centuries. The lower storey shows a slightly broken entablature and a pediment supported by six columns, while the upper storey represents a *tholos* surrounded on three sides by columns evoking a peristyle courtyard. As the front side had to be opened in order to offer an insight on the *tholos*, a deeply broken entablature and pediment was created. The whole monument is richly adorned with floral and figurative elements. Similar architectural comparisons to al-Khazna can be found on Roman wall paintings from Italy belonging to the so-called second Pompeiian style.[98] The question, then, is how to explain these similarities, and further, what is represented by these two categories of monuments, that is, al-Khazna on the one hand and the Roman wall paintings on the other? Although there are still attempts to explain the architectural wall paintings of the second Pompeiian style as representations of theatrical *scenae frontes* and, therefore, as not depicting real but fanciful architecture, it can be assumed on better evidence that they actually represent real architecture, especially the luxury architecture of Hellenistic palaces.[99] This, and the architectural comparisons referred to below, suggests that al-Khazna also reflects real architecture in a direct way and not, as sometimes supposed, a two-dimensional transformation of architectural complexes that in fact lie behind each other.[100]

If we are looking for comparisons showing similar architectural features like al-Khazna Faraoun, we have to consider first the 'Palazzo delle Colonne' in Ptolemais (Cyrenaica).[101] Lauter has shown that the basic elements of this monument go back to late Hellenistic times. Crucial for its dating are the capitals that can profitably be compared with examples especially from Italy and Alexandria,[102] as well as with figurative capitals from Messene, dating to the late Hellenistic period.[103] Almost all the architectural elements described above as being characteristic of al-Khazna are found at the 'Palazzo delle Colonne' as well. Examples include deeply broken entablatures and pediments in the upper storey of a façade facing an inner courtyard.[104] Especially interesting is a small rounded architectural fragment that most probably can be reconstructed as part of a *tholos*,[105] which could—according to its small size—very well be placed in the upper storey, as is the case at Petra. Since the works of Pesce and Lauter, several studies presenting more material from Egypt, especially from Alexandria, strengthen the hypothesis of a close connection between the 'Palazzo delle Colonne' and Ptolemaic palace architecture.[106]

This seems to be true as well for Qasr al-'Abd at 'Iraq al-Amir, situated between 'Amman and the Jordan Valley (Figure 11.17).[107] The monument, probably a kind of summer residence, can be attributed to Hyrcanus of the Tobiads' dynasty and is, therefore, dated to the years 182–175 BC. In this case, the affinities with al-Khazna can be seen in the general aspect of a lower, massive storey, with only a single entrance, combined with an upper storey with several windows and *intercolumnia*. Further, the positions of the figurative decorations show astonishing correspondences, although at 'Iraq al-Amir the lions are not placed on the top of the pediments. Of special interest for our further consideration is a particular feature of the Qasr al-'Abd. The entire building was placed in an artificial lake, accessible only by a path from the East.[108] The lake was filled by water from sources on the slopes surrounding the building and brought underground to two fountains, in the form of panthers, on the east side of the palace, close to its northeastern corner, and on the west side, close to its northwestern orner.[109] Therefore, the whole monument would have had the aspect of a ship floating in a lake that was filled by itself.[110] This makes it a good comparison for the luxury ship of Ptolemy IV, the *thalamegos*, which will be discussed below. An affiliation to Alexandria is also confirmed by the architectural elements, primarily by the capitals.[111] Alternatively, there are some characteristics in the general aspect of the façade that put Qasr al-'Abd in a close relationship with palace architecture that we know from Macedonia and Asia Minor.[112]

While not much has survived from the great royal palaces of the Ptolemies and Seleucids, we possess rather well-preserved archaeological remains of the palaces of the Hasmoneans[113] and of Herod the Great.[114] Of primary interest for our purpose is the so-called Northern Palace at Masada built c. 30–20 BC.[115] This monument shows a spectacular composition of three different building structures on three terraces of the steep northern slope of the citadel. The lower terrace bears a central rectangular hall surrounded by porticoes and some adjacent rooms. A huge *tholos*, offering a great view, stood on the middle terrace,

Figure 11.17. 'Iraq al-Amir (near 'Amman), Qasr al-'Abd (photo by S. Schmid).

while a semicircular balcony occupied the upper terrace with additional rooms at the back. One is especially struck by the similarity of the lower and middle terraces with the composition of al-Khazna when seen from the front,[116] proving again the connection between al-Khazna and (late) Hellenistic palace architecture.

Additional information may be gained from the description of the *thalamegos*, the famous riverboat of Ptolemy IV (Figure 11.18).[117] On the one hand, the recent discoveries in Macedonia make clear that the ship was strongly related to Macedonian luxury architecture.[118] On the other hand, it offers a few astonishing parallels for the façade of al-Khazna. We must first mention the general aspect of the façade with its *propylon*, flanked on both sides by colonnaded two-storied porticoes, which seems to be, according to the description, almost the same.[119] However, the most striking point is the mention of a *tholoid* temple (*monopteros*) of Aphrodite on the upper floor, flanked by dining rooms and bearing a marble statue of the goddess, and, therefore, very similar to the statue of a goddess, probably Isis-Tyche, in the *tholos* of al-Khazna.[120] If we imagine a section through the boat at the position of the *tholos* of Aphrodite, we would have exactly the façade of al-Khazna.[121] The description of details, such as capitals decorated with ivory and gold[122] and columns inlaid with precious Indian stones,[123] may very well be realistic and representative of Hellenistic palace architecture. This may also be seen in similar representations on the wall paintings at the *triclinium* in Oplontis or at the *oecus* of the Casa del Labirinto in Pompeii.[124]

In the light of recent discoveries in as-Siq at Petra, it is worth returning to the general aspect of al-Khazna once again. During cleaning and excavation works in as-Siq, much additional information on the water supply of ancient Petra was gained. Although not yet excavated, the square in front of the monument deserves special attention. The excavators of as-Siq, Bellwald and Keller, strongly suggest that the different water supplies leading towards the city's centre formed an artificial lake or pool just in front of al-Khazna. If this hypothesis is correct, the overall aspect of al-Khazna would be almost the same as that of the luxury ship of Ptolemy IV or of the Qasr al-'Abd, floating in an artificial lake. For the visitor stepping out of as-Siq, the huge façade reflected in the lake would have created the same impression.

Thus, we have demonstrated that al-Khazna Faraoun at Petra stands in a proper Hellenistic tradition, reflecting truly the palace architecture of the Hellenistic East. However, this is of no importance for a precise dating because Hellenistic elements could be conserved for quite a long time, especially in the East.[125]

Comparisons using the ornamental motifs of al-Khazna may be more promising. The floral ornaments of the capitals of al-Khazna (Figure 11.19)

Figure 11.18. The riverboat (*thalamegos*) of Ptolemy IV (after Nielsen 1994: fig. 71).

and the very similar capitals from the area around Qasr al-Bint (Figure 11.20) are usually compared with those of the Ara Pacis in Rome.[126] However, a closer look shows that the tendrils of the Ara Pacis are finer and intertwined in a complex way,[127] while the floral ornaments of al-Khazna and around Qasr al-Bint are more fleshy and straight in their composition and show a deeper relief. Better comparisons are offered by the painted tendrils from *oecus* 13 in the house of Augustus on the Palatine hill and ornaments in stucco from the upper floor of the same house, dated to c. 30 BC or immediately after.[128] In general, late republican Roman monuments come closer to the ornaments from al-Khazna, such as the ornamental decoration of a general's burial monument from the Via Appia, dating to c. 35 BC, may show.[129] The same stylistic characteristics can be found on several tendril fragments from Pergamon, which can be dated between c. 50 BC and the very early imperial years.[130] Indeed, relations between Petra and Pergamon have also been supposed on the basis of fragments of friezes from Petra that show garlands and weapons.[131]

We may therefore conclude that al-Khazna Faraoun was built in the second half, probably in the third quarter of the first century BC, that is, about the same time that most of the above-mentioned Roman walls were painted. As the architectural decoration of al-Khazna (e.g. the capitals) have close parallels from the area around Qasr al-Bint, we again have an indication that in the late first century BC manifold building activities

took place in Petra (cf. below). In this context, it has to be stressed that decoration in stucco showing broken and rounded pediments very similar to most of the monuments referred to so far, including al-Khazna, is applied on the exterior south wall of Qasr al-Bint.[132] If we bear in mind that a date in the late first century BC was convincingly proposed for that building, we would again have a good confirmation for the date of al-Khazna.[133] To this, we can add the broken pediment with figurative decoration (Figure 11.11) and thus we have, more or less, the contemporary appearance of this particular architectural feature in three areas: tomb façades, stucco decoration, and free-standing monuments, all dated to the late first century BC.

Following general models of evolution, it is believed that the simpler façades, showing stronger 'oriental' influences, as in Figure 11.21, would be older than the richly decorated 'western' ones like al-Khazna. After establishing a more or less precise date for al-Khazna, we should return to the dating of the simpler monuments. The inscriptions of the Hegra necropolis have already been mentioned. Starting with an inscription of the year 1 BC/AD, the inscriptions date a number of such 'oriental' monuments to the entire first century AD.[134] Another chronological indication is given by the theatre of Petra. As the theatre was built most probably in the first century AD and as the installation of its *cavea* destroyed several of these simpler tombs, they were clearly cut into the rock before the construction of the theatre.[135] The result

Figure 11.19. Petra, Al-Khazna Faraoun: three-quarter column- and quarter pilaster-capital from the lower storey (photo by S. Schmid).

Figure 11.20. Petra, tendril capital from the area around Qasr al-Bint (photo by S. Schmid).

of this short *tour d'horizon* is rather sobering: there is no good reason to believe that the simpler 'oriental' façades would predate such monuments as al-Khazna or the tomb of Hamrath. They could, at best, be contemporary. Therefore, we witness again a sudden appearance of a 'new' category of Nabataean material culture, clearly adopting Hellenistic elements.

Preliminary conclusions

The evidence listed in the previous sections shows that towards the end of the first century BC and the very beginning of the first century AD, Nabataean material culture passed to a stage of monumentalization. Houses, temples, rock-cut tomb façades, and

Figure 11.21. Petra, row of tomb façades south of the theatre (photo by S. Schmid).

architectural decoration are suddenly and prominently introduced into the repertoire of Nabataean artists and craftsmen. As with the sudden beginnings of the earliest remains of Nabataean material culture, coins and pottery, around 100 BC, one is surprised by the dynamic evolution. At least for modern scholars, it seems that these milestones could be related to an overall organization, since in both cases the sudden developments seem to come out of nowhere, that is, without any preceding 'test phase'. Just as the first pottery and coins fit perfectly into the Hellenistic *koiné* of the Near East, the same is true for the houses, the temples, and the tomb façades. In most cases, however, it is more difficult to point out direct prototypes.

The same scenario appears again in the general development of the city of Petra. For the buildings on the colonnaded street leading along Wadi Musa (no. 1 on Figure 11.4), as well as for the impressive pavement of as-Siq, including the water channels and dams, a date in the late first century BC can be established.[136] This shows that the entire city was indeed subject to a construction boom and must have been a huge building site for years, if not decades, until Qasr al-Bint, the temple of the winged lions, the 'South Temple', and so on were finished. The same seems to be true for smaller settlements. The first building phase of the Nabataean temple at Wadi Ramm is dated to around the turn of the eras and there, too, we see that at about the same time the entire infrastructure becomes rather monumental, including sophisticated water channels, some of them still *in situ*.[137] Maybe even this factor of monumentalization of public space can be considered in relation to the adoption of a kind of Near Eastern Hellenistic *koiné*, because it is precisely this aspect that marks Hellenistic cities, especially in Asia Minor.[138]

As indicated, there is some occasional Roman influence, for instance in the field of pottery production (Figure 11.3.6-8), but also most probably in the political alliterations of some of the relief-decorated monuments in the centre of Petra (Figures 11.11, 11.13). The stronger Roman influence appears with the taking over of the Seleucid empire by Pompey in 63 BC and of Ptolemaic Egypt by Octavian in 30 BC and his attempts to control directly long-distance trade in the Arabian peninsula as reflected by the expedition of Aelius Gallus.[139] It is clear that the increasingly stronger Roman influence in the Eastern Mediterranean since Pompey's campaigns in the 60s of the first century BC led also to new cultural elements in these regions. It is also during the early Roman imperial years that historical sources report direct contacts between Nabataeans and Romans.[140] Strabo not only witnesses the presence of many Romans and other foreigners in Petra by his time (*Geogr.* 16.4.21), but also notes houses of stone (16.4.26) as well as a plentiful supply of water and gardens in the city (16.4.21). He thus confirms entirely the picture obtained from archaeological evidence.

The later first century AD and the end of the Nabataean kingdom

The chronological basis: pottery and coins

Compared to his predecessors, the portraits of Rabbel II (AD 70–106) show a continuing dissolution of the naturalistic design in favour of a stronger ornamentalization. On most coins, Rabbel's face has a somewhat 'wooden' touch. This is due to the fact that only the basic lines occur in a heavily carved style. As a matter of fact, this seems to be a characteristic of several peripheral cultures of the wider Mediterranean area. The same evolution, from naturalistic to ornamental, may be traced within south and east Arabian coins, imitating Athenian and other Greek prototypes,[141] or within Celtic coins, imitating prototypes of Philip II of Macedonia and others.[142] Moreover, the culture of the Scythians and other peoples from the Black Sea area shows a similar approach to once naturalistic prototypes, mainly from the Greek world.[143] Within these cultures in particular we can find further confirmation for the model of settlement as outlined above for the Nabataeans. In these regions, where circumstances preserved more artifacts of organic material, it can be shown that a large part of the original material culture of these 'mounted' or nomadic peoples indeed consisted of carpets, wooden tools and the like.[144]

In the second and third quarters of the first century AD, the output of Nabataean pottery probably reaches its peak. Nabataean potters are now producing a great quantity of many different small forms, mainly jugs, juglets and *amphoriskoi*, many of them decorated with rouletting or stamped patterns (Figure 11.22.3-7). During the entire first and the beginning of the second centuries AD, the unpainted main forms are shallow plates with prominent vertical rims (Figure 11.22.1, 2). During the same period, the most prominent painted form is a flat open bowl, without any standing device, and a small, knobby rim (Figure 11.22.8). In the second half of the first century AD, a remarkable change in the decorative patterns of painted pottery can be observed. Up to this time, the Nabataean potters and painters preferred different small elements, such as quite realistic ivy and other leaves, covering the entire vessel (Figure 11.23). Now they use fewer but larger motifs, offering more space between them. Also, with the exception of the pomegranate, the floral elements lose much of their realism, becoming more abstract (Figures 11.22.8; 11.24). At about the same time, new, larger geometric motifs are introduced. The beginning of this new style falls almost exactly into the time of the accession to the throne of Rabbel II and can be dated to c. AD 70/80.[145] As we will see later, this development seems also to have some astonishing parallels in other categories.

Following the general evolution in the Mediterranean area, lamps found at Nabataean sites from the late first century BC and the beginning of the first century AD are either Roman imports or imitate Roman prototypes.[146] It is only towards the second quarter and, to a greater extent, in the second half of the first century AD that a characteristic Nabataean lamp type, the so-called Negev 1 type, makes its appearance.[147]

In general terms, one could say that until the third quarter of the first century AD there is a rather smooth development based on the previous phases as outlined above. From c. AD 70/80, a distinctive and different fashion appears.

Sculpture, reliefs and architectural decoration

As noted above, there is an early group of architectural decoration, both ornamental (Figures 11.19, 11.20) and figurative (Figures 11.11-11.15), that is clearly related to a stylistic background that could be called either late Hellenistic or early imperial, according to different parallels.

If we compare these sculptures and reliefs to the sculptural finds from Khirbat at-Tannur, we see obvious stylistic differences.[148] Their clearly less naturalistic and somewhat wooden faces and dresses reminded Glueck of Palmyrene and Parthian sculpture.[149] Furthermore, at Khirbat at-Tannur, several construction phases were observed, although their date was not easy to establish. A few reliefs, such as the so-called Atargatis panel (Figure 11.25) and a bust of Helios, could be attributed to an earlier phase, called phase II at Khirbat at-Tannur, while, for instance, the Nike supporting a zodiac (Figure 11.27) belongs to an intermediate phase IIA. The pieces from phase II and IIA show clear stylistic differences from the Petra sculptures discussed above. Glueck dated this phase to the years of Aretas IV and, therefore, contemporaneous with these reliefs from Petra. If this is correct, we would need to assume a provincial kind of style in order to explain the

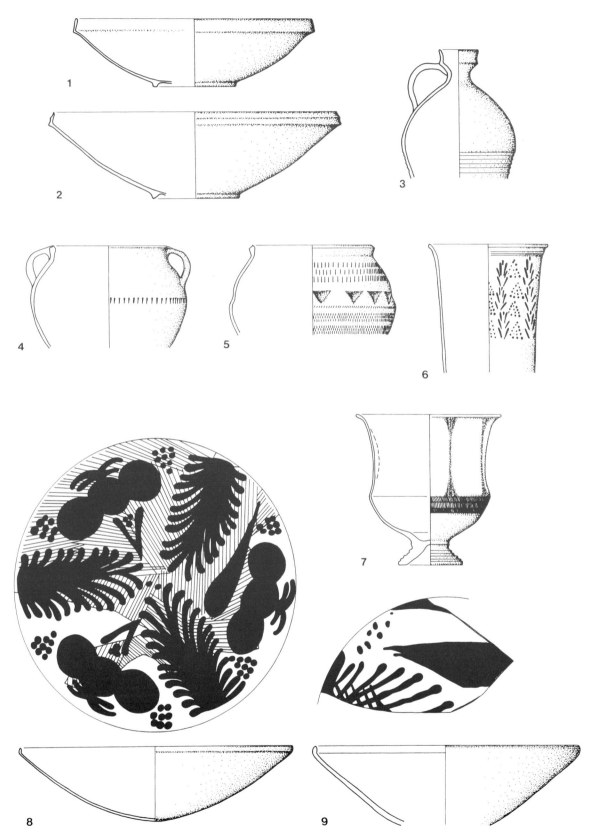

Figure 11.22. 1-8: Plain, rouletted, stamped and painted pottery of phase 3a-b (c. AD 20–100); 9: Painted bowl of the second/third century AD (drawings by S. Schmid). Scale 1:2.

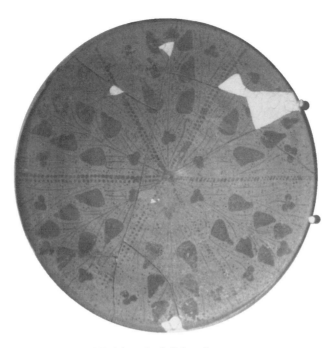

Figure 11.23. Painted bowl of phase 3a (c. AD 20–70/80) (photo by S. Schmid).

Figure 11.24. Painted bowl of phase 3b (c. AD 70/80–100) (photo by S. Schmid).

stylistic differences from Petra. Of further interest are the reliefs from phase III at Khirbat at-Tannur, including the famous dolphin and grain goddesses (Figure 11.28), which clearly show stronger stylistic differences from the earlier series of reliefs from Petra. Glueck suggested a date, without giving proof, in the early second century AD.[150] In addition, it needs to be mentioned that all the pottery illustrated in Glueck's reports on Khirbat at-Tannur and thought to be evidence for dating belongs to our phase 3c and would, therefore, confirm a date in the early second century AD.[151] Recently, the early date for phase II at Khirbat at-Tannur has been

Figure 11.25. 'Amman (Museum), Atargatis panel from Khirbat at-Tannur (photo by S. Schmid).

Figure 11.26. 'Amman (Museum), architectural element with a bust of Hermes/al-Khutbay from the temple at Khirbat adh-Dharih (photo by S. Schmid).

questioned and it seems that this phase belongs to the second half of the first century AD.[152]

Rather close in style to the sculptures from phases II and IIA at Khirbat at-Tannur is another series of relief blocks found in Petra in the destruction debris of the above-mentioned temenos gate.[153] A sample is the winged head with fleshy face and small curls (Figure 11.29) that Zayadine identified as Hermes/al-Khutbay.[154] To this piece may be added two heads of bearded men with a so-called Phrygian hat and strongly ornamental hairstyle, maybe pointing to Syrian or Mesopotamian influence.[155] These pieces, with their somewhat stiffly carved features, are not only related to the later pieces from Khirbat at-Tannur but are also clearly different from the earlier series from the same area (above, Figures 11.11-11.15). As this later phase of the temenos gate has a most probable *terminus post quem* of AD 76, this would confirm the observations made so far.[156] One of the closest parallels to the stylistic features of the sculptures from phase IIA at Khirbat at-Tannur was recently found at Khirbat adh-Dharih (Figure 11.26).[157] The relief slab, with the representation of Hermes/al-Khutbay, comes so close to the Atargatis on Figure 11.25 in the workmanship of the thick haircurls, the prominent setting of the eyebrows, the indication of the iris, the fleshy cheeks and so on, that we can probably conclude that it comes from the same workshop. Although not precisely dated due to its reuse in a Byzantine wall, the Hermes from Khirbat adh-Dharih is a very good example of Nabataean sculpture of the middle or second half of the first century AD.

The evolution in Nabataean sculpture follows, therefore, the same model as other categories of material culture, such as coins or pottery. It starts with a naturalistic style, clearly adopted from late Hellenistic and early imperial examples (Figures 11.11-11.15), developing into more abstract and stronger ornamentalizing representation (Figures 11.25-11.29). The zenith of this evolution is exemplified in the aniconic *baityloi* (Figure 11.30). Although they existed in earlier times, they too show considerable development in the latter part of the first century AD.[158] On the other hand, what is depicted on Figures 11.25, 11.26 and 11.29 could belong, according to their stylistic features, to

Figure 11.27. 'Amman (Museum), relief of a winged Nike supporting a zodiac (the upper part is at Cincinnati) from Khirbat at-Tannur (photo by S. Schmid).

Figure 11.28. 'Amman (Museum), relief of the so-called dolphin goddess from Khirbat at-Tannur (photo by S. Schmid).

an intermediate phase, distinguishing them both from the very naturalistic earlier group and from the stiffly carved later pieces.

Exactly the same general tendencies are seen in the floral decoration of capitals and other architectural features. The capitals and leaves decorating the temenos gate at the entrance to the area of Qasr al-Bint at Petra (Figure 11.31) show obvious and strong stylistic differences from the same elements of al-Khazna (Figure 11.19), from the temple of the winged lions and other earlier monuments (Figure 11.20). The capitals from the temenos gate are rather close to the ones from Khirbat at-Tannur,[159] making clear that they must belong to about the same period. Similar floral representation from the painted decoration of the Nabataean house on the north slope of az-Zantur may underline the late date of the temenos gate. In this case, a coin of Rabbel II (AD 70–106) was found in the plaster. Even when considering the difficulties in comparing painted with carved decoration, it is obvious that the style of the painting comes rather close to the floral decoration from the temenos (Figure 11.31).[160] This is further confirmed by the observation that the stylistic developments pointed out so far lead to the decorative elements of monuments such as the mausoleum of Qasr Nuweijis in the vicinity of 'Amman. Although not precisely dated, this funeral monument certainly belongs to the middle imperial years, that is, the second or third centuries AD, and shows similar ornamental details.[161] Comparisons with the architectural decoration of the Hauran region from the second and third centuries AD point in the same direction,[162] as do further comparisons with other Nabataean monuments.[163]

Tomb façades

This short overview on the Nabataeans and their culture is not the place to present a large number of Nabataean tomb façades, but just a few selected ones. The earlier monuments of this category stand in proper Hellenistic tradition. Here, a few later examples of about the second half of the first century AD are provided.

Figure 11.29. Petra, 'winged head', probably Hermes/al-Khutbay (photo by S. Schmid).

Figure 11.30. 'Amman (Museum), aniconic *baitylos* found in the temple of the winged lions (Petra), representing the goddess of Haiyan, son of Nayibat, according to the inscription (after Weber and Wenning 1997: fig. 81).

The Corinthian Tomb (Figure 11.32, second from left) is located in a prominent place among a row of façades on the north side of the al-Khubtha massif. It offers a full view of the city. The tomb's façade is of interest because its upper segment is an exact replica of al-Khazna, without, however, the sculptural décor. Its lower segment has eight semicolumns supporting a broken entablature.[164] Differences consist of two smaller entrances on the two *intercolumnia* on the left, one with a segmental pediment that is set back by two additional columns, and an architrave at the main entrance, bearing a strong resemblance to illusionistic architecture. In addition, there is a middle zone, an attica, with a central segmental pediment, different zones of broken miniature or dwarf entablatures, columns and a final broken pediment.

The lower and middle levels of the Corinthian Tomb have an almost exact parallel in the so-called Bab as-Siq triclinium just outside Petra.[165] This is important since an inscription attributed to Malichus II (AD 40/44–70) can date this tomb.[166]

The additional features of the Corinthian Tomb, such as miniature and illusionistic architecture, cannot help in dating the monument, since these features are also known from Hellenistic architecture. Some fragments of the Palazzo delle Colonne already prove the existence of miniature architecture in late Hellenistic times.[167] Illusionistic and false architectural elements, even with segmental pediments similar to those in Petra, can be found in Alexandrian architecture from the Hellenistic period onwards. These elements are usually thought to come from tombs from Alexandrian cemeteries, especially from the Mustafa Pascha necropolis.[168] However, the same elements occur in small altars and lockers for tomb *loculi* from different cemeteries dating from the Hellenistic period onwards.[169]

In general, the entire relief of the Corinthian Tomb is much flatter than that of al-Khazna and the whole monument shows less three-dimensional aspects. It can, therefore, be concluded that it should be dated somewhat later. The best evidence for dating is provided by the Corinthian capitals, being very similar

Figure 11.31. Petra, temenos gate, upper part of pilaster and capital (photo by S. Schmid).

Figure 11.32. Petra, west slope of al-Khubtha with row of tomb façades (photo by S. Schmid).

to those of the temenos gate, dated most probably to post-AD 76.[170] It is probably not incorrect to assume a date in the middle or in the third quarter of the first century AD for the Corinthian tomb.

Ad-Dayr (Figure 11.33) received its name, 'the monastery', from its use as a Christian church in late antiquity.[171] As with the Corinthian tomb, ad-Dayr too portrays considerably fewer three-dimensional elements in its lower storey compared to al-Khazna. This is due mainly to the missing propylon. The upper storey is again an almost exact parallel to al-Khazna. A difference consists in the fact that on both sides additional pillars and semicolumns have been added. The strongly broken entablature shows that this new feature

Figure 11.33. Petra, tomb façade of ad-Dayr ('the monastery') (photo by S. Schmid).

should be understood as running around the portico surrounding the *tholos*. Ad-Dayr has Nabataean capitals without floral ornaments and, therefore, does not offer evidence for dating.[172] In general, besides the Doric frieze in the upper storey, there is no additional decoration. However, in antiquity there were probably statues standing in the two niches of the lower and the three niches of the upper storey.[173] The façade of ad-Dayr appears much heavier and monumental, especially when compared to the general aspect of al-Khazna.

In the case of ad-Dayr, the additional pillars and semicolumns that frame the upper storey may give a chronological indication. Such elements, without any practical function in real architecture, are a distinguishing feature of Roman architecture of the last quarter of the first century AD onwards. These elements are a result of the interchange between traditional Roman architecture and Hellenistic luxury architecture, mainly of the East, such as the above-mentioned monuments.[174] The same elements can be found at the nymphaeum in Milet (c. AD 80) or at the library of Celsus in Ephesus (c. AD 113–117).[175] Comparable developments may also be observed in private architecture, mainly in villas of the second century AD.[176] Another good example is the monument of Philoppapos at Athens.[177] Until now, the façade of this tomb was usually considered as reflecting architectural elements of theatrical *scenae frontes*,[178] as formerly suggested for second style Pompeiian wall paintings.[179] In accordance with the evidence given above, it is much more likely that the monument of Philoppapos refers to luxury architecture as well, all the more so if one bears in mind the history of Philoppapos, being a descendant of the royal family of Commagene.[180] The date of AD 114–116 for Philoppapos's funeral monument corresponds perfectly to the above-outlined development of middle imperial architecture in the east.[181]

It seems, therefore, that a date not earlier than the last quarter of the first century AD may be proposed for ad-Dayr. Speculation about ad-Dayr being a monument erected during the reign of Rabbel II for the cult of the deified Nabataean king Obodas I or Obodas III should be treated with caution.[182] The hypothesis that ad-Dayr could have served as a cenotaph for the last Nabataean king, Rabbel II (AD 70–106),[183] is also speculative.

As far as the more complicated façades of al-Khazna and other monuments are concerned, it has been shown that they accurately reflect the very rich architecture of their times, that is, the late Hellenistic years in the case of al-Khazna or the years around AD 100 in the case of ad-Dayr.[184] Therefore, al-Khazna reflects the palace architecture of the Hellenistic Age, as do the rich private houses of Delos, for instance the 'House of the Hermes'.[185] Such characteristic features as the two (or more) storied peristyles of Hellenistic palace architecture are not only copied in the houses of the living

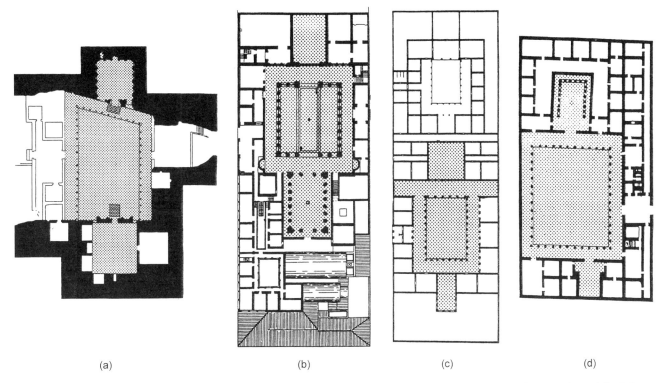

Figure 11.34. (a) Petra, Roman soldier tomb and triclinium 235 (after Schmidt-Colinet 1981: 78 fig. 19); (b) Ptolemais, Palazzo delle Colonne (after Nielsen 1994: 147 fig. 78); (c) Reconstruction of Vitruvius's description of the Greek house (after Nielsen 1994: 139 fig. 72); (d) Jericho, Herod's first winter palace (after Nielsen 1994: 196 fig. 105).

but also in the last resting places of the dead. This can be seen in examples from the Mustafa Pascha necropolis at Alexandria or from Nea Paphos on Cyprus.[186]

It is interesting to note that quite a few of the Nabataean tombs containing western elements show a strong tendency to evoke a third dimension, that is, at least the illusion of a peristyle courtyard, one of the most typical features of Hellenistic houses and palaces. Al-Khazna has a sort of small courtyard incorporated into its ground floor and evokes a peristyle in its upper storey (Figure 11.16). The Urn tomb (Figure 11.32, right) shows additional colonnaded *stoai* at its sides,[187] while the so-called Roman soldier tomb, together with the opposite *triclinium* 235 and structures between them, formed an entire complex (Figure 11.34a),[188] which is comparable to the most spectacular Hellenistic palace architecture known, such as the Palazzo delle Colonne (Figure 11.34b), the description of Ptolemy IV's *thalamegos*, or some of the Judaean palaces (Figure 11.34d). All of these are large-scale interpretations of Vitruvius's description of the Greek house (Figure 11.34c).[189] It is not only such richly decorated tombs that offer comparable structures, but rather 'traditional' monuments (but still with a pediment) too, like tomb 813.[190] Most probably, many more Nabataean tombs would prove to contain three-dimensional features if they had not been eroded or covered by sand and earth. This readily confirms the information gained from the Turkmaniye tomb's inscription, which mentions dining rooms, gardens and grottoes, all belonging to one complex.[191] We therefore see that much of the formerly enigmatic Nabataean funerary architecture in fact follows quite well the common shapes and forms of the Hellenized and Romanized Near East. In the extension towards a third dimension in the form of courtyards, *stoai* and peristyles, either in reality or evoked in relief, we obtain a new criterion for a classification of the Nabataean tombs. This is the case since these features clearly reflect the above-mentioned tendencies in Hellenistic palaces, houses and funeral monuments.[192]

Regarding the observed tendencies in pottery and architectural decoration, we witness a comparable evolution in the tomb façades, at least when comparing some richer examples like al-Khazna (Figure 11.16), the Corinthian tomb (Figure 11.32, second from left) and ad-Dayr (Figure 11.33). The huge and probably undecorated spaces of ad-Dayr's surface as shown on Figure 11.33 would, therefore, correspond to the large-scale ornamental decoration in Nabataean pottery of phase 3b and 3c (Figures 11.22.8; 11.24), that is the last quarter of the first and the beginning of the second centuries AD. Conversely, the very rich decoration of al-Khazna, covering and filling almost every available space (Figure 11.16), could very well be paralleled by the contemporary pottery of the late first century BC, where we see the same tendency towards a *horror vacui*, that is, the complete filling out of space available for decoration. This tendency also continues into the very early first century AD (Figures 11.3.5 on the right; 11.23).

As seen above, a good deal of the simpler tomb façades, showing stronger oriental influence in the form of steps on roofs, or evoking a rather tower-like aspect, are likely to date from the first century AD. It would, of course, fit well the general picture drawn so far, especially if we could see an increase of stronger Arabic elements for the *later* first century within the tomb façades too. This would fit perfectly with similar trends in sculpture or coin minting and even in pottery. Beside the usual stylistic arguments, there is another favourable element: logically the rich upper class of Petra would have tried to occupy topographically dominating places for their tombs. The best location for such façades is undoubtedly al-Khubtha north slope where the so-called kings' tombs, the Corinthian, Urn, and Palace, are placed (Figure 11.32; cf. the map Figure 11.4). All these tombs clearly show elements of elaborate Hellenistic architecture. Most of the orientalizing tomb façades are located in the different wadis leading away from the city centre (cf. Figure 11.4). One is tempted to posit that they were built as the city continued to grow and, therefore, not exclusively according to a social order but at least partially according to a relative chronological sequence.[193]

Renovation and annexation

The characteristic changes in Nabataean art and culture towards the last quarter of the first century, as outlined above, have been noted previously. This, along with some historical reflections, led to the hypothesis of a cultural and religious *renovatio* under the last Nabataean king, Rabbel II (AD 70–106).[194] It has been suggested that Rabbel II, who received the throne name 'the one who renewed and rescued his people', was responsible for a kind of official programme leading away from the previous realistic and 'western-oriented' representations of human figures towards a stronger 'oriental' traditional and abstract iconography as the prominent *baityloi* illustrate (Figure 11.30). Indeed, it is quite striking to see that, at least in cases where they can be more or less dated, the aniconic *baityloi* date to the later first century. In the earlier phases of Nabataean art, however, much stronger realistic features were used in order to represent humans and gods (Figures 11.11-11.15). Further evidence for a kind of struggle in Nabataean society about such features is seen in the fact that in the temple of the winged lions a kind of iconoclasm, which destroyed the earlier naturalistic decoration, seems to have taken place towards the later first century AD.[195]

It is, of course, very difficult to decide whether the observed changes in Nabataean art, as in reliefs or architectural decoration, do indeed fall together within a programme by Rabbel II. The difficulties come mainly from ignorance of the exact chronology of the first appearance of stronger ornamentalizing artistic features, for instance, in the representation of humans and gods on relief slabs (Figures 11.25-11.29) or *baityloi* (Figure 11.30).[196] However, there may be some relevant indications. As mentioned, the coins of Rabbel II do indeed show to a great extent less naturalistic features in the portrait of the king than did the coins of his predecessors. Additional information can be gained from the painted pottery, where the remarkable change from naturalistic, small-scale motifs to ornamental, large-scale decoration cannot only be paralleled with the development in other categories of Nabataean art, but has, above all, quite precise dates. Such pottery is found alongside characteristic forms of Eastern sigillata and coins of Rabbel II. In addition, Nabataean pottery on sites with a good historical chronological *terminus*—such as Masada—proves that this change happened around AD 70/80 and, therefore, precisely during the early years of Rabbel II's reign. As the conventionally applied term of *renovatio* for these new features within Nabataean culture of the later first century AD suggests, it is believed that the Nabataeans somehow returned to their indigenous roots with these new artistic expressions. This model demands two prerequisites: first, there was an earlier cultural and religious background to which they could return; and second, the intermediate-Hellenistic and Roman influenced period was only a superficial one.[197] If we stick to the evidence, this model necessarily encounters problems. First of all, about the entire 'pre-Hellenistic' period—not in chronological terms but in means of cultural aspects—we can only guess. This is because there is no written evidence that is peculiarly Nabataean that would enlighten us about the presedentarized Nabataeans. Moreover, the earliest material evidence already belongs to a completely

Hellenized form (cf. above). Further, the idea about *renovatio* was related to the assumption that both the 'western'-oriented programme in sculptural decoration and other elements around the turn of the eras, as well as the rather 'oriental' or 'traditional' programme under Rabbel II, were imposed by the ruling class, if not by the king himself.[198] As emphasized above, there may indeed be some allusions to a political programme in major categories of art, such as architectural sculpture. But what about 'smaller' categories, for example, coins and pottery? As for coin minting, we can suppose a strong programmatic influence by the ruler on how to design his portrait and on what attributes to choose. But can we imagine Obodas III indicating to the Nabataean potters which forms of the Roman thin-walled pots they should produce (Figure 11.3.6,7)? Or did Aretas IV chose the specific sigillata forms (Figure 11.3.8) to be copied by Nabataean workshops under his rule? Most probably these rulers did not, nor did Rabbel II decide on how to change the decorative patterns on painted pottery (Figures 11.22.8; 11.24). But as such 'unimportant' categories of crafts and workmanship are not only the best dated but also reflect the everyday culture of the common people, we can use them as a test of the model outlined above.

If we are dealing with a programme imposed by the authorities, we would expect at least some opposition or a contradictory tendency somewhere. The contrary, however, is the case. Pottery, coins, figurative decoration, and so on always seem to form a unity. Thus, if we do not want to conclude that the Nabataean kingdom was an 'Orwellian' state, we have to posit that both tendencies discussed here, that is, the 'western' and the 'oriental', were not imposed by the ruling class but were well based in the entire population, or at least the majority. As suggested above, the settlement around 100 BC was most probably the result of economic considerations combined with the increasing vacuum of power in the region by disinterest and weakness of at least one of the former supra-regional powers, the Seleucid empire. The archaeological evidence as well as the testimony of Strabo is proof that the new lifestyle was a complete success. Around the turning of the eras, Petra was a flourishing city of monumental public buildings and lavishly decorated residences. Huge funerary monuments were constructed or were under construction everywhere. Thus, a general economic wealth and welfare can be deduced.[199] But, early on, Strabo testifies to the importation of goods (*Geogr.* 16.4.26). Although he is referring to luxury items, this could very well be a clue for understanding further developments. As nomads, the Nabataeans had been, to a great extent, self-sufficient. Now, they became increasingly dependent on imported goods and trade.[200] As long as they had a wide selection of trading partners, the Nabataeans would have been in a rather strong position and would have been able to dictate prices and trading conditions. But with the increasing Roman political overweight in the region, they necessarily came under economic pressure too. Whether or not Rabbel II, indeed, tried to move his capital from Petra to Bosra,[201] the entire picture shows that the Nabataeans were struggling not only for political, but also for economic survival. It seems that under the reign of Rabbel II agricultural activities improved considerably.[202] This shows that the Nabataeans tried to turn again towards self-sufficiency in order to balance the loss in trade. We see therefore that the former advantage, that is, the better contacts, and so on, with the sedentarized infrastructure and increasing specialization, suddenly turned into a serious disadvantage.[203] As is evident from modern parallels, it must have been such elements of social and economic pressure that led to a more 'traditional' stream within Nabataean society and culture, rather than an imposed programme.[204]

In AD 106 the Roman emperor Trajan decided to incorporate the Nabataean kingdom into the Roman Empire.[205] The Roman annexation of the Nabataean kingdom and the creation of the province of Arabia is generally believed to have occurred peacefully. The movements of their forces show that the Romans were at least prepared for possible resistance. Not only had the III Cyrenaica from Egypt and probably the VI Ferrata from Syria sent considerable detachments into Arabia, but two additional *cohortes* (I Hispanorum and I Thebaeorum) were transferred to Judaea on the eve of the annexation.[206] Thus, the Nabataeans were hemmed in on three sides. Two Safaitic inscriptions, unfortunately undated, mention 'the year of the Nabataean war' and another one reports 'the year when the Nabataeans revolted against the people of Rome'.[207] These graffiti, together with the archaeological evidence, suggest that the annexation did not occur without military confrontation. According to archaeological results, it seems that places like Oboda, Moje Awad, Khirbat adh-Dharih, Dhiban and Sbaita were destroyed in the very early second century AD. In Petra, too, evidence of a contemporary destruction was reported on several occasions.[208]

A hostile component of the Roman takeover seems to be confirmed by a passage in Ammianus Marcellinus stating that, 'It [the Nabataean kingdom] was given the name of a province, assigned a governor, and compelled to obey our laws by the emperor Trajan, who, by frequent victories crushed the arrogance of its inhabitants...'[209] Furthermore, the promotion of a military tribune from the III Cyrenaica as well as the *ornamenta triumphalia* bestowed upon the commander of the annexation, A. Cornelius Palma, in AD 107, suggest military activities.[210] Bowersock observes a sort of an unofficial *damnatio memoriae* of the Nabataeans in the literary sources of the second century AD that could be very well interpreted as a reaction against their resistance to integration into the Roman Empire.[211] This would also explain why in the contemporary sources there is, besides the scanty mention in Dio Cassius, almost no report of the annexation, only in Ammianus Marcellinus, writing in the fourth century AD.

The annexation of the Nabataean kingdom has to be seen in a wider context within the planning of Trajan's further activities in the East. The annexation was probably not planned for 106 AD because at that time the Roman army was still involved in the war against the Dacians. The death of Rabbel II may have led to the decision in favour of immediate annexation.[212] After the annexation, the project most quickly realized in the new province was the building of the Via Nova Traiana, connecting the Gulf of al-'Aqaba with Syria.[213] It was built not simply with an economic but primarily with a military goal in mind because this road became one of the main routes for supplying the Roman army during its war against Parthia. It is generally believed that Trajan deliberately used the rich booty from Dacia and the income from the mines in Dacian territory to build new roads in the eastern provinces, especially for the planned campaign against the Parthians.[214] It seems that Trajan did not have too much trust in the loyalty of the Nabataeans and decided to integrate them into his empire, despite the not too favourable circumstances that a large part of the Roman army was tied down in Dacia.[215]

Reported contacts between the Parthians and the Nabataeans go back at least to the first century BC when in 38 BC the Romans fined the Nabataean king Malichus I for supplying the Parthian commander Pacorus during his invasion into Roman territories (41–38 BC).[216] Anthony's gift of Nabataean territories to Cleopatra may also be seen in this context.[217] Parthian influence on Nabataean policies can also be deduced from an episode during the Jewish throne quarrels, when Herod the Great asked Malichus I for refuge. The latter refused and argued that the Parthians had forbidden him to accept the refugees.[218]

Trajan could not therefore exclude the possibility of Nabataean supplies to the Parthians during his own campaign into Mesopotamia and Babylonia, for the Nabataeans would probably not have accepted the loss of their only potent trading partner in the Near East. That both events, the annexation of the Nabataean kingdom and the campaign against Parthia, were already seen as connected in antiquity may be deduced from the above-mentioned passage by Ammianus Marcellinus where he continues, 'It was given the name of a province, assigned a governor, and compelled to obey our laws by the emperor Trajan who, by frequent victories, crushed the arrogance of its inhabitants when he was waging glorious war with Media and the Parthians'.[219]

The factor of international trade in this confrontation was not only important for the Nabataeans but for the Romans as well.[220] This is shown by Trajan's—successful—efforts to control Spasinou Charax, the most important port in Mesopotamia, as well as his statement that he would like to travel to India if only he were younger.[221] The partners of the Far East trade quickly reacted to the new situation created by the provincialization of Nabataea. It is certainly no coincidence that in AD 107, immediately after the annexation, an Indian delegation made its way to Rome.[222]

Some final thoughts

When attempting to sum up the different aspects of Nabataean art, culture and history and put them in a wider context, we see a fascinating evolution during a relatively short period of time. In fact, from the first evidence of Nabataean material culture to the time of the Roman annexation, only two hundred years passed.

Thus, we see that there was in fact no continuous process of Hellenization, that is, a step-by-step adoption of what is considered Greek art and culture.[223] Indeed, the opposite was the case. The Nabataeans took over initially an almost completely Hellenized culture around 100 BC. Towards the second half of the first century BC and the first half of the first century AD, they were open to stronger Roman tendencies as well as to the first steps leading towards an identifiable Nabataean 'style'. A distinctive Nabataean culture is only attested during the next step, that is, during the second half of the first century AD, when their potters, sculptors and coin cutters consistently freed themselves from the earlier prototypes. It is difficult to decide whether this last step in development coincides with the accession to the Nabataean throne by Rabbel II, 'the one who renewed and rescued his people', or whether Rabbel was following an already existing general stream. Interestingly enough, it was only shortly before losing their political independence in AD 106 that the Nabataeans had fully developed their own material culture with the last of these phases.

When considering the evidence outlined above, one can conclude that Nabataean art and culture does show a comparable tripartite system of evolution—such can be applied to cultural development in general: a first phase of 'initialization'; a second phase of *akmé*, that is, the peak of the evolution; and, finally, a third phase, already leading towards decline. However, things seem to be more difficult in the case of the Nabataeans. The easiest case may be the pottery (cf. Figures 11.1, 11.3, 11.22), where indeed the subdivision into three phases—with some sub-phases—works quite well. In a first stage, the purely Hellenistic prototypes are taken over; in a second, the first tendency to a 'local' production is observed, combined with external (Roman) influence; and in the third step, the characteristic Nabataean style is achieved. Although of a similar general outline, other categories present different models of evolution. For instance, in sculpture and relief we also see a first phase of complete Hellenistic style (Figures 11.11-11.15), but with a chronological difference of about 100 years compared with the first appearance of the Hellenistic pottery! Yet another illustration is provided by temple architecture (Figures 11.8, 11.9), where from the beginning, that is, the end of the first century BC, there seem to be rather strong oriental elements, mixed with other influences from the Hellenized east.

This may be related to the particular case of the process of settlement as observed in the case of the Nabataeans. It may seem logical that in the first stage of settling down, people would not immediately start producing crafts, such as monumental architecture or sculpture that demanded a well-developed infrastructure and trained artists and craftsmen. But they would rather take their first steps in the field of 'smaller' categories, such as coin minting or pottery production.[224] This may explain why there is an important chronological difference, although they follow the same general outline, within pottery and numismatics, on the one hand, and sculpture and architecture, on the other. As for temple architecture, the impact of specific religious goals may play an important role, as stressed above, although lacking detailed written sources this must remain rather mysterious for the moment.[225]

Interestingly, most of the immediate geographical neighbours of the Nabataeans show a different model of cultural evolution. Most of them were settled for quite a long period, so that we find well-established local cultures that were increasingly influenced by Hellenistic and Roman features, until they became more or less completely Hellenized or Romanized and, at last, were incorporated into the Roman empire.

In a search for a parallel phenomenon to that found within the Nabataean culture, we must leave the Near East and turn towards Central Asia. There can be seen a comparable, suddenly increased input of western, that is, Hellenistic, culture that is due in these cases to the conquests of Alexander the Great.[226] The Greek colonists established, within a very short time, a strongly Hellenized culture, as can be seen chiefly in coins[227] and pottery,[228] while in the field of architecture is found a large amount of local forms and shapes beside the imported Greek elements.[229] Already during the first 150 years or so after the founding of these Greek settlements, there are some tendencies towards a similar evolution as observed in the case of the Nabataeans. The first coins of the Graeco-Bactrian and Indo-Greek kings show perfectly minted Greek coins with typical Hellenistic portraits of the ruler and Greek inscriptions. But around 200 BC at the latest, the first issues occur with a mixed legend, that is, in Greek on one side and in local script on the other.[230] Around the middle of the second century BC, nomads originating from regions in present-day Turkestan and China conquered the territories of most of these Greek kingdoms.[231] Therefore, the surviving Greek settlements were cut off from their direct connection to Mesopotamia and the Mediterranean and in the material culture a stronger tendency towards local elements can be observed. The coins now are consequently inscribed in local alphabets on one side and show at least partially non-Greek iconographic elements. Likewise, pottery develops its own style, although there are occasionally influences from the west. The last remnants of these former Greek cultural spots in Central Asia can be seen in the arts of the Kushan and the well-known Gandhara reliefs, incorporating both local and western elements.[232]

Although in general terms quite similar to the lines of development in Nabataean material culture, there are some crucial differences, mainly of a socio-political nature. The first Graeco-Bactrians were in fact Greeks, simply producing their own crafts in a place far away from their homelands. The first Nabataeans who produced their own material culture, on the contrary, were already 'at home' in central Jordan around 100 BC for more than two hundred years. They had, however, no preceding material culture that they could work on and, therefore, they adopted a foreign one, taking over the cultural *lingua franca* of the region. In a further development, the Graeco-Bactrians and Indo-Greeks continuously lost their former cultural roots because it became increasingly difficult for them to maintain a direct contact with the Mediterranean.[233] The Nabataeans, on the other hand, developed their own material culture in a surrounding that would have been extremely favourable to maintaining the former, that is, the Hellenistic one. Therefore, one could say that it was probably by distancing themselves from the predominant Hellenistic and later Roman cultural elements in their immediate neighbourhood that the Nabataeans became a sociocultural unity.[234]

The typical Nabataean culture did not entirely perish with the incorporation of the Nabataean kingdom into the Roman Empire. For instance, the characteristic Nabataean painting on pottery occurs at least into the fourth century AD as do the shapes of that pottery (Figure 11.22.9).[235] However, from AD 106 onwards no new elements can be found but rather a degeneration of the already established features. The shapes as well as the painting of the pottery could be said to be degenerated versions of the shapes and decorations as introduced in the last quarter of the first century AD (cf., for instance, Figure 11.22.9 with 11.22.8). Thus, it seems that with the Roman annexation, the Nabataeans not only lost their political independence, but also their innovative spirit.[236]

Acknowledgments

The author would like to thank (in alphabetical order) the following persons for their manifold practical and intellectual help: Ulrich Bellwald (Amman/Bern), Zbigniew T. Fiema (Helsinki), Alice K. Heyne (Basel), Daniel Keller (Amman/Schaffhausen), Bernhard Kolb (Basel), Martin Seyer (Athens/Vienna) and Fawzi Zayadine (Amman).

Notes

1. Exhibitions (selection): Schmitt-Korte 1976; Inoubliable Pétra 1980; Nabatäen 1981; Voie royale 1986; Königsweg 1987; Lindner and Zeitler 1991; Petra und die Weihrauchstrasse 1993. For publications on the Nabataeans or related to them, systematic research in the catalogue entries of the German Archaeological Institute at Rome, available on Dyabola (which, especially on peripheral cultures, is far from being complete), led to the following result: 1960–69: 34 entries; 1970–79: 60 entries; 1980–89: 111 entries; 1990–96: 119 entries. In other words, since 1960 output on the Nabataeans almost doubles every decade. Wenning gives the best overview on Nabataean archaeology (1987, 1990). Wenning (1987: 307-46, 1990: 367-77) gives a detailed bibliography on the subject. Therefore, in this chapter some of the older works are not quoted as they can be found easily in the contributions of Wenning. On the other hand, recently published works are emphasized.

2. For an attempt to put the available evidence on the Nabataeans in a wider perspective, see Wenning 1989; a concise overview is given by Dijkstra 1995: 34-80; quite inspiring, although rather superficial, is the chapter 'arte nabatea' in Garbini 1993: 211-19; strongly tending towards an ideological interpretation of Nabataean culture and, therefore, to be considered with caution is Patrich 1990; likewise, Mettinger 1995: 57-68.

3. Altorki and Cole 1989: 60-64.

4. See also below; for new research on as-Sela', see Lindner and Zeitler 1991: 180; Hart 1986; Hart and Falkner 1985 for an overall view of that survey. The question revolves not only on where was the stronghold of the Nabataeans, but also on 'al-Sela'' of the Edomites, mentioned in the Bible.

5. In favour of continuity: Bartlett 1979; against or at least sceptical: Bartlett 1990; Bienkowski 1990; Mattingly 1990; Hart and Falkner 1985; Hart 1987.

6. Graf 1990; Knauf 1986; Milik 1982; cf. also below notes 18 and 19.

7. See also Schmid 2001 and in press 2001.

8. Schneider 1996: 130-32, 141-42; see also Parr 1960: 135: 'There is evidence, in the form of Hellenistic pottery and coins of the third century BC, of settlement in Petra by that date; but so far no undisturbed levels of that period have been located'.

9. On the coinage of the Nabataeans, see Schmitt-Korte 1997: 101-04; 1991: 135-48; Weiser and Cotton 1996: 268-85; Schmitt-Korte and Price 1994; Peter 1993; Schmitt-Korte 1990; Schmitt-Korte and Cowell 1989; and Meshorer 1975.

10. Peter 1996: 91-93 and note 413 cat. nos. 1 and 2; Bowsher 1990; Meshorer 1975: 9-12, 85-86 pl. 1 cat. nos. 1-4. sup. 1.

11. Schmitt-Korte 1990: 125-26.

12. Weiser and Cotton 1996: 268 with n. 240; Kushnir-Stein and Gitler 1992–93.
13. Meshorer 1975: 12-16, 86-87 pl. 1 cat. nos. 5-8; Schmitt-Korte and Price 1994: 95-96; Augé 1990: 131-32. On the uncertainties about the beginning of the mint in Petra, cf. Schmitt-Korte 1997: 101-102 and fig. 105.
14. Knauf 1997: 15-16 and note 25 for the older references.
15. Schmid 1995: 637-38, 640-41, 1996a: 127-29, 1996b: 161-62, 164, 168, 1997b: 134, 2000b.
16. On this phenomenon see also Schmid 1996a: 130 and note 17; Hannestad 1983: 83-120; both have additional references.
17. The only distinctive characteristic is the infrequent simple painting. For a possible explanation, see Schmid 1996a: 130 and notes 18-22; cf. Schmid 2000b.
18. Boucharlat and Lombard 1985, 1991; Boucharlat and Mouton 1993a, 1993b; de Cardi 1984; Stevens 1994; Haerinck 1983: 98-100; cf. above note 6 on the origin of the Nabataeans.
19. Schmid 1996a: 130-34; 2000b: 118-21.
20. Zanoni 1996: 311-13, nos. 1 and 2.
21. This seems to be the case in other rather parallel developments from nomadism to sedentarism too, as can be seen for the Scythians and others; cf. below note 144. A small reflection of these 'archaic' features of the pre-settlement culture of the Nabataeans may be seen in some terracotta types, for example, the frequently occurring camels and the horse riders: Bignasca 1993: figs. 133-38; Parlasca 1997: 126-31 especially 130-31; Parlasca 1990a: 164-67 pl. 31, 1990b: 91-92 figs. 26-37. The chronology of Nabataean terracottas is still rather unsatisfactory. A first indication that the camels and horses should be rather early in date may be seen by the fact that from the Late Roman layers on az-Zantur only very few fragments were ascribed to these types, while in the earlier layers they are much more prominent: Bignasca 1996: 266-88.
22. Schmid 2001.
23. Wolski 1999 *passim* and especially pp. 83-116; Sherwin-White and Kuhrt 1993: 217-29; Fischer 1980, 1990a.
24. Drexhage 1988: 7-9. The passage, as well as the use of the monsoon winds, were known to the South Arabian people for a longer time, but they hid their knowledge from their rivals in trade; cf. Casson 1989: 11-12, 283-91; see also Kitchen 1994: 22-25 on the date of the *Periplus*.
25. Comparable phenomena, i.e., Arab tribes pushing into gaps of power at the periphery of the Mediterranean world and creating new political factors and eventually becoming sedentarized, occurred on several occasions in antiquity (Funke 1996).
26. Josephus, *Ant.* 13.13.5; *War* 1.4.4; Bowersock 1983: 23-24; Hammond 1973: 17.
27. Josephus, *Ant.* 13.15.1; *War* 1.4.7; Bowersock 1983: 24-25; Wenning 1993b: 30-31.
28. Josephus, *Ant.* 13.15.2; *War* 1.4.8; Bowersock 1983: 25-26; Hammond 1973: 18.
29. Wenning 1987: 22-24; Roche 1996: 73-99.
30. Stucky 1996: 14-17.
31. Nehmé 1997a; Nehmé 1997b: 66-70.
32. Schmid 2001.
33. Altorki and Cole 1989.
34. Altorki and Cole 1989: 67-82; however, this did not exclude the camel raiders from having houses in their home city.
35. Altorki and Cole 1989: 101-102, 208-31. The careful analysis in the case of 'Unayzah offers hope for reconstructing the process of settlement and social diversification among the Nabataeans. On this subject see Fiema 1996.
36. On this see Schmid 2001.
37. As is proven by the finds of Nabataean pottery at sites such as Jawf, Thâj, Hegra, Qaryat al-Fau, Gharrain, Marib, Qana and Khor Rori; on this see Schmid 2000b: 127-31. As this pottery apparently was not used as an export good, these finds point to Nabataean presence.
38. Schmid 2001.
39. Schmid 1996a: 132-33.
40. Schmid 1996a: 131 and notes 31-33; 2000b: 148-49.
41. Good pictures in Toynbee 1978: 155 figs. 307, 308.
42. Wreath: Meshorer 1975: 92 nos. 33, 35, 37, 39; no. 37 is interpreted by Meshorer as diademed but as Schmitt-Korte 1990: 110 nos. 21-22 shows wears a wreath; diademed: Meshorer 1975: 92 nos. 32, 34, 36, 38; Schmitt-Korte 1990: 111 no. 24; Schmid 1999.
43. Diademed: Meshorer 1975: 94-96 nos. 46, 47, 47A, 50, 52, 55; Schmitt-Korte 1990: 116-17 nos. 52-54; wreath: Meshorer 1975: nos. 48-49A.
44. Josephus, *Ant.* 15.10.2; 16.9.1-10.9; cf. Meshorer 1975: 32-33; Millar 1993: 39-40; Bowersock 1983: 49-54; Hammond 1973: 23-26.
45. Josephus, *Ant.* 16.9.2-3; on Syllaios see also Josephus, *Ant.* 16.7.6; Kokkinos 1998: 177-205 especially 182-84.

46. Stucky 1996: 17-21, 48-49 and note 195 referring to older excavations, especially at al-Khatute and al-Khubtha; Kolb 1997: 62-63.
47. Hoepfner *et al.* 1999: 515-24; Trümper 1998; Schmid in press.
48. A stronger oriental influence was deduced from the not so systematic orientation of the smaller units; see Stucky 1996: 48-49; Kolb 1997: 63. Parallels from Neo-Assyrian and Neo-Babylonian Mesopotamia show that even in such cases more extensive planning usually existed; see Miglus 1999: 254. The basic types of these Mesopotamian houses, showing one (Babylonia) or even two (Assyria) inner courtyards with the different rooms for 'official' and private use grouped around them, deserves special mention, see Miglus 1999: 133-75 (Assyria), 179-213 (Babylonia). As tempting as it is to consider such houses as the forerunners of the Nabataean ones, there are important chronological and geographical gaps yet to be filled.
49. See the plan by Andreae 1975: 77; in the villa at Boscoreale there is—beside the huge main peristyle—a small one (no. 15) at the periphery, showing porticoes only on three sides, while the fourth one contains only semicolumns against the wall; for Pergamon, see Pinkwart and Stamnitz 1984: 36-42, Pls. 51, 52.
50. Indeed, Hellenistic houses and palaces from Nippur and Seleucia show similar features: Kose 1996: 206 fig. 4, 208 fig. 6. As these monuments show considerable Hellenistic influence too, we may conclude that the houses so far excavated at Petra simply illustrate the *interpretatio nabataea* of the common Hellenistic house type; cf. in a broader context Colledge 1987. Taking into consideration also the remarks in note 48, it seems probable that the oriental variant of the Hellenistic house generally contained some older local elements.
51. Kolb *et al.* 1997, 1998; Kolb 1997.
52. The same is true for the houses on the northern slope of az-Zantur and on al-Khatute. From both places, an impressive view of the city's centre is possible. On the Ephesian houses, see Lang-Auinger *et al.* 1996 with an older bibliography; similar house types are reported from other areas in Asia Minor too, for instance, from Limyra: Seyer 1993; Seyer and Rasch 1997.
53. Kolb *et al.* 1998: 261 fig. 5.
54. On illusionistic paintings from Italy and built parallels from the Hellenistic world, see below notes 98, 99, 124.
55. Kolb *et al.* 1997: 234; on painting of phase 3a, see Schmid 1996b: 166 fig. 700 and here Figure 11.23.
56. See Kolb 1997: 65 fig. 69b for the painting of the so-called masonry style; cf. Rozenberg 1996 for parallels from Jericho; in general see Andreou 1989.
57. Rozenberg 1996; Fittschen 1996: *passim*, especially 150.
58. This is not the place to enter into detail on that subject. A detailed publication of these wall decorations, including the various interpretative aspects, is in preparation by B. Kolb (Basel).
59. On Dusares, see Arrian, *Anab.* 7.20.1; Wenning and Merklein 1997: especially 110; Zayadine 1990a; Gawlikowski 1990: 2662-65; Moutsopoulos 1990: *passim*, especially 53-54; Bowersock 1990a; Tran Tam Tinh 1990; Knauf 1990.
60. Nehmé 1997b: 1035-36.
61. On Qasr al-Bint, see Zayadine 1985; in 1999, IFAPO started new excavations on structures in front of and beside the temple. It is to be hoped that this will provide us with further evidence for the chronology of the monument.
62. Tholbecq 1997, 1998; Freyberger 1998; Freyberger 1997: 71-84.
63. To consult other research, not necessarily related to the Nabataeans, is normal practice. At an early stage, a rather simple model is usually proposed; it is replaced at a second stage by a very detailed differentiation; and at a third stage, it is often simplified again to some general outlines. This is, for instance, what usually happens with pottery studies all over the world, the work of the present writer not excluded.
64. In the case of the temple of al-Qasr (Figure 11.8a.1), it is not clear whether we have to reconstruct an inner subdivision in the form of a wall subdividing the rectangular building into a front and a rear part. If so, then al-Qasr would become an almost exact parallel to Qasr al-Bint (Figure 11.8a.6); see Altherr-Charon 1977 study on temples with a tripartite cella.
65. Freyberger 1998: 6-12; however, Freyberger stresses that the typically eclectic putting together of different elements may be a general characteristic of the Hellenistic Near and Middle East. This seems in general the best explanation for the phenomena observed in Nabataean culture as noted above in notes 50 and 56 relating to private architecture.
66. Tholbecq 1997: 1086-88, 1998: 248-52. One could also refer to the famous tent of Ptolemy II as

described by Kallixeinos *apud* Athenaeus, *Deipn.* 5.196-197. There, too, a central hall, surrounded on three sides by porticoes, is mentioned. This shows that such features were quite widespread in Ptolemaic Egypt; on the tent see also below notes 117-119.

67. See Freyberger 1998: 24; for the time being, we should be cautious in identifying the South Temple, also known as the Great Temple, as a temple. This is so since the discovery, during the 1997 season, of a theatre-like structure within it. Even its function as a temple has been questioned; see Joukowsky 1998. For the moment, it does not seem possible to distinguish the exact chronological relation among the *theatron*, the 'temple' proper and the temenos.
68. On the impact of Roman architecture on Near Eastern buildings and on interchanges between them, see Strocka 1988.
69. On Ptolemaic temple building in Egypt in general, see Aufrère *et al.* 1997: 215-61; Schloz 1994; Siegler 1969; Arnold 1999, especially 137-224; for a broader view on Egyptian temples, see the contributions in Gundlach and Rochholz 1994.
70. Zayadine 1986: 248.
71. On South Arabian temples in general, see now Schmidt 1997–98.
72. Freyberger 1998: 6-7; Zayadine 1985: especially 248-49, 1986: 238-48; if indeed, as Zayadine supposes, the temple of the late first century had a predecessor that could probably be dated to a generation before Obodas III (Zayadine 1985: 249), this would be the earliest cultic building known from the Nabataean area and would at least fill partially the chronological gap between their settlement and the first monumental buildings. Other possible early cult buildings are indicated by epigraphic evidence but not much is known about the architecture related to them (Wenning 1989: 249).
73. Tholbecq 1997: 1076-77, 1998: 243-45.
74. Freyberger 1998: 18-19; Hammond 1996: 5-14.
75. On pottery of phase 2b, see Schmid 1996b: 165-68, 173-74. For the pottery from the South Temple, see Joukowsky 1998: 313 and note 14. This pottery from trench 18 loci 3 and 10 (from the 1996 season) was identified by the author and the results were communicated in a letter to the excavator dated 27 January 1997. However, in 1997, in a trench parallel to trench 18 (trench 46 on fig. 1b in Joukowsky 1998: 295) apparently pottery of a much later date was found in layers below the pavement: the bowl on fig. 9 in Joukowsky 1998: 299 belongs not—as stated in the caption—to the second style, but to phase 3c, post-dating AD 100! If, indeed, the wrongly labelled pottery comes from below the pavement, this would be later than AD 100 and would, therefore, completely change the chronology of the entire building. According to a personal communication from the excavator (January 2001), later building activities were indeed attested in that area.
76. Tushingham 1972: 32-33; the latest element from the assemblages connected to the construction of the temple would be the painted bowl *ibid.* fig. 2, 59 of phase 2c; for a detailed analysis, see Schmid 2000b: 104-105.
77. On Nabataean sculpture in general, see McKenzie 1990; Lyttelton and Blagg 1990; McKenzie 1988, all with additional references.
78. Wright 1968; McKenzie 1988: 86-87; Freyberger 1998: 15-18.
79. On this sanctuary, see Tarrier 1990; Wenning 1989: 250 and no. 22.
80. Weber 1997: 118-19 fig. 130a; McKenzie 1988: 86-87 fig. 9; Zayadine 1986: 238-38 fig. 36; on the date of the Qasr al-Bint, see above.
81. On that, especially in connection with Nabataean sculpture, see Schmid 1999 and Smith 1991: 223-28, in a broader context. As a matter of fact, the first appearance of such classicizing features within Hellenistic sculptures goes back to the middle or even the first half of the second century BC. On the famous altar from Pergamon, the baroque style, as seen within the giants on the big frieze, already occurs together with the calm and fleshy faces on the small frieze: Smith 1991: 155-66. From then on, prominent examples like the group with Achilles and Penthesilea show that the classicizing style continues during the entire second and into the beginning of the first century BC; cf. Jaeggi and Schmid 1996: 23-24, 31-33.
82. Weber (1997: 118-19) points out that the above-mentioned bust of Helios from the Qasr al-Bint is another good example, and the fragment on Figure 11.14, despite its incomplete preservation, is of outstanding quality, both in technical terms as well as naturalistic styling.
83. This is true, for example, for such famous sculptural groups like the Laocoön or the sculptures from the Sperlonga cave (Smith 1991: figs. 143-46). While Laocoön himself is related predominantly to the baroque style of the high

Hellenistic period, his sons show rather the classicistic aspect. In Sperlonga, one would compare the beautiful baroque head of Ulysses with the fleshy face of Diomedes (Conticello and Andreae 1974: pls. 14. 37). On these aspects, see Jaeggi and Schmid 1996: 26-29, 32-33 with many additional references. Since then Despinis (1996: *passim*, especially 306-14) added some new information, relevant mostly to the presumed prototypes of the rape of the Palladion; on Laocoön and the Sperlonga groups, see now Lahusen 1999 and Strocka 1999.

84. See McKenzie 1988: 86-87; Lyttelton and Blagg 1990: 97 fig. 6.7.

85. Wenning 1989: 253-54. As pointed out above on the iconography of the royal coin portraits of the Nabataeans, we indeed see a stronger reference to Rome at this date. On the other hand, Thüroff suggests a very interesting hypothesis in an unpublished MA thesis at the University of Basel (Thüroff 1989: 96): therefore, the triton on a relief slab from Petra would be holding a torch, an otherwise rather unusual attribute for a sea creature, especially in connection with the weapons on the same frieze (tritons holding torches may occur on representations of Poseidon's wedding with Amphitrite). This could be understood as a reference to the Nabataean troops burning the ships of Cleopatra VII that she had prepared at a Red Sea port in order to flee Octavian after the battle at Actium (Dio Cassius 51.7.1; Plutarch, *Ant*. 69.3).

86. The most prominent example—but by far not the only one—for this thematic conjecture is the lower part of the outer frieze of the Ara Pacis at Rome, also showing a richly developed flora and fauna referring to Augustus's golden age; cf. on the Ara Pacis below notes 126-27.

87. The tomb façades of the Nabataeans in general, and the ones from Petra in particular, have been treated many times in the history of research. Concise overviews are given by McKenzie 1990; Zayadine 1989; Schmidt-Colinet 1981; see also the revised version in Weber and Wenning 1997: 87-94; Fedak 1990: 150-57; Lyttelton 1974: 61-83. The work of Matthiae 1991 gives a good introduction as well, although some interpretations, especially of the monuments treated here, seem difficult to substantiate; cf. below note 100. Special reference should be made to the seldom-used work of Qusus 1984; the basic works, especially for detailed descriptions, by Brünnow and von Domaszewski 1904: 137-428 and Bachmann *et al*. 1921: 8-28 are still irreplaceable; Schmid 2000a gives a brief introduction.

88. The covering with stucco is confirmed in most cases by the 'fixing' holes still *in situ* and, in rare cases, by fragments of conserved stucco. How spectacular these decorations could have been is shown by some gilded fragments related to the temple of Qasr al-Bint or even to a predecessor; see Zayadine 1985: 240. However, the rich fragments from private houses also confirm this general picture (cf. above).

89. Brünnow and von Domaszewski 1909: 98-101; Dentzer-Feydy 1985–86: 263-65; Fedak 1990: 86-87, 148-50.

90. In general, on Rhodian funerary monuments, see Fraser 1977, mostly dealing with smaller monuments; Mette 1992: 6-60; Lauter 1972, 1988; Fedak 1990: especially 83-87.

91. Lauter 1988: 155-56; Konstantinopoulos 1986: 229-31; Lauter 1972: 55-56; Fedak 1990: 85-87; on the Rhodian 'parks' with their grottoes and *triclinia*, see also Rice 1995.

92. The bibliography on the Mausoleum is legion; a good insight may be found in Jeppesen 1998, as well as in Higgs 1997; Jenkins *et al*. 1997; Jeppesen 1997; Walker and Matthews 1997; and Waywell 1997; see also Fedak 1990: 71-74.

93. Lauter 1986: 214, 1988: 156; Fedak 1990: 86-87. Lauter quotes parallels from Cyrene. This would, of course, fit well the often discussed Alexandrian influence in Nabataean tomb architecture. On tombs from Cyrenaica, see Stucchi 1975: 149-92.

94. Lauter 1988: 157 fig. 1; Konstantinopoulos 1973: 116 fig. 3.

95. Zayadine 1986: 217-21 with references to similar tombs around Jerusalem; on these see Bonato 1998. On similar monuments from Numidia, such as the Nabataean funeral architecture contesting the peripheral eclecticism between local tradition and Hellenistic influence, see Coarelli and Thébert 1988; Rakob 1979, 1983; and, in a broader framework, Camps 1995. It must be mentioned that a small tomb in the Bab as-Siq outside Petra is completely hewn out of the rock, i.e., free-standing. For other 'smaller' Nabataean monuments, strongly depending upon Hellenistic influence, see Weber 1997: 117-18.

96. Seigne and Morin 1995.

97. Brünnow and von Domaszewski 1904: 179-86, 223-31 no. 62. New excavations by Bellwald and

Keller in the square in front of al-Khazna will definitely increase our knowledge on the connection between the monument and its environs; cf. below.
98. See Schmid 2000a; McKenzie 1990: 85-101; Weber 1990: 108-109, 220-22 cat. nos. M 7-11; Tybout 1989; Lyttelton 1974: 17-25.
99. Tybout 1989: *passim*, especially 215-73, 325-52; Fittschen 1976a: 544-56; although the same author recently partially modified his theory about the cultural background of some eastern wall paintings (Fittschen 1996) this does not affect his and other reflections on what the architectural paintings of the second Pompeiian style represent; cf. also Schefold 1975.
100. This theory—for al-Khazna as well as for other façades—is mainly represented by Matthiae 1995, 1994, especially 284-85, 1991, especially 259-65, 1989, 1988.
101. Pesce 1950; Kraeling 1962: 83-89; Lauter 1971; Lyttelton 1974: 53-60; McKenzie 1990: 75-77; Nielsen 1994: 146-52, 284-86 cat. no. 22.
102. Lauter 1971: 151-52; cf. on Alexandrian capitals Pensabene 1991: 44-46 fig. 29-33; on their evolution and comparisons in general, see Hesberg 1978.
103. Lyttelton 1974: 58 pl. 63; Orlandos 1976: 36-38 fig. 37.
104. See the impressive reconstruction by Pesce 1950: pl. X; Lauter 1971: 163 fig. 15. It must be mentioned that all architectural elements used in that reconstruction were actually found, at least partially.
105. Pesce 1950: fig. 13 top left; Lauter 1971: 172 with n. 115.
106. First of all Pensabene 1993: 3-147, 195-216; McKenzie 1996; Schmidt-Colinet 1993; Pensabene 1991: *passim*, especially 49-55 fig. 42-62; Hesberg 1978; Lyttelton 1974: 40-52.
107. Will *et al.* 1991; especially on the connections with the Khazna, see Stucky 1990: 26; Nielsen 1994: 138-46.
108. Will *et al.* 1991: 37-38 with n. 157 and map 1-4.
109. Will *et al.* 1991: 211-18.
110. It does not matter whether or not the retaining walls for the lake were ever completed and, therefore, whether or not the lake was ever filled. The idea behind the installation is clear and it is this point that leads to further reflection.
111. Will *et al.* 1991: 149-208 especially 167-78; on the capitals see also Fischer 1990b: 7-11; for Alexandrian parallels cf. above nn. 102. 106.
112. Will *et al.* 1991: 277-86; cf. on related architecture Brands 1996; Siganidou 1996; Nielsen 1994: 81-99; Heermann 1986; Pandermalis 1976.
113. Netzer 1999: 3-31, 1996a, 1990: 37-42; Nielsen 1994: 155-63.
114. Netzer 1999: 32-34, 1996b, 1990: 42-50; Nielsen 1994: 181-208.
115. Foerster 1996: 58-63; Nielsen 1994: 189-95, 295-96 cat. no. 27; Netzer 1999: 80-87, 1991: 134-170, 575-88.
116. See the drawing by Netzer 1991: 583 plan 62, bearing in mind that the reconstruction of the roofs is hypothetical; cf. also Nielsen 1994: 193; Foerster 1996: 60-61.
117. Athenaeus, *Deipn.* 5, 204d-206c; the basic study is still Caspari 1916; cf. further Grimm 1981: 17; Nielsen 1994: 136-38; Pfrommer 1996a: 177-79; Pfrommer 1996b.
118. Pfrommer 1996b 102-104; cf. above n. 112.
119. Cf. the drawings by Pfrommer 1996a: 178 fig. 9; Pfrommer 1996b: 98 fig. 1; Nielsen 1994: 137 fig. 71.
120. Zayadine 1991: 300-306; Schmid 2000a.
121. Cf. the drawings by Pfrommer 1996a: 178 fig. 10; Pfrommer 1996b: 99 fig. 2.
122. Athenaeus, *Deipn.* 205c.
123. Athenaeus, *Deipn.* 205e; especially on these luxury features see Normann 1996: 153-61.
124. Strocka 1991: figs. 305, 308-11; for these paintings, see also above notes 98 and 99.
125. See below; in general, on the successors of Hellenism in the East, see still Schlumberger 1969; Bowersock 1990b, especially 29-40.
126. On the capitals from Petra see Ronczewski 1932; comparisons with the Ara Pacis: Schmidt-Colinet 1981: 62, 90-92; cf. also Freyberger 1997: 72, 1998: 6-7 who—although he points out the similarities with late republican ornaments—again returns to the comparison with the Ara Pacis.
127. In general on the Ara Pacis, see now Castriota 1995; cf. Settis 1988 and, for the tendrils, Sauron 1988.
128. Carettoni 1983: colour pl. 7, 15; on the dating 412-19; the paintings belong to the very late second (lower floor) and to the very early third Pompeiian style ('studiolo' in the upper floor).
129. Hölscher 1988: 363-64 cat. no. 199; Kraus 1976: 463-64 and fig. 8; Sydow 1974: especially 204-10; see also the very rich comparative material by Schörner 1995: 17-46 for the third quarter of the first century BC and 46-51 for the early

imperial years; see also the comparisons quoted by Freyberger 1998: 7 and note 58.
130. Börker 1973: 296-300 figs. 6-9; Rumscheid 1994: 292-94 pl. 114.5, 6 cat. no. 188.23-24.
131. Lyttelton and Blagg 1990: 95, 97; cf. Fittschen 1976b: 193 who thinks that the frieze slabs from Petra with representations of weapons are rather western/Roman in their composition and discusses the possibility of eastern influences on Augustean motifs.
132. Freyberger 1997: 74-76; Zayadine 1985: 246-47, 1986, 244-47.
133. It must be mentioned, however, that the stucco decoration could—theoretically—have been applied at a later time, although this does not seem very likely. At least the wall paintings from az-Zantur (above nn. 54-58 and Figure 11.6), with a well established *terminus post quem* of AD 20, show that such pediments continued to be used as decorative features into the first century AD.
134. On the Hegra tombs and their inscriptions, see McKenzie 1990: 11-31; Schmidt-Colinet 1987a, 1987b.
135. On the dating of the theatre, see McKenzie 1990: 35, 143-44. Hammond (1965: 55-65), the excavator, dates the theatre in the reign of Aretas IV (9 BC–AD 40).
136. Bellwald and Keller, personal communication.
137. Tholbecq 1998; cf. also above note 76 on Dhiban; other settlements could also be added.
138. Hoepfner *et al*. 1999: 445-46.
139. Strabo 16.4.21-24; Raschke 1978: especially 650-76; Sidebotham 1986a: 48-71, 113-30, 1986b and 1996 for a general overview; cf. further Eadie 1989; Romanis 1996: 19-21.
140. In the sense of increasing presence of Nabataeans in Italian ports and cities: Roche 1996: 86-95, 99.
141. Huth 1999; Haerinck 1998; Munro-Hay 1991, 1994: especially 192, 198-200, 1996; Arnold-Biucchi 1991; Naster 1983; in general terms, see Potts 1991.
142. Keller 1996; Sasianou 1993.
143. In general, on Scythian and related art, see Jacobson 1995; Schiltz 1994; Zazoff 1996. The tendency towards ornamentalization can be seen in an exemplary way within the so-called 'Rolltiere'. It was formerly believed that the Scythians arrived in the Black Sea area without a material culture of their own and started taking over and developing foreign prototypes; cf. Stucky 1976: especially 20-21 with the older references. As good as this theory is for a parallel to what can be observed in Nabataean culture, it now seems that the Scythians already had a culture of their own and the adoption of naturalistic (Greek) prototypes and their change towards ornamentalization was just one aspect within it; see Brentjes 1994: especially 159-61; on the manifold and complex aspects of Scythian culture, see also Cugunov 1998: especially 303-305; cf. on similar evolution and contacts within Celtic and Scythian art Guggisberg 1998.
144. See, for instance, Schiltz 1994: 248-89, 333-415 with many figs.; cf. above note 21.
145. On Nabataean pottery from the later first and the early second century AD and its historical context, see Schmid 1997a; on the date of the change from sub-phase 3a to 3b c. AD 70/80, see Schmid 1996b: 168, 173, 2000b.
146. Zanoni 1996: 316-19 nos. 10-13.
147. Zanoni 1996: 314-16 nos. 7-9. Interestingly, the Negev 1 type too seems to have close forerunners in Hellenistic types, mainly from Alexandria: Mlynarczyk 1998: 338-39, 341 fig. 12a, b.
148. On Khirbat at-Tannur, see Glueck 1965. On problems related to precise chronological sequences, see Wenning 1987: 77-81 and McKenzie 1988. Kader 1996: 130-31 and especially Freyberger 1998: 34-41 insist on a date in the reign of Aretas IV for all three phases and, therefore, for the entire sculptural decoration from Khirbat at-Tannur that appears impossible to the present writer, cf. also below notes 156 and 170.
149. Glueck 1965: 248-59 and others.
150. Glueck 1965: 138; cf. Wenning 1987: 77-81 and McKenzie 1988. Indeed, some of these sculptures bear some stylistic and technical similarities to a group of sculptures from the Hauran, dated to the second and third centuries AD: Diebner 1982; Bolelli 1985–86.
151. Glueck 1965: pls. 73a, 74a, 75a-b. All the other pottery—some of it earlier in date—illustrated in *Deities and Dolphins* comes from Petra and 'Amman and is not related to the temple of Khirbat at-Tannur. Glueck (1965: 139) refers to such pottery as being found under the pavement of phase II. Therefore, it would pre-date even the earlier sculpture. However, as Glueck himself notes, some of the pottery may have been intrusive from higher levels. Since there is no comprehensive study of the pottery from Khirbat at-Tannur relating it to layers and construction phases, it should be used only with care as dating evidence.

152. McKenzie 1988: *passim* and especially 88-89.
153. McKenzie 1990: 133-34 pls. 58-59; McKenzie 1988: 87-89.
154. Zayadine 1997: 618 no. 10; see also McKenzie 1988: 88, 91 no. 4 and fig. 12a.
155. The Department of Antiquities acquired one of the heads from the art market. It probably comes from Petra since a very close iconographic and stylistic parallel was found there; see Königsweg 1987: 224, 227 no. 210; Weber 1997: 116-17.
156. McKenzie 1990: 36, 132-34 pl. 47 b-d; the dating of the temenos gate is connected with a trench dug a few metres away: Parr 1970: especially 369-70; cf. Lyttelton and Blagg 1990: 92, 95; cf. above note 141 and below note 170.
157. Zayadine 1997: 618 no. 9; Villeneuve and Muheisen 1994: 745-46 fig. 5.
158. Wenning 1993a: 86-93; Wenning 1989: 257; Hammond 1981: 137-41.
159. See, for example, the pictures in McKenzie 1988: figs. 3, 7.
160. In addition, it may be noted that the painted floral motifs in the az-Zantur house have convincing parallels within the painted pottery of phase 3a (Figure 11.23), dated to c. AD 20–70/80 and, therefore, indicating a later date; cf. Schmid 1996b: 168, 173; Nabatäer 1981: pl. 79.
161. See Huff 1989: especially pl. 26.1, 2.
162. See Dentzer-Feydy 1985–86: 286-99.
163. For instance, the fragments with 'inhabited scrolls', in this case tendrils with panthers, from the temple of al-Qasr (Wenning 1990: pl. 21.7; cf. Tholbecq 1997: 1080-81), have good parallels within the wider region, confirming a date in the later first or early second century AD (Ovadiah and Turnheim 1994: 87-91 and others), although a chronological analysis is missing from the work of Ovadiah and Turnheim 1994.
164. See the drawing by Schmidt-Colinet 1981: 78 fig. 18 (right); Brünnow and von Domaszewski 1904: 168 fig. 192, 388-89 no. 766.
165. McKenzie 1990: pls. 126-27; Brünnow and von Domaszewski 1904: 172 fig. 197, 206-208 no. 34.
166. McKenzie 1990: 34 and notes 15-18 for the somewhat puzzling evidence.
167. Cf. the references above notes 101-106.
168. Adriani 1963–66: 107-83, especially 128-46, pls. 48-57 on the Mustafa Pascha necropolis; McKenzie 1990: 64-65.
169. See, for example, Pensabene 1993: 104-107 fig. 93 pl. 117.3; 133-35 pl. 117.7; 98 pl. XI 1 cat. no. 14 and many others.
170. On the capitals from the temenos gate, see Ronczewski 1932: 87-89 fig. 38; McKenzie 1990: 36. 132-34 pl. 47 b-d; on the chronology of the temenos gate, cf. above note 156. In a recent study, Kader (1996: 108-49) concludes that the gate and its capitals should be dated considerably earlier to the reign of Aretas IV. Kader 1996: 128-32 argues that the capitals are badly corroded and, therefore, were erroneously compared with later examples; Kader prefers comparisons with earlier parallels, such as the ones from al-Khazna, Qasr al-Bint and others (Figures 11.19, 20), and suggests a dating only slightly later than these, but still within the reign of Aretas IV. However, one should bear in mind that the corrosion of the capitals also reflects on the comparisons with earlier capitals and, therefore, should not be over-interpreted; cf. also above nos. 141 and 156.
171. On ad-Dayr, see Schmidt-Colinet 1981: 97-98; Zayadine 1989: 158; McKenzie 1990: 159-61 with additional bibliography; Brünnow and von Domaszewski 1904: 187 fig. 220, 331-35 no. 462.
172. On the capitals of ad-Dayr, see McKenzie 1990: 160-61; in general on Nabataean capitals and their possible origin in Ptolemaic Egypt, see Soren 1987: 206-12; Sinos 1990: 145-56, 227-29; Laroche-Traunecker 2000.
173. Also the painting of its façade may have contributed to the decoration of the monument.
174. Cf. above on the different palaces and other luxury architecture. Strocka (1988) has considered the exchange between traditional Roman architecture and Hellenistic luxury architecture that was especially fruitful in Asia Minor; cf. Schmidt-Colinet 1993: 6-13.
175. Strocka 1988: 294-97 and notes 10 and 16 for additional bibliography.
176. Mielsch 1997: 88.
177. Kleiner 1983.
178. Kleiner 1983: 78-79.
179. Cf. above note 99.
180. On the complex history of Philoppapos's family, see Baslez 1992.
181. For the chronology of Philoppapos's monument, see Kleiner 1983: 14-15.
182. Zayadine 1989: 158; Zayadine 1997: 53.

183. Rabbel II may have transferred his capital from Petra to Bosra, as suggested by some scholars: Milik 1958: 233-35; Peters 1977: 272-74; Bowersock 1983: 73, although others rejected this hypothesis: Wenning 1993a: 94-95. If the hypothesis is correct, Rabbel II could have died outside Petra and ad-Dayr could have been erected as his cenotaph. For the events connected with the annexation of Nabataea by the Romans in AD 106, see Schmid 1997a and below.
184. Schmid 2000a; McKenzie 1990: 119-20.
185. On the Hellenistic houses of Delos, see now Trümper 1998, on the 'House of the Hermes' ibid. 234-41 cat. no. 35; Kreeb 1988: 36-40, 200-215 cat. no. 24; both with the references to the excavation reports.
186. Mustafa Pascha: Adriani 1963–66: 128-46 pls. 48-57; McKenzie 1990: 64-65; Nea Paphos: Mlynarczyk 1990: 87-94, 223-32, 1996: especially 200-202 points out the close similarity between such peristyle tombs and the dwellings of the rich and wealthy. On the island of Rhodes there are, beside the above mentioned free-standing monuments, quite similar *hypogaea* as the ones from Alexandria and Nea Paphos: Konstantinopoulos 1986: 226-27 fig. 253.
187. McKenzie 1990: pls. 91 and 93.
188. McKenzie 1990: pls. 98, 100, 103, 104. The entire complex of the Roman soldier tomb and the triclinium 235 has not yet attracted the attention it deserves, although already Bachmann *et al.* 1921: 75-94 pointed out its importance; cf. further Schmidt-Colinet 1981: 77-82; Gagsteiger 1991: 59-61. With extensive cleaning or a small-scale excavation between these two structures, much additional information could be gained. See Schmid 2000c.
189. See above. On the connection between Vitruvius's description and reality as well as palace architecture, see Zoppi 1991–92; Raeder 1988; Reber 1988.
190. McKenzie 1990: pls. 163-64; Zayadine 1974, 1986: 229-37; Bockisch 1991: 89-97.
191. CIS 2.350; McKenzie 1990: 35 and note 30; Bachmann *et al.* 1921: 89-90 noted previously the close connection between what is mentioned on the Turkmaniye inscription and what is found in the Roman soldier complex.
192. This has already been noted (Schmidt-Colinet 1987a: especially 149-50) for the Urn tomb; in a broader sense on the same subject, see Schmidt-Colinet 1991. As interesting as is Schmidt-Colinet's hypothesis about the Urn tomb being the grave of Aretas IV, imitating the forum of Augustus at Rome, there is hardly any additional evidence supporting it since the Urn tomb can also be explained as a variation of a Hellenistic peristyle house or palace.
193. As Freyberger 1991 has pointed out, there is good evidence to date the tomb of Sextius Florentinus earlier than the usually adopted *terminus ad quem* of AD 129 (on this see McKenzie 1990: 33, 47, 165). However, if Freyberger's proposition of an Augustean date is correct, then we would have to suppose that the al-Khubtha north slope was already occupied at that time. This, however, seems rather improbable. The miniature architecture of the tomb of Sextius Florentinus shows strong similarities with the Palace tomb, the Corinthian tomb and the so-called Bab as-Siq triclinium. I would like therefore to propose an analogue date for all these monuments, related to the probable date of Bab as-Siq triclinium (McKenzie 1990: 34 and notes 15-18) and, therefore, around the third-quarter of the first century AD.
194. Wenning 1993a: *passim*, especially 86-93, 1989: *passim*, especially 257-58; Wenning and Merklein 1997: 110.
195. Freyberger 1998: 19 and note 231; Hammond 1986: 22 tentatively dates the reconstruction to the reign of Malichus II. In his later publications he is much less precise in dating and interpreting this phase: Hammond 1996: 11-12. For the temple in Wadi Ramm, a 'reorganization' of the monument including building activities was observed and tentatively dated to the time of Rabbel II (Tholbecq 1998: 245-47).
196. See above note 158.
197. See Wenning 1989: 255-59 using the metaphor of (Hellenistic) masks that were temporarily superposed on the indigenous Nabataean/Arabic features.
198. Wenning 1989: *passim*; Patrich 1990: *passim*, especially 165-66.
199. We can most probably see the not very flattering characterization of Obodas III (30–9 BC) as being weak and slack in the same context (Josephus, *Ant.* 16.7.6; Strabo, *Geogr.* 16.4.24). For all we know, there simply was no need to be more active or to handle the kingship more restrictively because the income from trade was more than sufficient. Most probably, Obodas's biggest mistake was to

count on the fact that after the incorporation of Egypt into the Roman Empire, the Nabataean position in the area would improve, all the more as the Nabataeans stood on the side of Octavian during his conflict with Anthony and Cleopatra (cf. above note 85). It seems that he did not take into consideration the Roman striving for control of long-distance trade. This miscalculation led to a series of conflicts with Rome, the first one being the unfortunate events during the campaign of Aelius Gallus into South Arabia; cf. above note 139.

200. As pointed out above, notes 36-38, settlement also led to a continuous specialization within Nabataean society.

201. See above note 183.

202. Bowersock 1983: 73 refers to irrigation installations in the Negev in order to optimize agricultural activities. The same may be true for the area around Petra. During a survey related to the Finnish Jabal Haroun Project (FJHP, see the first preliminary report by Frösén *et al.* 1998) in 1998 some barrages or terracing walls related to agricultural activities were investigated. The collected pottery was studied by the author during a seminar at Helsinki University in December 1998 and it turned out to be predominantly of phase 3b (cf. Figures 11.22.8; 11.24) and, therefore, from Rabbel II's time.

203. Interestingly, again the parallel examples from modern South Arabia seem to confirm this picture: Altorki and Cole 1989: 232-48; cf. above notes 34 and 35.

204. Of course, modern archaeology should be very sensitive and cautious not to over-interpret the connection between ancient and modern data in general. Some of the very tricky cases with many stimulating comments can be found in Meskel 1998.

205. See Schmid 1997a with further bibliography; Fiema 1987; Schmid 2000b: 143-46.

206. Bowersock 1983: 79-82; on the troops and their movements, see also Zayadine and Fiema 1986; Fiema 1987: 28 and notes 20 and 21; Kennedy 1980.

207. Sartre 1982: 131-32; 1985: 63-72 especially 68-69; Bowersock 1983: 80 and note 13; Wenning 1993a: 100-101; on these and other Safaitic inscriptions where Rome and Romans are mentioned, see Macdonald 1993: 328-34.

208. For detailed analysis as well as related bibliography, see Schmid 1997a, 2000b: 141-43.

209. Ammianus Marcellinus 14.8.13: 'Hanc provinciae imposito nomine, rectoreque adtributo, obtemperare legibus nostris Traianus compulit imperator, incolarum tumore saepe contunso...' (after Loeb Classical Library); Wenning 1993a: 100-103 and note 97 sees in this passage a proof for a Nabataean rebellion that justified the Roman intervention. For another translation of *tumor*, see Eadie 1985.

210. The tribune's promotion is discussed by Syme 1965: 353 and was—like the *ornamenta triumphalia* for Palma—not necessarily connected to the annexation of Arabia, although it is likely; the bestowing of the *ornamenta triumphalia* upon Palma is mentioned on an honorific statue in Rome (CIL 6.1386); the statue is also reported by Dio Cassius 68.16.2; cf. Wenning 1993a: 101 and note 107. There are only rare cases known where the *ornamenta triumphalia* were bestowed for non-military successes; cf. Zimmermann 1992: especially 295-96; Matteotti 1993: 194.

211. Bowersock 1988; Bowersock suggests that this suppression of the Nabataean *ethnikon* had been instigated by (among others) the cities of the Decapolis who had suffered economically under Nabataean competition (Bowersock 1988: 51-52). It seems unlikely that writers such as Ptolemaios of Alexandria and others would have been influenced by such local animosities. It is much more probable that they followed Roman policies. For Rome, on the other hand, there would have been no reason for treating the Nabataeans differently to other subject peoples, if they had not shown considerable opposition against their integration into the empire.

212. See also Starcky 1966: 920; Bowersock 1983: 82; Sartre 1985: 70; Wenning 1993a: 97-98 has doubts, arguing that Rabbel II could not have been very old in AD 106. In any case, the dynasty of the Nabataean kings did not cease with Rabbel II since, according to one of the Babatha papyri and one inscription from al-Khubtha, Rabbel II had two sons: RES 1434; Meshorer 1975: 87-88; Bowersock 1983: 80 with note 15; Graf 1988: 177 and note 39; for additional matters about the annexation, see Fiema 1987.

213. On the Via Nova Traiana, see Graf 1992: especially 256-59, and Freeman, this volume.

214. See, for instance, Axioti 1980: 189-90 and the references quoted in notes 20 and 21.

215. Fiema 1987: 28-89 is somewhat sceptical, thinking that Roman planning did not go that far.

On the other hand, Domitian had already made preparations for a campaign against the Parthians who were later conquered by Trajan; cf. Wolsky 1993: 176-78. This shows that the Roman emperors had freed themselves from the earlier republican step-by-step policies towards their provinces and clients and attempted to integrate them into future planning.

216. Dio Cassius 48.41.5; see also Debevoise 1938: 108; Wolsky 1976: 413-14, 1993: 41, 122-24 especially note 2.

217. Dio Cassius 49.32.5; Plutarch, *Ant.* 36.2; in the case of the Ituraean king Lysanias, his execution and the donation of Ituraean territories to Cleopatra are explicitly explained by his support of Pacorus.

218. Josephus, *Ant.* 14.14.1-2; Josephus, *War* 1.274-76; although Josephus claims this was just a pretext, this seemed plausible to Herod for he continued his flight to Cleopatra's Egypt.

219. Ammianus Marcellinus 14.8.13: 'Hanc provinciae imposito nomine, rectoreque adtributo, obtemperare legibus nostris Traianus compulit imperator, incolarum tumore saepe contunso, cum glorioso Marte Mediam urgeret et Parthos' (after Loeb Classical Library); see also Sartre 1985: 72.

220. On this see also Eadie 1986: especially 248-49.

221. Dio Cassius 68.28-29; cf. Delbrueck 1955–56: 245. The period from the conquest of Mesopotamia and Babylonia by Trajan (AD 115–117) until Hadrian abandoned them shortly thereafter is the only period in history when Rome directly controlled all starting points for the Far East trade—except the South Arabian ports.

222. Dio Cassius 68.15.1; the Indians were not the only delegation but they were probably the most exotic, which explains their special mention by Dio. It is, moreover, remarkable that the Chinese, who were very interested in contact with the West, withdrew completely, precisely in AD 107, from the Middle Asiatic Sea, their starting point for trade with the West; on this see Delbrueck 1955–56: 251, 264-69. It seems, therefore, that in the Middle and Far East, political and economic changes around the Mediterranean were sharply observed.

223. Such a process of Hellenization has been proposed for South Arabia by Pirenne 1965 with a first phase of a rather autochthonous culture and occasional western influence and a second phase with a strong western influence.

224. Also because, in the case of the first Nabataean coins, it seems that they simply used the well-established mint of Damascus.

225. Wenning 1989: 244 and *passim*.

226. In general see Litvinskij 1998; Harmatta *et al.* 1996; Invernizzi 1995; Errington and Cribb 1992; Arora 1991; Rapin 1990; Karttunen 1990; Holt 1988; Ozols and Thewaldt 1984, all with abundant additional references. Still not replaced is Schlumberger 1969.

227. Due to the circulation and publication of many coin hoards from Central Asia without precise archaeological context, the bibliography about this subject is much richer than in the cases of pottery or architecture: Smirnova 1995; Bopearachchi 1991, 1990; Guillaume 1991, 1987; Holt 1981; Kraay 1981; all with additional references.

228. Gardin 1990, 1985, 1984.

229. Pitschikjan 1996; Nielsen 1994: 124-28, 278-80 cat. no. 19; Bernard 1996: especially 110-16, 1990. The local elements are predominantly present in palace architecture, while other buildings, such as theatres and *gymnasia*, show purely Greek features as can be seen at Aï Khanoum. In general terms, the Hellenistic rulers of the East adopted much from their oriental forerunners, mainly because the forms of royal representation were not well developed in mainland Greece. The local tradition of the Bactrian palace buildings is obvious, as well as their connections to Babylon and the Achaemenid Empire: Sarianidi 1985.

230. These are not related to stronger non-Greek influence but to the fact that new territories in India were conquered by the Greeks and the coins, therefore, had to be understood also by the new subjects.

231. Enoki *et al.* 1996; Zadneprovskiy 1996; Bernard 1987.

232. Puri 1996; Pugachenkova *et al.* 1996; Nehru 1990. As Nehru correctly points out, the western influences in Gandharan sculpture are not the result of one single period of east–west contacts but of several such interactions from the Achaemenid to the Roman periods. However, the Hellenistic input may be considered the most important one; cf. also Taddei 1984; see now the bibliography on Gandhara by Guenée *et al.* 1998.

233. This shall not be understood as a negative development, because, while losing their former culture, they created a new one.

234. On this cf. also Bouzek 1995: 115: 'Es scheint also eine Regel zu sein, dass eigenständige Kunststile in der Alten Welt für ihre Herausbildung zwar den fruchtbaren Kontakt mit griechischen Vorbildern benötigten, aber gleichzeitig auch Distanz, damit die Kraft der griechischen Kunst nicht die Ausprägung der spezifischen Elemente unterdrücken konnte'.

235. On some fragments of this late painted pottery with a well-defined, late fourth- and early fifth-centuries AD context, see Schmid 1996b: 168 along with notes 614 and 615; Fellmann Brogli 1996: 240 fig. 844-49.

236. The fact that no new elements are introduced into Nabataean culture after AD 106 but older features are only repeated should also caution us from an interpretation that there was a continuing Nabataean political entity into the second to fourth centuries AD. In fact, the observations made above should finally offer a possible explanation why there are still remnants of Nabataean cultural elements long after the end of the Nabataean kingdom and resolve the *aporia* observed by Dijkstra 1995: 38-40 and others: cultural remnants, therefore, could exist without political entity.

References

Adriani, A.
1963–66 *Repertorio d'arte dell'Egitto greco-romano*. Serie C 1 and 2. Palermo: Banca di Sicilia.

Altherr-Charon, A.
1977 Origine des temples à trois cellae du bussin méditerranēen est: État de la question. *L'Antiquité classique* 46: 389-440.

Altorki, S. and D.P. Cole
1989 *Arabian Oasis City: The Transformation of 'Unayzah*. Austin: University of Texas.

Andreae, B.
1975 Rekonstruktion des grossen Oecus der Villa des P. Fannius Synistor in Boscoreale. In B. Andreae and H. Kyrieleis (eds.), *Neue Forschungen in Pompeji und den anderen vom Vesuvausbruch 79 n.Chr. verschütteten Städten*, 71-92. Recklinghausen: Bongers.

Andreou, A.
1989 *Griechische Wanddekorationen*. Michelstadt: Neuthor-Verlag.

Arnold, D.
1999 *Temples of the Last Pharaohs*. Oxford: Oxford University Press.

Arnold-Biuchi, C.
1991 Arabian Alexanders. In W.E. Metcalf (ed.), *Mnemata: Papers in Memory of Nancy M. Waggoner*, 99-115. New York: American Numismatic Society.

Arora, U.P. (ed.)
1991 *Graeco-Indica: India's Cultural Contacts with the Greek World*. New Delhi: Ramanand Vidya Bhawan.

Aufrère, S., J.-C. Golvin and J.-C. Goyon
1997 *L'Egypte restituée. I. Sites, et temples de Haute Egypte*. 2nd edn. Paris: Errance.

Augé, Ch.
1990 Sur la figure de Tyché en Nabatène et dans la province d'Arabie. In Zayadine 1990b: 131-46.

Axioti, K.
1980 Rōmaikoí drómoi tēs Aitōloakarnanías. *Archaiologikon Deltion* (Meletes) 35: 186-205.

Bachmann, W., C. Watzinger and Th. Wiegand
1921 *Petra*. Wissenschaftliche Veröffentlichungen des deutsch-türkischen Denkmalschutz-Kommandos 3. Berlin: Springer Verlag.

Bartlett, J.
1979 From the Edomites to the Nabataeans: A Study in Continuity. *Palestine Exploration Quarterly* 111: 53-66.
1990 From Edomites to Nabataeans: The Problem of Continuity. *Aram* 2: 25-34.

Baslez, M.
1992 La famille de Philopappos de Commagène. Un prince entre deux mondes. *Dialogues d'histoire ancienne* 18.1: 89-101.

Bernard, P.
1987 Les nomades conquérants de l'empire gréco-bactrien. Réflexions sur leur identité ethnique et culturelle. *Académie des inscriptions et belles lettres. Comptes rendus*: 758-68.
1990 L'architecture religieuse de l'Asie centrale à l'époque hellénistique. In Deutsches Archäologisches Institut (ed.), *Akten des XIII. internationalen Kongresses für klassische Archäologie, Berlin 1988*, 51-59. Mainz: von Zabern.
1996 The Greek Kingdoms in Central Asia. In Harmatta *et al*. 1996: 99-129.

Bienkowski, P.
1990 The Chronology of Tawilan and the 'Dark Age' of Edom. *Aram* 2: 35-44.

Bignasca, A.
1993 Die Terrakotten. In *Petra und die Weihrauchstrasse* 1993: 65-67.
1996 Terrakotten aus spätrömischen Befunden. In Ez Zantur 1 1996: 283-94.

Bockisch, Ch.
1991 Das Grab 813 aus archäologischer Sicht. In Lindner and Zeitler 1991: 89-97.

Bolelli, G.
1985–86 La ronde bosse de caractère indigène en Syrie du Sud. In J.-M. Dentzer (ed.), *Hauran I, Recherches archéologiques sur la Syrie du Sud à l'époque hellénistique et romaine*, I, 311-72. Paris: Geuthner.

Bonato, S.
1998 Les mausolées hellénistiques de la nécropole de Jérusalem, reflets de la propagation et de l'assimilation de l'hellénisme en Judée. *Annales d'histoire de l'art et d'archéologie* 20: 23-46.

Bopearachchi, O.
1990 Graeco-Bactrian Issues of Later Indo-Greek Kings. *Numismatic Chronicle* 150: 79-103.
1991 *Monnaies gréco-bactriennes et indo-grecques: Catalogue raisonné*. Paris: Bibliothèque nationale.

Börker, Ch.
1973 Neuattisches und pergamenisches an den Ara Pacis-Ranken. *Jahrbuch des deutschen archäologischen Instituts* 88: 283-317.

Boucharlat, R. and P. Lombard
1985 The Oasis of Al Ain in the Iron Age. Excavations at Rumeilah 1981–1983; Survey at Hili 14. *Archaeology in the United Arab Emirates* 4: 44-73. Al-Ain: Department of Antiquity and Tourism.
1991 Datations absolues de Rumeilah et chronologie de l'âge du fer dans la peninsule d'Oman. In K. Schippmann *et al.* (eds.), *Golf-Archäologie: Mesopotamien, Iran, Kuwait, Bahrain, Vereinigte Arabische Emirate und Oman*, 301-14. Göttingen: Leidorf.

Boucharlat, R. and M. Mouton
1993a Mleiha (3e s. avant J.-C. – 1er/2e s. après J.-C.). In U. Finkbeiner (ed.), *Materialien zur Archäologie der Seleukiden- und Partherzeit im südlichen Babylonien und im Golfgebiet*, 219-49. Tübingen: Wasmuth.
1993b Importations occidentales et influence de l'hellénisme dans la péninsule d'Oman. In Invernizzi and Salles 1993: 275-89.

Bouzek, J.
1995 Griechische Kunst in den Randkulturen: Ihre Rezeption, Funktion und Umbildung. In D. Rössler and V. Stürmer (eds.), *Modus in rebus: Gedenkschrift für Wolfgang Schindler*, 114-15. Berlin: Mann.

Bowersock, G.W.
1983 *Roman Arabia*. Cambridge, MA: Harvard University Press.
1988 The Three Arabias in Ptolemy's Geography. In Gatier, Helly and Rey-Coquais 1988: 47-53.
1990a The Cult and Representation of Dusares in Roman Arabia. In Zayadine 1990b: 31-36.
1990b *Hellenism in Late Antiquity*. Cambridge: Cambridge University Press.

Bowsher, J.M.C.
1990 Early Nabataean Coinage. *Aram* 2: 221-28.

Brands, G.
1996 Halle, Propylon und Peristyl—Elemente hellenistischer Palastfassaden in Makedonien. In Hoepfner and Brands 1996: 62-72.

Brentjes, B.
1994 Ortband, Rolltier und Vielfrass. Beobachtungen zur 'skythischen' Akinakes-zier. *Archäologische Mitteilungen aus Iran* 27: 147-64.

Brünnow, R.E. and A. von Domaszewski
1904, *Die Provincia Arabia auf Grund zweier in den Jahren 1897*
1909 *und 1898 unternommener Reisen und der Berichte früherer Reisender*, I (1904), III (1909). Strassburg: Trübner.

Camps, G.
1995 Modèle hellénistique ou modèle punique? Les destinées culturelles de la Numidie. In *Actes du IIIe Congrès international des études phéniciennes et puniques, Tunis 11–16 novembre 1991*, 235-48. Tunis: Institut national du patrimoine.

Cardi, B. de
1984 Survey in Ras al-Khaimah, U.A.E. In R. Boucharlat and J.-F. Salles (eds.), *Arabie orientale, Mésopotamie et Iran méridional de l'âge du fer au début de la période islamique*, 201-15. Paris: Editions recherche sur les civilisations.

Carettoni, G.
1983 La decorazione pittorica della Casa di Augusto sul Palatino. *Mitteilungen des deutschen archäologischen Instituts. Römische Abteilung* 90: 373-419.

Caspari, F.
1916 Das Nilschiff Ptolemaios IV. *Jahrbuch des deutschen archäologischen Instituts* 31: 1-74.

Casson, L.
1989 *The Periplus Maris Erythraei: Text with Introduction, Translation, and Commentary*. Princeton, NJ: Princeton University Press.

Castriota, D.
1995 *The Ara Pacis Augustae and the Imagery of Abundance in Later Greek and Early Roman Imperial Art*. Princeton, NJ: Princeton University Press.

Coarelli, F. and Y. Thébert
1988 Architecture funéraire et pouvoir: Réflexions sur l'hellénisme numide. *Mélanges de l'Ecole française de Rome. Antiquité* 100: 761-819.

Colledge, M.
1987 Greek and non-Greek Interaction in the Art and Architecture of the Hellenistic East. In A. Kuhrt and S. Sherwin-White (eds.), *Hellenism in the East: The Interactions of Greek and non-Greek Civilizations from Syria to Central Asia after Alexander*, 134-62. Berkeley: University of California Press.

Conticello, B., and B. Andreae
1974 Die Skulpturen von Sperlonga. *Antike Plastik* 14. Berlin: Mann.

Cugunov, K.V.
   1998   Der skythenzeitliche Kulturwandel in Tuva. *Eurasia antiqua* 4: 273-309.

Debevoise, N.C.
   1938   *A Political History of Parthia*. Chicago: The University of Chicago.

Delbrueck, R.
   1955–56   Südasiatische Seefahrt im Altertum. *Bonner Jahrbücher* 155–156: 8-58, 229-308.

Dentzer-Feydy, J.
   1985–86   Décor architectural et développement du Hauran dans l'antiquité (du Ier s. av. au VIIe s. de notre ère). In J.-M. Dentzer (ed.), *Hauran I: Recherches archéologique sur la Syrie du Sud à l'époque hellénistique et romaine*, I. 2: 261-309. Paris: Geuthner.

Despinis, G.I.
   1996   Studien zur hellenistischen Plastik II: Statuengruppen. *Mitteilungen des deutschen archäologischen Instituts. Abteilung Athen* 111: 299-336.

Diebner, S.
   1982   Bosra: Die Skulpturen im Hof der Zitadelle. *Rivista di archaeologia* 6: 52-71.

Dijkstra, K.
   1995   *Life and Loyalty: A Study in the Socio-Religious Culture of Syria and Mesopotamia in the Graeco-Roman Period Based on Epigraphical Evidence*. Leiden: E.J. Brill.

Drexhage, R.
   1988   *Untersuchungen zum römischen Osthandel*. Bonn: Habelt.

Eadie, J.W.
   1985   Artifacts of Annexation: Trajans' Grand Strategy and Arabia. In J.W. Eadie and J. Ober (eds.), *The Craft of the Ancient Historian: Essays in Honor of Ch. G. Starr*, 407-23. Lanham, MD: University Press of America.
   1986   The Evolution of the Roman Frontier in Arabia. In Ph. Freeman and D. Kennedy (eds.), *The Defence of the Roman and Byzantine East: Proceedings of a Colloquium held at the University of Sheffield in April 1986*, 243-52. British Archaeological Reports, International Series 297. Oxford: British Archaeological Reports.
   1989   Strategies of Economic Development in the Roman East: The Red Sea Trade Revisited. In D.H. French and C.S. Lightfoot (eds.), *The Eastern Frontier of the Roman Empire: Proceedings of a Colloquium held at Ankara in September 1988*, 1: 113-20. British Archaeological Reports, International Series 553. Oxford: British Archaeological Reports.

Enoki, K., G.A. Koshelenko and Z. Haidary
   1996   The Yüeh-Chih and their Migrations. In Harmatta *et al.* 1996: 171-89.

Errington, E. and J. Cribb (eds.)
   1992   *The Crossroads of Asia: Transformation in Image and Symbol*. Cambridge: Ancient India and Iran Trust.

Ez Zantur 1.
   1996   *Petra ez Zantur. I. Die Ergebnisse der schweizerisch-liechtensteinischen Ausgrabungen 1988–1992*. Terra archaeologica II. Monographien der schweizerisch-liechtensteinischen Stiftung für archäologische Forschungen im Ausland (SLSA/FSLA). Mainz: von Zabern.

Fedak, F.
   1990   *Monumental Tombs of the Hellenistic Age: A Study of Selected Tombs from the Pre-Classical to the Early Imperial Era*. Toronto: University of Toronto.

Fellmann Brogli, R.
   1996   Die Keramik aus den spätrömischen Bauten. In Ez Zantur 1 1996: 219-81.

Fiema, Z.T.
   1987   The Roman Annexation of Arabia: A General Perspective. *Ancient World* 15: 25-35.
   1996   Nabataean and Palmyrene Commerce—The Mechanisms of Intensification. In *Palmyra and the Silk Road*. Annales archéologiques arabes syriennes 42: 189-95.

Fischer, Th.
   1980   *Seleukiden und Makkabäer: Beiträge zur Seleukidengeschichte und zu den politischen Ereignissen in Judäa während der 1. Hälfte des 2. Jahrhunderts v.Chr.* Bochum: In Kommission beim Studienverlag N. Brockmeyer.
   1990a   Hasmoneans and Seleucids: Aspects of War and Policy in the Second and First Centuries B.C.E. In A. Kasher *et al.* (eds.), *Greece and Rome in Eretz Israel: Collected Essays*, 3-19. Jerusalem: Yad Izhak Ben-Zvi.

Fischer, M.L.
   1990b   *Das korinthische Kapitell im alten Israel in der hellenistischen und römischen Periode*. Mainz: von Zabern.

Fittschen, K.
   1976a   Zur Herkunft und Entstehung des 2. Stils—Probleme und Argumente. In Zanker 1976: 539-63.
   1976b   Zur Panzerstatue in Cherchel. *Jahrbuch des deutschen archäologischen Instituts* 91: 175-210.
   1996   Wall Decorations in Herod's Kingdom: Their Relationship with Wall Decorations in Greece and Italy. In Fittschen and Foerster 1996: 139-61.

Fittschen, K. and G. Foerster (eds.)
   1996   *Judaea and the Greco-Roman World in the Time of Herod in the Light of Archaeological Evidence*. Abhandlungen der Akademie der Wissenschaften in Göttingen. Philologisch-Historische Klasse. Dritte Folge 215. Göttingen: Vandenhoeck & Ruprecht.

Foerster, G.
   1996   Hellenistic and Roman Trends in the Herodian Architecture of Masada. In Fittschen and Foerster 1996: 55-72.

Fraser, P.M.
   1977   *Rhodian Funerary Monuments*. Oxford: Clarendon Press.

Freyberger, K.S.
1991 Zur Datierung des Grabmals des Sextus Florentinus. *Damaszener Mitteilungen* 5: 1-8.
1997 Blattranken, Greifen und Elephanten. In Weber and Wenning 1997: 71-84.
1998 *Die frühkaiserzeitlichen Heiligtümer der Karawanenstationen im hellenisierten Osten. Zeugnisse eines kulturellen Konflikts im Spannungsfeld zweier politischer Formationen.* Mainz: von Zabern.

Frösén, J. et al.
1998 The Finnish Jabal Harun Project: Report on the 1997 Season. *Annual of the Department of Antiquities of Jordan* 42: 483-502.

Funke, P.
1996 Die syrisch-mesopotamische Staatenwelt in vorislamischer Zeit. Zu den arabischen Macht- und Staatenbildungen an der Peripherie der antiken Grossmächte im Hellenismus und in der römischen Kaiserzeit. In B. Funck (ed.), *Hellenismus. Beiträge zur Erforschung von Akkulturation und politischer Ordnung in den Staaten des hellenistischen Zeitalters*, 217-38. Tübingen: J.C.B. Mohr.

Gagsteiger
1991 Die Architektur Petras. In Lindner and Zeitler 1991: 49-68.

Garbini, G.
1993 *Aramaica.* Studi semitici 10. Rome: Universita degli studi 'La Sapienza'.

Gardin, J.C
1984 Die Ursprünge der Kusana-Keramik. In Ozols and Thewaldt 1984: 110-26.
1985 Les relations entre la Méditerranée et la Bactriane dans l'antiquité, d'après des données céramologiques inédites. In J.-L. Huot *et al.* (eds.), *De l'Indus aux Balkans: Recueil à la mémoire de Jean Deshayes*, 447-60. Paris: Recherche sur les civilisations.
1990 La céramique hellénistique en Asie centrale: Problèmes d'interprétation. In Deutsches Archäologisches Institut (ed.), *Akten des 13. internationalen Kongresses für klassische Archäologie, Berlin 1988*: 187-93. Mainz: von Zabern.

Gatier, P.-L., B. Helly and J.-P. Rey-Coquais (eds.)
1988 *Géographie historique au Proche-Orient (Syrie, Phénicie, Arabie, grecques, romaines, byzantines): Actes de la Table Ronde de Valbonne, 16-18 septembre 1985.* Paris: Editions du centre nationale de la recherche.

Gawlikowski, M.
1990 Les dieux des Nabatéens. In H. Temporini and W. Haase (eds.), *Aufstieg und Niedergang der römischen Welt*, II. 18. 4: 2659-77. Berlin: W. de Gruyter.

Glueck, N.
1965 *Deities and Dolphins: The Story of the Nabataeans.* London: Farrar, Straus & Giroux.

Graf, D.F.
1988 Qura Arabiyya and Provincia Arabia. In Gatier, Helly and Rey-Coquais 1988: 171-211.
1990 The Origin of the Nabataeans. *Aram* 2: 45-75.
1992 Nabataean Settlements and Roman Occupation in Arabia Petraea. In A. Hadidi (ed.), *Studies in the History and Archaeology of Jordan*, IV: 253-60. Amman: Department of Antiquities.

Grimm, G.
1981 Orient und Okzident in der Kunst Alexandriens: In N. Hinske (ed.), *Alexandrien: Kulturbegegnungen dreier Jahrtausende im Schmelztiegel einer mediterranen Großstadt*, 13-25. Mainz: von Zabern.

Guenée, P. et al.
1998 *Bibliographie analytique des ouvrages parus sur l'art du Gandhara entre 1950 et 1993.* Mémoires de l'Académie des inscriptions et belles-lettres 16. Paris: de Boccard.

Guggisberg, M.
1998 'Zoomorphe Junktur' und 'Inversion'. Zum Einfluss des skythischen Tierstils auf die frühe keltische Kunst. *Germania* 76: 549-72.

Guillaume, O.
1987 *L'analyse des raisonnements en archéologie: Le cas de la numismatique gréco-bactrienne et indo-grecque.* Paris: Recherche sur les civilisations.
1991 *Graeco-Bactrian and Indian Coins from Afghanistan.* New York: Oxford University Press.

Gundlach, R. and M. Rochholz (eds.)
1994 *Ägyptische Tempel: Struktur, Funktion und Programm. Akten der ägyptologischen Tempeltagungen in Gosen 1990 und Mainz 1992.* Hildesheim: Gerstenberg.

Haerinck, E.
1983 *La céramique en Iran pendant la période parthe (circa 250 av. J. C. à circa 225 après J. C.): Typologie, chronologie et distribution.* Iranica antiqua Supplement 2. Leuven: Imprimerie Orientaliste.
1998 More Pre-Islamic Coins from Southeastern Arabia. *Arabian Archaeology and Epigraphy* 9: 278-301.

Hammond, Ph.C.
1965 *The Excavation of the Main Theater at Petra, 1961–1962: Final Report.* London: Quaritch.
1973 *The Nabataeans: Their History, Culture and Archaeology.* Studies in Mediterranean Archaeology 37. Gothenburg: Aström.
1981 Ein nabatäisches Weihrelief aus Petra. In Nabatüer 1981: 137-41.
1986 Die Ausgrabung des Löwen-Greifen-Tempels in Petra (1973–1983). In Lindner 1986: 16-30.
1996 *The Temple of the Winged Lions Petra, Jordan 1974–1990.* Arizona: Petra.

Hannestad, L.
1983 *The Hellenistic Pottery from Failaka. Ikaros. The Hellenistic Settlements 2.* Aarhus, Denmark: Jysk Arkaeologisk Selskab.

Harmatta, J. et al. (eds.)
1996 *History of Civilizations of Central Asia. II. The Development of Sedentary and Nomadic Civilizations:* 700 B.C. to AD 250. 2nd edn. Paris: UNESCO.

Hart, S.
1986 Sela': the Rock of Edom? *Palestine Exploration Quarterly* 118: 91-95.
1987 Five Soundings in Southern Jordan. *Levant* 19: 33-47.

Hart, S. and R. Falkner
1985 Preliminary Report on a Survey in Edom 1984. *Annual of the Department of Antiquities of Jordan* 29: 255-77.

Heermann, V.
1986 *Studien zur makedonischen Palastarchitektur*. Berlin: Papyrus-Druck.

Hesberg, H. von
1978 Zur Entwicklung der griechischen Architektur im ptolemäischen Reich. In H. Maehler and V.M. Strocka (eds.), *Das ptolemäische Ägypten: Akten des internationalen Symposions 27–29. September 1976 in Berlin*, 137-45. Mainz: von Zabern.

Higgs, P.
1997 A Newly Found Fragment of Free-standing Sculpture from the Mausoleum at Halicarnassus. In Jenkins and Waywell 1997: 30-34.

Hoepfner, W. and G. Brands (eds.)
1996 *Basileia: Die Paläste der hellenistischen Könige. Internationales Symposion in Berlin vom 16–20.12.1992*. Mainz: von Zabern.

Hoepfner, W. et al.
1999 Die Epoche der Griechen. In W. Hoepfner (ed.), *Geschichte des Wohnens. I. 5000 v.Chr. – 500 n.Chr.: Vorgeschichte, Frühgeschichte, Antike*, 123-608. Stuttgart: Deutsche Verlags-Anstalt.

Hölscher, T.
1988 Historische Reliefs. In Antikenmuseum Berlin (ed.), *Kaiser Augustus und die verlorene Republik*, 351-400. Mainz: von Zabern.

Holt, F.L.
1981 The Euthydemid Coinage of Bactria: Further Hoard Evidence from Ai Khanoum. *Revue numismatique* 23: 7-44.
1988 *Alexander the Great and Bactria: The Formation of a Greek Frontier in Central Asia*. Leiden: E.J. Brill.

Huff, D.
1989 Das Qasr Nuweijis bei Amman. *Istanbuler Mitteilungen* 39: 223-36.

Huth, M.
1999 An Important Hoard of Early South Arabian Coins from the Kingdom of Qataban. *Schweizerische numismatische Rundschau* 78: 37-51.

Inoubliable Pétra
1980 *Le royaume nabatéen aux confins du désert. Musées royaux d'art et d'histoire, 1 mars–1 juin 1980*. Brussels: Snoeck-Ducaju.

Invernizzi, A. (ed.)
1995 *In the Land of the Gryphons: Papers on Central Asian Archaeology in Antiquity*. Florence: Le lettere.

Invernizzi, A. and J.-F. Salles (eds.)
1993 *Arabia antiqua: Hellenistic Centers around Arabia*. Rome: Istituto italiano per il Medio ed Estremo Oriente.

Jacobson, E.
1995 *The Art of the Scythians: The Interpretation of Cultures at the Edge of the Hellenic World*. Handbuch der Orientalistik 8, 2. Leiden: E.J. Brill.

Jaeggi, O. and S.G. Schmid
1996 Beiträge zur Sammlung Lagunillas des Museo Nacional de Bellas Artes in Havanna (Kuba). *Antike Kunst* 39: 14-37.

Jenkins, I. and G.B. Waywell (eds.)
1997 *Sculptors and Sculpture of Caria and the Dodecanese*. London: British Museum.

Jenkins, I. et al.
1997 The Polychromy of the Mausoleum. In Jenkins and Waywell 1997: 35-41.

Jeppesen, K.
1997 The Mausoleum at Halicarnassus: Sculptural Decoration and Architectural Background. In Jenkins and Waywell 1997: 42-48.
1998 Das Maussoleion von Halikarnass, Forschungsbericht 1997. *Proceedings of the Danish Institute at Athens* 2: 161-231.

Joukowsky, M.S.
1998 Brown University 1997 Excavations at the Petra Great Temple. *Annual of the Department of Antiquities of Jordan* 42: 293-318.
1999 Brown University 1998 Excavations at the Petra Great Temple. *Annual of the Department of Antiquities of Jordan* 43: 195-222.

Kader, I.
1996 *Propylon und Bogentor: Untersuchungen zum Tetrapylon von Latakia und anderen frühkaiserzeitlichen Bogenmonumenten im Nahen Osten*. Damaszener Forschungen 7. Mainz: von Zabern.

Karttunen, K.
1990 Taxila: Indian City and a Stronghold of Hellenism. *Arctos* 24: 85-96.

Keller, D.
1996 Gedanken zur Datierung und Verwendung der Statere Philipps II. und ihrer keltischen Imitationen. *Schweizerische numismatische Rundschau* 75: 101-19.

Kennedy, D.L.
- 1980 *Legio VI Ferrata*: The Annexation and Early Garrison of Arabia. *Harvard Studies in Classical Philology* 84: 283-309.

Kitchen, K.A.
- 1994 *Documentation for Ancient Arabia. I. Chronological Framework and Historical Sources*. Liverpool: Liverpool University Press.

Kleiner, D.E.E.
- 1983 *The Monument of Philopappos in Athens*. Rome: Bretschneider.

Knauf, E.A.
- 1986 Die Herkunft der Nabatäer. In Lindner 1986: 74-86.
- 1990 Dushara and Shai' al-Qaum. *Aram* 2: 175-83.
- 1997 Der sein Volk liebt: Entwicklung des nabatäischen Handelsimperiums zwischen Stamm, Königtum und Klientel. In Weber and Wenning 1997: 14-24

Kokkinos, N.
- 1998 *The Herodian Dynasty: Origins, Role in Society and Eclipse*. Sheffield: Sheffield Academic Press.

Kolb, B.
- 1997 Petra–eine Zeltstadt in Stein. In Weber and Wenning 1997: 62-66.

Kolb, B., D. Keller and R. Fellmann Brogli
- 1997 Swiss-Liechtenstein Excavations at az-Zantur in Petra 1996: The Seventh Season. *Annual of the Department of Antiquities of Jordan* 41: 231-54.

Kolb, B., D. Keller and Y. Gerber
- 1998 Swiss-Liechtenstein Excavations at az-Zantur in Petra 1997. *Annual of the Department of Antiquities of Jordan* 42: 259-77.

Königsweg
- 1987 *Der Königsweg: 9000 Jahre Kunst und Kultur in Jordanien und Palästina*. Exhibition Cologne, Schallaburg, Munich.

Konstantinopoulos, G.
- 1973 Néa eurēmata ek Ródou kai Astupalaías. *Athens Annals in Archaeology* 6: 114-24.
- 1986 *Archaía Ródos: Episkópēsē tēs istorías kai tēs téchnēs*. Athens: Morphotiko Hidryme Ethnikes Trapezes.

Kose, A.
- 1996 Zur Säulenarchitektur im parthischen Mesopotamien. In E.-L. Schwandner (ed.), *Säule und Gebälk: Zu Struktur und Wandlungsprozess griechisch-römischer Architektur*, 203-20. Mainz: von Zabern.

Kraay, C.M.
- 1981 Demetrius in Bactria and India. *Numismatica e antichità classicue. Quaderni ticinesi* 10: 219-33.

Kraeling, C.H.
- 1962 *Ptolemais, City of the Libyan Pentapolis*. Chicago: The University of Chicago.

Kraus, Th.
- 1976 Überlegungen zum Bauornament. In Zanker 1976: 455-70.

Kreeb, M.
- 1988 *Untersuchungen zur figürlichen Ausstattung delischer Privathäuser*. Chicago: Ares.

Kushnir-Stein, A. and H. Gitler
- 1992–93 Numismatic Evidence from Tel Beer-Sheva and the Beginning of Nabataean Coinage. *Israel Numismatic Journal* 12: 13-20.

Lahusen, G.
- 1999 Bemerkungen zur Laokoon-Gruppe. In *Hellenistische Gruppen: Gedenkschrift für Andreas Linfert*, 295-305. Mainz: von Zabern.

Lang-Auinger, C., G. Forstenpointner and G. Lang
- 1996 *Hanghaus 1 in Ephesos: Der Baubefund*. Forschungen in Ephesos 8, 3. Vienna: Verlag der Österreichischen Akademie der Wissenschaften.

Laroche-Traunecker, F.
- 2000 Chapiteaux 'nabateéns', 'corinthiens inachevés' ou 'simplifiés'? Nouveaux exemples en Egypte. *Ktéma* 25: 207-13.

Lauter, H.
- 1971 Ptolemais in Libyen: Ein Beitrag zur Baukunst Alexandrias. *Jahrbuch des deutschen archäologischen Instituts* 86: 149-78.
- 1972 Kunst und Landschaft—ein Beitrag zum rhodischen Hellenismus. *Antike Kunst* 15: 49-59.
- 1986 *Die Architektur des Hellenismus*. Darmstadt: Wissenschaftliche Buchgesellschaft.
- 1988 Hellenistische Sepulkralarchitektur auf Rhodos. In S. Dietz and I. Papachristodoulou (eds.), *Archaeology in the Dodecanese*, 155-63. Copenhagen: National Museum of Denmark, Department of Near Eastern and Classical Antiquities.

Lindner, M. (ed.)
- 1986 *Petra: Neue Ausgrabungen und Entdeckungen*. Munich: Delp.
- 1989 *Petra und das Königreich der Nabatäer: Lebensraum, Geschichte und Kultur eines arabischen Volkes der Antike*. Munich: Delp.

Lindner, M. and J.P. Zeitler (eds.)
- 1991 *Petra: Königin der Weihrauchstrasse*. Fürth: UKA-Verlag.

Litvinskij, B.A.
- 1998 *La civilisation de l'Asie centrale antique*. Archäologie in Iran und Turan 3. Rahden: Leidorf.

Lyttelton, M.
- 1974 *Baroque Architecture in Classical Antiquity*. London: Thames & Hudson.

Lyttelton, M. and T. Blagg
- 1990 Sculpture in Nabataean Petra, and the Question of Roman Influence. In M. Henig (ed.), *Architecture and*

Architectural Sculpture in the Roman Empire, 91-107. Oxford: Oxford Committee for Archaeology.

Macdonald, M.C.A.
- 1993 Nomads and the Hawran in the Late Hellenistic and Roman Periods: A Reassessment of the Epigraphic Evidence. *Syria* 70: 303-413.

Matteotti, R.
- 1993 Zur Militärgeschichte von Augusta Rauricorum in der zweiten Hälfte des 1. Jahrhunderts n.Chr. Die Truppenziegel der 21. Legion aus Augst. *Jahresberichte aus Augst und Kaiseraugst* 14: 185-97.

Matthiae, K.
- 1988 Die Fassade des *Bab-es-Siq*-Trikliniums in Petra: Bemerkungen zu ihrer Gestaltung. *Zeitschrift des deutschen Palästina-Vereins* 104: 74-83.
- 1989 Die Fassade von Ed-Der in Petra. *Klio* 71: 257-79.
- 1991 Die nabatäische Felsarchitektur in Petra. *Klio* 73: 226-78.
- 1994 Der Platz vor der Khazne-Fassade in Petra: Beobachtungen zur Gestaltung der unteren Etage dieser Fassade. *Das Altertum* 39: 275-86.
- 1995 Die Tholos der *Hazne*-Fassade in Petra. *Zeitschrift des deutschen Palästina-Vereins* 111: 151-61.

Mattingly, G.L.
- 1990 Settlement on Jordan's Kerak Plateau from Iron Age IIC through the Early Roman Period. *Aram* 2: 309-35.

McKenzie, J.S.
- 1988 The Development of Nabataean Sculpture at Petra and Khirbet Tannur. *Palestine Exploration Quarterly* 120: 81-107.
- 1990 *The Architecture of Petra.* Oxford: Oxford University Press.
- 1996 Alexandria and the Origins of Baroque Architecture. In *Alexandria and Alexandrianism: Papers Delivered at a Symposium Organized by the J. Paul Getty Museum and The Getty Center for the History of Art and the Humanities and Held at the Museum April 22–25, 1993*, 109-25. Malibu: The J. Paul Getty Museum.

Meshorer, Y.
- 1975 *Nabataean Coins.* Qedem 3. Jerusalem: Institute of Archaeology, Hebrew University of Jerusalem.

Meskel, L. (ed.)
- 1998 *Archaeology under Fire: Nationalism, Politics and Heritage in the Eastern Mediterranean and Middle East.* London: Routledge.

Mette, B.D.
- 1992 Skulptur und Landschaft: Mythologische Skulpturengruppen in griechischer und römischer Aufstellung. Unpublished doctoral thesis, University of Cologne.

Mettinger, T.N.D.
- 1995 *No Graven Image? Israelite Aniconism in Its Near Eastern Context.* Stockholm: Almquist & Wiksell.

Mielsch, H.
- 1997 *Die römische Villa: Architektur und Lebensform.* Munich: Beck.

Miglus, P.A.
- 1999 *Städtische Wohnarchitektur in Babylonien und Assyrien.* Baghdader Forschungen 22. Mainz: von Zabern.

Milik, J.T.
- 1958 Nouvelles inscriptions nabatéennes. *Syria* 35: 227-51.
- 1982 Origines des Nabatéens. In A. Hadidi (ed.), *Studies in the History and Archaeology of Jordan*, I: 261-65. Amman: Department of Antiquities.

Millar, F.
- 1993 *The Roman Near East, 31 BC–AD 337.* Cambridge, MA: Harvard University Press.

Mlynarczyk, J.
- 1990 *Nea Paphos in the Hellenistic Period.* Nea Paphos 3. Warsaw: PWN-Editions scientifiques de Pologne.
- 1996 Palaces of Strategoi and the Ptolemies in Nea Paphos: Topographical Remarks. In Hoepfner and Brands 1996: 193-202.
- 1998 Terracotta Mould-Made Lamps in Alexandria (Hellenistic to Late Roman Period). In J.-Y. Empereur (ed.), *Commerce et Artisanat dans l'Alexandrie hellénistique et romaine*, 327-52. Bulletin de correspondance hellénique Supplément 33. Athens: Ecole française d'Athens.

Moutsopoulos, N.C.
- 1990 Observations sur les représentations du Panthéon nabatéen. In F. Zayadine (ed.), *Petra and the Caravan Cities*, 53-75. Amman: Department of Antiquities.

Munro-Hay, S.C.
- 1991 The Coinage of Shabwa (Hadramawt) and other Ancient South Arabian Coinage in the National Museum, Aden. *Syria* 68: 393-418.
- 1994 Coins of Ancient South Arabia. *Numismatic Chronicle* 154: 191-203.
- 1996 Coins of Ancient South Arabia, II. *Numismatic Chronicle* 156: 33-47.

Nabatäer
- 1981 *Die Nabatäer: Erträge einer Ausstellung im Rheinischen Landesmuseum Bonn, 24. Mai – 9. Juli 1978.* Cologne: Rheinland-Verlag/Bonn: Habelt.

Nabateërs
- 1971 *De Nabateërs, een vergeten volk aan de Dode Zee, 312 v.–106 n.Chr. Provincial gallo-romeins Museum, Tongeren (B), 19 februari tot 12 april 1971.* Tongeren: Provincial gallo-romeins Museum.

Naster, P.
- 1983 Remarques au sujet des imitations des monnaies d'Athènes dans la presqu'île arabique. In P. Naster, *Scripta Nummaria: Contributions à la méthodologique numismatique*, 141-46. Louvain-La-Neuve: Seminaire de numismatique Marcel Hoc.

Nehmé, L.
- 1997a L'habitat rupestre dans le bassin de Pétra à l'époque nabatéenne. In G. Bisheh, M. Zaghloul and I. Kehrberg (eds.), *Studies in the History and Archaeology of Jordan*, VI: 281-88. Amman: Department of Antiquities.
- 1997b L'espace cultuel de Pétra à l'époque nabatéenne. *Topoi* 7: 1023-67.

Nehru, L.
- 1990 Hellenism in Gandharan Sculpture. In Deutsches Archäologisches Institut (ed.), *Akten des XIII. internationalen Kongresses für klassische Archäologie, Berlin 1988*, 317-19. Mainz: von Zabern.

Netzer, E.
- 1990 Architecture in Palaestina prior to and during the Days of Herod the Great. In Deutsches Archäologisches Institut (ed.), *Akten des XIII. internationalen Kongresses für klassische Archäologie, Berlin 1988*, 37-50. Mainz: von Zabern.
- 1991 *The Buildings: Stratigraphy and Architecture. Masada 3*. Jerusalem: Israel Exploration Society.
- 1996a The Hasmonean Palaces in Palaestina. In Hoepfner and Brands 1996: 203-208.
- 1996b The Palaces Built by Herod — A Research Update. In Fittschen and Foerster 1996: 27-54.
- 1999 *Die Paläste der Hasmonäer und Herodes' des Grossen*. Antike Welt, special issue. Mainz: von Zabern.

Nielsen, I.
- 1994 *Hellenistic Palaces: Tradition and Renewal*. Aarhus, Denmark: Aarhus University.

Normann, A. von
- 1996 *Architekturtoreutik in der Antike*. Munich: tuduv.

Orlandos, A.K.
- 1976 Neóterai éreunai en Messéné. In U. Jantzen (ed.), *Neue Forschungen in griechischen Heiligtümern: Internationales Symposion in Olympia vom 10. bis 12. Oktober 1974*, 9-38. Tübingen: Wasmuth.

Ovadiah, A. and Y. Turnheim
- 1994 *'Peopled' Scrolls in Roman Architectural Decoration in Israel: The Roman Theatre at Beth Shean/Scythopolis*. Rome: Bretschneider.

Ozols, J. and V. Thewaldt (eds.)
- 1984 *Aus dem Osten des Alexanderreiches: Völker und Kulturen zwischen Orient und Okzident. Iran, Afghanistan, Pakistan, Indien*. Cologne: DuMont.

Pandermalis, D.
- 1976 Beobachtungen zur Fassadenarchitektur und Aussichtsveranda im hellenistischen Makedonien. In Zanker 1976: 387-97.

Parlasca, I.
- 1990a Seltene Typen nabatäischer Terrakotten: Östliche Motive in der späteren Provincia Arabia. In Ch. Börker and M. Donderer (eds.), *Das antike Rom und der Osten: Festschrift Klaus Parlasca zum 65. Geburtstag*, 157-74. Auslieferung Universitätsbibliothek Erlangen. Erlangen: Universitätsbund Erlangen-Nürnberg.
- 1990b Terrakotten aus Petra: Ein neues Kapitel nabatäischer Archäologie. In Deutsches Archäologisches Institut (ed.), Zayadine 1990b: 87-105.

Parlasca, Götter und Reittiere
- 1997 Religions- und kulturgeschichtliche Bedeutung nabatäischer Terrakotten. In Weber and Wenning: 126-131.

Parr, P.J.
- 1960 Excavations at Petra, 1958–59. *Palestine Exploration Quarterly* 92: 124-35.
- 1968 Découvertes récentes au sanctuaire du Qasr à Pétra, I: Comptes rendu des dernières fouilles. *Syria* 45: 1-24.
- 1970 A Sequence of Pottery from Petra. In J.A. Sanders (ed.), *Near Eastern Archaeology in the Twentieth Century: Essays in Honor of Nelson Glueck*, 348-81. Garden City, NY: Doubleday.

Patrich, J.
- 1990 *The Formation of Nabatean Art: Prohibition of a Graven Image among the Nabateans*. Jerusalem: Magnes.

Pensabene, P.
- 1991 Elementi di architettura alessandrina. In S. Stucchi and M. Bonanno Aravantinos (eds.), *Giornate di studio in onore di Achille Adriani*, 29-85. Studi Miscellanei 28. Rome: 'L'Erma' di Bretschneider.
- 1993 *Elementi architettonici di Alessandria e di altri siti egiziani: Repertorio d'arte dell'Egitto greco-romano*. Serie C 3. Rome: Bretschneider.

Pesce, G.
- 1950 *Il 'Palazzo delle colonne' in Tolemaide di Cirenaica*. Rome: 'L'Erma' di Bretschneider.

Peter, M.
- 1993 Münzprägung und Münzumlauf im Gebiet der Nabatäer. In Petra und die Weihrauchstrasse 1993: 18-20.
- 1996 Die Fundmünzen. In Ez Zantur 1 1996: 91-127.

Peters, F.E.
- 1977 The Nabataeans in the Hawran. *Journal of the American Oriental Society* 97: 263-77.

Petra und die Weihrauchstrasse
- 1993 *Petra und die Weihrauchstrasse*. Catalogue of an exhibition in Zurich and Basel. Zurich: Konzept Gut.

Pfrommer, M.
- 1996a Roots and Contacts: Aspects of Alexandrian Craftmanship. In *Alexandria and Alexandrianism: Papers Delivered at a Symposium Organized by the J. Paul Getty Museum and The Getty Center for the History of Art and the Humanities and Held at the Museum April 22-25, 1993*, 171-89. Malibu: The J. Paul Getty Museum.
- 1996b Fassade und Heiligtum: Betrachtungen zur architektonischen Repräsentation des vierten Ptolemäers. In Hoepfner and Brands 1996: 97-108.

Pinkwart, D. and W. Stamnitz
- 1984 *Peristylhäuses westlich der unterer Agora*. Altertümer von Pergamon 14. Berlin: W. de Gruyter.

Pirenne, J.
- 1965 Les phases de l'hellénisation dans l'art sud-arabe. In *Le rayonnement des civilisations grecque et romaine sur les cultures périphériques: Huitième congrès international d'archéologie classique (Paris 1963)*, 535-41. Paris: de Boccard.

Pitschikjan, I.
- 1996 Die Entwicklung des baktrischen Palast-Tempels. In Hoepfner and Brands 1996: 226-33.

Potts, D.T.
- 1991 *The Pre-Islamic Coinage of Eastern Arabia*. Copenhagen: Carsten Niebuhr Institute of Near Eastern Studies, University of Copenhagen/Museum Tusculanum.

Pugachenkova, G. et al.
- 1996 Kushan Art. In Harmatta *et al.* 1996: 331-95.

Puri, B.N.
- 1996 The Kushans. In Harmatta *et al.* 1996: 247-63.

Qusus, N.
- 1984 *Pétra: Ē nabataikē prōteúousa tes iordanikēs erēmou. Sumbolē stēn architektonikē kai tē morfologia tōn mnēmeíōn*. Thessaloniki: Marrogenis.

Raeder, J.
- 1988 Vitruv, de architectura VI 7 (aedificia Graecorum) und die hellenistische Wohnhaus- und Palastarchitektur. *Gymnasium* 95: 316-68.

Rakob, F.
- 1979 Numidische Königsarchitektur in Nordafrika. In H.G. Horn and Ch.B. Rüger (eds.), *Die Numider: Reiter und Könige nördlich der Sahara*, 119-71. Cologne: Rheinland-Verlag/Bonn: Habelt.
- 1983 Architecture royale numide. In *Architecture et société: De l'archaïsme grec à la fin de la république romaine. Actes du Colloque international organisé par le Centre national de la recherche scientifique et l'Ecole française de Rome (Rome 2–4 décembre 1980)*, 325-48. Paris: Le Centre national de la recherche scientifique.

Rapin, C.
- 1990 Greeks in Afghanistan: Aï Khanoum. In J.-P. Descoeudres (ed.), *Greek Colonists and Native Populations*, 329-42. New York: Oxford University Press.

Raschke, M.G.
- 1978 New Studies in Roman Commerce with the East. In H. Temporini and W. Haase (eds.), *Aufstieg und Niedergang der römischen Welt*, II. 9.2: 604-1076. Berlin: W. de Gruyter.

Reber, K.
- 1988 Aedificia graecorum: Zu Vitruvs Beschreibung des griechischen Hauses. *Archäologischer Anzeiger*: 653-66.

Rice, E.E.
- 1995 Grottoes on the Acropolis of Hellenistic Rhodes. *Annual of the British School at Athens* 90: 383-404.

Roche, M.-J.
- 1996 Remarques sur les Nabatéens en Méditerrané. *Semitica* 45: 73-99.

Rollin, S. and J. Streetly
- 1996 *Jordan*. Blue Guide. London: A&C Black/New York: Norton.

Romanis, F. de
- 1996 *Cassia, cinnamomo, ossidiana: Uomini e merci tra Oceano Indiano e Mediterraneo*. Rome: 'L'Erma' di Bretschneider.

Ronczewski, K.
- 1932 Kapitelle des El Hasne in Petra. *Archäologischer Anzeiger* 47: 38-89.

Rozenberg, S.
- 1996 The Wall Paintings of the Herodian Palace at Jericho. In Fittschen and Foerster 1996: 121-38.

Rumscheid, F.
- 1994 *Untersuchungen zur kleinasiatischen Bauornamentik des Hellenismus*. Mainz: von Zabern.

Sarianidi, V.
- 1985 Monumental architecture of Bactria. In J.-L. Huot *et al.* (eds.), *De l'Indus aux Balkans: Recueil à la mémoire de Jean Deshayes*, 417-32. Paris: Recherche sur les civilisations.

Sartre, M.
- 1982 *Trois études sur l'Arabie romaine et byzantine*. Collection Latomus 178. Brussels: Revue d'études latines.
- 1985 *Bostra: Des origines à l'Islam*. Bibliothèque archéologique et historique 117. Paris: Geuthner.

Sasianou, A.
- 1993 Copies and Imitations of Thasian Tetradrachms. In *Proceedings of the XIth International Numismatic Congress*, I: 123-31. Louvain-La-Neuve: Association Professeur Marcel Hoc pour l'encouragement des recherches numismatiques.

Sauron, G.
- 1988 Le message esthétique des rinceaux de l'Ara Pacis Augustae. *Revue archéologique*: 3-40.

Schefold, K.
- 1975 Der zweite Stil als Zeugnis alexandrinischer Architektur. In B. Andreae and H. Kyrieleis (eds.), *Neue Forschungen in Pompeji und den anderen vom Vesuvausbruch 79 n. Chr. verschütteten Städten*, 53-59. Recklinghausen: Bongers.

Schiltz, V.
- 1994 *Die Skythen und andere Steppenvölker: 8. Jahrhundert v.Chr. bis 1. Jahrhundert n.Chr.* Munich: Beck.

Schloz, S.
- 1994 Das Tempelbauprogramm der Ptolemäer: Die Darstellung eines Rekonstruktionsproblems. In Gundlach and Rochholz 1994: 281-86.

Schlumberger, D.
- 1969 Nachkommen der griechischen Kunst ausserhalb des Mittelmeerraums. In F. Altheim and J. Rehork (eds.),

Der Hellenismus in Mittelasien, 281-405. Darmstadt: Wissenschaftliche Buchgesellschaft.

Schmid, S.G.
- 1995 Nabataean Fine Ware from Petra. In K. 'Amr. F. Zayadine and M. Zaghloul (eds.), *Studies in the History and Archaeology of Jordan*, V: 637-47. Amman: Department of Antiquities.
- 1996a Die Feinkeramik der Nabatäer im Spiegel ihrer kulturhistorischen Kontakte. In M. Herfort-Koch, U. Mandel and U. Schädler (eds.), *Hellenistische und kaiserzeitliche Keramik des östlichen Mittelmeergebietes. Kolloquium Frankfurt 24–25. April 1995*, 127-45. Frankfurt am Main: Arbeitskreis Frankfurt und die Antike, archäologisches Institut der Johann Wolfgang Goethe-Universität.
- 1996b Die Feinkeramik. In Ez Zantur 1 1996: 151-218.
- 1997a Nabataean Fine Ware Pottery and the Destructions of Petra in the Late First and Early Second Century AD In G. Bisheh, M. Zaghloul and I. Kehrberg (eds.), *Studies in the History and Archaeology of Jordan*, VI: 413-20. Amman: Department of Antiquities.
- 1997b Eierschalendünne Tongefässe und grobe Waren. In Weber and Wenning 1997: 131-37.
- 1999 Un roi nabatéen à Délos? *Annual of the Department of Antiquities of Jordan* 43: 279-98.
- 2000a The 'Hellenistic' Tomb Façades of Nabataean Petra and their Cultural Background. In V. Christides and Th. Papadopoullos (eds.), *Proceedings of the Sixth International Congres of Graeco-Oriental and African Studies. Graeco-Arabica 7-8, 1999-2000*, 485-509. Nicosia: Archbishop Makarias III Cultural Centre.
- 2000b *Die Feinkeramik der Nabatäer. Typologie, Chronologie und kulturhistorische Hintergründe. Petra ez Zantur II 1. Ergebnisse der schweizerisch-liechtensteinischen Ausgrabungen (= Terra archaeologica IV)*. Mainz: von Zabern.
- 2000c The International Wadi al-Farasa Project (IWFP). Exploration Season 1999. *Annual of the Department of Antiquities of Jordan* 44: 335-354.
- 2001 The Impact of Pottery Production on the Sedentarisation of the Nabataeans. In J.R. Brandt and L. Karlsson (eds.), *From Huts to Houses. Transformations of Ancient Societies. Proceedings of an International Seminar Organized by the Norwegian and Swedish Institutes in Rome, 21-24 September 1997*, 427-436. Stockholm/Jonsered: Paul Aströms Förlag.
- in press The 'Hellenisation' of the Nabataeans: A New Approach. In Department of Antiquities (ed.), *Studies in the History and Archaeology of Jordan* VII. Amman: Department of Antiquities.

Schmidt, J.
- 1997–98 Tempel und Heiligtümer in Südarabien: Zu den materiellen und formalen Strukturen der Sakralbaukunst. *Nürnberger Blätter zur Archäologie* 14: 10-40.

Schmidt-Colinet, A.
- 1981 Nabatäische Felsarchitektur: Bemerkungen zum gegenwärtigen Forschungsstand. In Nabatäer 1981: 61-102.
- 1987a The Mason's Workshop of Hegra, its Relation to Petra, and the Tomb of Syllaios. In A. Hadidi (ed.) *Studies in the History and Archaeology of Jordan*, II: 143-50. Amman: Department of Antiquities.
- 1987b Zur nabatäischen Felsnekropole von Hegra, Medain Salih in Saudi-Arabien. *Antike Welt* 18, 4: 29-42.
- 1991 Exedra duplex: Überlegungen zum Augustusforum. *Hefte des archäologischen Seminars der Universität Bern* 14: 43-60.
- 1993 Der ptolemäische Eckgiebel: Ursprung und Wirkung eines Architekturmotivs. In Invernizzi and Salles 1993: 1-13.

Schmitt-Korte, K.
- 1990 Nabataean Coinage—Part II. New Coin Types and Variants. *Numismatic Chronicle* 150: 105-33.
- 1997 Das Münzsystem der Nabatäer. In Weber and Wenning: 101-104.

Schmitt-Korte, K. and M. Cowell
- 1989 Nabataean Coinage—Part I. The Silver Content Measured by X-Ray Fluorescence Analysis. *Numismatic Chronicle* 149: 33-58.

Schmitt-Korte, K. and M. Price
- 1994 Nabataean Coinage—Part III. The Nabataean Monetary System. *Numismatic Chronicle* 154: 67-131.

Schneider, Ch.
- 1996 Die Importkeramik. In Ez Zantur 1 1996: 129-49.

Schörner, G.
- 1995 *Römische Rankenfriese: Untersuchungen zur Baudekoration der späten Republik und der frühen und mittleren Kaiserzeit im Westen des Imperium Romanum*. Mainz: von Zabern.

Seigne J. and Th. Morin
- 1995 Preliminary Report on a Mausoleum at the Turn of the BC/AD Century at Jarash. *Annual of the Department of Antiquities of Jordan* 39: 175-91.

Settis, S.
- 1988 Die Ara Pacis. In *Kaiser Augustus und die verlorene Republik*. Catalogue of an exhibition, Berlin, 400-26. Mainz: von Zabern.

Seyer, M.
- 1993 Die Grabung in den Hanghgäusern von Limyra. In J. Borchardt and G. Dobesch (eds.), *Akten des II. Internationalen Lykien-Symposions*, I: 171-81. Vienna: Österreichische Akademie der Wissenschaften.

Seyer, M. and B. Rasch
- 1997 Die Grabung in der Nordweststadt. In J. Borchhardt (ed.), *Grabungen und Forschungen in Limyra aus den Jahren 1991–1996*, 338-48. Jahreshefte des Österreichischen archäologischen Institutes in Wien. Beiblatt 66.

Sherwin-White, S. and A. Kuhrt
    1993   *From Samarkhand to Sardis: A New Approach to the Seleucid Empire.* London: Duckworth.

Sidebotham, S.E.
    1986a   *Roman Economic Policy in the Erythra Thalassa 30 BC–AD 217.* Mnemosyne Supplement 91. Leiden: E.J. Brill.
    1986b   Aelius Gallus and Arabia. *Latomus* 45: 590-602.
    1996   Romans and Arabs in the Red Sea. *Topoi* 6: 785-97.

Siegler, K.G.
    1969   Die Tore von Klabsha. *Mitteilungen des deutschen archäologischen Instituts. Abteilung Kairo* 25: 139-53.

Siganidou, M.
    1996   Die Basileia von Pella. In Hoepfner and Brands 1996: 144-47.

Sinos, A.
    1990   *The Temple of Apollo Hylates at Kourian and the Restoration of its Southwest Corner.* Athens: Leventis Foundation.

Smirnova, N.
    1995   Bactrian Coins in the Pushkin State Museum of Fine Art. *Ancient Civilisations from Scythia to Siberia* 2: 335-52.

Smith, R.R.R.
    1991   *Hellenistic Sculpture.* London: Thames & Hudson.

Soren, E.
    1987   *The Sanctuary of Apollo Hylates at Kourion, Cyprus.* Tucson: University of Arizona.

Starcky, J.
    1966.   Pétra et la Nabatène. *Dictionnaire de la Bible, Supplément* 7, 886-1017. Paris: Letouzey et Ané.

Stevens, K.G.
    1994   Surface Finds from Qarn bint Sa'ud (Abu Dhabi Emirate—U.A.E.). *Mesopotamia* 29: 199-262.

Strocka, V.M.
    1988   Wechselwirkungen der stadtrömischen und kleinasiatischen Architektur unter Trajan und Hadrian. *Istanbuler Mitteilungen* 38: 291-307.
    1991   *Casa del Labirinto.* Häuser in Pompeji 4. Munich: Hirmer.
    1999   Zur Datierung der Sperlonga-Gruppen und des Laokoon. In *Hellenistische Gruppen: Gedenkschrift für Andreas Linfert,* 307-22. Mainz: von Zabern.

Stucchi, S.
    1975   *Architettura Cirenaica.* Rome: 'L'Erma' di Bretshneider.

Stucky, R.A.
    1976   Achämenidische Ortbänder. *Archäologischer Anzeiger,* 13-23.
    1990   Hellenistisches Syrien. In *Akten des XIII. Internationalen Kongresses für Klassische Archaeology, Berlin 1988,* 25-31. Mainz: von Zabern.
    1996   Die nabatäischen Bauten. In Ez Zantur 1 1996: 13-50.

Sydow, W. von
    1974   Die Grabexedra eines römischen Feldherrn. *Jahrbuch des deutschen archäologischen Instituts* 89: 187-216.

Syme, R.
    1965   Governors of Pannonia Inferior. *Historia* 14: 342-61.

Taddei, M.
    1984   Neue Forschungsbelege zur Gandhara-Ikonographie. In Ozols and Thewaldt 1984: 154-75.

Tarrier, D.
    1990   Baalshamin dans le monde nabatéen: A propos de découvertes récentes. *Aram* 2: 197-203.

Tholbecq, L.
    1997   Les sanctuaires des nabatéens: Etat de la question à la lumière de recherches archéologiques récentes. *Topoi* 7: 1069-95.
    1998   The Nabataeo-Roman Site of Wadi Ramm (Iram): A New Appraisal. *Annual of the Department of Antiquities of Jordan* 42: 241-54.

Thüroff, R.
    1989   Waffenfriese im Gebiet des östlichen Mittelmeers: Zur Entwicklung eines Friestyps in hellenistischer und frührömischer Zeit. Unpublished MA thesis, University of Basel.

Toynbee, J.M.
    1978   *Roman Historical Portraits.* London: Thames & Hudson.

Tran Tam Tinh, V.
    1990   Remarques sur l'iconographie de Dusares. In Zayadine 1990b: 107-14.

Trümper, M.
    1998   *Wohnen in Delos: Eine baugeschichtliche Untersuchung zum Wandel der Wohnkultur in hellenistischer Zeit.* Rahden: Leidorf.

Tushingham, AD
    1972   *The Excavations at Dibon (Dhibân) in Moab: The Third Campaign 1952-53.* Annual of the American School of Oriental Research 40. Cambridge, MA: American Schools of Oriental Research.

Tybout, R.A.
    1989   *Aedificiorum figurae: Untersuchungen zu den Architekturdarstellungen des frühen zweiten Stils.* Amsterdam: Gieben.

Villeneuve, F. and Z. al-Muheisen
    1994   Découvertes nouvelles à Khirbet edh-Dharih (Jordanie), 1991–1994. *Académie des inscriptions et belles lettres. Comptes rendus:* 735-57.

Voie royale
    1986   *La voie royale: 9000 ans d'art au royaume de Jordanie.* Musée du Luxembourg 26 novembre 1986–25 janvier 1987. Paris: Association française d'action artistique.

Walker, S. and K.J. Matthews
　1997　The Marbles of the Mausoleum. In Jenkins and Waywell 1997: 49-59.

Waywell, G.B.
　1997　The Sculptures of the Mausoleum at Halicarnassus. In Jenkins and Waywell 1997: 60-67.

Weber, M.
　1990　*Baldachine und Statuenschreine*. Rome: Bretschneider.

Weber, M.
　1997　Die Bildkunst der Nabatäer. In Weber and Wenning: 114-125.

Weber, Th. and R. Wenning (eds.)
　1997　*Petra: Antike Felsstadt zwischen arabischer Tradition und griechischer Norm*. Antike Welt, special issue. Mainz: von Zabern.

Weiser, W. and H.-M. Cotton
　1996　'Gebt dem Kaiser, was des Kaisers ist...' Die Geldwährungen der Griechen, Juden, Nabatäer und Römer im syrisch-nabatäischen Raum unter besonderer Berücksichtigung des Kurses von Sela'/Melaina und Lepton nach der Annexion des Königreiches der Nabatäer durch Rom. *Zeitschrift für Papyrologie und Epigraphik* 114: 237-87.

Wenning, R.
　1987　*Die Nabatäer—Denkmäler und Geschichte: Eine Bestandesaufnahme des archäologischen Befundes*. Novum Testamentum et Orbis Antiquus 3. Göttingen: Vandenhoeck & Ruprecht.
　1989　Maskierte Götter? Anmerkungen zum Aufeinandertreffen von Ost und West am Beispiel der arabischen Nabatäer. In K. Rudolph and G. Rinschede (eds.), *Beiträge zur Religion/Umwelt-Forschung*, I: 243-60. Geographia Religionum 6. Berlin: Dietrich Reimer.
　1990　Das Nabatäerreich: Seine archäologischen und historischen Hinterlassenschaften. In H.-P. Kuhnen, *Palästina in griechisch-römischer Zeit*, 367-415. Handbuch der Archäologie 2, 2 Munich: Beck.
　1993a　Das Ende des nabatäischen Königreiches. In Invernizzi and Salles 1993: 81-103.
　1993b　Eine neuerstellte Liste der nabatäischen Dynastie. *Boreas* 16: 25-38.

Wenning, R. and H. Merklein
　1997　Die Götter in der Welt der Nabatäer. In Weber and Wenning 1997: 105-10.

Will, E. *et al.*
　1991　*'Iraq al-Amir: Le château du Tobiade Hyrcan*. Paris: Geuthner.

Wolski, J.
　1999　*The Seleucids: The Decline and Fall of their Empire*. Krakow: Nakladem Polskiej Akademii Umiejetnosci.

Wolsky, J.
　1976　*Les Parthes et la Syrie*. Acta Iranica IIIème série 5. Louvain: Bibliothèque Pablavi.
　1993　*L'empire des Arsacides*. Acta Iranica IIIème série 18. Louvain: Peeters.

Wright, G.R.H.
　1968　Découvertes récentes au sanctuaire du Qasr à Pétra, II. Quelques aspects de l'architecture et de la sculpture. *Syria* 45: 25-40.

Zadneprovskiy, Y.A.
　1996　The Nomads of Northern Central Asia after the Invasion of Alexander. In Harmatta *et al.* 1996: 457-72.

Zanker, P. (ed.)
　1976　*Hellenismus in Mittelitalien: Kolloquium in Göttingen vom 5. bis 9. Juni 1974*. Abhandlungen der Akademie der Wissenschaften in Göttingen. Philologisch-Historische Klasse. Dritte Folge 97. Göttingen: Vandenhock & Ruprecht.

Zanoni, I.
　1996　Tonlampen. In Ez Zantur 1 1996: 311-36.

Zayadine, F.
　1974　Excavations at Petra (1973–1974). *Annual of the Department of Antiquities of Jordan* 19: 142-50.
　1985　Recent Excavation and Restoration at Qasr el Bint of Petra. *Annual of the Department of Antiquities of Jordan* 29: 239-49.
　1986　Tempel, Gräber, Töpferöfen: Ausgrabungen des Department of Antiquities in Petra. In Lindner 1986: 214-69.
　1989　Die Felsarchitektur Petras: Orientalische Traditionen und hellenistischer Einfluss. In Lindner 1989: 124-61.
　1990a　The Pantheon of the Nabataean Inscriptions in Egypt and the Sinai. *Aram* 2: 151-74.
　1991　L'iconographie d'Isis à Pétra. *Mélanges de l'Ecole française de Rome. Antiquité* 103: 283-306.
　1997　Hermes/al-Kutbay Lexicon Iconographicum Mythologiae Classicae (LIMC) Supplementum, 616-19. Zurich/Düsseldorf: Artemis.

Zayadine, F. (ed.)
　1990b　*Petra and the Caravan Cities*. Amman: Department of Antiquities.

Zayadine, F. and Z.T. Fiema
　1986　Roman Inscriptions from the Siq of Petra: Remarks on the Initial Garrison of Arabia. *Annual of the Department of Antiquities of Jordan* 30: 199-206.

Zazoff, P.
　1996　Bildchiffren der skytho-thrakischen Kunst. *Hamburger Beiträge zur Archäologie* 18 (1991): 167-80.

Zimmermann, B.
　1992　Zur Authentizität des 'Clemensfeldzuges'. *Jahresberichte aus Augst und Kaiseraugst* 13: 289-303.

Zoppi, C.
　1991–92　L'architettura abitativa in età ellenistica: Il modello vitruviano e i documenti superstiti. *Rendiconti della Accademia di archeologia, lettere e belle arti, Napoli* 63: 157-98.

12. Roman Jordan

Philip Freeman

Introduction

Jordan was the province of Arabia in the Roman period. However, the northern part of the province also extended into what is today Syria and where lay the 'capital' of Arabia, Bostra (modern Busra al-Sham) along with a number of adjacent towns. As such, they are omitted from this discussion. There is, however, a considerable modern literature that reports on recent work in the area (e.g. Dentzer 1986; MacAdam 1986b; Burns 1992). In contrast the 'Jordanian' cities of Gadara and Pella, once members of the informal pre-province league known as the Decapolis, were transferred to the province of Syria when the Nabataean kingdom was annexed in AD 106. Although part of another province in the Roman period, these particular cities are discussed here as they are now in Jordan. Finally, I have defined the Roman period as spanning the mid-first century AD to the end of the fourth century.

Writing about the Roman period is made difficult by the way that it is sandwiched between two more strongly defined periods. In its strictest sense, the Roman era in Jordan is relatively short. On the one hand it could be said to start c. 63 BC (with Pompey's 'eastern settlement') when it first came into intimate contact with Roman politics. Alternatively, it came into the full Roman sphere in AD 106 when the Nabataean monarchy was removed in preference to a full Roman province. This 'period' lasted down to sometime in the early fourth century when the Byzantine period is said to commence. This periodization represents a significant problem, since the start of the period is defined by a political event even though the region had been in contact with the Roman empire for at least 150 years. The end of the era is likewise thought to have been occasioned by modifications in the existing Roman administration of the region in the late third–early fourth century, or simply because the capital of the empire was shifted in the 320s. In both cases, whatever the criteria for defining the political/administrative transitions, they had minimal if no (immediate) implications for the economic, religious, social and cultural conditions of the region. The situation is compounded by the fact that for different researchers the term Roman means different things. Broadly speaking, British scholars have tended to regard the succeeding Byzantine period as commencing in the fifth and even the sixth century. On the other hand, American scholars tend to place the divide in the early fourth century (e.g. Parker 1986). But even then more discrepancies are evident. For instance, for the (Australian) Pella project, 'Early Roman' means 63 BC to AD 135 and Late Roman 135–324, a convention adopted by Khouri (1988). In contrast Homès-Fredericq and Hennessy define Early Roman as 63 BC–AD 106 and Late Roman as 106–324 (Homès-Fredericq and Hennessy 1989: 10), while, for example, the Hisban Survey saw its Early Roman period as 63 BC–AD 193, and its later equivalent spanning 193–365 (LaBianca 1989: 261-69). The start and end dates for the period as a whole are derived from historical sources. The sub-divisions owe as much to periods derived from the evidence of ceramics, and then from a limited number of excavated sites. In recent times Sauer's Hisban ceramics sequence has been extensively used to cross-date many surveys across all of Jordan (e.g. Parker 1986). In turn, it has led to the creation of the unrealistic Early Roman 1, 2, 3 and Late Roman 1, 2 and 3 subdivisions.

Moving away from the chronological issue, archaeology's great strength is its ability to look at humankind's cultural, technological and economic progress. It is less effective in exploring its political condition. But in examining Roman Arabia, we are dealing with a period that did not necessarily have an obvious cultural effect. For a political Roman Arabia does not equate with the on-going cultural situation. Irrespective of the fact that the region was part of the empire, culturally, socially and economically, Roman Arabia owes more to an earlier Hellenistic and Nabataean framework than to anything that could be described as distinctly Roman. While it would be rash to deny that changes did occur in these spheres, they came with time. That sense of cultural continuity exists in the equally arbitrarily defined Byzantine period, which owed as much to the earlier periods as it did to the consequences of the Roman presence. So, for instance, the urban framework of the province can now be shown clearly to pre-date the annexation. Where there is information, the same impression is apparent in the region's smaller communities as well as in the countryside. Likewise, in terms of material culture, there is the same clear sense of continuity from the Hellenistic period to the Roman and from the Roman to the Byzantine. In one respect, then, to talk of a specific archaeological period that is characteristically or purely Roman is misleading, if not erroneous.

Another inhibiting factor in attempting to synthesize the archaeology of Roman Jordan is the standard of fieldwork and its processing. While there are notable exceptions, it is disappointing to note how little substantial work can be reported for certain important themes about the Roman province. Stern *et al.*'s (1993) atlas of excavations in the Holy Land is site specific. It contains useful summaries of work at individual sites but fails to synthesize that data for the wider picture of Roman Arabia. In a similar way, Homès-Fredericq and Hennessy's (1989) synthesis of the archaeology of Jordan shows that, despite so many of the sites reported producing 'Roman' data (often recovered through excavation), little of that material is treated with satisfactory attention. The imbalances of treatment are frequently a consequence of the priorities that occasioned that work. A good example, but by no means unique, is the number of entries reporting work at Byzantine Christian sites. The majority of such sites have almost incidentally produced earlier Roman deposits and yet the emphasis in these entries is limited to discussions of architectural typologies and the style and the iconography of mosaics rather than consideration of the relationship of the pre-Byzantine to the Byzantine data. Another consequence of the explosion in the quantity of fieldwork, both in terms of survey and excavation, which is being undertaken in Jordan is that relatively little of that work has been written up and published as definitive statements. Instead there is a considerable literature of interim statements, results that can be easily rendered out-dated or liable to re-interpretation.

It is for these reasons that, despite the explosion in the scope and pace of research, it is surprising to find that, contrary to expectations, it is not an easy task to compose a synthesis outlining the current understanding of the archaeology and so the history of Roman Jordan. This situation is a consequence of a number of factors. Available discussions are either too dated, too limited in subject matter or paradoxically too general in scope. For instance, Bowersock's work on Roman Arabia (1983) is restricted to the period from 63 BC to the early third century and is written almost exclusively from the perspective of frontiers and Arabia as part of the wider empire. Parker's (1986) work on the province's military history is similarly restricted by its emphasis on a frontier system whose existence is debatable. Isaac's (1990) and Millar's (1993) studies on the army in the east and cultural milieu of the Roman Near East lose much in being generalized studies, and, in the case of the latter, does not employ the archaeological evidence to anything like its potential. Nor is there anything comparable to MacAdam's (1994) and Schick's (1994) surveys of the countryside in Byzantine Jordan.

Chronology and period parameters

The Romans first impinged politically on the region in the mid-first century BC as a consequence of Pompey's actions in Syria and then in Judaea. At this time the bulk of the region formed the kingdom of Nabataea. One of the consequences of Pompey's intervention was the creation of the league of cities known as the Decapolis. The precise form and function of this federation of Hellenistic cities remains unclear. One school of thought sees it as a cultural entity—islands of Greek culture in a Semitic ocean. More plausible is that it was a group of communities released from Seleucid, Hasmonean and Nabataean domination and made the responsibility of the Roman governor of Syria. 'Member' cities, although it is clear that the 'league' was never formally constituted, included Gerasa (known as Antioch Chrysorhoas in the second century AD: modern Jarash) (Figure 12.2), Pella (modern Tabaqat Fahl), Gadara (modern Umm Qays) (Figure 12.4), Abila (modern Quwayliba) (Figure 12.5), Philadelphia (= Rabbath-Ammon: modern 'Amman) (Figure 12.3), Capitolias (= ancient? Raphana, Arbela: modern Bayt Ras), Hippos (modern Qal'at al-Husn/Susiya) and Dium (location unknown) and to the north, in Syria, Damascus and Canatha (modern Qanawat). It is safe to assume that the Jordanian Decapolis cities must have enjoyed at least contiguous territories.

In subsequent Roman history the Nabataean kingdom played a marginal role in the Civil Wars. In ?6 BC, Augustus initiated a campaign, led by the legate Aelius Gallus, to Arabia Felix from Egypt. Bowersock's (1983: 45) hypothesis, on a reading of Strabo (16.4.21), that Nabataea became a (temporary) Roman possession in c. 2 BC is attractive. However, for the rest of the first century AD the region is almost totally anonymous in Greek and Latin sources. We are reliant on Nabataean

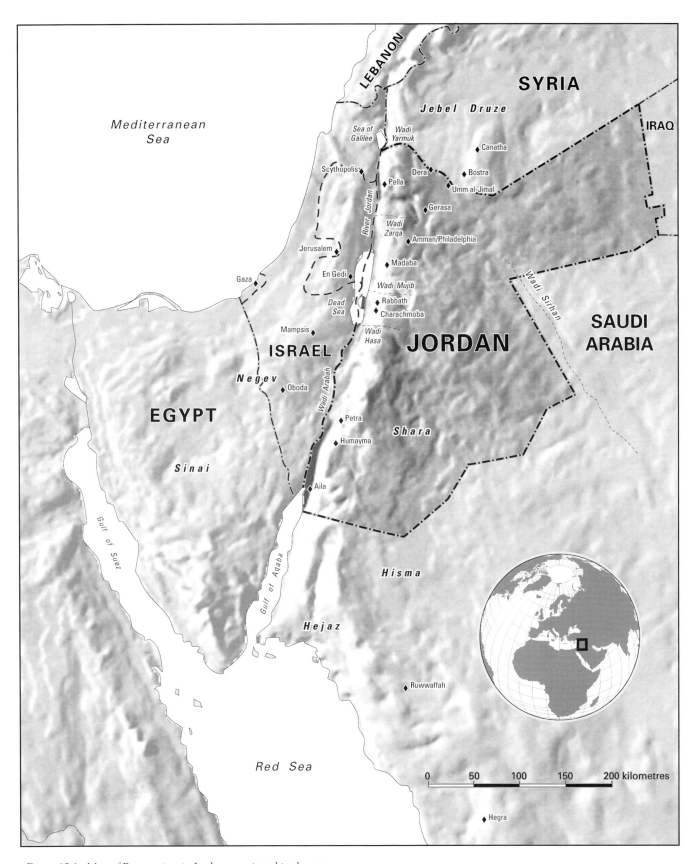

Figure 12.1. Map of Roman sites in Jordan mentioned in the text.

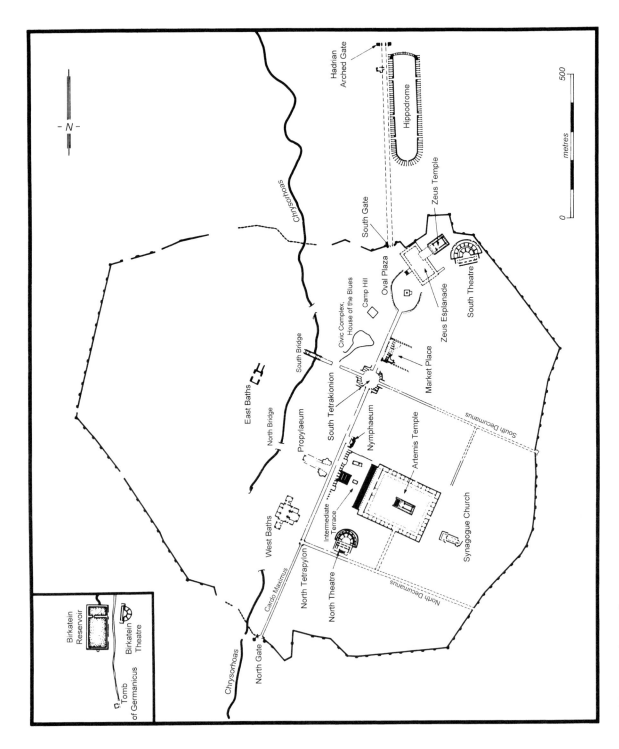

Figure 12.2. Gerasa of the Decapolis (Jarash).

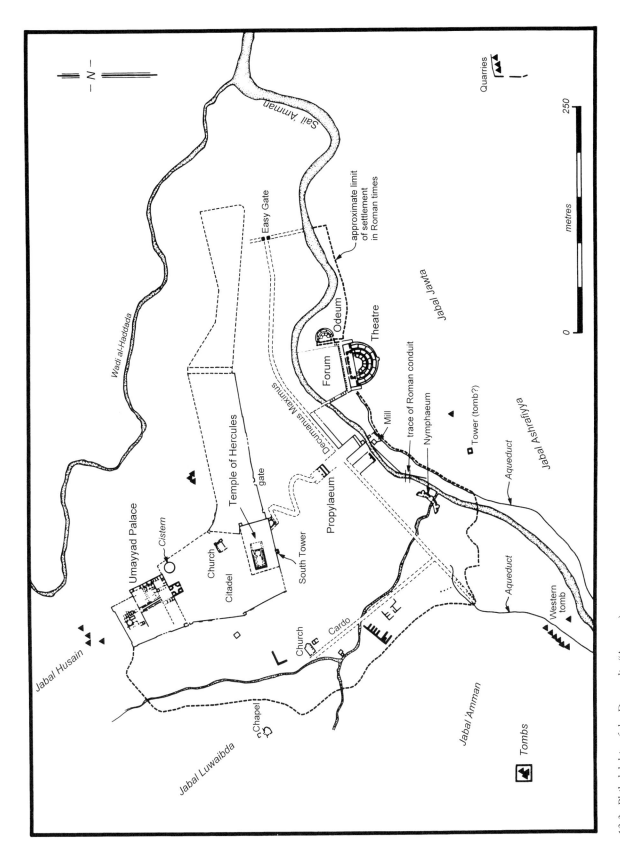

Figure 12.3. Philadelphia of the Decapolis ('Amman).

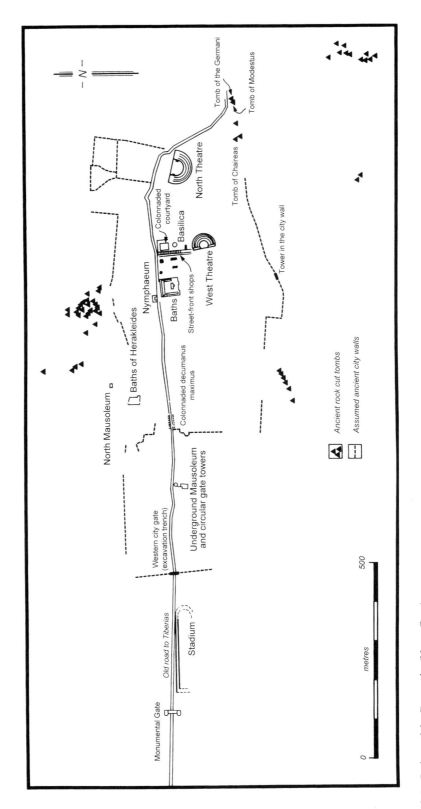

Figure 12.4. Gadara of the Decapolis (Umm Qays).

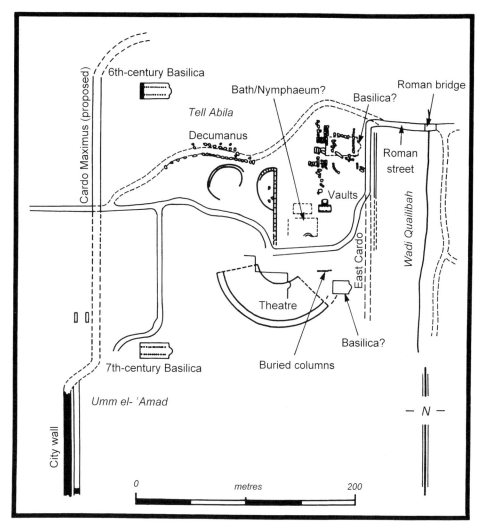

Figure 12.5. Abila of the Decapolis (Quwayliba).

inscriptions, fleshed out in part by Josephus's account of events in the Jewish world, for a chronology of events and developments in the region. In AD 106 the Nabataean kingdom was annexed as the new province of Arabia (Freeman 1996). While the circumstances behind the decision on Rome's part to annex the region following the death of Rabbel II, the last Nabataean king, are obscure, a number of developments have been dated to the immediate aftermath of the action. It appears, for instance, that the Decapolis was broken up with the transfer of Gadara and Pella to the province of Syria. The majority of the other communities became the responsibility of the governor of Arabia. At this time there is also the appearance of a new era for dating in Arabia which did not use the traditional regnal convention but had as its epoch the arrival of the Roman governor in Bostra. The annexation has been seen in some circles as inevitable; elsewhere as being the result of the force of circumstances. The only certain events that can be placed in this period are the commencement of the recruiting of Nabataean units into the Roman army from c. 113 and the erection of milestones on the Via Nova Traiana running from Aila to the borders of Syria. The tendency has been to regard this road as a coherent unit, one constructed to facilitate the process of annexation. Something, however, is lost in this characterization. The road, rather than being a fresh construction, may have been the consolidation of an earlier route known in biblical sources as the King's Highway (Num 20.17; 21.22, though see Miller 1989). Dated milestones from along its length reveal that sections of it were (re-)built at different times (e.g. the southern part, al-'Aqaba to Philadelphia in 111–112; the Bostra–Philadelphia section in 114). It is perhaps better then to see it more as a group of short stretches of roads linking communities than as a deliberate attempt to impose a strategic spine to the new province.

At about this time the city of Bostra is believed to have supplanted Petra as the 'capital' of Nabataea, if that change had not already occurred under Rabbel II. It was under the emperor Trajan that Bostra was renamed, presumably in recognition of its new status, as Nea Traiane Bostra, while Petra was elevated to the status of a *metropolis*. Because the governors of the province were based in Bostra, as was the main garrison (the legion III Cyrenaica), it is commonly asserted that the city became the administrative centre of the province. It is known, however, that the governors were never bound to remain at Bostra. They were expected to tour their province on a regular basis, not least to hear legal cases, and so on in its major communities. At the same time, the epigraphy of the region shows that the cities continued to undertake the same range of administration that they had executed before Nabataea was annexed. The origins of that administrative framework go back to the Hellenistic city model. Similarly, it would seem that Bostra was not the financial capital of the province. Judging from the number of inscriptions put up by imperial financial officers, the procurators, Gerasa seems to have become the financial focus of the region, which in turn must be a reflection of its commercial pulling power over the other cities in the province. It may not be coincidental then that the emperor Hadrian wintered in Gerasa in 129/130, as commemorated in an inscription and arch put up in that city. As the emperor proceeded on to Jerusalem, it is highly plausible that he would have already visited Bostra, and, perhaps later on, communities such as Rabbath-Moab and Petra. The region may have enjoyed the dubious privilege of another imperial tour in 199 when the emperor Septimius Severus toured the eastern provinces. It is known that it was Severus who decided to enlarge the Arabian province, with the transfer to it of the Jabal al-Druze and the region of the Le'ja (e.g. the territory of Canatha). A further indication of Severus's interests in the region is the way that a number of Arabs seem to have benefited from his patronage at the imperial court in Rome. The emperor Caracalla (211–218) raised Gerasa to the status of a *colonia*, with the name Aurelia Antoniniana, while the emperor Elagabalus (219–222) did the same for Petra. Bostra became a *metropolis* under Philip (244–249) who also accorded privileges to his previously insignificant hometown Shabha, now renamed Philipopolis. By c. 300 an additional small legionary fortress had been constructed at al-Lajjun in central Jordan in order to accommodate the legio IV Martia recently transferred from western Europe, while by the early fourth century legio X Fretensis had moved from Jerusalem to a new base at Aila. Under Gallienus, in 262, the governorship of Arabia passed from a pro-praetorian legate to an equestrian *praeses*. Because of the problems of the Roman empire at the end of the third century, direct control of Arabia passed to the Palmyrenes, under Zenobia and Vaballuthus. Roman control was reasserted in 272. In or after 295 and the reforms of Diocletian, the old province of Arabia was partitioned as part of a broader scheme of administrative reforms. The region south of Wadi al-Hasa, combined with the Sinai and Negev, became the new province of Palestina Salutaris (later called Palestina Tertia) with its capital at Petra. The now smaller original province retained the old name. Both provinces were controlled by *duces*.

Roman Arabia was subject to a number of earthquakes that at various times have inflicted considerable damage to the province's communities. For instance, major earthquakes have been identified in the early second century AD, 306 and 363, with the last of these, which destroyed much of Petra, being especially destructive (Russell 1985). Numerous references and allusions to low-scale warfare involving nomads, Arabs and Saracens along with Roman responses also punctuate the history of the province. It is difficult, however, to pin down these events. They are largely undated and geographically vague coupled with a laxity in terminological precision (Bowersock 1983; Shahid 1984, 1989).

The geography of Roman Arabia

At the time of its annexation, the northern boundary of the province would have extended into modern Syria, up to and beyond the ancient city of Bostra, the 'provincial' capital. The Jordan Valley determined the western boundary of the province. With respect to the province's eastern and southern borders, this is not so much of an issue, primarily because this was desert. The boundaries of the province were probably in the first place those of the Nabataean kingdom, and would thus, in the north and west, have been determined by the boundaries of the city-states in those areas. In north and northwest Arabia, some of the cities that had been part of the Decapolis since the mid-first century BC but nominally attached to the province of Syria were now added to Arabia. The communities affected included Philadelphia, Capitolias, Abila, Pella and Gadara. The boundaries of Pella and Gadara ran along the Jordan

Valley and more than likely would also have been the province's western boundaries. As there were no comparable city-states to the east and south, the boundaries here were ephemeral or non-existent, defined by tribal territories, themselves liable to be vague and disputed. Still, there has been debate about whether Roman authority extended as far south as the Hisma or even Madain Salih. There was a slight modification of the province's boundaries sometime after the early fourth century but by this time the region had begun to merge into its Byzantine phase.

The sources of evidence

As it lies in what might be called the historic era, any summary of the region under the Romans has to refer to the disparate historical sources: few, if any, originated from the region. This material tends to be written within the framework of diplomatic, military and political relations of the eastern Roman empire. There are relatively few ancient sources that write of Arabia in the Roman period *sensu stricto*. The early first-century AD geographical writer Strabo in book 16 of his *Geography* reports numerous ethnographic, historical and geographical anecdotes about the region. The country appears as a marginal area in Josephus's account of Jewish history down to the late first century AD in the *Jewish Antiquities* and *Jewish War*, but he does occasionally let slip some interesting incidental details. Those later historians that discuss the region, including the fourth-century Latin historian Ammianus Marcellinus, are inevitably limited in scope and subject matter. Almost an anomaly, the cities of Gadara and, to a lesser degree, Pella, are known from Greek, Latin and Jewish literary sources to have had reputations for artistic accomplishment as well as something akin to leisure resorts.

Potentially more valuable are cartographic sources (Bowersock 1983: 164-86). Ptolemy's *Geography*, composed in the mid-second century AD, is a register of place names which provides co-ordinates and lists of the significant places, regions and tribes of Arabia at the time the document was compiled. The Peutinger Table, also of the mid-second century AD, is a map that describes the road system of the Roman empire at various times. The Arabian section evidently illustrates elements of the pre-Roman and later Roman provincial network. The same applies to the Antonine Itinerary, while the early fourth-century *Notitia Dignitatum* is a summary, often not entirely reliable, of the military dispositions, by unit and location, of the Roman army in the region. The sixth-century AD Madaba Map in the Church of St George is a mosaic which depicts the Holy Land and lists 157 cities from Tyre and Sidon in the north to Egypt in the south, and from east to west the Mediterranean coast and the Syrian Desert. Close analysis of the places recorded on the mosaic suggests that it is based on the fourth-century *Onomasticon* of the Church historian Eusebius, who is an important source of information for the arrival of Christianity in the region and so indirectly about life in pagan Arabia.

Society, religion and the economy of the region rarely appear directly in the ancient sources, although the province's epigraphy offers a database which some scholars have started to exploit (MacAdam 1986a, 1990; Sartre 1982; Sartre 1986; Graf 1988, 1989; Macdonald 1993). Other than individual or small groups of inscriptions published in articles, Brünnow and von Domaszewski (1904–1909), Butler (1907–1919), and Kraeling (1938) have variously published the Latin and Greek inscriptions of the region. Subsequent survey and excavation projects have published inscriptions as they discover them. The bringing together of this evidence, a difficult task, has not enjoyed the most illustrious of histories (MacAdam 1986a, 1990). Work on the *Inscriptions grecques et latines de Syrie* (IGLS), which would have included Transjordan, began in 1905. In recent years, with the problems of the IGLS programme and the creation of a separate Jordanian state, this scheme has evolved into the *Inscriptions grecques et latines de Jordanie* (IGLJ), since modified to *Inscriptions de la Jordanie* (IJ). Five volumes of inscriptions and commentaries are planned of which volume 2, edited by Gatier, on the epigraphy of Central Jordan, including 'Amman, has been published (1986; and Gatier 1989: 54).

Finally, in recent years there has been published a collection of papyri from the caves above En Gedi at the southern end of the Dead Sea. These papyri are invaluable for understanding the condition of the region at the end of the first and early second century AD. The documents, the property of a Jewish widow who deposited them in the aftermath of the Bar Kochba rebellion of 132, are better known as the Babatha Archive. It includes a mish-mash of documents referring to aspects of a number of legal disputes that Babatha got into with the families of her deceased (two) husbands over the years 93/94–132. While not explicitly related to the province, the documents indicate aspects of life in the region around Petra at about the time of the Roman annexation of Nabataea (Lewis 1989; Isaac 1992). At the other end of the time scale, the Petra Papyri, recently excavated from a Byzantine church in the city, date to the later sixth century. Preliminary readings of these documents show that they contain indirect evidence of earlier (Roman) aspects of life in the city and countryside around Petra (Humphrey 1995).

An introduction to the history of fieldwork in Roman Jordan

The variable quality of research on the archaeology of (Roman) Jordan which has been noted is in part a consequence of the way that fieldwork in the region has evolved. Before attempting to summarize what is our present understanding of Roman Jordan, it is worth outlining what has conditioned the accumulation of that knowledge. In general terms, it is possible to isolate, over the 150 years or so that fieldwork has been conducted in the region, a number of broad shifts in research *mentalities*.

The earliest antiquarian-cum-archaeological explorations of Roman Jordan occurred in the early nineteenth century, although in truth these were not motivated purely by the cause of historical research (Geraty and Willis 1986; Khouri 1988). Here the emphasis was on visiting and recording the architectural splendours of what might be described as monumental sites, places such as Gerasa and Petra. The earliest (western) records of the archaeology of the region were the consequence of the exploits of explorers such as Seetzen's 1806–1807 visit to Gadara, Abila, Gerasa and Pella. Other explorers at this time include Burckhardt, whose original plan was to visit Mecca. One of the results of his sojourn in the region was the discovery of Petra in 1812 and a visit to Jarash. As the nineteenth century progressed, the pace of 'scientific' investigation accelerated. In this respect, research was increasingly driven by the support of institutions and academic organizations established to support exploration, especially of sites with biblical associations, as well as to promote, if not safeguard, national interests. Burckhardt operated under the auspices of the Royal Geographical Society of London. The Palestine Exploration Fund was established in 1865 to obtain and disseminate 'information respecting ancient and modern Syria (e.g. Syria, Lebanon, Jordan and Israel)...and the ancient and modern inhabitants thereof...the History, Literature...Ethnology, Mineralogy, Numismatics, Topography, Geography (physical and political), Geology, Zoology, Botany, Meteorology, Natural History and the manners and customs of the same (countries)'. American involvement in the region went through an even more tortuous process (MacAdam 1986a).

In addition to the part played by governments and learned societies, the support of academic institutions became increasingly significant from the late nineteenth century. In 1897–98 Brünnow and von Domaszewski traversed the northern and central parts of the country, describing, mapping and photographing the major (Classical and Byzantine) sites, forts, roads and other monuments. This phase of exploration culminated at the turn of the century in two major American expeditions, both led by Butler: the 1899–1900 American Archaeological Expedition to Syria and the 1904–1905 and 1909 Princeton Archaeological Expedition to Syria. The common feature of this period was that research continued to be city/site specific, recording architecture, monuments, inscriptions and the evidence of Christianity. Following on from the work of Brünnow and von Domaszewski and Butler *et al.* in the north, the popular perception is that systematic archaeological work on Roman and Byzantine Jordan, if not entirely drawing to a halt, then slowed to a trickle until a resumption in the late 1960s (Bowersock 1971). This impression is valid for the military dimension, although Stein and Poidebard did add much relevant and (as subsequently shown) irrelevant material to the picture by the use of air reconnaissance (Gregory and Kennedy 1985; Kennedy and Riley 1990). However, there was progress in other aspects of the Roman archaeology of the region.

As archaeology in general in the early twentieth century came to be equated with excavation, so in Transjordan the same perceptions shaped research, with the digging, often in the name of conservation and restoration, of the same types of sites that had attracted attention in the previous century. From the second decade of the twentieth century a new phase of exploration in the region becomes evident. From 1917 down to 1950 Transjordan was administered under a British Mandate. In this arrangement antiquities were the responsibility of a (British-run) Department of Antiquities, which naturally enough tended to focus research on themes of interest to British (and, by implication, European and American) scholarship. This partition meant the *de facto* exclusion of German work in the region after the First World War, as well as the effective curtailment of other European activity too. After Jordan gained its independence, the same interests and priorities continued to influence work in Jordan, not least because the earliest directors of Antiquities were Britons. From the mid-1920s various British teams, initially under the auspices of the British School at Jerusalem and then later the newly created Department of Antiquities, began programmes of clearance work and so limited excavation. Horsfield commenced such work at Gerasa in 1925. This became the route by which the Americans gained an entry to the site, culminating in the publication of Kraeling's *Gerasa* (1938). Work at the site continued with Lankester Harding clearing the 'Oval Piazza' and parts of the *cardo* and in the 1950s with Kirkbride's restoration of the South Theatre. In the meantime, Horsfield had moved to Petra where he commenced clearance work in 1929, which eventually culminated with Kirkbride, Parr and Wright each clearing a large part of the heart of the city in the 1950s and 1960s. Elsewhere, less romantically, field survey continued to be undertaken and was influential in initiating a new phase of research. Notable exponents at this time include the Americans Albright and Glueck. While their work may not have been to the highest methodological standards and both were more preoccupied with the identification of late (biblical) prehistoric sites, Glueck in particular located innumerable Roman and Byzantine sites, notably in his explorations in eastern Transjordan (Glueck 1934–1951; Parker 1976; Graf 1983).

From the late 1960s to the early 1970s the direction of archaeological work again shifted, with more extensive use of field survey. The reasons for the transition are numerous. Changes in research priorities, themselves a consequence of developments in archaeology in general, and recognition of the fact that sites are merely parts of larger (archaeological) landscapes are partly responsible. Equally significant is the way that the potential of commercial aviation made the region more accessible to Westerners. There was also the fact that politically the region was relatively stable. At about the same time the potential of (international) tourism was recognized by the Jordanian government who, with relative economic prosperity, encouraged foreign teams to explore the 'traditional' sites as preparatory to developing them for tourism. The major initiatives of this period were the concentrated programmes of excavation at Petra and Jarash. At the latter site was created, in 1983, the Jarash Archaeological Project, 'part of a comprehensive and long range programme of restoration and conservation of Jarash and its monuments. The Jordanian Government indeed recognises both Petra and Jarash as the country's chief tourist attractions. They are thus of enormous value to the economy and their conservation and protection are prerequisites for increasing this source of income' (Hadidi in Zayadine 1986: 6). Work at Petra and Gerasa was assisted not least because the two sites are relatively unencumbered with the over burden of subsequent reoccupation. The same cannot be said of the other major Roman-period sites. Gadara, which since the late 1970s has been largely a German monopoly under the auspices of the German Protestant Institute for Archaeology, is in part obscured by the late Ottoman settlement of Umm Qays (Weber 1990; Kerner 1994). Likewise Philadelphia is almost totally lost below 'Amman. In contrast Abila and Pella, while not built over, are covered by very substantial deposits of hillwash. Another of the consequences of this period of economic prosperity and the concomitant property boom was a refocusing of research priorities. With the Jordanian landscape under increased pressure from population growth, construction projects and agricultural and industrial exploitation, it was recognized by the late 1970s that part of the country's archaeological potential was diminishing before it was being identified. Field survey, now often used as a cheaper alternative to excavation, has increasingly been employed as a means of obtaining an overview of the archaeology of the region from which more intensive research might evolve. The culmination of these two tensions—academic research and the threat to Jordan's heritage—was the inception in 1987 of the Cultural Resources Management Programme (Tell *et al.* 1993). This is a joint Jordanian Department of Antiquities and American Center of Oriental Research (ACOR) initiative which is an umbrella title for a number of actions ranging from impact assessments in advance of construction work to statutory protection and exploration. The programme is also intimately linked with the JADIS (Jordan Antiquities Database and Information System), an initiative funded by the United States Agency for International Development (USAID), a computerized database for the registration and recording of archaeological sites from published information and archive files (Palumbo 1993).

The purpose of the preceding résumé is to explain the strengths and inadequacies of our understanding of Roman Jordan. Despite the massive upsurge in the range and quality of fieldwork in Jordan, there are still a number of problems which hinder efforts to synthesize the results of that work into a broader picture of the region under Roman administration. Excavation has long tended to concentrate almost exclusively on the 'big' urban complexes (notably Gerasa and Petra, and more recently Pella, Abila and Gadara) and to a lesser degree, Umm al-Jimal, Madaba and Humayma, and even then only on aspects of these communities. It remains a fundamental issue whether it is valid to construct a history of the region based on the interpretation of developments that in the first instance pertain to those communities. The situation is not assisted by the variable degree of publication of the results of some of this work. The 1980s phase of work at Gerasa is now reasonably accessible (Zayadine 1986, 1989), as are the 'Amman results (Northedge 1992; Kanellopoulos 1994). Work at Pella has tended to emphasize the pre-Classical and especially prehistoric deposits (McNicoll 1982, 1992). Its Roman levels have also been covered and/or destroyed by Byzantine activity. Compounding this is the fact that the (civic) core of the city would have lain in Wadi Jirm in which the hillwash from Tall al-Husn has accumulated. Abila too has produced more about the Byzantine and Umayyad phases of its history than the Roman. Synthesis of the Abila and Gadara material is much less advanced, although both sites have enjoyed the prompt publication of interim, but still raw, statements (Mare 1994; Weber 1990). The fact that work at the pair is on-going has in part hindered interpretation, as it has at Humayma and Petra. The results of nearly 20 years of fieldwork at Umm al-Jimal are promised (de Vries 1998).

With this background, where certain 'Roman' sites have attracted more attention and resources, gaps are now clearly apparent in other aspects of the region's 'Roman' archaeology. One major deficiency has been at the southern terminus of the major 'Roman' road through the country, ancient Aila, modern al-'Aqaba, a site that has long lost out to its Byzantine/early Islamic successor (Zayadine 1994). The search has, however, begun for its Roman progenitor. This project involves a regional environmental and archaeological survey (the Southeast Wadi 'Arabah Survey) and excavation of the Roman/Byzantine city for insights into its economy (Parker 1997). Elsewhere, recent fieldwork has also indicated more of the sort of 'large' towns that were often missed or ignored in the nineteenth-century surveys. For instance, At-Tuwana (Ptolemy's Thana/Thoana, Thoruna in the Peutinger Table) is actually a series of complexes. Originating in the Nabataean period, its urbanization commenced in the first century BC and continued down to the late Byzantine period. Its evident prosperity has been linked to long-distance trade, its proximity to Petra and the Via Nova (Fiema 1993). The excavators of Tall ash-Shuna in the Jordan Valley, while originally interested in the prehistory of the site, were struck by just how extensive were its Hellenistic, Roman and Byzantine deposits (Baird and Philip 1994). This was clearly a large site at this time, possibly as large as 25 ha, although little is currently known about the form and structure of this settlement. Unfortunately, understanding of the other communities in this hierarchy—the 'small towns'—remains poor, although with the on-going exploration of the likes of Rabbathmoba, Charachmoba (modern al-Karak), Hisban, Humayma, Walid, Umm ar-Rasas, Madaba and Udhruh there is potential. It is certain that more work at these sorts of sites will redress the balance in our understanding of settlements.

The greatest explosion of data has been in that amassed by field survey. But again there exist problems in relating the results of even adjacent surveys, let alone those spread across the country. Survey, although able to accumulate data rapidly and cheaply, tends to be preconditioned by the strategy that determines how and what it accumulates (Banning 1996). Different approaches can also be conditioned by the terrain to be explored as well as, for want of a better expression, what might be labelled different national traditions or approaches to field survey. This is evident when trying to correlate the results of British, American and French fieldwork in the region. Likewise surveys can work on the micro-regional level (e.g. as part of just one wadi, and so tend to be intensive in their data retrieval) or a macro-regional dimension (and so are more extensive and, therefore, less concerned with detail). The situation is compounded by the way that some surveys have been one season long, while others have been conducted over a number of years and may consequently have evolved their methodologies. Surveys too have a tendency, in spite of the best efforts of the project, to be initially, at least, especially interested in specific periods or themes. The Southeast 'Arabah Survey is interested in the Nabataean/Roman/Byzantine settlement pattern to the north of al-'Aqaba. The 1993 Madaba Plains Project was concerned mainly with inscriptions in a 5 km radius around Tall al-'Umayri. The Jarash Regional Survey, despite its title, was more concerned with the Neolithic–Early Bronze Age transition. The Wadi Faynan Project involves a collection of different (British) teams looking at various aspects of the geology and archaeology of a number of environmental zones in a wadi complex due south of the Dead Sea (Barker et al. 1997). This sort of project, despite its attempts to integrate the various results of the constituent projects, typifies the problems involved in correlating survey data. The variable geography, topography and climate of Jordan are bound to have had an effect on settlement patterns. It is, therefore, going to be just as difficult to extrapolate a nationwide perspective from regional surveys. A final problem which complicates the issue of field survey is the unevenness in the processing and interpretation of the data that each survey has accumulated. Indicative of this particular problem is the way that Alcock (1994), while acknowledging the considerable increase in fieldwork in Jordan, still had recourse to just the Wadi al-Hasa survey in her summary of the Jordanian countryside in the Hellenistic period. Part of her problem is that surveys can easily become what Banning (1986) has called the creation of inventories of sites.

These factors explain that in attempting to summarize what is known of the archaeology of Roman Jordan certain aspects of it are better understood than others. In particular, certain 'big' city sites predominate, where there has been considerable discussion about the form and evolution of their street plans and so the identification of building programmes (Barghouti 1982; Segal 1988, 1997). Histories of this limited sort can be constructed for Gerasa and Petra, into whose framework communities such as Pella, Gadara and Philadelphia can be reasonably accommodated. There is also scholarly attention on churches and Christianity, again because of the bias of western scholarship coupled with the fact that the evidence for them has been found in such cities.

Perhaps the greatest gap in our knowledge is what might have been occurring in the territories—the countryside—of these sorts of communities as well as in their larger counterparts. With the growth of field survey as a prospecting technique, this represents the biggest explosion of data available to the archaeologist. In its ancient sense the city was the smallest politically autonomous community, although in reality the degree of autonomy was restricted when the city was a possession of kings and empires. In this definition the city-state was the combination of the principal community (after which the city took its name) and satellite towns, villages and hamlets as well as the countryside (*chora*). The two elements, namely, of city and territory, were indivisible. And yet, for all the regional surveys that have been undertaken, it is disappointing to report just how little is known about these sorts of relationships in the Roman period. There have been undertaken a number of surveys of regions that must have lain within the territories of some of the major cities (Homès-Fredericq and Hennessy 1989: 13-97). For instance, Wadi al-Yabis lies between Pella and Gerasa and Wadi al-'Arab between Irbid and Gadara. Both wadis have been surveyed recently. However, such projects are frequently hindered or limited by the research biases noted above. In contrast, despite its title, the Greater 'Amman Survey (Abu Dayyah et al. 1991) is not really a systematic attempt to explore the territory of the city at any time in its past. Instead, it is more an exercise in trying to record sites as they emerge and/or are destroyed as the modern city grows. Its links with the Cultural Resource Management Programme are obvious. As a part of the excavation of the large late Roman, Byzantine and Umayyad town at Udhruh, Killick (1983) undertook a survey of some 800 km^2 of its hinterland. While not a 'classical' city by any stretch of the imagination, not least because very little is known of it, the Southeast 'Arabah Survey has been looking at the countryside around Nabataean and Roman al-'Aqaba (Smith et al. 1997). More specifically, only recently has work been directed towards exploration of the territories of Gerasa and Pella (Koucky in McNicoll et al. 1992; Watson 1996). The potential for the exploration of the territory of Gerasa has been increased immeasurably by the discovery of a series of marker stones which appear to delimit at least the eastern fringes of the city and so provide an indication of the resources that were available within its *territorium* (Seigne in press). This should complement earlier surveys that have been conducted in the region (Hanbury-Tenison 1987; Sapin 1987). Another aspect of the archaeology of the Roman countryside, one little explored, are the Roman deposits and architectural fragments scattered around many of the modern agricultural communities in the vicinity of Gerasa. Many of the (modern) communities in its vicinity, places such as Khirbat al-Qusayr southwest of the city, are generously scattered with Roman architectural fragments and other surface detritus and as such would almost certainly reward further investigation.

There is one more comment to be made about our relatively poor understanding of the archaeology of Roman Jordan. The range of field techniques has until recently been limited to a combination of traditional architectural studies, excavation and lately various forms of field survey. There has been the failure in the past, for understandable reasons, to use other forms of reconnaissance evidence. The obvious example in this respect is air photography, although the situation is beginning to change (Kennedy and Riley 1990; Kennedy 1998). Other forms of remote sensing technology, including Geographic Information Systems, are becoming increasingly used. The potential of these methods to provide a better understanding of the archaeology of Jordan as a whole is almost infinite.

A good example of the potential of new approaches to data collection with respect to the ancient countryside is to be found in the Umm al-Jimal report (de Vries 1998). While reporting on work in the town itself, that work concentrated mainly on excavation and survey of the main settlement. In doing so, it has identified the Nabataean antecedent to the visible Roman and early Byzantine site that attracts visitors today. However, exploration of the countryside around the site was not originally given a high profile in the project. More recently, Kennedy (1998) has used a set of aerial photographs taken in 1953 to identify sites on the swathe of land covered by the 1:50,000 K737 map of the Umm al-Jimal area, some 400 square kilometres. On these photographs were identified over 1050 'sites' of 22 different morphological types, making for an average site density of 2.5 per square kilometre. Of course the main problem with this sort of approach is that the air photographs fail to distinguish between different periods of activity, which could range from prehistoric to modern, but certain patterns can be identified, not least when the results are tested against ground proofing. At the very least, this particular exercise has served to provide an indication of just how intensively exploited was a region that had previously been imagined to be too desolate and desiccated to have been farmed in the past. Comparable studies of other parts of northern Jordan are promised. Assisting in the understanding of the importance of agriculture in Roman Jordan, field survey is constantly hinting at preserved ancient landscapes made up of field systems and clearance zones. Kennedy has identified what look to be extensive Hellenistic-Nabataean-Roman field systems in the Hauran region in the north of the country, while comparable systems have been seen in Wadi Faynan in the south (Kennedy, MacAdam and Riley 1986; Kennedy and Freeman 1995; Barker et al. 1997).

It is because of the absence of work on the relationship between town and country that what can be said of society and the economy in the Roman period is hampered. Our understanding of religion too is limited principally to urban-based cults, with some notable exceptions. However, one positive feature is that the situation with regard to the military history of the region is in some respects immeasurably better but in others similarly impaired. The evidence available is sufficient for the writing of a history of the defence of the province, but of only one dimension: namely its eastern (i.e. desert) margins. The issue of the relationship of the military to the city and towns (and so the countryside) remains neglected, if not ignored. Bearing in mind these comments, the following account is as much a summary of the current state of progress in the archaeology of Roman Jordan as a pointer to potential areas of research. Irrespective of these general comments, a number of recurring features in excavations and field surveys can be discerned.

Cities, towns and the countryside

As is well documented, there are a number of unmistakably Graeco-Roman cities in Jordan, namely, Gerasa (Figure 12.2), Philadelphia (Figure 12.3), Capitolias, Gadara (Figure 12.4) and Abila (Figure 12.5), as well as Petra in the south and Bostra in what is now Syria (and which is excluded from this discussion). Because of it being so far south and because of its unique location and architecture, there is a temptation to regard Petra as something of an exception to the picture in the northern cities that conversely often tend to be treated as a homogeneous group. And yet, in writing about all of these cities, there are a number of problems. We know next to nothing about their origins, other than that all these sites show evidence of prehistoric occupation (as far back as the Middle Palaeolithic), as revealed at Abila, Pella, Philadelphia, Gerasa and Petra. With the exception of Petra, they all appear to have been (re-)founded in the early Hellenistic period and to have thrived thereafter. On the other hand, little or no early Hellenistic occupation has so far been found at Petra. Indeed, there is relatively little 'Roman' data from this site, and that which exists is frequently problematic (Parr 1996; Fiema in press). Our next problem is that not much is known about the shape or form of these foundations. This is an important issue because it may have had a bearing on the development of the communities in the Roman period. Unfortunately, a combination of little work, post-Hellenistic destruction and the reutilization of those Hellenistic features and structures makes discerning this difficult. In the past it was common to believe that it was the Roman occupation of Arabia that provided the impetus or second wind for urbanization in Arabia. However, archaeological work is constantly demonstrating that this is not the case and that much of what has traditionally been attributed to the early second century can be traced back into the first century BC/AD. For instance, the street plan of Roman Gerasa (Figure 12.2), evidently redefined in the second century AD, owed much more to a Hellenistic antecedent than was once imagined before being remodelled in the mid-first century AD. The same applies to Bostra, where the framework of the city's layout is now known to have been created in the first century AD, under the Nabataeans, rather than as a consequence of the Roman occupation and the establishment of power there. Of Philadelphia, a Hellenistic (re-)foundation (Figure 12.3), it cannot be said if its plan was a Hellenistic or Roman adaptation. All of its extant buildings date to the second–third centuries AD, with the rest, along with its street plan, lost beneath the modern city.

The cumulative impression is that the layouts of the street plans of these cities show more than just a little care. The best example of this is Philadelphia where the city planners, whoever they were, were faced with an especially difficult topographic canvas. The site lies in a deeply incised valley (Wadi 'Amman), joined by subsidiary wadis, all of which dissect a limestone plateau that comprises a number of peaks (e.g. Jabal 'Amman). The wadi has a perennial stream, fed by the spring Ras al-'Ayn, 1 km southwest of the ancient town. Into this wadi was placed the *decumanus maximus*, probably built before 160, and the other civic accoutrements of a major Graeco-Roman community. The constricted location meant that it was not able to develop a full-blown street grid, which itself could be useful today for identifying the evolution of the city. It could only really spread to the southwest and the east. Unfortunately, with the way that 'down town' 'Amman has grown up in Wadi 'Amman, on present evidence it is not possible to identify any traces of a Hellenistic plan.

In contrast to Philadelphia, the topography of Gerasa is much simpler. The visible remains appear, superficially, to suggest that the public buildings of the city were constructed on the western bank of the River Chrysorohas. Admittedly, largely on negative evidence, it has been assumed by some that on the eastern side of the river, under the modern town of Jarash, would have been most of the city's domestic quarter. However, recent excavations have shown that private residences existed in the western sector of the site. It would also be speculative, if not foolhardy, to argue that there was ever a split between public and private organization of the city. The evidence from the western part of the city shows that much of the road network here originated in the late Hellenistic era, with minor readjustment of *insulae* in the Roman period, along with the attaching to it of new streets on largely the same alignments. Work commenced on the Temple of Zeus (Olympias) in at least AD 22/23 (although there must have been a Hellenistic foundation below it), at the South Theatre (90/91) and the 'Oval Piazza'. What is striking in all this work is that in the case of the former, at least, construction was funded in part by private donations. This sort of benefaction was common enough for citizens of such communities, but in this example the contribution was made by a Roman official at a time when the city was extra-provincial. Similar gifts from Roman officials and soldiers become increasingly common throughout the first century.

The two principal roads of the city are still visible at Gadara (Figure 12.4). It is difficult, however, to say much about the development of the site because there are very few extant structures in the lower part of the city (e.g. to the west of the acropolis) to project the arrangement of the subsidiary roads. The city was founded by the Ptolemies as a military colony in the Yarmuk Valley. Unfortunately the heart of that and the subsequent community would have been on the natural hill that became its acropolis and has always been attractive for settlement. However, recent excavations have discovered incontrovertible evidence of (early) Hellenistic deposits (Kerner 1994). The road from the west (Tiberias) Gate to the Jordan Valley was paved in c. 50–100, along with the appearance of a necropolis on both sides of the street. Also constructed at this time was the Tiberiade Gate (so-called because of its similarities with a comparable gate excavated at nearby Tiberias; Weber 1991a). This was a pair of circular towers forming a gate some 600 m beyond the city's west walls. Its purpose, outside a set of functioning city gates, is enigmatic. The excavators' view that it represented a customs or border post on the outskirts of the city is unconvincing. Excavations have exposed part of the southwest corner of the city wall on the south side of the acropolis, built sometime between 100 and 50 BC. This was evidently demolished some 100 years later (Josephus, *Ant.* 4.416). That this was the case is not difficult to understand. The first century BC/AD was, generally speaking, a period of stability, admittedly with the occasional blip. That period is reflected in the building work that has been dated to this period. Above all, it was Petra that seems to have prospered most at this time. During this time were constructed its theatre and a number of major civic temples, the *decumanus* and the temple known today as the Qasr al-Bint (McKenzie 1990; but see Parr 1996 for counter-arguments; Fiema in press). This prosperity and a sense of public-spirited community clearly continued, perhaps even accelerated, into the second century AD.

In all of this discussion of the Decapolis cities conspicuous by their absence are Capitolias, Abila and Pella, where the evidence is so poor that it is difficult even to estimate with any confidence the alignment of their *decumani* let alone what went with them. Pella, destroyed by Alexander Jannaeus in 83 BC, was still evidently in ruins by the time that Pompey passed through the region. Its rebirth as an urban complex seems to have been delayed to the late first century AD, when a new street layout with public buildings was established over the Hellenistic ruins in Wadi Jirm. Constructed after this were the *odeon*, public baths, a public courtyard (*parvis*) and *agora*. Coins of the city also suggest that a *nymphaeum* and at least one major temple still await discovery. The fact that so little is known about Roman Pella has led one commentator to observe that 'the Roman Decapolis city does not seem to have been as impressive as some of the other contemporary cities' (Smith and McNicoll 1992: 135). The failure to define Roman Pella is due to the overburden which now covers the site. It has been hypothesized that it may have also been a small community, perhaps affected by population decline at this time. The available evidence suggests that its main period of prosperity was in the fifth century (as it was in the other cities) although it must be noted that this may be more a reflection of the nature of the work that has been done at the site than historical fact (Smith and McNicoll 1992). Farther south, on-going excavations at Aila have found the Nabataean and Roman town and demonstrated that it was largely a creation of the first century BC. There does not appear to be a Hellenistic antecedent to the settlement (Parker 1997). The town continued into the Late Roman and Byzantine periods with its rise and success being attributed to the development of the trade route through the Gulf of al-'Aqaba.

Although, as has been noted, the first century AD witnessed significant building work at most of the major cities of Arabia, still the second century is striking for the quantity and quality of construction. The bulk of the public buildings at Philadelphia date to this century, perhaps as part of a coherent scheme. Those structures built include the *agora*, theatre, public baths and portico. In addition, inscriptions from the city point to the existence of a council house and a *gymnasium*. Elsewhere, at Gadara, the *nymphaeum* adjacent to the *decumanus maximus* preserves dedications of the second century. It is believed that the city expanded westward in the same century, away from the (Hellenistic) *acropolis* and heart of the original city. Once the city's west wall/gate had been established, there was then constructed a hippodrome and monumental arch farther west, and so beyond the west gate, in a fashion that also occurred at Gerasa. At Gerasa the central part of the *cardo* was widened in the second century. The South Theatre was reconstructed in the early part of the same century, and in the period 161–166, at a time when the Temple of Herakles/Hercules was being completed at Philadelphia, the Temple of Zeus Olympias was embellished further. In 129/130, in light of Hadrian's visit, the city's Triumphal Arch was erected. Standing so far south of the city's walls, it has been suggested that its position reflected aspirations to extend the city. This work was followed by the hippodrome (c. 140) and in c. 150 by the *propylaea* of the Artemis temple, which itself necessitated changes in the vicinity of the temple. Still later, the so-called 'Eastern Baths' and the North Theatre (or *odeon*) were added in 162–166. What is significant in this, as with the earlier building activity, is that it would have all been funded by private donations and benefactions (*euergetism*), as is reflected in the city's epigraphy.

There has been, in the past, a temptation to attribute those developments in Arabia that can be dated to the second century AD and after to direct Roman intervention, perhaps on the part of emperors or through their governors. However, the degree of imperial initiatives is debatable. Recent work is repeatedly showing that the towns and cities of the region prospered before the Roman annexation, although their success may be an indirect consequence of the relative stability that Roman suzerainty in the wider region brought. That stability and thus prosperity continued when Nabataea became a full Roman possession. In addition to this, the degree of intervention from the Roman authorities in local government was minimal. Communities appear to have retained their pre-Roman administrative frameworks and control of their resources, other than the taxation that they were expected to provide. Where the contribution of Romans to building work is preserved, this should be interpreted as a reflection of the sort of personal patronage that underscored social relations in the ancient world rather than a reflection of imperial policy. The one potential exception to this impression is Petra. Fiema (in press) has summarized the evidence for pre-Roman and Roman-period building activities in the city. Those that appear to post-date 106, he has tentatively linked with refurbishment following an earthquake of c. 113. This may have been assisted by relief from the imperial power. That said, he is as prepared to see them as the result of an influx of wealth and stability following the annexation.

The impression of prosperity in the major cities in the second and early third centuries is paralleled in some of their smaller counterparts. At Umm al-Jimal it has been found that an earlier, possibly Nabataean, 'rural' settlement was destroyed sometime in the third century, perhaps during the Palmyrene interlude, after which the focus of the community shifted some 300 m westwards to another site that appears to have been developing since the mid-second century AD. This settlement, the one that is visible today, prospered throughout the Byzantine period, defended by a *praetorium/castellum* of the late third century, built under the emperor Diocletian. This would imply that Umm al-Jimal was an important part in the later Roman defence systems (de Vries 1998). As a word of caution, it has to be said that this picture might not be universal. It may be the case that the broad picture is not consistent or uniform. For instance, at Humayma (ancient Auara), where there was a sizeable pre-Roman Nabataean community founded under Aretas III and an early (e.g. second/third century) Roman fort perhaps built on an earlier Nabataean one, there is at present little sign of the Roman period between the Nabataean and Byzantine horizons, other than a second-century bath house. The fort was probably abandoned in the late third century, then renovated for domestic occupation in the fourth century.

It is a truism, but another aspect of city life is that people lived in them. In the past, however, this simple fact lost out to grander research schemes. As regards domestic and private accommodation, as is to be expected where there has been excavation, there is good evidence for seamless continuity from pre-Roman to Roman-period housing. For instance, Late Roman houses have been found on top of late Nabataean dwellings at Petra. Unfortunately, these were subsequently buried under or incorporated into Byzantine structures (e.g. as with the recently excavated 'Petra church' constructed c. 363; Z. Fiema pers. comm.). Excavations by Zeitler (1990) revealed two complete buildings and part of a third in the east of the city. The earliest stone-built structure, a rectangular unit measuring 2.8 × 2.6 m, was erected 30/20 BC and abandoned at the start of the first century AD. In AD 50, after a period of dereliction, a second larger four room building (13 × 11.5 m) with a cistern was put up on the site. The third structure was placed beside this building. These two buildings were evidently destroyed (by an earthquake?) in the fourth century, after which they were further modified. Data recovered from the site revealed something of the diet of the occupants. The structures had their own bread kiln. Animal bones included much sheep/goat, but considerably less pig, dog, fish and bird. These dwellings complement those that have been excavated elsewhere in the city, at az-Zantur. Here the evidence of continuity spans the first four centuries AD (e.g. Stucky 1994). Houses were also found on the pre-second-century AD acropolis at Philadelphia. However, these structures were then levelled when the temple complex was prepared. At Gerasa, a group of houses has been excavated in the vicinity of the 'Oval Piazza' and along the southern *decumanus*. At Gadara, evidence of housing has been located perpendicular to the westward running *decumanus maximus*, in more-or-less rectangular blocks that are believed to date back to the Hellenistic period. Houses have also been reported under the third-century Roman bathhouse at Humayma. On a slightly larger scale, an archaeological park has been created at Madaba that has as its focal point the Roman street system using the *decumanus* and its adjacent buildings. Visible structures are mainly Byzantine churches built atop Roman-period domestic structures.

As already explained, what was occurring in the territories of these towns and cities at this time is difficult to discern. The situation is not helped by our inability to locate with any precision the territorial boundaries of the cities, which would in turn allow some estimations of their agricultural and other resource potential. The one exception to this is Gerasa, where recent discoveries of marker stones by Seigne (in press) have delineated at least parts of the eastern and southern extent of the *chora*. Parts of the territory of Petra have been explored by various survey projects but there has been little effort at correlating the results (Lindner 1989: 83). Otherwise the results of field survey remain too general, while there have been too few site-specific excavations. Despite this negative assessment, some interesting points are beginning to emerge. Broadly speaking, it seems that the Roman–Byzantine period was one of prosperity as reflected in higher settlement densities than previously had been the case. Survey of Wadi Ziqlab in the northern Jordan Valley showed that the most intensive settlement densities lay in the Roman–Byzantine period (Khouri 1988: 15-20). In turn, the survey argued that the decline in settlement in all periods can be linked with erosion and colluvation. This deterioration not only has implications for the visibility of sites during survey, but for the success of settlements in general over time (Banning 1996). This picture is paralleled to a certain degree in Wadi al-'Arab, where survey produced data that should be relevant to Gadara (and to a lesser extent Capitolias) and the upper Jordan Valley (Hanbury-Tenison *et al.* 1984). Topographically, this was a mixed region which was going to make large-scale arable farming difficult on the slopes between the valley but easier on the plateau upon which Gadara lies. Therefore, we might envisage an emphasis on olive and grape production in the hilly region and tillage on the plain. Again, field survey has created a picture of increasing settlement densities through the Iron Age, Classical and Byzantine periods in this region. The Wadi al-Yabis Survey found that in the Roman (and Byzantine) period settlement and agricultural activity reached an all-time peak (Palumbo *et al.* 1990). The South Hauran Survey in north Jordan has tentatively concluded that there was an increase in settlement densities in the Roman period compared to the preceding Iron Age and Hellenistic periods (Kennedy and Freeman 1995). Farther south, beyond the Dead Sea, exploration of the southern Ghors region and the northeast 'Arabah has demonstrated that they were extensively occupied in the Nabataean–Roman–Byzantine periods with clear continuity from the one to the other (MacDonald 1992: 83–95). The same trend was identified in Wadi al-Hasa, where Nabataean, Roman and, to a lesser degree, Byzantine sites proved to be the most common and where most of the sites explored were isolated farmsteads and small hamlets of a few households and corrals (MacDonald *et al.* 1988; Clark *et al.* 1994). Likewise, the Hisban Survey identified a progressive rise in rural settlement densities (LaBianca 1984 in Homès-Fredericq and Hennessy 1989: 261-69). In this particular case, there was also detected an accelerating trend towards 'urbanization' (= nucleation), a process that reached its height in the early Byzantine period. The Northwest 'Ard al-Karak Survey identified many hamlets, isolated farmsteads and campsites in a region that was extensively exploited under the Roman *pax*. Relative to the Wadi Faynan Project, the objectives of which have been noted above, the most substantial visible remains appear to be Roman and Byzantine in date (Barker *et al.* 1997). The Southeast 'Arabah Survey, looking at the southeast sector of Wadi 'Arabah and up to 7 km northeast of Aila, has suggested that the Nabataean–Early Roman periods were marked by significant occupation densities which began to fall away in the Late Roman–Byzantine periods. The rise in settlement density under the Romans is linked with the development of Aila and the rise of a regional road network in the first and second centuries AD (Smith *et al.* 1997).

Although there have been numerous regional field survey projects, few Roman-period rural sites have been excavated. Their data and interpretation, nevertheless, remains problematic. At Tall Faysal, located a few kilometres south of Gerasa and discovered in the early 1960s, a structure that was originally erected in the late second century AD and used on and off down to the twelfth or fourteenth centuries was excavated in 1991 (Palumbo *et al.* 1993). Because of its rectilinear shape it has been interpreted as a small fort or watchtower overlooking a crossing on the River az-Zarqa' on the road between Gerasa and Philadelphia. However, it might as easily be interpreted as a farmstead. At Khirbat Faris, survey of an area 3 × 3 km and then excavation of a specific building revealed something of a Classical (and later) agricultural settlement, which seems to have been especially prosperous in the first and second centuries AD (McQuitty and Falkner 1993). Recent excavations at Khirbat Salameh have revealed a 14-room building of the Iron Age to Hellenistic and then Late Roman to Byzantine periods. While there is as yet no firm explanation of its purpose, it looks to have been part of an agricultural complex (Bikai 1994a).

How one is to explain the rise in settlement density in the Roman period is problematic. Direct Roman (imperial) stimulus seems unlikely. More attractive is the explanation that it is an indirect consequence of a trickle-down effect of the peace and prosperity that the *pax Romana* brought to a process that is evident in parts of the country under the Nabataeans. After the burst of building activity in the towns and cities and the intensification in the exploitation of the countryside in the province, the momentum seems to have been lost. The amount and quality of civic building activity from the third century onwards declines significantly. Why the curtailment? Perhaps the momentum had reached a natural conclusion, where the cities now possessed most if not all the recognized features of major communities. For all of them exhibit the range of civic buildings expected of the typical Graeco-Roman city of the eastern Mediterranean: theatres (Abila, Philadelphia and Petra, along with three each at Gerasa and Gadara), basilicas/*stoas* (Abila and possibly Gadara), *odeons* (Abila, Pella and Philadelphia), *agora/fora/macella* (Gerasa, Philadelphia, Pella, Petra and Gadara), public baths (Pella and Gerasa), *nymphaea* (Pella, Philadelphia and Gadara), (extra-mural) monumental gates and arches (Gadara, Petra and Gerasa) and hippodromes (Gerasa and Gadara). The fact that at Gerasa and Gadara at least some building work seems to have been left incomplete at this time implies something more serious. Work on the hippodrome at Gerasa, itself the smallest known Roman example (at 245 m long), commenced sometime between the mid-second century and AD 209/212, although it ceased to act as a public arena for horse races some 100 years later. It was then modified solely for games at its northern end in the late third and fourth centuries, after which its buildings were reoccupied for industrial and domestic purposes as well as for residences and a quarry for dressed stone. Its initial decline was due to its extremely poor foundations that caused the southwestern section of the complex to collapse (Ostrasz 1989). The end of the working life of the hippodrome came in the seventh century following a series of earthquakes. At Gadara, construction again started in the later third century but, in this case, was never completed. Looking at the historical framework, the failure to complete or repair civic structures may have something to do with the problems of the wider region and the turmoil of the Roman Empire in general. This may have had a knock-on effect on the commercial prosperity of the region which could no longer fund these sorts of (privately funded) building programmes. An alternative explanation or else contributory factor may have been the proximity of these communities to one another coupled with their common (Hellenistic) origins. It is known that the rivalry between certain cities in the Greek East could be counter-productive. Pliny's (*Epistles* 10.39-40, 49-50, 70-71, 98-99) account of the mess that a number of cities in the province of Bithynia got into shows how the attempts of some to match and outstrip the facilities that neighbouring cities possessed could ruin them. It requires no stretch of the imagination to see something similar in Roman Jordan, judging from the way that most of the cities seem to possess the same array of civic buildings and that they appear to have been erected at roughly the same time.

The provision of public amenities in the cities of the region is important because, indirectly, it permits estimates about the size of their populations. Ancient city records rarely provide breakdowns of population. Where there are figures they usually relate to male citizens and exclude women, children, slaves and resident non-citizens. In trying to calculate the size of populations, the capacity of certain public amenities can provide rough indications of the potential size. Unfortunately, our knowledge of the public facilities of the cities of Arabia is limited. The best known are those at Gerasa (Figure 12.2). Here the South Theatre is thought to have been able to seat 4000 spectators. The later North Theatre could accommodate 1600. Perhaps more indicative of a minimum size of the community is the capacity of the hippodrome which could hold 15,000. What cannot be taken into account in calculating a minimum size for the population of the city is the numbers living in and on the countryside. The reasons for this inability have already been explained. As such, estimates that the population of Jarash was in the range 17,000-20,000 represent a minimum estimate for those living in just the city and not the city-territory. Similar estimates can be attempted with even less confidence for Philadelphia and Gadara.

The military dimension—the defence of the province

The military history of Roman Arabia is in part one of fluctuating diplomatic relations with the indigenous tribes rather than having to deal with any (major) external threats and invasions. In the first instance, there was the contact with the Nabataeans. Later on, when the Romans had moved into the region, the Ruwwafa inscription of 166/169 implies close ties between Rome and a Thamudic tribal confederation (Graf 1978; Parker 1986: 124). In the Palmyrene period, the Tanukh, an Arab tribe originally from the northeast Arabian peninsula, migrated northwest to avoid Sassanid control. This led them into the Hauran where they formed a tribal confederation which came to ally itself with Rome and which helped Rome in the destruction of Zenobia. In the later fourth century we learn more and more about the Arabs with implications for their organization. For instance, there was Mavia and her 'Saracens' in the 360/370s and the rise of the tribe of the Salih in the late fourth century, led by (Roman) *phylarchs*. It was these local potentates who came, in time, to assume responsibility for the defence of Arabia: the Kindites in the late fifth century and ultimately the Ghassanids in the later fifth–early sixth centuries (Graf 1988; Shahid 1984, 1989).

In the past, almost inevitably, the military occupation of the province has been written within the framework of the historical narratives, made tangible by those fort sites discovered by earlier field expeditions, notably those of Brünnow and von Domaszewski, Stein and Glueck. Added to this was the belief that the known activities of emperors elsewhere in the empire would have been paralleled by similar developments in Arabia. The military history has, therefore, usually been written in terms of frontier history. But for this approach to be valid, it has to: (1) identify who were the (frontier's) enemy, which in turn permits identification of (2) where the frontier was and (3) how it looked and functioned. Potential enemies who might have determined where the frontier was located have included the Parthians, the Arabs (also known as the Saracens), nomads and/or internal threats.

Credit for reawakening scholarly interest in the 'Roman frontier' in Arabia lies with Parker's *Limes Arabicus* Project which commenced work in 1975 and ceased in 1993 (Parker 1986, 1987). This project was a combination of field survey, test excavation at selected sites and a concerted programme of excavation at the site of the legionary fortress at al-Lajjun. From this work Parker was able to construct a narrative history of the frontier. For the Roman period, that history might be summarized as follows:

(1) Following the annexation of 106 the Nabataean defensive network provided the initial framework for the Roman occupation, although not all Nabataean sites were occupied. Added to this arrangement was the work on the Via Nova Traiana. That said, the second-century occupation of most forts in the region remains largely a blank.

(2) Under Septimius Severus the province's territory expanded northwards. The evidence of military activity indicates a major strengthening of the northern sector of the frontier, especially around Wadi Sirhan that implies, in Parker's opinion, trouble with nomads. There also occurred at this time a widespread repair to the road system. These developments are related to the aggressive frontier policy of Severus and in part to help his eastern campaigns in Mesopotamia. In contrast, the central and southern sectors appear to have been relatively quiet.

(3) Repairs to the road system, which was maintained until c. 250, occurred in the mid-third century, after which the Arabian garrison was probably denuded to bolster the frontiers to the north. This process may be detected in the archaeological record, although the system was maintained as the denuding of other garrisons continued in the Palmyrenian period.

(4) While the garrison in Arabia was restored under the emperor Aurelian, it is more likely that it was one of his successors, probably Diocletian, who was responsible for the actual building restoration of the frontier. This involved a number of elements: repair of the existing road network, new forts and repairs, especially in the central sector of the frontier's forts (including the construction of a new legionary fortress at al-Lajjun), and the creation of the *strata Diocletiana* to protect southern Syria and north Arabia from Persian and Arab raids. Behind the garrisons stationed on the province's frontier road system, a reserve of mobile cavalry units (*equites*) was kept in the towns.

(5) The fourth/fifth century represents the most heavily fortified period of the *Limes Arabicus*, although it was essentially that system created by Diocletian.

> a broad fortified zone in Transjordan from Bostra to Aila and a secondary zone of defence in southern Palestine from the Mediterranean to the Dead Sea. All the legions and the bulk of the *auxilia* were deployed in the outer zone. This zone was designed to contain the attacks of the Hijaz tribes and control movement through Wadi Sirhan…The inner zone was crucial for the security of Palestine (Parker 1986: 145).

(6) The late fifth century saw the decline of the Arabian frontier's defences, with the cessation of building activity and the abandonment of forts. This was probably occasioned by the weakening of the garrison in response to problems elsewhere. The evidence from the frontier's central sector (including that from al-Lajjun) is that this was abandoned in c. 550. Few sites remained occupied in the southern sector. Only in the north is there evidence of continued Roman military activity. This decline was matched by another in the quality of the troops in late fifth/early sixth-century Arabia.

Despite the neatness of Parker's explanation of the form and evolution of the Roman frontier in Arabia, there are a number of problems. Nearly all aspects of his work and interpretation have been subsequently challenged. These include his survey methodology, the accuracy of the ceramics sequence he used to date his sites and the presumption that datable occupation at his forts necessarily denotes military presence. However, the most aggressive debate concerns to what extent the 'frontier' was designed to halt a nomadic threat. Parker's position has long been that it was so designed. In contrast, others have challenged, with reference to other criteria, virtually every aspect of Parker's views. For instance, there was never a real nomadic threat and the frontier was never a frontier-in-depth. A central element to understanding the defence of the province has been widely seen as the Via Nova Traiana. Unfortunately, as we have already seen, the known facts about the road are actually quite slight. It was evidently constructed between c. 111 and 114, probably from the south northwards. In turn, there is some debate as to whether or not military sites were located with particular reference to the road. The lack of other information permits any number of hypotheses as to how the frontier might have worked. It may have functioned in any number of guises, none so far decisively identified. Parker failed to take into account the part played by urban-based garrisons (which were based far from the frontier proper). As an army of occupation, perhaps it was meant to look to internal problems. Indeed, there was a failure on the part of the project to search for new forts elsewhere (e.g. away from the frontier and along the Jordan Valley). Finally, non-military aspects of the frontier were not addressed. In conclusion then, in light of these many criticisms, perhaps Parker's most lasting achievement of his work is that it reinvigorated the subject, generating a debate, often vigorous, which is on-going (Isaac 1988, 1990; Kennedy 1992; Graf 1986, 1989, 1997; Fiema 1995 against Parker 1992).

The subject of the province's defences includes the issue of urban fortifications. While it is clear that a number of communities did possess walls and towers to defend themselves, the nature and date of these fortifications is problematic because the evidence has been so neglected. At Gerasa, for instance, the location of only four gates to the city is currently known, although there must have been more. The dating of the provision of defences is usually established by dating the gates themselves. On the available information, there seems to be no consistent explanation for when fortifications were thrown up, but if they were contemporary it would imply a common response to problems in the region. One might suggest that the erection of urban defences need not reflect a deteriorating security situation but could be another aspect of the sense of rivalry and emulation that existed between the major cities. At Gerasa the walls were 3500 m long with over 100 towers and enclosed an area of 850,000 square metres. The date of their erection is difficult to establish. They seem to have been built in AD 60–70. However, the North Gate replaced an earlier example in c. 115, a pattern that corresponds with the South Gate. In contrast at Gadara, the third-century necropolis on the west side of the city was robbed for the construction of the (west) gate and wall. The date of the monumental gate at Gadara is problematic as excavation revealed no significant stratified finds. The excavators, however, have argued for a dedication in the late third century, under the Severan emperors (Hoffman 1990). If the parallel with Gerasa holds, where it has been argued that the erection of the gate there was in anticipation of or to persuade the imperial authorities (e.g. Hadrian?) to patronize the extension of the city, then at Gadara one would be tempted to link it with Septimius Severus's visit to the east. Alternatively, under Philip the Arab in the mid-third century the region benefited from much imperial patronage. The Tiberiade Gate has been dated to the second century AD (Weber 1991a). The reasons for this fortification are unknown but they may be contemporary with developments at Gerasa. By this time the westward expansion of Gadara had ceased. There are no known fortifications around the lower city of Philadelphia, although the citadel/acropolis appears to have been fortified in the second century AD, as part of the initiative that saw the completion of the Temple of Herakles/Hercules. These walls evidently served as defences for the whole town, or so Ammianus would suggest. Other towns in Arabia were provided with fortifications but these appear to date to the Byzantine period (e.g. as at Umm ar-Rasas). The one exception is at Umm al-Jimal, where it has been argued on the basis of a now lost inscription that the town was fortified as early as the reign of Caracalla in the early third century. One site which does not fit entirely with the province's defences is that of Machaerus, a late Hellenistic–Early Roman palace and fortress 15 km north of Wadi al-Mujib on the eastern side of the Dead Sea. Used by the Jewish king Herod the Great, who also used the thermal springs at nearby Callirhoe (modern Zarka Ma'in or else at 'Ayn az-Zara), in the First Jewish Revolt it was besieged and sacked by the Romans, as shown by the impressive siege works that are still visible around it.

Another element of the archaeology of the province which is frequently associated with the military history of the province is the form and evolution of its road network, as emphasized by discussions about the Via Nova and the Strata Diocletiana in the north of the country. Two distinct areas of research might be highlighted: (a) in the north, the work of Bauzou (1986) on the Via Nova and Kennedy (1982, 1997) on the road system in the South Hauran region (and to a lesser degree by MacAdam 1986a); and (b) towards the south, Graf has been for a number of years exploring the network of roads which fed into the Via Nova around and south of Petra (Graf 1995, 1997).

The economy of the province: trade, production and agriculture

Measuring the economy of Roman Jordan covers a multitude of themes, ranging from the macro- to the micro-, from the local (e.g. the household) to the regional. Unfortunately, yet again, it is difficult to extrapolate a satisfactory interpretation of this evidence, derived as it is from a largely undigested database of field survey and limited excavation. One potential indicator of the economy of the region might be its coinage. Most of the cities before the Roman annexation minted their own coins. This privilege was frequently continued under the Romans. This is in itself a reflection of the often relaxed approach that they took to imperial government. It was, however, a privilege. Rome retained the sole authority to issue precious *specie* while at the same time curtailing and often removing the rights of cities to issue low denominations in their own name. The cities of the Decapolis are known to have minted coins well into the Roman period (Spijkerman 1978), but it is a contentious issue whether coinage is a direct or reliable indicator of the vitality of an economy. At the very least, there has been little study of the rise and fall in the circulation of coinage in Arabia that might in turn say something about the general state of the region at any particular time.

The strength of the province's economy has traditionally been measured by the vibrancy and quality of building work in its cities. In turn, the ability to initiate and fund this work is linked to the vitality of the (international) caravan long-distance trade that passed through the region, from the Far East to the Mediterranean (Miller 1969; Raschke 1978; Sidebotham 1986, 1991). It was the revenues of tax and the re-distribution and processing of these materials that is believed to have been the basis for the region's wealth. In turn, it is from this sort of trade that the evident prosperity of Petra in the first century BC/AD has been predicated. There may be something in the way that caravan trade indirectly funded building work. But just as fieldwork is constantly demonstrating that the countryside around these 'caravan' communities was densely occupied (and thus intensively exploited?), so the raw foundation of these communities was based on their agricultural prosperity. The relatively high cost of land transportation in antiquity is well documented, which meant that the cities had to rely on local sources for everyday items. For the caravan trade was never noted for the movement of such basics as grain and olive oil, but rather for luxuries and exotica. It may not be coincidental that the (Jordanian) Decapolis cities are to be found on the high plateau of northwest Jordan, with Pella at the western foothills. This region, although hilly and interspersed with valleys, is today agriculturally productive and must have been so in antiquity. Foodstuffs must then have had to be acquired and so transported locally.

The emphasis that has been given in the past to the part of the caravan trade is due in part to the lasting influence of Rostovtzeff's *Caravan Cities* (1932), an important work which drew extensively from the results of the sort of limited fieldwork that had been completed up until the 1930s. In addition to this, there was Rostovtzeff's (flawed) perception of the form and organization of the ancient economy, a model that consistently underplayed the importance of land. Nowadays the balance has been redressed as exploitation of land is now seen to be the basis of the ancient economy. Reflecting this change, with reference to the economy of Pella, Smith (1987) has argued that while engaged in trade at all periods of its history, it could never have maintained its viability solely on the basis of the caravan trade. It must have been dependent on other sources or activities. Such observations must have implications for other cities in the region. The same line of argument has been applied to the settlement at Umm al-Jimal. That the site comprises substantial and extensive standing remains today has misled scholars in the past to see the basis of the town's wealth as being derived almost exclusively from the profits from the caravans. However, the plainness of the town's architecture, coupled with appreciation of its domestic buildings (houses with enclosed courtyards and ground floor as barns and stables), implies, on modern parallels, households principally concerned with farming. This does not deny, however, that the community might have benefited from stop-over traffic passing through the region.

The two exceptions to this picture are Petra and perhaps Aila, where at the former the evidence of excavation suggests that there is little Roman data from a city that seems to have flourished in the first and second centuries AD, judging from the building work that occurred at this time. Why the city should have gone into an architectural (and thus economic?) stagnation in the third century has been attributed to a decline in the amount of commercial traffic that passed through it in the Roman period. This was a result of a shift in caravan routes, when the Roman occupation of ancient Syria and Arabia is supposed to have created the conditions for a better protected and more reliable route from the east than the preceding route through the Persian Gulf and the Arabian peninsula. The consequences of this shift were that the communities farther north, such as Gerasa, profited in turn. Parker's work at Aila is rather ambitiously trying to address the question: 'What was Aila's role in international trade between the Roman Empire and its neighbours?' (1997: 25). The evidence recovered so far shows that materials from the western and eastern Mediterranean and Egypt were passing through it. Literary sources also attest to its vibrant commercial links with the Far East.

Archaeology has also revealed something of the micro-regional production of everyday objects. But this is a far from complete picture. Shops have been discovered along the main roads of most of the major cities of the region. More formal market or commercial places (*agora*, *fora* and *macella*) are known at a number of sites. Recent excavations in Gerasa of an octagonal-shaped building at the southern end of the *cardo* and just north of the Oval Piazza exposed a *macellum* which was constructed c. 125 but which went through a substantial refurbishment in the late second–early third century (Martin-Bueno 1992). Segal (1997) has made the highly attractive suggestion that a similar-shaped building currently being excavated at Gadara may also have been a *macellum*. There is also the temptation to regard any large open spaces at the heart of cities and towns as *fora*. At least two such areas have been identified at Gerasa (e.g. the Oval Piazza and one adjacent to the North Theatre). At Petra, there are three suspiciously large terraces above the *cardo*. These have been seen as *fora* for the city, an idea that has its attractions, not least because they seem to be linked to a range of shops along the *cardo*. They are known today as the Upper Market. In addition to these commercial zones, we know quite a lot about particular industries in particular cities (e.g. the Jarash Bowl industry at late Gerasa — Watson 1989) but have difficulty in relating them to each other or to the wider economy of the individual cities. Metalworking, glass production and pottery kilns, as well as quarries, are now common enough discoveries at the major city sites. Parker's (1997) excavations at al-'Aqaba have exposed a second- to fourth-century complex of bread kilns and pottery workshops.

Religious life

Most of what can be said about religious life in the Roman province is derived yet again from what was occurring in its cities. In general terms, there has been a considerable failure to explore regional cult centres like that at Si' in southern Syria (Dentzer 1986). One exception is the Nabataean temple at Wadi Ramm in southern Jordan. Excavated in 1959, the complex seems to have been dedicated to the Nabataean goddess Allat in the first century BC or early first century AD. A similar complex is that at Khirbat at-Tannur, 23 km southwest of the Dead Sea, which was again founded in the first century BC and continued into the second century AD. This demonstrates yet another factor that has to be taken into account when considering religious life in Roman Jordan, namely, the longevity of religious sites and where the deity venerated could also evolve and/or be replaced by another. Likewise, there is the Roman propensity to adopt and manipulate regional deities to its own end. It should also be borne in mind that if the Roman occupation of Nabataean Arabia does not mark a cultural divide, we should have no problem in envisaging a continuity in indigenous religious practices. Pre-Roman deities are known to have continued in Gerasa where the Nabataean cult of Dushara (akin to the Greek Dionysius) is attested, as well as Pacides and another 'Arab' god (Welles 1938: Nos. 2, 17-22, 192). The cult of Maiuma at Gerasa, which was later associated with a Graeco-Roman festival celebrating the goddess Artemis, was probably originally a Phoenician cult that was later equated with the cult of Artemis. Its cult was later converted to a Christian 'harvest festival'. As elsewhere in the empire, the cities of the region played a part in the imperial cult along with the associated priesthood. Furthermore, a cult of Tiberius is known in Gerasa as early as 42/43.

It is inscriptions that refer to the existence of specific cults of a variety of deities in most of the region's cities and so indirectly to their worship at temples and shrines. The sites of most of these cults unfortunately remain to be located. However, an interesting feature common to Philadelphia, Pella, Gerasa, Abila and perhaps Gadara (Figures 12.2-12.5) is the way that each appears to have had substantial temple complexes in prominent, natural positions and that these positions were enhanced architecturally. Something similar may be visible at Tall ash-Shuna (Baird and Philip 1994). They all also show evidence of long occupation sequences that might imply a continuity in function. Each of these religious complexes was dedicated to different deities, which would serve to enhance the idea that these were specialist cult centres (e.g. of Herakles/Hercules at Philadelphia and Artemis at Gerasa).

At Philadelphia there is the so-called Great Temple, more commonly known as the Temple of Herakles/Hercules, set at the southern end of the acropolis and not the highest spot (Figure 12.3). This position made the temple more visible to the lower city and must be a reflection of its importance to the community. The acropolis was unfortunately heavily re-used in the early Islamic era which has meant that much of the detail of the temple is now lost (Bowsher 1992). Measuring 43.8 × 27.5 m, it was set on an artificial platform with a courtyard at its front. The entire complex was linked to the lower city by a monumental staircase which zig-zagged its way down the southern slope of the acropolis to a monumental *propylaeum* built on the northern side of the east–west *decumanus* which lay at the base of the acropolis. Inscriptions imply that the temple was dedicated between 161 and 167, which in turn provides a *terminus ante quem* for the *decumanus* (and thus the layout of the lower city). The temple was probably placed on the site, cleared in the early second century AD, of an earlier cult centre, perhaps the Ammonite deity Milkom, who in the Hellenistic period came to be equated with Herakles. The cult had its own festival in the Roman period. The Herakles temple, however, was not the only cultic establishment built on the acropolis, but was probably part of a scheme to accommodate a number of other temples. There is evidence for at least two other temples, namely, one dedicated to Tyche and the other known today as the Northern Temple. The Herakles site has more recently been subject to consolidation and restoration work carried out by the Department of Antiquities and the American Center of Oriental Research (Kanellopoulos 1994). Recently Bikai (1994b) has published the results of excavations on a sixth- to seventh-century church on the Jabal al-Luweibdeh. He has made the intriguing suggestion that this church incorporated material, if not the actual site, of an earlier Roman building, perhaps a temple. The likelihood is that the church was dedicated to St George, a saint who is known to have been equated with Herakles in the transition from paganism to Christianity.

While the Philadelphia temple is badly preserved, the best examples of cults placed on prominent sites that retained a spiritual significance over time are the Temples of Zeus (Olympias) and the Temple of Artemis and its Sanctuary at Gerasa (Figure 12.2). The Temple of Zeus, situated on a prominent rise at the south of the city, was placed on a site that shows occupation from the Middle Bronze and Iron Ages onwards. In 22/23 a new temple was built, although there had been a Hellenistic foundation on the site. It was further embellished, evidently under the patronage of a Roman governor, sometime between 161 and 167. This temple comprised a series of terraces that accommodated the hill's steep slope. It was presumably intimately linked with the South Theatre which was attached to the site. Judging by the size of the temple, the cult of Zeus at Gerasa was surpassed by another, that of Artemis, who was the community's protecting deity. The Temple of Artemis complex was an all-the-more substantial structure, deliberately and carefully planned to incorporate a number of mutually supportive elements (Kalayan 1982; Parapetti 1982). Five components can be highlighted in this scheme. From the eastern half of the city there was a *propylaeum* which entered another *propylaeum* on the western side of the *cardo*. Part of the temple complex, therefore, stretched over one of the city's principal roads. This second *propylaeum* led to the third element, an outer court before the temple court and finally the temple itself. This work involved complex engineering as it took in the slopes and hill terraces as part of the monumental approach to the temple. Work on the temple commenced in 79/80, with the addition of the *propylaea* in c. 150. This addition caused a change in the area immediately around the temple. Surprisingly, no pre-temple building is currently known at this site, although there must surely have been one. As already noted, a form of the Artemis cult, that of Maiuma, was celebrated elsewhere at Gerasa, at the Birketein complex, 1.2 km north of the city, which possessed two bathing/reservoir pools and a festival theatre. Finally, the so-called Great Cathedral at Gerasa, dedicated in c. 365, is believed to have re-used the impressive temple of an unnamed deity which was perhaps erected in the first century AD.

At Pella, the civic complex was positioned below Tall al-Husn, in Wadi Jirm. Coins of the city depict a large temple so far unlocated but presumably on the tell. Excavation in Wadi Jirm has revealed Roman buildings which, repaired in the fourth/fifth century, were possibly part of the *temenos* or *propylaeum* of a major Classical-period sanctuary. In the case of Abila, another site sandwiched between two principal hills, Tall Abila and Khirbat Umm al-'Amad, excavations on the former have exposed a substantial sixth-century basilica which in turn appears to have been placed on top of an earlier temple or church. The evidence for Roman Gadara is not so strong. Again, coins of the mid-second century AD show a sitting Zeus-type figure, placed in a tetrastyle temple. However, the city also venerated Tyche and the Three Graces. At Gadara, there was a bath and theatre complex at Hammeh Gader, comparable to Birketein, north of the city.

The similarities between these sites are too much of a coincidence. There seems to have been a tradition in the region to site (major) religious buildings on prominent natural locations and to embellish them further with architectural features and monumental elements. Why there should have been this phenomenon is difficult to explain. It might be a feature of religion in this area—the utilization and adaptation of particularly venerated cults in these communities. Equally, it might be indicative of the sort of private-funded but civic-encouraged competition which is better attested in regions such as Bithynia where civic pride had as much to do with religious devotion.

With respect to other sites, Petra exhibits a number of Greek-style temples. The 'South Temple' (or Great Temple) exhibits extensive Hellenistic influences (Joukowsky 1994). Its history is complex. The levelling and preparation of the site probably occurred in the mid-first century AD, with its precinct being modified sometime between the mid-second century and 330. It was then damaged by the earthquake of 303, 331 or 363, after which it was heavily modified but continued to function until the mid-sixth century, before being destroyed by an earthquake in 551. Another temple which has been excavated at Petra is the so-called Temple of the Winged Lions, built c. AD 25 and destroyed in the earthquake of 363 (see Parr 1996 for problems with the foundation date).

Traditionally, burials in the Roman period had to be placed outside the city limits. Extensive cemeteries have been reported around virtually all the major sites of Roman Jordan; there are Roman period rock-cut tombs at Petra (McKenzie 1990). The extensive collection of second- and third-century tombs at Marwa, 7 km northeast of Irbid, might be associated with the Decapolis city of Capitolias. A number of tombs have been excavated at Pella (Smith and McNicoll 1992). Gadara was surrounded by a number of large cemeteries and some particularly significant graves. The Tomb of the Germani and the Tomb of Sentius Modestus, both first century AD, are to be found outside the east gate. Modestus' funerary inscription describes him as a Holy Herald of the City that indicates that he was an organizer of sacred games and festivals. In addition to these tombs there are those of Chaireos (AD 90/91) and the North Mausoleum of an unknown family but dated to 355/356. There is also the so-called Underground Mausoleum that was extended in the fourth century but which was originally built outside the city walls. It should, therefore, be earlier than the third century when the walls were built. At Philadelphia, there are three main concentrations of burials (at the southern end of the Jabal al-Hussein, at the western end of Wadi 'Amman and along Jabal Ashrafiah and Jabal Jaufa to the south). There are also at least three noteworthy mausolea scattered around the city: at Quweismeh (Qasr al-Sab), at Qasr Nuweijis, and at Khirbat as-Souk, all dated on architectural criteria to the second or third century AD.

The other great theme in the archaeology of Jordan in the Roman period is the arrival of Christianity. This is manifested archaeologically by the appearance of churches, cemeteries and other iconographic representations. While the full flourishing of the religion and its archaeological evidence are features of the Byzantine period, its earliest manifestations date to the Roman period. As a generalization, it might be said that many of Jordan's Byzantine churches are actually based on earlier Roman structures, some of which were temples. For instance, at Tall Nimrin (near South Shunah in the Jordan Valley), excavations have produced a sixth- to eighth-century church, although the site appears in fourth- and fifth-century Christian sources as Bethnamaris (Piccirillo 1982). Likewise, Eusebius reports Christians and churches at Petra well before the dated archaeology. The earliest literary references in Arabia are to Christian refugees and later to martyrdoms under particular emperors. Refugees are reported at Pella at the end of the First Jewish Revolt (c. AD 70). Hagiographic sources imply that Christians lived in Philadelphia in the late third or early fourth century. At Gadara, martyrs are attested in the very early fourth century, while a bishop of the community is known to have attended the Council of Nicaea in 325. Gerasa was represented at the Council of Seleucia in 359. With respect to archaeological remains, at least 15 churches are known at Gerasa, although all of them date to the late fourth/fifth century. The Great Cathedral seems to have been dedicated as early as the mid-fourth century. This building can be associated with reports of a miracle at Gerasa. The Christian writer Bishop Epiphanus (c. 315–403) authored the *Panarion* or Medicine Chest, which offered advice for curing 80 heresies. In this work he refers to a *martyrium* and fountains at Gerasa where every year a spring ran with wine on the anniversary of the miracle of Cana, which was also the Feast of the Epiphany. The fact that there is a re-used first/second-century fountain in the Great Cathedral has prompted the suggestion that it was the location of this event. Finally, excavations by Parker (1997) at Aila have exposed a basilica-type structure. Initial and admittedly speculative interpretation of this building proposes it as an early fourth-century church, which, if it is, would make it one of the earliest known churches in Jordan to date. Christianity arrived in Aila by the early fourth century. The community there was represented by Bishop Petrus at the Council of Nicaea. Eusebius recounts that early Christians were sent to work in the copper mines at ancient Phaino (Khirbat Faynan in Wadi Faynan in southern Jordan). The mines are known to have been worked at this time (Hauptmann and Weisgerber 1992). At least three churches are preserved at the Faynan site, although they have not yet been securely dated. In time, Phaino was represented at a number of church councils in the fifth and sixth centuries.

Thanks to Josephus's account of events before the First and Second Jewish Revolts, we know that there were substantial Jewish communities in some of the Decapolis cities. The relationship between Jews and Gentiles in such cities at the outbreak of the First Revolt (66–70) does not appear to have been entirely friendly. Massacres in the 'Jordanian' Decapolis are specifically attested at Pella and Philadelphia. A more tolerant atmosphere is reported in Gerasa, where the population there offered protection to its Jewish community (*War* 2.458; 2.480). Architectural fragments from a synagogue there have been found in the triumphal arch built for Hadrian, while later still, one of the city's churches was built in c. 530 from a converted third/fourth-century synagogue. Finally, a synagogue was constructed in the first century AD at the bath complex of Hammeh Gader at Gadara.

Technology

Understanding of the exploitation of Roman Arabia's natural resources is similarly inhibited by a relative lack of fieldwork. However, two avenues of research might be noted here. With respect to Jordan's natural metal deposits, the clearest demonstration of their exploitation is to be found in the south of the country, in the Wadi 'Arabah region. Here, in the Wadi Faynan complex, fieldwork has shown that the deposits were exploited from the seventh millennium BC onwards. By the Roman period, the miners were largely reduced to reworking old deposits. Survey has also identified over 120 mines as well as over 70,000 tons of Roman-period slag (Khouri 1988: 121-27; Hauptmann and Weisgerber 1982; Hauptmann 1990). It seems that the ores from the mines scattered throughout the surrounding wadis were in the main smelted at the one site, one close to Khirbat Faynan. Extensive Roman (copper) mining activities have also been identified elsewhere in Wadi 'Arabah, north of Aila. More recently, complementing this industry, there is evidence of copper working in Aila itself (Parker 1997). Such was the scale of metal production and its need for charcoal and other resources, in at least the southern part of Arabia, that one can only imagine what were the environmental and floral consequences (cf. Barker *et al.* 2000: 44).

Roman-period stone quarries have been widely identified throughout the province, although the criteria for identifying them as such are often obscure. The scale of monumental building work in at least the cities and towns of the region would have ensured that this must have been a thriving industry. Unfortunately, little is known of its organization. Analysis of the marbles used for monumental architecture and funerary furniture across the province indicates that the Arabian cities imported this material from the usual centres of production in the eastern Mediterranean, Egypt and Asia Minor.

Another subject that is attracting increasing attention in the cities and towns of the country is that of the supply of water. Water-collection systems are known and have been studied at Gadara since 1992. Parts of what must have been a considerable system for the movement of water under the city have been recorded, with an upper channel of at least 380 m in length and lower one at over 360 m located directly under the city's acropolis (Weber 1991a, 1991b; Kerner 1994). Sources 3 km from the site, at 'Ayn Gadara, and 12 km east, at 'Ayn at-Terab, were evidently tapped. At least three (Roman) aqueducts are known at Abila: that from the 'Ayn Khureibah, 2.5 km away from the city, and two underground systems, one on top of the other under Umm al-'Amad, both certainly used in the Roman and Byzantine periods and perhaps dating back to the Hellenistic era. Gerasa too possessed a sophisticated system of water supply. The two reservoirs at the Festival complex at Birketein stored water to be piped into the city. Drawing in part on a natural spring, the pools measured 43.5×88.5 m and had a capacity of c. 1,500,000 litres. The flow of water into the city was regulated. It would have then fed at least the public bath houses, the nymphaea and public fountains of the city. In addition to this source of water, unpublished fieldwork to the north of the city has also discovered traces of pipe systems up to 5 km from the city that fed the Birketein pools.

With the subject of water supply comes consideration of the public and private use of the resource. In the case of the former, we are talking principally of baths and *nymphaea*. Two and perhaps three bathhouses have been identified at Gadara, with the Baths of Herakleides (third century AD) and another constructed into the very steep slope on the south side of the *decumanus maximus* in the early fourth century. *Nymphaea* or public fountains were located along the *decumanus* which dates to at least the late second century AD. In addition to these facilities, close to the city, some 7 km east of the Sea of Galilee in the Yarmuk Valley and 4 km north of Gadara is the bath and theatre resort of Hammeh Gader. This complex, which flourished in the second and third centuries, enjoyed something of a reputation (e.g. Strabo 16.2.45). Warm-water springs also existed at Pella, which likewise enjoyed something of reputation in rabbinic sources. At Gerasa, the baths were located close to the Chrysorohas, on the eastern side of that river. They remain undated. A *nymphaeum* was built adjacent to the Artemis complex. A well-preserved structure in downtown 'Amman has long been interpreted as a *nymphaeum* and dated to the second century AD. Recent restoration work on this structure has led to a better understanding of its form and evolution into the Byzantine and Islamic periods (Waheeb and Zubi 1994). A *nymphaeum* of the second century AD (?) along with a public bath lies on the main road through Petra, both of which imply a sophisticated hydraulic system.

Away from the larger sites, important work has been undertaken elsewhere. For instance, the current programme of work at Humayma is the result of an earlier survey of the hydraulic systems at the site (Oleson 1986, 1992; Oleson *et al.* 1993). Kennedy (1995) has summarized the evidence for the collection and management of water resources in a number of small northeastern 'Roman' villages. Utilizing a combination of survey data and air photographs, he has attempted to calculate the collection potential at a number of sites where considerable amounts of water were stored. With few if no signs that the water was for civic purposes, the conclusion has to be that it was for domestic uses as well as for animals. The results demonstrate just how sophisticated was the management of a vital resource, a skill that has been lost today. Meanwhile, the stepped dam at Wadi al-Jilat in northeast Jordan has recently been restudied (Politis 1993). While the date of its construction remains unknown, it has been compared to a Roman example at Mérida in Spain, although it has also been concluded that the Jordanian example might equally be Nabataean with a Byzantine reconstruction.

Conclusion

In summarizing the archaeology of Roman Jordan, the overall impression is detailed knowledge in a few limited spheres, a volume of largely undigested data in others, and a number of glaring gaps in certain fundamental areas. In the first category come aspects of the public buildings in the country's cities. In the second group should be included the processing of field survey data pertinent to the period, while the third represents the creation of better databases (e.g. for environmental evidence), understanding of the less monumental aspects of urban life, more detailed exploration of a range of rural sites, and finally the marrying of these various strands of evidence to create a broad impression of what it must have been like to live in a Roman province that had a long and sophisticated cultural, economic and political history.

Irrespective of these deficiencies, what can be said of Jordan under the Romans? From the evidence summarized in the preceding paragraphs, it seems clear that the second and early third centuries represented a period of considerable prosperity in Roman Jordan, if the scale and vitality of building work, settlement densities and agricultural activity are true indicators. This built on the accelerating prosperity of the first century BC and first century AD. Similarly, decline and depression are reflected in reduced building activity, as apparent in the late third and fourth centuries. These conditions appear to have been reversed from the late fourth century when, on the criterion of building activity (notably in the form of church building), the region seems to have enjoyed something of a renaissance, unless at least there was a general redistribution of how individuals and groups deployed their surplus resources, from work to benefit the community as whole to supporting the nascent church. This impression is derived from the towns. It does not seem to be reflected in the countryside, if settlement densities are an accurate reflection of prosperity. Indeed, increases in the distribution of what are only broadly dated Roman settlement patterns need not just be indicative of stability and success. They may in fact denote decline as farmers tried to re-establish themselves at new sites because of regional over-exploitation and/or climatic deterioration.

What explains these rises and declines? If the picture can be detected nationwide, then a single cause might be suggested. The stability Roman dominance brought to the Middle East as a whole in the first century BC/AD must surely have had positive consequences for the economy of the region. Conversely, an obvious example of the negative effects of Roman rule would be the general malaise that the political crises of the third century is supposed to have inflicted on most parts of the empire. The rivalry that is detectable in a number of aspects of the cities might have also burnt out their vitality. However, a nationwide interpretation may not necessarily be the only explanation. Regional aspects and factors could have had comparable consequences to the large-scale explanation. Bearing in mind how close was the relationship between town and country, environmental degradation and over-exploitation of the land in the later Roman period might have been a significant negative factor, as has been suggested in some regional case studies. Unfortunately, the quality of the evidence so far accumulated is not decisive either way. Finally, there is the possibility of localized factors, notably the effects of earthquakes and the consequent longer term dislocations.

The abiding impression of the archaeology of Roman Jordan is not so much what has been achieved as the incredible potential that remains. In spite of the long history of field work in the region and the immense threats to the surviving archaeology, the stable political conditions and climate of the region, coupled with what is still a vast untouched potential, will ensure that the region continues to contribute immeasurably to our understanding of what it must have been like to live in one of the frontier provinces of the Roman empire.

Acknowledgments

I am grateful to Graham Oliver, Zbigniew Fiema and Fiona Aiken for casting critical eyes at various stages over this text. They do not necessarily agree with the observations or conclusions made. Lorraine McEwan prepared the comparative illustration of the plans of Gerasa, Philadelphia, Gadara and Abila.

References

Abu Dayyah, A.S. et al.
1991 Archaeological Survey of Greater Amman Phase 1: Final Report. *Annual of the Department of Antiquities of Jordan* 35: 361-95.

Alcock, S.
1994 Breaking up the Hellenistic World: Survey and Society. In I. Morris (ed.), *Classical Greece: Ancient Histories and Modern Archaeologies*, 171-90. Cambridge: Cambridge University Press.

Baird, D. and G. Philip
1994 Preliminary Report on the Third Season (1993) of Excavations at Tell esh-Shuna North. *Levant* 27: 111-33.

Banning, E.B.
1986 Peasants, Pastoralists and *Pax Romana*: Mutualism in the Southern Highlands of Jordan. *Bulletin of the American Schools of Oriental Research* 261: 25-50.
1996 Highlands and Lowlands: Problems and Survey Frameworks for Rural Archaeology in the Near East. *Bulletin of the American Schools of Oriental Research* 301: 25-46.

Barghouti, A.N.
1982 Urbanization of Palestine and Jordan in Hellenistic and Roman Times. In A. Hadidi (ed.), *Studies in the History and Archaeology of Jordan*, I: 209-30. Amman: Department of Antiquities.

Barker, G. et al.
1997 The Wadi Faynan Project, Southern Jordan: A Preliminary Report on Geomorphology and Landscape Archaeology. *Levant* 29: 19-40.
2000 Archaeology and Desertification in the Wadi Faynan: The Fourth (1999) Season of the Wadi Faynan Landscape Survey. *Levant* 32: 27-52.

Bauzou, T.
1986 Les voies de communication dans le Hauran à l'époque romaine. In Dentzer 1986: 137-66.

Bikai, P.
1994a University of Jordan Excavations at Khirbet Salameh 1993. *Annual of the Department of Antiquities of Jordan* 38: 395-99.
1994b The Byzantine Church at Darat Al-Funun. *Annual of the Department of Antiquities of Jordan* 38: 401-15.

Bowersock, G.W.
1971 Report on Provincia Arabia. *Journal of Roman Studies* 61: 219-42.
1983 *Roman Arabia*. Cambridge, MA: Harvard University Press.

Bowsher, J.M.C.
1992 The Temple of Hercules: A Reassessment. In Northedge 1992: 129-38.

Browning, I.
1982 *Jerash and the Decapolis*. London: Chatto & Windus.
1989 *Petra*. London: Chatto & Windus.

Brünnow, R.E. and A. von Domaszewski
1904–1909 *Die Provincia Arabia auf Grund Zweier in den Jahre 1897 und 1898 unternommenen Reisen und der Berichte früherer Reisender*. 3 vols. Strassburg: Trübner.

Burns, R.
 1992 *Monuments of Syria: An Historical Guide*. London: I.B. Tauris.

Butler, H.C.
 1907–1919 *Publications of the Princeton University Archaeological Expedition to Syria in 1904–1905 and 1909*. Leiden: E.J. Brill.

Clark, G.A. *et al.*
 1994 Survey and Excavation in Wadi Al-Hasa: A Preliminary Report of the 1993 Field Season. *Annual of the Department of Antiquities of Jordan* 38: 41-55.

Dentzer, J.M. (ed.)
 1986 *Hauran I: Recherches archéologiques sur la Syrie du Sud à l' époque héllenistique et romaine*. Bibliotheque archéologique 124. Paris: Libraire Orientaliste Paul Geuthner.

de Vries, B.
 1998 *Umm el-Jimal: A Frontier Town and its Landscape in Northern Jordan*, I. Journal of Roman Archaeology Supplementary Series 26. Portsmouth, RI.

Fiema, Z.
 1993 Tuwaneh and the *via nova Traiana* in Southern Jordan: A Short Notice on the 1992 Season. *Annual of the Department of Antiquities of Jordan* 37: 549-50.
 1995 Military Architecture and the Defense 'System' of Roman-Byzantine Southern Jordan—A Critical Appraisal of Current Interpretations. In K. 'Amr, F. Zayadine and M. Zaghloul (eds.), *Studies in the History and Archaeology of Jordan*, V: 261-69. Amman: Department of Antiquities.
 2000 Byzantine Petra: A Reassessment. In T.S. Burns and J.W. Eadie (eds.), *Urban and Rural in Late Antiquity*, 111-131. East Lansing, MI: Michigan State University Press.

Freeman, P.
 1996 The Annexation of Arabia and Imperial Grand Strategy. In D.L. Kennedy (ed.), *The Roman Army in the East*, 91-118. Journal of Roman Archaeology Supplementary Series 18. Ann Arbor, MI.

Gatier, J.-P.
 1986 *Inscriptions de la Jordanie*. II. *Région central*. Paris: Geuthner.
 1989 Inscriptions of Jordan Survey. In Homes-Fredericq and Hennessy 1989: 54-55.

Geraty, L.T. and L.A. Willis
 1986 Archaeological Research in Transjordan. In L.T. Geraty and L.G. Herr (eds.), *The Archaeology of Jordan and Other Studies*, 3-75. Berrien Springs, MI: Andrews University.

Glueck, N.
 1934–1951 *Explorations in Eastern Palestine*. 4 vols. Annual of the American Schools of Oriental Research 14, 15, 18-19, 25-28. Philadelphia, PA and New Haven, CT: American Schools of Oriental Research.

Graf, D.
 1978 The Saracens and the Defense of the Arabian Frontier. *Bulletin of the American Schools of Oriental Research* 229: 1-26.
 1983 The Nabateans and the Hisma: In the Footsteps of Glueck and Beyond. In C. L. Meyers and M. O'Connor (eds.), *The Word of the Lord Shall Go Forth: Essays in Honor of David Noel Freedman in Celebration of his Sixtieth Birthday*, 647-64. Winona Lake, IN: Eisenbauns.
 1986 Review of Parker 1986. *Journal of Near Eastern Studies* 50.2: 151-53.
 1988 Qura 'Arabiyya and Provincia Arabia. In P.-L.Gatier (ed.), *Géographie-historique au Proche-Orient: Notes et Monographies Techniques* 23: 171-222. Paris: Editions du CNRS.
 1989 Rome and the Saracens: Reassessing the Nomadic Menace. In T. Fahd (ed.), *L'Arabie préislamique et son environment historique et culturel: Actes du colloque Strasbourg 24–27 Juin 1987*, 341-400. Leiden: E.J. Brill.
 1995 The *via nova Traiana* in Arabia Petraea. In J. Humphrey (ed.), *The Roman and Byzantine Near East: Some Recent Archaeological Research*, 241-65. Journal of Roman Archaeology Supplementary Series 14. Ann Arbor, MI.
 1997 The *via militaris* and the *limes Arabicus*. In W. Groenman-Waateringe *et al.* (eds.), *Roman Frontier Studies 1995*, 123-33. Oxbow Monographs 91. Oxford: Oxbow.

Gregory, S. and D.L. Kennedy
 1985 *Sir Aurel Stein's Limes Report*. British Archaeological Reports International Series 272. Oxford: British Archaeological Reports.

Hanbury-Tenison, J.W.
 1987 Jarash Region Survey. *Annual of the Department of Antiquities of Jordan* 31: 129-57.

Hanbury-Tenison, J.W. *et al.*
 1984 Wadi Arab Survey 1983. *Annual of the Department of Antiquities of Jordan* 28: 385-423.

Hauptmann, A.
 1990 The Copper Deposits of Feinan, Wadi Araba: Early Mining and Metallurgy. In S. Kerner (ed.), *The Near East in Antiquity: German Contributions to the Archaeology of Jordan, Palestine, Syria, Lebanon and Egypt*, I: 53-62. Amman: German Protestant Institute for Archaeology of the Holy Land/Al Kutba.

Hauptmann, A. and G. Weisgerber
    1992   Periods of Ore Exploitation and Metal Production in the Area of Feinan, Wadi Araba, Jordan. In M. Zaghoul et al. (eds.), *Studies in the History and Archaeology of Jordan*, IV: 61-66. Amman: Department of Antiquities.

Hoffman, A.
    1990   The Monumental Gate *extra muros* in Gadara. In S. Kerner (ed.), *The Near East in Antiquity: German Contributions to the Archaeology of Jordan, Palestine, Syria, Lebanon and Egypt*, I: 95-103. Amman: German Protestant Institute for Archaeology of the Holy Land/Al Kutba.

Homès-Fredericq, D. and J. B. Hennessy (eds.)
    1989   *The Archaeology of Jordan*, II/1-2. Akkadica Supplementum 7-8. Leuven: Peeters.

Humphrey, J.H.
    1995   Editor's Note (The Petra Papyri). In J.H. Humphrey (ed.), *The Roman and Byzantine Near East: Some Recent Archaeological Research*. Journal of Roman Archaeology Supplementary Series 14. Ann Arbor, MI.

Isaac, B.
    1988   Review of Parker 1986 and 1987. *Journal of Roman Studies* 78: 240-41.
    1990   *The Limits of Empire: The Roman Army in the East*. Oxford: Oxford University Press.
    1992   The Babatha Archive: A Review Article. *Israel Exploration Journal* 42: 62-75.

Joukowsky, M.
    1994   1993 Archaeological Excavations and Survey of the Southern Temple at Petra, Jordan. *Annual of the Department of Antiquities of Jordan* 38: 293-332.

Kalayan, H.
    1982   The Symmetry and Harmonic Properties of the Temples of Artemis and Zeus, and the Origin of Numerals as used in the enlargement of the South Terrace in Jerash. In A. Hadidi (ed.), *Studies in the History and Archaeology of Jordan*, III: 243-54. Amman: Department of Antiquities.

Kanellopoulos, C.
    1994   *The Great Temple of Amman: The Architecture*. American Center of Oriental Research Publications 2. Amman: American Center of Oriental Research.

Kennedy, D.L.
    1982   *Archaeological Explorations on the Roman Frontier in North-East Jordan*. British Archaeological Reports, International Series 134. Oxford: British Archaeological Reports.
    1992   The Roman Frontier in Arabia (Jordanian Sector). *Journal of Roman Archaeology* 5: 473-89.
    1995   Water Supply and Use in the Southern Hauran, Jordan. *Journal of Field Archaeology* 22.3: 75-90.
    1997   Roman Roads and Routes in North-East Jordan. *Levant* 29: 71-94.
    1998   The Area of Umm el-Jimal: Maps, Air Photographs and Surface Survey. In de Vries 1998: 39-90.

Kennedy, D.L., H.I. MacAdam and D.N. Riley
    1986   Preliminary Report on the Southern Hauran Survey, 1985. *Annual of the Department of Antiquities of Jordan* 30: 145-53.

Kennedy, D.L. and D.N. Riley
    1990   *Rome's Desert Frontier from the Air*. London: Batsfords.

Kennedy, D.L. and P.W. Freeman
    1995   Southern Hauran Survey 1992. *Levant* 27: 39-73.

Kerner, S.
    1994   The German Protestant Institute for Archaeology and Other German Projects in Jordan. In S. Kerner (ed.), *The Near East in Antiquity: Archaeological Work of National and International Institutions in Jordan*, IV: 49-64. Amman: German Protestant Institute for Archaeology of the Holy Land/Al Kutba.

Khouri, R.G.
    1988   *The Antiquities of the Jordan Rift Valley*. Amman: Al Kutba.

Killick, A.C.
    1983   Udruh—1980, 1981 Seasons: A Preliminary Report. *Annual of the Department of Antiquities of Jordan* 27: 231-44.

Kraeling, C.H.
    1938   *Gerasa: City of the Decapolis*. New Haven, CT: American Schools of Oriental Research.

LaBianca, Ø.S.
    1984   Objectives, Procedures and Findings of Ethnoarchaeological Research in the Vicinity of Heshbon in Jordan. *Annual of the Department of Antiquities of Jordan* 28: 269-88.
    1989   Hesban. In Homès-Fredericq and Hennessy 1989: 261-69.

Lewis, N. (ed.)
    1989   *The Documents from the Bar Kokhba Period in the Cave of Letters: Greek*. Jerusalem: Israel Exploration Society/The Hebrew University of Jerusalem/The Shrine of the Book.

Lindner, M.
    1989   Southern Jordan Survey. In Homès-Fredericq and Hennessy 1989: 84-90.

MacAdam, H.I.
    1986a   *Studies in the History of the Roman Province of Arabia: The Northern Sector*. British Archaeological Reports International Series 295. Oxford: British Archaeological Reports.
    1986b   *Bostra Gloriosa. Berytus* 34: 169-92.
    1990   The IGLS Series Then and Now (1905–1989). *Journal of Roman Archaeology* 3: 458-64.

1994 Settlements and Settlement Patterns in Northern and Central Transjordania, ca. 550–ca. 750. In G.R.D. King and A. Cameron (eds.), *The Byzantine and Early Islamic Near East. II. Land Use and Settlement Patterns*, 49-93. Princeton: Darwin.

MacDonald, B. et al.
1988 *The Wadi el Hasa Archaeological Survey 1979–82, West-Central Jordan*. Waterloo, ON: Wilfrid Laurier University.

MacDonald, B. et al.
1992 *The Southern Ghors and Northeast 'Arabah Archaeological Survey*. Sheffield Archaeological Monographs 5. Sheffield: J.R. Collis.

Macdonald, M.
1993 Nomads and the Hawran in the Late Hellenistic and Roman Periods. A Reassessment of the Epigraphic Evidence. *Syria* 70: 303-413.

McKenzie, J.
1990 *The Architecture of Petra*. British Academy Monographs in Archaeology 1. Oxford: Oxford University Press.

McNicoll, A.W. et al.
1982 *Pella in Jordan. I. An Interim Report on the Joint Expeditions of Sydney and the College of Wooster Excavations at Pella, 1979–1981*. Canberra: Australian National Gallery.

McNicoll, A.W. et al.
1992 *Pella in Jordan. II. The Second Interim Report of the Joint University of Sydney and College of Wooster Excavations at Pella, 1982–1985*. Mediterranean Archaeology Supplement 2. Sydney: Meditarch.

McQuitty, A. and R. Falkner
1993 The Faris Project: Preliminary Report on the 1988, 1989 and 1990 Seasons. *Levant* 25: 37-61.

Mare, W.H.
1994 The 1992 Seasons of Excavations at Abila of the Decapolis. *Annual of the Department of Antiquities of Jordan* 38: 359-78.

Martin-Bueno, M.
1992 The *Macellum* in the Economy of Gerasa. In M. Zaghoul et al. (eds.), *Studies in the History and Archaeology of Jordan*, IV: 315-20. Amman: Department of Antiquities.

Millar, F.
1993 *The Roman Near East 31 BC–AD 337*. Cambridge: Cambridge University Press.

Miller, J.
1969 *The Spice Trade of the Roman Empire*. Oxford: Oxford University Press.

Miller, J.M.
1989 Moab and the Moabites. In J.A. Dearman (ed.), *Studies in the Mesha Inscription and Moab*, 1-40. Atlanta, GA: Scholars Press.

Northedge, A.
1992 *Studies on Roman and Islamic Amman. I. History, Site and Architecture*. British Academy Monographs in Archaeology 3. Oxford: Oxford University Press.

Oleson, J.P.
1986 The Humayma Hydraulic Survey: Preliminary Report of the 1986 Season. *Annual of the Department of Antiquities of Jordan* 30: 253-60.

1992 The Water-Supply System of Ancient Auara: Preliminary Results of the Humeima Hydraulic Survey. In M. Zaghloul et al. (eds.), *Studies in the History and Archaeology of Jordan*, IV: 269-75. Amman: Department of Antiquities.

Oleson, J.P. et al.
1993 The Humeima Excavation Project: Preliminary Report of the 1991–1992 Seasons. *Annual of the Department of Antiquities of Jordan* 37: 461-502.

Ostrasz, A.
1989 The Hippodrome of Gerasa: A Report on Excavations and Research 1982–1987. In Zayadine 1989: 51-78.

Palumbo, G.
1993 JADIS (Jordan Antiquities Database and Information System): An Example of National Archaeological Inventory and GIS Applications. In J. Andresen et al. (eds.), *Computing the Past: Computer Applications and Quantitative Methods in Archaeology*. Computer Applications in Archaeology Conference, 1992. Aarhus: Aarhus University.

Palumbo, G. et al.
1990 The Wadi el-Yabis Survey: Report on the 1989 Season. *Annual of the Department of Antiquities of Jordan* 34: 95-118.

Palumbo, G. et al.
1993 Salvage Excavations at Tell Faysal. *Annual of the Department of Antiquities of Jordan* 37: 89-118.

Parapetti, R.
1982 The Architectural Significance of the Sanctuary of Artemis at Gerasa. In A. Hadidi (ed.), *Studies in the History and Archaeology of Jordan*, I: 255-60. Amman: Department of Antiquities.

Parker, S.T.
1976 Archaeological Survey of the *Limes Arabicus*. *Annual of the Department of Antiquities of Jordan* 21: 19-31.

1986 *Roman and Saracens: A History of the Arabian Frontier*. Winona Lake, IN: Eisenbauns.

1987 *The Roman Frontier in Central Jordan: Interim Report on the Limes Arabicus Project 1980–1985*. British Archaeological Reports International Series 340. Oxford: British Archaeological Reports.

1992 The Nature of Rome's Arabian Frontier. In V. Maxfield and M.J. Dobson (eds.), *Roman Frontier Studies 1989*, 498-504. Proceedings of the XVth International

Congress of Roman Frontier Studies. Exeter: University of Exeter.

1997 Preliminary Report on the 1994 Season of the Roman Aqaba Project. *Bulletin of the American Schools of Oriental Research* 305: 19-44.

Parr, P.
1996 The Architecture of Petra: Review Article. *Palestine Exploration Quarterly* 128: 63-70.

Piccirillo, M.
1982 A Church at Shunat Nimrin. *Annual of the Department of Antiquities of Jordan* 38: 335-42.

Politis, K.D.
1993 The Stepped Dam at Wadi el-Jilat. *Palestine Exploration Quarterly* 125: 43-49.

Raschke, M.G.
1978 New Studies in Roman Commerce with the East. In H. Temporini and W. Haase (eds.), *Aufstieg und Niedergang der römischen Welt*, II.9.2, 604-737. Berlin: W. de Gruyter.

Rostovtzeff, M.I.
1932 *Caravan Cities*. Oxford: Oxford University Press.

Russell, K.W.
1985 The Earthquake Chronology of Palestine and Northwest Arabia from the 2nd through the mid-8th Century AD. *Bulletin of the American Schools of Oriental Research* 260: 37-59.

Sapin, J.
1987 Jerash Region Survey. *Annual of the Department of Antiquities of Jordan* 21: 129-57.

Sartre, M.
1982 *Trois études sur l'Arabie romaine et byzantine*. Collection Latomus 178. Brussels: Revue d'études latins.
1986 Les peuplement et le développement du Hauran antique à la lumière des inscriptions grecques et latines. In Dentzer 1986: 189-204.

Schick, R.
1994 The Settlement Pattern of Southern Jordan: The Nature of the Evidence. In G.R.D. King and A. Cameron (eds.), *The Byzantine and Early Islamic Near East*. II. *Land Use and Settlement Patterns*, 133-54. Princeton: Darwin.

Segal, A.
1997 *From Function to Monument: Urban Landscapes of Roman Palestine, Syria and Provincia Arabia*. Oxbow Monograph 66. Oxford: Oxbow.
1988 *Town Planning and Architecture in Provincia Arabia*. British Archaeological Reports International Series 419. Oxford: British Archaeological Reports.

Seigne, J.
in press Les limites orientale et meridionale du territoire de Gerasa. *Syria*.

Shahid, I.
1984 *Rome and the Arabs: A Prolegomenon to the Study of Byzantium and the Arabs*. Washington, DC: Research History and Collections, Dumbarton Oaks.
1989 *Byzantium and the Arabs in the Fourth Century*. Washington, DC: Dumbarton Oaks Research Library.

Sidebotham, S.E.
1986 *Roman Economic Policy in the Eurythra Thalassa: 30 BC–AD 277*. Leiden: E.J. Brill.
1991 Ports of the Red Sea and the Arabia–India Trade. In V. Begley and R.D. de Puma (eds.), *Rome and India: The Ancient Sea Trade*, 12-37. Madison: University of Wisconsin.

Smith, A.M. *et al.*
1997 The South-East Araba Archaeological Survey: A Preliminary Report of the 1994 Season. *Bulletin of the American Schools of Oriental Research* 305: 45-72.

Smith, R.H.
1987 Trade in the Life of Pella of the Decapolis. In A. Hadidi (ed.), *Studies in the History and Archaeology of Jordan*, III: 53-58. Amman: Department of Antiquities.

Smith, R.H. and A. McNicoll
1992 The Roman Period. In McNicoll *et al.* 1992: 119-44.

Spijkerman, A.
1978 *The Coins of the Decapolis and Provincia Arabia*. Jerusalem: Franciscan Press.

Stern, E., A. Gilboa-Lewinson and B. Aviram (eds.)
1993 *The New Encyclopedia of Archaeological Excavations in the Holy Land*, I-IV. Jerusalem: Simon Schuster.

Stucky, R.A. *et al.*
1994 Swiss-Liechtenstein Excavations at Ez-Zantur in Petra 1993: The Fifth Campaign. *Annual of the Department of Antiquities of Jordan* 38: 271-92.

Tell, S. *et al.*
1993 The Cultural Resources Management Project in Jordan. *Annual of the Department of Antiquities of Jordan* 37: 67-164.

Waheeb, M. and Z. Zubi
1994 The Amman Nymphaeum Excavation Report: A Short Note on the 1992–93 Seasons. *Annual of the Department of Antiquities of Jordan* 38: 507-508.

Watson, P.
1989 Jerash Bowls: Study of a Provincial Group of Byzantine Decorated Fine Ware. In Zayadine 1989: 223-61.
1996 Pella Hinterland Survey 1994: Preliminary Report. *Levant* 28: 63-72.

Weber, T.
1990 One Hundred Years of Jordanian–German Fieldwork at Umm Qais (1890–1990). In S. Kerner (ed.), *The Near East in Antiquity: German Contributions to the Archaeology of Jordan, Palestine, Syria, Lebanon and Egypt*, I: 15-28.

Amman: German Protestant Institute for Archaeology of the Holy Land/Al Kutba.

1991a Gadara of the Decapolis: Tiberiade Gate, Qanawat el-Far'oun and Bait Rusan: Achievements in Excavation and Restoration at Umm Qais 1989–1990. In S. Kerner (ed.), *The Near East in Antiquity: German Contributions to the Archaeology of Jordan, Palestine, Syria, Lebanon and Egypt*, II: 123-34. Amman: German Protestant Institute for Archaeology of the Holy Land/Al Kutba.

1991b Gadara of the Decapolis. Preliminary Report on the 1990 Season at Umm Qais. *Annual of the Department of Antiquities of Jordan* 36: 223-31.

Welles, C.B.
1938 The Inscriptions. In kraeling 1938: 355-494.

Zayadine, F.
1986 *Jerash Archaeological Project 1981–1983*. Amman: Department of Antiquities.
1989 *Jerash Archaeological Project 1984–1988 2: Fouilles de Jerash 1984–1988*. Extrait de *Syria* 66. Paris: Libraire Orientalishe Paul Geuthner.
1994 Ayla-'Aqaba in the Light of Recent Excavations. *Annual of the Department of Antiquities of Jordan* 38: 485-505.

Zeitler, J.P.
1990 Houses, Sherds and Bones: Aspects of Daily life in Petra. In S. Kerner (ed.), *The Near East in Antiquity: German Contributions to the Archaeology of Jordan, Palestine, Syria, Lebanon and Egypt*, I: 39-52. Amman: German Protestant Institute for Archaeology of the Holy Land/Al Kutba.

13. The Byzantine Period

Pamela Watson

Terminology

The period in Western Asian history from the fourth to the mid-seventh centuries AD is variously referred to as 'Roman', 'Late Roman', 'Late Antiquity' or 'Byzantine'. The term 'Byzantine' is used here, following the refined period division of Levantine archaeologists to distinguish the last centuries of the eastern Roman empire before the Islamic conquest. There is no historically formal beginning for the period, the choice being one of scholarly convenience. The year AD 324, when Constantine I founded the new imperial capital of Constantinople, provides a logical marker. The eponymous site, at the junction between Asia and Europe, was the old Greek settlement of Byzantion. The focus of administration of the Roman empire moved from Rome to centre on the Balkans and the eastern Mediterranean. It is from here that a Christian and primarily Greek-speaking state evolved.

For Jordan and the general area of the Levant, the Byzantine period ended with the Muslim Conquest of 636/640, although the empire continued elsewhere until the mid-fifteenth century. Within this regional context only, there are broad chronological subdivisions into early Byzantine (fourth–fifth century) and late Byzantine (sixth–early seventh century).

'Classical' is the general term used to describe the period of Graeco-Roman political sway in the Levant, from the time of the conquest by Alexander the Great in 322 BC to the demise of the Byzantine empire in the southeast c. AD 640. It thereby encompasses the Hellenistic, Roman and Byzantine periods.

The modern state of Jordan does not coincide with state or provincial boundaries in antiquity, which were changeable. The term 'Transjordan', that is, the inhabited region east of the Jordan Rift Valley (the Jordan Valley, the Dead Sea and Wadi 'Arabah), is used as a neutral geographical location for the area.

Origins

There is no definable boundary between the Roman and Byzantine periods. Nevertheless, within an administrative, cultural and economic continuum, changes were occurring to the empire. At the beginning of the fourth century, the administrative and economic reforms of Diocletian had given it a new lease of life. In the east, this institutional and administrative structure remained largely intact until the Persian and Arab invasions of the early seventh century. Diocletian's successor Constantine I, responding to the growing importance of the eastern half of the empire, chose to administer it from an eastern centre, Constantinople. At the same time, he gave religious preference to a hitherto minor religion, Christianity, laying the foundations for its eventual dominance of the spiritual life of the state. The shift to the east and the rise of Christianity in the early fourth century represent a significant change in the character of the empire. Subsequently, the administration of the West and the East was separated and the West was later lost to the barbarians. Church hierarchy and theological conflicts influenced the affairs of state, while Christian mores affected social and cultural attitudes. The central bureaucracy was reinforced, court ceremonial permeated all aspects of life, and the army was restructured, giving prominence to the defensive forces. Provincial governors and their staff replaced municipal bodies, and military and civil powers were separated. The empire had evolved into a definably different entity to its precursor centred in Rome.

Sources and chronology

Chronological precision varies within the Byzantine period, according to the quality and availability of the sources at hand. Written sources may be biased, selective or inaccurate. The archaeological record is imperfectly preserved and its exploration through excavation and survey is uneven. The conjunction between the written and physical sources of information can be at times fortunately direct or unfortunately obscure. Written sources fall into a number of categories: ecclesiastical writers (documenting the history of the church and the lives of the saints); secular historians and panegyrics to the emperors; documentary material, for example, accounts of the church councils, the law codes of the emperors, and military

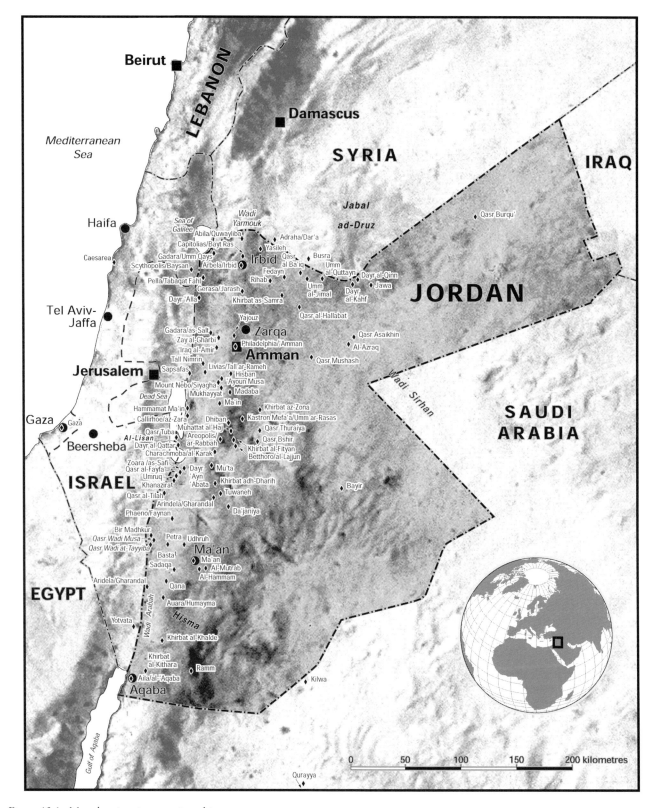

Figure 13.1. Map showing sites mentioned in text.

and administrative lists; and regional records such as papyrus archives. Some sources, such as later Arab and Byzantine chroniclers of the early seventh century, are not contemporary and rely on earlier records now lost. A comprehensive list of ancient sources and their editions is found in Jones (1973: 1464). For a discussion of the sources, see also Shahid (1984, 1989) and Kaegi (1992). Inscriptions, either dedicatory, public monumental, or funerary, are of primary importance in dating structures and events, providing names, genealogies, ranks, language usage and other glimpses of society. They are numerous but often fragmentary.

Environment

Jordan is climatically marginal, and small fluctuations in rainfall and temperature make a considerable difference to productivity. The land is particularly sensitive to human-induced degradation of the natural vegetation. Population peaked in the Byzantine period and land use was correspondingly intensive. It is not surprising to find indications of depleted natural vegetation as well as highly sophisticated land and water management strategies that were devised to maximize production from the existing resources.

There is evidence for deforestation in the highlands of northern Jordan during the Classical period (Sapin 1985: 225, 227). The ecological equilibrium maintained by the ancient village system was superseded by the demands of an urban socio-economy generated by the expanding cities of the region, such as Gerasa, Pella, Gadara, and Capitolias. Resources of wood and stone were heavily exploited and the Byzantine period witnessed an abrupt increase in deforestation. Agriculture was expanded to meet the needs of the cities, encroaching upon traditional pastoral zones that had protected low vegetation. The resulting degradation contributed to progressive erosion in this area (Beaumont 1985: 291-96).

Archaeobotanical research at Tall Hisban on the western edge of the central plateau indicates a change in the flora of the region during humankind's occupation. This was probably due to overgrazing which altered the selective advantage of certain species over others. It was neither a process of desertification nor the result of climate change. (LaBianca and Lacelle 1986: 123-40). During the Byzantine period, when land use was most intense, land that was previously pastoral was converted to cultivation, and elaborate water management works were constructed (Geraty and LaBianca 1985: 327-28).

Evidence from Pella in the north Jordan Valley documents a progressive shift from the use of forest species for fuel and timber to that of cultivated trees or naturally occurring wadi species. A species such as hackberry gradually disappears, becoming extinct by the Byzantine period. In the Umayyad period the population was reduced to lower-grade wadi plants for fuel, such as oleander and tamarisk (McNicoll et al. 1992: 256). Evidence from farther down the valley at Dayr 'Alla suggests a similar pattern of depleting resources of wood over time (Van Zeist 1985: 203).

In the southern copper mining area around Wadi Faynan in Wadi 'Arabah, the culling of local terrace and wadi growth provided the main source of fuel for the smelting operations. Exploitation of the southern forests had ceased by the Iron Age (Hauptmann 1986: 418). The extent to which progressive deforestation led to human-caused desertification in this area is still being researched.

The Byzantine period witnessed the climax of settlement and land use in Jordan, leading to the expansion of sedentary occupation and agriculture into marginal semi-desert areas. This is particularly apparent in the northeastern basalt area of the southern Hauran, along the steppic fringes of central Jordan, and to a lesser degree in the south. The viability of life in these marginal regions depended upon well-organized systems of water collection and distribution to support the growing of crops and to supply the needs of the villages and towns. At Umm al-Jimal, one of a network of towns in the Hauran, we find the remains of aqueducts bringing winter rainwater from the highlands in the north, a series of public and private reservoirs for long-term storage, and channels collecting local winter runoff. Traces of a complex hydraulic system for water diversion and storage exist in the surrounding countryside along with agricultural terracing, corrals and barns. Urban housing was designed specifically for the stabling and husbandry of animals as well as people (De Vries 1985). To the east, a network of fossilized field boundaries surround ancient villages such as Umm al-Quttayn (Kennedy 1982: 331-35).

In southern Jordan, the extensive ancient field systems of Wadi Faynan reveal systematic hydrological management for agricultural production. These fields continued to flourish in the Byzantine period, servicing the copper-mining communities (Barker *et al.* 1997). The urban centre of nearby Petra was sustained by the organization of sophisticated hydrological systems and the careful terracing and husbandry of all suitable wadis and plateaux within its mountainous environment (Gentelle 1984: especially 27).

Could a minor fluctuation towards a cooler/wetter climate favour and perhaps influence the unprecedented expansion of settlement in the Byzantine period? A variety of data sources for climate indicators has led to conflicting conclusions. Some analyses supported the cooler/wetter trend while others suggested that the Byzantine period was more arid than the Roman period. Whatever the case, minor undetectable climatic fluctuations as well as a range of human factors may also have had considerable influence on land use and demography. The prevailing political and economic conditions probably did more than climate to encourage population and settlement expansion at this time.

Historical background

In the late third century, the Emperor Diocletian instituted a major restructuring of the administration of the Roman empire. Provincial divisions were redrawn, and the military system was reorganized and strengthened. Constantine and his successors continued this process, establishing the administrative framework for the next two centuries. Jordan was divided between four Byzantine imperial provinces that also extended into Palestine/Israel and Syria (Palestine I, II, III, and Arabia; Figure 13.2).

After a prolonged period of struggle, Constantine achieved sole power, defeating his final rival Licinius in AD 324. He relocated imperial administration to Constantinople, a position more central to the geographical realities of the empire. Initially, the empire was run by joint rule, with the emperor delegating responsibility for certain areas to a junior Augustus and sometimes subordinate Caesars (Jones 1973: 325). On the death of Theodosius in 395, the empire was divided between his two sons Arcadius and Honorius, into the East and West respectively. This administrative division became a permanent reality when the West, which had been progressively declining, finally succumbed to external pressures in 476. The continuation of the Roman empire thereafter was carried by the eastern half alone.

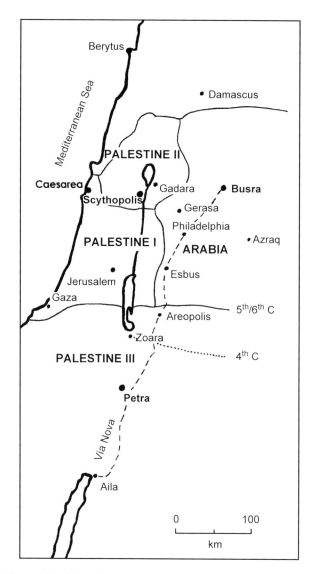

Figure 13.2. Map showing Byzantine provincial divisions.

The role of Christianity

Constantine's conversion to Christianity, a religion intensely persecuted by his predecessor Diocletian, marked another significant departure in the development of the empire. Christianity was legalized in 313, and, with a brief exception in the reign of Julian (361–63), grew rapidly under the patronage of the fourth-century emperors to become the official state religion. Constantine's energetic espousal of Christianity involved him in the doctrinal controversies of a developing church, setting a pattern of imperial involvement in ecclesiastical affairs that would continue for centuries. The provinces of Palestine, Arabia, and Syria, the geographical origin of the new religion, were affected by this favoured status.

The role of the Arab nomads

Economically, the provinces of Arabia and Palestine were of minor significance in the empire. They were relatively poor in resources, although the luxury trade by caravans through the area was a lucrative source of revenue. Their main strategic importance was geographical, as a land bridge between the rich provinces of Syria and Egypt. Bordered by a desert that was sparsely populated by nomads, the southeastern frontier was not considered as vulnerable to serious invasion as the frontiers with Persia and the West. Nevertheless, the nomadic Arab presence became an influential factor in the administration of the eastern provinces and Roman policy evolved to accommodate it. During Constantine's reign, a significant tribal group and its leader, 'Imru al-Qays, is attested in eastern Syria. The epitaph on his grave at Namara (between Busra and Damascus) describes him as 'king of the Arabs' (Isaac 1990: 239-40). The use of powerful Arab tribes as allies in their continuing conflicts becomes a feature of both Roman and Persian strategy in the buffer zone of the eastern desert from the fourth century onwards. The Lakhmids, based on the Euphrates, served the Persian cause. By the late fourth century, the Salih had become the dominant tribe in southern Syria and northern Transjordan. They are attested as Roman *phylarchs* (tribal commanders under the patronage of Byzantium) from the reigns of Valens to Anastasius at the beginning of the sixth century.

In the last decade of the fifth century, disruptive nomadic incursions were reported in the eastern provinces, probably reflecting competition between the major clans (Sartre 1982: 155-56). At this time, the Kinda tribe, moving out of central and northern Arabia, raided Palestine and southern Syria, an area hitherto dominated by the Salih. The Kindite war (498–502) revealed the weakness of the Byzantine defences in Transjordan and Palestine. The Emperor Anastasius eventually concluded a treaty with Harith the Kindite in 502, making him a formal Byzantine ally and *phylarch* of the Kindites in the area of Palestine III (Shahid 1989: 130). Another tribe, the Ghassanids, had filtered into Transjordan through Wadi Sirhan in c. 300. The Salih permitted them to settle in the Balqa' district south of 'Amman (Trimingham 1979: 178-79; Sartre 1982: 157). In 498, the leader of the Ghassan, Jabala, was reported to be raiding into Palestine and was curbed by the Byzantine *dux* of Palestine, Romanos. The Ghassanids proved their ascendancy over the Salih and the Byzantines gave them the *phylarchy* in the area of Arabia. Their position was enhanced in 529/530 when, after trouble with the Kindites and the Lakhmids, the Emperor Justinian appointed Harith the Ghassanid as supreme *phylarch*, patrician and king of all the Arab tribes allied to Byzantium (the *foederati* or federates). This co-ordinated federate force was formed in response to the effective power of the Persian Arab allies led by al-Mundhir, king of the Lakhmids. The Ghassanids thus became responsible for protection of the frontier from Palmyra to Wadi al-Hasa. The area of Palestine III was assigned to Harith's brother, Abu Karib. Justinian's policy was to reduce the role of the imperial frontier forces, the *limitanei*, shifting their responsibilities to the Arab federates unified under a single commander (see Defence).

Conflict with Persia

During the fourth century, several wars were fought against the Sassanian Persians to the northeast. This, however, had little impact on the southern areas. Constantine's son, Constantius, was engaged in a number of campaigns. His successor, the pagan emperor Julian, was killed on campaign in 363. The fifth century was a time of relative peace between the Persians and the Byzantines, although there were occasional lapses. Both parties were more concerned with internal problems and border pressures elsewhere.

During the reign of Justinian (527–65), there were three periods of conflict with the Persians, 527–32, 540–45, and 549–62, all occurring in Syria and beyond (Jones 1973: 269-94). Justinian's main concern was ensuring peace and security within his borders rather than conquest in the East. The focus of his aggression was in the opposite direction, that is, attempting to regain the lost western provinces. In 532, Justinian and Chosroes signed a treaty of Eternal Peace but this proved to be a misnomer. A 50-year peace was finally signed in 562. The Arab federates played a significant role as raiders, complementary to the activities of the main army. They were mentioned specifically in the peace treaty (Isaac 1990: 240-49).

The policy of Justinian's successors was less supportive of the Ghassanids and trouble ensued. Al-Mundhir, the Ghassanid leader, had problems with Justin II and then Tiberias, culminating in al-Mundhir withdrawing his forces during conflict with the Persians in 581. He was finally betrayed and exiled. In revenge, his son ravaged Syria, Palestine and Arabia over the next two years (Whitby 1995: 91). The success of this raiding illustrates the weakness of Byzantine control in the southeast provinces, with imperial forces concentrated east of the Euphrates.

The Persian invasion and occupation of the Levant between 613 and 628 was far more determined and acquisitive than anything seen in previous centuries. Instead of merely seeking booty and gaining an advantageous peace on the borders, as had been the pattern in the past, Chosroes II sought to conquer and retain. Besides pushing through Asia Minor to the Bosphorus, he overran Syria and Palestine, burning and sacking Jerusalem in 614 and removing the relic of the Holy Cross to Ctesiphon. He took Egypt in 619, cutting off the corn supply to the Byzantine capital. Energetic counterattacks by the emperor Heraclius (610–41) in Asia Minor and Persia eventually prevailed. Chosroes was deposed and murdered and his son sued for peace. The Persians evacuated all conquered territories in 628 and Heraclius returned the Holy Cross to Jerusalem in triumph in 630 (Schick 1992: 108-11).

The Islamic conquest

No sooner was the land recovered than an unexpected threat emerged from the remote southern regions of Arabia. The appearance of a small religious movement seeking a larger audience had almost gone unnoticed. Yet, within the space of six years, the Byzantines were pushed out of Palestine, Transjordan and Syria forever. They were victorious in their first encounter with the invading Muslim forces in 629, at Mu'ta, south of Wadi al-Mujib, but their triumph was short lived. Heraclius's decision to end subsidies to certain tribes in southern Palestine prompted them to assist the Muslim forces to gain access to Gazan territory in 633/34. Major defeats of the Byzantines occurred at Ajnadayn and then Fahl in 634, and Damascus was captured in 635. The decisive Muslim victory came at the Battle of Yarmuk in 636 and by 640 the final pockets of Byzantine resistance were removed from the region. Thus ended almost a thousand years of Graeco-Roman rule in the Levant (Kaegi 1992: 66-146).

Settlement patterns

Population and settlement intensity reached its peak in most of Jordan during the Byzantine period, as attested by recent excavations and surveys in many parts of the country. Numerous factors contributed to this situation, chief among them being the sense of general security that allowed settlement and food production to exist in safety. Healthy trade and economic networks encouraged an overall prosperity. An expanding religious community enjoyed the highest political support and attracted pilgrims as part of the Holy Land itinerary, albeit on the periphery.

Basic factors affecting settlement patterns are the geographical and climatic divisions of the landscape and the political organization. The following discussion is divided into natural geographical regions as they fall within the modern borders.

1. The northwest region, from the Yarmuk River to az-Zarqa' River, including the east side of the Jordan Valley and eastwards to the Via Nova Traiana on the edge of the steppe (roughly a line from az-Zarqa to Khirbat as-Samra to al-Mafraq).
2. The north-central plateau and westward wadis to the Dead Sea, from az-Zarqa' River to Wadi al-Mujib.
3. The south-central plateau (ancient Moab) from Wadi al-Mujib to Wadi al-Hasa, including the southern Ghors south of the Dead Sea.
4. The southern region below Wadi al-Hasa to the Gulf of al-'Aqaba.
5. The eastern steppe and desert areas, roughly eastwards from the line of the Via Nova, incorporating the southern Hauran and al-Azraq oasis.

Useful syntheses of information on settlement patterns in Jordan for the Byzantine period have been made by: Piccirillo (1985) concerning rural settlements throughout Jordan; MacAdam (1994) concerning northern and central Jordan; Parker (1992) for central Jordan; and Schick (1994) and Fiema (1991) for the southern half of Jordan. A register of surveys and site excavations up until the mid-1980s (with some omissions) is found in Homès-Fredericq and Hennessy (1986, 1989). King has published brief preliminary accounts of his overall survey of Byzantine and Islamic sites in Jordan (1982, 1983, 1987, 1989). The foundations for these investigations were laid by Glueck (1934–1951) in his extensive coverage of 'Eastern Palestine'. Primary information comes from more specific regional surveys and excavations, which vary in intensity, purpose and methodology. Consequently, there are gaps in the coverage and variation in the quality of information.

1. The northwest region and north Jordan Valley (within the provinces of Palestine II and Arabia): The climatically benign area of the northwestern

region, with its rolling wooded hills, lush valleys and perennial springs, supported the majority of the population. It accommodated members of the former classical league of Hellenized cities known as the Decapolis, such as Gerasa, Pella, Gadara, Abila, Capitolias, and the unidentified Dium. Another important urban centre existed at Arbela (Tall Irbid). These large centres were surrounded by a network of towns and villages, roads, agricultural installations, farmsteads and villas, reaching an occupational peak in the Byzantine period unsurpassed until the twentieth century. Eastwards, the land undulates more gently and rainfall decreases. Sites are more dispersed but the important north–south communication route of the Via Nova Traiana and its offshoots ensure a zone of occupation, supporting towns such as Khirbat as-Samra, Rihab, and Yasileh.

Excavations in the Decapolis cities have revealed concentrated and extensive occupation during the Byzantine period. Increased population and changing cultural requirements led to alterations in the metropolitan fabric. General tendencies, such as the encroachment upon public spaces and facilities by private enterprises, and the reuse or demolition of major institutional structures such as temples, theatres and hippodromes, are in evidence in these centres. New monumental architecture is dominated by the building of a multitude of churches in the fifth and sixth centuries. The grid-like order of Roman-town planning had been attempted at some sites such as Gerasa, and Gadara, but was not always achieved in other Hellenized cities such as Abila and Pella, where the hilly topography influenced street alignments. The Roman-period plans were often disturbed by the imposition or insertion of these new structures. A more organic approach to urban development, closer to the traditional oriental format, began to re-emerge. Archaeological evidence is detailed in numerous excavation reports from: Gerasa/Jarash (Kraeling 1938; Zayadine 1986, 1989; Kehrberg and Ostrasz 1997; Seigne 1992); Pella/Tabaqat Fahl (Smith 1973; Smith and Day 1989; McNicoll *et al.* 1982, 1992); Gadara/Umm Qays; Abila/Quwayliba; Capitolias/Bayt Ras; and Arbela/Irbid (most with preliminary reports in *ADAJ* from the 1980s).

2. The north-central plateau and south Jordan Valley/Dead Sea littoral (within the provinces of Palestine I and Arabia): The area lies between Wadi az-Zarqa' in the north and Wadi al-Mujib in the south. The heartland of this region is the rich agricultural plains of the Balqa' around Madaba, and the similarly productive Baq'ah Valley. The northern wooded hills and watered wadis of the western edge of the plateau give way to a more open, drier landscape in the south, and the eastern steppe merges into the desert. The old Decapolis city of Philadelphia ('Amman) dominated the north of this region but the agricultural lands to the south were serviced by important towns such as Hisban, Madaba, Umm ar-Rasas and Dhiban. Modern-day as-Salt (another ancient Gadara) had been the Roman capital of Peraea, set deep in a valley west of 'Amman, surrounded by densely wooded hills. The major wadis between the plateau and the Rift Valley operated as important link zones between the two extremes of elevation, providing access routes with monitoring facilities and pockets of agricultural settlement.

Surveys indicate the closely settled nature of this region. The area around 'Amman/Philadelphia contains many settlements, most of them first noted because of the presence of churches, such as Jubaiha, Swafiyeh, Quwaysma, Khirbat Salameh, Khilda, and Yajouz. Information on Byzantine Philadelphia and its territory is patchily known through survey and excavation (Glueck 1939; Abu-Dayyah *et al.* 1991; Northedge 1992: 39-45, 59-60, 155). The same pattern continues in the hilly region around as-Salt and north to az-Zarqa' River. Settlements tended to lie on hilltops, many now bearing modern villages. Very few of these sites continued into the Early Islamic periods.

The Baq'ah Valley, the rich terra rossa plain just northwest of 'Amman and adjacent to the above-described hills, is anomalous in showing little evidence of installations for this period. Perhaps, surrounded as it was by hilly settlements, it was devoted to large-scale agriculture. Intensive Byzantine settlement, dropping significantly in the early Islamic period, is attested in the hinterland of 'Iraq al-Amir in Wadi as-Sir, running westwards from 'Amman. Similarly, in the area of the Balqa' south of 'Amman, comprising the hinterland of Hisban and the Madaba plains, the Byzantine period was clearly the most populous, engaging in intensive food production throughout the landscape. Towns acted as thriving market centres, surrounded by scattered hamlets, farmsteads and related installations (cisterns, reservoirs, perimeter walls, terrace walls, mills, presses, roads). Settlement was not constrained by factors of defence or natural water supply, indicating a secure and highly organized environment (MacAdam 1994: 80-88, nn. 183, 184, 187 for references). Excavation has been undertaken at the town sites of Hisban, Tall al-'Umayri, Tall Jawa, al-Drayjat and Madaba (MacAdam 1994; also Bikai and Dailey 1996), and numerous Christian sites in the area, such as the monastery/church complex of the Memorial of Moses at Mount Nebo and its related town (Khirbat al-Mukhayyat) as well as hinterland chapels, monasteries and sanctuaries, at 'Ayn Musa, Ma'in, Massuh, and Abu Sarbut (Piccirillo 1993). The landscape is littered with Christian monuments and the remnants of intensive viticulture.

Connecting routes from this area down to the Jordan Valley north of the Dead Sea have been documented (e.g. Piccirillo 1987). Sites are fewer in the southern and drier half of the east Jordan Valley and mostly unexcavated. Work at Tall Nimrin near South Shuna suggests affluence in this period (Flanagan et al. 1994). There were more medium to small villages and farms than defensive or strategic types of sites. Many new agricultural sites on the valley floor were first settled in the Late Byzantine period. A pair of well-preserved Roman-Byzantine forts with extensive hydraulic facilities lie at the mouth of the nearby Wadi Kafrain. They probably controlled and protected the renowned warm springs below, popular with pilgrims, as well as the rich agricultural lands of the valley floor and a route to the plateau (Prag and Barnes 1996). The mountainous eastern shores of the Dead Sea contain isolated thermal springs with lush vegetation. The Roman bathing complex of 'Ayn az-Zara (Callirhoe) continued to be occupied in the Byzantine period, as part of a wider fertile area divided into farms with other farmsteads and villas, a possible fort and a paved road leading to the plateau ('Amr et al. 1996: 445).

The plateau between Wadi Walla and Wadi al-Mujib reveals a similar settlement pattern to that of the Madaba Plains, with a peak of activity in Roman and Byzantine times. The towns of Umm ar-Rasas/Kastron Mefa'a (Piccirillo 1989: 269-308) and Dhiban (Tushingham 1972) provided strong market and religious centres in the Byzantine period. Survey of the area between Hisban and al-Karak found a substantial number of Byzantine sites (just over half) but not as great as in the Roman period (Homès-Fredericq 1986: 81-100; MacAdam 1994: 80, 181). This variation in the settlement pattern, particularly for the area south of Wadi al-Mujib, is supported by other survey data from the al-Karak plateau, see below.

3. The south-central plateau and the southern Ghors (the provinces of Arabia and Palestine III): Different zones of the south-central plateau (the al-Karak plateau, or Central Moab) have been surveyed over time (Glueck 1939; Canova 1954; Parker 1987: chs. 3-4; Parker 1992; MacDonald et al. 1988; Miller 1991), as have the southern Ghors around the Lisan peninsular of the Dead Sea and south to the rise into Wadi 'Arabah (Rast and Schaub 1974; King et al. 1987; MacDonald et al. 1992).

Byzantine settlement patterns are less homogenous here than in the northern regions. On the plateau, the greatest density of population occurred in the Nabataean/Early Roman period, slumped somewhat in the Late Roman period and revived, but not quite to the same extent, in the early Byzantine period, declining again in the sixth century. The eastern line of fortifications was abandoned by the beginning of the sixth century and the eastern area in general was sparsely occupied (Parker 1992). The two main towns of the period were Areopolis (ar-Rabbah) and Charachmoba (al-Karak), although other towns and villages are known from ancient sources. These are largely unexcavated (Miller 1991: 12-13). The legionary base of Betthoro (al-Lajjun) reveals a functional decline from the fifth century (Parker 1986, 1987). Secondary Roman-Byzantine road systems have been documented through the agricultural hinterland west of the Via Nova Traiana down to the shores of the Dead Sea. They were associated with fortified isolated buildings and farms.

Byzantine settlement retains its ascendancy in the southern Ghors of the Rift Valley, where activity was widespread on the valley floor and up the tributary wadis, utilizing all available agricultural terrain. The area straddled important communication routes between southern Palestine and the northern Negev and the Moabite plateau. The largest concentration of settlement was around as-Safi, biblical Zoara, and the religious centre at Dayr 'Ayn 'Abata, the Monastery of Saint Lot (King 1987: 446-49; MacDonald et al. 1992: 104, 111-12; Politis 1997). The highest annual tax payment scheduled for Palestine III in the Beersheba Edict of the early sixth century was from Zoara, presumably an indicator of its economic standing. Both sites were depicted on the Madaba Map. The monasteries and hermitages in the vicinity indicate the religious importance of this area (Holmgren and Kaliff 1997; MacDonald et al. 1988: 242-44, 1992: 104). Military control of east–west routes across the northern 'Arabah continued through the occupation of sites such as Umm at-Tawabin, Qasr Fifa (the Praesidium of the *Notitia Dignitatum* and the Madaba Map), Umruq and Khanazir (King 1987: 449-51; Fiema 1991: 230; MacDonald et al. 1992: 104-05). The tomb remains of a Roman soldier who died around the end of the fourth/early fifth century were found at Haditha east of the Lisan. MacDonald et al. conclude that the dense sherd scatters on the valley floor are the only remnants of the many Byzantine village and town sites since subsumed under later activity (1992: 111).

4. The southern plateau, the Hisma and Wadi 'Arabah (Palestine III): Southern Jordan with its arid climate and remoteness from the centres of administration supported a lower population. Dry farming was possible on the western side of the plateau, but pastoralism remained the mainstay of the economy. Previously, in the time of Nabataean hegemony, the caravan trade from South Arabia was a primary source of wealth, and population and agriculture was at its height. Trade between the Hijaz and Syria continued to be important through to Islamic times. Byzantine settlement tended to cluster around a number of centres in the south, as isolated concentrations of activity rather than the integrated network that existed in Nabataean/Early Roman times.

Both textual sources and survey-excavation work in this area are limited (reviewed in Schick 1994; Fiema 1991). Recent surveys were conducted in the southern Wadi al-Hasa, around Petra, Dana, Wadi Faynan, the southern Wadi 'Arabah, and the Hisma (MacDonald et al. 1988; MacDonald 1996; Ruben et al. 1997; Smith et al. 1997; Graf 1992). A few major settlement sites such as Petra (Parr 1986; Fiema 1991, 1995; Patricia Bikai 1995), Udhruh (Killick 1986, 1987), Humayma (Oleson et al. 1992, 1994), Aila[1] (Parker 1996, 1996, 1998) and Gharandal/Arindela (Walmsley 1998) have been excavated, and a network of secondary road systems identified (Graf 1992).

Like the al-Karak plateau, the Edomite plateau was less populous in the Byzantine period than in the Roman period, declining in the sixth century towards a marked drop in early Islamic times. Byzantine activity concentrated around Gharandal/Arindela in symbiosis with the busily populated region of the southern Ghors and the copper-producing wadis flanking Wadi 'Arabah. Settlement decreases eastwards, with a clustering of sites around two routes from the south and southeast into Wadi al-Hasa (Wadi al-'Ali and entrances near Qal'at al-Hasa) but not in any other wadis. Sites were surprisingly sparse along the Via Nova itself and generally in the central plateau. This stretch of road seems little used or scarcely monitored at this time (MacDonald et al. 1988: 232, 246, 292-95; Fiema 1991: 223-25). Yet Khirbat adh-Dharih (c. 5 km west of the Via Nova in Wadi al-La'ban) has revealed Byzantine settlement that was not apparent in the original survey (MacDonald 1996: 326). Similarly Tuwaneh (Roman Thana/Thoana/Thornia, on the Via Nova south of adh-Dharih) flourished until its decline towards the end of the Late Byzantine period. However, a surveyed section of the Via Nova south of Wadi al-Hasa found no Late Byzantine sherds in the farthest watchtowers north or south from Tuwaneh.

The broad pattern of activity suggests a preferred use of the desert route for efficient north–south communications at the expense, but perhaps not the total demise, of the established central route, at least between Udhruh and Wadi al-Hasa. A secondary, undulating western route along the escarpment linked the more numerous western settlements (Fiema 1991: 283, Map 9).

Another focus of settlement was around Udhruh, attesting to the continued importance of this town. Udhruh paid the second highest amount of taxes recorded in the Beersheba Edict of the early sixth century (Killick 1983: 231). The Ghassanid, Jabal b. Harith, was reported to have rebuilt the town in the sixth century. There was a network of roads, small forts, watchtowers, farmsteads, large unwalled settlements, and agricultural systems, all dated to the Byzantine period (Killick 1983, 1986).

Recent excavations have altered the perception that Byzantine Petra was in terminal decay. The pattern of urban use changed in the same way as in the Decapolis cities of the north, and new churches were built or existing structures transformed. In the early sixth century Petra still had a city council (inferred from an inscription found at Elji), and the latest housing found so far dates to the middle and late sixth century. Textual references to bishops from Petra continue up to the late sixth century. Papyrus scrolls dating between 528 and 582 provide a mine of information on the daily life of the local inhabitants, presenting an image of social and economic vitality based now on land ownership rather than international trade (Pierre Bikai 1996). Surrounding settlements and buildings are mentioned, as well as key figures in the life of Petra. From the dating formulae used, it is apparent that imperial orders were received quickly and efficiently, confirming the continued status of Petra as an important regional administrative centre in the late Byzantine period. The recently deciphered Scroll 60 indicates that c. 540, Petra remained the dominant city in relation to Udhruh (Augustopolis) and that the fiscal administration of the imperial house was still active there (Kaimio and Koenen 1997).

The Nabataean town of Humayma (Auara) in the Hisma also retained its importance in Byzantine times although activity in the desert farther east was diminished. Four churches have now been identified and the occupation area doubled. The Roman fort ceased to function by the end of the fourth century, reducing the military character of the settlement. A third-century bathhouse was modified in the fifth/sixth century and, presumably, the extensive hydraulic systems were maintained.

The Roman road and control systems in the south continued through the early Byzantine period. Subsidiary roads radiating from the Via Nova were identified at Petra, Basta, Sadaqa, Qana and Humayma (Graf 1992). Only the Via Nova road forts of al-Khalde and Kithara towards Aila show evidence of Late Byzantine occupation. Farther north, the sixth-century occupation of Udhruh and Ayl may, like Humayma, represent civilian rather than military use (Parker 1986: 87-113).

The Roman-Byzantine town of Aila has recently been discovered within modern al-'Aqaba, northwest of Islamic Ayla. The site had long functioned as the gateway for seaborne and overland trade between India, Africa, South Arabia, Egypt, the Mediterranean and Syria, and maintained its importance through the Byzantine period (see Foreign Relations and Trade). Militarily, it was the base for the Tenth Fretensis Legion, possibly up to the sixth century (see Defence). Bishops from Aila attended ecclesiastical councils from the fourth-century onwards, and it was an important halt for pilgrims on the way to Mount Sinai (see Religion). Aila was fortified or refortified in the late fourth-fifth century by a city wall with towers, which fell into disrepair in the sixth/early seventh century. A mudbrick church, possibly the earliest known, and domestic areas of the Late Byzantine period have been exposed. Material remains attest to wide-ranging contacts in the Byzantine period (Parker 1996, 1998).

The number of sites declined in the arid hinterland of al-'Aqaba, the southern Wadi 'Arabah, after the early Byzantine period. These had been small way stations and forts, such as Gharandal/Aridela, Qasr Wadi at-Tayyiba, Qasr Wadi Musa, and Bir Madhkur, principally serving the east-west traffic across the valley along a limited number of routes. The recent discovery of a section of stone-paved road running north-south may alter our understanding of these connections, but dating is uncertain. Evidence from the west side of the 'Arabah suggests that a reduced traffic tended more to the west, from Yotvata to Gaza (Smith et al. 1997; Fiema 1991: 233-34). Sites are relatively sparse in the northern 'Arabah, principally comprising forts that control water sources, such as Qasr at-Tilah, Fifa, Khanazira and Umruq. Again, occupation fades away in the Late Byzantine period.

Passage between Aila and Wadi Faynan in the central 'Arabah must have remained feasible while it still produced and marketed copper. The extensive remains of ancient Phaeno include at least five churches, attesting to the legacy of the early Christian martyrs forced to work in the mines (Figure 13.3). The latest known bishop is recorded on a building inscription, dated 587-88 (Ruben et al. 1997). Two large cemeteries, an aqueduct, reservoir and mill

Figure 13.3. Khirbat Faynan, ancient Phaeno. View east showing building ruins covering the foreground and the tell behind.

facilities are located nearby. Inscriptions and grave goods from the South Cemetery date its use from the fifth up to the late sixth centuries (Findlater *et al.* 1998). Remnants of extensive field systems dominate the broad wadi to the west (Barker *et al.* 1997). There has been detailed investigation of the ancient mines and smelting centres in the vicinity, focusing on the geology and metallurgical developments (Hauptmann and Weisgerber 1992). In the Roman-Byzantine period, the remaining low-grade ores were worked at least until the end of the fifth century, and the town flourished. There is no evidence for Early Islamic activity, and the site seems to have been in decline during the Late Byzantine period.

5. The eastern steppe and desert (Province of Arabia): The only fertile areas in this region are the southern Hauran with its productive volcanic soil and the oasis of al-Azraq. Building on the early surveys of Butler (1919), the area has recently been examined by Kennedy (1982, 1997; Kennedy and Freeman 1995) and King (1983). This is supplemented by Parker's studies of the *limes* system (1986, 1987), excavations at Umm al-Jimal, Qasr Burqu', and Qasr al-Hallabat (De Vries 1985, 1993, 1998; Betts *et al.* 1991; Bisheh 1986). Work in the Syrian Hauran informs and complements all these studies. MacAdam summarized much of the evidence for settlement patterns in the region (1994: 53-68).

The southern Hauran comprises the basalt desert region of northeastern Jordan (the *harra*) in a marginal rainfall zone. The area extends into southern Syria and the lava slopes of the Jabal ad-Druze, with Busra, the capital of the province of Arabia, dominating the region. Systematic settlement on a large scale began in the Nabataean period and reached its apogee in the Byzantine period when settlements were closely scattered across the landscape. During the fourth and fifth centuries, a system of fortifications, watchtowers and signalling posts functioned to control and monitor these southeastern limits of the Byzantine empire, creating a wide area of civilian security. In the sixth century, military policy delegated this control to the Ghassanids, and the imperial garrisons disappeared. Civilian activities flourished and expanded in these times and many churches were built. At Umm al-Jimal, the Late Roman fort lost its military value during the fifth century, with its role assumed by the smaller Barracks complex. The fort was turned over to domestic (probably commercial) use. The construction of domestic housing reached its peak in the sixth century and 15 churches were built. Umm al-Jimal had become a prosperous rural

Figure 13.4. View across Umm al-Jimal.

Christian community within a similarly settled hinterland, a short day's journey from Busra (Figure 13.4).

Sophisticated systems of water collection and distribution supported cereal production as well as pastoral activities. A previously unknown Roman fort has been tentatively identified at Umm al-Quttayn, surrounded by domestic housing, extensive traces of field boundaries, and a Roman road leading towards Busra (Kennedy and Freeman 1995). These settlements are just two examples of the scatter of towns and villages, usually with churches or even monasteries, linked by a network of roads which facilitated communications beyond the line of the Via Nova (Kennedy 1997). Major trade routes straddled the region bringing associated economic incentives and profit to the inhabitants. Pre- and post-Islamic Arabic texts and traditions confirm the importance of the Hauran trade with the Hijaz and Mecca in the sixth and early seventh centuries (Sartre 1987: 160-62).

Farther east, occupation was restricted to isolated fortifications sometimes with small attached settlements. These were placed along the Strata Diocletiana, the easternmost zone of imperial control providing a communication route running north from al-Azraq to the Euphrates (Parker 1986: 15-36). The fourth–fifth century fort of Dayr al-Kahf monitored the southeastern Hauran, with a small surrounding settlement and traces of agricultural activity. Nearby is the Roman-Byzantine settlement of Dayr al-Qinn and an outlying watchtower overlooking Jawa (King et al. 1983: 415). The remote outpost of Qasr Burqu' is situated at an eastern water source along a minor route from the south (Betts et al. 1991; Parker 1986: 21) (Figure 13.5).

The large oasis of al-Azraq at the outlet of Wadi Sirhan was a pivotal location in the eastern desert. Besides the large fort and reservoir, it must have supported some settlement and agriculture. Subsequent occupation has destroyed most of the early evidence. Outlying sites comprise small forts and watchtowers along established routes. There is no surface evidence of Late Byzantine occupation at these sites. Farther west, Qasr al-Hallabat also had traces of field and terrace walls in the vicinity but later activity has obliterated any Roman-Byzantine evidence, save for the pottery. There are some traces of Byzantine activity in the steppe south of al-Hallabat, near the Umayyad agricultural villa of Qasr al-Mushash (King et al. 1983: 391-92). Byzantine settlement in these arid areas was minimal, associated with military installations where conditions allowed it. The remote Nabataean-Roman fort of Bayir was abandoned by the Byzantine period. As in the south, early Byzantine control seems consolidated within a zone further westwards.

Figure 13.5. Qasr Burqu'. View northeast when the seasonal lake is full (mid-ground), showing the surrounding basalt *harra*.

Material culture

The abundant remains of the Byzantine period in Jordan have yielded a wealth of cultural artifacts. Pottery is the most enduring and ubiquitous survivor of ancient material culture, but other hardy classes of material such as glass, metalwork, coins, worked bone, and carved stone are found, not to mention architectural decoration such as relief sculpture, wall painting and mosaics. Limited occupation in highly arid areas has allowed survival of some organic materials such as fabrics, wood, leather and matting that normally would have disintegrated.

Pottery

Pottery vessels, from the menial to the fine, were an essential part of daily life. Sites with extensive stratified deposits of occupation from the fourth to seventh centuries have aided documentation of the changes in pottery through time. In the north, excavations at Pella, Jarash, Umm al-Jimal and Hisban produced useful pottery sequences for this period (Smith 1973; Smith and Day 1989; McNicoll *et al.* 1992: 163-81; Watson 1992; Zayadine 1986, 1989: 85-97; De Vries 1998: ch. 11; Sauer 1973). The greater stability of settlement in the north, with large multi-period sites that attracted major archaeological investigations, has led to a bias in our current knowledge. The pottery sequence in the centre and south is less understood, although the situation is improving with the recent excavations at al-Lajjun, Humayma, Petra and al-'Aqaba.

The local pottery of Byzantine Jordan is regional in nature, with the northern sequence being distinct from the middle and south. The intra-regional relationships reflected in the pottery distribution patterns tend to be more east–west than north–south, broadly following provincial divisions of the time, as well as geographical ease of access. A few classes of pottery, mostly imported, range more widely. They are predominantly fine-ware bowls and amphorae. Some pottery forms are popular throughout Transjordan but are reproduced locally in different fabrics. The wares or fabrics tend to be well differentiated from previous Late Roman productions in the north, although this is less evident in the south. From the paler buffs, pinks and creamy-yellows, the favoured colouring of the clay becomes more red-orange and grey, with occasional creams. The cooking wares become grittier. Combed decoration, finger impressions and pie-crust decorations occur. At the end of the period, however, the production of many of the fabrics and forms continues with little change into the Umayyad period. Not until the Abbasid period is a marked change of direction noticeable. A few Umayyad wares are distinctive, as are developments in form and decoration, particularly in the eighth century. But the mid- to late seventh-century transition is difficult to identify without a large assemblage of material to evaluate (McNicoll et al. 1982: 100; Watson 1992).

A brief selection of some of the most common and typical Byzantine pottery types follows, as an example of the pottery of Byzantine Jordan. Of the ordinary daily wares, the so-called 'Palestinian bag-shaped amphora' is the most distinctive and common container recovered, presumably because it was used not only to transport liquids such as wine, but as the standard water carrier (Figure 13.7.2). In the north, it is generally grey (occasionally red or mottled) with white-painted swirling decoration on a ribbed body and two short loop handles on the shoulder. The fabric is hard and metallic. Farther east and south, the fabric changes to a red or buff. The ware of the common ribbed or plain cooking casserole with cut rim and lid also varies regionally, but the form is relatively constant (Figure 13.6.1). The basic cooking pot with spheroid body (mostly ribbed) and wide-mouthed short- or no neck, has a variety of diagnostic rim forms that change over time (Figure 13.6.2) and occurs in the same regional wares as the casserole. Very large hand-made storage jars with hole-mouth opening and folded collar rim continue to be used from the Roman period, generally in a pink to buff, hard fabric (changing regionally). In the north, it converts to grey ware by the Early Islamic period (Figure 13.7.1). Large hand-made basins have flat bases, flaring walls, and a variety of thickened rim forms. In the north, these were manufactured in a chaff-tempered buff-brown ware (Figure 13.7.3), and, in the Late Byzantine period, in the same fine hard grey ware as the later large storage jars, continuing into the Islamic period (Figure 13.7.4). Jarash is a known production centre for the grey ware. Jugs and juglets appear in many forms. In the Jarash red ware, the appearance of simple, white-painted swirls and single stripes on handles of jugs heralds the more exaggerated use of white painted decoration in the Early Islamic examples. Plain conical cups were a feature. The repertoire of fine ware bowls and plates is dominated by imported Phoenician and African red slip wares (from Asia Minor and North Africa respectively), and some Cypriot and Egyptian imitations (Hayes 1972). These were fine red wares, slipped and sometimes decorated with rouletting and stamps. Local manufacturers tended to imitate these forms, the most spectacular series being produced at Jarash (known as 'Jarash Bowls') and further embellished with pictorial decorations in white and red paint (Figure 13.6.5) (Watson 1992: 242). Another group of less distant imports that appeared widely is the bowls and jugs of 'Fine Byzantine' ware probably manufactured in Jerusalem (Figure 13.6.6; Watson 1992: 242; 'Ware K'). Common amphorae types of external origins were the 'Gaza' (Figure 13.8) (Peacock and Williams Class 49), the 'Antioch/Cypriot' type (Class 44), and the 'Abu Mina' (Egyptian Delta, Kellia Type 186), all attesting to dynamic relations with the Mediterranean world to the west (Watson 1992: 239-40, H2, H1, H4).

Byzantine oil lamps are mould-made, evolving out of the Hellenistic-Roman tradition. They vary in type and are usually regional in their distribution, with a pronounced difference between northern and southern Jordan. Popular forms in the early Byzantine period belong to the bow-shaped nozzle family and the bilanceolate candlestick family. The former develops into the broad-nozzle type, while the latter enlarges over time into the Late Byzantine 'candlestick' type, which continues into the Early Islamic period. A similar time span is observed for the 'Jarash' lamps, a distinctive group with tongue handles that develop into a zoomorphic form. The general tendency in the development of shape is a thickening of the nozzle until it becomes one piece with the body. There is a coarsening and simplification of the decoration and the size of the handle increases. The same handle form can appear on several body types. The most common Byzantine motif is a shoulder decoration of radiating lines (Zayadine 1986: 163-66, 367-75; 1989: 85-97; Da Costa forthcoming).

Two major pottery production centres have been identified in the north and the south, namely, at Jarash and Petra. The Jarash production ranged from Roman through to Abbasid times. The Byzantine pottery manufacturing area was concentrated in the disused hippodrome where, besides kilns producing many other types, a dump of unfired Jarash bowls was found (Kehrberg and Ostrasz 1994, 1997). A sixth-century potter's shop of deformed 'seconds' was unearthed in the retail area by the South Gate (Zayadine 1986: 71). Jarash pottery and their distinctive lamps had a wide distribution in northern Jordan. Doubtless, other centres existed in the north, for instance, a Late Byzantine pottery kiln was discovered in a salvage excavation on Wadi az-Zarqa' (Palumbo et al. 1993). Late Roman pottery kilns of the third/fourth centuries have recently been identified at both Abila and Pella, lending credence to the assumption that they also produced pottery in the Byzantine period.

The kilns at Petra-Zurrabah functioned from the Nabataean period through to the sixth century, the latest kiln proving the continued production of the

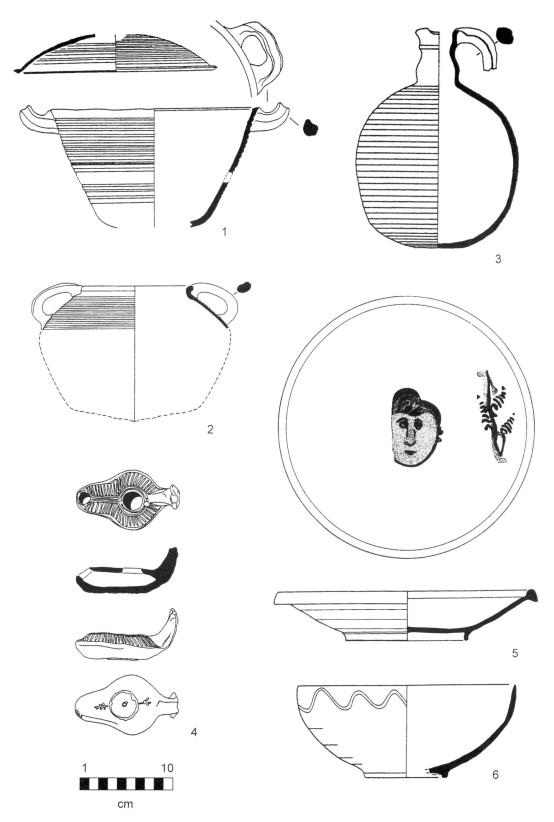

Figure 13.6. Byzantine pottery. 1. Cooking casserole and lid (Pella. Watson 1992: fig. 1.1, 2) 2. Cooking pot (Pella. Cat. No. 72) 3. Jug (Lajjun. Pub. No. 306) 4. Jarash lamp (Jarash. Rasson and Seigne, in Zayadine [ed.] 1989: 145, fig. 13.5) 5. Jarash bowl (Jarash. Watson, in Zayadine [ed.] 1989: 244, fig. 9.1) 6. 'Fine Byzantine' bowl (Amman. Koutsoukou *et al.* 1997: 107, no. 172).

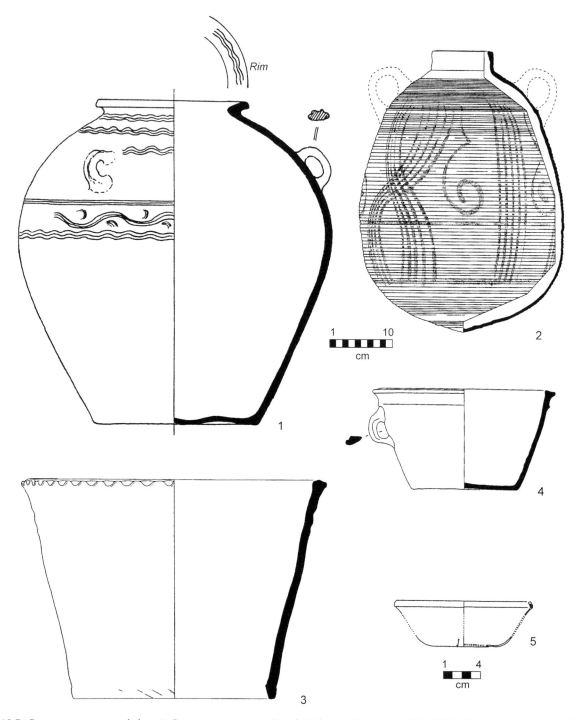

Figure 13.7. Byzantine pottery and glass. 1. Grey ware storage jar (Jarash. Fisher and McCown 1931: 35) 2. 'Palestinian bag-shaped' amphora (Pella. Cat. No. 10640) 3. Hand-made basin in chaff-tempered coarse ware (Pella. Cat. No. 2153) 4. Grey ware basin (Jarash. Rasson and Seigne, in Zayadine [ed.] 1989: 133, fig. 7.7) 5. Glass bowl (Jarash. Dussart 1989: 106, fig. 2b).

'Nabataean painted' ware into the Late Byzantine period. A predominance of cream surfaces was reported in the late phases, as well as a decline in craftsmanship generated by the requirements of mass production, leading to much coarser, poorly levigated clay ('Amr 1991). Kilns of the mid-seventh century at al-'Aqaba produced a typical Byzantine corpus, including the 'Ayla-Axum' amphora that appears as early as the fifth century. Kiln wasters and slag provide further evidence for production in the Byzantine period (Melkawi et al. 1994; Parker 1998: 388-89). The corpus from the seventh-century production is described az-Zarqa' River (Palumbo et al. 1993). Glass production is postulated at Abila, where a large quantity of glass fragments and lumps of glass slag was found in the theatre *cavea*. Most of the information on glass in Jordan comes from studies of glass found at Jarash, with minor contributions from other sites (such as Meyer 1988; Kraeling 1938: 505-46).

With the invention of moulds and, later, the technique of blowing, glass objects could be industrially produced in quantity, rather than being exclusive items of luxury. Byzantine glass is relatively abundant and diverse, with many new forms appearing. The colour, however, is standardized, being light blue, blue-green or green. Vessel walls are very thin, more so than the Roman product. Forms introduced include simple conical beakers (that are rare outside the area), the double unguentaria (double kohl tubes or flasks, characteristically 'Syro-Palestinian'), globular and piriform flasks with very long necks and flaring mouths, small bottles with narrow necks, and stemmed goblets. Glass lamps in the form of a goblet, with handles or stem, were introduced after the fourth–fifth centuries. They were especially popular in churches, eventually replacing ceramic lamps. Glass dishes and bowls were less common (Figure 13.7.5). Decorative techniques were limited to cutting, engraving and moulding, while ruffles and particularly trailed threads became characteristic decorative features of the region, commonly applied to bottles. Glass jewellery was popular, in the form of plain or twisted rings and bracelets of coloured glass.

Most of the Byzantine glass windowpanes found at Jarash and Pella are flat, and probably cast. In the early seventh century 'crown' panes, which permit more rapid manufacture than cast glass, prevail. The majority of windowpanes occur in the debris of churches, where glass mosaic tesserae used for wall and hemidome decoration are also common. Glass lamps also predominate in these contexts and the quantity of glass bottles and goblets suggests that they were used in church services. Apart from the generally fragmentary material provided by stratified excavations, tombs have produced the bulk of whole vessels. The advantage of intact survival in such contexts is diminished by their generally poor precision in dating, but they serve as useful reference parallels for the fragments.

Figure 13.8. Petra Church mosaic detail: man holding an amphora of 'Gaza' type (cf. McNicholl et al. 1992: pl. 116.8).

as 'southern Jordanian/Palestinian', with wares that are more 'sandy' and less hard and metallic than their northern counterparts, notably lacking in painted decoration. Strong associations with Egyptian products are noted.

Glass

Ancient 'Syria' was famous for its glass industry, and Syrian glass-blowers travelled far and wide. Recent excavations in Jordan suggest that some provincial centres had their own production facilities. At Jarash, numerous glass cakes were found in the 'Glass Court' near the Cathedral, possibly being reused for the manufacture of glass tesserae for mosaics. A thick dump of broken glass was found nearby, but no actual kilns have been located. A Byzantine pottery kiln that was converted to a glass furnace in the Early Islamic period was found by the Gerasa–Philadelphia road, near

Metalwork

The survival of metal in the archaeological record is haphazard, owing to the problems of corrosion. Apart from utilitarian objects such as nails, hinges and locks, occasional fine objects are recovered. Church ritual and embellishment encouraged the production of a wide variety of metal objects such as crosses, chandeliers, lamp holders, small bells and incense burners (see, for example, Zayadine 1986: 152, 155, 338; McNicoll *et al.* 1982: pl. 150, 2). A bronze repoussée plaque, from a fourth-century tomb at Pella, is remarkable for documentation of early Christian iconography. It apparently depicts Christ's entry into Jerusalem on one side and the twin Constantinian shrines of Golgotha and the Holy Sepulchre on the other (McNicoll *et al.* 1982: 100).

Coinage

After 324, a uniform monetary system was applied throughout the whole empire. It was based on a new gold unit, the *solidus*, valued at 72 to the pound. A gold half-*solidus*, the *semissis*, and the more obscure $1\frac{1}{2}$ *scripulum* were also introduced. The latter was replaced by the *tremissis* (third-*solidus*) during the reign of Theodosius I in 383. Silver coinage was revived under Constantine, with the *miliarensis* (one-eighteenth *solidus*) and the *siliqua* (one-twenty-fourth *solidus*), becoming widely used by his successors throughout the fourth century. Their production ceased around 400. As silver coinage became popular, the use of debased silver for smaller denominations was gradually abandoned, replaced by copper-alloy coins in the fourth and fifth centuries. These varied in size and weight but generally became smaller over time, reflecting a constantly changing and unstable standard. They are referred to as *nummi* or, more specifically, 'bronze' (Æ) 1, 2, 3, or 4, the Æ4 being the smallest and the most common by the fifth century. During the fifth century the most important coinage was gold. Very little silver was minted, and an inadequate quantity of copper coins was issued, causing a shortage of officially minted coins. The tiny base-metal coins were, therefore, imitated in large numbers by forgers. Many of them were needed to buy a commodity, hence their frequent discovery in large hoards.

The monetary reform of Anastasius in 498 introduced new larger copper coins. For numismatists, this marks the beginning of Byzantine, as distinct from Roman coinage. These new coins filled the awkward gap between the *tremissis* and the Æ4, where between 2400 and 4800 of the latter were needed to equal one of the former. The new copper denominations were of revolutionary design, with the most prominent feature being the use of a large Greek letter indicating the value in *nummi* on the reverse type: M (= 40 *nummi*) the *follis*; K (= 20 *nummi*) the half-*follis*; I (= 10 *nummi*) the *decanummium*; E (= 5 *nummi*) the *pentanummium*. This set a pattern for the next three centuries.

The designs on coins of the fourth century do not change suddenly from pagan to Christian imagery, as there was no official iconography to draw upon. The Christian *chi-rho* monogram appeared initially in a minor role as a mint-mark. The image of the 'hand of God' receiving the deified deceased emperor, or appointing a new emperor, was used in various forms. Such symbols became more prominent in the fifth century. The cross appeared as an attribute of the city of Constantinople or decorating the official sceptre. The standard design on the reverse of the eastern *solidi* from the fifth until the seventh centuries, was the cross with the figure of Victory. The traditional depiction of the emperor's head on the obverse, shown in profile to right, changed under Constantius II (337–61) to a three-quarter facing bust, helmeted, with spear and shield. This was regularly used until the sixth century under Justinian I, when the face became fully frontal, the characteristic 'Byzantine' representation. Later, a crown replaced the helmet. *Tremisses*, however, continued to use profile portraits. The post-reform *folli* show a transition on the obverse type from the typical Roman profile bust (standard under Anastasius) to the typical Byzantine facing bust introduced by Justinian, although the profile type lingered on occasionally until the seventh century. Justin II introduced the emperor and empress enthroned side-by-side, and double and even triple images became common in the seventh century.

After the mid-fourth century, the minting of gold coins was confined to the *comitatus*, the place of residence of the emperor (principally Constantinople and Thessalonica). In the east, base-metal coinage was produced separately at a number of state mints: Constantinople, Thessalonica, Heraclea, Nicomedia, Cyzicus, Antioch and Alexandria. Constantinople, Nicomedia and Cyzicus produced the bulk of the coinage. Before the Anastasian reform, most of these mints had collapsed, with only Constantinople and Thessalonica remaining in operation. After 498 and during the sixth century, they reopened, with the exception of Heraclea. Other mints were activated, such as Carthage and Constantine in Numidia. During the prosperous reign of Justinian (527–65), at least 15 mints were operational. The troubled reign of Heraclius in the early seventh century saw another reduction of Byzantine mints and short-lived operations at various new locations such as Alexandretta, Isaura, Seleucia (in Isauria), Jerusalem and Cyprus. Copper coins generally bore recognizable mint-marks unless they were too small, but, after Anastasius, silver and gold coins were rarely marked (Whitting 1973; Burnett 1987).

The most common Byzantine coinage recovered in Jordan and the Levant in general is that of Justinian I (527–65). His reign was long, and he needed money to conduct prolonged wars against Persia. His successor Justin II (565–78) is also well represented in finds. Apart from the *nummi*, fifth-century coinage is less common.

Organic remains

Textiles rarely survive in other than extremely arid climates. The carbonized remains of silks found in the Umayyad earthquake destruction of Pella are a unique find in Jordan, but the cloth was widely available in the region before this. The Persian control of the silk trade was broken in the mid-sixth century when silk moth eggs were brought to Constantinople. From here, under the patronage of the Byzantine emperors, the industry developed and regional workshops flourished. Syria produced silks at Tyre, Damascus, Aleppo and Antioch, and silk remains have been found in the Byzantine sites of Halabiyeh on the Syrian Euphrates as well as Avdat in the Negev (McNicoll *et al.* 1992: 257-65). Scythopolis, in the northwest Jordan Valley, was renowned for its manufacture of fine linen from the Late Roman period onwards. Remnants of shrouds have survived in Roman and Byzantine tombs in arid areas south of the Dead Sea (Politis 1998; Findlater *et al.* 1998). On the Lisan Peninsula of the Dead Sea, fragments of clothing, including wool and linen, ropes and plaited work were recovered from the excavation of the monastery of Dayr al-Qattar (Holmgren and Kaliff 1997: 324).

Leather, usually sandals, and wood also survive in such conditions, for example, at Faynan, in the form of small objects such as a spindle, a wooden kohl tube, jewellery (ring and bracelet), and a comb, as well as coffins (Findlater *et al.* 1998). Peculiar thermal conditions emanating through the bedrock and sealed within some tombs at Pella allowed coffins and even laurel leaves to survive in a much more humid climate (McNicoll *et al.* 1992: 141, 143).

Wall painting

Very little art in this genre has survived in Jordan, and for the Byzantine period, virtually none. Painted tombs of the Roman period have been found at Abila and Bayt Ras, and the renowned wall paintings of Qusayr 'Amra are Umayyad. Fragmentary remnants were found in the East Church at Pella, probably from the upper wall. Wealthier churches tended to be decorated with marble panels and mural mosaics, while the poorer churches would use plaster and paint, a far more fragile medium.

Sculpture

Monumental, free-standing sculpture was popular in the classical Greek and Roman artistic context. During the Byzantine period many of the significant religious and public buildings of the earlier era were abandoned, reused, altered and plundered for their useful components. Most Byzantine churches re-used columns, capitals, carved architraves and lintels from their pagan predecessors. More often than not, archaeologists have found major items of classical sculpture in the dump and fill deposits relating to the process of dismantling and reconstruction. Provincial sculptural expression during the Byzantine period was limited to relief carvings on tomb and building lintels, sarcophagi, church furniture such as chancel screens and altars, and small-scale figurines.

Architectural and furniture reliefs continue the design motifs of previous centuries with an additional overlay of Christian symbolism. The cross in its numerous manifestations was a dominant image, as well as the vine, the wreath, animals from paradise, the tree of life, the chalice, and others in common with the iconography found on the mosaics. Newly made column capitals break away from the traditional classical Corinthian form and depict a variety of the above images, including human figures (Piccirillo 1993: ills. 316, 321, 326-29, 727-30). They vary from simple elegant designs, recalling the format of their foliate predecessors, to busy overall patterning of details and images charged with meaning.

Production of pottery figurines continued from the Roman period, employing Christian iconography that easily transferred meaning from pagan imagery. For instance, the popular image of the 'Madonna and Child' grew from traditional Egyptian images of Isis with Horus and the Semitic figurines of Astarte (Bienkowski 1991: pl. 66). An interesting class of 'mirror plaques' in clay or plaster has been found in Byzantine tombs from the south to the north. They can be anthropomorphic or in the shape of fish, with painted details, and carry the inset for a small glass mirror or reflector. Possibly, they have a symbolic and prophylactic function (Bienkowski 1991: pl. 65; Humbert 1997: 55).

Social and political organization

Administrative structure

By the late fourth century AD, the area covered by modern Jordan was included in four Roman provinces: Palestine I (*Palaestina Prima*), Palestine II (*Palaestina Secunda*), Palestine III (*Palaestina Salutaris*, later *Palaestina Tertia*) and Arabia. They formed part of a diocese within the Prefecture of Oriens. The organizational details of this structure are unclear, but for the beginning of the fifth century we have the *Notitia Dignitatum*, an official 'List of Offices', which records the provinces, military placements and offices.

Starting from the northwest of Jordan, each province covered the following area (Figure 13.2): Palestine II extended from the west across the north Jordan Valley onto the western side of the plateau, including the cities of Pella, Capitolias (Bayt Ras), Arbela (Irbid), Abila (Quwayliba) and Gadara (Umm Qays). Scythopolis (Beth Shan) was the capital.

Palestine I extended from the west across the south Jordan Valley to the plateau, including Livias (Shunat Nimrin), Gadara (as-Salt), Amathus, and Bacatha(?). The capital was on the coast at Caesarea Maritima.

Palestine III initially extended south of Wadi al-Hasa to the Gulf of al-'Aqaba, with its capital at Petra. Sometime between 451 and 535 the northern boundary was moved north to Wadi al-Mujib.

The province of Arabia extended from the Hauran south to Wadi al-Hasa (later retracted to Wadi al-Mujib) and included Gerasa (Jarash), Philadelphia ('Amman), Esbus (Hisban), and Madaba. The capital was Busra, today in southern Syria. The southern and eastern boundaries fade into the adjacent deserts. Negative evidence suggests that earlier Roman penetration into the northern Hijaz was withdrawn in the Byzantine period (Fiema 1991: 128-33; Sartre 1982: 133).

A governor who was responsible for civil jurisdiction administered each province, while overall military authority was in the hands of the *dux* or military commander. In Arabia, the military performed civil functions as well, but inefficiently and for its own profit, prompting reorganization in the sixth century. Distribution of the military forces changed over the centuries, diminishing from the fifth to seventh centuries (see Defence). Ecclesiastical divisions seem to have followed the imperial administrative units. Jordan contained about 18 bishoprics, as attested by inscriptions in church mosaics (see Religion).

Society

The cities in northwestern Jordan had been under a strong Hellenizing influence since the conquests of Alexander the Great. Nevertheless, even here, the population retained its indigenous ties. Inscriptions and artifacts reveal personal names recorded in Semitic, Greek and Latin script. The proportion of Semitic to Graeco-Roman personal names increases in towns to the east such as Umm al-Jimal and Khirbat as-Samra, where the mix of cultures sees the same name inscribed in both Greek and Aramaean (Sartre 1985: 199). The strength of the local Aramaean communities is evident in the emergence of an Aramaean Christian scripture in a church dominated by Greek as the official language (Humbert and Desreumaux 1989: 121; MacAdam 1994: 74). The language of the Petra scrolls is overwhelmingly Greek, yet the Semitic character of the area is clearly evident in the transcription of personal names, toponyms and descriptive or designatory terms taken from Nabataean, Aramaic and Arabic. The papyri are economic and legal documents dealing with issues of real estate, inheritance, marriage, contracts, disputes, loans and sales. As well as ordinary people, administrators in the ecclesiastical, civilian, and military ranks are mentioned. The backbone of Petra society had shifted from engagement in long-distance trade to land-ownership and the maintenance of vineyards, orchards, sown lands, slaves and housing complexes (Bikai 1996; Kaimio and Koenen 1997).

Regional surveys show that the countryside in the watered areas was dotted with farmsteads in the Byzantine period. Their frequency and relatively unfortified nature suggests a peaceful and productive agricultural economy, as depicted in mosaics of the period. The hinterlands of towns in the more arid regions such as Umm al-Jimal, Phaeno, Petra or Humayma contain vestiges of careful water collection, storage and distribution for intensively worked fields (aqueducts, channels, cisterns, reservoirs and networks of field walls). The environment to the east was hostile in many ways, as the desert settlements, especially in the northeast, were fortified with continuous walls, towers and military posts. Farm workers and landholders must have formed a significant section of society, supporting the burgeoning towns and their economies.

Urban shopkeepers and artisans were organized into guilds (*collegia*), useful for tax collection, pricing, civic contributions and compulsory services of various kinds. Petra Scroll 60 refers to a *collegium* of tax collectors situated in Petra. One assumes that the potter's guild at Gerasa (attested in a third-century inscription) was maintained, as ceramic production continued as a major industry for the city. Slaves were integral to ordinary life, as personal servants, industrial and agricultural workers, and objects of commerce. The harsh copper mining centre of Phaeno in the foothills of Wadi 'Arabah was infamous for the exploitation of condemned Christians during the persecutions of Diocletian and later. They were the seed for a thriving Christian community as the mines continued working throughout the fourth and fifth centuries.

Changes to urban civic life reflect fundamental religious and philosophic shifts during these centuries. Major public buildings such as temples, theatres and hippodromes fell into disuse when their functions were no longer relevant or favoured. By the fourth century, drama had given way to mime, a kind of ballet with themes drawn from Greek mythology. This exacerbated Christian dislike of the theatre. Actors may have been popular idols but they were despised by Christians and excluded from the church. The hippodrome at Gerasa was poorly built and, therefore, had a short working life, while the hippodrome at Gadara was never finished. The changes at Gerasa are probably representative of the other Decapolis cities. The city was walled in the early fourth century and the abandoned hippodrome outside became an industrial centre for pottery and lime for the next three centuries. In the late fourth century, a 'cathedral' was built directly over a pagan temple. The sanctuary of the Temple of Artemis came to house a Christian complex in one area and another pottery production centre, operating from the fifth to eighth centuries. The North Theatre was out of use by the fifth century and Byzantine buildings encroached upon it. The main streets became encumbered by shop extensions within the colonnades, reducing their width and the original aesthetic of the streetscape. Nevertheless, the integrity of the basic urban structure survived. Churches were erected all over the city, often plundering the temples for their architectural elements. Attention was lavished on the interiors, with rich mosaics and marble panels (Kraeling 1938; Zayadine 1986, 1989; Seigne 1992; Kehrberg and Ostrasz 1997). Similar patterns of urban change have been recorded at Abila, Gadara, Pella, 'Amman and Petra. However, changes were incremental rather than absolute, and survival of the pagan ethos is noted at Gerasa, with the sporadic continuation of the controversial nautical festivals of Maiuma into the sixth century (Jones 1973: 1021). Bishops and their priests grew to have an important role in the community, reaching into the countryside with numerous village churches and monastic settlements. A strong sense of village solidarity and ecclesiastical involvement is gained from the acknowledgments in village church inscriptions on the Jordanian plateau.

Natural catastrophes affected the social and economic life of the Byzantine East during the fifth and sixth centuries. Numerous earthquakes left many urban centres in ruins, some with inadequate means of recovery. The bubonic plague appeared in 541 and recurred in a series of devastating epidemics lasting more than 200 years. The depopulation described by Procopius and John of Ephesus affected social and economic activities in the countryside and towns. Local Ghassanid poetry of the pre-Islamic *Jahiliya* expresses the fear it inspired. Archaeological evidence for hurried mass burials at sites such as Umm al-Jimal, Gerasa and Pella has been explained as the result of the plague (Conrad 1986; Cheyney 1995; Kehrberg and Ostrasz 1994; McNicoll *et al*. 1992: 218-19). The plague may have contributed to the decline of Byzantine society in this region in the late sixth and early seventh centuries.

Economy

Agriculture

The proliferation of towns and villages in the Byzantine period was sustained by intensive husbandry of precious food and water resources (see Environment). In the north, the rich plains of the western plateau continued to be important for cereals such as wheat and barley. As rainfall diminished towards the east, the settlements of the southern Hauran carefully tended their water resources to enable irrigated cultivation where today there is desert. The moister western plateau and the spring-fed wadis leading down to the Rift Valley permitted cultivation of a variety of products, particularly vines, olives, fruit and nut trees (such as fig and almond) and a range of horticultural products. Wine from the region of Capitolias and Gadara (Umm Qays) was praised in the sixth-century poetry of the pre-Islamic Arabs (Lenzen and Knauf 1987: 35-37).

The recent survey around Pella has found hillsides full of small wine presses and occasional larger complexes, attesting to a thriving local industry. The size and distribution of these presses suggests that this industry was predominantly small-scale, processed *in situ* as it was harvested, for local consumption rather than trade (Watson and O'Hea 1996). Surveys elsewhere in the northern hilly regions reveal a similar pattern for these installations. Fewer olive presses were found but they formed larger complexes. The more arid southern regions also produced cereals in the marginal rainfall zones. Irrigated agricultural activity was concentrated in spring-fed wadis, with the same dedication to water husbandry. Every town and village had its associated gardens. The ancient field and water systems recorded around Udhruh, Petra and Faynan are visible reminders of successful production in a harsh environment (Killick 1986; Gentelle 1984; Barker *et al.* 1997).

Pastoralism

Architectural remains attest to the herding of animals. The better-preserved basalt towns and villages in the Hauran have many examples of stables for larger animals such as cows and horses (Figure 13.9). Similar stabling facilities were excavated at Pella and Jarash (Figure 13.10) (see Architecture). Sheep and goats tend to be corralled rather than stabled, and remains of these are found in the countryside of the Hauran. Analysis of faunal remains from excavated sites of the period shows that sheep and goat formed a major part of the diet, followed by cattle, pig, camel, and other domestic animals, and hunted animals such as partridge and gazelle. Bucolic pastoral scenes on contemporary mosaics provide vivid depictions of such activities.

Industry

Knowledge of local industry in the Byzantine period is patchy. Study of the distribution pattern of pottery forms and wares reveals a strong regional pattern of domestic production overlaid by more widely travelled imported products linking the regions. Major centres of production were found in Jarash, Petra and Aila, with defined industrial areas and distribution markets beyond their immediate vicinity. The movement of goods tended to be more east–west rather than north–south, facilitated by the arrangement of administrative units. It is assumed, from the patterns of ware distribution, that there were many other centres (undiscovered) catering for local demand. The Hauran produced a distinctive ware, probably at Busra, where Islamic kilns have been identified. The Byzantine glazed pottery from Dayr 'Ayn 'Abata south of the Dead Sea (an accidental technological development that led nowhere) was another local production (Freestone *et al.* forthcoming). Late Roman pottery kilns found at Abila and Pella reinforce the impression of small-scale production at numerous sites, despite the close proximity of a dominant industrial centre such as Scythopolis was to Pella.

Evidence for the production of glass is sparse, with possible activity at Jarash and Abila (see Material Culture). The *macellum* (market place) at Jarash contained a tannery and dye works probably from the late Byzantine period, and limekilns were found in the area west of St Theodore's, and in the hippodrome. A limekiln has been identified in the domestic quarters at Gadara. Such industry represents localized production for local consumption.

Copper continued to be mined and smelted at Phaeno until the end of the fifth century when the ore bodies seem to have been exhausted. Until this time, the area of Wadi Faynan had been the major copper-production centre in the southern Levant. (Hauptmann and Weisgerber 1992). The trading port of Aila seems to have processed some of this ore, as metal slag was found in the recent excavations. Aila may have sustained a valuable fishing industry, as *garum* (fish sauce) from the Red Sea has been identified in the residue of a ceramic vessel found in a fifth-century context at Petra (Parker 1998: 11).

Trade

The region gained much of its commercial viability from its transit position on trade routes (see Foreign Relations and Trade). Aila continued to be commercially significant as a port of sea trade with India, Egypt, Africa and the Arabian Gulf. It was the terminus for the major trunk road linking Syria to the Red Sea, and the nexus for important caravan land routes to the Mediterranean across the Negev desert and to Egypt across the Sinai. The desert route from the Hijaz to Syria helped sustain the settlements in the southern Hauran.

Pilgrimage

The local economy was enhanced by a small amount of pilgrim traffic and the facilities to service it. While the

Figure 13.9. Umm al-Jimal. Mangers with stone troughs in a stable (house complex 49). Blocking of fenestrations is from later re-use.

Figure 13.10. Pella, Tall al-Husn. Mangers with flat sills in a domestic complex (XXXIVE). The bench behind belongs to a later phase of re-use.

majority of holy sites were located in western Palestine, pilgrims also travelled into Transjordan (see Religion). The main interest lay east and south of Jericho, at sites such as Sapsafas (on the east bank of the Jordan), Mount Nebo and the hot springs of Baaru (modern Hammamat Ma'in). The monastery of Saint Lot at the south end of the Dead Sea was also famous. Established routes crossed the Jordan, and Aila was an important staging post on the way to St Catherine's Monastery in Sinai.

Architecture

Churches

With the acceptance of Christianity, the construction of churches proliferated in the towns and countryside, reaching a peak in the sixth century. The dominant form was the basilica, consisting of a nave and two side aisles, sometimes expanded to four side aisles. They terminated in the east as a central apse flanked by two rooms, or three apses, with a larger central apse. The triapsidal type was more favoured in the sixth century. At the western end of the church was a square or rectangular colonnaded courtyard, the atrium. An antechamber or narthex sometimes separated the church from the atrium. The main altar was placed on a low dais in front of the central apse, separated from the body of the church by a chancel screen of decorated marble slabs set between marble posts. Chapels, baptisteries, sacristies and complexes for the clergy were often attached to the side of the main structure. Sometimes churches were built in clusters, as seen at Gadara and Gerasa.

Centralized churches were less common, but varied in form. The church at Gadara, for example, was planned as an octagon within a square (Figure 13.11), while the Church of St John the Baptist at Gerasa was based on a circle within a square. The now-destroyed Church of the Prophets, Apostles and Martyrs at Gerasa was a cross within a square, with a small projecting apse, while the Church of the Virgin Mary at Madaba has a round nave with an elongated apsed presbyterium (Piccirillo 1993).

The exterior of these churches was quite austere, with simple ornament restricted to doorjambs, lintels and cornices. Many of the architectural elements such as columns, capitals, and architraves were reused from earlier temples. However, the interiors were lavishly decorated with mosaics on the floors and walls (see Chapter 26), painted plaster walls, carved marble chancel screens and posts, marble facing, and patterned pavements.

Town Planning

Life in most towns continued within the framework of Roman-period urban planning and buildings. Adaptations were made as requirements changed. The Byzantine approach was more pragmatic than aesthetic, filling in and modifying open public spaces and monumental buildings to suit commercial and social requirements. Damage by intermittent earthquakes such as that of 363 must have aided this process of transformation. These changes are evident from excavations at Gerasa, Philadelphia, Pella, Gadara, Madaba and Petra. The colonnaded streets of Gerasa and Petra became cluttered with shops between the columns, as did the oval piazza in Gerasa. Some public buildings (temples, theatres and hippodromes) were re-used as industrial, commercial and religious centres, as well as 'quarries' for useful architectural elements. Pagan festivals and theatrical performances were progressively abolished from the fourth century onwards, and the process of transformation was well advanced by the fifth century. In Philadelphia, the vestibule of the Umayyad Palace based its cruciform plan on earlier Byzantine foundations dating no earlier than the fifth century. This plan closely resembles the praetorium or audience hall of the Ghassanids at Resafa in north Syria. However, the presence of the north and south doorways in the 'Amman structure suggests it may have been a monumental gate to the civic area (Almagro 1994).

In some parts of the east, where Hellenization had little impact, towns were not planned in the orthogonal system. They developed more organically, influenced by topography, existing structures, privately owned land and other practical considerations. Umm al-Jimal and associated settlements in the Hauran are good examples of this phenomenon. They have narrow winding streets, irregular open spaces, domestic and public agglomerations that are closed to the exterior but open internally onto courtyards, and an exterior circumference defensively closed by continuous walling between adjoining buildings. Their predominantly Arab population retained the traditional urban patterns documented from the earliest periods of urbanization in the Middle East, reflecting a social organization based on family and tribal authority rather than a civic municipal authority.

Domestic architecture

Our knowledge of domestic architecture of the period is relatively comprehensive due to the preservation of the Hauran settlements and to extensive excavations of domestic structures at Gerasa, Pella, and 'Amman. The availability of resources naturally led to regional differences in structural materials and solutions. The inhabitants of the Hauran, where wood was scarce,

Figure 13.11. Umm Qays. Columns (basalt) of the Octagonal Church with rectangular colonnaded atrium (in limestone) behind. View north towards Wadi Yarmouk and the Golan plateau.

resorted to ingenious methods of roofing using stone corbels and arches. This enabled the construction of sturdy multi-storey buildings, which have survived some 1500 years (Figure 13.4). Basalt was also used for doors, window grills and shutters. The wooded western hills, predominantly limestone country, facilitated the construction of stone buildings with wooden-framed roofs, either pitched or tiled, or flat with rush and pisé. Housing at Pella had ground floors of stone with an upper floor of mudbrick and pisé. As the snecked-stone construction was set in clay mortar that could erode in wet weather, the walls were probably plastered seasonally with pisé. This method of construction continued in use to the twentieth century. Finer buildings, such as the churches, used ashlar masonry, carefully fitted with no mortar or pebbles between the stones.

A common element in most houses was the internal courtyard, generally with an opening direct to the street, sometimes embellished by a vestibule. Classical houses have an atrium or peristyle, which have the formality of an interior room open to the sky, whereas this interior space functioned as a working yard. It was the place for domestic activities such as cooking, drawing water, the thoroughfare of animals and access to the upper stories or the roof. Native eastern traditions held sway over introduced western models. Many houses, particularly in the Hauran, have fenestrated internal walls on the ground floor, comprising a row of 'windows' with a flat sill or a sunken stone trough, resembling mangers in a stable (Figures 13.9 and 13.10). Cows and equids were caught in such rooms in the 749 earthquake at Pella. It seems likely that people and animals coexisted in many of these houses, with people inhabiting the upper floors (De Vries 1985; McNicoll *et al.* 1982: 136).

Military architecture

Most of the forts in Jordan were built in the late third or fourth centuries and continued in use through the early Byzantine period (Figure 13.12). Their plans fall into three main types: the quadriburgia (small forts with four projecting corner towers, such as Qasr Bshir, Qasr ath-Thuraiya), forts with projecting square interval and angle towers (Dayr al-Kahf, Qasr al-Azraq, Khirbat az-Zona, Khirbat al-Fityan, Da'janiya, Umm al-Jimal, Khirbat as-Samra), and forts with U-shaped and semi-circular external towers (al-Lajjun and probably Udhruh). Sometimes internal rooms were built against the curtain wall, but the larger forts tend to have

Figure 13.12. Qasr Bshir. View of the western (entrance) facade.

independent structures within the enclosure. Few forts are known to be constructed in the Byzantine period. The fifth-century barracks at Umm al-Jimal and Qasr al-Ba'iq have no projecting towers but were protected by internal towers. It seems that several different types of forts were built in the same period and region, perhaps motivated by different conditions or purposes (Parker 1995).

Foreign relations and trade

Byzantine Jordan was connected on three sides with imperial provinces, easing relations and trade with the immediate north, west and southwest. The arrangement of the three Palestinian provinces facilitated movement between Transjordan and the Mediterranean. Thus, the ceramic corpora from Transjordanian sites contain imports from North Africa, Egypt, coastal Palestine, Cyprus and coastal Anatolia (see Material Culture). At this time, the region was truly part of a Mediterranean culture.

While tensions fluctuated with the Persians to the northeast, interaction between the two great powers occurred well beyond the borders of Jordan and the effects were felt indirectly. This changed in the early seventh century with the Persian invasion of 613, although their 15-year occupation has left little or no trace in the archaeological record. The desert to the east and southeast was the domain of Arab tribes whose emergence during these centuries played a significant role in social and political developments (see History). The Salih, the Kinda and the Ghassanids were tribes who had migrated westwards into imperial territory. They were utilized and controlled by the Byzantines through alliances, subsidies and delegated military roles (see Defence).

The commercial significance of the region in Roman imperial times had been primarily as a transit route for trade between the Far East, India and the West, and between the more economically significant regions of Egypt and Syria. However, trading routes varied through the centuries according to political factors, changing the fortunes of towns along the way. By the sixth century, Petra was no longer an entrepôt for the caravan trade from South Arabia and beyond, as routes shifted north through the Persian Gulf and Syria, or south via the Red Sea. Some activity may have continued. Chinese annals, such as those of Liu-Sung (written c. 500, referring to the period 420–78) and the Wei Dynasty (written before 572, covering the period 386–556), may record a silk trade between China and Petra (eventually leading to Antioch). The sea route was longer but easier than the overland northern route, time was not so important, and the profits were high.

The Sassanian Persian policy in the northeast was to hinder Byzantine access to the overland route as much as possible and they controlled the traffic in silk through that area (Miller 1969: 132-39).

Excavations of the Roman/Byzantine port of Aila, at the head of the Gulf of al-'Aqaba, have revealed evidence for widespread contacts with the western Mediterranean, southern Arabia and beyond, throughout the Byzantine period (Parker 1998). Fine-ware pottery, oil, glass and metals passed through the port. The presence of Egyptian and Gaza amphorae implies the importation of wine and other agricultural products from Egypt and Palestine, and the importation of African red slip pottery is well attested. By the early fifth century, Aila seems to have been producing its own amphorae as transport containers (the 'Ayla-Axum' amphorae) that percolated the length of the Red Sea. Literary sources reveal the far-flung connections of the port. Eusebius in c. 290 notes that commercial traffic from India and Egypt passed through Aila (*Onomasticon* 6.17-21). The discovery of a freshly minted Byzantine-Sassanian hoard of coins and jewellery near Humayma attests to a significant Arabian–Persian trade coming through Aila in the early fifth century (Oleson, 'Amr and Schick 1992). In the late fifth century, Timotheus of Gaza mentioned that a native of Aila and a trader in Indian goods passed through Gaza bringing two giraffes and an elephant as gifts for the Emperor Anastasius. The importance of Red Sea commerce in c. 500 is reflected in the campaigns by Romanus, the *dux* of Palestine, to re-establish control of the Red Sea island of Iotabe where tariffs were levied on cargoes (see p. 490). The identification of Iotabe is controversial, but it is more likely to be Jezirat al-Far'un, an island c. 40 minutes sailing from Aila where Byzantine sherds are well represented and there is suitable anchorage (Glueck 1939: 11; Rothenberg 1970: 22; Fiema 1991: 175). Procopius of Caesarea, in the late sixth century, states that Roman vessels sailed from Aila into the Red Sea. Around 570, the Piacenza pilgrim records that it was eight staging posts from Mount Sinai to Aila and that shipping from India comes into port at Aila, bringing a variety of spices. Another text recounts an anecdote from an Egyptian monk Joseph, where a man from Aila offered to pay Joseph to pray for the successful return of a trading ship he had sent to Ethiopia (Parker 1996: 234-35; Schick 1994: 151).

The lands of Axum (Ethiopia) and Himyar (Yemen) flanked the gateway to the Red Sea. Byzantine diplomacy secured influence with these natural intermediaries for further trade to East Africa, India, Ceylon and China, thereby avoiding Persian customs duties. The principal source on eastern trade is Cosmas Indicopleustes, a sixth-century Egyptian merchant turned monk who describes the role of Ceylon (Taprobane) as a great emporium for trade between East and West. Silk, aloes, cloves, sandalwood, pepper, copper, musk and castor were traded (Bury 1958: II, 320-33). The Ethiopians acted as intermediaries for the Byzantines (although they did not carry silk), while the Persians traded directly, having their own commercial colony in Ceylon. Maritime traffic up the Red Sea headed for Clysma at the head of the Gulf of Suez, and Aila at the head of the Gulf of al-'Aqaba. When the Himyarite Dhu Nuwas attempted to disrupt this trade and to persecute Christians, the Byzantines supported the Ethiopian conquest of South Arabia in 525. Fifteen ships were sent from Aila. Fifty years later, South Arabia came under the control of the Sassanians and commercial and political power shifted to North Arabia.

Arabic sources record details of the terms of surrender of Aila to Muhammad in 630, where it was agreed to pay one dinar poll tax per adult, a total of 300 dinars. This indicates a considerable drop in population from the earlier Byzantine period when the legion was stationed there. Possibly, the legion was removed c. 530 when Justinian transferred the primary responsibility for defence of the frontier from the *limitanei* (the Roman frontier forces) to the allied Arab federates, the Ghassanids. Evidence for defortification in Arabia in the sixth century is widespread (Parker 1986: 149-52; see Defence). The number of sites in the hinterland of Aila and elsewhere in Wadi 'Arabah declined in the late Byzantine period, as did the facilities along the southern Via Nova. Goods from Aila were channelled through the Negev to Gaza and the Mediterranean, rather than through southern Transjordan (see Settlement Patterns).

There is increasing evidence that desert routes from the south, through Tabuk and Tayma to Bayir, al-Jafr and Ma'an, were developed in the sixth and early seventh centuries. Pre-Umayyad occupation of the 'desert castles' at al-Mushash and al-Hallabat, east of 'Amman, has been established. In the northern Hijaz, a Byzantine/early Islamic monastic hermitage existed at Kilwa, and a Byzantine presence continued at Qurayya until 600. The southern Via Nova was no longer important as a protected route, with most of its forts and caravanserais being abandoned by the late sixth century, yet the marginal land between Ma'an and Udhruh witnessed a substantial expansion of population in the late Byzantine period. Sites such as al-Mutrab, al-Hammam, Jabal al-Tahuna and Jarba had no previous occupation phases (Killick 1986: 236; Parker 1986: 100-102; Fiema 1991: 185). The commercial use of the caravan track from the northern Hijaz through Qurayya and al-Mudawwara to Maʻan may have been the sustaining factor behind these settlements.

At the same time, the more easterly connections were utilized between the Hijaz and Syria/Palestine through Wadi Sirhan and further east through Qasr Burquʼ (Sartre 1987; Fiema 1991: 178-91) (Figure 13.5). Burquʼ commands a lesser route between north Arabia and central Syria, running east of the *harra* from Wadi Sirhan via Wadi Muqat and Qaʻ Abu al-Hussein, where there is a small strategic site with Safaitic inscriptions and Roman-Byzantine sherds (King *et al.* 1983: 416). Historical information and material remains suggest Byzantine commercial presence in the northern Hijaz. Byzantine pottery occurs at sites in the central and northern Wadi Sirhan. Marginal settlements between al-Azraq and Busra expanded and flourished in this period, servicing the trade route and secured by forts and garrisons. Busra was renowned in the pre-Islamic Arabic sources as an artisanal and agricultural centre, and a redistribution market for trade with the Arabs and Persians. The trade south was not dependent on luxuries but on essentials such as cereals, wine, raisins, hides, leather products, arms and coarse cloths. During the sixth century Mecca rose to prominence as a dynamic trading centre, facilitated by the incessant wars in Mesopotamia between Sassanid Persia, Rome and their Arab clients, the Lakhmids and the Ghassanids. The Sassanian take-over of South Arabia in the late sixth century and the eclipse of Axum reinforced this process. Substantial caravans of 2500 camels were regularly organized from Mecca, bringing the spices and aromatics procured in South Arabia to Busra, where they were redistributed to the west and north. As yet, we have no evidence for the terms of this exchange. Thus, the Hauran flourished in the late Byzantine period, a time, for example, of unprecedented building and renovation of public structures and churches. Continuing trade with the south, combined with an ethnic affinity between the Arab populations of the Hauran and the Hijaz, reduced the impact of the Persian occupation and ultimately smoothed the transition from Byzantine to Islamic control. Surrounding areas of the empire, however, were unravelling under the strain of religious crises, military disorganization, the incursions of the Persians, earthquakes and recurrent plague.

Defence

The Transjordanian region formed the southeastern borders of the Byzantine Empire, protected farther east and southeast by hostile deserts. Security was more concerned with controlling the nomadic Arab tribes (called the *saraceni* by the Roman historians) who roamed these deserts and interacted with the sedentary agricultural population on the desert fringe. Constant friction with the Persians to the northeast was a less immediate problem, but was brought closer by the use on both sides of the nomadic tribes as allies, operating in a strategic buffer position between the two powers.

The Saracens existed both in conflict and symbiosis with Persia and Byzantium, who utilized their skills for reconnaissance, skirmishing and harassing an enemy army on the march. In return, they were paid subsidies in money or commodities in kind, and their leaders received titles and dignities. They were termed the *foederati* (federates) in the Byzantine system, serving as allies under their tribal leaders, headed by a *phylarch* (Parker 1986: 143-47; Mayerson 1989).

Early Byzantine defensive systems

During the late third and early fourth centuries, the Emperor Diocletian, followed by Constantine, reorganized the military structure of the empire, especially in the frontier zones. A system of forts, fortlets, watchtowers and roads was consolidated into a chain of military installations stretching from Egypt to the Euphrates. As part of this system, legionary bases were established or strengthened in the Transjordan at Udhruh, al-Lajjun (Betthoro), in southern Syria at Busra, and perhaps at Aila. Apart from al-Lajjun, these were all situated in towns on the major north–south road, the Via Nova Traiana. Al-Lajjun was sited to the east, at the primary water source for the area. With its associated watchtowers and fortlets, this fort covered the access routes from the east into Wadi al-Mujib. The Strata Diocletiana was established as a fortified road running from al-Azraq in northeast Transjordan, along the eastern slopes of the Jabal ad-Druze, to the Euphrates. Construction of the fort at Dayr al-Kahf is dated to c. 306 by an inscription, and the fort at Umm al-Jimal is dated to c. 300. The contemporary fort at al-Azraq seems to have been repaired in 326–33. In the south, there is no evidence of any military occupation in the Hisma east of the Via Nova and Wadi Ramm, or in the northern Hijaz, reflecting a retreat of influence beyond the marginal zone in this area. Attention to the frontier zone is matched by a withdrawal of legions from near the coast. No garrisons were posted on the road network within the populous agricultural heartlands or in their large urban centres, such as the cities of the Decapolis. Evidence from milestones along the Via Nova shows that the roads continued to be maintained in the fourth century (Parker 1986).

Early Byzantine occupation of the larger military sites in Wadi 'Arabah continued at sites such as Gharandal and Bir Madhkur. The abandonment of smaller defensive posts in this period may be due to the added security provided by the Tenth Fretensis Legion in Aila. A fourth-century road and milestones were identified in the western 'Arabah north of the fort at Yotvata (Smith *et al.* 1997: 67). The route from Aila up Wadi 'Arabah continued to be fully functional at this time, branching west and northwest across the Negev to the Mediterranean and the rest of Palestine. Like eastern Transjordan, the Negev was a monitored zone studded with forts and garrisons.

A number of Saracen attacks in the late fourth century underlines the need for this system of security. In 378 Mavia, queen of the Saracens, attacked the towns of the *limes* (frontier zone) of Palestine and Arabia and plundered the neighbouring provinces. Peace was restored through diplomacy. Around 383, another Saracen revolt is recorded, resulting in the Salih replacing the Tanukh as the dominant tribe (Parker 1987: 817-18).

In the fourth and fifth centuries, there were two basic categories of Roman troops: the mobile field army or *comitatenses*, elite troops under the direct command of the emperor or the *magistri militum*, and the regional garrisons or *limitanei*, under the command of the regional *duces*. Most auxiliary units were stationed in forts or towns between the legionary bases. The pattern is visible in the plateau area east of the Dead Sea where the hinterland of al-Lajjun has been extensively surveyed. A chain of forts spreads north and south of the legionary fortress, typically 10–15 km apart. From their size they must have held garrisons of c. 150 troops. The legion at al-Lajjun is estimated to have numbered 1000–2000 men. Fourth-century legions were smaller than their predecessors, which, in the second century, numbered around 6000 men. Numerous watchtowers were placed between the forts. Towers and forts cluster around the shallow eastern entrances to the wadis, which were the preferred routes of travel for the nomadic tribes. Within this secure system, settlement expanded significantly in the marginal areas. Survey of the Edomite plateau to the south has revealed a lattice of minor roads servicing an expanded population in the Byzantine period (Graf 1992: 259).

Use of the terms *limes* and *limitanei* has generated considerable debate. A reasonable consensus refers to the *limes* as a frontier district or zone rather than a defended border. It was under the military command of a *dux limitis*, and comprised a variety of military installations (forts, fortlets, police posts, watchtowers), protected water sources (such as the spring at al-Lajjun, or the reservoirs at al-Azraq), and communication routes. These bases served to contain and monitor tribal movements and behaviour, secure the passage of caravans and travellers, and protect the settled populations in the marginal lands.

In general, the *limitanei* were units under the regional command of a *dux limitis*, as distinguished from the field army or *comitatenses*. They were given tax-free land for their own cultivation and profit and received a salary (Parker 1987: 814). At the time of Justinian, Procopius reports that the pay fell into serious arrears, was later revoked, and eventually the *limitanei* were disbanded (*Secret History* 24.12-14). The accuracy of this account has been questioned, as elsewhere the *limitanei* are mentioned as operating throughout Justinian's reign. Payment arrangements, however, may have altered significantly (Isaac 1990: 210-11; Whitby 1995: 112-13). It should not be assumed that they were militarily less competent, being 'soldier-farmers', given that land ownership does not require farming in person.

Late Byzantine defensive system

In the early fifth century, Saracen incursions into Palestine and Syria (as recorded by Jerome in *Epistle* 126) may have encouraged a number of defensive measures. The barracks at Umm al-Jimal were built in 412/413, and the nearby fort of Qasr al-Ba'iq in 412. An inscription from the fort at Ziza, south of 'Amman, indicates some building activity in 410. The two forts east of Ma'an in Edom, al-Hammam and al-Mutrab, possibly date to the early fifth century (Parker 1986: 146). Inscriptions from Busra and Gerasa refer to refortification of the city walls in 440–41. However, the barracks complex at Umm al-Jimal represents a reduced military capacity as it replaced a large fourth-century fort that was converted into a market place in the early fifth century. Similarly, the fort at Humayma went out of use at the end of the fourth century (De Vries 1993: 452; Oleson *et al.* 1995: 324). It is apparent that fourth-century solutions were being revised in the fifth century, evolving a more decentralized imperial military authority.

In the reign of Leo (457–74), a tribal leader Amorkesus seized the island of Iotabe in the Gulf of al-'Aqaba. He levied tolls on the passing trade and applied successfully to Leo to be appointed *phylarch* of the 'Saracens around Petra'. Leo's acquiescence illustrates the weakness of Byzantine control in the area, but also his pragmatism in converting an outlaw into a law enforcement officer. It was not until 498 that this lucrative position was restored to the Byzantines. In the later fifth century, there is evidence for a decline in the quality of the *limitanei* and a continued depletion of military installations along the eastern frontier. The Theodosian law of 443 confirming the agricultural and land rights of the *limitanei* seems motivated by a need to redress a problem. The excavations at al-Lajjun suggest that life at the fort had run down, and the soldiers' families were now living within the precinct. A metalworking industry in the barracks, presumably maintaining weapons and equipment, ceased by the end of the century. The shoddy architectural restoration of the structure after the 502 earthquake implies a lack of resources. A number of military sites in the region were peacefully abandoned by c. 500: Khirbat al-Fityan, Rujm Beni Yasser and Qasr Bshir. No new forts were built after the early fifth century (Parker 1987: 818-20).

The continuity of the fortified zone was disrupted by the widespread abandonment of forts. Some concern is shown in the north, where an edict of Anastasius, found in fragments at Qasr al-Hallabat and elsewhere, declares that fines for illegal treatment of the *limitanei* should be used for restoration of the forts. Another inscription of 529 records the restoration of the fort at al-Hallabat. This is the last imperial military building inscription from Arabia (Parker 1987: 821). Nevertheless, the only other forts in the northern sector with sixth-century pottery are Ba'iq (near Umm al-Jimal) and Aseikhin (near al-Azraq). The evidence for al-Azraq is unclear, and the barracks at Umm al-Jimal may have been converted from military to monastic use. In the central sector, only three forts appear occupied after the early sixth century: Muhattat al-Haj (monitoring a crossing of Wadi al-Mujib), Khirbat az-Zona (monitoring the caravan route east along the edge of the desert) and al-Lajjun. Most military sites in Edom were abandoned by the mid-sixth century, other than Udhruh and Ail which may have been civilian by then. The fortlets of Khirbat al-Khalde and Qasr al-Kithara in the Hisma continued in use, but the remaining watchtowers were unoccupied. Weakened imperial protection probably led to the refortification of town walls recorded at Scythopolis in the 520s, Busra in 540, and Umm al-Jimal sometime in the sixth century.

The reduced commitment of imperial troops in the southeast frontier zone is associated with the increasing importance of the Arab federates, led in the sixth century by the Ghassanids (see History). Anastasius confirmed their leading role after the Kindite wars of c. 502/503, conferring the *phylarchy* of Palestine on the Kinda, and that of Arabia on the Ghassanids. Around 530, Justinian appointed the Ghassanid leader Harith as supreme *phylarch* of all the allied Arab tribes, responsible for protecting the eastern frontier from Palmyra to Wadi al-Hasa. They opposed in kind the Persian Arab allies united under the Lakhmid king Mundhir. Byzantine defence policy shifted from a network of imperial troops garrisoned in forts, to mobile troops of subsidized allies drawn from the local power base. It derived from a pragmatic assessment of the physical and economic realities of the time. Maintenance of the old fortified system was costly at a time when funds were needed to counter the Persians in Syria and Mesopotamia, and to reconquer the lost western provinces.

The reformed strategy worked well for the next half-century. However, emperors in the late sixth century progressively weakened the Ghassanids without taking any compensatory measures along the frontier. Justin II (565–78) cut off subsidies to some of the tribes. A major dispute erupted in 580–81 between the Byzantines and the Ghassanids, inflamed by intrigue and treachery on the part of the emperor Maurice. He was intensely hostile to the Monophysite creed followed by the Christian Ghassanids. After the capture and exile of the Ghassanid *phylarch* Mundhir, his sons defeated the *dux* of Arabia, laid siege to Busra, and for two years ravaged Syria, Palestine and Arabia. Eventually the Ghassanid confederation disintegrated into its constituent tribal elements (Sartre 1982: 189).

The eve of the Muslim Conquest

Following another peace with Persia in 591, many of the eastern troops were transferred to Europe to counter the Avars and the Slavs. Little is known about the Byzantine army in the southeast in the early seventh century. The paucity of historical sources reflects the declining economic and political significance of the area to the Byzantine state. Palestine and Arabia had not been invaded by regular armies since the time of

Zenobia in the third century, and their experience was limited to local security. In 613–14, the Persians under Chosroes II swept easily through Syria to Arabia and Palestine. Few details are known, but Arabia was probably overtaken prior to the siege and pillage of Jerusalem in 614. Busra and Adraha (Dera'a) were reportedly taken by assault. There is no clear evidence in the archaeological record of destruction of towns east of the Jordan, or of the following 14 years of occupation. The implied lack of resistance by local authorities foreshadows a similar reaction in two decades time. Trade between Busra and the Hijaz and Mecca continued without interruption, while inscriptions show that churches at Rihab and Samah were renovated and embellished (Schick 1992: 110-11).

The emperor Heraclius marshalled the imperial resources against the Persians and, after prolonged battles and the eventual murder of Chosroes, regained the region in 628. He had little time to re-establish control in the southeast or to reorganize the system of security. He seems to have continued parts of the previous administrative system, and old military units reappear (Kaegi 1992: 73-74).

The Byzantines were not expecting any concerted Arab invasion from the south, treating the initial conflicts as more of the same nomadic raids they had endured through recent centuries. They defeated the first Muslim advance at Mu'ta in 629. The tribal allies of the Byzantines initially opposed the Muslim invaders and a number of hostile encounters are recorded (Schick 1992: 113; Kaegi 1992: 66-111). Nevertheless, on the way to Mu'ta the raiders were able to stay two nights at Ma'an without any opposition. The fact that Mohammed in 630 accepted the surrender of Udhruh, al-Jarba or Aila, negotiated with the local bishop, implies that there were no regular garrisons stationed in the south or any imperial administration. The move to re-establish Byzantine control could not immediately cover the whole area and they may have set up an initial defensive line east of the Dead Sea (Parker 1986: 154). The concerted Muslim conquest of Syria had begun by early 634, with the capture of Areopolis (modern ar-Rabba) on the plateau south of Wadi al-Mujib. The Muslims then advanced across Wadi 'Arabah to western Palestine. Byzantine reliance on Arab allies was inevitable, and the decision to end subsidies to some of the tribes was a disaster. Disgruntled tribesmen aided the Muslims in defeating the Byzantines at Dathin near Gaza in early 634. As there were no garrisons in southern Palestine, a unit had to be dispatched from as far away as Caesarea. The battle of Ajnadayn, southwest of Jerusalem, was a more decisive victory, followed by Fihl (near Pella). Finally, the Byzantine army was annihilated at the Battle of Yarmuk in 636. Their commanders lacked unity and were ultimately out-manoeuvred. Syria was rapidly overtaken and by 640 the last of the opposition surviving behind the walls of 'Asqalan, Gaza, Jerusalem and Caesarea had been removed (Kaegi 1992: 112-46). Most towns submitted peacefully, perhaps mindful of the terms that had been offered to towns such as Aila, Tiberias and Pella, where security of life and property were guaranteed in return for payment of a poll tax (Schick 1995: 72-73). The conjunction of the Muslim invasion with the poorly re-established and vulnerable Byzantine recovery of Arabia and Palestine had proved disastrous for the Byzantines and opportune for the Muslims (Kaegi 1992: 236-87).

Burial customs

Burial practices in the fourth–seventh centuries continued in the traditions of the Roman period, with some minor adaptations for Christian sensibilities in decoration (the addition of crosses) and preferred orientation. Major towns were surrounded by cemeteries, often lining access routes or located in the rocky slopes of adjacent wadis. Interments inside sanctified structures were common. Mausolea and graves beneath the floors of churches have been documented at many sites. The sites of Abila, Umm Qays (Gadara), Yasileh, Umm al-Jimal, Pella, Jarash, 'Amman, Hisban and many others have all revealed tombs re-used over the centuries from the Roman period through the Byzantine period.

There were a variety of grave types, from the elaborate to the simple. The socio-economic position of the interred and geographical context seem to be the principal determinants of type. Elaborate monumental built tombs are found near urban centres such as Gadara, Jarash and 'Amman. The 'North Mausoleum' at Gadara is dated by inscription to the mid-fourth century. It was built in the form of a small temple, on a square podium with a staircase, and a crypt in the centre of the podium. The architect's name, 'Arabios', indicates his indigenous origins (De Vries 1973). The finely constructed subterranean Roman hypogaeum at the same site was expanded and altered in the Byzantine period to accommodate important burials associated with the Christian basilica built above it. Stone-lined shaft tombs (cists) were cut through the floor of the vaulted entrance hall to the west of the domed main chamber. The lead sarcophagus of a deaconess was placed in front of the barrel-vaulted passageway that girdled the main chamber. Two adjacent built tombs at Jarash, south of the Hadrianic Gate, were Roman constructions re-used in the Byzantine period (Zayadine 1986: 12-16, 1989: 201-18). They comprised vaulted chambers with a monumental superstructure (now destroyed), containing stone-lined cist graves as well as free-standing sarcophagi.

Most of these Transjordanian towns lie in rocky areas, suitable for rock-cut features and favouring cave formations. A popular tomb type of relatively high status is the chamber tomb with radiating loculi. The latter may be flat or have rectangular shafts cut into them. This type is common at Pella, for example, where there is no evidence for cremation or grave burials in the Roman/Byzantine period. Another type of chamber tomb contains *arcosolia*, arched openings off the main chamber. Both chamber types generally have an entrance passage (*dromos*), often approached by steps from above. The more elaborate façades have architraves, decorated lintels and stone doors with elaborate lock mechanisms. The remains of wooden coffins have been found in some of the loculi and chambers. Some chambers contain stone sarcophagi (McNicoll *et al.* 1992: 141-44). These types occur throughout the country, for instance, at Abila where single trough (cist) graves in chambers are also recorded. Two private tomb precincts with temenos walls are dated to the Byzantine period (Mare 1984). All these tombs contained multiple interments, of mixed gender and ages. When undisturbed, the skeletons are found supine, with legs outstretched and arms either by their side or laid across the chest. There is evidence for shrouds wrapping the bodies, and sometimes a suggestion of clothing and leather sandals. The peculiar thermal conditions at Pella have enabled the survival of laurel leaves in one tomb, presumably used to sweeten the air. Grave goods consisted of personal jewellery, glass vessels including unguentaria (for the dispensing of aromatics), spatulae and spoons, pottery lamps and vessels, plaster mirrors and make-up palettes. These can be found with the bodies or scattered around the chamber. Items peculiar to the interred may be included, such as the surgical tools of a medical practitioner recovered at Abila.

Significant cemeteries have been excavated in the east at Umm al-Jimal and Khirbat as-Samra (Brashler 1995; Nabulsi and Humbert 1996). They revealed simpler tomb types presumably for people of lower or poorer status. The seventh-century cemetery at Khirbat as-Samra consisted of trenches dug in the soil, with a shelf left c. 30-60 cm from the bottom, bearing the covering slabs of basalt flagstones. These were predominantly children's graves. Umm al-Jimal cemetery AA contained unlined pits with or without grave objects, well-constructed stone-lined cists for single occupants, and simple pits with coffins. Cemetery Z contained a number of types: loculi, cist and coffin-lined tombs, representing a range of socio-economic levels.

At Dhiban on the Moabite plateau, a late fourth-century cemetery revealed two types: a cist with earth floor, containing multiple burials, with head placed to the west; and simple inhumations. Early seventh-century cist graves were found inserted in the pavement of the disused Nabataean temple (Tushingham 1972: 86, 107-15).

In the large Byzantine cemetery at Faynan (Phaeno) in Wadi 'Arabah, the standard grave consists of a rectangular cut with rounded ends, oriented east–west, with a groove cut into the vertical sides (occasionally a ledge) to bear the covering slabs of roughly cut ovoid sandstone. These were laid perpendicular to the length of the grave, and sealed by a mixture of pebbles and mud. The grave contained a single body, laid supine with head to the west. Over 50% had remnants of shrouds. Generally, there were no grave goods other than personal jewellery. The majority of graves were marked with an orthostat stone at the western end, sometimes inscribed with a cross of varying design. One hundred and eighty out of 1200 orthostats have inscribed Christian crosses, but names are rare. The cemetery was well organized into rows of graves (Findlater *et al.* 1998).

A more rare and elaborate type of grave had a built stone superstructure. It comprised either cut and dressed sandstone blocks or more roughly cut blocks, laid on the ground surface or within a shallow foundation trench, forming a rectangular plan which was filled with rubble. Only one course is extant but the amount of rubble suggests that they may have stood up to 1 m high. While this type is not found elsewhere in Jordan, it has parallels with Coptic burials in North Saqqara of the late fourth and early fifth centuries, where the superstructure is of mudbrick. Another unusual type is represented by three examples, marked by a line of medium to large rocks following the outline of the grave cut. This is the standard type found in the Byzantine cemetery at Rehovot in the Negev. These links with

Egypt and southern Palestine accord with textual information, which reports that many of the early Christian martyrs forced to work in the mines of Phaeno came from Egypt and Gaza.

Vast Byzantine cemeteries in Ghor as-Safi have come to light through extensive plundering activities. Attempts to rescue the remaining information have recorded more than 300 funerary stelae, 90% of which are inscribed in Greek, the remainder in Aramaic (Politis 1999). The Monastery of Saint Lot at Dayr 'Ayn 'Abata overlooks the area. Disarticulated secondary burials were found in the venerated cave behind the church, and in one of the cisterns that was re-used as a communal burial chamber. Such burial customs are well known in early Christian monastic orders.

Examination of certain burials at Umm al-Jimal, Jarash and Pella has independently prompted speculation on the great plague which ravaged the region from the sixth to the eighth centuries (Conrad 1986: 146). Many of the Late Byzantine tombs at Pella had single loculi and sarcophagi containing up to five multiple interments of all ages and sexes. One coffin showed clear evidence of rapid multiple interment: cloth was laid over a body that had not yet decomposed, then plaster was poured over it to make a base for the next body. The jar in which the plaster was mixed was left nearby, suggesting carelessness and haste. Scant respect was often shown to the earlier occupants whose bones were simply scattered around the chamber (McNicoll *et al.* 1992: 143, 218-20). At Umm al-Jimal, a common-class cist grave was dug to hold a minimum of four individuals with a wide range of ages, buried close in time or contemporaneously with each other. The effort in digging such a grave is negligible and, therefore, the need to place them together is considered unusual (Cheyney 1995). More telling is the evidence from the hippodrome at Jarash. Two chambers were re-used for mass burial around the mid-seventh century, containing more than 100 bodies dumped without ceremony on top of mounds of discarded pottery. Prior to disintegration of the flesh, the stone-vaulted roof collapsed onto the bodies, probably due to the earthquake of 659 (Kehrberg and Ostrasz 1994).

Religion and ritual

The rise of Christianity

The fourth century is marked by the rapid rise of Christianity to a position of power and influence in the empire (see Historical Background). While Constantine I became an enthusiastic Christian he remained tolerant in his beliefs. On the one hand, he confiscated the treasures and endowments of the temples and banned pagan sacrifice; on the other, he founded two new pagan temples in Constantinople. There was a short interlude of pagan resurgence under the emperor Julian (361–63), but his early death prevented its consolidation. In 380, Christianity became the formal religion of the state. In 391, Theodosius I issued the first of a series of laws that progressively banned all pagan ceremonies. The privileges of the priestly class were withdrawn in 396. Temples were closed and many were demolished, their materials re-used for the repair of roads, bridges and aqueducts. Unauthorized attacks on temples and their conversion into churches had been tolerated for some time. In Jordan, disorders were recorded at Petra and Areopolis, where the strongly pagan local population had vigorously defended their temples. However, pagan beliefs were not banned outright, and the cult survived into the sixth century. Finally, Justinian forbade pagans to hold professorial chairs, to receive inheritances or bequests, or to testify in court. In 529, he ordered all pagans to accept baptism under penalty of confiscation and exile (Jones 1973: 167-69, 938-43; Trimingham 1979: 82).

Conversion of the nomads

The Arabs of the Syrian Desert began to be converted in the mid-fourth century, apparently through their encounters with priests who dwelt among them and the monks in retreat in the deserts. These ascetics, who eschewed urban forms of Christianity, were admired for their healing powers and ability to expel possessive spirits. Early in the reign of Valens, an Arab leader called Zokomos was able to beget a son through the intercession of a hermit and, in gratitude, he and his whole tribe became Christians. Barochius 'bishop of the Arabs' (nomads) is mentioned at the Council of Seleucia in 359, while Theotimus of the Arabs was present at the Synod of Antioch in 363. Bishops solely associated with nomads were not given town or territorial titles. Those based in towns, such as Germanus of Petra and Arabion of Adraa, as well as the bishops of Palestine III, also had relationships with the nomads. In the early fifth century, Aspebetos was the leader of a tribe that relocated from Persian to Roman territory. The monk Euthymius cured Aspebetos's son of paralysis and the whole tribe converted. They established an encampment (*parembole*) near Euthymius, west of the Dead Sea, and Aspebetos was made Bishop of the Encampments in 427 (Trimingham 1979: 94-120).

Christian administration and hierarchy

Bishoprics were established in centres of population, mostly in cities. However, in the province of Arabia, where the village was the normal unit of government, village bishoprics were common. In Palestine, the four regions of the Jordan Valley were independent bishoprics (Jericho, Amathus, Livias and Gadara/as-Salt).

The bishop of a metropolis or capital city of a province had the position of Metropolitan, with authority over the other bishops within his province. The boundaries of a bishopric or ecclesiastical province generally coincided with the civil administrative province. The bishoprics or episcopal sees of Madaba, Esbus, Philadelphia and Gerasa were within the province of Arabia under the Metropolitan of Busra. Palestine I, with Caesarea Maritima as its metropolis, incorporated the bishoprics of Livias, Gadara (as-Salt), Bacatha and Amathus. Palestine II, with Scythopolis as its metropolis, contained the bishoprics of Pella, Gadara (Umm Qays), Capitolias and Abila. In the south, Palestine III contained Areopolis (ar-Rabba), Charachmoba (al-Karak), Aila, Zoara, Phaeno, Arindela and Augustopolis, under the metropolitan city of Petra. The bishops of Arabia answered to the patriarch of Antioch, who had primacy over the area by tradition. In 431 the three provinces of Palestine came under the jurisdiction of the patriarch of Jerusalem. Mosaic inscriptions from the many churches in Jordan confirm or clarify these divisions, allegiances and hierarchical structure. Priests, deacons, deaconesses, lectors and others assisted the bishops in the towns, while an archpresbyter might assist in a village (Jones 1973: 873-94; Piccirillo 1993: 43-45).

A rich ecclesiastical geography had developed by the sixth century in Jordan. Churches and monasteries dotted the countryside, revealing the penetration of the new faith beyond the cities into the rural and nomadic communities. This was encouraged by the Emperor Justinian's promotion of an ecclesiastical building programme and the sympathetic security provided by the Christian Ghassanids.

Christian interpretation and schisms

Christianity was an evolving religion in this period, seeking consensus on interpretation and organization throughout the empire. Argument and division characterized its development as the faith of the establishment. A series of church councils attempted to deal with the variety of viewpoints on doctrinal issues and matters of order and authority. The Arian controversy, over the relationship of God the Son to God the Father, dominated the fourth century. During the fifth century, controversies concerning the nature of Christ reached a climax. The Nestorian doctrine was condemned, followed by the Monophysite interpretation of the single nature of Christ. A diophysite (dual nature) interpretation was promoted as the orthodox 'Chalcedonian' doctrine. These doctrinal conflicts and resolutions were as much an expression of political struggle as disputes over theology. Because of imperial support and the institutionalization of the Church, disagreements came to be heresies and liable to punishment by the state. Moreover, imperial religious policy tended to change with individual emperors, making it indecisive and inconsistent.

Syro-Palestinian Christians were divided into opposing communions. Antioch, the metropolis of orthodox Greek Christianity, was alien to the majority of Aramaean congregations in the region. The gap in communication was filled through the medium of the monastic movement that was predominantly Monophysite. Most of the people followed the monks rather than the bishops and clergy. However, Euthymius, the moulder of the Dead Sea monastic movement in the early fifth century, and his protégé the Bishop of Jerusalem, supported the Orthodox creed, creating conflict within the monastic movement. Eventually the Aramaean Monophysite element within the Palestinian monks was eliminated. In the province of Arabia, the Ghassanids became staunch Monophysites. It is impossible to determine, from the remains of its architecture or decoration, which rite an ancient church might have used. Thus, the affiliation of specific churches and, therefore, the pattern of affiliation, is still unknown (Trimingham 1979: 80-188; Jones 1973: 232-35, 929-42, 967).

In the early seventh century, Heraclius introduced an alternative interpretation to the religious debate, Monotheletism, in an attempt to resolve the doctrinal split. More controversy ensued and the effort was notably unsuccessful. At the time of the Muslim invasions, the eastern provinces were divided in their Christian allegiances, but the schism does not seem a decisive factor in determining loyalties. Many moderate Monophysites were prepared to unite with the

Orthodox Christians, and later Monophysite writers under Arab rule were not demonstrably anti-Byzantine (Schick 1995).

Monasticism and pilgrimage

The founder of the eremitic and monastic movements was the Egyptian Antony, who retired into the desert in the 270s. During the Great Persecution at the beginning of the fourth century, he organized the numerous disciples who had followed him into a loose-knit community. Such groups of hermits, who lived in separate cells and met only for common worship, were later known as *laurae*. A few decades later, another Egyptian, Pachomius, founded the first *coenobium*, where the monks led a communal life under strict discipline. Both forms of monasticism caught on rapidly in Egypt and soon spread to Palestine. In the late 350s, the movement spread to Syria (Jones 1973: 929-33; Trimingham 1979: 107-16, 183-84).

A number of monasteries, *laurae* and hermitages have been identified in Transjordan. In the north, the monastery at Fedayn, near al-Mafraq, is known from texts and excavation (Humbert 1989: 126). The 'barracks' complex at Umm al-Jimal was possibly converted into a religious institution in the sixth century (De Vries 1985: 251), and one interpretation of Qasr Burqu' is that it functioned as a reclusive refuge for a hermit, later expanded to a monastery (Betts *et al.* 1991: 15-16). An eremitic function is proposed for the isolated tower at Umm ar-Rasas (Figures 13.5 and 13.13). A partial list of the Monophysite monasteries in the province of Arabia, dating to c. 575, survives in the 'Letter of the Archimandrites'. There were 137 signatories, all heads of monasteries in the province (MacAdam 1994: 59).

Many of the monasteries in the fertile zones were working agricultural establishments: Zay al-Gharbi, west of 'Amman, Munyah, Mount Nebo, the nearby smaller monastery of al-Keniseh, and another near 'Ayn Keniseh. An inscription at Mukhayyat provides the first epigraphic evidence for a 'stylite' (an ascetic who lives on top of a pillar) in the Madaba region. The Memorial of Moses at Mount Nebo was a prominent sanctuary, servicing pilgrims. The remains of at least three small monastic complexes were found in the adjacent valley of 'Ayoun Musa, and a hermitage at 'Ayn Jammalah. A rock-cut *laura* has been identified at Muallaqah, in the desolate Wadi Jebara to the west of 'Iraq al-Amir (MacAdam 1994: 81 n. 185).

Mount Nebo was the main destination for most of the pilgrims venturing east of the Jordan, and the nearby desert area attracted monks and hermits. On the east bank of the Jordan River lies Sapsafas, connected with the baptism of Jesus and John the Baptist. A *laura*, known from historical records, has been identified at Wadi Kharrar in caves in the marl cliffs, as well as remains of a monastic-pilgrimage centre. The pilgrim road continues via Livias (Tall ar-Rama) towards Esbus (Hisban) on the plateau, turning off to the Springs of Moses and Mount Nebo. Ma'in lies further south, with its monastery of Khirbat Dayr and remains of an inn and baths. The nearby hot springs of Baaru were a therapeutic destination. A hermitage has been found in the vicinity at 'Ayn Qattara (Piccirillo 1987).

Isolation in a desert environment was the desirable context for an ascetic lifestyle. A number of remote hermitages have been identified in Wadi al-Mujib and Wadi al-Hasa, for example, at Hammam 'Afra (MacDonald 1980), and near the outlet to Ghor as-Safi. The southern end of the Dead Sea contains a cluster of hermitages and monasteries that link geographically with the Judaean desert and its monastic society. A 'bridge' is provided by a ford to the Lisan Peninsula, which contains a number of hermitages and a monastery. The major hermitage of Qasr at-Tuba was possibly the centre of a *laura*. The sixth-century monastery of Dayr al-Qattar was the focus for a large monastic community in the area (Holmgren and Kaliff 1997). The location of these monasteries may have been determined by the Byzantine tradition concerning the destroyed 'cities of the Plain' (Gen. 13.12), Sodom and Gomorrah. In Ghor as-Safi, overlooking ancient Zoara, the Sanctuary of Lot at Dayr 'Ain 'Abata comprised monastic and pilgrim facilities (Politis 1997). Many of these holy sites are depicted on the sixth-century mosaic floor known as the Madaba Map (see Mosaics).

Jabal Harun, southwest of Petra, was thought to be the burial site of Aaron, the brother of Moses. Monastic remains date back to the sixth century, according to surface pottery and textual references (Peterman and Schick 1996). A well-known hermitage lies above the processional route from Petra to ad-Dayr. The latter pagan monument was re-used as a Christian shrine, probably in the Byzantine period. A monastery called Dayr al-Qunfudh is reported at Aila. The monastic movement penetrated the northern Hijaz with a monastic hermitage identified at Kilwa (Fiema 1991: 185).

Figure 13.13. Umm ar-Rasas. View of the tower east of the town.

Churches

Eusebius states that churches had been built in many large cities before the Great Persecution at the end of the third century (*HE* 8.1.5), but none has survived. Church construction revived in the fourth century but most of these were replaced in the fifth and sixth centuries. Excavations at Aila have found a mudbrick basilica, believed to be a church, dated to the end of the third/beginning of the fourth century, and destroyed by an earthquake in the later fourth century (Parker 1998). This would make it the earliest known purpose-built church in the world.

Many towns had multiple churches, more than adequate for their population base. Umm al-Jimal had 15, Rihab and Khirbat as-Samra had 8 each, Umm ar-Rasas 16, Madaba 11, and so on. Their design gives no indication of differing tribal or doctrinal affiliations. Mosaic inscriptions reveal that many of the church donors in Jordan were individuals or groups of individuals, sometimes from one family. Sometimes a village was a donor. Donors possibly sought divine favour through such bequests, encouraging this proliferation.

Writing

Our knowledge of the written language for this period comes mainly from inscriptions and, to a lesser extent, literary references. A small proportion of the population was ethnically Greek, yet Greek was the primary language of written communication. It was the formal language of the Byzantine administration, of power and prestige and flourished particularly in the Hellenized northwest of Jordan. Latin, the other language of Roman administration, never dominated in the East and ceased to be used in inscriptions after the fourth century. The majority of the population was Semitic, speaking Arabic or Aramaic, despite any imposed cultural veneer. Syriac (eastern Aramaean) is mentioned as the language of the urban centres of Palestine I and II (Shahid 1984: 291). Many of the personal names found in Greek inscriptions from the Hauran and Moab are Arabic (Sartre 1985: 198-202; Canova 1954). Greek was the official language of the church used in dedicatory inscriptions, nevertheless, Syro-Palestinian Aramaic inscriptions are found in churches at Mount Nebo, 'Ayoun Musa, Quwaysma, and Khirbat al-Kursi. A cemetery at Khirbat as-Samra contained 46 funerary stelae with inscriptions in Greek and 65 stelae inscribed in Aramaean. The same names could appear in either script, attesting to the mixed cultural milieu. Arabic inscriptions, rare at this time, are found at the church of St George at Mukhayyat and the double church at Umm al-Jimal (Piccirillo 1993: 45; De Vries 1993: 448, 454; Humbert and Desreumaux 1989).

Further evidence for the use of Arabic as spoken language is found in the sixth-century papyrus scrolls from Petra. Roll 10, for example, reveals names used for places, districts, fields, orchards, buildings, houses, and even parts of houses, in Arabic or Aramaic, transcribed in Greek. It is clear that there was a scribal tradition in Petra in the sixth century that included knowledge of Arab philology (Bikai 1996). A few earlier inscriptions and references provide evidence for Arabic as the language of everyday life in the eastern and southern areas in the Byzantine period. The earliest dated inscription is the Namara text of 328, written in Arabic using the Nabataean Aramaic alphabet. Epiphanius, writing in the later fourth century, says that the language of the pagan liturgical celebrations of the feast of Venus in Petra was Arabic. Sozomen attests to the use of Arabic in the composition of poetry celebrating the victories of Queen Mavia in the later fourth century (Shahid 1984: 292). The Arabic script, the direct representation of the spoken language, developed out of Aramaic, the traditional lingua franca of the Semitic Middle East. The earliest dated inscription in this new script comes from Zabad (near Chalcis in northern Syria) c. 512, suggesting that the process of differentiation occurred during the fifth century (Shahid 1989: 409-19).

Writing and language at this time reflect a complex linguistic history. The public written language differed from the spoken languages, which themselves varied according to area. It is difficult to determine from the sources how closely the written language matched the language that was spoken, and we cannot assess the relative proportion of speakers of Greek, the different dialects of Aramaean, and Arabic.

Conclusion

The Byzantine period in Jordan marks a high point in the occupational history of the land, as abundantly attested by the archaeological record. A seasoned system of law and administration, from municipal to military and imperial levels, provided the security for the population to expand and flourish throughout the countryside. A healthy economy, utilizing the available resources and capitalizing on established trade routes that covered the region, supported this population and prospered within the zone of imperial security. A religion that had a local genesis became a prominent social and cultural force within the Byzantine empire, endowing the region with status and attention. The continuity of this society was undermined in later years by a number of factors. Administrative changes, determined by a variety of considerations and miscalculations, diluted imperial control. Christological controversies linked with political struggles affected regional cohesion. Natural catastrophes such as earthquakes and endemic plague made life more difficult and uncertain for ordinary people. When faced with a close sequence of external threats, first from Persia and then from northern Arabia, the region was ill-equipped to defend its physical territory or the status quo. Centuries of western dominance of the eastern Mediterranean were replaced with astounding rapidity by a focused political and religious force emerging from indigenous Arab society.

Note

1. This spelling of 'Aila' follows Parker's spelling and reasoning (1996: 232). A variety of spellings are attested in the sources, and 'Aila' for the Nabataean/Roman/Byzantine period distinguishes it from Islamic 'Ayla' which was at a different adjacent location.

References

'Amr, K.
   1991   Preliminary Report on the 1991 Season at Zurrabah. *Annual of the Department of Antiquities of Jordan* 35: 313-23.

'Amr, K. et al.
   1996   Archaeological Survey of the East Coast of the Dead Sea, Phase 1: Suwayma, az-Zara and Umm Sidra. *Annual of the Department of Antiquities of Jordan* 40: 429-49.

Almagro, A.
1994 A Byzantine Building with a Cruciform Plan in the Citadel of Amman. *Annual of the Department of Antiquities of Jordan* 38: 417-27.

Barker, G.W. et al.
1997 The Wadi Faynan Project, Southern Jordan: A Preliminary Report on Geomorphology and Landscape Archaeology. *Levant* 29: 19-40.

Beaumont, P.
1985 Man-induced Erosion in Northern Jordan. In A. Hadidi (ed.), *Studies in the History and Archaeology of Jordan*, II: 291-96. Amman: Department of Antiquities.

Betts, A. et al.
1991 The Burqu'/Ruweishid Project: Preliminary Report on the 1989 Field Season. *Levant* 23: 7-28.

Bienkowski, P.
1991 *Treasures from an Ancient Land: The Art of Jordan*. Stroud: Alan Sutton.

Bikai, Pierre
1996 Petra Church Project, Petra Papyri. *Annual of the Department of Antiquities of Jordan* 40: 487-89.

Bikai, Patricia
1995 Petra Ridge Church Project. *American Journal of Archaeology* 99: 530-31.

Bikai, Patricia and T.A. Dailey
1996 *Madaba Cultural Heritage*. Amman: American Center of Oriental Research.

Bisheh, G.
1986 Qasr al-Hallabat: A Summary of the 1984 and 1985 Excavations. *Archiv für Orientforschung* 33: 158-62.

Brashler, J.
1995 The 1994 Umm al-Jimal Cemetery Excavations: AREA AA and Z. *Annual of the Department of Antiquities of Jordan* 39: 457-69.

Burnett, A.
1987 *Coinage in the Roman World*. London: Seaby.

Bury, J.B.
1958 *History of the Later Roman Empire*. 2 vols. New York: Dover.

Butler, H.C.
1919 *Division II, Architecture: Section A, Southern Syria. Syria: Publications of the Princeton University Archaeological Expeditions to Syria in 1904–5 and 1909*. Leiden: E.J. Brill.

Canova, R.
1954 *Iscrizioni e monumenti protocristiani del Paese di Moab*. Rome: Vatican City.

Cheyney, M.
1995 Umm al-Jimal 1993: A Cist Burial. *Annual of the Department of Antiquities of Jordan* 39: 447-55.

Conrad, L.I.
1986 The Plague in Bilad al-Sham in Pre-Islamic Times. In M.A. Bakhit and M. Asfour (eds.), *Proceedings of the Symposium on Bilad al-Sham during the Byzantine Period II (English)*, 143-63. Amman: University of Jordan.

Da Costa, K.
forthcoming Byzantine and Early Islamic Lamps: Typology and Distribution. In E. Villeneuve and P. Watson (eds.), *Byzantine and Early Islamic Ceramics in Syria-Jordan (IVth–VIIIth Centuries)*. International Colloquium, Amman 1993. Beirut: IFAPO.

De Vries, B.
1973 The North Mausoleum at Umm Qeis. *Annual of the Department of Antiquities of Jordan* 18: 77.
1985 Urbanization in the Basalt Region of North Jordan in Late Antiquity: The Case of Umm el-Jimal. In A. Hadidi (ed.), *Studies in the History and Archaeology of Jordan*, II: 249-56. Amman: Department of Antiquities.
1993 The Umm al-Jimal Project, 1981–1992. *Annual of the Department of Antiquities of Jordan* 37: 433-56.

De Vries, B. (ed.)
1998 *Umm el-Jimal: A Frontier Town and its Landscape in Northern Jordan*, I. Journal of Roman Archaeology Supplementary Series 26. Portsmouth, RI: Journal of Roman Archaeology.

Dussart, O.
1989 Recherche sur le verre antique en Jordanie et Syrie du Sud. *Contribution française à l'archéologie jordanienne*, 105-108.

Fiema, Z.
1991 Economics, Administration and Demography of Late Roman and Byzantine Southern Transjordan. Unpublished PhD dissertation, Department of Anthropology, University of Utah.
1995 Culture History of the Byzantine Ecclesiastical Complex at Petra. *American Center of Oriental Research Newsletter* 7.2: 1-3.

Findlater, G. et al.
1998 The Wadi Faynan Project: The South Cemetery Excavation, Jordan 1996: A Preliminary Report. *Levant* 30: 69-83.

Fisher, C.S. and C.G. McCown
1931 Jerash-Gerasa 1930. *Annual of the American School of Oriental Research* 11: 1-62.

Flanagan, J., D. McCreery and K. Yassine
1994 Tell Nimrin: Preliminary Report on the 1993 Season. *Annual of the Department of Antiquities of Jordan* 38: 205-44.

Freestone, I.C., K.D. Politis and C.P. Stapleton
forth- The Byzantine Glazed Pottery from Deir 'Ain 'Abata,
coming Jordan. In E. Villeneuve and P. Watson (eds.), *Byzantine and Early Islamic Ceramics in Syria-Jordan (IVth-VIIIth Centuries)*. International Colloquium, Amman 1993. Beirut: IFAPO

Gentelle, P.
1984 Traces de champs autour de Petra. In F. Villeneuve (ed.), *Contribution française à l'archéologie jordanienne*, 24-27. Amman: IFAPO.

Geraty, L.T. and Ø. LaBianca
1985 The Local Environment and Human Food-Procuring Strategies in Jordan: The Case of Tell Hesban and its Surrounding Region. In A. Hadidi (ed.), *Studies in the History and Archaeology of Jordan*, II: 323-30. Amman: Department of Antiquities.

Glueck, N.
1934, *Explorations in Eastern Palestine*, I, II, III, IV. Annual of the
1935, American Schools of Oriental Research 14, 15, 18-19,
1939, 25-28. New Haven, CT: American Schools of Oriental
1951 Research.

Graf, D.
1992 Nabataean Settlements and Roman Occupation in Arabia Petraea. In M. Zaghloul et al. (eds.), *Studies in the History and Archaeology of Jordan*, IV: 253-60. Amman: Department of Antiquities.

Hauptmann, A.
1986 Archaeometallurgical and Mining-Archaeological Studies in the Eastern 'Arabah, Feinan Area, 2nd Season. *Annual of the Department of Antiquities of Jordan* 30: 415-19.

Hauptmann, A. and G. Weisgerber
1992 Periods of Ore Exploitation and Metal Production in Feinan. In M. Zaghloul et al. (eds.), *Studies in the History and Archaeology of Jordan*, IV: 61-66. Amman: Department of Antiquities.

Hayes, J.W.
1972 *Late Roman Pottery*. London: British School at Rome.

Holmgren, R. and A. Kaliff
1997 The 1995–1996 Excavation of Dayr al-Qattar al-Byzanti: A Preliminary Report. *Annual of the Department of Antiquities of Jordan* 41: 321-40.

Homès-Fredericq, D. and J.B. Hennessy (eds.)
1986, *Archaeology of Jordan*, I, II. Leuven: Peeters.
1989

Humbert, J.-B.
1997 Les figurines byzantines du cimetière de Samra. *Le Monde de la Bible* 104: 55.

1989 El-Fedein/Mafraq. In F. Villeneuve (ed.), *Contribution française à l'archéologie jordanienne*, 125-31. Amman: IFAPO.

Humbert, J.-B. and A. Desreumaux
1989 Khirbet es-Samra. In F. Villeneuve (ed.), *Contribution française à l'archéologie jordanienne*, 113-21. Amman: IFAPO.

Isaac, B.
1990 *The Limits of Empire: The Roman Army in the East*. Oxford: Clarendon Press.

Jones, A.H.M.
1973 *The Later Roman Empire 284–602: A Social, Economic and Administrative Survey*. Oxford: Basil Blackwell.

Kaegi, W.E.
1992 *Byzantium and the Early Islamic Conquests*. Cambridge: Cambridge University Press.

Kaimio, M. and L. Koenen
1997 Reports on Decipherment of Petra Papyri (1996/7). *Annual of the Department of Antiquities of Jordan* 41: 449-62.

Kehrberg, I. and A. Ostrasz
1994 Gerasa, Hippodrome. *American Journal of Archaeology* 98: 546-47.
1997 A History of Occupational Changes at the Site of the Hippodrome of Gerasa. In M. Zaghloul et al. (eds.), *Studies in the History and Archaeology of Jordan*, IV: 167-73. Amman: Department of Antiquities.

Kennedy, D.L.
1982 *Archaeological Explorations on the Roman Frontier in North-East Jordan*. British Archaeological Reports, International Series. Oxford: British Archaeological Reports.
1997 Roman Roads and Routes in North-East Jordan. *Levant* 29: 71-93.

Kennedy, D.L. and P. Freeman
1995 Southern Hauran Survey 1992. *Levant* 27: 39-73.

Killick, A.C.
1983 Udhruh, 1980, 1981, 1982 Seasons: A Preliminary Report. *Annual of the Department of Antiquities of Jordan* 27: 231-44.
1986 Udhruh and the Southern Frontier. In P. Freeman and D. Kennedy (eds.), *The Defence of the Roman and Byzantine East*, 431-46. British Archaeological Reports, International Series. Oxford: British Archaeological Reports.

King, G.R.D. et al.
1982, Survey of Byzantine and Islamic Sites in Jordan. *Annual*
1983, *of the Department of Antiquities of Jordan* 26: 85-95; 27:
1987, 385-436; 31: 439-60; 33:199-217.
1989

Koutsoukou, A. et al.
1997 *The Great Temple of Amman: The Excavations*. Amman: American Center of Oriental Research.

Kraeling, C.H.
1938 *Gerasa, City of the Decapolis*. New Haven, CT: American Schools of Oriental Research.

LaBianca, Ø.S. and L. Lacelle (eds.)
1986 *Environmental Foundations: Studies of Climatological, Geological, Hydrological, and Phytological Conditions in Hesban and Vicinity*. Berrien Springs, MI: Andrews University.

Lenzen, C.J. and E.A. Knauf
1987 Beit Ras/Capitolias. A Preliminary Evaluation of the Archaeological and Textual Evidence. *Syria* 64: 21-46.

MacAdam, H.I.
1994 Settlements and Settlement Patterns in Northern and Central Transjordania, ca 550–ca 750. In G. King and A. Cameron (eds.), *The Byzantine and Early Islamic Near East*. II. *Land Use and Settlement Patterns*. Princeton, NJ: Darwin.

MacDonald, B.
1980 The Hermitage of John the Abbot at Hammam 'Afra, Southern Jordan. *Liber Annuus* 30: 351-64.
1996 Survey and Excavation: A Comparison of Survey and Excavation Results from Sites of the Wadi al-Hasa and the Southern al-Aghwar and Northeast 'Arabah Archaeological Surveys. *Annual of the Department of Antiquities of Jordan* 40: 323-37.

MacDonald, B. *et al.*
1988 *The Wadi al-Hasa Archaeological Survey 1979–1983, West-Central Jordan*. Waterloo, ON: Wilfrid Laurier University.

MacDonald, B. *et al.*
1992 *The Southern Ghors and Northeast 'Arabah Archaeological Survey*. Sheffield Archaeological Monographs 5. Sheffield: J.R. Collis.

Mare, W.H.
1984 The 1982 Season at Abila of the Decapolis. *Annual of the Department of Antiquities of Jordan* 28: 39-54.

Mayerson, P.
1989 Saracens and Romans: Micro-Macro Relationships. *Bulletin of the American Schools of Oriental Research* 274: 71-79.

McNicoll, A.W., R.H. Smith and J.B. Hennessy
1982 *Pella in Jordan*, I. Canberra: Australian National Gallery.

McNicoll, A.W. *et al.*
1992 *Pella in Jordan*, II. Sydney: Meditarch.

Melkawi, A., K. 'Amr and D. Whitcomb
1994 The Excavations of Two Seventh Century Pottery Kilns at Aqaba. *Annual of the Department of Antiquities of Jordan* 38: 447-68.

Meyer, C.
1988 Glass from the North Theatre Byzantine Church, and Soundings at Jerash, 1982–1983. In W.E. Rast (ed.), *Bulletin of the American Schools of Oriental Research Supplement* 25, 175-222. Baltimore, MD: American Schools of Oriental Research.

Miller, J.I.
1969 *The Spice Trade of the Roman Empire, 29 BC–AD 641*. Oxford: Clarendon Press.

Miller, J.M.
1991 *Archaeological Survey of the Kerak Plateau*. Atlanta, GA: Scholars Press.

Nabulsi, A.J. and J.-B. Humbert
1996 Excavations in the Byzantine Cemetery at Khirbet as-Samra, Site B—1995. *Annual of the Department of Antiquities of Jordan* 40: 491-93.

Northedge, A. (ed.)
1992 *Studies on Roman and Islamic Amman*, I. Oxford: Oxford University Press.

Oleson, J.P., K. 'Amr and R. Schick
1992 The Humeima Excavation Project: Preliminary Report of the 1991 Season. *Echos du monde classique/Classical Views* 36, NS 11: 137-69.

Oleson, J.P. *et al.*
1995 Preliminary Report on the Humeima Excavation Project 1993. *Annual of the Department of Antiquities of Jordan* 39: 317-54.

Palumbo, G. *et al.*
1993 The Cultural Resources Management Project in Jordan: Salvage Excavations at Tell Faysal, Jarash. *Annual of the Department of Antiquities of Jordan* 37: 89-112.

Parker, S.T.
1986 *Romans and Saracens: A History of the Arabian Frontier*. Winona Lake, IN: Eisenbrauns.
1992 The Limes and Settlement Patterns in Central Jordan in the Roman and Byzantine Periods. In M. Zaghloul *et al.* (eds.), *Studies on the History and Archaeology of Jordan*, IV: 321-25. Amman: Department of Antiquities.
1995 The Typology of Roman and Byzantine Forts and Fortresses in Jordan. In K. 'Amr *et al.* (eds.), *Studies in the History and Archaeology of Jordan*, V: 251-60. Amman: Department of Antiquities.
1996 The Roman Aqaba Project: The 1994 Campaign. *Annual of the Department of Antiquities of Jordan* 40: 231-57.
1998 The Roman 'Aqaba Project: The 1996 Campaign. *Annual of the Department of Antiquities of Jordan* 42: 375-94.

Parker S.T. (ed.)
 1987 *The Roman Frontier in Central Jordan: Interim Report on the Limes Arabicus Project 1980–1985.* British Archaeological Reports, International Series. Oxford: British Archaelogical Report.

Parr, P.J.
 1986 The Last Days of Petra. In M. Bakhit and M. Asfour (eds.), *Proceedings of the Symposium on Bilad al-Sham during the Byzantine Period II (English Section)*, 192-205. Amman: University of Jordan.

Peterman, G. and R. Schick
 1996 The Monastery of Saint Aaron. *Annual of the Department of Antiquities of Jordan* 40: 473-80.

Piccirillo, M.
 1985 Rural Settlements in Byzantine Jordan. In A. Hadidi (ed.), *Studies in the History and Archaeology of Jordan*, II: 257-61. Amman: Department of Antiquities.
 1987 The Jerusalem–Esbus Road and its Sanctuaries in Transjordan. In A. Hadidi (ed.), *Studies in the History and Archaeology of Jordan*, III: 165-72. Amman: Department of Antiquities.
 1989 Chiese e i mosaici di Um er-Rasas—Kastron Mefaa. In M. Piccitillo (ed.) *Chiese e mosaici di Madaba*, 269-308. Jerusalem: Franciscan Printing Press.
 1993 *The Mosaics of Jordan.* Amman: American Center of Oriental Research.

Politis, K.D.
 1997 Excavations and Restorations at Dayr 'Ayn 'Abata 1995. *Annual of the Department of Antiquities of Jordan* 41: 341-50.
 1998 Survey and Rescue Collections in the Ghor as-Safi. *Annual of the Department of Antiquities of Jordan* 42: 627-34.

Prag, K. and H. Barnes
 1996 Three Fortresses on the Wadi Kafrain, Jordan. *Levant* 28: 41-61.

Rast, W. and T. Schaub
 1974 Survey of the Southeastern Plain of the Dead Sea, 1973. *Annual of the Department of Antiquities of Jordan* 19: 5-53.

Rothenberg, B.
 1970 An Archaeological Survey of South Sinai: First Season 1967/1968. *Palestine Exploration Quarterly* 102: 4-28.

Ruben, I., R.H. Barnes and R. Kana'an
 1997 Mapping and Preliminary Survey in Wadi Faynan South Jordan. *Annual of the Department of Antiquities of Jordan* 41: 433-52.

Sapin, J.
 1985 Prospection géo-archéologique de l'Ajlûn 1981–82: Example de recherche intégrante. In A. Hadidi (ed.), *Studies in the History and Archaeology of Jordan*, II: 217-27. Amman: Department of Antiquities.

Sartre, M.
 1982 *Trois études sur l'Arabie romaine et byzantine.* Revue des études latines. Brussels: Latomus.
 1985 Le peuplement et le développement du Hauran antique à la lumiere des inscriptions grècques et latines. In J.-M. Dentzer (ed.), *Hauran*, I: 189-202. Paris: Geuthner.
 1987 Le Hawran byzantin à la veille de la conquête musulmane. In M.A. Bakhit (ed.), *Proceedings of the Second Symposium on the History of Bilad al-Sham during the Early Islamic Period up to 40 A.H./640 AD*, I: 155-67. Amman: University of Jordan.

Sauer, J.A.
 1973 *Heshbon Pottery 1971.* Berrien Springs, MI: Andrews University.

Schick, R.
 1992 Jordan on the Eve of the Muslim Conquest AD 602-634. In P. Canivet and J.-P. Rey-Coquais (eds.), *La Syrie de Byzance à l'Islam, VII–VIII siècles: Actes du colloque international, Lyon–Paris, 1990*, 107-19. Damascus: IFAPO.
 1994 The Settlement Pattern of Southern Jordan: The Nature of the Evidence. In G. King and A. Cameron (eds.), *The Byzantine and Early Islamic Near East. II. Land Use and Settlement Patterns*, 133-54. Princeton, NJ: Darwin.
 1995 *The Christian Communities of Palestine from Byzantine to Islamic Rule: A Historical and Archaeological Study.* Princeton, NJ: Darwin.

Seigne, J.
 1992 Jérash romaine et byzantine: Développement urbain d'une ville provinciale orientale. In M. Zaghloul et al. (eds.), *Studies in the History and Archeology of Jordan*, IV: 331-41. Amman: Department of Antiquities.

Shahid, I.
 1984 *Byzantium and the Arabs in the Fourth Century.* Washington, DC: Dumbarton Oaks.
 1989 *Byzantium and the Arabs in the Fifth Century.* Washington, DC: Dumbarton Oaks.

Smith, A.M., M. Stevens and T.M. Niemi
 1997 The Southeast Araba Archaeological Survey: A Preliminary Report of the 1994 Season. *Bulletin of the American Schools of Oriental Research* 305: 45-71.

Smith, R.H.
 1973 *Pella of the Decapolis*, I. Wooster, OH: The College of Wooster.

Smith, R.H. and L.P. Day
 1989 *Pella of the Decapolis*, II. Wooster, OH: The College of Wooster.

Trimingham, J.S.
 1979 *Christianity among the Arabs in Pre-Islamic Times.* London: Longman.

Tushingham, A.D.
- 1972 *The Excavations at Dibon (Dhibân) in Moab: The Third Campaign 1952–53.* Cambridge, MA: American Schools of Oriental Research.

Walmsley, A.G.
- 1998 Gharandal in Jibal: First Season Report. *Annual of the Department of Antiquities of Jordan* 42: 433-41.

Watson, P.M.
- 1992 Change in Foreign and Regional Economic Links with Pella in the Seventh Century: The Ceramic Evidence. In P. Canivet and J.-P. Rey-Coquais (eds.), *La Syrie de Byzance à l'Islam, VII–VIII siècles: Actes du colloque international, Lyon–Paris 1990*, 233-47. Damascus: Institut français de Damas.

Watson, P.M. and M. O'Hea
- 1996 The Pella Hinterland Survey 1994: Preliminary Report. *Levant* 28: 63-76.

Whitby, M.
- 1995 Recruitment in Roman Armies from Justinian to Heraclius (ca. 565–615). In A. Cameron (ed.), *The Byzantine and Early Islamic Near East. III. States, Resources and Armies*, 61-124. Princeton, NJ: Darwin.

Whitting, P.D.
- 1973 *Byzantine Coins.* New York: G.P. Putnam's Sons.

Zayadine, F. (ed.)
- 1986 *Jerash Archaeological Project 1981–1983*, I. Amman: Department of Antiquities.
- 1989 *Jerash Archaeological Project 1984–1988*, II. Paris: Department of Antiquities of Jordan.

Zeist, W. van
- 1985 Past and Present Environments in the Jordan Valley. In A. Hadidi (ed.), *Studies in the History and Archaeology of Jordan*, II: 199-204. Amman: Department of Antiquities.

14. Umayyad and Abbasid Periods

Donald Whitcomb

The end of antiquity?

In a volume such as this on the archaeology of Jordan, it would not have been surprising in years past to find the articles ending with the 'coming of Islam'. The dominating model was that this momentous event sealed the past and began a new, fully historical Mediaeval period. This somehow implied that full historical documentation (would that such a thing existed) somehow obviated the need for archaeology; for scholars studying the period after this defining event, archaeological methodology was seen as subsidiary to historical, a scholarly tendency commonly encountered in the study of any society that left written sources. Although written records may provide details, thoughts, reasoning, and the like, broader trends, both temporal and spatial, frequently may be clarified only through the archaeological record.

In the past several decades the archaeology of Jordan in the early Islamic period (c. 600–1000) (Figure 14.2) has witnessed a concentration of research, both conceptual and in the field, unparalleled in previous years or in the experience of other countries of the Levant. Archaeologists working in the field have amassed an evidential base unrivaled in neighbouring regions of the Middle East (Figure 14.1). These excavations, combined with interpretative analysis relating this data to historical contexts, have made major advances illustrating, for example, the urban system of this region during the transition from Late Antiquity to early Islam.

Increased attention has been paid in recent decades to the concept of Late Antiquity as a research focus, often in combination with early Islam. They have been defined in various chronological periodizations (Figure 14.2). What is clear and significant is the abandonment of the assumption of the Fall of Rome in 476 as a pivotal turning point; as Cameron notes, no significant economic or social change seems to be associated with this event or date (1993: 33). Likewise, the idea that 636 was a turning point has given way to an appreciation of the continuity of material culture reflected in the cumulative value of archaeological research on individual sites and survey patterns (e.g. MacAdam 1994, and other articles in the second volume of *The Byzantine and Early Islamic Near East*, edited by King and Cameron).

A central facet of recent studies on Late Antiquity has been settlement patterns and in particular the study of urban systems. The fate of the Classical city has been explored, and it is widely recognized at this point that those cities had, in the immediately pre-Islamic Near East, developed into a type that was no longer 'classical'. Again, this realization is the result of careful analysis, the combination of textual and archaeological resources; the doctoral study of Schick, with its very useful gazetteer, is an exemplar of this trend (1995). His volume benefits from the intensive fieldwork undertaken in Jordan, usually reported in the *Annual of the Department of Antiquities of Jordan*. If one chronicles the occurrence of articles pertaining to Islamic archaeology, one observes only a rare report until the late 1970s and a curve of activity in the late 1980s; since the hiatus of the Gulf War, Islamic archaeology represents a fairly substantial portion of the archaeology of Jordan.

Jordan has in effect become a leading venue for re-evaluating historical assumptions and testing new ideas concerning the cultural ramifications of early Islam, a crucial period in the history of the Islamic world. Work now focuses on interpretations and historical synthesis; the next and subtler campaign needs to be directed toward delineation of cultural changes and their explanations. As current research in early Islamic history questions accepted perceptions of historical sources, it becomes more and more important to consider archaeological evidence in the framing and answering of historical questions. An example of archaeological applications to the investigation of the formation of early Islam, though somewhat controversial, is found in revisionism of evidence from the Negev (Whitcomb 1995: 500-501). The Negev and much of Jordan are peripheral regions where contemporary documentation is minimal (though one may note that anepigraphic sites such as Qasr al-Mushatta and Qasr al-Kharrana, discovered almost 100 years ago, have recently revealed inscriptions: Imbert 1995 and Imbert and Bacquey 1989; cf. Urice 1987). However,

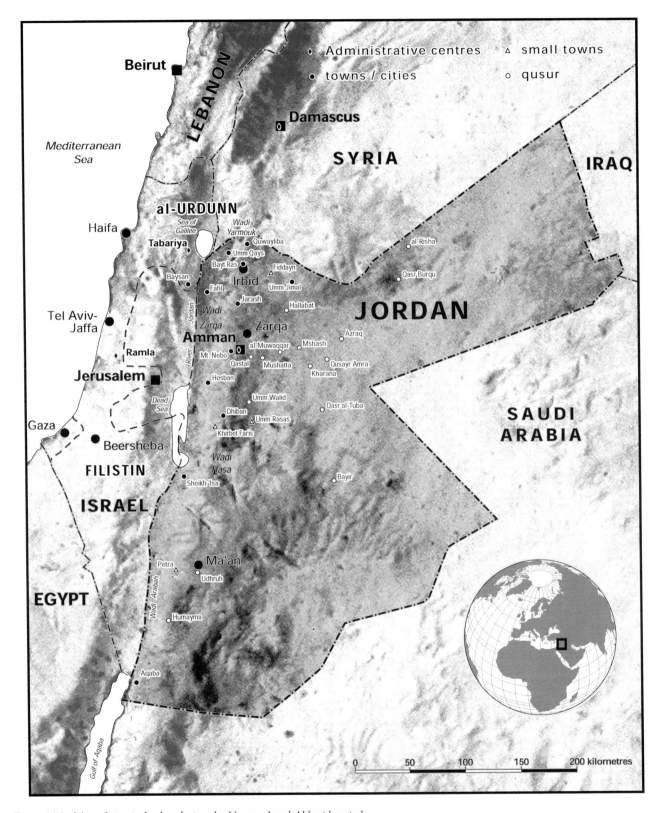

Figure 14.1. Map of sites in Jordan during the Umayyad and Abbasid periods.

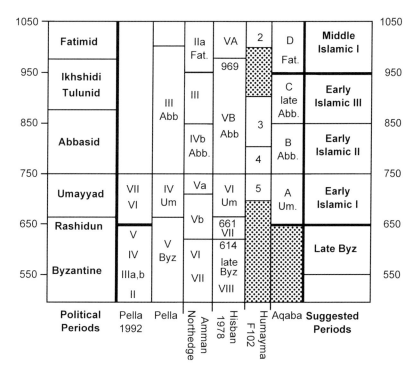

Figure 14.2. Chronological periodization of the early Islamic periods in Jordan.

The Umayyad state and Arab *Qusur*

The political structure of early Islamic Bilad ash-Sham (Greater Syria or the four modern Levantine states) was conceived in terms of *junds*, a term that is unusual and ambiguous. Following the lead of Shahid (1986), these political regions stretching 'from the sea to the desert' may be associated with the late Classical term 'theme' as indicative of a military district. These may further be associated, as is the word itself, with Persian toponymy, imposed during the reorganization during the Persian occupation (614 to 628; Whitcomb 1994a). The modern state of Jordan coincides with portions of three *junds*: Filastin (Filistia), al-Urdunn, and Dimashq, with corresponding capitals in Ramla, Tabariya, and Damascus. Jordan, or rather the Transjordanian segment of these *junds*, lay to the east and south of these political centres.

The structure of the *junds* has importance for our understanding of the role of archaeological sites in modern Jordan during the formative era of Islamic culture. Careful examination of Byzantine towns and monasteries shows, with increasing clarity, a continuation, often without apparent change, well into the Abbasid period (e.g. Madaba, Umm ar-Rasas, Zughar, and others). Likewise, cities such as 'Amman, Jarash, Pella, and Gadara also continued, although their structural organization and their relation to their dependencies bear some examination. The apparent innovations seem to lie on the eastern periphery of the *junds*, where extant settlements of earlier periods became the loci for new social relations (see Figure 14.1).

The desert had been crossed by roads and was the locus of settlements and fortifications in pre-Islamic times (usually, during the Classical period, under the rubric of the *limes*, however one may define the concept; Parker 1986). Such remains dating to Classical times must be seen in light of the Nabataean and Ghassanid political achievements which, while hardly confined to this periphery, indicate a strong role of Arab tribes in the pre-Islamic period (MacAdam 1994). An appreciation of this focus on the desert may begin with the remarkable but largely neglected study of Helms (1990). He begins with a study of Qasr Burqu, an apparently typical *qasr* (or 'desert castle', see below) of the Umayyad period; he expanded this field research by investigating the atypical site of ancient al-Risha, a collection of minimally preserved structures, individual buildings arranged in parallel lines with a mosque and large formal building. The settlement of al-Risha seems to be a sedentarized camp, but at least one of its buildings, Structure C, is parallel to the enclosures that are the core element for most of the *qusur* (as implied in the name, see Conrad 1981). Perhaps a better example of a settlement complex involves the well-excavated *qasr* at Umm al-Walid (Bujard and Haldimann 1997); the pattern of settlement, involving three *qusur*, seems to indicate a tension between unifying enclosure and social privacy dramatized in the subsidiary walls subdividing the common court.

Helms uses the morphology of al-Risha to reconstruct a crucial aspect of social organization, the process of sedentarization in the early Islamic period (a process important for many periods; Finkelstein 1995). He makes a strong case from ethnoarchaeological models of the bedouin camp for the *parembole nomadon* or *hadir* as developmental stages toward the *amsar* of the early Islamic period (Helms 1990: 39-41; Whitcomb 1994a). This model might be stronger with the term 'arab' rather than 'bedouin', since the later ethnic ramifications of 'arab' are crucial as this society develops beyond a socio-economic definition limited to the desert. Just as the architectural elements and ceramics of al-Risha come from urban sources, so the familiarity of Arab societies with urban society facilitated formation of the early Islamic state. The desert became the scene for a pastoral peasantry, that is, the affective regions of cities increased in a new or perhaps more intensive manner. Understanding 'this remarkable breakthrough in social organization' (Donner 1981: 269) has proven elusive from documentary sources. Donner's focus on nomads, or better, the tribal foundations of Arab Muslims, has been salutary but marred by neglect of the archaeological resources that might bear upon the subject.

Renewed archaeological research on the *qusur* has concentrated on broader contexts. In past decades, interpretations have shifted the balance from romantic explanations to economic factors (as caravanserai systems or desert exploitation, the latter close to Sauvaget's *latifundia*) or social organization (as interactive facilities between the settled and nomad/tribal, rather like *majlis* buildings). Thus Ghazi Bisheh excavated Qasr al-Hallabat for its stratigraphy and construction details and, more importantly, as a complex of mosque, agricultural enclosures, and quarries (1985, 1993; see also his work at Qasr Mashash 1989). On the contrary, the potential of Muwaqqar has not been realized in spite of important epigraphic and ceramic evidence (Hamilton 1946; Najjar 1989). Likewise, the efforts to record Qastal include a broader perspective, studying not only the *qasr* and mosque, but a bath, an agricultural barrage, and a cemetery (Carlier and Morin 1989; Qusayr 'Amra is a similar potential complex, Almagro et al. 1975). Another complex is the estate of the Abbasid family at Humayma (Foote 1994).

Consideration of function brings one back to more traditional approaches, including art historical analysis, for which one may note the new perspectives provided by Bacharach (1996). His construction of Marwanid political patronage relies heavily on traditional sources yet recognizes the importance of structural patterns, notably locational relationship of the palace, bath, and *dar al-imara* (aspects he develops more fully in broader contexts, 1991). The monuments concerned are recognized as set in a context of tents, that is, camps; further, the locations where they were constructed almost always had pre-Islamic structures, often ruins, psychological magnets for renewed occupation. The structural patterns of the *qusur* lead to the inescapable conclusion that these were intended to be urban, or, better, to bring an urban pattern to new regions. Further, the few decades of building occurred some 50 years after the conquest; like the Islamic urban pattern imposed on Jerusalem, this seems likely to have begun during the long rule of Mu'awiya (as governor and caliph, 640–80; Whitcomb 1995).

Abbasid revolution or devolution?

Study of archaeological sites in Jordan demonstrates how, until the mid-eighth century, this region played the role of an influential margin in the development of an imperial state. More clearly than in any other region of the early Islamic world, except potentially Arabia itself, one may observe the material remains of a new civilization that would dominate the entire Middle East for many centuries. The Abbasid revolution, and more precisely the period of changes directly following this political upheaval, produced instability expressed in material changes in the archaeological record in Jordan. It is still deceptively easy to see cultural discontinuities as regional devolution. Thus Kennedy may observe that 'there is no significant surviving architectural monument in Bilad ash-Sham from the period between the abandonment of the last Umayyad palace construction and the restoration of the Aqsa mosque by the Fatimids' (1992: 106; LaBianca [1990: 241], more dramatically, sees no settlement remains for the same period).

Oleg Grabar, who has made many fundamental contributions to the study of the *qusur*, discovered a historical fragment concerning the *qasr* or fort of Hisban (see below) and its role as a centre of Umayyad

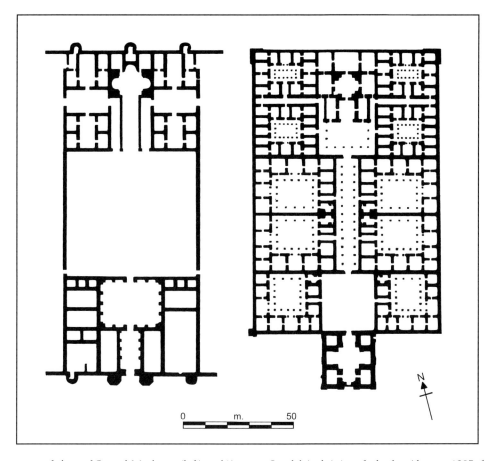

Figure 14.3. Comparison of plans of Qasr al-Mushatta (left) and 'Amman Citadel (right) (justified, after Almagro 1987: fig. 14).

partisans during the al-Faddayni revolt of 813 (1964; discussed in Cobb 1997). This historical fragment may reflect a richer settlement pattern than heretofore acknowledged for the earlier Abbasid period. The necessity to control viable Arab settlements in Jordan may have led the Abbasids to focus their attention on 'Amman and to establish this city as a regional capital and seat of a provincial governor. Following Umayyad attentions to the Balqa region, the Abbasids appointed three known governors (including Salih ibn Sulayman in 796, who may have restored the palace and citadel of 'Amman; see below). The Tulunid interim of 878–905 was followed by Abbasid restoration about 918, when the governor of 'Amman was Muhammad ibn Tughj (who later founded the Ikhshidid dynasty of Egypt in 935; Northedge 1992: 53).

To begin to understand 'Amman, one must consider the lower town, where Northedge has plausibly reconstructed a *jami'* or congregational mosque (1989). This mosque may be taken together with a putative *hammam* and *dar al-imara* to configure an early Islamic city within the remnants of Philadelphia (Whitcomb 2000: fig. 3); this urban complex was no doubt an Umayyad foundation, as Northedge suggests. Construction on the citadel is another matter; using stratigraphic elements presented by Northedge (1992; see Whitcomb 2000) and the structural elements excavated by Almagro, one may find an interpretation diametrically opposed to their Umayyad identification (Harding 1951). The focus of the 'Amman Citadel is the northern palace complex, which shows an axial orientation beginning with the Entrance Hall or *'iwan'*, then a court and street with nine residential units, and finally a court and Throne Room flanked again by four residential units. While the formation of the citadel is unlike most of the *qusur*, there is a striking resemblance to Qasr al-Mushatta. A careful comparison of Almagro's justified plan of 'Amman Citadel (1987: fig. 14) with Qasr al-Mushatta clearly shows striking parallels in size, orientation, and hierarchical arrangement of architectural elements (Figure 14.3). It may be significant that the only carved stone decoration in these structures is the façade panel at Qasr al-Mushatta and the analogous entrance pavilion at

'Amman. Grabar has argued a very plausible case on formal and stylistic grounds for the Abbasid foundation of Qasr al-Mushatta (1987).

The Entrance Hall may be seen reflexively as the north gate of the upper city, with a *hammam* to its east, transiting the palace into an urban complex. This is the startling result of Almagro and Arce's new excavations: a large court flanked by shops, more vernacular residences and streets (Northedge 1991), and, south of the court, a broad esplanade and arcade fronting a large *jami'* mosque. The combination of urban and palatial elements in a hierarchical arrangement clearly reflects the palace-city typical of Late Antiquity, as clearly described by Ćurčić (1993). This second *jami'* mosque, and indeed occupation of the citadel itself, are problematic for the Umayyad period but, along with the parallels with Qasr al-Mushatta, these features suggest that the entire 'Amman Citadel should be dated to the Abbasid period. The socio-cultural correlate of this architectural phenomenon may be found in the gathering of Umayyad estates in the hands of the Banu Salih, a family that Kennedy describes as a sub-dynasty maintaining the wealth and prosperity of Abbasid Bilad ash-Sham (1981: 74-75).

Both Qasr al-Mushatta and 'Amman Citadel have been considered in light of the so-called 'desert castles', but their more palatial character suggests they should also be studied with reference to contemporary urban entities. This leads to the necessity of examining the continuities of late Byzantine cities, particularly those of the Decapolis in modern Jordan which were typical of the Classical urban tradition and, more importantly, of transformations in Late Antiquity (Kennedy 1985). Jarash represents an important example of this phenomenon, where the temples, theatres, and baths of the Classical city had been transformed under Christian urban values exemplified in the construction of at least 15 churches, mainly in the sixth century. An early excavation, that of the bath of Placcus, noted later restorations which seem associated with numbers of Arabic coins (Kraeling 1938: 265-69). In a similar manner, the intensive archaeological research in the 1980s concentrated on reconstruction of the Classical city, recording, but not utilizing, the evidence of later Islamic occupations (Zayadine 1986, 1989).

Virtually everywhere in the city, the uppermost layers contained abundant evidence of kilns, indicating a ceramic industry and suggesting that the town was a regional production centre (Schaefer and Falkner 1986). Excavators consistently suggest that the apparently sudden cessation of production was linked with the 747/48 earthquake. An important exception to this interpretation is found in the study by Gawlikowski of a residential area near the North Decumanus that continued well into the Abbasid period, when it likewise began ceramic production. He boldly states that 'there is no evidence for the earthquake of 746–47 AD that destroyed Pella and which supposedly marked the end of Jerash as well' (1986: 115). The location of innumerable kilns, a notoriously undesirable industry, in theatres, temples, and other central locations indicates a metamorphosis in the urban structure of Jarash. The mosque found near the Cardo is not only small but also rather doubtful (Naghaweh 1982). The early Islamic city of Jarash may require examination of the wider cityscape, such as advanced by Parapetti (1993: fig. 35). The dual orientations he delineates suggest that the walled area east of the river, admittedly not fully occupied in earlier periods, may have been the *misr* or Islamic town in the Umayyad and Abbasid periods (as well as subsequent times).

Pella (Fihl) is another Classical city with a complex history (and archaeology) stretching into the early Islamic period. The excavations of Smith produced ample ceramic evidence of this continuity (1973), while detailing residential and civic institutions of the city. The excavations of the University of Sydney added a specific concern with the Islamic periods in 1979, investigating the residential sectors III, IV, VIII, and IX of the Umayyad period. Though other earthquake events are periodically cited (e.g. 659/60; note Russell's caveats on earthquakes, 1985), the destruction caused by the 747/48 catastrophe seems clear in the residential areas and in the Civic Complex. The Civic Complex is a particularly interesting urban centre with religious and commercial aspects, the latter dependent on the identification of a northern caravanserai (Smith and Day 1989; Walmsley 1992). Neither here nor in the Tall al-Husn complex (Watson 1992) is there clear evidence of major changes or the introduction of new 'Islamic' elements during the Umayyad period: nor was there an obvious continuation into the Abbasid period, as Walmsley noted:

> The delayed recognition of 'Abbasid settlement levels at Fihl is attributable to a number of factors. In particular there is a marked reluctance to acknowledge any permanent occupation of the site after the late Umayyad earthquake, a view that continues to find support from Smith... Furthermore both Sydney and Wooster have overlooked the remains of the successive Islamic settlements in the centre of the tell by concentrating on the east and west ends of the

mound. Finally as the 'Abbasid houses were probably demolished to make way for the Ayyubid/Mamluk village... the remains of this settlement are obscured by later human activity on the tell (Walmsley 1987: 114-15).

More recent research has uncovered expanded settlement, literally a suburb to the north of the main mound, Khirbat Fahl (Walmsley 1991). This settlement seems to be a new foundation associated with Abbasid period ceramics, an important key for defining this corpus (see below). While the function of these structures is not completely clear, the orthogonal plans with streets suggest an expansion of urban planning typical of other Abbasid foundations (Whitcomb 1995). Similar histories might be generated for Umm Qays (Gadara; Nielsen et al. 1993), Abila (Quwayliba; Fuller 1987), Umm al-Jimal (Knauf 1984), and Bayt Ras (Capitolias; Lenzen and Knauf 1987) in the north of Jordan (see Figure 14.1). In each case, one finds a provincial city of Late Antiquity continuing as a regional centre; structural and architectural changes began before the arrival of Islam, gradually unfolding into a still poorly defined Islamic city. The apparent absence of an Abbasid chapter in their urban histories may be a matter of self-fulfilling prophecy demanding renewed archaeological attention.

The definition of this urban tradition in the early Islamic period for other sites along the western edge of Jordan (Figure 14.1) must await further research and synthesis; though these sites are manifestly complex, the progress in understanding them has been significant. Within this context, the study of ceramic evidence was a fundamental technique for advances in chronological and cultural understandings. Perhaps the most important single accomplishment in this field was the contribution by Sauer (1973), which attempted a synthesis of ceramic diagnostics for the Islamic as well as earlier periods. His formulation of Islamic ceramics stood for a decade or longer, until evidence amassed from Hisban and other sites in Jordan led to necessary revisions in this first consensus. The heritage of earlier archaeological practices in Palestine, specifically the reliance on a pottery 'reader', has meant often-discarded evidence and hence unverifiable conclusions and idiosyncratic interpretations. Many of the excavations discussed here consciously addressed this problem, for example, Jarash, Pella, and al-'Aqaba (Ayla, see below). For the Umayyad and Abbasid periods, changes were connected with the problem of distinguishing Byzantine and earliest Islamic production, and secondly, the problem of identification of Abbasid wares. In both cases, misidentifications led to accentuation of the earlier period (i.e., Byzantine, Umayyad) and underestimation of settlement during the Abbasid and later periods for the large number of intensive regional surveys available in Jordan.

Research on the Pella ceramics has produced one of the clearest patterns for early Islamic ceramic changes. Walmsley builds upon the perceptive studies of Byzantine ceramics by Watson (1992) and acknowledges the termination of a number of types at the end of the sixth or early seventh centuries (1995: Wares 1-5). During the period 660 to 900, there are four stylistic phenomena:

1. Dark grey, metallic wares (Wares 11-14), some with white paint, continue throughout, associated with cooking pots and basins. These may be related to contemporary wares from Khirbat al-Mafjar (Whitcomb 1988b) as well as the manufactures of Jarash (Schaefer and Falkner 1986).
2. Cream wares appear around the mid-eighth century and may be an innovation of the Abbasid period (Wares 7 and 9); the Mahesh wares of the south (Whitcomb 1989c) may have influenced this introduction.
3. Red-painted wares (Ware 8) appear at about the same time as a regional Abbasid type; its apparent abundance at Jarash signals that site as a major production centre. Fine painted 'palace wares' and the 'fine Byzantine' cups (Ware 10) seem later elements within this tradition.
4. By the ninth century one sees the introduction of Abbasid ceramics, an early Syrian tradition of glazed ware, fine cream wares, and steatite imitations (Wares 15-19); these types form the classic definition of 'Mefjer wares', though the term needs to be used with caution. One may note that this understanding of Abbasid ceramics stands separate from the Samarra ceramic tradition, in part antecedent and in part a contemporary regional tradition of some enterprise.

The Seljuq/Fatimid epilogue

The archaeological story of Pella ends abruptly in the Abbasid period, only to pick up with the Mamluk mosque and village. Recognition of Abbasid settlement and sherds has significantly narrowed the divide between the Umayyad and the Ayyubid; as has been noted, 'there is more than enough [remaining] pottery to fill the gap...' (Whitcomb 1992). Within the past decade, the acceptance of Abbasid occupation layers and settlement achievements has resulted in the transfer of the gap, decline, abandonment or abatement to the late ninth and tenth centuries. As in the earlier periods, one cannot divorce the history of Transjordan from Palestine. This period marks the beginning of a new regional phenomenon, the hegemonic sequence of Egyptian dynasties; these are the Tulunids, Ikhshidids and Fatimids of the Early Islamic 3 and Middle Islamic 1 archaeological traditions (Figure 14.2). (These are followed by the Ayyubids and Mamluks, after the brief interruption.) Given the strength of these mature Islamic cultures, it is tempting to let peripheries decline to invisibility.

Al-'Aqaba (or Ayla, as it was known in this period) was by its very nature on this periphery but in a way that documents the interactions of Bilad ash-Sham, Egypt and Arabia. The excavations at Ayla have contributed to understanding aspects of the *misr* type of early Islamic urban development and early Islamic artifacts (Melkawi *et al.* 1994; Whitcomb 1989a). Ayla continued and indeed was revived during the Abbasid period; the latest periods (Phases C and D) document the late Abbasid and Fatimid occupations (Whitcomb 1994b). The Fatimid residence is an example of architectural transformation, where the crossing of the axial streets was marked by a tetrapylon that was converted into a large house (Whitcomb 1988a). The ceramics in this area and elsewhere duplicate Abbasid types outlined at Pella (although with clear regional variations, particularly the Mahesh and Coptic glazed wares; Whitcomb 1989b, 1989c). Since Ayla was a port, a number of Samarra ceramic wares are also present. The late Abbasid (period C, or the Ikhshidid and Tulunid periods) features ceramics of an interregional style, only recently being isolated at Alexandria, Askalon, and Caesarea. The glazed wares reflect commercial relationships and, on the other hand, the development of regional styles in Palestine (Whitcomb 1990–91). By the eleventh century, luxury trade in Chinese ceramics and Fatimid lustre wares are balanced with hand-made 'Tupper wares' in contexts of trash accumulation, suggesting a mixture of prosperity and decline.

The late ninth and tenth centuries witnessed transformations in religion and culture not unlike those seen in the sixth and early seventh centuries. As in the early transition, efforts at periodization are often misleading with no clear historical date. In general terms, by the beginning of the eleventh century, categories of material culture have changed enough to justify the beginning of a middle Islamic archaeological period. In terms of Ayla, the Qarmati revolt from 901 to 906 resulted in a battle fought near al-'Aqaba. Then the Qarmatians, allied with Jarrahids, captured and administered the southern region from Ramla in 968, and again they defeated a Fatimid army at al-'Aqaba in 982. Finally, the Jarrahids attacked and sacked Ayla in 1024, an action that may have caused the deposition of a hoard of Sijilmasa dinars (Whitcomb 1994b). This last example is perhaps clear archaeological evidence of events reflecting a broader political and economic instability. Once again, the spectre of thundering hoards of nomads may be overdrawn and suggestions of decline and collapse mask complex social and religious movements. The Qarmatians were apparently much more than successful desert warriors and the Jarrahids may have been tribal but not necessarily Bedouins. Once again, the history of Jordan may become an illuminating peripheral region through contributions of archaeological interpretation.

References

Amalgro, A.
    1987   Origins and Repercussions of the Architecture of the Umayyad Palace in Amman. In A. Hadidi (ed.), *Studies in the History and Archaeology of Jordan*, III: 181-91. Amman: Department of Antiquities.

Almagro, M. *et al.*
    1975   *Qusayr 'Amra: Residencia y banos Omeyad en el Desierto de Jordania.* Madrid: Ministerio de Asuntos Exteriores.

Bacharach, J.L.
    1991   Administrative Complexes, Palaces, and Citadels: Changes in the Loci of Medieval Muslim Rule. In I. Bierman *et al.* (eds.), *The Ottoman City and Its Parts: Urban Structure and Social Order*, 105-22. New Rochelle, NY: Caratzas.

    1996   Marwanid Umayyad Building Activities: Speculations on Patronage. *Muqarnas* 13: 27-44.

Bisheh, G.
    1985   Qasr al-Hallabat: An Umayyad Desert Retreat or Farm-land. In A. Hadidi (ed.), *Studies in the History and Archaeology of Jordan*, II: 263-65. Amman: Department of Antiquities.

    1989   Qasr Mshash and Qasr 'Ayn al-Sil: Two Umayyad Sites in Jordan. In M.A. Bakhit and R. Schick (eds.), *The Fourth International Conference on the History of Bilad al-Sham during the Umayyad Period*, II: 81-103. Amman: University of Jordan.

    1993   From Castellum to Palatium: Umayyad Mosaic Pavements from Qasr al-Hallabat in Jordan. *Muqarnas* 10: 49-56.

Bujard, J. and M.-A. Haldimann
1997 Umm al-Walid et Khan az-Zabib, cinq qusur omeyyades et leurs mosquées revisités. *Annual of the Department of Antiquities of Jordan* 41: 351-74.

Cameron, A.
1993 *The Mediterranean World in Late Antiquity,* AD 395–600. London: Routledge.

Carlier, P. and F. Morin
1989 Qastal al-Balqa': An Umayyad Site in Jordan. In M.A. Bakhit and R. Schick (eds.), *The Fourth International Conference on the History of Bilad al-Sham during the Umayyad Period*, II: 104-39. Amman: University of Jordan.

Cobb, P.M.
1997 White Banners: Contention in 'Abbasid Syria, 750-877. Unpublished PhD dissertation, The University of Chicago.

Conrad, L.I.
1981 The Qusur of Medieval Islam: Some Implications for the Social History of the Near East. *al-Abhath* 29: 7-23.

Čurčić, S.
1993 Late-Antique Palaces: The Meaning of Urban Context. *Ars orientalis* 23: 67-90.

Donner, F. McG.
1981 *The Early Islamic Conquests*. Princeton, NJ: Princeton University Press.

Finkelstein, I.
1995 *Living on the Fringe: The Archaeology and History of the Negev, Sinai and Neighboring Regions in the Bronze and Iron Ages*. Sheffield: Sheffield Academic Press.

Foote, R.
1994 The 'Abbasids and their Residence in Humeima. *al-'Usur al-Wusta* 6.1: 1-3.

Fuller, M.J.
1987 Abila of the Decapolis: A Roman-Byzantine City in Transjordan. Unpublished PhD dissertation, Washington University, St. Louis.

Gawlikowski, M.
1986 A Residential Area by the South Decumanus. In F. Zayadine (ed.), *Jerash Archaeological Project 1981–1983*, I: 107-36. Amman: Department of Antiquities.

Grabar, O.
1964 A Small Episode of Early 'Abbasid Times and Some Consequences. *Eretz-Israel* 7: 44*-47*.
1987 The Date and Meaning of Mshatta. *Dumbarton Oaks Papers* 41: 243-47.

Hamilton, R.W.
1946 An Eighth-Century Water-Gauge at al Muwaqqar. *Quarterly of the Department of Antiquities of Palestine* 12: 70-72.

Harding, G.L.
1951 Excavations on the Citadel, Amman. *Annual of the Department of Antiquities of Jordan* 1: 7-16.

Helms, S.
1990 *Early Islamic Architecture of the Desert: A Bedouin Station in Eastern Jordan*. Edinburgh: Edinburgh University Press.

Humbert, J.-B.
1989 El-Fedein/Mafraq. In F. Villeneuve (ed.), *Contribution française à l'archéologie jordanienne 1989*, 124-31. Amman: IFAPO.

Imbert, F.
1995 Inscriptions et espaces d'écriture au Palais d'al-Kharrana en Jordanie. In G. Bisheh (ed.), *Studies in the History and Archaeology of Jordan*, V: 403-16. Amman: Department of Antiquities.

Imbert, F. and S. Bacquey
1989 Sept graffiti arabes au palais de Musatta. *Annual of the Department of Antiquities of Jordan* 33: 259-67.

Kennedy, H.
1981 *The Early Abbasid Caliphate: A Political History*. London: Croom Helm.
1985 From Polis to Madina: Urban Change in Late Antique and Early Islamic Syria. *Past and Present* 106: 3-27.
1992 Nomads and Settled People in Bilad al-Sham in the Fourth/Ninth and Fifth/Tenth Centuries. In M.A. Bakhit and R. Schick (eds.), *Bilad al-Sham during the Abbasid Period*, 105-13. Amman: University of Jordan.

Knauf, E.A.
1984 Umm el-Jimal: An Arab Town in Late Antiquity. *Revue Biblique* 91: 578-86.

Kraeling, C.H. (ed.)
1938 *Gerasa, City of the Decapolis*. New Haven, CT: American Schools of Oriental Research.

LaBianca, Ø.S.
1990 *Sedentarization and Nomadization: Food System Cycles at Hesban and Vicinity in Transjordan*. Hesban I. Waltham, MA: Brandeis University.

Lenzen, C.J. and E.A. Knauf
1987 Beit Ras/Capitolias: A Preliminary Evaluation of the Archaeological and Textual Evidence. *Syria* 64: 21-46.

MacAdam, H.I.
1994 Settlements and Settlement Patterns in Northern and Central Transjordania, ca 550–ca 750. In G.R.D. King and A. Cameron (eds.), *The Byzantine and Early Islamic Near East*. II. *Land Use and Settlement Patterns*, 49-94. Princeton, NJ: Darwin.

Melkawi, A. et al.
1994 The Excavation of Two Seventh Century Pottery Kilns at Aqaba. *Annual of the Department of Antiquities of Jordan* 38: 447-68.

Naghaweh, A.
1982 An Umayyad Mosque at Jerash. *Annual of the Department of Antiquities of Jordan* 26: 20-22.

Najjar, M.
1989 Abbasid Pottery from el-Muwaqqar. *Annual of the Department of Antiquities of Jordan* 33: 305-22.

Nielsen, I. et al.
1993 *Gadara-Umm Qes. III. Die byzantinischen Thermen*. Wiesbaden: Otto Harrassowitz.

Northedge, A.
1989 The Umayyad Mosque of Amman. In M.A. Bakhit and R. Schick (eds.), *The Fourth International Conference on the History of Bilad al-Sham during the Umayyad Period*, II: 140-63. Amman: University of Jordan.
1991 The Citadel of 'Amman in the Abbasid Period. In M.A. al-Bakhit and R. Schick (eds.), *The Fifth International Conference on the History of Bilad al-Sham during the Abbasid Period*, 179-94. Amman: University of Jordan.
1992 *Studies on Roman and Islamic 'Amman. I. History, Site and Architecture*. Oxford: Oxford University Press.

Parapetti, R.
1993 Gerasa, Town Plan. *American Journal of Archaeology* 97: 500-501.

Parker, S.T.
1986 *Romans and Saracens: A History of the Arabian Frontier*. American Schools of Oriental Research Dissertation Series 6. Winona Lake, IN: Eisenbrauns.

Russell, K.W.
1985 The Earthquake Chronology of Palestine and Northwest Arabia from the 2nd through the mid-8th Century AD *Bulletin of the American Schools of Oriental Research* 260: 37-59.

Sauer, J.A.
1973 *Heshbon Pottery, 1971*. Berrien Springs, MI: Andrews University.

Schaefer, J. and R.K. Falkner
1986 An Umayyad Potters' Complex in the North Theater, Jerash. In F. Zayadine (ed.), *Jerash Archaeological Project 1981–1983*, I: 411-59. Amman: Department of Antiquities.

Schick, R.
1995 *The Christian Communities of Palestine from Byzantine to Islamic Rule: A Historical and Archaeological Story*. Princeton, NJ: Darwin.

Shahid, I.
1986 The Jund System in Bilad al-Sham: Its Origin. In M.A. Bakhit and M. Asfour (eds.), *Proceedings of the Symposium on Bilad al-Sham during the Byzantine Period*, II: 45-52. Amman: University of Jordan.

Smith, R.H.
1973 *Pella of the Decapolis. I. The 1967 Season of the College of Wooster Expedition to Pella*. Wooster, OH: The College of Wooster.

Smith, R.H. and L.P. Day
1989 *Pella of the Decapolis. II. Final Report on the College of Wooster Excavations in Area IX, the Civic Complex, 1979–1985*. Wooster, OH: The College of Wooster.

Urice, S.
1987 *Qasr Kharana in the Transjordan*. Durham, NC: American Schools of Oriental Research.

Walmsley, A.G.
1987 The Administrative Structure and Urban Geography of the Jund of Filastin and the Jund of al-Urdunn: The Cities and Districts of Palestine and East Jordan during the Early Islamic, 'Abbasid and Early Fatimid Periods. Unpublished PhD dissertation, University of Sydney.
1991 Architecture and Artifacts from 'Abbasid Fihl: Implications for the Cultural History of Jordan. In M.A. al-Bakhit and R. Schick (eds.), *The Fifth International Conference on the History of Bilad al-Sham during the Abbasid Period*, 135-59. Amman: University of Jordan.
1992 The Social and Economic Regime at Fihl (Pella) between the 7th and 9th Centuries. In P. Canivet and J.-P. Rey-Coquais (eds.), *La Syrie de Byzance à l'Islam VIIe–VIIIe siècles*, 249-61. Damascus: Institut français de Damas.
1995 Tradition, Innovation, and Imitation in the Material Culture of Islamic Jordan: The First Four Centuries. In G. Bisheh (ed.), *Studies in the History and Archaeology of Jordan*, V: 657-68. Amman: Department of Antiquities.

Watson, P.M.
1992 Change in Foreign and Regional Economic Links with Pella in the Seventh Century AD: The Ceramic Evidence. In P. Canivet and J.-P. Rey-Coquais (eds.), *La Syrie de Byzance à l'Islam*, VII–VIII siècles, 23-34. Damascus: Institut français de Damas.

Whitcomb, D.
1988a A Fatimid Residence in Aqaba, Jordan. *Annual of the Department of Antiquities of Jordan* 32: 207-24.
1988b Khirbet al-Mafjar Reconsidered: The Ceramic Evidence. *Bulletin of the American Schools of Oriental Research* 271: 51-67.
1989a Evidence of the Umayyad Period from the Aqaba Excavations. In M.A. Bakhit and R. Schick (eds.), *The Fourth International Conference on the History of Bilad al-Sham during the Umayyad Period*, II: 164-84. Amman: University of Jordan.

1989b   Coptic Glazed Ceramics from the Excavations at Aqaba, Jordan. *Journal of the American Research Center in Egypt* 26: 167-82.

1989c   Mahesh Ware: Evidence of Early Abbasid Occupation from Southern Jordan. *Annual of the Department of Antiquities of Jordan* 33: 269-85.

1990–91   Glazed Ceramics of the Abbasid Period from the Aqaba Excavations. *Transactions of the Oriental Ceramic Society* 55: 43-65.

1992   Reassessing the Archaeology of Jordan of the Abbasid Period. In S. Tell (ed.), *Studies in the History and Archaeology of Jordan*, IV: 385-90. Amman: Department of Antiquities.

1994a   Amsar in Syria? Syrian Cities after the Conquest. *ARAM* 6: 13-33.

1994b   *Ayla: Art and Industry in the Islamic Port of Aqaba*. Chicago: The Oriental Institute.

1995   Islam and the Socio-Cultural Transition of Palestine, Early Islamic Period (638–1099 C.E.). In T. Levy (ed.), *The Archaeology of Society in the Holy Land*, 488-501. London: Leicester University.

2000   Hesban, Amman, and Abbasid Archaeology in Jordan. In L.E. Stager, J.A. Greene, and M.D. Coogan (eds.), *The Archaeology of Jordan and Beyond: Essays in Honor of James A. Sauer*, 505-15. Winona Lake, IN: Eisenbrauns.

Zayadine, F. (ed.)

1986   *Jerash Archaeological Project 1981–1983*, I. Amman: Department of Antiquities.

1989   *Jerash Archaeological Project 1984–1988*, II. Amman: Department of Antiquities (= *Syria* 66: 1-260).

15. Fatimid, Ayyubid and Mamluk Jordan and the Crusader Interlude

Alan Walmsley

No longer hidden: a new archaeological view of middle Islamic Jordan

For over half a century archaeologists have generally assumed that the middle Islamic period of Jordan (roughly mid-ninth to fifteenth centuries) was, on the whole, historically inconsequential. To Jordan's pioneer archaeologist, G.L. Harding, it was a prolonged period of depressing decline:

> In the ninth century AD the conquering Abbasides transferred the capital to Baghdad, and Jordan began to be forgotten; not being on any trade route, or producing any natural wealth, the country was left to fall into decay. But it was still of sufficient importance for the Crusaders in the twelfth (sic) century to occupy part of it and build castles there, the chief of which were Shobak and Karak. After that its prosperity declined still further, and it was a country of small, poor villages, scraping a bare existence among the ruins of past splendour (Harding 1967: 52).

For apart from a few castles, and then mostly perceived as Crusader constructs, there was little monumental architecture to report and no major sites that could compete with those from earlier periods—that is, the equivalent of biblical Dhiban, Nabataean Petra, or Roman Jarash for instance.

Based more on unsubstantiated Eurocentric perceptions than any concrete evidence, this attitude grew out of biased and unchallenged nineteenth-century attitudes: a belief in the true magnificence of Classical and early Christian civilization and the 'desolation of Islam'. Writing of the Hauran, the American explorer Selah Merrill lamented the decimation of an early Christian society in a historical reconstruction sounding more like an imaginative Hollywood script than impartial scholarship: 'The crosses and other Christian symbols which exist on the buildings here indicate that the houses remain as they were left in the early centuries, when the Moslems swept away the inhabitants of this region, leaving the cities and the land in desolation' (Merrill 1986 [1881]: 22).

This theme of social and urban devastation in the wake of an Arab/Muslim/Bedouin invasion, embraced with enthusiasm by Merrill apparently after reading De Vogüé (Merrill 1986 [1881]: 60-70), dominates much nineteenth-century writing (Johns 1994: 1). Few explorers, it seems, could resist its attention-grabbing appeal; and likewise neither could many archaeologists of the early twentieth century. At Jarash C.S. Fisher was oblivious to any major Islamic presence, even though there was overwhelming archaeological evidence (including numismatic) to the contrary. From this point the belief in the collapse of civilization with the 'Arab conquest' easily entered popular literature, for instance the frequently reprinted *Pleasure of Ruins* by Rose Macaulay:

> Entering some Arab village of squalid hovels, we are in a Roman colony, among temple columns, triumphal arches, traces of theatres and baths which no one has had the intellect or the cleanliness to use since the Arabs expelled the civilized Graeco-Roman-Syrian inhabitants and squatted among their broken monuments, stabling their horse in the nave of a Christian basilica, their camels in a richly carved pagan temple (Macaulay and Beny 1977 [1953]).

Clearly it was going to be difficult to discard this biased, indeed bigoted, viewpoint and allow for the appropriate recognition of middle Islamic Jordan.

The concept of Muslim 'thundering hoards' was finally superseded in the academic world (but not, unfortunately, in popular belief) by a new and more reasoned theory on the socio-urban collapse of late antique Jordan. This argued for the continuation of urban communities (mostly Christian) after the Islamic Conquest, but their dispersion in the mid-eighth century due to a combination of natural factors (notably earthquakes) and the relocation of the capital of the Islamic empire from Damascus to Iraq. This was the chronology for the breakdown of urban society in Jordan presented in brief by Harding in his seminal *Antiquities of Jordan* (above). Only with the fall of the Umayyads, it was now believed, did Jordan become politically inconsequential and a cultural backwater. Another change saw a developing awareness of widespread rural settlement in Mamluk times, seen in some way as connected to the active market economy of Mamluk Egypt and the sugar industry. This view, however, did not allow for a major urban component in the settlement of Mamluk Jordan.

Archaeological research of the 1970s and 1980s largely bypassed the middle Islamic period, with a few notable exceptions. Consequently there was little motive or information to question prevailing ideas, and Harding's explanation for the supposed social and urban disintegration of Islamic Jordan developed, by default, into a three-stage model. It presented a most gloomy scenario.

Stage 1: urban collapse and rural depopulation

It was widely accepted that by the Fatimid period the absence of strong central government after the fall of the Umayyads had precipitated a rapid collapse of the urban infrastructure and growing political chaos. This had opened up the region to predatory Bedouin incursions, resulting in rural depopulation and economic recession, especially in the south. The 'desert' had won at the expense of the 'sown'.

Stage 2: crusader reassertion

In the twelfth century the Crusaders re-imposed central authority over south Jordan, constructing imposing castles at ash-Shawbak, Wadi Musa and, later, al-Karak in particular. This resulted in an improvement in the level of rural settlement. As a zone of conflict, however, the north was sparsely populated except for military outposts of both sides, notably the castles at 'Ajlun and as-Salt.

Stage 3: mamluk revival

The expulsion of the Crusaders after the Battle of Hattin (1187) and the eventual reunification of greater Syria (Bilad ash-Sham) with Egypt under the Mamluks in the late thirteenth century heralded a new era of moderate prosperity based on a rejuvenated rural economy, especially a flourishing sugar industry in the tropical climate of the Jordan Valley. The widespread occurrence of brightly coloured glazed pottery on the surface of many sites in Jordan attests, it is often suggested, a population increase due to stable administration until the invasion of ash-Sham by Timur-Leng and growing political division and economic impoverishment at the onset of the fifteenth century.

This three-stage model was convenient, seemingly authoritative, and popular, especially as it dealt quickly with a bothersome period of little concern to archaeologists whose primary interest (and that of their funding bodies) was biblical, Classical and early Christian Jordan. Reflections of this model have emerged, unwittingly it seems, in overviews of the settlement history of many sites, for instance Pella (Smith 1973; Smith and Day 1989) and Hisban (LaBianca 1990; but note the review by McQuitty 1993), and also in more general accounts of Islamic history (for instance Kennedy 1986: 293).

It is unlikely that many contemporary archaeologists would be brave enough to offer such a simplistic scenario for middle Islamic Jordan, especially as single-cause explanations for social transformations in the early Islamic period have been effectively discarded in the last few years. In a recent review Bienkowski wisely adopted a neutral stance, although mention was made of the Mamluk revival and subsequent decline (Bienkowski 1991: 28). A major concern of this chapter is to challenge the three-stage model outlined above by drawing upon available textual and archaeological data. Recent archaeological work has gone some way to reassessing the character and extent of Fatimid, Seljuq, Crusader, Ayyubid and Mamluk settlement in Jordan, although these periods are grossly underrepresented in current research. Significant contributions have been made by, among others, Brown (1984, 1992), Johns (1992, 1994), Kareem (1993), Lenzen and Knauf (1987), Schick (1997) and Whitcomb (1995a, 1997b), while the author has researched Fatimid and Mamluk Jordan (Walmsley 1987, 1992b). The objective is to depict a society in the throes of major cultural and economic transformations that saw it progressively integrated, in uneven stages, into the Arab-Muslim world of Egypt and ash-Sham, perhaps in such a manner and to such an extent that it became almost imperceptible to modern research. Yet it can be confidently maintained that the major centres and the rural population—agricultural and pastoral—can be identified through textual and archaeological sources. The assumed hiatus in the settlement history of middle Islamic Jordan can now be sufficiently filled so as to seriously question that assumption, and permanently consign to the rubbish bin the eighteenth–nineteenth-century European 'romance' of a Bedouin influx at the expense of urban and settled life.

Between history and archaeology: the contribution of written sources

The failure of past scholarship

Past scholarship has, with only rare exceptions, shown scant interest in promoting the archaeological study of the post-antique world in the east Mediterranean (Whitcomb 1995a: 496). Archaeology 'properly' terminated with the Islamic Conquest, or even earlier with Constantine's empire; the Middle Ages belonged to historical studies and art history. One major reason for the poor showing of Islamic archaeology in Jordan was the obvious discrepancy between the few archaeological reports available and written sources—historical, biographical and geographical. While the sources mentioned towns and an active rural sector, archaeologists reported urban decline and a depopulation of the countryside. It comes as no surprise, therefore, that Islamic historians have happily ignored obviously flawed archaeological results.

This rather pessimistic picture has changed in the last 10 years, mostly through the actions of a handful of archaeologists. It would now be inconceivable to consider the progress of archaeological research into Fatimid, Seljuq, Crusader, Ayyubid and Mamluk Jordan without reference to written material, especially the detailed works of geographers and travellers. The information preserved in these sources will serve as a starting point for this study, and offers an informative complement to the archaeological studies available.

The historical view

Since the time of Ahmad Ibn Tulun (d. A.H. 270/ AD 84), if not long before, most of Bilad ash-Sham came within the political and cultural orbit of Egypt, and the extension of Fatimid control over ash-Sham in 359/970 represented a straightforward historical continuation of this arrangement. Control extended to the region east of the Jordan Rift Valley through garrisons and tribal alliances, and it seems that the region played a continuous role in the social and economic life of Fatimid Palestine-Jordan. This is particularly apparent from the detailed account of the geographer al-Maqdisi (al-Muqaddasi, d. 1000), a native of Jerusalem (Bayt al-Maqdis) who wrote about 985 (Muqaddasi [al-Maqdisi] 1994). According to al-Maqdisi, the area of Jordan north of Wadi az-Zarqa', called Jabal Jarash, was included in the

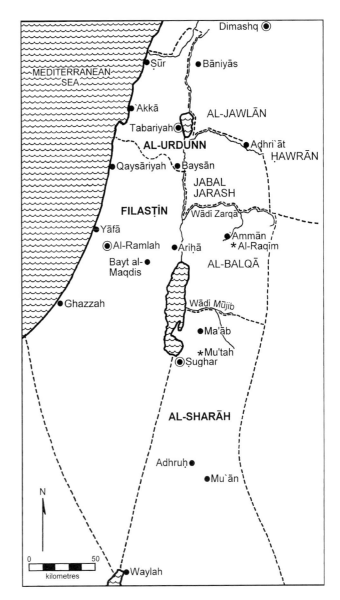

Figure 15.1. The geo-political features of tenth-century Jordan, based on the account of the Arab geographer al-Maqdisi.

district of al-Urdunn and had Adhri'at (Dar'a) as its chief town (Figure 15.1). The area was populated with many villages and produced olives, various fruits including grapes, and honey. This agricultural productivity was such, reports al-Maqdisi, that it contributed to the wealth of Tabariyah (Tiberias), al-Urdunn's capital. Between Wadi az-Zarqa' and Wadi al-Mujib was the al-Balqa' area, which belonged to the district of Filastin (Palestine). Around its principal town of 'Amman were villages, farms, grain-fields, mills and flocks, and nearby was ar-Raqim where pilgrims could visit the Cave of the Sleepers. Perhaps the most interesting aspect of al-Maqdisi's account is his identification of a separate district south of Wadi al-Mujib called ash-Sharah. The capital was located at Sughar (Zughar, Zoar) in the Ghawr as-Safi, south of the Dead Sea, and had six other townships including Ma'ab (Rabbah, Areopolis), Mu'an (Ma'an), Udhruh and Waylah (Ayla, Aila). Ma'ab, the principal town in the mountains, had many productive villages and the holy site of Mu'tah with the tombs of the Prophet's companions. Ayla and Zughar, strategically located at either end of Wadi 'Arabah, were thriving commercial centres. The well-populated Red Sea port of Ayla served Filastin and the Hijaz, with roads heading northwards to ar-Ramlah (six marches) and Zughar (four marches). From Zughar, roads continued onto Jerusalem, Nablus, and 'Amman (Figure 15.2). Another important road described by al-Maqdisi was the Hajj or pilgrimage route from Damascus. This passed through Adhri'at and az-Zarqa' to 'Amman, which was a major assembly place for pilgrims from Palestine, and continued south by way of two watering stops to Mu'an and then Tabuk, three stops later.

From al-Maqdisi's account it is difficult to avoid the conclusion that Fatimid Jordan supported a minimum of eight major population centres (and probably more; Maqdisi's description appears incomplete especially for north Jordan) through an active rural economy and considerable inter- and intra-regional trade. In particular, southern Jordan came to benefit from the revival of the Red Sea trade routes under the Fatimid caliphs of Egypt, which found expression in the political events of the later tenth and early eleventh centuries when the tribal shaykhs of south Palestine and Jordan played an active, and sometimes counter-productive, role in the affairs of Filastin, either as allies of or rivals to Fatimid domination (Kennedy 1991: 307, 320-40). Facing often formidable opposition, the Jarrahids' strength lay in a stable and prosperous base in ash-Sharah and the northern Hijaz, held through a number of strongholds including al-Karak (Schick 1997).

Following the imposition of Seljuq control over Bilad ash-Sham (1071–79), the Arabic sources continued to refer to the old districts of al-Balqa', Ma'ab, al-Jibal and ash-Sharah, but within these districts political power had already shifted to other population centres, particularly Wadi Musa (for Sharah) and al-Karak (for Ma'ab). Accordingly, Crusader sources identify Wadi Musa, ancient Petra (McKenzie 1991), as the principal object of military activity south of the Dead Sea during the first quarter of the twelfth century (Schick 1997). The first expedition took place in 1100 when Baldwin

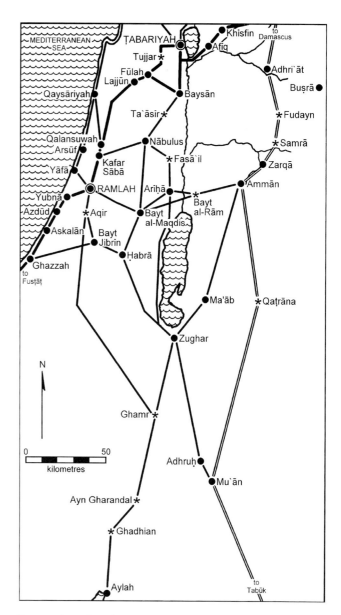

Figure 15.2. The pilgrimage route (divided line) and other major roads of Jordan in the tenth and eleventh centuries.

I (r. 1100–18) led a campaign across the Jordan, reaching Wadi Musa by way of Zughar south of the Dead Sea (Figure 15.3), causing the (Muslim) inhabitants of both localities to flee. After staying three days and visiting the monastery of St Aaron on Jabal Harun, Baldwin returned to Jerusalem by the same route (Fulcher of Chartres 1969: 145-47; William of Tyre 1976 [1941]-a: 427). Another campaign to Wadi Musa was undertaken in 1106–1107. This succeeded in evicting a Seljuq force under al-Ispahbad sent from Damascus to regain control over ash-Sharah, al-Jibal, Ma'ab and al-Balqa', after which Baldwin returned to Jerusalem by a route north of the Dead Sea. Together al-Ispahbad's mission,

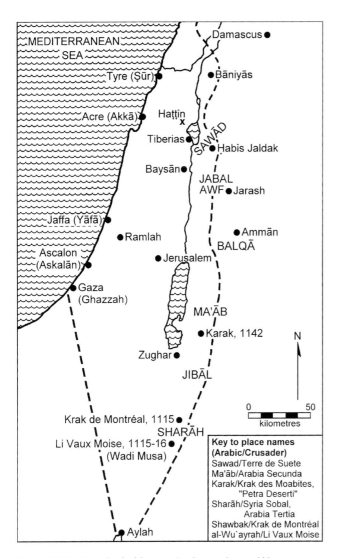

Figure 15.3. Crusader holdings in Jordan in the twelfth century.

but that by 1111 the Crusaders under Baldwin had deeply penetrated this territory and established a permanent presence, probably with the tacit support of sections of the local Christian population.

The Latin kings of Jerusalem now set about consolidating their control over Jordan, and were particularly successful in the south. In 1115 Baldwin built a castle at ash-Shawbak, which he named Krak de Montréal (Fulcher of Chartres 1969: 215; Kennedy 1994: 23-25; William of Tyre 1976 [1941]-a: 506-507). This ensured Crusader domination of ash-Sharah, the Syria Sobal and Arabia Tertia of the Crusader sources (earlier in 1105 Baldwin had failed in a similar attempt to secure the Sawad). In the following year Baldwin extended control as far as Ayla (al-'Aqaba) on the Red Sea (Fulcher of Chartres 1969: 215-16; William of Tyre 1976 [1941]-a: 513), occupying or possibly building a fortress there. At about this time Christians from other presumably less protected regions (perhaps al-Balqa' and the Jarash region) were encouraged to migrate and settle in Jerusalem (William of Tyre 1976 [1941]-a: 507-508). Soon after (in 1120), Tughtigin built a castle at Jarash to enforce his interests in Jabal Awf, but this was destroyed in the following year by a Crusader raid under Baldwin II (r. 1118–31). Fulcher (d. 1127) is the principal source on the castle:

> The inhabitants of the area called this fortress Jarash. It was inside a city wondrously and gloriously built in ancient times on a strong site. The castle was built of large squared stones. But when the king considered that he had taken the place with difficulty and that it would be hard to provide it with the necessary men and provisions, he ordered the castle destroyed and all his men to return home (Fulcher of Chartres 1969: 235; see also William of Tyre 1976 [1941]-a: 538-39).

In the south, control of the strategically located settlement of Wadi Musa was not without its problems for the Latin kings. Baldwin II suppressed a revolt there in 1127 (Ibn al-Qalanisi 1967 [1932]: 182), and, although fortified and garrisoned with a formidable castle (Li Vaux Moise, al-Wu'ayra, built by 1115–16), the town again rebelled against Crusader domination in 1144. The townsfolk managed to capture the castle with the help of a small Seljuq force (William of Tyre 1976 [1941]-b: 144-45), but the youthful Baldwin III (r. 1143–62) besieged and eventually captured the fortress through brutal economic warfare. Perhaps at this time the Crusader outpost at al-Habis in the heart of ancient Petra was erected to monitor the activities of the cave-dwelling Arabs and the Wadi 'Arabah routes. To boost Crusader domination over south Jordan another

Baldwin's response and his return route indicate the considerable spread of Crusader control as far north as al-Balqa'. Also in the same year, the Crusaders campaigned in the Sawad (Crusader Terre de Suete, essentially the Jawlan and north Jordanian plains) and Jabal Awf (the 'Ajlun district), thereby gaining (or regaining) a measure of control over north Jordan. This de facto situation was soon formalized by the signing of share agreements with Tughtigin, Atabeg of Damascus, in 1109 and 1111. The agreements covered nearly all Jordan, from the Sawad to ash-Sharah in the south, although the status of 'Amman and al-Balqa' is unclear (Ibn al-Qalanisi 1967 [1932]: 74-75, 81-82, 92, 113; William of Tyre 1976 [1941]-a: 429-30, 1976 [1941]-b: 470). The agreements show that at the time of the arrival of the Crusaders (and presumably before) Jordan was under the jurisdiction of the *atabeg* of Damascus, castle, this time on a massive scale, was constructed in 1142 at al-Karak in Ma'ab (Crusader Arabia Secunda, William of Tyre 1976 [1941]-b: 127). This site, the former Jarrahid stronghold (above), was endowed with formidable natural defences and controlled the profitable Dead Sea routes. The move completed Crusader domination over the fertile wheat-growing and pastoral lands south of the strategic Wadi al-Mujib and heralded the establishment of the great Lordship of Oultrejourdain based at al-Karak.

While providing a clear outline of Crusader activity in Jordan, the Latin sources also preserve a valuable glimpse of socio-economic conditions in the early twelfth century. Dates grew at Zughar, villages lined the route to Wadi Musa, itself a valley 'very rich in the fruits of the earth' with 'luxuriant olive groves which shaded the surface of the land like a dense forest' and mills on its streams, while around ash-Shawbak was 'fertile soil, which produces abundant supplies of grain, wine, and oil'. In other areas were 'many Christians living in villages' who left for Jerusalem with their 'flocks and herds', while camels and asses were common booty from raids (Fulcher of Chartres 1969: 146-47; William of Tyre 1976 [1941]-a: 506-508; 1976 [1941]-b: 145). Although only scattered references, Fulcher and William nevertheless depict Jordan as having a mixed settled and nomadic population actively engaged in agricultural and pastoral pursuits.

The Crusader presence east of the Jordan was short-lived, for Oultrejourdain including the still disputed lands in the north were all lost in the second half of the twelfth century through the actions of Nur ad-Din (d. 1174) and the famous Salah ad-Din b. Ayyub or Saladin (d. 1193). Unlike the fortified south, Crusader claims over north Jordan were always weak, consisting of an agreement to share the crops (wine, grain, olives) and revenues with the rulers of Damascus. This could only last as long as the Damascus *atabegs* agreed, and Nur ad-Din—who gained control of Damascus in 1154—certainly did not. Continuing the *jihad* against the Franks begun by his father Zangi, *atabeg* of Aleppo and Mosul (d. 1146), Nur ad-Din besieged the Crusader outpost of al-Habis Jaldak on the south bank of Wadi Yarmuk, but after battle with Baldwin III retreated to Damascus (1158). Towards the end of his rule, Nur ad-Din, with the ambiguous support of Salah ad-Din (at this point governor of Egypt), penetrated Oultrejourdain with the intention to take al-Karak and ash-Shawbak, but without success. Salah ad-Din returned in 1173, causing 'everything found outside the fortresses to be burned, bushes and vines to be cut down, and villages to be destroyed' (William of Tyre 1976 [1941]-b: 272-73, 387-90, 470).

The real losses, however, came with the rise of Salah ad-Din, and culminated in the decisive Muslim victory at Hattin in northern Palestine (1187). Although it took Salah ad-Din 12 years after the death of Nur ad-Din to effect the political unification of Syria and Egypt, he was active during much of this period. In 1182 he encamped at Jarba and conducted raids around ash-Shawbak before continuing to Damascus, while other Muslim forces took al-Habis Jaldak, only to lose it a few months later. In 1183 and 1184 al-Karak was besieged for months on end, and although the castle withstood Salah ad-Din's armies the lower town was looted with great loss of property: 'all their household possessions, all their furniture and utensils of every quote, were seized by the enemy' (Ibn Jubayr 1952: 301, 311, 313-14; William of Tyre 1976 [1941]-b: 467-72, 482-85, 498-504). The brutish and rash lord of Oultrejourdain, Reynald de Châtillon (r. 1177–87), attracted Salah ad-Din's rage by raiding the pilgrimage routes from Damascus and Cairo, attempting to attack the holy sites of Makkah and Madinah, and his barbarous treatment of prisoners (Maalouf 1984: 186-88). Once Salah ad-Din had brought Syria and Egypt under his control (1183), he was able to draw on vast resources to oppose the Latin Kingdom. In the following year he entered Crusader territory north of Tabariyah (Tiberias) and took the town. The Crusaders responded and, as was their practice, brought together all available manpower to repel the Muslim force. In the resulting battle (July 1187) the Crusader force was totally defeated with great loss of life. Of the captured leaders, Salah ad-Din's greatest wrath was reserved for 'the most malicious, evil and treacherous' Reynald, who was promptly executed (Holt 1986; Maalouf 1984: 190-94). The Crusader disaster at Hattin left Oultrejourdain, as with the whole Latin Kingdom, largely undefended, so in 1188 al-Karak and ash-Shawbak capitulated with little resistance to al-'Adil (d. 1218), Salah ad-Din's brother and his viceregent in Egypt. By 1193, the year of Salah ad-Din's death, the Arabic sources reveal that Jordan was divided between two administrations. The area south of Wadi az-Zarqa', including as-Salt and al-Balqa', was governed from Egypt by al-'Aziz Uthman, one of Salah ad-Din's sons, while Jabal Awf ('Ajlun) and the Sawad in north Jordan belonged to al-Afdal 'Ali, Salah ad-Din's eldest son based in Damascus (Humphreys 1977: 63-64, 75-77, 83).

For the remainder of the Ayyubid period (to 1263 in Jordan) and in the subsequent Mamluk period (1263–1517) Jordan and its castles, of both Crusader and Muslim origin, continued to play a leading role in the political and social history of Egypt and Bilad ash-Sham. In addition to the existing Crusader constructs in south Jordan, Muslim castles were built at 'Ajlun (between 1188 and 1192), al-Azraq (1236–37) and as-Salt (1220), which had become the chief town of al-Balqa'. These were intended to oversee communication routes and the local population, being built after the Crusader threat to Jordan had been largely neutralized. Most frequently mentioned in the Ayyubid-Mamluk historical sources was the immensely strong and strategically located castle of al-Karak (see in general Humphreys 1977; Irwin 1986; Thorau 1987 [1992]: esp. 134-41). As in earlier times, this castle controlled the Red Sea–Wadi 'Arabah trade routes to the west and the Hajj road from Damascus to the east, but most significantly it dominated movement between Cairo and Damascus, while the Crusaders continued to hold sections of coastal Palestine. For a while in the mid-thirteenth century al-Karak remained independent of both Damascus and Cairo, becoming particularly active in political affairs under the Ayyubid prince al-Mughith Umar (d. c. 1264), the great grandson of al-'Adil. Although al-Mughith backed Baybars during the turbulent first decade of the Mamluk sultanate (1250–60), soon the essentially centralized nature of the Syro-Egyptian Mamluk administration as constituted by Baybars (r. 1260–77) could not allow an independent-acting al-Karak, and al-Mughith was treacherously deposed by his former confederate (1263). Al-Karak, with ash-Shawbak, now became one of six Syrian provinces, but it often served as a distant outpost of Egypt, a place of refuge, exile and sometimes rebellion. Its most celebrated residents, willing or not, were:

- *as-Sa'id Muhammad Baraka Khan* (d. 1280), the deposed son of Baybars, along with his brothers and the queen-mother (in al-Karak 1279-96);
- *an-Nasir Muhammad* (d. 1341) on two occasions, firstly as a child (1297–99) then in 1309–10 in a successful ploy to regain the sultanate;
- *an-Nasir Ahmad* (d. 1344), son of an-Nasir Muhammad who grew up in al-Karak and, after becoming sultan, removed himself and much of the treasury back there before being deposed (1342). Seven campaigns were required before al-Karak was taken and an-Nasir Ahmad captured;
- *Barkuk* (d. 1399), the first of the Circassian Mamluk sultans, who gained support while exiled in al-Karak during 1389–90 to reclaim the sultanate.

Geographical works and the journals of travellers offer a particularly valuable source of information on socio-economic conditions in Ayyubid and Mamluk Jordan. There is much material from the thirteenth to fifteenth centuries: the encyclopaedic work of Yaqut (d. 1229) with numerous and often extensive entries on places and regions arranged alphabetically, the geography of Dimashqi (d. c. 1327), a sometimes confusing mixture of brief first-hand and copied accounts, also that of Abu 'l-Fida (d. 1331) extensively drawn from personal observations, the travelogue of Ibn Battuta (d. 1377) and al-Qalqashandi's authoritative encyclopaedic work (d. 1418). In these works are described the administrative divisions, places and localities in Jordan, although the quality and detail of the accounts vary considerably. The mostly original fourteenth- and fifteenth-century works emphasize the castle towns (especially al-Karak, ash-Shawbak and 'Ajlun), holy sites (Mu'tah/Mazar, Amata, ar-Raqim) and the stopping points along the pilgrimage route between Damascus and the Hijaz. The evidence from the written sources suggests that these three settlement types dominated Mamluk Jordan. The sources also pay attention to the vibrant rural economy in which pastoral industries and agriculture, both subsistence and cash crops, played an equally important role.

Jordan in the Mamluk period was divided between two administrative divisions called *mamlakah*, 'kingdoms': al-Karak and Damascus. As in Ayyubid times, much of Jordan was administered from al-Karak. The Mamlakat al-Karak included the castle-towns of al-Karak and ash-Shawbak, and extended northwards to Zizia (al-Jiza), westwards to include Zughar, Wadi 'Arabah and the Sinai peninsula, and into the northern Hijaz (Figure 15.4). Other places mentioned include Ma'an, Mutah, al-Lajjun, al-Hasa and Wadi Musa, plus the districts of al-Jibal and ash-Sharah. While Dimashqi also placed 'Amman, as-Salt, az-Zarqa', al-Azraq and the Balqa' in the Mamlakat al-Karak, this probably reflected the earlier Ayyubid structure. He again lists the Balqa' and the towns of as-Salt, 'Amman and az-Zarqa' as part of the Mamlakat Dimashq, adding Hisban (which was the capital of the Balqa', according to Abu 'l-Fida). Certainly included in Dimashq were Jabal Awf, the district of Jarash with its castle-town of 'Ajlun, the Sawad, the

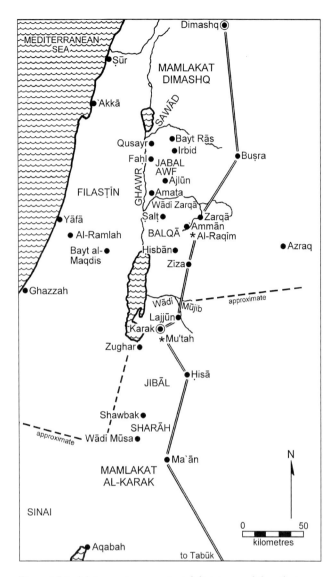

Figure 15.4. Major centres, provincial divisions and the pilgrimage route in Mamluk Jordan.

importance of individual sites had changed considerably. The north was dominated by 'Ajlun, its strong castle and fruit-bearing lands being repeatedly mentioned in the sources (see Le Strange 1965 [1890]: 388-89). Also important were populous as-Salt, then as now renowned for its pomegranates, and Hisban with (as Abu 'l-Fida relates) mills and fields lining the wadi to the Ghor (Le Strange 1965 [1890]: 456, 529-30). In the south, the sources emphasize the castle-towns of al-Karak and ash-Shawbak, both at the centre of a thriving arboricultural industry (apricots, pears and pomegranates) and largely populated with Christians (Gibb 1958: 159-60; Le Strange 1965 [1890]: 479-80, 536). Also given major treatment in the thirteenth- to fifteenth-century works was a number of holy sites. Leading these were: Mu'tah/Mazar with the tombs of three Companions of the Prophet including Ja'afar ibn Abi Talib, the brother of 'Ali, slain by a Byzantine force in 629 at the Battle of Mutah; Amata, burial place of Abu Ubaydah (d. 639), conqueror and first governor of ash-Sham; ar-Raqim, the reputed site of the Cave of the Sleepers; Wadi Musa and its spring (Qur'an 2.60); Jabal Harun, burial place of the prophet Aaron, brother of Moses; and Irbid with the tombs of Moses' mother and four sons of Jacob (on these see further below). Even the major commercial centres of the Ghor, namely Qusayr Mu'in (ash-Shunah) and Zughar, were credited with religious traditions (Gibb 1958: 82-83; Le Strange 1965 [1890]: 73-74, 274-92, 393, 457-58, 548-49). A third important group of settlements were those positioned on the Darb al-Hajj, the pilgrimage route from Damascus to the Hijaz (Figure 15.4). Ibn Battuta lists these as, after Busra, Zizia, al-Lajjun, al-Karak and Ma'an, which was 'the last town in Syria' (Gibb 1958: 159-60). 'Amman was bypassed, for it was in a state of ruin (*dimnah*) according to Dimashqi. The pilgrimage route probably also passed through az-Zarqa' and al-Hasa, both mentioned by Dimashqi in his description of the *mamlakah*.

The historical, geographical and biographical literary works provide a valuable source of information on socio-economic conditions in Jordan from the Fatimid to the Mamluk period. They reveal a number of major developments, which together had a major impact on the social history of middle Islamic Jordan. In essence these were:

- the political and probably social distinctness of Jordan south of Wadi al-Mujib, perhaps of long standing but first very apparent in the Fatimid period;

districts of Bayt Ras and Fihl, and the upper and middle Ghor and their respective towns of Qusayr (North Shunah) and Amata (Abu Ubaydah). The fourteenth- and fifteenth-century sources show that the boundary between Mamluk al-Karak and Dimashq ran along (or just north of) Wadi al-Mujib/Wala, then the Dead Sea and Wadi 'Arabah, which meant that Jordan was again divided along a long-established politico-cultural boundary that reappeared in the Fatimid period (compare Figures 15.1 and 15.4).

The entries in Dimashqi, Abu 'l-Fida, Ibn Battuta and al-Qalqashandi leave little doubt that living conditions in the towns and countryside had largely recovered from the military and political upheavals of the twelfth and thirteenth centuries, although the relative

- the dynamic economy of Fatimid Jordan, embracing agriculture and arboriculture, pastoral industries, and trade and commerce;
- the geographical extent of Crusader control over much of Jordan, which was achieved by treaty (for the north) or direct rule (the lordship of Oultrejourdain) south of Wadi al-Mujib/Wala catchment;
- a permanent shift in settlement from sites along the *badia* fringe (e.g. 'Amman, Ma'ab) to a line of defensive sites (e.g. al-Karak, 'Ajlun) on high points along the Jordan Valley scarp. This development first appeared in the Fatimid period but intensified in Crusader and Ayyubid times. A parallel development is the greater importance of Holy Places (e.g. Mu'tah, ar-Raqim, Abu Ubaydah), again beginning in the Fatimid period but increasing in Ayyubid and Mamluk times;
- the social and economic reconstruction of Jordan after the Crusader experience, concentrating on arboriculture and commercial crops such as sugar.

Site settlement: an archaeological survey

The archaeology of tenth- to sixteenth-century Jordan is beset with difficulties, for the paucity of research means that there is no established starting point and the quality of data is patchy. Few excavations have been undertaken with the specific purpose of investigating the nature and extent of site occupation, and little attention has been paid to determining the range, characteristics and development of material culture in middle Islamic Jordan. Likewise, regional surveys have frequently cursorily, sometimes recklessly, treated the usually abundant architectural and ceramic evidence for the occupation of sites in the Islamic periods, and a disinterested attitude to the middle Islamic periods has introduced considerable distortion to settlement profiles as a result. Archaeologically, Fatimid and Seljuq Jordan are almost invisible, the twelfth century solely evidenced by Crusader and Muslim castles, while Ayyubid and Mamluk Jordan (the two periods, covering 250 years, are commonly clumped together) are typified by the many small village sites identified in regional surveys.

Due to this haphazard approach, it becomes exceedingly difficult to build up an archaeological profile that faithfully represents the full extent of social and economic life in the formative middle Islamic period.

Inevitably, then, given the current state of research, this study ends up concentrating on the architectural features of surviving monuments—predominantly castles, mosques and shrines—and on the distribution of mostly poorly recorded village and hamlet sites located by regional surveys. Only a scattering of middle Islamic sites, large and small, has been excavated (and then usually unintentionally) and little research has been done on urban sites, many of which are the major towns of modern Jordan. The absence of reliable, well-contexted and sufficiently extensive archaeological data imposes considerable limitations on its usefulness in accurately detailing the form and development of socio-economic structures and the emergence of modern Jordan out of the middle ages into a key player in the Middle East today.

Sites of the Fatimid and Seljuq periods (tenth and eleventh centuries)

Archaeological evidence for settlement during the tenth and eleventh centuries is particularly sparse, which is at variance with al-Maqdisi's positive account of urban and rural conditions in the late tenth century. Only a few sites—'Amman and Ayla (al-'Aqaba) are the important examples—have clearly identified Fatimid occupation levels, and then usually associated with structures of an earlier Islamic date. Other major sites with reported Fatimid occupation include: Pella, Tall Abu Qa'dan, Nimrin and Khirbat Shaykh 'Isa in the Rift Valley; and Hisban, Dhiban, Bayt Ras, Faris and Qam in the mountains.

At 'Amman, excavations between 1975 and 1979 by Bennett and Northedge on behalf of the Department of Antiquities produced clear evidence for settlement on the Citadel in the Fatimid period (eleventh century), with occupation concentrated in the Umayyad Palace complex and adjacent domestic structures (Northedge 1992: 159). Stratum III of the excavations, dated to the eleventh century, was the last major period of occupation on the citadel, and represented an unbroken continuation of the previous stratum (IVb, ninth–tenth centuries). In Area C, two parallel lines of squares sectioned the west slope to the Umayyad fortification wall, revealing two streets and four building complexes. Although probably constructed in the Abbasid period, these buildings were substantially modified in Fatimid times. Their sudden and simultaneous destruction was perhaps due to the earthquake of 1068. Dating is provided by two coins, a dinar of al-Hakim (407/1016–17) and a worn dirham of az-Zahir (411-27/1021–36), the latter found embedded in the terminal floor surface of Building 3. A similar sequence is found in Areas B and D, where Umayyad-period houses were continuously utilized and modified until the Stratum III destruction. In Area B five buildings and a street were excavated, although one unit (Building C) produced some evidence of Ayyubid-period reuse. The Palace also continued to be occupied into Fatimid times but was subdivided into smaller units; these were residential according to Northedge, indicating a reduced or non-official role for the palatial complex by the eleventh century. Most notably, the reception hall was converted into a self-contained building, which required the insertion of a wooden roof over the central court (Northedge 1992: 78, 81, 83-84, 88). The fortification wall surrounding the upper terrace of the citadel had fallen out of use long before the eleventh century as it had been built over by Abbasid-Fatimid domestic structures in Areas C and D (Wood 1992: 124-25).

The archaeological evidence on the nature of Fatimid and Seljuq settlement at 'Amman is paralleled by the discoveries at Islamic Ayla, although the role of Ayla, which was primarily commercial, was quite different from the political and military functions served by 'Amman. At Ayla, the Fatimid and Seljuq periods coincide with the end of the early Islamic II and the first half of the middle Islamic I periods, as defined by Whitcomb. In this last phase of occupation, which ended with the Crusader conquest of 1116, the originally Umayyad-'Abbasid structures showed continued use and modifications, although construction standards were poorer. Excavations, directed by Whitcomb, of the town gates and wall towers, the Central Pavilion building in the middle of the town, the Egyptian street (one of four axial roadways that led from the gates to the Central Pavilion), the beach-side suq along the south wall, and the congregational mosque have detailed an uninterrupted sequence from late Abbasid into Fatimid times. Particularly revealing is the later occupation of the Pavilion Building, excavated in 1986 and 1987 (Whitcomb 1988a and 1988b). Originally set up as a public monument at the junction of Ayla's four axial streets, the structure in its last manifestation was a well-appointed residence of eight rooms around a central courtyard, onto which opened a deep *iwan* (porch). The courtyard, *iwan* and adjacent rooms formed the domestic nucleus of the house. In a 4.2 m^2 corner room, entered from a doorway in the west wall of the *iwan*, a polychrome fresco-decorated wall was uncovered. This room and another on the east produced an astounding collection of luxury and everyday ceramics as well as other objects, including a dirham of al-Hakim (996–1020). A bath, latrine and kitchen were exposed to the north of the courtyard, while to the east the piers of a staircase to the roof or a second storey were identified. Entrance to the house was from Egypt Street to the west through an anteroom, with doorways positioned off-axis to protect the privacy of the household. Whitcomb's excavations at Ayla have revealed dramatic evidence for a major urban dislocation in the mid-eleventh century, involving 'slumping and possibly massive subsidence of walls and buildings of the city' and due, in all likelihood, to the powerful earthquake of 1068 (Whitcomb 1995b: 505). There is no archaeological evidence for a comprehensive rebuilding program after the earthquake, suggesting that the town never recovered from this major setback and hence presented an easy target for Baldwin's forces.

The archaeological work at 'Amman and Ayla reveals a shared trend in their urban history: a severe reduction, if not end, to their political and administrative functions, probably in the second half of the tenth century but no later than the early eleventh century. The clearest evidence for the change is the usurpation and conversion of public buildings into private dwellings, notably the palace on the 'Amman Citadel and the Central Pavilion at Ayla. While construction techniques were generally mediocre, levels of personal wealth were still good as illustrated by the quality and quantity of objects recovered from these structures. A similar downgrading of urban functions can be documented for the towns of north Jordan, although the evidence from Fihl/Pella, Bayt Ras and Jadar suggests that this transformation happened more than a century earlier, about the middle of the ninth century (Walmsley 1992a).

The evidence for Fatimid and Seljuq occupation at other sites in Jordan is patchy and undistinguished when compared with 'Amman and Ayla, but nevertheless suggests a settlement profile similar to what probably were the two major towns in tenth–eleventh-century Jordan. Bayt Ras Phase IVb, dated 900–1100, saw the continued utilization of urban structures including a line of vaults in Area A, thought to have been a market place of the Byzantine and early Islamic town (Lenzen 1995; Lenzen and Knauf 1987). During Phase IVb, the original Byzantine-period tessellated pavement of Phase VI (300–600) was substituted with flagstones. However, a continuing public role for the vaults appears unlikely especially as the church opposite the vaults had seemingly fallen out of use in the preceding Phase V (600–900). Elsewhere, eleventh-century ceramics attest continued occupation of Bayt Ras during the Fatimid period. At Pella, the Abbasid town centre has revealed domestic and industrial occupation continuing into the early Fatimid period (Walmsley 1995). Continuity of occupation throughout the Fatimid and Seljuq periods at Zughar, modern Khirbat Shaykh 'Isa, has been established by two surveys of the site (Whitcomb 1992a: 115-17).

A continuing use of existing stone buildings and ambiguous ceramic typologies for Fatimid and Seljuq Jordan could well account for the absence of reported tenth–eleventh-century occupation at other major archaeological sites in Jordan. Any major hiatus in occupation at historically important sites such as Hisban and Dhiban is difficult to accept. Sauer correctly recognizes early Fatimid pottery at both sites (Sauer 1982: 333; on ceramics, see further below), but the architectural evidence from the Dhiban excavations for Fatimid and Seljuq occupation on the tell, while tantalizing, is impossibly confused (Tushingham 1972: 77-83). Also contributing to the shortage of data is the lack of excavation or publication of a number of sites including some of the major centres mentioned in al-Maqdisi, notably Ma'an (unexcavated), Ma'ab and Udhruh (both partially excavated but not published). In other cases the substantial redevelopment of leading eleventh-century centres in later times would have largely destroyed the evidence for Fatimid and Seljuq-period settlement: al-Karak is the obvious example. Admittedly, this reasoning relies on mostly negative evidence, but then neither does the absence of available evidence demonstrate site abandonment. The material from 'Amman, Ayla, Bayt Ras, Pella and Zughar suggests a continuing settlement history into Fatimid times for many of the major towns of Jordan, although their politico-administrative and hence urban role was considerably diminished or even extinguished during the ninth and tenth centuries.

Fatimid and Seljuq occupation at other sites is not widely reported, although there are some notable examples, especially in the Jordan Valley. Soundings at Tall Abu Qa'dan (Gourdan), next to Tall Dayr 'Alla, produced 19 phases (A to T, omitting I), all of Islamic date (Franken and Kalsbeek 1975). The soundings came across a long sequence of courtyard surfaces over mudbrick collapse but few other features. Unlike the other sites considered so far, the buildings at Tall Abu Qa'dan were of mudbrick, and, as they had a much shorter life, a more representative early to middle Islamic sequence has been preserved. Chronology, however, is a problem, as dating the finds (mostly ceramics) was not the purpose of the excavations and no firm dating evidence (namely coins) was found. A reinterpretation of the pottery (in progress) indicates that, while Phase A is possibly eighth-century Umayyad, the sequence properly begins with the early Abbasid period (B-C = ninth century) and continues with late Abbasid and early Fatimid occupation (Phases D-G = tenth–eleventh centuries). The stratigraphic and pottery evidence very strongly suggests that the major 'destruction' at the end of Phase E can be equated with the end of the Abbasid town centre at Pella, maybe early in the Fatimid period. If so, Phases F and G would be firmly Fatimid, and specifically of eleventh-century date. Phases H-T continue the sequence into the Ayyubid and Mamluk periods, making eight centuries of settlement that, contrary to the view of Franken and Kalsbeek, would not allow for any significant gap in site occupation. A similar picture emerges from the important excavations at Tall Qudsiyah, 15 km south of North Shuna in the north Jordan Valley (Kareem 1987: 92-123). Again, the main building material was mudbrick, and by trimming bulldozer cuttings and excavating soundings Kareem has identified a major unbroken sequence extending from late antiquity to the Ottoman period. The Fatimid levels, consisting of ash, soil and mudbrick layers (courtyards?), were dated by lustre and other glazed wares and also decorated cream wares.

In the mountains, the use of stone as the primary building material has resulted in evanescent Fatimid remains, yet the identification of eleventh–twelfth-century levels has not been an impossible obstacle for problem-driven projects such as Khirbat Faris (Johns et al. 1989; McQuitty and Falkner 1993). Excavations in and around House 2 have recorded the continuous use of this stone structure from its construction (first century) until today with Stage 3, made up of a series of surfaces 60-70 cm deep, ranging across the ninth/tenth–twelfth centuries. Interestingly, this time span was barely accounted for in the surface sherding survey, with only five out of the 246 squares surveyed producing material identified as ninth–twelfth centuries. The lesson is obvious, often stated, but bears repeating: surface collections, even when correctly read, do not accurately represent either the periods of settlement or extent of occupation of a site.

The Faris project demonstrates unequivocally that little weight can be given to the results of regional surveys based upon surface collection (note the comments of Johns 1994: 3-9). Thus the low representation of Fatimid and Seljuq occupation in many surveys, in both the north and south of the country, reflects neither the nature nor extent of urban and rural settlement and land use in the tenth–twelfth centuries. Clearly, there were important differences with the situation in the Byzantine and Umayyad periods, but simply to stipulate a settlement 'gap' or site abandonment in the Fatimid and Seljuq periods based on the 'presence/absence' of very imperfectly known ceramic types (especially those of the middle Islamic period) avoids a fuller consideration of a more complex series of human land-use issues (Brown 1991: 229-32; Johns 1994; Miller 1991: 19-20). A number of related reasons can be postulated to explain the poor showing of Fatimid and Seljuq sites in surveys: the imperfect and cursory nature of survey techniques; the substantial refurbishment of towns, villages and holy sites after the expulsion of the Crusaders (and hence the virtual absence of major structures such as shrines and mosques, see below); different exploitation strategies; and, as already noted, a very poor understanding of the material culture of the period, especially ceramics. In those areas where middle Islamic ceramics are better known, for instance the north Jordan Valley, survey results have been considerably more successful in identifying widespread settlement (e.g., Kareem 1987: 450-52).

While it is probably unfair to single out any one survey, for the error is widespread, the high-profile Hisban survey offers a clear example of unrepresentative survey results and their skewed interpretation. From a Byzantine high of 126 inhabited sites in the Hisban region, only 33 Umayyad and seven Abbasid period sites were identified; no Fatimid and Crusader occupation was recorded. From these figures it has been proposed that the population plunged from an all-time high of over 70,000 in 635 (the Islamic Conquest) to as few as 1200 in 900. Concurrently, intensive land-use strategies based on cereals, gardens and orchards were replaced by extensive pastoral/nomadic strategies until there was no sedentary occupation at or around Hisban during the three centuries of Fatimid, Seljuq and Crusader domination (Geraty and LaBianca 1985; LaBianca 1990). As McQuitty (1993: 168-69) points out: 'LaBianca does not offer any different interpretations of the settlement history of Jordan. He fits the food-system concept into the conventional models of the last fifty years despite tantalizingly touching on the necessity of discussing the *type* of land-use and settlement if the complexity of each period is to be characterized.' Johns (1994: 8-9) precisely summarizes the problem plaguing many surveys: often early Islamic wares, originating in a Byzantine tradition, are misassigned to the earlier period, while the origin of many Ayyubid-Mamluk types (especially hand-made wares) in the Fatimid and Seljuq periods has yet to be recognized, leading to similar mistakes in periodization; and the situation is probably worse south of Wadi al-Mujib. As Whitcomb (1992b: 388) has noted, 'there are more than enough ceramics published as Umayyad and Ayyubid/Mamluk to fill in the Abbasid/Fatimid period and make it quite a respectable occupation in Jordan's history'. Not surprisingly, given the differences of approach and interpretation, we seem to be in completely different worlds when reading LaBianca on Hisban and Johns on the Ard al-Karak during the early to middle Islamic periods.

Sites of the Crusader period (twelfth century)

The archaeology of Crusader-period Jordan (1100–88) suffers from all the problems associated with Fatimid and Seljuq-period archaeology. Apart from the few monumental castles at ash-Shawbak, al-Karak and Petra, where later Ayyubid and Mamluk additions have often obscured the original Crusader structures (see Chapter 27), the twelfth century is virtually invisible in archaeological terms. The castle at 'Ajlun, now thought to date from between 1188 and 1192, is an Ayyubid construct and will be considered with the thirteenth-century castles of as-Salt and al-Azraq in the following architecture section.

Few excavated sites have produced identifiable twelfth-century occupation, but one major exception is al-Wu'ayra, the Crusader castle of Li Vaux Moise that dominated the entrance to Petra (Brown 1987; Kennedy 1994: 25-27; Vannini and Tonghini 1997). Excavations by Brown during 1987 in two areas of the castle identified two distinct architectural and depositional phases. Phase I was Crusader and dated 1108/1116–88, while Phase II was early Ayyubid and dated to the late twelfth to early thirteenth centuries. In Square 4, set in front of the northeast corner tower, a ceramic-rich sequence of occupational levels was excavated. The earliest was Phase IA, a series of courtyard levels, above which came Phase IB characterized by a floor and domestic occupation levels, and finally Phase II when a secondary enclosure was erected for domestic occupation. At the east tower, Phase II was also identified by the erection of secondary structures, again for domestic purposes. More recent work at al-Wu'ayra has expanded Brown's original conclusions (Vannini and Tonghini 1997; Vannini and Vanni Desideri 1995). Phase I represents the original Crusader castle, built between 1107 and 1116; Phase II the substantial remodelling of the castle defences in the 1140s; Phase III Crusader occupation levels; Phases IIIa and IV a short span of non-military use in the Ayyubid period (roughly equivalent to Brown's Phase II); and four major collapse levels (V-VIII). Crusader to early Ayyubid material, similar to the al-Wu'ayra finds, was also recovered by Brown from Phase I of the Ayyubid Palace at ash-Shawbak (Brown 1988: 230-32). The work at al-Wu'ayra and ash-Shawbak, by successfully isolating twelfth- to early thirteenth-century occupational strata, carries major implications for other middle Islamic sites with reputedly 'Ayyubid-Mamluk' levels, for instance, Dhiban where the Ayyubid-period material includes ceramics of the previous century; hence twelfth-century occupation at Dhiban should be suspected.

Other evidence for twelfth- to early thirteenth-century settlement in Jordan comes from Jarash where two occupation levels were identified in the vaulted passages of the Zeus complex, terminating with a 1202 destruction level containing copious amounts of pottery. Considering the degree of occupation around the Zeus temple and the absence of comparable material at the Temple of Artemis, Seigne has quite plausibly suggested that Tughtigin's stronghold was located in the Zeus area, probably at the south theatre (cf. the Seljuq-Ayyubid fortress at Busra). One site that seems to have experienced a considerable erosion of its position at this time was 'Amman, as the excavations on the citadel, unwalled since the Abbasid-Fatimid period, produced no significant occupation between the 1068 earthquake and the late twelfth century at the earliest (Northedge 1992: 161). Similarly Whitcomb suggests that the early Islamic town of Ayla, badly damaged in the earthquake of 1068, was abandoned with Baldwin's expedition of 1116 (1997a). As no archaeological work has been undertaken around the Mamluk fort to the east where the Crusader stronghold (if it existed) may have been located, any evidence for twelfth-century Ayla (now al-'Aqaba) has yet to be found.

Regional surveys make scant mention of the twelfth century as this period is equally, if not more, invisible as the Fatimid and Seljuq centuries. Of particular hindrance has been the amorphous nature of Hand-Made Geometrically Painted Ware (HMGPW), generally but inaccurately identified as a hallmark of a generalized 'Ayyubid-Mamluk' phase of occupation. However, current understanding of HMGPW does not recognize any differentiation within a 300-year-plus period nor, until recently, a pre-Ayyubid (i.e., late Fatimid-Crusader) variety of the hand-made pottery (on HMGPW, see more below). Furthermore, Ayyubid-Mamluk redevelopment of settlements after the Crusader occupation would have also obscured the twelfth century, and, as with the Fatimid and Seljuq periods, few reliable conclusions can be drawn from survey data on settlement density in the twelfth century.

Sites of the Ayyubid and Mamluk periods (thirteenth–sixteenth centuries)

The range of Ayyubid and Mamluk-period sites in Jordan is extensive, and includes examples with significant architectural features. The major categories range from holy sites adorned with mosques or shrines to fortified localities, typically town-based castles and forts/khans, and from agricultural and industrial establishments to innumerable villages, sometimes with identified cemeteries. The contrast with the earlier Islamic periods, between the Abbasid and Fatimid-Seljuq, is immediately apparent: this was an age of reconstruction and building after the Crusader debacle and has left, in the archaeological record, a more tangible and high-profile presence. Consequently, the number of excavated and surveyed sites with identifiable Ayyubid-Mamluk occupation is considerably greater, although the period is often poorly reported, particularly when the recovery of Islamic remains is not the primary intention of a project.

Important work on sites with major architectural remains includes the controlled excavations at al-Karak and ash-Shawbak by Brown (1988, 1989a, 1989b, 1989c); a study of the castle at 'Ajlun (Johns 1932; Minnis and Bader 1988); the exposure of a bathhouse complex at Hisban (de Vries 1986, 1994), the survey of Mamluk forts at az-Zarqa', Zizia and al-'Aqaba (Glidden 1952; Petersen 1991); soundings in a Mamluk khan at Dhra' al-Khan (Kareem 1997); and the excavation of village mosques at Pella and al-Lahun (Whitcomb in Homès-Fredericq 1989; Walmsley 1992b). The architectural component of these and other sites will be considered in the next section (see also Chapter 27). The intentional (compared to accidental) excavation of settlements with

Ayyubid-Mamluk occupation includes Khirbat Faris, Bayt Ras, Khirbat 'Ayn Jenin (Hart 1987: 45), Nakhal (Schick et al. 1994), Tall Abu Qa'dan and Tall Abu Sarbut (Haas et al. 1989; 1992), Tall Qudsiyah and Pella. Sites with evidence (usually ceramic) for an Ayyubid-Fatimid presence figure prominently in large-scale regional surveys. Usually the sites are multi-period with evidence for earlier settlement periods but in other instances sites appear to have only an Ayyubid-Mamluk presence, perhaps an indication of settlement expansion to some degree in the thirteenth–fourteenth centuries.

Up to 1994, excavations and surveys in Jordan had identified Ayyubid and Mamluk occupation at 795 out of a total of 3843 historical sites (i.e. excluding prehistoric and unknown-period sites), or 20.69 percent, with 442 of these sites (55.6 percent) also recording Umayyad-Fatimid occupation (Palumbo 1994). While this latter figure is almost certainly too low for the reasons already elucidated, these figures do suggest two aspects of Ayyubid-Mamluk settlement in Jordan: a reasonably high proportion of site occupation and considerable and largely unbroken continuity from the earlier Islamic periods. The JADIS figures, however, when treated by individual sector (Table 15.1), also indicate that the density of site occupation was much higher in the north of the country (range 20-40 percent, average 28.19 percent) while south of al-Karak the density abruptly fell to within a low 3-12 percent range (average 4.84 percent). As far as it is possible to tell, the figures show greater uniformity across the different environmental zones of the Jordan Valley, the elevated mountains and plains, and *badia*.

While this rather severe dichotomy between the north and south of Jordan may be a product of both a poor understanding of regional ceramic traditions in the thirteenth–sixteenth centuries and the research objectives of survey teams, it may alternatively indicate a relatively lower density of sedentary population in the south due to the greater resources of the north with the intensification of a village-based, cash-crop economy. Unfortunately, developments during the three centuries of Ayyubid-Mamluk hegemony in Jordan cannot be chronologically isolated due to an elementary knowledge of the pottery. Similarly, as it is impossible to isolate early Ottoman ceramics, the number of northern sites attributed to the Ayyubid-Mamluk period could have been inflated by the inclusion of sites with late sixteenth–seventeenth-century occupation (Johns 1994: 19-22). Thus, any settlement processes that produced this differential cannot be identified on the basis

JADIS sector	Ayyubid-Mamluk	Total no. historic sites	% of total	zone
1	187	531	35.22	Irbid/N. Jordan Valley
2	176	525	33.52	'Ajlun/Jarash
3	29	135	21.48	al-Mafraq
4	121	530	22.83	'Amman/S. Jordan Valley
5	38	111	34.23	Kharraneh
6	28	178	15.73	Dhiban
7	11	44	25.00	Qasr al-Tuba
8	8	20	40.00	al-Azraq/Ruweished
9	93	362	25.69	al-Karak
10	29	298	9.73	Ader
11	16	306	5.23	At-Tafila
12	49	439	11.16	Wadi al-Hasa
13	0	2	0.00	Bayir
14	10	340	2.95	al-'Aqaba/Petra
15	0	22	0.00	Ma'an/al-Jafr

Table 15.1. Total number of Ayyubid-Mamluk sites and percentage of total number of historical-period sites by JADIS sectors.

of these figures alone, although they raise tantalizing issues of variable site density, changing land use and divergent economies.

A relatively good record of regional surveys and excavations in the al-Karak district has permitted an unusually reliable and particularly illuminating survey of Ayyubid-Mamluk occupation in the towns and the countryside of the Ard al-Karak (Brown 1984, 1992; Johns 1994: 20-22; Miller 1991). The excavations by Brown at al-Karak and ash-Shawbak castles have been particularly decisive in identifying Ayyubid and Mamluk phases in a continuous site history from the Crusader period to the recent past. At ash-Shawbak, Brown (1988, 1989c) conducted excavations in the Ayyubid Palace complex, concentrating on the Reception Hall, the North Palace and the East Palace. Four major phases were identified: Phase I (Ayyubid), the construction of the Palace complex; Phase II (later Ayyubid), represented by additions and rebuilds in the North Palace; Phase III (Mamluk, perhaps fourteenth–fifteenth century), which saw major spatial changes with the abandonment of the Reception Hall and alterations to the North and East palaces; and Phase IV (Ottoman). The phases were determined by architectural changes. Few chronological indicators were recovered except a limited number of ceramics; unfortunately their broad dating makes them particularly unsuitable for the task. Interestingly, however, HMGPW was largely absent from the otherwise overwhelming Mamluk assemblage of Phase III, and was seen by Brown as a reflection of the higher social status of the castle occupants. A small-scale excavation at al-Karak also concentrated on the Islamic Palace, a large building complex on the upper court originally interpreted as Crusader private apartments (Brown 1989a, b). A sounding in the south *iwan* (recess) of the Palace Reception Hall identified two major occupational phases: Phase I (Mamluk), which produced ceramic and numismatic evidence for a fourteenth-century construction date (indicating that this complex should be equated with the Palace of Sultan as-Nasr Muhammad, built in 1311); and Phase II, Ottoman. As at ash-Shawbak, HMGPW and local monochrome glazed wares were scarce in the Mamluk levels whereas imported glazed and wheel-made cream wares dominated the corpus. Again this was seen as evidence for the bifurcated nature of Mamluk society in Jordan, divided between the urban ruling elite and the rural peasantry.

Stemming from a detailed analysis of the results of Miller's al-Karak Plateau survey, Brown (1992) has argued for a major demographic shift on the Ard al-Karak from the highly productive arable plains in the centre of the plateau to the more broken and isolated terrain located on the south and southwest. As a zone of perennial springs but limited cultivable land, food production became more intensive, revolving around irrigated fields and orchards. Brown suggests that this shift took place after the fifteenth century as a result of competition for resources with nomadic groups (the Bedouin), but Johns (1994: 22) has firmly and convincingly argued for a seventeenth- or eighteenth-century date for this change and rejected the theory of competition for land with pastoral groups.

Architecture: religious and secular, public and private

The architectural heritage of middle Islamic Jordan is modest and almost exclusively Crusader, Ayyubid and Mamluk in date. This is not to say that nothing was built in the Fatimid period. Very possibly, the Shi'ite Fatimids enlarged or embellished the shrine at Mazar to commemorate the three Companions of the Prophet including Ja'afar ibn Abi Talib, brother of the Orthodox Caliph 'Ali most favoured by the Shi'ites. Tombstones and other finds certainly attest a strong Fatimid interest in Mazar. The burial places of other Companions, for instance, that of Abu Ubaydah in the Jordan Valley, could have received similar attention. However, these and any other possible Fatimid structures have not survived due to the extent of Ayyubid-Mamluk reconstruction. The holy sites of Palestine and Jordan generally attracted a higher profile in the ninth and especially tenth centuries, and quite probably the erection of a new Ayyubid-Mamluk mosque or shrine was undertaken at the expense of an existing (but perhaps derelict) Fatimid, or Fatimid-augmented, structure.

Castles, forts and other fortifications

One of the most tangible expressions of the Crusader presence in south Jordan is the presence of impressive, if ruined (to varying degrees), castles at al-Karak, ash-Shawbak and Petra (detailed in Chapter 27). The Ayyubid and, after them, Mamluk elites that replaced the Crusader lords at al-Karak and ash-Shawbak repaired and extended the fortifications of both castles, building towers and walls. At ash-Shawbak, the Ayyubids inserted a palace in an Islamic style, and at al-Karak the Mamluks did likewise; these palaces parallel examples at Raqqa, Cairo and Jerusalem (Brown 1988, 1989a, 1989b, 1989c).

The Ayyubid princes similarly secured their domination over north Jordan following the Crusader defeat at Hattin (1187) by constructing castles at 'Ajlun (between 1188 and 1192) and as-Salt (1220), a watch tower over 'Amman, a fortlet at az-Zarqa', and a more substantial fort at al-Azraq (by Izz ad-Din Aybeg, Lord of Salkhad in 1236–37).

The watch tower at 'Amman was built on the south edge of the Citadel, overlooking the lower town where the Ayyubid and Mamluk settlement would have been located. Its plan is a 9.45 by 7.55 m rectangle, with a single internal chamber measuring 4.8 by 3.1 m and entered through a doorway in the north wall (Wood 1992: 113-14, 125). As with other Ayyubid towers, this chamber would have been vaulted. Arrow slits in rectangular recesses penetrate the west, south and east walls, while a staircase within the north wall once led to the roof. The tower was constructed out of large limestone blocks, many re-used from the nearby Roman-period Temple of Hercules. Of particular note are two column drums laid horizontally in the south wall and wedged in place with triangular-shaped stones. Horizontally laid column drums were a feature of Ayyubid military architecture (cf. the theatre/fort at Busra, with towers dated 1202–28), and the discovery of an Ayyubid *fals* from Damascus sandwiched between two floor levels confirms an Ayyubid date (late twelfth or early thirteenth century) for the tower at 'Amman.

Of probable similar date, based on architectural comparisons, is the small fort at az-Zarqa' named Qasr Shabib (Petersen 1991). The fort sits on an elevated wadi spur above the az-Zarqa' River, and is an uncomplicated structure. Made of large limestone blocks, some of which are clearly reused, the plan is a plain 13.75 m square with walls about 3 m thick. The single entrance is located in the north wall, and consists of a pointed-arch doorway set back within a 3.4 m high double-voussoired arch similarly pointed. A recess above the doorway once held an inscription (cf. the al-Azraq castle, below), while behind the outer arch and above the doorway was a concealed machicoulis (cf. the outer gateway at 'Ajlun, below). The doorway leads into a barrel-vaulted chamber and a single room, from the northeast corner of which a staircase led to the roof. Arrow slits, now blocked, in pointed-arch recesses are visible in the west, south and east walls. The building once stood much higher than its surviving 8 m, but the upper storey is considerably ruined. Nevertheless, its plan and architectural features (doorway, arrow slits) suggest a thirteenth-century date for Qasr Shabib.

The Ayyubid castle at as-Salt, now largely destroyed, was built on a hill top to the north of the town, and boasted stout towers of undressed stone and rock-cut dry moats. That as-Salt was equipped with a castle and 'Amman, it seems, only with a watch tower represents the changing fortunes of these two centres in the twelfth–thirteenth centuries, for by this date as-Salt, as already noted, had replaced 'Amman as the chief town of the Balqa'.

Also to benefit from its strategic position was the village of 'Ajlun in the Jabal Awf, immediately above which was built a small castle on a hilltop overlooking the central Jordan Valley and Wadi Kafrinji, a major route into the Jordanian highlands from the valley. Constructed by the Amir Izz ad-Din Usama between 1188 and 1192 (and probably not the commonly accepted date of 1184; see Humphreys 1977: 77), the castle centred on an unevenly shaped quadrilateral, determined by the shape of the hilltop, with four square towers at each corner and a single entrance in the east wall (Figure 15.5). An outer gate was positioned between the northeast and southeast towers, and, while seemingly part of the original structure (it bonds with the southeast tower), the doorway proper might be a later reconstruction. At a lower level on the east and south sides were two baileys, of which the eastern was the larger and better equipped having a tower at either end. All the towers and intervening walls were periodically pierced by plain arrow slits. Water was provided by five rock-cut cisterns under the castle. In the early thirteenth century, after the castle was captured from Izz ad-Din Usama, the defences were strengthened by the construction of a massive tower on the south side and the original structure given added height to match the new tower. To accommodate these changes, two new gates were added, the inner one of which was decorated with birds and provided with a machicoulis accessed from the first floor (Figure 15.6). An inscription on the south tower dates these additions to 1214–15, and they were undertaken by Izz ad-Din Aybeg, Lord of Salkhad, for the Ayyubid prince al-Mu'azzam (d. 1227), son of al-'Adil and his successor in Damascus (1218). The dry moat that surrounds the castle must have been cut, or at least completed, as part of these additions for it makes allowance for the projection of the south tower. Today the castle at 'Ajlun, named Qal'at ar-Rabad, shows evidence of later repairs but the plan is as Izz ad-Din Aybeg left it, and hence it is a fully Ayyubid structure.

The al-Azraq fort was also constructed by Izz ad-Din Aybeg, but later, in 1236–37 as recorded in an inscription over the entrance (Figure 15.7). Perhaps originally Diocletianic in date, the solely basalt structure approaches a square (c. 79 by 72 m) with projecting rectangular corner towers and intervening towers of irregular spacing and number (Kennedy 1982: 75-96; Figure 15.8). The basic plan conforms to a late Roman type, and the original masonry is identifiable in places, especially the outer walls. Differing construction styles indicate that the thirteenth-century work involved considerable reconstruction of the towers, walls and internal features, even taking into account Druze additions in the twentieth century. Entry to the central courtyard of the Ayyubid fort was through a projecting tower set off-centre in the south wall (the original entrance, now gone, was probably between the two towers in the east wall). The tower defended a right-angled entrance with a doorway set back under a slightly pointing arch and equipped with heavy basalt door leaves, the left of which is a recent restoration. Above the arch is the inscription, an arrow slit (one of four in the room above the entrance, each set in pointed arches) and a machicoulis. The tower once had a third storey, now largely destroyed. Inside the fort are rooms of varying dimensions opening out onto a spacious courtyard. The originally three-storied group in the middle of the west wall had three large windows in the courtyard wall of the second storey, and has been appropriately called the 'praetorium' for it was clearly the nucleus of the fort.

Figure 15.5. Ajlun castle: plan showing the original structure (1188–1192) with gateway at (A), and early thirteenth-century additions with an inner (B) and outer (C) gateway (based on C.N. Johns 1932: plate 21).

Large rectangular arched rooms along the north wall probably functioned as stables. Water could be accessed through a subterranean room on the east wall. Very possibly, these parts of the fort served similar functions in both the late Roman and the Ayyubid periods, as the structures appear to form part of the original fort yet retained their essential architectural integrity with Izz ad-Din Aybeg's rebuild. In the northeast quadrant of the courtyard is a small free-standing mosque, which probably dates to the Ayyubid garrisoning of the fort (see below).

The extent of military building in the later twelfth and thirteenth centuries reveals that the Ayyubid princes and their governors paid particular attention to securing the regions of north and south Jordan. The Crusaders, although critically weakened by the loss at Hattin, still remained firmly entrenched on coastal Palestine, which ensured the continued prominence of the inland Jordanian routes between Cairo and Damascus. Existing Crusader castles in the south were restored and improved while new strongholds were built in the north to defend the main routes and monitor the local population, some of whom had spent nearly three generations under Crusader influence if not rule.

In the following Mamluk period, especially after the final expulsion of the Crusaders from Palestine, attention shifted to improving security on the main roads especially those that passed between Damascus, the Hijaz and Egypt. On the *Darb al-Hajj* (Pilgrimage route) from Damascus to Makkah (Mecca) a small but solid fort was built at the strategic site of Jiza (az-Zizia), where a large reservoir provided water for travellers especially pilgrims and armies (Petersen 1991; see Chapter 28). Although modified in more recent times by the addition of second-storey windows, the plan is original and consists of a main rectangular structure 21.5 (N–S) by 12.2 (E–W) m and a smaller square annex on the east side which continues the line of the southern facade of the fort. Walls of large squared limestone blocks average 2.5 m thick, and were originally perforated by slit windows. The doorway, positioned flush with the south facade, has a flat lintel and is surmounted by a slit window then a

Figure 15.6. Ajlun castle: the inner gateway to the additions of 1214–15 (at B on the plan), featuring a broad machicoulis and an arch decorated with two birds. View to the northwest (photo by A. Walmsley).

Figure 15.7. Azraq: entrance to the fort, 1236–37. View to the northwest (photo by A. Walmsley).

graceful triple-arched window. Inside are two storeys. The ground floor is a large chamber with a slightly pointed barrel vault, from which a staircase leads to the upper floor in the thickness of the east wall. Upstairs is a complex of four rooms around a courtyard, all of later (probably nineteenth-century) construction. The doorways have charming rosette-decorated flat lintels surmounted by a relieving arch and a circular vent in the spandrels. Other structures once existed near the fort, including a mosque, but these are no longer extant. To the west of az-Zizia at Hisban soundings have identified a large and well-built early Mamluk structure, suggested to be a khan, southwest of the acropolis (de Vries 1994: 163). A khan at Hisban is to be expected; the town served as a major road station and functioned as the capital of al-Balqa' for a while.

A fortified khan was also constructed at al-'Aqaba which, although substantially modified inside, has a well-preserved gateway of the late Mamluk period (Glidden 1952). The current fort seemingly replaced a Crusader or, more likely, Ayyubid structure around which a new settlement had arisen in place of the ruined Ayla to the west (Whitcomb 1997a). Abundant fresh water and the reinstitution of the pilgrimage route around the north shore of the Red Sea ensured the town's importance throughout the Mamluk period. The fort is constructed of small squared blocks laid in even courses of alternating colours. It was equipped with stout round corner towers and an unpretentious gateway, consisting of an arched entrance between two half-round towers, mid-point in the north wall. An inscription spanning both sides of the doorway declares that the fort was constructed in the time of Sultan Qansawh al-Ghawri (1501–16), the penultimate Mamluk sultan and a great builder. The doorway, which was reduced in size at a later date, leads into a passageway roofed with a pointed barrel vault. Beyond stands a large open courtyard, which once was flanked with stables, storehouses and sleeping quarters for travellers and pilgrims. The style and construction of the al-'Aqaba fort confirms a late Mamluk date, although

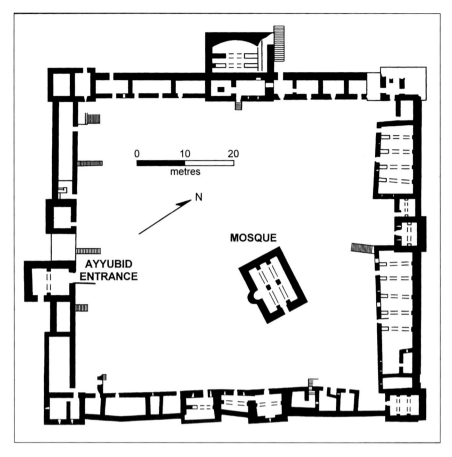

Figure 15.8. Azraq: plan of the fort, with the mosque in the centre (based on Kennedy 1982: 74, fig. 14).

al-Ghawri's contribution may have been limited to a substantial refurbishment around the gateway rather than a complete rebuilding of the fort (cf. the Citadel of Aleppo). The fort played a key role during the Arab Revolt in the drive for independence from the Ottoman empire, as recalled by the Hashemite coat of arms above the fort entrance.

Road stations were also constructed on the route known as the Darb al-Quful that passed from Damascus into Palestine through Irbid and the north Jordan Valley in the Mamluk period. At Dhra' al-Khan, at the foot of the pass to Irbid in the Jordan Valley, excavations have exposed a major caravanserai, rectangular in plan (64 by 54 m) with 1.2 m thick buttressed walls constructed from large limestone blocks (Kareem 1992, 1997). Rooms were arranged around a central courtyard, entered by a gateway in the middle of the northern wall. Coins suggest a construction date under the Mamluk sultan Qalawun (1280–90) and a major reconstruction under an-Nasr Muhammad (third reign, 1310–41). Occupation continued into the Ottoman period. To support these developments Baybars, Barquq and other Mamluk sultans had bridges built or rebuilt over the Jordan river at a number of locations between Damiyeh (in 1265 under Baybars) and Jisr al-Majamia' to connect north Jordan with Baysan (Beth Shan), Tabariyah and Nablus.

Mosques and shrines

The Ayyubid and Mamluk reconstruction of Jordan following the Crusader interregnum was especially directed towards restoring and improving the religious infrastructure of the region. Two categories of mosque were built: commemorative mosques (*mashhad*) at holy sites usually constructed at state expense, and congregational mosques (*masjid*) in towns and villages financed by the local community. At the Muslim holy sites new mosques replaced earlier and, especially in the former Crusader regions, probably neglected structures, including Mazar and Mu'ta south of al-Karak, at the tombs of Abu Ubaydah at Amata and Mu'adh bin Jabal near North Shuna in the Jordan Valley, and at ar-Raqim near 'Amman. Most notably a shrine was built atop Jabal Harun near Wadi Musa (Petra), for seemingly the Christian monastery encountered by Baldwin I in

1100 did not continue after the mid-thirteenth century. Numerous mosques were also built to meet the needs of towns and villages, and these are known from the modest specimens from al-Lahun, Tabaqat Fahl and al-Azraq.

The commemorative shrine on the summit of Jabal Harun was constructed over the reputed tomb of Aaron, the brother of Moses (Musa) and also a prophet in Islam (Qur'an 19.53, 'Miriam'). Later Muslim traditions recount the death of Aaron in a cave after he and his brother were attracted to enter by a bright light, and, seeing there a golden throne, Aaron was drawn to sit upon it at which moment the Angel of Death struck. Moses was subsequently accused by the Israelites of killing his brother, but miraculously his innocence was proven. The present structure, which at c. 1350 m above sea level is visible from a great distance, is dated by an inscription to the year 1338–39 in the time of the third reign of the Sultan an-Nasr Muhammad. Constructed from local stone, much of it re-used, the shrine measures 12.5 (N–S) by 9 (E–W) m and has a flat stone roof with a prominent dome in the southeast corner. The single chamber, roofed with low vaults resting on a central pier, is entered by an arched doorway located at the south end of the west wall. Inside by the south wall stands the cloth-draped cenotaph of Aaron, positioned midway between the doorway and a *mihrab*. Above the *mihrab* stands the dome. At the opposite (northern) end of the chamber in the west corner a descending staircase leads to the sepulchre of Aaron, a rock-cut chamber with 19 loculi and clearly a tomb of much earlier date (Nabataean or Roman?). The mid-fourteenth-century Muslim shrine partially rests on the foundations of an earlier and larger memorial, almost certainly a Byzantine-period church, which in 1217 was still inhabited by two Greek monks (Peterman and Schick 1996). That only fragments of the church were incorporated in the Muslim shrine, including marble screen and column pieces, suggests that the summit church was abandoned for a period before the construction of the current structure. On a plateau to the west of the summit are the remains of a large monastic complex some 60 (N–S) by 50 (E–W) m in area, consisting of rooms, a chapel/church, courtyards and cisterns. Conceivably, this was the monastery 'joyously found' by Baldwin I, although this reference could equally apply to the summit church.

Another holy site to receive renewed attention in the Mamluk period was the Cave of the Sleepers (Ahl al-Kahf ar-Raqim), located a short distance from the Hajj route southeast of 'Amman. Although the Christian legend of the sleepers locates the cave near Ephesus in Turkey, the al-Balqa' cave of ar-Raqim is specifically linked to the Qur'anic version of the story, for instance by two tenth-century geographical sources (al-Istakhri, al-Maqdisi) and al-Yaqut in the thirteenth century. In the Qur'an (18.9-26; the whole chapter is named 'al-Kahf' after the tale) the youths, variously given as three, five or seven in number, and their dog hide in a cave to preserve their faith in one God, for others had taken to believing in other gods. They sleep for 309 years, the youths turning occasionally (as one does in sleep), with their dog, forelegs outstretched (as dogs can sleep, yet remain alert, with their heads resting on their forelegs), at the doorway. Upon waking, they reckon that they have been asleep no more than one day. It is a miraculous story, conveying the promise of resurrection and extolling the virtues of Islamic monotheism in the face of persecution and rejecting compromise—precisely the challenges facing the fledgling Muslim community in Makkah at the time this chapter was revealed to Muhammad.

The inspiration to build a mosque over the cave would have come directly from the Qur'an (18.21):

> Thus did We make their case known to the people, that they might know that the promise of God is true, and that there can be no doubt about the Hour of Judgement. Behold, they dispute among themselves as to their affair. (Some) said, 'Construct a building over them': Their Lord knows best about them: those who prevailed over their affair said, 'Let us surely build a place of worship over them' (Yusufali translation).

There are, however, two mosques at al-Kahf, a square structure with a courtyard and minaret above the cave, and a rectangular building with a long east–west axis in front of the cave. The latter mosque conforms to a Mamluk style in Jordan and in all probability is thirteenth or fourteenth century (Figures 15.9, 15.10); architecturally the mosque above the cave appears earlier as one would expect from the Qur'anic account. The walls of the upper mosque survive to 1.5 m or less in height, and are made from large limestone blocks up to a metre in length. A deep *mihrab* in the south wall confirms that this building functioned as a mosque, but the large blocks suggest the conversion of an existing Roman or Byzantine structure (if *in situ*) or are re-used. Two rows of columns and wall pilasters suggest that the mosque roof was originally supported by

Fatimid, Ayyubid and Mamluk Jordan and the Crusader Interlude

Figure 15.9. Al-Kahf: view of the Mamluk mosque, looking south. Cave entrance located at bottom left-hand corner (photo by A. Walmsley).

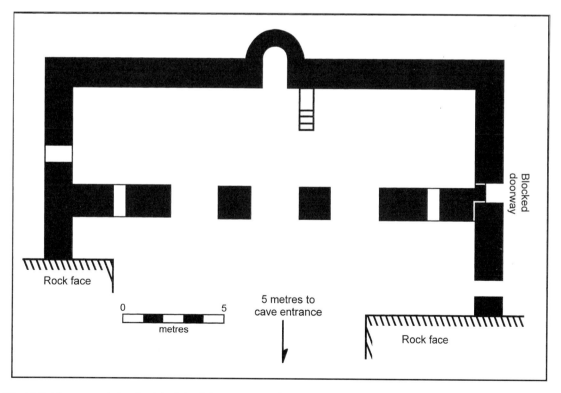

Figure 15.10. Al-Kahf: measured sketch of the Mamluk mosque, south to top.

535

arches. A doorway in the east wall opened out onto a portico as evidenced by two columns, and was flanked on the north by a minaret and staircase, of which four steps remain. The date of this mosque is uncertain, but the square plan, side entrance and deep *mihrab* complies with the small mosque style of Umm al-Walid, Khan az-Zabib and Jabal Usays, usually attributed to the Umayyad period by association with larger adjacent complexes (King *et al.* 1983: 399-405). Interestingly, as with the mosque at Umm al-Walid, the upper structure at al-Kahf was mistakenly interpreted as a temple (Brünnow and von Domaszewski 1905: 195-206). The lower mosque sits immediately in front of the cave entrance and, as side walls butt up to the rock face, access to the cave entrance was only through two doorways, one later blocked, in the west wall (Figure 15.10). The open doorway gave entry into an irregular courtyard positioned between the mosque and the rock face. From here entry to the mosque, measuring c. 19.6 (E–W) by 4.65 (N–S) m, was through three broad (2.37 m) doorways in the north wall, probably originally arched and permanently open. In the centre of the south wall was a 2 m-deep *mihrab* and, to its right (west), a *minbar* built of stone blocks with four steps remaining. As the layout of this mosque is very similar to the village mosques of the thirteenth and fourteenth centuries at Tabaqat Fahl and al-Azraq (below), a late Ayyubid or early Mamluk date is very likely. The cave, the centre of all this activity, is a much earlier (late Roman or Byzantine) construct. A doorway in the decorative facade leads down into a rock-cut chamber flanked with three deep arcosolia, the two lateral arcosolia having three graves apiece. Apart from its legendary associations, the tomb has no exceptional features.

As already discussed, particular attention was paid to embellishing the burial places of the Companions of the Prophet in the Ayyubid and Mamluk periods, notably Abu Ubaydah in the Jordan Valley and the tombs of Ja'afar ibn Abi Talib, Zayd ibn al-Harithah and Abd Allah bin Rawahah at Mazar. Accordingly, these sites figure prominently in the sources of the thirteenth–fifteenth centuries. Accompanying these developments was the recognition of many burial places belonging to a local 'saint' or *wali*. One example is the shrine of Abu Sulayman ad-Dirani near ash-Shawbak (Khammash 1986: 81-83). Of probable Mamluk date, the shrine is rectangular in shape (17 by 8.2 m) with a dome emphasizing the burial place at the west end (Figure 15.11). A single doorway in the north wall gives entry to a rectangular, tunnel-vaulted room with a recessed, pointed-arched *mihrab* in the south wall opposite the entrance. To the west, the tomb is marked by a headstone and foot stone of marble within an oval of stones, above which is the partly collapsed dome. The tomb chamber is a modest but elegant structure with a second, plain *mihrab* in the centre of the south wall. The remains of the dome rest on an octagon supported by four low-slung pointed arches that spring from piers in each corner of the room. The octagon, with four corner decorative squinches and four small windows at the apex of the arches, cleverly and decoratively acts as a transition between the square room and the circular base of the dome. Khammash (1986: 83) records an interesting account of the ceremonies at the shrine:

> In weekly ceremonies at the gravesite, the villages would pay tribute by drumming and lighting oil lamps or candles. They would bring new fabric to cover the grave and would tear off small pieces of the old fabric to keep for good fortune. Prayers for the sick and other supplications were addressed to the wali. Women dyed their hands and their hair with henna, which was also applied to the walls as an offering. 'They bring incense, henna and sacrifices. Women walk about the shrine. Before, it was said that there used to be ostrich eggs hanging in the dome in a net. There were lamps, some of glass and some of clay,' explained the old caretaker.

Similar, if not the same, ceremonies have probably been enacted at this shrine, and many others like it, since Ayyubid-Mamluk times (and possibly long before), as local holy men served an important function especially in rural communities. Country women in particular were attracted to local shrines as access to the congregational mosque of the town proved geographically and socially difficult.

The restoration of the social infrastructure of Jordan after the expulsion of the Crusaders equally required the erection of community mosques in towns and villages. Serving both a social and political purpose, this activity was widespread with numerous examples known and many more being identified as archaeological and anthropological work intensifies in Jordan. At Tibnah near Irbid Khammash has documented the 'old mosque' (*al-jami'a al-qadim*) of the village which he suggests is of Ottoman date (eighteenth century). Built over a cave, the mosque is 'simply and powerfully proportioned, ... a perfect square, 12.5 by 12.5 meters, approximately 5 meters high' (Khammash 1986: 60). Walls are solidly built in even courses of roughly dressed stone blocks averaging 24 by 34 cm,

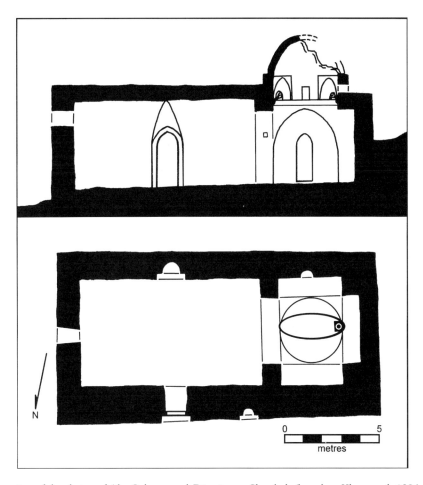

Figure 15.11. Plan and section of the shrine of Abu Sulayman al-Dirani near Shawbak (based on Khammash 1986: 81).

with a double-arched doorway offset to the east in the north wall. Inside is a single chamber with a *mihrab* opposite the doorway in the south wall and a central pier to support the roof, cross-vaulted except for the southeast quadrant where a dome once stood over the *mihrab*. The similarities with the Shrine of Harun are clear (vaulted chamber, dome over *mihrab*, cave underneath), suggesting a Mamluk date for the Tibnah mosque (conversely, it could be argued that the Harun shrine is an Ottoman rebuild and the inscriptions date an earlier structure, but this seems less likely).

Recent excavations at Tabaqat Fahl and al-Lahun have firmly identified the Mamluk-period village mosques for these two centres. The mosque at Tabaqat Fahl was excavated by Bisheh as part of the University of Sydney Excavations at Pella in 1982 (Walmsley 1992b). The plan is very similar to the Mamluk mosque at al-Kahf, consisting of a plain rectangle measuring 15.8 (N–S) by 6.85 (E–W) m formed by two-faced, rubble core stone walls with a deep *mihrab* midpoint in the south (*qiblah*) wall (Figure 15.12). Engaged columns, clearly re-used, were built into the internal face of the *qiblah* wall either side of the *mihrab*. Immediately to the right (west) of the *mihrab* the lower section of a minbar was uncovered with three stone steps remaining. Opposite the *mihrab* in the north wall was the only doorway into the mosque, although three elongated windows either side of the entrance would have admitted plenty of light. The roof of the mosque, probably of perishable materials, was supported by two rows of evenly spaced columns from which, at about a metre above the floor level, once sprung high-pointed arches, the voussoirs of which were found fallen on the packed earth floor of the mosque. Floor to ceiling height would have been around 3.85 m. Short buttresses carried the thrust of the arches on the end columns to the outside walls, the resultant alcoves being hidden by the construction of thin dividing walls. Some care was taken in the building of the mosque, with re-used squared blocks being used to strengthen the corners of the building. All but the south wall were built in 35 cm-deep foundation trenches, while the load-bearing columns were

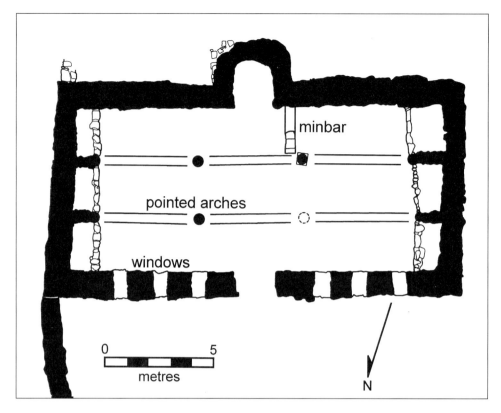

Figure 15.12. Tabaqat Fahl: plan of the Mamluk mosque (drawing: Hart/Pella Excavations).

sunk 50–75 cm below ground level. Ceramics and coins suggest a fourteenth- to fifteenth-century date for the construction and use of this mosque. Village houses contemporary with the mosque have been located to the north, and excavations in this area, while limited in extent, have uncovered a well-stratified sequence of Mamluk ceramics.

The similarities in overall plan between the mosques of Tabaqat Fahl and al-Azraq are immediately apparent. The al-Azraq mosque is a free-standing structure of local basalt in the courtyard of the fort, one imagines built as part of Izz ad-Din Aybeg's reconstruction in the mid-thirteenth century. Essentially rectangular in plan, the long axis of the mosque runs parallel to the *qiblah* wall, as does a double line of high pointed arches that rest on deeply buried columns. However, at al-Azraq the mosque is not as wide as the example at Tabaqat Fahl, with two instead of three arches per row, while the arches terminate on pilasters, not engaged pillars. Also different is the location of the doorway, which at al-Azraq is in the short east wall, and the inclusion of only two elevated windows in the *qiblah* wall makes the mosque interior very dark. In spite of the differences, these two mosques clearly belong to a common tradition—the arch spans are within 5 cm of each other—that can be dated to the late Ayyubid and early Mamluk period (thirteenth–fourteenth centuries) in Jordan.

The small mosque at al-Lahun, investigated by Whitcomb as part of the Belgian Committee of Excavations in Jordan (D. Homès-Fredericq, Director), is rectangular in shape with walls of rough field stones. In the short south wall is a deep *mihrab*, and facing it in the north wall is the sole doorway. Thus, the long axis of the mosque is orientated N–S. No windows or *minbar* are present. Pottery and a coin recovered from within the building suggest a late Mamluk date (c. fifteenth century) for the al-Lahun mosque. Houses belonging to a small village have been identified nearby.

Other secular architecture: urban and rural, domestic and industrial

Evidence for secular architecture in the Fatimid and Seljuq periods is meagre due, in part, to the extensive building programs of the Crusader, Ayyubid and Mamluk elites. From the eleventh century the emphasis in secular architecture is on military or quasi-military architecture (castles, forts and khans as described above). Furthermore, in the highlands, stone-built domestic structures, often of Roman date, show continuity of use from earlier periods, with superimposed floors and internal modifications to the structures the only evidence for ongoing occupation. Better preserved is the domestic and industrial architecture of the Jordan Valley. Construction was widespread due to a prosperous agricultural economy, while at village sites the use of unbaked brick and the frequency of earthquakes has resulted in the preservation of superimposed levels of domestic architecture.

The few sites with reported Fatimid-Seljuq domestic occupation indicate an easy continuity from the Abbasid period; structures were modified or rebuilt, depending on requirements. As described earlier, the central monument of al-'Aqaba was redesigned in the Fatimid period as an aristocratic residence, featuring a central courtyard with a porch and eight rooms (Whitcomb 1988b). Similarly, on the 'Amman Citadel, the Umayyad Palace, including the Reception Hall, was extensively converted into dwellings and the nearby stone houses of Umayyad and Abbasid construction were substantially modified before their destruction in the mid-eleventh century (Northedge 1992: 159 and above).

The improved economic environment after the expulsion of the Crusaders had a deep but varied impact on the new generation of domestic architecture in the Ayyubid and Mamluk periods. In many places, especially in the uplands, the erection of completely new houses was unnecessary as existing stone structures simply continued to be used and, where necessary, were altered to meet domestic requirements. This reutilization of architectural resources is clear from work at Khirbat Faris. Structures in the House 2 area show an unbroken sequence of occupation, represented by a series of surfaces, through the twelfth/thirteenth century until perhaps as late as the sixteenth century (McQuitty and Falkner 1993). This sequence followed on without major interruption from earlier Fatimid and Seljuq settlement outlined earlier. Yet between about the eighth and nineteenth centuries, no major construction work was undertaken, as existing structures, when re-roofed and given new floors, made perfectly adequate quarters. While investigations at other upland sites do not match the detail of the Faris work, instances of existing architecture being continuously used or reutilized in the Ayyubid and Mamluk periods are recorded from Bayt Ras, 'Amman, Umm al-Quttayn (after a proposed hiatus from the mid-eighth century), Umm ad-Dananir, Qastal and Udhruh. Of particular interest are the reoccupied sites of Balu', al-Lahun and 'Ara'ir, where the Mamluk village was constructed within, and by re-using elements from, the derelict Iron Age houses.

Surveys and excavations have turned up evidence for new Ayyubid and Mamluk domestic architecture at a number of sites, for instance Gharandal, Nakhal, Hisban, Dhiban and Khirbat 'Ayn Janin. However, detailed information is scarce, as sites with thick Islamic deposits have been assiduously avoided in the past, and most discoveries have been incidental to wider research programmes. Consequently, the archaeological evidence may favour sites with continuous or resettled histories, and overlook developments and innovations in the domestic architecture of new establishments. At Dhiban part of a large house and other structures were uncovered (Tushingham 1972: 83-84). The house, made from re-used stone, comprised a courtyard and large rooms (c. 6 by 7 m), each spanned with closely spaced transverse arches. Bins, ovens and a cistern were found in associated occupation levels. However, elsewhere at Dhiban pre-existing structures, obviously still extant at the time, were converted into dwellings. At the multi-period sites of Gharandal and Nakhal domestic structures were built within the remains of Byzantine churches. The Gharandal structures, part of a larger settlement (Figure 15.13), incorporated existing walls built in a continuous sequence from the ninth century, when the ecclesiastical function of the church had apparently ceased. Well reported are the thick Mamluk strata at Hisban (Str. 3, 1260–1400; Str. 2, 1400–56), where a housing quarter was excavated to the west of the acropolis complex of bath and reception hall (de Vries 1994: 161-65). These were new stone structures, haphazardly built on a freshly cut terrace, with each unit typically consisting of a number of originally vaulted rooms around a court. The Hisban regional survey located many outlying sites with Ayyubid and Mamluk occupation (52, or 35 percent of all sites) with a village architecture distinguished by arches, vaulted buildings, semi-subterranean rooms and caves (Ibach 1987: 191-94).

Differences in domestic architectural styles to those in the uplands can be observed in the Jordan Valley. A major variation was in the building materials utilized, with most houses on the valley floor being made of sun-dried mudbrick, not stone (which is scarce in the valley), during the middle Islamic period. Where necessary, houses were replaced as they became derelict or collapsed, resulting in deep, stratigraphically distinct sites made up of a long sequence of surfaces and mudbrick collapse as uncovered at Tall Abu Qa'dan

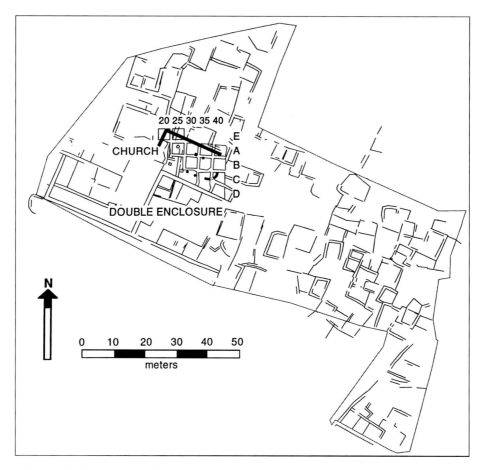

Figure 15.13. Gharandal, near Tafila: plan of the Department of Antiquities area showing Mamluk structures within and around earlier Nabataean and Byzantine monuments (Barnes/Gharandal Archaeological Project).

and Tall Abu Sarbut, both village sites near Dayr 'Alla (Franken and Kalsbeek 1975; Haas *et al.* 1992). At Tall Abu Sarbut walls of mudbricks were built on foundations of field stones before both the walls and floor were covered with a yellowish plaster.

Overall domestic constructions of the Ayyubid and Mamluk periods were basic and practical, representing personal undertakings rather than state or community projects. Houses were generally one to three adjacent single-storied rooms with independent doorways to a common walled courtyard. In stone-built houses, transverse arches were employed in larger rooms to support the roof. Mudbrick rooms were generally small enough to be bridged by wooden beams, and while domes could have been employed, surviving walls do not appear thick enough.

Surviving public architecture in an urban context is rare for middle Islamic Jordan as most of the important sites (e.g., as-Salt, al-Karak), being major centres in modern Jordan, have undergone considerable urban renewal especially in the last half century and the evidence is lost or hidden. The written sources record the construction of a mosque, bath, khan, hospital and *madrasah* (school) at al-Karak besides the palace (Brown 1989b). Similarly at Hisban, the capital of the Balqa' region for a time in the Mamluk period, the urban accruements of a mosque and/or palace, khan and bath are attested (de Vries 1994). The bath, excavated in 1973 by Van Eldreren (de Vries 1986), is a modest 14.5 by 4.0–5.5 m in size, constructed of stone and originally roofed with barrel vaults. The layout consisted of an entry hall, a dressing room with storage niches and two elevated platforms where bathers could rest, a lateral transitional corridor to the bath room with a rest bench, and the bathing room with a water tank, benches either side and a heated floor. Beyond, and not accessible from the bath, were the water tanks above a furnace, reached by way of a service room. The Hisban bath is very similar in layout to the Hammam al-'Ayn in Jerusalem, although on a smaller scale. The Hammam al-'Ayn, a typical Mamluk public bathhouse (*hammam*), forms part of the large Suq al-Qattanin ('Market of the Cotton Merchants') complex west of the Haram ash-Sharif and built in 1336–37 during the sultanate of an-Nasir Muhammad (Burgoyne 1987: 273-98). However, the bath itself appears to have been built a little earlier, as it was in existence by 1330. The Hammam al-'Ayn begins with an architecturally appealing change-room featuring a domed roof, central octagonal fountain and raised platforms in pointed-arched recesses, very similar to (although grander than) the dressing room at Hisban. A doorway, as at Hisban, gives entrance to a transverse corridor but (properly) with a small cold room at the far end, not simply a bench. From the corridor a doorway leads directly into the warm and then hot rooms, unusually six in number; at Hisban the functions of these rooms were condensed into a single hot room. While considerably smaller than Hammam al-'Ayn, the Hisban bathhouse conforms to a standard early Mamluk type and probably dates to sometime in the fourteenth century.

One significant architectural legacy of the economic restructuring of rural post-Crusader Jordan was the erection of water mills for sugar cane crushing and the grinding of grain. In the face of declining agricultural production in fourteenth- and fifteenth-century Egypt, the Mamluk elite directly invested in developing the resources of the Jordan Valley including the general utilization of water energy through the construction of mills (Rogan 1995). Along the edge of the many valleys that open out into the east bank of the Jordan Valley stand the ruins of numerous stone-built mills, but their construction dates are problematical. While a Mamluk date can be (and often is) plausibly offered for the widespread construction of water-driven mills in the Jordan Valley, many of the mills as they exist today appear to be substantially mid-nineteenth to early twentieth century, in some (perhaps many) instances the result of major restorations of derelict Mamluk-period mills. For instance, 10 water mills were identified along the upper Wadi Hisban during the Hesban Regional Survey, but Ayyubid/Mamluk pottery was found at only one of these (Ibach 1987: 194). Numerous mills were located during the East Jordan Valley Survey in 1975, 'often with millstones *in situ* and often associated with large quantities of "sugar pots"' (Ibrahim *et al.* 1976: 182-83, 202-203). However, the excavation of a horizontal-wheel flour mill in the fast-flowing Wadi Jirm near Tabaqat Fahl produced a firm late Ottoman date with no evidence of an earlier Mamluk-period structure or occupation. Nevertheless, mills are widely reported in the written sources for the Mamluk period and earlier, specifically Fatimid, times. The earlier Islamic mills were, perhaps, mostly of the vertical wheel variety, for at Tabaqat Fahl a heavy pottery vase recovered from the mid-eighth-century deposits was originally intended for roping to a vertical water wheel (Walmsley 1982: 149, pl. 145.4).

The economy, trade, pilgrimage and routes

The dearth of archaeological data on the Fatimid and Seljuq periods in Jordan is especially apparent when attempting to describe economic activity in the tenth and eleventh centuries. The prevailing view for the highlands, typified by the conclusions of the Heshbon Expedition (LaBianca 1990: 218), is for a shift to 'low intensity configuration' of land use in the Abbasid period, that is from plough agriculture to pastoral activities, and a 'retreat' from a sedentary lifestyle to nomadism by the Fatimid and Seljuq periods. However, this view does not rest comfortably with the textual evidence, particularly as the tenth-century geographers Ibn Hawqal and al-Maqdisi depict Jordan as agriculturally productive. North Jordan was populated with villages producing olives, honey and fruits especially grapes, the Balqa' was an area of villages, fields and water mills with grain, fruit and honey as the main produce, the Ma'ab region was renowned for its grapes and especially almonds, while Jibal and ash-Sharah were also prosperous. These sources further remark upon the cultivation of tropical crops, specifically dates, indigo, rice and bananas, in the Jordan Valley. Both writers also note that around 'Amman and farther south the herding of sheep and goats played a significant role in the economy. As mentioned earlier, the Crusader sources of the early twelfth century likewise describe Jordan as widely settled with villages. The crops are unchanged: grain, grapes (for wine), dates, olives and oil, again with sheep and goat herding. Just before the Muslim reconquest of south Jordan Ibn Jubayr (1952: 301) could still describe the al-Karak region as 'the choicest part of the land ... it being said that the number of villages reaches four hundred'. Together the tenth-century Islamic geographies and the twelfth-century Crusader-period sources depict Jordan as supporting a mixed settled and nomadic population engaged in mutually beneficial agricultural and pastoral activities.

The rise of the Fatimid dynasty in Egypt stimulated the Indian Ocean trade, with ships reaching as far as China, Sumatra and Java (Chaudhuri 1985: 34-62).

Ceramic evidence from Fustat indicates that the volume of Red Sea trade to China increased markedly after about AD 1000 (Scanlon 1970). At this time, al-'Aqaba, strategically positioned at the eastern head of the Red Sea, was described by al-Maqdisi as Palestine's port on the 'Chinese Sea', and the excavations of Whitcomb have vividly illustrated the international outlook of the town in the Fatimid period. Of particular note is the glazed pottery from the Pavilion Building (Whitcomb 1988b). Two large blue-green jars with barbotine decoration seemingly originated from Basra in southern Iraq (Figure 15.15.5), lustre wares came from Iraq and perhaps Egypt, while fine celadons and porcelains originated in China. A hoard of 32 gold dinars of the early eleventh century and a collection of glass and bronze weights further attest the importance of the Indian Ocean and Far East trade to Fatimid Ayla. In addition to Palestine, land routes from Ayla linked Udhruh, Mu'an, Ma'ab and 'Amman to the Red Sea trade networks.

In part, the Crusader occupation of Wadi Musa and Ayla was an attempt by Baldwin I to gain some control over the Red Sea trade, thereby restricting Fatimid domination and weakening the economic benefits to Egypt. Under Reynald, Lord of al-Karak, raiding reached the Holy Cities of the Hijaz, but in general Crusader activity in the Red Sea was limited and ended altogether with the recovery of south Jordan by Salah ad-Din in 1188. However, within the Crusader realms there was a flourishing trade in agricultural produce and minerals. Wheat, oil, wine, dates, sugar, bitumen and salt in particular were shipped in boats across the Dead Sea from the southern Ghor and the highlands of Oultrejourdain to Jericho and Jerusalem (Johns 1994: 10-14). Annual fairs were held in the Sawad and al-Karak, while shipping and caravan traffic was subject to taxes.

The return of all Jordan to Islamic hegemony and its economic reintegration into the market systems of Damascus and Cairo revived the many traditional agricultural practices temporarily disrupted, but clearly not completely destroyed, by the political and social partitioning of Jordan in the Crusader period. The critical destructive factor was the protracted period of warfare between Salah ad-Din and Reynald, which appears to have inflicted considerable damage on the village economy of south Jordan (Johns 1994: 12-13). Conditions did not improve substantially until well into the thirteenth century. In the rural context, the return to normality resulted in the restoration of a village-based economy, founded upon conventional agricultural pursuits (grain, legumes and aboriculture), pastoral activities and a revival of cash crops, especially sugar and indigo. The produce was destined for the large urban markets of Damascus, Cairo, and ar-Ramlah, locally to al-Karak, as-Salt and other centres, and for export to more distant locations.

Sugar cane cultivation was introduced to the Jordan Valley and coastal Palestine perhaps as early as the seventh century, and had become a major crop by the tenth century (Galloway 1989: 23-47; Watson 1983). Arab-Islamic society of the seventh and eighth centuries provided a favourable political and economic environment for the growth of the sugar industry. Irrigation technology, a requisite for water-thirsty cane, was improved and diffused throughout the Islamic realm, enforceable legal frameworks controlled the distribution of water, investment capital was plentiful to finance the expensive cane processing factories, and a burgeoning urban population— increasingly wealthy and sophisticated—was keen to adopt new tastes in food. After the expulsion of the Crusaders these factors re-emerged to promote a revival of the sugar industry, which flourished until the mid-fourteenth century. Arabic sources and travel accounts concur that cane was cultivated in the Jordan Valley wherever water was available in copious quantities, including east and south of the Dead Sea (Ashtor 1981: 92-93). Al-Karaki sugar was highly regarded in Europe. The process of obtaining sugar from the cane was laborious, inefficient and costly (Galloway 1989: 37-40). The juice was extracted by milling and pressing, usually through a combination of both techniques, which required the cane to be cut into short lengths. Mill technology differed little from that used to grind wheat or extract liquids from olives or grapes. Two types of mills were used in Palestine and Jordan: the upper rotating stone variety with either a flat or conical millstone, and the 'edge-runner' in which a vertically fixed round stone was rolled on a pivot in a circle over the cane. They were powered by flowing water, animals or humans. Once extracted, the juice was boiled to thicken it, and then poured into thick pottery cones to crystallize. The cones were set upon bag-shaped jars to collect molasses. Repeated boiling improved the quality of the sugar, but consumed great quantities of fuel. Olive waste, used in earlier times for firing pottery kilns, was an obvious source.

Sherds from sugar cones and jars are common discoveries at Mamluk sites in the Ghor (Figure 15.17.10-13), and numerous stone-built mills have been identified along the east bank including south of the Dead

Sea (see, for instance, Abu Dalu 1995; Greene 1995; Ibach 1987: 194; King et al. 1987; McQuitty 1995; Whitcomb 1992a: 114-18). As already noted, a Mamluk date for many of these mills is not assured; indeed, an Ottoman date is more likely for the majority of them.

The decline of the Mamluk sugar industry in the Jordan Valley and the rest of Syria-Palestine was just as sudden and complex as its thirteenth-century resurgence (Ashtor 1981). A fall off in cultivation can be detected in the fourteenth century, but by the sixteenth century the sugar industry had disappeared. The principal factors behind the failure were economic and technological. The industry was labour and energy intensive and became increasingly expensive to operate following a major decline in the population, perhaps by a third, due to a series of plagues and warfare during the fourteenth and fifteenth centuries. In addition, the policies of the Mamluk elite perpetuated inefficiencies and corruption, while mills were not maintained or technologies improved due to the prohibitive costs involved (Rogan 1995). The arrival of cheap and better sugar from the Americas in the sixteenth century precipitated the final collapse.

Other primary and secondary industries in Jordan thrived under the direct or indirect patronage of the Mamluk elites. The horses, camels and sheep of the al-Karak region were highly prized in Cairo for transport, wool and meat (Irwin 1986: 185-86). Iron was mined and smelted at Mugharat al-Wardah in the 'Ajlun mountains throughout the Ayyubid and Mamluk periods. Alongside the mine-smelting furnaces, an ore-roasting oven, slag and ash were identified and dateable by the associated pottery (Coughenour 1989). Copper and perhaps iron mining and smelting was resumed at Faynan in the 'Arabah, and copper finger rings and earrings were commonly worn for personal adornment. Many local industries flourished at a village level. For instance, glass may have been manufactured into bracelets and other items at Dayr 'Alla in the Jordan Valley, and cottage potting industries producing the ubiquitous Hand-Made Geometric Painted Ware (see below) must have been a feature of every village of note. Unquestionably, other village industries such as weaving, woodworking, basketry and mat making would have equally prospered, but these are rarely recorded as they are archaeologically 'invisible' and were of no great interest to scholars or travellers.

A major boost to the local economy would have been the passage of the annual pilgrimage to the Holy City of Makkah in the Hijaz. The pilgrimage route changed a number of times during the period under consideration, mostly in response to the Crusader occupation of south Jordan in the eleventh century, hence different centres benefited at different times. During the Fatimid period, the path followed a set route through the *badia* (eastern steppe lands) of Jordan (Figure 15.2), and is described in some detail in the geographical accounts, especially al-Maqdisi (Musil 1926: 326-31 and Appendix 15; 1927: 516-20). Going southwards from Adhri'at (Dar'a in Syria), az-Zarqa' was reached after a day, then another day led to the major assembly point of 'Amman where pilgrims from surrounding districts could join the caravan. After 'Amman two unspecified watering places (most likely al-Qatrana and al-Hasa) were needed to reach Mu'an (Ma'an) where provisions could be obtained, after which the route continued south to Tabuk and the Hijaz.

The *badia* route continued in use during the Crusader occupation of Jordan, but while this route avoided Crusader territory it still left the caravans liable to attack. In 1180 a caravan of merchants was looted on its way to Makkah by Reynald de Châtillon, Lord of al-Karak. More seriously, in 1186 merchants and pilgrims were attacked and summarily executed or captured in blatant violation of a truce with Salah ad-Din. All protestations were ignored, leading Salah ad-Din to revive the *jihad* against the Crusaders, and which saw victory—and the immediate execution of Reynald—within the year (Maalouf 1984: 186-94).

As already described, the prolonged struggle between the Crusaders and Muslims of Syria and Egypt brought to prominence the fortress towns of Jordan, especially al-Karak. By the Mamluk period al-Karak served as a major stop for the Damascene pilgrimage before entering the more physically challenging desert territory to the south. Two sources of the fourteenth century present a comparatively detailed picture of the route and conditions on the journey (Abu'l-Fida 1983; Ibn Battuta 1958; Peters 1994: 79-86). Ibn Battuta gives a full account of the route he followed in 1326 (Figure 15.4). From Damascus the caravan made two stops before Busra, where a stop of four nights allowed latecomers to catch up and the pilgrims to buy provisions. From Busra the caravan travelled to the Pool of az-Zizia. The most likely route for the caravan was through az-Zarqa', as suggested by the construction of a small fort in the thirteenth century (described above). 'Amman, seemingly not repopulated to any extent since Crusader times, held no attraction so was bypassed. From az-Zizia, where there was also a Mamluk fort (above), the caravan continued to al-Lajjun, where water was plentiful, and then reached al-Karak. After a stop of four days at al-Karak the pilgrims set out for Ma'an, almost certainly by way of al-Hasa where there was a toll station, and then Tabuk. The brief account of Abu' l-Fida shows that the wealthy travelled by horseback to al-Karak, sending their baggage and servants in advance, and from there went by camel (1983: 65). The Mamluk caravans were always large. In 1432 some 3000 camels returned to Damascus; in 1503 35,000 camels, 40,000 people and a guard of 60 Mamluks set out from the same city (Peters 1994: 80-85). The figures are obviously inaccurate guesses, but nevertheless reveal the considerable size of the annual pilgrimage.

Other roads transversed Jordan, linking the settlements on the Hajj route to places to the west and south. In the Fatimid period, a major road joined 'Amman with Jerusalem by way of Hisban, Bayt ar-Ram and Jericho (Figure 15.2). A second route from 'Amman ran down the highland range to Ma'ab, probably by way of Hisban, and then to Zughar in the 'Arabah. From here, the road travelled down the 'Arabah through Ghamr (where another road left to ar-Ramlah), 'Ayn Gharandal and Ghadhian to Ayla. From Zughar, a second road ascended the scarp to Udhruh and Mu'an, probably by way of Wadi Musa. As Ibn Jubayr (1952: 300-301) makes clear, roads remained open despite warfare between the Muslims and Crusaders, with taxes being widely imposed on travellers by both sides. An important route ran from Busra and Adhri'at to Tiberias and Safad by way of Bayt Ras and Jadar (Figure 15.3), and both Crusader and Muslim armies marched along this road in the battle for Tiberias and the north Jordan Valley (Mershen and Knauf 1988: 132-36). Later a toll station was placed in the village of Jadar, resulting in a name change to *mkis* (from which the Umm Qays of today).

The Ayyubid and especially Mamluk road system was extensive in its coverage, consisting of the official *barid* (postal service) roads, which doubled as trade routes and were usually well maintained, and a network of connecting routes. The Darb al-Quful, mentioned earlier, crossed into north Jordan near Baysan (Beth Shan) at Jisr al-Majamia', and continued eastwards through al-Qusayr (Shuna) before ascending the scarp to Irbid. As an official state road, the Darb al-Quful remained popular until the late fifteenth century, when Safad rose to prominence and the northern crossing at Jisr Binat Ya'qub gained greater popularity. Another important road was the southern route from Cairo to ash-Shawbak and al-Karak, transversed by Baybars in June 1276 in only 11 days (Zayadine 1985). The first five days were spent crossing the Sinai to Wadi Musa (Petra), entering from Wadi 'Arabah. Once watered, he continued onto ash-Shawbak and al-Karak, where a rebellion was suppressed. From al-Karak the link to Damascus followed the Hajj route through az-Zizia and az-Zarqa', while another route crossed at Zughar to Khalil (Hebron) and Jerusalem. A third route ran up the mountain range between Dhiban, Hisban (link to Jericho and Jerusalem), as-Salt, 'Ajlun (link to Nablus) and Irbid, finally joining the Darb al-Quful at Umm Qays. Farther east, a circuit route through the *badia* joined Petra and al-'Aqaba with Ma'an, then to Bayir and al-Azraq (link to az-Zarqa'), and northwards to the Hauran. As already mentioned, most of these strategically placed nodules in the communications network were progressively fortified in Ayyubid and Mamluk times. Other roads crossed between the major routes and centres and very possibly many were restored in the Mamluk period and marked out with edging stones. Frequently, one suspects, these roads are mistakenly classified as Roman constructs, especially as many would have traced old Roman routes.

Material culture

At first glance, the more than five hundred years between the late tenth century and the start of the sixteenth century coincided with significant changes, in keeping with neighbouring Palestine and Syria, in the material culture of Jordan. These changes have often been seen as indicative of, and explained by, broader trends in the socio-economic regime of Islamic Jordan. The view has been largely negative: the collapse of urban life, political disintegration, external domination, the end of specialized manufacturing industries (wheel-thrown pottery, for instance), a short rural boom especially in the Jordan Valley, and a rustication of the population. Many of these judgments were ostensibly supported by an assessment of the relative worth of wheel-made and hand-made ceramics, based on the assumption that the latter—of coarse fabric and soft firing—were definitely inferior. The hand-made wares were interpreted as representing an easy if misleading picture of small, isolated communities surviving at little more than a subsistence level. Likewise, the large and heavy bowls and jars used in the sugar industry came to portray equally disadvantaged migrant labour in the cane fields of the Jordan Valley.

The material culture of middle Islamic Jordan clearly did experience major shifts in the tenth to sixteenth centuries, including the rising popularity of hand-made wares, especially Hand-Made Geometric Painted Ware (or HMGPW), and the ubiquitous appearance of sugar pots in the Jordan Valley, but also the proliferation of local and imported glazed wares. A re-evaluation of all the material, not simply ceramics (or one class of ceramics), presents a more complicated scenario based on the exploration of new economic opportunities, the prescription of socio-political identities, and the status of elites in middle Islamic Jordan.

The repertoire of surviving material finds from middle Islamic sites in Jordan is standard for a historical period, and includes ceramics, glass, metals (including coins), stone artifacts, including architectural decoration, and ivory or bone. The characteristics of each group will be summarized before entering into discussion on the socio-economic implications of the material.

Ceramics

As numerous authors have already observed, pottery traditions are mostly unaffected by political events and dynastic changes. Other factors, social and economic, promoted change, and thus ceramics can be a valuable tool for the historian as they reflect developments over a wide cross-section of society, especially the otherwise invisible lower socio-economic groups (Walmsley 1994). In this necessarily brief treatment of middle Islamic pottery from Jordan, the material will be classified by type: plain and decorated wheel-made, mould-made, glazed (wheel- and mould-made), and hand-made wares including HMGPW. The development of each type will be traced over time with the intention of avoiding dynastic labels for particularly distinctive ceramic types. The treatment will largely concentrate on Jordanian material, and the referenced reports should be consulted if further information is required on dates and parallels from Palestine and Syria.

Pottery dated to the eleventh to early twelfth centuries

A reliable starting point for eleventh-century pottery comes from the excavations at the 'Amman Citadel and al-'Aqaba. On the Citadel of 'Amman, excavations by Bennett of two buildings in Area C recovered a limited range of ceramics from an eleventh-century destruction at the end of Stratum III, while in a house in Area B an abandonment deposit, also of Phase III date, and a pit produced further eleventh-century material (Northedge 1984: 275-81, 1992: 143-47, figs. 137, 141, 150-51). The pottery is largely utilitarian, and features:

- cooking jars in a red fabric with broad strap handles, often with applied clay lugs to body and a glazed interior (Figure 15.14.1);
- thin-bodied, cream-ware jars featuring flaring necks and turban handles (for this ware in Palestine and Jordan, see Walmsley forthcoming);
- glazed bowls, slipped with yellow, blue and green glaze (Figure 15.14.3);
- plain 'earthenwares', a general grouping of wheel-made jars and bowls in a reddish-yellow to light brown fabric, mostly undecorated although one example has white painted wavy lines (Figure 15.14.4);
- 'bag-shaped' jars (*zirs*), light-brown fabric with body-combing (Figure 15.14.6);
- moulded 'arcade' lamps with bunched grape decoration (Figure 15.14.4).
- large hand-made, thick-rimmed basins and storage jars with loop handles, fired light-brown to reddish-yellow and sometimes decorated with combing and thumb impressions (Figure 15.14.5);
- *Kerbschnitt* (engraved) bowls;
- coarse hand-made ware of coil manufacture.

As Northedge observes, the corpus shows considerable continuity from the preceding Stratum IVb (Abbasid); likewise, it represents a development of the tenth-century material from Pella (Walmsley 1991). Noticeably different are the glazed cooking pots and the first appearance of coarse hand-made ware. Northedge dates this corpus to the end of Stratum III, which he suggests was caused by the earthquake of 1068, although elements of the corpus may suggest a date at the start of the twelfth century. Hence, rather than an earthquake destruction, the end of this phase may represent the evacuation of Seljuq forces from 'Amman under Crusader pressure.

The pottery from the Pavilion Building at al-'Aqaba ranges from plain to glazed wares of local and imported origin (Whitcomb 1988b). Like the 'Amman Citadel material, the Pavilion Building corpus may spill over into the early twelfth century, but the absence of distinctive twelfth-century Chinese imports suggests a mostly eleventh-century date with some tenth-century material persisting. Major varieties include:

- glazed monochrome and polychrome lustre ware bowls of Iraqi or Egyptian manufacture (Figures 15.15.1, 3);
- monochrome, splash-decorated, Fayyumi and sgraffiato glazed bowls of Iraqi, Egyptian,

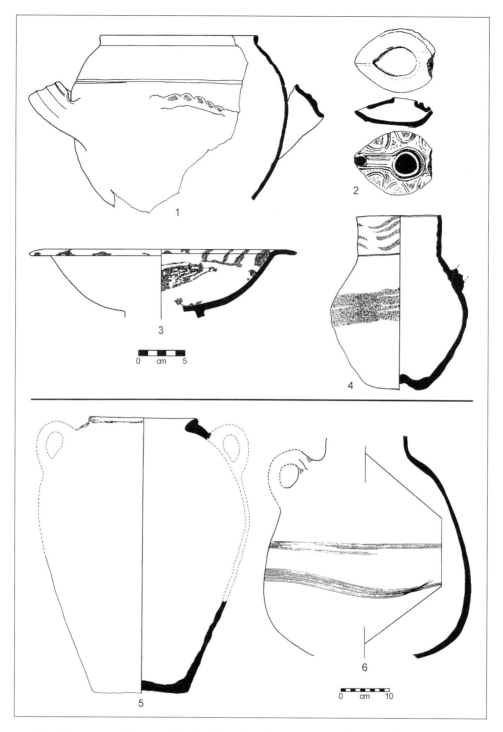

Figure 15.14. 'Amman Citadel: pottery of the second half of the eleventh century: 1. red ware cooking jar; 2. moulded lamp with grape decoration; 3. white-slipped bowl with blue-green glaze; 4. red fabric jar with white paint; 5. pale brown storage jar; 6. light brown bag jar (based on Northedge 1992: figs. 137, 141, 150).

north Syrian and local manufacture (Figure 15.15.2);
- Chinese celadons and porcelains, most commonly fine Qingbai porcelains including comb decorated variety (Figure 15.15.4);
- large 'Arab-Sassanid' jars, with incised and applied decoration and a heavy blue-green glaze (Figure 15.15.5);
- common earthenwares, including jars, juglets and bowls in red to orange-red and tan fabric and

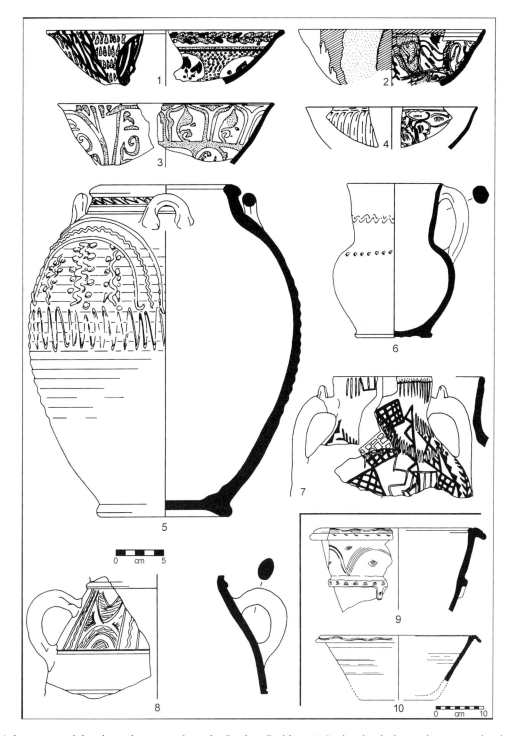

Figure 15.15. Ayla: pottery of the eleventh century from the Pavilion Building: 1, 3. glazed polychrome lustre ware bowls; 2. polychrome glazed sgraffiato bowl; 4. Qingbai porcelain bowl; 5. blue-green glazed jar with applied and incised decoration; 6. buff-coloured incised juglet; 7. black painted and incised hand-made jar; 8. red to buff-coloured jar with incised decoration; 9-10. buff-orange to cream incised bowls, 10 with dark green glaze on interior (based on Whitcomb 1988b).

with instances of combed and incised decoration (Figure 15.15.6, 8-10);
- large storage jars with fattened rims and loop handles in a red to red-orange gritty fabric, sometimes

with applied decoration below neck (similar to the Citadel types);
- hand-made bowls, basins, cups and jars with sand temper (no chaff), fired buff, orange to

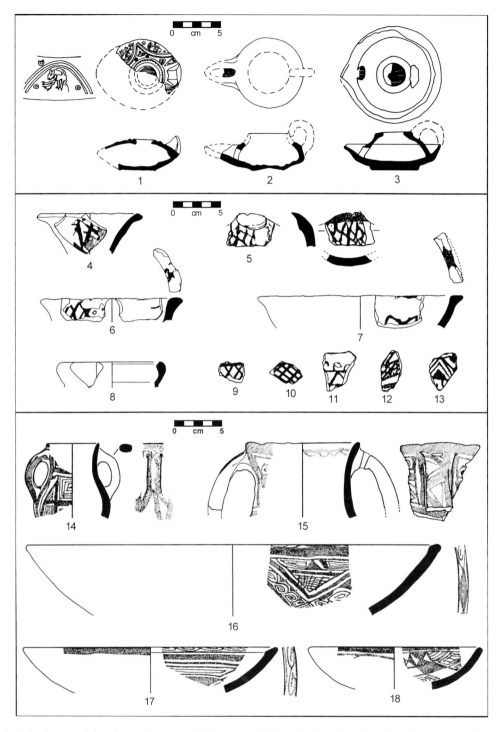

Figure 15.16. 1-3. Ayla, lamps of the eleventh century (Whitcomb 1994); 4-13. Shawbak, hand-made, red-painted coarse wares of the twelfth–early thirteenth centuries (Brown 1988); 14-18. Tabaqat Fahl (Pella), dark red to purplish painted jars and bowls in Hand-Made Geometric Painted Ware of the second half of the fourteenth to fifteenth centuries (Walmsley 1992b and unpublished).

grey and sometimes decorated with red paint (Figure 15.15.7).

It needs to be noted that most of the recovered pottery was common earthenwares (around 60 percent of the count). Glazed wares were always rarer (about 10 percent) while the coarse hand-made wares only become common in the last period.

Another significant eleventh-century change at al-'Aqaba was the appearance of two quite different types of lamps. Islamic lamps in Jordan were traditionally almond-shaped with moulded decoration of arcades, vine tendrils, grape bunches, animals and palmettes, as with the 'Amman Citadel example (above). Glazed examples are known from the ninth and tenth centuries (Figure 15.16.1). Both of the new eleventh-century forms at al-'Aqaba are notably different in shape and manufacture. The first has a wheel-made enclosed bowl with a deep channel spout, raised loop handle and a green or yellow glaze (Figure 15.16.2). The form is most likely of Egyptian origin, with parallels at al-Fustat. The second type, first attested in the later eleventh century, is the saucer lamp, also wheel-made with a loop handle and central reservoir (Figure 15.16.3). Both types develop in the following centuries; most notably the central reservoir of the saucer lamp diminishes in size to almost vestigial proportions by the fifteenth century.

The 'Amman Citadel and al-'Aqaba present a reliable ceramic horizon for Jordan in the eleventh century (with perhaps elements from the very beginning of the twelfth century). The material ranges from imported and local glazed wares to utilitarian storage vessels and gritty-fabric cookers. At al-'Aqaba new lamp types, conventionally considered Ayyubid-Mamluk in date, were in circulation, while at both sites coarse hand-made wares had made an appearance. Some of the hand-made pottery was decorated with paint, and may be a precursor for the ubiquitous Hand-Made Geometric Painted Wares of the thirteenth–fourteenth centuries and later. Fortunately, a twelfth-century link has now been established.

Pottery dated to the twelfth and early thirteenth centuries

Excavations in the Crusader castles of al-Wu'ayra and ash-Shawbak have identified a limited corpus of twelfth- to early thirteenth-century pottery for Jordan, but overall this period is poorly represented. In a preliminary treatment of the ceramics from the 1987 excavations at al-Wu'ayra, Brown (1987) noted the preponderance of hand-made wares with only a few wheel-thrown *zirs* and glazed bowls. In the earlier ('Crusader', twelfth century) phase, the hand-made coarse wares were decorated with linear designs and dots in red paint, while in Phase II ('Ayyubid', late twelfth to early thirteenth centuries) the linear-painted ware persisted but geometric painting in red to grey was more prevalent. Wheel-made bowls glazed in yellow and green, common in later assemblages (below), first appeared in Brown's Phase II. More recent excavations at al-Wu'ayra by Vannini and Tonghini (1997) confirmed the predominance of hand-made wares, but without being able to duplicate the sequence from linear to geometric painted. However, they did observe that most Syrian gritwares of Phase III ('Crusader occupation levels' of the later twelfth century) gave way to monochrome glazed wares by Phase IV ('Ayyubid'). Overall, glazed wares constituted a small percentage of recovered pottery. Unglazed pottery made on a wheel was also attested, including cream ware with stamped decoration from Phase IV. One interesting feature of the twelfth-century pottery from al-Wu'ayra is its overwhelmingly local character. To Brown (1987: 284), there was 'absolutely nothing specifically "Crusader"' about the pottery corpus she recovered at al-Wu'ayra, but Vannini and Tonghini (1997: 382) recognized a Crusader presence in the Phase III north Syrian fritwares.

At ash-Shawbak castle, Brown (1988) was again able to identify a discrete twelfth- to early thirteenth-century date for early hand-made coarse wares in the constructional Phase I ('Ayyubid') of the Ayyubid Palace complex. The coarse wares included linear-painted types, often with criss-cross patterns (Figure 15.16.4-13). The assemblage also contained a few examples of HMGPW and monochrome glazed, as would be expected of a corpus that ran into the thirteenth century. By Phase III ('Mamluk', fourteenth–fifteenth centuries) monochrome glazed bowls, underglaze painting, celadons, wheel-thrown cream wares and HMGPW had become typical (see below).

Admittedly the twelfth- and early thirteenth-century pottery groups from al-Wu'ayra and ash-Shawbak are limited in size, and caution must be exercised in their use. However, some support for Brown's thesis of an early series of hand-made coarse wares including linear red painting comes from the excavations at Khirbat Faris, where a similar hand-made pottery painted in red line and dot patterns was recovered in a ninth- to twelfth-century context (Falkner in McQuitty and Falkner 1993: 53-54). Similarly, a preliminary analysis of the middle Islamic pottery from new excavations at Gharandal, near at-Tafilah, has identified a precursor for HMGPW. The Gharandal material is thin-walled and heavily tempered with chaff, unlike al-'Aqaba where there was no chaff but much sand. These initial impressions suggest hand-made wares exhibited considerable local variability in fabric and decoration between the eleventh and early thirteenth centuries, whereas by the fourteenth century HMGPW had, in relative terms, attained a much greater degree of regional standardization. Generally, the extent of ceramic variability in the eleventh–thirteenth centuries, and hence the absence of an easily identified corpus, would account for the poor showing of these centuries in regional surveys. Understandably, the tendency would be to lump the early coarse wares into the Mamluk, or even Ottoman, periods, thereby greatly under-representing over two centuries of settlement in Jordan. The extent of the error is demonstrated by a reconsideration of the stratigraphy and pottery from a number of sites in Jordan, for instance Dhiban and Tall Abu Qa'dan, in which strata and material of the eleventh–thirteenth centuries can be successfully identified (and is the subject of a separate study in progress).

Farther north at Tall Hisban, a discrete 'Ayyubid' (c. 1174–1263) corpus has been isolated and dated by coin evidence by Sauer (1994: 268-70). HMGPW was 'abundant' and decorated with set designs in dark purplish red and brown paint. Glazed wares are rarer, and included glazed cookers. No earlier twelfth-century coarse hand-made wares could be identified, and both the Hisban coins and the al-Wu'ayra and ash-Shawbak pottery suggest that this Ayyubid corpus is mostly mid-thirteenth century and after. Finally, excavations at Rujm al-Kursi have produced three early thirteenth-century lamps from a coin-dated context. They were hand made in two parts from a coarse gritty fabric, and feature a central filling hole, handle and pierced wick hole.

Pottery dated between the later thirteenth and fifteenth centuries

The equation between certain pottery types and Mamluk-period occupation in Jordan has almost exclusively concentrated on two distinct varieties: slipped wheel-made bowls supporting a yellow or green glaze and, more usually, HMGPW. They have the advantage of being easily identified during surveys and are a relatively common feature of the archaeological landscape in Jordan, in that they represent the last major pre-modern phase in the occupation of many sites. As with earlier periods, the material is generally poorly published and dating imprecise, and similarly this assessment of pottery from the later thirteenth to fifteenth centuries relies upon the more reliable reports.

The excavation of the mosque on the main mound at Tabaqat Fahl (Pella) produced a representative range of ceramics that can be dated, substantially on coin evidence, to the second half of the fourteenth and early fifteenth centuries (Walmsley 1992b). The major pottery varieties were:

- Hand-Made Geometric Painted Ware (HMGPW), slipped and sometimes burnished and liberally decorated with painted monochrome or bichrome geometric designs in dark red, purple and black. Jars, jugs, bowls and basins were the common shapes (Figures 15.16.14-18, 15.17.1-2).
- 'Cooking pot' ware; coarse with numerous inclusions and used in the hand manufacture of large, splayed-necked cooking jars with solid handles in the shape of a pointed ear (Figure 15.17.15-16).
- Plain ware, wheel-made elongated jars and straight-sided bowls in a soft, light yellow-brown to reddish-yellow ware and linked with the sugar industry (Figure 15.17.10-13).
- Thin-walled painted ware, wheel-made and reddish-yellow to pink in colour with limestone grits, freely painted in reddish-brown to brown-black on the rim, handles and body.
- Underslipped glazed ware bowls, internally slipped or decorated with brush strokes and over-glazed with brilliant yellow or green glazes over a wheel-made bowl (Figure 15.17.3, 5-7).
- Underglaze painted ware with very fine bichrome and polychrome painted designs under a clear glaze (Figures 15.17.8-9).
- 'Silhouette' ware, with black paint under turquoise, blue or green glaze.
- Sgraffiato ware, white-slipped and incised with a yellow to green glaze (Figure 15.17.41).

HMGPW, sugar-pot ware and underslipped glazed ware bowls, all seen as local products, formed the largest groups in the corpus. Only a few sherds of the last three glazed varieties, most likely imports from north Syria, were recovered. They are conventionally dated to the fourteenth and fifteenth centuries, which neatly complies with the numismatic evidence.

Also in the Jordan Valley, the neighbouring sites of Tall Abu Qa'dan and Tall Abu Sarbut produced large quantities of pottery of about the same date as the material from Tabaqat Fahl. At Tall Abu Qa'dan a long sequence of thirteenth- to fifteenth-century pottery was recovered, although there are considerable problems with stratigraphy and dating (Franken and Kalsbeek 1975: 107-203; but note the review of Sauer 1976). The different pottery types as identified by Franken and Kalsbeek are:

Fatimid, Ayyubid and Mamluk Jordan and the Crusader Interlude

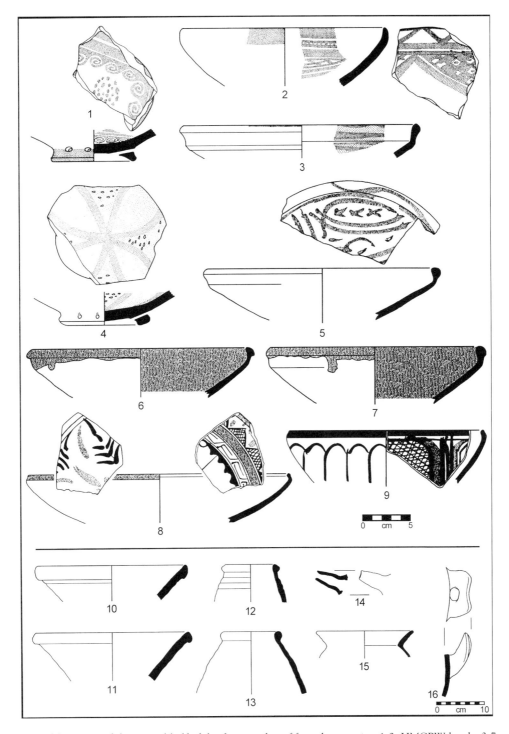

Figure 15.17. Tabaqat Fahl, pottery of the second half of the fourteenth to fifteenth centuries: 1-2. HMGPW bowls; 3-7. green and yellow glazed bowls with brushed (3, 5) or full slip (6-7) and some sgraffiato (4); 8-9. bichrome and polychrome underglaze painted wares; 10-13. light brown to reddish-yellow sugar pots; 14-16. heavy hand-made coarse wares including cooking jars (Walmsley 1992b and unpublished).

- HMGPW, found in profusion. It appears in a developed form, typically with a burnished surface and decorated with care at first. Later forms are slipped and the decoration poorer.

- Glazed wares, predominantly slipped wheel-made bowls and including sgraffiato decoration. A major change in firing technology occurred in the fourteenth century when bowls were stacked

right-side up (previously upside down) in the kiln.
- Faience bowls decorated in blue and black.
- Hand-made cooking pots with large pierced handles.
- Sugar pots, both jars and bowls, mass produced in large kilns roughly 5 m in height. This type of pottery was completely absent before Period 2.
- Red-firing wheel-made pottery, often with a salt bloom, used in the manufacture of cooking pots, bowls, jars and jugs, some of the latter with coarse and fine sieve necks and spouts.
- Light red, hard-firing jars (equivalent to thin-walled painted ware at Tabaqat Fahl).
- White wares, but very scarce (and perhaps remnants from earlier levels).

Franken considered HMGPW unrelated to the prevailing ceramic traditions in Islamic Jordan and Palestine, and decided upon a North African inspiration for the ware (Franken and Kalsbeek 1975: 213-22; Homès-Fredericq and Franken 1986: 239-44). However, the identification of eleventh- and twelfth-century precursors with linear decoration (above) would argue that HMGPW was an entirely local development. Volumes of wheel-made wares at Tall Abu Qa'dan were much less frequent than HMGPW, and were mostly confined to Phase 2 (which is more satisfactorily dated to the later thirteenth and fourteenth centuries). Franken also saw these thrown wares as sufficiently different as to indicate an independent tradition, but this view is also difficult to accept.

Few sites in the highlands have produced distinct later thirteenth- to fifteenth-century ceramics from stratigraphically sound and coin datable levels. The most accessible material has come from Hisban, Rujm al-Kursi, Khirbat Faris, the al-Karak plateau survey, ash-Shawbak and, most interestingly from a socio-economic aspect, al-Karak. Based on the coin evidence, Sauer has differentiated two pottery groups dating to the later thirteenth and fourteenth centuries at Tall Hisban. These were labelled 'Early Mamluk I' (EMI), dated c. 1263-1300, and 'Early Mamluk II' (EMII), c. 1300-1401, and mostly originate from a well-stratified cistern (Sauer 1994: 270-73). In EMI brown became the preferred paint colour on HMGPW, while glazed wares were more common. In EMII the paint had become darker brown or black and the pot surface lighter (light yellowish to greenish); glazed wares continued. Sauer also suggested that a 'Late Mamluk' (c. 1401–1516) stratum was distinguishable in which HMGPW, featuring a darker paint and narrower lines, was present but glazed wares were scarce. In a combined study of all HMGPW groups at Hisban, Sauer (1973: 55-56) identified 13 major designs: squares, waves, single and double zigzags, hourglass, diamond nets, semi-spirals, interlocking triangles, dotted lines, fine-lined lozenges, lined dots, nets and wavy lines. A single vessel often was decorated with a combination of designs, which currently do not appear to have chronological significance. At Rujm al-Kursi the mid-thirteenth-century corpus of HMGPW, dated by coins, was slipped and decorated with reddish-brown paint in 10 design categories (Khadija 1992). In addition to seven of Sauer's 13 groups, Khadija identified three new categories of motifs: diamonds, triangles and stars. A selection of HMGPW has also been published from Khirbat Faris, although no firm dates are offered (McQuitty and Falkner 1993). However, the comment is made that some doubt still exists as to a terminal date of HMGPW, and a post-sixteenth- to seventeenth-century date is not unreasonable as HMGPW sometimes appears with tobacco pipes, which cannot be dated earlier than the seventeenth century.

A very valuable assessment of thirteenth- to fifteenth-century pottery in Jordan was undertaken by Brown as part of an archaeological survey of the al-Karak plateau directed by Miller (Brown 1991: 232-41). The study was based on pottery picked up during the survey that, from this period, was mostly HMGPW with some monochrome glazed wheel-thrown bowls. Very rare were:

- unglazed wheel-thrown wares, both household vessels (e.g., strainer jugs) and industrial (sugar pots), and ranging in colour from buff to pink and red; glazed bowls with sgraffiato decoration;
- glazed moulded vessels of pink, white and red fabric;
- a hard, metallic grey ware used in the production of small moulded conical vessels, probably containers for mercury or other precious liquids;
- imported glazed wares.

Brown provides a full evaluation of each category, referencing widely published and unpublished material from sites in Jordan, neighbouring Syria and Palestine, and beyond. While the range of sites with comparative material is extensive, no new conclusions on the chronology of forms and decoration were possible as all the sherds came from unprovenanced surface collections.

Based upon her investigation of al-Karak castle, Brown has been able to draw a sharp distinction between the ceramic profile in the countryside, as identified by the al-Karak plateau survey, and the pottery recovered from within the castle confines (Brown 1992). Unlike the countryside, where monochrome glazed wares and especially HMGPW predominated, the castle ceramics included many fine imported wares and, in proliferation, wheel-made pale-bodied glazed bowls and many types of plain vessels. Imported glazed wares included Chinese celadons and imitation celadons, polyglaze sgraffiato, underglaze painted in blue and black, and 'Rusafa' ware. Apart from the celadons, these glazed varieties demonstrate wide inter-regional contacts between al-Karak, north Syria and Egypt in the period. HMGPW, by way of contrast, was uncommon even though it was readily available locally. Brown argues convincingly that the contrast between the rural and urban assemblages at al-Karak reflected the deep socio-economic gap that existed between the town of al-Karak and the surrounding countryside. The Mamluk elite of al-Karak, which (as has been seen) included high-profile exiles from the ruling class of Cairo, enjoyed a lifestyle considerably better than that in rural areas, where fine and imported wares—indicative of a wider consumption of luxury goods—were exceedingly rare. In Brown's analysis, the ceramics reveal that urban al-Karak and rural Ard al-Karak were located at the opposite ends of the socio-economic scale under Mamluk rule.

Non-ceramic finds

Non-ceramic finds manufactured from materials including stone, glass, ivory, bone and metals have only been sporadically mentioned and erratically described, as most writers with any interest in middle Islamic Jordan have concentrated on pottery in an attempt to establish a basic ceramic chronology for the eleventh to fifteenth centuries. Accordingly, no general survey can be offered at this time. However, specific mention should be made of select categories of non-ceramic finds, especially in the few instances where these are reported in sufficient detail.

A significant collection of stone, glass, metal and ivory artifacts, dating to the later tenth and eleventh centuries, was recovered from the upper levels at al-'Aqaba (Whitcomb 1994). Outstanding were the many ivory decorative plaques, probably furniture inlays, depicting enigmatic scenes military and heroic in tone, some seemingly of quite ancient inspiration. Other Fatimid-period objects from al-'Aqaba included steatite bowls from western and southern Arabia (also popular in earlier periods), and a collection of eleventh-century glass and bronze weights used to measure precious-metal coins. The weights are testimony to the expanding commercial role of al-'Aqaba in the Fatimid period, a development further supported by the discovery of a hoard of 32 dinars just inside the north gate. All but three (which are standard Fatimid issues) originated far to the west in Morocco, and were minted under the authority of the Spanish Umayyad Caliph Hisham II of Cordoba (r. 976–1009). They travelled as a group, probably with a merchant, until their concealment near the mosque of al-'Aqaba. Also datable to the Fatimid period is a group of engraved stone tombstones at Mazar, including a refined marble panel in decorative Kufic from the tomb of Ja'afar ibn Abi Talib, brother of 'Ali and a favourite of the Fatimid caliphs. The tenth- and eleventh-century finds from south Jordan, while still extremely limited in scope, are further evidence for the prevailing influence of Egypt, the Hijaz and south Arabia on the cultural matrix of the region under Fatimid hegemony.

The non-ceramic finds from the twelfth to fifteenth centuries betray the richness of the period, but little has been adequately reported. At Hisban, a wide range of finds were recovered, mostly domestic, commercial and personal objects manufactured from stone, clay, bone, ivory, metals and glass. Likewise at Tabaqat Fahl, the partial excavation of a fourteenth- and fifteenth-century settlement and cemetery has produced a corpus of copper finger rings, earrings and bangles for arms and ankles. Also present were bangles of glass, mostly made by twisting and joining multi-coloured canes. Similar glass bracelets were recovered at Tall Abu Sarbut and the cemetery on Tall Dayr 'Alla, with some evidence for a local manufacturing centre. Copper coins struck under the Ayyubids and Mamluks are a relatively common discovery at larger sites, although no Ayyubid or Mamluk mints operated in Jordan. At Tabaqat Fahl, 11 Mamluk copper coins were found, all of fourteenth-century date and spanning the reigns of al-Mansur Abu Bakr (1341) to az-Zahir Barquq (second reign, 1390–99). The positively identified mints were Damascus and Cairo, with others possibly originating in Hamah, Alexandria and Tarablus. The geographical spread of mints reveals long-distance contacts with districts in Egypt and ash-Sham. From Hisban, a total of 39 copper and one silver Ayyubid-period coins were recovered, increasing to 65 copper and 71 silver in the Mamluk period. Most of the Mamluk silver coins were found as a hoard of 66 concealed in a lamp. A larger hoard of 2244 early Mamluk silver coins was found at al-Karak in a ceramic pot. All were struck in the short period between 1257 and 1296, but as many were worn they were probably concealed after the end of the thirteenth century.

While the evidence is tantalizing for an abundant and varied cultural horizon in Ayyubid and Mamluk Jordan, the available evidence continues to be patchy and incomplete. Until large-scale, systematic excavations are undertaken on a major Ayyubid and Mamluk site in Jordan (which probably requires working in a currently occupied town), and while existing excavations remain unpublished, the prevailing narrow view of 'Ayyubid-Mamluk Jordan equals only rough hand-made pottery' will persist unchallenged. The Hisban results—its architecture, coins, ceramics and the many other finds—indicate the great promise of any future project.

Settlement patterns and population: towards a social history of middle Islamic Jordan (a parting note)

The Arabic and Crusader written sources, archaeological results (what little exists) and, most critically, the activities of the Crusaders east of the Jordan Rift valley point to a sustained level of socio-economic activity in the eleventh and twelfth centuries. There were at least eight major population centres reliant on Jordan, supported by a vibrant rural sector, trade and commerce. The importance of trade, especially in the eleventh century, is demonstrated by the finds at Ayla (al-'Aqaba), but the benefits would have spread up the main communication routes. There is growing evidence, at Khirbat Faris and Dhiban for instance, that many sites with easily recognized Ayyubid and Mamluk occupation were also settled in Fatimid, Seljuq and perhaps Crusader times, the later material often disguising earlier evidence (a problem especially apparent in surveys). If so, the supposed 'settlement gap' between the early Abbasid and Mamluk periods quickly disappears. Settlement levels were probably not significantly less than in the Ayyubid and Mamluk periods.

The emergence of new local power nodules, especially the Jarrahids, in eleventh-century Jordan had a profound impact on urban and rural settlement, especially the relative importance of the major population centres. In the mountains of south Jordan, where the trend is very clear, the traditional political and social functions of Ma'ab were taken over by al-Karak, strategically placed on major communication routes and easier to defend. Likewise Wadi Musa, astride routes leading into the 'Arabah, took the place of Udhruh. In the north, a similar trend can be recognized, although the evidence is more circumstantial. Probably during the eleventh century, the traditional socio-political roles of 'Amman and Jarash were transferred to as-Salt and 'Ajlun, for again both sites could be defended and were strategically placed on Jordan Valley routes. Equally, the ascendancy of Irbid at the expense of Bayt Ras probably began at this time. Geographically, the overall shift was westwards, away from the open *badiyah* to strategic, defensible locations on high points along the east Jordan rift—to citadels in other words. In the gestation of these strongholds three important points need emphasizing:

- The rise of citadels was in response to the political uncertainties of the eleventh and twelfth centuries, and represents the political aspirations of local shaykhdoms, usually of Bedouin origin, and their opposition to expanding Fatimid domination.
- They did not originate in a 'desert and sown' conflict. Indeed many of the shaykhdoms gained powerful backing from the urban populations.
- This movement marked the final phase in the urban eclipse of many of the once dominant Classical-period centres, for instance Udhruh, Ma'ab, 'Amman, Jarash and Bayt Ras. They were still occupied, obviously, but served no major politico-military role.

The changes appear to have happened quite rapidly and may have been accompanied by a severe, if relatively brief, impact on overall occupation levels. Hence the rise of citadels in Jordan, as strong points for the ruling elite, was well under way before the arrival of the Crusaders in the early twelfth century.

Seen in this light, the Crusader occupation of south Jordan and their attempts in the north represent an intensification of a process already underway. Major castles were built at ash-Shawbak and Wadi Musa in 1115–16, and al-Karak in 1142 (replacing the Jarrahid stronghold). In the north, the Muslims built a fort at Jarash in 1120, but it was soon destroyed. The Crusader–Muslim struggle brought fortified towns to the fore in the twelfth century, a process that continued into the Ayyubid period. Further castles were built by the Ayyubids at 'Ajlun (1188–92), as-Salt (1220) and al-Azraq (1236–37). In addition, al-Karak and its ancillary but equally impregnable castle of ash-Shawbak were both substantially renovated under the Ayyubids and early Mamluks. Walls were strengthened, towers added and palaces built to house the elite in suitable style, although all this activity essentially ceased after the rise of the Circassian Mamluks in 1382. In general, the Mamluks were more interested in securing the major roads, especially the Darb al-Hajj and the postal routes between major centres. A number of forts and caravanserais were erected at major passage points, especially where there was water, and toll stations were established. The Mamluks also directed their attention to the Muslim holy sites of Jordan, for instance Mu'tah-Mazar, Amata, Jabal Harun and ar-Raqim. The Fatimids had taken particular interest in developing these holy sites, notably Mazar, and in part the Mamluk intention was to reduce the Fatimid-Shi'ite presence by reasserting Muslim orthodoxy. The holy sites represent another important example of social continuity from the Fatimid period into Mamluk times.

In the fields of settlement, society and culture, the threads of continuity are demonstrably very strong between the Fatimid and Mamluk periods although, as has been seen, change is undeniably apparent. The ceramics further reflect this dichotomy. Hand-Made Geometric Painted Wares were clearly present in the later eleventh to twelfth centuries, but in these early stages it displayed considerable regional variation in fabric and manufacture. By the thirteenth century there was greater standardization across Jordan, Palestine and Syria, with the wares continuing into the fifteenth century and probably beyond. The rise of HMGPW took place at the expense of wheel-made utilitarian pottery, although it can still be found in urban contexts. The ceramics at al-Karak, especially the fine glazed and wheel-made wares, argue for a high degree of social bifurcation in the Mamluk period, the final outcome of more than 300 years of socio-cultural development in south Jordan.

The tone of this chapter has been deliberately 'upbeat'. It has tried to redress the mistakenly negative view of Jordan under Fatimid, Seljuq, Crusader, Ayyubid and Mamluk hegemony. The available historical and archaeological information has been interpreted within a positive and assertive framework, using the limited data at hand to develop an image of Islamic Jordanian society in the throes of deep and permanent political, cultural and economic change in the immediately pre-modern age. Only when middle Islamic Jordan is evaluated in this manner, freed of irrelevant concepts of state and society devised from Classical and biblical archaeology, will a more honest assessment become possible, and the modern Middle East more accessible to the Western mind.

Acknowledgments

I am very grateful to Ahmad Shboul for assisting with the Qur'anic passages, to Lawrence Pontin for research in 'Amman and to Peter Magee for library work in Sydney. Many thanks are also due to Nigel Oram and John Couani at the Arts Faculty IT Unit for assisting with the illustrations. A Mamluk bibliography provided over the World Wide Web by the Middle East Documentation Center at the University of Chicago gave useful pointers (http://www.lib.uchicago.edu/e/su/mideast/MamBib.html).

The research for and writing up of this paper was completed while the author was an Australian Research Fellow in the Department of Semitic Studies, University of Sydney, Australia, through funding provided by the Australian Research Council.

References

Abu Dalu, R.
    1995   The Technology of Sugar Mills in the Jordan Valley during the Islamic Periods (in Arabic). In K. 'Amr, F. Zayadine and M. Zaghoul (eds.), *Studies in the History and Archaeology of Jordan*, V: 37-48. Amman: Department of Antiquities.

Abu'l-Fida
    1983   *The Memoirs of a Syrian Prince*. Wiesbaden: Steiner.

Ashtor, E.
    1981   Levantine Sugar Industry in the Late Middle Ages: A Case of Technological Decline. In A.L. Udovitch (ed.), *The Islamic Middle East, 700–1900*, 91-132. Princeton, NJ: Darwin.

Bienkowski, P. (ed.)
    1991   *Treasures from an Ancient Land: The Art of Jordan*. Liverpool: National Museums and Galleries on Merseyside.

Brown, R.M.
    1984   Late Islamic Settlement Patterns on the Kerak Plateau, Trans-Jordan. Unpublished MA thesis, State University of New York, Binghamton.
    1987   A 12th century A.D. Sequence from Southern TransJordan: Crusader and Ayyubid Occupation at el-Wu'eira. *Annual of the Department of Antiquities of Jordan* 31: 267-88.

1988 Summary Report of the 1986 Excavations: Late Islamic Shobak. *Annual of the Department of Antiquities of Jordan* 32: 225-45.

1989a Excavations in the 14th Century AD Mamluk Palace at Kerak. *Annual of the Department of Antiquities of Jordan* 33: 287-304.

1989b Kerak Castle. In D. Homès-Fredericq and J.B. Hennessy (eds.), *Archaeology of Jordan. II/1. Field Reports: Surveys and Sites A-K*, 341-47. Akkadica Supplementum 7. Leuven: Peeters.

1989c Shaubak. In D. Homès-Fredericq and J.B. Hennessy (eds.), *Archaeology of Jordan. II/2. Field Reports: Sites L-Z*, 559-66. Akkadica Supplementum 8. Leuven: Peeters.

1991 Ceramics from the Kerak Plateau. In J.M. Miller (ed.), *Archaeological Survey of the Kerak Plateau*, 168-279. Atlanta, GA: Scholars Press.

1992 Late Islamic Ceramic Production and Distribution in the Southern Levant: A Socio-economic and Political Interpretation. Unpublished PhD dissertation, State University of New York, Binghamton.

Brünnow, R.E. and A. von Domaszewski
1905 *Die Provincia Arabia*. Strassburg: Trübner.

Burgoyne, M.H.
1987 *Mamluk Jerusalem: An Architectural Study*. London: British School of Archaeology in Jerusalem/World of Islam Festival Trust.

Chaudhuri, K.N.
1985 *Trade and Civilisation in the Indian Ocean: An Economic History from the Rise of Islam to 1750*. Cambridge: Cambridge University Press.

Coughenour, R.A.
1989 Mugharat el Wardeh. In D. Homès-Fredericq and J.B. Hennessy (eds.), *Archaeology of Jordan. II/2. Field Reports: Sites L-Z*, 386-90. Akkadica Supplementum 8. Leuven: Peeters.

De Vries, B.
1986 The Islamic Bath at Tell Hesban. In L.T. Geraty and L.G. Herr (eds.), *The Archaeology of Jordan and Other Studies Presented to Siegfried H. Horn*, 223-35. Berrien Springs, MI: Andrews University.

1994 Hesban in the Ayyubid and Mamluk periods. In D. Merling and L.T. Geraty (eds.), *Hesban after 25 Years*, 151-166. Berrien Springs, MI: Andrews University.

Franken, H.J. and J. Kalsbeek
1975 *Potters of a Medieval Village in the Jordan Valley*. Leiden: North Holland.

Fulcher of Chartres
1969 *A History of the Expedition to Jerusalem 1095-1127*. Knoxville: University of Tennessee.

Galloway, J.H.
1989 *The Sugar Cane Industry: An Historical Geography from its Origins to 1914*. Cambridge: Cambridge University Press.

Geraty, L.T. and Ø.S. LaBianca
1985 The Local Environment and Human Food-procuring Strategies in Jordan: The Case of Tell Hesban and its Surrounding Region. In A. Hadidi (ed.), *Studies in the History and Archaeology of Jordan*, II: 323-30. Amman: Department of Antiquities.

Gibb, H.A.R.
1958 *The Travels of Ibn Battuta*. Cambridge: Cambridge University Press/The Hakluyt Society.

Glidden, H.W.
1952 The Mamluk Origin of the Fortified Khan at al-'Aqabah, Jordan. In G.C. Miles (ed.), *Archaeologica Orientalia: In Memoriam Ernst Herzfeld*, 116-18. New York: Augustin.

Greene, J.A.
1995 The Water Mills of the 'Ajlun-Kufranja Valley: The Relationship of Technology, Society and Settlement. In K. 'Amr, F. Zayadine and M. Zaghloul (eds.), *Studies in the History and Archaeology of Jordan*, V: 757-65. Amman: Department of Antiquities.

Haas, H. de, H.E. LaGro and M. Steiner
1989 First Season of Excavations at Tell Abu Sarbut, 1988: A Preliminary Report. *Annual of the Department of Antiquities of Jordan* 33: 323-26.

1992 Second and Third Season of Excavations at Tell Abu Sarbut, Jordan Valley (Preliminary Report). *Annual of the Department of Antiquities of Jordan* 36: 333-43.

Harding, G.L.
1967 *The Antiquities of Jordan*. New York: Praeger.

Hart, S.
1987 Five Soundings in Southern Jordan. *Levant* 19: 33-47.

Holt, P.M.
1986 *The Age of the Crusades: The Near East from the Eleventh Century to 1517*. London: Longman.

Homès-Fredericq, D.
1989 Lehun. In D. Homès-Fredericq and J.B. Hennessy (eds.), *Archaeology of Jordan. II/2. Field Reports: Sites L-Z*, 349-59. Akkadica Supplementum 8. Leuven: Peeters.

Homès-Fredericq, D. and H.J. Franken
1986 *Pottery and Potters—Past and Present: 7000 Years of Ceramic Art in Jordan*. Tübingen: Attempto.

Humphreys, R.S.
1977 *From Saladin to the Mongols: The Ayyubids of Damascus, 1193–1260*. New York: State University of New York.

Ibach, R.D.
1987 *Archaeological Survey of the Hesban Region: Catalogue of Sites and Characterisation of Periods*. Berrien Springs, MI: Andrews University.

Ibn al-Qalanisi
1967 *The Damascus Chronicle of the Crusades*. London: Luzac.
(1932)

Ibn Battuta
1958 *The Travels of Ibn Battuta*, A.D. *1325–1354*. Cambridge: Cambridge University Press.

Ibn Jubayr
1952 *The Travels of Ibn Jubayr*. London: Jonathan Cape.

Ibrahim, M., J.A. Sauer and K. Yassine
1976 The East Jordan Valley Survey, 1975. *Bulletin of the American Schools of Oriental Research* 222: 41-66.

Irwin, R.
1986 *The Middle East in the Middle Ages: The Early Mamluk Sultanate 1250–1382*. London: Croom Helm.

Johns, C.N.
1932 Medieval 'Ajlun. *Quarterly of the Department of Antiquities of Palestine* 1: 21-33.

Johns, J.
1992 Islamic Settlement in Ard al-Karak. In M. Zaghloul *et al.* (eds.), *Studies in the History and Archaeology of Jordan*, IV: 363-68. Amman: Department of Antiquities.

1994 The *Longue Durée*: State and Settlement Strategies in Southern Transjordan across the Islamic Centuries. In E.L. Rogan and T. Tell (eds.), *Village, Steppe and State: The Social Origins of Modern Jordan*, 1-31. London: British Academic Press.

Johns, J., A. McQuitty and R.K. Falkner
1989 The Faris Project: Preliminary Report upon the 1986 and 1988 Seasons. *Levant* 21: 63-95.

Kareem, J.M.H.
1987 Evidence of the Umayyad Occupation in the Jordan Valley as Seen in the Jisr Sheikh Hussein Region. Unpublished MA thesis, Yarmouk University.

1992 Darb al-Quful in the Light of Archaeological and Historical Evidence. *Newsletter of the Institute of Archaeology and Anthropology, Yarmouk University* 13: 3-8.

1993 The Settlement Patterns in the Jordan Valley in the Mid- to Late Islamic Period. Unpublished PhD dissertation, Freie Universität Berlin.

1997 The Site of Dhra' al-Khan: A Main Caravanserai on Darb al-Quful. In G. Bisheh, M. Zaghloul and I. Kerberg (eds.), *Studies in the History and Archaeology of Jordan*, VI: 365-69. Amman: Department of Antiquities.

Kennedy, D.L.
1982 *Archaeological Explorations on the Roman Frontier in North East Jordan*. Oxford: British Archaeological Reports.

Kennedy, H.
1986 *The Prophet and the Age of the Caliphates: The Islamic Near East from the Sixth to the Eleventh Century*. London: Longman.

1991 Nomads and Settled People in Bilad ash-Sham in the Fourth/Ninth and Fifth/Tenth Centuries. In M.A. Bakhit and R. Schick (eds.), *Proceedings of the Fifth International Conference on the History of Bilad ash-Sham: Bilad ash-Sham during the Abbasid Period (English and French Section)*, 105-13. Amman: History of Bilad ash-Sham Committee.

1994 *Crusader Castles*. Cambridge: Cambridge University Press.

Khadija, L.
1992 Designs on Painted Ayyubid/Mamluk Pottery from Rujm al-Kursi, 1990 Season. *Annual of the Department of Antiquities of Jordan* 36: 345-56.

Khammash, A.
1986 *Notes on Village Architecture in Jordan*. Lafayette: University of Southwestern Louisiana.

King, G., C.J. Lenzen and G.O. Rollefson
1983 Survey of Byzantine and Islamic Sites in Jordan: Second Season Report, 1981. *Annual of the Department of Antiquities of Jordan* 27: 385-436.

King, G.R.D. *et al.*
1987 Survey of Byzantine and Islamic Sites in Jordan. Third Season Preliminary Report (1982): The Southern Ghor. *Annual of the Department of Antiquities of Jordan* 31: 439-59.

LaBianca, Ø.S.
1990 *Sedentarization and Nomadization: Food System Cycles at Hesban and Vicinity in Transjordan*. Hesban I. Berrien Springs, MI: Andrews University.

Le Strange, G.
1965 *Palestine under the Moslems*. Beirut: Khayats.
(1890)

Lenzen, C.J.
1995 From Public to Private Space: Changes in the Urban Plan of Bayt Ras/Capitolias. In K. 'Amr, F. Zayadine and M. Zaghoul (eds.), *Studies in the History and Archaeology of Jordan*, V: 235-39. Amman: Department of Antiquities.

Lenzen, C.J. and E.A. Knauf
1987 Beit Ras/Capitolias: A Preliminary Evaluation of the Archaeological and Textual Evidence. *Syria* 64: 21-46.

Maalouf, A.
1984 *The Crusades through Arab Eyes*. London: Al Saqi Books/American University in Cairo.

Macaulay, R. and R. Beny
1977 *The Pleasure of Ruins*. London: Thames & Hudson.
(1953)

McKenzie, J.
1991 The Beduin at Petra: The Historical Sources. *Levant* 23: 139-45.

McQuitty, A.
1993 Book Review of Ø.S. LaBianca, *Sedentarization and Nomadization*, Hesban I. *Palestine Exploration Quarterly* 125: 167-69.

1995 Water-mills in Jordan: Technology, Typology, Dating and Development. In K. 'Amr, F. Zayadine and M. Zaghloul (eds.), *Studies in the History and Archaeology of Jordan*, V: 745-51. Amman: Department of Antiquities.

McQuitty, A. and R.K. Falkner
1993 The Faris Project: Preliminary Report on the 1989, 1990 and 1991 Seasons. *Levant* 25: 37-61.

Merrill, S.
1986 *East of the Jordan: A Record of Travel and Observation in*
(1881) *the Countries of Moab, Gilead, and Bashan*. London: Darf.

Mershen, B. and E.A. Knauf
1988 From *Gadar* to *Umm Qais*. *Zeitschrift des deutschen Palästina-Vereins* 104: 128-45.

Miller, J.M. (ed.)
1991 *Archaeological Survey of the Kerak Plateau*. Atlanta, GA: Scholars Press.

Minnis, D. and Y. Bader
1988 A Comparative Analysis of Belvoir (Kawkab al-Hawa) and Qal'at al-Rabad ('Ajlun Castle). *Annual of the Department of Antiquities of Jordan* 32: 255-63.

Muqaddasi (al-Maqdisi)
1994 *The Best Divisions for the Knowledge of the Regions (Ahsan al-Taqasim fi ma'rifat al-Aqalim)*. Reading: Garnet.

Musil, A.
1926 *The Northern Hegaz: A Topographical Itinerary*. New York: American Geographical Society.
1927 *Arabia Deserta*. New York: American Geographical Society.

Northedge, A.E.
1984 Qal'at 'Amman in the Early Islamic Period. Unpublished PhD dissertation, University of London.

Northedge, A.
1992 *Studies on Roman and Islamic 'Amman: The Excavations of Mrs C.-M. Bennett and Other Investigations*. Oxford: Oxford University Press.

Palumbo, G. (ed.)
1994 *JADIS The Jordanian Antiquities Database and Information System: A Summary of the Data*. Amman: Department of Antiquities of Jordan/American Center of Oriental Research.

Peterman, G.L. and R. Schick
1996 The Monastery of Saint Aaron. *Annual of the Department of Antiquities of Jordan* 40: 473-80.

Peters, F.E.
1994 *The Hajj: The Muslim Pilgrimage to Mecca and the Holy Places*. Princeton, NJ: Princeton University Press.

Petersen, A.
1991 Two Forts on the Medieval Hajj Route in Jordan. *Annual of the Department of Antiquities of Jordan* 35: 347-59.

Rogan, E.L.
1995 Reconstructing Water Mills in Late Ottoman Transjordan. In K. 'Amr, F. Zayadine and M. Zaghloul (eds.), *Studies in the History and Archaeology of Jordan*, V: 753-56. Amman: Department of Antiquities.

Sauer, J.A.
1973 *Heshbon Pottery 1971: A Preliminary Report on the Pottery from the 1971 Excavations at Tell Hesban*. Berrien Springs, MI: Andrews University.
1976 Pottery Techniques at Tell Deir 'Alla (review article). *Bulletin of the American Schools of Oriental Research* 224: 91-94.
1982 The Pottery of Jordan in the Early Islamic Period. In A. Hadidi (ed.), *Studies in the History and Archaeology of Jordan*, I: 329-37. Amman: Department of Antiquities.
1994 The Pottery at Hesban and its Relationship to the History of Jordan: An Interim Hesban Pottery Report, 1993. In D. Merling and L.T. Geraty (eds.), *Hesban after 25 Years*, 225-281. Berrien Springs, MI: Andrews University.

Scanlon, G.T.
1970 Egypt and China: Trade and Imitation. In D.S. Richards (ed.), *Islam and the Trade of Asia: A Colloquium*, 81-95. Oxford: Bruno Cassirer.

Schick, R.
1997 Southern Jordan in the Fatimid and Seljuq Periods. *Bulletin of the American Schools of Oriental Research* 305: 73-85.

Schick, R., H. Mahasneh and S. Ma'ani
1994 Nakhal. *American Journal of Archaeology* 98: 553-54.

Smith, R.H.
1973 *Pella of the Decapolis* I. *The 1967 Season of the College of Wooster Expedition to Pella*. Wooster, OH: College of Wooster.

Smith, R.H. and L.P. Day
1989 *Pella of the Decapolis* II. *Final Report on the College of Wooster Excavations in Area IX, the Civic Complex, 1979–1985*. Wooster, OH: College of Wooster.

Thorau, P.
1987 *The Lion of Egypt: Sultan Baybars I and the Near East in*
(1992) *the Thirteenth Century*. London: Longman.

Tushingham, A.D.
1972 *The Excavations at Dibon (Dhiban) in Moab: The Third Campaign 1952–53*. Annual of the American Schools of Oriental Research 40. Cambridge, MA: American Schools of Oriental Research.

Vannini, G. and C. Tonghini
1997 Medieval Petra: The Stratigraphic Evidence from Recent Archaeological Excavations at al-Wu'ayra.

In G. Bisheh, M. Zaghloul and I. Kehberg (eds.), *Studies in the History and Archaeology of Jordan*, VI: 371-84. Amman: Department of Antiquities.

Vannini, G. and A. Vanni Desideri
- 1995 Archaeological Research on Medieval Petra: A Preliminary Report. *Annual of the Department of Antiquities of Jordan* 39: 509-40.

Walmsley, A.G.
- 1982 The Umayyad Pottery and its Antecedents. In A. McNicoll, R.H. Smith and B. Hennessy (eds.), *Pella in Jordan I. An Interim Report on the Joint University of Sydney and The College of Wooster Excavations at Pella 1979–1981*, 143-57. Canberra: Australian National Gallery.
- 1987 The Administrative Structure and Urban Geography of the *Jund* of Filastin and the *Jund* of al-Urdunn: The Cities and Districts of Palestine and East Jordan during the Early Islamic, 'Abbasid and Early Fatimid Periods. Unpublished PhD dissertation, University of Sydney.
- 1991 Architecture and Artifacts from Abbasid Fihl: Implications for the Cultural History of Jordan. In M.A. Bakhit and R. Schick (eds.), *Proceedings of the Fifth International Conference on the History of Bilad ash-Sham: Bilad ash-Sham during the Abbasid Period (English and French Section)*, 135-59. Amman: History of Bilad ash-Sham Committee.
- 1992a Fihl (Pella) and the Cities of North Jordan during the Umayyad and Abbasid Periods. In K. 'Amr et al. (eds.), *Studies in the History and Archaeology of Jordan*. IV: 377-84. Amman: Department of Antiquities.
- 1992b The Islamic Period: The Later Islamic Periods. In A.W. McNicoll et al. (eds.), *Pella in Jordan II: The Second Interim Report of the Joint University of Sydney and College of Wooster Excavations at Pella 1982–1985*, 188-98. Sydney: Mediterranean Archaeology.
- 1994 Ceramic Analysis and Social Processes: Pottery and Society in Antiquity. In C.C. Sorrell and A.J. Ruys (eds.), *Proceedings of the International Ceramics Conference, Austceram94* (International Ceramic Monographs Volume 1 issues 1), 10-15. Sydney: Australian Ceramic Society.
- 1995 Tradition, Innovation, and Imitation in the Material Culture of Islamic Jordan: The First Four Centuries. In K. 'Amr, F. Zayadine and M. Zaghoul (eds.), *Studies in the History and Archaeology of Jordan*, V: 657-68. Amman: Department of Antiquities.
- forthcoming Turning East: The Appearance of Islamic Cream Wares in Jordan—The End of Antiquity? In E. Villeneuve and P.M. Watson (eds.), *Byzantine and Early Islamic Ceramics in Syria–Jordan (IVth–VIIIth Centuries): Actes du colloque international*. Beirut: Institut français d'archéologie du Proche-Orient (IFAPO).

Watson, A.M.
- 1983 *Agricultural Innovation in the Early Islamic World: The Diffusion of Crops and Farming Techniques, 700–1000.* Cambridge: Cambridge University Press.

Whitcomb, D.S.
- 1988a *Aqaba: 'Port of Palestine on the China Sea'.* Chicago: Al Kutba.
- 1988b A Fatimid Residence at Aqaba, Jordan. *Annual of the Department of Antiquities of Jordan* 32: 207-23.
- 1992a The Islamic Period as seen from Selected Sites. In B. MacDonald et al., *The Southern Ghors and Northeast 'Arabah Archaeological Survey*, 113-18. Sheffield Archaeological Monographs 5. Sheffield: J.R. Collis.
- 1992b Reassessing the Archaeology of Jordan of the Abbasid Period. In K. 'Amr et al. (eds.), *Studies in the History and Archaeology of Jordan*, IV: 385-90. Amman: Department of Antiquities.
- 1994 *Ayla: Art and History in the Islamic Port of Aqaba.* Chicago: The Oriental Institute, University of Chicago.
- 1995a Islam and the Socio-Cultural Transition of Palestine—Early Islamic Period (638–1099 CE). In T.E. Levy (ed.), *The Archaeology of Society in the Holy Land*, 488-501. London: Leicester University Press.
- 1995b A Street and the Beach at Ayla: The Fall Season of Excavations at 'Aqaba, 1992. *Annual of the Department of Antiquities of Jordan* 39: 499-507.
- 1997a The Town and Name of Aqaba: An Inquiry into the Settlement History from an Archaeological Perspective. In G. Bisheh, M. Zaghloul and I. Kehrberg (eds.), *Studies in the History and Archaeology of Jordan*, VI: 359-63. Amman: Department of Antiquities.
- 1997b Mamluk Archaeological Studies: The State of the Art. *Mamluk Studies Review* 1: 97-106.

William of Tyre
- 1976 (1941)-a *A History of Deeds Done beyond the Sea*, I. New York: Octagon.
- 1976 (1941)-b *A History of Deeds Done Beyond the Sea*, II. New York: Octagon.

Wood, J.
- 1992 The Fortifications. In A. Northedge, *Studies on Roman and Islamic 'Amman: The Excavations of Mrs C.-M. Bennett and Other Investigations*, 105-27. Oxford: Oxford University Press.

Zayadine, F.
- 1985 Caravan Routes between Egypt and Nabataea and the Voyage of Sultan Baibars to Petra in 1276. In A. Hadidi (ed.), *Studies in the History and Archaeology of Jordan*, II: 159-74. Amman: Department of Antiquities.

16. The Ottoman Period[1]

Alison McQuitty

Introduction

For most of the span of Ottoman rule in Jordan (AD 1516–1918) the archaeological evidence for settlement, society and economy within the country is both inconclusive and invisible. Apart from the Hajj forts (see Petersen, this volume) and associated sherd dumps, a sprinkling of tobacco pipes, railway stations and a few mid- to late-nineteenth-century buildings, there is little material culture to indicate the Ottoman nature of the country. Jordan was an area of rural settlement, predominantly ahistorical, certainly tribal and definitely 'the Other' as far as the authors of the historical documentation available were concerned, that is, the Ottoman administration and nineteenth-century European travellers. In addition, the Antiquities Law in Jordan does not yet protect material post-AD 1700 and, until recently, these late periods have not aroused much interest or attention from archaeologists (Palumbo et al. 1993: 69-84; Silberman 1989). All of the above factors have contributed to a patchy and vague picture of the settlement and economy of Jordan in the Ottoman period. The following is a consideration of the strands of evidence available which can be used to weave a more complete impression of the period: historical (official, tribal, and oral), archaeological, architectural (official and vernacular), and the material culture (Table 16.1).

History

Before considering the historical evidence in more detail, a brief outline of Ottoman administration and the terminology involved is presented.[2]

Ottoman administration (terms in Turkish/Arabic)

At the head of the provincial administrative hierarchy was the *vilayet/wilayah* (province) where the *vali/wali* (governor) resided. The new territories of the province were divided into *sancaks/liwas* (sub-provinces), which again were sub-divided into *kazas/qadas* (districts). The leader of the *sancak/liwa* was the *mutasarrif* and of the *kaza/qada* the *kaymakam* (district governor). The smallest unit of administration, several of which comprised the *kaza*, was the *nahiye* headed by a *mudir* (Hütteroth and Abdulfattah 1977: 17-20; Rogan 1994: 37). The *nahiye* was made up of a collection of villages and nomads/tribes. These administrative units were both military and fiscal in purpose—they served as recruiting areas for the provision of troops for the Ottoman army and taxes were levied according to the economic potential of the *nahiye*. It is the description at the *nahiye* level that offers the greatest information to the archaeologist.

Historical sources and their use
(Table 16.1)

Archaeologists fortunate enough to work in historical periods often rely too heavily on interpretations based on documents that were not primarily compiled to describe the nature of settlements, their interaction with each other, their burial customs and their material culture. They try to 'fit' their data to the historical overview. As Davis summarizes it: 'historians have tended to focus on the big pictures, while archaeological investigations by their very nature are more likely to generate information that pertains to little pictures of the past' (1991: 133).

The treasury of sixteenth-century Ottoman tax records, so comprehensively studied and analysed by Hütteroth and Abdulfattah, gives the scholar a checklist of the agricultural produce from each area. The records give an insight into the way in which the Ottoman administration collected taxes, provide a terminology for the settlement hierarchy, and pinpoint the geographical position of some types of settlements. The records do not reveal the material culture correlates, for example, what a *qarya* (village) or *mazra'a* (farm with no permanent settlement) looks like and if they can be distinguished archaeologically. Nomads are recorded but no light is shed (or can be shed from this collection of documents) on their interaction with more sedentary populations. Above all, the flexibility and variety of human adaptation to the

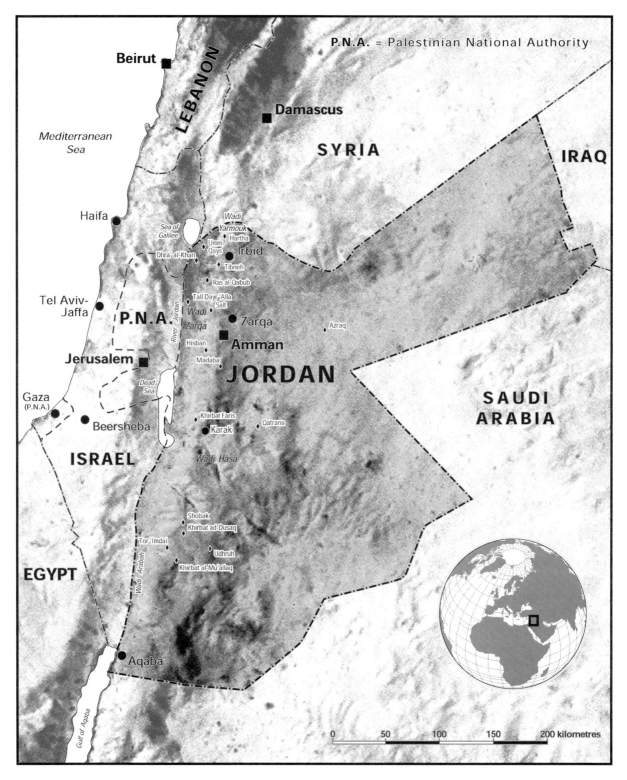

Figure 16.1. Map of Jordan showing sites mentioned in text.

environment is not conveyed by the static categorization that bureaucracies such as the Ottoman tax collectors demand. Nineteenth-century land records or *tapu* documents reveal much about the relationship of the administration to its 'subjects' although there is debate about whether the regional diversity shown can be explained by the lack of registration by the Ottoman authorities or regional diversity

Historical source	Related bibliography	Date
defter i mufassal	Hütteroth and Abdulfattah 1977	16th century
tapu (land) registers and court records	Mandaville 1966; Mundy 1996; Rogan 1991, 1994, 1995	19th and 20th centuries
Papers of Church Missionary Society	Held at University of Birmingham (UK) library	19th and 20th centuries
Archives of the Latin Patriarchate of Jerusalem		19th and 20th centuries
ethnography and tribal/oral history	Abu Jaber 1989; Lewis 1987 and all European travellers' accounts	17th–20th centuries
various photographic archives—only UK locations mentioned	Palestine Exploration Fund; Gertrude Bell Archive, University of Newcastle; St Antony's College, Oxford	19th and 20th centuries

Table 16.1. Historical sources for the Ottoman period in Jordan.

per se (Mundy 1996). For the archaeologist, the value of these documents lies in background information that can be added about population centres and the agricultural diversity that existed in the country as a whole.

Tribal and oral history give a vivid view of the movements of peoples and the way-of-life as remembered before 'modernization'.[3] The accounts of nineteenth-century European travellers are invaluable for casting light and detail on the lives, interactions and material culture of a population—the people history usually forgets—of a rural landscape. Two caveats in using their information are: (1) these travellers were often motivated by a desire to see traces of biblical pastoral scenes or classical ruins which certainly coloured what they reported (Silberman 1982: 242); and (2) there is a real danger of extrapolating back from the nineteenth century to characterize the seventeenth and eighteenth centuries which are largely missing from the historical record.

Historical evidence for Jordan in the Ottoman period (Table 16.2)

In the control of its provinces, the Ottoman empire, after a glorious and firm beginning, exerted less and less control which, by the eighteenth century, had become nominal in several cases. In many areas, the history of government was the history of inter-tribal conflicts and the local *sheikh* or tribal leader was the link between the Ottoman administration and the province (Findley 1986: 4). As an imperial authority, the Ottomans were notable for their ability to incorporate local realities and translate them into the Ottoman administrative system. Jordan became Ottoman in AD 1516 after the Ottomans had defeated the local Mamluk governor of Aleppo. By early 1517, the Ottomans had proceeded through Bilad ash-Sham to Cairo and the local chiefs had all paid homage to their new rulers (Pitcher 1972: 104-106). Throughout the history of Ottoman hegemony in Jordan, the priority always lay with keeping open the Hajj route from Damascus to the Holy Places of Mecca and Medina. It is here that Ottoman architecture and material culture is seen.

The country lay within the *vilayet/wilayah* (province) of Damascus. At the *sancak/liwa* (sub-province) level, Jordan consisted of the *Qada Hawran* and *Liwa 'Ajlun*, itself sub-divided, as far as the destination of

Selim I	1512–1520
Suleiman I	1520–1566
Selim II	1566–1574
Murad III	1574–1595
Mehmed III	1595–1603
Ahmed I	1603–1617
Mustafa I	1617–18; 1622–23
Osman II	1618–1622
Murad IV	1623–1640
Ibrahim	1640–1648
Mehmed IV	1648–1687
Suleiman II	1687–1691
Ahmed II	1691–1695
Mustafa II	1695–1703
Ahmed III	1703–1730
Mahmud I	1730–1754
Osman III	1754–1757
Mustafa III	1757–1774
Abdul Hamid I	1774–1789
Selim III	1789–1807
Mustafa IV	1807–1808
Mahmud II	1808–1839
Abdul Mejid	1839–1861
Abdul Aziz	1861–1876
Murad V	1876
Abdul Hamid II	1876–1909
Mehmed V	1909–1918
Mehmed VI	1918–1922

Table 16.2. Ottoman sultans.

tax revenues were concerned, between 'Ajlun and al-Karak, Jabal al-Karak and ash-Shawbak (Hütteroth and Abdulfattah 1977: 18). The sixteenth-century *defter-i mufassal* was arranged according to this framework but probably reflected the countryside, administrative divisions and agricultural products of the preceding Mamluk administration (Hütteroth and Abdulfattah 1977: 18).

The *defter* is a detailed register of the villages within each *nahiye* compiled for taxation purposes. It contains: the village names and number of households in each village; the names of the tribes that occupy land; the types, value and proportions of agricultural produce within the various environmental zones of Jordan; the types of settlement varying from *mazra'a* to small town; non-agricultural activities carried out in the *nahiye*; and the destinations of the taxation. A fiscal landscape can be drawn which also gives valuable insight into the agricultural landscape of Jordan. Models for the type of agriculture practised can be suggested on the basis of this *defter*. What cannot be revealed is whether *mazra'a* and *qaraya* are materially distinguishable; whether areas registered as under the unit of the tribe (*'ashira* or *jama'a*), and yet producing agricultural products, were necessarily barren of settlement; and how long this snapshot of late sixteenth-century rural life continued. It is not until the 1930s, under the British Mandate, that a comparative exercise was carried out throughout Jordan (Fishbach 1994). However, land registration had been proceeding since the mid-nineteenth century, and nineteenth-century travellers' accounts, the 1881 Palestine Exploration Fund map of a portion of Jordan, the 1899 Schumacher map of 'Ajlun and the 1889 Conder map of Moab do give some comparison.

The sixteenth-century *defter* demonstrates variation in economic strategies within Jordan; nineteenth-century land registration documents and court records show variation in the relationships between region and government. No variation in material culture can be shown by these documents and here one must turn to archaeology.

Environment

Climate, vegetation and wildlife

While the climate in Jordan has not dramatically changed for the last 4000 years, there have been long-term wet and dry cycles which can have major effects on the rural economy and settlement patterns (see discussion below on drought and Chapter 17, this volume). It has been suggested that one such cool, wet cycle affected the Middle East in the seventeenth–nineteenth centuries, bringing more favourable conditions for agriculture (Koucky 1987: 19: Faroqhi 1994: 467). More evidence on the micro-level (e.g. dendrochronology) is needed before such a conclusion can be sustained or rejected. Certainly the local variability in climatic conditions is not in doubt and this has had an impact on the type of economic strategy followed. It is ironic that, if correct, such an optimum period for the agricultural economy seems to equate with a period of low settlement figures.

Chapter 1 of this volume (Macumber) gives an overview of the potential of the vegetation of Jordan. By the Ottoman period, it seems from historical accounts that most of the natural vegetation cover had been removed in the course of agriculture, grazing and for the collection of firewood and building material. The increased demand for wood during the building and running of the Hijaz railway is inevitably cited as the reason behind the denuding of Jordan's forests. The fact is that the demand for wood and pressures on forested land always existed and by the late Ottoman period there were probably not many more trees to cut down. Nineteenth-century travellers' accounts mention the wide-open pastures on the plateau, the fields of corn and thickets in the wadis but not great stands of woodland. The analysis of botanical remains from archaeological sites gives a checklist of the kinds of wild fauna and flora found in a localized area. At Khirbat Faris, faunal remains attest to the presence and/or hunting of gazelle, oryx, fox and porcupine (Rielly 1989). At Hisban many more wild animals were recorded including gazelle, wild boar, ibex, hare, fox and hyena (LaBianca and Lacelle 1986: 68).

Natural disasters: plague, drought and earthquakes

The cycle of natural disasters that affected Jordan throughout this and preceding periods has often been used as an explanation for the comparative paucity of rural settlement. Along with general insecurity, plague, drought and earthquakes are taken to account for 'the flight of the peasant from the land', the decline in sedentary agriculture and permanent settlement (Brown 1984; Lewis 1987: 13). While all these natural disasters are the backdrop against which the period should be understood, they are not always phenomena detectable in the archaeological record.

Plague

The plague epidemic that struck Egypt and Syria in the preceding Mamluk Period was undoubtedly the devastating pneumonic plague which led to a definite demographic decline (Dols 1979). However, the strain of plague[4] affecting this area in the sixteenth–nineteenth centuries is not identified in the historical literature and its impact on population figures cannot be assessed. Added to this is the question of to what extent the plague would have spread rapidly in rural and often remote districts. Rogan suggests that while the plague may not have spread rapidly in rural districts, it would have caused depopulation of the countryside as the population migrated into the plague-ravaged towns of Jerusalem and Damascus (pers. comm.).

Drought

Throughout the Ottoman period, agriculture in Jordan was largely dependent on rainfall. In years of no rain, the impact on agriculture is devastating. In a more recent period, the droughts of the 1940s–50s in central Jordan led to far-reaching changes as the village of al-Qasr was more permanently occupied to take advantage of the government water and food-distribution centre while the cattle of the nearby Beni Hamaida tribe were slaughtered (Kana'an and McQuitty 1994; Lancaster and Lancaster 1995: 116). Such periods are not recorded in history on this local level but their occurrence may account for radical shifts in local settlement type and pattern as well as economic strategy. The effect of droughts on pastoralism is equally dramatic as herds are slaughtered or die for want of food and water, and nomads and their herds move farther into cultivated lands in search of water and grazing.

Earthquakes

Earthquakes and their effects, notwithstanding that the response to earthquake damage is often to clear it up rather than to abandon the site, are the favourite response of archaeologists to levels of destruction and abandonment within their excavations. However, one major earthquake, in 1588, is historically attested for this period with its epicentre in the Northern Hijaz (Ambraseys and Melville 1989: 1279). There were regular smaller and localized shocks and tremors throughout the Ottoman Period (Ghawanmeh 1992: 53-60) but the effects of the earthquakes are not known outside urban areas.

Land-use and economy

Throughout history, the main economic strategy followed in Jordan has been mixed farming. Rather than seeing a dichotomy between nomadic pastoralism and settled agriculture, it is considered more useful to conceive of a continuum with specialized nomadic pastoralism at one end and intensive sedentary agriculture at the other (Johns *et al.* 1989). The proportion of pastoralism and agriculture to each other varies between communities exploiting the same environmental zone and through time. There are additional variations in the type of agriculture practised, whether intensive or extensive (Davis 1991: 138-39). Mixed farming can be practised from a permanently built settlement or a cave or a tent or a combination of the three, that is, the population can be sedentary or semi-nomadic and the archaeologist cannot always be certain that lack of permanent settlement means nomadic pastoralism.[5] In short, the simple term 'mixed farming economy' conceals a wealth of complexity.

The *defter*s give a fine overview of the type of rural economy extant in the sixteenth century and reveal the economic variation in the country with its balance of agriculture and pastoralism (Figure 16.2). In the northern hills, a concentration of mixed-farming villages is found. Olives, fruit-trees and pulses were cultivated on terraces with extensive cultivation of wheat and barley taking place on flatter land. To the east, in the southern reaches of the Hauran, the concentration of agricultural production is on wheat, with summer-crops and barley providing a sizeable fraction. The area to the south of Wadi az-Zarqa' is recorded in the *defter*s as nomadic[6] with a large number of *mazra'a* producing a fixed amount of taxes, that is, it is not clear what the balance of produce is. On the al-Karak plateau the balance is also characteristic of a mixed-farming economy, although different from that farther north, with wheat/barley/summer crops and livestock being in almost equal quantity. This balance remains true for the hilly area of ash-Shawbak where fruit trees including vines and olives replace summer crops. Livestock form part of the produce proportion of all areas but do show concentrations on the al-Karak plateau and on the eastern edge of the Jordan Valley—they are almost entirely absent from the southern Hauran plain. Singer

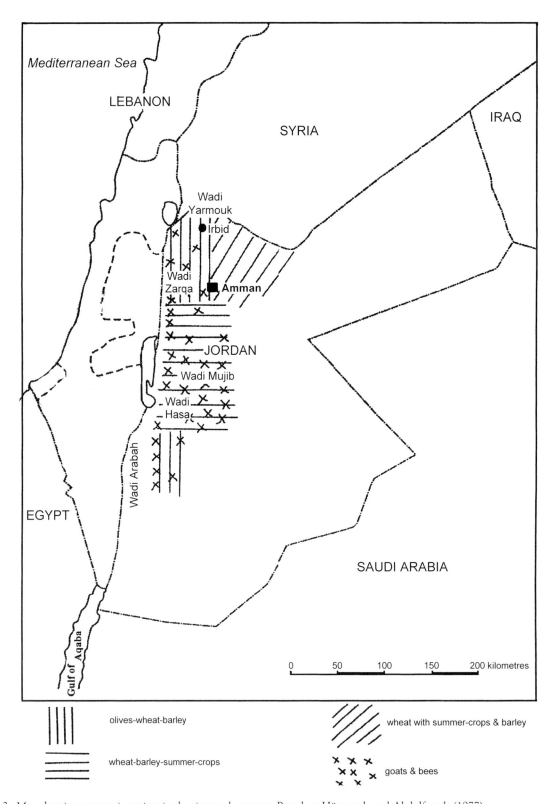

Figure 16.2. Map showing economic variety in the sixteenth century. Based on Hütteroth and Abdulfattah (1977).

notes that in the *defter* for Palestine the term 'goat' may well include sheep as well (1994: 51). The area south of ash-Shawbak lay in the *vilayat* of *Misr* and the taxable produce for these areas is not recorded in this *defter*.

The archaeological information to compare with this situation is simply not available: archaeology can provide the localized view but rarely the grand sweep. Two sites that have produced adequate information for the domesticated plants and animals consumed at the sites are Khirbat Faris (Rielly 1989) and Hisban (LaBianca and von den Driesch 1995: 72). In the early Ottoman period at Khirbat Faris, the environmental evidence shows a mixed-farming community herding overwhelmingly sheep/goat, with cattle being raised as a significant minority (for an account of this practice in the same area in recent times see Lancaster and Lancaster 1999: 202-203). Over time, the proportion of goats to sheep changes, with goats becoming more numerous. Sheep exist better with cultivation and goats fare better in degraded conditions (Köhler-Rollefson 1989: 91). The age profile shows that these cattle were mature animals, suggesting that they were kept for their milk and work potential. Other domesticated animals included donkey, horse, camel, pig, dog, cat, dove and chicken. Salt-water fish are present, suggesting trade with the Red Sea. Botanical evidence from the same period suggests the cultivation of barley, wheat, millet, oat and possibly sorghum with other legume crops, for example, lentil, bitter vetch and common pea being grown alongside. Stones of cultivated fruit/nut trees were also found on the site, including grape, date, fig, pistachio and olive (Charles 1992). The botanical remains analysed in detail seem to be crop by-products that have ultimately been used as fuel for clay ovens/*tawabeen*, whether directly or as animal fodder which, in dung form, was used as fuel. The faunal evidence from Hisban is interpreted as belonging to a later period, namely, 1870–1976. The range of domesticated species is similar to that at Khirbat Faris: sheep, goat, cattle, pig, camel, ass, horse, dog and cat. There is a slight preponderance of goat and the cattle bones have butcher-marks, indicating that they were used for meat (LaBianca and von den Driesch 1995: 103). The faunal information is based on a few samples and shows cultivation of wheat and barley with legumes and a few olive stones (Gilliland 1986: 125).

Trade

The provisioning of the Hajj caravan with its demand for transport and baggage animals, grain and fresh meat provided a major annual market for the Jordanian countryside (Landau 1971: 59). The traffic in grain was also carried on westwards to Palestine and the markets in Jerusalem and al-Khalil/Hebron. The Bedouin also sent plant-ash to Nablus for the soap industry (Lancaster and Lancaster 1995: 118) while bitumen was collected from the Dead Sea for use as fuel and in veterinary remedies. Indigo and sulphur were traded from the Jordan Valley (Hütteroth and Abdulfattah 1977: 83). Urban merchants from the towns of Palestine and Syria came to Jordan to trade manufactured goods such as pottery, textiles, tobacco pipes and glass but for the most part the economy was subsistence-based where barter rather than money-exchange was the rule. The trade was between the raw materials of Jordan and the manufactured goods of Syria and Palestine. Exchange undoubtedly took place between the various life-style groups, that is, pastoralists exchanging animals and dairy products with the grain of agriculturists. Such things, however, are not recorded in history or visible to archaeologists.

Archaeological framework

Archaeological syntheses of Ottoman settlement are few and far between and mainly based on survey data. The framework with which this survey data is interpreted varies from using a 'core–periphery' model (Brown 1984, 1992) to analysis based on a 'food-systems' model which takes into account the fluctuations in the balance between pastoralism and agriculture, between production and productivity (LaBianca 1990). As with other Islamic periods, the pitfalls of using dynastic terms to characterize settlement patterns and material culture are very real (Johns 1992a: 363; Whitcomb 1992: 386). In addition, there are severe difficulties in distinguishing Ottoman material culture. While some scholars use dynastic terms to divide periods of history (Sauer 1986), various alternatives for a more archaeologically oriented terminology have been suggested (Whitcomb 1992: 386). For the purposes of this chapter, organized in dynastic terms, this writer prefers the solution adopted by MacDonald *et al.* (1992: 75) who refer to two main periods within this date range: Mamluk/Ottoman (thirteenth–end sixteenth centuries) and Ottoman/Modern (mid-nineteenth–early twentieth centuries). Given the present state of archaeological knowledge, much of what is identified as Mamluk is likely to be Early Ottoman while the Late Ottoman period overlaps with the Modern. In the Mamluk/Ottoman period archaeologists are dealing with information from excavations, surveys and a few standing buildings. In the Ottoman/Modern period standing buildings are the overwhelming bulk of evidence. The gap in-between, that is, the elusive seventeenth–early nineteenth centuries, is what remains to be detailed.

Surveys

In spite of the wealth of archaeological activity in Jordan over the past 20 years, the country is still relatively poorly covered in terms of survey, particularly in the hilly areas (see Chapter 21, this volume). While the Eastern Desert and the Jordan Valley, the al-Karak plateau, Wadi al-Hasa and the Madaba Plains have been relatively well covered, other areas have received partial but intensive coverage, for example, concentration in northern Jordan has been on the east–west wadi systems: Wadi al-Yabis, Wadi Ziqlab. Coupled with the lack of ability to distinguish Ottoman material culture, this has led to an underestimation of settlement numbers and lack of clarity in distinguishing a settlement hierarchy. Many surveys do not differentiate between Ayyubid/Mamluk and Ottoman settlements (Ibrahim *et al.* 1976: 61; Mabry and Palumbo 1988) or totally ignore the Ottoman period.[7]

Numbers alone of sites attesting Ottoman presence do not add to the picture of settlement and land-use: it is more crucial to distinguish site types and site sizes and to quantify the data on which conclusions are based (Finkelstein 1998). This was done for the Ayyubid/Mamluk settlements recorded from the Hisban Survey and attempted for the Ottoman period (Ibach 1987: 191-95). Analysis of the size, location and quantity of ceramics used to date the sites suggests that the settlement pattern of that period was dominated by large nucleated settlements on the plateau. It seems very likely that this settlement pattern continued into the Early Ottoman period and scholars are justified in drawing cautious conclusions about the nature of Early Ottoman settlement from this. The al-Karak Plateau Survey has been one of the most comprehensive surveys in Jordan and has received a considerable amount of analytical attention. The survey methodology was purposive although collection on sites was systematic. Twenty-four of the 27 sites identified as Ottoman were identified on the basis of less than five sherds. Of these same 27 sites, 10 sites are considered to be settlements of considerable size, probably villages, two are modern towns, 11 are individual structures with associated cisterns, two are religious structures, and two are sherds unassociated with structures.

In their Southern Ghors and Northeast 'Arabah Survey, MacDonald *et al.* (1987) report 14 Ottoman sites, eight of which are cemeteries/individual tombs. Although the settlement distribution maps of the Ottoman period of the Wadi al-Hasa survey are difficult to use—they are divided into Ottoman and Late Ottoman, Ottoman/Modern and Late Ottoman/Modern—and it is not clear on what grounds the distinctions have been made, the conclusions provide a useful framework for considering the settlement of the period. MacDonald *et al.* report that the major sites that are still occupied villages lie on the western edge of the plateau. The remainder of sites towards the east and along Wadi al-Hasa appear to be more ephemeral sites, that is, tent emplacements, animal corrals and caves, which MacDonald *et al.* consider to be the traces of a semi-nomadic/nomadic population (1988: 270-77).

Settlement pattern

However patchy, the survey information shows that by the Ottoman period settlement in Jordan was overwhelmingly rural. Settlement included deserted villages—probably dated Ayyubid/Mamluk on surveys but continuing into the Early Ottoman period—villages that are still occupied today or were so in the recent past, plus traces of the trappings of agriculture and pastoralism. Based on the historical and survey evidence, it appears that areas in which olives and fruit trees were part of the agricultural produce seem to have retained their permanent village settlement into the nineteenth century, while areas of grain and summer-crop production seem to have become seasonally occupied by the nineteenth and perhaps were already so in the sixteenth century.

Knowledge about the area north of 'Amman and south of Wadi al-Hasa is less reliable because archaeological work in those areas is incomplete. In the area north of 'Amman, the archaeological information is not yet available to conjecture about settlement patterns at that date. In most cases, the settlements that in earlier times had been large towns, for example, Jarash and Umm Qays, have produced evidence for Ottoman occupation of some kind. Walmsley suggests that in the ninth–twelfth centuries AD there seems to have been a change from urban to rural settlements in the north (1992: 382). It can be suggested that this rural settlement continued throughout later centuries although the balance between agriculture and pastoralism changed. The results from the Hisban survey have been detailed above: analysis of the size, location and quantity of ceramics used to date the sites suggests that the settlement pattern of that period is dominated by large nucleated settlements on the plateau. Several of these settlements continue as modern villages and the town of Madaba. For the central plateau of Jordan, a picture of settlement distribution can be built up: larger and, in many cases, continuously occupied villages occupying the plateau edge and the 300 m spring-line along the western plateau edge and more ephemeral sites in the hinterland and farther east towards the desert. It is tempting to see the sites, and particularly the cemetery sites, on the floor of the Ghor as the traces of a nomadic population that practised transhumance between the valley floor and the plateau and buried their dead in the Ghor. In areas where survey work has been relatively intensive, for example, al-Karak plateau, a settlement pattern relating to the continuously occupied villages can be discerned which probably relates to a different economic strategy that the population followed. The continuously occupied villages are clustered on the western edge of the plateau with access to the wadis and the plateau hinterland that was suitable for growing wheat/barley and grazing animals (Miller 1991; Brown 1984). In the immediately preceding period, the pattern suggests that settlements continued round to the eastern edge of the plateau.

Site types

Since much of the evidence for site types comes from standing architecture, the two topics are considered together. The site types/architecture of the Ottoman period can be divided into 'examples of Ottoman architecture' and 'architecture of the Ottoman period' that is regional, even within Jordan, and bears little relation to anything Ottoman. Apart from the architecture, very little is known of the associated material culture. Most of the sites are those recorded by surveys, as described above, or in studies of vernacular architecture. Apart from Hisban and Khirbat Faris, both on the grain-producing plateau of central Jordan, little excavation of Ottoman occupation levels has been done. Even distinctly Ottoman sites, such as the Hajj forts, have not been investigated by excavation.

Ottoman architecture

Hajj forts and associated installations

The Hajj forts themselves are considered separately in this volume (see Chapter 28). Related to the Hajj route are various installations including the bridge and the associated road at al-Hasa, built in 1730–33 on the orders of Aydinilli Abdullah Pasha. During these periods, fortified sites such as 'Ajlun, al-Karak and ash-Shawbak Castles were also used (Brown 1988: 240), and it may be that sites such as Khirbat ad-Dusaq,[8] near ash-Shawbak, should be considered as part of Ottoman guardianship of communication routes. Brünnow and von Domaszewski recorded Khirbat ad-Dusaq as 'eine Karawanserai aus sarazenische Zeit' (1904: 98; Glueck 1934: 76) and produced a plan showing three structures built on an enclosing courtyard wall. Wallin (cited in Brünnow and von Domaszewski 1904: 463) had visited the site earlier in 1854 when it was abandoned. His Bedouin guide told him that the site, known as 'Khan Alzebib', was a Hajj route station built by Sultan Suleiman. Now all that remains are two parallel barrel-vaulted galleries with no evidence for connecting walls. The construction is of stone with simple stone surround to the windows.

Udhruh fort also seems to be a candidate guarding both a communication route and a water source. Its construction date is unknown but it is similar in form to the sixteenth-century fort at al-Qatrana. The elevation and *khanat* plan show several phases of re-building. The facade is plain with a machicolation over the doorway and arrow-slits along the top storeys (Killick 1983: 115).

It is the Hajj route, not surprisingly, which again provides a good example of nineteenth-century Ottoman architecture, standardized throughout the empire. The Hijaz railway, running from Istanbul to Makkah (Mecca), was completed in 1908. Railway stations took the place of Hajj forts and were standardized in design from north to south. The simple rectangular two-storey structures were roofed with a pitched, tiled roof supported by iron I-beams and represent a typically Ottoman implant into the local architectural tradition (Daher 1995: 341-49) (Figure 16.3).

Other Ottoman buildings

As the Ottomans reasserted their control over the province south of Damascus, a school was built at al-Karak.[9] It was built in 1893 on the orders of Sultan Abdul Hamid II and is a two-storey building arranged around a central courtyard. It is still in use today as the al-Karak Boys' Secondary School. One other government building that is still in use is the *serai* in Irbid, converted from the *kaymakam* residence to the town's prison to the town's museum. This building was constructed in the mid-nineteenth century and is very similar in detail to the urban architecture of that period. Its plan is akin to the earlier Hajj forts and caravanserai being square surrounding a central courtyard. The entrance, which would have contained the *kaymakam*

Figure 16.3. Al-Qatrana railway station.

Figure 16.4. The *serai* (governor's residence) at Irbid.

residence, consists of two storeys and is located opposite an *iwan* wing (Figure 16.4).

Towns

In the *defters*, size was no criterion for determining whether a settlement was a town (Hütteroth and Abdulfattah 1977: 23). The term 'town' applied to settlements that included non-agricultural activities and that were subject to different taxes, that is, its function not its form. Until the mid-nineteenth century, little is known about the physical aspect of settlements that in the sixteenth-century Ottoman tax-records are called towns. Hubras, near Irbid, is mentioned in the *defters* as a town. It is now a large village and there are few traces of pre-nineteenth-century

Figure 16.5. Nineteenth-century house of Haj Guweida Suleiman Obeidat in Hartha.

buildings. One exception may be the mosque that is more reminiscent, however, of a building serving a village community rather than a town. It was not until the mid-nineteenth century that settlements recognizable as towns or urban centres, for example, as-Salt and Irbid, started to appear in Jordan. Until that time the urban centres lay north and west in Damascus, Jerusalem and al-Khalil, while seasonal markets were regularly provided by the Hajj pilgrimage. Al-Karak, which, due to its strategic situation, is an exception, commanded both the routes north–south and westwards to the Jordan Valley and the fertile hinterland of the al-Karak plateau, and is a natural candidate for a market centre.

The architecture of the town of as-Salt stands out in Jordan; the revival of as-Salt and its transformation from a small rural centre to an urban settlement started in 1867 when the *vali* of Damascus re-established order and Ottoman control. New styles and building materials were introduced and a large number of trading families moved east from Nablus bringing new ideas and architects with them. A Middle Eastern regional style, recognizable from Jaffa to Nablus to Alexandria, developed and is evident in the facades of the merchants' buildings. In 1907 the Small Mosque, a building that with its pitched tile roof and large windows would not look out of place in any Anatolian town, was erected in as-Salt (Khatib and al-Asir 1995). The town of Madaba is another excellent example of a settlement that includes buildings of regional urban style (Denton and St. Laurent 1996) and even Irbid boasted houses with intricate facades. Again al-Karak is notable for not being part of the same trend. By the early twentieth century, urban style was becoming more common in some of the villages, mainly of the north. In the village of Hartha, the house of Hajj Guweida Suleiman Obeidat, a member of the dominant tribe in the district, shows ornate treatment of the entrance facade which would be more in place in towns such as as-Salt or Irbid (Figure 16.5).

Architecture of the Ottoman period

'Villages'

For the purposes of this volume, the term 'village' includes those settlements defined in the *defters* as *qarya*, *mazra'a* and *khirbah* (Hütteroth and Abdulfattah 1977: 22-56). *Qarya* refers to a permanently inhabited rural settlement of any size. The normal translation of *mazra'a* is 'an agricultural area with no permanent settlement'; however, Hütteroth and Abdulfattah acknowledge the complexities of that term and, based on work in Syria, Rafeq (1984) and Schilicher (1991: 186) have suggested that the term should be interpreted as defining the *type* of agricultural produce, that

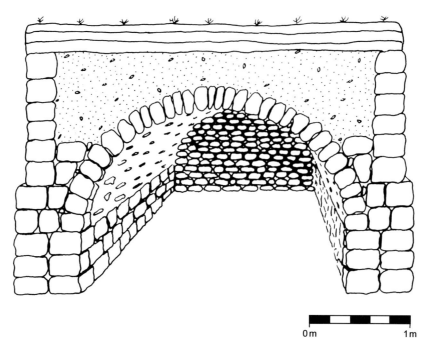

Figure 16.6. Construction method of barrel-vaulted structure.

is, grain, rather than the type of occupation. *Khirbah* is generally translated as 'ruins' but the term may stay in use as part of a populated settlement name. These fiscal categories are not distinguishable archaeologically: *mazra'a* are almost definitely the sites found on various surveys, many of which are considered to be Ayyubid/Mamluk but which should probably be considered Mamluk/Early Ottoman, and which are not mentioned by name in the *defter*s. In the survey of the al-Karak plateau, Miller (1991: 171-72) records 77 'deserted village' settlements of the Middle/Late Islamic period. This can be compared with the 23 villages recorded in the *defter* for the same area, Nahiya Kerak. In addition, for the same area, 20 percent of the taxation revenue is indicated as coming from *mazra'a*: it seems reasonable to assume that some of the 77 sites that Miller recorded represent *mazra'a* (Hütteroth and Abdulfattah 1977: 96, 172).

The villages found on surveys are described variously but a comprehensive generalization is that given by Ibach: 'One mark of such a site is the sharply undulating surface of the ground. Small mounds are interspersed with depressions and cave entrances. These mounds and depressions are caused by several architectural features of the period, namely, arches, vaulted buildings, semi-subterranean rooms and caves' (1987: 191-92).

For the Early Ottoman period, little is known about the individual structures within these rural settlements. It is likely that the buildings of the Early Ottoman period continued with the construction techniques and spatial arrangement of earlier periods. Excavated examples of rural buildings are few and far between. At Hisban a complex of barrel-vaulted structures surrounding a central courtyard was excavated (LaBianca 1990: 220-21). An identical complex was excavated at Khirbat Faris. At both sites, construction of these complexes can be dated to the Mamluk period, but it is probable that their use did not stop immediately with the coming of the Ottomans. Subsequently at Khirbat Faris, these structures appear to have been extensively re-used—as stables, or as oven-houses. It is not yet clear how long the period of re-use continued and whether the re-use existed at the same time as occupation in tents or occupation in houses that have not yet been excavated. Banning *et al.* (1989: 54) excavated a nearly complete barrel-vaulted structure containing Late Ottoman artifacts (see discussion below) at Ras al-Qabub in northern Jordan. The barrel-vault was finished with a lime-plaster (1989: 54). The writer at Hmoud (Figure 16.6) recorded an early twentieth-century barrel-vaulted structure that was used for dwelling and storage.

A recently excavated settlement at Khirbat al-Mu'allaq has shown re-use of earlier Islamic and Edomite buildings in the construction of two houses during the Ottoman period (Lindner *et al.* 1996: 116 and 133).[10] The type of occupation suggested by the excavation results is very reminiscent of the post-Mamluk levels at Khirbat Faris: ovens constructed within well-built corners of larger and earlier buildings.

The situation in the continuously occupied villages or those recently abandoned is clearer. These villages are a common sight throughout Jordan and have been the subject of study by archaeologists, architects, anthropologists, historians and developers (Biewers 1992, 1997; Biewers and Kana'an 1993; Fakhoury and Sweiss 1995; Kana'an and al-Rifai 1988; Kana'an and McQuitty 1994; Khammash 1986; Merschen 1992; Noca 1985). These settlements include structures that can be traced back to the mid-nineteenth century.

The individual structure plans show variation from north to south but are based on a similar construction method. Apart from the Jordan Valley, all construction is in stone. The house is a simple rectangular shape with minimal openings. Arches support a flat roof of beams, brushwood and earth. In the area north of Wadi az-Zarqa' and west of the modern Irbid–'Amman highway, examples can be found that closely mirror those common in Palestine. Humans, agricultural stores and animals are all housed under one roof and the division in use of space is marked by differences in height. These buildings are very rarely vaulted but the village of Umm Qays is an important exception (Shami 1989). Here many of the houses were domed and cross-vaulted using the skill of professional masons brought from Palestine. Few of the villages include structures of more than one storey. A notable exception is the village of Tibna in which several of the village houses possess two storeys.

The more common Jordanian type of house, found both in the north and as far south as Ras an-Naqb, is not marked with height divisions. Agricultural storage facilities are built into the fabric of the building, as grain-bins between the arches (*rawiyat*), as ranges of mud-brick bins (*kawair*), or as separate 'rooms' within the basic building (Biewers 1997). The antecedents for this type of construction are unclear. Stone-arched houses are common in the Classical period and were re-used in the Islamic periods. However, these arches sprang directly from the house wall. The nineteenth-century examples are of houses whose roof is supported by arches that spring from an arch-wall—a wall built perpendicular from the house-wall (Kana'an and McQuitty 1994: 142) (Figure 16.7). Examples of this type of construction have been reported from northern Syria dating from the fourteenth (Fuller and Fuller 1987-88: 279) and eighteenth centuries (Aurenche 1990: 46). This type of structure is the core of the nucleated 'traditional' village seen today. These houses are, in some cases, part of courtyard complexes which are themselves the spatially organizing feature of the village; in others cases, the spatial pattern is more dispersed and the courtyards are a late introduction (Kana'an and McQuitty 1994: 149).

Caves

Often these buildings incorporated caves or there are examples of caves themselves forming nucleated settlements. LaBianca has quite rightly emphasized the importance of the cave as a dwelling house and its importance in the site hierarchy. Failure to take caves into account leads to a significant under-estimation of settlement figures. Habitation caves typically had a masonry entrance and smoke-blackened ceilings from the use of domestic fires (LaBianca 1991: 227-28; MacDonald *et al.* 1988: 272). Perhaps the best-documented example of caves being used for habitation comes from Petra where, until recently, the Bedul occupied both Nabataean tombs and caves (Bienkowski and Chlebik 1991). Temporary use of rock-shelters in the Ottoman period has been archaeologically recorded at Tur Imdai (Russell and Simms 1987).

Individual structures

Individual structures which seem to be particular to regions where grain production and storage was of major importance are recorded on surveys. Perhaps these buildings are the physical expression of the *mazra'a*. Such buildings do not appear to develop into the nucleus of villages (but see below). The structure is similar to the Jordanian village houses described above, that is, arches supporting a flat roof with grain-bins placed in between the arches. These houses, however, are larger, frequently enclosing $30 \, m^2$, and the arch-walls are longer, allowing greater space for the grain-bins. The impression that these buildings were primarily for storage rather than habitation is overwhelming. Examples of such structures, dubbed the 'arch and grain-bin'- type because of their most salient features, have been studied in detail at Khirbat Faris and the modern villages of Smakieh and al-Qasr, all on the al-Karak plateau. The two standing 'arch and grain-bin' houses at Khirbat Faris date from the end of the nineteenth century (Figure 16.8). Excavation within the houses has revealed that the side-walls of the house stood directly on earlier walls on the ground while the arch-walls were founded in a deep foundation trench. Ceramics from the foundation trench are

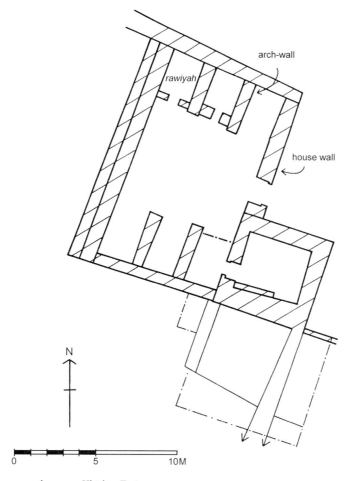

Figure 16.7. Plan of nineteenth-century house at Khirbat Faris.

Figure 16.8. Interior of nineteenth-century house at al-Qasr.

hand-made, painted and would normally be termed Ayyubid/Mamluk. Obviously these ceramics are either residual or their production continued far into later periods. The floors of both houses were cobbled. Prior to and at the same time as grain-storage in such houses, grain was stored and hidden in cisterns. The term for these buildings is *qasr* and it seems likely that a semi-nomadic population used them in conjunction with tents (Kana'an and McQuitty 1994: 149; Lancaster and Lancaster 1999: 263). Mithqal Pasha, from the Beni Sakhr tribe, while continuing to live in his tent, constructed one such house, which was part store-house, part family and tribal headquarters, at the centre of his territory in Umm al-'Amad near 'Amman (Lewis 1987: 231). LaBianca suggests that these buildings can develop into '*qusur* villages', being clusters of fortified farms/substantial agricultural compounds (1992: 224).

Tribal strongholds

One type of building that falls between urban and rural settlement and seems to be a grander version of the above 'individual structures' is the family or tribal stronghold, scattered north and south and exemplified in sites such as Tibneh, Yadoudeh and Muhei. The Shraydeh family from the late eighteenth century dominated Tibneh and the surrounding villages. The family's traditional residence, *al-Alali*, crowns the hill on which the village is situated. It is a two-storey groin-vaulted construction that stands out from the surrounding village of flat-roofed and arched houses. A large pointed arch distinguishes the entrance and there is ribbed stone decoration around the windows. A prayer-room is included in the upper floor. It is further distinguished from the surrounding village houses of agriculturalists since no courtyard is attached (Khammash 1986: 62-66). Muhei is an occupied village midway between al-Karak and the Desert Highway or Hajj route. Again, a large vaulted building dominates the skyline that was the traditional residence of the local sheikhly family. In both cases, these buildings were constructed in the late nineteenth century and, in their architectural decoration, hint at the increasing trend of urban style creeping into essentially non-urban buildings.

Pastoral sites

Several surveys record sites that are almost certainly related to nomadic pastoralist and semi-nomadic populations (MacDonald *et al.* 1988: 255-80; Miller 1991). These sites not only include caves but also tent sites and animal pens. Modern studies of tent sites have shown that the same location is used repeatedly for seasonal exploitation of the area and that the traces of tent emplacements and their hearths can be easily distinguished (Banning and Köhler-Rollefson 1992). These sites are difficult to date but they must be considered as part of the settlement hierarchy if the complexity of Ottoman land-use is to be realized. It is much easier to detect these sites on the fringes of cultivated land but that does not mean that their occupants were confined to these fringes: all trace of these ephemeral sites on land now cultivated will have been erased over time.

Water-mills

Water-mills for grinding grain were an integral part of the Jordanian rural economy and are commonly attributed to the Ottoman period. There are many historical references to mills in the sixteenth-century *defter* since they were taxed and nineteenth-century European travellers also comment frequently on their occurrence. Conder reports: 'At Sumieh and below 'Ain Hesban are the mills, which were erected by Dhiab of the Adwan in the year 1191 of the Hejirah (AD 1777)' (1889: 129). Again the problem is one of dating. By calculating the rate of tufa accumulation in the water-channel leading to the penstock, Gardiner and McQuitty suggest an eighteenth-century date for mills in the Wadi al-'Arab in north Jordan (1987: 28). Ottoman land registers of the late nineteenth century record their registration, sale and restoration (Rogan 1995). However, these mills probably had earlier antecedents (McQuitty 1995): the same technology was used for the earlier fourteenth-century sugar mills in the Jordan Valley (Abu Delu 1995).

The mills recorded on the Wadi al-Hasa survey are almost definitely wrongly assigned to the Roman/Byzantine period (MacDonald *et al.* 1988: 284-85). They are of the arubah penstock/horizontal-wheeled type and usually comprise a single tower attached to what would have been a mill-house (Figure 16.9). The mills are often part of a chain of mills fed by a water-channel taken from higher up in the wadi. The masonry of the tower, the chute and its immediate channel were of good quality. The techniques required for building the arubah penstock and the wheel-chamber were specific and it seems likely that

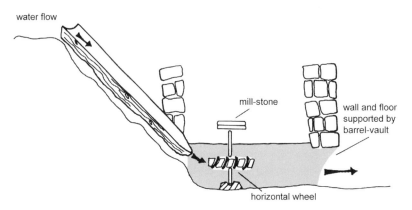

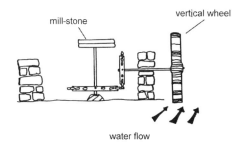

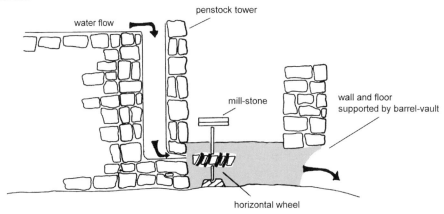

Figure 16.9. Diagram to show watermill technology.

there were groups of masons specializing in building the mechanisms of water-mills.

The only exception to this horizontal-wheeled type mill are the five vertical undershot and possible overshot mills at al-Lajjun. These mills relate to the early twentieth-century stationing of a Turkish garrison at the spring of al-Lajjun and are a totally alien import into Jordanian mill technology (de Vries pers. comm.).

Material culture

Identifying Ottoman material culture remains the major challenge for archaeologists working in this period. Distinctions within material culture are traditionally used in Near Eastern archaeology as the main interpretative tool for survey and excavation results. The material culture may not itself be well dated, as is the case in the Ottoman period. Too often the distinctions become the basis for identifying chronological divisions when differences may be reflecting social divisions, for example, between pastoral and village sites, and between urban and village sites.[11] In addition, scant attention has been paid to 'late' levels that are often cleared rather than being stratigraphically excavated. The situation in the Late Ottoman period is clearer because collections of ethnographic material are available for consultation.

Ceramics

There were several trends that continued into Ottoman and Early Modern times from the preceding periods:

1. An increasing division in terms of material culture assemblages between site types within regions, that is, a rural material assemblage as opposed to an urban material assemblage. Excavations of Mamluk/Early Ottoman Khirbat Faris, a rural site, have produced a handful of glazed sherds. Al-Karak Castle, a major site of the Mamluk Sultanate and only 20 km away, produced large amounts of glazed and unglazed wheel-made ware that probably continued into the Early Ottoman period. Petrographic analysis has suggested its sources to be both local and from Damascus and Jerusalem (Mason and Milwright 1998).
2. An increasing tendency for the production of hand-made rather than wheel-made coarse ware. By the Ottoman period, the overwhelming bulk of ceramics were hand-made and their chronological position is largely distinguished by stratigraphic means, that is, their date is not readily identified on survey. This fact too is subject to variation within site types and between regions: excavations at Ti'innik in Palestine include a far greater percentage of wheel-made wares in the assemblages of later periods than those recognized in Jordan (Ziadeh 1995).
3. A decrease in the use of ceramics in the household assemblage and their replacement by wooden vessels, metal and animal skins (Oleson and 'Amr et al. 1993: 479). This statement is subject to variation—it was made with regard to southern Jordan: the situation in the hill-villages around 'Ajlun may have been quite different.

Fine-wares

Greater emphasis has been laid, in all periods, on the identification of fine-wares and the potential these sherds have for pointing out chronological divisions and trading patterns—the time period and place of manufacture of these ceramics is usually known. The sites that have been excavated in Jordan and that have produced ceramics likely to be Ottoman are sites unlikely to produce fine pottery. Ottoman fine-ware is confined to museum collections and heirlooms and is almost always unprovenanced. The sites that might produce fine-ware have not been excavated, that is, the Hajj forts, and the excavator must look outside Jordan for information on such ceramics (Gilmors et al. 1984; Pringle 1986; Miles 1985).

Equally, candidate sites for producing fine-wares have been extensively re-used and successively cleared so that the material culture of recent centuries has no chance of surviving. Al-Karak Castle was used by the locally pre-eminent tribes as a seat of 'government' and by the Ottomans to garrison their troops until the beginning of the last century. Brown excavated in the Mamluk Palace which, according to the elders of the town, had been used in the early twentieth century as a prison (1989: 295). Since that time most of the vaults have been cleared out for various purposes, most recently to become part of the museum.

The occurrence of grass-green glazed ware over a white-slipped reddish or brown body is an exception to the lack of knowledge of fine-wares. These wares have been reported both on survey (Kareem 1987: 455) and excavation (McQuitty and Falkner 1993: 42) and appear to continue from the fourteenth into the sixteenth centuries (Figure 16.10).

Coarse wares

Hand-made as opposed to wheel-made coarse wares constitute the overwhelming majority of ceramics found both on survey and on excavation in late levels. It is not possible to distinguish between Mamluk/Early Ottoman and Late Ottoman/Modern except on the basis of stratigraphy and then the problem of these sherds being mixed with residual, earlier examples clouds the picture. Pottery, like site-types, does not change with dynasty. It is thus likely that the ceramics identified as Mamluk continued into the sixteenth century. Indeed, based on the excavations at Ti'innik in Palestine, Ziadeh suggests that this so-called Mamluk, hand-made, geometrically painted

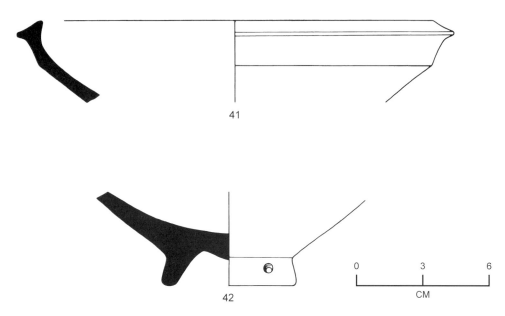

Figure 16.10. Sixteenth-century pottery from Khirbat Faris: (41) Glazed wheel-turned bowl. Sandy orange fabric with translucent grits. Green glaze over white slip. (42) Glazed base of a wheel-turned bowl. Fabric as 41. Hole bored in ring-base.

ware became dominant only between the sixteenth and nineteenth centuries (1995: 210; Johns 1998). There is good evidence to suppose, as the later discussion on tobacco-pipes shows, that many assemblages interpreted by excavators as Mamluk may in fact be later. Subsequent to the Mamluk period, relative to excavated sites in Jordan, there seems to be a trend for painted decoration to continue, although on more crudely made fabric. Brown reports, from ash-Shawbak, decorated jar/jug forms with painted geometric decoration on crudely fashioned bodies with heavily chaff-pocked surfaces. Cooking-pots of hard grey ware with impressed raised bands on the exterior for decoration come from the same context. A base with a chaff-pocked, red slipped surface was also found (Brown 1988: 240) (Figure 16.11). It is this feature of a chaff-pocked, red-slipped surface that Sauer focuses on as distinguishing Ottoman ceramics from Hisban (1993: 273). From the Negev, Schaeffer reports the same type of hand-made cooking pots containing calcite and organic temper with a red burnished-slipped exterior and records the ceramics in all levels at the Islamic site of Tel Jemmeh (1989: 42). He also records wheel-made grey-ware jars that were known by Late Ottoman times to come from Gaza. Such ceramics seem to have had initially a distribution in Jordan and north Palestine but by the Mamluk/Ottoman period this had spread to central and southern Palestine (Schaeffer 1989: 43). At Tur Imdai, a total of 16 pots was found, all but two of them hand-made. The fabric contained chaff but the vessels were not slipped or burnished. The wheel-made vessels were storage jars with a very dark grey exterior and dark brown core, perhaps Schaeffer's Gaza Ware. All of these ceramics are dated on the basis of C^{14} to the eighteenth century (Russell and Simms 1987). In almost all of the major urban sites of earlier periods, for example, Umm Qays, Jarash, Pella, Humayma, and the continuously occupied villages throughout the country, ceramics have been found that are probably Ottoman but that have no context.

The continuation of the ceramic tradition in Jordan has been extensively documented in the last few years (Franken and Kalsbeek 1975; London and Sinclair 1991; Merschen 1985). In these ethnographic studies, the potters are always women producing hand-made pottery as part of the agricultural year, that is, seasonally. The forms are usually jars and jugs as well as large storage and water vessels that are decorated with applications of 'Tree of Life' motifs or simulated rope bands, sherds of glazed pottery, fragments of mirror (Figure 16.12). The decoration does not include paint or slip. The pots are fired in a clamp or bonfire and the basic fuel is dung.

The context is vital in establishing the chronology of Ottoman ceramics—apart from the fine ware, much of the pottery may never be recognizable standing on its own. The excavators of the ceramics at Tur Imdai report that they could easily be mistaken for prehistoric examples were it not for the fact that their dating is

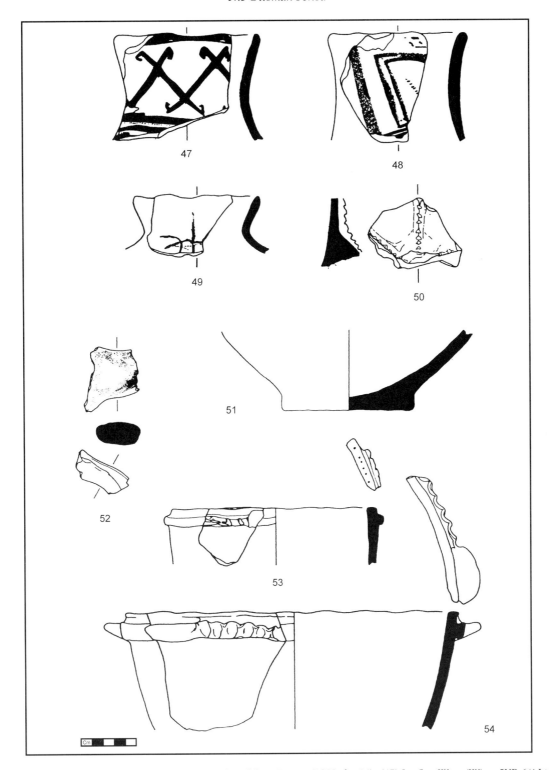

Figure 16.11. Ottoman pottery from ash-Shawbak. Reproduced from Brown (1988: fig. 14): (47) Jug/Jar: Ware (W) = 5YR 6/4 Light Reddish Brown; Exterior Slip (ES) 10R 5/6 Red; Exterior Paint (EP) = 10R 3/1 Dark Reddish Gray; Handmade (H). (48) Jug/Jar: W = 2.5YR 6/6 Light Red; ES = 10R 5YR 8/4 Pink; EP = 10R 5/4 Weak Red, 5YR 3/1 Dark Gray H. (49) Jug/Jar: W = 5YR 8/4 Pink; Self Slip; H. (50) Handle: W = 5YR 7/4 Pink; ES = 10YR 8/3 Very Pale Brown; H. (51) Base: W = 5YR 7/4 Pink; ES = 10R 5/4 Weak Red; H. (52) Handle: W = 2.5YR 6/6 Light Red; ES = 10R 5/6 Red; EP = 10R 4/1 Dark Reddish Gray; H. (53) Cooking Pot: W = 5YR 5/1 Gray; H. (54) Cooking Pot: W = 5YR 5/1 Gray; H.

Figure 16.12. Twentieth-century *habia* from north Jordan.

fixed by radiocarbon. The functions suggested by the forms are that ceramics were used for cooking, food containers and storage.

Lamps

Lamp-types, traditionally called Mamluk, are variations of the saucer lamp, very similar to Bronze/Iron Age examples. Some were made with a loop handle reaching from the middle of a small saucer formed at the centre of the reservoir to the rim of the lamp and many were glazed (Kareem 1993: 220). Gichon and Linden suggest, based on stratigraphic evidence from an excavated bath-house/shrine in Palestine, that these open saucer lamps (without the handle) date from the eighteenth century (1984: 166). These lamps are not glazed: Gichon and Linden consider glaze and decorated closed shapes to be an indication of a Mamluk date. Little is known in Jordan of post-Mamluk lamps and suggestions for examples of later Ottoman lamps are usually based on value judgments, for example, 'crude manufacture'.

A lamp was found in abandonment rubble of the latest occupation phase at Khirbat Faris. It is hand-made from two pieces of clay. The height of the top section is more decorative than functional and the oil reservoir is exceptionally small[12] (Figure 16.13).

Tobacco-pipes (Figure 16.13)

Tobacco pipes are the most reliable indicator of Ottoman material culture. Tobacco smoking was first introduced into the Ottoman empire in the early seventeenth century and became widespread in the latter part of the century. Any contexts containing pipe fragments are automatically dated to this century and later. Many previous excavations have erroneously reported assemblages containing pipes as being Mamluk (e.g., Riis 1957).

Istanbul, Diyabaker and Luleburgaz near Edirne in Turkey, as well as Assiut in Egypt, are known to have been centres of pipe production (Robinson 1985: 152; Simpson 1990) and, closer to Jordan, Nazareth and Jerusalem (Scholch cited in Ziadeh 1995: 211). Robinson comments that 'Probably every town of any size had at least one pipe-maker. In lesser villages perhaps the potter would turn out a few pipes; he may have made his own moulds from a pipe or pipes brought from a major center' (1985: 153).

The Eastern or Turkish pipe consisted of a ceramic bowl, the most frequent find on survey and excavation, into which the stem was inserted. The bowls were made of specially washed and filtered fine clay in two-part moulds of stone or metal (Robinson 1985: 157). The earliest pipes were of white/grey clay, probably imitating European examples, but by the nineteenth century the most popular fabric was red clay which could be obtained in Egypt, the Lake Van area, and Bulgaria. When this red clay was not available, the alternatives would often be covered with a red slip (Robinson 1983: 266). Early pipe-bowls were small because of the high price and luxury nature of tobacco but by the end of the century the pipe bowls and stem opening were becoming larger. Their chronology and development have been well illustrated and analysed for urban sites in the Western Ottoman empire: the Sarachane, Istanbul (Hayes 1980), Corinth and the Athenian Agora (Robinson 1985), and the more rural Kerameikos, near Athens (Robinson 1983), but as yet not in the Eastern Ottoman empire. Their distribution in the Eastern Ottoman empire is conveniently summarized in the *Newsletter of the Society of Clay Pipe Research* (Simpson 1990, 1994). Pipes are, unfortunately, sometimes totally unreported or are surface finds. On excavations, they are often from the upper levels in situations of poor stratigraphic control. In Jordan, pipes have been reported from Jarash, Hisban, Dhra' al-Khan, Ras al-Qabub, Tawilan, Udhruh and Khirbat Faris. In Jarash, four bowl fragments from the upper levels of the re-occupation in the northern Roman Theatre are recorded (Clark *et al.* 1986: 266). At Hisban pipes 'may have been near the surface in the village part of the site' (Sauer 1993: 273; Wimmer 1978: 150). At Dhra' al-Khan, stratified tobacco pipes were found in levels dated by the excavator to the fourteenth–fifteenth centuries according to the ceramic type. These fragments of pipe were 'dusky-red' (Kareem 1993: 216, 259-60). At Ras al-Qabub, fragments of small pipe bowls were found in association with wheel-made pottery in a barrel-vaulted structure (Banning *et al.* 1989: 52). At Tawilan, a pipe fragment was found apparently out of context, associated with Nabataean/Roman pottery (Bennett and Bienkowski 1995: 92 and fig. 9.41.7). At Udhruh, pipe fragments appear to come from the southwest corner tower of the Roman fort (Killick 1983: 129). At Khirbat Faris, only three of the 11 fragments are stratified. One of these is, however, from a well-stratified context associated with the construction of an arched room.

Ovens (McQuitty 1993–94)

Another form of material culture frequently found on settlement sites and in modern villages is the clay oven, which is used for many forms of baking, not just bread. Like the pottery of the recent past, the women of the village make these ovens as part of the agricultural year. Most of the ovens found in late contexts on excavations and in use today are of the *tabun/tawabeen* type. Their form varies in detail from the north to the south of Jordan but the basic method of manufacture and use is the same. The ovens are made of locally collected clay to which grit temper, animal hair and perhaps calcite is added. The ovens are coil-made and highly burnished before being left to dry in the sun. They are then fired in position in the oven house, which may be a purpose-built structure, an abandoned house or cave, but this type of oven is never used in the open in a courtyard. The oven is heated up for baking by lighting a fast, hot fire. The heat is retained when the oven is not in use by covering the mouth and then clamping the whole with animal dung, straw and/or the remains of olive pressing (*jift*).[13] Several of the ceramic finds from archaeological sites in this late period seem to be clay oven covers (McQuitty 1993–94: 56; Lindner *et al.* 1996: 124-25). A curved line of stones to contain the ash heap when it is cleared out is a frequent feature near the oven house.

Glass

Little to nothing is known from excavation about Ottoman vessel glass. It seems probable that the fifteenth-century Mamluk glass continued into the Ottoman period. By the nineteenth century, ethnographic accounts by European travellers take up the story and known glass-production centres, for example, Hebron, are well documented. One of the features of glass, namely, its recyclability, may also account for its apparent absence in the archaeological record. In addition, as with ceramics, the type of site excavated is extremely significant for the type of assemblage that will be found.

Glass bracelets, in contrast, are an extremely common find on surveys and on both settlement and cemetery sites. They were used in conjunction with metal bracelets at the cemetery of Tall Dayr 'Alla, in which many examples were found in female graves dating from the past 500 years (Steiner 1995: 538). Eight seventeenth–eighteenth-century female graves excavated at Umm Qays contained glass, metal and studded leather bracelets (Merschen 1991: 137). They are known from the Roman period but, based on a recent survey of excavated sites and museum collections, Spaer appears to have isolated types that are both Mamluk/Ottoman and Ottoman (1992: 44-62). The bracelets are almost exclusively of the seamless type, that is, a small piece of molten glass was pierced by a metal rod after which a larger ring was formed by rotating it on the rod with the help of a second tool (Steiner 1995: 537). The glass is both translucent and opaque, both single and multi-coloured, although the use of opaque blue, green, red and white glass seems to occur in the Late Ottoman period. The bracelets are also decorated in a myriad of patterns although some decorative styles are exclusively Ottoman, for example, 'eye ornamentation', and others appear to be pre-Ottoman, for example, the speck-pattern (Spaer 1992: 51). The cross-section of the bracelet is semi-circular, flat, evenly pointed or obliquely pointed. Spaer (1992) comments that the latter type, the obliquely pointed bracelet, is unlikely to have been used before the Mamluk period and some variations are of the Late Ottoman period. She also posits that there was a huge increase in the production

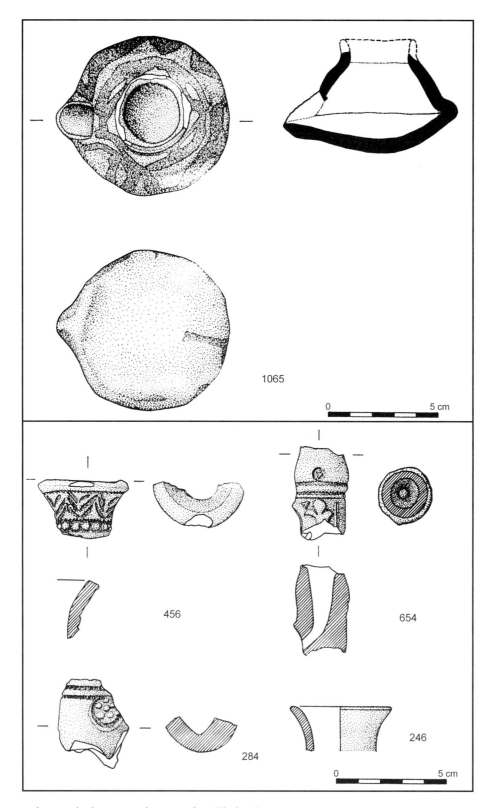

Figure 16.13. Ottoman lamp and tobacco pipe fragments from Khirbat Faris.

of these bracelets throughout the Ottoman period that suggests they were being mass-produced. The quality of the bracelets, however, seems to have declined as the quantity increased.

Beads and jewellery

As well as glass and metal bracelets, the excavations of the Abu an-Naml cemetery in Umm Qays produced large numbers of beads and perforated coins that had been sewn onto textile to compose headdresses, bead necklaces, rings, an earring, cosmetic utensils and rosaries from bone-beads (Merschen 1991: 137). Many of the beads are recognizable from ethnographic contexts and are still in use today for particular medicinal or magical properties. The materials used range from stone, bone, amber, shell and coral to glass. Most useful for dating purposes are the easily recognizable trade-glass beads that were produced in industrial centres in Italy, the Netherlands, Czechoslovakia and elsewhere and distributed all over the world. Based on this evidence and that from the perforated coins used in the jewellery, Merschen was able to date this cemetery to the seventeenth–eighteenth century.

Coinage and money

The most distinctive feature of Ottoman coinage is the *tughra*, or Ottoman sultan's name, which appears on the coins as well as the information relating to the denomination, year of minting and mint. Each sultan had his own *tughra* which is often unreadable but always recognizable as each *tughra* is distinct. The *tughra* was included 'as a sign of common allegiance to the Sunni Muslim state and its leader, the Sultan' (Darley-Doran 1988: 90).

When considering Ottoman coinage and money-exchange, the key words are variability and regionality, both chronologically and geographically. The two main points to take into account are:

1. the extent to which Ottoman coinage was *produced* for the provinces; and
2. the extent to which Ottoman coinage was *used* in the provinces.

Lamp and Pipe Description (descriptions by Madeline Sarley)

(1065) Lamp: Core = 10R 6/6 Light Red; Exterior Surface slipped = 5YR 7/6 Pink and much effected by lime concretions. Interior Surface = 10YR 7/1 Light Grey. Oxidized, hard, even firing. Plaster, basalt, shell and mica inclusions. Decoration = Weak Red 10R 4/4 paint unsteadily applied with wide zig-zag pattern.

(246) Tobacco Pipe bowl rim and body. Core = 10R 6/8 Light Red; Exterior and Interior = 10R 5/1 Reddish Grey. Exterior Surface highly burnished, 10R 4/3 Weak Red. Interior Surface 10R 4/3 Weak Red, with no burnishing. Reduced firing, very hard, very smooth. Plaster, grog and basalt inclusions. Bell shaped bowl with everted rim. No decoration present. Ceramic spot date: c. 1850 onwards. Parallels: Hayes VIII.

(284) Tobacco Pipe-stem. Core = 10YR 6/1 Light Brownish Grey; Exterior Surface burnished (10R 4/4) Weak Red. Oxidized, very hard, very smooth. Dense fabric. Flint inclusions and voids. Decoration: moulded then stamped; moulded serrated double lines around the stem, over-stamped by a rosette in a circle. Ceramic spot date: c. 1850 onwards. Parallels: form like #654.

(456) Tobacco Pipe bowl rim and body. Core = 10YR 7/1 Light Grey; Exterior Surface highly burnished = 10R 3/6 Dark Red; Interior Surface = 10R 4/3 Weak Red. Oxidized, very hard, very smooth. Shale (?) inclusions and voids. Bell shaped bowl. Decoration: moulded then cut. Plain rim, with a moulded serrated line below. Under the serrated line is a wide area left plain by the moulding, which has had wedges cut from it to create a tulip leaf pattern. Lastly, a heavily moulded line of pearls, high-lighted along their top edge by three knife point incisions. Ceramic spot date: c. 1850 onwards. Parallels: Hayes XV.

(654) Tobacco Pipe stem and socket. Core = 10YR 8/1 White; Exterior Surface highly burnished = 7.5YR N3/0 Very Dark Grey. Oxidized, very hard, very smooth. No added inclusions to the dense, fine fabric. Decorated by moulding and then stamping.
Stem: Base has a wide, plain band decorated with parallel lines, one each side, running from the socket. Around the stem side of the socket are two parallel lines, the first thick and serrated, the second is plain. On each side of the stem, are moulded, stylized leaves. The left is a *fleur de lis* and on the right, acanthus leaves.
Socket: Plain, except for a stamp towards the stem, on the left. This is faint and could be a backwards looking deer. Ceramic spot date: c. 1850 onwards. Parallels: Form is like Hayes XVI, but no mention of black burnishing.

Just as Ottoman administration incorporated local realities into its execution, so too did the coinage. When the Arab provinces were conquered in the sixteenth century, there were already well established local Muslim currencies in use. Darley-Doran comments, 'the Ottomans made no concerted effort to displace them because they wished to avoid economic disruption and popular unrest ... In Egypt and Mamluk Syria the standard silver coin continued to be the *medin*, which originally began as a half dirham under the Burji Mamluks' (1988: 89-90), that is, Mamluk coinage was not immediately supplanted by Ottoman coinage. In the seventeenth and early eighteenth centuries, Ottoman coinage started disappearing and the use of earlier coinage was supplemented by the use of foreign coinage. In addition, the actual mint activity declined (Pamuk 1994: 962). A new monetary unit, the *kurus* (piaster), was introduced at the end of the seventeenth century as an attempt by the Ottoman government to reassert control over the economy. This coinage was, however, not minted anywhere in Syria while the mint in Baghdad was used only occasionally (Pamuk 1994: 966). The examples found in Jordan probably came either from Istanbul or Cairo. It was not until the mid-nineteenth century that the Ottomans were able to gain more control over the production of their coinage. It was also at this time that imitation coins started to be produced for jewellery purposes—to adorn female head-dresses and necklaces. The imitations often bore a countermark or included an incorrect date in order to differentiate them from genuine coins. Prior to this, it seems that old coins were hammered and perforated for jewellery purposes.

It is no easy matter to determine the use of coinage in Jordan throughout the Ottoman period. The agricultural taxation referred to in the *defters* was presumably gathered in kind (Hütteroth and Abdulfattah 1977: 89), although at some point this would have been converted into money if the taxation destination was far from the point of production, that is, Istanbul rather than Damascus. In Palestine, it appears to have been given in cash form (Singer 1994: 56). Surplus agricultural produce would have been sold at local or Hajj markets. However, it seems likely that much local exchange was carried out on a barter basis and that the Ottoman economy in Jordan was *not* a monetary economy. The absence of Ottoman coins on archaeological excavations does not reflect the absence of an Ottoman population per se: coins were not part of the rural material culture assemblage. However, there have been a few stratified and many surface finds of Ottoman coins.

At Khirbat Faris, a possible sixteenth-century coin was found in abandonment rubble and a coin used for necklace decoration was found on an occupation surface under abandonment rubble.

Ethnography and material culture

One of the most useful aspects of the nineteenth-century European travellers' and ethnographers' accounts is the insight that they give into the everyday life and material culture of the inhabitants of Jordan and their material culture. Engravings and early photographs from the period give hints of the way things were used and the general environment of the nineteenth century. In the twentieth century, Ottoman material culture became the focus of collections and catalogues: in particular, the jewellery and dress of that period, although the everyday and agricultural items have also been collected, for example, cooking vessels, animal harnesses, threshing sleds, Circassian carts (Volger *et al.* 1987). In addition, very important studies of the 'Way of Life' and the technology of both urban and rural populations have been carried out, mainly in Palestine (Canaan 1933, 1934; Dalman 1932–33; Granquist 1965). There is a treasury of information about material culture of the recent past in Jordan and it should not be divorced from the information gained from archaeology.

Religion and ritual

Throughout the Ottoman period, Jordan was overwhelmingly Sunni Muslim with a minority of Orthodox Christians (Hütteroth and Abdulfattah 1977: 40), who began to convert to Catholicism and Protestantism in small numbers during the late nineteenth century.

Mosques and churches

There are no remains of mosques dating from the sixteenth–nineteenth centuries found on excavation although examples dated as Mamluk may have continued into the sixteenth century, for example, al-Lahun and Pella. Many mosques that are in use today have incorporated earlier structures into their fabric, for example, al-Karak Congregational Mosque which was built in the late nineteenth century (Dowling 1896: 330), but a few standing examples of early mosques survive. In al-Azraq, a mosque, very similar in plan to the example from Pella, survives in the centre of the castle courtyard. The Druze most recently used it in the early twentieth century but it is believed to have been built in the early fourteenth century. In Tibneh, there are three traditional mosques: one is a simple rectangular structure with an exterior staircase leading to part of the roof which was used by the muezzin for the call to prayer; another was on the first floor of the *al-Alali* referred to above; and the third, the largest, constructed of different stone from the houses, rendering it very distinctive within the village. The main Tibneh mosque is a perfect square, 12.5 by 12.5 m. It comprised three cross-vaulted units supporting a flat roof while a fourth, in front of the mihrab, probably supported a dome (Khammash 1986: 60). The stone facades of the mosque are simple and undecorated although a symbolic *mihrab* protrudes from the *qibla* wall and the east wall. A staircase is built into the thickness of the west wall giving access to the roof from which the muezzin made the call to prayer. Khammash concludes that this mosque is undoubtedly Ottoman and may be eighteenth century (1986: 61). The plan and proportions are certainly different from the examples known to be earlier and more akin to eighteenth-century examples from Palestine (Petersen 1996). As mentioned previously, the Small Mosque in as-Salt, dating to 1907, is an excellent example of Ottoman architecture that owes little to regional tradition. However, the style of classic Ottoman minarets became that which was followed throughout Jordan in the early-mid twentieth century.

A common but practically undatable (within the Islamic centuries) type of mosque found on survey is 'the desert mosque' or *musalla*. This type comprises the outline of a mosque and *mihrab* traced with a rough stone wall about two courses high. These mosques usually measure c. 4 m east–west and only 1–2 m north–south. Inscriptions, both of Arabic and earlier scripts, for example, Safaitic, often distinguish the stones of the *mihrab*. These structures are referred to as 'desert mosques' because they are easily recognized within desert environments, although they occur anywhere where there is temporary occupation of a site and a group of Muslims who require a facility for communal worship: the author saw a 'desert mosque' that had just been constructed in the proximity of a group of tents before the Holy Month of Ramadan. As mosques are associated with built shrines (see discussion below), these 'desert mosques' are associated with tombs (Lancaster and Lancaster 1993: 153).

Although Christianity survived in Jordan throughout the Mamluk and Ottoman centuries, practically nothing is known of the associated buildings, burial practices and rituals until the influx of Christian missionaries in the mid-nineteenth century. It is not until this time that the building of churches is reported. The most famous example is St George's Church in Madaba, constructed in 1898 on the foundation of a sixth-century basilica that included the Madaba Map (see Chapter 26, this volume). The Orthodox Christians of al-Karak worshipped in the Church of St George, built in 1849 (Dowling 1896: 330) while a church 'building-boom' took place in as-Salt as the town expanded.

Shrines

There are numerous shrines, both Islamic and Christian, in Jordan: shrines for the Companions of the Prophet, for prophets mentioned in the Q'uran as in the Old Testament, and for local wise and holy men (Turab 1996). These shrines are visited by pilgrims, both local and from afar, for example, Mazar. Often the shrines become the nucleus for cemeteries. The *defter i mufassal* does not mention these shrines specifically nor do *waqf*[14] payments go towards their upkeep. In Jordan few taxes went for *waqf* payments—the exceptions are in the Jordan Valley—(Hütteroth and Abdulfattah 1977: Map 4). In most cases, the construction of the shrines, usually over a grave and with a small mosque or at least a *mihrab* associated, is undated. It is assumed that many of the buildings date to the Ayyubid/Mamluk period but no real architectural study or analysis has been done to verify or refute this. To date, only the shrine of Abu Sliman al-Dirany, near ash-Shawbak, has been documented in detail (Khammash 1986: 82–83). Khammash cautiously dates this, on the basis of the ceramics found in the vicinity, to the Mamluk period. But as the section on 'Material Culture' in this chapter suggests, the ceramics could equally well be Early Ottoman. Considerably more analysis has been done on the shrines of Palestine, many of which appear to incorporate Ottoman construction into their fabric (Toombs 1985: 31-32; Petersen 1996).

Graves, tombs and cemeteries

The stipulations for Islamic burial are well known and documented.

> Burial itself is in a grave deep enough to conceal odor and prevent abuse of the body by animals. Within the grave a niche is dug on the *qiblah* side of the grave or else a smaller trench is dug on the floor of the grave. Into this the body is placed without a coffin, lying on its right side, facing the *qiblah*. The cheek is bared

and placed on a stone. The niche or trench is then sealed with bricks or stones, and earth is replaced in the grave and rounded slightly above ground level (Reinhart 1995: 34-35).

Simplicity is the keyword: 'a simple headstone may mark the grave, but anything more elaborate is to be avoided' (Denny 1995: 35-37). These are the stipulations: what is the practice?

Again, the nineteenth-century travellers and twentieth-century ethnographers give an insight into the ritual surrounding burial, including the construction of graves (Granquist 1965: 55-56).

Most archaeological surveys record graves, obviously Islamic because of their orientation, which stand out as a heap of raised earth and stones surrounded by a kerb of small stones. Invariably these graves are recorded as Bedouin or Ottoman/Modern. Often these graves are within a cemetery group. They may well be the graves of nomadic or semi-nomadic pastoralists—a fact that can be suggested by their location. They need not necessarily be Ottoman/Modern—as yet archaeologists know little of grave types in the Islamic periods and it may be that an Umayyad nomadic pastoralist grave looks much like an Ottoman one! One type of individual and very distinctive grave, associated with Bedouin, is the stone tomb/cairn that is a common sight on the prominent headlands of the basalt desert, the *harra*. One group of these, recently built by the Ahl al-Gabal group of tribes, has been extensively described by Lancaster (1993: 151-69). These elaborate tombs consist of a cairn, marking the tomb itself, flanked by two 'wings' or stone walls about 1 m high. Near the tomb is a stone platform on which the body was washed prior to burial and occasionally a 'desert mosque', usually south of the tomb, is found. The symbolism of these structures is considered to be the representation of a tent (the 'wings'). From the interviews conducted by Lancaster and Lancaster with the Ahl al-Gabal, it appears that 'none of these elaborate tombs were built for politically influential figures in the conventional sense' (1993: 157). Some were built for healers who 'were (are?) important for the political health of the community'. While the particular tombs recorded by Lancaster and Lancaster are modern, this type of distinctive 'desert tomb' must be considered in the repertoire of Ottoman grave-types.

Graves/cemeteries associated with settlements follow the same pattern. The superstructure of the grave, however, is often more elaborate. Some are marked by headstones which are obviously Ottoman (Figure 16.14).

Figure 16.14. Nineteenth-century grave.

Excavation of Muslim cemeteries in Jordan is not common. Many projects, however, come across Muslim graves in the course of their work. These graves excavated are often dated, not always on firm grounds, to the Mamluk period. Two cemeteries have been partially excavated in the northern town of Umm Qays. In 1987, Andersen and Strange reported 16 disturbed graves located within an earlier outdoor structure, possibly a courtyard. Some of the graves were dated to the Mamluk period and all were of the stone-cist type, that is, the dug grave was lined with stones and the body was covered with flat cap-stones. Two graves included glass and iron bangles (1987: 78-100). This grave-type was repeated, also in Umm Qays, at the cemetery of Abu an-Naml that appears to have clustered around the shrine of Abu an-Naml, which has since disappeared. Merschen dates these 15 graves to the seventeenth–eighteenth century (see discussion under 'beads and jewellery'). Steiner reports similar burials from Tall Dayr 'Alla (1995: 537) while Parker reports Bedouin graves from his excavations at the Roman fort of al-Lajjun (1987: 260). At Tall al-Hesi in

Grave Types		
Type	Description	Notes
I	grave cut neither capped nor lined with stones	
II	grave cut capped but not lined with stones	majority
III	grave cut both capped and lined with stones	
IV	grave cut capped but not lined with stones	uncommon

Type Frequency			
Type	Field I	Field V	Field VI/IX
I	46%	18.6%	21.2%
II	36%	79.1%	71.5%
III	14.5%	0	6.6%
IV	2.7%	0	0.3%

After Toombs 1985: 74

Table 16.3. Grave types. Based on Eakins (1993).

Palestine, a major cemetery of the Mamluk/Ottoman period has been excavated and this allows a fuller inventory of grave types. A total of 861 burials was excavated, none marked by any kind of superstructure but nucleated around a shrine. The graves were located in three discrete areas, Fields I, V and VI/IX, and show variety between the areas, in grave type (Table 16.3), and the frequency of objects found within the graves. However, the excavator was struck by the 'sameness of the burials' (Eakins 1993: 71), and while it is tempting to correlate different types of burial with different dates within the Mamluk/Ottoman period, it is not possible to draw definite conclusions. Toombs suggested that the cemetery was used first in Field VI/IX, then Field V and finally Field I. As Table 16.3 shows, this would imply an overwhelming early date for Grave Type II (Toombs 1985: 116). However, the dating evidence is so poor that a conclusion that these cemetery areas simply represent discrete communities with different burial traditions may be a more acceptable option.

Conclusions

Perhaps more than most periods, our impression of Ottoman settlement within Jordan suffers from preconceptions. The Ottoman centuries are characterized as a time of decline and devastation: of lawlessness when wild Bedouin preyed on defenceless peasants, savaged the crops and converted arable land into pastures for their flocks. Continuing on these lines, the argument goes that by the seventeenth century central Ottoman control had waned and, in the face of this lack of security, villages and agricultural life retreated to the security of the mountains or were totally abandoned in favour of nomadism and pastoral life or flight to the cities of Syria and Palestine. To be sure, there are elements of truth in this: historical and tribal records report great population movements, tribal raiding and a retreat of the area under cultivation (Abujaber 1989; Lewis 1987).

But this historical narrative based on the perceived influence of external powers (or rather lack of influence) is not the whole story. The more enduring factor is that the overwhelmingly rural population of Jordan continued to practise its mainly subsistence economy with elements of pastoralism and agriculture. Life carried on in the countryside not the towns. Tribalism, and not the Ottoman State, was the major force in the land until the late nineteenth century.[15] By no means was life a rural idyll but Jordan did not simply grind to a halt in the sixteenth century only to be revived in the nineteenth century. The challenge for archaeologists is to examine the evidence for settlement, society and economy during those centuries objectively and to characterize the Ottoman period on that basis.

Notes

1. I am extremely grateful to Dr Eugene Rogan (Middle East Centre, Oxford University) for commenting on an earlier draft of this contribution. His suggestions were invaluable but any omissions and mistakes remain my own. I would like to acknowledge and thank the several people who assisted in the completion of illustrations for this contribution: Fridtjof Eykenduyn, Kevin Hicks, Isabelle Ruben, Madeleine Sarley-Pontin and Ilka Schacht.
2. For the sake of simplicity the terminology given below relates to the nineteenth century. Different units and terms could be used in previous centuries.
3. 'Modernization' is an extremely value-laden concept that implies progress or improvement. That is not meant here; the term is used as short-hand for the changes in economy, society's values and customs taking place in Jordan during the Late Ottoman period.
4. Pneumonic plague is almost always fatal. Bubonic plague is 'one of the least contagious epidemic diseases and causes few deaths' (Dols 1979: 169).

5. The more compelling question is to ask why populations choose to build settlements at times if the economic strategy followed is basically the same. Johns suggests a sequence for the process of settlement related to this mixed farming continuum (1992: 366-67).
6. In many cases it may be that the term 'nomadic' is interchangeable with the term 'tribal'—see discussion in Lancaster and Lancaster 1995: 121.
7. A frontal attack on the surveys, which have been conducted in Jordan, is not the intention of the author. It is appreciated that surveys are not always conducted to retrieve multi-period evidence. In addition, it is better to be cautious than over-differentiate on the basis of ill-understood material culture. However, the lack of consideration of Ottoman settlements in surveys contributes as much to the 'invisibility' of this period as does the lack of knowledge regarding its material culture. A particular note that illustrates perfectly the problem of the visibility of Ottoman occupation is that neither al-Karak Castle nor Khirbat Tadun (Khirbat Faris), both of which have historically and archaeologically attested Ottoman occupation, produced ceramics on survey that could definitely be assigned to the Ottoman period (Miller 1991: 49, 89).
8. Brünnow and von Domaszewski and Glueck spelled the toponym Khirbat ed-Doshak. The spelling used here is that of Palumbo 1995: site 2099024.
9. After the Ottoman Education Law of 1869, which tried to regulate more efficiently the organization of education, a number of state schools were set up, mainly for the Muslim population (Landau 1975: 499). Ottoman education was free; the schools were supported by *waqf* endowments. Turkish was the language of instruction for a curriculum that included religious instruction (Luke and Keith-Roach 1930: 236). The foundation of these schools was a direct case of instilling Ottoman values in previously excluded tribal populations and thus 'incorporating the periphery' (Rogan 1991).
10. Indeed, the ceramics from the earlier Islamic period appear very Ottoman (1996: figs. 12 and 21) although this level was dated AD 785–1015 by C^{14} (Lindner et al. 1996: 132).
11. See Brown 1989: 297-98, 1992: 232-41 for a full discussion of this phenomenon.
12. I am very grateful to the Khirbat Faris Project ceramicist Madeline Sarley for this description as well as for the description of the pottery and tobacco pipes.
13. This may be a reason for the scarcity of olive stones in archaeo-botanical samples: if the olive waste, including the stones, is used for fuel it will not appear in the archaeological record.
14. *Waqf* refers to the practice of endowing religious institutions with the proceeds from taxation, that is, rather than the taxation going to the Ottoman government in Istanbul it went towards the upkeep of a religious institution.
15. Lancaster and Lancaster suggest that 'attention should be focussed on *hadhari* (civilized urban society) and tribal rather than village and nomadic ... (being) more consistent with the concepts of the region' (1995: 121). This is a very attractive and logical proposition: the problem for archaeologists and students of material culture is that they deal with concrete manifestations of these concepts which often invite categorization based on settlement and economy.

References

Abu Dalu, R.
1995 The Technology of Sugar Mills in the Jordan Valley during the Islamic Periods (in Arabic). In K. 'Amr, F. Zayadine and M. Zaghloul (eds.), *Studies in the History and Archaeology of Jordan*, V: 37-48. Amman: Department of Antiquities.

Abujaber, R.S.
1989 *Pioneers over the Jordan: The Frontier of Settlement of Transjordan 1850–1914*. London: I.B. Tauris.

Ambraseys, N.N. and C.P. Melville
1989 Evidence for Intraplate Earthquakes in Northwest Arabia. *Bulletin of the Seismological Society of America* 79(4): 1279-81.

Andersen, F.G. and J. Strange
1987 Bericht über drei Sondagen in Umm Qes, Jordanien, im Herbst 1983. *Zeitschrift des deutschen Palästina-Vereins* 103: 78-100.

Aurenche, O.
1990 Habitat de nomades et habitat de sédentaires en Syrie et en Jordanie: Etude de cas. In H.F. Frankfort (ed.), *Nomades et sédentaires en Asie centrale*, 31-48. Paris: CNRS.

Banning, E.B. et al.
1989 Wadi Ziqlab Project 1987: A Preliminary Report. *Annual of the Department of Antiquities of Jordan* 33: 43-58.

Banning, E.B. and I. Köhler-Rollefson
- 1992 Ethnographic Lessons for the Pastoral Past: Camp Locations and Material Remains Near Beidha, Southern Jordan. In O. Bar-Yosef and A. Khazanov (eds.), *Pastoralism in the Levant*, 181-204. Monographs in World Archaeology 10. Madison, WI: Prehistory.

Bennett, C.-M. and P. Bienkauski
- 1995 *Excavations at Tawilan in Southern Jordan*. Oxford: Oxford University Press.

Bienkowski, P. and B. Chlebik
- 1991 Changing Places: Architecture and Spatial Organisation of the Bedul in Petra. *Levant* 23: 147-80.

Biewers, M.
- 1992 Occupation de l'espace dans le village traditionnel de 'Aima: Approche ethnoarchéologique. In S. Tell *et al.* (eds.), *Studies in the History and Archaeology of Jordan*, IV: 397-402. Amman: Department of Antiquities.
- 1997 *L'habitat traditionnel à 'Aima: Enquête ethnoarchéologique dans un village jordanien*. Maison de l'orient méditerranéen/British Archaeological Reports, International Series 662. Oxford: Archaeopress.

Biewers, M. and R. Kana'an
- 1993 Vernacular Architecture Survey of Kurkuma, 316-19. In G. Palumbo *et al.* (eds.), The Wadi el-Yabis Survey and Excavations Project: Report on the 1992 Season. *Annual of the Department of Antiquities of Jordan* 37: 307-24.

Boling, R.G.
- 1989 Site Survey in the el-'Umeiri Region. In L.T. Geraty *et al.* (eds.), *Madaba Plains Project*, I: 98-188. Berrien Springs, MI: Andrews University.

Brown, R.M.
- 1984 Late Islamic Settlement on the Kerak Plateau, Trans-Jordan. Unpublished MA dissertation, State University of New York at Binghamton.
- 1988 Summary Report of the 1986 Excavations: Late Islamic Shobak. *Annual of the Department of Antiquities of Jordan* 32: 225-45.
- 1989 Excavations in the 14th Century AD Mamluk Palace at Kerak. *Annual of the Department of Antiquities of Jordan* 33: 287-304.
- 1992 Late Islamic Ceramic Production and Distribution in the Southern Levant: A Socio-Economic and Political Interpretation. Unpublished PhD dissertation, State University of New York at Binghamton.

Brünnow, R.E. and A. von Domaszewski
- 1904 *Die Provincia Arabia*, I. Strassburg: Trübner.

Burkhardt, J.L.
- 1992 *Notes on the Bedouins and Wahabis*. London: Garnet Publishing.

Canaan, T.
- 1932 The Palestinian Arab House. *Journal of the Palestine Oriental Society* 12(4): 223-47.
- 1933 The Palestinian Arab House. *Journal of the Palestine Oriental Society* 3: 183.

Charles, M.
- 1992 *Charred Plant Remains: Khirbat Faris*. Unpublished archive report.

Clark, V.A., J.M.C. Bowsher and J.D. Stewart
- 1986 The Jerash North Theatre. In F. Zayadine (ed.), *Jerash Archaeological Project*, I: 205-202. Amman: Department of Antiquities.

Conder, C.R.
- 1889 *The Survey of Eastern Palestine*. London: Palestine Exploration Fund.

Daher, R.
- 1995 The Resurrection of the Hijaz Railroad Line. In K. 'Amr, F. Zayadine and M. Zaghloul (eds.), *Studies in the History and Archaeology of Jordan*, V: 341-50. Amman: Department of Antiquities.

Dalman, G.
- 1932–33 *Arbeit und Sitte in Palästina*. Gütersloh: Bertelsmann.

Darley-Doran, R.E.
- 1988 An Alternative Approach to the Study of Ottoman Numismatics. In *A Festschrift Presented to Ibrahim Arturk on the Occasion of the 20th Anniversary of the Turk Numismatik Dernagi*, 87-90. Istanbul: Turk Numismatik Dernegi Yayinlari.

Davis, J.L.
- 1991 Contributions to a Mediterranean Rural Archaeology: Historical Case Studies from the Ottoman Cyclades. *Journal of Mediterranean Archaeology* 4/2: 131-216.

Denny, F.M.
- 1995 Modern Practice. In J.L. Esposito (ed.), *The Oxford Encyclopedia of the Modern Islamic World*, 35-37. New York: Oxford University Press.

Denton, B. and B. St. Laurent
- 1996 Early Twentieth Century Architecture. In P.M. Bikai and T.A. Dailey (eds.), *Madaba: Cultural Heritage*, 47-89. Amman: American Center for Oriental Research.

Dols, M.W.
- 1979 The Second Plague Pandemic and its Recurrences in the Middle East: 1347-1894. *Journal of the Economic and Social History of the Orient* 22(2): 162-89.

Dowling, T.E.
- 1896 Kerak in 1896. *Palestine Exploration Fund Quarterly Statement*: 327-32.

Eakins, J.K.
- 1993 *Tell el Hesi: The Muslim Cemetery in Fields V and VI/IX (Strata II)*. Winona Lake, IN: Eisenbrauns.

Fakhoury, L. and R. Sweiss
1995 Tayyiba 'A Thriving Village'. In K. 'Amr, F. Zayadine and M. Zaghloul (eds.), *Studies in the History and Archaeology of Jordan*, V: 361-74. Amman: Department of Antiquities.

Faroqhi, S.
1994 Crisis and Change. In H. Inalcik and D. Quataert (eds.), *An Economic and Social History of the Ottoman Empire*, 411-636. Cambridge: Cambridge University Press.

Findley, C.F.
1986 The Evolution of the System of Provincial Administration as Viewed from the Center. In D. Kushner (ed.), *Palestine in the Late Ottoman Period*, 3-29. Leiden: E.J. Brill.

Fishbach, M.R.
1994 British Land Policy in Transjordan. In E.L. Rogan and T. Tell (eds.), *Village, Steppe and State: The Social Origins of Modern Jordan*, 58-79. New York: British Academic Press.

Franken, H.J. and J. Kalsbeek
1975 *Potters of a Medieval Village in the Jordan Valley*. North-Holland Ceramics Studies in Archaeology 3. Amsterdam: North-Holland.

Fuller, M. and N. Fuller
1988 Tell Tuneir on the Khabur: Preliminary Report on Three Seasons. *Les Annales archéologiques arabes syriennes* 37: 279-90.

Gardiner, M. and A. McQuitty
1987 A Water Mill in Wadi el Arab, North Jordan and Water Mill Development. *Palestine Exploration Quarterly* 119(2): 24-32.

Ghawanmeh, Y.
1992 Earthquake Effects on Bilad ash-Sham Settlements. In S. Tell *et al.* (eds.), *Studies in the History and Archaeology of Jordan*, IV: 53-59. Amman: Department of Antiquities.

Gichon, M. and R. Linden
1984 Muslim Oil Lamps From Emmaus. *Israel Exploration Journal* 34(2): 156-69.

Gilliland, D.R.
1986 Palaeoethnobotany and Palaeoenvironment. In Ø.S. LaBianca and L. Lacelle (eds.), *Environmental Foundations: Hesban 2*, 123-42. Berrien Springs, MI: Andrews University.

Gilmors, M., S. al-Hiwah and I. Resseeni
1984 Dhrub al Hajj Architectural Documentation Program: Preliminary Report on the Architectural Survey of the Northern Pilgrimage Routes. *Atlal* 8: 143-61.

Glueck, N.
1934 Explorations in Eastern Palestine I. *Annual of the American Schools of Oriental Research* 14: 1-113.

Granquist, H.
1965 *Muslim Death and Burial: Arab Customs and Traditions Studied in a Village in Jordan*. Commentationes Humanorum Litterarum. Helsinki.

Hayes, J.W.
1980 Turkish Clay Pipes: A Provisional Typology. In P. Davey (ed.), *The Archaeology of the Clay Tobacco Pipe*, IV: 3-10. British Archaeological Reports, International Series 92. Oxford: British Archaeological Reports.

Herr, L.G. *et al.* (eds.)
1991 *Madaba Plains Project 2*. Berrien Springs, MI: Andrews University.

Humphrey, J.W.
1990 The Turkish Clay Smoking Pipes of Mytilene. *Society for Clay Pipe Research Newsletter* 26: 2-9.

Hütteroth, W.-D. and K. Abdulfattah
1977 *Historical Geography of Palestine, Transjordan and Southern Syria in the Late 16th Century*. Erlangen: Frankischen geographischen Gesellschaft.

Ibach, R.D.
1987 Archaeological Survey of the Hesban Region. In Ø.S. LaBianca (ed.), *Hesban 5*, 191-95. Berrien Springs, MI: Andrews University.

Ibrahim, M., J.A. Sauer and K.N. Yassine
1976 The East Jordan Valley Survey 1975. *Bulletin of the American Schools of Oriental Research* 222: 41-66.

Inalcik, H. and D. Quartaert (eds.)
1994 *An Economic and Social History of the Ottoman Empire*. Cambridge: Cambridge University Press.

Jaussen, A.
1948 *Coutumes des Arabes au pays de Moab*. Paris: Librairie d'Amérique et d'Orient.

Johns, J.
1992 Islamic Settlement in Ard al-Kerak. In S. Tell *et al.* (eds.), *Studies in the History and Archaeology of Jordan*, IV: 363-68. Amman: Department of Antiquities.
1998 The Rise of Middle Islamic Hand-Made Geometrically-Painted Ware in Bilad al-Sham (11th–13th Centuries AD). In R.-P. Gayraud (ed.), *Colloque international d'archéologie islamique*, 65-93. Textes arabes et études islamiques 36: Institut française d'archéologie orientale.

Johns, J., A. McQuitty and R. Falkner
1989 The Faris Project: Preliminary Report upon the 1986 and 1988 Seasons. *Levant* 21: 63-95.

Kana'an, R. and T. al-Rifai
1988 *'Iraq al-Amir wa al-Bardon*. Amman: University of Jordan.

Kana'an, R. and A. McQuitty
1994 The Architecture of al-Qasr on the Kerak Plateau: An Essay in the Chronology of Vernacular Architecture. *Palestine Exploration Quarterly* 126: 127-51.

Kareem, J.
1987 Evidence of the Umayyad Occupation in the Jordan Valley as seen in the Jisr Sheikh Hussein Region. Unpublished MA thesis, Yarmouk University, Jordan.
1993 The Settlement Patterns in the Jordan Valley in the Mid- to Late-Islamic Period. Unpublished PhD dissertation, Freie Universität, Berlin.

Khalidi, T. (ed.)
1984 *Land Tenure and Social Transformation in the Middle East.* Beirut: American University.

Khammash, A.
1986 *Notes on Village Architecture in Jordan.* Lafayette, LA: University Art Museum.

Khatib, R.Y. and H.S. al-Asir
1995 The Golden Era of as-Salt (1870–1950): Urban and Architectural Development. In K. 'Amr, F. Zayadine and M. Zaghloul (eds.), *Studies in the History and Archaeology of Jordan,* V: 351-60. Amman: Department of Antiquities.

Killick, A.
1983 Udhruh: The Frontier of an Empire: 1980 and 1981 Seasons, A Preliminary Report. *Levant* 15: 110-31.

Köhler-Rollefson, I.
1987 Ethnoarchaeological Research into the Origins of Pastoralism. *Annual of the Department of Antiquities of Jordan* 31: 535-39.

Koucky, F.L.
1987 The Regional Environment. In T.S. Parker (ed.), *The Roman Frontier in Central Jordan,* 11-40. British Archaeological Reports 340i. Oxford: British Archaeological Reports.

LaBianca, Ø.S.
1990 *Sedenterization and Nomadization: Hesban 1.* Berrien Springs, MI: Andrews University.
1991 A Note on Seasonally Occupied Cave Villages. In L.G. Herr *et al.* (eds.), *Madaba Plains Project 2,* 353-55. Berrien Springs, MI: Andrews University.

LaBianca, Ø.S. and A. von den Driesch (eds.)
1995 *Faunal Remains: Hesban 13.* Berrien Springs, MI: Andrews University.

LaBianca, Ø.S. and L. Lacelle (eds.)
1986 *Environmental Foundations: Hesban 2.* Berrien Springs, MI: Andrews University.

Lancaster, W. and F. Lancaster
1993 Graves and Funerary Monuments of the Ahl al-Gabal, Jordan. *Arabian Archaeology and Epigraphy* 4(3): 151-69.
1995 Land Use and Population in the Area North of Karak. *Levant* 27: 103-24.
1999 *People, Land and Water in the Arab Middle East,* II. Studies in Environmental Anthropology. Amsterdam: Harwood Academic.

Landau, J.M.
1971 *The Hejaz Railway and the Muslim Pilgrimage: A Case of Ottoman Political Propaganda.* Detroit, MI: Wayne State University Press.

Lewis, N.N.
1987 *Nomads and Settlers in Syria and Jordan 1800–1980.* Cambridge: Cambridge University Press.

Lindner, M., E.A. Knauf and J.P. Zeitler
1996 An Edomite Fortress and a Late Islamic Village Near Petra (Jordan): Khirbat al-Mu'allaq. *Annual of the Department of Antiquities of Jordan* 40: 111-35.

London, G.A. and M. Sinclair
1991 An Ethnoarchaeological Survey of Potters in Jordan. In L.G. Herr *et al.* (eds.), *Madaba Plains Project 2,* 420-28. Berrien Springs, MI: Andrews University.

Luke, H.C. and E. Keith-Roach (eds.)
1930 *The Handbook of Palestine and Transjordan.* London: Macmillan.

Mabry, J. and G. Palumbo
1988 The 1987 Wadi el-Yabis Survey. *Annual of the Department of Antiquities of Jordan* 32: 275-306.

MacDonald, B. *et al.*
1987 Southern Ghors and Northeast 'Arabah Archaeological Survey 1986, Jordan: A Preliminary Report. *Annual of the Department of Antiquities of Jordan* 31: 391-418.
1988 *The Wadi el Hasa Archaeological Survey 1979–1983, West-Central Jordan.* Waterloo, ON: Wilfrid Laurier University.
1992 Settlement Patterns along the Southern Flank of Wadi al- Hasa: Evidence from 'The Wadi al-Hasa Archaeological Survey'. In S. Tell *et al.* (eds.), *Studies in the History and Archaeology of Jordan,* IV: 73-76. Amman: Department of Antiquities.
1992 *The Southern Ghors and Northeast 'Arabah Archaeological Survey.* Sheffield Archaeological Monographs 5. Sheffield: J.R. Collis.

Mandaville, J.E.
1966 The Ottoman Court Records of Syria and Jordan. *Journal of the American Oriental Society* 86(3): 311-19.

Mason, R.B. and M. Milwright
1998 Petrography of Middle Islamic Pottery from Karak. *Levant* 30: 175-90.

McQuitty, A.
1986 Architectural Study of Bait Ras. *Archiv für Orientforschung* 33: 153-55.
1993– Ovens in Town and Country. *Berytus* 41: 53-76.
94
1995 Water-Mills in Jordan: Technology, Typology, Dating and Development. In K. 'Amr, F. Zayadine and M. Zaghloul (eds.), *Studies in the History and Archaeology of Jordan,* V: 745-52. Amman: Department of Antiquities.

McQuitty, A. and R. Falkner
1993 A Preliminary Report on the Khirbat Faris Project: The 1989, 1990 and 1991 Seasons. *Levant* 25: 37-62.

McQuitty, A.M. and C.J. Lenzen
1989 An Architectural Study of the Irbid Region with Particular Reference to a Building in Irbid. *Levant* 21: 119-28.

McQuitty, A. *et al.*
1997–1998 Mamluk Khirbat Faris. *Aram* 9-10: 1-47.

Merschen, B.
1985 Recent Hand-Made Pottery from Northern Jordan. *Berytus* 33: 75-87.
1990 The Islamic Cemetery of Abu an-Naml. In T. Weber and A. Hoffman (eds.), Gadara of the Decapolis: Preliminary Report of the 1989 Season at Umm Qeis. *Annual of the Department of Antiquities of Jordan* 34: 331-33.
1991 The Islamic Cemetery of Abu en-Niml. In S. Kerner (ed.), *The Near East in Antiquity* 2: 135-41. Amman: Al Kutba.
1992 Settlement History and Village Space in Late Ottoman Northern Jordan. In S. Tell *et al.* (eds.), *Studies in the History and Archaeology of Jordan*, IV: 409-16. Amman: Department of Antiquities.

Miles, G.C.
1985 In A.D. Tushingham (ed.), *Excavations in Jerusalem 1961–7*, I: 176-77. Toronto, ON: Royal Ontario Museum.

Miller, J.M.
1991 *Archaeological Survey of the Kerak Plateau*. Atlanta, GA: Scholars Press.

Mundy, M.
1996 Qada' 'Ajlun in the Late Nineteenth Century: Interpreting a Region from Ottoman Land Registers. *Levant* 28: 77-95.

Noca, L.
1985 *Smakieh: Un village de Jordanie*. Ecole d'architecture de Lyon: Travail personnel de troisième cycle, Lyon.

Oleson, J.P. and K. 'Amr *et al.*
1993 The Humeima Excavation Project: Preliminary Report of the 1991–1992 Seasons. *Annual of the Department of Antiquities of Jordan* 37: 461-502.

Palmer, C.
1998 'Following the Plough': The Agricultural Environment of Northern Jordan. *Levant* 30: 129-66.

Palumbo, G. *et al.*
1993 Cultural Resources Management in Jordan. *Annual of the Department of Antiquities of Jordan* 37: 69-84.

Palumbo, G. (ed.)
1995 *The Jordan Antiquities Database and Information System*. Amman: Department of Antiquities and the American Center of Oriental Research.

Pamuk, S.
1994 Money in the Ottoman Empire, 1326–1914. In H. Inalcik and D. Quataert (eds.), *An Economic and Social History of the Ottoman Empire*, 947-85. Cambridge: Cambridge University Press.

Parker, S.T. (ed.)
1987 *The Roman Frontier in Central Jordan*. British Archaeological Reports 340i. Oxford: British Archaeological Reports.

Petersen, A.
1996 A Preliminary Report on Three Muslim Shrines in Palestine. *Levant* 28: 97-113.

Pitcher, D.E.
1972 *An Historical Geography of the Ottoman Empire from Earliest Times to the End of the Sixteenth Century*. Leiden: E.J. Brill.

Pringle, D.
1986 *The Red Tower*. British School of Archaeology in Jerusalem Monographs, 1. London: British School of Archaeology in Jerusalem.

Rafeq, A.-K.
1984 Land Tenure Problems and Their Social Impact in Syria around the Middle of the Nineteenth Century. In T. Khalidi (ed.), *Land Tenure and Social Transformation in the Middle East*, 371-96. Beirut: American University.

Reinhart, A.K.
1995 Legal Foundations. In J.L. Esposito (ed.), *The Oxford Encyclopedia of the Modern Islamic World*, 34-35. New York: Oxford University Press.

Reilly, K.
1989 Khirbat Faris 1989 Excavation Season: Interim Animal Bone Report. Unpublished archive report.

Riis, P.J. and V. Poulsen
1957 *Hama. Fouilles et recherches de la Fondation Carlsberg 1931–1938*. IV. 2. *Les verreries et poteries médiévales*. Copenhagen: Danish National Museum.

Robinson, R.
1983 Clay Tobacco Pipes from the Kerameikos. *Mitteilungen des deutschen archäologischen Instituts* 98: 265-84.
1985 Tobacco Pipes of Corinth and of the Athenian Agora. *Hesperia* 14: 147-203.

Rogan, E.L.
1991 Incorporating the Periphery: The Ottoman Extension of Direct Rule over South-Eastern Syria (Transjordan), 1867–1914. Unpublished PhD dissertation, Harvard University.
1994 Bringing the State Back: The Limits of Ottoman Rule in Jordan, 1840–1910. In Rogan and Tell 1994: 32-57.

1995   Reconstructing Water Mills in Late Ottoman Transjordan. In K. 'Amr, F. Zayadine and M. Zaghloul (eds.), *Studies in the History and Archaeology of Jordan*, V: 753-56. Amman: Department of Antiquities.

Rogan, E.L. and T. Tell (eds.)
1994   *Village, Steppe and State: The Social Origins of Modern Jordan*. New York: British Academic Press.

Royal Scientific Society
1990   *Salt: A Plan for Action*. 3 vols. Amman: Salt Development Corporation.

Russell, K.W. and S.R. Simms
1997   Tur Imdai Rockshelter: Archaeology of Recent Pastoralism in Jordan. *Journal of Field Archaeology* 24/4: 459-72.

Sauer, J.A.
1986   Umayyad Pottery from Sites in Jordan. In L.T. Geraty and L.G. Herr (eds.), *The Archaeology of Jordan and Other Studies Presented to Siegfried H. Horn*, 301-30. Berrien Springs, MI: Andrews University.
1993   The Pottery at Hesban and its Relationship to the History of Jordan: An Interim Hesban Pottery Report, 1993. In D. Merling and L.T. Geraty (eds.), *Hesban after 25 Years*, 225-81. Berrien Springs, MI: Andrews University.

Schaeffer, J.
1989   Archaeological Remains from the Medieval Islamic Occupation of the Northwest Negev Desert. *Bulletin of the American Schools of Oriental Research* 274: 33-60.

Schilcher, L.
1991   The Grain Economy of Late Ottoman Syria and the Issue of Large-Scale Commercialization. In C. Keyder and F. Tabak (eds.), *Landholding and Commercial Agriculture in the Middle East*, 173-228. Albany: State University of New York.

Shami, S.
1989   Settlement and Resettlement in Umm Qeis: Spatial Organisation and Social Dynamics in a Village in North Jordan. In J.P. Bourdier and N.A. Sayyad (eds.), *Dwellings, Settlements and Tradition*, 451-76. New York: American Universities.

Silberman, N.A.
1982   *Digging for God and Country: Exploration, Archaeology and the Secret Struggle for the Holy Land 1799–1917*. New York: Knopf.
1989   *Between Past and Present: Archaeology, Ideology and Nationalism in the Modern Middle East*. New York: Holt, Reinhart & Winston.

Simpson, St. John
1990   Ottoman Clay Pipes from Jerusalem and the Levant: A Critical View of the Published Evidence. *Society for Clay Pipe Research Newsletter* 28: 6-16.
1993   Turkish Clay Pipes: A Review. *Society for Clay Pipe Research Newsletter* 39: 17-23.
1994   Near Eastern Pipe News. *Society for Clay Pipe Research Newsletter* 44: 14-15.

Singer, A.
1994   *Palestinian Peasants and Ottoman Officials*. Cambridge: Cambridge University Press.

Spaer, M.
1992   The Islamic Glass Bracelets of Palestine: Preliminary Findings. *Journal of Glass Studies* 34: 44-62.

Steiner, M.
1995   Glass Bracelets from Tell Abu Sarbut. In K. 'Amr, F. Zayadine and M. Zaghloul (eds.), *Studies in the History and Archaeology of Jordan*, V: 537-40. Amman: Department of Antiquities.

Toombs, L.E.
1985   *Tell el-Hesi: Modern Military Trenching and a Muslim Cemetery in Field I, Strata I-II*. Waterloo, ON: Wilfrid Laurier University.

Turab
1996   *The Holy Sites of Jordan*. Amman: Turab.

Tushingham, A.D.
1985   *Excavations in Jerusalem 1961–67*. Toronto: Royal Ontario Museum.

Volger, G., K. von Welck and K. Hackstein (eds.)
1987   *Pracht und Geheimnis*. Cologne: Rautenstrauch-Joest-Museum.

Walmsley, A.
1992   Fihl (Pella) and the Cities of North Jordan during the Umayyad and Abbasid Periods. In S. Tell *et al.* (eds.), *Studies in the History and Archaeology of Jordan*, IV: 377-84. Amman: Department of Antiquities.

Whitcomb, D.
1992   Reassessing the Archaeology of Jordan of the Abbasid Period. In S. Tell *et al.* (eds.), *Studies in the History and Archaeology of Jordan*, IV: 385-90. Amman: Department of Antiquities.

Wimmer, D.H.
1978   Area G.4, 13, 15. In R.S. Boraas and L.T. Geraty (eds.), *Heshbon 1976*, 149-56. Andrews University Monographs 10. Berrien Springs, MI: Andrews University.

Ziadeh, G.
1995   Ottoman Ceramics From Ti'innik, Palestine. *Levant* 27: 209-45.

17. Climatic Changes in Jordan through Time

Burton MacDonald

Introduction

It is generally agreed that archaeological data does not suffice on its own as an indicator of palaeoclimates. Other evidence, for example, palynology, palaeomorphology, faunal remains, and dendrochronology, must also be taken into consideration (see, e.g., Henry 1985; Henry and Turnbull 1985; Rosen 1995). In fact, the interpretation of palaeoclimates should be based on multidisciplinary studies. It must be noted, however, that natural evidence, such as pollen and palaeomorphology, does not always record pure climatic events. Natural evidence may also suffer from human interference such as deforestation and agriculture (Frumkin *et al.* 1994: 315).

There has been a great deal of study done on the palaeoclimates of Israel, especially in the southern part of the country and in the area of the Dead Sea (see, e.g., Horowitz 1971, 1974, 1978, 1979; Marks 1976; Goldberg 1981, 1984, 1986, 1995; Goldberg and Bar-Yosef 1982; Issar and Bruins 1983; el-Moslimany 1994; Goodfriend and Magaritz 1988; Goodfriend 1990; Frumkin *et al.* 1991, 1994; Bar-Yosef and Kra 1994). The same amount of work has not been done on the palaeoclimates of Jordan. It is often therefore necessary to extrapolate from information about palaeoclimates west of the 'Arabah, the Dead Sea, and the Jordan river to make judgments concerning climatic conditions in Jordan. This being the case, the work of Henry (1985) and Clark *et al.* (1997) in southern Jordan is, nevertheless, a contribution to the study of the palaeoenvironments of Jordan.

In studies of palaeoclimates, scholars generally define the precipitation/evaporation ratio higher than today's as 'moist' and lower than today's as 'dry' (Frumkin *et al.* 1994: 321). It is therefore necessary to begin with a description of the present-day temperature and precipitation of Jordan in order that we may be able to compare this with palaeoclimates.

Present-day climate and rainfall

The Middle East year may be divided into three seasons: (1) summer, from mid-June to mid-September; (2) rainy season in the cooler half of the year extending roughly from mid-October to mid-April; and (3) transitional seasons which comprise the remainder of the year (Anonymous 1984: iv; see also Shehadeh 1985: 29-30). There are, normally, five continuous months, from the first week in May to the first week in October, without any rain at all (Baly 1974: 43). The transitional seasons are characterized by two important phenomena, namely, desert storms and the sirocco (Baly 1974: 51), which the Arabs refer to as *khamasin* (Shehadeh 1985: 29-30).

The main characteristics of Jordan's climate reflect the transitional location of the country between the Mediterranean climate to the west and arid climates in the east and in the south (Anonymous 1984: iv; Shehadeh 1985: 25). There is, in fact, a great variety in Jordan's climate from one morphological unit to another. For the present purposes, however, it is sufficient to follow Bender and divide the country into three climatic zones, namely, Mediterranean, semi-arid, and arid (1974: 11-15; see also Ahmad 1989: 6). This classification does justice to the transitional location of the country.

Mediterranean climate, which ranges from sub-humid (> 600 mm precipitation per year; average January temperature = 3°C, average August temperature = 27-33°C) to semiarid (precipitation between 300 and 600 mm a year; average January temperature = 3-7°C, average August temperature = 30-35°C) (Bender 1974: 187), dominates in the highlands at the eastern rim of Wadi 'Arabah-Jordan Graben as far south as ash-Shawbak. From April till October this climate has dry summers with an average maximum annual temperature of 38.8°C; from November until March it is winter with an average minimum temperature of 0.5°C.[1]

There is a wide transition zone, east and south of the areas with Mediterranean climate, to the fully arid climate that characterizes the desert areas of east and south Jordan. In this semi-arid transition zone, the average annual maximum temperature is +40°C, and the average annual minimum temperature is −1.6°C; the average annual rainfall is between 50 and 300 mm (Baly 1974: 61). The segment of this zone which is immediately east of the highlands and at times close to the fields and farming villages is often referred to as the 'steppeland' and is more often pastoral than agricultural. In times of drought, the Bedouin moved from the desert to these steppes, in which dry grass provides some kind of pasture land (Hütteroth and Abdulfattah 1977: 62).

The arid or desert regions of Jordan receive less than 50 mm precipitation per year. Moreover, the distribution is extremely erratic from year to year, and there is a strong tendency for much of it to come in heavy storms during the transitional seasons (Baly 1974: 61-62).

About 90% of Jordan receives less than 200 mm of precipitation per year, which is totally dissipated by evaporation. Only about 14% of the mean annual rainfall is exploitable: 70% drains towards the Jordan Valley, Dead Sea, and Wadi 'Arabah; the remaining 30% can only be utilized by drilling wells (Anonymous 1986: i). Relative to this fact, it is important to point out that at least 300 mm of rainfall per year is considered necessary for sustainable agriculture (Issar 1995: 351) (Figure 17.1).

Climatic changes through the ages

There is little firm evidence relative to the climates of the Lower Palaeolithic period of Jordan (see Chapter 2, this volume). It is, thus, necessary to begin our discussion with the Middle Palaeolithic period.

The Middle Palaeolithic begins with the deposition of massive, well-rounded gravels. These gravels are associated with Middle Palaeolithic artifacts. Goldberg suggests that this points to higher and more sustained discharges during a climatic regime, between 85,000 to 75,000 bp, which was substantially wetter than today's (1981, 1984, 1995: 46). He (1986: 237) supports his interpretation with radiometrically dated fossil travertines associated with Mousterian industries in the Negev and Judaean desert regions (see Schwarcz et al. 1979, 1980) and by a relatively high percentage of arboreal pollen from sediments in Mousterian sites in the Central Negev (see also Horowitz 1979). Moreover, in Goldberg's opinion, the widespread occurrence of early Middle Palaeolithic sites in the Negev and Sinai is a good case for wetter conditions at this time (1986: 241). These conditions, in the opinion of Clark et al. (1997), appear to be similar for west-central Jordan.

The above-described moist period was followed by an erosional episode during the latter part of the Middle Palaeolithic or roughly between 70,000 to 45,000 years

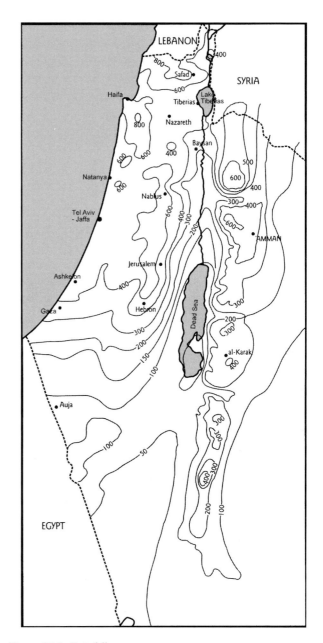

Figure 17.1. Rainfall map.

ago. During this period, that is, between the early Mousterian gravels and the following deposition of finer sediments of the Upper Palaeolithic, there are very few Middle Palaeolithic sites. Goldberg has interpreted this to represent arid conditions, similar to today's (1981, 1984). The Lisan Formation is dated by radiocarbon and Uranium-series methods to 63,000–15,000 or 17,000 years ago (Goldberg 1995: 44). Lake Lisan, the predecessor of the Dead Sea, reached an altitude of −180 m asl (Goldberg 1995: 44).[2] This fact is another indication of moister conditions. It must be noted, however, that there was fluctuation of the lake bed over the last 30,000 years of its existence (Begin et al. 1974).

Study of both the diatoms and the sediments demonstrates an increase in salinity through time. Goldberg interprets this to represent a general trend toward climatic aridity, with greater precipitation/evaporation ratios at the base of the Lisan Formation, some 63,000 years ago (1995: 44).

Wetter conditions are generally posited for the early Upper Palaeolithic, that is, between c. 40,000 to c. 25,000 years ago (Henry 1985: 75; Goodfriend and Magaritz 1988; Goldberg 1995: 46; see also Chapter 2, this volume). Goodfriend and Magaritz (1988) base this conclusion on their study of paleosols (dark paleosols indicate a wet climate). Goldberg, moreover, found Upper Palaeolithic sites associated with massive accumulations of fine-grained sediments. He interprets this as indicative of 'gentle, sustained rainfall in a climatic regime noticeably wetter than today's and producing perhaps perennial stream flow' (1986: 239). Furthermore, pollen analyses from Upper Palaeolithic sites in the Nahal Zin region of Israel show a higher percentage of arboreal pollen for this period than for the present but less than for the Middle Palaeolithic (Horowitz 1979). Issar and Bruins (1983) speculate 50-100% more precipitation than at present.

Archaeologically, for the Upper Palaeolithic between 40,000 to 22,000 bp, sites abound in the Sinai as well as in the Nahal Zin region of the central Negev (Marks 1976, 1977, 1983; Goldberg 1986: 241). On the other hand, there are very few sites for the late Upper Palaeolithic, that is, from c. 22,000 to c. 18,000 bp (Goldberg and Bar-Yosef 1982; Goldberg 1986: 241; for Jordan, see Henry 1985: 75, 76: table 4). This evidence is in line with the fact that, 'with the exception of the Besor/Bi'r as-Saba' sequence and possibly that of nearby Nahal Sekher, most areas show a period of erosion from c. 22,000 to 15,000 bp' (Goldberg 1986: 240). This phase, which falls within the maximum of the last glacial period, has been interpreted as representing arid conditions similar to today's (Goldberg 1986: 240; see Chapter 2, this volume) (and a hyper-arid phase is associated with the Late Glacial maximum at c. 20,000–18,000 bp). Evidence for this comes from the records of sedimentation, pollen and faunal analyses, and stable isotope variations in foraminifera preserved in deep-sea cores (Fontugne et al. 1994: 75). During this time, the vegetation and climate developed in a similar manner in southern Europe and the Near East. A sagebrush semi-desert, dominated by *Artemisia* with interspersed *Pinus*, was present at all elevations. Such vegetation is characteristic of regions with low minimum precipitation (100-350 mm/year) and very cold winters with numerous freezing days ($< -15°C$). Summer mean temperatures are estimated to have been between 27° and 35°C in the Near East (Fontugne et al. 1994: 76-78).

There is conflicting evidence on the climate relative to the Epipalaeolithic period. Fontugne et al. (1994: 76-78) claim that the period from 18,000 to 13,000 bp was dry. Goldberg cites geomorphic and palynological evidence that suggests that climatic conditions were wetter during the Geometric Kebaran and Early Natufian, and became increasingly arid during the Late Natufian (1995: 45). He is also of the opinion that wetter conditions prevailed between c. 14,500 to c. 12,500 years ago (1986: 240-41). Goodfriend and Magaritz's analysis of paleosols also confirmed that 13,000 bp was wet (1988). Henry's conclusions (1985) concerning the climate for southern Jordan for the period between 18,000 and 11,000 bp are, in part, at variance with the above scenario. He posits, on pollen and faunal evidence, a moist phase (probably equivalent to the Kebaran) that was replaced by drier conditions (equivalent to the Geometric Kebaran) that, in turn, were replaced by a moist interval during the Early Natufian (1985: 75; see also Henry and Turnbull 1985: 60-61). These conflicting positions on the climate of the Epipalaeolithic may be explained, at least in part, by the fact that there were probably shorter climatic cycles at work and/or there were climatic cycles within the period that have not yet been accounted for.

Epipalaeolithic sites (including the Geometric Kebaran and Mushabian) are widespread both in upland and lowland localities from Lebanon to the Suez and as far south as Wadi Feiran (Goldberg and Bar-Yosef 1982). The Negev Kebaran, however, has a more limited distribution suggesting increased aridity c. 12,000–11,000 years ago (Goldberg 1986: 241).

The beginning of the Holocene period c. 10,000 bp is marked by the expansion of deciduous oak and the masking of herbaceous vegetation in the pollen record. This reforestation occurred along with an increase in humidity, following the meltwater discharge into the global ocean (Bard et al. 1987). This also coincides with the beginning of a period of high freshwater discharge into the eastern Mediterranean from the Nile and other rivers (Fontugne et al. 1994: 78). Moreover, el-Moslimany (1994) found that during the almost three-thousand-year period between 9500 and 6750 bp, there was a high percentage of arboreal pollen at Ali Kosh and Tepe Sabz, implying increased rainfall. Sauer (1996), making use of an Arabian Sea core sample, also concludes that 9000–7000 bp was a damp period. Kafafi (1996), on the basis of palaeobotanical and geomorphic evidence from PPNB levels at Beulha, posits that the eastern desert of Jordan was very moist from 8800 to 8250 bp.

Frumkin *et al.* (1991), through a study of salt caves at Mt Sedom along the southwestern shore of the Dead Sea, determined that 7000 bp was a wet period. They support this conclusion on the basis of their humidity indicators: abundance of wood (more trees means more water); cave width (wider indicates more humidity, since there is more water to erode the caves); and height of outlets (higher outlets mean the water was high enough to erode through at that point which is indicative of more humidity).

All the above evidence is in line with wetter conditions in the Jordan Valley (Schuldenrein and Goldberg 1981; Goldberg and Bar-Yosef 1982) and the southern and eastern parts of the area that were probably influenced by rains derived from the south, probably the African monsoon (Goldberg 1986: 241-42). Such is not the case, however, for the western Negev as compared to the Jordan Rift Valley, southern Sinai, and northern Israel indicating that arid conditions continued in this western area (Goldberg 1986: 241).

A drying trend within the Holocene was interrupted during the Chalcolithic by the deposition of 3-4 m of fluvial silts at Shiqmim, in the northern Negev desert, and the accumulation of a 30-40 cm thick loess mantle on the slopes adjacent to the site (loess deposition is also recorded for the Har Harif area) (Goldberg 1986: 240). Goldberg infers a wetter climate for the Chalcolithic on the basis of the hypotheses that dust accumulation and alluviation tend to occur during wetter periods (Goldberg 1986: 240). Moreover, the pollen record indicates a humid episode at this time (Goldberg 1986: 240; see also Henry 1985: 76, table 4). Archaeologically, there is an increase in the number of sites in Sinai as well as in the southern and western Negev at this time (Goldberg 1986: 242).

Frumkin *et al.* have concluded, based on research done on the salt caves of Mount Sedom, that there was a moist period which was roughly synchronous with the Early Bronze I–III period (1994: 325-26). This was, in their opinion (1994: 329), the moistest period during the last 6000 years (see also Shehadeh 1985: 27). This was followed by a desiccation that started during the EB III period and continued into the Middle Bronze Age (Frumkin *et al.* 1994: 326, 329). Weiss *et al.* (1993) attribute the collapse of third-millennium BC Mesopotamian civilization, at least in part, to increased aridity beginning at 2200 BC (see also Sauer 1994: 370).

Nicholson and Flohn (1980) reviewed various palaeoclimate data for dozens of sites in the Sahara and eastern Africa. They also concluded that wetter-than-present conditions prevailed in these areas until c. 4000 bp.

Rosen concludes, on the basis of geomorphologic studies from the Negev desert, southern foothills (Shephela) and the coastal plain of Israel, that there was 'widespread alluvial activity and aggrading floodplains in the Chalcolithic and Early Bronze Age' (1995: 31). This suggests a moister climate with a higher water table than at present (Rosen 1986). Rosen's study also indicates, however, fluctuating moist conditions in the Early Bronze Age (1995: 31).

The moist period of the Early Bronze period was followed by unstable conditions during MB I and desiccation beginning around MB II (Frumkin *et al.* 1994: 328-29; see also Chapter 8, this volume). Palynological information from a core in the Hula basin of Israel suggests climate deterioration at c. 4500 bp from warm and humid vegetation to the present Mediterranean-type vegetation. Moreover, there is evidence, in the form of pollen data, for the shrinkage of Lake Hula, in the north of Israel, at this time (Frumkin *et al.* 1994: 326).

There appears to be a correlation between post-Early Bronze desiccation and similar evidence from the Sahara and eastern Africa. This is in the form of high lake levels until c. 4000 bp when their levels fell rapidly until they reached their present levels (Gasse 1980).

Sauer (1996) and Rosen (1995) agree that 3550–3200 bp was a dry period. Lake levels in Switzerland and Central Europe declined in 3400–3300 bp. The Near East may have been dry as well since there seems to be a correlation between moisture trends in the two areas. Thin tree rings dated to 3300–3100 bp indicate that the climate was cold and dry during this time. Danin's (1995) study of the Dead Sea sediment levels indicates that this body of water dropped to –400 m asl. Rosen (1995) points to the extinction of oak in Galilee as indicating dryness during the Iron Age period.

Goodfriend (1990) holds that 3000 bp marked a period of low rain in the Negev. Evidence for this condition includes short-distance transport, associated with an arid period, in the northern Negev and the southern Shephela during the Iron Age (Goldberg 1995: 47).

A wetter phase is one of the explanations for the occurrences of widespread silts from Qadesh Barnea in the south to the central Shephela region in the north, during the Byzantine and Early Islamic periods, that is, between roughly 1700 and 600 bp (Goldberg 1986: 240-41; see also Horowitz 1978: fig. 1). The widespread silts, however, could have been caused by the influence of human activity on the landscape in the form, for example, of deforestation and over-grazing (see Chapter 13, this volume). Archaeology supports the hypothesis of a wetter climate since the area was densely occupied by large Byzantine settlements. It is hard to envisage how such a large population could survive under the regime of today's arid conditions. Goldberg (1986: 240-41) points out that independent verification of this hypothesis is needed, for example, from palynology.

A more arid phase begins at least around AD 1400, if not earlier, in late Mamluk times (Ghawanmeh 1995). This decline continues until Modern times. The present climate is 'substantially drier than most previous Holocene periods, indicating that current conditions in the Near East are not reflective of earlier periods in its history' (Sauer 1994: 378).

Conclusions

It must be kept in mind that caution is advised about all of the above. There is need for continual revision of the above-described climatic conditions in the light of new information (Goldberg 1995: 53). This is especially true in the present case since the data is taken from Israel, for the most part, and not directly from east of the Jordan, the Dead Sea, and the 'Arabah. The data may, thus, not be pertinent in all cases.

Notes

1. Some average annual rainfalls in the Mediterranean zone: in the area of Quwayliba/Abila, between 350 and 450 mm; at as-Salt, 700 mm; at Jabal 'Amman (the part of the plateau in the centre of the city), 375 mm; in the Hisban region, 400 mm; at Na'ur, 525 mm; in the al-Karak region, 350 mm; at Mazar, 375 mm; and at ash-Shawbak, 300 mm.

    It is interesting to compare the average annual rainfall of 268 mm at the 'Amman airport at Marka (having the longest rainfall records in Jordan going back to 1923) to that of Jabal 'Amman.
2. The present Dead Sea level is c. − 410 m asl.

References

Ahmad, A.A.
1989   *Jordan Environmental Profile: Status and Abatement.* Amman: United States Agency for International Development.

Anonymous
1984   *National Atlas of Jordan. I. Climate and Agroclimatology.* Amman: National Geographic Centre.
1986   *National Atlas of Jordan. II. Hydrology and Agrophodrology.* Amman: Royal Jordanian Geographic Centre.

Baly, D.
1974   *The Geography of the Bible.* Rev. edn. New York: Harper & Row.

Bar-Yosef, O. and R.S. Kra (eds.)
1994   *Late Quaternary Chronology and Paleoclimates of the Eastern Mediterranean.* RADIOCARBON. Tucson: The University of Arizona.

Bard, E. et al.
1987   Retreat Velocity of the North Atlantic Polar Front during the Last Deglaciation Determined by 14C Accelerator Mass Sectrometry. *Nature* 328: 791-94.

Begin, Z.B. et al.
1974   Lake Lisan, the Pleistocene Precursor of the Dead Sea. *Geological Survey of Israel Bulletin* 63: 1-30.

Bender, F.
1974   *The Geology of Jordan.* Trans. M.K. Khdeir. Berlin: Gebrüder Borntraeger.

Clark, G.A. et al.
1997   Chronostratigraphic Contexts of Middle Paleolithic Horizons at the Ain Difla Rockshelter (WHS 634), West-Central Jordan. In H.G. Gebel, Z. Kafafi and G. Rollefson (eds.), *The Prehistory of Jordan. II. Perspectives from 1996,* 77-100. Studies in Near Eastern Production, Subsistence, and Environment. Berlin: Ex oriente.

Danin, A.
1995   Man and the Natural Environment. In T.E. Levy (ed.), *The Archaeology of Society in the Holy Land,* 25-37 London: Leicester University Press.

Frumkin, A. et al.
1991   The Holocene Climatic Record of the Salt Caves of Mount Sedom, Israel. *The Holocene* 1.3: 191-200.
1994   Middle Holocene Environmental Change Determined from the Salt Caves of Mount Sedom, Israel. In Bar-Yosef and Kra 1994: 315-32.

Fontugne, M. et al.
1994   Paleoenvironment, Sapropel Chronology and Nile River Discharge during the Last 20,000 Years as Indicated by Deep-Sea Sediment Records in the Eastern Mediterranean. In Bar-Yosef and Kra 1994: 75-88.

Gasse, F.
1980 Late Quaternary Changes in Lake-Levels and Diatom Assemblages on the Southeastern Margin of the Sahara. *Palaeoecology of Africa* 12: 333-50.

Goldberg, P.
1981 Late Quaternary Stratigraphy of Israel: An Eclectic View. C.N.R.S. Colloque No. 598, *Préhistoire du Levant*, Lyons 1980, 55-66.
1984 Late Quaternary History of Qadesh Barnea, Northeastern Sinai. *Zeitschrift für Geomorphologie* 28/2: 193-217.
1986 Late Quaternary Environmental History of the Southern Levant. *Geoarchaeology: An International Journal* 1/3: 225-44.
1995 The Changing Landscape. In T.E. Levy (ed.), *The Archaeology of Society in the Holy Land*, 40-54. London: Leicester University Press.

Goldberg, P. and O. Bar-Yosef
1982 Environmental and Archaeological Evidence for Climatic Change in the Southern Levant and Adjacent Areas. In J.L. Bintliff and W. van Zeist (eds.), *Palaeoclimates, Palaeoenvironments and Human Communities in the Eastern Mediterranean Region in Later Prehistory*, 399-414. British Archaeological Reports S-133. Oxford: British Archaeological Reports.

Goodfriend, G.A.
1990 Rainfall in the Negev Desert during the Middle Holocene, Based on a 13c of Organic Matter in Land Snail Shells. *Quarternary Research* 28: 374-92.

Goodfriend, G.A. and M. Magaritz
1988 Paleosols and Late Pleistocene Rainfall Fluctuations in the Negev Desert. *Letters to Nature* 332: 144-46.

Ghawanmeh, Y.
1995 The Effect of Plague and Drought on the Environment of the Southern Levant during the Late Mamluk Period. In A. Hadidi (ed.), *Studies in the History and Archaeology of Jordan*, II: 315-22. Amman: Department of Antiquities.

Henry, D.O.
1985 Late Pleistocene Environment and Paleolithic Adaptions in Southern Jordan. In A. Hadidi (ed.), *Studies in the History and Archaeology of Jordan*, II: 67-77. Amman: Department of Antiquities.

Henry, D.O. and P.F. Turnbull
1985 Archaeological and Faunal Evidence from Natufian and Timnian Sites in Southern Jordan with Notes on Pollen Evidence. *Bulletin of the American Schools of Oriental Research* 257: 45-64.

Horowitz, A.
1971 Climatic and Vegetational Development in Northeastern Israel during Upper Pleistocene-Holocene Times. *Pollen et Spores* 13: 244-78.
1974 Preliminary Palynological Indications as to the Climate of Israel during the Last 6000 Years. *Paléorient* 2: 407-14.
1978 Human Settlement Patterns in Israel. *Expedition* 20.4: 55-58.
1979 *The Quaternary of Israel*. New York: Academic Press.

Hütteroth, W.-D. and K. Abdulfattah
1977 *Historical Geography of Palestine, Transjordan and Southern Syria in the Late Sixteenth Century*. Frankische geographische Gesellschaft. Erlangen: Palm und Enke.

Issar, A.
1995 Climatic Change and the History of the Middle East. *American Scientist* 83: 350-55.

Issar, A. and H. Bruins
1983 Special Climatological Conditions in the Deserts of Sinai and the Negev during the Latest Pleistocene. *Palaeogeography, Palaeoclimatology, Palaeoecology* 43: 63-72.

Kafafi, Z.
1996 The Environment Impact on the Neolithic Settlement Patterns in Jordan. A paper submitted to the International Seminar 'The Syrian Peninsula: Cultural Heritage and Contacts' held at Deir ez-Zor, Syria, April 22-25.

Marks, A.E.
1977 The Upper Paleolithic Sites of Boker and Boker Tachtit: A Preliminary Report. In A.E. Marks (ed.), *Prehistory and Paleoenvironments in the Central Negev, Israel*, II: 61-80. Dallas: Southern Methodist University.
1983 The Sites of Boker and Boker Tachtit: A Brief Introduction. In A.E. Marks (ed.), *Prehistory and Paleoenvironments in the Central Negev, Israel*, III: 15-37. Dallas: Southern Methodist University.

Marks, A.E. (ed.)
1976 *Prehistory and Paleoenvironments in the Central Negev, Israel*, I. Dallas: Southern Methodist University.

el-Moslimany, A.P.
1994 Evidence of Early Holocene Summer Precipitation in the Continental Sequences in Israel. In Bar-Yosef and Kra 1994: 121-30.

Nicholson, E. and H. Flohn
1980 African Environmental and Climatic Changes and the General Atmospheric Circulation in Late Pleistocene and Holocene. *Climatic Change* 2: 313-48.

Rosen, A.M.
1986 *Quaternary Stratigraphy and Paleoenvironments of the Shephela, Israel*. Geological Survey of Israel, Reports 25/86. Jerusalem: Geological Survey of Israel.
1995 The Social Response to Environmental Change in Early Bronze Age Canaan. *Journal of Anthropological Archaeology* 14: 26-44.

Sauer, J.A.
- 1994 A New Climatic and Archaeological View of the Early Biblical Traditions. In M.D. Coogan, J.C. Exum and L.E. Stager (eds.), *Scripture and Other Artifacts: Essays on the Bible and Archaeology in Honor of Philip J. King*, 366-98. Louisville, KY: Westminster/John Knox.
- 1996 The River Runs Dry. *Biblical Archaeology Review* 22/4: 52-55, 64.

Schuldenrein, J.S. and P. Goldberg
- 1981 Late Quaternary Paleo-Environments and Prehistoric Site Distributions in the Lower Jordan Valley: A Preliminary Report. *Paléorient* 7: 57-71.

Schwarcz, H.P. *et al.*
- 1979 Uranium Series Dating of Travertine from Archaeological Sites, Nahal Zin, Israel. *Nature* 277: 558-60.
- 1980 Uranium Series Dating of Archaeological Sites in Israel. *Israel Journal of Earth-Science* 29: 157-65.

Shehadeh, N.
- 1985 The Climate of Jordan in the Past and Present. In A. Hadidi (ed.), *Studies in the History and Archaeology of Jordan*, II: 25-37. Amman: Department of Antiquities.

Weiss, H. *et al.*
- 1993 The Genesis and Collapse of Third Millennium North Mesopotamian Civilization. *Science* 261: 995-1004.

18. Water Supply in Jordan through the Ages

John Peter Oleson

Introduction

Like much of the eastern Mediterranean area, Jordan has a dry climate with an annual rainy season, usually from October through April (Sanlaville 1981; Shehadeh 1985; Dutton *et al.* 1998; MacDonald, 'Climate', in this volume). There is enormous variation in the amount of annual precipitation from one part of Jordan to another; the high northern and western hills have a significant annual rainfall, totaling 495 mm at Irbid, 600 mm at as-Salt, and 488 mm at at-Tayiba but the figures fall off steeply to the south and east, to 70 mm at al-Azraq, 44 mm at Ma'an, and 50 mm at al-'Aqaba. The greater part of Jordan in an average year receives less than 250 mm of precipitation, the lower limit at which drought farming can be undertaken (Natural Resources Authority 1977). Rainfall not only is meagre in the desert region, but it is unevenly distributed in space and time, and the total amount that actually falls in a given area can vary dramatically from year to year (Shehadeh 1985). In addition, the rates of evapo-transpiration are very high. Since local water resources are seldom sufficient to sustain life in this region, settled populations have always had to manipulate the natural hydrology to the best of their ability.

Nothing in ancient Jordan equaled the idyllic circumstances of Egypt and Mesopotamia, where great rivers brought an abundant exogenous water supply that sustained an enormous population (van Laere 1980; Bonneau 1993). Only the Jordan Valley, which in Gen. 13.10 is compared to both Egypt and the Mesopotamian Eden, provided a similar, though smaller, resource. The Jordan river even managed its own inundation, like the Nile (Josh. 3.15), although at an inconvenient time, like the Tigris and Euphrates.

Although the earliest evidence is lost, humans, from at least the early Neolithic period, must have found it advantageous to enhance the flow of springs, to divert the local runoff from precipitation, to conduct and store water, and to tap the flow of occasional or perennial streams. This range of activities was assisted by water works: structures, tools, and procedures designed to manipulate the flow of water for human benefit. The structures involved, depending on the situation and human or technological resources, could be spring houses, cisterns, reservoirs, dams, wells, enhanced runoff fields, agricultural terraces, wadi barriers, conduits, canals, and water-lifting devices. A wide variety of procedures evolved alongside these structures. The importance of a secure water supply and the vulnerability of most water works stimulated the appearance in the ancient Near East of law codes regulating, among other things, the management of water (Smith 1976: 10; Métral and Métral 1982). Such codes formalized age-old, informal rules of ownership, access, and maintenance. The hierarchy of applications usually applied to water in antiquity included drinking, watering livestock, washing, industrial or craft processes, and irrigation. The great importance of water in this desert landscape is reflected in the positive symbolism accorded to flowing water in the Jewish, Christian and Islamic holy books (cf. Jones 1928).

Sources

Source material for the history of water supply in ancient Jordan is varied. The archaeological evidence is the most extensive, including the remains of many types of structures from all periods. Literary sources, such as the Bible (Oleson 1992a) or the works of Greek and Roman technical writers and historians (such as Herodotus, Diodorus of Sicily, Vitruvius, Frontinus, and Pliny the Elder), Moabite, Nabataean, Greek, and Latin inscriptions, and a few papyrus documents, all provide critical data on the design, construction, use, and social context of water works. Although they came from a different cultural background, the Greeks and Romans occupied much of the region from the fourth century BC to the seventh century AD, and the written comments and sophisticated structures they left behind testify to the special character of problems of water supply in the ancient Near East. Unlike the case of Egypt and Mesopotamia, no visual representations survive in Jordan of such structures or implements in use.

For the most part, the development of ancient water-supply technology in the region was incremental in character, successive cultures building on the accomplishments of their predecessors. Naturally, the developments recorded for sites in modern Jordan often were part of a larger regional cultural expression that can only be touched upon here.

Springs

What were the beginnings of hydraulic technology? Most city sites in the ancient Near East were located over or near springs, but the earliest stages of development have nearly always been lost (cf. Drower 1954: 525). Pools of water attract significant human and animal traffic, and unimproved springs soon become muddy and polluted. This condition, a symbol of wickedness in Prov. 25.26, concerned Roman writers on water supply, who emphasize the need to keep springs and wells free of reeds, vegetation, and mud (Vitruvius 8.4.2; Pliny, *Nat.* 31.36-37; Frontinus, *Aq.* 2.90). One common procedure was excavation of a small pool in the marsh below a spring to receive the flow and keep the surrounding area dry. Although there is no evidence for improvement of the spring at al-Azraq, Palaeolithic and Neolithic strata around this and other springs in the region contain significant quantities of lithic material and the bones of game animals, testifying to the attractions of the location (Garrard et al. 1985). Settlement remains around 'Ayn Jamam in Edom indicate that this spring also served as a centre for Neolithic occupation in the region (Waheeb 1996).

As urban centres developed in the Early Bronze Age, the ground level around such springs gradually rose, the walls of the original pool were reinforced with stone blocks, and steps were provided to facilitate access to the water. This sort of structural protection, including paving and the provision of basins, was repeated countless times throughout the Near East. At Tall as-Sa'idiyya, by the Late Bronze Age the spring issued from the bottom of a walled pit well below ground level, but was accessible from the top of the tall by means of a monumental stone stairway (Miller 1988).

Spring houses appeared later to collect the outflow, protect it from the sun and pollution, and to provide convenient access. Paving and runoff channels were essential. The use of carefully engineered conduits allowed freer placement of the spring house, often at some distance from the source. There are two ornate fountains of the second century BC at the Hasmonean palace of 'Iraq al-Amir, where water poured through the mouths of panthers cut in low relief in stone slabs set into the facade (Lapp 1993: 647). In the first century

Figure 18.1. 'Ayn Shelaleh conduit (photo: J. P. Oleson).

AD the Nabataean King Rabb'ell II reworked the 'Ayn Shelaleh, connected it to an aqueduct system serving a nearby sanctuary, and left an inscription recording his accomplishments (Figure 18.1) (Savignac 1933: 407-11). We know little about the administration of these water sources, but a Nabataean inscription from Khirbat at-Tannur of 8/7 BC mentions an official in charge of a nearby spring: 'resh 'Ain La'ban' (Glueck 1965: 512-13). Greek and Roman urban planners built architecturally impressive facades for fountain houses. The most elaborate structure of this type was the second-century AD nymphaeum at Gerasa, combining two stories of rich architectural decoration with sculpture, jets of water, and a large pool (Browning 1982: 143-47) (Figure 18.2). Another has been identified at Jericho (Dorrell 1993).

Wells

The settled, agricultural character of Neolithic life meant that communities could no longer move to ensure their water supply, as they had once followed

Figure 18.2. Gerasa nymphaeum, view (photo: J.P. Oleson).

animal and plant food sources. The concomitant restriction of pastoral nomadic groups to increasingly marginal land meant that they, too, were hard-pressed to find sufficient water (Helms 1982). If springs were not accessible, the easiest method of obtaining water was to tap an aquifer through excavation. The presence of wells may explain the flourishing early Neolithic herding and agricultural communities at al-Bayda and Wadi Dhobai (Miller 1980: 331-33). The earliest wells must have been irregular, unimproved pits dug in wadi beds. Early improvements included the provision of a wood, stone, or baked-brick lining to prevent collapse of the shaft, construction of a well-head to keep people and animals from falling in, a drinking trough for animals, and possibly shelter against the sun.

Vitruvius (8.1.1-5, 6; 6.12-13; cf. Pliny, *Nat.* 31.43-49) provides detailed instructions on the selection of a suitable site for a well and the proper procedures for safe excavation and lining of the shaft. These procedures are based on long experience, and many must already have been developed in the Neolithic Near East; they involve the observation of topography, soil, surface moisture, and vegetation. Because of the difficulty of digging wells and the importance of their water, this type of supply was generally privately owned and access guarded jealously. During the exodus, Moses asked for permission for his followers to travel through Edom on the King's Highway, first promising the local king that they would not drink water from any well, then (more realistically) offering to pay for any water the people or livestock consumed (Num. 20.17-19; cf. Deut. 2.27-28).

Once found, water had to be lifted from wells by means of buckets or jars on ropes. Most wells in the Middle East in antiquity were provided only with a sturdy well-head or curb with circular opening through which the container was lowered directly on a rope. In drawing up the heavy, full container, the rope often was allowed to rub on the sides of the stone curb, leaving sets of abrasion grooves. Pulley wheels or rollers were known after the ninth century BC as aids for lifting water from wells or deep cisterns (van Laere 1980: 33), but the windlass with crank is not attested until the early Mediaeval period.

Cisterns

Even where wells and springs were available, cisterns were also needed to provide emergency drinking water or low-quality water for irrigation and craft processes. In some households, cistern water also replaced well or spring water that was difficult of access, controlled by hostile families, or too expensive. The authorities could also become involved. In the inscription on the Moabite Stone, set up in Dhiban around 830 BC, King Mesha commemorates his regulation of this aspect of the local water-supply system: 'Since there was no cistern inside the city ... I said to all the people,

"Each of you make himself a cistern in his house!'" (Ullendorff 1958). Cisterns are mentioned frequently in the Egyptian papyri of the Hellenistic and Roman periods, primarily in the context of irrigation, but also in use for stock-raising and various craft processes (Bonneau 1993). Some of the recently discovered Petra papyri treat similar concerns, mentioning agricultural plots fitted out with cisterns and water channels to allow irrigation (Koener 1996; Gagos and Frösén 1998).

Even in extremely arid regions of Jordan, runoff water from local precipitation constituted a substantial resource that was often tapped for filling cisterns. Bedrock or stony slopes could be altered by removing or piling up stones in order to increase runoff to small fields, cisterns and reservoirs. The Nabataeans were particularly adept at this type of water-harvesting (Evenari et al. 1982: 95-119, 127-47), and it is mentioned in a Dedanite inscription from Wadi Ramm (Farès-Drappeau 1995). In the Negev, the runoff fields on the hills above sometimes were enhanced by heaping up the small stones on the surface in long rows or regular patterns of piles (Evenari et al. 1982: 127-47). This technique is rare at Nabataean sites in Jordan, but it can be seen at Hawara/Humayma (Figure 18.3) (Oleson 1986: 258, pl. 46.2). The practice probably developed much earlier. In 2 Kgs 3.25 the Israelites are said to destroy fields in the hill country of Moab by piling stones on them—possibly a reference to interference with runoff fields, since the action is linked with the stopping up of springs. Where runoff was not available, even very large cisterns could be filled by means of donkey-trains or porters, as seems to have been the case for some of the Herodian cisterns at Machaerus and Masada (Netzer 1993: 975).

The numerous cisterns of varying periods and designs that surround most Near Eastern sites testify to the importance of this type of water source. Ancient examples often have remained in use up to the present. Built cisterns were not only less common but also less likely to survive. Cutting the cisterns into the bedrock both saved on building materials and ensured the stability of the structure. Where possible, as at the Edomite mountain stronghold of Jabal al-Qseir near Petra (Lindner et al. 1996) or at the Muqawwar cascades in the Hisma (Jobling 1989), natural basins were used to collect water, with or without modification. Since the limestone, sandstone, gravel, or sand constituting most of the landscape of Jordan are very porous, the application of an impermeable clay or plaster lining facilitated the construction of waterproof cisterns and consequently the spread of habitation to locations distant from springs and wells. At Tall 'Ai, the early Iron Age inhabitants cut their cisterns in deposits of a chalk that became impermeable automatically when wet (Callaway 1993: 45).

The simplest and earliest form of cistern in the entire ancient Near East is bottle-shaped: a deep, round, rock-cut reservoir that tapers upwards to a small entrance hole. This design and its variations, made waterproof with clay or plaster, appears in the Early Bronze Age at Bab adh-Dhra' and Ta'annek (Miller 1980: 337-38), in the Middle Bronze Age at Hazor, and down into the Iron Age at many other sites, such as as-Sela' (Glueck 1939: 26-29). The Augustan historian Diodorus Siculus describes the same type of cistern in use among the Nabataeans of his time (19.94.6-8). At Petra, runoff was the only possible source of water for many areas of the settlement, and in consequence there are numerous cisterns around the hilly site (al-Muheisen and Tarrier 1997). There are so many cisterns around the Iron Age settlement on the high acropolis that the Bedouin named it Umm al-Biyara, 'Mother of Cisterns'.

During the Hellenistic and Roman periods, cisterns tended to take on a regular square or rectangular shape that allowed larger dimensions and easier excavation and access. The plaster lining frequently was made more durable by the addition of crushed potsherds. A design first made popular in this region by the Nabataeans was a deep square or rectangular basin cut in the rock or built of blocks and roofed with long flat slabs carried on arches that crossed the cistern on its short axis. This is a design well suited to the treeless regions of the desert. It was probably borrowed from the Greeks, since examples from the Hellenistic period appear at Delos and Klaros, visited by Nabataean merchants (Oleson 1995: 717). Nabataean examples can be seen in large numbers at Petra, Hawara/Humayma (Figure 18.4), and Umm al-Jimal (Kennedy 1998: 67). At Jewish sites, special cisterns, filled at least in part with rain water that had not been dipped or poured in a broken stream, were used as *mikva'ot*, or ritual baths. Cisterns of this type, provided with a stairway to allow immersion, have been found in large numbers at Jerusalem, Qumran, and Herod's palaces at Herodion and Machaerus.

Despite its importance as a reserve supply, cistern water had a poor reputation among the Greeks and Romans with regard to taste, clarity and healthiness (e.g. Pliny, *Nat.* 31.34). Judging from the relative lack of complaint in biblical texts, ancient cultures in the

Figure 18.3. Humayma: run-off field with stone piles (photo: J.P. Oleson).

Figure 18.4. Humayma: cistern with arch and slab roof (photo: J.P. Oleson).

Middle East had to be less fastidious about their water, although Jer. 2.13 implies the superiority of spring water. Most ancient cisterns were supplied with a settling tank, but the sudden intensity and infrequent occurrence of rainstorms in the Middle East meant that significant amounts of sediment could enter a cistern each year. The provision of a roof over open tanks prevented further pollution, the proliferation of animal and plant life, and loss of the contents through evaporation. In contrast to the supply of water from springs and wells, that from cisterns is frequently associated in biblical passages with agricultural activities as well as stock-raising (e.g. Deut. 6.10-11; 2 Kgs 18.31; Isa. 36.16; Neh. 9.25; 2 Chron. 26.10; and Eccl. 2.4-6).

Terracing

Runoff water could be held back by terraces and wadi barriers so that it would sink into the soil and sustain crops. At least by the Early Bronze Age, terraces appeared in the hilly areas of Jordan at sites such as Tall al-Handaquq (Mabry 1989), and this approach to soil and water conservation has continued in use up to the present, making it difficult to date terracing accurately (Hopkins 1997: 27-28). Plato (*Laws* 761 a-b) describes the practice and its long-term effect on the landscape. The Nabataeans were particularly skilled at terracing, and farms of this type are common throughout Jordan (e.g. near Umm al-Jimal; Kennedy 1998: 67-80).

Reservoirs and dams

Reservoirs appear in the archaeological record as early as cisterns. While cisterns tended to be small and private, reservoirs were large, usually unroofed pools that stored quantities of water for public use or for applications that required significant volume. Because of their size, reservoir support walls usually were at least partly built rather than cut in the rock (Figure 18.5). Reservoirs might be filled by springs, as at Birketein near Gerasa, or by runoff water. In the late fourth millennium BC, settlers at Jawa built a system of reservoirs with a volume of 75,000 cubic metres, designed to be filled by runoff. The water was meant for human consumption, for livestock, and for irrigation in an otherwise almost waterless region (Helms 1981). A similar system appeared in the Early Bronze Age at Tall al-Handaquq, where long dams built of boulders both diverted wadi flow and shaped a pool 1.5 ha in area (Mabry 1989). The inscription on the Moabite Stone from Dhiban mentions that King Mesha himself built 'the retaining walls of the reservoir ... inside the city' (Ullendorff 1958). Numerous enormous rain-fed reservoirs were built in Jordan by the Nabataeans and Romans: for example, at Dayr al-Kahf, Umm al-Quttayn, Umm al-Jimal, Qasr Hallabat, al-Jiza, Petra, Hawara/Humayma and Quweira (Kennedy 1995; al-Muheisen and Tarrier 1997).

Nabataean and Roman dams were built of cut stone blocks or mortared rubble. Very substantial Nabataean dams for diverting or pooling the flow of wadis survive at Petra (Parr 1967; Lindner 1987) and Hawara/Humayma (Figure 18.6) (Oleson 1991a). Large Roman dams, sometimes based on Nabataean predecessors, were built along the Arabian frontier to provide drinking and irrigation water for forts and settlements, as at Qasr Bshir and al-Lajjun (de Vries 1987). The chronology and function of a large dam at Wadi al-Jilat is uncertain (Politis 1993).

Conduits

The earliest water conduits undoubtedly were simple earth channels leading from a spring or stream to an irrigated field. As long as the soil is not too porous and the stream velocity not high enough to cause serious erosion, earth conduits are satisfactory for most purposes. Even the great irrigation systems of Egypt and Mesopotamia depended on earth conduits (van Laere 1980; Bonneau 1993). At Jawa in the late fourth millennium BC, the conduits of the urban water system supplied by runoff were excavated in the stony soil (Helms 1981). Much later, at Nabataean and Byzantine Umm al-Jimal, the main water-supply conduit was a simple earth channel (de Vries 1998: 93, 99).

The main drawbacks of water conduits of more than local significance are the cost of labour and materials required for their construction, the need for constant maintenance, and their vulnerability. These characteristics meant that in the pre-Roman Middle East, as in pre-Roman Greece, long aqueducts were extremely rare. The small and vulnerable states of this period simply did not have the resources or stability to support aqueducts on the scale later made customary by the Romans. The Romans built aqueduct systems at Jarash, Umm Qays (Kerner *et al.* 1997), and many other cities in the region. Such water-supply systems were expensive but could completely change the natural hydrology of a city.

Where erosion or the porosity of the soil was a problem, conduits were built of stone slabs, waterproofed with clay or mortar, or constructed from a series of blocks carved with a longitudinal channel, set end to end. Both types are found already in the fourteenth century BC at Ugarit, and they continued to appear throughout the Near East into the Late Roman period. This type of conduit is characteristic of Nabataean water-supply systems varying in scale from a single house to a whole settlement. At Hawara/Humayma in the Hisma, 26 km of conduit carried the water from three springs to the habitation centre (Eadie and Oleson 1986) (Figure 18.7). Alternatively, a conduit

Figure 18.5. Petra: reservoir (photo: J.P. Oleson).

Figure 18.6. Humayma: dam (photo: J.P. Oleson).

could be built of mortared rubble covered with a layer of waterproof plaster, a design in use throughout the Middle East from the early Iron Age on. Particularly extensive systems have been found serving the Hasmonean and Herodian palaces at Jericho and Machaerus (Netzer 1977), at as-Sadeh (Lindner *et al.* 1990: 215), and at al-Lajjun (de Vries 1987).

Short ceramic pipelines appear during the late second millennium BC in Mesopotamia, and by the Late Bronze Age the innovation had spread throughout the eastern Mediterranean. Such pipelines enjoyed great popularity because they were very efficient, inexpensive to manufacture, relatively easy to lay, and provided a clean water supply with the added

Figure 18.7. Humayma: aqueduct conduit (photo: J.P. Oleson).

convenience and security of a subterranean course. Even in early imperial Rome, where lead pipes had become common, Vitruvius notes that terracotta pipelines were preferable for the purity of their water and the ease of repair (8.6.10-11). A long Nabataean terracotta pipeline carried water through the Siq at Petra.

Water in open conduits can flow only downhill, while tightly sealed metal, stone, or terracotta pipelines can be pressurized to bring the water back up almost as high as its source. As long as the pipe sections and joints could withstand the increase in pressure at the low points, this procedure eliminated the need for long detours around valleys and made it possible to provide a supply of running water to an isolated acropolis. Pressurized descending sections (termed invented siphons) appeared during the Roman period in the water-supply systems of Jerusalem (Issar 1974) and Susita. The technology is described by Vitruvius (8.6.6-9). Small-scale pressurized systems also allow the use of valves to turn the water on and off, but only one typically Roman bronze stop-cock has been found in Jordan, at Hawara/Humayma (Figure 18.8) (Oleson 1988: 163, pl. 30).

The *qanat* is an exception to the local scale of water conduits in the pre-Roman Middle East. This system consists of a gently sloping tunnel tapping the aquifer of a debris slope at the foot of a mountain range. The course of the channel and the proper water level were found by sinking a line of vertical access shafts at intervals of 30 to 50 m (Goblot 1979). These shafts allowed excavation to proceed from two directions at once, provided a means for removing debris, and facilitated maintenance. The *qanat* could extend from several kilometers up to 20 or 30 km in length, depending on the depth of the source, surface topography, and the distance from source to intended point of use. The *qanat* was probably developed in Urartu, where their use is alluded to for the first time in the late eighth century BC by Sargon II, who was busy destroying them (Forbes 1964: 156-57; Goblot 1979). The devices became characteristic of Persia and spread probably from there westward, appearing in the Jordan Valley (Ionides 1939: 349; Glueck 1951: 361-63; Kennedy and Riley 1990: 70-76), at Umm al-Jimal (Kennedy 1998: 67), Udhruh (Graf 1990: 137; Killick 1987), and at Qasr al-Mushash (Homès-Frederick and Hennessy 1989: 391-96) in the Roman or early Islamic period. The fact that they were cut in the subsoil rather than built protected *qanats* to an extent from the problems of materials and vandalism that affected conventional aqueducts, but they did require constant maintenance.

Water-lifting devices

The only water-lifting device in use before the Hellenistic period was the *shaduf*, a swing beam with counterweight at the short end and a pole or rope suspended from the other to lift a bucket or skin bag from a well, cistern, river or water channel. This simple but effective device first appears in Mesopotamian art in the mid-third millennium BC and in Egyptian art and literature slightly later. It is still in use today essentially unchanged. Irrigation accounts in ancient archives throughout the Middle East record its use in lifting water for irrigation, livestock, craft processes, and human consumption (van Laere 1980; Drower 1954: 522-25; Oleson 1984, 2000).

The other mechanical water-lifting devices of the ancient eastern Mediterranean region were probably invented by scholars or engineers associated with the Museum in Ptolemaic Alexandria: the wheel with compartmented rim or body, the water-screw,

Figure 18.8. Humayma: Roman bronze stopcock (photo: J.P. Oleson).

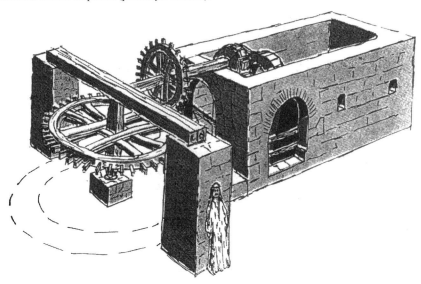

Figure 18.9. Qusayr 'Amra: reconstruction of *saqiya* arrangement (Schiøler 1973: 94, fig. 64) (reproduced with the kind permission of Odense University Press).

the bucket-chain and the force pump (Oleson 1984: 291-385, 2000). The water-screw, a rotating tube with interior helix of wooden or copper vanes to lift the water, is said to have been invented by Archimedes in the late second century BC to solve certain irrigation problems in the Nile delta. All of these devices except the water-screw could be driven by animals by means of a large angle gear device now called a *saqiya*. Although it is likely that most of these water-lifting devices were used in ancient Jordan, particularly to assist irrigation in the Jordan Valley, the only evidence testifying to their presence is an eighth-century AD *saqiya*-driven bucket-chain installation at Qusayr 'Amra (Figure 18.9) (Schiøler 1973: 92, 148, 155).

References

Bonneau, D.
   1993   *Le régime administratif de l'eau du Nil dans l'Egypte grecque, romaine et byzantine*. Leiden: E.J. Brill.

Browning, I.
   1982   *Jerash and the Decapolis*. London: Chatto & Windus.

Callaway, J.A.
1993 Ai. In E. Stern (ed.), *New Encyclopedia of Archaeological Excavations in the Holy Land*, I: 39-45. New York: Simon & Schuster.

Dorrell, P.G.
1993 The Spring at Jericho from Early Photographs. *Palestine Exploration Quarterly* 125: 95-115.

de Vries, B.
1987 The el-Lejjun Water System. In S.T. Parker (ed.), *The Roman Frontier in Central Jordan*, part 1: 399-428. British Archaeological Reports, International Series 340 (i). Oxford: British Archaeological Reports.

Drower, M.S.
1954 Water-Supply, Irrigation, and Agriculture. In C. Singer (ed.), *History of Technology*, I: 520-57. New York: Oxford University Press.

Dutton, R.W., J.I. Clarke and A. Batticki (eds.)
1998 *Arid Land Resources and their Management: Jordan's Desert Margin*. London: Kegan Paul.

Eadie, J.W. and J.P. Oleson
1986 The Water-Supply Systems of Nabataean and Roman Humayma. *Bulletin of the American Schools of Oriental Research* 262: 49-76.

Evenari, M., L. Shanan and N. Tadmor
1982 *The Negev: The Challenge of a Desert*. 2nd edn. Cambridge, MA: Harvard University Press.

Farès-Drappeau, S.
1995 L'inscription de type dedanite de Abu ad-Diba'/Wadi Ramm. Une nouvelle lecture. *Annual of the Department of Antiquities of Jordan* 39: 493-97.

Forbes, R.J.
1964 *Studies in Ancient Technology*, I. Leiden: E.J. Brill.

Gagos, T. and J. Frösén
1998 Petra Papyri. *Annual of the Department of Antiquities of Jordan* 42: 473-83.

Garrard, A. *et al.*
1985 The Environmental History of the Azraq Basin. In A. Hadidi (ed.), *Studies in the History and Archaeology of Jordan*, II: 109-15. Amman: Department of Antiquities.

Glueck, N.
1939 *Explorations in Eastern Palestine*, III. Annual of the American Schools of Oriental Research 18-19 (1937–1939). New Haven, CT: American Schools of Oriental Research.
1951 *Explorations in Eastern Palestine*, IV. Annual of the American Schools of Oriental Research 25-28 (1945–1949). New Haven, CT: American Schools of Oriental Research.
1965 *Deities and Dolphins: The Story of the Nabataeans*. New York: Farrar, Straus & Giroux.

Goblot, H.
1979 *Les Qanats: Une technique d'aquisition de l'eau*. Paris: Mouton.

Graf, D.
1990 Arabia during Achaemenid Times. In H. Sancisi-Weerdenburg and A. Kuhrt (eds.), *Achaemenid History. IV. Centre and Periphery*, 131-48. Leiden: Nederlands Instituut voor het Nabije Oosten.

Helms, S.W.
1981 *Jawa: Lost City of the Black Desert*. Ithaca, NY: Cornell University Press.
1982 Paleo-Beduin and Transmigrant Urbanism. In A. Hadidi (ed.), *Studies in the History and Archaeology of Jordan*, I: 97-113. Amman: Department of Antiquities.

Homès-Fredericq, D. and J.B. Hennessy (eds.)
1989 *Archaeology of Jordan. II. Field Reports: Surveys and Sites*. Akkadica Supplementum 7-8. Leuven: Peeters.

Hopkins, D.C.
1997 Agriculture. In E. Meyers (ed.), *Oxford Encyclopedia of Archaeology in the Near East*, I: 22-30. New York: Oxford University Press.

Ionides, M.G.
1939 *Report on the Water Resources of Transjordan and their Development*. London: Government of Transjordan.

Issar, A.
1974 Ancient Jerusalem: Its Water Supply and Population. *Palestine Exploration Quarterly* 106: 33-51.

Jobling, W.J.
1989 Aqaba-Ma'an Archaeological and Epigraphical Survey. In D. Homès-Fredericq and J.B. Hennessy (eds.), *Archaeology of Jordan. II/1. Field Reports: Surveys and Sites A–K*, 16-24. Akkadica Supplementum 7. Leuven: Peeters.

Jones, T.J.
1928 *Quelle, Brunnen und Zisterne im alten Testament*. Morgenländische Texte und Forschungen 1.6. Leipzig: Pfeiffer.

Kennedy, D.L.
1995 Water Supply and Use in the Southern Hauran, Jordan. *Journal of Field Archaeology* 22: 275-90.
1998 The Area of Umm el-Jimal: Maps, Air Photographs, and Surface Survey. In B. de Vries (ed.), *Umm el-Jimal: A Frontier Town and its Landscape in Northern Jordan*, I: 39-90. Portsmouth, RI: JRA.

Kennedy, D.L. and D. Riley
1990 *Rome's Desert Frontier from the Air*. Austin: University of Texas.

Kerner, S., H. Krebs and D. Michaelis
1997 Water Management in North Jordan: The Example of Gadara Umm Quays. In G. Bisheh *et al.* (eds.), *Studies in the History and Archaeology of Jordan*, VI: 265-70. Amman: Department of Antiquities.

Killick, A. (ed.),
1987 *Udhruh: Caravan City and Desert Oasis*. Romsey, Hampshire: A.C. Killick.

Koener, L.
1996 The Carbonized Archive from Petra. *Journal of Roman Archaeology* 9: 177-88.

Lapp, P.
1993 'Iraq el-Emir. In E. Stern (ed.), *New Encyclopedia of Archaeological Excavations in the Holy Land*, II: 646-49. New York: Simon & Schuster.

Lindner, M.
1987 Nabatäische Talsperren. In G. Garbrecht (ed.), *Historische Talsperren*, 147-74. Stuttgart: Wittwer.

Lindner, M. *et al.*
1990 Es-Sadeh—A Lithic-Early Bronze-Iron II (Edomite)-Nabataean Site in Southern Jordan: Report on the Second Exploratory Campaign, 1988. *Annual of the Department of Antiquities of Jordan* 34: 193-237.
1996 Jabal al-Qseir: A Fortified Iron II (Edomite) Mountain Stronghold in Southern Jordan. *Annual of the Department of Antiquities of Jordan* 40: 137-66.

Mabry, J.
1989 Investigations at Tell el-Handaquq, Jordan (1987–88). *Annual of the Department of Antiquities of Jordan* 33: 59-95.

Métral, F. and J. Métral (eds.),
1982 *L'Homme et l'eau en Méditerranée et au Proche Orient. II. Aménagements hydrauliques, état et législation*. Lyon: Maison de l'Orient.

Miller, R.
1980 Water Use in Syria and Palestine from the Neolithic to the Bronze Age. *World Archaeology* 11: 331-41.
1988 The Water System (Tell es-Sa'idiyeh). *Levant* 20: 84-88.

al-Muheisen, Z. and D. Tarrier
1997 Ressources naturelles et occupation du site de Pétra. In G. Bisheh *et al.* (eds.), *Studies in the History and Archaeology of Jordan*, VI: 143-48. Amman: Department of Antiquities.

Natural Resources Authority
1977 *National Water Master Plan of Jordan*. Amman: Natural Resources Authority, in cooperation with the German Agency for Technical Cooperation, Federal Republic of Germany.

Netzer, E.
1977 The Winter Palaces of the Judean Kings at Jericho at the End of the Second Temple Period. *Bulletin of the American Schools of Oriental Research* 228: 1-13.
1993 Masada. In E. Stern (ed.), *New Encyclopedia of Archaeological Excavations in the Holy Land*, III: 973-85. New York: Simon & Schuster.

Oleson, J.P.
1984 *Greek and Roman Mechanical Water-Lifting Devices*. Toronto: University of Toronto.
1986 Humayma Hydraulic Survey: Preliminary Report of the 1986 Season. *Annual of the Department of Antiquities of Jordan* 30: 253-59.
1988 Humayma Hydraulic Survey: Preliminary Report of the 1987 Season. *Annual of the Department of Antiquities of Jordan* 32: 157-69.
1991a Eine nabatäische Talsperre in der Nähe von Humeima (das antike Auara) in Jordanien. In G. Garbrecht (ed.), *Historische Talsperren*, II: 65-72. Stuttgart: Wittwer.
1991b Aqueducts, Cisterns, and the Strategy of Water Supply at Nabataean and Roman Auara (Jordan). In E.T. Hodge (ed.), *Future Currents in Aqueduct Studies*, 45-62. Bristol: Cairns.
1992a Water Works. In D.N. Freedman (ed.), *Anchor Bible Dictionary*, VI: 883-93. New York: Doubleday.
1992b The Water-Supply System of Ancient Auara: Preliminary Results of the Humeima Hydraulic Survey. In G. Bisheh (ed.), *Studies in the History and Archaeology of Jordan*, IV: 269-76. Amman: Department of Antiquities.
1995 The Origins and Design of Nabataean Water-Supply Systems. In K. 'Amr, F. Zayadine and M. Zaghloul (eds.), *Studies in the History and Archaeology of Jordan*, V: 707-19. Amman: Department of Antiquities.
2000 'Irrigation' and 'Water-Lifting'. In O. Wikander (ed.), *Handbook of Ancient Water Technology*, 183-302. Leiden: E.J. Brill.

Parr, P.J.
1967 La date du barrage du Siq à Pétra. *Revue biblique* 74: 45-49.

Politis, K.
1993 The Stepped Dam at Wadi el-Jilat. *Palestine Exploration Quarterly* 125: 43-49.

Sanlaville, P.
1981 Réflexions sur les conditions générales de la quête de l'eau au proche orient. In J. Métral and P. Sanlaville (eds.), *L'Homme et l'eau*, I: 9-21. Lyon: Presses Universitaires.

Savignac, M.R.
1933 Le sanctuaire d'Allat àIram. *Revue biblique* 42: 405-22.

Schiøler, T.
1973 *Roman and Islamic Water-Lifting Wheels*. Odense: Odense University.

Shehadeh, N.
- 1985 The Climate of Jordan in the Past and Present. In A. Hadidi (ed.), *Studies in the History and Archaeology of Jordan*, II: 25-37. Amman: Department of Antiquities.

Smith, N.A.F.
- 1976 *Man and Water: A History of Hydro-Technology*. London: Peter Davies.

Ullendorff, E.
- 1958 The Moabite Stone. In D.W. Thomas (ed.), *Documents from Old Testament Times*, 195-98. London: Nelson.

van Laere, R.
- 1980 Techniques hydrauliques en Mésopotamie ancienne. *Orientalia lovaniensia periodica* 11: 11-53.

Waheeb, M.
- 1996 Ain al-Jammam. *American Journal of Archaeology* 100: 514-16.

19. Pastoralism

Alison Betts

The term 'pastoralism' refers to the practice of grazing or pasturing herds or flocks of domestic animals such as sheep, goats, cattle and camels. Pastoralism is frequently associated with a nomadic way of life since land too arid or poor for agriculture often supports sufficient grazing to sustain herds of animals. In Jordan, however, many settled villagers combine agriculture with the herding of small flocks of sheep and goat. Camel herds, though, need fairly extensive areas of grazing. Since they are more adapted to arid conditions, they are usually kept out in the steppe where they are managed by people who live in tents, and who move with the herds as they seek new pastures.

A somewhat arbitrary line that defines the land suitable for dry farming, normally around the 200–250 mm isohyets, divides Jordan, like much of the Near East. Steppe is classified as land where the rainfall is too low to support dry farming, but sufficient to sustain drought-tolerant vegetation and an annual growth of grasses and flowering plants.

Geology as well as rainfall divides the steppe lands of Jordan. The eastern steppe, the *hammad*, consists largely of open rolling gravel plains with occasional outcrops of limestone. From the oasis of al-Azraq eastwards for roughly 100 km the limestones are overlain by a thick bed of ancient lava, the *harra*, producing a landscape strewn with basalt boulders, cut by deeply incised wadis interspersed with stretches of white mudflats. In the south, the sandstones of the Petra region stretch out to the east, creating the dramatic scenery of Wadi Ramm with its huge cliffs and mesa-like rock formations.

Each area of the steppe lands has its own peculiar features. The *harra*, because of its rough and difficult terrain, was traditionally a place of refuge and a home for brigands. Its deep wadis hold water well into the dry season and provide shelter from the winter winds for herds of sheep and goat. However, its rocky ground is not very suitable for camels. The open plains of the *hammad* are simpler to traverse on foot or with camels, and east of the *harra* a series of wadi systems drain down into the Euphrates valley, providing 'grazing corridors' along which herders and their flocks could move relatively easily. In the south, the sandstone country provides shelter and limited water supplies for sheep and goat in winter and early summer, while the sandy floors of the canyons are also suited to camels.

Pastoralism is as old as the domestication of animals. By the seventh millennium BC, the first farmers of the Early Neolithic period kept sheep and goats that they probably herded in the uncultivated lands around their settlements. By the later Neolithic period, at the beginning of the sixth millennium BC, there is archaeological evidence for the first nomadic pastoralism in the eastern steppe, where prehistoric peoples combined the herding of sheep and goat with hunting and gathering wild plants (Köhler-Rollefson 1992; Betts 1993a; Garrard *et al.* 1994). The camel was probably first domesticated somewhere in South Arabia or in Iran in the third to second millennium BC (Compagnoni and Tosi 1978), but was not adopted as a major source of livelihood for nomadic pastoralists in North Arabia until the first millennium BC. From then until the recent past, sheep and goat were herded in the good steppe and in uncultivated lands around the villages, while camels were kept mostly in the dry steppe/desert regions. Sheep and goats usually need to be watered at least once a day, particularly in the hotter months, while camels can go for several days without water. Flocks of the former could, therefore, never be herded far from available water sources. Today, however, with the introduction of trucks to the desert areas, water can be transported to remote locations, and sheep and goats have replaced camels as the main herd animals of the eastern steppic regions (Russell 1988; Lancaster and Lancaster 1991).

Archaeological evidence for pastoralism comes from various sources. Nomadic people have little in the way of material possessions and, since they move so frequently, rarely leave much behind by which the archaeologist can identify them. However, the remains of sheep, goat and other herd animals in the faunal record of archaeological sites indicate that their inhabitants probably kept flocks as part of their subsistence basis. The short-term encampments of nomadic pastoralists

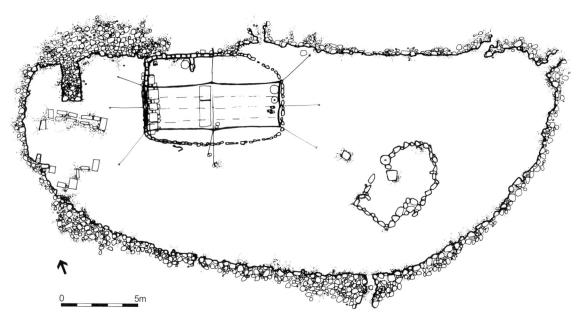

Figure 19.1. Modern bedouin encampment: a Bedul family tent in Petra showing organization of space (from Bienkowski and Chlebik 1991: fig. 16).

leave few traces but, because they frequently live in remote regions, their camps often remain undisturbed for many hundreds of years. A row of stones, an enclosure or a rock-lined hearth may indicate an ancient camp, while a cleared area in the gravel *hammad* with an associated scatter of potsherds shows that a tent was pitched there in antiquity (Simms 1988; Helms 1990; Cribb 1991; Banning and Köhler-Rollefson 1992; Betts 1993b). Some pastoralists, like the Bedul of the Petra region, lived in caves where their material remains are better preserved for the archaeologist (Banning and Köhler-Rollefson 1983; Bienkowski and Chlebik 1991) (Figure 19.1). In addition to archaeological evidence, pastoralists appear in the historical records, often as a source of trouble to the authorities as their mobile lifestyle kept them partially beyond the reach of tax administrators and military control (Matthews 1978; Briant 1982; Eph'al 1984).

Although nomadic herders are hard to identify in the archaeological record, pastoralists and their flocks have always featured as an important element in the economy of the ancient Near East. Nomadic pastoralists depended on urban and village markets to sell the produce of their herds: meat, hides, wool, cheese, butter, cloth and other items. In the same markets, they also purchased metal tools and containers, various foodstuffs, fine cloth and other basic items that they could not produce for themselves. With the introduction of the camel, pastoralists entered the caravan trade, providing transport and taking goods across the desert routes. Because of the uncertain nature of the steppe, where the availability of pasture and rainfall can vary dramatically from place to place and from year to year, nomads have never relied wholly on their flocks and herds, but have always availed themselves of other opportunities to increase their income, provided that the activities involved were in keeping with their mobile lifestyle. Thus young men would often take up arms as mercenaries or as regular troops, or become involved in trade and transport (Barth 1973; Lancaster 1981; Bocco 1984). Among modern pastoralists, social, economic and political organization is bound up with marriage patterns, familial relations and, on a larger scale, with tribal affiliations (Lancaster 1981). It is likely, although not certain, that similar patterns existed in the past. Throughout prehistoric and historic times, the nomadic pastoralists of the Syro/Jordanian steppe lands have shown themselves capable both of individual autonomy and cohesive political action.

The first Neolithic pastoralists in Jordan's eastern steppe left a record of their passing in the form of stone shelters, chert implements and occasionally exchange items such as beads made from exotic stones or marine shells, which showed that they had some form of economic contact with the villagers of the fertile lands to the west (Betts *et al.*1991; McCartney 1992; Garrard *et al.* 1994). Traces of Neolithic pastoral herding camps have been found all across the steppe from the eastern heights of the Jordan Rift Valley to the Euphrates. It is not certain whether these hunter/herders left their villages in the fertile lands to pasture their flocks in the steppe for the winter months or whether they were a separate population who lived an almost exclusively nomadic life. By the Chalcolithic period, the hunter/herders of the eastern steppe were using pottery that had probably been obtained in village markets (Betts 1992; Henry 1995). By this time also, around the fourth to the early third millennium BC, the appearance of walled towns in the fertile zones and up to the margins of the good steppe implies the development of more complex economic networks in which nomadic pastoralists may well have played a role (Helms 1981; Betts 1991). By the early second millennium BC, donkey caravans were transporting goods over long distances across the ancient world (Larsen 1976) and it is possible that nomadic groups were involved in similar trade in Jordan, although no clear evidence has yet come to light on this. A little later, however, by the second half of the first millennium BC, one of the greatest trading peoples of the Orient, the Nabataeans, established the capital of their commercial kingdom at Petra in southern Jordan. The success of the Nabataeans was based on their nomadic heritage and knowledge of the desert routes southwards into Arabia which enabled them to bring spices, incense and other exotic merchandise up to the Mediterranean coast (Glueck 1965: 3-45; Hellenkemper Salies 1981). The extensive historical record for the Hellenistic and Roman periods in Jordan gives a more detailed picture of the role of nomads in the economic and political spheres of the time (Isaac 1990). Jordan's eastern deserts are strewn with rocks bearing inscriptions in a number of languages and scripts, including Nabataean, Safaitic and Thamudic, used primarily by nomadic peoples. The inscriptions and accompanying sketches of camels, wild animals and scenes of warfare and hunting (Figure 19.2) give a tantalizing glimpse into the life of the nomadic herders of the period (Macdonald 1993). In the Byzantine period, the steppic tribes also served a more formal function. The Romans, in their longstanding conflict with the Persians, had neither the resources nor the wish to fight in the desert and so employed the Ghassanids to protect their eastern borders in the sixth and early seventh centuries AD. The Persians (Sassanians), in similar fashion, enlisted the support of nomads of the Euphrates region, the Lakhmids, so that the imperial frontier battles were fought largely by client tribesmen (Nöldeke 1887;

Figure 19.2. Safaitic carving from eastern Jordan (photo: Alison Betts).

Sartre 1982; Helms 1990). As these imperial struggles drew to a close in the first half of the seventh century AD, a new movement was developing in the south, the power of the Islamic faith, which was to be carried by Arabs, many of nomadic pastoralist heritage, across vast portions of the ancient world (Hitti 1970; Glubb 1978).

References

Banning, E.B. and I. Köhler-Rollefson
- 1983 Ethnoarchaeological Survey in the Beidha Area, Southern Jordan. *Annual of the Department of Antiquities of Jordan* 27: 375-83.
- 1992 Ethnographic Lessons for the Pastoral Past: Camp Locations and Material Remains near Beidha, Southern Jordan. In O. Bar-Yosef and A. Khazanov (eds.), *Pastoralism in the Levant: Archaeological Materials in Anthropological Perspectives*, 181-204. Monographs in World Archaeology 10. Madison, WI: Prehistory Press.

Barth, F.
1973 A General Perspective on Nomad-Sedentary Relations in the Middle East. In C. Nelson (ed.), *The Desert and the Sown: Nomads in the Wider Society*, 11-22. Berkeley: University of California, Institute of International Studies.

Betts, A.V.G.
1992 Tell el-Hibr: A Rock Shelter Occupation of the Fourth Millennium B.C.E. in the Jordanian Badiya. *Bulletin of the American Schools of Oriental Research* 287: 5-23.
1993a The Neolithic Sequence in the East Jordan Badia: A Preliminary Overview. *Paléorient* 19/1: 43-53.
1993b The Burqu'/Ruwayshid Project: Preliminary Report on the 1991 Field Season. *Levant* 25: 1-11.

Betts, A.V.G. (ed.)
1991 *Excavations at Jawa 1972–1986*. Edinburgh: Edinburgh University Press.

Betts, A. et al.
1991 The Burqu'/Ruweishid Project: Preliminary Report on the 1989 Field Season. *Levant* 23: 7-28.

Bienkowski, P. and B. Chlebik
1991 Changing Places: Architecture and Spatial Organisation of the Bedul in Petra. *Levant* 23: 147-80.

Bocco, R.
1984 Nomadismo pastorale e società beduina in Giordania: Un orientamento bibliografico. *Studi per l'ecologia del quaternario (Italy)* 6: 129-34.

Briant, P.
1982 *Etat et pasteurs au Moyen-Orient ancien*. Editions de la Maison des sciences de l'homme. Cambridge: Cambridge University Press.

Compagnoni, B. and M. Tosi
1978 The Camel: Its Distribution and State of Domestication in the Middle East During the Third Millennium B.C. in the Light of Finds from Shar-i Sokhta. In R.H. Meadow and M.A. Zeder (eds.), *Approaches to Faunal Analysis in the Middle East*, 91-103. Peabody Museum Bulletin 2. Cambridge, MA: Peabody Museum.

Cribb, R.L.D.
1991 Mobile Villagers: The Structure and Organisation of Nomadic Pastoral Campsites in the Near East. In C.S. Gamble and W.A. Boismier (eds.), *Ethnoarchaeological Approaches to Mobile Campsites*, 371-94. International Monographs in Prehistory, Ethnoarchaeological Series 1. Ann Arbor: International Monographs in Prehistory.

Eph'al, I.
1984 *The Ancient Arabs*. Jerusalem: Magnes Press.

Garrard, A. et al.
1994 Prehistoric Environment and Settlement in the Azraq Basin: An Interim Report on the 1987 and 1988 Excavation Seasons. *Levant* 26: 73-109.

Glubb, J.B.
1978 *A Short History of the Arab Peoples*. London: Quartet.

Glueck, N.
1965 *Deities and Dolphins*. London: Cassell.

Hellenkemper Salies, G.
1981 *Die Nabatäer: Erträge einer Ausstellung im Rheinischen Landesmuseum Bonn, 24. Mai–Juli 1978*. Bonn: Habelt.

Helms, S.W.
1981 *Jawa: Lost City of the Black Desert*. Ithaca, NY: Cornell University Press.
1990 *Early Islamic Architecture of the Desert*. Edinburgh: Edinburgh University Press.

Henry, D.O.
1995 *Prehistoric Cultural Ecology and Evolution: Insights from Southern Jordan*. New York: Plenum.

Hitti, P.K.
1970 *History of the Arabs from the Earliest Times to the Present*. 10th edn. London: Macmillan.

Isaac, B.
1990 *The Limits of Empire*. Oxford: Clarendon Press.

Köhler-Rollefson, I.
1992 A Model for the Development of Nomadic Pastoralism on the Transjordanian Plateau. In O. Bar-Yosef and A. Khazanov (eds.), *Pastoralism in the Levant: Archaeological Materials in Anthropological Perspectives*, 11-18. Monographs in World Archaeology 10. Madison, WI: Prehistory.

Lancaster, W.
1981 *The Rwala Bedouin Today*. Cambridge: Cambridge University Press.

Lancaster, W. and F. Lancaster
1991 Limitations on Sheep and Goat Herding in the Eastern Badia of Jordan: An Ethno-archaeological Enquiry. *Levant* 23: 125-38.

Larsen, M.T.
1976 *The Old Assyrian City-State and its Colonies*. Copenhagen Studies in Assyriology 4. Copenhagen: Akademisk Forlag.

Macdonald, M.C.A.
1993 Nomads and the Hawran in the Late Hellenistic and Roman Periods: A Reassessment of the Epigraphic Evidence. *Syria* 70: 303-413.

Matthews, V.
   1978   *Pastoral Nomadism in the Mari Kingdom (ca. 1830–1760)*. American Schools of Oriental Research Dissertation Series 3. Cambridge, MA: American Schools of Oriental Research.

McCartney, C.
   1992   Preliminary Report of the 1989 Excavations at Site 27 of the Burqu'/Ruweishid Project. *Levant* 24: 33-54.

Nöldeke, T.
   1887   *Die Ghassanischen Fürsten aus dem Hause Gafna's*. Berlin: Königlichen preussischen Akademie der Wissenschaften zu Berlin.

Russell, K.W.
   1988   *After Eden: The Behavioural Ecology of Early Food Production in the Near East and North Africa*. British Archaeological Reports, International Series 391. Oxford: British Archaeological Reports.

Sarte, M.
   1982   *Trois études sur l'Arabie romaine et byzantine*. Collection Latomus 178. Brussels: Latomus.

Simms, S.
   1988   The Archaeological Structure of a Bedouin Camp. *Journal of Archaeological Science* 15: 197-211.

20. Traditional Agriculture

Carol Palmer

> We journeyed through wide valleys, treeless, uninhabited, and almost uncultivated... A generation or two hence it will be deep in corn and scattered over with villages... I shall not be there to see. In my time the uplands will still continue to be that delectable region of which Omar Khayyam sings: 'the strip of herbage strown that just divides the desert from the sown'... (Bell 1907: 23).

Gertrude Bell, the famous traveller, passed through the land that is now modern Jordan, criss-crossing between cultivated and desert lands. Archaeological evidence demonstrates the extent to which cultivation of these lands has changed through time. At the time Bell was travelling through what was then Ottoman Syria, government authority was re-establishing itself and the social and political conditions for a settled agricultural mode of existence were becoming increasingly favourable and, at the same time, improving conditions for travellers, Hajj pilgrims and traders alike. The archaeological monuments Bell recorded, such as those in the Balqa' district to which the above quote relates, demonstrate that communities once thrived beyond the then limits of agriculture. Today, that 'strip of herbage' is, indeed, 'scattered over with villages' and even some 'desert' regions are 'sown'.

Over 90 percent of contemporary Jordan is considered too arid for rain-fed cereal production (Duwayri 1985; el-Hurani 1989). Even in areas suitable for agriculture, annual rainfall is highly variable such that yields cannot be assured. Agriculture involves risk, the chance of failure, and uncertainty such that farmers do not really know in advance whether years will be good or bad (Halstead and O'Shea 1989). Through time, good yields have been closely correlated with abundant rainfall (Arabiat et al. 1983). Lower annual average rainfall is associated with greater chances of failure. In areas receiving in the range of 200-500 mm average annual precipitation, rainfall varies on average between 25 and 50 percent per year. Between 100 and 300 mm, variability is 50-100 percent, and above 500 mm it is less than 25 percent (Dixon et al. 1989). Even though Irbid, in the north, receives comparatively 'good' precipitation, about 455-81 mm (Stewart 1989), approximately one year in five rainfall is less than 300 mm and wheat yields are very poor or fail (Qasem and Mitchell 1986). There is also extreme variation and series of drought years occur, as happened in the 1930s. Current average wheat yields in Jordan are approximately 650-700 kg/ha (ICARDA 1984), a figure which is the same as estimates from earlier this century (Pinner 1930). Comparatively low yields are partly due to this variable pattern of rainfall, but also to the persistence of traditional cultivation practices (see below).

During the period of waning Ottoman authority, between the seventeenth and nineteenth centuries, agriculture as a way of life receded (Lewis 1987). The two main reasons for the abandonment of settlements and associated depopulation are the environmental vulnerability of the south and east (modern-day Jordan) to drought combined with the fact that the area was the most exposed to Bedouin attack. For the sixteenth century, there are tax records documenting significant agricultural productivity for parts of what was then southeastern Syria. Records for the district of 'Ajlun, the district immediately north of the Balqa', reveal a high level of tax collection and the district appears to have been the most economically important area east of the Jordan river (Hütteroth and Abdulfattah 1977). It is in this area that settled village-based agriculture continued, albeit on a much reduced scale (Burckhardt 1822; Lewis 1987). 'Ajlun was the first area where Ottoman authority was reinstated and where titles to land were first and, ultimately, most comprehensively registered (Mundy 1992; Rogan 1994). In the Balqa' district, as-Salt was the only permanent settlement in the mid-nineteenth century due to the ravages of Bedouin incursions (Rogan 1994). Even in the hilly areas in 'Ajlun, where some 80 villages survived, Jarash, to the southeast, was only permanently re-inhabited in the late nineteenth century with the arrival of the Circassian community (Lewis 1987).

The nomadic lifestyle that so typified the 'desert' and the more sedentary peasant modes that were the 'sown' represent complementary, dependent modes of existence that are not mutually exclusive. Plant and animal husbandry are interdependent activities.

Furthermore, although some groups strongly identify themselves as Bedouin, as typified by wealthy, powerful camel-herding nomads, and others as peasants, or *fellaheen* (literally, 'cultivators'), there also exists a continuum between the two. Many less wealthy 'Bedouin' in the Balqa' district practised both animal and plant husbandry. Similarly, the Bedul Bedouin of Petra grew crops, as well as ranging their animals over large distances (Simms and Russell 1997). Even in the district of 'Ajlun, where settled agriculture remained, *fellaheen* husbanded their own animals and these had to be grazed, so some members of the community followed a lifestyle more similar to nomadic pastoralists (that is, they were transhumant)—following the availability of grazing—taking their animals to the Ghor (Jordan Valley) in the west or *badia* (arid steppe) to the east and south during the winter. Some villages also had an association with poorer *'arab* (Bedouin) where the *'arab* looked after village flocks in return for a proportion of the new-born goats or sheep (usually a third or a quarter) and some grain. Farther south, the *fellaheen* of Dana moved cyclically with their animals between grazed and cultivated areas. The flexibility between the two modes of existence is now often used in archaeological interpretation to explain settlement abandonment and nucleation (e.g. LaBianca 1990). Relations between people who identified themselves as Bedouin and *fellaheen* were, nevertheless, often hostile. In northern Jordan, the strong Bedouin tribes extracted a share of the crop, a tax known as the *khuwwa*, from village communities in return for protection (Rogan 1994). Older *fellaheen* tell stories about this period, the brutalities of Bedouin raids and the extortionate *khuwwa*.

Lack of year-round settlement does not represent absence of agriculture. A wheat crop is planted between November and January and is ready for harvest in June—in the Jordan Valley even earlier. Thus, people can plant a crop and leave after the harvest or even go away between sowing and harvest, on occasion hiring a guard to protect crops. Particularly in wadi beds in the dry areas, where barley may be grown opportunistically, crops are left unattended after sowing.

Starting in the nineteenth century, local *fellaheen* expanded from older settlements, refugees from the Caucasus re-settled villages such as Jarash, 'Amman and az-Zarqa', and Bedouin tribes established plantation villages using mostly Palestinian and Egyptian labour (Rogan 1994). Al-Yaduda farm (Bell 1907; Abujaber 1989), south of 'Amman, is a particularly well-recorded Bedouin plantation village. During the British Mandate period (1921–46) and into the early years of the Hashemite Kingdom, agricultural land was permanently partitioned into individual plots with legal title. In villages where *musha'*, a system of communal land tenure, had prevailed that entailed regular redistribution of shares, this was a very important change. It broke down social control of land and cultivation, which meant the state could intrude into the lives of villagers, and allowed land to be mortgaged or sold to outsiders (Fischbach 1994). Since the break-up of communal land tenure, kinship ties weakened and the household became the focus of economic and social activity (Antoun 1972). This has had a profound effect on land management practices as farmers began to base decision-making on their ability to fulfil household requirements, first, and broader familial alliances, second, in the process reducing the number of people involved in agricultural labour and the amount of available common village grazing land (Palmer 1999).

Traditional farming is rapidly disappearing as Jordan becomes part of the global economy (see Palmer 1998 for a discussion of the term 'traditional' as applied to contemporary farmers and farming practices). Yet, in some ways, old practices still persist. This is partly because older small-scale farmers, usually over 55 years old (el-Hurani 1988), dominate the agricultural scene and maintain former practices. Consequently, this also means that many traditional practices will soon disappear. It should be noted, however, that some traditional farming practices are linked to risk avoidance. For example, farmers are unwilling to invest in expensive fertilizers or the latest improved crop types because they know that if rainfall is not good, all will be lost. There are some major differences, however, in the contemporary rural economy: few farmers now depend on agriculture for their income (Mundy and Smith 1990; el-Hurani 1989) and labour is less readily available than in the previous kin-based system. As many farmers are less dependent upon farming for subsistence, what are traditionally perceived as good farming practices, such as hand-weeding and good seed bed preparation, are eliminated. This is especially true when family labour is not available; grown-up family members often have paid employment and children are at school (Aydin 1990). Hired non-local male labour is increasingly common, which discourages women from participating in agricultural production. Most farmers in Jordan now use tractors, although these may be the main, or only, aspect of modern farming practice adopted (el-Hurani 1989; Aydin 1990). Traditional *ards*, or scratch-ploughs, are still made (Figure 20.1) and used (Figures 20.2 and 20.3) (Palmer and Russell

Figure 20.1. Adzing an oak beam to form the stilt/sole of an *ard*.

Figure 20.2. Close-up of an *ard* during spring tillage (see Figure 20.3). The iron share is completely submerged and the ploughman is holding an ox goad in his right hand while guiding the *ard* with his left.

1993), particularly on inaccessible or stony ground and under tree crops. Non-capitalist relations and forms of production have persisted in Middle Eastern rural society, despite the expansion of capitalism (Glavanis and Glavanis 1983, 1990). Rural farmers, or peasants, cannot simply be defined as 'simple commodity producers' (see Friedmann 1980; Ellis 1988), but operate on a number of levels (e.g. household

Figure 20.3. Dual animal tillage in the spring using a traditional *ard* near the village of al-Mazar, northern Jordan. The second man is trickling chick-peas into the furrow created by the share.

economies, risk aversion, drudgery aversion) while at the same time integrating with the market and the state.

Today, the total cultivated area of Jordan is estimated at approximately 4 percent of land area with 93% of this considered dryland and 7 percent partially or completely irrigated (el-Hurani 1989). The lowest annual precipitation needed for barley and wheat production is 200 mm and 250 mm respectively (Arnon 1972), the two crops predominating in the area today and in the past. Large-scale fruit, vegetable and cereal irrigation farming in the Jordan Valley is comparatively recent; the East Ghor Canal Project was opened only in 1963. Field crops, 92% of which are wheat and barley, account for 80% of the total cultivated dryland area (el-Hurani 1989). Durum wheats predominate. Legumes such as lentil, vetches (bitter-vetch and common vetch) and faba bean are also important field crops. All have a long history of cultivation in the region (Zohary and Hopf 1994). Cereals are winter crops sown at the start of the rainy season. The legumes are usually planted later as they tend to be susceptible to frost. Important summer crops (planted in spring or early summer) that are grown without irrigation include chick-pea, sesame, tobacco, okra, maize (which is replacing Durra sorghum) and broom sorghum. A range of vegetables and fruits such as melon, tomato and cucumber are also grown under unirrigated and irrigated conditions. With the settlement of land title, increased political stability and decline in the availability of labour for agricultural tasks, tree crops are increasingly grown. Olives, in particular, are considered easy to maintain and yield a good return. In high mountainous areas, vines are successful. Figs, almonds and pomegranates are traditional tree crops with enduring popularity, the latter usually grown under irrigation.

Cultivation does take place in areas receiving low levels of rainfall. The pattern of rainfall means that, when they occur, heavy winter rains collect naturally in wadi beds. People in the past collected runoff water on a large and small scale (see Chapter 18, this volume), and a number of ancient systems are currently in the process of formal study (e.g. Barker *et al.* 1997). Today, local Bedouin may grow cereals in wadi beds where water collects naturally, usually as a supplement to animal feed, or may clear, scratch-out and till small fields at breaks in an incline (Figure 20.4). The field shown in Figure 20.4 is used for growing *heeshee*, a local tobacco with mild psychotropic effects. Similar fields, like this one in Wadi Siagh below Petra, overlay the remains of a vast water collection and irrigation system from the Nabataean period. Stone banks and terraces help to

Figure 20.4. Field fed by runoff water on a break in slope in the Wadi Siagh below Petra. Note the channel in the centre-bottom portion of the photograph directing water to the field and the direction of furrows allowing water to be spread.

trap water and stabilize soil, preventing erosion. Terraces are particularly used for tree cultivation, though there are distinctly fewer ancient terraces east of the Jordan river than on the West Bank.

There are important times in the agricultural year when tasks must be completed; failure to sow or harvest at optimum times results in yield losses. The following is a brief description of the traditional rain-fed agricultural practices for a variety of cultigens, based on field observations by the author. (For fuller details on practices in northern Jordan, see Palmer [1998], earlier surveys [Pinner 1930; Dalman 1932, 1933; Antoun 1972], and recent agronomic literature [e.g. Arabiat et al. 1983; Duwayri 1985; el-Hurani 1988; Jaradat 1988].) Land is prepared for sowing by tilling before the onset of the rains that helps moisture penetrate the soil surface. Grain crops are broadcast sown (Figure 20.5), although legumes were formerly planted using a tube attached to the rear of a traditional *ard*. Where an *ard* is still used, the land to be sown is divided into smaller units in order to help distribute the seed evenly while preventing seed loss. Cereals and winter legumes are usually sown between November and February, depending upon the onset of rains. Cereals are always planted first. Some small-scale farmers still hand weed wheat fields in the late winter. In areas where they can be grown, spring crops are planted on fallow areas in the late winter and early spring (Figure 20.3).

Faba bean, vetches and lentil are harvested first, followed by barley and wheat. Legumes are always harvested by uprooting, but cereals are uprooted (Figure 20.6) and sickle-harvested depending upon the density of the crop and local soil conditions. It is more difficult to uproot cereals that are densely sown or planted in heavy clay-rich soils. The final crop to be harvested is sesame in September. The olive harvest takes place in October, sometimes extending into November. In the past, the summer months were spent on the threshing floor, processing and cleaning each crop after it was harvested. Hand-processing crops is now extremely rare (Figure 20.7) due to the immense time-saving incurred with the use of mechanized threshing machines. Many farmers hand-harvest cereal crops which could be harvested by combine-harvesters because hand-harvesting enables harvesting closer to the ground, thereby allowing them to obtain more better quality straw.

Farmers have to rotate crops in order to maintain yields and, in the Mediterranean and Near East, cultivated (ploughed or bare) fallow is commonly used to restore soil moisture and fertility, particularly in the case of cereal farming. The benefits of cultivated fallow as a moisture-saving strategy are optimum when

Figure 20.5. Broadcast-sowing wheat at Mu'ta, near al-Karak, before tractor ploughing.

Figure 20.6. Harvesting wheat by uprooting in northern Jordan.

a dry cropped year follows a wet fallow year (Harris et al. 1991). The moisture saved in the fallow, even though this may be only 10 percent of rainfall received in the previous year, can prevent crop failure. Average annual rainfall levels provide a broad guideline for the crop rotation regimes that can be practised. Cereal-fallow is the main rotation regime (87 percent) used in areas receiving less than 300 mm of rainfall annually (el-Hurani 1988). Above 300-350 mm, the area being fallowed can be planted with summer crops.

Figure 20.7. An up-turned threshing sledge at rest. Note the basalt pieces wedged into holes to break the straw and cereal ears during threshing. The woman in the background is beating fresh, roasted chick-pea pods to release the seeds.

Wheat-lentil rotation is also possible, but only a small minority (5.4 percent in el-Hurani 1988) practises it. Above 400 mm, three-course rotations are possible and predominate on the plains around Irbid. Interestingly, in the lowest rainfall ranges, some 42% of wheat growers (el-Hurani 1988) practise continuous cereal cropping, representing an opportunistic strategy where the crop is probably grown as fodder.

There are a number of possible strategies to minimize losses caused by rainfall variability and offset the effects of uncertainty. There is, for example, a relationship between the onset of rainfall and total rainfall; six out of 10 high rainfall seasons have onsets in November (Stewart 1989). Farmers are aware of this and delay the start of planting until onset and may only finish in January or early February when it is clear how much rain is expected, but before any delay would incur serious yield losses. Accordingly, they adjust the area planted with wheat in nearly perfect accordance with actual rainfall, planting the maximum area when high rainfall is expected and the minimum area when lower levels portend. The drier the area, the later the date of planting. Farmers may also change to crops requiring lower levels of precipitation, taking into account economic returns. Lentils are the most valuable of the legumes, but chickpeas are more liable to succeed with late onset and low water expectations (Stewart 1989). Sowing rates are higher in higher rainfall and irrigated areas. Sowing rates for wheat are about 80 kg/ha at 'Amman and about 60 kg/ha around al-Karak (Figure 20.5) (Arabiat et al. 1983) but as high as 150-200 kg/ha on the Irbid plains. Because summer crops are planted after most of the rainfall has fallen, farmers know precisely which crops to plant, if any, and how they should be spaced.

Other strategies to offset the effects of variability include diversification in the range of crops grown, exchange (including forcible exchange such as the *khuwwa* and feasting) and storage, both of crops and crops 'stored' in animals (a 'walking larder') (Halstead and O'Shea 1989). Many traditional foods, such as *kishk*—dried, fermented milk and bulgar wheat—have a long storage capacity. Storage is less important today than in the recent past and many traditional foods, and the knowledge to make them, are disappearing (Basson 1981). It is evident from recent history that another important coping strategy is to alter subsistence regime or to move on. Political and social conditions have a large part to play in the adoption of all these strategies and, consequently, also to the shifting boundary between 'the desert and the sown'.

In much of the anthropological literature, the terms 'farmer' and 'peasant' have been used interchangeably (Forbes 1989) and, indeed, have both been used here. The term 'peasant' does, however, have certain disadvantages with its frequent uncomplimentary connotations and sense of subordination. This is particularly unfortunate in the case of the Middle East where anthropological inquiry was for a long time framed in a functionalist-orientalist paradigm (Eickelman 1981; Glavanis and Glavanis 1990). This approach sees rural life as static and self-contained with change coming principally from outside. The term 'farmer', on the other hand, does seem to imply a high degree of commercialization and integration into a market economy that is not entirely justified. In the archaeological literature, the term peasant is largely ignored, yet so many interpretations using an agrarian ecological paradigm clearly rely on analogies drawn from the contemporary study of peasants. The avoidance of the term is perhaps justified because archaeologists wish to disavow the complex social connotations of the term as scholars of feudalism employ it. Instead, terms such as 'subsistence' and 'small-scale' are preferred. And yet, in concentrating on the ecological adaptations of rural life, the social and historical dynamic implicit in treatments using the term peasant often remains under-appreciated in archaeological interpretations.

References

Abujaber, R.S.
- 1989 *Pioneers over Jordan: The Frontier of Settlement in Transjordan, 1850–1914*. London: I.B. Tauris.

Antoun, R.T.
- 1972 *Arab Village: A Social Structural Study of a Trans-Jordanian Peasant Community*. Bloomington, IN: Indiana University Press.

Arabiat, S., D. Nygaard and K. Somel
- 1983 *Factors Affecting Wheat Production in Jordan*. Aleppo: International Center for Agricultural Research in the Dry Areas.

Arnon, I.
- 1972 *Crop Production in Dry Regions*. II. *Systematic Treatment of the Principal Crops*. London: Hill.

Aydin, Z.
- 1990 Agricultural Labour and Technological Change in Jordan. In D. Tully (ed.), *Labour and Rainfed Agriculture in West Asia and North Africa*, 185-208. Dordrecht: Kluwer.

Barker, G.W. *et al.*
- 1997 The Wadi Faynan Project, Southern Jordan: A Preliminary Report on Geomorphology and Landscape Archaeology. *Levant* 29: 19-40.

Basson, P.
- 1981 Women and Traditional Food Technologies: Changes in Rural Jordan. *Ecology of Food and Nutrition* 11: 17-23.

Bell, G.
- 1907 *The Desert and the Sown*. London: Virago Press edition (1985) reproduced from the William Heinemann original.

Burckhardt, J.L.
- 1822 *Travels in Syria and the Holy Land*. London: Murray.

Dalman, G.
- 1932 *Arbeit und Sitte in Palästina*. II. *Der Ackerbau*. Gütersloh: Bertelsmann.
- 1933 *Arbeit und Sitte in Palästina*. III. *Von der Ernte zum Mehl*. Gütersloh: Bertelsmann.

Dixon, J.A., D.E. James and P.B. Sherman
- 1989 *The Economics of Dryland Management*. London: Earthscan.

Duwayri, M.
- 1985 Farm Systems in Rain-Fed Areas. In A.B. Zahlan (ed.), *The Agricultural Sector of Jordan: Policy and Systems Studies*, 126-58. London: published for the Abdul Hameed Shoman Foundation, Amman by Ithaca.

Eickelman, D.
- 1981 *The Middle East: An Anthropological Approach*. Englewood Cliffs, NJ: Prentice-Hall.

Ellis, F.
- 1988 *Peasant Economics: Farm Households and Agrarian Development*. Cambridge: Cambridge University Press.

Fischbach, M.R.
- 1994 British Land Policy in Transjordan. In E.L. Rogan and T. Tell (eds.), *Village, Steppe and State: The Social Origins of Modern Jordan*, 80-107. London: British Academic Press.

Forbes, H.A.
- 1989 Of Grandfathers and Grand Theories: The Hierarchised Ordering of Responses to Hazard in a Greek Rural Community. In Halstead and O'Shea, 1989: 87-97.

Friedmann, H.
- 1980 Household Production and the National Economy: Concepts for the Analysis of Agrarian Formations. *Journal of Peasant Studies* 7(2): 158-84.

Glavanis, K. and P. Glavanis
- 1983 The Sociology of Agrarian Relations in the Middle East: The Persistence of Household Production. *Current Sociology* 31(2): 1-109.

Glavanis, K. and P. Glavanis (eds.)
- 1990 *The Rural Middle East: Peasant Lives and Modes of Production*. London: Zed Books and Birzeit University.

Halstead, P. and J. O'Shea (eds.)
- 1989 *Bad Year Economics: Cultural Responses to Risk and Uncertainty*. Cambridge: Cambridge University Press.

Harris, H.C. et al.
1991 The Management of Crop Rotations for Greater WUE under Rainfed Conditions. In H.C. Harris, P.J.M. Cooper and M. Pala (eds.), *Soil and Crop Management for Improved Water Use Efficiency in Rainfed Areas*, 237-50. Aleppo: International Center for Agricultural Research in the Dry Areas.

el-Hurani, M.H.
1988 *Report on the Wheat Baseline Data Survey*. Amman: Jordan Highland Agricultural Development Project.
1989 Analysis of Agricultural Policy in the Jordan Drylands. In C.E. Whitman *et al.* (eds.), *Soil, Water, and Crop/Livestock Management Systems for Rainfed Agriculture in the Near East Region*, 36-56. Washington, DC: United States Department of Agriculture, United States Agency for International Development, International Center for Agricultural Research in the Dry Areas.

Hütteroth, W.-D. and K. Abdulfattah
1977 *Historical Geography of Palestine, Transjordan and Southern Syria in the Late 16th Century*. Erlangen: Erlanger geographische Arbeiten.

ICARDA
1984 *A Report on the Jordan Cooperative Cereal Improvement Project*. Aleppo: International Center for Agricultural Research in the Dry Areas.

Jaradat, A.A. (ed.)
1988 *An Assessment of Research Needs and Priorities for Rainfed Agriculture*. Jordan: The United States Agency for International Development.

LaBianca, Ø.S.
1990 *Sedentarization and Nomadization: Food System Cycles at Hesban and Vicinity in Transjordan*. Hesban 1. Berrien Springs, MI: Institute of Archaeology and Andrews University.

Lewis, N.N.
1987 *Nomads and Settlers in Syria and Jordan, 1800–1980*. Cambridge: Cambridge University Press.

Mundy, M.
1992 Shareholders and the State: Representing the Village in the Late Nineteenth Century Registers of the Southern Hawran. In T. Philipp (ed.), *The Syrian Land in the Eighteenth and Nineteenth Century*, 217-38. Berliner Islamstudien 5. Stuttgart: Steiner.

Mundy, M. and R.S. Smith (eds.)
1990 *Part-time Farming: Agricultural Development in the Zarqa River Basin, Jordan*. Irbid: Institute of Archaeology and Anthropology, Yarmouk University.

Palmer, C.
1998 'Following the Plough': The Agricultural Environment of Northern Jordan. *Levant* 30: 129-65.
1999 Whose Land is it Anyway? An Historical Examination of Land Tenure and Agriculture in Northern Jordan. In C. Gosden and J. Hather (eds.), *The Prehistory of Food: Appetites for Change*, 288-305. London: Routledge.

Palmer, C. and K.W. Russell
1993 Traditional Ards of Jordan. *Annual of the Department of Antiquities of Jordan* 37: 37-53.

Pinner, L.
1930 Wheat Culture in Palestine. *Bulletin of the Palestine Economic Society* 5(2).

Qasem, S. and M. Mitchell
1986 The Problems of Rainfed Agriculture. In A. Burrell (ed.), *Agricultural Policy in Jordan*, 30-40. London: Ithaca.

Rogan, E.L.
1994 Bringing the State Back: The Limits of Ottoman Rule in Jordan, 1840–1910. In E.L. Rogan and T. Tell (eds.), *Village, Steppe and State: The Social Origins of Modern Jordan*, 32-57. London: British Academic Press.

Simms, S.R. and K.W. Russell
1997 *The Ethnoarchaeology of the Bedul Bedouin of Petra: Implications for the Food Producing Transition, Site Structure, and Pastoralist Archaeology*. Utah State University Contributions to Anthropology 22. Logan, UT: Utah State University.

Stewart, J.I.
1989 Response Farming for Improvement of Rainfed Crop Production in Jordan. In C.E. Whitman *et al.* (eds.), *Soil, Water, and Crop/Livestock Management Systems for Rainfed Agriculture in the Near East Region*, 288-306. Washington, DC: United States Department of Agriculture, United States Agency for International Development, International Center for Agricultural Research in the Dry Areas.

Zohary, D. and M. Hopf
1994 *Domestication of Plants in the Old World*. 2nd edn. Oxford: Clarendon Press.

21. Archaeological Survey in Jordan

Edward Banning

Regional archaeological survey, which has goals and methods quite different from excavation, has a long history in Jordan. It began with the explorations of Europeans, who had preserved knowledge of Transjordanian place names from biblical and classical sources, but had not visited these places from the end of the Crusades until the early nineteenth century. Notable among such early explorers was Burckhardt, 'discoverer' of Petra. After nearly a century of research on identifying cities and towns mentioned in the ancient sources, archaeologists, armed with the new tool of pottery typology, began to record much larger numbers of sites in the early twentieth century, and found that regional survey provided unique perspectives on ancient settlement patterns, demography, agricultural exploitation, and even political boundaries, that excavations at a few key sites could not. By the late 1970s, survey had become a systematic and more scientific component of archaeological research in Jordan because it offered insights into cultural patterns at a regional, rather than site-specific, scale. It was also becoming increasingly important as a tool in cultural resource management as the rapid pace of Jordan's development caused exponential increase in the discovery, and often destruction, of archaeological sites. There is not room here for a comprehensive history of archaeological surveys in Jordan, but the following will introduce some of the turning points in the design and results of surveys to illustrate their potential and methodology.

The early archaeological exploration of Jordan was mainly by individual foreign travellers who published itineraries of their journeys. Among descriptions of the landscape, vegetation and contemporary culture, these itineraries offered descriptions of Mediaeval castles and classical ruins, such as Petra and Jarash, as well as speculations on the identification of other ruins based principally on similarities between modern Arabic place names and classical or biblical ones. Consequently it is convenient to describe this as the toponymic stage of archaeological survey in Jordan. As most of these travellers had no archaeological training, and especially as there were as yet no independent means for dating archaeological remains except for those with datable inscriptions, it was impossible for authors to infer changes in the distribution or material culture of sites over the millennia, and there was virtually no reference to sites dating earlier than the Bronze Age. Prominent toponymic surveyors include Burckhardt (1822, 1829), Buckingham (1821, 1825), Merrill (1881), Oliphant (1880) and Robinson (1841, 1856).

Partway through the toponymic stage of exploration, organized expeditions began to supplement individual explorations. Funded by such organizations as the Palestine Exploration Fund, these expeditions carried out topographical and geographical surveys, published detailed maps, took thousands of photographs, and spent many months or years in the field instead of the few weeks typical of some individual efforts. Conder and Kitchener's *Survey of Eastern Palestine* (Conder 1883, 1889) and Brünnow and von Domaszewski's (1904, 1905) research on Roman Arabia are classic examples of this work. Some surveys continued to be carried out more casually by individuals, but now some of these, such as Tristram (1897) and Schumacher, were living in the regions they studied and accumulating years of familiarity with local landscapes and their antiquities. Determining the routes of Roman roads and the distribution of Roman forts was now a common research goal.

Schumacher's studies of the antiquities of northern Jordan are in many ways a turning point, as his methods and results really anticipated the better-known work by Glueck (Schumacher 1886, 1890; Steuernagel 1925, 1926). Both these researchers accumulated enough detailed information about their archaeological landscapes to begin to transcend toponymic identifications and to draw inferences about the history and late prehistory of whole regions.

But the thing that did most to expand the usefulness of archaeological survey in the region was the growing awareness that careful typology of pottery from stratified sites, as Albright pioneered at Tall Beit Mirsim (Albright 1932, 1933, 1938, 1943), allows us to order archaeological sites in time. Small surveys by Albright (1926, 1929) in the 1920s, and Glueck's much more extensive surveys (Glueck 1934, 1935, 1939, 1946, 1951), benefited enormously from the use of ceramic chronology. Albright was now able to make inferences about the routes of caravans and Chedorlaomer's army in the Bronze Age, regional abandonments in the Late Bronze Age, and the clearing of forest in Iron Age Gilead (Albright 1929). Glueck used his much larger database to promote the view that much of Jordan was unoccupied in the Middle and Late Bronze Ages, and particularly that there were no sites in what would become the kingdoms of Edom and Moab until the early Iron Age (Glueck 1934, 1939). The methods of these surveys, however, remained fairly casual survey on horseback or by vehicle, generally following roads and tracks between modern villages, and occasionally relying on local informants for advice on where to find ruins. Some later surveys (e.g. Mittmann 1970; Miller 1991) continued in this tradition.

For the next two decades there was less archaeological survey in Jordan, but something new was the increasing attention to prehistoric sites. Field's drives around the Jordanian deserts from 1925 to 1950 had resulted in the discovery of many scatters of Palaeolithic and later stone tools (Field 1960), but most prehistoric sites were being found and reported by geologists, not archaeologists. From 1953, the Jordanian Department of Antiquities undertook the country's first archaeological impact study, the Point IV Irrigation Project, under which de Contenson and Mellaart surveyed and carried out soundings on Neolithic and Chalcolithic, as well as later sites, in the Jordan Valley (de Contenson 1960, 1964; Leonard 1992). This led to small-scale explorations by other prehistorians, such as Kirkbride (1956, 1960), but not many new sites were discovered in this period. Typically, small excavations at selected sites followed small and fairly casual survey.

The pace of survey picked up again in the 1970s, and more of these surveys began to search for prehistoric sites and to incorporate explicit, systematic prospecting methods or statistical sampling. The Jordan Valley Survey, for example, employed systematic fieldwalking by larger crews (Ibrahim *et al.* 1976; Yassine *et al.* 1988). The Heshban regional survey was the first in Jordan to attempt regional sampling methods (Ibach 1976, 1978). The Judayid Basin survey and later surveys in the Ras an-Naqb area employed systematic pedestrian survey (Henry 1979, 1982, 1995). The Wadi al-Hasa Survey used regional sampling, systematic fieldwalking and provided estimates of survey coverage (Banning 1985, 1988). In spite of its claim to have detected only a small percentage of Wadi al-Hasa's sites, the survey was still much more intensive than the older surveys and recorded more than 1000 sites from 1979 to 1982, orders of magnitude more than old surveys had done (MacDonald *et al.* 1988). This was partly due to a broader definition of what constituted a 'site'. Older surveys had mainly recorded relatively large settlements, such as tells, and sites with visible architecture, such as stone towers. The newer surveys recorded small scatters of flaked stone and pottery, as well as petroglyphs, fragments of aqueducts, terrace walls, and stone cairns. These surveys were particularly successful at increasing the number of known prehistoric sites.

Often these surveys sampled a population that consisted of a geometric grid, typically consisting of 1 km squares, and then crew members walked back and forth across each square in the grid with a spacing between transects ranging from 25 to 500 m, depending on the survey's intensity (Figures 21.1, 21.2). Often these surveys only sampled about 20 percent of the squares within the whole grid. Technically the sample consists of the set of grid squares examined, although most practitioners proceeded as though they had sampled a population of sites. This practice is called cluster sampling (Mueller 1975), and can be effective if the sampled units approximate a microcosm of the whole region of interest. Consequently, most of the surveys' designers attempted to ensure that their transects intersected different kinds of topography and different vegetation zones.

The 1980s saw a flurry of surveys of this type, including surveys of Wadi Ziqlab (Banning 1982; Banning and Fawcett 1983), Wadi al-Yabis (Mabry and Palumbo 1988, 1992), Wadi al-'Arab (Hanbury-Tenison 1984), the Jarash region (Hanbury-Tenison 1987), Wadi Bayir (Rolston and Rollefson 1982), the upper Wadi az-Zarqa (Simmons and Kafafi 1992), the eastern desert (Betts 1984, 1988, 1992), and the southern Ghor (MacDonald *et al.* 1992). Other less systematic, but relatively intensive, surveys also took place with focus on particular types of antiquities, such as Thamudic and Safaitic petroglyphs (Macdonald 1983, 1992) or Roman hydraulic systems (Oleson 1990). Virtually all of these surveys relied on surface scatters of artifacts for the identification of sites, and the main thing that distinguished them was that it was now much more feasible to estimate differences in the density, size and distribution of sites over time. This led to further scrutiny of Albright's and Glueck's hypotheses about fluctuating

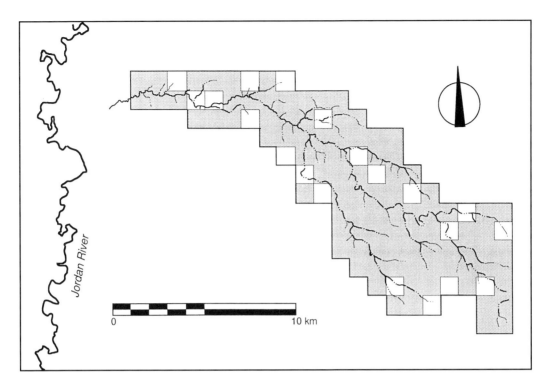

Figure 21.1. Map of Wadi Ziqlab's drainage basin showing the boundaries of the sampling frame and the quadrats that were sampled (white squares).

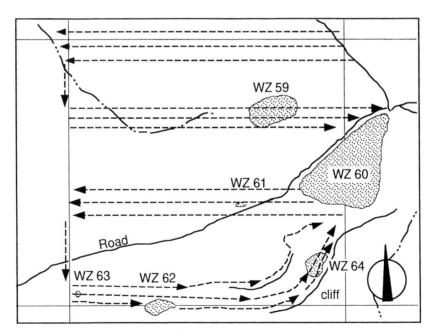

Figure 21.2. Map of pedestrian transects (dashed lines) in the quadrat bordering site WZ 60 in the southeastern Wadi Ziqlab basin on 8 December 1981. Stippled areas indicate sites detected, and the quadrat (hairlines) is 1 km².

settlement and fairly ambitious attempts to estimate regional population sizes, to study nomad-sedentary interactions and fluctuations (Banning 1985; LaBianca 1986), and to infer the economic factors that led to changes in site location (Banning 1985).

The probability of these surveys detecting archaeological sites is highly sensitive to 'edge effects', site size, obtrusiveness and visibility (Plog et al. 1978; Schiffer et al. 1978; Banning 1986, 1988). Because sites are not points, and occupy varying amounts of space, they can be discovered at the edges of a sample unit when their centres actually lie outside the sample (Figure 21.3). This tends to exaggerate the number of large sites. Meanwhile, because most of these surveys had crew members walking roughly parallel transects fairly widely spaced (such as 100 m or even more), sites that are small or consist only of low-density scatters of artifacts can 'slip through the net' and be missed, so the number of such sites tends to be underestimated (Figure 21.4). Consequently the 'coverage', or probability of finding sites, varies with the size and obtrusiveness of those sites (Figure 21.5). Obtrusiveness is the inherent character of sites that makes it possible to see them. Standing ruins and tells are more obtrusive than sherd scatters. Variations in visibility, caused by vegetation, soil cover and other things that can obscure the view of a site, are actually a greater challenge for these surveys.

Increasing awareness that geomorphologic and other processes that affect visibility have distorted the picture of how archaeological materials are distributed (e.g. Bar-Yosef and Goren 1980; Brooks *et al.* 1982) led the Wadi Ziqlab Project to supplement these kinds of survey with a sub-surface sample. One of the differences between this and other surveys is that the population being sampled is not a grid of geometric spaces, but a set of landforms, in this case a set of stream terraces where recent colluvium can be expected to have buried archaeological materials. At each stream terrace in this set, team members excavate one or two trenches, each 3 m² in area and up to 2 m deep, to see if any buried sites can be detected. Although the sub-surface survey is only conducted as a small adjunct to excavations at particular sites, testing of 13 stream terraces has led to the discovery of buried Neolithic, Chalcolithic and Early Bronze Age sites (Banning 1996).

One set of techniques that has not had much application in regional archaeological survey in Jordan is remote sensing. These are surveys that depend on contrasts between archaeological materials and their environment in their magnetic, electrical, seismic, optical or chemical properties and on instrumentation able to measure those properties. Unfortunately the materials—mud and limestone—which predominantly make up archaeological sites in Jordan usually do not differ very much from their environments, making these techniques difficult to apply at the regional scale. Aerial photography and satellite imagery, however, are types of remote sensing that do have some application, at least for the most optically obtrusive sites and

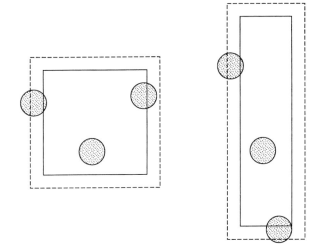

Figure 21.3. Edge effects on site discovery by transects (long rectangles) and square quadrats. Because some sites will be discovered whose centres lie outside the sample unit, the effective area surveyed (dashed rectangle) is actually larger than the area of the units themselves (inner rectangle). Because the ratio of edge to surface area is greater for transects than for squares, transects tend to detect disproportionately more sites (after Plog *et al.* 1978).

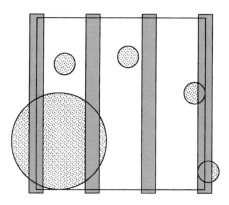

Figure 21.4. The effect of site size on the probability of site detection by pedestrian transects (grey rectangles). Note that some small sites (small circles) will lie completely between transects, while larger sites will always be intersected by at least one transect (after Plog *et al.* 1978).

regions of good visibility, such as the desert (e.g. Kennedy 1998a and 1998b).

Nor have Jordanian surveys gone very far yet in the use of 'non-site' or 'inter-site' approaches that attempt to take the entire distribution of artifacts on the landscape into account, and not just those that are

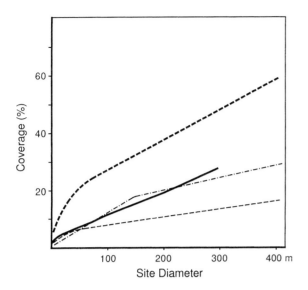

Figure 21.5. Estimates of survey coverage as a function of site diameter for the Wadi Ziqlab Survey 1981 (solid curve), Wadi al-Hasa Survey 1979 (dots and dashes), Wadi al-Hasa 1982 (heavy dashed curve) and Wadi al-Hasa stratified random sample (light dashed curve) (after Banning 1985, 1988). Because of the effects noted in Figures 21.3 and 21.4, larger sites are more likely to be discovered.

clustered into 'sites'. These approaches are increasingly important in surveys in many parts of the world, including North America (e.g. Ebert 1992; Stafford 1995), Europe (Neustupny 1998) and the Middle East (Wilkinson 1989). Jordanian surveys often record 'isolated finds' in addition to 'sites' but are only beginning to adopt fully non-site approaches.

Survey is an evolving method that provides insights into the spatial dimensions of ancient culture at the regional scale. Although not without difficulties, it is better suited than site-focused excavations at revealing regional settlement patterns and the demographic, economic and ecological contexts that helped to shape them.

References

Albright, W.F.
    1926   The Jordan Valley in the Bronze Age. *Bulletin of the American Schools of Oriental Research* 6: 13-74.
    1929   New Israelite and Pre-Israelite Sites: The Spring Trip of 1929. *Bulletin of the American Schools of Oriental Research* 35: 1-14.
    1932   *The Excavation of Tell Beit Mirsim I*. Annual of the American Schools of Oriental Research 12. New Haven, CT: American Schools of Oriental Research.
    1933   *The Excavation of Tell Beit Mirsim IA*. Annual of the American Schools of Oriental Research 13. New Haven, CT: American Schools of Oriental Research.
    1938   *The Excavation of Tell Beit Mirsim II*. Annual of the American Schools of Oriental Research 17. New Haven, CT: American Schools of Oriental Research.
    1943   *The Excavation of Tell Beit Mirsim III*. Annual of the American Schools of Oriental Research 18. New Haven, CT: American Schools of Oriental Research.

Banning, E.B.
    1982   The Research Design of the Wadi Ziqlab Survey, 1981. *American Schools of Oriental Research Newsletter* 8: 4-8.
    1985   Pastoral and Agricultural Land Use in the Wadi Ziqlab, Jordan: An Archaeological and Ecological Survey. Unpublished PhD dissertation, University of Toronto.
    1986   Peasants, Pastoralists and *Pax Romana*: Mutualism in the Southern Highlands of Jordan. *Bulletin of the American Schools of Oriental Research* 261: 25-50.
    1988   Methodology. In B. MacDonald et al., *The Wadi el Hasa Archaeological Survey 1979–1983, West-Central Jordan*, 13-25. Waterloo, ON: Wilfrid Laurier University.
    1996   Highlands and Lowlands: Problems and Survey Frameworks for Rural Archaeology in the Near East. *Bulletin of the American Schools of Oriental Research* 301: 25-45.

Banning, E.B. and C. Fawcett
    1983   Man–Land Relationships in the Ancient Wadi Ziqlab: Report of the 1981 Survey. *Annual of the Department of Antiquities of Jordan* 27: 291-309.

Bar-Yosef, O. and N. Goren
    1980   Afterthoughts Following Prehistoric Surveys in the Levant. *Israel Exploration Journal* 30: 1-16.

Betts, A.
    1984   Black Desert Survey, Jordan: Second Preliminary Report. *Levant* 16: 25-34.
    1988   The Black Desert Survey. Prehistoric Sites and Subsistence Strategies in Eastern Jordan. In A.N. Garrard and H.G. Gebel (eds.), *The Prehistory of Jordan: The State of Research in 1986*, 369-91. British Archaeological Reports, International Series 396. Oxford: British Archaeological Reports.
    1992   Eastern Jordan: Economic Choices and Site Location in the Neolithic Periods. In M. Zaghloul et al. (eds.), *Studies in the History and Archaeology of Jordan*, IV: 111-14. Amman: Department of Antiquities.

Brooks, I.A., L.D. Levine and R.W. Dennell
    1982   Alluvial Sequence in Central West Iran and Implications for Archaeological Survey. *Journal of Field Archaeology* 9: 285-99.

Brünnow, R.E. and A. von Domaszewski
   1904   *Die Provincia Arabia auf Grund Zweier in den Jahren 1897 und 1898 unternommenen Reisen und der Berichte früherer Reisender.* I. *Die Römerstrasse von Madeba über Petra und Odruh bis El-'Akaba.* Strassburg: Trübner.
   1905   *Die Provincia Arabia auf Grund Zweier in den Jahren 1897 und 1898 unternommenen Reisen und der Berichte früherer Reisender.* II. *Die äussere Limes und die Römerstrassen von el-Ma'an bis Bosra.* Strassburg: Trübner.

Buckingham, J.S.
   1821   *Travels in Palestine, through the Countries of Bashan and Gilead, East of the River Jordan: Including a Visit to the Cities of Geraza and Gamala in the Decapolis.* London: Longman, Hurst, Rees, Orme, and Brown.
   1825   *Travels among the Arab Tribes Inhabiting the Countries East of Syria and Palestine, Including a Journey from Nazareth to the Mountains beyond the Dead Sea, and from thence through the Plains of Hauran to Bozra, Damascus, Tripoly, Lebanon, Baalbeck, Antioch, and Aleppo.* London: Longman, Hurst, Rees, Orme, Brown, and Green.

Burckhardt, J.L.
   1822   *Travels in Syria and the Holy Land.* London: Murray.
   1829   *Travels in Arabia.* London: Colburn.

Conder, C.R.
   1883   *Heth and Moab: Explorations in Syria in 1881 and 1882.* London: Bentley.
   1889   *The Survey of Eastern Palestine, Memoirs of the Topography, Orography, Hydrography, Archaeology, etc.,* I. London: Committee of the Palestine Exploration Fund.

de Contenson, H.
   1960   Three Soundings in the Jordan Valley. *Annual of the Department of Antiquities of Jordan* 4/5: 12-98.
   1964   The 1953 Survey in the Yarmuk and Jordan Valleys. *Annual of the Department of Antiquities of Jordan* 8-9: 30-46.

Doughty, C.M.
   1936   *Travels in Arabia Deserta.* London: Cape.

Ebert, J.I.
   1992   *Distributional Archaeology.* Albuquerque, NM: University of New Mexico.

Field, H.
   1960   *North Arabian Desert Archaeological Survey 1925–1950,* II. Papers of the Peabody Museum of Archaeology and Ethnology 45. Cambridge, MA: Harvard University Press.

Glueck, N.
   1934   *Explorations in Eastern Palestine* I. Annual of the American Schools of Oriental Research 14. Philadelphia: American Schools of Oriental Research.
   1935   *Explorations in Eastern Palestine* II. Annual of the American Schools of Oriental Research 15. New Haven, CT: American Schools of Oriental Research.
   1939   *Explorations in Eastern Palestine* III. Annual of the American Schools of Oriental Research 18-19. New Haven, CT: American Schools of Oriental Research.
   1946   Some Chalcolithic Sites in Northern Gilead. *Bulletin of the American Schools of Oriental Research* 104: 12-20.
   1951   *Explorations in Eastern Palestine* IV. Annual of the American Schools of Oriental Research 25-28. New Haven, CT: American Schools of Oriental Research.

Hanbury-Tenison, J.W.
   1984   Wadi Arab Survey 1983. *Annual of the Department of Antiquities of Jordan* 28: 385-423.
   1987   Jerash Region Survey 1984. *Annual of the Department of Antiquities of Jordan* 31: 129-59.

Henry, D.O.
   1979   Palaeolithic Sites within the Ras en-Naqb Basin, Southern Jordan. *Annual of the Department of Antiquities of Jordan* 23: 93-99.
   1982   The Prehistory of Southern Jordan and Relationships with the Levant. *Journal of Field Archaeology* 9: 417-44.
   1995   *Prehistoric Cultural Ecology and Evolution: Insights from Southern Jordan.* New York: Plenum.

Ibach, R.J.
   1976   Archaeological Survey of the Hesban Region. *Andrews University Seminary Studies* 14: 119-26.
   1978   Expanded Archaeological Survey of the Hesban Region. *Andrews University Seminary Studies* 16: 201-14.

Ibrahim, M., J. Sauer and K. Yassine
   1976   The East Jordan Valley Survey, 1975. *Bulletin of the American Schools of Oriental Research* 222: 41-66.

Kennedy, D.
   1998a   Aerial Archaeology in Jordan. *Levant* 30: 91-96.
   1998b   Gharandal Survey 1997: Air Photo Interpretation and Ground Verification. *Annual of the Department of Antiquities of Jordan* 42: 573-85.

Kirkbride, D.
   1956   A Neolithic Site at Wadi el Yabis. *Annual of the Department of Antiquities of Jordan* 3: 56-60.
   1960   A Brief Report on the Pre-Pottery Flint Cultures of Jericho. *Palestine Exploration Quarterly* 92: 114-19.

LaBianca, Ø.S.
   1986   The Return of the Nomad: An Analysis of the Process of Nomadization in Jordan. *Annual of the Department of Antiquities of Jordan* 30: 251-54.

Leonard, A., Jr
   1992   *The Jordan Valley Survey, 1953: Some Unpublished Soundings Conducted by James Mellaart.* Annual of the American Schools of Oriental Research 50. Winona Lake, IN: Eisenbrauns.

Mabry, J. and G. Palumbo
   1988   The 1987 Wadi al-Yabis Survey. *Annual of the Department of Antiquities of Jordan* 32: 275-305.

1992 Environmental, Economic and Political Constraints on Ancient Settlement Patterns in the Wadi al-Yabis Region. In M. Zaghloul et al. (eds.), *Studies in the History and Archaeology of Jordan*, IV: 67-72. Amman: Department of Antiquities.

MacDonald, B. et al.
1988 *The Wadi el Hasa Archaeological Survey 1979–1983, West Central Jordan*. Waterloo, ON: Wilfrid Laurier University.
1992 *The Southern Ghor and Northeast 'Arabah Archaeological Survey*. Sheffield Archaeological Monographs 5. Sheffield: J.R. Collis.

Macdonald, M.C.A.
1983 Inscriptions and Rock-Art of the Jawa Area, 1982. *Annual of the Department of Antiquities of Jordan* 27: 571-76.
1992 The Distribution of Safaitic Inscriptions in Northern Jordan. In M. Zaghloul et al. (eds.), *Studies in the History and Archaeology of Jordan*, IV: 303-07. Amman: Department of Antiquities.

Merrill, S.
1881 *East of the Jordan: A Record of Travel and Observation in the Countries of Moab, Gilead, and Bashan*. London: Bentley.

Miller, J.M. (ed.)
1991 *Archaeological Survey of the Kerak Plateau*. Atlanta, GA: Scholars Press.

Mittmann, S.
1970 *Beiträge zur Siedlungs- und Territorialgeschichte des nördlichen Ostjordanlandes*. Abhandlungen des deutschen Palästinavereins. Wiesbaden: Otto Harrassowitz.

Mueller, J.W.
1975 Archaeological Research as Cluster Sampling. In J.W. Mueller (ed.), *Sampling in Archaeology*, 33-41. Tucson: University of Arizona.

Neustupny, E.
1998 *Space in Prehistoric Bohemia*. Prague: Institute of Archaeology, Academy of Sciences of the Czech Republic.

Oleson, J.P.
1990 Humeima Hydraulic Survey, 1989: Preliminary Field Report. *Echos du monde classique/Classical Views* 34(2): 145-63.

Oliphant, L.
1880 *The Land of Gilead*. London: Blackwood.

Plog, S., F. Plog and W. Wait
1978 Decision-Making in Modern Surveys. *Advances in Archaeological Method and Theory* 1: 384-421.

Robinson, E.
1841 *Biblical Researches in Palestine, Mount Sinai and Arabia Petraea: A Journal of Travels in the Year 1838*. Boston: Crocker.
1856 *Later Biblical Researches in Palestine and Adjacent Regions: A Journal of Travels in the Year 1852*. London: Murray.

Rolston, S. and G.O. Rollefson
1982 The Wadi Bayir Palaeoanthropological Survey, a First Season Report. *Annual of the Department of Antiquities of Jordan* 26: 211-19.

Schiffer, M.B., A.P. Sullivan and T.C. Klinger
1978 The Design of Archaeological Surveys. *World Archaeology* 10: 1-28.

Schumacher, G.
1886 *Across the Jordan*. London: Bentley.
1890 *Northern 'Ajlun, 'Within the Decapolis'*. London: Watt.

Simmons, A. and Z. Kafafi
1992 The 'Ain Ghazal Survey: Patterns of Settlement in the Greater Wadi az-Zarqa Area, Central Jordan. In M. Zaghloul et al. (eds.), *Studies in the History and Archaeology of Jordan*, IV: 77-82. Amman: Department of Antiquities.

Stafford, C.R.
1995 Geoarchaeological Perspectives on Paleolandscapes and Regional Subsurface Archaeology. *Journal of Archaeological Method and Theory* 2 (1): 69-104.

Steuernagel, D.C.
1925 Der Adschlun. *Zeitschrift des deutschen Palästina-Vereins* 48: 1-144, 201-392.
1926 Der Adschlun. *Zeitschrift des deutschen Palästina-Vereins* 49: 1-167.

Tristram, H.B.
1897 *Bible Places: The Topography of the Holy Land*. London: Society for Promoting Christian Knowledge.

Wilkinson, T.J.
1989 Extensive Sherd Scatters and Land-Use Intensity: Some Recent Results. *Journal of Field Archaeology* 16 (1): 31-46.

Yassine, K., J.A. Sauer and M. Ibrahim
1988 The East Jordan Valley Survey, 1976 (Part Two). In K. Yassine (ed.), *Archaeology of Jordan: Essays and Reports*, 189-207. Amman: University of Jordan.

22. The Analysis of Chipped Stone

Douglas Baird

Introduction

The aim of this chapter is to shed light on the manner in which chipped-stone studies have been used to help in the interpretation of past human behaviour in Jordan. Chipped stone has been used in a manner analogous to other artifact categories in interpretations of past behaviour, but it has certain strengths and weaknesses by comparison. In particular, it is worth outlining those areas where effective insights are gained and the basis of that effectiveness.

As with other artifact categories, distinctive chipped-stone tools and technologies have been used to provide a date for sites and/or deposits on sites relative to others. In particular, in periods (Palaeolithic–Aceramic Neolithic) and areas (eastern Jordan—where ceramic manufacture and use is much more restricted, e.g. post-Aceramic Neolithic, than farther west) where chipped stone is the dominant artifact category, numerically speaking, this has been an important role for chipped-stone studies. While absolute dating, particularly C14 dating, is seen as important, it remains the case, given the expense of C14 dating, the demands for particular sorts of samples and the need to date survey assemblages, that relative dating using lithics will remain important.

The aim of chipped-stone production is predominantly the making of tools used in cutting, scraping, and piercing tasks, and as elements of projectile armatures. The elements of the chipped stone assemblage used for such tasks, where they can be recognized as being used for specific activities, potentially stand to tell us much about activities carried out by a society, the ways in which such activities were carried out and their relative importance in terms of frequency of particular activities. Further, we can document contrasts between sites in the range of activities present, and understand something of the use of structures and areas on sites, both through relatively primary artifact distributions and the nature of artifact disposal/deposition. Opportunities to strengthen this functional approach exist through micro-wear studies (studies of microscopic alterations to the character of the surface of the chipped stone, which can indicate the relative hardness and softness of material worked, the direction and use area of the tool, and perhaps identify the specific materials worked). Much debate exists about the effectiveness of micro-wear studies. One thing is clear, however. The greater the specificity of a claim for the function of a particular tool, the higher the potential inaccuracy of the claim. The most convincing micro-wear-based claims tell us about the general nature of the use of tools, rather than very specific tasks or materials worked (Levi Sala 1996: 69).

The manner of chipped-stone manufacture, presumably handed down from artisan to apprentice, embodies distinctive practices indicative of restricted networks of communication and aspects of group identity. Distinctive features of lithic manufacture preserved on the chipped stone, when collated and comprehended, are, therefore, likely to be diagnostic of such identity. Equally, features of the final product visible to others may have been deliberately singled out by manufacturers and users as emblematic of their group affiliations. Technological and morphological features of chipped stone have been examined in an attempt to establish an indication of distinct groups at various levels.

Chipped-stone tool manufacture, unlike some other manufacturing processes, produces abundant quantities of waste. This material, apart from indicating the manufacturing procedures, may also point to the location and intensity of production within and between sites and thus help in site interpretation. In addition, such information on locales, intensity, complexity, and skill levels involved in production may give some indication of the socio-economic role and/or status of knappers. Was knapping practised at the household level, a domestic mode of production, or within only a few households, pointing towards more specialized divisions of labour? Was it likely to be a full-time, economically based specialization or a status-based pursuit, perhaps practised less intensively?

Since all stages of manufacture are preserved and the locations of those different stages can be documented, we can easily chart how raw materials travelled from sources and the processing they were subjected to along the way. The organization of production and distribution, and potentially exchange, can be studied relatively easily using chipped stone.

Olszewski (Chapter 2, this volume) outlines the basic procedures of flint knapping and some of the typical products: blades, flakes, bladelets, cores and microliths. Generally, chipped-stone analysts make a distinction between tools and debitage. Tools are removals (blades, flakes, bladelets or other) from cores that have been further modified by use or deliberate 'retouch', that is, the removal of small flakes from edges and surfaces. Tools can also be specifically shaped cores (e.g. bifaces). Debitage is removals that do not show evidence of such modification. This does not mean that such debitage could not have been used without retouching or retouch developing. So while all tools have been used or intended for use as such, some debitage may have been so used as well.

There are different aspects of chipped-stone production technology. A valuable distinction can be made between the strategies used in producing particular forms of debitage used as blanks for tools, and the techniques used in such strategies (Baird 1994: 525; Newcomer 1975: 97-98). Here, strategy means the way in which the core is shaped to produce certain sorts of removals and the order involved in the shaping and removals (Figure 22.1.8). Technique refers to the way in which force is applied to a core, including potentially highly idiosyncratic features like the force of the blow delivered on a core, instrument used, angle of blow, speed of blow, and point of impact in relation to the edge of the striking platform on a core. Clearly, these techniques can vary quite distinctively from the production strategy.

Epipalaeolithic

Chronological indicators

Microliths (Figure 22.1.2) are tools (see above) manufactured on small, elongated elements of debitage called bladelets, a frequent conventional definition that includes pieces less than 5 cm long and 1.2 cm wide. These are retouched, frequently into distinctive shapes, and often with very steep/abrupt retouch called backing (Figure 22.1.2). Different shaped microliths were employed with variable frequency at different periods. There is both chronological and spatial patterning in their use. It is clear that non-geometric types, for example, elongated bladelets with retouch that hardly modify the form of the bladelet, occur earlier in the Epipalaeolithic and after 16,000 bp (uncalibrated) microliths with geometric shapes predominate (Figure 22.1.2). After 12,000 bp (uncalibrated), lunates (Figure 22.1.2), a geometric microlith with a curved back, are typical. Statistical analysis of C14 dates confirms that an early Natufian can be distinguished by larger lunates with 'Helwan' retouch (Figure 22.1.2), that is, retouch cutting onto both faces of the microlith, creating an acute angled or sharp back. At later Natufian sites, however, smaller lunates, with bipolar abrupt retouch, that is, retouch that creates a steep blunt back by retouching from both faces, predominate (Byrd 1989: 163-67). Caution is necessary in the use of variation in the proportional occurrence of such microliths for chronological purposes because of evidence indicating the diversity and complexity of the factors influencing the use of such tools. The al-Azraq Project's work in eastern Jordan indicates that in some later Epipalaeolithic sites (al-Jilat 8 and 10) there are very low frequencies of geometric microliths, probably reflecting the distinctive nature of the activities carried out at those sites (Byrd 1988). Evidence from Jordan has, thus, disrupted simple views of the development of Epipalaeolithic industries and what might be chronological markers. A standard sequence as understood from west of the Rift Valley is unlikely to apply straightforwardly. A further good example of this is the micro-burin technique (Figure 22.1.1), a distinctive way of segmenting bladelets believed once to be a feature of the later Epipalaeolithic but now documented in eastern Jordan from the al-Azraq Basin as early as 20,000 bp (uncalibrated) (Garrard 1998: 143) and equally early in the Qalkhan of southern Jordan (Henry 1995: 225). Given that assemblage groups with chronological significance are mainly diagnosed upon the basis of the proportions of differently shaped geometrics, one might also ask whether it is always the case that there are clear distinctions between the shapes of such tools and how much triangles might grade into lunates, lunates into trapezes, trapezes into rectangles and where lies the overlap with backed and truncated bladelets and scalene bladelets. Furthermore, it may be difficult, in practice, to differentiate systematically between different shape categories. For example, when is a lunate with a flattened end a trapeze (Neeley and Barton 1994)?

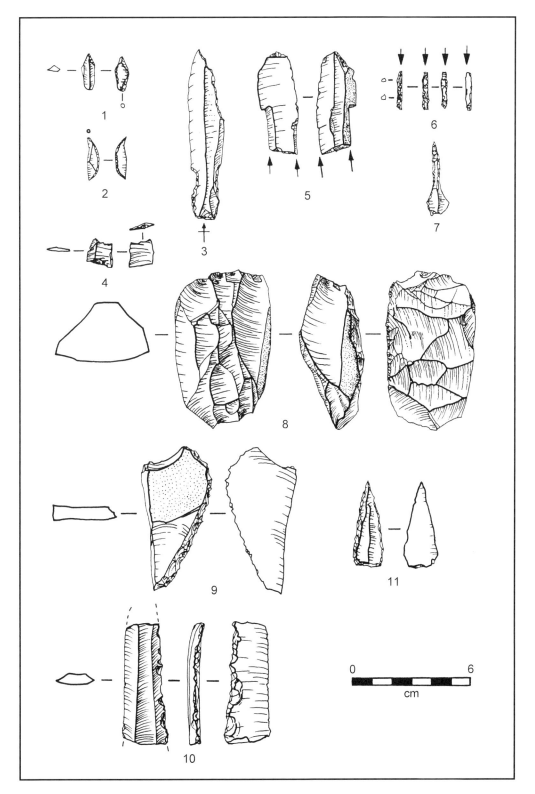

Figure 22.1. Chipped stone types from Jordan: 1. Microburin; 2. Lunate; 3. Al-Jilat knife; 4. Hagdud truncation; 5. Truncation burin; 6. Spall drill; 7. Bladelet piercer; 8. Naviform core; 9. Tabular scraper; 10. Canaanean blade; 11. Shuna point.

Technological trends

Why should this period be distinguished by the use of microliths and was there a particular advantage to the manufacture and use of microliths? Production of microliths would require relatively small amounts of raw material to generate a cutting or piercing edge. Microliths would have been part of composite tools made up of wooden hafts and shafts and would have provided great flexibility in tool design as they could be hafted in very diverse fashions to produce a large variety of tools employing the same basic components. Such tools could be easily refurbished by the replacement of one or two small damaged components. The tool elements themselves were highly portable, many easily carried in a pouch, bag or pocket. Such technology suits mobile groups for whom it would be advantageous to reduce the quantity and weight of material goods carried either on specific expeditions or camp shifts. It would also suit groups for whom it would be advantageous to refurbish tools rapidly, for example, on hunting and gathering expeditions where game might be available for only specific periods or where specific crops might be available for only a short time. In such cases, it would be inappropriate to spend much time on refurbishing or making new tools.

The nature of their sites suggests that many of the groups using microliths would have been relatively mobile. In the Natufian, however, some sites with substantial architecture appear, such as Wadi Hammeh 27 in Jordan (Edwards 1991), suggesting longer periods of occupation of sites. Nevertheless, microliths remained in use throughout the period. This is probably because they remained advantageous, in part, because mobility was a significant element in societies' activity, whether of whole communities in more arid areas (for example, al-Bayda' in the south or al-Azraq 18 in eastern Jordan) or components of communities in the Mediterranean zone. Further, task groups from longer-term settlements may regularly have exploited the landscape quite extensively. In addition, hunting clearly remained a significant activity whether in economic or status terms. Furthermore, it is likely that such a 'traditional' technology was deeply embedded in society. The issue then becomes why microlithis were abandoned rather than retained (see below).

Settlement patterns

The identification of the length of occupation and range of activities on sites has been considered important in understanding hunter-gatherer settlement patterns. This is because groups will occupy a range of settings in different ways, quite possibly at different seasons, thus exploiting different resources as part of a mobile settlement pattern. As a result, no one site will be representative of a community's overall behaviour.

Attempts have been made to establish a basic distinction between sites that are longer-term residencies from which task groups might have operated, the sites of such task groups, and sites that are more uniform in their occupational characteristics, suggesting regular moves of similar sorts of groups (Binford 1980). Similarly, attempts have been made to establish whether different sorts of groups congregated at different seasons and in different sorts of locations, as suggested by the study of modern hunter-gatherers (Henry 1995: 422).

Lithics have been used in the above-described studies. Byrd (1989: 185-86), for example, in looking at lithic variation on Natufian sites, pointed out the varying ratios between cores, debitage, microliths and other tools, and correlated patterning in such aspects with thickness of deposit and density of structures on sites. He suggested three groups of sites with greater and lesser degrees of permanence of occupation on this basis. He correlated this finding with the environmental zones occupied, combined with different functions of sites.

In the later Epipalaeolithic of southern Jordan (Hamran), sites on the Ras an-Naqab plateau show marked distinctions in terms of lithics with broadly contemporary sites in the lower elevations of Wadi Hisma (Henry 1995: 279-83). Densities of chipped stone on lowland sites are higher per cubic metre, directly reflecting more manufacture and discard of tools. Evidence for greater degrees of both manufacture and discard is likely to indicate longer occupations, particularly when associated with greater areas of occupation and depths of cultural deposit on the lowland sites (Henry 1995: 281, table 10.6). In addition, there were considerably higher proportions of microliths and perforating tools on lower-lying sites compared to higher proportions of varied retouched pieces in upland sites. These latter are more likely to have been manufactured on an *ad hoc* basis to meet diverse needs. This may reflect, Henry suggests, more variability in activities carried out between different, shorter-term occupations than between longer-term occupations. At lowland sites, there may have been more refurbishment and manufacture of microlith-mounted tools, possibly for hunting, but also possibly for harvesting plant materials. In addition, the high proportion of perforating tools presumably represents not just the greater frequency of manufacture of these tools, but also greater use of these for other manufacturing activities including possibly the manufacture of tools, items of clothing and ornamentation, and site facilities. Henry further interprets such differences as reflecting summer occupation of the piedmont and winter occupation of the lowlands. There is no direct evidence, however, of seasonality (through flora or fauna, for example) and other interpretations than the implied vertical transhumance are possible. These could include, of course, the greater mobility of communities and different resources exploited in higher elevations, whatever the season. Henry (1995: 330) interprets the much higher density of chipped stone at an early Natufian site in this area as indicative of longer occupation periods.

Activities

Chipped stone can also indicate the specialized nature of certain sites. Olszewski (Chapter 2, this volume) has already indicated the low proportions of microliths and high proportions of al-Jilat knives (Figure 22.1.3) on the site of al-Jilat 22. Micro-wear on these knives suggests cutting activities (Garrard 1998: 143). At late Epipalaeolithic al-Jilat 10, microliths are extremely rare, while large blades and blade tools are extremely common. In conjunction with faunal evidence, this suggests a limited butchering/carcass processing and possibly consumption event (Byrd 1988). High proportions of non-geometric microliths in early Epipalaeolithic sites in the al-Azraq Basin suggests sites related mainly to hunting activities. More diverse tools and variability of the type just discussed at al-Jilat 10 and 22 suggest a greater range of, and variation in, activities on sites and between sites in the late Epipalaeolithic, namely, some longer-term camps combined with the camps of task groups (Byrd 1988). Byrd's (1989) analysis of Natufian chipped-stone assemblages suggests that tools can distinguish three groups of sites: (1) where plant processing activities are important, which include, in Jordan, sites in the Mediterranean zone such as Wadi Hammeh 27; (2) where a range of larger tool types point to diverse activities; for example, al-Bayda' and the Black Desert sites of Khallat Anaza and Jabal as-Subhi fall into this category in the steppic areas of Jordan; and (3) in steppe/desert areas, sites dominated by geometric microliths, probably represent hunting encampments, namely, 'Ayn as-Saratan = al-Azraq 18 and sites in Wadi Judayid.

Assemblage groups: group identity or adaptation

Various researchers concerned with the Epipalaeolithic of the southern Levant have recognized geographically contiguous groups of sites with assemblages that have a number of common characteristics. As Olszewski (Chapter 2, this volume) notes, some might represent ethnically distinct populations. For Jordan, Henry (1995: 342-43) has been a major exponent of this view. It may be that adaptations by groups to different environments might play a role in explaining this spatial variability. In addition, Neeley and Barton (1994) have suggested, for example, that microlith form may relate to aspects of raw material acquisition and the nature of the manufacturing process, use life, and discard of the tools that included the microliths, as much as to ethnic affiliation. In the southern Jordanian area that Henry studied, an interesting situation exists where two of these assemblage groups, representative of wider contrasts, are found in the same area. Features of Henry's Middle–Late Hamran broadly parallel the Geometric Kebaran of the Mediterranean zone of the western southern Levant and his Madamaghan (in a different sense to that defined by Olszewski, see Chapter 2, this volume), the Mushabian of the southern arid areas. In the Madamaghan, elongate, non-geometric retouched bladelets dominate (straight-backed, arched-backed, and scalene bladelets) (Henry 1995: 301) and evidence of the micro-burin technique (Figure 22.1.1) is common (Henry 1995: 305). Geometrics are rarer, but consist of the trapeze rectangles and then Helwan (Figure 22.1.2) and other lunates typical of Mediterranean zone industries on slightly later sites. The Hamran assemblages have high frequencies of non-geometrics, consisting of straight-backed, bladelet segments, often truncated. In addition, trapeze-rectangle shapes (Henry 1995: 341) dominate geometrics. Late Hamran assemblages see the appearance of lunates, especially Helwan lunates (Figure 22.1.2). The micro-burin technique is absent (Henry 1995: 341). The essential differences relate to the documented use of the micro-burin technique, the proportion of geometrics, and the shape of the non-geometrics. Since the latter could be related largely to the use of the micro-burin technique, the use or lack of use of this technique might explain a number of the other contrasts. It is plausible that variations in the geometric microliths are as likely to relate to the type of tools used and the nature of their refurbishment as to be indicators of ethnic affiliations. In this situation, such structured, spatial variation could be explained by varying activities carried out in different settings and/or seasons. Equally, the type of non-geometric microlith desired might relate to the type of tools used and thus influence the employment of the micro-burin technique (Figure 22.1.1). On the other hand, the micro-burin technique itself might well be a culturally embedded technique only communicated through specific and limited networks of learning and thus a good reflection of ethnicity, at least at specific periods. These alternatives, currently difficult to resolve, reflect some of the problems of the interpretation of variation between groups of lithic assemblages. However, experimentation and replication might demonstrate the effectiveness of different bladelet-segmentation processes for producing different products and allow us to understand this variability better.

The late-final Hamran assemblages with lunates and microburin technique are thus like the Madamaghan and Natufian and one wonders whether these assemblages are functional variants created by closely related groups spanning the immediately pre-Natufian into the period of the earliest classically Natufian assemblages. Given that there are few absolutely distinctive tool types that currently characterize these assemblages and assemblage groups, but rather proportions of certain sorts of microliths and the use of a distinctive bladelet-segmenting method, it may be that detailed analysis of the retouch on the microliths and its relation to microlith morphology might be more clearly diagnostic of reflections of group identity, if such exist in the chipped-stone record of the Epipalaeolithic. The restricted use of very specific tool types, such as the al-Jilat knives (Figure 22.1.3), may point to that possibility. Clearly, these were used at a specific time period, as far as evidence allows at the moment, in a restricted area of eastern Jordan.

Neolithic

Technological trends

There is undoubtedly considerable significance in the disappearance of microlithic technology during the transition to sedentary-farming communities, which occurred from the Natufian through the Pre-Pottery Neolithic A (PPNA) into the Pre-Pottery Neolithic B (PPNB). During this period, the chronological status of sites can be established by reference to the projectile point types recovered (cf. Chapter 3: Figure 3.2). The points of the Neolithic contrast morphologically with microlithic predecessors because of their larger individual size (cf. Chapter 3: Figure 3.2). They are generally elongated bladelets and blades, usually retouched at one end to form a pointed tip, or utilize blades with naturally pointed tips, and with retouched 'tang' areas opposite the relatively thin, sharp point (cf. Chapter 3: Figure 3.1). They may not all invariably have served as the tips of projectiles, but, almost certainly, many did during much of the Neolithic. Microliths do not disappear quickly or completely. Many PPNA assemblages have debitage/blanks where small blades and bladelets are important. The PPNA Khiam points are small. Hagdud truncations (Figure 22.1.4), which are clear microliths (short bladelet segments truncated at either end), are used from PPNA into Early PPNB. It seems plausible that the use of large numbers of microlith-fitted tools declined in conjunction with the degree, frequency and scale of mobility and, possibly, the relative economic importance of hunting. Hunting, nevertheless, probably remained significant, first economically, as reflected by the large proportions of gazelle in eighth-millennium BC (uncalibrated) faunal assemblages from the southern Levant generally (Bar-Yosef 1995: fig. 3) and, secondly, socially (Kensinger 1989), as reflected by the presence of the 'projectile' points themselves. Whatever their other uses, these were almost certainly regularly used as projectile tips, largely arrowheads given their weight and tang size (Roodenberg 1986: 32-36), presumably mainly for hunting, although we should not discount the possibility of their use in warfare.

Assemblage groups: group identity and adaptation

The change in the nature of projectile armatures may well reflect a change in the status of hunting as an activity. The large size of these projectile points relative to microliths and the elaboration and variation witnessed through time and across space in their design (cf. Chapter 3: Figure 3.2) hint at the possibility of emblematic style, indicative of group or more personal identity. This would be associated with differing attitudes to, and the different role of, hunting in a more sedentary, grain-harvest-dominated society. The more ostentatious points perhaps indicate an enhanced social status of hunting and hunters (Kensinger 1989:

25) in a world where its contribution to subsistence may actually have declined. An example of the reflection of such identities may be provided by a combination of observations made by Gopher (1994: 249-51), Rollefson (Chapter 3, this volume), and the present writer. These observations suggest that large Jericho points with large wings/barbs may be a typical feature of the Jordan Valley and immediately adjacent areas in the Middle PPNB of the early seventh millennium BC (uncalibrated). However, at contemporary sites to the east and south, Jericho points are rarer, smaller and have much smaller barbs and in the east, at least, Amuq and Byblos points (cf. Chapter 3: Figure 3.2) form parts of distinct Middle PPNB assemblages. This may marry with evidence for distinct production techniques in eastern Jordan (Baird 1994).

In the introduction to this chapter, a distinction was made between production strategy and technique. The latter refers to the more idiosyncratic features that relate to how the core may be held or immobilized during knapping, the force, angle, and placement of the blow, and the nature of the hammers used. Very specific techniques are highly likely to be learned, and thus shared, through direct face-to-face interaction. Indications that techniques occur as part of regional traditions may indicate learning networks typical of closely affiliated communities. Work on technique indicators of groups in the al-Azraq Basin suggests significant contrasts with knappers to the west and southwest at broadly contemporary sites such as 'Ayn Ghazal, Basta and Jericho during the PPNB (Baird 1994).

Such identities are also reflected by a variety of habitual practices as represented by lithics, which effectively document the way that 'functional' adaptations and cultural identities are tightly entwined within the material record. Rollefson (Chapter 3, this volume) outlines the characteristics of the 'burin sites' of northeastern Jordan, particularly the northeastern steppe. Burins (Figure 22.1.5) are blades or flakes with elongated slivers or splinters, termed 'spalls' (Figure 22.1.6), removed from along the edge of the blank. This produces a blade or flake with a robust edge and a spall, either of which are suitable for further use as a tool. Varieties of burins are defined by the location of spall removal and the sort of platform the spall was removed from, for example, a truncation (Figure 22.1.5) or previous spall removal (dihedral) burin. Sites in Wadi al-Jilat have large proportions of the truncation (Figure 22.1.5) and angle burins typical of burin sites, alongside a range of other tools, and evidence of intensive but still relatively short-term occupations compared to moister areas. This suggests that pure burin sites with few other tools represent yet shorter or more specialized episodes of occupation, in which a low range of activities were undertaken, at least in relation to those reflected in the preserved material culture. These activities would include manufacture of the blanks for removal of burin spalls, spall removal, manufacture of drill bits (Figure 22.1.6) from the burin spalls, manufacture of drills themselves, and stone (possibly shell and wood) bead manufacture using the hafted spall drills (Figure 22.1.6). Evidence from some sites indicates that at least, on occasion, the burins themselves were used, so we may include such activities employing robust-edged burins (Figure 22.1.5). These activities presumably fitted neatly into the hunting, cultivating, gathering and, later on, herding round of groups exploiting the eastern steppe. Thus, they give us some indication of activities and the place of those activities in these eastern societies. Given the frequency with which burin sites are encountered, this was an important place. However, other groups in the southern Levant also engaged in bead manufacture on a large scale. The site of Fidan A, now excavated by Adams and Levy as Fidan 1, is a good example where huge quantities of piercing tools are found (personal observation and Raikes 1980: figs. 6-7). The additional important factor about the burin sites is that they indicate not just a particular adaptation/set of activities but a particular culturally enshrined technology. The manufacture of piercing tools from burin spalls (Figure 22.1.6) represents a distinct practice compared to the manufacture of piercing tools from blade and bladelet elements (Figure 22.1.7). The latter has its own distinct features in the case of the Fidan sites where in the PPNB at Fidan A/1 small bladelets were produced from naviform cores which had been reduced unusually small for this purpose (personal observation). The economic, technological and scheduling activities of the groups in northeastern Jordan were clearly bound together as part of a distinctive cultural identity. This identity may also be reflected by techniques of manufacture specific to northeastern Jordan and contrasted with neighbouring areas (see above).

The burin sites also provide us with important information relating to continuity of populations in at least northeast Jordan throughout a period of economic and social change. Rollefson (Chapter 3, this volume) has described the introduction of herding (to the northeastern steppe at c. 6000 BC [uncalibrated]) (Garrard 1998: 145) and some of the changes in the moister area from the PPN to Pottery Neolithic, including the decreasing elaboration reflected in the material culture (Baird 1997: 372-74). The work of the al-Azraq project clearly demonstrates that the burin sites span these changes. The inter-connected economic, scheduling, and technological practices they reflect commenced in the Middle PPNB, c. 6700 BC (uncalibrated), and run to at least 5000 BC (uncalibrated), encompassing the introduction of herding.

Specialization and exchange

Other equally important issues during the Neolithic can be approached through the study of chipped stone. There is evidence for much social complexity in the Neolithic of Jordan. Such complexity involves the diversification of social and economic roles and the related status of individuals and families (Baird 1997: 375-76). Specialization and exchange may indicate one direct measure of such economic diversification. Since lithic tools allow one to approach manufacture and, more importantly, the context of manufacture, they offer approaches to understanding the emergence of craft specialization. During the PPNB, elongated blades used for many tools were produced from so-called naviform cores, that is, large cores with two opposed, angled platforms and an elongated narrow and relatively flat removal surface between the platforms, and a crest opposite the main removal surface (Figure 22.1.8). This strategy requires careful preparation of the core and a certain amount of skill to perfect (Wilke and Quintero 1994; Gebel 1996: 268). Skill levels are difficult to judge and, therefore, are not an adequate basis on which to suggest specialization when considering ancient societies, where many of the skills may have been practised from infancy. However, the fact that this was a time-consuming production strategy, with specific raw material requirements, may beg the question whether specialization emerged. One way to document this, that allows us to approach not simply the question of whether any specialization may have developed but also the nature of that specialization, is to look at the distribution of the evidence of manufacture on sites and the evidence of particular modes of production. Thanks to the work of Wilke and Quintero (1994), we can document the different stages of naviform production effectively.

Thus on the eighth–seventh millennium BC (uncalibrated) sites of Wadi al-Jilat and the al-Azraq area, naviform cores (Figure 22.1.8) and the full range of production debris from them are found in and around all structures. In the sixth millennium BC (uncalibrated), when naviform core use disappears, a similar situation persists with evidence for all phases of production, using other strategies, located in and next to structures. In some cases, this production debris has been dumped from neighbouring areas; in others, it relates to production within the structures. It seems clear that there were a number of knappers at work within each group and probably at least one per co-habitational unit. Given the sizes of sites and structures, it is clear that these groups, and indeed co-residential units, were of small size. We might therefore deduce that a significant proportion of these groups were involved in knapping. In these contexts, it would appear that naviform production is not to be seen as necessarily a specialist craft. This does not mean it was not so in other contexts, and the low proportion of naviform cores (Figure 22.1.8) and non-uniform distribution of naviform production waste may hint at a different situation in the large settlements in the moister areas (Gebel 1996: 268). But whether specialization was economic and went beyond the provisioning of extended family units by individuals is another matter that only detailed contextual analysis in large settlements might resolve. Indeed, it may be that the structuring of activities in these large communities was quite distinct from that in smaller communities. Naviform core reduction may have occurred in spatially discrete areas (Gebel 1996: 268-69) and on an episodic basis to fit in with other, possibly seasonally structured, activities. In such a scenario naviform production evidence would only be found in certain limited settings in settlements or on the edge of settlements (Gebel 1996: 269), a function of the new organizational demands of life in the relatively large agricultural communities of the Middle to Late PPNB.

Technological trends from Pre-Pottery to Pottery Neolithic

Changes from PPN to Pottery Neolithic have already been indicated (Baird 1997: 372-74). PPNC lithic assemblages seem to be direct ancestors of those of the Pottery Neolithic. Indeed, on desert sites where pottery is rare, it is currently impossible to distinguish between assemblages contemporary with the PPNC and those contemporary with the earlier Pottery Neolithic in the absence of C14 dates, hence the use of the term Late Neolithic in those areas. Significant changes occur from the seventh to sixth millennium BC (uncalibrated). The frequency of projectile points in assemblages declines in the Mediterranean zone and its immediate fringes. Projectile points also become smaller (cf. Chapter 3:

Figure 3.2 top row). The naviform strategy disappears, to be replaced by more expedient technologies, at the transition from PPNB to PPNC. Associated with this is a decline in the use of finer-grained flint in large tabular- or flat-cobble form, procured from specific settings. Wadi and colluvial cobbles are used more frequently. Biface tools become more common. The density of lithics on Mediterranean zone sites seems to decline throughout the Pottery Neolithic. Because of these changes and partly because of the reduction in the production of long blades, the range of relatively standardized tools becomes more limited into the Pottery Neolithic B and Wadi Rabah related assemblages. Basically, only sickles, transverse arrowheads, tabular scrapers (Figure 22.1.9) and a range of woodworking tools show much standardization by the Pottery Neolithic B and period of the Wadi Rabah assemblages. Ad hoc tools, that is, a variety of simply or hardly modified blades and flakes, become common, often referred to as an expedient tool kit (Banning and Siggers 1997: 325). These developments might plausibly be associated with a number of trends. Hunting clearly becomes less important in the Mediterranean zone as indicated by changes in faunal assemblages. By the sixth millennium BC (uncalibrated), there is a clear decline in hunted animals in the Mediterranean zone faunal assemblages. In the seventh millennium BC (uncalibrated), the degree to which faunas represent hunting as opposed to herding is variable and far from clear. This change might account not just for a decline in the frequency of projectile points, but also a decrease in the range of standardized tools and change in character of tools as butchery and hide processing became less common, particularly if animal husbandry came to involve exploitation of secondary products. With these shifts, the context of butchery may well have changed. More butchery may have taken place on or close to settlements rather than at game kill sites and, thus, a less standardized and curated tool kit may have been required for this purpose. Before pottery was introduced, skin and wooden containers would have had a particular importance. Once pottery was introduced, these items may not have been so varied, frequent, or symbolically loaded, although their use would have certainly continued. The cumulative result of these changes was a reduction in the range, standardization, and diversity of uses of chipped stone tool kits. This meant not just a quantitative reduction in chipped stone use, but also a major change in the way tool kits might be designed and assembled (Banning and Siggers 1997: 329). These processes were lengthy and cumulative and should not all be associated directly with the more specific changes from PPNB to PPNC, which may well have had a social dimension and probably reflect an additional factor in these changes, that is, the status of the knapper in society (Baird forthcoming).

Chalcolithic and Early Bronze Age

Technological trends

The trends characterizing the development of lithic production technology and tool kits from earlier to later Neolithic continue with the introduction of metal tools. A consideration of the development of chipped stone manufacture and use, alongside the use of metal tools for similar purposes, can be instructive. In the southern Levant, it is during the fifth millennium BC (uncalibrated) that metal tools appear in significant but low numbers in the archaeological record (Levy and Shalev 1989). One suspects the readily recyclable nature of metal means that the low numbers of tools and other items we find masks their true frequency, whatever that was, as hinted by the Nahal Mishmar hoard. Intriguingly, in the Chalcolithic of the southern Levant, copper and flint tools apparently designed for similar purposes have a lengthy period of contemporary use. Certainly, from the middle of the Chalcolithic to the beginning of the Early Bronze Age, it would appear that chisels, axes and adzes were frequently made out of both flint and copper. Equally, awls/piercers are found in both media. This may reflect limited differences in performance between these media. Or it may mean that significantly better copper tools were less readily available, more 'costly' to procure or associated with high status and accessible only through certain social networks. If we see a thousand-year coexistence we must ask why these flint tools disappear when they do. It would appear that flint woodworking tools disappear at the beginning of or during early EB I as evidence from the long ash-Shuna sequence and from 'Ayn Shadud might suggest (Rosen 1997: 98). We might then speculate that it was at this point that metal equivalents were superior enough or readily available enough to complete the substitution. Since the basic design and composition of metal adzes, axes and chisels changes little from Chalcolithic to EB I, it seems likely that these metal tools became more readily available at this point. This indicates substantial changes in the social aspects of the organization of metal and lithic production at the transition from Chalcolithic to EB I. Interestingly, perforating tools continue to be common in the EB I period, suggesting flint retained its value for these purposes.

Specialization, exchange, contextual analysis and symbolic significance

It seems likely that metal production involved a certain degree of specialization. Is there evidence within the context of lithic production for the appearance of specialization during these periods? Certainly by the EB I period, from 3600 BC (calibrated) onwards, there is such evidence. This takes the form of the appearance of Canaanean blades (Figure 22.1.10) on sites. At many sites, there is no evidence of Canaanean blade production in the form of cores from which such blades could have been manufactured. The material used for such blade production was usually distinctive and often relatively fine grained. It may well be that Canaanean blade cores were reworked for the production of other sorts of tools. At some sites, there is evidence for tools made on similar material; on others, however, there is not, for example, Tall ash-Shuna. On many sites, production debris in such material is not present either. In such circumstances, it is likely that production settings for Canaanean blades (Figure 22.1.10) were quite restricted to certain areas of sites or to only certain sites. The methods used to produce such blades, although not necessarily complicated, require a certain degree of sophisticated knowledge. Thus, it seems likely that Canaanean production involved a limited number of 'specialists' supplying much of society, for example, those involved in agricultural production, on an exchange basis.

Evidence from Anatolia suggests that Canaanean blade production appears earlier in the north. At Hacenebi, it is documented in Phase A in the first centuries of the fourth millennium BC, before the beginnings of EB I in the south (Stein et al. 1996). The technology is thus likely to have been adopted in the southern Levant from the north. However, the social context and organization of Canaanean blade production (Fig. 22.1.10) in the south may well have been different from that in the north, as Canaanean blade cores or evidence of manufacture appear more frequently in northern (see, for example, Behm-Blancke 1992; Hanbury-Tenison 1983; Stein et al. 1996) than in southern sites. It may be that Canaanean blade producers were more commonly located in settlements in the north, possibly because demand was higher in larger northern settlements. It is important to appreciate, however, that similar artifacts and production technologies do not automatically imply that production is organized in the same way. The appearance of this new technology almost certainly represents a change in the organization and technology of production of sickle blades rather than in the hafting and use of the artifacts concerned. This new technology and associated organizational changes may not have been adopted uniformly. There is evidence, again from ash-Shuna, that 'Chalcolithic' production technologies manufacturing 'Chalcolithic' style backed and truncated blades continued alongside Canaanean manufacture for some time in EB I. Certainly, a number of Canaanean blades are backed and truncated in a manner that suggests that they would have been hafted like their 'Chalcolithic style' predecessors and possible contemporary items. In addition, some southern areas of Jordan and Israel (Rosen 1997: 58) produce local variants of non-Canaanean backed sickle blades (the Uvda, Bab adh-Dhra') alongside Canaanean blades. The latter are documented in the south at Bab adh-Dhra', Numayra and Wadi Fidan 4 (Adams 1997: 653).

At Tall ash-Shuna in EB I, we see another distinctive tool type in large numbers, but barely documented elsewhere. As caches and individual items, large numbers of Shuna points (Figure 22.1.11) were recovered at the site (Baird 1987: 477). Their character suggests that they might well be projectiles. If so, such numbers might indicate warfare given the very limited evidence for hunted fauna at the site. Their highly standardized size, evidence for 'mass' production and absence of evidence indicating production within the excavation areas at ash-Shuna suggest again specialized and limited production episodes.

Evidence for specialized production of chipped stone in the Chalcolithic is less conclusive. Sickle blades are the most common formal tool, along with axes and chisels in most Chalcolithic settlements. Detailed studies of the Early Chalcolithic levels of Tall ash-Shuna in the north Jordan Valley suggest chipped-stone manufacture was common in most of the architectural complexes and open areas at the site. This probably included manufacture of blade blanks for sickle blades. Indeed, in some cases, manufacture of the tools themselves in these areas at ash-Shuna is indicated. At Chalcolithic Abu Hamid, a number of locations for chipped stone manufacture or dumped debris therefrom have been found in open areas around domestic structures, although here manufacture involved mainly flake production for a range of flake tools and possibly axe resharpening, without clear evidence for sickle-blade manufacture (Barberan 1997).

Other aspects of the chipped-stone distribution at ash-Shuna well illustrate how such studies can help in the understanding of these settlements. In the Early Chalcolithic, the distribution of axes, adzes, and chisels, as well as sickle segments, indicates that all of the major building complexes exposed have evidence, not just for the manufacture of chipped stone, but for the storage, refurbishment and/or discard of chipped-stone agricultural and woodworking tools. This points to a situation where the Early Chalcolithic households documented at ash-Shuna are all engaged in agriculture and the manufacture of tools to support this activity. The manufacture of the axe class is another matter. It may be that these were the products of specialist manufacture as suggested by limited evidence for biface production on most sites. The occurrence of specific and limited settings for the manufacture of such agricultural and craft tools is likely, whether or not their manufacture was in the hands of a limited number of specialists. This is because of its episodic character and the fact that it is likely to have been related to activity during specific seasons, for example, manufacture of sickle blades as part of gearing up for harvest. Such activities could well be seen as involving groups of producers carrying out production in specific and potentially off-site settings.

The Chalcolithic and Early Bronze Age tabular scrapers (Figure 22.1.9) are an informative chipped-stone item. In studying material from the western part of the southern Levant, Rosen (1997: 75) documented a decrease in frequency of these items from north to south and from arid areas to the Mediterranean zone. He suggested a pattern of down-the-line exchange. Evidence for manufacture comes from sites in the Negev, but not farther north. It seems that these items were also manufactured in eastern Jordan where in the area of al-Jilat, for example, blocks from which such tabular scrapers have been removed are extensively scattered (personal observation), as in the al-Jafr basin. At a Chalcolithic site in the al-Azraq basin, al-Jilat 27 tabular scrapers are common. Tabular scrapers (Figure 22.1.9) are also common in Henry's (1995: table 15.2) Timnian sites of the Hisma area, dated to the Chalcolithic and Early Bronze Age and in Faynan/Fidan (Adams 1997: 653) and in Early Bronze levels at Dayr 'Ayn 'Abata in the northeast 'Arabah (personal observation). It would seem that they are common in the arid areas of southern and eastern Jordan and less common in the Mediterranean zone and northern sites like ash-Shuna where, for example, they are only 0.78 percent of the Early Chalcolithic tool assemblage and 4.25 percent of the EB I assemblage from the 1983–84 excavations (Baird 1987). Again, some movement of materials or finished products is suggested, in this case probably the latter, since tabular scraper cores seem only documented in the arid areas. Their frequency in the arid areas may relate to the presence of suitable raw material since tabular material is common in eastern Jordan, for example, it also probably relates to the specific uses to which they were put. Henry (1995: 372) suggests that their association with communities to whom herding was undoubtedly important might indicate that they were for shearing wool. Others have suggested that they might be butchery items (McConaughy 1979). Occasional examples have silica sheen that Henry (1995: 372) suggests might result from shearing. His experiments, however, were unable to document this. Tabular scrapers were likely used for a variety of purposes, with a variety of significance depending partly on the communities using them and how they were acquired. While plausibly items of exchange, in those communities that used them regularly in the arid zones their frequency and the scatter of cores across the al-Jilat landscape do not suggest specialist production. This accords with Rosen's view of manufacture in the Negev (Rosen 1997: 109). The low frequencies of tabular scrapers in Mediterranean zone sites suggest movement from communities where use was common to those where it was not. In such a process, the significance and function of these items may well have changed. That it may have changed through time is suggested by the fact that some Early Bronze Age tabular scrapers have incised cortex in contrast to Chalcolithic predecessors. We can look at the evidence of the contexts from which tabular scrapers have been recovered on sites in Jordan in order to gain some idea of their role/function. These contexts are predominantly domestic in nature. Tabular scrapers are found alongside other household tools on both arid zone and moister zone sites, for example, in the Jordan Valley at Chalcolithic and Early Bronze Age ash-Shuna, Early Bronze Age Tall Abu al-Kharaz, and in the arid zone at Chalcolithic al-Jilat 27 and the Hisma sites. These domestic and functional roles should not be taken to exclude a simultaneous symbolic significance that might attach to exotic items in areas to which they were imported, but it does indicate the household settings of their use and suggests a functional role as well. At Dayr 'Ayn 'Abata, tabular scrapers may be associated with Early Bronze Age mortuary practices or domestic use of the cave; it is unclear which. They may then have had different meanings in different contexts, sometimes utilitarian and functional as suggested for many sites in Jordan, possibly also symbolically significant in other settings, for example, at Dayr 'Ayn 'Abata or on sites like Mizpe Shalem and in the Biq'at Uvda (Rosen 1993: 44).

Conclusions

This chapter has attempted to indicate the varied ways in which the evidence from chipped stone is being used to provide insights into important aspects of human behaviour in the archaeology of Jordan. Chipped-stone studies have been used to provide a relative dating tool, to assess the nature and reasons for variations in activities carried out by communities at different sites or on sites, to investigate the broadest behavioural distinctions in identity between contemporary communities that might relate to what we understand as ethnicity, to study issues of specialization and exchange and understand the significance of technological development for society and economy.

References

Adams, R.
- 1997 On Early Copper Metallurgy in the Levant: A Response to Claims of Neolithic Metallurgy. In H.G. Gebel, Z. Kafafi and G. Rollefson (eds.), *The Prehistory of Jordan. II. Perspectives from 1997*, 651-56. Berlin: Ex oriente.

Baird, D.
- 1987 A Preliminary Analysis of the Chipped Stone from the 1985 Excavations at Tell ash-Shuna North. *Annual of the Department of Antiquities of Jordan* 31: 461-80.
- 1994 Chipped Stone Production Technology from the Azraq Project Neolithic Sites. In H.G. Gebel and S. Kozlowski (eds.), *Neolithic Chipped Stone Industries of the Fertile Crescent*, 525-42. Berlin: Ex oriente.
- 1997 Goals in Jordanian Neolithic Research. In H.G. Gebel, Z. Kafafi and G. Rollefson (eds.), *The Prehistory of Jordan. II. Perspectives from 1997*, 371-82. Berlin: Ex oriente.
- forthcoming Explaining Technological Change from the 7th to the 6th Millennium BC in the South Levant. In I. Caneva (ed.), *Beyond Tools: Reconsidering Definitions, Counting and Interpretation in Lithic Assemblages*.

Banning, E. and J. Siggers
- 1997 Technological Strategies at a Late Neolithic Farmstead in Wadi Ziqlab, Jordan. In H.G. Gebel, Z. Kafafi and G. Rollefson (eds.), *The Prehistory of Jordan. II. Perspectives from 1997*, 319-31. Berlin: Ex oriente.

Barberan, C.
- 1997 Trois ateliers de Taille de Silex à Abu Hamid (Jordanie). In H.G. Gebel, Z. Kafafi and G. Rollefson (eds.), *The Prehistory of Jordan. II. Perspectives from 1997*, 383-94. Berlin: Ex oriente.

Bar-Yosef, O.
- 1995 Earliest Food Producers—Pre Pottery Neolithic (8000–5500). In T. Levy (ed.), *The Archaeology of Society in the Holy Land*, 190-204. London: Leicester University Press.

Behm-Blancke, M. (ed.)
- 1992 *Hassek Höyük naturwissenschaftliche Untersuchungen und lithische Industrie*. Tübingen: Wasmuth.

Binford, L.
- 1980 Willow Smoke and Dogs' Tails: Hunter-Gatherer Settlement Systems and Archaeological Site Formation. *American Antiquity* 45(1): 4-20.

Byrd, B.
- 1988 Late Pleistocene Settlement Diversity in the Azraq Basin. *Paléorient* 14: 257-64.
- 1989 The Natufian: Settlement Variability and Economic Adaptations in the Levant at the End of the Pleistocene. *Journal of World Prehistory* 3/2: 159-97.

Edwards, P.
- 1991 Wadi Hammeh 27: An Early Natufian Site at Pella, Jordan. In O. Bar-Yosef and F. Valla (eds.), *The Natufian Culture in the Levant*, 123-48. International Monographs in Prehistory. Ann Arbor, MI: Prehistory.

Garrard, A.
- 1998 Environment and Cultural Adaptations in the Azraq Basin: 24,000–7,000 BP. In D. Henry (ed.), *The Prehistoric Archaeology of Jordan*, 139-48. British Archaeological Reports, International Series 705. Oxford: British Archaeological Reports.

Gebel, H.G.
- 1996 Chipped Lithics in the Basta Craft System. In S. Kozlowski and H.G. Gebel (eds.), *Neolithic Chipped Stone Industries of the Fertile Crescent and their Contemporaries in Adjacent Regions*, 261-70. Berlin: Ex oriente.

Gopher, A.
- 1994 *Arrowheads of the Neolithic Levant*. Winona Lake, IN: Eisenbrauns/American Schools of Oriental Research.

Hanbury-Tenison, J.
- 1983 The 1982 Flaked Stone Assemblage at Jabal Aruda, Syria. *Akkadica* 33: 27-39.

Henry, D. (ed.)
- 1995 *Prehistoric Cultural Ecology and Evolution: Insights from Southern Jordan*. New York: Plenum.

Kensinger, K.
- 1989 Hunting and Male Domination in Cashinahua Society. In S. Kent (ed.), *Hunters as Farmers*, 18-26. Cambridge: Cambridge University Press.

Levi Sala, I.
- 1996 *A Study of Microscopic Polish on Flint Implements*. British Archaeological Reports, International Series 629. Oxford: British Archaeological Reports.

Levy, T. and S. Shalev
- 1989 Prehistoric Metalworking in the Southern Levant: Archaeometallurgical and Social Perspectives. *World Archaeology* 20/3: 352-72.

McConaughy, M.
- 1979 Formal and Functional Analysis of Chipped Stone Tools from Bab edh Dhra. Unpublished PhD dissertation, Ann Arbor, MI: University Microfilms.

Neeley, M. and C. Barton
- 1994 A New Approach to Interpreting Late Pleistocene Microlith Industries in Southwest Asia. *Antiquity* 68: 275-88.

Newcomer, M.
- 1975 Punch Technique and Upper Palaeolithic Blades. In E. Swanson (ed.), *Lithic Technology*, 97-102. The Hague: Mouton.

Raikes, T.
- 1980 Notes on Some Neolithic Sites in the Wadi Araba. *Levant* 12: 40-60.

Roodenberg, J.
- 1986 *Le mobilier en pierre de Bouqras*. Istanbul: Nederlands historisch-archaeologisch instituut te Istanbul.

Rosen, S.
- 1993 Metals, Rocks, Specialization, and the Beginning of Urbanism in the Northern Negev. In A. Biran and J. Aviram (eds.), *Biblical Archaeology Today 1990: A Pre-Congress Symposium Supplement—Population, Production and Power*, 41-56. Jerusalem: Israel Exploration Society.
- 1997 *Lithics after the Stone Age*. London: Altamira.

Stein, G. *et al.*
- 1996 Uruk Colonies and Anatolian Communities: An Interim Report on the 1992–1993 Excavations at Hacinebi, Turkey. *American Journal of Archaeology* 100: 205-60.

Wilke, P. and L. Quintero
- 1994 Naviform Core and Blade Technology: Assemblage Character as Determined by Replicative Experiment. In H.G. Gebel and S. Kozlowski (eds.), *Neolithic Chipped Stone Industries of the Fertile Crescent*, 33-60. Berlin: Ex oriente.

23. Pottery Technology

Henk Franken

Definition and purpose

Pottery technology, in archaeology, is the study of ancient techniques of pottery making. It also includes anything that is connected with the life and human conditions of the potters and with the trade of pottery. It thus has a strong anthropological component. The study of ancient techniques of pottery production has great potential for the discovery of cultural traits like the level of technological knowledge and inventiveness, as well as of markets and interregional trade of pottery (Orton *et al.* 1993). It also replaces the subjective elements in comparative pottery studies by creating a more objective base for selecting relevant pottery attributes for typological work, while revealing the problems of the existing practice of using pottery for the purpose of dating archaeological levels.

Pottery technology does not replace the description of the shape of pottery but adds an explanation of that shape. It tries to analyse the production of the ceramic vessel by identifying the type of clay and other raw materials used. It describes the construction of the pot step-by-step, the tools and the firing technique used. It identifies the conditions that determine the technical level of the work and its history, and, as a result, can often explain how and why a given shape was made.

The requirements for a technological study of ancient pottery are knowledge of, and experience with, the production of pottery from natural clays (in contrast to factory-composed clays), and a working knowledge of clay chemistry as well as facilities for testing clays for their properties of workability, for example their degree of plasticity. This also includes mixing clays with other clays or non-plastics, and making firing tests under various conditions such as reduced flow of oxygen during the kiln's cooling off period.

This type of research does not focus primarily on the chemical composition of the clays used in antiquity (Rice 1987: 51). Analysis of the chemical composition of clays is required, however, if the research aims at identifying the origin of the clays used by the potters in antiquity. Such analyses are also often useful for the understanding of the reaction of the clay to firing at various temperatures. However, without an exact chemical analysis it is possible, in most cases, to make reliable statements about the production process and the raw materials used. This is quite fortunate because when adequate skill and knowledge are available, large samples of sherds can be processed at little cost with the aid of comparatively cheap equipment (van As 1992: 1-3).

Clays used by the ancient potters

At the very beginning, when a potter's family settled on a suitable site to set up a business, the potter must have done some preliminary search for and testing of the local clays. He must have found clay that had the right properties for his pots. What the potter could make largely depended on what his father had taught him, including the ability to recognize the type of clay that suited his purpose and his technical skill. In the case of a village population—where the women used to make the pots for their own households—the same applies. One would expect that this coarse kitchenware could be made of any soil containing clay but this is not the case. Mixing the clay with non-plastics may produce a very easy-to-handle substance, but observation of potters in present-day communities generally shows that much care is taken to achieve the right substance. The production method the potters employ is closely related to the properties of the clay substance they prepare. The test of good workable clay comes when the finished pot is put aside to dry or when it is fired in the kiln. Pots should not crack during firing; but how does one prevent this cracking? The potters' answer is invariably the same: we do exactly what we have been told to do or not to do by our fathers and/or mothers. The firing of pots is the test of sound work: cracked pots indicate bad work.

Pottery technology can identify the mistakes that lead to failures. This is done in two steps. First of all, a number of constant features of clays and the behaviour of clays and clay mixtures are described in the literature. Secondly, tests can be applied to detect aspects particular to a specific tradition of pot making. To the constants belongs a whole series of distinctive features classifying the nature and the properties of potter's clays: primary or secondary clay (clay that has been transported by water from the site of its origin), plastic or lean clay, a thrower's clay or not a thrower's clay, an alkaline clay or an acid clay, and so on. The second step is an analysis of the pottery-making techniques to replace such general and often misleading descriptions like 'hand-made' or 'wheel-made'. Instead of the latter, a description of the steps taken by the potter in the construction of the vessels is given. It can easily be demonstrated that there are countless ways of making pottery shapes. But potters from a common tradition will use the same methods and apply them to their entire production.

Study of the temper materials

Analysis of the non-plastics, whether of organic or inorganic origin, that the potter added to the clay plays an important part in the study of temper materials. This is so since non-plastics can change the properties of a clay body and turn it into a better product. For instance, if organic material such as dung is added to a clay several months before it is to be processed, it will, through bacterial action, make better thrower's clay. Otherwise, manure from herbivores, dried and mixed with clay and used without delay, will also strongly affect the properties of the clay. It will make the pot porous, reduce its weight and save fuel because it burns inside the pot wall. When the firing period is short, one may find carbonized organic matter in the core of the sherd. A short firing time will always leave the core of the sherd black, because burning out the carbon takes time. One other important effect is that shrinkage of the clay body during drying and firing is reduced by organic matter, and right through the ages one finds pottery with inorganic temper but critical parts, for example the handle and the base, tempered with organic temper. As a result, handles will not shrink away from the pot wall and bases will not crack during drying.

The study of inorganic temper is a rich source of information that is relevant for the understanding of the work of the potter but also for determining the provenance of the pots. When looking for the provenance of the raw materials, it is much easier to locate the rock or the sand, after the temper that was used by the potters has been identified by microscopic analysis, than to find the original clay bed by sampling clay for chemical analysis. Clay, found in or near an excavated potter's workshop, may have lost components during an interval of hundreds of years and thus may no longer be suitable in the production process for which it was originally intended. Rainwater, seeping through from the time the site existed to the time of excavation, will transport lime and even iron down into the subsoil. Furthermore, during evaporation, the soluble salts will be extracted from the clay and transported to the surface. But lime rock, calcite, fine quartz sand or quartz silt or basalts may give a hint as to the provenance of pots. The majority of the pottery will usually reflect the local geological circumstances. However, when pots with basalt sand as a temper are found in an area devoid of volcanic rock one must think in terms of interregional transportation of the pots. There is another possibility, namely, that potters originally working in a basaltic region moved to another area and, being dissatisfied with the sand that was available locally, used donkey caravans to bring basalt sand to their new site. In some cases, there must have been trade in quartz sand to places where only lime sand was available. But the sand is not the only possible indication of provenance. For it is often a combination of elements seen through a binocular microscope, using thin sections and making other observations, that throws light on the origins of pottery.

Reactions to kiln firing

The difference in the use of fine quartz sand and lime sand, as a tempering material, can be rather significant for the production process. Lime sand adds to the porosity of the sherd, while quartz sand does not, and is prohibitive for kiln temperatures well over 800°C. At that temperature, it dehydrates and begins to turn into quick lime. However, quartz sand, provided it is very fine, permits high temperatures. Another disadvantage of lime temper is that it may add salts to the body of the pot that can form scum on the surface during firing and blur painted decoration. Exact temperatures are known for the decomposition of lime but normally natural lime is not pure lime and the process may start at lower temperatures, even close to 800°C. Normally, one will not find sherds tempered with coarse quartz sand because this does not even allow for a temperature of c. 600°C, unless one has low-fired pots that were made under rather primitive conditions. As a rule, very fine, dust-size temper allows for higher temperatures. Yet in the Bronze and Iron Ages, very little pottery was ever fired over 800-900°C. Even afterwards, during Hellenistic and Roman times, potters did not go much beyond 850°C. This means that pottery from those periods is never hard and always very porous. Some of the

Chalcolithic pottery from Tulaylat al-Ghassul seems to have been fired to near stoneware temperatures close to 1200°C, but there are also ways and means of making very hard pottery without really creating stone-ware temperatures. One such method is to let a fire in which the pottery is placed continue for several days without much oxygen and use an enormous amount of fuel. It has been estimated that keeping a fire at the same temperature level for 20 hours has the same effect on pottery as raising the temperature by 100°C. On the whole, the clays in Jordan have a short melting trajectory, that is, the pot holds its shape to a certain temperature, but if one goes a little above that temperature the clay will suddenly melt and become liquid. This simply means that there was no clay from which a kiln could be built to fire pottery to well over 1100°C, not to mention the technological know-how required to construct such a kiln, normally a down-draft one. All this does not mean that there were no high temperature firings, but these were of short duration. Quick firing, using high temperatures for a short time, gave the pots a hard and reddish coloured surface but left the core much softer and dark. This was economic in terms of fuel consumption, not too hard on the kiln, and resulted in the production of satisfactory utility wares. Lime grains in or near the surface of the pot would then cause blisters only at the surface, as can be observed quite frequently on Jordanian pottery of all periods. This would happen, for the most part, to pots stacked close to the fire chamber. It is assumed that the potters in most periods used updraft kilns as they do today. The difference in temperature in an updraft kiln between the floor of the pottery chamber and the vent hole in the top of the kiln can be as much as 200°C.

Temper and pottery typology

Another aspect of the study of temper materials is the grouping of pottery according to temper for typological purposes. No matter how much two pots seem to be identical in shape, they should not be lumped under one heading if they have different non-plastics as a temper. If both have coarse lime temper but the matrix of one of the sherds has a considerable amount of microfossils and that of the other none, then one will normally find that the lime temper itself is different in each case, one being fossiliferous and the other not. This indicates a difference in origin and may point to local trade and different workshops.

Tradition

At this point, the influence of a tradition of pottery making becomes relevant for ceramic research since the idea of tradition is an indispensable axiom in the interpretation of our observations. It is, however, based on the observations of cultural anthropologists in the study of third-world potters as well as on the documented lives of potters in small workshops in western Europe until well into the twentieth century. The potters are, and were, totally dependent on the teachings of their predecessors, be they their parent-potters or masters. They would do exactly as they had been told, even in what appear to be very minor details of the production process.

The importance of this strict regime for success can easily be seen by comparing fine ware of high quality such as Roman *terra sigillata* with the ordinary pottery produced in the Levant. The clays, so-called illitic clays, used in the production of *terra sigillata* wares are very difficult to work with. It is easy to see the difference between the good and not so good pieces, and this even holds for Petra wares. These clays demand great skill on the part of the potters, and a good understanding of processes that take place during kiln firing. For good results, a lot of intelligence and knowledge is required and precise control over the entire production line. There is thus a kind of general rule: beautiful pottery can only be made from clays that are difficult to handle and require highly skilled potters.

Such is too much to expect from the potters who through history provided the communities with pottery for daily use. Nevertheless, the same chemical laws apply to their clays and slips, and the only difference is that the potters searched for what may be called easy-going clays or mixtures that accepted temperature differences of 200°C in kiln firing and still survived as a tolerable product. A good example of this is Mediaeval glazed pottery, where the chemical and physical processes that take place during the heating of the colouring oxides used in glazes are extremely sensitive to colour, shine, or to shrinkage and cracking and many other faults. Even a difference in temperature of no more than 5°C may change the result. The common Mediaeval glazed wares in Jordan often show a whole range of such mistakes and merely indicate that it was impossible for the potters to make a uniform product such as we expect when buying glazed plates or cups. Persian glazed wares are an example of how it was possible, with great skill, to make glazed-art pottery, using the same glazes. Potters accustomed to work with 'easy' or undemanding clays will be able to produce pottery but never become master potters. The non-glazed wares show the same range of faults if one makes uniformity of product a criterion. Thus, the tolerance of the materials that were used was wide. One step beyond the limits, however, might mean loss of the product, usually consisting of a kilnload of cracked and broken pots.

What potters make depends primarily on the nature of the clays that are available and on the skill that is required to manipulate them. This explains the difference in quality of pottery from the Greek mainland or Cyprus and those from Palestine and Jordan.

One would expect that potters in antiquity were conservative people who followed 'correct' procedures. These potters were probably not unlike many present-day, traditional potters and were very anxious to keep the capricious spirit of the kiln happy, as well as the other demons that may have been thought to own the clay bed or roam about the drying room. Potters working in a certain tradition will not change anything, either in their use of raw materials or in the construction of their pots, unless forced to do so by outside circumstances.

In order to demonstrate the above, one must be able to survey a tradition over a period of several centuries, and, most importantly, have a statistically reliable sample of pottery for each sub-period at one's disposal. In both instances that the writer studied the early Iron Age from Tall Dayr' Alla and the Iron Age II from Jerusalem, it was clear that every shape that was found in the beginning could still be found, and was thus still produced, at the end. In the case of the Jerusalem pottery, it was the amount of pots of each 'type' that might diminish or increase but, certainly, there was no change in the shape between the ninth and the seventh centuries BC. Noticeable changes in the production are caused by influences from outside. It may, then, be rather difficult for the archaeologist to find out what actually happened. But the constants in the production process are the important aspects and make it possible to define traditions and productions.

To distinguish between locally made pottery and regional trade in pottery, to distinguish between pottery workshops, or to find out about the existence of pottery markets and pottery peddlers or other means of transport, we need to posit the axiom of the conservatism of ancient potters as well as study the composition of the non-plastics and clays. In combination with an analysis of the manufacturing techniques, one gets an impression of the type of pottery production and the way in which the pots were used.

The history of a tradition

The technological study of the pottery from one phase or stratum of a site is only one step towards reconstructing the history of a tradition of pot making. Here the study addresses such questions as the continuity and discontinuity of the pottery production at 'transitional' periods. For instance, the relationship between Late Bronze Age and Iron Age pottery, which has to be approached from a technological angle. Likewise, the explanation of the invention of pot-making in Neolithic times can only be given by understanding the technical problems involved. To be able to survey the history of a tradition of pot-making from its beginning to its decline and end, including its spread over an area, and to tie this up with the economic and social history of that period, creates a fresh understanding of the cultural processes. It does away with a lot of imaginative but unrealistic explanations often encountered in pottery studies.

Technological research

Technological study can be performed with the aid of fairly simple tools, provided the ceramist, apart from having been trained in the chemistry of clays, also has experience in working with clays that are not factory-made clay mixtures. In other words, one has to be able to test local clays for their workability. There are standard tests for clays to measure all the aspects that play a role in the construction of pots. It is often only in the attempt to make replicas that one discovers the advantages and the drawbacks of the clays and the techniques of pottery making used in antiquity. The study of non-plastics requires making thin sections of sherds and using a binocular or petrographic microscope. The microscope is first used to look at a large number of sherds on a fresh break. One then selects sherds to make thin sections of each temper group found. A small electric test kiln must be used to measure shrinkage or to find changes of colour at various temperatures and other features of ancient sherds. Test bars can be made from fresh clays taken from the area surrounding the site under study. Sampling for suitable clays that may have been used in antiquity requires some experience and training as well as the advice of a geologist. Simple field tests must be used, however, to ensure unsuitable soils are discarded before the samples are sent to the home base.

Knowledge of the chemistry, combined with a lot of knowledge from practical experiments with raw materials found in nature, goes a long way in explaining the attributes that can be observed in ancient ceramics. For exact measurements of clay elements, one must use Neutron Activation (NAA) or X-Ray Diffraction (XRD) analyses or similar laboratory techniques. However, in order to select the right samples for such tests, it is important first to undertake the groundwork by selecting test pieces, using the binocular microscope. This reduces the chance of the sample being non-representative for the main ware or wares that make up the bulk of the pottery.

A sketch of the technical development of the pottery production in Jordan is found in Bienkowski (1991: 62-85).

References

As, A. van
    1992   Pottery Technology: The Bridge between Archaeology and Laboratory. *Newsletter, Department of Pottery Technology, Leiden University* 9/10: 1-3.

Bienkowski, P. (ed.)
    1991   *Treasures from an Ancient Land: The Art of Jordan*. Stroud: Alan Sutton.

Hendrix, R.E., P.R. Drey and J.B. Storfjell
    1997   *Ancient Pottery of Transjordan: An Introduction Utilizing Published Whole Forms—Late Neolithic through Late Islamic*. Berrien Springs, MI: Institute of Archaeology/Horn Archaeological Museum, Andrews University.

London, G.
    1999   Central Jordanian Ceramic Traditions. In B. MacDonald and R.W. Younker (eds.), *Ancient Ammon*, 57-102. Studies in the History and Culture of the Ancient Near East 17. Leiden: E.J. Brill.

Orton, C., P. Tyers and A. Vince
    1993   *Pottery in Archaeology*. Cambridge Manuals in Archaeology. Cambridge: Cambridge University Press.

Rice, P.M.
    1987   *Pottery Analysis: A Source Book*. Chicago: University of Chicago Press.

24. Writing and Texts

Alan Millard

Writing began in Babylonia before 3000 BC and was commonly done on clay tablets in the cuneiform script. Clay tablets are quite durable and have survived in large numbers, but can break and suffer from dampness or erosion, so are often incomplete. With the diffusion of Sumero-Babylonian culture, the writing system spread and reached north Syria by 2300 BC, where it is abundantly represented at Ebla. A single tablet from Byblos and texts from Babylonia referring to the city show that writing was known there about 2000 BC. A scatter of examples attests its use at Hazor two or three centuries later and thereafter throughout Canaan.

Influenced by Babylonia, the hieroglyphic and hieratic scripts developed in Egypt. The latter was mainly written on papyrus, sometimes on leather, or, for ephemeral notes, on potsherds. Since papyrus and leather disintegrate in damp soil, documents survive only in very arid or otherwise unusual conditions. Ostraca can last a long time, although the ink may easily be washed away. Egyptian inscriptions on stone vases reached Byblos by the Second Dynasty (c. 2800 BC).

As yet no examples of writing from the third millennium BC have come to light on either side of the Jordan, apart from some Egyptian pharaonic names on pots and hieroglyphic cylinder seal impressions from the far south of Palestine (e.g. Arad, 'Ayn Besor). Indeed, in this and other aspects, the region appears as a backwater compared with Syria. This may be due to the fact that trade passed from Byblos to Egypt by sea rather than by land.

One form of identification mark deserves notice, although it is not actually writing. Early Bronze Age jars impressed before firing with designs made by roughly cut cylinder seals are found from eastern Iraq through northern Mesopotamia, Syria and into Palestine and Jordan (Ben-Tor 1978, 1992; Esse 1990: 32*; Collon 1988: 20-24).

Middle Bronze Age sites in Jordan have yielded cylinder seals and scarabs, but at present it seems unlikely that the inscriptions on them were understood as writing, rather they served as magical signs.

The first examples of writing appear in Jordan in the Late Bronze Age. Two tiny fragments of cuneiform tablets were unearthed at Pella' and the El-Amarna Letters, found in Egypt (Moran 1992), include a reference to Pella in a way that indicates that someone there could read (Letter 256). That letter and one other (no.197) refer to Ashtartu (Tall 'Ashtara) and another (no. 204) may have been sent from Qanawat (Qanu). Since the senders of several letters have not been identified, there may have been other towns in Jordan with scribes writing and reading cuneiform in the Late Bronze Age. The fact that no tablets have been found does not mean they did not exist. That observation applies more acutely to Egyptian papyri. None have come to light anywhere in the area, but clay bullae that once sealed them—the imprints of the fibres remain on the backs—attest their former presence (e.g. at Tall as-Sa'idiyya; Tubb 1990: 28). Egyptian inscriptions do occur in the Late Bronze Age on cliffs, on stelae, such as that of Ramesses II at Sheikh Sa'id in the Hauran (Kitchen 1979: 223), and on small objects in Sinai and Canaan. Others may appear in Jordan. Scarabs, with hieroglyphic signs conveying good luck rather than specific messages, are common. The stela from al-Balu' has faint traces of script above its carved relief that may be Egyptian signs. If so, they remain undeciphered (Vincent and Horsfield 1932).

The second millennium BC in the Levant was an age of experiment in writing from which emerged the Canaanite alphabet, the mother of all alphabets, and its imitation in cuneiform, known principally from Ugarit but spread south as far as Beth-Shemesh. Since this alphabet was written mainly with ink on papyrus or leather, only a few specimens of the early stages of the alphabet survive, for example, notes and names that happened to be scratched or painted on stone, pottery or metal.

Although no Late Bronze Age specimens of Canaanite alphabetic writing have yet appeared in Jordan, the site of Tall Dayr 'Alla uniquely attests one abortive scribal experiment. Several clay tablets were found in the ruins of a shrine. Apart from crushed and fragmentary pieces, seven have small holes made with a sharp point in their surfaces and five have signs made of scratched lines and holes. A vertical line separates groups of signs, marking sense boundaries. The signs themselves have not been satisfactorily related to any other script and thus the contents are unknown (Franken 1965, 1992: 57-59, 64-65, 67-70; Ibrahim and van der Kooij 1994).

As the Iron Age states of Ammon, Edom and Moab established themselves, each adopted forms of the alphabet for writing their West Semitic languages that had developed in Late Bronze Age Canaan. In Ammon, the influence of Damascus was strong, giving the script Aramaic traits. In Moab, however, the Israelite presence east of the Jordan doubtless led to the stronger Hebrew appearance of the letters in the ninth century. Aramaic features, nevertheless, entered Moabite and also had an impact on the poorly represented Edomite script (Naveh 1982: 100-11; van der Kooij 1987). There are public displays of writing on stone from Ammon and Moab recording military success and building works (e.g. Citadel Inscription from 'Amman, Moabite Stone or Mesha Stela from Dhiban) or dedicating statues (at 'Amman and perhaps at al-Karak). The impression of an Edomite king's seal from Umm al-Biyara displays his name and title, 'Qosgabbar, king of Edom'. The national god Qos, forming part of the personal name, points to the area of origin, even without the title, as do Chemosh for Moab, Yahweh for Israel and Judah and, more rarely, Milcom for Ammon in other personal names. Such distinctive names on seals help in identifying local scripts, although some characteristics overlapped as engravers may have moved from kingdom to kingdom. The seals, some bearing designs as well as letters, usually give the owner's name, the name of his father, and sometimes a title. Impressions of the seals on clay, that is, bullae, have come to light at some Iron Age sites. The imprints of papyrus fibres on their backs testify to documents—legal deeds, letters, wills—that are now lost.[1]

Graffiti are meagre relics of everyday use of writing. They usually consist of personal names, scratched on pottery sherds, that is, ostraca, which are written in ink and convey brief messages or notes. Examples are known from several sites, including 'Amman, Hisban, Busayra and Umm al-Biyara. Peculiar is the small bronze bottle from Tall Siran with a commemorative text of a seventh-century Ammonite king engraved on the outside.

Seals were stamped on pots to mark ownership or origin. Nearly two dozen stamped jars have been found at Tall al-Khalayfi.

Letters of the alphabet also served as guides for craftsmen assembling objects made in several pieces. Well known on carved ivories, once decorating wooden furniture from Samaria and other sites, the same custom appears in the case of the eyes inlaid in stone heads found at 'Amman, each eye having a letter engraved on the back. (For the Ammonite texts, see Aufrecht 1989, 1999; Hübner 1992: 15-129; for Moabite, see Timm 1989: 158-302.)

No literary texts have been found, although hints of literary composition may be seen in the Moabite Stone, with one outstanding exception, the 'Balaam' text of Tall Dayr 'Alla (Hoftijzer and van der Kooij 1976, 1991). Written in black and red ink on the plaster of a wall late in the ninth century BC and using an Aramaic cursive script, this fragmentary narrative, in a local dialect of Aramaic, tells of a vision the gods gave to Balaam, son of Beor (known from Num. 22–24). The writing seems to imitate a column of a scroll, implying such books were current in Iron Age towns. It is the only example of Aramaic literature surviving in the Levant in a copy earlier than the Dead Sea Scrolls.

No cuneiform inscriptions from the period of the Assyrian empire have as yet been found in Jordan. However, carved high on a remote cliff at as-Sela' is a rock carving of Mesopotamian style. It probably represents the neo-Babylonian or Chaldaean king, Nebuchadrezzar of Babylon (605–562 BC), or his last successor Nabonidus (556–539 BC). The accompanying cuneiform text is not legible (Dalley and Goguel 1997).

A solitary cuneiform tablet from the first millennium BC was excavated at Tawilan. It is a legal document drawn up in Harran probably in 521, but perhaps in 423, between people with Aramaic names and one Edomite, Qos-shama, guarantor for two sheep. Apparently, its Edomite owner had taken it home (Dalley 1995).

The Persian empire adopted the Aramaic language and alphabet as its administrative tool and its effects are apparent in Jordan as elsewhere. No papyrus or leather documents have survived, but a few ostraca from various sites, for example, Tall al-Mazar, Hisban, and as far south as Tall al-Khalayfi (DiVito 1993), demonstrate Aramaic in bureaucratic and private use. Occasional examples are found on other materials, such as seal impressions of an official of Ammon on jar handles from Tall al-'Umayri (Herr 1992), and an owner's name on a bronze bowl from Umm Udhayna.

Alexander the Great's conquests introduced Greek as the language for monuments and official documents. Evidence for this is the discovery in Egypt of papers of Zenon, secretary to an Egyptian minister under Ptolemy II (282–246 BC), who had to deal with affairs in Palestine and Jordan. Among his documents are records of slave trading at Birtha of Ammon ('Iraq al-Amir; Millar 1987: 118-19). However, apart from rare weights and stamps on wine-jars imported from Rhodes, little Greek is seen in Jordan before the Roman period when numerous memorials appear in the Decapolis cities (Jarash, 'Amman, etc.), and their coins continue to bear Greek legends. Latin occurs, notably on milestones and occasionally on monuments, most famously the tomb of Sextus Florentinus in Petra (for some texts see Gatier 1986), wherever the Roman army was. The continuance of both Greek and Latin for legal deeds is demonstrated by the hoard of carbonized papyri, written in the sixth century and mostly in Greek, found in a ruined church at Petra (Bikai 1994; Peterman 1994). In daily life, however, the Aramaic script maintained its position and was occasionally prominent, as with the name 'Tobia' boldly carved on the cliff by Qasr al-'Abd ('Iraq al-Amir). The Hebrew form, that is, the 'square script', of Aramaic letters was normal where there were Jewish settlements. The Dead Sea Scrolls and other documents from caves near the Dead Sea give evidence of this practice. So far, however, no specimens have been found in Jordan.

Accordingly, it was the Aramaic script that the Nabataean Arab tribes adopted when they settled in the former Edom and northern Sinai. Among the oldest examples is the inscription of Aslah from Bab as-Siq outside Petra of c. 95 BC. These examples display a few of the features that came to characterize the Nabataean script of the next centuries. While the formal letters were developed for epitaphs, dedications, legends on coins and stamps on pottery lamps, a more cursive style was written with ink on papyrus, now revealed by documents of the late first and early second centuries AD discovered in caves west of the Dead Sea. The continuance of the script can be traced through innumerable graffiti scribbled by travelers, pilgrims and herdsmen on rock faces and boulders after the Roman conquest of Petra in AD 106. It can also be traced throughout the second and third centuries to the memorial for 'Imru'lqais, king of all the Arabs, at Namara in the Hauran (AD 328/29), which is an Arabic composition written in developed Nabataean letters.

Texts from the fourth to sixth centuries are rare. Three in Arabic, however, from the sixth century exhibit the first stages of the Arabic script branching off from cursive Nabataean (Gruendler 1993; Healey 1993).

Another alphabetic writing, an offshoot of the Canaanite alphabet, developed in southern Arabia and letters of this script were found scratched on jars of the Iron Age at Tall al-Khalayfi. Several centuries later, Arab herdsmen in the deserts of southern Syria, Jordan and northern Arabia, used two forms of this Old South Arabic alphabet, known as Safaitic and Thamudic. Their texts are almost all graffiti on rocks and stones, scratched as they travelled or watched their flocks, giving names and genealogies, epitaphs and short messages. Dated between the first century BC and the fourth century AD, they attest a surprisingly free use of writing in an apparently non-literary environment, without descendants. (For a convenient conspectus see Macdonald 1992.)

Until the appearance of the Nabataean, Thamudic and Safaitic graffiti, writing was not a normal part of the skills ordinary people practised. It was the monopoly of the scribes in the Middle and Late Bronze Ages and, even when the alphabet made the art of reading and writing much simpler, it was not a necessity for most people, although anyone who really wanted to learn to write could probably have done so. Nevertheless, it should be noted that the amount of written material surviving from ancient Jordan is only a very small proportion of what existed there in the ages of antiquity. (For a slightly more extensive survey, see Millard 1991.)

Note

1. The authenticity of a 'Moabite' papyrus with a bulla published by Bordreuil and Pardee (1990) remains questionable.

References

Aufrecht, W.E.
   1989   *A Corpus of Ammonite Inscriptions*. Lewiston, NY: Edwin Mellen Press.
   1999   Ammonite Texts. In B. MacDonald and R.W. Younker (eds.), *Ancient Ammon*, 163-88. Studies in the History and Culture of the Ancient Near East 17. Leiden: E.J. Brill.

Ben-Tor, A.
   1978   *Cylinder Seals of Third-Millennium Palestine*. Bulletin of the American Schools of Oriental Research Supplement Series 22. Cambridge, MA: American Schools of Oriental Research.
   1992   New Light on Cultic Cylindrical Seal Impressions from the Early Bronze Age in Eretz Israel. *Eretz Israel* 23: 38-44 (Hebrew), 146* (English summary).

Bikai, P.M.
　1994　The Petra Papyri. *Annual of the Department of Antiquities of Jordan* 38: 509-11.

Bordreuil, P. and D. Pardee
　1990　Le papyrus du marzeah. *Semitica* 38: 49-68.

Collon, D.
　1988　*First Impressions*. London: British Museum.

Dalley, S.
　1995　The Cuneiform Tablet. In C.-M. Bennett and P. Bienkowski, *Excavations at Tawilan in Southern Jordan*, 67-68. Oxford: Oxford University Press.

Dalley, S. and A. Goguel
　1997　The Selaʻ Sculpture: A Neo-Babylonian Rock Relief in Southern Jordan. *Annual of the Department of Antiquities of Jordan* 41: 169-76.

DiVito, R.A.
　1993　The Tell el-Kheleifeh Inscriptions. In G. Pratico, *Nelson Glueck's 1938–1940 Excavations at Tell el-Kheleifeh: A Re-appraisal*, 51-63. Atlanta: Scholars Press.

Esse, D.L.
　1990　Early Bronze Age Cylinder Seal Impressions from Beth Yerah. *Eretz Israel* 21: 27*-34*.

Franken, H.J.
　1965　Clay Tablets from Tell Deir ʻAlla. *Vetus Testamentum* 14: 377-79.
　1992　*Excavations at Tell Deir ʻAlla: The Late Bronze Age Sanctuary*. Louvain: Peeters.

Gatier, P.L.
　1986　*Inscriptions de la Jordanie*. II. *Région centrale*. Paris: Geuthner.

Gruendler, B.
　1993　*The Development of the Arabic Scripts: From the Nabataean Era to the First Islamic Century According to Dated Texts*. Atlanta, GA: Scholars Press.

Healey, J.F.
　1993　Nabataean to Arabic: Calligraphy and Script Development among the Pre-Islamic Arabs. *Manuscripts of the Middle East* 5: 41-52.

Herr, L.G.
　1992　Two Stamped Jar Impressions of the Persian Province of Ammon from Tell El-ʻUmeiri. *Annual of the Department of Antiquities of Jordan* 36: 163-66.

Hoftijzer, J. and G. van der Kooij
　1976　*Aramaic Texts from Deir ʻAlla*. Leiden: E.J. Brill.
　1991　*The Balaam Text from Deir ʻAlla Re-evaluated*. Leiden: E.J. Brill.

Hübner, U.
　1992　*Die Ammoniter: Untersuchungen zur Geschichte, Kultur und Religion einer transjordanischen Volkes im 1. Jahrtausend v. Chr*. Abhandlungen des deutschen Palästinavereins 16. Wiesbaden: Otto Harrassowitz.

Ibrahim, M.M. and G. van der Kooij
　1994　Excavations at Deir ʻAlla 1994. *Newsletter of the Institute of Archaeology and Anthropology, Yarmouk University* 16: 16-20.

Kitchen, K.A.
　1979　*Ramesside Inscriptions*, II. Oxford: Blackwell.

Macdonald, M.C.A.
　1992　Inscriptions, Safaitic. In D.N. Freedman (ed.), *The Anchor Bible Dictionary*, III: 418-23. New York: Doubleday.

Millar, F.
　1987　The Problem of Hellenistic Syria. In A. Kuhrt and S. Sherwin-White (eds.), *Hellenism in the East*, 110-33. London: Duckworth.

Millard, A.R.
　1991　Writing in Jordan: From Cuneiform to Arabic. In P. Bienkowski (ed.), *Treasures from an Ancient Land: The Art of Jordan*, 133-49. Stroud: Alan Sutton.

Moran, W.L.
　1992　*The Amarna Letters*. Baltimore, MD: The Johns Hopkins University Press.

Naveh, J.
　1982　*Early History of the Alphabet*. Jerusalem: Magnes Press.

Peterman, G.L.
　1994　Discovery of Papyri in Petra. *Biblical Archaeologist* 57: 55-57.

Timm, S.
　1989　*Moab zwischen den Mächten: Studien zu historischen Denkmälern und Texten*. Ägypten und Altes Testament 17. Wiesbaden: Otto Harrassowitz.

Tubb, J.
　1990　Preliminary Report on the Fourth Season of Excavations at Tell es-Saʻidiyeh in the Jordan Valley. *Levant* 22: 21-42.

van der Kooij, G.
　1987　The Identity of the Trans-Jordanian Alphabetic Writing in the Iron Age. In A. Hadidi (ed.), *Studies in the History and Archaeology of Jordan*, III: 107-21. Amman: Department of Antiquities.

Vincent, L.H. and G. Horsfield
　1932　Une stèle Egypto-Moabite au Balouʻa. *Revue Biblique* 41: 417-44.

25. The Bible, Archaeology and Jordan

Burton MacDonald

Introduction

The Bible may be viewed as a collection of historical, theological and instructional materials which were produced, utilized, preserved and eventually 'canonized' (declared authoritative) by ancient Israel and the early Christian Church. It is actually a library of books numbering 24 for Jews, 66 for Protestants, and 72 and 73 for Eastern Orthodox and Roman Catholics respectively.

There are various views on the nature of the Bible. Moreover, there are several approaches to its study. Such being the case, the archaeologist would do well to become familiar with the latest developments in 'biblicals criticism', an approach that seeks to understand the biblical materials within the context of the original communities that preserved and used them and within the broader social, cultural, and religious context of Near Eastern life and history. For the biblical critic, the Bible is to be read and studied in exactly the same way as any other collection of ancient documents. The initial concern of the critic is to understand the biblical materials in their original historical and cultural context. The first concern is to answer the question: What did the text mean at the time it was composed?

A biblical text must not always be viewed as being contemporaneous with the events, peoples and places with which it deals. For example, the biblical narrative of Israel's defeat of Sihon king of Heshbon (Num. 21.21-26; Deut. 1.4; etc.) may or may not recount an actual historical event (see below). The biblical critic must, therefore, attempt to determine the date of the origin of the narrative and the biblical writer's intention in using it. Archaeologists are not generally expected to be also biblical critics. However, they would do well to be familiar with the results of such study when using a particular text.

The Bible contains a great deal of information on ancient Jordan between the Iron Age and the Roman period inclusively. In fact, it is the best literary source for anyone attempting to write the history and understand the culture of Ammon, Moab and Edom, three nations/states/peoples of ancient Jordan.

Most of the information in the Bible about Jordan comes from the Old Testament/Hebrew Scriptures. There is little about the area in the New Testament.

The Bible refers to the territory to the east of the Jordan River as 'beyond the Jordan' (Gen. 50.10, 11; Deut. 3.8, 20, 25; Mt. 4.25; Mk 3.8; 10.1; *et passim*), 'to the east' (Num. 32.19), 'beyond the Jordan toward the east' (Josh. 12.1), 'east of the Jordan' (1 Kgs 17.3), 'from the Jordan eastward' (2 Kgs 10.33), 'on the east side of the Jordan' (Deut. 4.41; 1 Chron. 6.78), and 'across the Jordan' (Jn 1.28; 10.40). The name Perea appears in variant readings of Lk. 6.17 for the region that the Gospels refer to as 'beyond the Jordan'. The modern expression 'Transjordan' comes from these biblical expressions.

Some scholars think that the biblical writers may not have been very clear about the geography and the topography of the territory to the east of the Jordan. In their opinion, the biblical writers may have written about the area as outsiders rather than as people with first-hand knowledge (Baly 1987: 124; Noort 1987: 125; Bartlett 1989: 101). Narratives about the routes, namely, Num. 20.14-20, 22 (and 21.4), 21.10-20, 33.1-49, Deut. 2.1-26 and Judg. 11.14-26, supposedly taken by the Israelites during their travels through and/or around the territories of Edom and Moab, support this conclusion (Miller 1989a).

Approximately 160 biblical sites and regions are located in the area that is presently called the Hashemite Kingdom of Jordan. Many are still to be identified with certainty (MacDonald 1993, 2000).

Old Testament/Hebrew scriptures and Jordan

When Lot, due to a shortage of pastureland, separated from his uncle Abraham (Gen. 11.31), he settled in the plain of the Jordan (Gen. 13.10-11). It was in this area that the 'cities of the Plain' (Gen. 13.12), namely, Sodom, Gomorrah, Admah, Zeboiim and Bela (that is, Zoar) (Gen. 14.8) were located, probably in the area along the southeast shores of the Dead Sea (Rast 1987; Neev and Emery 1995). It was from Zoar that Lot fled and took refuge in a cave and subsequently became the father of the Ammonites and the Moabites (Gen. 19.30-38). The narrative thus indicates a close relationship, at least in the mind of the biblical writer, between the Israelites and two of their neighbours east of the Jordan.

It was in the area of the Jabbok (= Wadi az-Zarqa') that Jacob met his twin brother Esau (Gen. 25.25-26), the eponymous ancestor of the Edomites (Gen. 36.1, 8, 9, 19), on the former's return from Haran to the land of Canaan. Esau, whom the biblical writers portray in a poor light (Gen. 25.29-34; 26.34-35; 27.40), occupied the territory of Seir (Gen. 32.3; 26.8, 9; Deut. 2.8; Judg. 5.4), that is, the hill country south of Moab and both east and west of Wadi 'Arabah.

When the Israelites were unable to enter the Promised Land from the south (Num. 13–14), they set out from Kadesh (Num. 20.1) in the Negev and either detoured around the lands of the Edomites (Num. 20.21; 21.4) and the Moabites (Judg. 11.17-18) or passed through the land of Edom (Deut. 2.1-8) before camping in the plains of Moab, that is, the southern extremity of the east Jordan valley between Wadi Nimrin and the Dead Sea, across from Jericho (Num. 22.1; 33.48). It was close by on Mount Nebo (= Ras al-Siyagha) that Moses viewed the whole land (Deut. 34.1), died, and was buried in the land of Moab (Deut. 34.5-6).

Yahweh, the god of Israel, gave the Israelites their inheritance east of the Jordan. Similarly, Israel's god is attributed with giving Mount Seir to Esau (Deut. 2.5—Judg. 5.4 associates Yahweh with Seir). The same god is said, furthermore, to have given Moab and the Ammonites, the descendants of Lot, their territorial possessions (Deut. 2.9, 19).

The Israelites settled east of the Jordan (Num. 21.31) after they had defeated the Amorite king Sihon of Heshbon 'and took possession of his land from the Arnon to the Jabbok, as far as to the Ammonites' (Num. 21.24; see also Deut. 2.31-37) (the site of Heshbon will be treated later). They next defeated another Amorite king, Og of Bashan (Num. 21.33-35; Deut. 3.1-6), whose territory some consider to have been located both south and north of the Yarmuk River. As a consequence, the territory beyond the Jordan from Wadi Arnon (= Wadi al-Mujib) to Mount Hermon was said to belong to Israel (Deut. 3.8). It became the possession of the Reubenites, the Gadites, and the half-tribe of Manasseh (Num. 32; Deut. 3.12-17; Josh. 13.8-11).

Israelite possessions east of the Jordan would, of course, have varied from time to time. They would have depended, at least in part, on the development of the nations/states of Ammon, Moab and Edom (see, e.g., 2 Kgs 3.1-27). Moreover, the encroachment of the Aramaeans (see, for example, 2 Kgs 8.28-29) and later the Assyrians (2 Kgs 17.1-6), especially in the northern part of the county, would have affected Israelite possessions (Dion 1995: 1285).

The Ammonites are said to have inhabited the territory to the east of the above-described Israelite territory in the vicinity of the upper region of the Jabbok (Deut. 2.37; 3.16). It appears, however, that the Ammonites were interested in extending their territory, especially to the north, since they came into conflict with the Israelites (Judg. 10.6–12.7; 1 Sam. 10.27; Amos 1.13).

The biblical writers locate Moab to the south of the Arnon (Num. 21.13; 22.36). There are indications, however, that the territory of Moab included land north of the Arnon since part of this region is said to have been once in its possession (Num. 21.26) and part of it is referred to as the plains of Moab (Num. 22.1; 26.3; 32.12; 33.48; 33.50; 36.13; Josh. 13.32).

The area east of the Jordan was a place of refuge for Israelites. Ishbaal, one of Saul's sons, was set up as the head of a government in refuge at Mahanaim (2 Sam. 2.8, 12), probably located at Tulul adh-Dhahab al-Gharbiyya on the Jabbok (Coughenour 1989). David, who fled from Saul, took his parents across the Jordan to protect them (1 Sam. 22.3-4). Moreover, when Absalom usurped the throne of his father David (2 Sam. 16.1-23), the latter was forced to cross the Jordan and came to Mahanaim (2 Sam. 17.24, 27; 19.32; 2 Kgs 2.8). In the same vein, the cities of Bezer, Ramoth in Gilead, and Golan, from the tribes of Reuben, Gad and the half-tribe of Manasseh respectively (Deut. 4.41-43), are said to be cities of refuge to which a homicide, that is, someone who unintentionally killed another person, could flee (Deut. 19.3-4) (only the first two of these cities are located in Jordan; Golan is located to the north of the Yarmuk River). Much later in history, the land of Jordan became a place of refuge for Israelites fleeing Babylon's might (Obad. 14) and, later still, for Jews driven as fugitives from their homeland (2 Macc. 4.26; 5.7).

The Levites received 48 cities with the surrounding land for grazing (Josh. 21.1-42). Several of these cities are located in Jordan: (1) Bezer, Jahzah, Kedemoth and Mephaath (= Umm ar-Rasas), out of the tribe of Reuben (Josh. 21.36-37; 1 Chron. 6.78-79); and (2) Ramoth in Gilead, Mahanaim, Heshbon and Jazer, out of the tribe of Gad (Josh. 21.38-39; 1 Chron. 6.80-81).

It was during the Israelite siege of Rabbath-Ammon (= Jabal al-Qal'a/'Amman Citadel) that Uriah, the husband of Bathsheba (2 Sam. 11.3), was killed, at David's request (2 Sam. 11.15). Uriah's death cleared the way for David to take Bathsheba, who in time became the mother of Solomon (2 Sam. 12.24), as one of his wives.

David is said to have subdued Edom, Moab and the Ammonites (2 Sam. 8.12). He received tribute from the Moabites (2 Sam. 8.2; 1 Chron. 18.2) and put garrisons in Edom and all the Edomites became his servants (2 Sam. 8.14; 1 Chron. 18.13). He also fought against Rabbath Ammon and took it, as well as all the cities of the Ammonites (2 Sam. 12.29-31; cf. 1 Chron. 20.3).

Solomon built a fleet of ships at Ezion-geber, which is near Eloth on the shore of the Red Sea in the land of Edom (1 Kgs 9.26). From there, the ships went to Ophir (probably southern Arabia or east Africa) and brought back gold for Solomon (1 Kgs 9.28).

Three of Solomon's 12 districts were located east of the Jordan. Their headquarters were at Ramoth-Gilead (1 Kgs 4.13), Mahanaim (1 Kgs 4.14) and in the land of Gilead (1 Kgs 4.19).

Solomon loved many foreign women, some of whom were Moabites, Ammonites and Edomites (1 Kgs 11.1). One of these women, Naamah, an Ammonite, was the mother of Rehoboam (1 Kgs 14.21, 31; 2 Chron. 12.13), Solomon's son and successor.

The biblical writers viewed the gods of the non-Israelites east of the Jordan as national gods. For example, they have Chemosh, who is not able to save Moab (Jer. 48.13, 46), going into exile along with the people (Jer. 48.7).

Relations between the Israelites and the nation/states of Jordan were less than amicable also during the time of Israel's divided monarchy (925–586 BC). It appears that the Ammonites were in constant conflict with the Israelites (2 Chron. 20; 24; 26; 27). Moab, moreover, rebelled against Israel after the death of Ahab (2 Kgs 1.1; 3.4-5) and Edom, for its part, came into conflict with the kings of Judah (1 Kgs 22.47; 2 Kgs 8.20, 22; 14.7). There was probably constant struggle between Judah and Edom for control of the ports on the Gulf of al-'Aqaba (2 Kgs 14.22; 16.6).

The prophets of Israel have little good to say about the Ammonites (Amos 1.13-15; Zeph. 2.8-11; Jer. 49.1-6; Ezek. 21.28-32; 25.2-5), Moab (Amos 2.1-3; Isa. 15–16; Zeph. 2.8-11; Jer. 48.1-47; Ezek. 25.8-11) and Edom (Amos 1.11-12; Jer. 49.7-22; Ezek. 25.12-14; 35.5, 14-15; Mal. 1.3-4).

The book of Obadiah, written soon after the Babylonian destruction of Jerusalem in 587/586 BC, is generally an indictment against Edom for its stance when its neighbour and kin, that is, Israel, was in danger. It depicts Edom as failing to come to his brother in need and even accuses Edom of gloating and rejoicing over Israel's misfortune. Tobiah, an Ammonite, was one of those opposed to the rebuilding and/or repairing of the walls of Jerusalem after the Israelites returned from exile in Babylon after 538 BC (Neh. 2.10; 4.3, 7-9). Neh. 13.1 states that Ammonites and Moabites are excluded in perpetuum from the congregation of Israel. This was probably due to their opposition to the Jews returning from Babylon.

New Testament and Jordan

The area east of the Jordan river is not prominent in the New Testament. There are, nevertheless, several mentions of it in its books.

John the Baptist is said to have baptized at Bethany across the Jordan (Jn 1.28; 10.40). It is to this same place that Jesus went when 'the Jews took up stones again to stone him' (Jn 10.31, 40).

The place of John the Baptist's imprisonment and death at the hands of Herod Antipas (Mt. 14.1-12; Mk 6.14; Lk. 9.7-9) is not mentioned in the New Testament. Josephus, however, locates it at Machaerus (= Mukawir), east of the Dead Sea (*Ant.* 18.109-19).

Jesus is said to have visited the cities of the Decapolis region (Mk 5.20; 7.31). This would have brought him into the area of northern Jordan since all but one of these cities is located east of the Jordan river. Moreover, the narrative of Jesus' cure of the demoniac is frequently located at either Gadara (= Umm Qays) or Gerasa (= Jarash) (Mt. 8.28; Mk 5.1; Lk. 8.26-37), two of the Decapolis cities (Fitzmyer 1981: 736-37).

Mark 10 has Jesus making his way from Galilee up to Jerusalem by way of the area to the east of the Jordan. Crowds gathered around him and he taught them (v. 1).

Arabia, the place to which Paul went following his encounter with Jesus on the road to Damascus (Acts 9.3-6), is generally located in the Nabataean kingdom (Gal. 1.17). Petra was, of course, the capital of this kingdom.

Biblical archaeology and Jordan

Archaeology reconstructs material culture and cultural changes. It can also be instrumental in the verification, illumination, and supplementation of the written sources that, for the present purposes, are the books of the Bible.

With models taken from the social sciences, the archaeological record can be utilized now to reconstruct the social organization of Jordan during the biblical period. In fact, the history of Jordan during the Iron Age, Persian and Roman periods can be written only with the help of both biblical and archaeological studies. For it is the combination of material evidence derived through archaeology and of textual data provided, to a large extent, by the Bible that makes such a history possible. In this respect, it is important to emphasize that in the relationship between archaeology and the Bible, the central concern is that both scientific biblical study and scientific archaeological investigation inform one another in constructive ways (Meyers 1984: 39).

A prime example of combining the information provided by the biblical text and artifactual evidence, along with the study of toponyms, is in the effort to locate biblical sites in Jordan. In such a study, three factors come into play: (1) topographical and historical information derived from the ancient written sources, particularly the Bible; (2) analysis of the site's name, its development, and its preservation in the area (the study of toponyms); and (3) the artifactual evidence recovered by excavation and/or survey (Grollenberg 1956: 19-20; Aharoni 1979: 124; Miller 1983: 119-20). The results of such work can provide a mass of information on the life, economy and resources available to the peoples about whom the Bible speaks.

The site of Heshbon may be used as an illustration of the attempt to identify a biblical site east of the Jordan. Heshbon, mentioned 38 times in the Bible, can be located in a general manner on the central, Transjordanian plateau (Deut. 3.10; 4.43; Josh. 13.16) east of the Jordan River between Wadi Arnon in the south and Wadi Jabbok in the north (Deut. 2.24; Josh. 12.2). It is west of both the territory of the Ammonites (Josh. 12.2) and 'the wilderness of Kedemoth' (Deut. 2.26). It is related to such well-known and confidently identified sites as Medeba (= Madaba) and Dibon (= Dhiban) (Num. 21.30). From the biblical sources, nothing more definite can be said about its location.

From a toponymic point of view, there is no doubt that the modern village and associated tell of Hisban bears the biblical name. The question is whether the biblical name has remained at the same site down through the centuries or has it migrated to modern-day Tall Hisban from some near-by or, even, distant location? Eusebius places Heshbon, which in his day was called Esbus, in the mountains of Gilead c. 20 Roman miles from the Jordan across from Jericho (*Onomasticon* 84.1-6). The Talmud locates it at Housban/Hisban (Neubauer 1868: 21).

Tall Hisban (elevation 895 m asl) is situated in a rolling plain. It is c. 9 km north of the modern town of Madaba and c. 20 km southwest of 'Amman. Andrews University excavated Tall Hisban for five seasons from 1968 to 1976. Baptist Bible College continued the excavation of a Byzantine church at the site in 1978. The excavators uncovered no remains, other than some Late Bronze Age sherds, earlier than the Iron Age I period when there was probably a small, unfortified village at the site, dated to the twelfth–eleventh centuries BC and dependent on an agrarian-pastoral economy (Geraty 1992: 182, 1997: 20). Although there is evidence of the site's habitation during the tenth–eighth centuries BC, the best-preserved Iron Age remains date to the seventh–sixth centuries BC. They indicate 'a general prosperity and continued growth, probably clustered around a fort' (Geraty 1997: 20-21; see also Geraty 1992: 182). This settlement may have come to a violent end (Geraty 1992: 182, 1997: 21). There is no evidence for occupation during the Persian period but the site was reoccupied in the Late Hellenistic period. Habitation at Hisban continued throughout the Roman and Byzantine periods (when it reached its zenith) (LaBianca 1989: 264-67; Geraty 1992: 182-83, 1997: 21).

A problem with the location of biblical Heshbon at Tall Hisban is the apparent discrepancy between the biblical account of the Israelite capture of the site from the Amorite king Sihon (Num. 21.21-35; see also Deut. 2.26-35) and the archaeological evidence. The site's conquest, as narrated in the Bible, would have taken place, according to traditional dating, around the end of the thirteenth and the beginning of the twelfth centuries BC. As noted above, however, the archaeological evidence does not support the location of an Amorite capital city at Tall Hisban in either the Late Bronze or Early Iron Ages.

A solution to the above-identified problem can be found in the fact that the majority of literary critics agree that the prose segment of Num. 21.21-35 belongs to a late, Deuteronomistic stratum of the Pentateuch while the poetic portion (vv. 27b-30) originally had nothing to do with the conquest of Heshbon (Miller 1983: 124). Thus, from a literary-critical point of view, the narrative is either legendary (Van Seters 1980: 117-19; Timm 1989a: 94-95, 1989b: 175) or anachronistic (Miller 1983: 124).

In conclusion, most commentators, with very few exceptions, identify Tall Hisban with biblical Heshbon (Noth 1935: 248, 1944: 51, 53, 1968: 240; Simons 1959: 117, 121, 298, 449; Van Zyl 1960: 92; Abel 1967, 2: 348-49, 424; Ottosson 1969: 86; Baly 1974: 233 n. 11; Peterson 1980: 622-24; Aharoni 1979: 436; Boling 1985: 25; Geraty and Running 1989: ix; Miller 1989b: 28; Timm 1989b: 175; Knauf 1990; Lemaire 1992: 68* [with a question mark]; MacDonald 2000: 93). There are, however, a few scholars who look for biblical Heshbon at another site, for example, Jalul (Horn 1976, 1982: 10, 11; Ibach 1978; Boling 1988: 47), or understand Heshbon as more than the name of a 'city' but primarily as the name of a region (Merling 1991). Despite these dissenting voices, it appears almost certain that the biblical site of Heshbon is to be identified with Tall Hisban. This certainty is based on textual, toponymic and archaeological grounds.

Dornemann's work on Jabal al-Qal'a (= 'Amman Citadel) is another example of the relationship between the Bible and archaeology. As a result of his soundings north of the upper terrace, he posits that his findings, which span much of the tenth century BC and continue on into the next century, must be understood in the context of the biblical reference to the time of David and Solomon (1983: 172). On the basis of his work (1983: 172, 1997: 99), he thinks that there is evidence for the rebuilding of the city wall, which he dates to c. 980 BC, after its capture by David (2 Sam. 12.29).

Another example of the relationship between the Bible and archaeology comes from the writer's work in southern Jordan, that is, ancient Edom (MacDonald et al. 1988, 1992; MacDonald 1994). The archaeological evidence on eastern Edom, namely, the area east of Wadi 'Arabah and south of the Dead Sea, causes one to question many of the biblical texts relative to Edom and the Edomites. It provides little support for a major Edomite presence in the area at the end of the Late Bronze and/or the beginning of the Iron Age. It is, thus, hard to envisage, on the basis of the present archaeological evidence, the biblical portrayal of a strong fighting force in Edom during the above-mentioned periods (the traditional date for the exodus of the Israelites from Egypt and the period of their 'wanderings') that would have deterred a large group from passing through—and forced the group to detour around—this territory. Moreover, a large population in the same region would have been necessary to account for the numbers killed by David and/or Joab (1 Kgs 11.15-16). Evidence for such is not yet available (MacDonald 1994: 243). Finally, archaeology has not brought to light firm support for a Solomonic seaport on the Red Sea (MacDonald 1994: 243).

Conclusions

The Bible does provide important data on early peoples dwelling east of the Jordan river, especially during the first millennium BC. However, it needs to be assessed in the light of textual, literary and historical criticism. It must, moreover, be studied in relation to the latest scientific archaeological findings. Only by the wedding of scientific, biblical and archaeological information can a plausible portrayal of first-millennium BC Jordan be painted.

References

Abel, F.-M.
    1967   *Géographie de la Palestine*. I. *Géographie physique et historique*; II. *Géographie politique des villes*. 3rd edn. Paris: Gabalda.

Aharoni, Y.
    1979   *The Land of the Bible: A Historical Geography*. 2nd rev. edn. Trans. and ed. A.F. Rainey. London: Burns & Oates.

Baly, D.
    1974   *The Geography of the Bible*. Rev. edn. New York: Harper & Row.
    1987   The Pitfalls of Biblical Geography in Relation to Jordan. In A. Hadidi (ed.), *Studies in the History and Archaeology of Jordan*, III: 123-24. Amman: Department of Antiquities.

Bartlett, J.R.
    1989   *Edom and the Edomites*. Journal for the Study of the Old Testament, Supplement Series 77. Sheffield: JSOT Press.

Boling, R.G.
    1985   Levitical Cities: Archaeology and Texts. In A. Kort and S. Morschauser (eds.), *Biblical and Related Studies Presented to Samuel Iwry*, 23-32. Winona Lake, IN: Eisenbrauns.
    1988   *The Early Biblical Community in Transjordan*. Sheffield: Almond Press.

Coughenour, R.A.
    1989   A Search for Mahanaim. *Bulletin of the American Schools of Oriental Research* 273: 57-66.

Dion, P.E.
    1995   Aramaean Tribes and Nations of First-Millennium Western Asia. In J.M. Sasson (ed.), *Civilizations of the Ancient Near East*, II: 1281-94. New York: Charles Scribner's Sons.

Dornemann, R.H.
- 1983 *The Archaeology of the Transjordan in the Bronze and Iron Ages*. Milwaukee, WI: The Milwaukee Public Museum.
- 1997 Amman. In E.M. Meyers (ed.), *The Oxford Encyclopedia of Archaeology in the Near East*, I: 98-102. New York: Oxford University Press.

Fitzmyer, J.A.
- 1981 *The Gospel According to Luke (I–IX): Introduction, Translation, and Notes*. The Anchor Bible 28. New York: Doubleday.

Geraty, L.T.
- 1992 Heshbon. In D.N. Freedman (ed.), *The Anchor Bible Dictionary*, III: 181-84. New York: Doubleday.
- 1997 Hesban. In E.M. Meyers (ed.), *The Oxford Encyclopedia of Archaeology in the Near East*, III: 19-22. New York: Oxford University Press.

Geraty, L.T. and L.G. Running (eds.)
- 1989 *Historical Foundations: Studies of Literary References to Hesban and Vicinity*. Hesban 3. Berrien Springs, MI: Andrews University.

Grollenberg, L.A.
- 1956 *Atlas of the Bible*. Trans. and ed. J.M.H. Reid and H.H. Rowley. London: Nelson.

Horn, S.H.
- 1976 Heshbon. In K. Crim (ed.), *The Interpreter's Dictionary of the Bible, Supplementary Volume*, 410-11. Nashville: Abingdon Press.
- 1982 *Hesban in the Bible and Archaeology*. Occasional Papers of the Horn Archaeological Museum 2. Berrien Springs, MI: Horn Archaeology Museum, Andrews University.

Ibach, R.D.
- 1978 An Intensive Surface Survey at Jalul. *Andrews University Seminary Studies* 16: 215-22.

Knauf, E.A.
- 1990 Hesbon, Sihons Stadt. *Zeitschrift des deutschen Palästina-Vereins* 106: 135-44.

LaBianca, Ø.S.
- 1989 Hesban. In D. Homès-Fredericq and J.B. Hennessy (eds.), *Archaeology of Jordan. II/1. Field Reports: Surveys and Sites A–K*, 261-69. Akkadica Supplementum 7. Leuven: Peeters.

Lemaire, A.
- 1992 Hesbon=Hisban? In E. Stern and T. Levi (eds.), *Eretz-Israel: Archaeological, Historical and Geographical Studies (Abraham Biran Volume)*, 23: 64*-70*. Jerusalem: The Israel Exploration Society/Hebrew Union College-Jewish Institute of Religion.

MacDonald, B.
- 1993 *Ammon, Moab and Edom: Early States/Nations of Jordan during the Biblical Period (End of the 2nd and the 1st Millennium BC)*. Amman: Al Kutba.
- 1994 Early Edom: The Relation between the Literary and Archaeological Evidence. In M.D. Coogan, J.C. Exum and L.E. Stager (eds.), *Scripture and Other Artifacts: Essays on Archaeology and the Bible in Honor of Philip J. King*, 230-46. Louisville, KY: Westminster/John Knox Press.
- 2000 *'East of the Jordan': Territories and Sites of the Hebrew Scriptures*. Boston, MA: American Schools of Oriental Research.

MacDonald, B. et al.
- 1988 *The Wadi el Hasa Archaeological Survey 1979–1983, West-Central Jordan*. Waterloo, Ontario: Wilfrid Laurier University.
- 1992 *The Southern Ghors and Northeast 'Arabah Archaeological Survey 1985–1986, Southern Jordan*. Sheffield Archaeological Monographs 5. Sheffield: J.R. Collis.

Merling, D.
- 1991 Heshbon: A Lost City of the Bible. *Archaeology in the Biblical World* 1/2: 10-17.

Meyers, E.M.
- 1984 The Bible and Archaeology. *Biblical Archaeologist* 47/1: 36-40.

Miller, J.M.
- 1983 Site Identification: A Problem Area in Contemporary Biblical Scholarship. *Zeitschrift des deutschen Palästina-Vereins* 99: 119-29.
- 1989a The Israelite Journey through (around) Moab and Moabite Toponymy. *Journal of Biblical Literature* 108/4: 577-95.
- 1989b Moab and the Moabites. In A. Dearman (ed.), *Studies in the Mesha Inscription and Moab*, 1-40. Atlanta: Scholars Press.

Neev, D. and K.O. Emery
- 1995 *The Destruction of Sodom, Gomorrah, and Jericho: Geological, Climatological, and Archaeological Background*. New York: Oxford University Press.

Neubauer, A.
- 1868 *La géographie du Talmud*. Paris: Michel Lévy Frères.

Noort, E.
- 1987 Transjordan in Joshua 13: Some Aspects. In A. Hadidi (ed.), *Studies in the History and Archaeology of Jordan*, III: 125-30. Amman: Department of Antiquities.

Noth, M.
- 1935 Studien zu den historisch-geographischen Dokumenten des Josuabuches. *Zeitschrift des deutschen Palästina-Vereins* 58: 185-255.
- 1944 Israelitische Stamme zwischen Ammon und Moab. *Zeitschrift für die alttestamentliche Wissenschaft* 60: 11-57.
- 1968 *Numbers: A Commentary*. Trans. J.D. Martin. Philadelphia: Westminster Press.

Ottosson, M.
1969 *Gilead: Tradition and History.* Trans. J. Gray. Lund: C.W.K. Gleerup.

Peterson, J.L.
1980 A Topographical Surface Survey of the Levitical 'Cities' of Joshua 21 and I Chronicles 6: Studies on the Levites in Israelite Life and Religion. Unpublished PhD dissertation, Chicago Institute of Advanced Theological Studies and Seabury-Western Theological Seminary, Evanston, IL.

Rast, W.E.
1987 Bab edh-Dhra and the Origin of the Sodom Saga. In L.G. Perdue, L.E. Toombs and G.L. Johnson (eds.), *Archaeology and Biblical Interpretation: Essays in Memory of D. Glen Rose*, 185-201. Atlanta: John Knox Press.

Simons, J.
1959 *The Geographical and Topographical Texts of the Old Testament: A Concise Commentary in XXXII Chapters.* Leiden: E.J. Brill.

Timm, S.
1989a *Moab zwischen den Machten: Studien zu historischen Denkmalern und Texten.* Ägypten und Altes Testament 17. Wiesbaden: Otto Harrassowitz.
1989b Die Ausgrabungen in Hesban als Testfall der neueren Palästina-Archäologie. *Nederduits gereformeerde teologiese tydskrif* 30: 167-78.

Van Seters, J.
1980 Once Again the Conquest of Sihon's Kingdom. *Journal of Biblical Literature* 99: 117-19.

Van Zyl, A.H.
1960 *The Moabites.* Leiden: E.J. Brill.

26. The Mosaics of Jordan

Michele Piccirillo

Introduction

The oldest mosaic of Jordan belongs to the Herodian era (the last decades of the first century BC) and consists of a fragment of the floor of the *apoditerium* (changing room) in the baths of the fortified palace built by the king of Judaea at Machaerus in the desert on the southern border of the oriental province of Peraea.

A mosaic floor of the Roman period, second–third centuries AD, was discovered accidentally at Jarash in 1907. The remains consist of the external border with the personifications of the Seasons represented in the corners and of busts of the Muses and portraits of Greek poets and writers surrounded by a garland of fruit and flowers supported by a naked putti. The main motifs of the mosaic are dedicated to the cycle of Dionysus and Herakles.

After more than a century of accidental findings and discoveries from research carried out by professional archaeologists all over Jordan, the mosaic art in the region is attested chronologically from the first century BC to the second half of the eighth century AD. Such discoveries cover the whole Transjordanian territory from north to south. The main centres attested for mosaic production are the northern regions around Jarash/Irbid, Philadelphia-'Amman, and Madaba.

The great majority of the mosaic floors are dated to the sixth century and decorated the floors of ecclesiastical buildings. These mosaics must, therefore, be seen and studied in the context of Byzantine art contemporary with the spread of Christianity in the region during a period of a flourishing economy at the time of the Emperor Justinian and his immediate successors on the imperial throne of Constantinople.

The recent discovery of mosaic floors in Umayyad palaces and in contemporary churches demonstrates better than in other regions the continuity of mosaic art as an uninterrupted tradition from the Hellenistic to the Arabic period.

The mosaics of secular buildings

The discovery, in 1982 and 1985 respectively, in Madaba of the Hippolythus Hall under the church of the Virgin Mary and of the mosaic floors of the Burnt Palace, dated by the archaeological context to the Byzantine period, have clarified historically and functionally some mosaic floors known previously, for example, the Bacchich Procession, still *in situ* in the Archaeological Museum, the mosaic of Achilles, and the mosaic of the Seasons.

The mythological scenes depicted in the mosaics are not due to the pagan faith of the Madabites, as concluded by the first scholars, but to the classical renaissance that flourished in the fifth century, culturally nourished by the reading of Greek-Roman literature.

The mosaic of Hippolythus, found under the church of the Virgin Mary in the northern quarter of the city, is the most significant example of the mythological subjects used in the Byzantine mosaics of Jordan (Figure 26.1). The figurative motifs of the central and eastern panels are inspired by Euripides' Hippolythus. Beyond the border, on the eastern side, are the enthroned personifications of the Tychai of the christianized cities of Rome, Gregoria and Madaba, each holding a cross on a long staff. This mosaic remains the best example to date of the skill achieved by the mosaicists of the Madaba School.

The church mosaics: outstanding works

There is no doubt that the best works of the Byzantine mosaicists working in Jordan have been found in Madaba and its vicinity. The evidence for this assertion is found in the masterpieces that paved the floors of the churches of Madaba, of Mount Nebo and Ma'in to the west, and of Kastron Mefa a/Umm ar-Rasas to the east. These works are products of the so-called Madaba Mosaic School. In the vast repertoire of floor mosaics of the Late Roman–Byzantine period, the sixth-century

Figure 26.1. Mosaic of Hippolythus, Madaba. The figurative motifs are inspired by Euripides' Hippolythus. Beyond the border (top left) are the enthroned Tychai of Rome, Gregoria and Madaba (photo courtesy of M. Piccirillo).

mosaics of Jordan are a homogeneous historical and artistic group that witness the classical renaissance that found its maximum expression and diffusion in Justinian's time.

To this classical renaissance belong thematically the agricultural, pastoral and hunting scenes, the Nilotic motifs, the portraits of the benefactors, the personifications of the Earth with the young Carpoforoi on the sides, the Sea-Thetis-Abyssos, the Seasons, and the Foliate Masks. The Rivers of Paradise are often used as an erudite reference to the classical world and the Bible as are the allegory of sacrifice (explained by Ps. 50.21) in the Theotokos Chapel on Mount Nebo, and the allegory of the messianic kingdom of Emmanuel, with the citation from the prophet Isaiah (65.25b), at Ma'in. Geographical and topographical motifs have a special place in this new interest in classical art and culture. The most representative examples of this are the Madaba Map, the topographical border of Ma'in, and the city plans in the church of St Stephen at Umm ar-Rasas. The most direct references to the classical world are the mythological scenes used to decorate the secular houses of Madaba.

The mosaicists of Madaba, and more generally the mosaicists of Jordan, preferred miniaturized compositions. They divided the floors, normally using vine or acanthus scrolls in which they represent the single figures of a more complex scene, into smaller geometrical compartments.

From the inscriptions added to their compositions, we know the names of a few mosaicists working in the area, such as the teams composed of Soel, Kaium and Elia, who signed their work in the Memorial of Moses on Mount Nebo (AD 530); Naum, Kiriacos and Toma, who signed their work in the church of St George in the village of Nebo (AD 536); Salaman, who signed the central medallion of the Sea in the church of the Apostles at Madaba (AD 578); Staurachios from Esbus; and Euremios who signed the upper mosaic floor of the sanctuary in the church of St Stephen at Umm ar-Rasas (AD 756).

Among the mosaics of the sixth century, the period best represented in the mosaics of Jordan, we can highlight some outstanding works that decorated the churches of the city of Madaba and its territory, starting with the Madaba Map, the most famous of the mosaics of Jordan.

The Madaba Map (Figure 26.2) decorates the floor of the basilica near the north gate of the city. It is oriented towards the east. Vignettes and captions indicate localities. From a geographical point of view, the nearly 150 captions of localities added to the map refer mainly to the territory of Palestine from the region of Tyre and Sidon on the north, to the Egyptian Delta

Figure 26.2. Mosaic map of the Holy Land from Madaba, with Jerusalem at its centre (photo courtesy of M. Piccirillo).

on the south, and from the Mediterranean Sea to the desert. The mosaic reaches its figurative high-point in the vignette of Jerusalem, which in a way is the ideal centre of the composition even if it is not the precise physical centre.

The mosaicists of Madaba favoured topographical representations. This is true also of the mosaicists of Jarash and of Jordan in general. Besides the Madaba Map, two churches of Jarash, the church of St John and the church of the Holy Apostles Peter and Paul, were decorated with city plans. Moreover, walled cities are depicted on Mount Nebo, both in the Memorial of Moses and in the church of the Holy Martyrs Lot and Procopius at Khirbat al-Mukhayyat, in the church on the acropolis of Ma'in, on Jabal 'Ajlun, at Zay near as-Salt, in the lower church of al-Quwaysmah-'Amman, in the church of St John at Khirbat as-Samra, and in the churches of Kastron Mefa a-Umm ar-Rasas. However, the Madaba Map, more than a century after its discovery, remains unique and is unsurpassed by new discoveries.

The publication of the Madaba Map in 1897 was not only an international event which focused the attention of scholars from all over the world upon Madaba, but became the starting point of new important discoveries. In the year of its discovery, Fr Giuseppe Manfredi, the Latin Priest of Madaba, excavated a number of churches: three along the Roman Road;

the church of the Prophet Elias with its crypt; the eastern church, known as the church of the Sunna' Family; the church of al-Khadir, now known as the church of the Holy Martyrs; the 'cathedral' church; and the Apostles' church on the southern slope of the tell. The pavement of the latter preserves one of the most interesting works of the mosaicists of Madaba. The decorative programme rotates around the central medallion with the personification of Thalassa (the Sea), represented as Thetis, a woman, with a rudder as a standard, coming out of the waves amid jumping fish and sea monsters (Figure 26.3).

There is preserved on Mount Nebo, in the Sanctuary of Moses and the surrounding monastic buildings atop Siyagha, in the churches of the village of Khirbat al-Mukhayyat (Holy Martyrs Lot and Procopius), and in the valleys of 'Uyun Musa and 'Ayn al-Kanisah, a concentration of mosaic works that are fundamental for understanding the development of Byzantine-Umayyad mosaic art in the territory of Madaba from the fourth to the second half of the eighth century.

The main part of the mosaics of Mount Nebo, except for the mosaics of the basilica of Moses, were not affected by the iconoclastic crisis. On Mount Nebo an extraordinary figurative repertoire has, therefore, been preserved that is a very important reference for the re-reading of the iconographic interpretation of less

Figure 26.3. Mosaic medallion with the personification of the Sea. Madaba, Church of the Apostles (photo courtesy of M. Piccirillo).

fortunate contemporary works discovered at Madaba and Umm ar-Rasas, the other two centres of major activity for the artisan mosaicists in the territory of the diocese.

The mosaics of the Umayyad period in Jordan

The themes and their development, as well as the decorative motifs of sacred and secular buildings, find their natural result in the mosaics of the churches and palaces of the Umayyad epoch. These mosaics are the last works of an uninterrupted tradition of workmanship. Floor mosaics found recently in the Umayyad palaces of Qasr al-Hallabat, Qusayr 'Amra, Qastal, al-Muwaqqar, and on the 'Amman Citadel are few but, nevertheless, representative of the secular art of the period.

The best preserved examples come from Qasr al-Hallabat. In Room 4 of this palace a large and elaborate border of looped circles surrounds the geometric composition with lobed squares enclosed by a band of lotus buds along the wall and a continuous series of black waves on the interior. Flowers and fruits, such as pomegranates, lemons and a variety of citrus, fill the circles of the border. Animals decorate the compartments of the inner composition of the carpet: a standing onyx moves among flowers; two partridges facing a flower; a wolf; a hare nibbling grapes; a bull; a fish; two birds confronting one another; and a leopard.

The floor mosaics of excavated churches dated to the Umayyad period in Jordan are a real novelty. They are an addition to knowledge of mosaic art by modern archaeological research. To date in Jordan we have at least nine mosaic floors of churches dated to the Umayyad period. They are found in the church of the Virgin Mary in Madaba (AD 767), in the church on the acropolis of Ma'in (719), in the lower church at al-Quwaysmah (718), in the church of St Stephen at Umm ar-Rasas (756), in the nave and the two aisles of the same church (eighth century), in the nave of the church of St Lot at Ghor as-Safi in the Jordan Valley (691), in the upper mosaic floor of the double church at Khilda-'Amman (687), in the upper mosaic floor of the church of Shunah al-Janubiyya in the Jordan Valley (eighth century), and the added mosaic in the chapel of the Theotokos Monastery at Wadi 'Ayn al-Kanisah on Mount Nebo (762).

Of all the above, the outstanding work is the mosaic floor of the church of St Stephen at Umm ar-Rasas (Kastron Mefaa) (Figure 26.4). Its major artistic interest focuses on the double geographical frame depicting on the north, eight cities of Palestine starting with the Holy City of Jerusalem and ending with the city of Gaza, on the south, seven cities and villages of Jordan, beginning with the double city-plan of Kastron Mefa'a, identified by the main inscription with the ruins of Umm ar-Rasas, and going down to Charach Mouba, plus 10 cities of Egypt beginning with Alexandria.

On the basis of the method of writing common to the three contemporary mosaics at Umm ar-Rasas in the sanctuary of the church of St Stephen, in the church of the Virgin Mary at Madaba, and at 'Ayn al-Kanisah, one can suggest a single author, Staurachios from Hisban and his team, of which Euremios was a member. These, according to present evidence, were the

Figure 26.4. Mosaic floor of the Church of St Stephen, Umm ar-Rasas, showing cities and villages of Jordan, Palestine and Egypt (photo courtesy of M. Piccirillo).

last mosaicists of the Madaba region. They were still active during the second century of the Muslim Era (Hegira).

Iconoclasm in the mosaics of Jordan

Except for the mosaics of the church of the Virgin Mary at Madaba and the upper mosaic in the sanctuary of St Stephen at Umm ar-Rasas, other mosaics of this period show the use of figures. They show at the same time traces of iconoclastic damage, which is, therefore, subsequent to their dates. The date of the mosaics with figurative composition gives a *terminus post quem* for the iconoclastic damage, presently dated to 719/20, the date of the mosaics of Ma'in, Quwaysmah and, possibly, of St Stephen. All three of these mosaics have figures. Therefore, scholars agree in dating iconoclasm to some time in the eighth century.

In most cases the damage was restored with care using the same tesserae of the removed figure, or patched up with larger tesserae. In some cases the mosaicists in charge of the work made new motifs. The additions below are, therefore, chronologically related to the restoration of the iconoclastic damages and dated to that period.

The most famous of such additions is found in the mosaic of the north chapel of the church of Ma'in. The original composition with a lion facing a bull, of which one can see the tail, two paws, a hoof, a hump and the points of two horns, was replaced with a bush to the left and a jar to the right. In the church of St Stephen at Umm ar-Rasas, the rectangular panel in front of the sanctuary had been decorated with seven portraits of benefactors. A leaf, a circle with florets and a diamond replaced the first benefactor, whose name was partly obliterated. The second, carrying a lamb whose head was spared, was replaced with diamonds and circles. A branch with leaves of the same genre replaced a benefactor carrying a jar on his shoulder. A short branch with leaves among three florets replaced the fourth figure. The same motif, with the addition of a cross of florets and a circle, replaced the fifth figure carrying a peacock, the tail of which was not damaged. A fluteplayer (the flute is visible near the tree on the right) was replaced by a branch between two florets.

The question still remains unanswered as to who was responsible for the order to destroy the images of humans and animals in the mosaics of the churches of Jordan.

Historically, the dated mosaics of the church of St Stephen at Umm ar-Rasas (AD 756), that of the chapel of the Theotokos on Mount Nebo (762) and the mosaic of the church of the Virgin at Madaba (767) witness to a Christian community in existence in the second half of the eighth century. This community was well organized with a bishop, presbyters and deacons, peacefully carrying on church affairs without interference and having economic resources and artistic skills to decorate their churches with such sophisticated mosaic floors.

References

Avi-Yonah, M.
    1954   *The Madaba Map*. Jerusalem: Israel Exploration Society.

Bagatti, B.
    1957   Il significato dei mosaici della scuola di Madaba (Transgiordania). *Rivista di archeologia cristiana* 33: 139-60.

Kraeling, C.H.
    1938   *Gerasa, City of the Decapolis*. New Haven, CT: American Schools of Oriental Research.

Piccirillo, M.
    1981   *Chiese e mosaici della Giordania settentrionale*. Jerusalem: Franciscan Printing Press.
    1989   *Chiese e mosaici di Madaba*. Studium Biblicum Franciscanum, Collectio Maior 34, Documentazione grafica a cura di E. Alliata. Jerusalem: Franciscan Printing Press.
    1993   *The Mosaics of Jordan*. Eds. P.M. Bikai and T. Dailey. Amman: American Center of Oriental Research.

Piccirillo, M. and E. Alliata
    1994   *Umm al-Rasas—Mayfa'ah I: Gli scavi del complesso di Santo Stefano*. Jerusalem: Franciscan Printing Press.
    1998a   *Mount Nebo: New Archaeological Excavations 1967–1997*. Jerusalem: Franciscan Printing Press.
    1988b   *The Madaba Map Centenary 1897–1997: Travelling through the Byzantine–Umayyad Period*. Jerusalem: Franciscan Printing Press.

Saller, S.
    1941   *The Memorial of Moses on Mount Nebo*, I-II. Jerusalem: Franciscan Printing Press.

Saller, S. and B. Bagatti
    1949   *The Town of Nebo (Khirbet el-Mukhayyat)*. Jerusalem: Franciscan Printing Press.

27. Crusader Castles in Jordan

Denys Pringle

Having captured Jerusalem on 14 July 1099 and Tiberias in September of the same year, the Crusaders soon began extending their control east of the Jordan Valley. The importance of the area to the newly arrived westerners was essentially threefold. In the first place, there was a strategic need to secure defensible frontiers, to confront the Muslim strongpoint of Damascus, and to control the line of communications between Damascus and Egypt. Secondly, there were economic attractions in being able to control the caravan routes between Syria and the Jazira in the north and the Hijaz and Egypt in the south, besides exploiting the agricultural and mineral resources of the land itself. Thirdly, there was the ideological imperative which underpinned the whole Crusading enterprise, namely to extend Christendom to its former limits and to sustain the Christian communities of the east.

In an area corresponding to what is today northern Jordan and southern Syria, around the valley of the Yarmuk (the area known as the Sawad, or terre de Sueth), a number of villages that had belonged to the church in earlier times were granted by Tancred, prince of Galilee, to the Benedictine abbey of Mount Tabor as early as 1100 (Röhricht 1893: 5-6, no. 36). In 1105, the Franks from Tiberias built a castle at al-'Al, in the southern Jaulan, to protect this region; but within a year it had been taken by Tughtagin, *atabeg* of Damascus, and the garrison killed (Ibn al-Qalanisi 1932: 71-72). Although this castle has been identified with the promontory site of Qasr Bardawil (cf. Deschamps 1939: 100-101), recent investigations have found no indication of occupation there later than Middle Bronze II (see Pringle 1997: 117, no. R14). It seems, therefore, that the Crusader castle may have been located elsewhere.

In 1109, King Baldwin I concluded a treaty with Tughtagin, which effectively split the agricultural resources of the Sawad and the Jabal 'Awf between them, with a third share being left to the Bedouin chiefs of the region (Ibn al-Qalanisi 1932: 92). But although Crusader control was later to extend as far east as Muzairib and Dar'a (cité Bernard d'Étampes) and political influence to Basra and Salkhad, Frankish settlement in this area was slight. Castles, however, were an essential means of enforcing military and hence economic and political control. Thus, in 1109 Baldwin I established a second castle in a Byzantine monastic cave complex on the southern side of the Yarmuk gorge at 'Ayn al-Habis. The Muslims occupied this castle, which was known as the Cava de Suet, or Habis Jaldak, in 1111–13 and again briefly in 1118 and 1182; otherwise, it remained under Crusader control until it fell to Saladin in 1187. Much of the cave complex has collapsed since the twelfth century, although the site has been investigated by Schumacher in 1913–14, by Horsfield in 1933–35, and more recently by Nicolle (1988; cf. Deschamps 1933, 1935, 1939: 99-116, 211-13, pl. XXXI; Pringle 1993: I, 26, 1997: 18, no. 10).

Farther south in the Jabal 'Awf or Jabal 'Ajlun, which the Franks made little attempt to control directly, the only castles that we hear of are Muslim ones. In Jarash, for example, Tughtagin of Damascus 'at great expense had caused a fortress of immense hewn stones to be erected' around the Temple of Artemis; Baldwin II captured it in 1121 but then dismantled it, since it would have been impractical to maintain (William of Tyre 1941: I, 539, 1986: 565-66; cf. Harding 1974: 98). In 1139, Thierry of Alsace, count of Flanders, also stormed another cave castle between as-Salt and Jarash, near to Mount Gilead (*iuxta Mons Galaad*: Jabal Jal'ad; William of Tyre 1941: II, 102-05, 1986: 681-84; Deschamps 1933: 47, 48 n. 2).

It was in the southern part of the country, however, that Crusader settlement, and hence castle building, was more intensive. In November and December 1100, Baldwin I, following his accession to the throne, led an expedition of 140 knights and 500 infantry from Hebron southeast around the southern end of the Dead Sea as far as Wadi Musa and Jabal Harun, before returning to Bethlehem. In 1107, the king returned to destroy a castle that the Damascenes had recently built in Wadi Musa. He evacuated to the west of the Jordan the Christians whom he encountered there. This suggests that permanent occupation of the area was not yet contemplated (Deschamps 1939: 40-41; Mayer 1987: 199, 1990: 19-20, 29-30).

This policy was overturned, however, in 1115, when Baldwin founded the castle of ash-Shawbak, or 'Mount Royal' (Montreal, grid ref. 2037.9932). The site of this castle was evidently chosen to allow the knights stationed in it to control the desert road. However, it is also clear from the accounts of its foundation that it was intended to act as a centre for Frankish settlement (Prawer 1980: 109-11, 476). William of Tyre, for example, says that 'after the completion of the work, [Baldwin] placed as inhabitants both knights and foot-soldiers, granting them extensive possessions', while the Old French version of his chronicle describes these inhabitants as 'knights, sergeants and villeins' (William of Tyre 1941: I, 506-507, 1986: 534-35; Pringle 1993: II, 304). Later on, Montreal possessed its own court of burgesses.

The castle of ash-Shawbak occupies the summit of an oval hill at the head of a valley (Figure 27.1). William of Tyre describes the defences as comprising 'a wall, towers, an outer wall and ditch', strengthened by arms and machines. Since the castle has been added to by the Ayyubids and Mamluks, it is now difficult to identify precisely those elements belonging to the Crusader phase. The castle would appear, however, to have had a double wall, with an outer gate (possibly with a bent-entrance) leading to an inner one on the south. Although it is likely that there was also a *faubourg* extending down the valley below the castle, it would appear that the nucleus of the Crusader settlement lay inside these defences, as William of Tyre implies. This is indicated both by references in charters to private houses located within it, and by the fact that the castle contains the remains of two churches: one of them a small barrel-vaulted chapel lying between the two gates and possibly intended for the use of the local Orthodox Christians, the other a fine three-aisled building situated in the central area and probably to be identified as the Latin parish church. An inscription from above the door of the larger church gives the date 1118, and appears to have referred to a viscount named Hugh (Pringle 1993: II, 304-14, nos. 229-30, 1997: 75-76, no. 157).

Soon after constructing this castle, and certainly before his death in 1118, Baldwin granted it and the lordship dependent on it to Roman of Le Puy. However, in the 1130s, Roman and his son Ralph took part in a revolt against King Fulk and lost the lordship to the king's steward, Pagan (Mayer 1987: 201, 1990: 102-14). In 1142/43, Pagan began to build a new castle to the north at al-Karak, in Moab, to replace ash-Shawbak as the centre of his lordship. Construction work on al-Karak was continued by his successors, Maurice and Philip of Milly (Deschamps 1939: 46; Mayer 1990: 115-34). It was Maurice who, in 1152, assigned the defence of the barbican, or lower ward, of this castle to the Hospitallers in return for various concessions within the lordship (Delaville le Roulx 1894: I, 160, no. 207; cf. Röhricht 1893: 71, no. 279).

The castle of al-Karak occupies a rocky spur at the southern end of the plateau on which the town now stands (see Müller-Wiener 1966: 47-48, pls. 23-27; Deschamps 1939: 80-93, pls. IV-XXVII; Kennedy 1994: 45-52, pls. 15-19; Pringle 1993: I, 286, 1997: 59-60, no. 124) (Figures 27.2 and 27.3). In the twelfth century the town was still unwalled, relying for its defence on the natural cliffs that surrounded it. The castle is roughly triangular in shape, some 220 m long and 40–110 m wide, and is detached from the town by a 30 m–wide ditch. The walls are strengthened by projecting rectangular towers. A lower outer ward, some 30 m wide, is on its west side. Although mostly Mamluk in its present form, this appears to represent the barbican, which is described in 1152 (see above) as extending from the tower to the left of the gate as far as the Tower of St Mary. This suggests that the main entrance to the castle lay through the barbican and that St Mary's Tower may have been the predecessor of the present Ayyubid keep, at the southern end of the inner ward; quite possibly, it was badly damaged in the siege of 1183, when Saladin bombarded it with two siege engines placed on the hill to the south (cf. Deschamps 1939: 64).

James of Vitry, who was bishop of Acre in the thirteenth century, records that there were seven very strong fortresses dependent on Crusader al-Karak and ash-Shawbak in the twelfth century; however, he does not name them. Deschamps has suggested that they were: Ahamant ('Amman), Taphila (at-Tafila), Hurmus (Khirbat al-Hurmuz), li Vaux Moïse (al-Wu'ayra), Sela (al-Habis, Petra), Aila (al-'Aqaba) and the île de Graye (Jazirat Fara'un) in the Gulf of al-'Aqaba (1939: 39 n.1). However, excavations undertaken at Jazirat Fara'un in the 1980s demonstrated conclusively that the castle was Ayyubid (see Pringle 1997: 117, no. R9). If the number given by James of Vitry is correct, it would, therefore, seem more plausible to identify the castles, from north to south, as: 'Amman, at-Tafila, Khirbat as-Sil', Khirbat al-Hurmuz, al-Wu'ayra, al-Habis and al-'Aqaba.

The existence of a Crusader castle in 'Amman is suggested by a charter of 31 July 1161, by which Baldwin III gave Philip of Milly everything

Figure 27.1. Ash-Shawbak (Montreal): the castle from the west (photo D. Pringle).

Figure 27.2. Al-Karak (Crac): the castle from the southwest (photo D. Pringle).

Figure 27.3. Al-Karak (Crac): the northeast tower of the barbican, which covered the principal gate into the castle (photo D. Pringle).

in Transjordan, 'that is to say, Mount Royal and Karak (Crach) and 'Amman (Ahamant), with all their appurtenances…between Zerca and the Red Sea' (Strehlke 1869: 3, no. 3; Röhricht 1893: 96-97, no. 366). Philip later joined the Templars, and in 1166 Amalric I granted 'Hamam with all its territory' to the order, together with half of what Philip had held in the Balqa' (Belcha) (Delaville le Roulx 1905–1908: 184; cf. Deschamps 1935: 42-47, 1939: 38-48). A masonry tower, some 8 m high with plain arrow-slits, which stands on the Citadel of 'Amman, was at one time thought to be Crusader but has now been shown by excavation to date more probably from the thirteenth century (Wood 1993; Northedge 1983; Northedge et al. 1993: 105-27). The site of the Crusader castle has, therefore, still to be verified (cf. Pringle 1997: 112-13, no. P3).

The 'most renowned castle of Tafila' (*castrum nobilissimum apud Traphyla*) is mentioned around 1239 in a list of former Crusader castles then in the hands of the sultan (Deschamps 1943: 96); one of its former lords, Martin (*dominus Martinus de Taphilia*), may perhaps be identified in the witness list to a charter of 1177 (Röhricht 1893: 144, no. 542). It seems likely that the castle occupied the same site as the present rectangular redoubt of the Ottoman period (Johns 1937: 96; Lawrence 1939: pl. 103; Pringle 1997: 98, no. 214).

A castle called al-Sala' was captured from the Franks by the Ayyubid amir, Sa'd al-Din Kamshabah al-Asadi, along with al-Karak and ash-Shawbak in Ramadan 584 H/AD 1188–89 ('Imad al-Din 1972: 99; al-Maqrizi 1980: 88); and the Latin list of castles in Muslim hands around 1239 refers to it as *Celle* (Deschamps 1943: 96-97). Although Yaqut mentions al-Sala' around 1225 as a 'fort in Wadi Musa' (Le Strange 1890: 528), causing some to identify it with al-Habis (e.g., Deschamps 1939: 19, 38; Horsfield and Horsfield 1938: 5), it seems unlikely to be identifiable with either of the known Crusader castles in Wadi Musa (see below). A more plausible location is Khirbat as-Sil', a natural rock castle some 10 km south of at-Tafila, which is sometimes proposed as Sela of 2 Kgs 14.7. Here Mediaeval occupation is attested by surface finds of pottery (Zayadine 1985: 164-66, fig. 9).

Hurmus is a castle mentioned as having fallen to Malik al-'Adil, Saladin's brother, in 1187 ('Imad al-Din 1972: 99; Abu Shama 1896: IV, 382). Although the likeliest site for this probably lies at or near

Khirbat al-Hurmuz, some 10 km north of Petra (Deschamps 1939: 38; Johns 1937: 30), definitive archaeological confirmation is still lacking and other possible candidates have also been suggested (cf. Kob 1967; Brooker and Knauf 1988: 187).

Although Baldwin I had attacked a Muslim castle in Wadi Musa in 1107, the first reference to a Frankish castle there is in 1144 when, hearing that it had been captured by the Muslims, Baldwin III set about retaking it (William of Tyre 1941: II, 144-45, 1986: 721-22). The date of construction of the *castrum...Vallis Moysi* (al-Wu'ayra) is unknown, although it seems possible that it had been built in 1127/31, following a punitive raid on the area by Baldwin II (Mayer 1990: 97-99, 130, 159). In 1158, it endured an eight-day siege by the Egyptians. In 1161, when it was granted to Philip of Milly along with al-Karak and ash-Shawbak, its previous holders are identified as Ulric, viscount of Nablus, and his son Baldwin (Strehlke 1869: 3, no. 3). It fell to Saladin in 1188.

Al-Wu'ayra castle lies northeast of Petra, on the north side of the Wadi Musa (see Musil 1907: II, 65-70, figs. 20-33; Savignac 1903; Kennedy 1994: 24-27, 30, 45, pls. 2-3; Vannini and Desideri 1995; Pringle 1993: II, 373-77, 1997: 105-106, no. 230). It has a roughly quadrangular plan, measuring some 90 m north–south by 60-80 m east–west, and is surrounded on all sides by natural ravines, that on the east having been artificially deepened leaving a rock pinnacle to support the entrance bridge. The walls are defended by rectangular towers and by linear outworks. The Christian nature of this castle is demonstrated by the existence of a chapel, very similar in design and in the detail of its apse moulding to that in the outer ward at ash-Shawbak (Pringle 1993: II, 375-76, no. 277). To judge from the size of the castle, it is likely that in addition to a lordly residence it also contained a dependent civilian settlement. Baldwin II's attempt to encourage Christian settlement in the area is also attested by the mention around 1160 of a village in the plain below Wadi (or 'Jabal') Musa, called Hara, which contained a church of St Moses (Mayer 1990: 98-99; Pringle 1993: II, 376-77, no. 278).

James of Vitry's sixth castle may be identified as that occupying al-Habis, in Petra (Figure 27.4). This comprises a tower, some 8.4 by 5.3 m, standing on the summit and surrounded by discontinuous and irregular lines of walls, pierced by embrasures, which form two enclosures lower down (Hammond 1970; Brooker and Knauf 1988: 186-87; Kennedy 1994: 28-30, 33, 121, pls. 4-6; Marino *et al.* 1990: 4-5; Pringle 1997: 49-50, no. 97). The castle is identified as Crusader on account of its masonry, although its Crusader name is unknown. It may be identified, however, as the castle of al-Aswit that al-Nuwayri mentions as having been visited by Sultan Baybars in 1276 (Zayadine 1985: 173; cf. 164-66).

Al-'Aqaba, or Aila, was occupied by the Franks in 1116 and remained theirs until 1170, when the fortress, attacked by sea and land, fell to Saladin (Deschamps 1939: 54, n.1). The Crusader castle evidently lay on the mainland and not on Jazirat Fara'un, whose castle post-dates 1170. Excavations have shown that the early Muslim fort of al-'Aqaba was deserted around 1100, leading one to suppose that its Crusader replacement must have been located elsewhere, possibly on the site later occupied by the castle that is credited to al-Nasir Muhammad, about 1320 (Khouri and Whitcomb 1988: 33; Pringle 1997: 113, no. P5). The Department of Antiquities has provided some support for this theory by its recent excavations on the castle.

In 1181 the lord of al-Karak, Reynald of Châtillon, raided the Hijaz and followed this over the winter of 1182–83 by a sea-borne *razzia* along the Arabian coast. In retaliation for this affront to Islam, Saladin besieged, unsuccessfully, the castle of al-Karak in 1183 and again in 1184. On 4 July 1187, however, the sultan personally beheaded Reynald after the battle of the Horns of Hattin, above Tiberias. This battle effectively destroyed the Frankish royal army and the king, Guy of Lusignan, was taken prisoner. Al-Karak fell eventually to Malik al-'Adil in October/November 1188 and ash-Shawbak in April/May 1189, having held out for almost two years after the battle. Although the Franks were to regain much of the Levantine coastlands from 1192 onwards, they never again controlled lands east of the Jordan.

During the Ayyubid counter-Crusade that led up to Saladin's victory, other Muslim castles had been built in Jordan. In addition to Jazirat Fara'un, already mentioned, in 1184–85 Saladin's amir 'Izz al-Din Usama began building Qal'at ar-Rabad, near 'Ajlun, in direct response to the construction of Crusader Belvoir (Kawkab al-Hawa), which faces it across the Jordan Valley (Johns 1932). The Mediaeval castles at as-Salt and az-Zarqa' (Qasr Shabib) may also be Ayyubid in origin, although both places had been briefly under Frankish control in the twelfth century. It is largely because the castles of al-Karak and ash-Shawbak were rebuilt by the Ayyubids and Mamluks, however, that they have survived to this day as two of the most impressive Mediaeval fortresses of the Middle East.

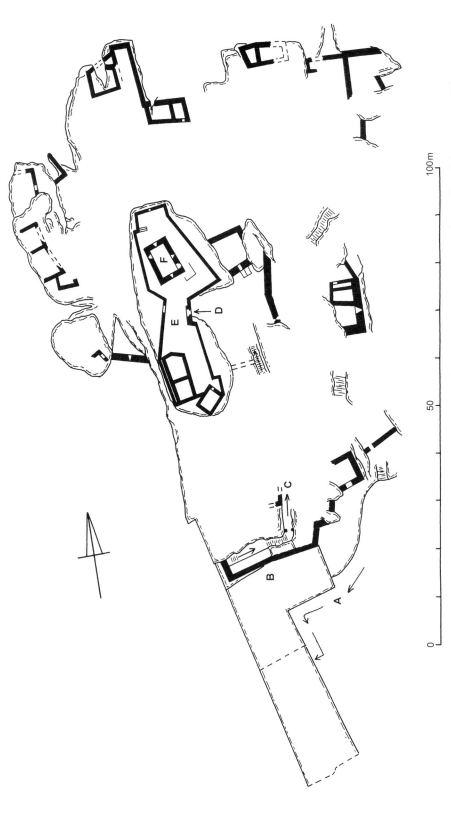

Figure 27.4. Al-Habis, Petra (identified as al-'Aswît): plan of the castle (redrawn after Hammond 1970). Key: (A) access route from valley; (B) outer gate; (C) entrance gate and ramp; (D) gate to inner ward; (E) inner ward; (F) keep or *donjon*.

References

Abu Shama
- 1896 *Le livre des deux jardins*. Ed. and trans. A.C. Barbier de Meynard. In *Recueil des historiens des croisades: Historiens orientaux*, 4-5. Paris: Academie des inscriptions et belles-lettres.

Brooker, C.H. and E.A. Knauf
- 1988 Notes on Crusader Transjordan. *Zeitschrift des deutschen Palästina-Vereins* 104: 184-88, pls. 14-15.

Delaville le Roulx, J.
- 1905–1908 Chartes de Terre Sainte. *Revue de l'Orient latin* 11: 181-91.

Delaville le Roulx, J. (ed.)
- 1894–1906 *Cartulaire général de l'Ordre des Hospitaliers de Saint-Jean de Jérusalem (1100–1310)* 4 vols. Paris E. Leroux.

Deschamps, P.
- 1933 Deux positions stratégiques des croisés à l'est du Jourdain: Ahamant et el Habis. *Revue Historique* 172: 42-57.
- 1935 Une grotte-forteresse des croisés au delà du Jourdain: El Habis en Terre de Suète. *Journal Asiatique* 227: 285-99.
- 1939 *Les châteaux des croisés en Terre-Sainte*. II. *La défense du royaume de Jérusalem*. 2 vols. Bibliothèque historique, géographique et monumentale, 34. Paris: Haut-Commission de la République française en Syrie et au Liban, Service des antiquités.
- 1943 Etude sur un texte latin énumérant les possessions musulmanes dans le royaume de Jérusalem vers l'année 1239. *Syria* 23: 86-104.
- 1964 *Terre Sainte romane*. La nuit des temps 21. La Pierre-qui-Vire: Zodiaque.

Hammond, P.C.
- 1970 *The Crusader Fort on El-Habis at Petra: Its Survey and Interpretation*. Middle East Center, University of Utah, Research Monograph 2. Salt Lake City: University of Utah.

Harding, G.L.
- 1974 *The Antiquities of Jordan*. Rev. edn. London: Lutterworth/Amman: Jordan Distribution Agency.

Horsfield, G. and A. Horsfield
- 1938 Sela-Petra, the Rock, of Edom and Nabatene [I-II]. *Quarterly of the Department of Antiquities in Palestine* 7: 1-42.

Ibn al-Qalanisi
- 1932 *The Damascus Chronicle of the Crusades*. Trans. H.A.R. Gibb. University of London Historical Series 5. London: Luzac.

'Imad al-Din al-Isfahani
- 1972 *Conquête de la Syrie et de la Palestine par Saladin (al-Fath al-qussî fî l-fath al-qudsî)*. Trans. H. Massé. Documents relatifs à l'histoire des croisades 10. Paris: Geuthner.

Johns, C.N.
- 1932 Medieval 'Ajlun, I: The Castle (Qal'at ar-Rabad). *Quarterly of the Department of Antiquities in Palestine* 1: 21-33. (Reprinted in D. Pringle [ed.], *Pilgrims' Castle ['Atlit], David's Tower [Jerusalem] and Qal'at ar-Rabad ['Ajlun]: Three Middle Eastern Castles from the Time of the Crusades*. Aldershot: Variorum [1997]).
- 1937 *Palestine of the Crusaders: A Map of the Country on Scale 1:350,000 with Historical Introduction and Gazetteer*. Jaffa: Survey of Palestine.

Kennedy, H.
- 1994 *Crusader Castles*. Cambridge: Cambridge University Press.

Khouri, R.G. and D. Whitcomb
- 1988 *Aqaba: Port of Palestine on the China Sea*. Amman: Al Kutba.

Kob, K.
- 1967 Zur Lage von Hormoz: Ein territorialgeschichtliches Problem der Kreuzfahrerzeit. *Zeitschrift des deutschen Palästina-Vereins* 83: 136-64.

Lawrence, T.E.
- 1939 *Oriental Assembly*. Ed. A.W. Lawrence. London: Williams and Norgate.

Le Strange, G.
- 1890 *Palestine under the Moslems*. New York: Houghton, Mifflin.

al-Maqrizi
- 1980 *A History of the Ayyubid Sultans of Egypt*. Trans. R.J.C. Broadhurst. Library of Classical Arabic Literature 5. Boston: Twayne.

Marino, L. *et al.*
- 1990 The Crusader Settlement in Petra. *Fortress* 7: 3-13.

Mayer, H.E.
- 1987 The Crusader Lordship of Kerak and Shaubak: Some Preliminary Remarks. In A. Hadidi (ed.), *Studies in the History and Archaeology of Jordan*, III: 199-203. Amman: Department of Antiquities.
- 1990 *Die Kreuzfahrerherrschaft Montréal (Sobak): Jordanien im 12. Jahrhundert*. Abhandlungen des deutschen Palästina-Vereins 14. Wiesbaden: Otto Harrassowitz.

Müller-Wiener, W.
- 1966 *Castles of the Crusaders*. Trans. J. Maxwell Brownjohn. London: Thames & Hudson.

Musil, A.
- 1907–1909 *Arabia Petraea*. 3 vols. Vienna: Hölder.

Nicolle, D.
- 1988 'Ain al Habis: The Cave de Sueth. *Archéologie Médiévale* 18: 113-40.

Northedge, A.
- 1983 The Fortifications of Qal'at Amman ('Amman Citadel): Preliminary Report. *Annual of the Department of Antiquities of Jordan* 27: 437-59.

Northedge, A. et al.
- 1993 *Studies on Roman and Islamic Amman. I. History, Site and Architecture.* British Academy Monographs in Archaeology 3. Oxford: Oxford University Press.

Prawer, J.
- 1980 *Crusader Institutions.* Oxford: Clarendon Press.

Pringle, D.
- 1993 *The Churches of the Crusader Kingdom of Jerusalem: A Corpus.* 3 vols. (in progress). Cambridge: Cambridge University Press.
- 1997 *Secular Buildings in the Crusader Kingdom of Jerusalem: An Archaeological Gazetteer.* Cambridge: Cambridge University Press.

Röhricht, R. (ed.)
- 1893 *Regesta Regni Hierosolymitani.* Innsbruck: Libraria Academica Wagneriana.

Savignac, R.
- 1903 Ou'aïrah. *Revue Biblique* 12: 114-20.

Strehlke, E. (ed.)
- 1869 *Tabulae ordinis theutonici ex tabularii regii berolinensis codice potissimum.* Berlin: Weidmann. (Reprinted with preface by H.E. Mayer, Toronto: University of Toronto, 1975.)

Vannini, G. and A.V. Desideri
- 1995 Archaeological Research on Medieval Petra: A Preliminary Report. *Annual of the Department of Antiquities of Jordan* 39: 509-40.

William of Tyre
- 1941 *A History of Deeds Done beyond the Sea.* Trans. A.A. Babcock and A.C. Krey. 2 vols. Records of Civilization, Sources and Studies 35. New York: Columbia University.
- 1986 *Chronicon.* Ed. R.B.C. Huygens. In *Corpus christianorum, continuatio mediaeualis,* 63-63a. Turnhout: Brepols.

Wood, J.
- 1993 The Fortifications of Amman Citadel, Jordan. *Fortress* 16: 3-15.

Zayadine, F.
- 1985 Caravan Routes between Egypt and Nabataea and the Voyage of Sultan Baybars to Petra in 1276. In A. Hadidi (ed.), *Studies in the History and Archaeology of Jordan,* II: 159-74. Amman: Department of Antiquities.

28. Ottoman Hajj Forts

Andrew Petersen

From the beginning of the Islamic period the territory comprising the modern Kingdom of Jordan was of significance because of its position on the Syrian Hajj route (Muslim pilgrimage route) from Damascus to Mecca. This was one of three main routes that formed the principal overland access to Mecca from the Arab world. The other major routes were from Iraq (Baghdad) and from Cairo. The relative importance of each route was dependent on political factors. Thus during the ninth century (Abbasid period) the Iraqi route (*Darb Zubayda*) was the most significant and attracted considerable investment with halting places on most of the route (see Petersen 1994). Between the thirteenth and fifteenth centuries the political importance of the Mamluk regime was reflected in the development of facilities along the Egyptian Hajj route. One of these structures survives in al-'Aqaba in southern Jordan where the Egyptian route passed round the northern shores of the Red Sea (Glidden 1952; Khouri and Whitcomb 1988: 33-35). However, the majority of structures associated with pilgrimage in Jordan belong to the Syrian Hajj route (*Darb al-Hajj al-Shami*), a continuation of pre-Islamic trade routes that ran from north to south Arabia (for description and history of this route, see Tresse 1937). During the seventh and early eighth centuries the Umayyads established this as the principal road connecting Mecca with their capital at Damascus. Structures associated with this period have been identified at a number of sites in Jordan including al-Qastal, Khan al-Zabib and Humayma. With the fall of the Umayyad caliphate in AD 750, the significance of the Syrian route declined and did not recover until the sixteenth century when the Ottoman conquest of Syria made it part of a direct route between Mecca and the imperial capital at Constantinople (modern Istanbul).

Before the Ottoman period the Hajj followed a more westerly route along the King's Highway that follows the eastern edge of the highlands overlooking the Jordan Valley. The three fortresses of 'Ajlun, al-Karak and ash-Shawbak protected this route. During the reign of Suleyman the Magnificent the route was changed so that it now lay along the edge of the desert. This new road was known as *Tariq al-Bint* after the daughter of Selim I (1512–1520) who had complained of bandits on the King's Highway and caused this new route to be adopted. This route is now the one followed by the modern Desert Highway.

The renewed importance of the Syrian Hajj route under the Ottomans was manifested in a number of structures and installations that line the road (Petersen 1989). To the north of Damascus, a number of khans were built to provide accommodation for pilgrims (Sauvaget 1937). In Damascus itself, the official starting point of the Hajj, a number of pilgrim facilities were built including a mosque, soup kitchen and camping ground (Goodwin 1971: 256-57). There were no caravanserais to the south of Damascus and pilgrims were expected to camp. At each halting place there were water supplies either in the form of wells or cisterns. As the life of the Hajj caravans depended on these water resources a number of forts were built to protect them from the Bedouin or other potential attackers. The forts were not intended to provide protection or accommodation for the pilgrims although, where possible, the caravan would camp in front of the fort (cf. Tresse 1937: 229).

There are nine forts within the boundaries of Jordan stretching from the Syrian frontier in the north to the border of Saudi Arabia at al-Mudawwara in the south-east. The oldest forts, Jiza/Zizia and Qasr Shabib at az-Zarka', date from the Mamluk period (1250–1517) although both were extensively re-furbished during the Ottoman period (Petersen 1991). The earliest Ottoman forts date from the sixteenth century and include the structures at al-Mafraq (destroyed c. 1950), al-Qatrana, Ma'an and al-'Unaiza. However, it was not until the eighteenth century that the majority of stops were fortified with forts built at al-Hasa, al-Fassu'a and al-Mudawwara. The forts continued in use into the nineteenth century although the advent of steamships, which were able to navigate the Red Sea, meant that the numbers of pilgrims using this route declined. For a few years at the beginning of the twentieth century the land route again became important with the construction of the Hijaz railway (Dahir 1995).

During the nineteenth century travellers such as Burckhardt (1882) and Doughty (1926) described the forts as run-down but functioning institutions. By the early years of the twentieth century, they were abandoned and were described as archaeological relics, first by Brünnow and von Domaszewski (1904–1905) and later by Jaussen and Savignac (1909). Today the majority of the buildings are abandoned structures vulnerable to decay and vandalism although the fort at al-Qatrana has recently been restored with a grant from the Turkish Government.

The forts all have the same basic square plan, 20 by 20 m average, and are built of local materials, generally either limestone or basalt although plate flint and sandstone are also used. Where suitable, the stone is finely dressed as at al-Qatrana, Ma'an and al-Mudawwara, although where this was not possible, plaster was used to provide a smooth finish to architectural features such as the gateway at al-'Unaiza or the *mihrab* at al-Fassu'a. The forts are entered through a gateway that leads into a central courtyard opening on to stables and storerooms. There was an opening in the centre of the courtyard that led either into a subterranean cistern or a well. Access to the upper floor is generally by twin staircases located either side of the main entrance. The upper floor usually contained accommodation for the garrison, a *mihrab* or mosque and sometimes a stairway leading to a parapet.

There are significant differences between the sixteenth-century forts and those built in the eighteenth century. The sixteenth-century ones tended to have more decorative features. Thus, the fort at Ma'an has two carved stone rosettes above the entrance and the fort at al-Qatrana has three carved stone balls above the entrance. Other distinctive features include stepped crenellations, carved rainspouts and arrow-slits with splayed bases. All of these features are also found in the gateways and towers of the sixteenth-century walls of Jerusalem. The arrow-slits with splayed bases are of particular significance as they represent a particular phase in the development of artillery when small, low cannon was used in conjunction with bows. In the latter forts, the arrow loops are replaced by more numerous, less visible, slits that are suitable for handheld firearms. A related development is the use of projecting corner towers. The sixteenth-century forts had projecting box machicolations above the gateways although the corners were left undifferentiated. In the eighteenth century there were projecting corner turrets at first-floor level that allowed those inside the fort to protect the base of the structure. The latter two developments probably reflect the acquisition of handheld firearms by Bedouin raiders (Petersen 1995).

There is usually a graveyard and a reservoir or cistern outside the fort. The graves are marked with rough unmarked headstones or kerbs made of fieldstones. The water resources range from large Roman reservoirs such as those at Jiza/Zizia and al-Qatrana to small rectangular open cisterns such as those at al-Mudawwara. It is notable that the sixteenth-century forts are built next to existing water resources whereas those of the eighteenth century had specially constructed reservoirs. The cisterns were generally fed by water diverted from wadis although at al-Fassu'a rainwater appears to have been used.

Although the primary function of the forts was to protect the water resources of the Hajj caravan, they had a number of related functions including their use as post stations, police posts and supply depots. Set within the wider context of Ottoman policy, the Hajj forts may be seen as part of an attempt to gain greater control over Arabia and to protect the newly acquired province of Syria. Similar considerations led in 1585 to the construction of the Ottoman fortress of Qal'at al-Sai in the Nile valley 650 km south of Aswan (Alexander 1997: 16-20). Certainly the strategic value of the Hajj route and its facilities were not lost on Muhammad 'Ali Pasha who in 1825 commissioned a survey of the condition of forts and water resources on the road (Abujaber 1995: 741). The Hajj forts, as strategic outposts, had always been part of a wider network that covered Jordan and Palestine. The outer limits of this network were marked on the one side by the Via Maris which ran from Cairo to Damascus through the coastal plain and on the east by the Hajj road (cf. Petersen 1998: 109-11). Within Jordan, the fortresses on the King's Highway ('Ajlun, al-Karak and ash-Shawbak) were all repaired and used to supply troops to the smaller outposts on the Hajj route.

The size of the Hajj fort garrisons varied depending on size and the availability of manpower. Some idea of the relative strengths of each garrison may be gained from the fact that the Hajj forts usually had about 20 *mustahfizan* (fortress soldiers) whilst the fortress of ash-Shawbak had 68 soldiers. The troops were generally jannisaries from the province of Damascus although Ma'an and al-Qatrana were manned by imperial jannisaries. There were, however, constant problems of undermanning

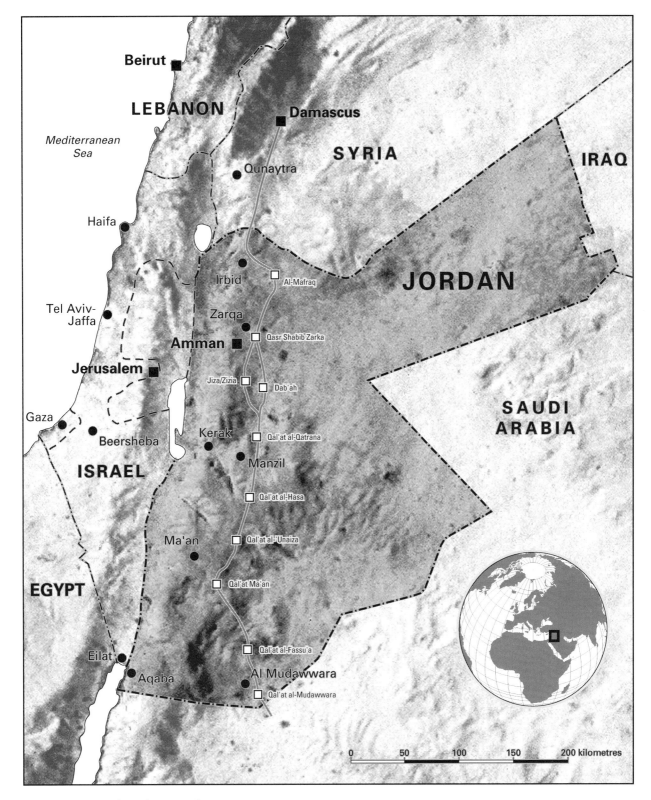

Figure 28.1. Location of Hajj forts in Jordan.

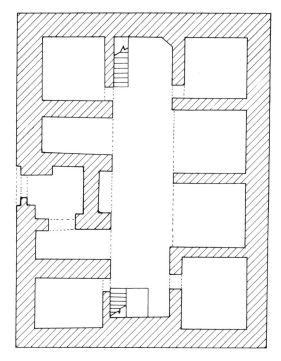

Figure 28.2. Qal'at al-Qatrana: ground floor plan.

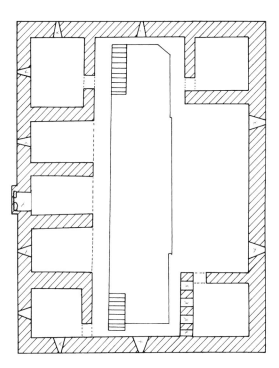

Figure 28.3. Qal'at al-Qatrana: first/upper floor plan.

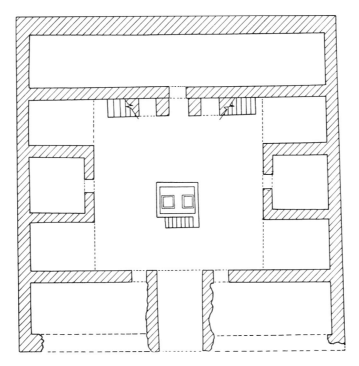

Figure 28.4. Qal'at al-Hasa: ground floor plan.

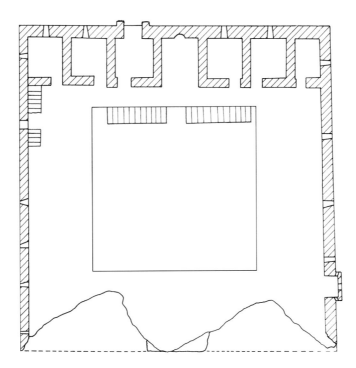

Figure 28.5. Qal'at al-Hasa: first/upper floor plan.

Figure 28.6. Qal'at al-Fassu'a: entrance and projecting corner towers.

and the substitution of jannisaries with local Arab tribesmen. The function of the garrisons was purely to protect the water resources and aid communications along the route. Military protection of the Hajj caravan was assigned to the *Cerde* or military escort that in the sixteenth century comprised 100 cavalry and 200 janissaries equipped with camels, guns and cannon. The main form of protection, however, was the *Sürre* (purse) that was a sum of money distributed to tribes along the length of the route to guarantee the safety of the caravan.

Summary of Hajj forts in Jordan
(Figure 28.1)

1. Al-Mafraq, also known as Tall Fa'un: Hajj fort built in the early sixteenth century and destroyed in 1957 for construction of the Trans-Arabian Pipeline.
2. Qasr Shabib Zarka: Mediaeval fort visited by Ibn Battuta in the fourteenth century. It is a single-room structure covered with cross vaults.
3. Jiza/Zizia: Mediaeval fort standing next to a Roman reservoir that could hold 20,000 cubic litres of water. Remains of the fort, which was extensively refurbished during the nineteenth century, consist of single large barrel vault and smaller later rooms. Ibn Battuta visited the fort during a pilgrimage in the fourteenth century.
4. Dab'ah, also known as Qal'at al-Balqa: small square fort (25 by 25 m) with central courtyard and solid corner towers. Instead of the normal twin staircases, there is a single set of stairs providing access to the first floor. The fort may have older, possibly Byzantine, origins. It was rebuilt in the sixteenth and eighteenth centuries.
5. Qal'at al-Qatrana (Figures 28.2, 28.3): sixteenth-century (1531) fort recently restored by Turkish Government. Rectangular plan (22.20 by 17.35 m) with central courtyard stands next to Roman reservoir with settling tank.
6. Qal'at al-Hasa (Figures 28.4, 28.5): eighteenth-century (1757–74) fort with square plan (24.20 by 23.70 m) with a central courtyard, projecting towers, and a well in the centre. The north wall and interior are mostly in ruins though the mosque (vault with *mihrab*) is still visible. Nearby there is a bridge and paved road that were also constructed in the eighteenth century. The road is made of basalt and flint cobbles edged with limestone blocks and stretches for a distance of more than 1.5 km in the vicinity of the fort. There are drains at irregular intervals that run beneath the road. The bridge, which has the same cobbled surface as the road, is supported on two large arches with a smaller arch to the north.

7. Qal'at al-'Unaiza: late sixteenth-century (1576) fort built of basalt with square plan (28.00 by 28.85 m). It is in a ruinous condition with the corners robbed out. In the entrance there are traces of decorative plasterwork. Behind the fort to the west there are traces of reservoirs (now filled in).
8. Qal'at Ma'an: sixteenth-century (1531) fort built of ashlar limestone with square plan (23.60 by 23.60 m). The fort has been re-modelled in recent times for use as a local prison (now disused). Arrowslits and machicolations are still visible above the original doorway (now blocked). Immediately outside the entrance to the fort there is a large open cistern, a well and a *qanat* system that supplies the numerous walled gardens in the vicinity. As an oasis, Ma'an was the largest settlement on the Hajj route (in Jordan) and pilgrims spent two days here resting and buying supplies for the next stage of the journey.
9. Qal'at al-Fassu'a, also referred to as Zahr al-'Aqaba: eighteenth-century fort built of flint with limestone details. It has a large central courtyard with projecting corner towers (Figure 28.6). There are two reservoirs nearby that are fed by rainwater and a wadi.
10. Qal'at al-Mudawwara, also known as Jughayman: eighteenth-century (1730–33) fort built of sandstone. There is evidence of its reuse in the early twentieth century. Wells and two cisterns are located 80 m to the south. Nearby are the remains of a metalled road made of flint cobbles edged with sandstone similar to that near Qal'at al-Hasa.

References

Abujaber, R.S.
  1995   Water Collection in a Dry Farming Society. In K. 'Amr, F. Zayadine and M. Zaghloul (eds.), *Studies in the History and Archaeology of Jordan*, V: 737-44. Amman: Department of Antiquities.

Alexander, J.
  1997   Qalat Sai, the Most Southerly Ottoman Fortress in Africa. *Sudan and Nubia* 1: 16-19. (The Sudan Archaeological Research Society Bulletin.)

Burckhardt, J.L.
  1882   *Travels in Syria and the Holy Land*. London: Murray.

Brünnow R.E. and A. von Domaszewski
  1904   *Die Provincia Arabia*, I. Strasbourg: Trübner.

Dahir, R.
  1995   The Resurrection of the Hijaz Railroad Line. In K. 'Amr, F. Zayadine and M. Zaghloul (eds.), *Studies in the History and Archaeology of Jordan*, V: 341-50. Amman: Department of Antiquities.

Doughty, C.M.
  1926   *Travels in Arabia Deserta*. London: Cape & The Medici Society.

Glidden, H.W.
  1952   The Fortified Mamluk Khan at al-'Aqabah, Jordan. In G.C. Miles (ed.), *Archaeologia in Memoriam Ernst Herzfeld*, 116-18. New York: Augustin.

Goodwin, G.
  1971   *Ottoman Architecture*. London: Thames & Hudson.

Jaussen, J. and R. Savignac
  1909   *Mission archéologique en Arabie*, I. Paris: La Société française des fouilles archéologiques.

Khoury, R.G. and D. Whitcomb
  1988   *Aqaba: Port of Palestine on the China Sea*. Amman: Al Kutba.

Petersen, A.D.
  1989   Early Ottoman Forts on the Darb al-Hajj. *Levant* 21: 97-118.
  1991   Two Forts on the Medieval Hajj Route in Jordan. *Annual of the Department of Antiquities of Jordan* 35: 347-60.
  1994   The Archaeology of the Syrian and Iraqi Hajj Routes. *World Archaeology* 26/1: 47-56.
  1995   The Fortification of the Pilgrimage Route during the First Three Centuries of Ottoman Rule (1516–1757). In K. 'Amr, F. Zayadine and M. Zaghloul (eds.), *Studies in the History and Archaeology of Jordan*, V: 299-305. Amman: Department of Antiquities.
  1998   Qal'at Ras al-'Ayn: A Sixteenth-Century Ottoman Fortress. *Levant* 30: 97-112.

Sauvaget, J.
  1937   Les caravansérails syrien du Hadjdj de Constantinople. *Ars Islamica* 4: 98-121.

Tresse, R.
  1937   *Le pélerinage syrien aux villes saintes de l'Islam*. Paris: Imprimerie Chaumette.

Index of Place Names and Sites Mentioned in the Text

Editors' Note

Every attempt has been made to standardize various transliterations of the Arabic place names in this index, but where several variants occur in the text, they are cited as alternative spellings in the index. Sites beginning with Khirbat, Tall, Tel, Qasr and the Arabic definite article in all variants have been listed under the primary name of the site. Geographical place names (Wadi, 'Ayn, Jabal, Har, etc.), have been listed in alphabetical order following the geographical feature (i.e. Wadi-al Hasa, 'Ayn Besor). Russell Adams would like to acknowledge the assistance of Anna Piskorowski-Adams for assistance in compiling this index.

'Abd, Qasr al- 386, 387f, 661
Abila (Quwayliba) 272f, 280, 296f, 297, 303, 311, 323, 324, 325, 328, 334, 349, 350f, 428, 433f, 434, 436, 437, 438, 440, 442, 444, 449, 450, 452, 462f, 467, 474, 479, 480, 481, 482, 491, 494, 504f, 509, 599
Abila, Tall *see* Abila
Abu Alayiq, Tall 164f, 204
Abu al-Kharaz, Tall 113, 114, 164f, 169, 171, 180, 181f, 182t, 183, 188, 204, 206f, 208, 210, 211f, 212, 235f, 241, 272f, 276, 296f, 298, 303, 303, 307, 309, 311, 325, 328, 332, 334, 649
Abu Alubah, Tall (Abu Aluba) 43f, 47, 241
Abu an-Naml 583, 586
Abu an-Ni'aj, Tall 235f, 236, 238, 240, 241, 241f, 245, 246, 252, 253, 257, 258, 259, 261, 262
Abu Gosh 86
Abu Habil North, Tall 241
Abu Habil, Khirbat 235f
Abu Habil 107, 113, 114
Abu Hamid 68f, 95, 109, 111, 112t, 113, 113t, 114, 118, 119, 123, 124f, 125, 126, 127, 128, 133, 134, 142, 144, 145, 147, 148, 150, 151, 648
Abu Hawam, Tall 309
Abu Hudhud 68f, 70
Abu Hureyra, Tall 99
Abu Ishribsheh 244, 251
Abu Matar/Safadi 147
'Abu Mina' 474
Abu Nijrah, Tall 298
Abu Nusayr 334
Abu Qarf, Khirbat 235f, 241
Abu Sarbut, Tall 468, 528, 540, 550, 553
Abu Sharer 43f, 47
Abu Sliman al-Dirany 585
Abu Snesleh 112t, 114, 115, 119, 128, 129f, 272f, 280, 285
Abu Thawwab, Tall 68f, 92, 93, 95, 112t
Abu Ubaydah 523
Adeimeh 149, 164f, 200, 201
Ader 164f, 235f, 239, 243, 245, 251, 257, 258, 262, 334, 528
Adir 175, 193
Admah 663
Adoni-Nur, tomb of 350f, 351
Adraha (Adhruh, Adhri'at, Dar'a) 462f, 491, 517, 517f, 518f, 518, 543, 544
Adwan 575
Afiq 518f
Africa 15, 17, 33, 60, 470, 482, 598, 665
Afula 274
Ahl al-Kahf ar-Raqim 534, 535f, 536, 537
Ai (at-Tall) 164f, 177, 192, 206, 245, 606
'Ain La'ban 604
'Ajjul, Tall al- 235f, 247, 254
Ajlun 164f, 523, 564, 569, 577, 621, 622, 681, 685, 686
'Ajlun 19, 191, 202, 330, 335, 516, 521, 522, 522f, 526, 527, 528, 529, 530, 531f, 543, 544, 554, 555
Ajnadayn 491
Akko (Acre, Akka) 309, 517f, 519f, 522f, 678
'Al, al- 677
Aleppo 479, 520. 533
Alexandretta 479
Alexandria 412n, 478, 510, 553, 571, 610, 674
Ali Kosh 597
'Alla, Tall 235f, 241
Amarna, Tall al- 314, 659
Amata (Abu Ubaydah) 521, 522, 522f, 533, 555
Amathus 480, 494
'Amman (Philadelphia, Rabbath-Ammon) 255, 272f, 297, 306, 309, 310, 315, 323, 324, 325, 329, 332, 333, 334, 335, 336, 337, 348, 351, 361, 386, 394, 410n, 428, 431f, 433, 434, 435, 438, 439, 440, 441, 443, 444, 445, 449, 449, 452, 462f, 464f, 467, 475f, 477, 480, 481, 484, 487, 489, 491, 494, 495, 507, 517f, 518, 518f, 519, 519f, 521, 523, 524, 525, 527, 528, 529, 530, 533, 534, 539, 543, 544, 545, 554, 568, 575, 622, 627, 660, 661, 663, 664, 665, 666, 671, 673, 674, 675, 678, 680
'Amman Airport 291, 303, 304, 306, 307, 308, 309, 310, 313, 314, 522, 522f, 599
'Amman Citadel (al-Qal'ah) 235f, 242, 280, 285, 331, 333, 338, 350f, 351, 507, 507f, 508, 524, 529, 539, 545, 546f, 549, 665, 667, 674, 680
Ammon 306, 315, 329, 331, 334, 335, 336, 337, 340, 347, 348, 360, 361, 660, 661
'Amra, Qasr 479, 504f, 506, 611, 674
Anatolia 211, 214, 253, 254, 486
Antioch 478, 479, 486, 493, 494
'Aqabah region, al- 323

'Aqabah, al- (Ayla, Aila, Waylah) 16f, 361, 433, 434, 438, 439, 444, 446, 448, 449, 452, 462f, 464f, 469, 470, 473, 476, 482, 483, 487, 488, 489, 491, 494, 495, 496, 497, 509, 510, 518, 518f, 519, 519f, 522f, 523, 524, 525, 527, 528, 532, 539, 542, 544, 545, 547f, 548f, 549, 553, 554, 603, 562f, 665, 678, 681, 685
Aqir 518f
Ara'ir 235f, 239, 240, 243, 245, 257, 262, 539
Arab'in, Tall al- 164f, 191t
Arabia 309, 348, 367, 368, 370, 371, 379, 385, 390, 401, 413n, 414n, 427, 428, 433, 434, 435, 440, 442, 445, 446, 447, 449, 451, 452, 465, 466, 467, 468, 469, 470, 471, 480, 486, 487, 488, 489, 490, 491, 494, 495, 510, 553, 615, 617, 631, 665, 685, 686
Arabian Sea 15, 482, 597
Arad 116, 164f, 179, 183, 206, 212, 213, 659
Aram 334
Arbela *see* Irbid
Ard al-Karak 526, 528, 553
Areopolis (ar-Rabbah, Ma'ab, Arabia Secunda) 462f, 464f, 468, 493, 494, 517f, 519f, 520, 523, 525, 544, 554
Ariha 517f, 518f
Arindela 494
Arqub ad-Dhahir 164f, 208
Arslantepe 146
Arsuf 518f
Asaikhin, Qasr 462f
Ashkelon 272f, 276, 518f, 519f
Ashtara, Tall (Ashtartu, Ashtaroth) 294, 659
Asia Minor 390
Assiut 580
Assouad, Tall 92
Assur 294
Assyria 294, 335, 406n
Aswan 686
Aswit, al- 681
'Ara'ir, Khirbat 298, 323, 324, 325, 330, 331, 332, 333
'Arbid 334, 338
'Argadat, Tall al- 298
'Artal, Tall 235f, 253, 263
'Asqalan 491
Athenian Agora 580
Athens 580
Atlit Yam 86
at-Tall, *see* Ai
Augustopolis 494
Ayl 470
Ayla (Aila) *see* 'Aqaba, al-
'Ayn al-Assad 37, 38
'Ayn al-Assawir 199
'Ayn al-Bassah 298, 299
'Ayn al-Bayda 10
'Ayn al-Buhira 43f, 45
'Ayn al-Habis 677
'Ayn-Hanatziv 235f, 253
'Ayn al-Kanish 673, 674
'Ayn al-Saratan 49f, 50f, 53, 57
'Ayn as-Saratan 643
'Ayn at-Terab 452
'Ayn az-Zara (Callirhoe) 447, 468

'Ayn Besor 659
'Ayn Difla 37, 39, 40, 41
'Ayn Gadara 452
'Ayn Ghazal 68f, 71, 72, 72f, 73, 74, 74f, 75, 76, 77, 78, 78f, 79, 80, 80f, 81, 82, 83, 83f, 84f, 85, 85f, 86, 86f, 87, 88, 88f, 89, 89f, 90, 90f, 91, 92, 92f, 93, 93f, 94, 97, 98, 109, 148, 272f, 274, 298, 645
'Ayn Hesban 575
'Ayn Jammalah 495
'Ayn Jammam 68f, 79, 81, 82, 85, 90, 96, 604
'Ayn Jenin, Khirbat 528, 539
'Ayn Keniseh 495
'Ayn Khureibah 452
'Ayn Mallaha 53
'Ayn Musa 462f, 468, 495, 497
'Ayn Qattara 495
'Ayn Rahub 49f, 50f, 57, 59, 68f, 92, 93, 97
'Ayn Shadud 179, 204, 647
'Ayn Shelaleh 604, 604f
'Ayn Soda, C-Spring 37, 38
Azdud (Ashdod) 518f
Azeimeh, Tall al- 113
Azraq Basin, al- 2, 3f, 4, 9, 10, 11, 12, 13, 14, 17, 18, 19, 24, 25t, 26, 27, 31, 32f, 36, 38, 42, 44, 46, 47, 48, 51, 52, 53, 54, 55, 57, 60, 70, 81, 192, 193, 584, 603, 604, 615, 640, 643, 645, 646, 649
Azraq Shishan, al- 81
Azraq site 17, al- 43f, 44, 49f, 50f, 53
Azraq site 18, al 642, 643
Azraq site 31, al- 9f, 10, 12
Azraq, al- Marsh 9, 9f, 47
Azraq, al- 12, 18, 462f, 464f, 466, 471, 472, 488, 489, 490, 504f, 521, 522f, 526, 529, 530, 532f, 533f, 534, 536, 538, 544, 555, 562f
Azraq, Qasr al- 485

Ba'iq, Qasr al- 486, 489, 490
Ba'ja III 334, 338
Ba'ja 68f, 79, 87
Bab adh-Dhra 4f, 5, 149, 164f, 170, 171, 172, 175, 178, 180, 182t, 183, 184, 185, 187, 188, 197, 198f, 199, 199f, 200, 203, 205, 205f, 206, 206f, 207, 208, 209, 210, 210f, 211, 212, 212f, 213, 214, 214f, 215, 215f, 216, 216f, 233, 235f, 240, 243, 245, 246, 247, 251, 257, 258, 262, 263, 280, 606, 647
Babylon 309, 329, 336, 347, 414n, 660, 665
Babylonia 294, 347, 406n, 414n, 659
Bacatha 480, 494
Badia 2, 13, 14, 17, 27
Baghdad 515, 584
Balqa' district, al- 465, 467, 517, 518, 519, 519f, 521, 522f, 532, 534, 540, 541, 621, 622
Balu', al- 296f, 298, 300, 306, 330, 331, 332, 333, 334, 338, 539, 659
Baniyas 517f, 519f
Baq'ah Valley 293, 296f, 298, 300, 303, 306, 309, 310, 311, 324, 336, 351, 467
Baqawiyya Fault 10
Baraqis, South Arabia 381
Bardawil, Qasr 677

Barqat al-Hattiye 164f, 195, 208
Barru 495
Basra 542
Basta 68f, 78, 79, 80, 81, 82, 85, 87, 91, 462f, 645
Batash, Tel 337
Bawwab al Ghazal 68f
Bayda Fault, al- 10
Bayda, al- 49f, 50f, 57 ,58, 59, 68f, 71, 72, 73, 75, 76, 79, 82, 84, 86, 88, 89, 97, 98, 605, 642
Bayir 462f, 487, 528, 544
Baysan (Bashan) 183, 504f, 517f, 518f, 519f, 533, 664
Bayt al-Maqdis 517f, 518f, 522f
Bayt al-Ram 518f, 544
Bayt Jibrin 518f
Bayt Ras (Capitolias, Raphana) 109, 428, 434, 440, 442, 451, 462f, 463, 467, 479, 480, 481, 494, 504f, 509, 522f, 522, 523, 524, 525, 528, 539, 544, 554
Be'er Resisim 235, 235f, 244
Beisamoun 99
Beit Mirsim, Tall 233, 235f, 254, 631
Bela 663
Belvoir Castle (Kawkab al-Hawa) 681
Beni Hasan 263
Beqa'a Valley (Lebanon) 108, 109, 280
Berytus 464f
Besor 597
Beth Shan 164f, 171, 177, 251, 285, 295, 309, 544
Beth Shemesh 659
Bethany 665
Bethlehem 677
Beulha 597
Bezer 664
Bint, Qasr al- (Petra) 374, 375f, 377, 379, 382, 383, 388, 389f, 390, 395, 406n, 407n, 408n, 411n, 441
Biq'at Uvda 650
Bir Madhkur 462f, 470, 489
Bire, Tel 20f, 21
Birketein 451
Bithynia 445, 451
Black Desert 9, 14, 31, 32f, 51, 53, 59, 60, 110, 643
Black Sea 391, 410n
Bostra (Bosra, Busra al-Sham) 310, 348, 401, 427, 429f, 433, 434, 440, 446, 462f, 464f, 465, 471, 472, 480, 488, 489, 490, 491, 494, 518f, 522, 522f, 527, 543, 544
Bshir, Qasr 462f, 485, 485f, 490, 608
Burqu', Qasr 462f, 471, 472, 473, 488, 495, 504f, 505
Busayra 323, 332, 333, 334, 338, 339, 340, 347, 348, 350f, 351, 352, 355, 356f, 357f, 358, 367, 660
Byblos 108, 659

Caesarea Maritima (Qaysariyah) 462f, 464f, 480, 487, 491, 510, 518f
Cafer Hoyuk 98
Cairo 521, 529, 531, 542, 543, 544, 553, 563, 584, 685, 686
Canaan 659
Canatha *see* Qanawat
Capitolias *see* Bayt Ras
Carthage 478
Çatal Hüyük 145, 146
Cava de Suet 677

Central Asia 414n
Central Negev 236
Ceylon 487
Chagar Bazar 112t
China 486, 487, 541, 542
Chrysorohas River 441
Coele Syria 348
Constantinople (Istanbul) 478, 569, 580, 584, 685
Cordoba 553
Corinth 580
Cyprus 274, 348, 399, 479, 486, 656
Cyrene 408n
Cyzicus 478

Da'janiya 462f, 485
ad-Dabab 164f, 191t
Dab'ah 687f, 690
Dacia 402
Dahariya, Khirbat 68f
Dahr 'Ayn Abata 272f, 280, 462f, 469, 482, 493, 495, 649, 650
Dahr al-Kahf 462f, 472, 485, 488, 608
Dahr al-Medineh 272f
Dahr al-Mudayna 280, 285
Dahr al-Qattar 462f, 479, 495
Dahr al-Qinn 462f, 472
Dahr al-Qunfudh 495
Damascus (Dimashq) 294, 309, 330, 332, 348, 368, 414n, 428, 464f, 466, 479, 505, 517f, 518, 519, 519f, 520, 521, 522, 522f, 529, 530, 531, 533, 542, 543, 544, 553, 563, 569, 571, 577, 584, 660, 665, 677, 685, 686, 687f
Damascus Basin 140
Damiyah 164f, 200, 201, 533
Dana 469, 622
Dar'a 677
Darb al-Hajj 522, 555
Darb al-Quful 544
ad-Dayr (Petra) 397, 398, 398f, 399, 411n, 412n, 495
Dayr 'Alla, Tall 272f, 276, 278, 285, 291, 292, 293, 295, 296f, 298, 299, 300, 303, 304, 308, 309, 314, 324, 325, 326, 328, 329, 330, 331, 332, 334, 335, 336, 349, 350f, 352, 355f, 359, 360, 462f, 525, 540, 543, 553, 562f, 581, 586, 656, 659, 660
Dayr, Khirbat 495
Dead Sea Basin 3, 293
Dead Sea Plain 276, 280
Dead Sea 2, 3, 4, 5, 8, 21, 22, 24, 27, 36, 47, 69, 109, 110, 111, 113, 130, 149, 152, 153, 170, 176, 201, 251, 280, 291, 298, 299, 338, 348, 435, 438, 444, 446, 447, 449, 461, 466, 467, 479, 482, 483, 489, 491, 494, 495, 518, 520, 522, 542, 543, 567, 595, 596, 598, 599, 661, 663, 664, 665, 667, 677
Degania 235f, 247
Delos 371, 374, 412n, 606
Dera'a 310
Dhahret Umm al-Marar 235f, 241, 245, 252, 258
adh-Dharih, Khirbat 95, 116, 394, 394f, 401, 462f, 469
Dhiab 575
Dhiban (Dibon) 326, 330, 331, 332, 333, 334, 351, 382, 401, 462f, 467, 468, 492, 504f, 515, 523, 525, 527, 528, 539, 544, 550, 554, 605, 608, 660, 666

695

Dhra' 67, 68f, 69, 95, 113, 116,
Dhra' al-Khan 527, 533, 562f, 581
Dium 428, 467
Diyabaker 580
Doshak, Khirbat ed- 588
Drayjat, al- 334, 335, 336, 350f, 351, 354, 355f, 358, 468
ad-Dusaq, Khirbat 562f, 569

East Africa 487
Ebla 233, 238, 254, 263, 659
Edfu 380
Edirne 580
Edom 294, 306, 323, 329, 331, 338, 339, 340, 347, 348, 349, 361, 489, 490, 604, 605, 632, 660, 661, 663, 664, 665, 667
Egypt 111, 142, 144, 150, 163, 165, 186, 214, 216, 238, 244, 254, 271, 278, 294, 337, 348, 361, 371, 379, 390, 401, 407n, 413n, 414n, 428, 449, 452, 465, 470, 482, 487, 488, 493, 516, 517, 518, 520, 521, 531, 541, 542, 543, 553
Ein Gedi 4f, 130, 132, 133, 149, 429f, 435
Eloth see 'Aqaba
En Shadud 164f
Enan 235f, 254, 255
Enoqiyya 11, 43f
Ephesus 374, 398, 534
Esbus 464f, 494, 495, 666
Ethiopia (Axum) 487
Euphrates River 488, 603, 615
Ezion-geber 665

Fa'un, Tall 690
Fadayn (Fedein, Fiddayn, al-Faddayni) 330, 331, 462f, 495, 504f, 507, 510, 565, 584, 603, 608, 659, 660, 667, 674, 677
Fahij 504f
Fahl, Khirbat 125, 509
Far'a North, Tall al- 127, 164f, 183, 205, 207
Fardakh 348
Faris, Khirbat 332, 333, 444, 504f, 523, 525, 528, 539, 549, 552, 554, 562f, 564, 567, 569, 572, 573, 574f, 577, 580, 581, 582f, 584, 588
Fasa'il 518f
Fassu'a, al- 685, 686
Fayfa, Qasr al- 462f, 469
Faynan, Khirbat (Phaino) 452, 462f, 470, 471, 479, 480, 481, 482, 492, 493, 494
Faysal, Tall 444
Feidat ed-Dihikiya 9f, 10
Fendi, Tall 113, 114
Feqeiqes 235f, 243, 251, 258
Fidan 4, Wadi 109, 110, 164f, 195, 213, 231, 648
Fidan site A, Wadi 645
Fihl 491, 524, 508
Fityan, Khirbat al- 462f, 485, 490
Fjaje 10, 37, 39
Fukhar, Tall al- 292, 293, 295, 296f, 297, 303, 303f, 305f, 306f, 306, 307, 308, 309, 310, 313, 324, 326, 328, 349, 350f, 352, 361
Fulah 518f

Gadara see Umm Qays
Galilee 140, 200, 247, 251, 275, 598, 665
Gaza 339, 348, 429f, 462f, 464f, 470, 474, 487, 491, 493, 517f, 518f, 519f, 522f, 578, 674
Gerasa see Jarash
Get, Qasr al- 379
Ghadhian 518f, 544
Ghamr 518f, 544
Ghanam, Tall 113
Gharandal 19, 462f, 469, 470, 489, 518f, 539, 540f, 544, 549
Gharrain 405n
Ghassul 68f, 108, 109, 112t, 113t, 118, 119, 123, 125, 130, 133, 135, 136, 138, 140, 141, 142, 144, 145, 146, 147, 148, 149, 150, 151, 152, 153
Ghazaleh, Tall 298
Ghazzeh Site H, Wadi 233
Ghor ad-Dhra 164f, 176
Ghor al-Katar 4, 7
Ghor as-Safi 493, 495, 518f, 674
Ghrareh 334, 339
Ghrubba 107, 113, 114
Gilat 112t, 113t, 130, 133, 147, 164f, 175
Gilead 348, 632, 664, 665, 666
Golan (Sawad, Terre de Suete) 112t, 140, 185, 186, 200, 247, 263, 519, 520, 521, 522f, 542, 664, 677
Gomorrah 495, 663
Grar 144
Great Rift Valley 3, 4, 5, 7, 8, 10, 15, 16, 19, 21, 27, 46, 47, 67, 196
Greece 368, 608
Gregoria 671
Gulf of Aden 3
Gulf of Aqaba 3, 16, 16f, 19, 21, 331, 339, 402, 442, 466, 480, 487, 490, 665, 678
Gulf of Suez 3, 487

Habis (Petra), al- 519, 678, 680, 681, 682f
Habis Jaldak, al- 519f, 520, 677
Habra 518f
Habuba Kabira 113t
Haditha 469
Hajjar, Khirbat al- 326, 328, 332, 333, 334, 350f, 351
Halat Ammar Formation 17
Halif, Tall 211
Hallabat, Qasr al- 462f, 471, 472, 487, 490, 504f, 506, 674
Hama region 255, 259
Hamad 13
Hamah 553
Hammam 'Afra 495
Hammam, Khirbat 68f, 80, 85
Hammam, Tall al- 164f, 174, 182t, 191t, 462f, 488, 489
Hammamat Ma'in (Baaru) 462f, 483
Hammeh Gader 451, 452, 453
Hammeh sites (26, 27, 34-44, 49, 53-54), Wadi al- 6f, 7f, 8, 43f, 47, 49f, 50f, 53, 55, 642, 643
Hammeh sites (31-32, 51-52), Wadi al- 6f, 43f, 47, 49f, 50f, 55
Hammeh sites (50), Wadi al- 6f
Hammeh Tomb 80E, Wadi 249
Hammeh, Tall al- 252

Hamra Ifdan, Khirbat 164f, 212, 263
Handaquq North, Tall al- 172, 174, 182t, 184, 185, 200, 204, 208, 608
Handaquq South, Tall al- 171, 174, 182t, 190, 209,
Handaquq, Tall al- 113, 164f, 235f, 241
Har Harif 598
Haram ash-Sharif 541
Harbush 113
Hari, Khirbat al- 335
Harra 13, 14, 681
Harran 348, 544, 660
Hartha 562f, 571, 571f
Hartuv 164f, 176, 179, 197, 201
Hasa site 1021, Wadi al- 49f, 58
Hasa site 1065, Wadi al- 22, 23, 24f
Hasa site 618, Wadi al- 22, 23, 24f
Hasa site 621, Wadi al- 22, 23, 24f
Hasa site 623X, Wadi al- 43f
Hasa site 634, Wadi al- 22
Hasa, al- 521, 522, 522f, 543, 569, 685
Hassan, Tall 10
Hatti 294
Hattin (Horns of Hattin) 519, 520, 529, 531 681
Hauran 187, 188, 204, 410n, 444, 445, 447, 466, 471, 472, 480, 481, 482, 484, 485, 488, 497, 515, 517f, 565, 659, 661
Haurina (Transjordan) 348
Hawar 606, 608, 610
Hayyat, Tall al- 235f, 236, 238, 240, 241, 252, 259, 262, 272f, 276, 278, 279, 279f, 285
Hazeva 338, 339, 340
Hazor 659
Hebron region 262, 544, 677
Hebron 581
Hegra 405
Hemamieh (Egypt) 113t
Heraclea 478
Heraclius 479
Herodian 606
Hesban *see* Hisban, Tall
Heshbon *see* Hisban, Tall
Hesi, Tall al- 587
Hierakonpolis Site 29 113t
Hierakonpolis T.100 145
Hijaz 446, 469, 472, 480, 482, 487, 488, 491, 495, 518, 521, 522, 531, 542, 543, 553, 565, 569, 677, 681
Hippos *see* Qal'at al-Husn
Hisban, Tall (Hesban, Heshbon) 296f, 324, 326, 328, 330, 331, 332, 333, 334, 335, 337, 348, 350f, 351, 352, 360, 438, 444, 462f, 463, 467, 468, 473, 480, 491, 504f, 507, 516, 521, 522, 522f, 523, 525, 526, 527, 532, 539, 540, 541, 544, 550, 552, 553, 554, 632, 562f, 564, 567, 568, 569, 572, 578, 581, 660, 663, 664, 666, 667, 674
Hisma 435, 469, 470, 488, 490, 608, 649
Hmoud 572
Horvat Beter 113
Housban 666
Hubras 570
Hujayrat al-Ghuzlan 130
Huleh Valley 254, 261
Humayma (Auara) 438, 443, 453, 469, 470, 473, 480, 489, 504f, 506, 578, 606, 607f, 608, 609f, 610, 610f, 611f, 685
Humayma J405, Wadi 49f, 50f, 56
Humayma J406, Wadi 49f, 50f, 56, 57
Humayma J407, Wadi 56
Hurmus 678, 680
Hurmuz, Khirbat al- 678, 681
Husn, Tall al- 172, 173, 182, 251, 277, 296f, 330, 331, 332, 438, 450, 508

Idumaea *see* Edom
Iktanu, Tall 164f, 204, 210, 211, 235f, 238, 240, 241, 245, 246, 247, 251, 257, 258, 262, 263, 334, 350f
Ile de Graye 678
'Imru al-Qays 465
India 470, 482, 486, 487
Indian Ocean 542
Iran 615
Iraq 542
'Iraq ad-Dubb 67, 68f, 69
'Iraq al-'Amir 110, 326, 328, 334, 335, 352, 361, 386, 387f, 462f, 467, 495, 604, 661
Irbid plains 627, 671
Irbid, Tall (Arbela) 109, 272f, 280, 285, 297, 310, 312, 324, 326, 328, 330, 331, 332, 439, 451, 467, 480, 522f, 528, 533, 536, 544, 570, 570f, 571, 687f
Iskander, Khirbat 164f, 176, 235f, 236, 238, 240, 243, 243f, 245, 246, 251, 258, 259, 262, 263
Israel 77, 86, 95, 97, 98, 233, 249, 251, 252, 257, 273, 274, 285, 329, 330, 331, 334, 436, 464, 597, 598, 599, 648, 660, 663, 664, 665

Jabal 'Ajlun 673, 677
Jabal Abu al-Khas 334
Jabal Abu Thawwab 94, 95, 97
Jabal ad-Druze 281, 471, 488
Jabal al 'Arab 174
Jabal al-Bishri 234
Jabal al-Druze 10, 11, 434
Jabal al-Hussein 451
Jabal al-Jill 110, 129, 130f, 194
Jabal al-Karak 564
Jabal al-Luweibdeh 450
Jabal al-Qsayr 334, 338, 606
Jabal al-Tahuna 488
Jabal Amman 441, 599
Jabal ar-Rahil 235f, 240, 241, 242f, 246f, 257
Jabal ar-Reheil; 164f, 192
Jabal Ashrafiah 451
Jabal as-Subhi 49f, 50f, 57, 59, 643
Jabal Awf 519, 519f, 520, 522f, 530, 677
Jabal el-Hamra 16f
Jabal el-Jill 17f
Jabal Fatma 49f, 50f, 56
Jabal Hamra 17, 19, 49f, 50f, 56, 57
Jabal Harun 495, 518, 522, 533, 534, 555, 677
Jabal Humayma 43f, 46
Jabal Jarash 517
Jabal Jaufa 451
Jabal Mishraq 49f, 50f, 56, 57
Jabal Mueisi 17, 19

Jabal Muheimi 49f, 50f, 56, 57
Jabal Mutyawwaq 172, 179, 201, 201f
Jabal Na'ja 68f, 96
Jabal Qa'aqir 235, 235f, 238, 250
Jabal Qalkha 17, 18, 19, 46
Jabal Queisa J-24 67, 68f, 69, 70, 129
Jabal Ram 17
Jabal Sahm 17
Jabal Sartaba 109, 113, 125, 128
Jabal Sheikh 207
Jabal Usays 536
Jabbok River see Wadi az-Zarqa'
Jadar 524, 544
Jaffa (Yaffa) 518f, 519f, 522f
Jafr Basin, al- 2, 3f, 9, 10, 16f, 19, 24, 26, 32f, 60
Jafr, al- 487, 528, 649
Jahzah 664
Jalul, Tall 164f, 174, 193, 272f, 280, 281, 283, 298, 324, 331, 332, 333, 334, 335, 336, 350f, 351, 352, 354, 358, 667
Jarash (Gerasa) 109, 125, 164f, 192, 272f, 280, 296f, 297, 306, 310, 324, 326, 332, 361, 385, 430f, 434, 436, 437, 438, 439, 440, 441, 442, 443, 444, 445, 447, 449, 450, 451, 452, 453, 462f, 463, 463, 464f, 467, 473, 474, 475f, 476f, 477, 480, 481, 482, 484, 489, 491, 492, 493, 494, 504f, 505, 508, 509, 515, 519, 519f, 568, 581, 605f, 608, 621, 631, 632, 661, 665, 671, 673, 677, 678
Jarba, al- 488, 491, 520
Java 541
Jawa South, Tall 324, 332, 333, 334, 335, 336, 468
Jawa 14, 164f, 171, 172, 174, 179, 180, 180f, 182t, 184, 192, 205, 272f, 281, 285, 462f, 472, 608
Jawf 405n
Jazayi, al- 298
Jazer 664
Jazira 677
Jemmeh, Tel 578
Jericho 3, 4f, 68f, 72, 86, 87, 95, 98, 108, 110, 114, 164f, 206, 208, 233, 235f, 238, 239, 246, 249, 250, 251, 254, 256, 257, 258, 259, 263, 275, 399f, 406n, 483, 494, 542, 544, 604, 609, 645, 666
Jerusalem 329, 337, 340, 408n, 429f, 466, 474, 478, 479, 491, 494, 517, 518, 519, 519f, 529, 542, 544, 567, 571, 577, 580, 606, 610, 656, 665, 677, 686
Jezirat al-Far'un 487, 678, 681
Jezreel Valley 5, 207, 236, 251, 261, 285
Jibal, al- 518, 519f, 521
Jilat 7-1, Wadi al- 70, 73
Jilat site 10, Wadi al- 12, 49f, 50f, 53, 648, 643
Jilat site 22, Wadi al- 49f, 50f, 53, 55
Jilat site 22, Wadi al- 643
Jilat site 26, Wadi al- 73
Jilat site 27, Wadi al- 649
Jilat site 6, Wadi al- 12, 49f, 50f, 53, 54
Jilat site 8, Wadi al- 12, 49f, 50f, 53, 648
Jilat site 9, Wadi al- 12, 43f, 44
Jirs al-Majamia 533, 544
Jisr Binat Ya'qub 544
Jiza, al- (Zizia, az-) 521, 522, 522f, 527, 531, 532, 543, 544, 608, 609, 685, 686, 687f, 690
Jordan River 3, 4f, 95, 272f, 276, 278, 280, 282, 283, 285, 313, 495, 603, 621, 666, 667
Jordan Valley 2, 3, 5, 8, 19, 21, 22, 26, 27, 72, 75, 77, 86, 95, 97, 98, 99, 107, 108, 109, 110, 111, 115, 116, 117, 118, 119, 128, 139, 144, 147, 150, 151, 152, 170, 171, 172, 174, 184, 185, 186, 190, 192, 193, 201, 202, 204, 205, 206, 207, 208, 234, 235, 240, 245, 247, 251, 257, 258, 259, 260, 261, 274, 275, 276, 278.280, 281, 282, 283, 284, 284t, 285, 291, 294, 295, 298, 299, 306, 309, 310, 314, 323, 324, 325, 328, 330, 334, 335, 336, 337, 351, 386, 434, 435, 441, 443, 444, 447, 451, 461, 466, 467, 468, 479, 480, 494, 516, 522, 525, 528, 530, 533, 536, 541, 542, 543, 550, 565, 567, 568, 573, 575, 585, 671, 596, 598, 603, 610, 611, 622, 632, 640, 645, 648, 664
Jreyyeh 235f, 241, 245
Judaea 428
Judaean Desert 596
Judah 329, 336, 337, 340, 347, 348, 360, 660, 665
Judayid Basin 17, 17f, 19
Judayid site J2, Wadi 18, 51
Judayid site J26, Wadi 18
Judayid site J31, Wadi 18

Kadesh 664
Kafar Saba 518f
Kalabsha (Egypt) 380
Kalavassos 309
Karabeh, Tall al- 298
Karak (Charachmoba, Crac), al- 164f, 294, 348, 350f, 462f, 468, 494, 515, 516, 518, 519f, 520, 521, 522, 522f, 523, 525, 526, 527, 528, 529, 533, 540, 541, 542, 544, 553, 554, 555, 562f, 564, 565, 569, 571, 575, 577, 584, 588, 626f, 627, 680, 681, 679f, 680, 680f, 685, 686, 687f
Karak Plateau, al- 110, 116, 175, 187, 193, 194, 195, 215, 283, 284t, 323, 328, 330, 333, 338, 352, 468, 469, 543, 552, 553, 568, 569, 571, 572, 573, 599, 660, 678
Karka Ma'in 447
Katarit-as Samra' 109, 113, 298, 299, 303, 311
Kedemoth 664
Keniseh, al- 495
Keraimeh, Tall al- 298
Kerak, Khirbat 164f, 171, 208, 208f
Kerameikos, Athens 580
Khalayfi, Tall al- 323, 331, 332, 334, 335, 338, 339, 340, 348, 350f, 351, 360, 361, 660, 661
Khalde, Khirbat al- 462f, 470, 490
Khalil (Hebron) 544, 567, 571
Khallat 'Anaza 49f, 50f, 51, 57, 59, 643
Khan az-Zabib 536, 685
Khanazir Fault 3
Khanazir, Khirbat (Khanazira) 235f, 243, 244, 462f, 469, 470
Kharaysin 68f, 85, 98
Kharrana 4, Qasr al- (Kharraneh) 9f, 12, 49f, 50f, 53, 54, 55, 56
Kharrana, Qasr al- (Kharraneh) 11, 503, 504f, 528
Khatute, al- 374, 406n
Khazna Faraoun (Petra), al- 385, 386, 387, 388, 389, 389f, 395, 396, 397, 398, 399, 400, 409n, 411n
Khilda 334, 336, 350f, 351, 352, 359, 360, 467, 674
Khisfin 518f
Khor Rori 405n

698

Khubtha (Petra), al- 396, 400, 406n, 413n
Kilwa 487, 495
Kirmil, Khirbat 235
Kitan, Tall 178
Kithara, Khirbat al- (Qasr al-Kithara) 462f, 470, 490
Klaros 606
Kos 371
Krak de Montreal see ash-Shawbak
Kufr Abil 263
Kursi, Khirbat al- 497

Lahun, al- 296f, 298, 324, 326, 328, 330, 331, 332, 333, 335, 338, 527, 534, 537, 538, 539, 584, 608, 609
Lajjun (Palestine) 518f
Lajjun, al- 164f, 174, 175, 193, 235f, 243, 244f, 245, 251, 434, 446, 462f, 468, 473, 475f, 485, 488, 489, 490, 521, 522, 522f, 543, 576, 586
Lake al-Hasa 4, 22, 23, 24, 27, 44, 45, 52
Lake Hula 598
Lake Lisan 3, 4, 5, 7, 8, 13, 22, 24, 26, 36, 596
Lake Samra 3, 5
Lake Tiberias 19
Lake Van (Bulgaria) 580
Lebanon 233, 294, 436, 597
Leilan, Tall 112t
Li Vaux Moise (al-Wu'ayrah, Wadi Musa) 519, 519f,, 526, 527, 678
Limestone Desert 9
Lisan Peninsula 3, 5, 479, 596, 597
Livias (Shunat Nimrin) 480, 494
Livias (Tall ar-Rama) 495
Liwa 'Ajlun 563
Luleburgaz 580
Luxor 314

Ma'an (Mu'an) 19, 79, 462f, 487, 488, 489, 491, 518, 522, 522f, 525, 543, 544, 603, 685, 686, 687f
Ma'in (Arabia) 348
Ma'in 462f, 468, 671, 672, 673, 674, 675
Ma'ab see Areopolis
Maadi 113t
Mabrak, al- 307
Macedonia 386
Machaerus 606, 608, 665, 671
Madaba Plain 193, 194, 237, 284t, 330, 335, 468
Madaba site 11 496
Madaba, Tall 193
Madaba 201, 280, 296f, 297, 309, 324, 326, 328, 330, 331, 332, 333, 335, 350f, 351, 438, 443, 435, 467, 468, 469, 480, 484, 494, 495, 505, 562f, 568, 571, 585, 666, 671, 672, 672f, 673, 673f, 674, 675
Mafjar, Khirbat al- 509
Mafraq, al- 296f, 297, 330, 331, 466, 495, 528, 685, 687f, 690
Magass, Tall 110, 114, 115, 116, 129, 131f, 142, 144, 164f, 195, 214, 216
Mahanaim 664, 665
Malih, Tall al- 113
Mampsis 429f
Marazzah North East 43f, 47
Marazzah North 43f, 47

Mardikh, Tall 176, 254, 263
Mari 234, 238
Marib (South Arabia) 381, 405n
Marwa 451
Masada 386, 400, 606
Masharia 1 6f, 8, 37, 39
Masharia 2 6f, 8, 37, 38
Masharia 3 6f, 8
Masharia 4 6f, 8, 37, 38
Masharia 5 6f, 8, 37, 38
Massuh 468
Mazar, Tall al- 298, 327, 328, 330, 331, 332, 334, 335, 337, 349, 350f, 351, 352, 358, 359, 360, 521, 522, 529, 536, 553, 555, 585, 599, 660
Mecca (Makkah) 488, 491, 520, 531, 534, 543, 563, 569, 685
Media 402, 520
Medina 563
Mediterranean 4, 15, 150
Megiddo 164, 176, 183, 235f, 246, 255, 259, 309
Mephaath 664
Meqabelein (tomb), al- 350f, 351, 352, 359, 362
Meqbereh, Tall al- 298
Mérida (Spain) 453
Merimde 112t, 113t
Mesopotamia 150, 151, 163, 211, 214, 234, 236, 238, 281, 299, 309, 313, 402, 403, 406n, 414n, 446, 488, 490, 603, 608, 609, 659
Miletus 371
Minshat Abu Omar 216
Mishash 504f
Mitanni 294
Mizpe Shalem 650
Moab 294, 298, 301, 306, 314, 315, 323, 329, 331, 333, 338, 340, 347, 348, 349, 361, 468, 564, 632, 660, 663, 664, 665, 678
Moje Awad 401
Mosul 520
Motza 235f, 255
Mount Carmel 34
Mount Gilead 677
Mount Hermon 664
Mount Nebo 272f, 280, 330, 331, 332, 333, 335, 462f, 468, 483, 495, 497, 504f, 664, 671, 672, 673, 674
Mount Sedom 598
Mount Seir 664
Mount Sinai 470, 487
Mount Tabor 677
Mu'allaq, Khirbat 335, 338, 562f, 572
Mu'ta 462f, 491, 517f, 521, 522, 522f, 523, 555, 626f
Mu'amariyeh 235f, 257, 263
Muallaqah 495
Mudawwara, al- 17, 19, 488, 685, 686
Mudaybi' 335
Mudayna 'Alia 327, 328, 330
Mudayna Mu'arraja 327, 328, 330
Mudayna Thamad 332, 335, 338
Mudiq, al- 112t
Mugayyir, Tall al- 349
Mugayyit, Tall al- 350f
Mugharat al-Wardah 543

Muhattat al-Haj 462f, 490
Muhei 575
Muheisen 48
Mukawir 665
Mukhattat 462f
Mukhayyat, Khirbat al- 468, 495, 497, 673
Mukhayzin 335
Munhatta 86
Munyah 495
Musa, Qasr Wadi 462f, 470
Mushash, Qasr al- 462f, 472, 487, 506, 610
Mushatta, Qasr al- 503, 504f, 507, 507f, 508
Mustah, Tall 164f, 191t, 298
Mutrab, al- 462f, 488, 489
Muwaqqar, al- 504f, 506
Muzairib 677

Naaharina 294
Nabataea 402
Nablus 518, 518f, 533, 544
Nahal Hemar 68f, 82, 98
Nahal Lavan 109 72
Nahal Mishmar 113t, 130, 139, 147, 149, 647
Nahal Oren 1-11 72
Nahal Qanah 112t, 149
Nahal Sekher 597
Nahal Zin 597
Nahariya, Tel 310
Nahiya Kerak 572
Nakhal 528, 539
Namara 465, 661
Nami, Tall 309, 310
Naqada KH3 113t
Naqada KHI 113t
Naqada NT 113t
Naur 109, 599
Nazareth 580
Nea Paphos 399, 412n
Negev 14, 39, 40, 41, 45, 48, 51, 52, 53, 59, 95, 107, 111, 114, 117, 130, 138, 147, 150, 179, 194, 211, 212, 244, 250, 253, 257, 259, 260, 263, 338, 391, 413n, 434, 468, 482, 487, 489, 492, 503, 578, 596, 597, 598, 606, 649, 664
Nekhiel South, Tall an- 235f, 257, 258, 298
Neve Ur 126
Nicomedia 478
Nile Delta 259, 274, 299, 611
Nile River 13, 15, 597, 603
Nile Valley 686
Nimrin, Tall 152, 276, 278, 280, 285, 296f, 297, 327, 329, 330, 331, 332, 334, 335, 350f, 351, 360, 451, 462f, 468, 523,
Nippur 406n
Nizzanim 112t
North Africa 338, 486
Northern Basalt Plateau 3f
Numayra (Nuwayri, al-) 4f, 164f, 169, 172, 174, 180, 181, 182t, 185, 209, 210, 211, 212, 214, 235f, 245, 272f, 275, 648, 681
Numidia 478
Nuweijis, Qasr 395, 451

Oboda 401, 429f
Oman 2, 13, 15

Palestine (Filastin) 139, 211, 212, 215, 233, 237, 247, 249, 253, 254, 255, 259, 263, 282, 285, 291, 292, 293, 294, 310, 311, 312, 313, 315, 329, 347, 348, 349, 358, 359, 361, 446, 464, 465, 466, 467, 468, 469, 480, 483, 485, 487, 488, 489, 490, 491, 494, 495, 497, 517, 518, 520, 529, 531, 533, 542, 543, 544, 552, 505, 510, 566, 567, 573, 577, 578, 580, 584, 585, 587, 656, 659, 672, 674, 686
Palmyra 490
Parthia 402
Pella (Tabaqat Fahl, Fahl, Fihl) 2, 3f, 4, 5, 6, 6f, 7, 8, 13, 22, 25t, 27, 31, 32f, 36, 38, 39, 47, 48, 53, 57, 60, 109, 110, 111, 112t, 113, 114, 118, 119, 125, 126, 126f, 127, 133, 134, 142, 144, 145, 147, 149, 151, 164f, 172, 182t, 201, 205, 212, 213, 272f, 276, 277, 277f, 278, 283, 285, 291, 294, 295, 296f, 298, 299, 300, 303, 307, 308, 309, 310, 311, 313, 314, 324, 327, 328, 330, 331, 332, 334, 337, 427, 428, 433, 434, 435, 436, 437, 438, 439, 440, 442, 448, 449, 450, 451, 452, 453, 462f, 463, 466, 467, 473, 474, 475f, 476f, 477, 479, 480, 481, 482, 483f, 484, 485, 491, 492, 493, 494, 505, 508, 509, 510, 515, 516, 522, 522f, 523, 524, 525, 527, 528, 534, 536, 537, 538, 538f, 541, 545, 550, 551f, 553, 568, 584, 659
Peraea 467, 663
Pergamon 388, 407n
Persia 347, 348, 479, 488, 490, 610
Persian Gulf 449, 485
Petra 31, 53, 56, 58, 338, 348, 350f, 367, 368, 371, 374, 375f, 377, 377f, 379, 380, 382, 382f, 384f, 385f, 387, 388, 389f, 390, 390f, 391, 393, 394, 395, 396, 396f, 397f, 398f, 399f, 401, 404, 405n, 406n, 408n, 409n, 410n, 411n, 412n, 413n, 434, 436, 437, 438, 439, 440, 441, 442, 443, 444, 448, 449, 451, 453, 462f, 469, 470, 473, 474, 480, 481, 482, 484, 485, 493, 494, 495, 497, 526, 528, 529, 533. 544, 504f, 573, 606, 608, 610, 615, 616f, 617, 624, 625f, 631, 661, 665, 678, 681, 682f
Petra-Zurrabah 474
Phaino *see* Faynan, Khirbat
Philadelphia see Amman
Phoenicia 347
Plain of Dhra' 280
Poleg, Tel 272f, 273
Priene 371
Ptolemais 399f

Qa' al-'Umayri 9f, 13
Qa' al-Azraq 10, 11
Qa' al-Jafr 10
Qa' Salab 18, 49f, 50f, 57
Qa'dan South, Tall 298
Qa'dan, Tall 298, 523, 525, 528, 539, 550, 552
Qa' Abu al-Hussein 488
Qada Hawran 563
Qadesh Barnea 599
Qafzeh 40
Qaisiyeh Fault 10
Qal'at ar-Rabad 530
Qal'at Ma'an 687f, 691

700

Qal'at al-Balqa 690
Qal'at al-Fassua' 687f, 690f, 691
Qal'at al-Hasa 469, 687f, 689f, 690, 691
Qal'at al-Husn (Hippos, Susiya) 428
Qal'at al-Mudawwara 687f, 691
Qal'at al-Qatrana 687f, 688f, 690
Qal'at al-Sai 686
Qal'at al-Unaiza 687f, 691
Qal'at ar-Rabad 681
Qalansuwah 518f
Qam 523
Qana 405n, 462f, 470
Qanawat (Qanu, Canatha) 428, 429f, 659
Qar al-Ba'ig 462f
Qarnini 348
Qaryat al-Fau 405n
Qasr, al- (Qusayr, Khirbat al-) 406n, 411n, 439, 565, 573, 574f
Qastal al-Muwaqqar 674
Qastal, al- 504f, 506, 539, 685
Qatif 3 112t
Qatrana, al- 518f, 543, 562f, 569, 570, 685, 686
Qitmit 339, 340
Qos, Tall al- 164, 191t
Qudsiyah, Tall 525, 528
Quirma Formation 10
Qumran 606
Qurayya 487, 488
Qurrein North, Khirbat 110, 114, 115
Qusayr Mu'in (ash-Shunah) 522
Quwaylibah *see* Abila
Quwayra Depression, al- 15, 16
Quwayra, al- (al-Quweirah) 16, 16f, 19, 49f, 50f, 57, 608
Quwaysmeh (Qasr al-Sab) 451, 467, 497, 673, 674, 675

Rabbah, ar- *see* Areopolis
Rabbath-Ammon *see* Amman
Rabbath-Moab 434
Rabi', Tall ar- 298
Rahub, Khirbat ar- 164f, 190
Rameh (Livias), Tall ar- 462f
Ramlah, al- 504f, 505, 510, 517f, 518f, 519f, 522f, 544
Ramm 462f
Ramoth 664, 665
Ramtha, ar- 109, 310
Raphana *see* Bayt Ras
Raqim, ar- (Raqim, al) 517f, 521, 522, 522f, 523, 555
Raqqa 529
Ras al-'Ayn 441
Ras al-Qabub 562f, 572, 581
Ras an-Naqab Scarp 16f, 17f, 26
Ras an-Naqab site 440 43f, 47
Ras an-Naqab 9, 15, 16f, 17, 21, 31, 32f, 36, 39, 40, 41, 42, 44, 45, 46, 47, 51, 52, 53, 54, 56, 57, 58, 60, 69, 97, 573, 632, 642
Ras Hamid, Tall 235f, 263
Ras Shamra 112t, 276
Rasm Harbush 115
Red Sea Rift 19
Red Sea 3, 142, 150, 216, 334, 482, 486, 487, 518, 521, 532, 542, 567, 665, 667, 680, 685
Reqa, Khirbat ar- 235f, 263
Reseifeh, Khirbat ar- 235f, 241, 242
Rhodes 371, 412n, 661
Rihab 462f, 467, 491, 496
Risha, al- 504f, 505, 506
Rome 373, 388, 408n, 412n, 413n, 414n, 488, 671
Rujm al Kursi 550, 552
Rujm al-Abd 300, 313
Rujm al-Henu 324, 334, 335, 336, 349
Rujm al-Henu West 349, 350f
Rujm al-Hiri 263
Rujm al-Malfuf North 336
Rujm al-Malfuf South 327, 328, 335
Rujm Beni Yasser 490
Rukeis, Tall 272f, 281
Rumeith, ar- 330, 332
Rumman, ar- 164f, 192
Rusaifah 245
Ruweished 528
Ruwwafa 445

Sa'idiyya Tahta 107, 113, 114
Sa'idiyya, Tall as- 164f, 177, 177f, 182t, 183, 185, 186, 188, 191t, 272f, 276, 292, 295, 296f, 298, 299, 300, 303, 306, 307, 308, 309, 310, 312, 324, 325, 327, 328, 330, 331, 332, 334, 335, 336, 349, 350f, 352, 354f, 355f, 358, 359, 360, 361, 362, 604, 659
Sabra 1 67, 68f, 69
Sabra 4 43f, 47
Sadaqa 462f, 470
Safad 544
Safadi 112t, 113t
Safi, as- (Zoar, Zoara, Zughar, Sughar) 4f, 164f, 170, 200, 201, 462f, 464f, 468, 469, 494, 495, 517f, 519f, 520, 521, 522, 522f, 525, 544, 663
Safut, Tall 272f, 280, 298, 310, 313, 315, 324, 327, 328, 332, 333, 335, 336
Sahab, Tall 109, 110, 114, 115, 118, 119, 128, 128f, 133, 144, 151, 272f, 280, 285, 296f, 297, 298, 306, 309, 310, 312, 315, 324, 327, 328, 330, 331, 332, 333, 335, 336
Sahara 598
Sahem 312, 324
Sakhineh, Tall as- 164f, 191t, 298
Sal 114, 140
Salameh, Khirbat (Salama) 335, 444, 467
Salkhad 677
Salt, as- (Gadara) 462f, 467, 480, 494, 516, 521, 522, 522f, 526, 529, 530, 540, 542, 544, 554, 555, 562f, 571, 585, 603, 621, 673, 677, 681
Samah 491
Samaria 337, 347, 660
Samra, Khirbat as- 19, 20, 20f, 21, 462f, 466, 467, 480, 485, 492, 496, 497, 518f, 673
Sapsafas 462f, 483
Saqqara 492
Sarachane 580
Saudi Arabia 2, 9f, 10, 13, 14, 15, 17, 19
Sawad *see* Golan
Sbaita 401

Scythopolis (Baysan, Beth Shan) 429f, 462f, 464f, 480, 494
Sea of Galilee 5, 453
Sedeh, as- 609
Seir 664
Sela', as- 335, 338, 340, 367, 370f, 371, 404, 606, 678, 680
Seleucia 406n, 479
Seyyeh, as- 68f, 92, 93
Shabib, Qasr 530, 681, 685, 687f, 690
Sham, ash- 553
Shammah plateau, ash- 14
Sharah, ash- (Syria Sobal) 518, 519, 519f, 521
Sharah 519f
Shawbak, ash- (Shobak, Krak de Montreal) 10, 36, 39, 515, 516, 519, 519f, 520, 521, 522, 522f, 526, 527, 528, 529, 536, 537f, 544, 549, 555, 562f, 564, 565, 566, 569, 578, 579f, 585, 595, 599, 678, 679f, 680, 681, 685, 686
Shaykh 'Isa, Khirbat 523, 525
Shechem 272f, 276
Sheikh 'Ali, Khirbat 86, 99
Sheikh 'Isa 504f
Sheikh Mohammed, Tall- 235f, 263
Sheikh Sa'ade 294
Shephela 599
Shiqmim 110, 112t, 113t, 130, 147, 149, 598
Shiyab, al- 94
Shuna (ash-Shamaliyya), Tall ash- (Shuna North, al-Qusayr) 107, 108, 109, 110, 111, 113, 113t, 114, 118, 119, 127, 127f, 133, 134, 139, 144, 149, 151, 164f, 169, 170, 171, 178, 178f, 179, 179f, 183, 185, 186, 186t, 187, 189, 191, 191t, 197, 203, 203f, 204, 204f, 205, 206, 207, 208, 208f, 214, 215, 217, 291, 438, 449, 522, 522f, 525, 533, 544, 647, 648, 649
Shuna South 4f, 278, 451, 468
Shunah al-Janubiyya 674
Si' 449
Sicily 603
Sidon 329, 337, 435, 672
Sifiya, as- 68f, 79, 80, 85, 90, 97
Sihab, Tall al- 294, 296f
Sil', Khirbat as- 660, 678, 680
Sinai Peninsula 16f, 41, 42, 48, 51, 52, 59, 194, 211, 212, 216, 244, 259, 339, 434, 482, 484, 521, 522f, 597, 598, 661
Siran, Tall 335, 660
Smakieh 573
Sodom 495, 663
Soleb 294
Souk, Khirbat as- 451
Southern Ghor 194, 206, 209, 210, 212, 213, 215, 247, 263, 299, 444, 466, 468, 568
Souweida 385
Sperlonga cave 406n, 407n
St Catherine's Monastery 484
Suez 597
Sukas 112t
Sukhna formation 21
Sukhne, Tall as- 164f, 192, 208, 215f
Sumatra 541
Sumieh 575
Sur 517f
Susita 610

Susiya *see* Qal'at al-Husn
Swafiyeh 467
Sweimeh 140
Syria 14, 99, 111, 139, 150, 176, 181, 192, 211, 214, 215, 233, 234, 236, 253, 254, 255, 259, 261, 263, 274, 278, 281, 293, 294, 299, 309, 347, 348, 359, 361, 371, 379, 401, 402, 427, 428, 433, 434, 436, 440, 446, 449, 464, 465, 466, 469, 471, 477, 479, 480, 482, 484, 486, 488, 489, 490, 491, 495, 516, 520, 522, 543, 544, 552, 553, 565, 567, 571, 573, 584, 587, 621, 659, 661, 677, 685, 686

Ta'annek, Tall (Ti 'innik) 577, 606
Ta'asir 518f
Tabaqa 49f, 50f, 57, 58, 59
Tabaqat Fahl *see* Pella
Tabariyah 504f, 505, 517f, 518f, 533
Tabuk 487, 518, 543, 544
Tadun, Khirbat 34, 588
Tafila, at- 323, 352, 528, 540f, 549, 678, 680
Tahune, Tall al- 298
Tannur, Khirbat at- 379, 382, 391, 393, 394, 394f, 395f, 410n, 411n, 449, 604
Tarablus 553
Tawilan 323, 324, 335, 338, 339, 348, 350f, 351, 360, 581, 660
Tayma 487
Tayyiba, Qasr Wadi at- 462f, 470, 603
Tepe Sabz 597
Terqa 238
Terre de Suete *see* Golan
Thâj 405n
Thalab al-Buhira 45
Thamayil 335, 338
Thebes 263
Thessalonica 478
Thuraiya, Qasr ath- 462f, 485
Tiberias (Tabariyah) 491, 517, 519f, 520, 544, 677, 681
Tibneh (Tibna) 562f, 573, 575, 585
Tigris 603
Tilah, Qasr al- 462f, 470
Timna 150, 235f, 253, 261, 294
Tor 'Imdai 562f, 573, 578
Tor Aeid 43f, 46, 47
Tor al-Tareeq 49f, 50f, 54, 55, 56
Tor Faraj 18, 37, 40, 41, 42, 46
Tor Fawaz 47
Tor Hamar 18, 43f, 46, 48, 49f, 50f, 56
Tor Sabiha 18, 37, 40, 41, 42, 46
Tor Sadaf 43f, 45
Tor Sageer; 43f, 55
Tsaf 112t
Tuba, Qasr at- 462f, 495, 504f, 528
Tubeiq Highlands 9, 15
Tubna 142
Tujjar 518f
Tulay 1 120, 143
Tulay 3 120, 123, 125, 131, 147
Tulay 5 130
Tulay 6 121
Tulaylat al-Ghassul 107, 108, 109, 110, 111, 113, 114, 116, 117, 118, 119, 120, 121f, 122f, 123, 125, 126, 127, 129,

130, 132f, 133, 135, 136, 139, 148, 151, 152, 164f, 185, 655
Tulul adh Dhahab al-Gharbiyya 664
Turkey 534
Turkman, at- 112t
Tuwana, at- (Thana/Thoana, Tuwaneh) 438, 462f, 469
Tyre 329, 348, 435, 479, 519, 519f, 520, 522f, 672, 677, 678

Ubeidiya 1, 4, 5
Udruh (Udhruh) 438, 439, 462f, 469, 470, 482, 485, 487, 488, 490, 491, 504f, 518, 525, 539, 544, 554, 562f, 569, 581, 610
Ugarit 294, 313, 608, 659
'Umayri, Tall al- 10, 17, 164f, 171, 180, 182t, 188, 190, 200, 201, 209, 210, 215, 216, 235f, 248f, 252, 252f, 255, 256, 256f, 272f, 280, 281, 283, 285, 298, 315, 324, 325, 325f, 326f, 327f, 328, 331, 332, 333, 335, 336, 337, 347, 349, 350f, 351, 352, 353f, 354, 358, 359, 360, 361, 362, 438, 468, 660
Umbashi, Khirbat 164f
Umm ad-Dananir, Tall 297, 298, 304, 306, 307, 309, 313, 314, 315, 324, 335, 349, 350f, 352, 539
Umm al-'Amad, Khirbat see Abila
Umm al-'Amad 452, 575
Umm al-Biyara 323, 335, 338, 339, 340, 367, 375f, 606, 660
Umm al-Ghozlan, Khirbat 235f, 247, 249f
Umm al-Jimal 429f, 438, 440, 443, 447, 448, 462f, 463, 471, 473, 480, 481, 483f, 484, 485, 486, 488, 489, 490, 491, 492, 493, 495, 496, 497, 504f, 505, 509, 606, 608, 610, 664, 671, 672
Umm al-Qanafah 297
Umm al-Quttayn 462f, 463, 472, 539, 608
Umm al-Walid 536
Umm ar-Rasas (Kastron Mefa'a) 438, 447, 462f, 467, 468, 495, 496, 496f, 671, 673, 674, 675, 675f
Umm as-Sedeirah 235f, 251
Umm at-Tawabin 469
Umm Bighal 235f, 263
Umm Hammad (Ash Sherqi), Tall 107, 109, 113, 114, 164f, 170, 179, 183, 190, 191t, 203, 204, 205, 205f, 206, 206f, 207, 211, 235f, 239, 240, 245, 246, 258, 262
Umm Hammad al-Gharbiyya, Tall 252, 253, 257, 258, 263
Umm Qays (Gadara) 361, 427, 428, 432f, 433, 434, 435, 436, 437, 438, 439, 440, 441, 442, 443, 444, 445, 447, 449, 451, 453, 462f, 463, 464f, 467, 480, 481, 482, 484, 485f, 491, 494, 504f, 505, 509, 544, 562f, 568, 573, 578, 581, 583, 586, 608, 665
Umm Rujm, Khirbat 235f, 257
Umm Udhayna 350f, 351, 352, 359, 360, 660
Umruq 462f, 469, 470
'Unaiza, al- 685, 686
Urartu 610
Uvda 648
Uwaynid site 14, Wadi al- 12, 13, 18, 49f, 50f, 53, 54
Uwaynid site 18, Wadi al- 12, 13, 18, 49f, 50f, 53, 54
Uwaynid 149f
'Uyun Musa 673

Wada 48
Waddum du-Masma'im (South Arabia) 381

Wadi 'Ajlun 113, 142
Wadi 'Arabah 2, 3, 4f, 10, 15, 16f, 19, 39, 51, 108, 109, 194, 212, 244, 283, 294, 339, 444, 452, 461, 463, 468, 469, 470, 481, 487, 489, 491, 492, 518, 519, 521, 522, 543, 544, 554, 566f, 595, 596, 599, 649, 664, 667
Wadi 'Ayn Umm Hudayb 113
Wadi Abu an-Naml 280
Wadi Aghar 43f, 46
Wadi al-'Ali 22, 24, 39, 41, 469
Wadi al-'Arab 109, 113, 272f, 283, 284t, 439, 444, 575, 632
Wadi al-'Ajib 281
Wadi al-Ahmar 58
Wadi al-Ajib 110
Wadi al-Bustan 10, 36
Wadi al-Ghadaf 9f
Wadi al-Hammam 6f
Wadi al-Hammeh 2, 5, 6, 6f, 7, 7f, 8, 19, 21, 22, 57, 235f, 272f, 283
Wadi al-Hasa (Zered) 2, 3f, 4, 4f, 5, 10, 13, 15, 19, 21, 22, 23, 24, 25t, 26, 27, 31, 32f, 39, 42, 44, 46, 48, 51, 52, 53, 54, 55, 56, 58, 60, 70, 85, 109, 194, 244, 263, 284, 291, 294, 296f, 298, 299, 314, 330, 331, 339, 434, 439, 444, 465, 466, 469, 480, 490, 495, 528, 566f, 568, 575, 632, 635f
Wadi al-Hibr 110, 114
Wadi al-Himar 6f
Wadi al-Janab 9f
Wadi al-Jilat 9f, 25t, 44, 57, 81, 110, 164f, 194, 453, 608, 645, 646, 649
Wadi al-Kafrayn 352
Wadi al-Kerak 4f, 10, 19, 85, 113, 184, 272f
Wadi al-La'ban 469
Wadi al-Mujib (Arnon) 4f, 10, 19, 79, 85, 97, 296f, 298, 299, 333, 447, 466, 467, 468, 480, 488, 490, 491, 495, 517, 518, 520, 522, 523, 526, 566f, 664, 666
Wadi al-Musa 309, 390, 516, 518, 519, 520, 521, 522f, 522, 533, 542, 544
Wadi al-Qattar 110, 114, 115, 116, 118
Wadi al-Quwayra 5, 19
Wadi al-Uwaynid 9f, 12, 37, 38, 57
Wadi al-Yabis 69, 109, 113, 192, 201, 247, 263, 272f, 283, 284, 284t, 297, 352, 439, 444, 568, 632
Wadi Amman 441, 451
Wadi ar-Rajib 298
Wadi ar-Rama/Djaffara 113
Wadi as-Sirhan Basin 2, 9, 10, 13, 14, 19
Wadi as-Sirhan Fault 10
Wadi as-Sirhan 446, 465, 472, 488
Wadi ath-Thamad 335
Wadi at-Tafila 85
Wadi Azeimeh 113
Wadi az-Zarqa' (Jabbock) 5, 19, 20, 21, 25t, 26, 85, 90, 98, 108, 110, 113, 190, 192, 193, 195, 201, 205f, 207, 208, 291, 298, 323, 335, 467, 474, 517, 520, 521, 522f, 565, 566f, 573, 632, 664, 666
Wadi Bayir 632
Wadi Beersheva 164f, 184, 213, 597
Wadi Butm 11
Wadi Dana 85
Wadi Dhobai 605
Wadi Dhulail 19

703

Wadi Ed-Dabi 9f
Wadi Faynan 68f, 79, 96, 98, 109, 110, 112t, 113t, 113, 114, 115, 144, 199, 206, 208, 213, 235f, 244, 253, 254, 259, 261, 293, 294, 296f, 299, 339, 340, 438, 440, 444, 452, 463, 469, 470, 543, 649
Wadi Faynan, Tall 110
Wadi Feiran 597
Wadi Fidan 645
Wadi Ghwair 68f, 79, 91, 98
Wadi Hisban 8, 25t, 109, 200, 201, 237, 283, 284t, 541, 599, 606, 642
Wadi Hisma 2, 3f, 4, 5, 13, 15, 16, 16f, 17, 18, 19.21, 25t, 27, 114, 115, 116, 119, 128, 194,
Wadi Humayma 49f, 50f, 57, 58, 59
Wadi Huwan 6f
Wadi Huweijir 71, 86
Wadi Jashsha el-Adla 9f
Wadi Jebara 495
Wadi Jirm al-Moz 5, 6f, 7, 19, 22, 113, 172, 276, 277, 438, 442, 450, 541
Wadi Judayid 16, 16f, 17f, 18, 49f, 50f, 57, 58, 59, 110, 632, 643
Wadi Kafrinji 530
Wadi Khalid 263
Wadi Kharrar 495
Wadi Kufrein 109, 113, 299, 468
Wadi Kufrinjeh 109, 113
Wadi La'ban 95
Wadi Ma'ayn 4f
Wadi Madamagh 48, 49f, 50f, 56
Wadi Muqat 488
Wadi Musa 85, 383, 385f, 664, 677, 680, 681
Wadi Nimrin 109, 113, 299
Wadi Qalkha 16, 16f, 18, 37, 39, 46, 47
Wadi Queisa 110, 114
Wadi Rabah 95, 96, 647
Wadi Rajil 9f, 10, 110, 174
Wadi Ramm 16, 19, 379, 390, 412n, 449, 488, 606, 615
Wadi Ramman 16
Wadi Rattama 11
Wadi Sarar 109, 113
Wadi Shellale 310
Wadi Shu'ayb 68f, 71, 72, 73, 75, 76, 78, 79, 80, 85, 89, 90, 91, 92, 93, 94, 95, 97, 98, 278
Wadi Siagh 624, 625f
Wadi Thallaga 16
Wadi Wala 335, 468, 522, 523
Wadi Walid 504f, 506
Wadi Yarmuk 296f, 520, 566f
Wadi Yutm 16, 16f, 19
Wadi Ziqlab 68f, 95, 96, 97, 109, 272f, 283, 443, 568, 632, 633f, 634, 635f
Wahiba Sands 15
Walid 438
Wu'ayra, al- 678, 681

Yadoudeh 575
Yajouz 462f, 467
Yarmouk Valley 441, 453
Yarmouth, Tel 164f, 172, 176, 187

Yarmuk River 19, 283, 291, 295, 296f, 298, 466, 664, 677
Yasileh 462f, 467, 491
Yehudiyeh, Tall al- 274
Yemen (Himyar) 487
Yiftahel 86, 164f, 178, 179, 204, 213
Yotvata 462f, 470, 489
Yutil Abu Kharaf 43f
Yutil al –Hasa 43f, 45, 49f, 50f, 55, 57, 58

Zabad 497
Zaharat adh-Dhra 280
Zahr al-'Aqaba 691
Zahrat adh-Dhra' site 1 272f, 280
Zantur, az- (Petra) 374, 375f, 376f, 377f, 395, 406n, 410n, 411n, 443
Zara, az- (Callirhoe) 462f
Zara, az- site 25 113
Zaraqun, Khirbat az- 164f, 171, 172, 173, 173f, 174, 175, 176, 177, 181, 182t, 183, 186, 186t, 187, 188, 190, 208, 209, 209f, 210f, 214, 215, 235f, 240, 241, 262
Zarqa', az- 2, 3f, 19, 238, 278, 466, 467, 518f, 522, 522f, 527, 528, 529, 530, 543, 544, 622, 681
Zarqa' River, az- 19, 239, 240, 242, 283, 444, 466, 477, 518, 530
Zay al-Gharbi 462f, 495
Zay 673
Ze'aze'iyeh 235f, 241, 258, 263
Zeboim 663
Zib Atuf (Petra) 377, 377f
Ziglab site 200, Wadi 111, 112t, 113
Ziza 489,
Zoar (Zoara) *see* as-Safi
Zona, Khirbat az- 462f, 485, 490